STRENGTH OF MATERIALS

SMTS-I

STRENGTH OF MATERIALS

By

Dr. B.C. PUNMIA

Formerly,
Professor and Head, Deptt. of Civil Engineering, &
Dean, Faculty of Engineering
M.B.M. Engineering College
Jodhpur

Er. ASHOK KUMAR JAIN
Director
Arihant Consultants
Jodhpur

Dr. ARUN KUMAR JAIN
Assistant Professor
M.B.M. Engineering College
Jodhpur

TENTH EDITION
(THOROUGHLY REVISED AND ENLARGED)

LAXMI PUBLICATIONS (P) LTD
(An ISO 9001:2015 Company)
BENGALURU • CHENNAI • GUWAHATI • HYDERABAD
KOCHI • KOLKATA • LUCKNOW • RANCHI
NEW DELHI

SMTS–I STRENGTH OF MATERIALS

Printed and bound in India
Typeset at : Goswami Associates, Delhi
First Edition : July 1964, *Tenth Edition* : June 2013, 2017, 2018, 2019, 2024, *Edition* : 2025
ISBN : 978-81-318-0925-9

Branches

Branch	Telephone
Bengaluru	080-26 75 69 30
Chennai	044-24 34 47 26
Guwahati	0361-254 36 69
Hyderabad	040-27 55 53 83
Kochi	0484-405 13 03
Kolkata	033-40 04 77 79
Lucknow	0522-430 36 13
Ranchi	0651-224 24 64

PUBLISHED IN INDIA BY

Laxmi Publications (P) Ltd.
(An ISO 9001:2015 Company)
113, GOLDEN HOUSE, GURUDWARA ROAD, DARYAGANJ,
NEW DELHI - 110002, INDIA
Telephone : 91-11-4353 2500, 4353 2501
www.laxmipublications.com
info@laxmipublications.com

C—00000/024/011
Printed at: Harsh Printing Press, Delhi.

Preface to the Tenth Edition

In the Tenth Edition of the Book, the subject matter has been thoroughly revised, updated and rearranged. Many chapters have been re-written, incorporating latest literature on the subject. Many tables have been revised/updated corresponding to the latest Editions of some Indian Standards. All the diagrams have been redrawn using computer graphics. The book has been typeset in bigger format, using two colours, keeping in pace with modern trend.

The book contains 28 chapters, divided into five sections. Section 1, containing six chapters deals with various aspects of stresses and strains, including theories of failures. Section 2 on 'Beams and Bending' contain ten chapters dealing with various aspects of strength and stiffness of beams. Section 3, containing three chapters, deals with compression members, including various masonry structures, such as Retaining Walls, Dams and Chimneys. Section 4 is devoted to shafts, springs and cylinders and has four chapters. Lastly, section 5 contains miscellaneous topics including Space Frames, Riveted Joints and Welded Joints.

Each chapter begins with clear statement of pertinent definitions, theories and basic principles. These are further supplemented with numerous solved examples. Each Chapter also contains *Advanced Problems* as well as problems from various *Competitive Examinations*. Account has been taken throughout of the suggestions offered by many users of the book and grateful acknowledgement is made to them.

'Sthapana'
4th April, 2011

B.C. Punmia
Ashok Kumar Jain
Arun Kumar Jain

Contents

Section 1: Stresses and Strains

Section 2 : Beams and Bending

Section 3 : Compression Members

Section 4 : Shafts, Springs and Cylinders

Section 5 : Miscellaneous Topics

SECTION 1 STRESSES AND STRAINS

CHAPTERS

1. Simple Stresses
2. Elastic Constants
3. Compound Stresses
4. Analysis of Strain
5. Strain Energy, Resilience and Impact Loading
6. Theories of Failure

1

Simple Stresses

1.1. MECHANICAL PROPERTIES OF SOLIDS

From engineering stand point, the most important properties of engineering materials are : elasticity, plasticity, ductility, brittleness, malleability, toughness and hardness. A particular material cannot possess all of these simultaneously because several of these are in opposition.

Elasticity. An external system of forces acting on a body tends to deform the body while the molecular forces acting between the molecules offer resistance against deformation. The deformation or displacement of the particles continues till full resistance to the external forces is set up. If the forces are now gradually diminished, the body will return, wholly or partly, to its original shape. *The property by which a body returns to its original shape after the removal of external load is called elasticity.* If the body regains completely its original shape, it is said to be *perfectly elastic*. A material is said to be *isotropic* if it is equally elastic in all directions. Steel, copper, aluminium, stone etc., may be considered to be perfectly elastic, within certain limits. For any particular material, a critical value of the load, known as the *elastic limit*, marks the partial breakdown of elasticity beyond which removal of the load results in some degree of permanent deformation.

Plasticity. Plasticity is the property of a material by which no strain disappears when it is relieved from the stress. Plasticity is thus the converse of elasticity. Most of the materials become plastic under the action of large forces, and behave in a manner similar to a viscous fluid.

Ductility. Ductility is the property of a material by which it can be drawn out by tension to a small section. In a ductile material, therefore, large deformation is possible before the absolute failure or rupture takes place. During ductile extension, a material shows a certain degree of elasticity, together with a considerable degree of plasticity. This property is utilised in wire-drawing.

Brittleness. A material is said to be brittle when it cannot be drawn out by tension to smaller section. Brittleness is, therefore, lack of ductility. In a brittle material, failure takes place with a relatively small deformation. This property is generally considered to be highly undesirable.

Malleability. Malleability is the property by which a material can be uniformly extended in a direction without rupture. A malleable material possesses a high degree of plasticity. The property is utilised in many operations, such as forging, hot rolling, drop stamping etc.

Toughness. Toughness is the property of a material which enables it to absorb energy without fracture. This property is very desirable in components subject to cyclic or shock loading.

Hardness. Hardness is the ability of a material to resist indentation or surface abrasion.

1.2. STRESS

When a body is acted by external force or load, internal resisting forces are set up and it is then said to be in a *state of stress*.

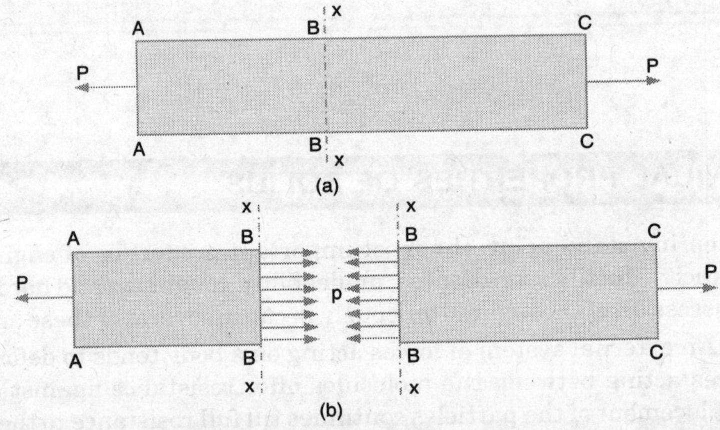

Fig. 1.1

Stress is the resistance offered by the body to deformation. Stress is usually measured by the intensity of force per unit area.

Consider a bar acted upon by a force P along its axis, as shown in Fig. 1.1. If we consider the equilibrium of a section XX, the internal resistance offered by the molecules against the external force will be assumed to be uniformly distributed over the whole area.

Thus, $$p = \frac{P}{A} \qquad \text{...(1.1)}$$

where p = Internal resistance = intensity of stress, or simply, *stress.*

A = Area of cross-section normal to its axis.

Saint Venant's Principle. The principle states that except in the region of the extreme ends of a bar carrying direct loading, the stress distribution over the cross-section is uniform.

Fig. 1.2 (*a*) shows a bar of square-section, subjected to an external force P. The stress distribution at section 1-1, at a distance equal to $b/2$ from the end, is shown in Fig. 1.2 (*b*), where the maximum stress is found to be equal to about 1.387 times the average stress. At a section 2-2, distant b from the end [Fig. 1.2 (*c*)], the maximum stress is very near to the average stress. If a section 3-3 is considered, at a distance greater than b from one end, the maximum stress and average stress are found to be equal for all practical purposes. This illustrates the famous Saint Venant's Principle of rapid dissipation of localised stresses. In practically all the stress analysis in this book, this principle will be followed, and a stress distribution given by Eq. 1.1 and Fig. 1.2 (*d*) will be assumed.

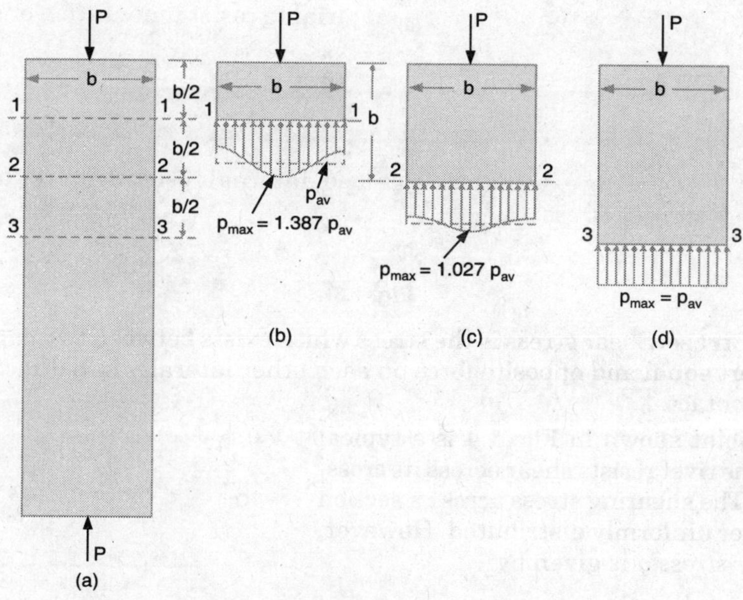

Fig. 1.2

The state of stress is *simple* when the stress acts in one direction only. The state of stress is compound when two or more stresses are acting in different directions.

There are five kinds of stress :

(*a*) Tensile Stress
(*b*) Compressive Stress } Direct stresses
(*c*) Shear Stress or Tangential Stress or Transverse Stress
(*d*) Bending Stress or Longitudinal Stress
(*e*) Twisting or Torsional Stress

(*a*) **Tensile Stress.** Tensile stress is the stress which exists between two parts of a body when each draws the other towards itself. Fig. 1.1 illustrates a tensile stress, where

$$p_{av} = p = \frac{P}{A} \quad \text{...[1.1(a)]}$$

(*b*) **Compressive Stress.** Compressive stress is the stress which exists between two parts of a body when each pushes the other from it. If in Fig. 1.3, a push P is applied to a bar AC, the internal resistance, or the compressive stress p will be given by

$$p_{av} = p = \frac{P}{A} \quad \text{...(1.2)}$$

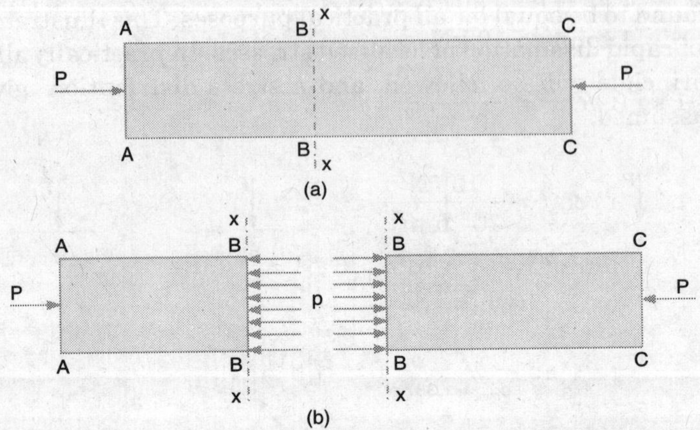

Fig. 1.3

(*c*) **Shear Stress.** Shear stress is the stress which exists between two parts of a body when the two parts exert equal and opposite force on each other laterally in a direction tangential to their surface in contact.

A riveted joint shown in Fig. 1.4 is a typical example, where the rivet resists shear across its cross-sectional area *A*. The shearing stress across a section is practically never uniformly distributed. However, the *average shear stress q* is given by

$$q = \frac{Q}{A} \quad \text{...(1.3)}$$

Bending and torsional stresses will be considered later.

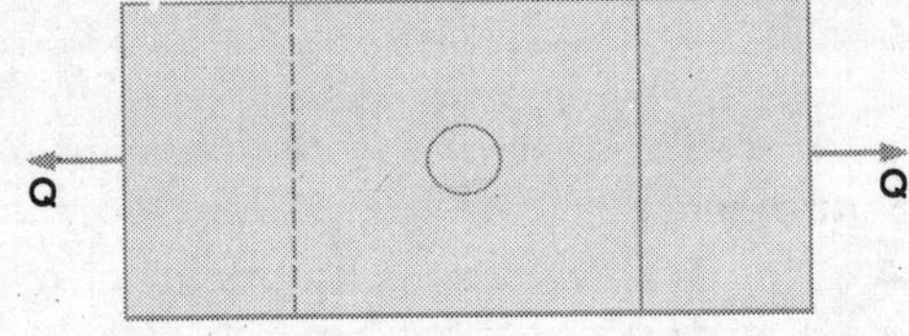

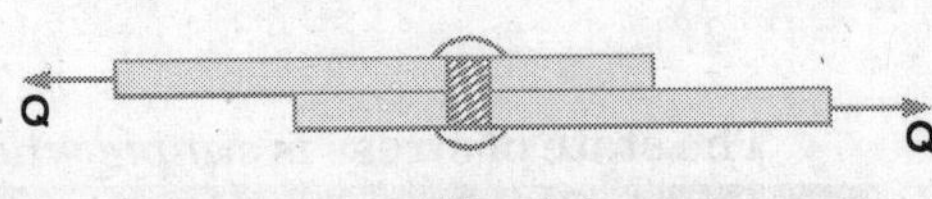

Fig. 1.4

Units of Stress

In M.K.S. system, the unit of force is in the *gravitation unit*, *i.e.*, kilogramme force kg (f), commonly expressed as kg only. When the force is large, it is expressed as tonnes, where 1 t = 1000 kg. The unit of stress is usually expressed as kg/cm^2.

In SI units, the unit of force is *newton* expressed by the symbol N. Since 1 newton is small, force is generally expressed in terms of kilo-newton and mega-newton. Thus,

1 kilo-newton (kN) = 10^3 N

1 mega-newton (MN) = 10^6 N

1 giga-newton (GN) = 10^9 N.

The stress is expressed in units of N/m^2 (Pascal). However, since this is very small, stress is generally expressed in terms of kN/m^2, MN/m^2, GN/m^2 and N/mm^2. A more common form of unit for stress is in terms of N/mm^2 (or MPa).

From the above, it is clear that the SI system has discarded the gravitational unit of force (and stress) followed in MKS system.

The following inter-relationships will be found to be useful

1 kg (f) = g newtons

Taking $g = 9.807$

$$1 \text{ kg (f)} = 9.807 \text{ N} \simeq 10 \text{ N}$$

Also, $1 \text{ tonne} \simeq 10 \text{ kN}$

$$1 \text{ kg (f)/cm}^2 = \frac{10 \text{ N}}{10^{-4} \text{ m}^2} = 10^5 \text{ N/m}^2 = 10^5 \text{ Pascals}$$

or $$1 \text{ kg (f)/cm}^2 = \frac{10^5 \text{ N}}{10^6 \text{ mm}^2} = 0.1 \text{ N/mm}^2$$

$$1 \text{ N/mm}^2 = 10^6 \text{ N/m}^2 = 10^6 \text{ Pa} = 1 \text{ MPa}$$

(where Pa = Pascal)

1.3. STRAIN

Strain is the measure of the deformation caused due to external loading. If a load P is applied to a body, the body deforms. Let Δ be the deformation of the body and L be its original length. The strain e is given by

$$e = \frac{\Delta}{L} \qquad \text{...(1.4)}$$

Thus, *strain is the ratio of deformation to the original length and is a dimensionless number.*

If the stress is tensile, the corresponding strain is known as *tensile strain*.

If the stress is compressive, the corresponding strain is known as *compressive strain*.

If the stress is shear, the corresponding strain is known as *shear strain*.

1.4. HOOKE'S LAW

Hooke's law states that within the elastic limits, stress is proportional to strain. Thus

$$\text{Stress} \propto \text{Strain}$$

or $$\frac{\text{Stress}}{\text{Strain}} = \text{constant}$$

or $$\frac{p}{e} = \text{constant} = E \qquad \text{...(1.5)}$$

The constant, known as the *modulus of elasticity, is defined* as *the intensity of stress which causes unit strain*. If p is the direct stress (tensile or compressive), the constant is called Young's Modulus of Elasticity and is denoted by E.

Thus, $$\frac{p}{e} = E$$

Now $$p = \frac{P}{A} \quad \text{and} \quad e = \frac{\Delta}{L}$$

$$\therefore \quad \frac{PL}{A\Delta} = E \quad \text{or} \quad \Delta = \frac{PL}{AE} \qquad \ldots(1.6)$$

Values of Young's modulus of elasticity (E)

Table 1.1 gives the values of E for some common materials.

TABLE 1.1. VALUES OF MODULUS OF ELASTICITY

S. No.	*Material*	*Modulus of elasticity E (kN/mm² or GPa)*
1.	Aluminium (Pure)	70
2.	Aluminium alloys	70 – 79
3.	Brass	90 – 110
4.	Bronze	96 – 120
5.	Cast iron	83 – 170
6.	Concrete (compression)	18 – 30
7.	Copper (Pure)	110 – 120
8.	Rubber	0.0007 – 0.004
9.	Steel	190 – 210
10.	Wrought iron	190

1.5. TENSILE TEST DIAGRAM (STRESS-STRAIN DIAGRAM)

If a piece of ductile material is subjected to a gradually increasing pull in a tension test machine, it is found that the resultant strain is directly proportional to the corresponding stress upto a certain limit only and beyond that the relation is not linear. In investigating the mechanical properties of materials beyond this limit, the relationship between the strain and the corresponding stress is usually represented graphically by a *tensile test diagram*. During the test, the loads are measured on the main dial of the machine while the elongations are measured with the help of extensometers.

Fig. 1.5 represents a stress-strain diagram in which the stress (calculated on the basis of original area) is plotted against strain. The straight line from O to A represents that part of the test where the extension remains proportional to the load and is termed the *line of proportionality*. Hooke's law is valid for this range. The stress at A is known as the *proportional limit*. If the load is increased beyond this, elongation increases more rapidly and the diagram becomes as represented by A to C. At a point B, between A and C, the elastic limit is reached and after that permanent set or deformation takes place. C is the point of sudden large extension, known as the *yield point*. After the yield point stress is reached, the *ductile extensions* take place, the strains increasing at an accelerating rate as represented by C to D. During the ductile extension, the area of cross-section decreases in practically the same proportion that the length increases. The material becomes *perfectly plastic* in the region CD. For mild steel, the elongation in this region is 10 to 15 times the elongation between O and A. If the load is further increased, the steel begins to *Strain harden*. During strain hardening region, the material appears to regain some of its strength and offers more resistance thus requiring increased tensile load for further deformation, because of changes in its atomic and crystalline structure.

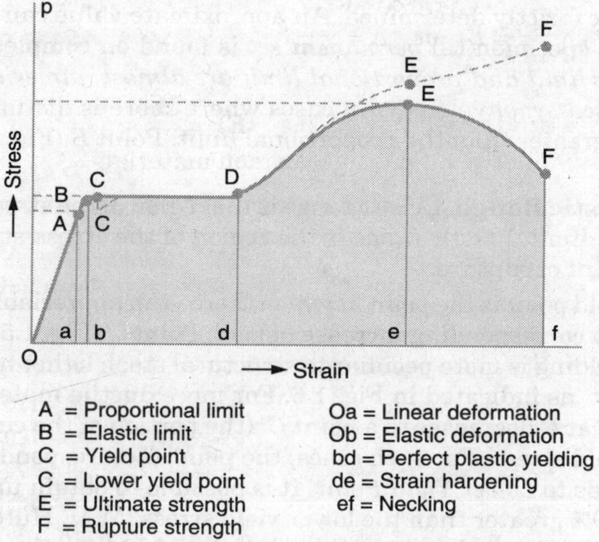

Fig. 1.5. Tensile test diagram (not to scale)

After D, with further increases in load and extension, the point E of the *maximum load or ultimate strength* is reached. Up to the maximum load, the bar extends uniformly over its parallel length but, if straining is continued, a local reduction in cross-section occurs (*i.e.,* formation of waist) and as the load is concentrated at this reduced area, a considerable local extension also takes place, till the failure or rupture takes place at F. If the load is divided by the original area of cross-section, the result is the *nominal stress* which is less at the rupture load than at the maximum load, as indicated by points F and E respectively. However, the breaking load divided by the reduced area of section (*i.e., actual stress intensity*) is greater than at the maximum load. The dashed portion of the curve represents relation between *actual stress intensity* and the strain, from which it is evident that the actual stress at rupture (represented by F') is greater than the stress at maximum load (represented by E').

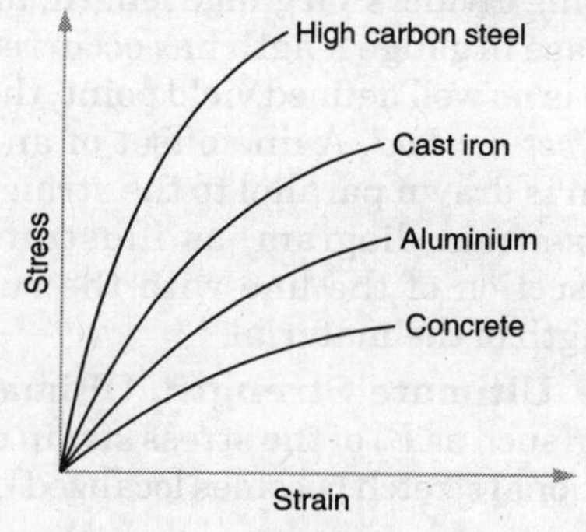

Fig. 1.6

Fig. 1.6 shows the comparative stress-strain curves for different materials.

Gauge Length. Gauge length is the failure length of the parallel portion of the test piece over which extensions are measured. For cylindrical test pieces, the gauge length used for calculation of percentage elongation is a function of the cross-sectional area or the diameter of the test bar.

Limit of Proportionality. It is the stress at which a stress-strain diagram ceases to be a straight line, *i.e.,* that stress at which extensions cease to be proportional to the corresponding stresses. The stress corresponding to point A (Fig. 1.5) represents the proportional limit. Robert Hooke's famous law *"Ut tensio sic vis"*, *i.e., "As strain, so force"*, related strain to stress and did not recognize a limit to this proportionality.

Elastic Limit. Elastic limit represents the maximum stress that may be developed during a simple tension test such that there is no permanent or residual deformation after the removal of the load. This cannot be exactly determined. An approximate value can be obtained by loading and unloading the test specimen till *permanent set* is found on complete removal of load. *For many materials, elastic limit and proportional limit are almost numerically the same, and the terms are sometimes used synonymously*. In cases where there is distinction between the two, elastic limit is always greater than the proportional limit. Point *B* (Fig. 1.5) corresponds to the elastic limit.

Elastic and Plastic Range. Elastic range is the region of the stress-strain curve between the origin to the elastic limit. Plastic range is the region of the stress-strain curve between the elastic limit and the point of rupture.

Yield Point. Yield point is the point at which there is an appreciable elongation or yielding of material without any corresponding increase of load. Point *C* (Fig. 1.5) shows the yield point. The phenomenon of yielding is more peculiar to structural steel. Other materials do not possess well defined yield point, as indicated in Fig. 1.6. For more ductile materials like annealed low carbon steel, the stress at *C* decreases to a point *C′* (the portion of the curve shown by dots) and then finally attains yield stress. For such cases, the point *C* corresponds to *Higher Yield Point* and point *C′* corresponds to *Lower Yield Point*. It is possible to obtain upper yield point in mild steel of the order of 100% greater than the lower yield stress. M.M. Hutchinson obtained upper yield point at 73,000 lb/sq. in. and lower yield point at 37,000 lb/sq. in. from a tensile test on a 0.038 in. diameter mild steel wire after proper annealing and careful preparation of the test piece.

Yield Strength. The yield strength, closely associated with the yield point, is defined as the lowest stress at which extension of the test piece increases without increase in load. It is determined by certain indications given by the testing machine and checked by using dividers on gauge length, to see that a visible permanent increase in gauge length has occurred. For brittle material, where there is no well defined yield point, the yield strength is determined by *offset method*. A line offset of an arbitrary amount of 0.2% of strain is drawn parallel to the straight line portion of the original stress-strain diagram, as illustrated in Fig. 1.7. The point of intersection of the line with the curve corresponds to the yield strength of the material.

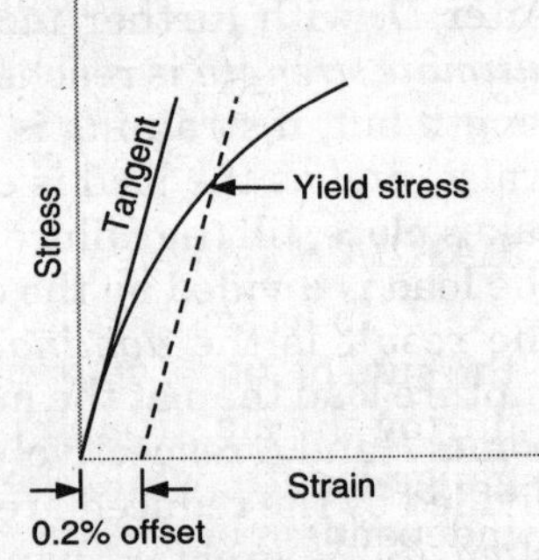

Fig. 1.7

Ultimate Strength. Ultimate stress or ultimate strength corresponds to the highest point (such as *E*) of the stress strain curve. At the point *E* corresponding to the ultimate strength, additional stretch becomes localized (resulting in the formation of *neck*) and then the load decrease.

$$\text{Ultimate strength} = \frac{\text{Maximum load}}{\text{Original area of cross-section}}.$$

Rupture Strength. The rupture strength is the stress corresponding to that at failure. In ductile materials, the rupture strength (represented by point *F*) is lower than the ultimate strength because the rupture strength is calculated by the failure load divided by the original area of cross-section. Actually, due to neck formation, the cross-sectional area of the test piece reduces considerably and the *actual rupture strength* (obtained by dividing the breaking load by the area at the time of rupture) is much higher than the *actual ultimate strength*. In Fig. 1.5, point *E′* corresponds to the *actual ultimate strength* while *F′* corresponds to the actual *rupture strength* which is higher. However, the ultimate strength (corresponding to point *E*) is commonly taken as the maximum stress of the material. For brittle materials, no neck formation takes place and the ultimate strength and the rupture strength are the same.

Proof Stress. Proof stress is the stress necessary to cause a non-proportional or permanent extension equal to a defined percentage of gauge-length. Alternatively, it can be expressed as the stress at which the stress-strain diagram departs by a specified percentage of the gauge length from the produced straight line of proportionality. If the specified percentage is 0.1% of the gauge length, the corresponding proof stress is designated as 0.1% proof stress. If a certain value for proof stress is specified for a material, and, after loading to that stress and then unloading, the permanent extension is less than the specified percentage of the gauge length, then the material is considered in respect of the minimum proof stress requirement.

Percentage Elongation. The percentage elongation is the percentage increase in length of the gauge-length. If L_o is the original gauge length and L_f the final gauge-length (measured after fracture),

$$\text{Percentage elongation} = \frac{L_f - L_o}{L_o} \times 100 \qquad \text{...(1.7)}$$

Working Stress and Factor of Safety. Working stress or safe stress is taken within the elastic range of the material. For brittle materials, it is taken equal to the ultimate strength divided by a suitable factor of safety. However, if there is a definite yield point, it is also taken as equal to yield point divided by a factor of safety. Thus,

$$\text{Working stress} = \frac{\text{Ultimate stress}}{\text{Factor of safety}} = \frac{p_u}{n} \qquad \text{...(1.8)}$$

$$= \frac{\text{Yield stress}}{\text{Factor of safety}} = \frac{p_y}{n'} \qquad \text{...[1.8 (a)]}$$

BARBA'S LAW

In a tensile test, the total elongation of the specimen of a certain gauge-length is equal to the sum of *uniform extension* and a large *local extension* due to waisting. The uniform extension during the elastic and plastic range (upto point E in Fig. 1.5) is found to be proportional to the gauge-length, while the local extension (*i.e.* deformation between point E and F, in Fig. 1.5) is independent of the *gauge-length*. This makes it important to specify the *gauge-length* in a tension test ; otherwise, if the gauge-length is increased, the effect of local extension would decrease the percentage elongation.

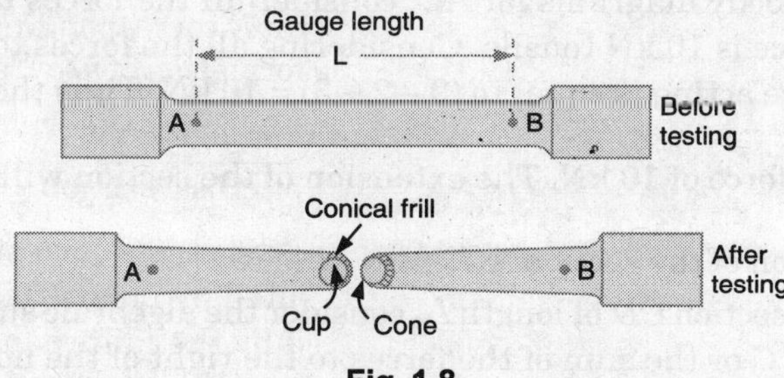

Fig. 1.8

Prof. Unwin has given the following expression for total extension :

$$\Delta_T = aL_o + b\sqrt{A} \qquad \text{...(1.9)}$$

where Δ_T = total extension of the gauge length, before fracture

L_o = gauge length

A = original area of cross-section of the specimen

a, b = constants.

Based on experiments, Barba has given the following law :

Geometrically similar specimens of the same material deform similarly if they are so proportioned that $L_o/\sqrt{A}$ *is constant.*

1.6. PRINCIPLE OF SUPERPOSITION

When a number of forces are acting on a body, the resulting strain, according to the principle of superposition, will be the algebraic sum of the strains caused by the individual forces.

Free Body Diagram. If an elastic body is subjected to a number of direct forces (tensile or compressive) at different sections along the length of the body, the deformation of individual sections can be very easily found if the free body diagrams are drawn for the individual sections. The total deformation of the body will then be equal to the algebraic sum of deformations of the individual sections. Thus, in Fig. 1.9 (*a*), the bar *AB* is acted upon by several forces. The free body diagrams for *AC*, *CD* and *DB* have been drawn in Fig. 1.9 (*b*), (*c*) and (*d*) respectively.

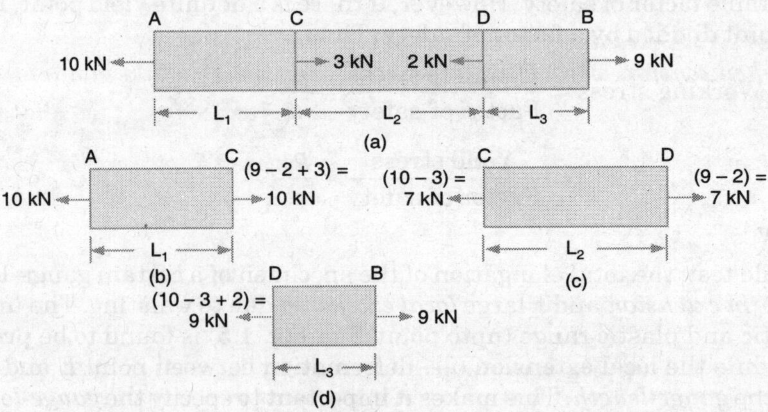

Fig. 1.9

To draw the free body diagrams for *AC*, consider all the forces to the left of the normal section at *A*. The net force is 10 kN tensile. Considering all the forces to the right of the normal section at *C*, the net force acting is equal to (9 – 2 + 3) = 10 kN. Thus the section *AC* of length L_1 is subjected to a tensile force of 10 kN. The extension of the section will be $\Delta_1 = \dfrac{10L_1}{A_1E}$, where A_1 is the area of cross-section of the section *AC*.

Similarly, for the section *CD* of length L_2 consider the algebraic sum of the forces to the left of the normal section at *C* or the sum of the forces to the right of the normal section *D*. The net force on *CD* = (10 – 3) = 7 = (9 – 2) = 7 kN. The extension will be

$$\Delta_2 = \frac{7L_2}{A_2E}$$

Similarly, the section *DB* will be subjected to a tensile force of 9 kN and its extension will be

$$\Delta_3 = \frac{9L_3}{A_3E}.$$

The total extension Δ of the bar AB will be given by

$$\Delta = \Delta_1 + \Delta_2 + \Delta_3.$$

1.7. BARS OF VARYING SECTIONS

If an axial force P (tensile or compressive) is acting on a bar of varying areas of cross-section, the total elongation or contraction of the bar will be equal to the sum of elongation or contraction of each section under the action of axial force P. Thus, referring to Fig. 1.10, the total elongation Δ will be given by $\Delta = \Delta_1 + \Delta_2 + \Delta_3$, where Δ_1, Δ_2 and Δ_3 are the elongations of the three portions.

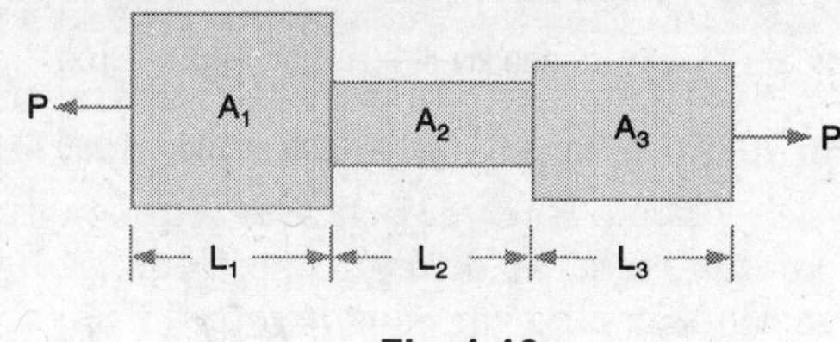

Fig. 1.10

Thus $$\Delta = \frac{P}{E}\left(\frac{L_1}{A_1} + \frac{L_2}{A_2} + \frac{L_3}{A_3}\right) \quad ...(1.10)$$

Example 1.1. *A circular rod of steel 10 mm diameter is tested for tension and it was observed that when tension was 11 kN, the total extension on a 300 mm length was 0.20 mm. Find the value of E.*

Solution.

Stress, $$p = \frac{P}{A} = \frac{11 \times 1000}{\frac{\pi}{4}(10)^2} = 140.06 \text{ N/mm}^2$$

Strain, $$e = \frac{\Delta L}{L} = \frac{0.2}{300} = 6.667 \times 10^{-4}$$

$\therefore$ $$E = \frac{\text{Stress}}{\text{Strain}} = \frac{p}{e} = \frac{140.06}{6.667 \times 10^{-4}} = \mathbf{2.1 \times 10^5 \text{ N/mm}^2 = 210 \text{ kN/mm}^2.}$$

Example 1.2. *The piston of a steam engine is 300 mm in diameter and the piston rod is of 50 mm diameter. The steam pressure is 1 N/mm². Find the stress in the piston rod and the elongation of a length of 800 mm, taking E = 205 kN/mm², when the piston is on the in-stroke.*

Solution. (Fig. 1.11)

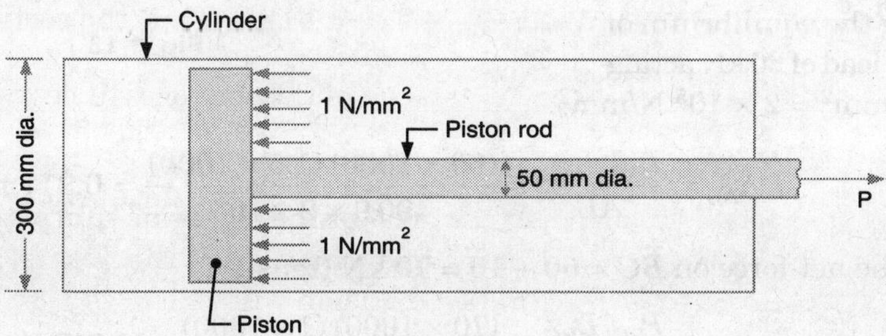

Fig. 1.11

Net area of piston $$= \frac{\pi}{4}(300^2 - 50^2) = 70686 - 1963 = 68723 \text{ mm}^2$$

The load P on piston rod = 68723 × 1 = 68723 N

Area of piston rod $= A = 1963 \text{ mm}^2$

$$\therefore \quad \Delta = \frac{PL}{AE} = \frac{68723 \times 800}{1963 \times 205 \times 10^3} = \mathbf{0.137 \text{ mm.}}$$

Example 1.3. *A rod as shown in Fig. 1.12 is subjected to a pull of 800 kN on the ends. Taking E = 2.05 × 10⁵ N/mm², find the extension of the rod.*

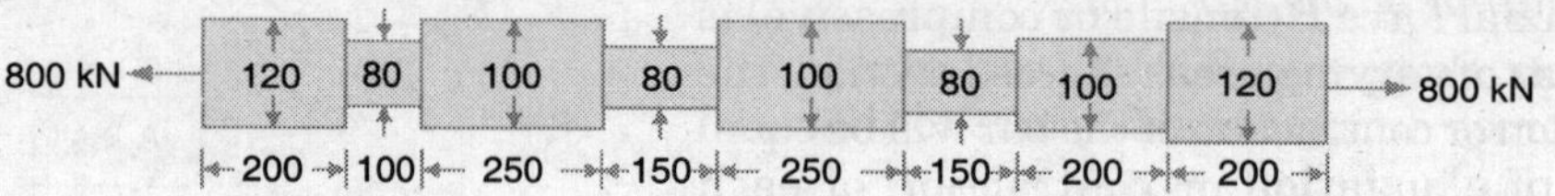

Fig. 1.12

Solution. $$\Delta = \frac{P}{E}\left[\frac{L_1}{A_1} + \frac{L_2}{A_2} + \frac{L_3}{A_3} \ldots\ldots\right]$$

$$A_1 = \frac{\pi}{4}(120)^2 = 11310 \text{ mm}^2 \; ; A_2 = \frac{\pi}{4}(80)^2 = 5027 \text{ mm}^2$$

$$A_3 = \frac{\pi}{4}(100)^2 = 7854 \text{ mm}^2.$$

$$\therefore \quad \Delta = \frac{800 \times 1000}{2.05 \times 10^5}\left[\frac{200+200}{11310} + \frac{(100+150+150)}{5027} + \frac{(250+250+200)}{7854}\right]$$

$$= \frac{8}{2.05}[0.03537 + 0.07957 + 0.08913] = \mathbf{0.796 \text{ mm.}}$$

Example 1.4. *A steel bar of 25 mm diameter is acted upon by forces shown in Fig. 1.13. Determine the total elongation of the bar.*

Take E = 200 kN/mm² (GN/m²).

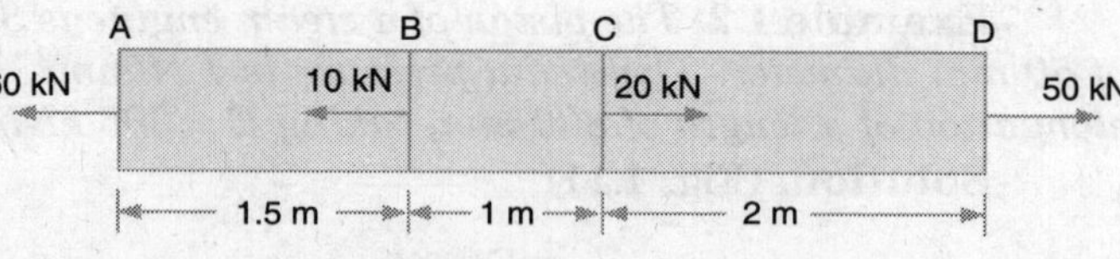

Fig. 1.13

Solution. Area of cross-section

$$= \frac{\pi}{4}(25)^2 = 490.9 \text{ mm}^2$$

Considering the equilibrium of AB, we get a tensile load of 60 kN acting on it. $E = 200 \text{ kN/mm}^2 = 2 \times 10^5 \text{ N/mm}^2$.

$$\therefore \quad \Delta_{AB} = \frac{P_{AB}\, L_{AB}}{AE} = \frac{(60 \times 1000)(1.5 \times 1000)}{490.9 \times 2 \times 10^5} = 0.917 \text{ mm}$$

Similarly, the net force on $BC = 60 + 10 = 70$ kN (tensile)

$$\therefore \quad \Delta_{BC} = \frac{P_{BC}\, L_{BC}}{AE} = \frac{(70 \times 1000)(1 \times 1000)}{490.9 \times 2 \times 10^5} = 0.713 \text{ mm}$$

And the net force on $CD = 50$ kN (tensile)

$$\therefore \quad \Delta_{CD} = \frac{P_{CD}\, L_{CD}}{AE} = \frac{(50 \times 1000)\,(2 \times 1000)}{490.9 \times 2 \times 10^5} = 1.019 \text{ mm}$$

$$\therefore \quad \Delta = \Delta_{AB} + \Delta_{BC} + \Delta_{CD} = 0.917 + 0.713 + 1.019 = \mathbf{2.649\ mm.}$$

Example 1.5. *A steel bar ABCD of varying sections is subjected to the axial forces as shown in Fig. 1.14. Find the value of P necessary for equilibrium. If E = 210 kN/mm², determine the total elongation of the bar.*

Solution. For the equilibrium of the bar, equate total forces acting towards right to the total forces acting to the left.

$\therefore$ 50 + 100 = 80 + P, from which

P = **70 kN**

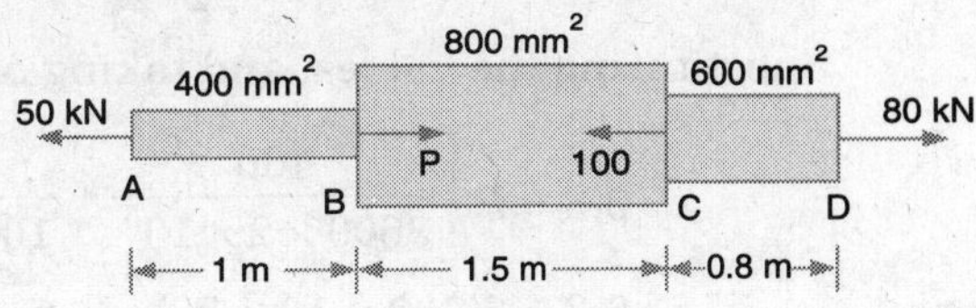

Fig. 1.14

Now consider equilibrium of each of the parts AB, BC and CD, as shown in Fig. 1.15.

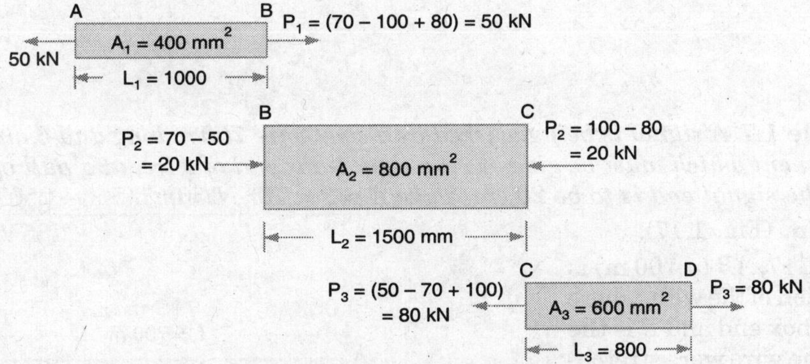

Fig. 1.15. Free body diagrams.

Hence
$$\Delta_{AB} = \frac{P_1 L_1}{A_1 E} = \frac{50 \times 1000 \times 1000}{400 \times 2.1 \times 10^5} = 0.595 \text{ mm (Extension)}$$

$$\Delta_{BC} = \frac{P_2 L_2}{A_2 E} = \frac{20 \times 1000 \times 1500}{800 \times 2.1 \times 10^5} = 0.179 \text{ mm (Compression)}$$

$$\Delta_{CD} = \frac{P_3 L_3}{A_3 E} = \frac{80 \times 1000 \times 800}{600 \times 2.1 \times 10^5} = 0.508 \text{ mm (Extension)}$$

$\therefore$ Total elongation = 0.595 – 0.179 + 0.508 = **0.924 mm.**

Example 1.6. *A member is formed by connecting a steel bar to an aluminium bar as shown in Fig. 1.16. Assuming that the bars are prevented from buckling side ways, calculate the magnitude of axial force P that will cause the total length of member to decrease by 0.3 mm. The values of elastic moduli for steel and aluminium are 2 × 10⁵ N/mm² and 0.7 × 10⁵ N/mm² respectively.*

Solution. Use suffix 1 for steel and 2 for aluminium bar

Area of steel bar = A_1 = 60 × 60 = 3600 mm²

Area of aluminium bar $= A_2 = 100 \times 100 = 10000 \text{ mm}^2$

$L_1 = 400$ mm ; $L_2 = 500$ mm ;

$E_1 = 2 \times 10^5$ N/mm^2 ; $E_2 = 0.7 \times 10^5$ N/mm^2

The total deformation of the member is

$$\Delta = P\left[\frac{L_1}{A_1E_1} + \frac{L_2}{A_2E_2}\right]$$

Substituting the values, and taking $\Delta = 0.3$ mm, we get

$$0.3 = P\left[\frac{400}{3600 \times 2 \times 10^5} + \frac{500}{10000 \times 0.7 \times 10^5}\right]$$

or $\quad 0.3 = 6.269 \times 10^{-7}\, P$

From which $P = 478481$ N $=$ **478.481 kN.**

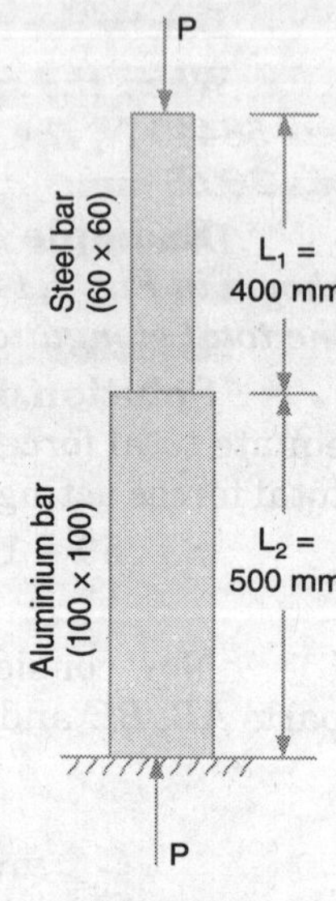

Fig. 1.16

Example 1.7. *A signal is being worked by a steel wire 700 m long and 6 mm in diameter. Find the movement which must be given to the signal-box end of wire, at a pull of 1.6 kN, if the movement at the signal end is to be 20 cm. Take $E = 2 \times 10^5$ N/mm^2.*

Solution. (Fig. 1.17).

In Fig. 1.17, AB (= 700 m) is the initial position of the wire, where A is the signal box end and B is the signal end. If the wire were stretch-proof, the movement of end A at the signal box would have been equal to the movement of 200 mm at the end B. But the wire extends due to a tensile force $P = 1.6$ kN. Let end A be given a movement δ (*i.e.* $AA' = \delta$) to bring end B to B' through 200 mm. If Δ is the extension of the wire under the pull P, we have

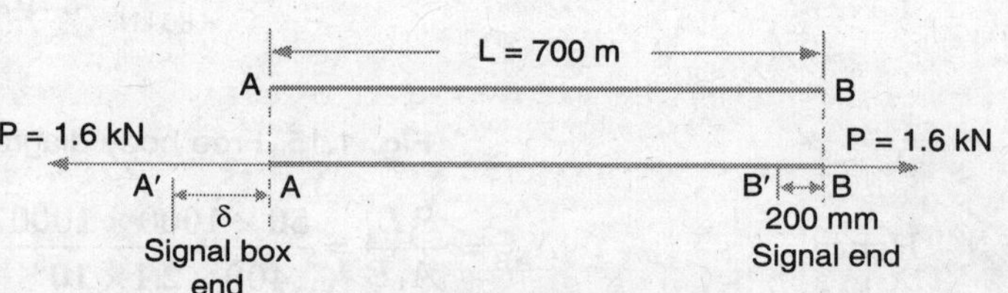

Fig. 1.17

$$\delta = \Delta + 200 \qquad ...(i)$$

Now, from Hooke's law, $\Delta = \dfrac{PL}{AE} = \dfrac{1.6 \times 1000\ (700 \times 1000)}{\dfrac{\pi}{4}(6)^2 \times 2 \times 10^5} = 198$ mm.

Hence $\quad \delta = 198 + 200 =$ **398 mm.**

Example 1.8. *In a tensile test on a specimen of mild steel 12.5 mm diameter, gauge length 200 mm, the following readings were recorded :*

Load (N)	5000	10000	15000	20000	25000	30000	35000	40000
Extension (mm)	0.039	0.0778	0.117	0.156	0.1945	0.235	0.273	0.3125

Determine the value of Young's modulus.

When the specimen was afterwards tested to destruction, the maximum load recorded was 58440 N, the diameter of the neck was 7.37 mm and the length between the gauge marks was 260.5 mm.

Determine the ultimate tensile strength, percentage reduction of area and percentage elongation.

Solution.

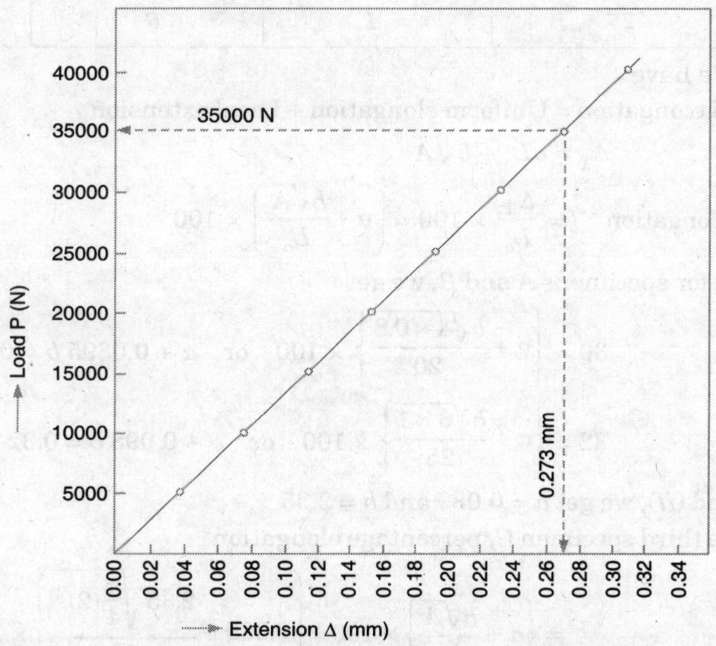

Fig. 1.18

The graph between load and extension is shown in Fig. 1.18. Slope of the graph

$$= \frac{P}{\Delta} = \frac{35000}{0.273} = 128200 \text{ N/mm.}$$

Area of cross-section of specimen $= \frac{\pi}{4}(12.5)^2 = 122.72 \text{ mm}^2$

$$E = \frac{PL}{A\Delta} = \frac{L}{A} \times \frac{P}{\Delta} = \frac{200}{122.72}(128200) = 2.089 \times 10^5 \text{ N/mm}^2$$

$$= \mathbf{208.9 \text{ kN/mm}^2 = 208.9 \text{ GN/m}^2}.$$

Ultimate tensile strength $= \frac{\text{Ultimate (maximum) load}}{\text{Original area of cross-section}} = \frac{58440}{122.72} = \mathbf{476.2 \text{ N/mm}^2}$

Percentage reduction in area $= \frac{122.72 - \frac{1}{4}\pi(7.37)^2}{122.72} \times 100 = \mathbf{65.24\%}.$

Percentage elongation $= \frac{\text{Increase in gauge length}}{\text{Original gauge length}} \times 100 = \frac{260.5 - 200}{200} \times 100 = \mathbf{30.25\%}.$

Example 1.9. *Table below shows the results of tensile tests (to destruction) on two specimens A and B of the same material. Estimate the percentage elongation of a third specimen C of the same material, having a gauge length of 15 cm and a diameter 2 cm.*

Specimen	*Gauge length (cm)*	*Thickness (cm)*	*Width (cm)*	*Elongation (%)*
A	*20*	*0.8*	*4*	*30*
B	*25*	*1*	*6*	*32*

Solution. We have

Total elongation = Uniform elongation + Local extension

$$\Delta_T = aL_o + b\sqrt{A} \quad \text{...(Eq. 1.9)}$$

$$\text{Percentage elongation} = \frac{\Delta_T}{L_o} \times 100 = \left(a + \frac{b\sqrt{A}}{L_o}\right) \times 100$$

Substituting for specimens A and B, we get

$$30 = \left\{a + \frac{b\sqrt{4 \times 0.8}}{20}\right\} \times 100 \quad \text{or} \quad a + 0.0895\,b = 0.3 \quad \text{...}(i)$$

and

$$32 = \left\{a + \frac{b\sqrt{6 \times 1}}{25}\right\} \times 100 \quad \text{or} \quad a + 0.098\,b = 0.32 \quad \text{...}(ii)$$

Solving (*i*) and (*ii*), we get $a = 0.089$ and $b = 2.35$.

Hence for the third specimen C, percentage elongation

$$= \left\{a + \frac{b\sqrt{A}}{L_o}\right\} \times 100 = \left(0.089 + \frac{2.35\sqrt{\frac{\pi}{4}(2)^2}}{15}\right) \times 100 = \mathbf{36.6\%}.$$

1.8. BAR OF TAPERING SECTION

Let us now consider the case of a circular bar of tapering section under an action of axial force P. In Fig. 1.19, the bar AB tapers uniformly from a diameter d_1 to diameter d_2 in length of L. Considering a very short section of length δx at a distance x form A, the diameter of the section will be

$$d_x = d_1 + \frac{d_2 - d_1}{L} x$$

$$\therefore \quad \text{Extension of the short length} = \delta\Delta = \frac{P\,\delta x}{\frac{\pi}{4}\left\{d_1 + \frac{d_2 - d_1}{L} x\right\}^2 E}$$

For the whole length of the rod

$$\Delta = \sum_{x=0}^{x=L} \delta\Delta = \int_0^L \frac{4P\,dx}{\pi(d_1 + kx)^2 E}, \text{ where } k = \frac{d_2 - d_1}{L}.$$

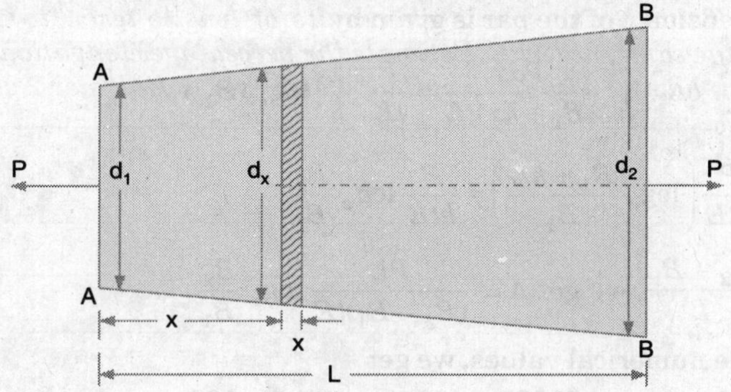

Fig. 1.19

$$\therefore \qquad \Delta = -\frac{4P}{\pi E}\frac{1}{k}\left\{\frac{1}{d_1 + kx}\right\}_0^L = -\frac{4PL}{\pi E(d_2 - d_1)}\left\{\frac{1}{d_1 + d_2 - d_1} - \frac{1}{d_1}\right\}$$

or
$$\Delta = \frac{4PL}{\pi E(d_2 - d_1)}\left\{\frac{1}{d_1} - \frac{1}{d_2}\right\} = \frac{4PL}{\pi E d_1 d_2} \qquad ...(1.11)$$

When $d_2 = d_1 = d$, $\Delta = \frac{4PL}{\pi E d^2} = \frac{PL}{AE}$ which is the same as Eq. 1.6.

Example 1.10. *A round bar of steel tapers uniformly from a diameter of 2.5 cm to 3.5 cm in length of 50 cm. If an axial force of 60000 N is applied at each end, determine the elongation of the bar. Take E = 205 kN/mm².*

Solution. From Eq. 1.11, $\Delta = \frac{4PL}{\pi E d_1 d_2}$

Here $L = 50$ cm $= 500$ mm ; $d_1 = 25$ mm ; $d_2 = 35$ mm

$$\therefore \qquad \Delta = \frac{4 \times 60000 \times 500}{\pi \times 205 \times 10^3 \times 25 \times 35} = \mathbf{0.213\ mm.}$$

Example 1.11. ***Plate of trapezoidal section.*** *A flat steel plate is of trapezoidal form of uniform thickness of 10 mm and tapers uniformly from a width 50 mm to 100 mm in a length of 400 mm. Determine the elongation of the plate under an axial force of 50 kN at each end. Take $E = 2.05 \times 10^5$ N/mm².*

Solution. Consider a small section of length δx at a distance x from the width B_1 (Fig. 1.20).

The width at that section $= B_x$

$$= B_1 + \frac{B_2 - B_1}{L}x = B_1 + kx,$$

where $k = \frac{B_2 - B_1}{L}$

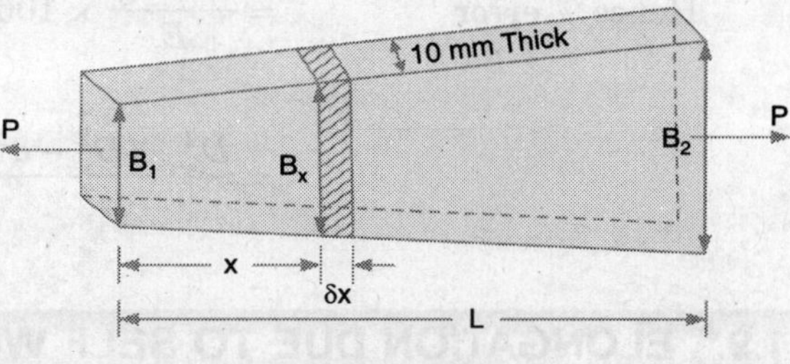

Fig. 1.20

Area $= (B_1 + kx)\, t$

$$\therefore \qquad \delta\Delta = \frac{P(\delta x)}{(B_1 + kx)tE}, \text{ where } \delta\Delta \text{ is the extension of a short length } \delta x.$$

Hence total extension Δ of the bar is given by

$$\Delta = \sum_{x=0}^{x=L} \delta\Delta = \int_0^L \frac{P dx}{(B_1 + kx)tE} = \frac{P}{tE} \cdot \frac{1}{k} \left[\log_e (B_1 + kx)\right]_0^L$$

or $$\Delta = \frac{P}{ktE}\left(\log_e \frac{B_1 + kL}{B_1}\right) = \frac{P}{ktE} \log_e \frac{B_2}{B_1} \qquad \text{...[1.11(a)]}$$

Putting $k = \dfrac{B_2 - B_1}{L}$, we get $\Delta = \dfrac{PL}{(B_2 - B_1)tE} \log_e \dfrac{B_2}{B_1}$ ...[1.11(b)]

Substituting the numerical values, we get

$$\Delta = \frac{50000 \times 400}{(100 - 50) \times 10 \times 2.05 \times 10^5} \log_e \frac{100}{50} = \mathbf{0.1352\ mm.}$$

Example 1.12. *If a tension test bar is found to taper from* (D + a) *diameter to* (D – a) *diameter, prove that the error involved in using the mean diameter to calculate the Young's modulus is* $\left(\dfrac{10a}{D}\right)^2$ *percent.*

Solution. Larger diameter $d_2 = D + a$; Smaller diameter $d_1 = D - a$

Let E = actual Young's modulus ; E_0 = modulus computed by mean dia.

Δ = measured extension of the bar

From Eq. 1.11, $$E = \frac{4PL}{\pi d_1 d_2 \Delta} = \frac{4PL}{\pi (D - a)(D + a)\Delta}$$

or $$E = \frac{4PL}{\pi\Delta(D^2 - a^2)} = \frac{K}{D^2 - a^2} \quad \left(\text{where } K = \frac{4PL}{\pi\Delta}\right) \qquad \text{...(1)}$$

Also, on the basis of mean diameter, we have

$$\Delta = \frac{PL}{AE_o} = \frac{PL}{\frac{\pi}{4} D^2 . E_o}$$

From which $$E_o = \frac{4PL}{\pi\Delta D^2} = \frac{K}{D^2} \qquad \text{...(2)}$$

Hence % error $$= \frac{E - E_o}{E} \times 100 = \frac{\dfrac{K}{D^2 - a^2} - \dfrac{K}{D^2}}{\dfrac{K}{D^2 - a^2}} \times 100$$

$$= \frac{D^2 - (D^2 - a^2)}{D^2} \times 100 = \frac{100a^2}{D^2} = \left(\frac{10a}{D}\right)^2 \text{ percent.}$$

1.9. ELONGATION DUE TO SELF WEIGHT

(*a*) **Bar of uniform section.** Consider a bar *AB* hanging freely under its own weight (Fig. 1.21). Considering a small section δ*x*, its deformation δΔ will be given by

$$\delta\Delta = \frac{W_x \delta x}{AE}$$

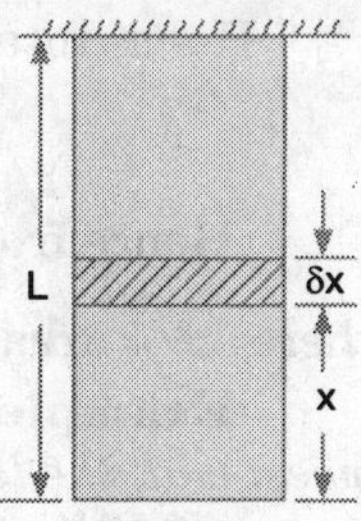

Fig. 1.21

where W_x = weight of portion below the section = $Ax\lambda$, where λ is the *specific weight* or *unit weight* of the material. In M.K.S. units, λ is expressed in kg (f)/m^3, while in SI units, $\lambda = \rho.g$ where ρ is the *specific mass* or *unit mass* or *density* of the material (expressed in kg/m^3) and g is the acceleration due to gravity which may be taken as 9.81 m/ sec^2. In SI units, λ is expressed in N/cm^3 or kN/m^3, since it is gravitational weight developed by the body per unit volume.

$$\therefore \quad \delta\Delta = \frac{(Ax\lambda)\delta x}{AE} = \frac{x\lambda\delta x}{E}$$

$$\therefore \quad \text{Total deformation of the rod } \Delta = \sum_{x=0}^{x=L} \delta\Delta = \int_0^L \frac{x\lambda}{E}\,dx$$

or
$$\Delta = \left[\frac{x^2\lambda}{2E}\right]_0^L = \frac{\lambda L^2}{2E} = \frac{\rho g L^2}{2E} \quad \text{...(1.12)}$$

If W = total weight of the bar, we have $\lambda = \dfrac{W}{AL}$

$$\therefore \quad \Delta = \frac{WL}{2AE} \quad \text{...[1.12(a)]}$$

Thus, the deformation of the bar under its own weight is equal to half the deformation if the body is subjected to the direct load equal to the weight of body.

(*b*) **Bar of tapering section** (Fig. 1.22).

Consider an elementary section XX of length δx, at a distance x from the lower end. The force acting on it will be equal to the weight of the portion below it.

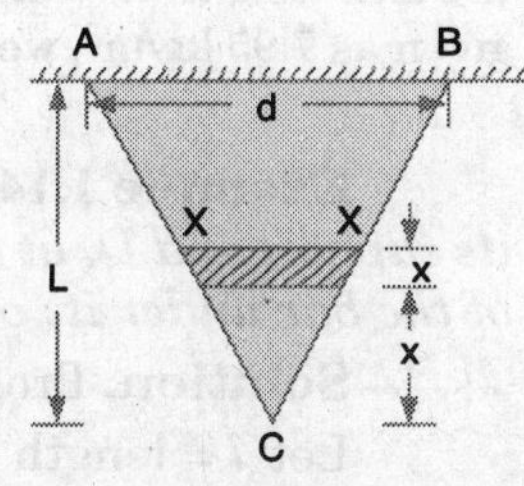

Fig. 1.22

If A_x is the area of cross-section at XX, its extension is given by

$$\delta\Delta = \frac{W_x \delta x}{A_x E}$$

where W_x = weight of portion below the section = $\dfrac{1}{3} A_x \,.\, x \,.\, \lambda$

where λ = specific weight or unit weight of the material.

$$\therefore \quad \delta\Delta = \frac{1}{3}\frac{A_x \,.\, x \,.\, \lambda \,.\, \delta x}{A_x E} = \frac{x\lambda\delta x}{3E}$$

Hence total extension of the whole bar is given by

$$\Delta = \sum_{x=0}^{x=L} \delta\Delta = \int_0^L \frac{x\lambda\delta x}{3E}$$

$$\therefore \quad \Delta = \frac{\lambda L^2}{6E} = \frac{\rho g L^2}{6E} \quad \text{...(1.13)}$$

where ρ is the density of the material.

Let W = total weight of the bar = $\dfrac{1}{3}\left(\dfrac{\pi}{4} d^2\right) . L . \lambda$

where d = diameter of the bar at AB.

$$\therefore \quad \lambda = \frac{12W}{\pi d^2 L}$$

Hence from Eq. 1.13, $\Delta = \dfrac{2WL}{\pi d^2 E} = \dfrac{WL}{2AE}$ ...[1.13 (*a*)]

where A = area of section at support AB.

Example 1.13. *A steel wire of 6 mm diameter is used for lifting a load of 1.5 kN at its lowest end, the length of the wire hanging vertically being 160 metres. Taking the unit weight of steel = 78 kN/m³ and E = 2 × 10⁵ N/mm², calculate the total elongation of the wire.*

Solution. Area of cross-section of wire $= \dfrac{\pi}{4}(6)^2 = 28.27 \text{ mm}^2$

Extension of wire due to load $= \Delta_1 = \dfrac{PL}{AE} = \dfrac{(1.5 \times 1000)(160 \times 1000)}{28.27 \times 2 \times 10^5} = 42.45$ mm

Extension of wire due to self weight $= \Delta_2 = \dfrac{\lambda L^2}{2E}$

where λ = gravitational force developed by the rod per unit volume

$= \text{specific weight} = 78 \text{ kN/m}^3 = \dfrac{78 \times 1000}{(1000)^3} \text{ N/mm}^3 = 78 \times 10^{-6} \text{ N/mm}^3$

$$\therefore \quad \Delta_2 = \frac{(78 \times 10^{-6})(160 \times 1000)^2}{2 \times 2 \times 10^5} = 4.99 \text{ mm}$$

$\therefore$ Total elongation = 42.45 + 4.99 = **47.44 mm.**

Note. In the above example, the *unit weight* λ is directly given. If, instead, the *density* ρ were given, the *unit weight* (λ) will be computed by multiplying the unit mass (ρ) by g. Thus, if the *density* of steel is given as 7.95 kg/m³, we have

$$\rho = 7.95 \text{ kg/m}^3 \quad \text{and} \quad \lambda = \rho . g = 7.95 \times 9.81 \simeq 78 \text{ kN/m}^3.$$

Example 1.14. *A solid bar is in the form of a truncated cone having D_1 as the diameter at its support and D_2 at its free end, and is hanging as shown in Fig. 1.23. Determine the elongation of the bar under its own weight.*

Solution. Prolong AD and BC to meet at O.

Let l = length of the truncated cone $ABCD$

L = length of the cone ABO ;

l' = length of the cone DCO

From Eq. 1.13, extension of the full cone under its own weight is given by

$$\Delta_1 = \frac{\lambda L^2}{6E} \quad ...(1)$$

Similarly, extension of the cone CDO under its own weight is given by

$$\Delta_2 = \frac{\lambda l'^2}{6E} \quad ...(2)$$

Let W' = weight of cone CDO

$$= \frac{1}{3}\left(\frac{\pi}{4} D_2{}^2\right) l' . \lambda = \frac{\pi}{12} D_2{}^2 l' \lambda \quad ...(3)$$

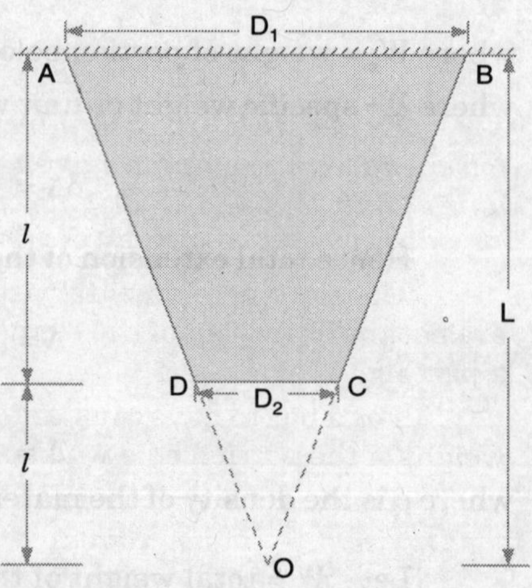

Fig. 1.23

From Eq. 1.11, extension of a tapering bar of length l under an axial load W' is given by,

$$\Delta' = \frac{4W'l}{\pi E D_1 D_2} = \frac{4l}{\pi E D_1 D_2}\left[\frac{\pi}{12} D_2{}^2 l' \lambda\right] = \frac{\lambda}{3}\frac{ll'}{E}\frac{D_2}{D_1} \quad ...(4)$$

Now extension of $ABCD$ = (Extension of ABO) – (Extension of CDO)

– Extension of $ABCD$ due to weight W' of CDO)

$$\therefore \quad \Delta = \Delta_1 - \Delta_2 - \Delta' = \frac{\lambda L^2}{6E} - \frac{\lambda l'^2}{6E} - \frac{\lambda l l'}{3E}\frac{D_2}{D_1} \quad ...(5)$$

In the above expression, let us eliminate L and l', since they are not directly known. From the geometry of the cone,

$$\frac{D_1}{D_2} = \frac{L}{l'} = \frac{l+l'}{l'} = \frac{l}{l'} + 1$$

$$\therefore \quad \frac{l}{l'} = \frac{D_1 - D_2}{D_2} \quad \text{or} \quad l' = \frac{D_2 l}{D_1 - D_2} \quad ...(i)$$

and

$$L = \frac{D_1}{D_2} l' = \frac{D_1}{D_2} \times \frac{D_2 l}{D_1 - D_2} = \frac{D_1 l}{D_1 - D_2} \quad ...(ii)$$

Substituting these values in (5), we get

$$\Delta = \frac{\lambda}{6E}\left(\frac{D_1 l}{D_1 - D_2}\right)^2 - \frac{\lambda}{6E}\left(\frac{D_2 l}{D_1 - D_2}\right)^2 - \frac{\lambda}{3ED_1}\cdot\frac{D_2{}^2 l^2}{D_1 - D_2}$$

$$= \frac{\lambda l^2}{6E}\left\{\frac{D_1{}^3 - D_1 D_2{}^2 - 2D_2{}^2 (D_1 - D_2)}{D_1 (D_1 - D_2)^2}\right\}$$

or

$$\Delta = \frac{\lambda l^2}{6E}\left[\frac{D_1{}^3 + 2D_2{}^3 - 3D_1 D_2{}^2}{D_1 (D_1 - D_2)^2}\right] \quad ...(1.14)$$

which is the required expression.

If $D_2 = 0$, and $D_1 = D$ (*i.e.*, a case of Fig. 1.22), we get

$$\Delta = \frac{\lambda l^2}{6E}\left[\frac{D^3}{D^3}\right] = \frac{\lambda l^2}{6E}$$ which is the same as Eq. 1.13.

Example 1.15. *Determine the shape of the bar hanging freely subjected to a pull W at its lower end, to have uniform strength. The area at the free end is A, the permissible stress is f and unit weight (or specific weight) of the material of the bar is λ.*

Solution. Refer Fig. 1.24.

Consider an elementary section of length δx, at a distance x from the lower end. Let the area at the lower end of the elementary section be equal to A_x and that at its upper end (*i.e.* at δx a part) be $A_x + \delta A_x$.

For a bar to have constant strength, the stress at any section, due to external load and the weight of the portion below it should be constant.

Thus, $f_x = f$ = constant.

$$f = \frac{W}{A} = \frac{W_x}{A_x} = \frac{W_x'}{A_x + \delta A_x} \quad ...(1)$$

where $W_x = W$ + Weight of the bar below the elementary section = $W + W_1$ (say)

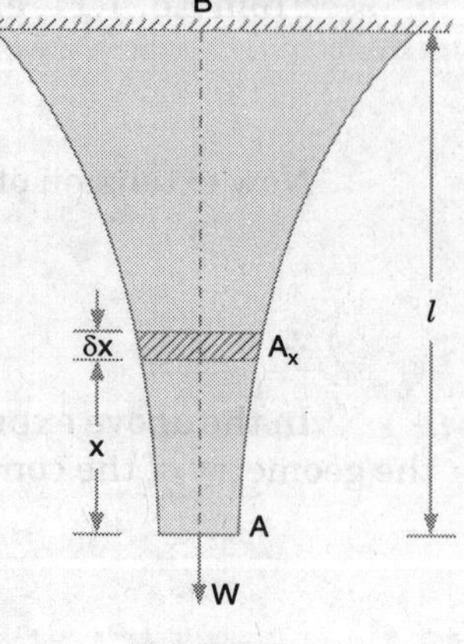

Fig. 1.24

and $W_x' = W_x + \delta W_x$ where δW_x is the weight of the elementary section itself

or $W_x' = W + W_1 + A_x.\lambda.\delta x$

Hence from (1), we get

$$fA = W \qquad ...(2)$$

$$fA_x = W + W_1 \qquad ...(3)$$

and $$f(A_x + \delta A_x) = W + W_1 + A_x.\lambda.\delta x \qquad ...(4)$$

where λ = specific weight (or unit weight) of the material = ρg

ρ = specific mass (or unit mass) of the material

Subtracting (3) from (4), we get

$$f.\delta A_x = A_x.\delta x.\lambda$$

or $$\frac{\delta A_x}{A_x} = \frac{\lambda}{f}.\delta x$$

Integrating, $$\int_A^{A_x} \frac{dA_x}{A_x} = \int_0^x \frac{\lambda}{f}\, dx \quad \text{or} \quad \log_e \frac{A_x}{A} = \frac{\lambda}{f}\, x$$

from which $$\frac{A_x}{A} = e^{\frac{\lambda.x}{f}} \quad \text{or} \quad A_x = Ae^{\frac{\lambda.x}{f}} = Ae^{\frac{\rho g x}{f}} \qquad ...(1.15)$$

The above expression gives the area of section A_x at any distance x from the lower end. Hence knowing A, λ and f, the shape of the bar can be determined from this equation.

Example 1.16. *A pile of uniform section is embeded in soil by a depth H. The pile supports a structural load P at its top, which is transferred to the soil entirely by friction, shown in Fig. 1.25 (a). The variation of friction (f) along the depth of the pile is given by $f_y = ky^2$, where y is the elevation above the bottom of the pile. Determine the total shortening of the pile.*

Solution. Fig. 1.25 (*b*) shows the variation of frictional resistance f with y. In order to determine the value of constant k, let us find the total frictional resistance for the full height and equate it to the external load P (neglecting the self weight of the pile).

$$\therefore \quad \text{Frictional resistance} = \int_0^H ky^2\,.\,dy = \frac{kH^3}{3}$$

$$\therefore \quad P = \frac{kH^3}{3} \quad \text{or} \quad k = \frac{3P}{H^3} \qquad ...(i)$$

Now the compressive force on the pile will also vary with the depth. At any height y above the bottom of the pile, the *total compressive force* will be equal to the *frictional resistance* of clay for the height y.

$$\text{Total compression} \quad P_y = \int_0^y ky^2\,.\,dy = \frac{ky^3}{3} = \frac{3P}{H^3}\,.\,\frac{y^3}{3} = \frac{Py^3}{H^3} \qquad ...(ii)$$

The variation of P_y is shown in Fig. 1.25 (*c*).

The shortening of small length dy of the pile is = $\dfrac{P_y\,.\,dy}{AE}$, assuming that the compressive force is constant for a length dy.

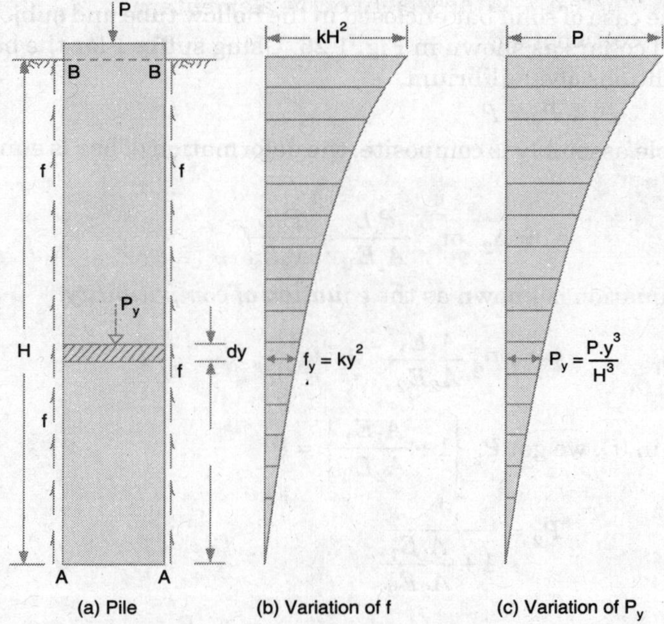

Fig. 1.25

$$\therefore \quad \text{Total shortening of pile} = \sum_{y=0}^{y=H} \frac{P_y \cdot dy}{AE} = \int_0^H \frac{Py^3}{H^3} \cdot \frac{dy}{AE} = \frac{P}{AE\,H^3} \int_0^H y^3 \cdot dy$$

$$= \frac{P}{AE\,H^3}\left(\frac{H^4}{4}\right) = \frac{PH}{4AE}.$$

Note. If the pile were a *point bearing pile* (*i.e.* transferring the load to the solid strata at its end), the shortening of the pile would have been equal to $\frac{PH}{AE}$.

1.10. COMPOSITE SECTIONS

When a bar consists of two different materials, it is said to be composite. Since there are two unknowns, two equations will be required. The conditions of equilibrium will provide one equation for the stresses in the individual sections. The other equation can be obtained from the consideration of the deformation of the whole structure.

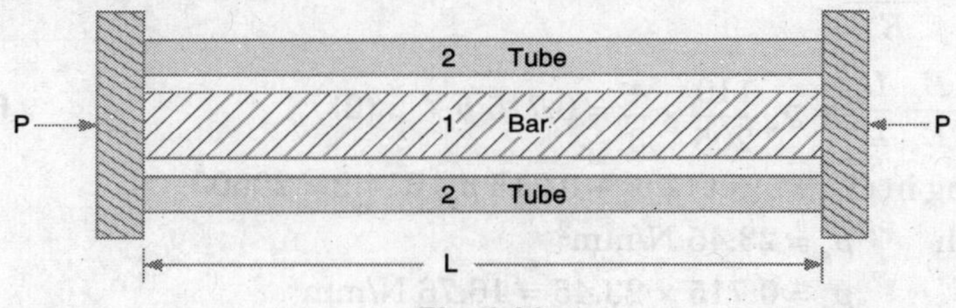

Fig. 1.26

Let us take the case of solid bar enclosed in the hollow tube and subjected to a compressive force P through rigid collars as shown in Fig. 1.26. Using suffix 1 for the bar and 2 for the tube, we get from the conditions of equilibrium,

$$P_1 + P_2 = P \qquad ..(i)$$

Since the whole assembly is composite, the deformation of bar is equal to the deformation of the tube. Thus,

$$\Delta_1 = \Delta_2 \text{ or } \frac{P_1L}{A_1E_1} = \frac{P_2L}{A_2E_2} \qquad ...(ii)$$

The second equation is known as the *equation of compatibility.*

Now from (*ii*), $$P_1 = P_2\frac{A_1E_1}{A_2E_2} \qquad ...(iii)$$

Substituting in (*i*), we get $P_2\left\{1+\dfrac{A_1E_1}{A_2E_2}\right\} = P$

or $$P_2 = \frac{P}{1+\dfrac{A_1E_1}{A_2E_2}} \qquad ...(1.16)$$

Substituting the value of P_2 in (*ii*), we get $P_1 = \dfrac{P}{1+\dfrac{A_2E_2}{A_1E_1}}$ ...[1.16 (*a*)]

Example 1.17. *A load of 30 kN is jointly supported by three equidistant vertical rods in the same vertical plane as shown in Fig. 1.27, the rods being so adjusted as to share to load equally. Calculate the final stresses if a further load of 20 kN is added. Take E_c = 110 kN/mm² and E_s = 205 kN/mm². Neglect any possibility of lateral buckling.*

Solution. Area of each rod

$$= A = \frac{\pi}{4}(20)^2 = 314.2 \text{ mm}^2$$

Initial stress in each rod

$$= \frac{30\times1000}{3\times314.2} = 31.83 \text{ N/mm}^2$$

Due to the addition of 20 kN load, let p_s be the stress in steel and p_c that in copper.

We have, thus, $(2p_s + p_c)A = 20000$ ...(1)

Also, deformation of steel = deformation of copper

$$\therefore \quad \frac{p_cL_c}{E_c} = \frac{p_sL_s}{E_s}$$

or $$p_c = p_s\frac{E_c}{E_s}\frac{L_s}{L_c} = p_s\frac{110}{205}\times\frac{4}{3} = 0.715\,p_s \qquad ...(2)$$

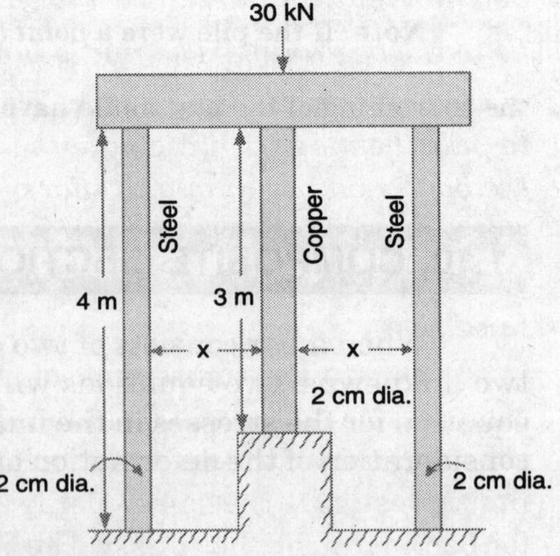

Fig. 1.27

Substituting in (1), we get $(2\,p_s + 0.715\,p_s)\,314.2 = 20000$

From which $p_s = 23.45$ N/mm²

Hence $p_c = 0.715 \times 23.45 = 16.76$ N/mm².

Thus the final stresses are :

(1) In copper rod, $p = 31.83 + 16.76 =$ **48.59 N/mm²**

(2) In steel rod, $p = 31.83 + 23.45 =$ **55.28 N/mm²**.

Example 1.18. *A copper rod, 25 mm in diameter is encased in steel tube 30 mm internal diameter and 35 mm external diameter. The ends are rigidly attached. The composite bar is 500 mm long and is subjected to an axial pull of 30 kN. Find the stresses induced in the rod and the tube. Take E for steel = 2 × 10⁵ N/mm² and E for copper as 1 × 10⁵ N/mm².*

Solution. (Fig. 1.28)

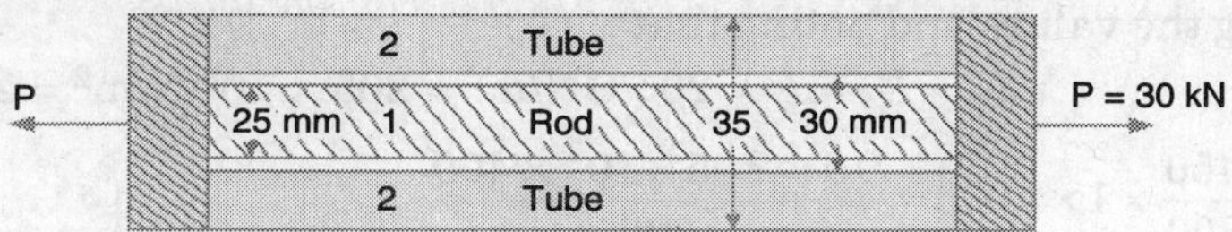

Fig. 1.28

Let us use suffix 1 for rod (copper) and 2 for tube (steel)

$$A_1 = \frac{\pi}{4}(25)^2 = 490.9 \text{ mm}^2 \; ; A_2 = \frac{\pi}{4}[(35)^2 - (30)^2] = 255.3 \text{ mm}^2$$

Now $\qquad p_1A_1 + p_2A_2 = P = 30000 \text{ N}$ ...(1)

Also, $\qquad \dfrac{p_1L}{E_1} = \dfrac{p_2L}{E_2} \quad \text{or} \quad p_1 = p_2\dfrac{E_1}{E_2} = \dfrac{1\times 10^5}{2\times 10^5}p_2 = 0.5p_2$ ...(2)

Hence from (1), $\quad 0.5p_2(490.9) + p_2(255.3) = 30000$

From which $\quad p_2 =$ **59.92 N/mm²** and hence $p_1 = 0.5p_2 =$ **29.96 N/mm²**.

Example 1.19. *Two steel plugs fit freely into the ends of steel tubular distance piece 300 mm long and are drawn together by a steel bolt (450 mm long) and nut, the nut being tight fit in the beginning. The nut is further tightened by $\frac{1}{4}$ turn to draw the piece together, the pitch of the bolt thread being 2 mm. The pieces A and B are then subjected to forces of 60 kN tending to pull them apart. Calculate the stresses in the bolt and the tube. The area of cross-section of the bolt is 600 sq. mm and that of the tube is 400 sq. mm. E for steel 205 GN/m².*

Solution. (Fig. 1.29).

(*a*) ***When there is no external load :***

Due to the tightening of the nut, compressive stress will be induced in the tube and tensile stress in the bolt. Let us first investigate the same. Using suffix 1 for bolt and 2 for the tube, we have, from statics

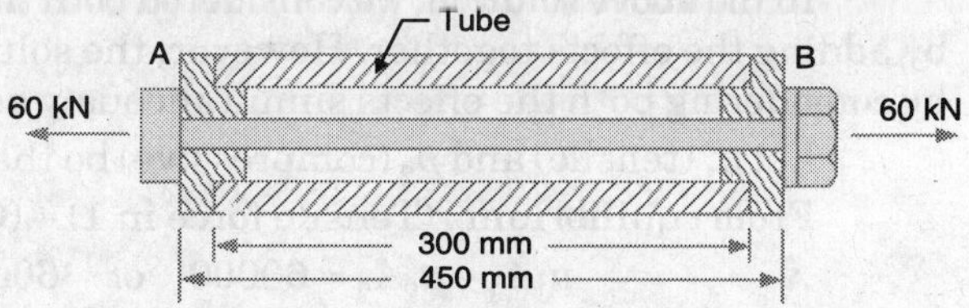

Fig. 1.29

$$p_1A_1 = p_2A_2 \qquad ...(i)$$

Since the final length on the bolt and tube is the same,

Shortening of tube + Extension of bolt = Movement of nut.

or $$\frac{p_2L_2}{E_2}+\frac{p_1L_1}{E_1}=\frac{1}{4}\times 2=0.5 \qquad ...(ii)$$

or $$p_2=\left(0.5-\frac{p_1L_1}{E_1}\right)\frac{E_2}{L_2}$$

Substituting these values in (*i*), we get

$$p_1A_1=\left(0.5-\frac{p_1L_1}{E_1}\right)\frac{E_2}{L_2}A_2 \quad \text{or} \quad p_1\left\{A_1+\frac{L_1}{L_2}\frac{E_2}{E_1}A_2\right\}=0.5\frac{E_2A_2}{L_2}$$

Substituting the values and noting that

$$E_2=E_1=E=205\text{ GN/m}^2=205\times 10^9\text{ N/m}^2=2.05\times 10^5\text{ N/mm}^2$$

$$p_1\left(600+\frac{450}{300}\times 1\times 400\right)=\frac{0.5\times 2.05\times 10^5\times 400}{300}$$

From which $p_1 = 113.9$ N/mm^2 (tensile)

and hence $$p_2=p_1\cdot\frac{A_1}{A_2}=113.9\times\frac{600}{400}=170.8\text{ N/mm}^2\text{ (compressive).}$$

(*b*) ***Considering the effect of external load :***

Let p_1' and p_2' be the additional stresses (tensile) due to external load.

$$\therefore \quad p_1'A_1+p_2'A_2=60000 \qquad ...(iii)$$

and $$\frac{p_1'L_1}{E_1}=\frac{p_2'L_2}{E_2} \text{ (compatibility equation)}$$

or $$p_2'=p_1'\frac{L_1}{L_2}\cdot\frac{E_2}{E_1}=p_1'\frac{450}{300}\times 1=1.5\,p_1' \qquad ...(iv)$$

Substituting in (*iii*), we get

$$600p_1'+(1.5\,p_1')\,400=60000$$

From which $p_1' = 50$ N/mm^2 (tensile) and $p_2' = 1.5 \times 50 = 75$ N/mm^2 (tensile)

Hence total stress in bolt = 113.9 + 50 = **163.9 N/mm^2 (tensile)**

Total stress in tube = 170.8 – 75 = **95.8 N/mm^2 (compressive).**

Alternative Solution

In the above solution, we considered both the actions separately, and then found the stresses by adding the effects together. However, the solution can also be obtained directly, and *in short*, by considering both the effects simultaneously as under.

Let p_1 (tensile) and p_2 (compressive) be the *final* stresses in the bolt and tube respectively.

From equilibrium : (Tensile force in 1) – (Comp. force in 2) = External tensile force

$$\therefore \quad p_1A_1-p_2A_2=60000 \quad \text{or} \quad 600\,p_1-400\,p_2=60000$$

or $$p_1-\frac{2}{3}\,p_2=100 \qquad ...(1)$$

From compatibility :

(Total extension of bolt) + (Total compression of tube) = movement of nut

or $$\Delta_1+\Delta_2=\text{movement of nut}$$

$$\therefore \quad \frac{p_1l_1}{E_1}+\frac{p_2l_2}{E_2}=\frac{1}{4}\times 2 \quad \text{or} \quad p_1\frac{450}{2.05\times 10^5}+p_2\frac{300}{2.05\times 10^5}=0.5$$

$$\therefore \qquad p_1 + \frac{2}{3} p_2 = \frac{0.5 \times 2.05 \times 10^5}{450} = 227.78 \qquad ...(2)$$

Adding (1) and (2), we get

$$p_1 = \mathbf{163.9\ N/mm^2\ (tensile)}$$

Hence $\quad p_2 = \frac{3}{2}(p_1 - 100) = \frac{3}{2}(163.9 - 100) = \mathbf{95.8\ N/mm^2\ (compressive).}$

Example 1.20. *A reinforced concrete column 50 cm diameter has four steel rods of 30 mm diameter embedded in it and carries a load of 850 kN. Find the stresses in steel and concrete. Take E for steel = 2.04 × 10⁵ N/mm² and E for concrete = 0.136 × 10⁵ N/mm².*

Find also the adhesive force between the steel and the concrete.

Solution. Area of column $= \frac{\pi}{4}(500)^2 = 196350\ \text{mm}^2$

Area of steel $= A_s = 4\left(\frac{\pi}{4} 30^2\right) = 2827\ \text{mm}^2$

Area of concrete $A_c = 196350 - 2827 = 193523\ \text{mm}^2$

Now $\quad p_s.A_s + p_c.A_c = 850 \times 1000$ N (from equilibrium) ...(1)

and $\quad \frac{p_s}{E_s} = \frac{p_c}{E_c}$ (from compatibility)

or $\quad p_s = p_c \frac{E_s}{E_c} = \frac{2.04}{0.136} p_c = 15 p_c \qquad ...(2)$

Fig. 1.30

Substituting in (1), we get $15p_c(2827) + p_c(193523) = 850000$

From which p_c = **3.603 N/mm²** and hence $p_s = 15 \times 3.603$ = **54.042 N/mm²**

Again, average compressive stress $= \frac{850000}{196350} = 4.329\ \text{N/mm}^2$

Average load carried by concrete $= 4.329 \times 193523 = 837762$ N

Actual load carried by concrete $= 3.603 \times 193523 = 697263$ N

$\therefore$ Adhesive force $= 837762 - 697263 = 140499\ \text{N} \simeq$ **140.5 kN.**

Example 1.21. *A rigid bar AC is supported by three rods in the same vertical plane and equidistant. The outer rods are of brass and of length 40 cm and diameter 2 cm. The central rod is of steel of 60 cm length and of 2.5 cm diameter. Calculate the forces in the bars due to an applied force P, if the bar AC remains horizontal after the load has been applied. Take $E_s/E_b = 2$.*

Solution. (Refer Fig. 1.31).

Let suffix 1 be used for brass and 2 for steel.

$$A_1 = \frac{\pi}{4}(2)^2 = 3.14\ \text{cm}^2\ ;$$

$$A_2 = \frac{\pi}{4}(2.5)^2 = 4.9\ \text{cm}^2$$

From statical equilibrium :

$$2P_1 + P_2 = P \qquad ...(i)$$

From compatibility equation :

$$\Delta_1 = \Delta_2$$

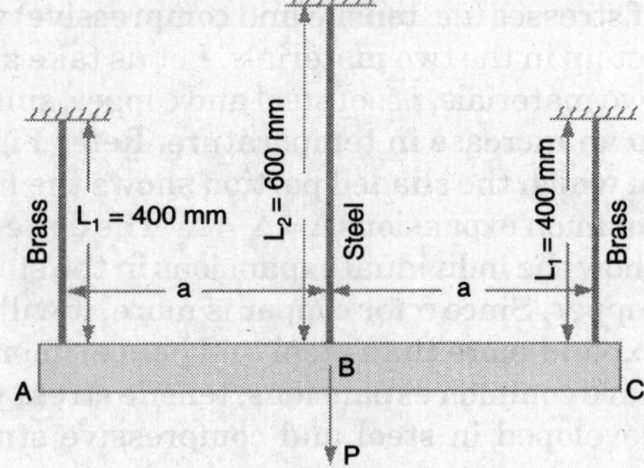

Fig. 1.31

or $$\frac{P_1L_1}{A_1E_1} = \frac{P_2L_2}{A_2E_2}$$

or $$P_2 = P_1\,\frac{A_2E_2L_1}{A_1E_1L_2} = P_1\,\frac{4.9}{3.14} \times \frac{2}{1} \times \frac{40}{60}$$

$\therefore$ $P_2 = 2.08\ P_1$ ...(*ii*)

Substituting in (*i*), we get $2P_1 + 2.08\ P_1 = P$

or $P_1 = \frac{P}{4.08}$ = **0.245 P**

Hence $P_2 = 2.08(0.245P)$ = **0.51 P.**

1.11. TEMPERATURE STRESSES

When a material undergoes change in temperature, its length is varied and if the material is free is do so, no stresses are developed. If the material is constrained so that no change in length is permissible, stresses are developed in the material. The stress so developed is known as *temperature stress* and may be tensile or compressive depending upon whether the contraction is checked or extension is checked.

Let us consider a bar of length L, snugly fitting between two walls or supports. If the temperature is increased through $t°$, the bar will be increased in length by an amount $\Delta = L\alpha t$, where α is the co-efficient of thermal expansion. Due to the external conditions, the bar is not free to expand by the amount Δ and, therefore, from Hooke's law,

$$\Delta = \frac{pL}{E}, \text{ where } p \text{ is the temperature stress}$$

or $$p = \frac{\Delta E}{L} = L\alpha t\,\frac{E}{L} = \alpha\, tE \quad ...(1.17)$$

Similarly, if the temperature falls, the tensile stress is αtE.

Again, let us consider the case when the ends yield by an amount a. In this case, the total amount of expansion checked = $\Delta - a$.

Therefore, $$p = (\Delta - a)\,\frac{E}{L} = (L\alpha t - a)\,\frac{E}{L} \quad ...(1.18)$$

Composite Section

If the bar consists of two materials and is subjected to temperature change, opposite kinds of stresses (*i.e.* tensile and compressive) will be set up in the two materials. Let us take a bar of two materials, *i.e.* of steel and copper, subjected to an increase in temperature. Refer Fig. 1.32 in which the shaded portion shows the final or common expansions $\Delta = \Delta_c = \Delta_s$. The dotted lines show the individual expansions in the steel and copper. Since α for copper is more, it will try to expand more than steel and hence, in order to have common expansions, tensile stress will be developed in steel and compressive stress in copper.

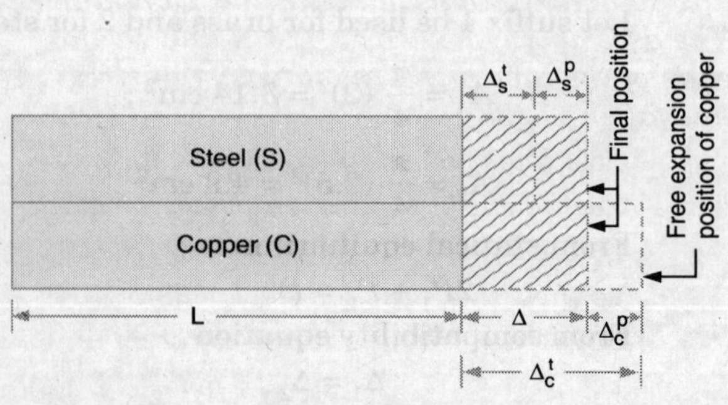

Fig. 1.32

From the stress equation, we get

Tensile force in steel = Compressive force in copper

or $$P_s = P_c = P \quad ...(i)$$

From the compatibility equation, Extension of steel = Extension in copper

$$\Delta_s = \Delta_c = \Delta \text{ (say)} \quad ...(ii)$$

Also from Fig. 1.32,

$$\Delta_s = \Delta_s^t + \Delta_s^p \text{ and } \Delta_c = \Delta_c^t - \Delta_c^p \quad ...(iii)$$

where $\Delta_s = \Delta_c = \Delta$ = Final expansion

Δ_s^t = Free expansion of steel due to temperature

Δ_c^t = Free expansion of copper due to temperature

Δ_s^p = Expansion of steel due to temperature stress p

Δ_c^p = Compression of copper due to temperature stress p.

Substituting these values in (*ii*), we get

$$\Delta_s^t + \Delta_s^p = \Delta_c^t - \Delta_c^p \quad \text{or} \quad \Delta_s^p + \Delta_c^p = \Delta_c^t - \Delta_s^t$$

or $$\Delta_s^p + \Delta_c^p = L.t(\alpha_c - \alpha_s) \quad ...(1.19)$$

or $$PL\left(\frac{1}{A_s E_s} + \frac{1}{A_c E_c}\right) = Lt(\alpha_c - \alpha_s) \quad ...(iv) ...[1.19\ (a)]$$

or $$P = \frac{t(\alpha_c - \alpha_s)}{\dfrac{1}{A_s E_s} + \dfrac{1}{A_c E_c}} \quad ...[1.19\ (b)]$$

Knowing the value of P, the individual stresses in the material will be $p_s = \dfrac{P}{A_s}$ and $p_c = \dfrac{P}{A_c}$.

Equation (*iv*) can also be written in the form,

$$\frac{P}{A_s E_s} + \frac{P}{A_c E_c} = t(\alpha_c - \alpha_s)$$

or $$\frac{p_s}{E_s} + \frac{p_c}{E_c} = t(\alpha_c - \alpha_s) \quad ...(1.20)$$

or $$e_s + e_c = (\alpha_c - \alpha_s)\, t \quad ...[1.20\ (a)]$$

which rolates the strains in the two materials.

Example 1.22. *A gun metal rod screwed at the ends passes through a steel tube. The tube has 30 mm external diameter and 25 mm internal diameter. The diameter of the rod is 22 mm. The assembly is heated to 423 K and the nuts on the rod are then screwed tightly home on the ends of the tube. Find the intensity of stress in rod and in tube when the common temperature has fallen to 293 K.*

Coefficient of expansion per K for steel = 12 × 10^{-6}

Coefficient of expansion per K for gun metal = 20 × 10^{-6}

Young's modulus for gun metal = 0.915 × 10^5 *N/mm*^{2} *(91.5 kN/mm*^{2}*)*

Young's modulus for steel = 2.05 × 10^5 *N/mm*^{2} *(205 kN/mm*^{2}*).*

Solution. Using suffix 1 for steel and 2 for gun metal, we get

$$p_1 A_1 = p_2 A_2 \quad ...(i)$$

$$\Delta_1 = \Delta_2 = \Delta = \text{net reduction in length.}$$

Since the temperature falls by (423 – 293) = 130 K, gun metal will try to shrink more, while steel will shrink less. Hence compressive stress will be produced in steel and tensile stress in gun metal.

Hence from Eq. 1.20,

$$\frac{p_1}{E_1} + \frac{p_2}{E_2} = (\alpha_2 - \alpha_1)t \quad ...(ii)$$

Now $A_1 = \frac{\pi}{4}(30^2 - 25^2) = 216 \text{ mm}^2 \text{ ; } A_2 = \frac{\pi}{4}(22)^2 = 380 \text{ mm}^2$

Hence from (*i*), $p_2 = p_1 \frac{A_1}{A_2} = p_1 \times \frac{216}{380} = 0.568\, p_1$

Substituting this in (*ii*), we get

$$p_1\left[\frac{1}{E_1} + \frac{0.568}{E_2}\right] = (20 - 12)\, 10^{-6} \times 130$$

or $p_1\left[\frac{1}{2.05} + \frac{0.568}{0.915}\right]\frac{1}{10^5} = 1040 \times 10^{-6}$

or $p_1\,[0.488 + 0.621] = 104$, which gives $p_1 = \frac{104}{1.109} = 93.78 \text{ N/mm}^2$

Hence $p_2 = 0.568 \times 93.78 = 53.27 \text{ N/mm}^2$

Hence intensity of stress in rod = **53.27 N/mm² (tensile)**

Intensity of stress in tube = **93.78 N/mm² (compressive).**

Example 1.23. *A railway is laid so that there is no stress in the rails at 10°C. Calculate (a) the stress in the rails at 60°C if there is no allowance for expansion, (b) the stress in the rails at 60°C if there is an expansion allowance of 10 mm per rail, (c) the expansion allowance if the stress in the rail is to be zero when the temperature is 60°C, (d) the maximum temperature to have no stress in the rails, if the expansion allowance is 13 mm per rails. Take $\alpha = 12 \times 10^{-6}$ per 1°C and $E = 2 \times 10^5$ N/mm². The rails are 30 m long.*

Solution. Rise of temperature = $t = 60 - 10 = 50°\text{C}$

(*a*) Stress = $\alpha t E = 12 \times 10^{-6} \times 50 \times 2 \times 10^5 =$ **120 N/mm².**

(*b*) $\frac{pL}{E} = \Delta = (L\alpha t - 10) = (30000 \times 12 \times 10^{-6} \times 50 - 10) = 18 - 10 = 8 \text{ mm}$

$\therefore \quad p = \frac{2 \times 10^5}{30000} \times 8 =$ **53.3 N/mm².**

(*c*) If the stresses are to be zero, the expansion allowance

$= L\alpha t = 30000 \times 12 \times 10^{-6} \times 50 =$ **18 mm.**

(*d*) If the stresses are to be zero, we have

$$p = \frac{E}{L}(L\alpha t - 13) = 0$$

$\therefore \quad L\alpha t = 13$ or $t = \frac{13}{L\alpha} = \frac{13}{12 \times 10^{-6} \times 30000} = 36°$

Allowable temperature = 10 + 36 = **46°C.**

Example 1.24. *Two parallel walls 8 m apart, are stayed together by a steel rod 25 mm in dia. passing through metal plates and nuts at each end. The nuts are screwed upto the plates while the bar is at a temperature of 423 K. Find the pull exerted by the bar after it has cooled to 288 K (a) if the ends do not yield, (b) if the total yielding at the two ends is 6 mm. Co-efficient of thermal expansion of steel = 12 × 10^{-6} per K and E = 2.05 × 10^5 N/mm².*

Solution.

Fall in temperature $= 423 - 288 = 135$ K.

Area of cross-section of rod, $A = \frac{\pi}{4}(25)^2 = 490.9 \text{ mm}^2$

(*a*) ***If the ends do not yield,***

Stress $p = \alpha t E = 12 \times 10^{-6} \times 135 \times 2.05 \times 10^5 = 332.1 \text{ N/mm}^2$

∴ Pull = 332.1 × 490.9 = 163030 N = **163.03 kN**

(*b*) ***If the walls yield by 6 mm***, we have

$$\frac{pL}{E} = \Delta = (L\alpha t - 6)$$

or
$$p = \frac{E}{L}(L\alpha t - 6) = E\alpha t - \frac{6E}{L} = 332.1 - \frac{6 \times 2.05 \times 10^5}{8 \times 1000} = 178.35 \text{ N/mm}^2$$

∴ Pull exerted on walls = 178.35 × 490.9 = 87550 N = **87.55 kN.**

Example 1.25. *A circular section tapered bar is rigidly fixed at both the ends as shown in Fig. 1.33. If the temperature is raised by 30°C, calculate the maximum stress in the bar. Take E = 2 × 10^5 N/mm² and α = 12 × 10^{-6} per 1°C.*

Solution.

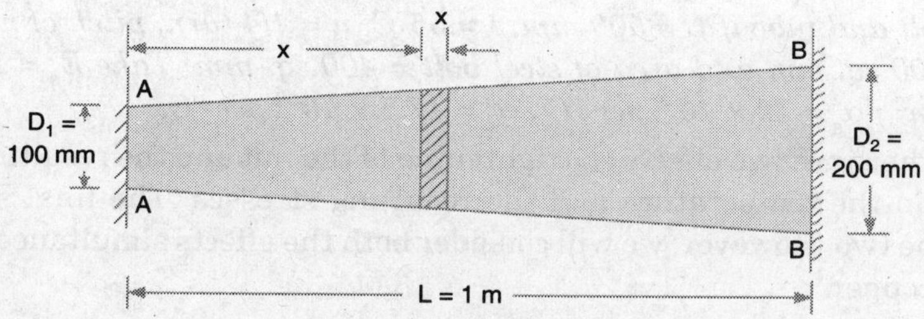

Fig. 1.33

When the temperature is raised by *t*°C, a compressive force *P* is induced. Since the force is the same for all cross-sections, maximum stress will be at section *AA*.

If the bar were free to expand, its expansion would be

$$\Delta = L\alpha t = 1000 \times 12 \times 10^{-6} \times 30 = 0.36 \text{ mm} \quad ...(1)$$

The force induced will be a force *P* which is required to prevent a free expansion of 0.36 mm.

For an element *dx*, the expansion is $\delta\Delta = \frac{Pdx}{A_x E}$

∴ Total extension $\Delta = \int_0^L \frac{P\,dx}{A_x E}$

But $A_x = \frac{\pi}{4}\left(D_1 + \frac{D_2 - D_1}{L}x\right)^2 = \frac{\pi}{4}(D_1 + kx)^2$, where $k = \frac{D_2 - D_1}{L}$.

$$\therefore \quad \Delta = \frac{4P}{\pi E}\int_0^L \frac{dx}{(D_1 + kx)^2} = -\frac{4P}{\pi kE}\left[\frac{1}{D_1 + kx}\right]_0^L$$

$$= -\frac{4PL}{\pi E(D_2 - D_1)}\left[\frac{1}{D_1 + D_2 - D_1} - \frac{1}{D_1}\right] = \frac{4PL}{\pi ED_1D_2} \quad ...(2)$$

Equating (1) and (2), we get

$$L\alpha t = \frac{4PL}{\pi ED_1D_2}$$

From which $$P = \frac{\pi E\alpha t}{4}D_1D_2 \quad ...(1.21)$$

Maximum stress induced is given by

$$p = \left(\frac{\pi}{4}E\alpha tD_1D_2\right) \div \left(\frac{\pi}{4}D_1^2\right) = E\alpha t\,\frac{D_2}{D_1} \quad ...(1.22)$$

when $D_1 = D_2$, $p = E\alpha t$, which is the same as Eq. 1.17.

Substituting the numerical values in Eq. 1.22, we get

$$p = (2 \times 10^5)(12 \times 10^{-6})(30)\left(\frac{200}{100}\right) = \mathbf{144\ N/mm^2.}$$

Example 1.26. *A steel bolt of length L passes through a copper tube of the same length, and the nut at the end is turned up just snug at room temperature. Subsequently, the nut is tightened up by a number of turns and the entire assembly is raised by a temperature t. Calculate the stress in the bolt and tube if L = 500 mm, t = 55°C, n = 1/4 turn, pitch of nut = 2 mm ; area of copper tube = 500 sq. mm and area of steel bolt = 400 sq. mm. Take* $E_s = 2 \times 10^5$ *N/mm²* ; $E_c = 1 \times 10^5$ *N/mm²* ; $\alpha_s = 12 \times 10^{-6}$ *per °C* ; $\alpha_c = 17.5 \times 10^{-6}$ *per °C.*

Solution. There are two effects : (*i*) tightening of the nut and the resulting stresses due to this, and (*ii*) rise in the temperature and the resulting stresses. The final stress will be the algebraic sum of the two. However, we will consider both the effects simultaneously. Use suffix *s* for steel and *c* for copper.

Since there is no external loading,

Tensile force in steel = compressive force in copper

$$\therefore \quad P_s = P_c = P \quad ...(1)$$

Since final length of bolt and tube is the same,

(total extension of bolt) + (total compression of tube) = (Movement of nut)

$$\therefore \quad \Delta_s + \Delta_c = np, \text{ (where } p \text{ is the pitch of bolt threads)} \quad ...(2)$$

Now $$\Delta_s = \frac{PL}{A_sE_s} + \alpha_sLt \text{ (extension) and } \Delta_c = \frac{PL}{A_cE_c} - \alpha_cLt \text{ (compression)}$$

Hence from (2), we get

$$\left(\frac{PL}{A_sE_s} + \alpha_sLt\right) + \left(\frac{PL}{A_cE_c} - \alpha_cLt\right) = np$$

from which $$P\left[\frac{1}{A_sE_s}+\frac{1}{A_cE_c}\right]=(\alpha_c-\alpha_s)\,t+\frac{np}{L} \quad ...(1.23)$$

Substituting the numerical values,

$$P\left[\frac{1}{400\times2\times10^5}+\frac{1}{500\times1\times10^5}\right]=(17.5-12)\,10^{-6}\times55+\frac{1}{4}\times\frac{2}{500}$$

From which $P = 40000$ N

$$\therefore \quad p_s=\frac{40000}{400}=\mathbf{100\ N/mm^2}\text{ and } p_c=\frac{40000}{500}=\mathbf{80\ N/mm^2}.$$

Example 1.27. *A composite bar made up of aluminium and steel is held between two supports as shown in Fig. 1.34. The bars are stress free at a temperature of 40°C. What will be the stresses in the two bars when the temperature drops to 20°C, if (a) the supports are unyielding, (b) the supports come nearer to each other by 0.1 mm. Take E for aluminium as 0.7×10^5 N/mm² and that for steel as 2.1×10^5 N/mm². The coefficient of thermal expansion for aluminium and steel are 23.4×10^{-6} per °C and 11.7×10^{-6} per °C respectively. The cross-sectional area of steel bar is 2 cm² and that of aluminium bar is 3 cm².*

Solution.

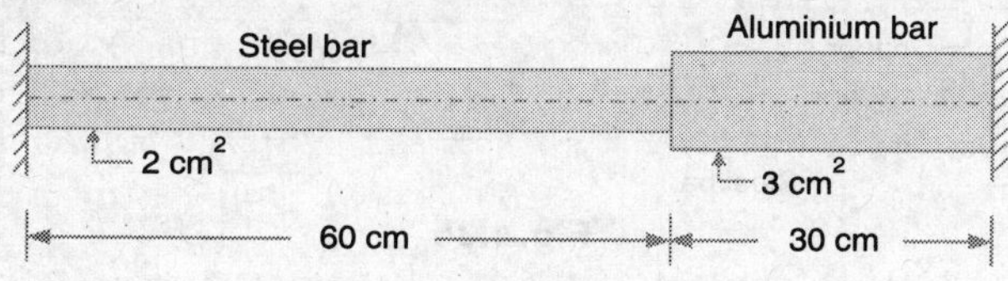

Fig. 1.34

Let us use suffix *s* for steel and *a* for aluminium bar.

Free contraction $\Delta = L_s\alpha_s t + L_a\alpha_a t$

$$= 600 \times 11.7 \times 10^{-6} \times (40-20) + 300 \times 23.4 \times 10^{-6}\,(40-20) = 0.2808 \text{ mm.}$$

Since the free contraction is checked, either fully or partly, tensile stresses will be set up in both steel as well as aluminium bars.

Since the force in both aluminium and steel bars is the same,

$$A_s.p_s = A_a.p_a$$

$$\therefore \quad 2p_s = 3p_a \quad\text{or}\quad p_s = 1.5p_a \quad ...(1)$$

$$\Delta_s^p = \text{elongation of steel bar due to temperature stress} = \frac{p_s}{E_s}.L_s=\frac{600}{2.1\times10^5}\,p_s$$

$$\Delta_a^p = \text{elongation of aluminium bar due to temperature stress} = \frac{p_a}{E_a}.L_a=\frac{300}{0.7\times10^5}\,p_a$$

Case (*a*) ***Supports are non-yielding***

$$\Delta_s^p + \Delta_a^p = \text{Free contraction } \Delta$$

$$\therefore \quad \frac{600}{2.1\times10^5}\,p_s+\frac{300p_a}{0.7\times10^5}=0.2808 \quad ...(2)$$

Substituting, $p_s = 1.5p_a$ from (1) and solving, we get

p_a = **32.76 N/mm²** (tensile) and hence $p_s = 1.5\,p_a$ = **49.14 N/mm²** (tensile)

Case (*b*) ***Supports yielding***

$$\Delta_s^p + \Delta_a^p = \Delta - 0.1$$

$$\therefore \quad \frac{600}{2.1\times 10^5}\,p_s + \frac{300\,p_a}{0.7\times 10^5}\,p_a = 0.2808 - 0.1 = 0.1808$$

Substituting $p_s = 1.5\,p_a$, we get

p_a = **21.09 N/mm²** (tensile) and hence p_s = **31.64 N/mm²** (tensile).

Example 1.28. *A composite bar is made up by connecting a steel member and a copper member rigidly fixed at their ends as shown in Fig. 1.35. The cross-sectional area of steel member is A cm² while that of the copper member is 2A cm² for half the length and A cm² for the other half of the length. The coefficients of expansion of steel and copper are α and 1.25 α respectively while the elastic moduli are E and 0.5 E respectively. Estimate the stress induced in the members due to a temperature rise of T degrees.*

Solution.

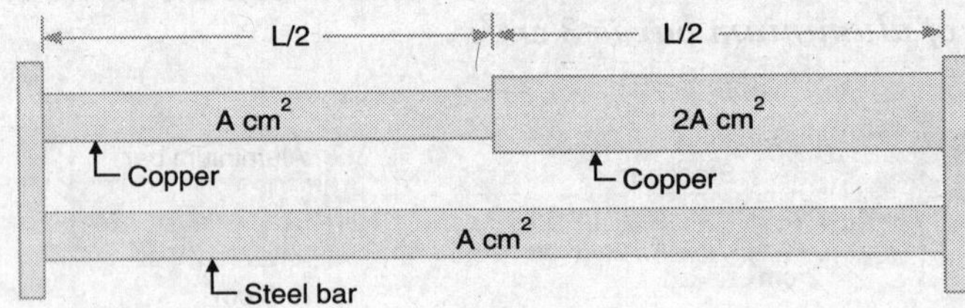

Fig. 1.35

Due to rise in temperature, copper will try to expand more than steel and hence compressive stress will be induced in copper and tensile stress in steel.

Let p_c = compressive stress in copper bar of A cm² area of cross-section.

$\therefore$ Stress in copper bar for the portion of $2A$ cm² area $= \dfrac{p_c A}{2A} = 0.5\,p_c$

$\therefore \quad \Delta_c^p$ = Compression of copper bar due to temperature stress

$$= \frac{p_c}{E_c}\cdot\frac{L}{2} + \frac{0.5\,p_c}{E_c}\,\frac{L}{2} = \frac{3}{4}\,\frac{p_c L}{E_c} = \frac{3}{2}\,\frac{p_c L}{E}$$

Δ_s^p = expansion of steel bar due to temperature stress $= \dfrac{p_s L}{E}$

Now from Eq. 1.19

$$\Delta_s^p + \Delta_c^p = LT\,(\alpha_c - \alpha_s)$$

$$\therefore \quad \frac{p_s L}{E} + \frac{3\,p_c L}{2E} = LT\,(1.25\alpha - \alpha) \quad \text{or} \quad p_s + 1.5\,p_c = 0.25\,E\,\alpha\,T \qquad \text{...(1)}$$

Since there is no external force acting,

$$p_c.A = p_s.A \quad \text{or} \quad p_c = p_s \qquad \text{...(2)}$$

Hence from (1) and (2), we get

$$p_c = p_s = \mathbf{0.1\,E\,\alpha\,T.}$$

Example 1.29. *A copper bar 30 mm diameter, is completely enclosed in a steel tube, 30 mm internal diameter and 50 mm external diameter. A pin, 10 mm in diameter is fitted transversely to the axis of the bar near each end, to secure the bar to the tube. Calculate the intensity of shear stress induced in the pins when the temperature of the whole is raised by 50 K. Take* $E_c = 1 \times 10^5$ *N/mm*^{2} *;* $E_s = 2 \times 10^5$ *N/mm*^{2} *;* $\alpha_c = 17 \times 10^{-5}$ *per K ;* $\alpha_s = 11 \times 10^{-6}$ *K.*

Solution. Area of cross-section of copper bar $= A_c = \dfrac{\pi}{4}(30)^2 = 706.9 \text{ mm}^2$

Area of cross-section of steel tube $= A_s = \dfrac{\pi}{4}(50^2 - 30^2) = 1964 - 706.9 = 1257.1 \text{ mm}^2$

Now
$$\frac{p_s}{E_s} + \frac{p_c}{E_c} = (\alpha_c - \alpha_s)T$$

$$\therefore \quad \frac{p_s}{2 \times 10^5} + \frac{p_c}{1 \times 10^5} = (17 - 11)\, 10^{-6} \times 50 \quad \text{or} \quad p_s + 2p_c = 60 \qquad \text{...(1)}$$

Since there is no external force, $p_s . A_s = p_c . A_c$

$$\therefore \quad p_s = p_c \cdot \frac{A_c}{A_s} = p_c \frac{706.9}{1257.1} = 0.562\, p_c \qquad \text{...(2)}$$

Substituting in (1) and solving, we get

$$p_c = 23.42 \text{ N/mm}^2$$

Hence force P between tube and copper bar

$$= p_c \times A_c = 23.42 \times 706.9 = 16550 \text{ N}$$

Area of cross-section of pin $= \dfrac{\pi}{4}(10)^2 = 78.54 \text{ mm}^2$

Since the pin is fitted transversely, and passes through the tube and the rod, it will be in *double shear*.

$$\therefore \quad \text{Shear stress in pin} = \frac{16550}{2 \times 78.54} = \mathbf{105.4 \text{ N/mm}^2}.$$

1.12. EQUIVALENT MODULUS OF A COMPOUND BAR

Let us consider a compound bar of length L consisting of two materials of modulus E_1 and E_2 and having areas of cross-section A_1 and A_2 as shown in Fig. 1.36 (*a*). Fig 1.36 (*b*) shows an *equivalent bar* of cross-sectional area $A(= A_1 + A_2)$ and length L, made of a material having modulus E, such that its extension under the given load is equal to that of the compound bar.

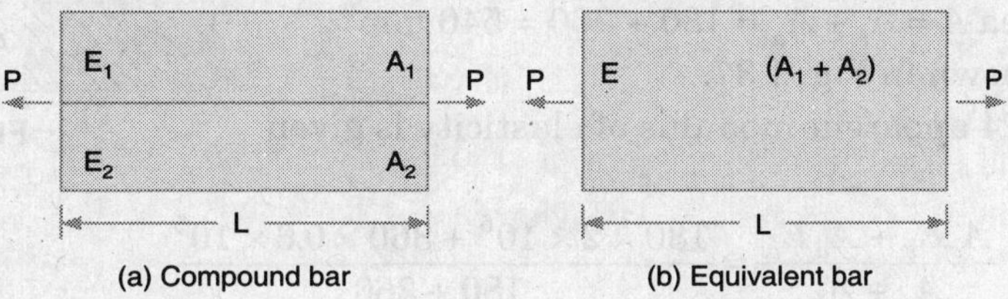

(a) Compound bar (b) Equivalent bar

Fig. 1.36

For the equivalent bar

Extension $\Delta = \dfrac{PL}{AE}$, where $A = (A_1 + A_2)$

$$\therefore \quad (A_1 + A_2)E = \frac{PL}{\Delta} \quad ...(1)$$

For the compound bar

Since the strain in both the materials is equal

$$\frac{p_1}{E_1} = \frac{p_2}{E_2} \quad \text{or} \quad \frac{p_1}{p_2} = \frac{E_1}{E_2} \quad ...(2)$$

Also $\quad p_1A_1 + p_2A_2 = P$

$$\therefore \quad \left(p_2 \frac{E_1}{E_2}\right) A_1 + p_2A_2 = P \quad \text{or} \quad p_2 = \frac{P}{\dfrac{E_1}{E_2}A_1 + A_2} \quad ...(3)$$

Now, common extension Δ is given by $\Delta = \dfrac{p_2L}{E_2}$

Substituting the value of p_2 from (3),

$$\Delta = \frac{PL}{A_1E_1 + A_2E_2} \quad \text{or} \quad \frac{PL}{\Delta} = A_1E_1 + A_2E_2 \quad ...(4)$$

Equating (1) and (4), we get

$$(A_1 + A_2)E = A_1E_1 + A_2E_2 \quad \text{or} \quad E = \frac{A_1E_1 + A_2E_2}{A_1 + A_2} \quad ...(1.24)$$

which is the required expression for the *equivalent* or *apparent modulus of elasticity*.

Example 1.30. *A composite bar is made by fastening one flat bar of steel between two similar bars of aluminium alloy. The dimensions of each bar are 30 mm wide × 6 mm thick, so that the cross-section of the composite bar measures 30 mm × 18 mm. If E for steel = 2 × 10⁵ N/mm², and E for the alloy = 0.6 × 10⁵ N/mm², find the apparent value of E for the composite bar when loaded in tension. If the respective elastic limits are 230 N/mm² and 50 N/mm², find the elastic limit of the compound bar.*

Solution. Let us use suffix *s* for steel and suffix *a* for alloy.

Area of steel, $A_s = 30 \times 6 = 180 \text{ mm}^2$

Area of alloy, $A_a = 30 \times 12 = 360 \text{ mm}^2$

$\therefore$ Total area $A = A_s + A_a = 180 + 360 = 540 \text{ mm}^2$

The bar is shown in Fig. 1.37.

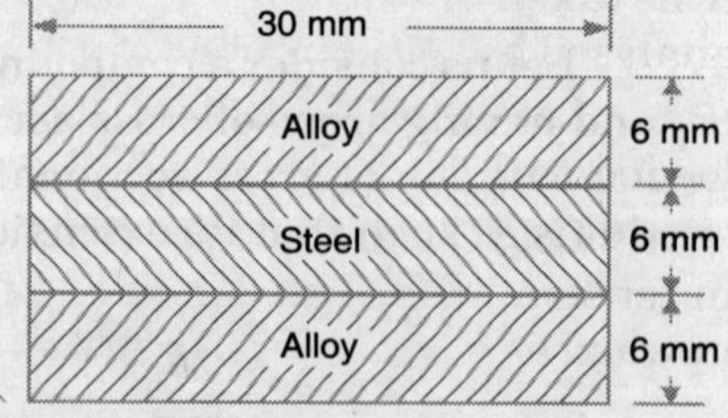

Fig. 1.37

From Eq. 1.24 apparent modulus of elasticity is given by

$$E = \frac{A_sE_s + A_aE_a}{A_s + A_a} = \frac{180 \times 2 \times 10^5 + 360 \times 0.6 \times 10^5}{180 + 360}$$

$$= \mathbf{0.1067 \times 10^6 \ N/mm^2.}$$

The elastic limit of the compound bar is the stress at which one of the members, either steel or alloy, is stressed to its elastic limit.

Let P be the load at the elastic limit and p_a and p_s be the corresponding stresses in alloy and steel respectively.

Since strains in both the materials are equal,

$$\frac{p_s}{E_s} = \frac{p_a}{E_a} \quad \text{or} \quad p_s = p_a \frac{E_s}{E_a} = p_a \frac{2}{0.6} = \frac{10}{3} p_a \qquad \text{...(1)}$$

Also, $\quad p_s.A_s + p_a.A_a = P$

$$\therefore \quad \frac{10}{3} p_a.A_s + p_a.A_a = P$$

or $\quad p_a \left[\frac{10}{3} \times 180 + 360 \right] = P \quad \text{or} \quad 960\, p_a = P$

From which $\quad p_a = \frac{P}{960} \quad \text{and} \quad p_s = \frac{10}{3} \times \frac{P}{960} = \frac{P}{288} \qquad \text{...(2)}$

Let f = elastic limit of the compound bar

$$\therefore \quad f(A_a + A_s) = P \quad \text{or} \quad f \times 540 = P \qquad \text{...(3)}$$

But $P = 960\, p_a$ from Eq. (2). Hence $540\, f = 960\, p_a$

or $\quad f = \frac{960}{540} p_a = \frac{960}{540} \times 50 = 88.89 \text{ N/mm}^2 \qquad \text{...(I)}$

Also, from Eq. (2), $P = 288\, p_s$. Hence from Eq. (3), $540\, f = 288\, p_s$

$$\therefore \quad f = \frac{288}{540} p_s = \frac{288}{540} \times 230 = 122.67 \text{ N/mm}^2 \qquad \text{...(II)}$$

The elastic limit is the lower of the two values of f given by (I) and (II). Hence the elastic limit of the compound bar will be **88.89 N/mm²**.

1.13. SIMPLE STATICALLY INDETERMINATE PROBLEMS

Statically indeterminate problems are those which cannot be solved by the simple equations of statics alone. To solve such problems, the deformation characteristics of the structure will have to be taken into account. Here we will consider some simple combinations which can be solved or analysed by the equations of compatibility, along with equations of statical equilibrium.

Let us first consider the case of bar of uniform cross-section, held between rigid plates and loaded with an axial force P (Fig. 1.38). The axial forces in AC and CB will be different.

Let R_1 and R_2 be the reactions at ends. A little consideration will show that bar AC will be in tension, while bar CB will be in compression. Hence both reactions R_1 and R_2 will be acting upward as shown. From statics,

$$R_1 + R_2 = P \qquad \text{...(1)}$$

Again, since the bar is rigidly held in position, extension in AC will be equal to compression of CB.

Tensile force in AC will be equal to R_1.

Hence $\quad \Delta_{AC} = \frac{R_1 a}{AE}$

Compressive force in CB will be equal to $R_2 (= P - R_1)$

Hence $$\Delta_{CB} = \frac{R_2 b}{AE}$$

Equating the two, we get

$$\frac{R_1 a}{AE} = \frac{R_2 b}{AE} \qquad ...(2)$$

From which $$R_1 = R_2 \frac{b}{a}$$

Substituting in (1), we get $R_2 \dfrac{b}{a} + R_2 = P$ or $\dfrac{R_2(a+b)}{a} = P$

$$R_2 = \frac{Pa}{a+b} = \frac{Pa}{L} \quad \text{and} \quad R_1 = R_2 \cdot \frac{b}{a} = \frac{Pa}{L} \times \frac{b}{a} = \frac{Pb}{L}.$$

Thus, the reactions and hence the axial forces are known.

The extension of $AC = \dfrac{R_1 a}{AE} = \dfrac{Pab}{LAE}$.

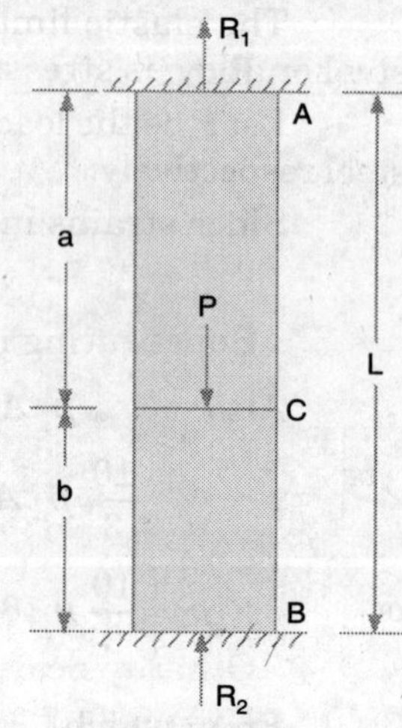

Fig. 1.38

Another example may be taken of a rigid bar AB supported by the equidistant rods in the same vertical plane (Fig. 1.39). All the three rods are of different materials. Three equations will be required to find the three unknowns. Using suffix 1, 2 and 3 for the rods, we have from statics,

$$P_1 + P_2 + P_3 = P \qquad ...(i)$$

Taking moments about B, $(P_2 a + P_3 2a) = P.c$ ...(*ii*)

The dotted lines show the deformed position of the structure. From the compatibility equation,

$$\Delta_2 = \frac{\Delta_1 + \Delta_3}{2} \quad \text{or} \quad \frac{P_2}{A_2 E_2} = \frac{P_1}{2A_1 E_1} + \frac{P_3}{2A_3 E_3} \qquad ...(iii)$$

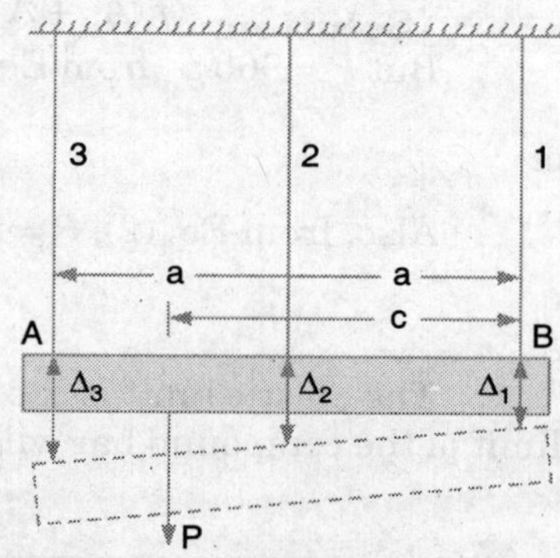

Fig. 1.39

From equations (*i*), (*ii*) and (*iii*) the three unknowns P_1, P_2 and P_3 can be easily calculated.

Next example may be taken of the system shown in Fig. 1.40 in which the outer rods are identical and are of the same material, and the inner rod is of different material.

From statics, we get

$$2P_1 \cos\theta + P_2 = P \qquad ...(1)$$

The dotted lines show the deformed shape of the structure. Assuming that the angle θ does not change appreciably, we get the compatibility equation

$$\Delta_1 = \Delta_2 \cos\theta$$

or $$\frac{P_1 L_1}{A_1 E_1} = \frac{P_2 L_2}{A_2 E_2} \cos\theta \qquad ...(2)$$

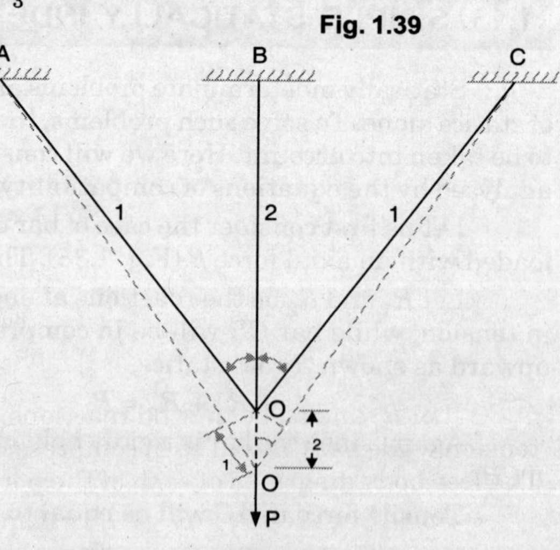

Fig. 1.40

But $$\frac{L_2}{L_1} = \cos\theta$$

$$\therefore \quad P_1 = P_2 \frac{A_1E_1}{A_2E_2} \cos^2\theta, \text{ from (2)}$$

Substituting in (1), we get $2P_2 \dfrac{A_1E_1}{A_2E_2} \cos^3\theta + P_2 = P$

or $$P_2 = \frac{P}{1 + \dfrac{2A_1E_1}{A_2E_2}\cos^3\theta} \quad \text{...(1.25)}$$

Knowing P_2, P_1 can be calculated.

Other typical combinations have been illustrated in the solved examples.

Example 1.31. *A circular bar ABCD is rigidly fixed at A and D and is subjected to axial forces as shown in Fig. 1.41 (a). Determine forces in each portion of the bar and the displacements of point B and C. Take E = 200 kN/mm².*

Solution.

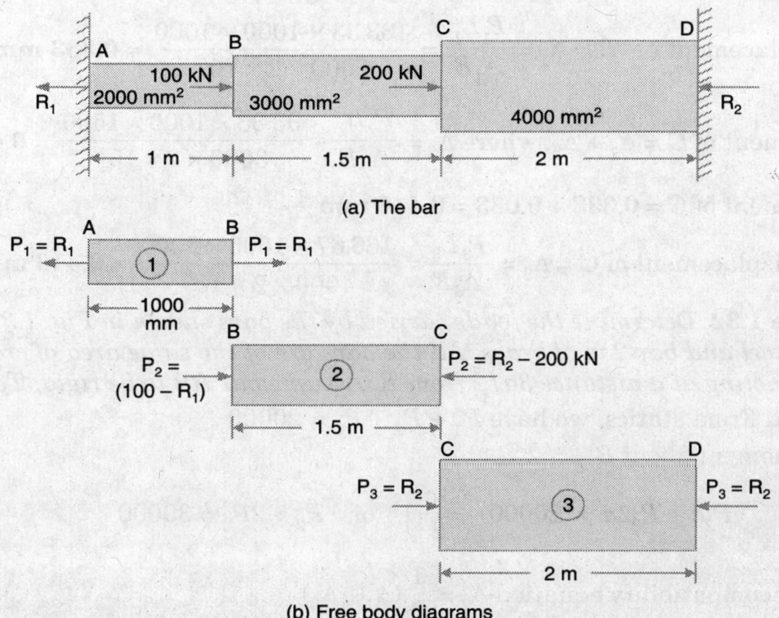

Fig. 1.41

Let R_1 and R_2 be the end reactions. A little consideration will reveal that bar *AB* will be in tension while bar *CD* will be in compression. Hence the directions of R_1 and R_2 will be as marked. The free body diagrams of each of the portions *AB*, *BC* and *CD* are shown in Fig. 1.41 (*b*).

For AB, Force $P_1 = R_1$; $L_1 = 1$ m $= 1000$ mm ; $A_1 = 2000$ mm².

For CD, Force $P_3 = R_2$; $L_3 = 2$ m $= 2000$ mm ; $A_3 = 4000$ mm².

For BC, let the axial force be P_2 (compressive). Considering forces to the left, $P_2 = 100 - R_1$. Similarly, considering forces to the right, $P_2 = R_2 - 200$.

Thus, $100 - R_1 = R_2 - 200 \quad \text{or} \quad R_1 + R_2 = 100 + 200 = 300$ ...(1)

The above equation could also be obtained from statics.

Now, from compatibility, extension of AB = compression of BC and CD

$$\therefore \quad \Delta_1 = \Delta_2 + \Delta_3 \quad \text{or} \quad \frac{P_1L_1}{A_1E} = \frac{P_2L_2}{A_2E} + \frac{P_3L_3}{A_3E}$$

Substituting the values, we get

$$\frac{R_1\,(1000)}{2000} = \frac{(100 - R_1)\,(1500)}{3000} + \frac{R_2\,(2000)}{4000}$$

or $$R_1 = (100 - R_1) + R_2 \quad \text{or} \quad 2R_1 - R_2 = 100 \qquad ...(2)$$

From (1) and (2), we get

$$R_1 = \frac{400}{3} = 133.33 \text{ kN} \quad \text{and} \quad R_2 = 300 - 133.33 = 166.67 \text{ kN}$$

Hence $P_1 = R_1 =$ **133.33 kN** (tensile)

$P_2 = 100 - R_1 = 100 - 133.33 = -33.33 =$ **33.33 kN** (tensile)

$P_3 = R_2 =$ **166.67 kN** (compressive)

$$\text{Now displacement of } B = \Delta_1 = \frac{P_1L_1}{A_1E} = \frac{133.33 \times 1000 \times 1000}{2000 \times 2 \times 10^5} = \mathbf{0.333\ mm} \rightarrow$$

$$\text{Displacement of } C = \Delta_1 + \Delta_2, \text{ where } \Delta_2 = \frac{P_2L_2}{A_2E} = \frac{33.33 \times 1000 \times 1500}{3000 \times 2 \times 10^5} = 0.083 \text{ mm} \rightarrow$$

Displacement of $C = 0.333 + 0.083 =$ **0.416 mm** →

$$\textit{Check.}\ \text{Displacement of } C = \Delta_3 = \frac{P_3L_3}{A_3E} = \frac{166.67 \times 1000 \times 2000}{4000 \times 2 \times 10^5} = \mathbf{0.416\ mm} \rightarrow$$

Example 1.32. *Determine the loads carried by the bars shown in Fig. 1.39 in which bars 1 and 3 are of steel and bar 2 is of brass. All the bars are of the same area of cross-section. The load P = 20 kN acting at a distance 3a/2 from B. Assume bar AB to be rigid. Take* $E_s/E_b = 2$.

Solution. From statics, we have $P_1 + P_2 + P_3 = 20000$ ...(*i*)

Taking moment about B,

$$P_2a + P_32a = (20000)\left(\frac{3a}{2}\right) \quad \text{or} \quad P_2 + 2P_3 = 30000 \qquad ...(ii)$$

From the compatibility equation $\Delta_2 = \dfrac{1}{2}(\Delta_1 + \Delta_3)$

or $$\frac{P_2}{A_2E_2} = \frac{1}{2}\left(\frac{P_1}{A_1E_1} + \frac{P_3}{A_3E_3}\right)$$

But $A_1 = A_2 = A_3 \quad \text{and} \quad E_1 = E_3 = 2E_2$

$$\therefore \quad \frac{P_2}{E_2} = \frac{P_1}{4E_2} + \frac{P_3}{4E_2}$$

or $$P_2 = \frac{P_1}{4} + \frac{P_3}{4} \quad \text{or} \quad P_1 + P_3 = 4P_2 \qquad ...(iii)$$

From (*ii*), $P_2 = 30000 - 2P_3$

Substituting in (*iii*), $P_1 + P_3 = 4(30000 - 2P_3) = 120000 - 8P_3$

or $$P_1 = 120000 - 9P_3$$

Substituting in (*i*), $(120000 - 9P_3) + (30000 - 2P_3) + P_3 = 20000$

or $$10P_3 = 150000 - 20000 = 130000$$

$\therefore$ $P_3 = 13000$ N = **13 kN**

$P_2 = 30000 - 2 \times 13000 = 4000$ N = **4 kN**

$P_1 = 120000 - 9 \times 13000 = 3000 =$ **3 kN.**

Example 1.33. *Determine the forces acting in the bars of the system represented in Fig. 1.40 in which P = 20 kN. All the bars have the same area of the cross-section. The outer bars are of steel and the central bar is of brass. Take θ = 30° and $E_s/E_b = 2$.*

Solution. From statics, $2P_1 \cos\theta + P_2 = 20$ ...(*i*)

And from compatibility, $\dfrac{P_1 L_1}{A_1 E_1} = \dfrac{P_2 L_2}{A_2 E_2} \cos\theta$, $\left(\text{where } L_1 = \dfrac{L_2}{\cos\theta}\right)$ ...(*ii*)

From (*i*) and (*ii*), we get $$P_2 = \frac{20}{1 + \dfrac{2A_1E_1}{A_2E_2}\cos^3\theta}$$

But $A_1 = A_2$; $E_1/E_2 = 2$; $\cos^3\theta = \cos^3 30° = 0.65$; $\cos^2\theta = \cos^2 30° = 0.75$

$\therefore$ $$P_2 = \frac{20}{1 + (2 \times 2 \times 0.65)} = \frac{20}{3.60} = 5.556 \text{ kN}$$

$\therefore$ $$P_1 = P_2 \frac{E_1}{E_2} \cos^2\theta = 5.556 \times 2 \times 0.75 = 8.333 \text{ kN}$$

Load taken by each steel rod = **8.333 kN** ; Load taken by the brass rod = **5.556 kN.**

Example 1.34. *Three identical pin-connected bars are arranged as shown in Fig. 1.42 and support a load P. Find the axial force in each bar and also the vertical displacement of the point of application of load. Neglect any possibility of lateral buckling of the bar. All the bars are of the same length and same area of cross-section.*

Solution. (Fig. 1.42).

Use suffix 1 for *OA* and *OB*, and 2 for *OC*.

From statics, we get $2P_1 \cos 60° + P_2 = P$

or $$2P_1 \times \frac{1}{2} + P_2 = P \quad \text{or} \quad P_1 + P_2 = P \qquad ...(i)$$

The dotted lines show the deformed shape of the structure. Assuming that the angles between the rods do not change appreciably, we get, from elastic deformations,

$$\Delta_1 = \Delta_2 \cos 60° = \tfrac{1}{2}\Delta_2 \qquad ...(ii)$$

From (*i*), $P_2 = P - P_1$ ...(*iii*)

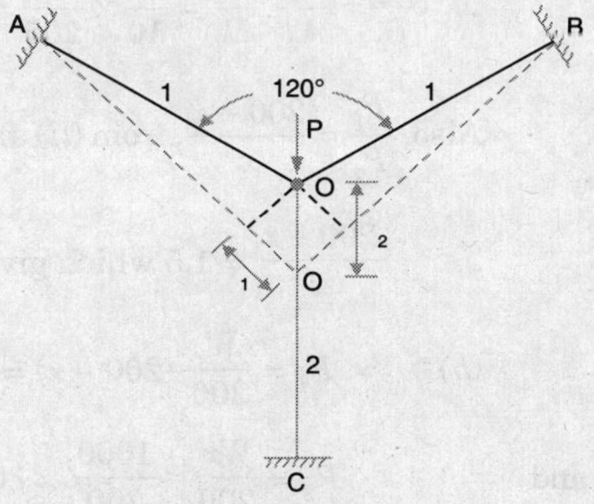

Fig. 1.42

From (*ii*), $2\Delta_1 = \Delta_2$

or $$\frac{2P_1L}{AE} = \frac{P_2L}{AE} \quad \text{or} \quad 2P_1 = P_2 \qquad ...(iv)$$

Solving (*iii*) and (*iv*), we get $P_1 = \frac{P}{3}$ and $P_2 = \frac{2}{3}P$

Hence $$\Delta_2 = \frac{P_2L}{AE} = \frac{2PL}{3AE}. \quad \textbf{Ans.}$$

Example 1.35. *Two vertical wires are suspended at a distance apart of 200 mm as shown in Fig. 1.43. Their upper ends are firmly secured and their lower ends support a rigid horizontal bar which carries load W. The left hand wire has a cross-sectional area of 25 mm² and is made of copper and right hand wire has an area of 10 sq. mm and is made of steel. Both wires initially are exactly 5 m long.*

(*a*) *Determine the position of the line of action of W, if due to W both wires extend by the same amount, and*

(*b*) *Determine the load, stress and extension of each wire if W = 1000 N. Neglect the weight of bar and take E for steel = 205 kN/mm² and E for copper = 123 kN/mm².*

Solution. (*a*) Let suffix 1 stand for copper wire and 2 for steel wire

From statics, $P_1 + P_2 = W \qquad ...(i)$

Taking moment about A,

$$P_1 = \frac{W(200 - x)}{200} \qquad ...(ii)$$

Taking moment about B,

$$P_2 = \frac{Wx}{200} \qquad ...(iii)$$

Since AB remains horizontal,

$$\Delta_1 = \Delta_2 \quad \text{or} \quad \frac{P_1}{A_1E_1} = \frac{P_2}{A_2E_2} \qquad ...(iv)$$

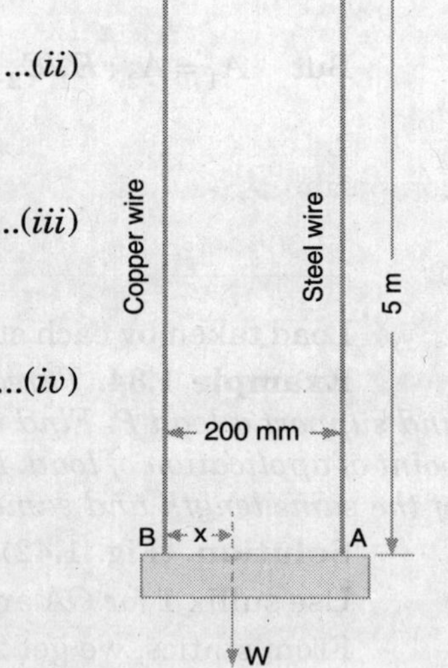

Fig. 1.43

$$\therefore \quad \frac{P_1}{P_2} = \frac{A_1}{A_2}\cdot\frac{E_1}{E_2} = \frac{25}{10} \times \frac{123}{205} = 1.50$$

Also, $\frac{P_1}{P_2} = \frac{200 - x}{x}$, from (*ii*) and (*iii*)

$$\therefore \quad \frac{200 - x}{x} = 1.5 \text{ which gives } x = \mathbf{80\ mm.}$$

(*b*) $$P_1 = \frac{W}{200}(200 - x) = \frac{1000}{200} \times 120 = 600 \text{ N}$$

and $$P_2 = \frac{Wx}{200} = \frac{1000}{200} \times 80 = 400 \text{ N}$$

$$p_1 = \frac{600}{25} = \mathbf{24\ N/mm^2}\ \text{(tension)}\ ; \quad p_2 = \frac{400}{10} = \mathbf{40\ N/mm^2}\ \text{(tension)}$$

$$\Delta = \frac{p_1 L}{E_1} = \frac{24 \times 5000}{123 \times 10^3} = \mathbf{0.976\ mm.}$$

Example 1.36. *For the bar shown in Fig. 1.44, calculate the reaction produced by the lower support on the bar. Take E = 2 × 10⁵ N/mm². Find also the stresses in the bars.*

Solution. Let R_2 = Reaction at the lower support when the bar touches it.

R_1 = Reaction at the upper support.

If the bar *BC* finally rests on the lower support, we have

$$R_1 + R_2 = 50000 \qquad \text{...(i)}$$

For bar *AB*, the total force = $R_1 = 50000 - R_2$ (tensile)

For bar *BC*, the total force = R_2 (comp.)

$$\text{Hence } \Delta_1 = \text{Extension of } AB = \frac{(50000 - R_2)\,1000}{100 \times 2 \times 10^5}\ \text{mm}$$

$$= \frac{50000 - R_2}{2 \times 10^4}\ \text{mm}$$

$$\Delta_2 = \text{contraction of } BC = \frac{R_2 \times 2000}{200 \times 2 \times 10^5}\ \text{mm}$$

$$= \frac{R_2}{2 \times 10^4}\ \text{mm}$$

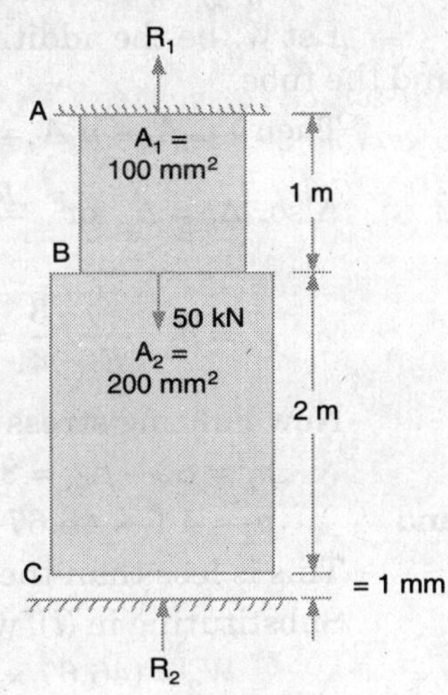

Fig. 1.44

In order that *C* rests on the lower support, we have from compatibility equation, $\Delta_1 - \Delta_2 = \Delta = 1$ mm. ...(ii)

$$\text{or} \quad \frac{50000 - R_2}{2 \times 10^4} - \frac{R_2}{2 \times 10^4} = 1 \quad \text{or} \quad 50000 - 2R_2 = 1 \times 2 \times 10^4 = 20000$$

$$\therefore \quad R_2 = \frac{30000}{2} = 15000\ \text{N} = \mathbf{15\ kN}\ \text{and}\ R_1 = 50000 - 15000 = 35000\ \text{N} = \mathbf{35\ kN}$$

$$\text{Stress in } AB = \frac{R_1}{A_1} = \frac{35000}{100} = \mathbf{350\ N/mm^2}\ \text{(tensile).}$$

$$\text{Stress in } BC = \frac{R_2}{A_2} = \frac{15000}{200} = \mathbf{75\ N/mm^2}\ \text{(compressive).}$$

Example 1.37. *Fig. 1.45 shows a round steel rod supported in a recess and surrounded by a co-axial brass tube. The upper end of rod is 0.1 mm below that of the tube and an axial load is applied to a rigid plate resting on the top of the tube.*

(a) Determine the magnitude of the maximum permissible load if the compressive stress in the rod is not to exceed 110 N/mm² and that in the tube is not to exceed 80 N/mm².

(b) Find the amount by which tube will be shortened by the load if the compressive stress in tube is the same as that in the rod. For steel, take E = 200 kN/mm² and that for brass 100 kN/mm².

Solution. Let suffix b be for brass and s for steel.

$$A_s = \frac{\pi}{4}(30)^2 = 706.9 \text{ mm}^2 ; A_b = \frac{\pi}{4}[(46)^2 - (40)^2] = 405.3 \text{ mm}^2$$

Part (a). The initial compressive stress p_{bo} in the tube before the steel rod is compressed

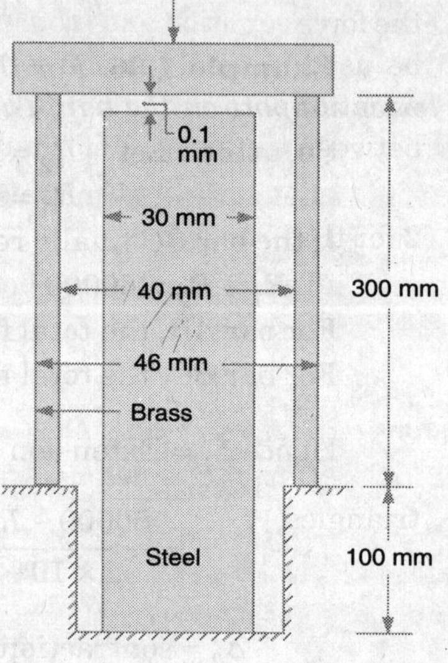

Fig. 1.45

$$= \frac{\Delta E_b}{L_b} = \frac{0.1 \times 1 \times 10^5}{300} = 33.33 \text{ N/mm}^2$$

$W_1 = p_{bo}.A_b = 33.33 \times 405.3 = 13500 \text{ N} = 13.50 \text{ kN}$

Let W_2 be the additional load to compress both the bar and the tube.

Then $\quad p_b A_b + p_s A_s = W_2 \quad ...(i)$

Also, $\Delta_b = \Delta_s$ or $\dfrac{p_b \times 300}{E_b} = \dfrac{p_s \times 400}{E_s}$

$$\therefore \quad p_s = p_b \cdot \frac{E_s}{E_b} \cdot \frac{3}{4} = p_b \frac{200 \times 10^3}{100 \times 10^3} \times \frac{3}{4} = 1.5\, P_b \quad ...(ii)$$

Now limiting stress in brass = 80 N/mm²

$\therefore \quad p_b = 80 - p_{bo} = 80 - 33.33 = 46.67 \text{ N/mm}^2$

and $\quad p_s = 1.5 \times 46.67 = 70 \text{ N/mm}^2$.

This is less than the limiting stress in steel.

Substituting in (i), we get

$$W_2 = (46.67 \times 405.3) + (70 \times 706.9) = 68400 \text{ N} = 68.4 \text{ kN}$$

$\therefore$ Total max. load = 13.50 + 68.4 = **81.9 kN**

Part (b)

Stress in brass $= \dfrac{E_b \,.\, \Delta_b}{L_b} = \dfrac{100 \times 10^3}{300}(0.1 + \Delta_b)$, where Δ_b = additional shortening = Δ_s.

Stress is steel $= \dfrac{E_s\, \Delta_s}{L_s} = \dfrac{200 \times 10^3}{400} \times \Delta_b$

Equating both the stresses, $\dfrac{100 \times 10^3}{300}(0.1 + \Delta_b) = \dfrac{200 \times 10^3}{400}\Delta_b$

This gives $\Delta_b = 0.2$ mm

$\therefore$ Total deformation of brass = $\Delta_{bt} = 0.1 + 0.2 =$ **0.3 mm.**

Example 1.38. *A rigid bar AB is hinged at A and supported by a bronze rod 2 m long and a steel rod 1 m long. A load of 8 tonnes is applied at the end B as shown in Fig. 1.46. Using the data in the following table, calculate the stress in each rod and the reaction at A.*

	Area (mm²)	*E (N/mm²)*
Steel	*900*	*2.05 × 10⁵*
Bronze	*600*	*0.82 × 10⁵*

Solution. There are three unknowns, *i.e.*, reaction R_A at A, force P_1 in the steel rod and force P_2 in bronze rod. To solve the problem, the two equations of static equilibrium (*i.e.*, the force equation and moment equation) must be used in combination with the additional equation obtained from the geometric relations between the elastic deformations.

Fig. 1.46

Let suffix 1 stand for the steel rod and 2 for the bronze rod.

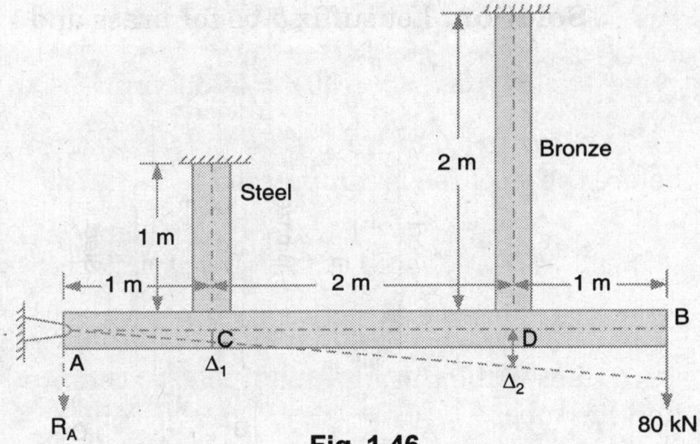

Taking moments about A, we get

$$\Sigma M_A = 0$$

$$= (P_1 \times 1) + (P_2 \times 3) - (80000 \times 4)$$

or $$P_1 + 3P_2 = 320000 \quad ...(1)$$

If Δ_1 is the deformation of the steel rod and Δ_2 that of bronze rod we have, from similar triangles,

$$\frac{\Delta_1}{1} = \frac{\Delta_2}{3} \quad \text{or} \quad \frac{P_1 L_1}{A_1 E_1} = \frac{1}{3}\frac{P_2 L_2}{A_2 E_2}$$

or $$\frac{P_1 \times 1000}{900 \times 2.05 \times 10^5} = \frac{1}{3}\frac{P_2 \times 2000}{600 \times 0.82 \times 10^5}$$

from which $$P_1 = 2.5\,P_2 \quad ...(2)$$

Substituting the value of P_1 in (1), we get $2.5\,P_2 + 3P_2 = 320000$

From which, P_2 = **58182 N** and hence P_1 = **145455 N**

For finding the reaction at A, use the force equation (*i.e.*, $\Sigma P = 0$). Assuming R_A to act downward,

$$P_1 + P_2 - R_A - 80000 = 0$$

$$\therefore \quad R_A = 145455 + 58182 - 80000 = \mathbf{123637\ N}$$

The positive sign to R_A shows that the assumed direction of R_A is correct.

Example 1.39. *A rigid block ABC, weighing 200 kN is supported by three rods symmetrically placed, as shown in Fig. 1.47 (a). Assuming the block to remain horizontal, determine the stress in each rod after a temperature rise of 20°C. The lower ends of the rods are assumed to have been at the same level before the block was attached and the temperature changed. Given : Area of steel rod 900 sq. mm, area of bronze rod 1500 sq. mm, E for steel 205 kN/mm², E for bronze 82 kN/mm² ; α for steel 12 × 10⁻⁶ per °C ; α for bronze 20 × 10⁻⁶ per °C. What will be the stress in each rod if the weight of the block is 100 kN only ?*

Solution. (*a*) Use suffix 1 for steel rods and 2 for the bronze rod. Fig. 1.47 (*b*) shows the extensions for the rods at different stages. Before increase in temperature and attachment of the block, the ends of all the three rods are at the same level. With the block *detached,* let the

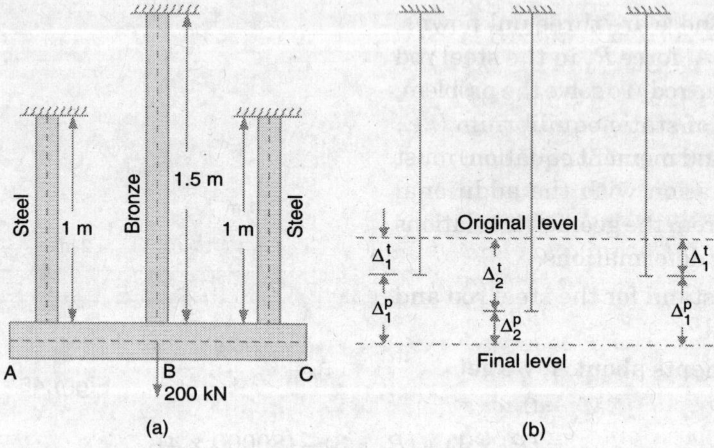

Fig. 1.47

temperature rise. The extensions due to temperature are represented by Δ_1^t and Δ_2^t respectively for the steel and bronze, Δ_2^t being greater than Δ_1^t. After the temperature has increased, if the block is attached to the rods, the expanded ends of the rods will have to be pulled through the load deformations Δ_1^p and Δ_2^p so that finally, all the ends are at the same level. It is clear, therefore, that the sum of the two deformations of steel rod is equal to the sum of the deformations of the bronze rod. Hence, we get

$$\Delta_1^t + \Delta_1^p = \Delta_2^t + \Delta_2^p \quad \text{or} \quad L_1\alpha_1 t + \frac{P_1 L_1}{A_1 E_1} = L_2\alpha_2 t + \frac{P_2 L_2}{A_2 E_2}$$

where, P_1 is the load carried by each of the steel rods and P_2 is the load carried by bronze rod.

Substituting the values, we get

$$(1000 \times 12 \times 10^{-6} \times 20) + \frac{P_1 \times 1000}{900 \times 205 \times 10^3} = (1500 \times 20 \times 10^{-6} \times 20) + \frac{P_2 \times 1500}{1500 \times 82 \times 10^3} \quad \text{...(1)}$$

or
$$P_1 - 2.25\, P_2 = 66420$$

Another equation is obtained from statics :

$$2P_1 + P_2 = 200 \times 1000 \quad \text{...(2)}$$

Solving (1) and (2), we get $P_1 = 93895$ N and $P_2 = 12210$ N

Hence the stresses are

$$p_1 = \frac{93895}{900} = \mathbf{104.33\ N/mm^2}\ \text{(tension) and } p_2 = \frac{12210}{1500} = \mathbf{8.14\ N/mm^2}\ \text{(tension)}$$

(*b*) If the weight of the block is 100000 N only and not 200000 N, the equation from statical equilibrium will be

$$2P_1 + P_2 = 100000 \quad \text{...(3)}$$

Solving (1) and (3), we get $P_1 = 52985$ N and $P_2 = -5971$ N.

The minus sign with P_2 indicates that the bronze bar will be in compression. The stresses are

$$p_1 = \frac{52985}{900} = \mathbf{58.87\ kN/mm^2}\ \text{(tension) and } p_2 = \frac{5971}{1500} = \mathbf{3.98\ N/mm^2}\ \text{(compression).}$$

Example 1.40. *A rigid beam AB is hinged at A and connected to the aluminium alloy bar BC at the other end. The beam is also supported on a steel post DE through a rigid bearing at D as shown in Fig. 1.48. At no load, the beam is horizontal and there is an initial clearance of δ= 0.1 mm between the beam and rigid bearing. Determine the maximum permissible load W which may be applied at B if the axial stresses are not to exceed 100 N/mm² for the steel and 120 N/mm² for the aluminium alloy. All the connections may be treated as smooth pins. For steel E = 2 × 10⁵ N/mm² and for aluminium alloy, E = 0.8 × 10⁵ N/mm².*

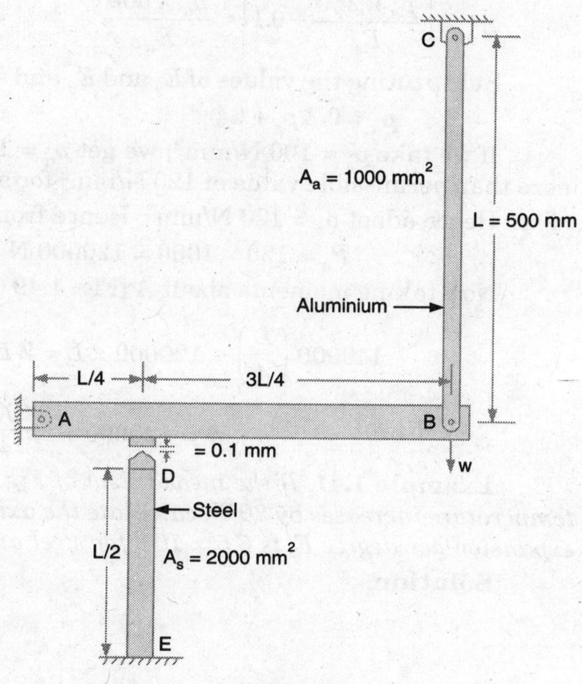

Fig. 1.48

Solution. Use suffix *a* for aluminium alloy and *s* for steel.

The free body diagram in Fig. 1.49 (*a*) shows the four unknown forces R_A, P_s, P_a and *W*. Let p_s and p_a be the axial stresses in steel post and aluminium bar respectively. One of the axial stresses will be equal to the permissible value (which is given), while the other one will be lesser than the allowable value.

Fig. 1.49 (*b*) shows the deformed position of the bars. From compatibility of deformation, we have

$$\frac{\Delta_s + \delta}{L/4} = \frac{\Delta_a}{L} \quad \text{or} \quad 4(\Delta_s + 0.1) = \Delta_a$$

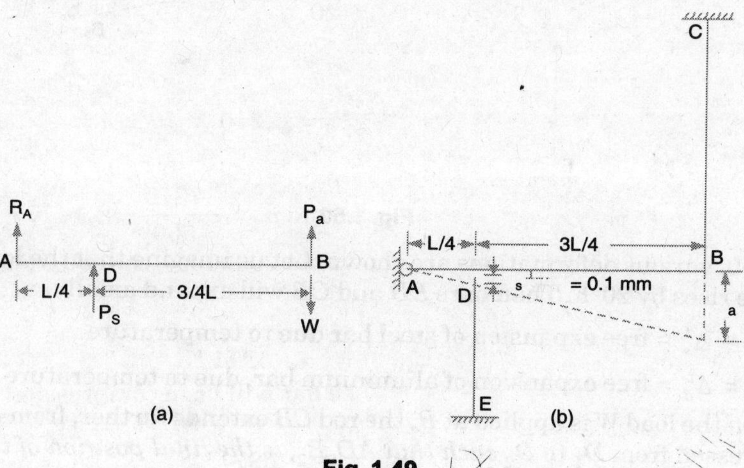

Fig. 1.49

$$\therefore \quad 4\left[\frac{p_s \times 250}{E_s} + 0.1\right] = \frac{p_a \times 500}{E_a}$$

Substituting the values of E_s and E_a and simplifying, we get

$$p_a = 0.8\, p_s + 64 \qquad \text{...(1)}$$

If we take p_s = 100 N/mm², we get p_a = 144 N/mm² from the above expression, which is more than permissible value of 120 N/mm² for aluminium alloy.

Hence adopt p_a = 120 N/mm². Hence from (1), p_s = 70 N/mm²

$$\therefore \quad P_a = 120 \times 1000 = 120000 \text{ N and } P_s = 70 \times 2000 = 140000 \text{ N.}$$

Now taking moments about A [Fig. 1.49 (*a*)], we get

$$140000\left(\frac{L}{4}\right) + 120000 \times L = WL$$

$$\therefore \quad W = 120000 + \frac{140000}{4} = 155000 \text{ N} = \mathbf{155\ kN.}$$

Example 1.41. *If the mechanism of Fig. 1.48 is loaded with a force W = 50 kN and the temperature increases by 20°F, calculate the axial stresses in BC and DE. Coefficient of thermal expansion per degree F is 6.5 × 10⁻⁶ for steel and 13 × 10⁻⁶ for aluminium alloy.*

Solution.

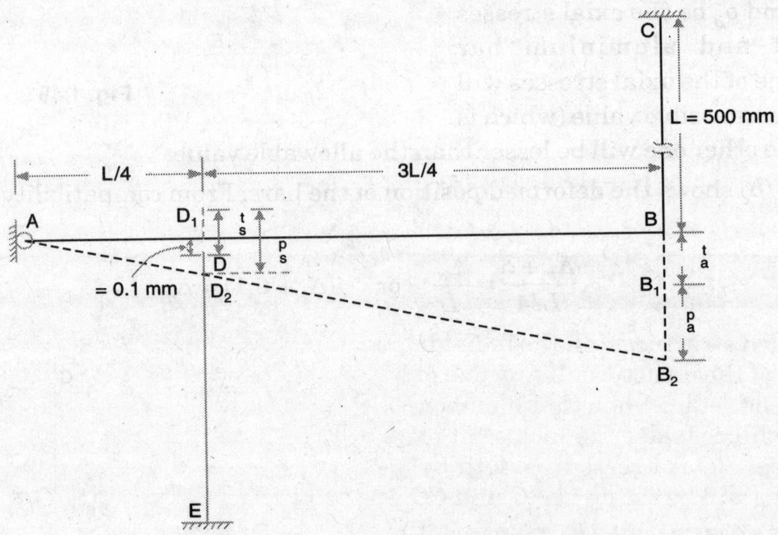

Fig. 1.50

In Fig. 1.50, various deformations are shown. Let us imagine that the load W does not act, and temperature rises by 20°F. Then bars ED and CB will expand axially.

Let $DD_1 = \Delta_s^t$ = free expansion of steel bar due to temperature

$BB_1 = \Delta_a^t$ = free expansion of aluminium bar, due to temperature

Now, when the load W is applied at B, the rod CB extends further, from B_1 to B_2, while the bar DE is compressed from D_1 to D_2 *such that* AD_2B_2, *is the final position of the rigid bar*.

Let $D_1D_2 = \Delta_s^p$ = contraction of steel bar, due to external load

$B_1B_2 = \Delta_a^p$ = expansion of aluminium bar due to external load.

The compatibility equation may be written as follows :

$$\frac{\Delta_s^p - \Delta_s^t + 0.1}{L/4} = \frac{\Delta_a^t + \Delta_a^p}{L} \quad \text{or} \quad 4(\Delta_s^p - \Delta_s^t + 0.1) = (\Delta_a^t + \Delta_a^p) \qquad ...(i)$$

Now $\Delta_s^p = \dfrac{p_s L_s}{E_s} = \dfrac{p_s(250)}{2 \times 10^5}$

$$\Delta_s^t = L_s \alpha_s t = 250 \times 6.5 \times 10^{-6} \times 20 = \frac{32500}{10^6}$$

$$\Delta_a^p = \frac{p_a L_a}{E_a} = \frac{p_a(500)}{0.8 \times 10^5}$$

$$\Delta_a^t = L_a \alpha_a t = 500 \times 13 \times 10^{-6} \times 20 = \frac{130000}{10^6}$$

Substituting these in (*i*), we get

$$4\left[\frac{250 p_s}{2 \times 10^5} - \frac{32500}{10^5} + 0.1\right] = \left[\frac{500 p_a}{0.8 \times 10^5} + \frac{130000}{10^6}\right]$$

which on simplification gives : $p_s = 1.25\, p_a - 28$ ...(1)

Also taking moments about A, we get [Fig. 1.49 (*a*)]

$$P_s\left(\frac{L}{4}\right) + P_a L = WL \quad \text{or} \quad (2000 p_s)\frac{L}{4} + (1000 p_a)L = 50000L$$

which gives $p_s + 2p_a = 100$ ...(2)

From (1) and (2), we get

p_a = **39.38 N/mm²** (tensile) and p_s = **21.33 N/mm²** (compressive).

1.14. STRESSES DUE TO LACK OF FIT

In framed structures, all the individual members are to be of definite length as required by the geometry of the structure. If any one of the members is not of correct length, stress in other members will be set up when that member is fitted. Such stress is known as *stress due to lack of fit*. Here, we will consider the lack of fit of a simple structure shown in Fig. 1.40, in which the central member BO is excess in length by an amount λ (say). When the member OB is fitted, tensile stresses will be set up in AO and CO and compressive stress in OB.

From statics, we get $2P_1 \cos\theta = P_2$ ...(*i*)

From compatibility, $\Delta_1 = \Delta_2 \cos\theta$ ...(*ii*)

$$\Delta_1 = \frac{P_1 L_1}{A_1 E_1} = \frac{P_1 L_2}{A_1 E_1 \cos\theta} \text{ and } \Delta_2 = \lambda - \frac{P_2 L_2}{A_2 E_2}$$

Hence, substituting in (*ii*), we get

$$\frac{P_1 L_2}{A_1 E_1} = \left(\lambda - \frac{P_2 L_2}{A_2 E_2}\right)\cos^2\theta \quad \text{or} \quad P_1 = \frac{A_1 E_1}{L_2}\left(\lambda - \frac{P_2 L_2}{A_2 E_2}\right)\cos^2\theta$$

Substituting in (*i*), $2\dfrac{A_1 E_1}{L_2}\left(\lambda - \dfrac{P_2 L_2}{A_2 E_2}\right)\cos^3\theta = P_2$

From which
$$P_2 = \frac{\frac{2A_1E_1}{L_2}\lambda \cos^3 \theta}{1+\frac{2A_1E_1}{A_2E_2}\cos^3 \theta} = \frac{\lambda A_2 E_2}{L_2\left(1+\frac{A_2E_2}{2A_1E_1 \cos^3 \theta}\right)} \quad ...(1.26)$$

Knowing P_2, P_1 can be calculated. It should be noted that P_1 will be of different nature than P_2.

Temperature Stresses. If in the place of lack of fit λ, the central bar of the system is heated through a temperature $t°$, the stresses will be found exactly in the same manner as explained above, except that $\lambda = L\alpha t$.

Example 1.42. *Determine the forces in all the three members represented in Fig. 1.40 if the central rod is 2 mm excess in length. All the three rods are having cross-section of 100 sq. mm and are of the same material. Take $\theta = 30°$, $L_2 = 200$ mm and $E = 2 \times 10^5$ N/mm².*

Solution. From Eq. 1.26, we get $P_2 = \dfrac{\lambda A_2 E_2}{L_2\left(1+\frac{A_2E_2}{2A_1E_1 \cos^3 \theta}\right)}$

where $\lambda = 2$ mm ; $A_1 = A_2 = 100$ mm² ; $L_2 = 200$ mm

$$\frac{A_2}{A_1} = 1 \; ; \; \frac{E_2}{E_1} = 1 \text{ and } \cos^3 \theta = \cos^3 30° = 0.65$$

$$\therefore \quad P_2 = \frac{2 \times 100 \times 2 \times 10^5}{200\left(1+\frac{1}{2 \times 0.65}\right)} = 113 \times 10^3 \text{ N} = \mathbf{113\ kN} \text{ (compressive)}$$

$$\therefore \quad P_1 = \frac{P_2}{2 \cos \theta} = \frac{113}{2 \cos 30°} = \mathbf{65.3\ kN} \text{ (tensile).}$$

Example 1.43. *In the system of bars represented in Fig. 1.40, the central rod is heated through 100°F. Calculate the stress in the rods if the outer rods are 200 sq. mm area and of steel, and the central rod is of 300 sq. mm area and of brass. The length of central bar is 1000 mm and the angle $\theta = 30°$. Take $E_s = 2.04 \times 10^5$ N/mm², $E_b = 1.02 \times 10^5$ N/mm² and $\alpha_b = 10.0 \times 10^{-6}$ per °F.*

Solution.

$$\lambda = L\alpha t = 1000\,(10 \times 10^{-6}) \times 100 = 1 \text{ mm}$$

$$\therefore \quad P_2 = \frac{\lambda A_2 E_2}{L_2\left(1+\frac{A_2E_2}{2A_1E_1 \cos^3 \theta}\right)} = \frac{1 \times 300 \times 1.02 \times 10^5}{1000\left(1+\frac{300}{2 \times 200} \times \frac{1.02}{2.04} \times \frac{1}{\cos^3 30°}\right)}$$

$$= \frac{306 \times 10^5}{1000\left(1+\frac{3}{8} \times \frac{1}{0.65}\right)} = \mathbf{19405\ N} \text{ (compressive)}$$

$$P_1 = \frac{P_2}{2 \cos \theta} = \frac{19405}{2 \times 0.866} = \mathbf{11204\ N} \text{ (tensile).}$$

1.15. SHRINKING ON

When a thin tyre of steel or any other material is to be shrunk on to a wheel, the diameter of the tyre is to be slightly smaller than that of the wheel so that it does not come out easily. Let D be the diameter of the wheel and d be the diameter of the tyre. The temperature of the tyre is to be increased by $t°$ so that it increases in diameter from d to D. When the tyre is slipped on to the wheel and the temperature falls, the steel tyre will try to come to its original diameter and, in doing so, the hoop stress (tensile) will be set up. The tensile strain imposed in the tyre will be

$$e = \frac{\pi D - \pi d}{\pi d} = \frac{D - d}{d}$$

∴ Circumferential tensile stress or *hoop stress* developed

$$= eE = \left(\frac{D-d}{d}\right) E \qquad ...(1.27)$$

Example 1.44. *A thin steel tyre is to be shrunk on to a rigid wheel of 1 m diameter. If the hoop stress is limited to 100 N/mm², calculate the internal diameter of the tyre. Find also the least temperature to which tyre tyre must be heated above that of the wheel before it could be slipped on. Take α for the tyre = 12 × 10⁻⁶ per °C and E = 2.04 × 10⁵ N/mm².*

Solution. Hoop stress $= \dfrac{E(D-d)}{d}$

or $\left(\dfrac{D-d}{d}\right) = \dfrac{100}{2.04 \times 10^5} = 4.9 \times 10^{-4}$ or $\dfrac{D}{d} = 1 + 4.9 \times 10^{-4}$

or $\dfrac{d}{D} = (1 + 4.9 \times 10^{-4})^{-1} \simeq (1 - 4.9 \times 10^{-4})$ from which $d = (1000 - 0.49) =$ **999.51 mm**.

Also $\pi D = \pi d(1 + \alpha t)$ or $\alpha t = \dfrac{D-d}{d} = 4.9 \times 10^{-4}$

∴ $t = \dfrac{4.9 \times 10^{-4}}{12 \times 10^{-6}} =$ **40.85°C.**

1.16. ADDITIONAL EXAMPLES USEFUL FOR COMPETITIVE EXAMINATIONS

Example 1.45. *A rigid bar AB, 9 m long, is suspended by two vertical loads at its ends A and B, and hangs in a horizontal position by its own weight. The rod at A is brass. length 3 m, cross-sectional area 1000 mm² and modulus of elasticity 1 × 10⁵ N/mm². The rod at B is steel, length 5 m, cross-sectional area 445 mm², modulus of elasticity 2 × 10⁵ N/mm². At what distance x from A may a vertical load P be applied if the bar is to remain horizontal after the load is applied ?*

Solution. (Fig. 1.51).

The three governing equations are

$$P_B + P_S = P \qquad ...(1)$$

$$9P_S = P.x \qquad ...(2)$$

and
$$\frac{P_B.L_B}{A_B E_B} = \frac{P_S.L_S}{A_S.E_S}$$

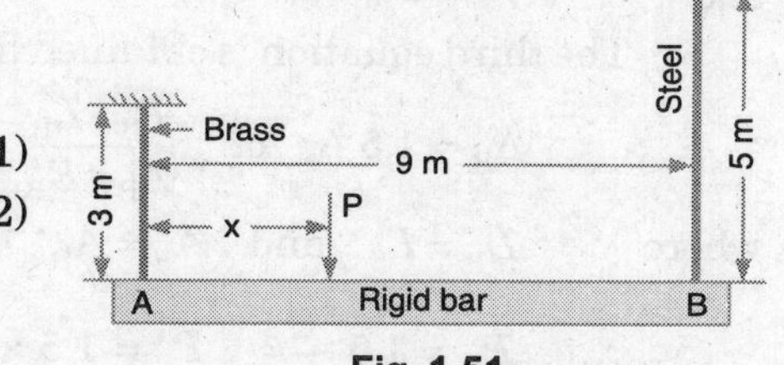

Fig. 1.51

or $$P_B = P_S\left[\frac{A_B}{A_S}\cdot\frac{E_B}{E_S}\cdot\frac{L_S}{L_B}\right]$$

$$= P_S\left[\frac{1000}{445}\times\frac{1\times10^5}{2\times10^5}\times\frac{5}{3}\right] = 1.873\ P_S \qquad ...(3)$$

From (1), $$P_S = \frac{P}{1+1.873} = 0.348\ P$$

From (2), $$x = \frac{9\times0.348\ P}{P} = \mathbf{3.133\ m.}$$

Example 1.46. *A hollow steel cylinder of length L = 300 mm, inside diameter d = 150 mm and uniform wall thickness t = 3 mm is filled with concrete and compressed between two rigid parallel plates at the ends by a load P = 500 kN. Find the compressive stress in each material and the total shortening of the cylinder if Young's modulus of steel, $E_S = 2\times10^5$ N/mm² and Young's modulus for concrete, $E_C = 2\times10^4$ N/mm². Assume that both the materials obey Hooke's law.*

Solution. (Refer Example 1.18)

$$A_S = \frac{\pi}{4}[(150+2\times3)^2-(150)^2] = 1442\ \text{mm}^2$$

$$A_C = \frac{\pi}{4}(150)^2 = 17671\ \text{mm}^2$$

$$\frac{p_S}{E_S} = \frac{p_S}{E_C} \text{ or } p_S = p_C\cdot\frac{E_S}{E_C} = p_C\ \frac{2\times10^5}{2\times10^4} = 10\ p_C \qquad ...(1)$$

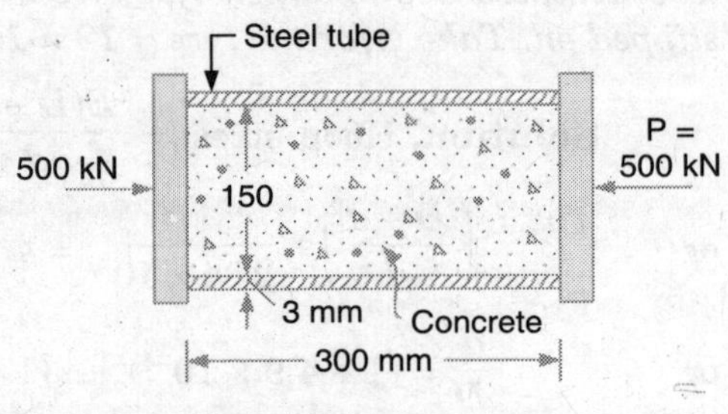

Fig. 1.52

Also, $$p_S A_S + p_C\cdot A_C = P \qquad ...(2)$$

or $(10\ p_C)\ 1442 + p_C\ (17671) = 500{,}000$

From which p_C = **15.581 N/mm²** and hence $p_S = 10\ p_C$ = **155.81 N/mm²**

Also, $$\Delta_S = \Delta_C = \frac{p_S\ .\ L_S}{E_S} = \frac{155.81\times300}{2\times10^5} = \mathbf{0.234\ mm.}$$

Example 1.47. *Two vertical bars, one of steel and another of bronze, each 2 m long and of equal area of cross-section hung vertically with 700 mm distance between their axes. A rigid member is connected to the lower ends of the bars. Calculate at what distance from the steel bar should a concentrated load of 70 kN is applied on the rigid member so that the elongation of bronze bar will be 1.5 times that of steel bar. E for steel is 2 × 10⁵ N/mm². E for bronze is 1 × 10⁵ N/mm².*

Solution. (Refer Example 1.35)

$$P_S + P_B = 70\ \text{kN} \qquad ...(i)$$

and $$0.7\ P_S = 70(0.7 - x) \qquad ...(ii)$$

The third equation is obtained from the compatibility condition that

$$\Delta_B = 1.5\ \Delta_S \text{ or } \frac{P_B\ .\ L_B}{A_B\ .\ E_B} = 1.5\frac{P_S\ .\ L_S}{A_S\ .\ E_S}$$

where $L_B = L_S$ and $A_B = A_S$

$$\therefore \quad P_B = 1.5\ \frac{E_B}{E_S}\ .\ P_S = 1.5\times\frac{1\times10^5}{2\times10^5}\ P_S = 0.75\ P_S \qquad ...(iii)$$

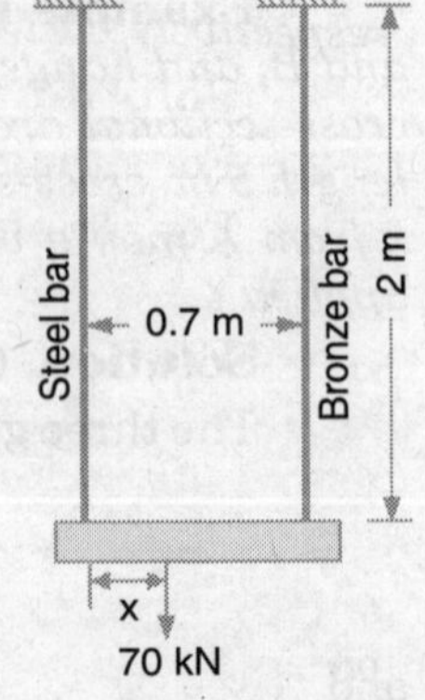

Fig. 1.53

Hence from (*i*), $P_S = \dfrac{70}{1 + 0.75} = 40$ kN

Again, from (*ii*), $0.7 \times 40 = 70(0.7 - x)$, from which $x =$ **0.3 m.**

Example 1.48. *A rigid beam ABC is hinged at A and suspended at two points B and C as shown in Fig. 1.54, by two bars BD and CE made of aluminium and steel respectively. The beam ABC carries a load of 2 kN at the mid-point of BC. The cross-sectional area of aluminium bar BD is 6 mm² and that of steel bar CE is 4 mm². Determine the loads taken up by the two bars BD and CE. Also, find the stresses developed in each bar. Modulus of elasticity is 0.7 × 10⁵ N/mm² for aluminium and 2 × 10⁵ N/mm² for steel.*

Solution. (Refer Example 1.38).

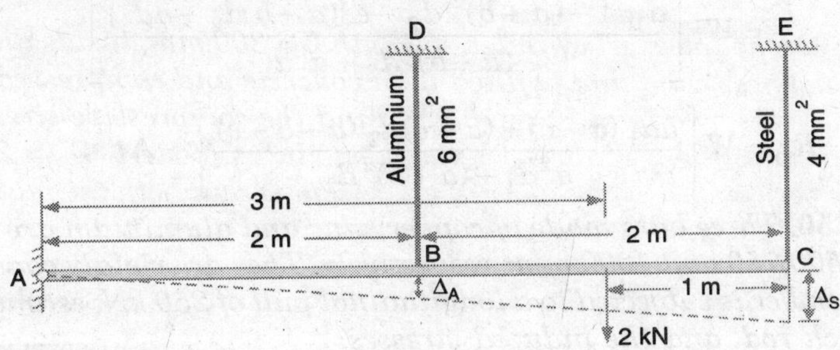

Fig. 1.54

Taking moments about hinge *A*, $4P_S + 2P_A = 2 \times 3$ ...(1)

From compatibility, $\dfrac{\Delta_S}{4} = \dfrac{\Delta_A}{2}$

$\therefore \quad \dfrac{1}{4} \cdot \dfrac{P_S}{(2 \times 10^5)4} = \dfrac{1}{2} \cdot \dfrac{P_A}{(0.7 \times 10^5)6}$ or $P_S = 3.8095\, P_A$ ...(2)

Hence from (1), $P_A = \dfrac{6}{(4 \times 3.8095 + 2)} \triangleq$ **0.348 kN** and $P_S \triangleq 3.8095 \times 0.348 =$ **1.326 kN**

$\therefore \quad p_A = \dfrac{0.348 \times 1000}{6} =$ **58 N/mm²** and $p_S = \dfrac{1.326 \times 1000}{4} =$ **331.4 N/mm².**

Example 1.49. *A rigid beam ABCD is hinged at D and supported by two springs at A and B as shown in Fig. 1.55. The beam carries a vertical load W at the point C at a distance c from the hinged end. The flexibilities (deflection/unit load) of the springs at A and B are d_1 and d_2 respectively. Determine the forces in the springs and also the reaction at the hinged support.*

Solution.

For spring *A*, extension $\Delta_A = F_A . d_1$;

For spring *B*, extension $\Delta_B = F_B . d_2$

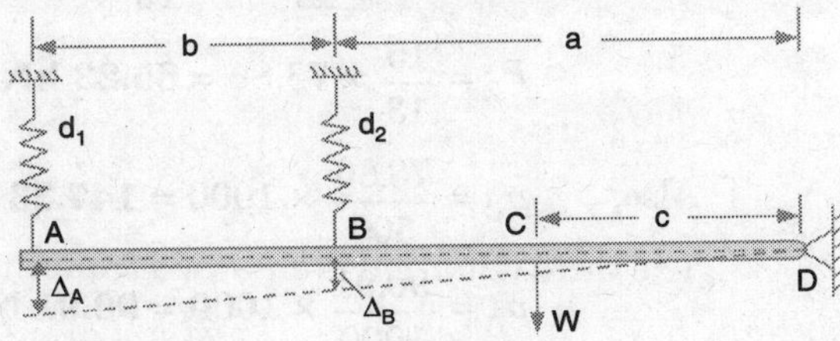

Fig. 1.55

From compatibility, $\dfrac{\Delta_A}{a + b} = \dfrac{\Delta_B}{a}$

$\therefore \quad \dfrac{F_A . d_1}{a + b} = \dfrac{F_B . d_2}{a}$

or $\quad F_A = F_B \cdot \dfrac{d_2}{d_1} \cdot \dfrac{a + b}{a}$ ...(*i*)

Also, by taking moments about the hinge D,

$$F_A(a+b)+F_B.a = W.c \qquad ...(ii)$$

Solving (i) and (ii), we get

$$F_A = \frac{W.c(a+b)\,d_2}{d_2(a+b)^2+d_1a^2}.\ \textbf{Ans.}\ \text{and}\ F_B = \frac{W.c.ad_1}{d_2(a+b)^2+d_1a^2}.\ \textbf{Ans.}$$

Reaction $\quad R_D = W - F_A - F_B$ (From statics)

$$\therefore \quad R_D = W - \frac{W.c}{d_2(a+b)^2+d_1a^2}\,[(a+b)d_2+ad_1]$$

$$= W\left[\frac{d_1a^2+(a+b)^2\,d_2 - c\,\{(a+b)d_2+ad_1\}}{(a+b)^2d_2+a^2d_1}\right]$$

or

$$R_D = W\left[\frac{ad_1\,(a-c)+(a+b)d_2\,(a+b-c)}{a^2d_1+(a+b)^2\,d_2}\right].\ \textbf{Ans.}$$

Example 1.50. *Three bars made of copper, zinc and aluminium are of equal length and have cross-section 500,750 and 1000 mm² respectively. They are rigidly connected at their ends. If the compound member is subjected to a longitudinal pull of 250 kN, estimate the proportion of load carried on each rod, and the induced stresses. Take,* $E_C = 1.3 \times 10^5$ *N/mm²,* $E_Z = 1 \times 10^5$ *N/mm² and* $E_A = 0.8 \times 10^5$ *N/mm².*

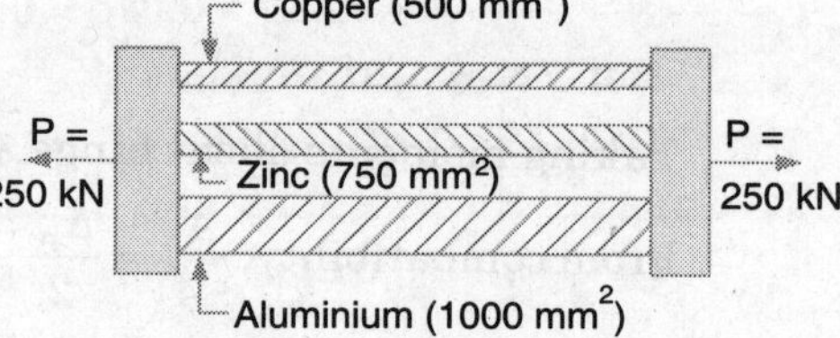

Fig. 1.56

Solution. From statics $P_C + P_Z + P_A = P = 250 \qquad ...(1)$

Since all the rods are rigidly connected at their ends, their deformations will be equal.

$$\therefore \quad \frac{P_C.L}{A_C.E_C} = \frac{P_Z.L}{A_Z.E_Z} = \frac{P_A.L}{A_A.E_A}$$

Hence $\quad P_Z = P_C \cdot \frac{A_Z}{A_C} \cdot \frac{E_Z}{E_C} = \frac{750}{500} \times \frac{1\times10^5}{1.3\times10^5}\,P_C = \frac{15}{13}\,P_C \qquad ...[2\,(a)]$

and $\quad P_A = P_C \cdot \frac{A_A}{A_C} \cdot \frac{E_A}{E_C} = \frac{1000}{500} \times \frac{0.8\times10^5}{1.3\times10^6}\,P_C = \frac{16}{13}\,P_C \qquad ...[2\,(b)]$

$\therefore$ From (1), $P_C + \frac{15}{13}P_C + \frac{16}{13}P_C = 250$ from which $P_C = \frac{250\times13}{44} = \mathbf{73.86\ kN}$

$$P_Z = \frac{15}{13} \times 73.86 = \mathbf{85.23\ kN}\ ;\ P_A = \frac{16}{13} \times 73.86 = \mathbf{90.91\ kN}$$

Also, $\quad p_C = \frac{73.86}{500} \times 1000 = \mathbf{147.33\ N/mm^2}\ ;\ p_Z = \frac{85.23}{750} \times 1000 = \mathbf{113.64\ N/mm^2}$

$$p_A = \frac{90.91}{1000} \times 1000 = \mathbf{90.91\ N/mm^2}.$$

Example 1.51. *A rigid bar LO is pinned at L and attached to the bars MN and OP, as shown in Fig. 1.57. The whole system is initially stress free and the weight of all bars are negligible. The temperature of the bar MN is lowered by 25 K and that of bar OP is raised by 25 K. Neglecting any possibility of lateral buckling, find the normal stresses in bar MN and*

OP. For MN which is brass, assume E = 0.9 × 10⁵ N/mm², α = 20 × 10⁻⁶ per K, and for OP, which is steel, take E = 2 × 10⁵ N/mm² and α = 12 × 10⁻⁶ per K. The cross-sectional area of MN is 500 mm², of OP is 600 mm².

Solution. The expansion of steel bar due to temperature,

$$\Delta_S^t = 250\,(12 \times 10^{-6}) \times 25$$
$$= 0.075 \text{ mm (increase)}$$

Free contraction of brass bar due to temperature,

$$\Delta_B^t = 300\,(20 \times 10^{-6})\,25$$
$$= 0.15 \text{ mm (decrease)}$$

For no temperature stress to be developed,

$$\frac{\Delta_B^t}{250} \text{ should be equal to } \frac{\Delta_S^t}{600}$$

Hence $\Delta_B^t = \Delta_S^t \times \dfrac{250}{600} = 0.417\ \Delta_S^t$

$$= 0.417 \times 0.075 = 0.03125 \text{ mm}$$

But $\Delta_B^t = 0.15$ mm.

Fig. 1.57

Hence in the equilibrium position attained due to rigidity of bar *LO*, tensile stress will be developed in the brass bar, and consequently, tensile stress will also be developed in steel bar.

If Δ_S and Δ_B are the *final deformations* in the two bars, at equilibrium position, we have

$$\frac{\Delta_B}{250} = \frac{\Delta_S}{600} \quad \text{or} \quad \Delta_B = \frac{250}{600}\,\Delta_S = 0.417\ \Delta_S \qquad ...(i)$$

where
$$\Delta_B = \Delta_B^t - \Delta_B^p = L_B \,.\, \alpha_B t_B - \frac{p_B L_B}{E_B} = 0.15 - \frac{p_B L_B}{E_B}$$

and
$$\Delta_S = \Delta_S^t + \Delta_S^p = L_S \,.\, \alpha_S \,.\, t_S + \frac{p_S \,.\, E_S}{E_S} = 0.075 + \frac{p_S \,.\, L_S}{E_S}$$

Substituting in (*i*), we get

$$\left(0.15 - \frac{p_B \times 300}{0.9 \times 10^5}\right) = 0.417\left(0.075 + \frac{p_S \times 250}{2 \times 10^5}\right)$$

or $\quad 52.125\, p_S + 333.33\, p_B + (3127.5 - 15000) = 0 \quad$ or $\quad p_S + 6.395\, p_B = 227.8 \qquad ...(ii)$

Taking moment about *L*,

$$(p_B A_B)\,250 - p_S A_S \times 600 = 0 \quad \text{or} \quad p_B \times 500 \times 250 - p_S \times 250 \times 600 = 0$$

$$\therefore \qquad p_B = 1.2\, p_S \qquad ...(iii)$$

From (*ii*) and (*iii*),

$$p_S = \frac{227.8}{1 + 6.395 \times 1.2} = \mathbf{26.26\ N/mm^2} \text{ and } p_B = 1.2 \times 26.26 = \mathbf{31.51\ N/mm^2}.$$

Example 1.52. *A rigid bar shown in Fig. 1.58 is hinged at A and supported by a steel rod at B. Determine the largest load P that can be applied at C if the stress in the rod is limited to 140 N/mm² and the vertical movement of end C is not to exceed (a) 2 mm, (b) 2.5 mm. Take E = 2 × 10⁵ N/mm² for steel rod.*

Solution. (Fig. 1.58)

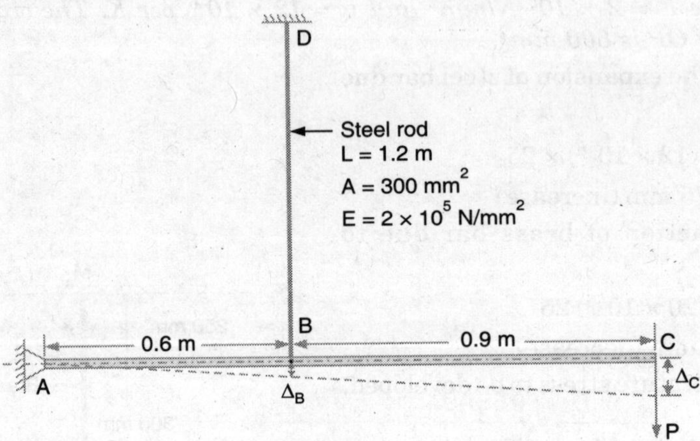

Fig. 1.58

(*a*) For Δ_C not to exceed 2 mm, $\dfrac{\Delta_B}{0.6} = \dfrac{\Delta_C}{1.5}$

or $$\Delta_B = \frac{0.6}{1.5}\,\Delta_C = \frac{0.6}{1.5} \times 2 = 0.8 \text{ mm}$$

But $$\Delta_B = \frac{p_B L}{E} \quad \text{or} \quad p_B = \frac{E\Delta_B}{L} = \frac{2 \times 10^5 \times 0.8}{1200} = 133.3 \text{ N/mm}^2 < 140.$$

$$\therefore \quad P_B = p_B.A = 133.3 \times 300 = 40000 \text{ N}$$

$$\therefore \quad P = P_B \times \frac{0.6}{1.5} = 40000 \times \frac{0.6}{1.5} = 16000 \text{ N} = \mathbf{16\ kN}.$$

(*b*) For Δ_C not to exceed 2.5 mm, $\Delta_B = \dfrac{0.6}{1.5} \times 2.5 = 1$ mm

$$\therefore \quad p_B = \frac{2 \times 10^5 \times 1}{1200} = 167 \text{ N/mm}^2 > 140 \text{ N/mm}^2 \quad \therefore \quad (p_B)_{max} = 140 \text{ N/mm}^2$$

$$\therefore \quad P_B = (p_B)_{max} \times 300 = 140 \times 300 = 42000 \text{ N} = 42 \text{ kN}$$

$$\therefore \quad P = P_B \times \frac{0.6}{1.5} = \frac{42 \times 0.6}{1.5} = \mathbf{16.8\ kN.}$$

Example 1.53. *A rigid bar ACBD, of negligible mass is pinned at C and is attached to two vertical rods BE and DF. Rod BE is of steel having E = 2 × 10⁵ N/mm² and rod DF is of bronze, having E = 0.83 × 10⁵ N/mm². Assuming that the rods were initially stress free, what maximum load P can be applied without exceeding stresses of 150 N/mm² in steel rod and 70 N/mm² in bronze rod.*

Solution. Let p_S = Stress in steel rod (limited to 150 N/mm²) ; p_B = Stress in brass rod (limited to 70 N/mm²)

For the equilibrium of the rod, $1.2\,P = 0.9\,p_S A_S + 1.8\,p_B.A_B$ ...(1)

Also, from compatibility, $\dfrac{\Delta_S}{0.9} = \dfrac{\Delta_B}{1.8}$ or $\Delta_S = \dfrac{0.9}{1.8}\,\Delta_B = 0.5\,\Delta_B$

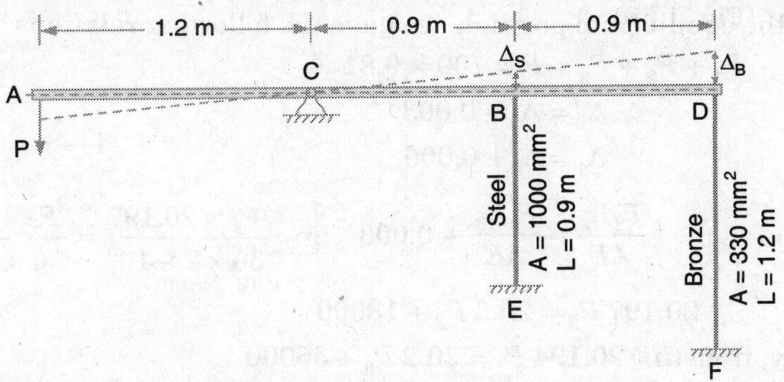

Fig. 1.59

$$\therefore \quad \frac{p_S L_S}{E_S} = \frac{0.5\, p_B L_B}{E_B}$$

or $$p_S = 0.5\,\frac{L_B}{L_S}\cdot\frac{E_S}{E_B}\cdot p_B = 0.5 \times \frac{1.2}{0.9} \times \frac{2}{0.83}\, p_B = 1.606\, p_B \quad \text{...[2 (a)]}$$

or $$p_B = 0.6225\, p_S \quad \text{...[2 (b)]}$$

If p_B is kept equal to its maximum permissible value of 70 N/mm², we get p_S = 1.606 × 70 = 112.42 N/mm², which is less than maximum permissible value of 140 N/mm². On the other hand, if p_S is kept equal to its maximum permissible value of 140 N/mm², we get p_B = 0.6225 × 140 = 87.15 N/mm² which exceeds the permissible value of 70 N/mm².

Hence the governing value of stress is p_B = 70 N/mm² and the corresponding value of p_S = 112.42 N/mm². Substituting these in Eq. (1), we get

$$1.2P = (0.9 \times 112.42 \times 1000) + (1.8 \times 70 \times 330)$$

From which $$P = 118965 \simeq \mathbf{118.97\ kN.}$$

Example 1.54. *As shown in Fig. 1.60, three steel wires, each 30 mm² in area are used to lift a load W = 700 kg. Their unstressed lengths are 20.194 m, 20.197 m and 20.200 m. (a) What stress exists in the longest wire ? (b) Determine the stress in the shortest wire if W = 235 kg. Take E = 2 × 10⁵ N/mm².*

Solution. (*a*) When the stretching in the shortest wire reaches 0.003 m, load taken by it is,

$$P = \frac{\Delta AE}{L} = \frac{0.003 \times 30 \times 2 \times 10^5}{20.194} = 891.35 \text{ N} = 90.86 \text{ kg} < 700 \text{ kg}.$$

Hence the second wire will also be stretched. Thus, when the second wire get stretched by an amount 0.003 m and the first wire now by an amount = 0.003 + 0.003 = 0.006 m, then Δ_1 = 0.006 m and Δ_2 = 0.003 m.

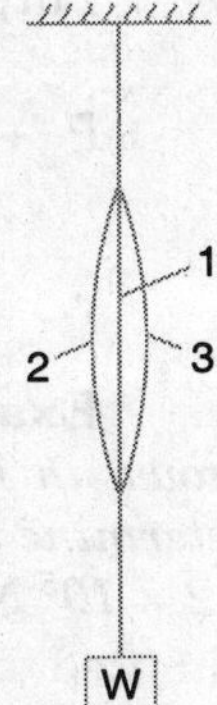

Fig. 1.60

$$\therefore \quad P_1 = \frac{0.006 \times 30 \times 2 \times 10^5}{20.194} = 1782.7 \text{ N} = 181.72 \text{ kg}$$

and $$P_2 = \frac{0.003 \times 30 \times 2 \times 10^5}{20.197} = 891.22 \text{ N} = 90.84 \text{ kg}$$

$\therefore \; P_1 + P_2$ = 181.72 + 90.84 = 272.56 kg < 700 kg.

Thus, the third wire also gets stretched.

Now, at the equilibrium position, we have the following relations :

$$P_1 + P_2 + P_3 = P = 700 \times 9.81 \text{ N} \quad ...(i)$$

$$\Delta_2 = \Delta_3 + 0.003 \quad ...(ii)$$

$$\Delta_1 = \Delta_3 + 0.006 \quad ...(iii)$$

From (*ii*), $\dfrac{P_2 L_2}{AE} = \dfrac{P_3 L_3}{AE} + 0.003$ or $\dfrac{P_2 \times 20.197}{30 \times 2 \times 10^5} = \dfrac{P_3 \times 20.200}{30 \times 2 \times 10^5} + 0.003$

or $$20.197\, P_2 = 20.2\, P_3 + 18000 \quad ...(iv)$$

Similarly, from (*iii*) $20.194\, P_1 = 20.2\, P_3 + 36000$ ...(*v*)

Putting these values of P_1 and P_2 in eqn. (*i*), we get

$$\frac{20.2\, P_3 + 36000}{20.194} + \frac{20.2\, P_3 + 18000}{20.197} + P_3 = 700 \times 9.81$$

or $$P_3 = 1397.69 \text{ N and } p_3 = \frac{1397.69}{30} = 46.59 \text{ N/mm}^2.$$

(*b*) When $W = 235$ kg is applied, then as the second wire gets stretched by an amount of 0.003 m,

$$P_1 + P_2 = 272.56 \text{ kg} > 235 \text{ kg}.$$

Hence the second wire does not get fully stretched. Thus, the final equilibrium equations are

$$P_1 + P_2 = 235 \times 9.81 \text{ N} \quad ...(vi)$$

and $$\Delta_1 = \Delta_2 + 0.003 \text{ m} \quad ...(vii)$$

From eqn. (*vii*), $\dfrac{P_1 \times 20.194}{30 \times 2 \times 10^5} = \dfrac{P_2 \times 20.197}{30 \times 2 \times 10^5} + 0.003$

or $$20.194\, P_1 = 20.197\, P_2 + 18000$$

or $$P_2 = \frac{20.194\, P_1}{20.197} - \frac{18000}{20.197}$$

Putting this value of P_2 in eqn. (*vi*),

$$P_1 + \frac{20.194\, P_1}{20.197} - \frac{18000}{20.197} = 235 \times 9.81 \quad \text{or} \quad P_1 = 1598.28 \text{ N}$$

$$\therefore \quad p_1 = \frac{1598.28}{30} = \mathbf{53.27 \text{ N/mm}^2}.$$

Example 1.55. *A light rigid bar ABCD pinned at B and connected to two vertical rods, is shown in Fig. 1.61. Assuming that the bar was initially horizontal and the rods stress free, determine the stress in each rod after the load P = 1000 N is applied at end D. Take E for steel = 2 × 10⁵ N/mm² and E for aluminium as 1 × 10⁵ N/mm².*

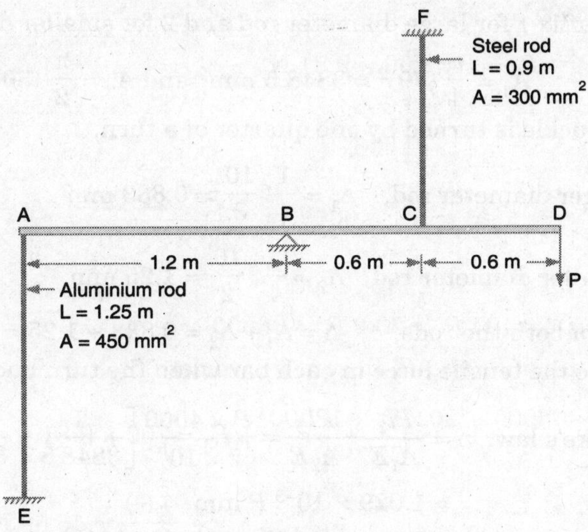

Fig. 1.61

Solution. (Fig. 1.62).

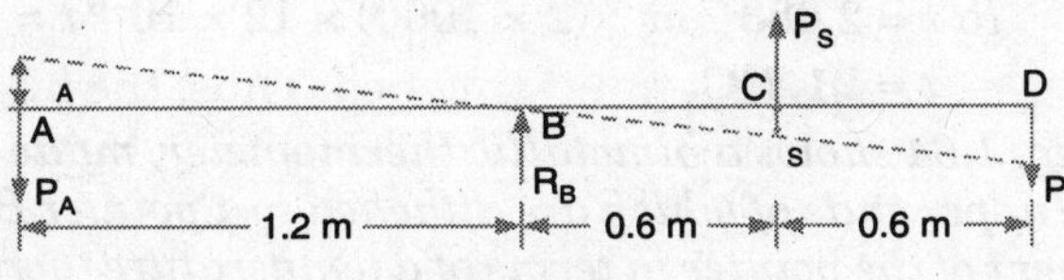

Fig. 1.62

Let P_A and P_S be the forces in the aluminium and steel bars, respectively, as shown in Fig. 1.62. At the equilibrium condition, we have

$$P_A \times 1.2 + P_S \times 0.6 = P \times 1.2 \quad \text{or} \quad 2P_A + P_S = 2P = 2000 \text{ N} \quad ...(i)$$

Also,
$$\frac{\Delta_A}{1.2} = \frac{\Delta_S}{0.6} \quad \text{or} \quad \Delta_A = 2\Delta_S$$

or
$$\frac{P_A \times 1.25}{450 \times 1 \times 10^5} = \frac{2 \times P_S \times 0.9}{300 \times 2 \times 10^5} \quad \text{or} \quad P_A = 1.08\, P_S \quad ...(ii)$$

Hence, from eqn. (*i*), P_A = 632.9 N and hence P_S = 586.03 N

$$\therefore \qquad p_A = \frac{632.9}{450} = \mathbf{1.41\ N/mm^2}\ ;\ p_S = \frac{586.03}{300} = \mathbf{1.95\ N/mm^2}.$$

Example 1.56. *Two steel rods, one of 70 mm diameter and the other of 50 mm diameter and joined end to end by means of a turn buckle, as shown in Fig. 1.63. The other end of each rod is rigidly fixed and there is initially a little tension in the rods. The effective length of each rod is 4 m. Calculate the increase in the tension when the turn buckle is tightened by one quarter of a turn if there are 2 threads per cm on the bigger diameter and 3 threads per cm on the other end. Neglect the extension of turn buckle. Also, find what rise in temperature could nullify the increase in tension. Take E = 2 × 10⁵ N/mm² and α = 12 × 10⁻⁶ per °C.*

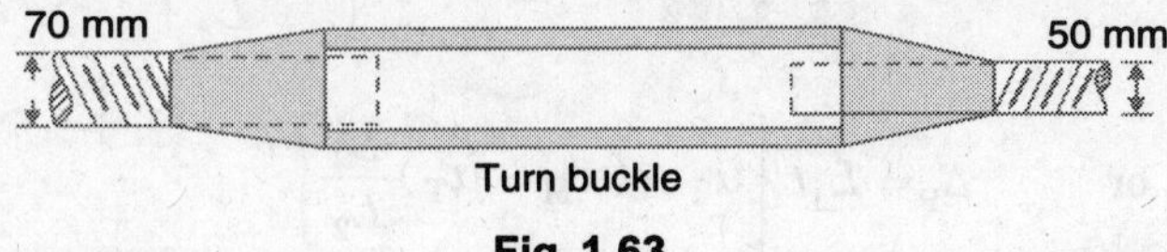

Fig. 1.63

Solution. Use suffix 1 for large diameter rod and 2 for smaller diameter rod.

$\therefore \quad A_1 = \frac{\pi}{4}(70)^2 = 3848.5 \text{ mm}^2$ and $A_2 = \frac{\pi}{2}(50)^2 = 3927 \text{ mm}^2$

When the turn-buckle is turned by one quarter of a turn,

Extension of larger diameter rod, $\Delta_1 = \frac{1}{4} \cdot \frac{10}{3} = 0.833$ mm

Extension of smaller diameter rod $\Delta_2 = \frac{1}{4} \cdot \frac{10}{2} = 1.25$ mm.

Total extension of both the rods, $\Delta = \Delta_1 + \Delta_2 = 0.833 + 1.25 = 2.083$ mm ...(*i*)

Let P Newtons be the tensile force in each bar when the turn buckle is tightened.

Hence from Hooke's law, $\Delta = \frac{Pl_1}{A_1E} + \frac{Pl_2}{A_2E} = \frac{P \times 4000}{2 \times 10^5}\left[\frac{1}{3848.5} + \frac{1}{3927}\right]$

$= 1.029 \times 10^{-5} P$ mm ...(*ii*)

Equating (*i*) and (*ii*), we get $P = 202434$ N = **202.434 kN**

Now, in order to nullify this tension, the rise of temperature of the two rods must be such that the total thermal expansion of the two rods is equal to 2.083 mm.

$\therefore \quad l\alpha t = 2.083 \quad$ or $\quad (2 \times 4000) \times 12 \times 10^{-6}\, t = 2.083$

From which t = **21.7°C.**

Example 1.57. *Fig. 1.64 shows a bimetallic thermometer, made of a tungsten bar AB and a magnesium bar CD the upper ends of which are attached to a pointer BDP. Derive an expression for the vertical displacement of the pointer in terms of a temperature increase t and the dimensions shown.*

Solution. The coefficient of thermal expansion of magnesium is more than that of tungsten. Hence the pointer P will move up (*i.e.*, +ve vertical displacement) for rise of temperature and vice-versa.

Now $\Delta_B = L_1\, \alpha_T\, t$ and $\Delta_D = L_1\, \alpha_M\, t$

$\therefore \quad (\Delta_D - \Delta_B) = L_1 t\,(\alpha_M - \alpha_T)$

Fig. 1.64 (*b*) shows the displacement diagram, from which :

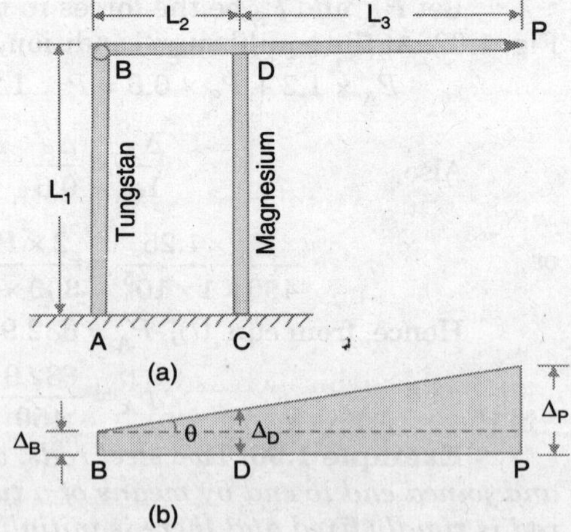

Fig. 1.64

$$\Delta_P = \Delta_B + \frac{\Delta_D - \Delta_B}{L_2}(L_2 + L_3)$$

$$= L_1\alpha_T t + \frac{L_1\, \alpha_M t - L_1\, \alpha_T t}{L_2}(L_2 + L_3)$$

or $$\Delta_P = L_1\alpha_T t + L_1 t\,(\alpha_M - \alpha_T)\left(1 + \frac{L_3}{L_2}\right)$$

or $$\Delta_P = L_1 t\left[\alpha_T + (\alpha_M - \alpha_T)\frac{L_3}{L_2}\right].$$

Example 1.58. *A solid cylinder of steel is placed inside a copper tube. The assembly is compressed between rigid plates by forces P. Find the value of increase in temperature so that all the load is carried by the copper tube.*

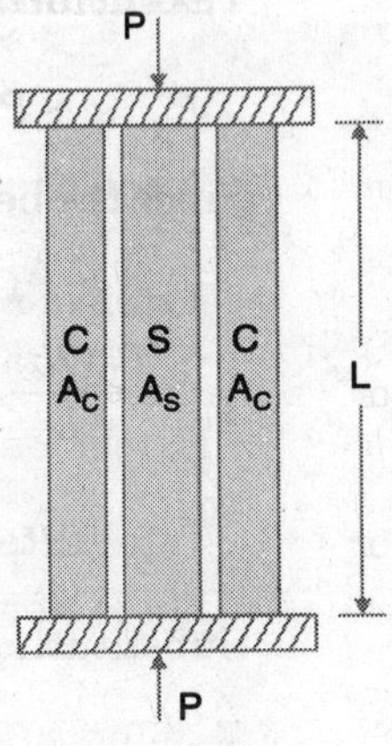

Fig. 1.65

Solution. Let the increase in temperature be t. Due to this increase, copper will expand more than steel.

Difference between the two expansions $= \Delta_t = L\,(\alpha_c - \alpha_s)\,t$

If the load P is to be borne entirely by the copper tube, the shortening (Δ_c) of the copper tube due to P should be equal to Δ_t

$$\therefore \qquad \Delta_c = \Delta_t$$

or
$$\frac{PL}{A_c E_c} = L\,(\alpha_c - \alpha_s)\,t$$

or
$$t = \frac{P}{A_c E_c\,(\alpha_c - \alpha_s)}.$$

Example 1.59. *A steel bar AB of length L is fixed at the ends and is subjected to a non-uniform increase in temperature represented by the expression $t_x = t \cdot \dfrac{x^2}{L^2}$, as shown in Fig. 1.66. Determine the value of compressive stress set up in the bar.*

Solution. Consider a short length dx, at a distance x from end A.

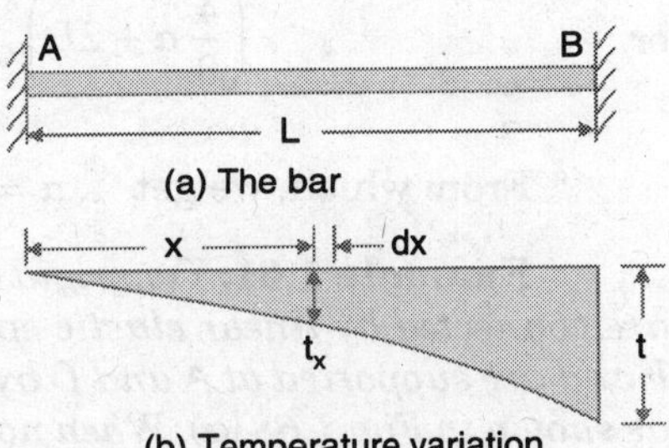

Fig. 1.66

Temperature $\quad t_x = t\,\dfrac{x^2}{L^2}$

$\therefore \quad$ Increase in length of dx due to temperature rise

$$= \Delta_{tx} = dx \cdot \alpha t_x = \alpha t\,\frac{x^2}{L^2}\,dx$$

Hence total increase in length $= \Delta_t$

$$= \Sigma\Delta_{t\,x} = \int_0^L \alpha t\,\frac{x^2}{L^2}\,dx = \frac{\alpha t}{L^2} \cdot \frac{L^3}{3} = \frac{\alpha t L}{3}$$

Since this increase in length is restricted by end supports, compressive stress will be set up in the bar, the value of which is given by.

$$p_c = \Delta_t \cdot \frac{E}{L} = \frac{\alpha t L}{3}\left(\frac{E}{L}\right) = \frac{E\,\alpha t}{3}.$$

Example 1.60. *Three steel bars A, B and C, having the same axial rigidity EA support a horizontal rigid beam ABC as shown in Fig. 1.67. Determine the distance a between bars A and B in order that the rigid beam will remain horizontal when a load P is applied at its mid-point.*

Solution. Let us use suffix 1, 2 and 3 for bars A, B and C respectively.

From statics, $\quad P_1 + P_2 + P_3 = P \qquad$ ...(1)

Taking moments about A,

$$P_2 \,.\, a + P_3 L = P \,.\, \frac{L}{2} \quad ...(2)$$

Since the beam ABC remains horizontal,

$$\Delta_1 = \Delta_2 = \Delta_3$$

or
$$\frac{P_1\,(2b)}{AE} = \frac{P_2\,(1.5b)}{AE} = \frac{P_3\,.\,b}{AE}$$

or
$$2P_1 = \frac{3}{2}\,P_2 = P_3 \quad ...(3)$$

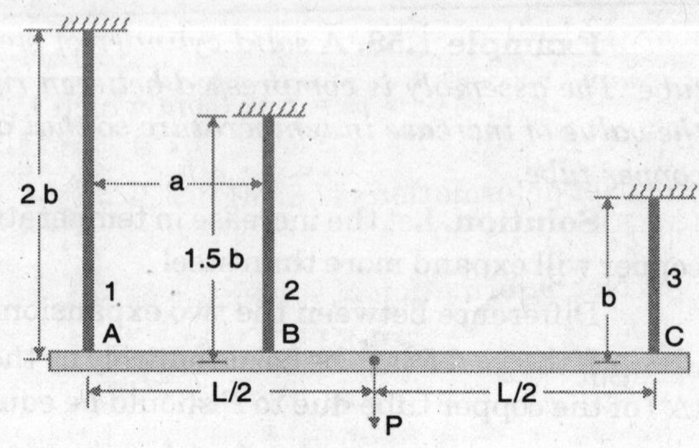

Fig. 1.67

Substituting in (1), we get

$$P_1 + \frac{4}{3}\,P_1 + 2P_1 = P, \text{ from which } P_1 = \frac{3}{13}\,P \quad ...(4)$$

Also, from (2), $\left(\frac{4}{3}\,P_1\right)\,a + 2P_1 L = P \,.\, \frac{L}{2}$

or
$$\left(\frac{4}{3}\,a + 2L\right)\,P_1 = P \,.\, \frac{L}{2}$$

or
$$\left(\frac{4}{3}\,a + 2L\right)\,\frac{3}{13}\,P = \frac{PL}{2}$$

From which, we get $\quad a = \frac{L}{8}$.

Example 1.61. *Two rigid bars AB and CD are connected by linear elastic springs of stiffness k and are supported at A and D by hinged supports as shown in Fig. 1.68 (a). When no loads are acting, the bars are horizontal and the springs are unstressed. Determine the vertical deflection at point C when the point load P is applied.*

Solution. Let P_1 be the force in spring 1 at C and P_2 be the force in spring 2. Fig. 1.68 (*b*) shows the deformation diagram while Fig. 1.68 (*c*) shows the free body diagrams for bars AB and CD. A careful study of these diagrams will reveal that spring 1 (at C) carries a tensile force (P_1) and spring 2 (at B) carries a compressive force P_2.

Taking moments about A [Fig. 1.68 (*c*)], we get $P_1 = 3P_2$.

Taking moments about D [Fig. 1.68 (*c*)], we get $(P_1 - P)\,3a = P_2 a$

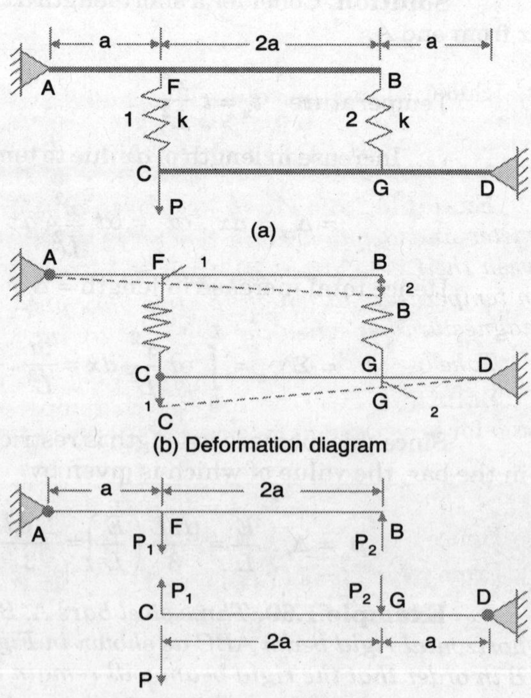

Fig. 1.68

or $$(P_1 - P)3 = \frac{P_1}{3}, \text{ from which } P_1 = \frac{9}{8} P \text{ and hence } P_2 = \frac{3}{8} P$$

Net deformation of spring 1 = $\delta_1 - \delta_1' = \Delta_1 = \frac{P_1}{k}$ (extension)

Net deformation of spring 2 = $\delta_2' - \delta_2 = \Delta_2 = \frac{P_2}{k}$ (compression)

But $$\frac{\delta_1}{\delta_2} = \frac{3a}{a}, \text{ from which } \delta_1 = 3\delta_2 \quad ...(i)$$

Also, $$\frac{\delta_2'}{\delta_1'} = \frac{3a}{a}, \text{ from which } \delta_2' = 3\delta_1' \quad ...(ii)$$

Now $$\frac{P_1}{k} = \delta_1 - \delta_1' = \delta_1 - \frac{\delta_2'}{3}$$

or $$\frac{3P_1}{k} = 3\delta_1 - \delta_2' \quad ...(a)$$

and $$\frac{P_2}{k} = \delta_2' - \delta_2 = \delta_2' - \frac{\delta_1}{3}$$

or $$\frac{P_2}{k} = -\frac{\delta_1}{3} + \delta_2' \quad ...(b)$$

Eliminating δ_2' from (*a*) and (*b*), we get $\delta_1 = \frac{3}{8k}(3P_1 + P_2)$ ...(I)

Substituting the values of P_1 and P_2, we get

$$\delta_1 = \frac{3}{8\,k}\left(3 \times \frac{9}{8} P + \frac{3}{8} P\right) = \frac{45\,P}{32\,k}.$$

Example 1.62. *Fig. 1.69 shows a bimetallic thermal control, made of a brass bar of 8 mm diameter and a magnesium bar of 12 mm diameter. The bars are so arranged that the gap between their ends are 0.1 mm at room temperature. Find the temperature increase above the room temperature at which the two bars just come in contact. Also find the compressive stress in magnesium bar when the temperature increase in 320° F.*

Take $\alpha_b = 10 \times 10^{-6}$ *per° F*, $\alpha_m = 15 \times 10^{-6}$ *per° F*; $E_b = 100\ kN/mm^2$ *and* $E_m = 40\ kN/mm^2$.

Solution. Let t be the rise in temperature. Using suffix b for brass and m for magnesium, we have

$$\Delta_{bt} = L_b\,\alpha_b\,t = 25 \times 10 \times 10^{-6}\,t$$

and $$\Delta_{mt} = L_m\,\alpha_m\,t = 40 \times 15 \times 10^{-6}\,t$$

Hence $\Delta_{bt} + \Delta_{mt} = 0.1 = 250 \times 10^{-6}\,t + 600 \times 10^{-6}\,t$

From which t = **117.65°F**

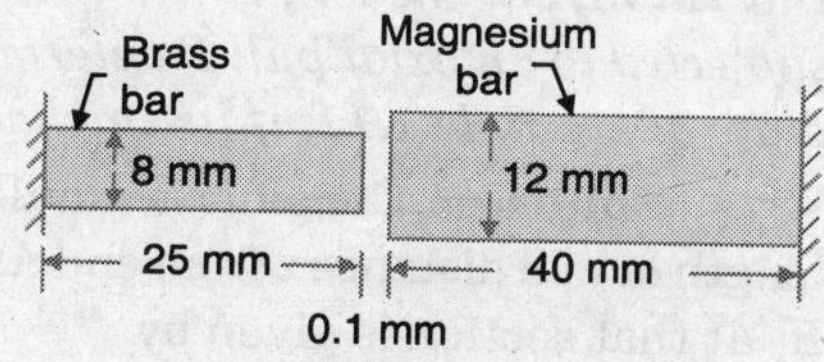

Fig. 1.69

$$A_b = \frac{\pi}{4}(8)^2 = 50.27 \text{ mm}^2 \text{ and}$$

$$A_m = \frac{\pi}{4}(12)^2 = 113.1 \text{ mm}^2$$

When $t = 320°$ F, $\Delta_{bt} = 25 \times 10 \times 10^{-6} \times 320 = 0.08$ mm

and $\Delta_{mt} = 40 \times 15 \times 10^{-6} \times 320 = 0.192$ mm

Now expansion restricted, $\Delta_t = (\Delta_{bt} + \Delta_{mt}) - 0.1 = (0.08 + 0.192) - 0.1 = 0.172$ mm ...(*i*)

If *P* is the compressive force induced, compression (Δ_c) of the two bars is

$$\Delta_c = \frac{PL_b}{A_bE_b} + \frac{PL_m}{A_mE_m} = P\left[\frac{25}{50.27 \times 100} + \frac{40}{113.1 \times 40}\right] = 0.0138\,P \quad ...(ii)$$

Equating (*i*) and (*ii*), we get $0.0138\,P = 0.172$ from which $P = 12.45$ kN

$$\therefore \quad p_m = \frac{P}{A_m} = \frac{12.45}{113.1} \times 1000 = \mathbf{110.1\ N/mm^2}\ \text{(comp.)}$$

Example 1.63. *A vertical bar ABC is attached to a rigid bar BDF, as shown in Fig. 1.70 (a). Find the ratio of loads P_2 and P_1 so that the vertical deflection of point C is zero.*

Solution. From the moment equilibrium of the rigid bar *BDF*, we find that a vertical force *P*

(thrust) $= P_2 \times \dfrac{1.5\,L}{L} = 1.5\,P_2$ is transmitted to the vertical bar, through pin *B*. Hence the vertical bar *ABC* is subjected to forces as shown in Fig. 1.70 (*b*). The reaction $R_A = P - P_1$. Thus, the portion *AB* is subjected to a compressive force $= R_A\ (= P - P_1)$ and portion *BC* is subjected to a tensile force P_1.

Hence $\delta_c = \Sigma \dfrac{PL}{AE} = \dfrac{P_1(L)}{aE} - \dfrac{(P - P_1)(2L)}{2aE}$

Substituting the value of *P* and noting that δ_c has to be zero, we have

$$\frac{P_1L}{aE} - \frac{(1.5\,P_2 - P_1)\,2L}{2aE} = 0$$

or $\quad P_1 - 1.5\,P_2 + P_1 = 0 \quad$ or $\quad 2P_1 = \dfrac{3}{2}P_2$

From which $\quad \dfrac{P_2}{P_1} = \dfrac{4}{3}$

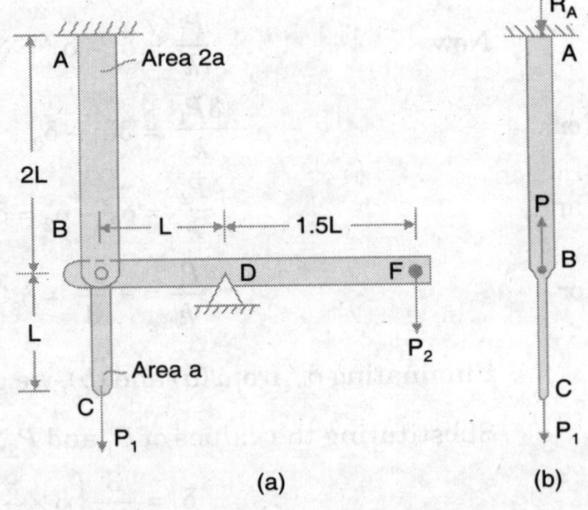

Fig. 1.70

Example 1.64. *The width of a trapezoidal section bar AB of length L tapers from B_1 at end A to B_2 at the end B. Its thickness also tapers from t_1 at end A to t_2 at the end B. If it is subjected to an axial pull P determine its elongation. Hence find the value of the extension of the plate of example 1.11, if its thickness varies from 8 mm at one end to 12 mm at the after end.*

Solution. Consider a small section of length δx at a distance *x* from end *A*. The width B_x at that section is given by

$$B_x = B_1 + \frac{B_2 - B_1}{L}\,x$$

or $\quad B_x = (B_1 + kx)$

where $\quad k = (B_2 - B_1)/L$

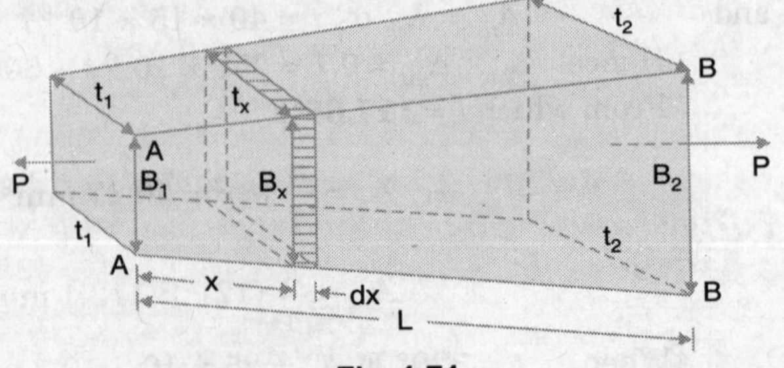

Fig. 1.71

Also, the thickness t_x at that section is given by

$$t_x = t_1 + \frac{t_2 - t_1}{L}\, x = (t_1 + k'x) \text{ where } k' = \frac{t_2 - t_1}{L}$$

Hence area A_x at that section $= B_x \cdot t_x = (B_1 + kx)\,(t_1 + k'x)$

$$\therefore \qquad \delta\Delta = \frac{P\,\delta x}{A_x E} = \frac{P\delta x}{(B_1 + kx)\,(t_1 + k'x)\,E}$$

where $\delta\Delta$ is the extension of the short length δx.

Hence the total extension Δ of the whole bar is given by

$$\Delta = \sum_{x=0}^{x=L} \delta\Delta = \int_0^L \frac{P\,dx}{(B_1 + kx)\,(t_1 + k'x)\,E}$$

or

$$\Delta = \frac{P}{E(kt_1 - k'B_1)} \left[\log_e \frac{B_1 + kx}{t_1 + k'x}\right]_0^L = \frac{P}{E(kt_1 - k'B_1)} \log_e \frac{t_1\,(B_1 + kL)}{B_1\,(t_1 + k'L)} \qquad \text{...*[1.11 (a)]}$$

when $\quad t_1 = t_2 = t, \quad k' = (t_2 - t_1)/L = 0$

$$\therefore \qquad \Delta = \frac{P}{Ekt} \log_e \frac{B_1 + kL}{B_1}, \text{ which is same as Eq. 1.11 } (a).$$

Substituting B_1 = 50 mm, B_2 = 100 mm, t_1 = 8 mm, t_2 = 12 mm and L = 400 mm, P = 50 kN

$k = (100 - 50)/400 = 0.125$; $k' = (12 - 8)/400 = 0.01$ and $E = 2.05 \times 10^5$, we get

$$\Delta = \frac{50000}{2.05 \times 10^5\,(0.125 \times 8 - 0.01 \times 50)} \log_e \frac{8\,(50 + 0.125 \times 400)}{50\,(8 + 0.01 \times 400)} = \mathbf{0.1403\ mm.}$$

Note that the elongation with a uniform thickness of 10 mm was 0.1352 mm in example 1.11.

1.17. EXAMPLES FROM COMPETITIVE EXAMINATIONS

Example 1.65. *A rigid block ABC weighing 200 kN is supported by three rods symmetrically placed as shown in the figure (Fig. 1.72). The lower ends of the rods are at the same level before the block is attached and the temperature changed. Assuming the block to remain horizontal, determine the stress in each rod after temperature rise of 20°C.*

(UPSC Engg. Services Exam., 1993)

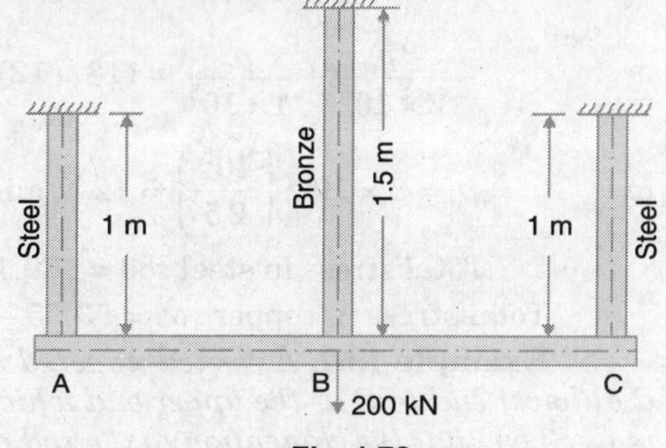

Fig. 1.72

Solution. *This problem was set exactly from Example 1.39 of this book. For solution, see Example 1.39.*

*Note that $\displaystyle\int_0^x \frac{dx}{(a_1 + b_1x)\,(a_2 + b_2x)} = \frac{1}{a_2b_1 - a_1b_2} \log_e \frac{a_2\,(a_1 + b_1x)}{a_1\,(a_2 + b_2x)}$.

Example 1.66. *A steel rod of 20 mm diameter passes centrally through a tight fitting copper tube of 40 mm external diameter. The tube is closed with the help of a rigid washer of negligible thickness and nuts threaded on the rod. The nuts are tightened, till the compressive load on the tube is 50 kN. Determine the stresses in the rod and the tube when the temperature of the assembly falls by 50°C. Take E for steel as 200 × 10³ N/mm² and for copper as 100 × 10³ N/mm² and the coefficient of expansion α for steel as 12 × 10⁻⁶ per °C and for copper as 18 × 10⁻⁶ per °C.* (UPSC Engg. Services Exam. 1994)

Solution. See example 1.26.

Area of rod $= A_s = \frac{\pi}{4}(20)^2 = 100\,\pi\text{ mm}^2$; Area of tube $= A_c = \frac{\pi}{4}(40^2 - 20^2) = 300\,\pi\text{ mm}^2$

(*a*) ***Effect of tightening of the nut.*** When the nut is tightened, compressive load (= 50 kN) is induced in tube while tensile load (= 50 kN) is induced in steel rod.

$$\therefore \quad p_s A_s = p_c A_c \quad \text{or} \quad p_s = \frac{A_c}{A_s}\,p_c = \frac{300\,\pi}{100\,\pi}\,p_c = 3p_c \qquad \ldots(i)$$

But $\quad p_c = \frac{P_c}{A_c} = \frac{50 \times 10^3}{300\,\pi} = 53.05\text{ N/mm}^2$ (compressive)

Hence $\quad p_s = 3p_c = 3 \times 53.05 = 159.15\text{ N/mm}^2$ (tensile).

(*b*) ***Effect of fall in temperature.*** When the temperature falls, copper tube will shrink more than the steel rod. Hence tensile stress will be induced in copper and compressive force will be induced in steel rod. Since total compressive force in steel rod = total tensile force in copper tube,

$$p_s A_s = p_c A_c \quad \text{or} \quad p_s = \frac{A_c}{A_s}\,p_c = \frac{300\,\pi}{100\,\pi}\,p_c = 3p_c$$

Now $\quad \Delta_s = \alpha_s L t + \frac{p_s}{E_s} L \quad \text{and} \quad \Delta_c = \alpha_c L t - \frac{p_c}{E_c} L$

Since $\Delta_s = \Delta_c$ at equilibrium position, we have

$$\alpha_s L t + \frac{p_s}{E_s} L = \alpha_c L t - \frac{p_c}{E_c} L$$

or $\quad 12 \times 10^{-6} \times 50 + \frac{p_s}{2 \times 10^5} = 18 \times 10^{-6} \times 50 - \frac{p_c}{1 \times 10^5}$

or $\quad \frac{3p_c}{2 \times 10^5} + \frac{p_c}{1 \times 10^5} = (18 - 12)\,10^{-6} \times 50 = 3 \times 10^{-4}$

or $\quad p_c = 3 \times 10^{-4}\left(\frac{10^5}{2.5}\right) = 12\text{ N/mm}^2$ (tensile) and $p_s = 3 \times 12 = 36\text{ N/mm}^2$ (comp.)

∴ Total stress in steel rod = 159.15 – 36 = **123.15 N/mm²** (tensile)

Total stress in copper rod = 53.05 – 12 = **41.05 N/mm²** (comp.)

Example 1.67. *A vertical tapered rod of length L has its diameter varying linearly from d at lower end to D at the upper end which has fixed support. Young's modulus of the material is E. Show that the elongation to the rod at its lower end when subjected to a longitudinal force F is given by* $\delta = \frac{4FL}{\pi\,dDE}$. (UPSC Engg. Services Exam. 1996)

Solution. ***This problem has been solved in § 1.8 of the book.***

Example 1.68. *A prismatic bar is fastened between two rigid walls at A and B subjected to loads as shown [Fig. 1.73(a)]. Determine the reactions at the supports.*

(UPSC Engg. Services Exam. 2001)

Solution. Refer Example 1.31.

Let R_A and R_B be the reactions at A and B, in the directions marked in Fig. 1.73 (*b*). The free body diagrams for AC, CD and DB are shown in Fig. 1.73 (*c*), (*d*) and (*e*) respectively

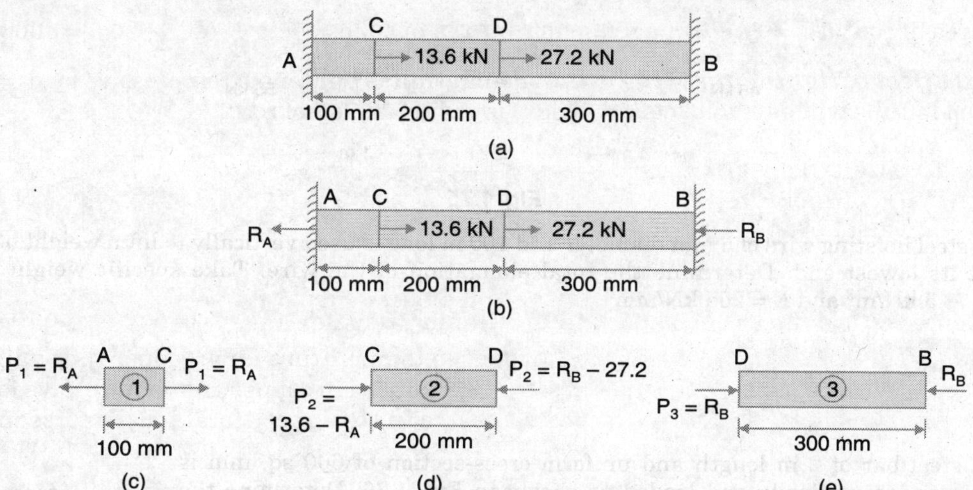

Fig. 1.73

From equilibrium $R_A + R_B = 13.6 + 27.2 = 40.8$ kN ...(1)

From compatibility, extension of AC = compression of CD and DB.

or $\Delta_1 = \Delta_2 + \Delta_3$ or $\dfrac{P_1L_1}{AE} = \dfrac{P_2L_2}{AE} + \dfrac{P_3L_3}{AE}$

$\therefore$ $R_A(100) = (13.6 - R_A)\,200 + R_B\,(300)$

Substituting the value of R_B from (1), we get

$R_A = (13.6 - R_A)\,2 + 3\,(40.8 - R_A)$, from which R_A = **24.933 kN**

Hence $R_B = 40.8 - 24.933 =$ **15.867 kN.**

PROBLEMS

1. A circular rod of steel 12 mm diameter is tested in a testing machine and it is found that when the tension is 16 kN the total extension on a 20 cm length is 0.14 mm. Find the value of E.
2. The piston of a steam engine is 40 cm diameter and the piston rod 6 cm diameter. The steam pressure is 1.05 N/mm^2. Find the stress in the piston rod and the elongation of a length of 75 cm, taking E = 205 kN/mm^2 when the piston is on the in-stroke.
3. A rod as shown in Fig. 1.74 is subjected to a pull of 500 kN on the ends. Taking E = 205 kN/mm^2, find the extension of the rod.

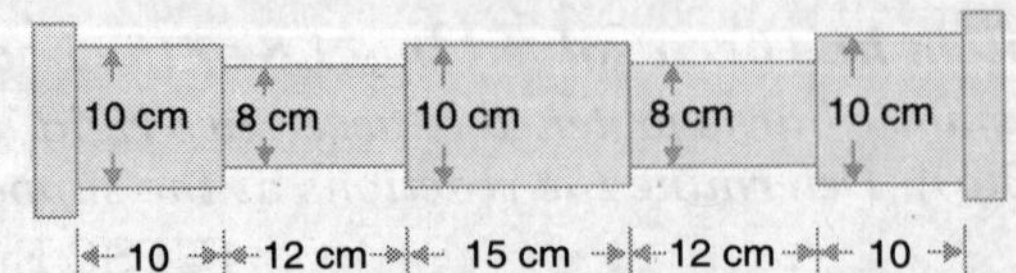

Fig. 1.74

4. A steel bar 3.2 m long is acted upon by forces as shown in Fig. 1.75. Determine the elongation of the bar. Take E = 205 kN/mm^2.

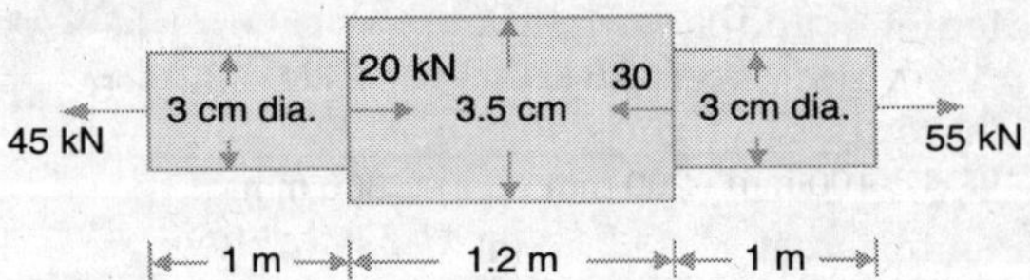

Fig. 1.75

5. A steel hoisting wire of 5 mm diameter and 100 m long is used vertically to lift a weight of 2000 N at its lowest end. Determine the total elongation of the wire. Take specific weight of steel = 78.5 kN/m^3 and E = 205 kN/mm^2.

6. A steel bar of 2 m length and uniform cross-section of 600 sq. mm is suspended vertically and loaded as shown in Fig. 1.76. Determine the total elongation of the bar. Take $E = 2.05 \times 10^5$ N/mm^2.

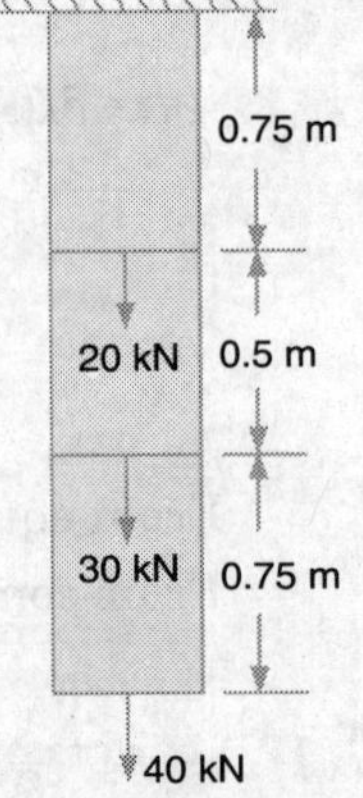

Fig. 1.76

7. A flat steel plate is of trapezoidal form. The thickness of the plate is 15 mm and it tapers uniformly from a width of 60 mm to 10 mm in a length of 30 cm. If an axial force of 100 kN is applied at each end, determine the elongation of the plate. Take E = 204 kN/mm^2.

8. A conical bar tapers uniformly from a diameter of 4 cm to 1.5 cm in a length of 40 cm. If an axial force of 80 kN is applied at each end, determine the elongation of the bar. Take $E = 2 \times 10^5$ N/mm^2.

9. A solid conical bar tapers uniformly from a diameter of 6 cm to 2 cm in a length of 100 cm. It is suspended vertically at the 6 cm diameter, the 2 cm diameter end being downward. Calculate the elongation of the bar due to self-weight. Take unit weight of the bar material as 78.5 kN/m^3 and E = 204 kN/mm^2.

10. A reinforced concrete column 45 cm × 45 cm has four steel rods 25 mm diameter embedded in it. Find the stresses in steel and concrete when the total load on the column is 1000 kN. Find also the adhesive force between the steel and concrete. E_s = 205 kN/mm^2 ; E_c = 13.6 kN/mm^2.

11. Calculate the stresses that will be produced in a steel bolt and a copper tube by $\frac{1}{4}$ of a turn of the nut if the length of the bolt (and that of the tube) is 50 cm, the pitch of the bolt being 3 mm. The area of cross-section of the bolt is 650 sq. mm and that of tube is 1300 sq. mm.

If a tensile load of 15 kN is applied at the ends, calculate the resulting stress in the bolt and the tube. Take E = 205 kN/mm^2 for steel ; E = 110 kN/mm^2 for copper.

12. A load of 20 kN is jointly supported by three vertical rods equal in length, each 15 mm diameter, and equidistant in a vertical plane. Initially, the rods are so adjusted as to share the load equally. The outer rods are of steel and the middle one of copper. Calculate the total stresses if a further load of 15 kN is added. Take E_c = 110 kN/mm^2 ; E_s = 205 kN/mm^2.

13. A steel rod 20 mm diameter and screwed at the end passes through a copper tube of 23 mm internal diameter and 30 mm external diameter. The temperature of the whole assembly is raised to 395 K and the nuts are tightened. Calculate the stresses if the common temperature falls to 310 K.

Coefficient of expansion for steel $= 11 \times 10^{-6}$ per K

Coefficient of expansion for copper $= 19 \times 10^{-6}$ per K

E for steel $= 2.05 \times 10^5$ N/mm^2

E for copper $= 1.10 \times 10^5$ N/mm^2.

14. Two walls 10 metres apart are stayed together by a 30 mm diameter steel rod with metal plates and nuts screwed at each end. The nuts are screwed when the bar is heated to a temperature of 420 K. Calculate the pull exerted by the bar on cooling to a temperature 300 K if (*a*) the ends do not yield, (*b*) if they yield by 3 mm. Take α for steel $= 11.5 \times 10^{-5}$ per K and $E = 2.05 \times 10^5$ N/mm^2.

15. A load of 20 kN rests on a rigid beam and is being taken equally by three vertical struts, each of 20 mm diameter and equi-spaced in a vertical plane. The central strut is of gun metal and the two outer ones are of steel. The temperature is then raised through 30 K. Calculate the stress in each strut, before and after the rise of temperature. Take E_s = 204 kN/mm^2 ; E_g = 92 kN/mm^2 ; $\alpha_s = 11 \times 10^{-5}$ per K and $\alpha_g = 19 \times 10^{-6}$ per K.

16. A rigid bar is supported by three rods, the outer one of steel and the central one of copper. The cross-sectional area of each steel rod is 300 mm^2 and of the copper rod is 1000 sq. mm. The three rods are equally spaced and the loads of 50 kN are each applied midway between the rods. Determine the forces in each of the vertical bars if the rigid bar remains horizontal after the loads have been applied. Neglect the weight of the rigid bar. Take E_s = 205 kN/mm^2 and E_g = 110 kN/mm^2.

17. A rigid bar AC is supported by three rods of the same material and of equal diameter as shown in Fig. 1.31. Calculate the stress in the bars due to an applied force P, if the bar AC remains horizontal after the load has been applied. Neglect the weight of the rigid bar. Take $L_2 = 2L_1$.

18. A railway is laid so that there is no stress in the rails at 289 K. Calculate (*a*) the stress in rails at 333 K if there is no allowance for expansion, (*b*) the stress in the rails at 333 K if there is an expansion allowance of 10 mm per rail, (*c*) the expansion allowance if the stress in the rail is to be zero when the temperature is 333 K, (*d*) the maximum temperature to have no stress in the rails, if the expansion allowance is 12.5 mm per rail. The rails are 25 m long. Take $\alpha = 11.8 \times 10^{-5}$ per K and E = 240 kN/mm^2.

19. Three pin connected bars OA, OB and OC are arranged as in Fig. 1.42 and support a load P. Find the axial force in each bar and also the vertical displacement of the point of application of the load. Bar OA and OB are of steel having area of cross-section a each and length L. Bar OC is of copper having area of cross-section $2a$ and length 1.5 L. Neglect the possibility of lateral buckling. Take $E_s/E_c = 1.9$.

20. Determine the forces acting in the bars of the system represented in Fig. 1.40 if a load of 30 kN is applied at O. The outer rods OA and OC are of steel and of area of cross-section 150 sq. mm. The central rod OB is of copper and area of cross-section 200 sq. mm. The angle AOC is 90°. Take $E_s/E_c = 1.85$.

21. A load of 10 kN is applied at O in the system represented in Fig. 1.40. Calculate the final stress in the rods if the central bar is heated through 55 K. All the three rods are of steel and of the same cross-sectional area of 20 sq. mm. The central bar is 40 cm long. The angle AOC is 90°. Take E = 200 kN/mm² and $\alpha = 11.8 \times 10^{-6}$ per K.

22. For the steel bar shown in Fig. 1.77 (a) determine the longitudinal force P and the stresses p at all cross-sections. Determine also the vertical displacement Δ of all cross-sections of the bar. Represent the result graphically by plotting P, p and Δ diagrams. Take $E = 2 \times 10^5$ N/mm².

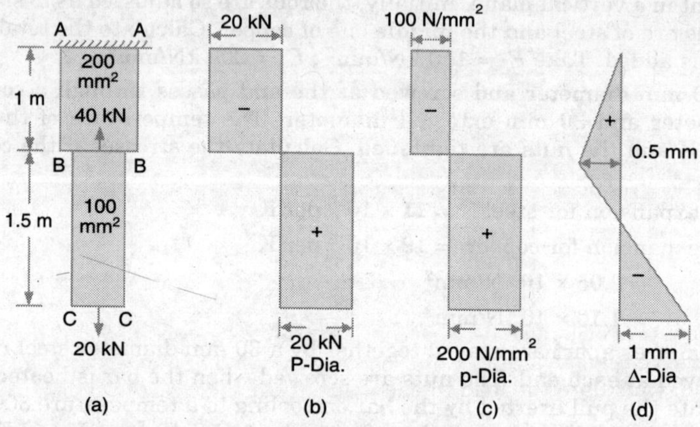

Fig. 1.77

23. A rigid bar ABC carries a load of 100 kN and is supported as shown in Fig. 1.78. Calculate the stresses in the rods and the reactions at A. Given : E_s = 205 kN/mm² ; E_b = 82 kN/mm².

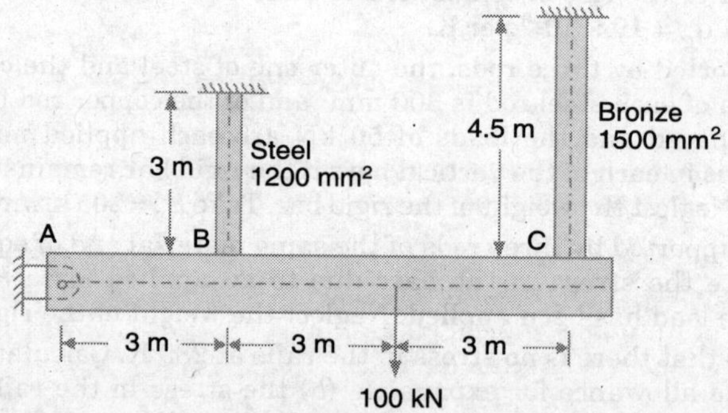

Fig. 1.78

24. A solid steel bar, 50 cm long and 7 cm diameter is placed inside an aluminium tube having 7.5 cm inside diameter and 10 cm outside diameter. The aluminium cylinder is 0.15 mm longer than the steel cylinder. An axial load of 600 kN is applied to the bar and cylinder through rigid cover plates as shown in Fig. 1.26. Find the stresses developed in the steel bar and aluminium tube. Assume for steel E = 220 kN/mm² and for aluminium E = 70 kN/mm². (*A.M.I.E. May 1964*)

25. A rigid bar *AB* is hinged at *A* and supported by a bronze rod *GD* of length $2L$ and a steel *FC* of length L. A load P is applied at the end *B* as shown in Fig. 1.79. Calculate the load carried by each rod and the reaction at *A*. Area of steel rod is equal to 1.5 times the area of the bronze rod. The modulus of elasticity for steel is 2.5 times that of bronze.

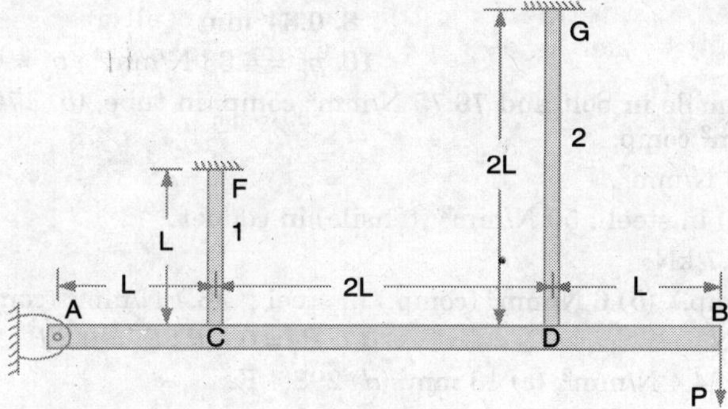

Fig. 1.79

26. A rigid member *ABC* of length $2L$ is supported at ends *A* and *C* and at its mid-point *B* by three vertical wires *DA*, *EB* and *FC* (Fig. 1.80). The wire *DA*, *EB* and *FC* extend α_1, α_2 and α_3 respectively under unit load. Calculate the load in each wire if the member *AC* weighs w per unit length.

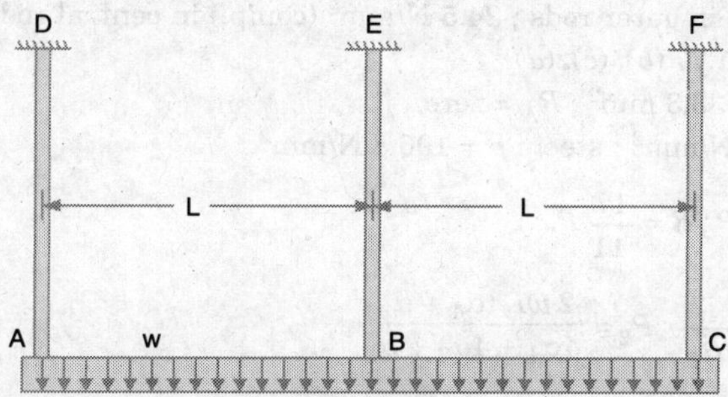

Fig. 1.80

27. A rigid bar *AB* is supported by three identical bars as shown in Fig. 1.81. All the three bars are equispaced and are in the same vertical plane. Calculate the load carried by the vertical bars, due to a load P applied to the rigid bar at a distance of 1.5 a from bar 1. The outer rods are of steel and the inner rod of brass. Take $E_s/E_b = 2$.

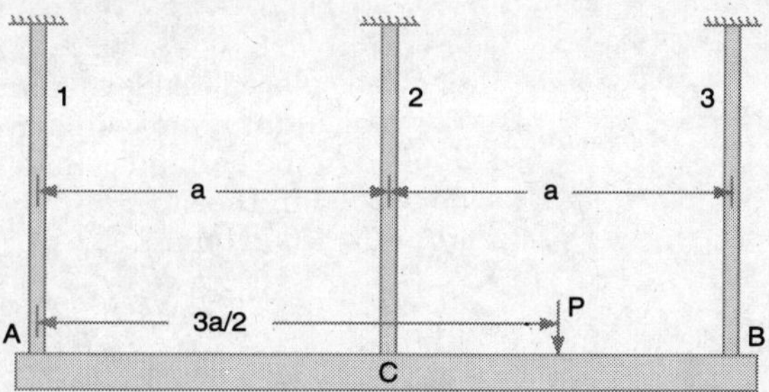

Fig. 1.81

ANSWERS

1. $E = 202$ kN/mm^2.
2. 45.6 N/mm^2 ; 0.168 mm.
3. 0.226 mm.
4. 0.845 mm.
5. 51.7 mm.
6. 1.075 mm.
7. 0.351 mm.
8. 0.34 mm.
9. 1.07×10^{-4} mm.
10. $p_c = 4.33$ N/mm^2 ; $p_s = 65.2$ N/mm^2 ; 119 kN.
11. (*a*) 159.5 N/mm^2 tensile in bolt and 79.75 N/mm^2 comp. in tube, (*b*) 170.6 N/mm^2 tensile in bolt and 73.77 N/mm^2 comp.
12. 71.2 N/mm^2 ; 55.75 N/mm^2.
13. 46.2 N/mm^2 (comp.) in steel ; 50 N/mm^2 (tensile) in copper.
14. (*a*) 197 kN, (*b*) 153.7 kN.
15. (*a*) 21.2 N/mm^2 (comp.), (*b*) 6 N/mm^2 (comp.) in steel ; 36.4 N/mm^2 (comp.) in gun metal.
16. 26.40 kN ; 47.20 kN.
17. $P_1 = 0.4\,P$; $P_2 = 0.2\,P$.
18. (*a*) 106 N/mm^2, (*b*) 24.4 N/mm^2, (*c*) 13 mm, (*d*) 298.4 K.
19. 0.416 *P* in steel bars : 0.584 *P* in copper bar.

$$\Delta = 0.438\,\frac{PL}{aE_c}$$

20. 10.5 kN in steel ; 15.1 kN in copper.
21. 52.9 N/mm^2 (tensile) in outer rods ; 24.5 N/mm^2 (comp.) in central rod.
22. Represented in Fig. 1.77 (*b*), (*c*), (*d*).
23. $p_s = 41.7$ mm^2 ; $p_b = 33.3$ mm^2 ; R_A = zero.
24. Aluminium : $p = 55$ N/mm^2 ; steel : $p = 106.8$ N/mm^2.
25. $P_1 = \frac{20}{11}P$; $P_2 = \frac{8}{11}P$; $R = \frac{17}{11}P$.
26. $P_1 = P_2 = \frac{4wL\alpha_2}{\alpha_1 + 4\alpha_2 + \alpha_3}$; $P_2 = \frac{2wL(\alpha_1 + \alpha_3)}{4\alpha_1 + 4\alpha_2 + \alpha_3}$.
27. $P_1 = \frac{3}{20}P$; $P_2 = \frac{1}{5}P$; $P_2 = \frac{13}{20}P$.

2

Elastic Constants

2.1. LATERAL STRAIN : POISSON'S RATIO

In the previous chapter, we have discussed the longitudinal strain of a bar or rod, under the action of axial or normal force. We have now to consider the effect of the axial force on the lateral dimensions.

If an axial tensile force is applied to a bar, the length of the bar will increase and its lateral dimensions (*i.e.* either the diameter or the width and breadth) will be decreased. *Thus, any direct stress produces a strain in its own direction and an opposite kind of strain in every direction at right angles to this.* The ratio of the lateral strain to the longitudinal strain is constant for a given material and is known as *Poisson's Ratio*.

Thus, $\dfrac{\text{Lateral strain}}{\text{Longitudinal strain}} = \text{Constant} = \dfrac{1}{m} = \mu$

or

$$\text{Lateral strain} = \frac{1}{m} \times \text{Longitudinal strain} \qquad \ldots(2.1)$$

Poisson's ratio is denoted by $\dfrac{1}{m}$ or μ, and its value lies between 0.25 to 0.42 for most of the metals. For rubber, its value ranges from 0.45 to 0.50.

Values of Poisson's ratio (μ or 1/m)

Table 2.1 gives the value of Poisson's ratio for some common materials. It should be noted that the Poisson's ratio is the same in tension and compression.

TABLE 2.1. VALUES OF POISSON'S RATIO

S.No	*Material*	*Poisson's Ratio* $\frac{1}{m}$ *or* μ
1.	Aluminium (Pure)	0.33
2.	Aluminium alloys	0.33
3.	Brass	0.34
4.	Bronze	0.34
5.	Cast iron	0.2 – 0.3
6.	Concrete (compression)	0.1 – 0.2
7.	Copper (Pure)	0.33 – 0.36
8.	Rubber	0.45 – 0.50
9.	Steel	0.27 – 0.30
10.	Wrought iron	0.3

Let us now find the values of lateral and longitudinal strains, under the direct stresses in different directions.

2.2. SINGLE DIRECT STRESS ALONG THE LONGITUDINAL AXIS

Let p be the stress (tensile).

Longitudinal strain e_1 (say) $= \frac{p}{E}$ (tensile).

Lateral strain $e_2 = -\frac{1}{m} e_1 = -\frac{1}{m}\frac{p}{E} = \frac{p}{mE}$ (compressive)

For a rectangular bar of length L, width b and thickness d, we have

$$V = L.b.d$$

or

$$\frac{\delta V}{V} = \frac{\delta L}{L} + \frac{\delta b}{b} + \frac{\delta d}{d}$$

Here

$$\frac{\delta L}{L} = e_1 = \frac{p}{E};\ \frac{\delta b}{b} = e_2 = -\frac{p}{mE};\ \frac{\delta d}{d} = e_3 = -\frac{p}{mE}$$

$\therefore$ Volumetric strain $e_v = \frac{\delta V}{V} = \frac{p}{E} - \frac{1}{m}\frac{p}{E} - \frac{1}{m}\frac{p}{E} = \frac{p}{E}\left(1 - \frac{2}{m}\right)$ ...(2.2)

If the bar is circular, of diameter d, we get

$$V = \frac{\pi}{4} d^2 L \quad \text{or} \quad \frac{\delta V}{V} = \frac{\delta L}{L} + \frac{2\delta d}{d}$$

where

$$\frac{\delta L}{L} = e_1 = \frac{p}{E};\ \frac{\delta d}{d} = e_2 = -\frac{p}{mE}$$

$\therefore$

$$e_v = \frac{\delta V}{V} = \frac{p}{E} - \frac{2p}{mE} = \frac{p}{E}\left(1 - \frac{2}{m}\right).$$

2.3. STRESS p_1 ALONG THE AXIS AND p_2 AND p_3 PERPENDICULAR TO IT

Generalised Hooke's Law

Let p_1, p_2 and p_3 be all tensile.

Longitudinal strain, $$e_1 = \frac{p_1}{E} - \frac{p_2}{mE} - \frac{p_3}{mE} \quad \text{...[2.3 (a)]}$$

Similarly, $$e_2 = \frac{p_2}{E} - \frac{p_3}{mE} - \frac{p_1}{mE} \quad \text{...[2.3 (b)]}$$

and $$e_3 = \frac{p_3}{E} - \frac{p_1}{mE} - \frac{p_2}{mE} \quad \text{...[2.3 (c)]}$$

Eq. 2.3 are known as the *generalised equations of Hooke's Law*.

If some of the stresses are of opposite sign, necessary change in the algebraic signs of the above expressions will have to be made.

2.4. UPPER LIMIT OF POISSON'S RATIO

Adding the three expressions of equation 2.3, we get

$$e_1 + e_2 + e_3 = \left(1 - \frac{2}{m}\right)\left(\frac{p_1 + p_2 + p_3}{E}\right) \quad \text{...(2.4)}$$

The sum of the three strains on the left is known as *dilatation* and, for small strains, it represents change in volume per unit volume.

In the case of hydrostatic tension, we have $p_1 = p_2 = p_3 = p$ and the resulting strains are $e_1 = e_2 = e_3 = e$.

Hence $$E \,.\, e = p\left(1 - \frac{2}{m}\right) \quad \text{...(2.5)}$$

In the above expression, $E \,.\, e$ and p are positive numbers. Hence $\left(1 - \frac{2}{m}\right)$ must also be positive. This limits $\frac{1}{m}$ to a maximum of 0.5. No material is known to have a higher value for Poisson's ratio, although $\frac{1}{m}$ for materials like rubber approaches this value.

Example 2.1. *A steel bar of rectangular cross-section 20 mm × 10 mm is subjected to a pull of 20 kN in the direction of its length. Taking E = 0.2 × 10⁶ N/mm² and m = 10/3, find the length of the sides of the cross-section and percentage decrease of area of cross-section.*

Solution. The strain in the direction of pull is

$$e_1 = \frac{p}{E} = \frac{P}{AE} = \frac{20000}{20 \times 10 \times 0.2 \times 10^6} = 5 \times 10^{-4}$$

$$\text{Lateral strain} = -\frac{e_1}{m} = -\frac{3}{10} \times 5 \times 10^{-4} = 1.5 \times 10^{-4}$$

Hence 20 mm side is decreased by $20 \times 1.5 \times 10^{-4} = 0.0030$ mm

10 mm side is decreased by $10 \times 1.5 \times 10^{-4} = 0.0015$ mm

New area of cross-section = (20 – 0.003) (10 – 0.0015) = (200 – 0.06) approx.

$\therefore$ % decrease of area of cross-section $= \dfrac{0.06}{200} \times 100 = \mathbf{0.03}.$

Example 2.2. *A piece of steel 20 mm long and 20 mm × 20 mm cross-section is subjected to a tensile force of 40 kN in the direction of its length. Calculate the change in volume. Take 1/m = 0.3, and E = 2.05 × 10⁵ N/mm².*

Solution. $e_1 = \dfrac{40000}{(20 \times 20)(2.05 \times 10^5)} = 4.88 \times 10^{-4}$;

$$e_2 = e_3 = -\frac{e_1}{m} = -4.88 \times 0.3 \times 10^{-4} = -1.464 \times 10^{-4}$$

$$\frac{\delta V}{V} = e_1 + e_2 + e_3 = [4.88 - (1.464 \times 2)]\, 10^{-4} = 1.952 \times 10^{-4}$$

$$V = 200 \times 20 \times 20 = 80000 \text{ mm}^3; \quad \therefore \quad \delta V = 1.952 \times 80000 \times 10^{-4} = \mathbf{15.62 \text{ mm}^3}.$$

Example 2.3. *The plates of a cylindrical boiler 2 m diameter and 3 metre long are under a tensile stress of 80 N/mm² in the direction of circumference and a tensile stress of 40 N/mm² in the axial direction. Taking 1/m = 0.3 and E = 200 kN/mm², find the increase in the internal capacity. Neglect the compressive stress due to steam pressure on the inner surface.*

Solution. $E = 200 \text{ kN/mm}^2 = 2 \times 10^5 \text{ N/mm}^2$

Let x be the suffix in the circumferential direction.

Let y be the suffix in the axial direction.

$$e_x = \frac{p_x}{E} - \frac{p_y}{mE} = \frac{80}{E} - \frac{0.3 \times 40}{E} = +\frac{68}{E}$$

and

$$e_y = \frac{p_y}{E} - \frac{p_x}{mE} = \frac{40}{E} - \frac{0.3 \times 80}{E} = +\frac{16}{E}$$

Now, the diameter of the boiler increases or decreases in direct proportion to the circumference. Hence e_x also denotes the diametrical strain. Hence the strains are $16/E$ parallel to axis and $68/E$ along any two perpendicular radii.

$$\therefore \quad \frac{\delta V}{V} = e_1 + e_2 + e_3 = \frac{16}{E} + \frac{68}{E} + \frac{68}{E} = \frac{152}{E}$$

$$V = \frac{\pi}{4} \times 4 \times 3 \text{ m}^3 = 9.425 \text{ m}^3 = 9.425 \times 10^9 \text{ mm}^3$$

$$\delta V = (9.425 \times 10^9)\frac{152}{2 \times 10^5} = 7163 \times 10^3 \text{ mm}^3 = \mathbf{7163 \text{ cm}^3 = 7.163 \text{ litres.}}$$

2.5. MODULUS OF RIGIDITY

Modulus of Rigidity or Modulus of Transverse Elasticity or Shearing Modulus expresses the relation between shear stress and shear strain.

Thus $$\frac{q}{\phi} = N \qquad ...(2.6)$$

where N = Modulus of Rigidity and ϕ = Shear strain, in radians.

2.6. COMPLEMENTARY SHEAR STRESS

"A stress in a given direction cannot exist without a balancing shear stress of equal intensity in a direction at right angles to it."

To prove the above rule of Complementary Shear Stress, let us take an infinitely small block *ABCD*, under shear stress of intensity q [Fig. 2.1 (a)]. The shear stress acting on the block will have a tendency to rotate the block in the clockwise direction. The considerations of statical equilibrium show that since there is no other force acting on the block, equilibrium can be attained only if a couple is applied, contrary to the previous one. This can be attained by having shear stress intensity q' on faces AD and CB as shown in Fig. 2.1 (b).

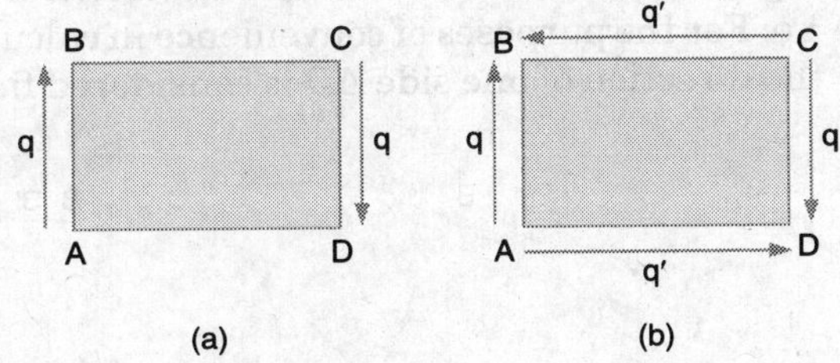

Fig. 2.1

The moment of given couple = Force × Lever arm = $q(AB)\,AD$

(Considering unit thickness perpendicular to the figure)

Similarly, the moment of the balancing couple = $(q'\,AD)\,AB$

For equilibrium, equating the two, we get

$$q'\,AD.\,AB = q\,AB\,.\,AD \quad \text{or} \quad q' = q \qquad ...(2.7)$$

Thus, the intensity of shearing across two planes at right angles are equal. Stress q' is known as *complementary shear stress.*

2.7. STATE OF SIMPLE SHEAR

The state of stress shown in Fig. 2.1 (b) is known as *state of simple shear*, as no other kind of stress is acting. Let us study a square block *ABCD* under the state of simple shear (Fig. 2.2).

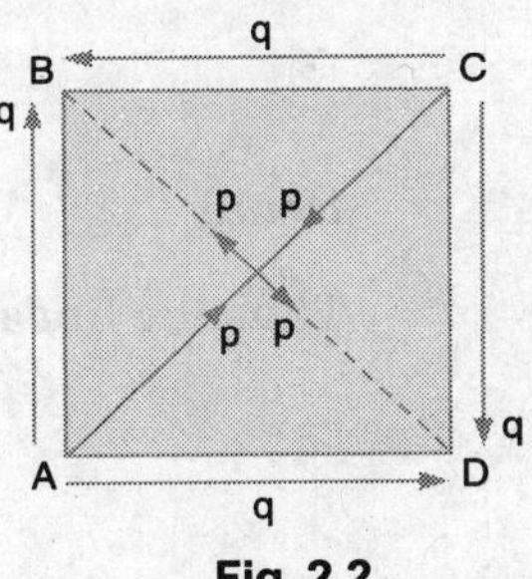

Fig. 2.2

Let the side of the square be b. The length of diagonal $AC = \sqrt{2}\,b$. Consider unit thickness perpendicular to the block.

Considering the equilibrium of the piece ABC, the resolved sum of q perpendicular to the diagonal

$$= 2(q \times b \times 1) \cos 45° = \frac{2qb}{\sqrt{2}} = \sqrt{2}\,q.b$$

If p is the tensile stress so produced on the diagonal, we have

$$p(AC \times 1) = \sqrt{2}\, q.b \quad \text{or} \quad p(\sqrt{2}\, b) = \sqrt{2}\, q.b$$

Hence $$p = q \text{ (tensile)} \qquad ...(2.8)$$

Resolving the shear stress along the diagonal, the resultant stress is zero. Similarly, it can be shown that the intensity of compressive stress on plane BD is numerically equal to q.

Hence a state of simple shear produces pure tensile and compressive stresses across planes inclined 45° to those of pure shear, and the intensities of these direct stresses are each equal to the intensity of the pure shear stress.

2.8. SHEAR STRAIN

Let us consider a square block $ABCD$, subjected to a state of simple shear q, in Fig. 2.3 (*a*). Fig. 2.3 (*b*) shows the resultant distortion of the block, the total change in the corner angles being $\pm\phi$. For the purposes of convenience in calculation, Fig. 2.3 (*c*) shows the distorted shape in which the direction of one side AD is considered fixed, the shearing strain ϕ still remaining the same.

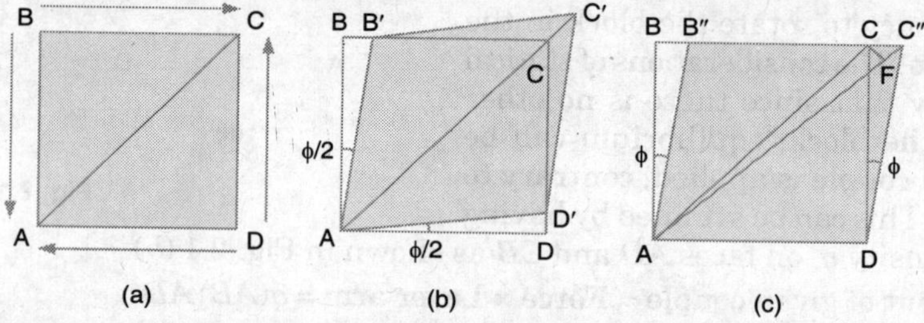

Fig. 2.3

Since ϕ is extremely small, we can assume BB'' as an arc with A as centre and AB as radius.

$$\therefore \qquad \phi = \frac{BB''}{AB} = \frac{CC''}{CD}$$

The elongation of the diagonal AC can be nearly taken equal to FC''.

Linear strain of the diagonal

$$e = \frac{FC''}{AC} = \frac{CC'' \cos 45°}{CD \sec 45°}$$

$$\therefore \qquad e = \frac{CC''}{2CD} = \frac{1}{2}\phi, \quad \text{where} \quad \phi = \frac{q}{N} \qquad ...(2.9)$$

Hence $$e = \frac{1}{2}\frac{q}{N} \qquad ...(2.10)$$

Thus the linear strain of the diagonal is equal to half the shear strain ϕ.

Example 2.4. *A lever is keyed to a shaft of 80 mm diameter. The width of the key is 10 mm and the length 50 mm. If the shear stress in the key is not to exceed 80 N/mm², find the load that can be applied at a radius of 1 metre.*

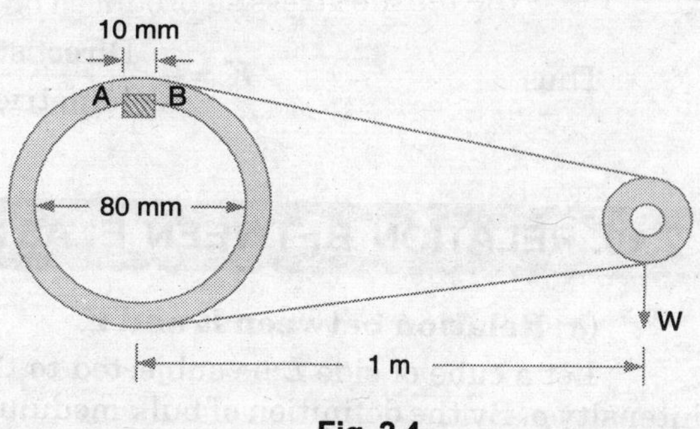

Fig. 2.4

Solution. Let W N be the load.

Torque produced at the shaft

$$= 1000\ W \text{ N-mm}$$

Shear force in the plane AB

$$= \frac{1000\ W}{40} = 25\ W \text{ N}$$

Area of key = 10 × 50 = 500 mm²

$$\text{Shear stress in key} = \frac{25\ W}{500} = 0.05\ W \text{ N/mm}^2.$$

But this is not to exceed 80 N/mm².

$$\therefore \quad 0.05\ W = 80 \quad \text{or} \quad W = \frac{80}{0.05} = \mathbf{1600\ N.}$$

Example 2.5. *A steel punch can be worked to a compressive stress of 800 N/mm². Find the least diameter of hole which can be punched through a steel plate 10 mm thick if its ultimate shear strength is 350 N/mm².*

Solution. Let the diameter be d.

Let q be the shear stress and p be the compressive stress.

Compressive force exerted = $\frac{\pi}{4} d^2 p$

Shear force required = $(\pi d t)\, q$ (where t is the thickness of the plate)

Equating the two, $\frac{\pi}{4} d^2 p = \pi d t q$

$$\therefore \quad d = \frac{4tq}{p} = \frac{4 \times 10 \times 350}{800} = \mathbf{17.5\ mm.}$$

Example 2.6. *A shaft is subjected to a twisting moment which produces a shearing stress at the surface of 110 N/mm² in planes perpendicular to the axis of the shaft. A small square is scratched on the surface of the shaft with two of its sides parallel to the axis of the shaft. Taking N = 85 kN/mm², find the change in the angle of the corners of the square.*

Solution. Within the limits of proportionality, $q = N\phi$

$$\therefore \quad \phi = \frac{q}{N} = \frac{110}{85 \times 1000} = 1.295 \times 10^{-3}$$

But 1 radian = 206265 seconds and hence ϕ in seconds = $1.295 \times 10^{-3} \times 206265 = 267$

$\therefore$ Change in angle $= \phi = \mathbf{4'\ 27''.}$

2.9. BULK MODULUS

When a body is subjected to three mutually perpendicular like direct stresses of the equal intensity, the ratio of the direct stress p to the volumetric strain is defined as the bulk modulus K of the body.

Thus $$K = \frac{\text{Direct stress}}{\text{Volumetric strain}} = \frac{p}{e_v} \qquad ...(2.11)$$

2.10. RELATION BETWEEN ELASTIC CONSTANTS

(*a*) Relation between E and K

Let a cube of side L be subjected to three mutually perpendicular like stresses of equal intensity p. By the definition of bulk modulus,

$$\frac{p}{e_v} = K \quad \text{or} \quad e_v = \frac{\delta V}{V} = \frac{p}{K} \qquad ...(i)$$

The total linear strain of each side $= e = \frac{p}{E} - \frac{p}{mE} - \frac{p}{mE}$

$$\therefore \quad \frac{\delta L}{L} = e = \frac{p}{E}\left(1 - \frac{2}{m}\right) \qquad ...(ii)$$

Now $V = L^3$ and hence $\delta V = 3L^2.\delta L$

$$\therefore \quad \frac{\delta V}{V} = e_v = \frac{3\delta L}{L} = 3e = \frac{3p}{E}\left(1 - \frac{2}{m}\right) \qquad ...(iii)$$

Equating (*i*) and (*iii*), $\frac{p}{K} = \frac{3p}{E}\left(1 - \frac{2}{m}\right)$

or $$E = 3K\left(1 - \frac{2}{m}\right) \qquad ...(2.12)$$

This gives the relation between E, K and m.

(*b*) Relation between E and N

We have seen in § 2.7 that the linear strain e of the diagonal AC [Fig. 2.3 (*c*)]

$$= e = \tfrac{1}{2}\phi = \frac{q}{2N} \qquad ...(i)$$

We have also seen in § 2.7 that a state of simple shear produces tensile and compressive stresses along diagonal planes and the value of such direct stresses is numerically equal to q. Thus, on diagonal AC [Fig. 2.3 (*c*)], there is a tensile stress p, and on diagonal BD there is a compressive stress p.

Hence, from § 2.1, the linear strain e of the diagonal AC, due to these two mutually perpendicular direct stresses, is given by

$$e = \frac{p}{E} - \left(-\frac{p}{mE}\right) = \frac{p}{E}\left(1 + \frac{1}{m}\right) \quad (\text{where } p = q) \qquad ...(ii)$$

$$\therefore \quad e = \frac{q}{E}\left(1 + \frac{1}{m}\right) \qquad ...(iii)$$

Equating (*i*) and (*iii*), we get $\frac{1}{2}\frac{q}{N} = \frac{q}{E}\left(1 + \frac{1}{m}\right)$

or $$E = 2N\left(1+\frac{1}{m}\right) \quad ..(2.13)$$

which gives the relation between E, N and $1/m$

We have already proved the relation between E and K

i.e. $$E = 3K\left(1-\frac{2}{m}\right) \quad ...(\text{Eq. } 2.12)$$

Equating the two, we get

$$E = 2N\left(1+\frac{1}{m}\right) = 3K\left(1-\frac{2}{m}\right) \quad ...(2.14)$$

Eliminating E from Eq. 2.14, we get $\dfrac{1}{m} = \dfrac{3K-2N}{6K+2N}$ ...(2.15)

Eliminating m, we get $E = \dfrac{9KN}{N+3K}$ ...(2.16)

Example 2.7. (*a*) *Determine the percentage change in volume of a steel bar 50 mm square in section and 1 m long when subjected to an axial compressive load of 20 kN.* (*b*) *What change in volume would a 100 mm cube of steel suffer at a depth of 5 km in sea water ?*

Take E = 2.05 × 10⁵ N/mm² and N = 0.82 × 10⁵ N/mm².

Solution. (*a*) Volume of the bar = $V = a^2L$

$$\therefore \quad \frac{\delta V}{V} = 2\frac{\delta a}{a} + \frac{\delta L}{L} = 2\,e_a + e_L$$

or $$\frac{\delta V}{V} = \left(-\frac{2}{m}\cdot\frac{p}{E}\right) + \frac{p}{E} = \frac{p}{E}\left(1-\frac{2}{m}\right)$$

(reduction, since p is compressive) ...(1)

Also $$E = 2N\left(1+\frac{1}{m}\right) \quad (\text{Eq. } 2.13) \quad ...(2)$$

or $$\left(1+\frac{1}{m}\right) = \frac{E}{2N} = \frac{2.05}{2\times 0.82} = 1.25$$

$$\therefore \quad \frac{1}{m} = 0.25 \quad ...(3)$$

Hence from (1) $$\frac{\delta V}{V} = \frac{p}{E}(1-0.50) = \frac{0.5\times 20000}{2.05\times 10^5\times 2500} = 1.95\times 10^{-5}$$

$\therefore$ Percentage reduction in volume = $\dfrac{\delta V}{V}\times 100 = 1.95\times 10^{-5}\times 100 =$ **0.00195.**

(*b*) On the 100 mm cube of steel

$p = wh$, where $w = 10080$ N/m³ for sea water and $h = 5$ km = 5000 m

$\therefore$ $p = 10080 \times 5000 = 50.4\times 10^6$ N/m² = 50.4 N/mm²

From Eq. 2.12, we have

$$K = \frac{E}{3\left(1-\frac{2}{m}\right)} = \frac{E}{3(1-0.5)} = \frac{2}{3}E = \tfrac{2}{3}\times 2.05\times 10^5 = 1.365\times 10^5 \text{ N/mm}^2.$$

Now $\frac{\delta V}{V} = \frac{p}{K}$; hence $\delta V = \frac{Vp}{K} = \frac{(100)^3 \times 50.4}{1.365 \times 10^5}$ = **369 mm³.**

Example 2.8. *If a bar is stretched in such a manner that all the lateral strain is prevented, what is the value of the modified modulus of elasticity and modified Poisson's ratio ? Take μ as the Poisson's ratio.*

Solution. Let p_1 be the axial stress applied.

Since e_2 and e_3 are each zero, the corresponding stresses p_2 and p_3 will be induced in the lateral directions and will be of the same sign as that of p_1. We have, thus,

$$e_1 = \frac{p_1}{E} - \frac{\mu}{E}(p_2 + p_3) \quad ...(1)$$

$$e_2 = \frac{p_2}{E} - \frac{\mu}{E}(p_3 + p_1) = 0 \quad ...(2)$$

and
$$e_3 = \frac{p_3}{E} - \frac{\mu}{E}(p_1 + p_2) = 0 \quad ...(3)$$

Adding (2) and (3), we get $p_2 + p_3 = 2p_1 \frac{\mu}{1-\mu}$

Substituting in (1), we get

$$e_1 = \left(\frac{p_1}{E} - \frac{\mu}{E}\frac{2p_1\mu}{1-\mu}\right) = \frac{p_1}{E}\left(1 - \frac{2\mu^2}{1-\mu}\right) = \frac{p_1}{E}\left(\frac{1-\mu-2\mu^2}{1-\mu}\right) = \frac{p_1}{E}\frac{(1+\mu)(1-2\mu)}{(1-\mu)}$$

$$\frac{p_1}{e_1} = E' = \text{modified modulus} = \frac{E(1-\mu)}{(1+\mu)(1-2\mu)} \quad \textbf{Ans.}$$

Also, $\mu' = \text{modified Poisson's ratio} = \frac{e_3}{e_1} = \text{zero.}$

Example 2.9. *A metal specimen is subjected to an axial stress of p_1. No constraint is exerted on one pair of sides but on the other, external pressures restrict the lateral strain to one-half of what it would be if there were no restraint. Calculate the modified modulus of elasticity. Take μ as Poisson's ratio.*

Solution. Let the two lateral directions be represented by suffix 2 and 3.

Let p_3 be the stress in the direction of restraint. p_3 will have same sign as that p_1. Since there is no restraint in the second direction, $p_2 = 0$. We have, thus,

$$e_1 = \frac{p_1}{E} - \frac{\mu p_3}{E} \quad ...(1)$$

$$e_2 = -\frac{\mu}{E}(p_1 + p_3) \quad ...(2)$$

and
$$e_3 = \frac{p_3}{E} - \frac{\mu}{E}p_1 \quad ...(3)$$

If there were no restraint in the third direction, $e_3' = -\frac{\mu}{E}p_1$ ...(4)

As per given condition, $e_3 = \frac{1}{2}e_3'$

Hence $\frac{p_3}{E} - \frac{\mu}{E} p_1 = -\frac{1}{2}\frac{\mu}{E} p_1$ or $p_3 = \frac{\mu p_1}{2}$ ...(5)

Substituting in (1), we get $e_1 = \frac{p_1}{E} - \frac{\mu}{E}\left(\frac{\mu p_1}{2}\right) = \frac{p_1}{E}\left(1 - \frac{\mu^2}{2}\right) = \frac{p_1}{2E}(2 - \mu^2)$

$$\therefore \quad \frac{p_1}{e_1} = E' = \text{modified modulus of elasticity} = \frac{2E}{2 - \mu^2}.$$

Example 2.10. *A bar 30 mm in diameter was subjected to tensile load of 54 kN and the measured extension on 300 mm gauge length was 0.112 mm and change in diameter was 0.00366 mm. Calculate Poisson's ratio and the values of three moduli.*

Solution.

Stress $= \frac{54 \times 1000}{\frac{\pi}{4} \times 900} = 76.4$ N/mm^2 ; Linear strain $= \frac{0.112}{300} = 3.73 \times 10^{-4}$

$$E = \frac{\text{Stress}}{\text{Strain}} = \frac{76.4}{3.73 \times 10^{-4}} = 2.046 \times 10^5 \text{ N/mm}^2 = \mathbf{204.6\ kN/mm^2}$$

Lateral strain $= \frac{\delta d}{d} = \frac{0.00366}{30} = 1.22 \times 10^{-4}$; But Lateral strain $= \frac{1}{m} e = \frac{1}{m}(3.73 \times 10^{-4})$

$$\therefore \quad \frac{1}{m}(3.73 \times 10^{-4}) = 1.22 \times 10^{-4} \quad \text{or} \quad \frac{1}{m} = \frac{1.22}{3.73} = \mathbf{0.326.}$$

Again $N = \frac{E}{2\left(1 + \frac{1}{m}\right)} = \frac{204.6}{2(1 + 0.326)} = \mathbf{77.2\ kN/mm^2.}$

$$K = \frac{E}{3\left(1 - \frac{2}{m}\right)} = \frac{204.6}{3(1 - 0.652)} = \mathbf{196\ kN/mm^2.}$$

Example 2.11. *The modulus of rigidity for a material is 0.51 × 10^5 N/mm^2. A 10 mm diameter rod of the material was subjected to an axial pull of 10 kN and the change in diameter was observed to be 3 × 10^{-3} mm. Calculate Poisson's ratio and the moduli of Elasticity.*

Solution. $A = \frac{\pi}{4}(10)^2 = 78.54$ mm^2 ; $p = \frac{10 \times 1000}{78.54} = 127.3$ N/mm^2

Lateral strain $= \frac{\delta d}{d} = \frac{3 \times 10^{-3}}{10} = 3 \times 10^{-4}$; But lateral strain $= \frac{p}{mE}$

$$\therefore \quad \frac{p}{mE} = 3 \times 10^{-4} \quad \text{or} \quad mE = \frac{p}{3 \times 10^{-4}} = \frac{127.3}{3 \times 10^{-4}} = 42.44 \times 10^4$$

Also, $E = 2N\left(1 + \frac{1}{m}\right)$ or $mE = 2N(m + 1)$

$$\therefore \quad 42.44 \times 10^4 = 2 \times 0.51 \times 10^5 (m + 1) \quad \text{or} \quad (m + 1) = \frac{42.44 \times 10^4}{2 \times 0.51 \times 10^5} = 4.16$$

$\therefore \qquad m = 3.16 \quad \text{or} \quad \frac{1}{m} = \mathbf{0.316}$

$$E = \frac{42.44 \times 10^4}{3.16} = \mathbf{1.343 \times 10^5\ N/mm^2}$$

$$K = \frac{E}{3\left(1 - \frac{2}{m}\right)} = \frac{1.343 \times 10^5}{3(1 - 0.632)} = \mathbf{1.216 \times 10^5\ N/mm^2.}$$

2.11. ADDITIONAL EXAMPLES USEFUL FOR COMPETITIVE EXAMINATIONS

Example 2.12. *The square frame ABCD (Fig. 2.5) consists of five steel bars of 600 mm² cross-sectional area and is submitted to the action of two forces P = 60 kN in the direction of the diagonal. Determine the changes of the angles at A and C due to deformation of the frame.*

Solution. Because of pin joints, the diagonal DB will take the complete load P. Let us assume that hinge D and the direction of the diagonal are stationary.

Then the displacement of the hinge B in the direction of the diagonal will be $\delta = PL/AE$, where L is the length of the diagonal.

In Fig. 2.5, C' is the new position of the hinge C. Draw CC_1 and $C'C_1$ parallel to DB and DC respectively. From the small right triangle CC_1C', $CC' = \delta/\sqrt{2}$

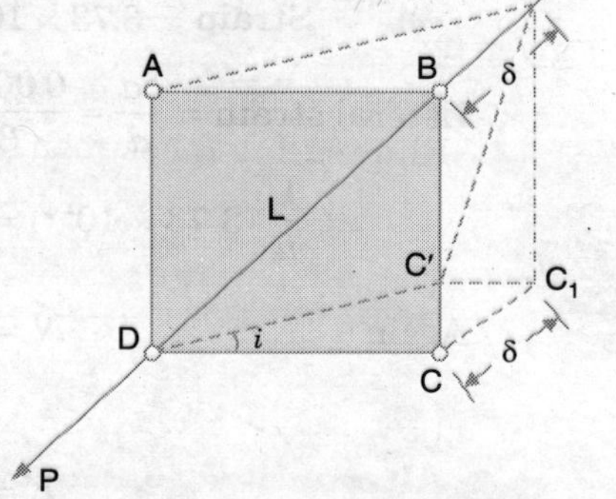

Fig. 2.5

Hence the angle of rotation of the bar DC due to the deformation of the frame is,

$$i = \frac{CC'}{DC} = \frac{\delta}{\sqrt{2}} \div \frac{L}{\sqrt{2}} = \frac{\delta}{L}$$

Substituting the value of δ, $\quad i = \frac{PL}{AE} \cdot \frac{1}{L} = \frac{P}{AE} = \frac{60 \times 1000}{600 \times 2 \times 10^5} = \frac{1}{2000}$ radian

Hence increase in the angle at C will be $= 2 \times \frac{1}{2000} = \mathbf{\frac{1}{1000}}$ **radian.**

Example 2.13. *A rectangular parallelopiped is subjected to tension in two perpendicular directions as shown in Fig. 2.6. Find the unit elongation ε in OC direction.*

Solution. Let ε_x and ε_y be the unit elongations in x and y directions respectively. Let a and b the co-ordinates of point C. These co-ordinates, after deformation, will be $a(1 + \varepsilon_x)$ and $b(1 + \varepsilon_y)$. Hence length $\overline{OC}$ after deformation will be

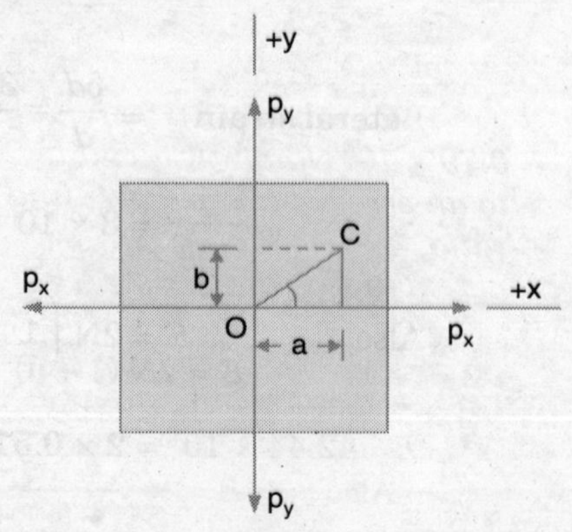

Fig. 2.6

$$(\overline{OC})' = \sqrt{a^2(1+2\varepsilon_x)+b^2(1+2\varepsilon_y)} \simeq \sqrt{a^2+b^2}\left(1+\frac{a^2\varepsilon_x}{a^2+b^2}+\frac{b^2\varepsilon_y}{a^2+b^2}\right)$$

Initial length of $\overline{OC}$ is $\overline{OC} = \sqrt{a^2+b^2}$

$$\therefore \qquad \frac{(\overline{OC})' - \overline{OC}}{\overline{OC}} = \varepsilon = \frac{a^2\varepsilon_x}{a^2+b^2}+\frac{b^2\varepsilon_y}{a^2+b^2}$$

But $\qquad \dfrac{a^2}{a^2+b^2} = \cos^2\alpha$ and $\dfrac{b^2}{a^2+b^2} = \sin^2\alpha$

$\therefore \qquad \varepsilon = \varepsilon_x \cos^2\alpha + \varepsilon_y \sin^2\alpha$, **which is the required expression.**

Example 2.14. *Show that if E is assumed to be correct, an error of 1% in the determination of N will involve an error of about 5% in the calculation of Poisson's ratio when its correct value is 0.25.*

Solution. Let the correct values of the elastic constants be E, N and μ.

We have $\qquad E = 2N(1+\mu) \qquad$ ...(1)

Due to error 1%, let N be increased to 1.01 N. The calculated value of Poisson ratio μ' is given by

$$E = 2 \times 1.01\, N(1+\mu') \qquad ...(2)$$

Eliminating E between (1) and (2), we get

$$N(1+\mu) = 1.01\, N(1+\mu') \quad \text{or} \quad 1+\mu = 1.01 + 1.01\,\mu'$$

or $\qquad \mu' - \mu = -0.01 - 0.01\,\mu' = -0.01\,(1+\mu') \qquad$ ...(3)

Hence % error in μ is

$$\frac{\mu'-\mu}{\mu} \times 100 = -0.01\,\frac{1+\mu'}{\mu} \times 100 \simeq -\frac{1+0.25}{0.25} \simeq -\mathbf{5\%.}$$

Alternative Solution

$$E = 2N(1+\mu)$$

Differentiating partially, and noting that $\delta E = 0$ since E does not vary, we have

$$0 = 2\,\delta N(1+\mu) + 2N\,\delta\mu \quad \text{or} \quad \delta\mu = -\frac{\delta N}{N}(1+\mu)$$

$$\therefore \qquad \%\ \text{error} = \frac{\delta\mu}{\mu} \times 100 = -\frac{\delta N}{N}\cdot\frac{1+\mu}{\mu} \times 100$$

$$= -\frac{1}{100} \times \frac{1+0.25}{0.25} \times 100 = -5\%\ \textbf{Ans.}$$

Example 2.15. *A determination of E and N gives values of 2 × 10⁵ N/mm² and 0.79 × 10⁵ N/mm². Calculate Poisson's ratio and the bulk modulus. If both the moduli are liable to an error of ± 2 percent, find the maximum percentage error in the derived value of Poisson's ratio.*

Solution.

$$E = 2N(1+\mu) \quad \text{or} \quad \mu = \frac{E}{2N} - 1 = \frac{2\times10^5}{2\times0.79\times10^5} - 1 = \mathbf{0.266}$$

Also, $\qquad K = \dfrac{E}{3(1-2\mu)} = \dfrac{2\times10^5}{3(1-2\times0.266)} = \mathbf{1.423 \times 10^5\ N/mm^2}$

Again, $\mu = \dfrac{E}{2N} - 1$

Let E be + 2 percent in error and N be – 2 percent in error.

Then incorrect value of $\mu = \dfrac{(2 \times 10^5) \times 1.02}{2(0.79 \times 10^5 \times 0.98)} - 1 = 0.3175$

$\therefore$ % error in $\mu = \dfrac{0.3175 - 0.266}{0.266} \times 100 = \mathbf{19.36\%}$.

Example 2.16. *A vertical rod of length L and diameter d is fixed at its upper end and carries a load W at its lower end. The extension of the rod due to W is δ. If a torque T_1 applied in a horizontal plane at the lower end of the rod gives an angle of twist θ radians, show that Poisson's ratio is given by*

$$\mu = \left(\frac{Wd^2.\theta}{16T.\delta} - 1\right).$$

Solution. As a vertical rod, the extension δ is given by

$$\delta = \frac{WL}{AE} = \frac{WL}{\frac{\pi}{4}d^2E} = \frac{4WL}{\pi d^2 E} \quad \text{or} \quad E = \frac{4WL}{\pi d^2 \delta} \qquad \text{...(1)}$$

Also, as a shaft, $\dfrac{T}{J} = \dfrac{N\theta}{L}$, where J = polar moment of inertia = $\dfrac{\pi d^4}{32}$

$$\therefore \quad N = \frac{TL}{J\theta} = \frac{TL}{\frac{\pi}{32}d^4\theta} = \frac{32TL}{\pi d^4 \theta} \qquad \text{...(2)}$$

Now, $E = 2N(1 + \mu)$

$$\therefore \quad 1 + \mu = \frac{E}{2N} = \frac{4WL}{\pi d^2 \delta} \cdot \frac{1}{2} \times \frac{\pi d^4 \theta}{32TL} = \frac{Wd^2\theta}{16T\delta}$$

Hence $\mu = \left(\dfrac{Wd^2\theta}{16T\delta} - 1\right)$. **Proved.**

Example 2.17. *A bar of steel is 38 mm in diameter and 450 mm long. A tensile load of 100 kN is found to stretch the bar by 0.20 mm. The same bar, when subjected to a torque of 1.27 kN-m is found to twist through 1.922°. Find the values of the four elastic constants.*

Solution. As a bar, $\delta = \dfrac{WL}{AE} = \dfrac{4WL}{\pi d^2 E}$

$$\therefore \quad E = \frac{4WL}{\pi d^2 \delta} = \frac{4 \times 100 \times 10^3 \times 450}{\pi(38)^2 \times 0.2} = \mathbf{1.984 \times 10^5\ N/mm^2}.$$

As a shaft, $N = \dfrac{TL}{J\theta} = \dfrac{32TL}{\pi d^4 \theta}$, where $\theta = \dfrac{\pi}{180} \times 1.922 = 0.033545$ radians.

$$\therefore \quad N = \frac{32 \times (1.27 \times 10^6) \times 450}{\pi(38)^4 \times 0.033545} = \mathbf{0.832 \times 10^5\ N/mm^2}.$$

Also, $E = 2N(1 + \mu)$

From which $\mu = \dfrac{E}{2N} - 1 = \dfrac{1.984 \times 10^5}{2 \times 0.832 \times 10^5} - 1 = \mathbf{0.1923.}$

And, $K = \dfrac{E}{3(1-2\mu)} = \dfrac{1.984 \times 10^5}{3(1 - 2 \times 0.1923)} = \mathbf{1.075 \times 10^5\ N/mm^2.}$

Example 2.18. *A bar 40 mm in diameter is subjected to a tensile force of 400 kN. The extension of the bar measured over a gauge length of 200 mm was 0.318 mm. The decrease in diameter was found to be 0.02 mm. Calculate the values of Young's modulus of elasticity and modulus of rigidity of the material.*

Solution. Stress $p = \dfrac{400 \times 1000}{\frac{\pi}{4}(40)^2} = 318.31\ \text{N/mm}^2$

Longitudinal strain, $\varepsilon_L = \dfrac{0.318}{200} = 1.59 \times 10^{-3}$; Lateral strain, $\varepsilon_d = \dfrac{0.02}{40} = 0.5 \times 10^{-3}$

$$\therefore \quad \mu = \frac{\epsilon_d}{\epsilon_L} = \frac{0.5 \times 10^{-3}}{1.59 \times 10^{-3}} = 0.3145$$

$$E = \frac{p}{\epsilon_L} = \frac{318.31}{1.59 \times 10^{-3}} \simeq \mathbf{2 \times 10^5\ N/mm^2}$$

$$N = \frac{E}{2(1+\mu)} = \frac{2 \times 10^5}{2(1 + 0.3145)} = \mathbf{0.76 \times 10^5\ N/mm^2.}$$

Example 2.19. *A bar of elastic material is subjected to a direct compressive stress of p_1 in the longitudinal direction. Suitable lateral compressive stress p_2 is applied along the other two lateral directions to limit the net strain in each of the lateral directions to half the magnitude that could be under p_1 acting alone. Find the magnitude of p_2 and the net strain in the longitudinal direction.*

Solution. $\varepsilon_2' = \varepsilon_3' = \dfrac{p_2}{E} - \dfrac{\mu p_2}{E} - \dfrac{\mu p_1}{E}$; $\varepsilon_2 = -\dfrac{\mu p_1}{E}$

$$\therefore \quad \frac{p_2}{E} - \frac{\mu p_2}{E} - \frac{\mu p_1}{E} = -\frac{1}{2}\frac{\mu p_1}{E}$$

Hence $p_2 = \dfrac{1}{2}\dfrac{\mu}{1-\mu}\, p_1.$ **Ans.**

Net strain in the longitudinal direction :

$$\varepsilon_1' = \frac{p_1}{E} - \frac{\mu p_2}{E} - \frac{\mu p_2}{E} = \frac{p_1}{E}\left[1 - 2\mu\,\frac{\mu}{2(1-\mu)}\right]$$

or $\varepsilon_1' = \dfrac{p_1}{E}\left(\dfrac{1 - \mu - \mu^2}{1-\mu}\right)$ compressive. **Ans.**

Example 2.20. *A rectangular block 250 mm × 100 mm × 80 mm is subjected to axial load as follows :*

480 kN tensile in the direction of its length,

1000 kN compressive on the 250 mm × 100 mm faces, and 900 kN tensile on 250 × 80 mm faces.

Assuming Poisson's ratio as 0.25, find in terms of modulus of elasticity of the material E, the strain in the direction of each force.

If E = 2 × 10⁵ N/mm², find the values of the modulus of rigidity and bulk modulus for the material of the block. Also, calculate the change in volume of the block due to loading specified above.

(Based on Cambridge University)

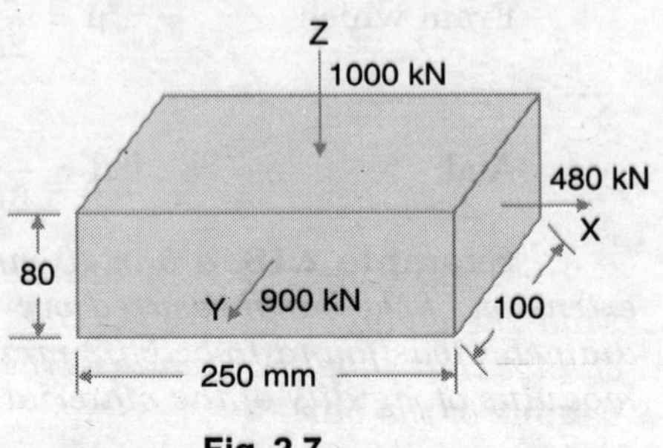

Fig. 2.7

Solution. Volume $V = 250 \times 100 \times 80 = 2 \times 10^6$ mm³ ; $\dfrac{1}{m} = 0.25$; $m = 4$

(a) ***Strains in each direction***

Stress in x-direction, $p_x = \dfrac{P_x}{A_x} = \dfrac{480 \times 10^3}{100 \times 80} = 60$ N/mm² (tension)

Stress in y-direction, $p_y = \dfrac{P_y}{A_y} = \dfrac{900 \times 10^3}{250 \times 80} = 45$ N/mm² (tension)

Stress in z-direction, $p_z = \dfrac{P_z}{A_z} = \dfrac{1000 \times 10^3}{250 \times 100} = 40$ N/mm² (comp.)

Taking tension as +ve and compression as negative, we have

$$e_x = \frac{p_x}{E} - \frac{p_y}{mE} + \frac{p_z}{mE} = \frac{60}{E} - \frac{45}{4E} + \frac{40}{4E} = \mathbf{\frac{58.75}{E}}$$

$$= \frac{58.75}{2 \times 10^5} = \mathbf{2.9375 \times 10^{-4}}.$$

$$e_y = \frac{p_y}{E} + \frac{p_z}{mE} - \frac{p_x}{mE} = \frac{45}{E} + \frac{40}{4E} - \frac{60}{4E} = \mathbf{\frac{40}{E}}$$

$$= \frac{40}{2 \times 10^5} = \mathbf{2 \times 10^{-4}}$$

and

$$e_z = -\frac{p_z}{E} - \frac{p_x}{mE} - \frac{p_y}{mE} = -\frac{40}{E} - \frac{60}{4E} - \frac{45}{4E} = -\mathbf{\frac{66.25}{E}}$$

$$= -\frac{66.25}{2 \times 10^5} = \mathbf{-3.3125 \times 10^{-4}}$$

(b) ***Change in volume***

$$\frac{\delta V}{V} = e_x + e_y + e_z = 2.9375 \times 10^{-4} + 2 \times 10^{-4} - 3.3125 \times 10^{-4} = +1.625 \times 10^{-4}$$

$$\therefore \quad \delta V = +1.625 \times 10^{-4} \times (2 \times 10^6) = \mathbf{+325\ mm^3}$$

(c) ***Modulus of rigidity***

$$N = \frac{mE}{2(m+1)} = \frac{4 \times 2 \times 10^5}{2(4+1)} = \mathbf{0.8 \times 10^5\ N/mm^2}$$

(d) ***Bulk Modulus***

$$K = \frac{mE}{3(m-2)} = \frac{4 \times 2 \times 10^5}{3(4-2)} = \mathbf{1.333 \times 10^5\ N/mm^2.}$$

Example 2.21. *A metallic bar 250 mm × 100 mm × 50 mm is loaded as shown in Fig. 2.8. Find the change in volume. Take E = 200 kN/mm² and Poisson's ratio = 0.25. Also find the change that should be made in the 4000 kN load, in order that there should be no change in the volume of the bar.* (Based in U.L.)

Solution.

$V = 250 \times 100 \times 50 = 1250 \times 10^3\ \text{mm}^3$;

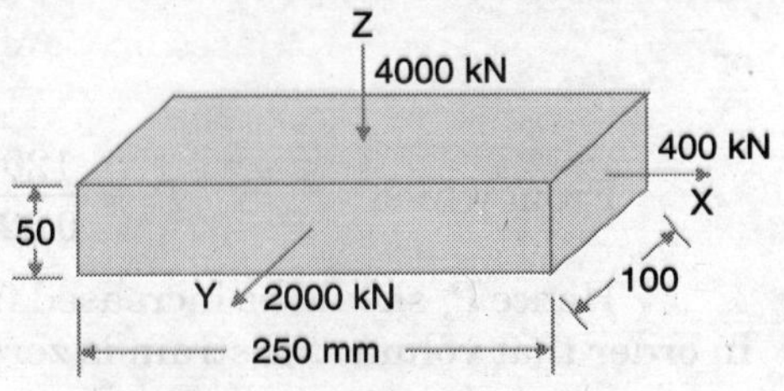

Fig. 2.8

$\frac{1}{m} = 0.25$; $m = 4$

(a) ***Change in volume***

$$p_x = \frac{P_x}{A_x} = \frac{400 \times 10^3}{100 \times 50} = 80\ \text{N/mm}^2\ \text{(tension)}$$

$$p_y = \frac{P_y}{A_y} = \frac{2000 \times 10^3}{250 \times 50} = 160\ \text{N/mm}^2\ \text{(tension)}$$

$$p_z = \frac{P_z}{A_z} = \frac{4000 \times 10^3}{250 \times 100} = 160\ \text{N/mm}^2\ \text{(comp.)}$$

Considering tension as positive and compression as negative, we have

$$e_x = \frac{p_x}{E} - \frac{p_y}{mE} + \frac{p_z}{mE} = \frac{80}{E} - \frac{160}{4E} + \frac{160}{4E} = \frac{80}{E}$$

$$e_y = \frac{p_y}{E} + \frac{p_z}{mE} - \frac{p_x}{mE} = \frac{160}{E} + \frac{160}{4E} - \frac{80}{4E} = \frac{180}{E}$$

$$e_z = -\frac{p_z}{E} - \frac{p_x}{mE} - \frac{p_y}{mE} = -\frac{160}{E} - \frac{80}{4E} - \frac{160}{4E} = -\frac{220}{E}$$

Now $\frac{\delta V}{V} = e_x + e_y + e_z = \frac{80}{E} + \frac{180}{E} - \frac{220}{E} = \frac{40}{E}$

$\therefore$ $\delta V = \frac{40}{E} \times V = \frac{40}{200 \times 1000} \times 1250 \times 10^3 = \mathbf{250\ mm^3}$

(b) ***Change in 4000 kN load***

Let P_z kN be the load, in place of 4000 kN load.

$\therefore$ $p_z = \frac{P_z \times 10^3}{250 \times 100} = 0.04\ P_z\ \text{N/mm}^2$ (comp.)

Hence $e_x = \frac{p_x}{E} - \frac{p_y}{mE} + \frac{p_z}{mE} = \frac{80}{E} - \frac{160}{4E} + \frac{0.04\ P_z}{4E} = \frac{40}{E} + \frac{0.01\ P_z}{E}$

$$e_y = \frac{p_y}{E} + \frac{p_z}{mE} - \frac{p_x}{mE} = \frac{160}{E} + \frac{0.04\,P_z}{4E} - \frac{80}{4E} = \frac{140}{E} + \frac{0.01\,P_z}{E}$$

and
$$e_z = -\frac{p_z}{E} - \frac{p_x}{mE} - \frac{p_y}{mE} = -\frac{0.04\,P_z}{E} - \frac{80}{4E} - \frac{160}{4E} = -\frac{0.04\,P_z}{E} - \frac{60}{E}$$

In order that $\frac{\delta V}{V} = e_x + e_y + e_z$ should be equal to zero, we have

$$e_x + e_y + e_z = 0$$

$$\therefore \quad \left(\frac{40}{E} + \frac{0.01\,P_z}{E}\right) + \left(\frac{140}{E} + \frac{0.01\,P_z}{E}\right) - \left(\frac{0.04\,P_z}{E} + \frac{60}{E}\right) = 0$$

or
$$\frac{120}{E} - \frac{0.02\,P_z}{E} = 0$$

From which $P_z = \frac{120}{0.02} = \mathbf{6000\ kN}$

Hence P_z should be increased from its present value of 4000 kN to a new value of 6000 kN, in order that volumetric strain is zero.

Example 2.22. *If two pieces of materials A and B have the same bulk modulus, but the value of E for B is 1% greater than that for A, find the value of N for the material B in terms of E and N for material A.*

Solution. Given $K_A = K_B = K\,;\ E_B = 1.01\,E_A$

From Eq. 2.16, we have
$$E = \frac{9KN}{N + 3K}$$

or
$$EN + 3EK = 9KN$$

or
$$3K\,(3N - E) = EN, \text{ from which } K = \frac{EN}{3(3N - E)}$$

Hence
$$\frac{E_A N_A}{3\,(3N_A - E_A)} = K = \frac{E_B N_B}{3\,(3N_B - E_B)}$$

$$\therefore \quad E_A N_A (3N_B - E_B) = E_B N_B (3N_A - E_A)$$

or
$$3\,E_A N_A N_B - E_A N_A E_B = 3E_B N_B N_A - E_A E_B N_B$$

or
$$N_B\,(3E_A N_A - 3E_B N_A + E_A E_B) = E_A N_A E_B$$

From which
$$N_B = \frac{E_A N_A E_B}{3E_A N_A - 3E_B N_A + E_A E_B} = \frac{1.01E_A N_A E_A}{3E_A N_A - 3 \times 1.01E_A N_A + 1.01E_A . E_A}$$

or
$$N_B = \frac{1.01E_A N_A}{1.01E_A - 3\,(1.01 - 1)\,N_A} = \frac{\mathbf{101}E_A N_A}{\mathbf{101}E_A - \mathbf{3}N_A}.$$

Example 2.23. *A bar of elastic material is subjected to a direct compressive stress of* p_1 *in the longitudinal direction. Suitable lateral compressive stress* p_2 *is applied along the other two lateral directions to limit the net strain in each of the lateral directions to one third the magnitude that could be under* p_1 *acting alone. Find the magnitude of* p_2 *and the net strain in the longitudinal direction.*

Solution. When p_2 is applied along the two lateral directions,

$$e_2' = e_3' = \frac{p_2}{E} - \frac{\mu p_2}{E} - \frac{\mu p_1}{E}$$

However, if p_2 is not applied,

$$e_2 = -\frac{\mu p_1}{E}$$

$$\therefore \quad \frac{p_2}{E} - \frac{\mu p_2}{E} - \frac{\mu p_1}{E} = -\frac{1}{3}\frac{\mu p_1}{E}$$

Hence $$p_2 = \frac{\mathbf{2}}{\mathbf{3}}\frac{\mu}{\mathbf{1}-\mu}\,\boldsymbol{p_1}. \quad \textbf{(Ans.)}$$

Net strain in the longitudinal direction :

$$e_1' = \frac{p_1}{E} - \frac{\mu p_2}{E} - \frac{\mu p_2}{E} = \frac{p_1}{E}\left[1 - 2\mu \times \frac{2}{3}\frac{\mu}{1-\mu}\right]$$

or $$e_1' = \frac{p_1}{E}\left[\frac{3 - 3\mu - 4\mu^2}{3(1-\mu)}\right] \text{ compressive.}$$

Substituting $\mu = \frac{1}{m}$, the above expression reduces to

$$e_1' = \frac{\boldsymbol{p_1}}{\boldsymbol{E}}\left[\frac{\mathbf{3}\boldsymbol{m}^2 - \mathbf{3}\boldsymbol{m} - \mathbf{4}}{\mathbf{3}\boldsymbol{m}(\boldsymbol{m}-\mathbf{1})}\right].$$

2.12. EXAMPLES FROM COMPETITIVE EXAMINATIONS

Example 2.24. *A cylindrical piece of steel 80 mm diameter and 120 mm long is subjected to an axial compressive force of 50000 kg. Calculate the change in the volume of the piece if bulk modulus = 1.7 × 10^6 kg/cm² and Poisson's ratio = 0.3.* (UPSC Engg. Services Exam. 1997)

Solution. Volume of cylinder $V = \frac{\pi}{4} d^2 L$

$$\therefore \quad \frac{\delta V}{V} = \frac{\pi L}{4} \cdot \frac{2d\,\delta d}{\frac{\pi}{4} d^2 L} + \frac{\pi d^2}{4} \cdot \frac{\delta L}{\frac{\pi}{4} d^2 L}$$

or $$\frac{\delta V}{V} = e_v = \frac{2\delta d}{d} + \frac{\delta L}{L} = -2\mu\frac{p}{E} + \frac{p}{E} = \frac{p}{E}(1 - 2\mu)$$

Here, $$p = \frac{P}{A} = \frac{50000}{\frac{\pi}{4}(8)^2} = 994.72 \text{ kg/cm}^2 \,; \; V = \frac{\pi}{4}(8)^2 \times 12 = 603.186 \text{ cm}^3$$

$$E = 3K(1-2\mu) = 3 \times 1.7 \times 10^6 (1 - 2 \times 0.3) = 2.04 \times 10^6 \text{ kg/cm}^2$$

$$\therefore \quad \delta V = \frac{Vp}{E}(1-2\mu) = \frac{603.186 \times 994.72}{2.04 \times 10^6}(1 - 2 \times 0.3) = 0.1176 \text{ cm}^3 = \mathbf{117.6 \text{ mm}^3.}$$

Example 2.25. *Fig. 2.9 shows three metal cubes A, B and C, of side 100 mm indirect contact, resting on rigid base and confired in x-coordinate direction between two rigid end-plates.*

If the upper face of the centre cube (cube B) is subjected to uniform compressive stress of 0.5 kN/mm², compute for the cube B, the following :

(i) The direct stress in the x-direction (σ_x)

(ii) The direct strains in the three coordinate directions x, y and z

(iii) Volumetric strain.

The elastic properties for the three cubes A, B and C are given in Fig. 2.9. State all the assumptions made. (UPSC Engg. Services Exam. 1999)

Solution. When compressive stress σ_z is applied on cube *B*, it will try to expand in the *x*-direction. However, due to contraints applied by cubes *A* and *C* in the lateral direction, a compressive stress σ_x will be induced in the *x*-direction. The free body diagrams for the three cubes are shown in Fig. 2.10. Since cube *B* is free to expand in *y*-direction, σ_y will be zero. Let us use +ve sign for compressive stress and compressive strain.

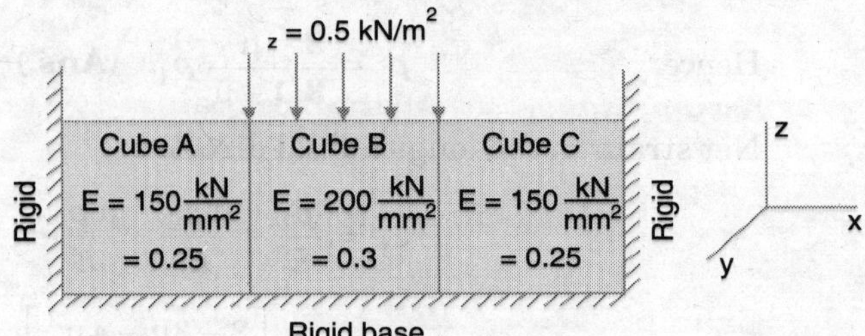

Fig. 2.9

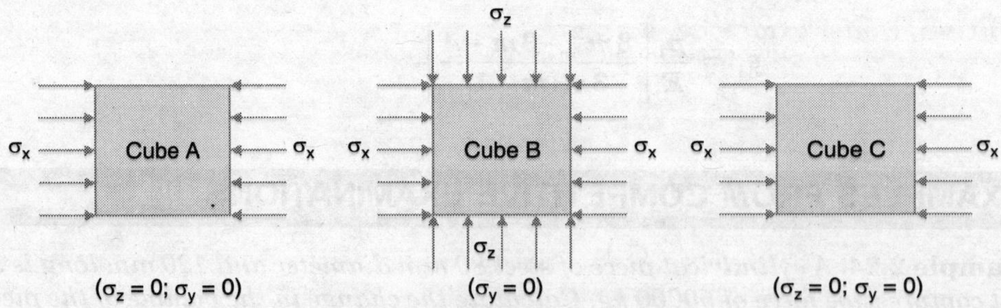

Fig. 2.10. Free body diagrams.

For cube B $\quad e_{xb} = \dfrac{\sigma_x}{E_b} - \dfrac{\mu_b \sigma_z}{E_b}$ (compression) $= \dfrac{\mu_b \sigma_z}{E_b} - \dfrac{\sigma_x}{E_b}$ (expansion)

For cubes A and C $\quad e_{xa} = \dfrac{\sigma_x}{E_a}$ (compression) and $e_{xc} = \dfrac{\sigma_x}{E_c}$ (compression)

From compatibility, expansion e_{xb} of cube B = compression e_{xa} + compression e_{xc}

$$\therefore \quad \frac{\mu_b \sigma_z}{E_b} - \frac{\sigma_x}{E_b} = \frac{\sigma_x}{E_a} + \frac{\sigma_x}{E_c} \text{ (where } E_a = E_c)$$

or

$$\frac{2\sigma_x}{E_a} + \frac{\sigma_x}{E_b} = \frac{\mu_b \sigma_z}{E_b}$$

$$\therefore \quad \sigma_x\left[\frac{2}{150}+\frac{1}{200}\right]=\frac{0.3\times 0.5}{200}, \text{ from which } \sigma_x = \mathbf{0.0409\ kN/mm^2}.$$

Positive sign shows that the assumed direction of σ_x is correct, *i.e.* σ_x is compressive.

Strains in cube B

$$e_{zb}=\frac{\sigma_z}{E_b}-\frac{\mu_b\sigma_x}{E_b}=\frac{0.5}{200}-\frac{0.3\times 0.0409}{200}=\mathbf{+\,2.4386\times 10^{-3}}$$

$$e_{xb}=\frac{\sigma_x}{E_b}-\frac{\mu_b\sigma_z}{E_b}=\frac{0.0409}{200}-\frac{0.3\times 0.5}{200}=\mathbf{-\,5.4545\times 10^{-4}}$$

$$e_{yb}=-\frac{\mu_b\sigma_z}{E_b}-\frac{\mu_b\sigma_x}{E_b}=-\frac{0.3\times 0.5}{200}-\frac{0.3\times 0.0409}{200}=\mathbf{-\,8.1135\times 10^{-4}}$$

$$\therefore \quad e_{vb}=e_{xb}+e_{yb}+e_{zb}=-\,5.4545\times 10^{-4}-8.1135\times 10^{-4}+2.4386\times 10^{-3}$$

$$= \mathbf{+\,1.0818\times 10^{-3}}. \ (\textit{i.e.}\ \text{decrease})$$

Assumptions. **(*i*) All the blocks are of homogeneous and isotropic materials**

(*ii*) All the deformed faces remain plane

(*iii*) The contact surfaces are frictionless

(*iv*) The end faces and base are rigid and non-yielding.

Example 2.26. *A steel specimen of 12 mm diameter extends by 6.31 × 10^{-2} mm over a gauge length of 150 mm when subjected to an axial load of 10 kN. The same specimen undergoes a twist of 0.5° on a length of 150 mm under a twisting moment of 10 Nm. Using the above data, determine the elastic constants E, μ, G and k.* (UPSC Engg. Services Exam. 2001)

Solution. *Under axial load,* $\delta=\dfrac{PL}{AE}=\dfrac{4PL}{\pi d^2 E}$

$$\therefore \quad E=\frac{4PL}{\pi d^2\delta}=\frac{4\times 10000\times 150}{\pi\,(12)^2\times 6.31\times 10^{-2}}=\mathbf{2.102\times 10^5\ N/mm^2}$$

Under torque $\dfrac{T}{J}=\dfrac{N\theta}{L}$ or $N=\dfrac{TL}{J\theta}=\dfrac{32TL}{\pi d^4\theta}$

where $\theta=\dfrac{\pi}{180°}\times 0.5°=8.7266\times 10^{-3}$ radians

$$\therefore \quad N=\frac{32\times 10\times 10^3\times 150}{\pi(12)^4\times 8.7266\times 10^{-3}}=\mathbf{0.844\times 10^5\ N/mm^2}$$

Also, $$\mu=\frac{E}{2N}-1=\frac{2.102\times 10^5}{2\times 0.844\times 10^5}-1=\mathbf{0.245}$$

and $$K=\frac{E}{3(1-2\mu)}=\frac{2.102\times 10^5}{3(1-2\times 0.245)}=\mathbf{1.374\times 10^5\ N/mm^2}.$$

PROBLEMS

1. A piece of steel 1 metre long and 3 cm × 3 cm cross-section is subjected to a tensile force of 120 kN in the direction of its length. Calculate the change in volume. Take $\frac{1}{m} = 0.29$ and $E = 205$ kN/mm^2.
2. A metal specimen is compressed in the direction of its axis and means are employed to reduce lateral expansion to one-third of what it would be if free to expand. Calculate the modified value of the elastic constant. Prove that its value will be $\frac{9}{8}$ E if $m = 4$.
3. A bar of elastic material is subjected to direct stress in a longitudinal direction, and its strains in the two directions at right angles are reduced to one-half and one-third respectively to those which normally occur in an ordinary tension member. If $E = 200$ kN/mm^2 and $m = 4$, what is value of the elastic constant ?
4. A lever is keyed to a shaft of 6 cm diameter. The width of the key is 1 cm and length 6 cm. What load can be applied at a radius of 60 cm without causing the shear stress is the key to exceed 75 N/mm^2 ?
5. A lever is secured to a shaft by a taper pin as shown in Fig. 2.11. The diameter of the shaft is 4 cm and the mean diameter of the pin is 1 cm. What torque can be applied to the lever without causing the shear stress in the pin to exceed 70 N/mm^2 ?
6. A shaft is subjected to shearing stress of 85 N/mm^2 at its surface. Calculate the change in the angle of square, scratched on its surface. Take $N = 80$ kN/mm^2.

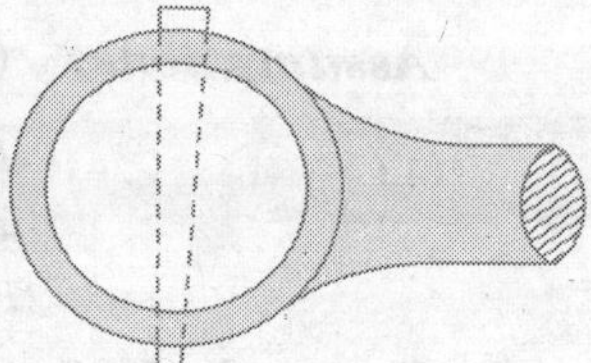
Fig. 2.11

7. A steel punch can be worked to a compressive stress of 1100 N/mm^2. Find the least diameter of the hole which can be punched through a steel plate 12.5 mm thick if its ultimate shear stress is 360 N/mm^2.
8. Rivet holes 20 mm diameter are punched in a steel plate 10 mm thick. The shearing strength of plate is 350 N/mm^2. Find the compressive stress in the punch at the time of punching.
9. A bar of steel 2 cm diameter was subjected to a tensile load of 25 kN and the measured extension on a 20 cm gauge length was 7.8×10^{-3} cm and the change in a diameter was 2.32×10^{-4} cm. Calculate the Poisson's Ratio and the values of the three moduli.
10. Derive the relationship between Modulus of Elasticity, Modulus of Rigidity and Poisson's Ratio of an elastic body.

    A bronze specimen has a Modulus of Elasticity of 120 kN/mm^2 and a Modulus of Rigidity of 47 kN/mm^2. Determine the Poisson's ratio of the material. *(A.M.I.E. Nov. 1974)*
11. Tensile stresses f_1 and f_2 act at right angles to one another on an element of isotropic elastic material. The strain in the direction of f_1 is twice that in the direction of f_2. If E for the material is 120 kN/mm^2 and $\mu = 0.3$, find the ratio of f_1 to f_2.

ANSWERS

1. 246 mm^3.
2. $E' = \frac{3m(m-1)}{3m^2 - 3m - 4} E$.
3. 221 kN/mm^2.
4. 2250 N.
5. 220 Nm.
6. 3′39″.
7. 16.4 mm.
8. 700 N/mm^2.
9. $\frac{1}{m} = 0.298$; $E = 204$ kN/mm^2 ; $N = 78.5$ kN/mm^2 ; $K = 169.5$ kN/mm^2.
10. $\frac{1}{m} = 0.277$.
11. 1.438.

3

Compound Stresses

3.1. INTRODUCTION

Upto this stage, we have studied the behaviour of materials under the action of either an axial stress (*i.e.* tension or compression) or a shear stress. In actual practice, all the three-types of stresses may act simultaneously on appropriate planes passing through a point in the strained material. We shall, therefore, investigate such general stress system and find the resultant stress on inclined planes, which may carry greater stress than the applied ones. In the treatment that we shall be doing now, it will be seen that the most complicated stress system is equivalent to a combination of two mutually perpendicular direct stresses.

3.2. STRESS COMPONENTS ON INCLINED PLANES

(*a*) **Stresses induced by direct stress**

In Fig. 3.1, a bar is subjected to tensile force P. It is required to find the stress components along and across an inclined plane AB the normal to which makes an angle θ with the direction of P.

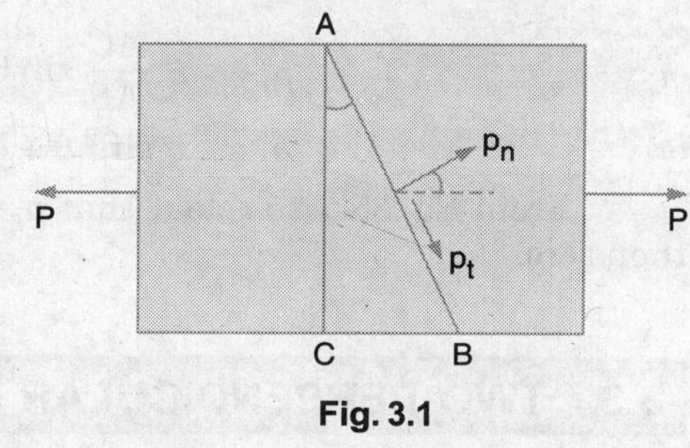

Fig. 3.1

Let A = Area of cross-section of normal section AC

$\therefore$ Area of section of $AB = A \sec \theta$.

$$\text{Stress Normal to } AB = p_n = \frac{\text{Force Normal to } AB}{\text{Area of section of } AB}$$

$$= \frac{P \cos \theta}{A \sec \theta} = \mathbf{p \cos^2 \theta} \quad ...(3.1)$$

$$\text{Stress tangential to } AB = p_t = \frac{P \sin \theta}{A \sec \theta} = p \sin \theta \cos \theta = \frac{p}{2} \sin 2\theta \quad ...(3.2)$$

The resultant stress $p_r = \sqrt{p_n^2 + p_t^2} = \sqrt{p^2 \cos^4 \theta + p^2 \sin^2 \theta \cos^2 \theta} = p \cos \theta$ ...(3.3)

The variations of p_n, p_t and p_r with θ are shown in Fig. 3.2. Evidently p_t reaches the maximum value of $p/2$ when θ = 45°. Hence all surfaces inclined at 45° to *AC* (and therefore to the axis of pull) are subjected to maximum shear stress.

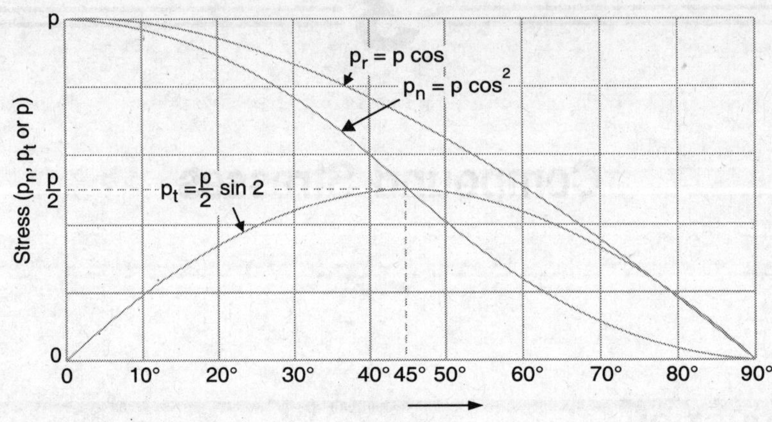

Fig. 3.2

(*b*) **Stresses induced by state of simple shear**

In Fig. 3.3, there is a state of simple shear. It is required to find the normal and tangential stresses for the plane *AB*.

Area of *BC* = *A* ; Area of *AB* = *A* sec θ

Considering unit thickness perpendicular to the diagram, we have

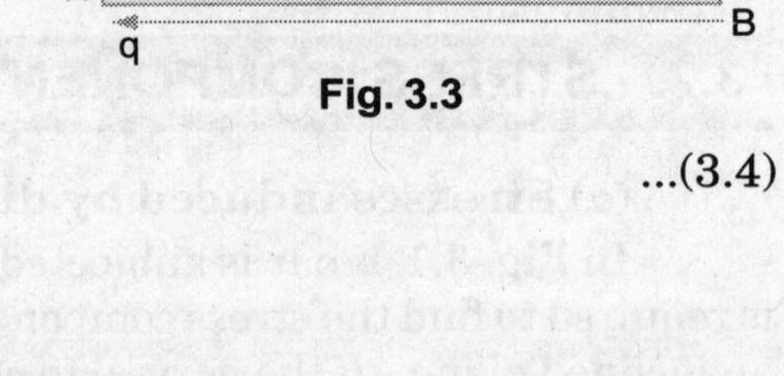

Fig. 3.3

$$p_n AB = q\, BC \sin \theta + q\, AC \cos \theta$$

or $$p_n = q\,\frac{BC}{AB} \sin \theta + q\,\frac{AC}{AB} \cos \theta$$

$$= q \cos \theta \sin \theta + q \sin \theta \cos \theta = q \sin 2\theta \quad \text{...(3.4)}$$

Similarly, $$p_t AB = -\,q\, AC \sin \theta + q\, BC \cos \theta$$

or $$p_t = -\,q\,\frac{AC}{AB} \sin \theta + q\,\frac{BC}{AB} \cos \theta$$

or $$p_t = -\,q \sin^2 \theta + q \cos^2 \theta = q\,(\cos^2 \theta - \sin^2 \theta) = q \cos 2\theta \quad \text{...(3.5)}$$

From Eq. 3.4, it is clear that p_n has its maximum value of q when θ = 45° and that p_t is then zero.

3.3. TWO PERPENDICULAR NORMAL STRESSES

(*a*) **When both the stresses are like** (Fig. 3.4). Let us take the case when two perpendicular like stresses p_1 and p_2 act on the material.

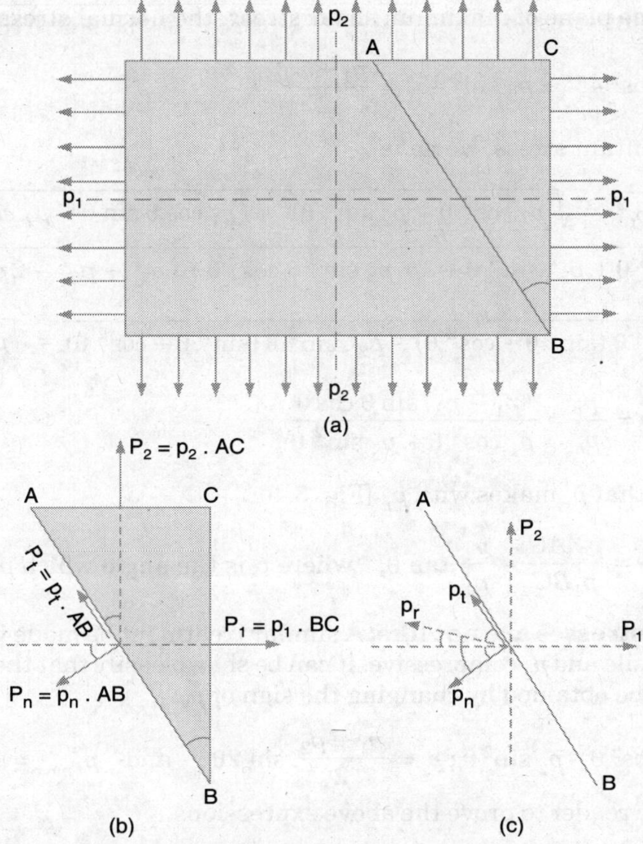

Fig. 3.4

To find the normal and tangential stress components on plane AB, let us consider the equilibrium of the block ABC.

Taking unit thickness of the block and resolving forces perpendicular to AB, we have

$$P_n = P_1 \cos\theta + P_2 \sin\theta \quad \text{or} \quad p_n AB = p_1 BC \cos\theta + p_2 AC \sin\theta$$

$$\therefore \quad p_n = p_1 \frac{BC}{AB} \cos\theta + p_2 \frac{AC}{AB} \sin\theta = p_1 \cos^2\theta + p_2 \sin^2\theta \quad \text{...[3.6(a)]}$$

or $$p_n = \tfrac{1}{2}(p_1 + p_2) + \tfrac{1}{2}(p_1 - p_2)\cos 2\theta \quad \text{...[3.6(b)]}$$

Similarly, resolving forces along AB, we have

$$P_t = P_1 \sin\theta - P_2 \cos\theta$$

or $$p_t AB = p_1 BC \sin\theta - p_2 AC \cos\theta \quad \text{or} \quad p_t = p_1 \frac{BC}{AB} \sin\theta - p_2 \frac{AC}{AB} \cos\theta$$

or $$p_t = p_1 \cos\theta \sin\theta - p_2 \sin\theta \cos\theta = \frac{p_1 - p_2}{2} \sin 2\theta \quad \text{...(3.7)}$$

Maximum shear stress will be obtained when $\theta = 45°$

Then $$(p_t)_{max} = \frac{p_1 - p_2}{2}$$

Across the same plane of maximum shear stress, the normal stress

$$p_n = p_1 \cos^2 45° + p_2 \sin^2 45° = \frac{p_1 + p_2}{2}.$$

If p_r is the resultant stress, we have

$$p_r = \sqrt{p_n^2 + p_t^2} = \sqrt{(p_1 \cos^2 \theta + p_2 \sin^2 \theta)^2 + (p_1 \cos \theta \sin \theta - p_2 \sin \theta \cos \theta)^2}$$

$$= \sqrt{p_1^2 \cos^4 \theta + p_2^2 \sin^4 \theta + 2p_1p_2 \sin^2 \theta \cos^2 \theta + (p_1^2 + p_2^2 - 2p_1p_2) \sin^2 \theta \cos^2 \theta}$$

$$= \sqrt{p_1^2 \cos^2 \theta (\sin^2 \theta + \cos^2 \theta) + p_2^2 \sin^2 \theta (\sin^2 \theta + \cos^2 \theta)} = \sqrt{p_1^2 \cos^2 \theta + p_2^2 \sin^2 \theta}$$

$$\tan \phi = \frac{p_t}{p_n} = \frac{(p_1 - p_2) \sin \theta \cos \theta}{p_1 \cos^2 \theta + p_2 \sin^2 \theta}$$

where ϕ is the angle that p_r makes with p_n [Fig. 3.4(*c*)]

$$\tan \alpha = \frac{P_2}{P_1} = \frac{p_2\, AC}{p_1 BC} = \frac{p_2}{p_1} \tan \theta,$$ where α is the angle which p_r makes with P_1.

(*b*) **When the stresses are not like.** A similar treatment be made when p_1 and p_2 are not like *i.e.* p_1 may be tensile and p_2 compressive. It can be shown easily that the different expressions for p_n, p_t, p_r etc., can be obtained by changing the sign of p_2.

Thus $p_n = p_1 \cos^2 \theta - p_2 \sin^2 \theta$; $p_t = \dfrac{p_1 + p_2}{2} \sin 2\theta$; and $p_{t\,max} = \dfrac{p_1 + p_2}{2}$.

It is left for the reader to prove the above expressions.

3.4. TWO PERPENDICULAR NORMAL STRESSES ACCOMPANIED WITH STATE OF SIMPLE SHEAR

Let us now take the most general case of stress system when two perpendicular normal stresses (both like) are accompanied with a state of simple shear as illustrated in Fig. 3.5 (*a*).

Consider the equilibrium of the prism *ABC* of unit length perpendicular to the diagram [Fig. 3.5 (*b*)].

Resolving perpendicular to *AB*, we get

$$p_n\, AB = p_1\, AC \cos \theta + p_2\, BC \sin \theta + q\, AC \sin \theta + q\, BC \cos \theta$$

or

$$p_n = p_1 \frac{AC}{AB} \cos \theta + p_2 \frac{BC}{AB} \sin \theta + q \frac{AC}{AB} \sin \theta + q \frac{BC}{AB} \cos \theta$$

$$= p_1 \cos^2 \theta + p_2 \sin^2 \theta + q \cos \theta \sin \theta + q \sin \theta \cos \theta$$

$$= p_1 \cos^2 \theta + p_2 \sin^2 \theta + q \sin 2\theta \quad \text{...[3.8 (a)]}$$

Eq. 3.8 (*a*) can also be written as follows :

$$p_n = \tfrac{1}{2}(p_1 + p_2) + \tfrac{1}{2}(p_1 - p_2) \cos 2\theta + q \sin 2\theta \quad \text{...[3.8(b)]}$$

Resolving along the face *AB*

$$p_t\, AB = p_1 AC \sin \theta - p_2 BC \cos \theta - q\, AC \cos \theta + q\, BC \sin \theta$$

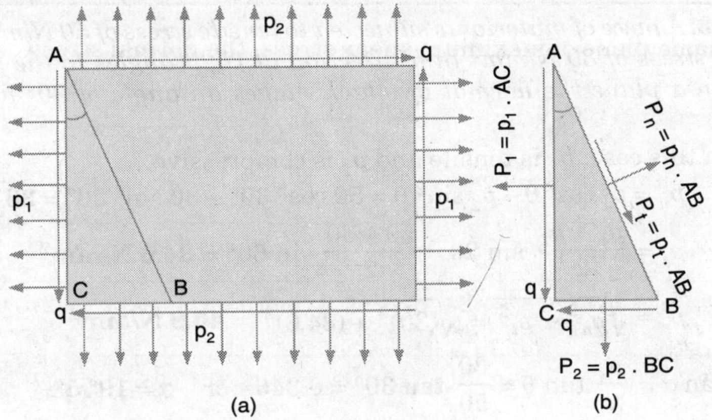

Fig. 3.5

or $$p_t = p_1 \frac{AC}{AB} \sin\theta - p_2 \frac{BC}{AB} \cos\theta - q \frac{AC}{AB} \cos\theta + q \frac{BC}{AB} \sin\theta$$

$$= p_1 \cos\theta \sin\theta - p_2 \sin\theta \cos\theta - q \cos^2\theta + q \sin^2\theta$$

$$= \frac{p_1 - p_2}{2} \sin 2\theta - q \cos 2\theta \qquad \text{...(3.9)}$$

If p_1 and p_2 are not like, similar expression for p_n and p_t will be obtained by changing the sign of p_2. Thus, we have

$$p_n = p_1 \cos^2\theta - p_2 \sin^2\theta + q \sin 2\theta \quad \text{and} \quad p_t = \frac{p_1 + p_2}{2} \sin 2\theta - q \cos 2\theta.$$

Example 3.1. *A tie bar is subjected to a uniform tensile stress of 80 N/mm². Find the intensity of shear stress, normal stress and resultant stress on a plane the normal to which is inclined 30° to the axis of the bar.*

Solution. $p_n = p \cos^2\theta = 80 \cos^2 30° =$ **60 N/mm²**

$$p_t = \frac{p}{2} \sin 2\theta = 40 \sin 60° = \mathbf{34.6\ N/mm^2}$$

$$p_r = \sqrt{p_n{}^2 + p_t{}^2} = \sqrt{(60)^2 + (34.6)^2} = \mathbf{69.2\ N/mm^2}.$$

Example 3.2. *A piece of material is subjected to tensile stresses of 60 N/mm² and 30 N/mm² at right angles to each other. Find fully the stresses on a plane the normal of which makes an angle of 40° with the 60 N/mm² stress.*

Solution. From equation 3.6(*b*), we have

$$p_n = p_1 \cos^2\theta + p_2 \sin^2\theta = 60 \cos^2 40° + 30 \sin^2 40° = \mathbf{47.5\ N/mm^2}$$

Similarly, from equation 3.7, we have

$$p_t = \frac{p_1 - p_2}{2} \sin 2\theta = \frac{60 - 30}{2} \sin 80° = \mathbf{14.8\ N/mm^2}$$

The resultant stress $p_r = \sqrt{p_n{}^2 + p_t{}^2} = \sqrt{(47.5)^2 + (14.8)^2} = 49.6$ N/mm²

Angle α of p_r with the direction of p_1 is given by

$$\tan\alpha = \frac{p_2}{p_1} \tan\theta = \frac{30}{60} \tan 40° = 0.42 \quad \text{or} \quad \alpha = \mathbf{22° 30'}.$$

Example 3.3. *A piece of material is subjected to tensile stress of 50 N/mm² in one direction and a compressive stress of 30 N/mm² in a direction at right angles to the previous one. Find fully the stresses on a plane the normal of which makes an angle of 30° with the 50 N/mm² stresses.*

Solution. In this case, p_1 is tensile and p_2 is compressive.

Hence, $$p_n = p_1 \cos^2 \theta - p_2 \sin^2 \theta = 50 \cos^2 30° - 30 \sin^2 30° = \mathbf{30\ N/mm^2}$$

and $$p_t = \frac{p_1 + p_2}{2} \sin 2\theta = \frac{50+30}{2} \sin 60° = \mathbf{34.6\ N/mm^2}$$

$$p_r = \sqrt{p_n^2 + p_t^2} = \sqrt{(30)^2 + (34.6)^2} = \mathbf{45.9\ N/mm^2}$$

$$\therefore \quad \tan \alpha = \frac{p_2}{p_1} \tan \theta = \frac{30}{50} \tan 30° = 0.346 \quad \text{or} \quad \alpha = \mathbf{19° 5'}.$$

Example 3.4. *A piece of material is subjected to tensile stresses of p_1 and p_2 at right angles to each other ($p_1 > p_2$). Find the plane across which the resultant stress is most inclined to the normal. Find the value of this inclination and the resultant stress when $p_1 = 70\ N/mm^2$ and $p_2 = 50\ N/mm^2$ (both tensile).*

Solution. Referring to Fig. 3.4 (*c*), let ϕ be the inclination of the resultant with the normal. We have

$$\tan \phi = \frac{p_t}{p_n} = \frac{(p_1 - p_2) \sin \theta \cos \theta}{p_1 \cos^2 \theta + p_2 \sin^2 \theta} \qquad \text{...(1)}$$

For ϕ to be maximum, $\tan \phi$ is also maximum, and $\dfrac{d(\tan \phi)}{d\theta} = 0$

Hence, differentiating and dividing out common factors

$$(p_1 \cos^2 \theta + p_2 \sin^2 \theta) \cot 2\theta + (p_1 - p_2) \sin \theta \cos \theta \times \sin 2\theta = 0$$

or $$p_n \cos 2\theta + p_t \sin 2\theta = 0$$

Hence $$\tan 2\theta = -\frac{p_n}{p_t} = -\cot \phi = \tan\left(\frac{\pi}{2} + \phi\right) \quad \text{or} \quad 2\theta = \frac{\pi}{2} + \phi \qquad \text{...(2)}$$

Substituting the value of θ in (1), we get

$$\tan \phi = \frac{(p_1 - p_2)\cos \phi}{p_1 (1 - \sin \phi) + p_2 (1 + \sin \phi)}$$

or $$\frac{p_2}{p_1} = \frac{1 - \sin \phi}{1 + \sin \phi} \quad \text{or} \quad \sin \phi = \frac{p_1 - p_2}{p_1 + p_2} \qquad \text{...(3)}$$

Hence $$\theta = \tfrac{1}{2}\left(\frac{\pi}{2} + \sin^{-1} \frac{p_1 - p_2}{p_1 + p_2}\right)$$ which gives the position of the required plane.

Substituting the value of p_1 and p_2, we get

$$\theta = \tfrac{1}{2}\left(90° + \sin^{-1} \frac{20}{120}\right) = \tfrac{1}{2}(90° + 9° 34') = 49° 47'$$

$$\phi = \sin^{-1} \frac{p_1 - p_2}{p_1 + p_2} = \sin^{-1} \frac{20}{120} = \mathbf{9° 34'}$$

$$p_r = \sqrt{(70 \cos 49° 47')^2 + (50 \sin 49° 47')^2} = \mathbf{58.7\ N/mm^2}.$$

3.5. CIRCULAR DIAGRAM OF STRESS : MOHR CIRCLE

In § 3.2, § 3.3 and § 3.4, we have seen how we can calculate the values of p_n and p_t if p_1, p_2 and q are given. The circular stress diagram, which we shall study now, offers an easy solution to some problems which might offer difficulties in mathematical manipulation. The circular stress diagram is a simple geometrical construction used by Rankine. It is known as 'Mohr circle' or 'stress circle'.

Let us first take the case when the material is subjected to two like stresses p_1 and p_2, $(p_1 > p_2)$. We know from equation 3.6 (*b*) that

$$p_n = p_1 \cos^2 \theta + p_2 \sin^2 \theta$$

which may be re-written as $p_n = \frac{1}{2}(p_1 + p_2) + \frac{1}{2}(p_1 - p_2) \cos 2\theta$ ...(3.10)

Also, from equation 3.7, $p_t = \frac{(p_1 - p_2)}{2} \sin 2\theta$

It is clear from the above equations that p_t and part of p_n vary with angle θ and may, therefore, be represented by the two rectangular projections of a radius vector of length $\frac{1}{2}(p_1 - p_2)$ and making an angle 2θ with the reference axis. This immediately suggests a simple geometrical construction for finding p_t and p_n.

MOHR CIRCLE CONSTRUCTION FOR LIKE STRESSES

(Refer Fig. 3.6)

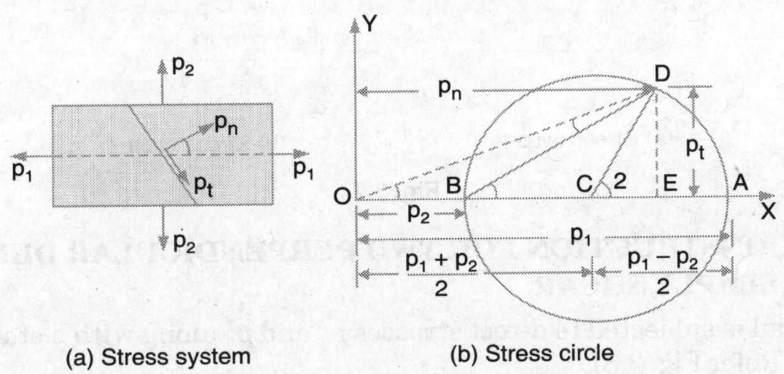

(a) Stress system (b) Stress circle

Fig. 3.6

On an axis *OX*, measure *OA* and *OB* equal to p_1 and p_2 respectively, to some scale. Bisect *AB* at *C*. With *C* as centre and *CA* or *CB* as radius, draw a circle. At the centre *C*, draw a line *CD* at an angle 2θ, in the same direction as the normal to the plane makes with the direction of p_1. In Fig. 3.6 (*a*), which represents the stress system, the normal to the plane makes an angle θ in the anticlockwise direction with the direction of p_1. The line *CD* has a perpendicular *DE* on the axis *OX*. Thus *DE* will represent p_t and *OE* will represent p_n.

From the stress diagram, $CD = CA = \frac{1}{2}(p_1 - p_2)$

$$DE = CD \sin 2\theta = \frac{1}{2}(p_1 - p_2) \sin 2\theta = p_t \quad \text{(From Eq. 3.7)}$$

Similarly, $OE = OC + CE = \frac{1}{2}(p_1 + p_2) + \frac{1}{2}(p_1 - p_2) \cos 2\theta = p_n$ ...(Eq. 3.10)

Hence it is proved that the two rectangular projections of the radius vector give p_t and p_n.

Also, from the stress circle, p_t is maximum when $2\theta = 90°$, or $\theta = 45°$ and $p_{t\,max} = \frac{1}{2}(p_1 - p_2)$, which is same as obtained earlier.

In order to mark p_t in the stress system, we will take the clockwise shear as negative and anti-clockwise shear as positive. Negative values of p_t will be above the axis and positive values below the axis.

Similarly, if θ is in the anti-clockwise direction, the radius vector will automatically be above the axis and θ will be reckoned positive. If θ is in the clockwise direction, it will be negative and the radius vector will be below the axis.

Tensile stress will be reckoned positive and will be plotted to the right of the origin O. Compressive stress will be reckoned negative and will be plotted to the left of the origin O.

MOHR CIRCLE CONSTRUCTION FOR UNLIKE STRESSES

If p_1 and p_2 are not like, the same procedure will be followed, except that p_1 and p_2 will be measured to the opposite sides of the origin O. The construction is clear from Fig. 3.7.

It is clear that the direction of p_n will depend upon the position of point E. If it is to the right O, the direction of p_n will be the same as that of p_1.

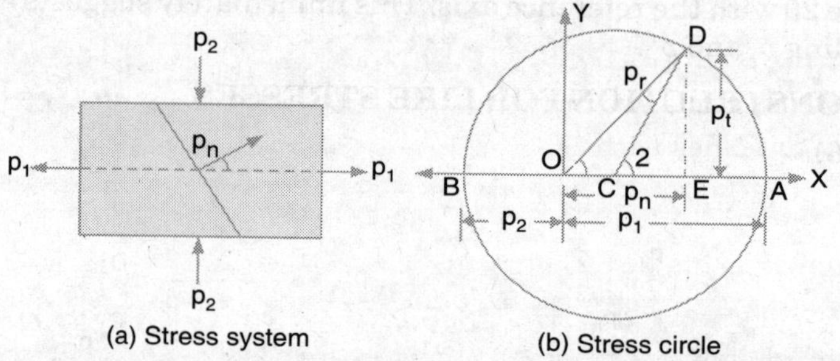

(a) Stress system (b) Stress circle

Fig. 3.7

MOHR CIRCLE CONSTRUCTION FOR TWO PERPENDICULAR DIRECT STRESSES WITH STATE OF SIMPLE SHEAR

If the material is subjected to direct stresses p_1 and p_2 along with a state of simple shear, proceed as follows (Refer Fig. 3.8).

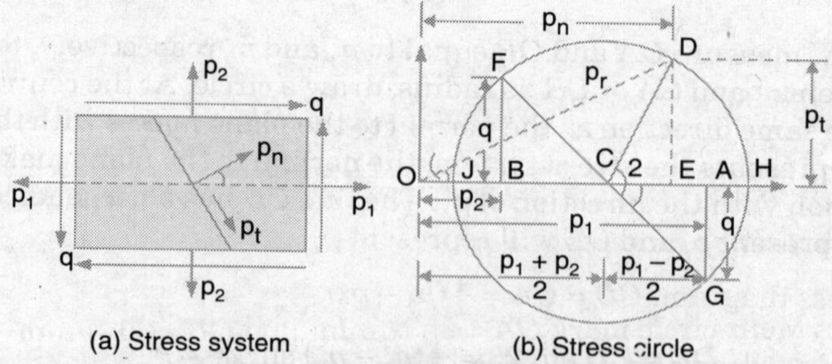

(a) Stress system (b) Stress circle

Fig. 3.8

Along the axis OX, measure $OA = p_1$ and $OB = p_2$ to some scale. At A, draw AG perpendicular to OX, and equal to q. AG has been drawn downward (as per sign conventions adopted) because q

is acting up with respect to the plane across which p_1 is acting, tending to rotate it in the anti-clockwise direction and is positive. Similarly, make BF perpendicular to OX and equal to q, but above OX. Join FG to cut the axis in C. With C as centre and CF or CG as radius, draw a circle. At C make CD at angle 2θ with CG in the anti-clock-direction. Draw DE perpendicular to the axis. DE will give p_t while OE will give p_n and OD will give p_r.

Proof. Let R = radius of stress circle

$$= \sqrt{CA^2 + AG^2} = \sqrt{\left(\frac{p_1 - p_2}{2}\right)^2 + q^2}$$

$$R \cos \beta = CA = \frac{p_1 - p_2}{2} \; ; R \sin \beta = AG = q$$

Now $$OE = OC + CE = OC + R \cos (2\theta - \beta)$$

$$= OC + R \cos \beta \cos 2\theta + R \sin \beta \sin 2\theta$$

$$= \tfrac{1}{2}(p_1 + p_2) + \frac{(p_1 - p_2)}{2} \cos 2\theta + q \sin 2\theta = p_n \quad \text{[as per equation 3.8 (a)]}$$

Similarly, $$DE = R \sin (2\theta - \beta) = R \cos \beta \sin 2\theta - R \sin \beta \cos 2\theta$$

$$= \tfrac{1}{2}(p_1 - p_2) \sin 2\theta - q \cos 2\theta = p_t \quad \text{(as per equation 3.9)}$$

Conclusions

From the stress circle, we draw the following conclusions :

(1) When D coincides with H, p_n attains maximum value :

$$p_{n.max} = OC + CH = \tfrac{1}{2}(p_1 + p_2) + \sqrt{\tfrac{1}{4}(p_1 - p_2)^2 + q^2}$$

$p_{n.max}$ is known as major principal stress σ_1. (see § 3.7)

$$p_t = 0 \; ; p_{r.max} = p_{n.max}$$

$$\tan 2\theta = \tan \beta = \frac{q}{\frac{p_1 - p_2}{2}} = \frac{2q}{(p_1 - p_2)} \quad \text{(see § 3.7)}$$

(2) When D coincides with J, p_n attains minimum value :

$$p_{n.min.} = OJ = OC - CJ = \tfrac{1}{2}(p_1 + p_2) - \sqrt{\tfrac{1}{4}(p_1 - p_2)^2 + q^2}$$

$p_{n.min.}$ is known as minor principal stress σ_2. (see § 3.7)

$$p_t = 0 \; ; p_{r.min.} = p_{n.min} \; ; \theta = 90° + \frac{\beta}{2}.$$

When $2\theta = (\beta + 90°)$, p_t attains maximum value $p_{t.max} = \sqrt{\tfrac{1}{4}(p_1 - p_2)^2 + q^2} = \dfrac{\sigma_1 - \sigma_2}{2}$

When $2\theta = [\beta + 270°]$, $p_{t.max} = -\sqrt{\tfrac{1}{4}(p_1 - p_2)^2 + q^2}$.

Example 3.5. *Solve example 3.2 graphically by Mohr circle.*

Solution. (Fig. 3.9)

To construct Mohr circle mark OA = 60 N/mm² and OB = 30 N/mm² to the same side of O. Bisect AB to get C and draw the circle with C as centre and CB as radius. Draw CD at 2θ = 80° to CA. Drop perpendicular DE. The Mohr circle is shown in Fig. 3.9 (*b*).

From Fig. 3.9 (*b*), we get the following :

$$p_t = DE = 14.8 \text{ N/mm}^2 \ ;$$
$$p_n = OE = 47.5 \text{ N/mm}^2 \ ;$$
$$p_r = OD = 49.6 \text{ N/mm}^2 \ ;$$
$$\phi = 17° \ ; \ \alpha = 22° \ 30'.$$

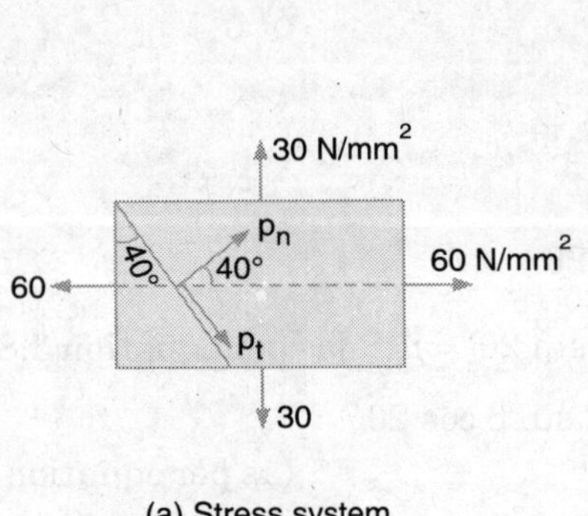

(a) Stress system

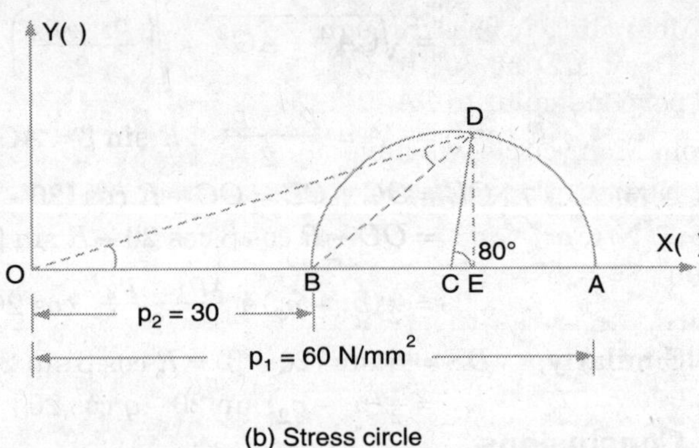

(b) Stress circle

Fig. 3.9

Example 3.6. *Draw the Mohr stress circle for direct stresses of 60 N/mm² (tensile) and 40 N/mm² (comp.) and estimate the magnitude and direction of the resultant stresses on planes making angles of 25° and 70° with the plane of the first principal stress. Find also the normal and tangential stresses on these planes.*

Solution. Plot OA = 60 N/mm² to one side and OB = 40 N/mm² to the other side and bisect AB to get C, as shown in Fig. 3.10. Draw the Mohr circle.

Draw CD_1 at 50° to CA and CD_2 at 140° to CA. Draw D_1E_1 and D_2E_2 perpendicular to AB.

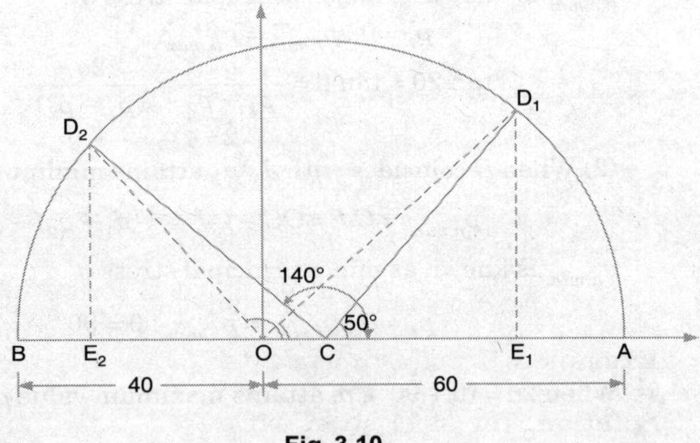

Fig. 3.10

For the first plane,

$p_t = D_1E_1 =$ **38 N/mm²** ;

$p_n = OE_1 =$ **42 N/mm²** (tensile)

$p_r = OD_1 =$ **57 N/mm²** ;

$\phi =$ **42°.**

For the second plane,

$$p_t = D_2E_2 = \mathbf{32 \ N/mm^2} \ ; \ p_n = OE_2 = \mathbf{28.5 \ N/mm^2} \text{ (comp.)}$$
$$p_r = OD_2 = \mathbf{42.5 \ N/mm^2} \ ; \ \phi' = \mathbf{131.5°}.$$

Example 3.7. *At a point in a bracket, the stresses on two mutually perpendicular planes are 40 N/mm² (tensile) and 20 N/mm² (tensile). The shear stress across these planes is 10 N/mm². Find using Mohr stress circle, the magnitude and direction of the resultant stress on plane making an angle of 30° with the plane of the first stress. Find also the normal and tangential stresses on this plane.*

Solution. (Fig. 3.11).

Draw OA = 40 N/mm² and OB = 20 N/mm². Erect perpendiculars BF and AG, each 10 N/mm² as shown in Fig. 3.11 (*b*).

Join F and G to get C and draw the Mohr circle to pass through F and G. Draw CD at 60° to CG. Draw DE perpendicular to OA.

From Mohr circle, we have

$p_t = DE = 3.7$ N/mm² ; $p_n = OE = 43.7$ N/mm² ; $p_r = OD = 43.8$ N/mm² ; $\phi \simeq 5°$.

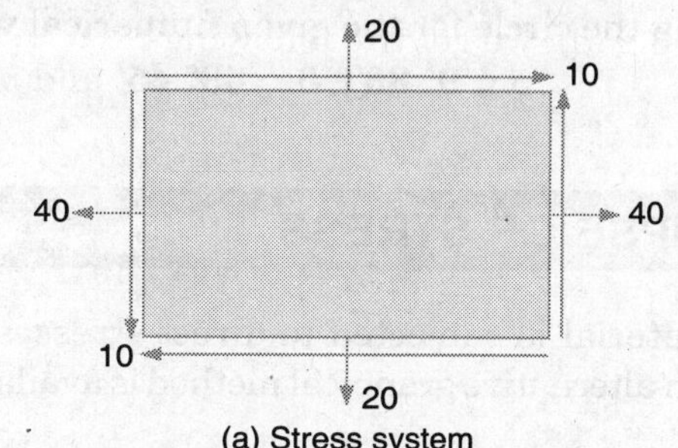

(a) Stress system

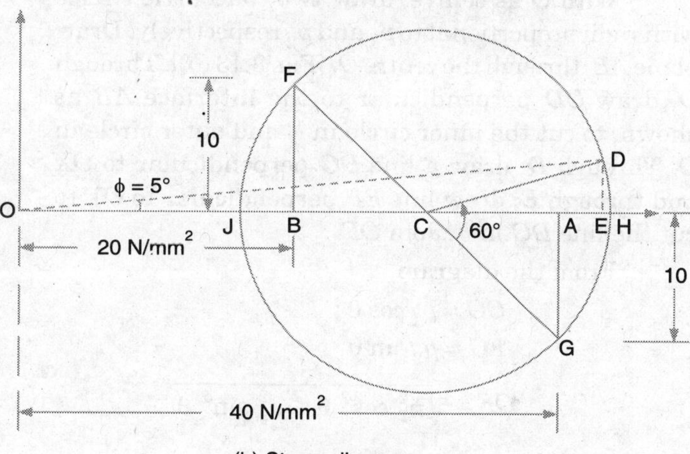

(b) Stress diagram

Fig. 3.11

Note. From the theoretical solution, the corresponding values will be as under :

$$p_n = \tfrac{1}{2}(p_1 + p_2) + \tfrac{1}{2}(p_1 - p_2)\cos 2\theta + q \sin 2\theta$$

$$= \tfrac{1}{2}(40 + 20) + \tfrac{1}{2}(40 - 20)\cos 60° + 10 \sin 60°$$

$$= 30 + 5.0 + 8.67$$

$$= 43.67 \text{ N/mm}^2$$

$$p_t = \tfrac{1}{2}(p_1 - p_2)\sin 2\theta - q\cos 2\theta$$

$$= \tfrac{1}{2}(40 - 20)\sin 60° - 10\cos 60° = 8.67 - 5.0 = 3.67 \text{ N/mm}^2$$

$$p_r = \sqrt{p_n^2 + p_t^2} = \sqrt{(43.67)^2 + (3.67)^2} = 43.82 \text{ N/mm}^2$$

and $$\phi = \tan^{-1}\left(\frac{p_t}{p_n}\right) = \tan^{-1}\left(\frac{3.67}{43.67}\right) = 4.8°.$$

Example 3.8. *Solve Example 3.4 using Mohr stress circle.*

Solution. The Mohr stress circle can be drawn as usual and is shown in Fig. 3.12. To obtain the maximum obliquity of the resultant stress, draw OD tangential to the circle. Thus from the Mohr circle, we get

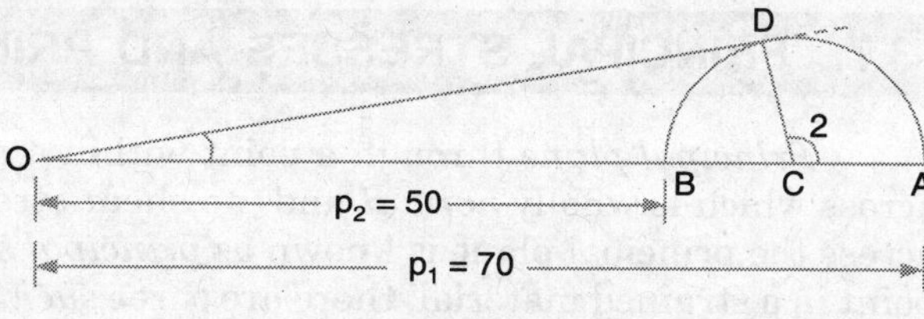

Fig. 3.12

$$\sin \phi = \frac{DC}{OC} = \frac{\tfrac{1}{2}(p_1 - p_2)}{\tfrac{1}{2}(p_1 + p_2)} = \frac{(p_1 - p_2)}{(p_1 + p_2)}$$

and $$\theta = \tfrac{1}{2}(90° + \phi)$$

Plotting the circle for the given numerical values, we get

$$\phi \simeq \mathbf{9° \ 30'} \ ; \ \theta \simeq \mathbf{49° \ 45'} \text{ and } p_r = \mathbf{58.7} \text{ N/mm}^2.$$

3.6. ELLIPSE OF STRESS

If a material is subjected to direct stresses p_1 and p_2 only, an alternative graphical method is available (Fig. 3.13).

With O as centre, draw two concentric circles with radii proportional to p_1 and p_2 respectively. Draw plane AB, through the centre O. [Fig. 3.13 (*b*)]. Through O, draw OD perpendicular to the interface AB as shown, to cut the inner circle in E and outer circle in D. Through D, draw a line DG perpendicular to OX and through E, draw line EF perpendicular to OY to cut the line DG in F. Join OF.

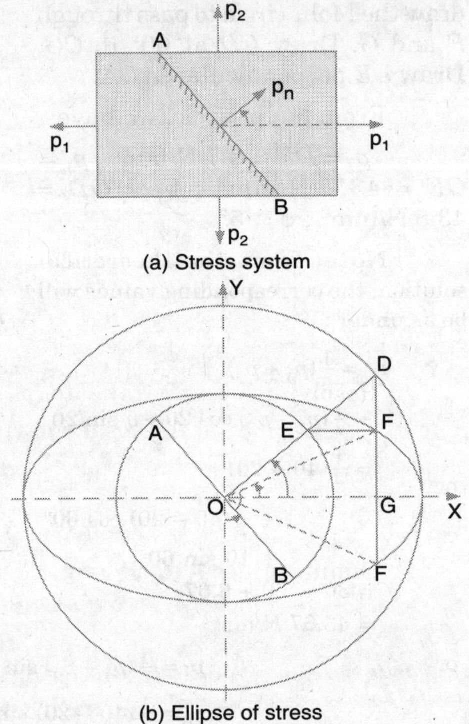

(a) Stress system

(b) Ellipse of stress

Fig. 3.13

From the diagram,

$$OG = p_1 \cos\theta \ ;$$

$$FG = p_2 \sin\theta$$

$$\therefore \quad OF = \sqrt{p_1^2 \cos^2\theta + p_2^2 \sin^2\theta}$$

$$= p_r$$

Hence OF gives the resultant stress p_r for that plane, and $\tan\alpha = \dfrac{p_2 \sin\theta}{p_1 \cos\theta} = \dfrac{p_2}{p_1}\tan\theta$, as earlier.

For different value of θ, the point F may be established. The locus of F will evidently be an ellipse. The diagram is, therefore, known as *ellipse of stress*.

3.7. PRINCIPAL STRESSES AND PRINCIPAL PLANES

Principal plane through a point within a material under stress is that plane the stress across which is wholly *normal* and no shear stress exists along this plane. The normal stress across the principal plane is known as *principal stress*. It may be stated in general, that at any point in a strained material, there are three *such* planes, mutually perpendicular to each other, carrying direct stresses only and no tangential stress. Out of the three, the plane carrying the maximum normal stress is called the *major principal plane* and the stress is called the *major principal stress*. The plane carrying the minimum normal stress is known as *minor principal plane* and the stress is known as *minor principal stress*. Here we shall be discussing only those cases where third principal stress is zero. We shall be denoting the major principal stress by σ_1 and minor by σ_2.

In order to find the principal stresses σ_1 and σ_2 under the action of p_1, p_2 and q (most general case), refer Fig. 3.14.

Let plane AB be the principal plane. Let us consider the equilibrium of the wedge ABC, in which the plane AB is having the normal stress (or principal stress) σ [Fig. 3.14 (b)].

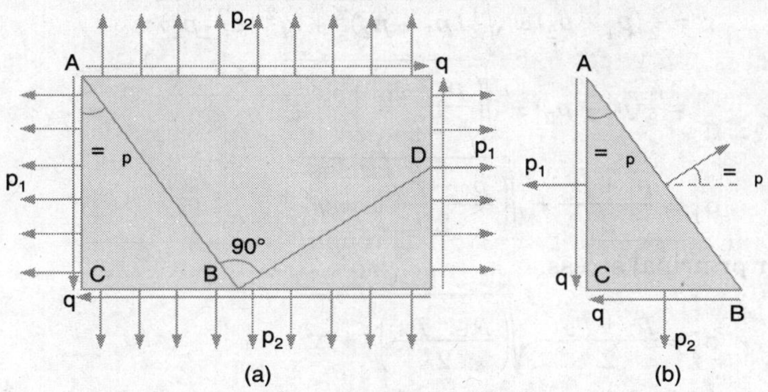

Fig. 3.14

Resolving forces parallel to BC, we get

$$\sigma AB \cos\theta = p_1 AC + qBC$$

or
$$\sigma \cos\theta = p_1 \frac{AC}{AB} + q\frac{BC}{AB} = p_1 \cos\theta + q \sin\theta$$

or
$$(\sigma - p_1)\cos\theta = q\sin\theta \quad \text{or} \quad (\sigma - p_1) = q\tan\theta \qquad ...(i)$$

Similarly, resolving along AC, we get

$$\sigma AB \sin\theta = p_2 BC + qAC$$

or
$$\sigma \sin\theta = p_2 \frac{BC}{AB} + q\frac{AC}{AB} = p_2 \sin\theta + q\cos\theta$$

$$(\sigma - p_2)\sin\theta = q\cos\theta \quad \text{or} \quad (\sigma - p_2) = q\cot\theta \qquad ...(ii)$$

Subtracting (i) from (ii), we get

$$(p_1 - p_2) = q(\cot\theta - \tan\theta) = \frac{2q}{\tan 2\theta}$$

Hence $\tan 2\theta = \tan 2\theta_p = \dfrac{2q}{p_1 - p_2}$ ...(3.11)

where θ_p is the angle θ which the principal plane makes with the plane of p_1

Equation 3.11 can be represented diagrammatically by Fig. 3.14(c) in which

$$\tan 2\theta_p = \frac{q}{\frac{(p_1 - p_2)}{2}} \left(= \frac{2q}{p_1 - p_2} \right)$$

$\therefore$
$$\cos 2\theta_p = \frac{p_1 - p_2}{2R} \qquad ...[3.11(a)]$$

and
$$\sin 2\theta_p = \frac{q}{R} \qquad ...[3.11(b)]$$

where
$$R = \sqrt{\left(\frac{p_1 - p_2}{2}\right)^2 + q^2} \qquad ...[3.11(c)]$$

$R = \sqrt{\left(\frac{p_1 - p_2}{2}\right)^2 + q^2}$, q, $2\theta_p$, $\frac{p_1 - p_2}{2}$

Fig. 3.14 (*c*)

Again, multiplying (*i*) and (*ii*), we get

$$(\sigma - p_1)(\sigma - p_2) = q^2$$

or $\sigma^2 - p_1\sigma - \sigma p_2 + p_1 p_2 - q^2 = 0$ or $\sigma^2 - \sigma(p_1 + p_2) + (p_1 p_2 - q^2) = 0$

which is quadratic equation for σ.

$$\therefore \quad \sigma = \tfrac{1}{2}(p_1 + p_2) \pm \sqrt{\tfrac{1}{4}(p_1 + p_2)^2 + (q^2 - p_1 p_2)}$$

$$= \tfrac{1}{2}(p_1 + p_2) \pm \sqrt{\left(\frac{p_1 - p_2}{2}\right)^2 + q^2} \qquad \text{...(3.12)}$$

$$\therefore \quad \sigma_1 = \frac{p_1 + p_2}{2} + \sqrt{\left(\frac{p_1 - p_2}{2}\right)^2 + q^2} \qquad \text{...[3.12(a)]}$$

which is the major principal stress,

$$\text{and} \quad \sigma_2 = \frac{p_1 + p_2}{2} - \sqrt{\left(\frac{p_1 - p_2}{2}\right)^2 + q^2} \qquad \text{...[3.12(b)]}$$

which is the minor principal stress.

From § 3.3 the maximum value of shear stress is given by

$$p_{\max} = \frac{\sigma_1 - \sigma_2}{2} = \sqrt{\left(\frac{p_1 - p_2}{2}\right)^2 + q^2} \qquad \text{...(3.13)}$$

The planes carrying maximum shear stress will be inclined at 45° to those of principal streses.

If p_1 and p_2 are not like, the sign for p_2 in equations 3.11, 3.12 and 3.13 will be changed.

Mohr Circle Method

Principal stresses and principal planes can either be found analytically from equations 3.11 and 3.12 or Mohr circle method can be used, as described below (Fig. 3.15).

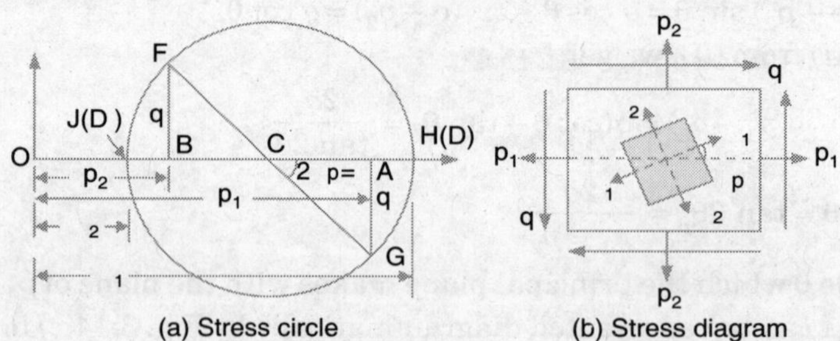

(a) Stress circle (b) Stress diagram

Fig. 3.15

Mark *OA* and *OB* proportional to p_1 and p_2. At *A* and *B*, erect perpendiculars *AG* = *BF* proportional to *q* in appropriate directions. Join *FG*, intersecting the axis in *C*. Since $p_t = 0$, *CH* represents the major principal plane, *D* coinciding with *H*. Similarly *CD′* represents minor principal plane, *D′* coinciding with *J*. Thus

$$OH = OC + CH = \tfrac{1}{2}(p_1 + p_2) + R, \text{ where } R = \text{radius of circle.}$$

$$= \tfrac{1}{2}(p_1 + p_2) + \sqrt{\left(\frac{p_1 - p_2}{2}\right)^2 + q^2} = \sigma_1$$

Similarly,

$$OJ = OC - CJ = \tfrac{1}{2}(p_1 + p_2) - \sqrt{\left(\frac{p_1 - p_2}{2}\right)^2 + q^2} = \sigma_2$$

$$\tan \beta = \frac{AG}{AC} = \frac{q}{\dfrac{p_1 - p_2}{2}} = \frac{2q}{p_1 - p_2} = \tan 2\theta_p \text{ (where } \beta = 2\theta_p).$$

Thus, the normal to the principal plane makes an angle θ_p in the clockwise direction with FG. The positions of principal planes have been located in the stress diagram of Fig. 3.15 (*b*).

Example 3.9. *At a point in an elastic material under strain, there are normal stresses of 50 N/mm² and 30 N/mm² respectively at right angles to each other with a shearing stress of 25 N/mm². Find the principal stresses and the position of principal planes if*

(a) 50 N/mm² is tensile and 30 N/mm² is also tensile.

(b) 50 N/mm² is tensile and 30 N/mm² is compressive.

Find also the maximum shear stress and its plane in both the cases.

Solution. (*a*) ***Analytically,***

$$\sigma = \frac{p_1 + p_2}{2} \pm \sqrt{\left(\frac{p_1 - p_2}{2}\right)^2 + q^2}$$

$$= \frac{50 + 30}{2} \pm \sqrt{\left(\frac{50 - 30}{2}\right)^2 + (25)^2}$$

$$= 40 \pm 26.9$$

$\therefore$ σ_1 = **66.9 N/mm²** (tensile) and

σ_2 = **13.1 N/mm²** (tensile)

$$\tan 2\theta_p = \frac{2q}{p_1 - p_2} = \frac{50}{50 - 30} = 2.5$$

$\therefore$ $2\theta_p = 68°\ 12'$

$\therefore$ θ_{p_1} = **34°6′** and θ_{p_2} = **124° 6′**

$$p_{t\,max} = \frac{66.9 - 13.1}{2} = \mathbf{26.9\ N/mm^2}$$

at (45° + 34° 6′) = **76° 6′.**

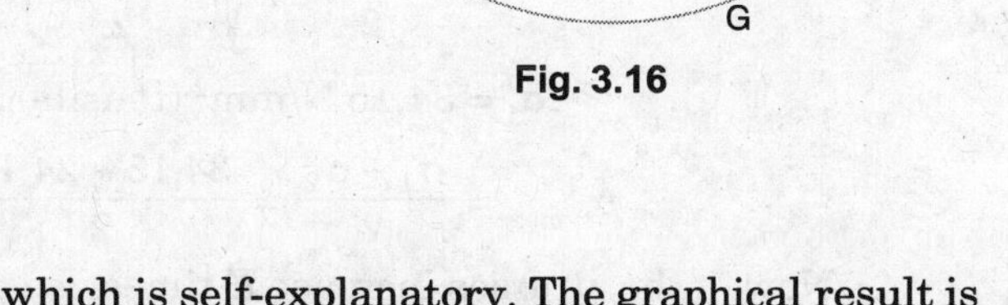

Fig. 3.16

Fig. 3.16 shows the graphical solution, which is self-explanatory. The graphical result is shown in the figure itself.

(*b*) ***Analytically,***

$$\sigma = \frac{50 - 30}{2} \pm \sqrt{\left(\frac{50 + 30}{2}\right)^2 + (25)^2}$$

$$= 10 \pm 47.1$$

$\therefore$ σ_1 = **57.1 N/mm²** (tensile) and

σ_2 = **– 37.1 N/mm²** (*i.e.*, comp.)

$$\tan \theta_p = \frac{2 \times 25}{50 + 30} = \frac{50}{80} = 0.625$$

$\therefore \quad 2\theta_p = 32°$

or $\quad \theta_{p_1} = \mathbf{16°}$ and $\theta_{p_2} = \mathbf{106°}$

$$p_{t\,max} = \frac{57.1 - (-37.1)}{2}$$

$$= \mathbf{47.1\ N/mm^2}$$

at (45° + 16°) = **61°** inclination with the plane of p_1.

Fig. 3.17 shows the graphical solution.

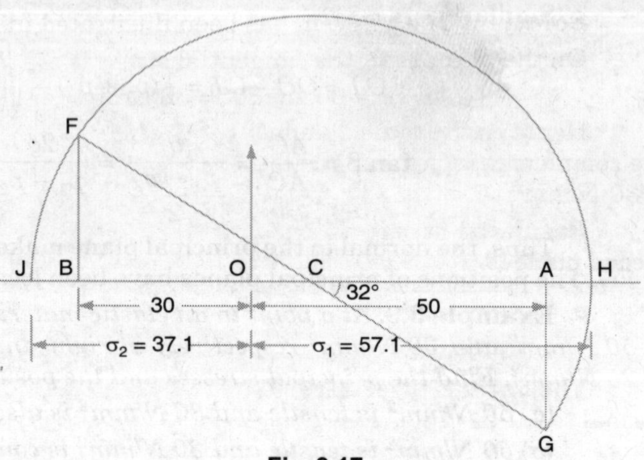

Fig. 3.17

Example 3.10. *At a certain point in a piece of elastic material there are normal stresses of 30 N/mm², tension and 20 N/mm², compression on two planes at right angles to one another, together with shearing stresses of 15 N/mm² on the same planes. If the loading on the material is increased so that the stresses reach values of K times those given, find the maximum permissible value of K if the maximum direct stress in the material is not to exceed 80 N/mm², and maximum shear is not exceed 50 N/mm².*

Solution. Working from the fundamentals, it can be shown that

$$\sigma = \frac{p_1 + p_2}{2} \pm \sqrt{\left(\frac{p_1 - p_2}{2}\right)^2 + q^2}\ , \text{ where } p_1 \text{ and } p_2 \text{ are alike}$$

Substituting the numerical values,

$$\sigma = \frac{30 - 20}{2} \pm \sqrt{\left(\frac{30 + 20}{2}\right)^2 + (15)^2} = 5 \pm 29.15$$

$\therefore \quad \sigma_1 = 34.15$ N/mm² (tensile) and $\sigma_2 = 24.15$ N/mm² (compressive)

$$p_{t,\,\max} = \frac{\sigma_1 - \sigma_2}{2} = \frac{34.15 + 24.15}{2} = 29.15 \text{ N/mm}^2$$

Now, if the stresses increase K times, we get

(*a*) Maximum principal stress (or direct stress) = 34.15 K which is not to exceed 80 N/mm².

$$\therefore \quad 34.15\,K = 80 \quad \text{or} \quad K = \frac{80}{34.15} = 2.34.$$

(*b*) Maximum shearing stress = 29.15 K, which is not to exceed 50 N/mm².

$$\therefore \quad 29.15\,K = 50 \quad \text{or} \quad K = \frac{50}{29.15} = 1.715$$

Hence the permissible value of K = **1.715.**

Example 3.11. *At a point in a material under stress, the intensity of the resultant stress on a certain plane is 60 N/mm² (tensile) inclined 30° to normal of that plane. The stress on a plane at right angles to this has a normal tensile component of intensity 40 N/mm². Find fully (1) the resultant stress on the second plane, (2) the principal planes and stress, (3) the plane of maximum shear and its intensity.*

Solution. The problem has been illustrated in Fig. 3.18 (*a*).

On the first plane, the tangential stress is

$$q = 60 \sin 30° = 30 \text{ N/mm}^2$$

Hence on the second plane the complementary shear stress is 30 N/mm².

Resultant stress on the second plane is

$$p_r = \sqrt{40^2 + 30^2}$$
$$= \mathbf{50 \text{ N/mm}^2}$$

The intensity of stress normal to the first plane is

$$= 60 \cos 30° = 52 \text{ N/mm}^2$$

The final stress system is represented in Fig. 3.18 (*b*), from which the principal stresses are :

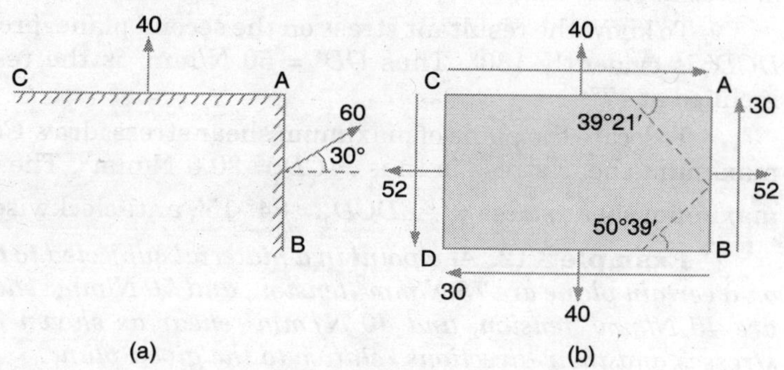

Fig. 3.18

$$\sigma = \frac{p_1 + p_2}{2} \pm \sqrt{\left(\frac{p_1 - p_2}{2}\right)^2 + q^2} = \frac{52 + 40}{2} \pm \sqrt{\left(\frac{52 - 40}{2}\right)^2 + (30)^2}$$

$$= 46 \pm 30.6$$

$\therefore \quad \sigma_1 = 76.6 \text{ N/mm}^2$ (tensile) and $\sigma_2 = 15.4 \text{ N/mm}^2$ (tensile)

$$\tan 2\theta_p = \frac{2q}{p_1 - p_2} = \frac{2 \times 30}{52 - 40} = \frac{60}{12} = 5$$

$\therefore \quad 2\theta_p = 78° \, 42'$ or $\theta_{p_1} = \mathbf{39° \, 21'}$ and $\theta_{p_2} = \mathbf{129° \, 21'}$

$$p_{t,\, max} = \frac{76.6 - 15.4}{2} = \mathbf{30.6 \text{ N/mm}^2}.$$

Mohr Circle Method :

To solve the problem graphically by Mohr Circle, refer Fig. 3.19.

On the plane *AB*, p_n is positive (being tensile) and p_t is positive being in the anticlockwise direction. Hence draw *OD* at an inclination of 30° to *OX* in the clockwise direction and make *OD* = 60 N/mm². Draw *DE* perpendicular to the axis *OX*. Set off *OE′* = 40 N/mm² to the right of *O* (since the stress on plane *AC* is tensile and is positive). Bisect *EE′* to get *C*. With *C* as the centre and *CD* as the radius, draw the Mohr circle.

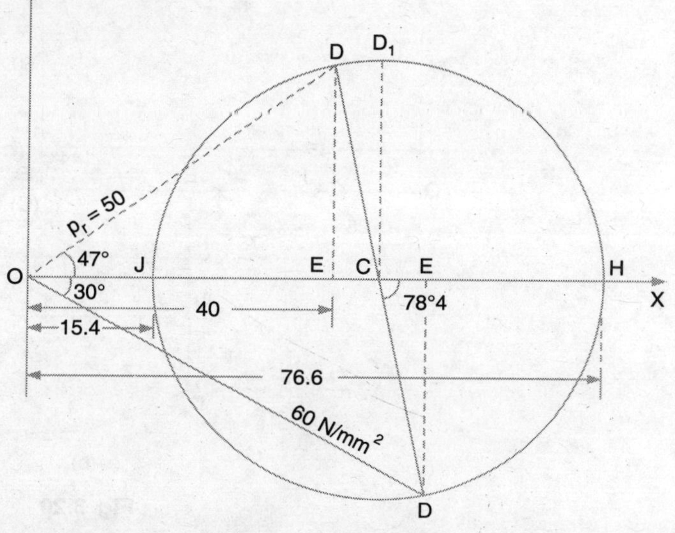

Fig. 3.19

From Mohr circle

$OH = \sigma_1 = 76.6$ N/mm^2 ; $OJ = \sigma_2 = 15.4$ N/mm^2 ; $\angle HCD = 78° \ 42' = 2\theta_p$.

The plane AB [Fig. 3.18 (*b*)] makes angles of 39° 21′ with the principal plane in the clockwise direction. Thus, the principal plane is inclined at an angle 39° 21′ in the anti-clockwise direction.

To know the resultant stress on the second plane, prolong DC to cut the circle in D'. Angle DCD' is evidently 180°. Thus $OD' = 50$ N/mm^2 is the resultant stress on the second plane, inclined at 47°.

To locate the plane of maximum shear stress, draw CD_1 perpendicular to the axis OC. The maximum shear stress is thus $= CD_1 = 30.6$ N/mm^2. The angle of inclination of the plane of maximum shear stress $= \frac{1}{2}\angle DCD_1 = 84° \ 15'$, anti-clockwise with line CD (*i.e.*, with plane AB).

Example 3.12. *At a point in a material subjected to two dimensional stresses, the stresses on a certain plane are 75 N/mm² tension, and 50 N/mm² shear and on another plane the stresses are 45 N/mm² tension, and 40 N/mm² shear as shown in Fig. 3.20 (a). Find the principal stresses and their directions relative to the given plane.*

Solution. (Fig. 3.20).

For the plane XY

$$p_n = +75 \text{ N/mm}^2 \text{ (being tensile)} ; \ p_t = -50 \text{ N/mm}^2 \text{ (being clockwise)}$$

$\therefore$ Make $OE = +75$ N/mm^2 and $ED = -50$ N/mm^2 (*i.e.*, in upward direction)

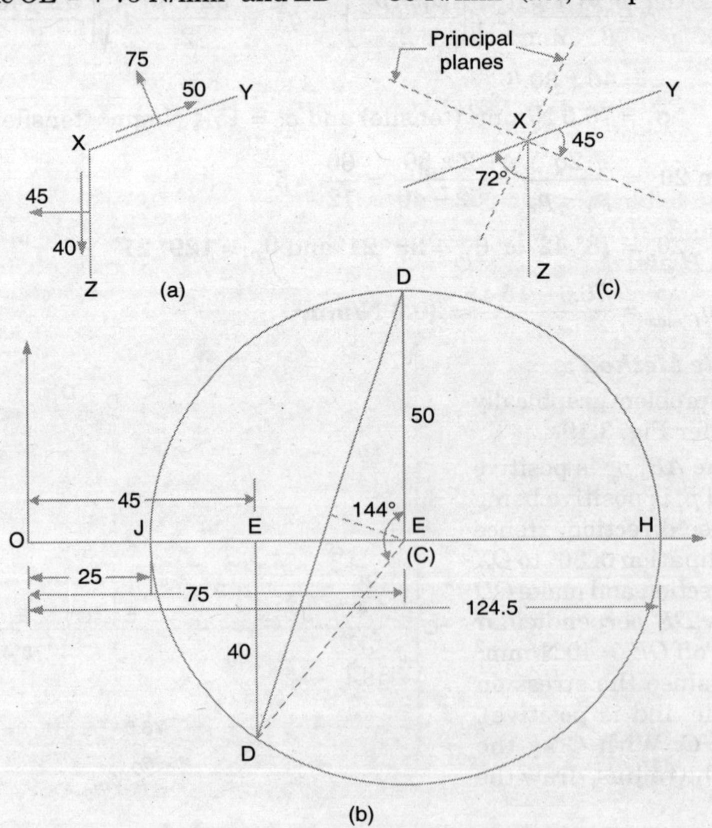

Fig. 3.20

For the plane XZ

$p_n = +\,45$ N/mm² (being tensile) ; $p_t = +\,40$ N/mm² (being anti-clockwise)

Make $OE' = +\,45$ and $E'D' = +\,40$ (*i.e.* downwards) as shown in Fig. 3.20 (*b*).

The Mohr circle should now pass through *D* and *D′*. Bisect *DD′* and draw the perpendicular bisector to cut the axis *OX* in *C*. In this special case, *C* coincides with *E*. Join *ED′*. $\angle DCH = 90°$.

The plane *XY*, is, therefore, inclined at 45° to the principal plane in anti-clockwise direction. Thus the major principal plane is inclined at 45° (clockwise) the plane *XY*. The position of the principal planes has been shown in Fig. 3.20 (*c*).

$$\sigma_1 = OH = \mathbf{+\,124.5\ N/mm^2}\ ;\ \sigma_2 = OJ = \mathbf{+\,25\ N/mm^2}\ ;\ \angle DCD' = 144°$$

∴ Angle between the planes *XY* and *XZ* = **72°** (anti-clockwise).

Example 3.13. *At a point in a material under stress, the resultant intensity of stress across a certain plane is 80 N/mm² (compressive) inclined at 30° to its normal. On another plane, the intensity of resultant stress is 60 N/mm² (tensile) inclined at 45° to its normal as shown in Fig. 3.21. Find the principal stresses and their directions to the given plane. Find also the angle between the planes.*

Solution. *For plane XY*, $p_r = 80$ N/mm²

p_n is negative and p_t is positive.

Draw $OD' = 80$ N/mm², at 30° (anti-clockwise) as shown in Fig. 3.22 so that p_n is negative and p_t is positive.

For plane XZ : $p_r = 60$ N/mm². Hence p_n is positive and p_t is negative.

Draw $OD = 60$ N/mm² at 45° with *OX* so that p_n is positive and p_t is negative.

The circle is to pass through *D* and *D′*. Join *DD′* and draw its perpendicular bisector to cut the axis in *C*. With *C* as centre and *CD′* as radius draw Mohr circle cutting the axis in *H* and *J*.

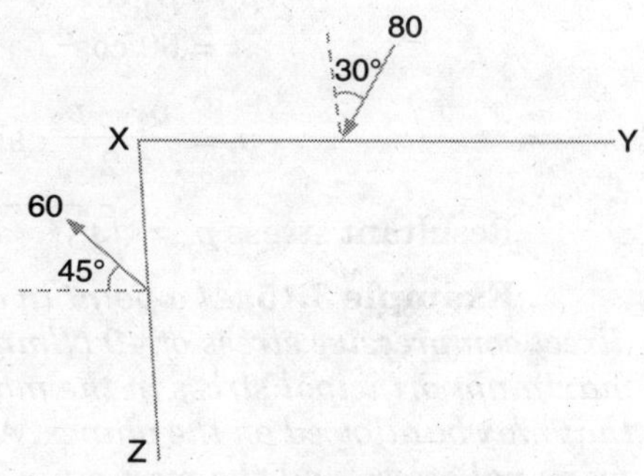

Fig. 3.21

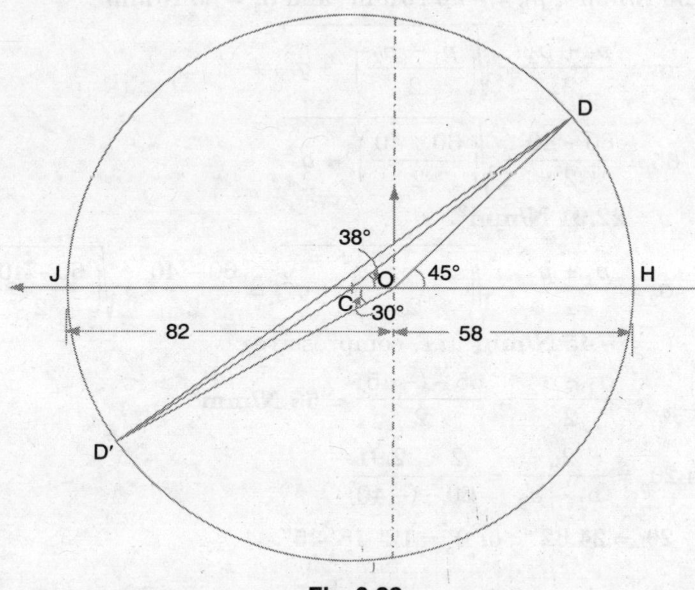

Fig. 3.22

Thus $OJ = \sigma_1 = -$ **82 N/ mm²** (*i.e.* compressive) ; $OH = \sigma_2 =$ **+ 58 N/mm²** (*i.e.* tensile)

$\angle HCD = 38°$

Inclination of plane of $\sigma_2 = 38°/2 = 19°$, clockwise with the plane XZ.

Inclination of plane of $\sigma_1 = (90° + 19°)$, clockwise with the plane XZ.

Angle between the planes $= \frac{1}{2} \angle D'CD = \frac{1}{2} \times 178 =$ **89°.**

3.8. ADDITIONAL EXAMPLES USEFUL FOR COMPETITIVE EXAMINATIONS

Example 3.14. *At a point in a loaded specimen, the principal stresses acting on two mutually perpendicular planes are 60 N/mm² and 40 N/mm², both being compressive. Determine the resultant stress acting on a plane inclined at 60° measured clockwise to the plane on which the larger normal stress is acting.*

Solution. Refer Fig. 3.4 (*a*). Here θ is in the clockwise and is therefore *negative*

$$p_n = p_1 \cos^2 \theta + p_2 \sin^2 \theta \qquad \text{...(Eq. 3.6)}$$

$$= 60 \cos^2 (-60°) + 40 \sin^2 (-60) = 45 \text{ N/mm}^2$$

$$p_t = \frac{p_1 - p_2}{2} \sin 2\theta = \frac{60 - 40}{2} \sin (-2 \times 60°) = -8.66 \text{ N/mm}^2$$

Resultant stress $p_r = \sqrt{45^2 + (-8.66)^2} =$ **45.83 N/mm².**

Example 3.15. *At a point in an elastic material, a direct tensile stress of 60 N/mm² and a direct compressive stress of 40 N/mm² are applied on planes at right angles to each other. If the maximum principal stress in the material is to be limited to 65 N/mm², find out the shear stress that may be allowed on the planes. Also, determine the magnitude and direction of the minimum principal stress and the maximum shear stress.*

Solution. (Refer Fig. 3.14).

Given $p_1 = +60$ N/mm² ; $p_2 = -40$ N/mm² and $\sigma_1 = 65$ N/mm².

$$\sigma_1 = \frac{p_1 + p_2}{2} + \sqrt{\left(\frac{p_1 - p_2}{2}\right)^2 + q^2}$$

or
$$65 = \frac{60 - 40}{2} + \sqrt{\left(\frac{60 + 40}{2}\right)^2 + q^2}$$

From which $q =$ **22.91 N/mm².**

Also,
$$\sigma_2 = \frac{p_1 + p_2}{2} - \sqrt{\left(\frac{p_1 - p_2}{2}\right)^2 + q^2} = \frac{60 - 40}{2} - \sqrt{\left(\frac{60 + 40}{2}\right)^2 + (22.91)^2}$$

$$= \mathbf{-45 \text{ N/mm}^2} \ (\textit{i.e. } \text{compressive})$$

$$p_{t,\,max} = \frac{\sigma_1 - \sigma_2}{2} = \frac{65 - (-45)}{2} = \mathbf{55 \text{ N/mm}^2}$$

$$\tan 2\theta_p = \frac{2q}{p_1 - p_2} = \frac{2 \times 22.91}{60 - (-40)}$$

$$\therefore \qquad 2\theta_p = 24.62° \text{ ; or } \theta_p = \mathbf{12° \, 18' \, 35''}.$$

Thus the major principal plane is oriented at 12° 18′ 35″ in anticlockwise direction with the plane on which p_1 is acting. The minor principal plane is inclined at 12° 18′ 35″ in clockwise direction with the plane on which p_1 is acting. The plane of maximum shear is inclined at 45° – 12° 18′ 35″ = **32° 41′ 25″** with the plane of p_1.

Example 3.16. *At a point in a piece of elastic material, there are three mutually perpendicular planes on which the stresses are as follows : tensile stress 50 N/mm² and shear stress 40 N/mm² on one plane ; compressive stress 35 N/mm² and complimentary shear stress 40 N/mm² on the second plane ; no stress on the third plane. Find (a) the principal stresses and the positions of the planes on which they act, (b) the positions of the planes on which there is no normal stress.*

Solution. (Fig. 3.23).

Mark off $OA = 50$ and $AG = 40$

$OB = -35$ and $BF = 40$.

Join F and G, cutting AB in C. Draw the circle with C as centre and CG (or CF) as radius.

Then

$\sigma_1 = OH = \mathbf{65.9\ N/mm^2}$ (tensile) ;

$\sigma_2 = OJ = \mathbf{50.9\ N/mm^2}$ (compressive)

$2\theta_p = 43°$ or $\theta_p = \mathbf{21°\ 30'}$.

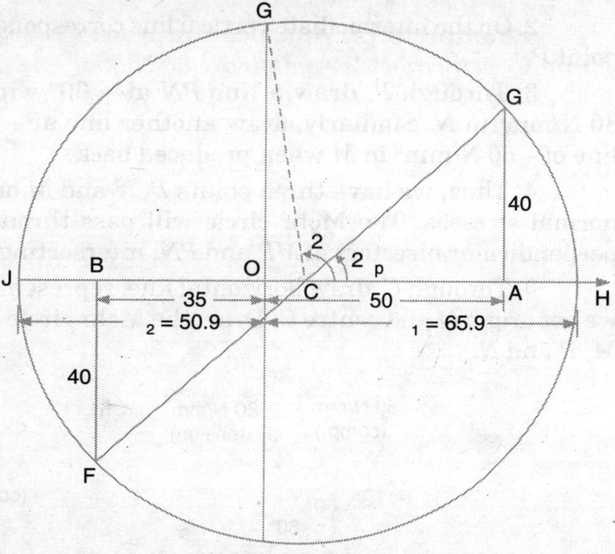

Fig. 3.23

Analytically

$$\sigma = \frac{50-35}{2} \pm \sqrt{\left(\frac{50+35}{2}\right)^2 + (40)^2}$$

$$= 75 \pm 58.36$$

$\therefore \quad \sigma_1 = \mathbf{65.86\ N/mm^2}$ (tensile)

and $\quad \sigma_2 = \mathbf{-\ 50.86\ N/mm^2}$ (*i.e.* compressive)

$$2\theta_p = \tan^{-1}\frac{2q}{p_1 - p_2} = \tan^{-1}\frac{2 \times 40}{50+35} = 43°\ 16'$$

$\therefore \quad \theta_p = \mathbf{21°\ 38'}$.

The relative positions of the planes are shown in Fig. 3.24.

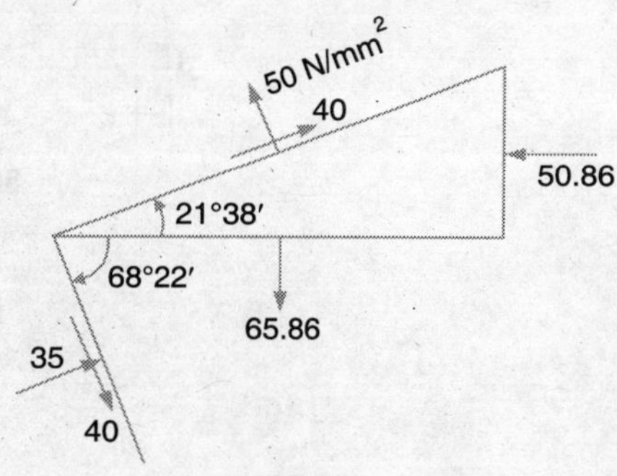

Fig. 3.24

If there is no normal stress on a plane, then for that plane, O and A will coincide. The radius vector for that plane will be CG' in Fig. 3.23, for which $2\theta' \simeq 97°$ (graphically).

Analytically,

$$2\theta' = 180° - \cos^{-1}\frac{OC}{CG'}$$

where $OC = OA - CA = 50 - \frac{1}{2}(50 + 35) = 7.5$

$$CG' = \text{Radius} = CG = \sqrt{AG^2 + CA^2}$$

$$= \sqrt{(40)^2 + (42.5)^2} = 58.36$$

$\therefore \quad 2\theta' = 180° - \cos^{-1}\dfrac{7.5}{58.36} = 180° - 82°\ 37' = 97°\ 23'$

$\therefore \quad \theta' = 48°\ 41.5'$ to the principal plane.

Example 3.17. *Fig. 3.25 (a) shows the direct stresses in three coplanar directions differing by 60°, at a particular point. Find the magnitude and directions of principal stresses.*

Solution. The problem can best be solved by Mohr circle construction, shown in Fig. 3.25 (*b*). The construction is done in the following steps :

1. Draw vertical line representing *y*-axis and set off three vertical lines at + 80, + 30 and – 50 distances from the *y*-axis.

2. On the intermediate vertical line corresponding to + 30 N/mm^2 stress, select any arbitrary point *P*.

3. Through *P*, draw a line *PN* at + 60° with vertical intersecting the vertical line of 80 N/mm^2 in *N*. Similarly, draw another line at + 120° with vertical, intersecting the vertical line of – 50 N/mm^2 in *M* when produced back.

4. Thus, we have three points *P*, *N* and *M* on three vertical lines corresponding to three normal stresses. The Mohr circle will pass through *P*, *N* and *M*. To locate its centre, draw perpendicular bisectors of *MP* and *PN*, intersecting at *C*.

5. Through *C*, draw horizontal line, representing *x*-axis, intersecting the *y*-axis in *O*. Thus, we get origin *O* and centre *C*. Draw the Mohr circle with radius *CM*. The circle will pass through *M*, *P* and *N*.

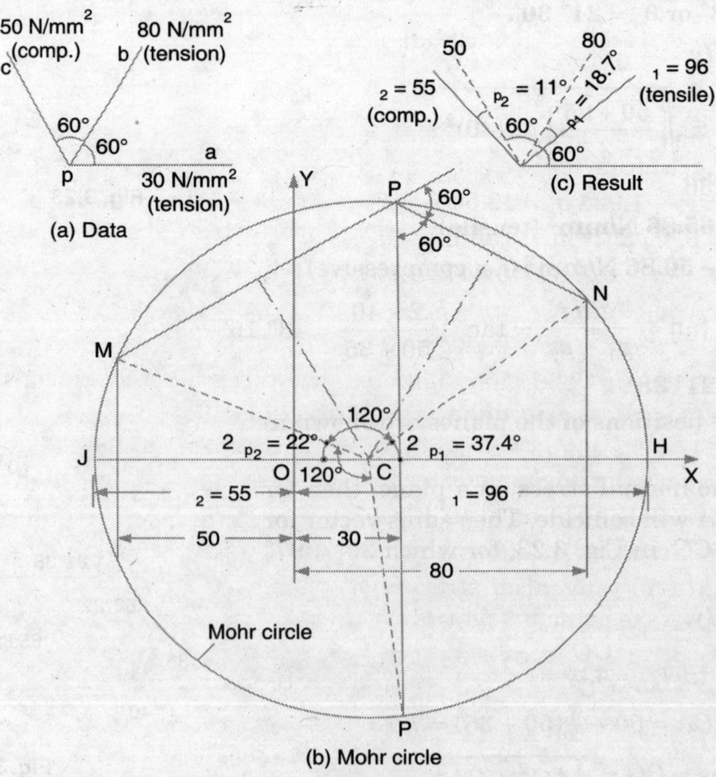

Fig. 3.25

6. The stress conditions on the three given planes are represented by points M, N and P', where P' is the vertical line through P. The justification of the construction lies from the fact that the angles between the three radii vectors CM, CN and CP' are 120°, which is twice the angle between each pair of given direct stresses.

7. The principal stresses are given by

$$\sigma_1 = OH = +96 \text{ N/mm}^2 \text{ (tension) and } \sigma_2 = OJ = -55 \text{ N/mm}^2 \text{ (i.e. compression).}$$

8. σ_1 is inclined at $\theta_{p_1} = \frac{1}{2}\angle NCH = \frac{1}{2} \times 37.4° = 18.7°$ to the 80 N/mm² stress and σ_2 is inclined at $\theta_{p_2} = \frac{1}{2}\angle MCJ = \frac{1}{2} \times 22° = 11°$ to the – 50 N/mm² stress.

The directions of σ_1 and σ_2 relative to the given normal stresses are given in Fig. 3.25(*c*).

Example 3.18. *A piece of material is subjected to three perpendicular tensile stresses. The strains in the three directions are in the ratio 3 : 4 : 5. If Poisson's ratio is 0.286, find the ratio of the stresses and their values if the greatest is 60 N/mm².*

Solution. Let the stresses be σ_1, σ_2 and σ_3 and the corresponding strains are 3ε, 4ε and 5ε.

$$\therefore \quad 3\varepsilon E = \sigma_1 - 0.286(\sigma_2 + \sigma_3) \qquad ...(i)$$

$$4\varepsilon E = \sigma_2 - 0.286(\sigma_3 + \sigma_1) \qquad ...(ii)$$

$$5\varepsilon E = \sigma_3 - 0.286(\sigma_1 + \sigma_2) \qquad ...(iii)$$

Subtracting (*i*) from (*iii*), we get

$$2\varepsilon E = (\sigma_3 - \sigma_1) - 0.286(\sigma_1 - \sigma_3) \text{ From which } \sigma_3 - \sigma_1 = 2\varepsilon E/1.286 \qquad ...(iv)$$

or

$$1.286\,\sigma_3 - 1.286\,\sigma_1 = 2\varepsilon E \qquad ...(a)$$

Again, from (*iii*)

$$\frac{\sigma_3}{0.286} - \sigma_1 - \sigma_2 = \frac{5\varepsilon E}{0.286} \qquad ...(v)$$

and from (*ii*), $\sigma_2 - 0.286\,\sigma_3 - 0.286\,\sigma_1 = 4\varepsilon E$ ...(*vi*)

Adding (*v*) and (*vi*), $3.21\,\sigma_3 - 1.286\,\sigma_1 = 21.5\varepsilon E$ ...(*b*)

Subtracting (*a*) from (*b*), we get

$$1.924\,\sigma_3 = 19.5\varepsilon E \quad \text{or} \quad \sigma_3 = 10.14\varepsilon E$$

Hence $\sigma_1 = 8.48\varepsilon E$ (from *iv*) and $\sigma_2 = 9.34\varepsilon E$ (from *ii*)

$$\sigma_1 : \sigma_2 : \sigma_3 = 0.847 : 0.921 : 1$$

Taking $\sigma_3 = 60$ N/mm²,

$$\sigma_1 = 0.847 \times 60 = \mathbf{50.8\ N/mm^2} \text{ and } \sigma_2 = 0.921 \times 60 = \mathbf{55.3\ N/mm^2}.$$

Example 3.19. *Fig. 3.26 shows normal and tangential stresses on two planes. Determine the principal stresses. Also, determine shear stress along CB.*

Solution. Let the plane AB make an angle θ, in the anticlockwise direction, with the principal plane. Hence the angle made by plane BC with the principal plane will be $\theta + 40°$.

Let σ_1 and σ_2 be the principal stresses.

Thus, there are three unknowns : σ_1, σ_2 and θ.

The normal and tangential stresses on plane AB, due to stresses σ_1 and σ_2 on the principal planes are given by

$$(p_n)_{AB} = 110 = \frac{\sigma_1 + \sigma_2}{2} + \frac{\sigma_1 - \sigma_2}{2}\cos 2\theta \qquad ...(1)$$

and

$$(p_t)_{AB} = -30 = -\frac{\sigma_1 - \sigma_2}{2}\sin 2\theta$$

or

$$30 = \frac{\sigma_1 - \sigma_2}{2}\sin 2\theta \qquad ...(2)$$

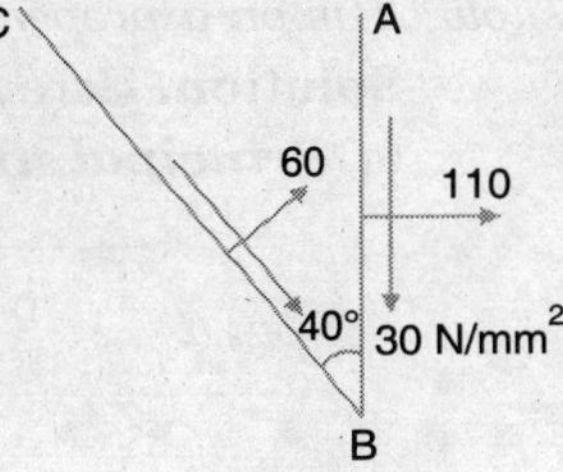

Fig. 3.26

Similarly, the normal stress on plane BC, due to stresses σ_1 and σ_2 on the principal planes is given by

$$(p_n)_{BC} = 60 = \frac{\sigma_1 + \sigma_2}{2} + \frac{\sigma_1 - \sigma_2}{2} \cos (2\theta + 80°) \qquad ...(3)$$

Subtracting Eq. (3) from Eq. (1),

$$\frac{\sigma_1 - \sigma_2}{2} [\cos 2\theta - \cos (2\theta + 80°)] = 50 \qquad ...(4)$$

Dividing Eq. (4) by Eq. (2),

$$\frac{\cos 2\theta - \cos (2\theta + 80°)}{\sin 2\theta} = \frac{50}{30} = 1.6667$$

or $\cos 2\theta - \cos 2\theta \cos 80° + \sin 2\theta \sin 80° = 1.6667 \sin 2\theta$

or $\cos 2\theta - 0.1736 \cos 2\theta = 1.6667 \sin 2\theta - 0.9848 \sin 2\theta$

or $0.6819 \sin 2\theta = 0.8264 \cos 2\theta$

or $\tan 2\theta = 1.2119$

From which $2\theta = 50.48°$ or $\theta = 25.24° = 25°14'$

Now from (2), $\dfrac{\sigma_1 - \sigma_2}{2} = \dfrac{30}{\sin 2\theta} = 38.89$

or $\sigma_1 - \sigma_2 = 77.79 \qquad ...(i)$

From (i), $\dfrac{\sigma_1 + \sigma_2}{2} + \dfrac{\sigma_1 - \sigma_2}{2} \cos 2\theta = 110$

or $\dfrac{\sigma_1 + \sigma_2}{2} + \dfrac{77.79}{2} \cos 50.48 = 110$

From which $\sigma_1 + \sigma_2 \simeq 170.5 \qquad ...(ii)$

From (i) and (ii) we get σ_1 = **124.1 N/mm²** and σ_2 = **46.4 N/mm²**

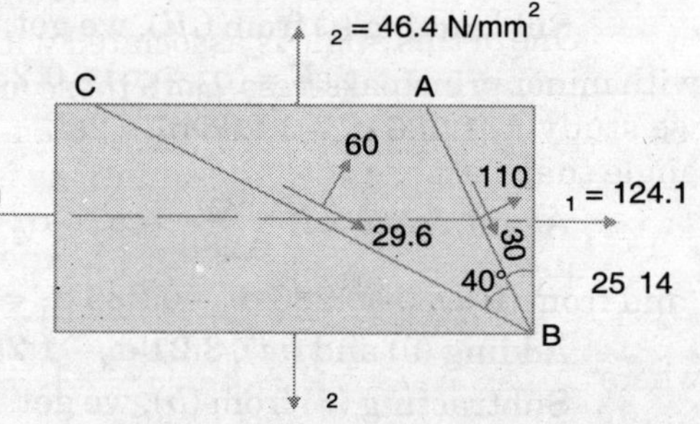

Fig. 3.27

Also, $(p_t)_{BC} = -\dfrac{\sigma_1 - \sigma_2}{2} \sin (2\theta + 80°)$

$$= -\frac{77.79}{2} \sin (50.48° + 80°)$$

≈ -29.6 N/mm² (*i.e.*, 29.6 in the direction marked in Fig. 3.27)

The position of various planes are marked in Fig. 3.27.

Example 3.20. *An element in plane stress is subjected to stresses p_1 = 120 N/mm², p_2 = – 40 N/mm² and q = – 45 N/mm², as shown in Fig. 3.28(a). Determine* (a) *the principal stresses and show them on a sketch of a properly oriented element.* (b) *the maximum shear stresses and show them on a properly oriented element. Consider only the in-plane stresses.*

Solution. Here p_1 = + 120 N/mm² ; p_2 = – 40 N/mm² and q = – 45 N/mm²

(a) ***Principal stresses***

$$\sigma = \frac{p_1 + p_2}{2} \pm \sqrt{\left(\frac{p_1 - p_2}{2}\right)^2 + q^2} = \frac{120 + (-40)}{2} \pm \sqrt{\left(\frac{120 + 40}{2}\right)^2 + (-45)^2}$$

$$= 40 \pm 91.79$$

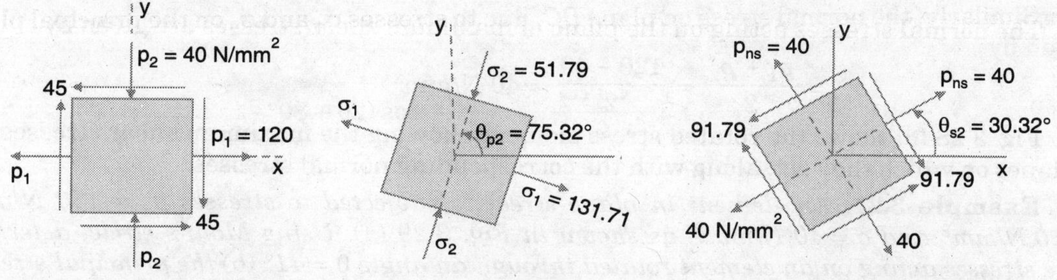

(a) Element in plane stress (b) Principal stresses (c) Maximum shear stresses

Fig. 3.28

From which σ_1 = 131.79 N/mm^2 (tensile) and σ_2 = – 51.79 N/mm^2 (*i.e.* compressive)

$$\tan 2\theta_p = \frac{2q}{p_1 - p_2} = \frac{2 \times (-45)}{120 + 40} = -0.5625, \text{ from which } 2\theta_p = -29.36°$$

Since $2\theta_p$ varies in the range from 0° to 360°, we have

$$2\theta_p = 360° - 29.36° = 330.64° \text{ and } \theta_p = 165.32°$$

or

$$2\theta_p = 180° - 29.36° = 150.64° \text{ and } \theta_p = 75.32°$$

One of this angle is associated with the major principal stress while the other is associated with minor principal stress. Both these angles differ by 90°. However, in order to remove confusion, we study Eqs. 3.11 (*a*, *b*) and Fig. 3.14 (*c*), correlating principal angles and stresses. The only angle that satisfies both the equations is θ_{p1}. Hence we can rewrite these equation as follows :

$$\cos 2\theta_{p1} = \frac{p_1 - p_2}{2R} \quad \text{and} \quad \sin 2\theta_{p1} = \frac{q}{R}$$

where

$$R = \sqrt{\left(\frac{p_1 - p_2}{2}\right)^2 + q^2} = 91.79 \text{ (found earlier)}$$

$$\therefore \quad \cos 2\theta_{p1} = \frac{120 + 40}{2 \times 91.79} = 0.8716 \text{ from which } 2\theta_{p1} = 29.36° \text{ (or } 360° - 29.36° = 330.64°)$$

and

$$\sin 2\theta_{p1} = -\frac{45}{91.79} = -0.4905, \text{ from which } 2\theta_{p1} = -29.30° \text{ (or } 360° - 29.36°) = 330.64°$$

The only angle between 0° and 360°, satisfying both of these conditions is $2\theta_{p1}$ = 330.64°. Hcncc θ_{p1} **= 165.32°.**

The other angle θ_{p2} is 90° larger or smaller than θ_{p1}

Hence θ_{p2} = 165.32° – 90° = **75.32′**

Fig. 3.28 (*b*) shows the rotated stress element showing the principal planes and principal stresses.

(*b*) ***Maximum shear stresses***

$$p_{t,\max} = \sqrt{\left(\frac{p_1 - p_2}{2}\right)^2 + q^2} = \sqrt{\left(\frac{120 + 40}{2}\right)^2 + (-45)^2} = 91.79$$

Also $\theta_{s1} = \theta_{p1} - 45° = 165.32° - 45° = 120.32°$

Negative shear stress acts as the plane for which θ_{s2} = 120.32° – 90°

= 30.32° (or $\theta_{s2} = \theta_{p2} - 45° = 75.32° - 45° = 30.32°$)

The normal stresses acting on the plane of maximum shear stresses are given by

$$p_{ns} = \frac{p_1 + p_2}{2} = \frac{120 - 40}{2} = 40 \text{ N/mm}^2$$

Fig. 3.28 (*c*) shows the rotated stress element showing the maximum shear stresses and the planes on which they act, along with the corresponding normal stresses.

Example 3.21. *An element in plane stress is subjected to stresses p_1 = 150 N/mm², p_2 = 50 N/mm² and q = 40 N/mm², as shown in Fig. 3.29 (a). Using Mohr's circle, determine (a) the stresses acting on an element rotated through an angle θ = 41° (b) the principal stresses, and (c) the maximum shear stresses. Show all results on sketches of properly oriented elements.*

Solution. The Mohr's circle is shown in Fig. 3.29 (*b*), the construction of which is self explanatory. From the stress circle of Fig. 3.29 (*b*), we get the following values :

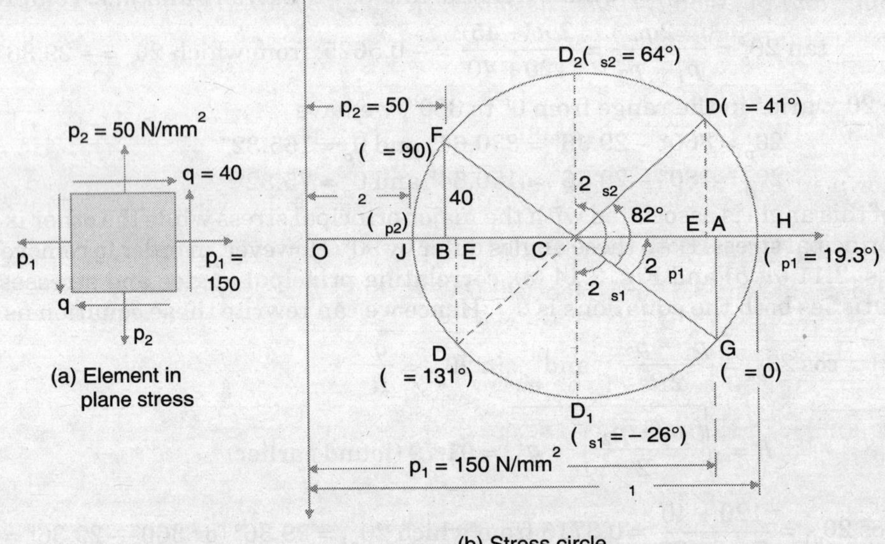

Fig. 3.29

$$(p_n)_D = OE = +\,147 \text{ N/mm}^2 \,;\; (p_t)_D = ED = -\,44 \text{ N/mm}^2 \,;\; (p_n)_{D'} = OE' = 54 \text{ N/mm}^2$$

$$\sigma_1 = OH = +\,164 \text{ N/mm}^2 \,;\; \sigma_2 = OJ = 36 \text{ N/mm}^2$$

$$2\theta_{p1} = +\,38.6° \,;\; \theta_{p1} = 19.3°$$

$$(p_t)_{max} = CD_1 = +\,64 \text{ N/mm}^2 \,;\; (p_t)_{min} = CD_2 = -\,64 \text{ N/mm}^2$$

$$2\theta_{s1} = -\,52° \,;\; \theta_{s1} = -\,26° \,;\; 2\theta_{s2} = +\,128° \,;\; \theta_{s2} = +\,64°$$

$$(p_n)_{D1} = \frac{p_1 + p_2}{2} = p_{av} = 100 \text{ N/mm}^2 = (p_n)_{D2}$$

Fig. 3.30 (*a*) shows the stresses on stress element at θ = + 41°

Fig. 3.30 (*b*) shows the stress element containing principal planes on which only the principal stresses act and there are no shearing stresses. Fig. 3.30 (*c*) shows the orientation of stress element containing the planes on which maximum shear stresses act. The corresponding normal stresses on these planes are evidently equal to the average stress.

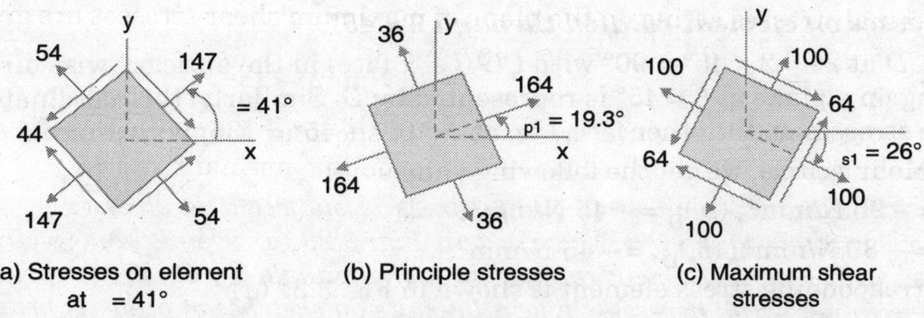

(a) Stresses on element at = 41° (b) Principle stresses (c) Maximum shear stresses

Fig. 3.30

Example 3.22. *An element in plane stress is subjected to stresses p_1 = – 75 N/mm², p_2 = 15 N/mm² and q = – 60 N/mm², as shown in Fig. 3.31 (a). Using Mohr's circle, determine (a) the stresses acting on an element rotated through an angle = 45° in the anticlockwise direction, (b) the principal stresses and (c) the maximum shear stresses. Show all results on sketches of properly oriented elements.*

Solution. Given : $p_1 = -75$ N/mm² ; $p_2 = 15$ N/mm² and $q = -60$ N/mm². The Mohr's circle is shown in Fig. 3.31 (*b*). The centre *C* of the circle has *x*-coordinate = $p_{av} = \dfrac{p_1 + p_2}{2}$ $= \dfrac{-75+15}{2} = -30$ N/mm². The coordinates of point $G(\theta = 0)$ are given by the stresses on the *x*-face of the element [Fig. 3.31 (*a*)]. Hence mark *OA* = – 75 and *AG* = – 60 (*i.e.* upwards). Similarly, the co-ordinates of point $F(\theta = 90°)$ are given by the stresses on the *y*-face of the element of Fig. 3.31 (*a*). Hence make *OB* = + 15 and *BF* = + 60 (*i.e.* downwards). Join *G* and *F*, to intersect *x*-axis on *C*, giving the centre of the circle. With *C* as centre, and *CG* (or *CF*) as radius *R* $(= \sqrt{(45)^2 + (60)^2} = 75$ N/mm²) draw the circle. The circle cuts the *x*-axis in points *H* and *J*.

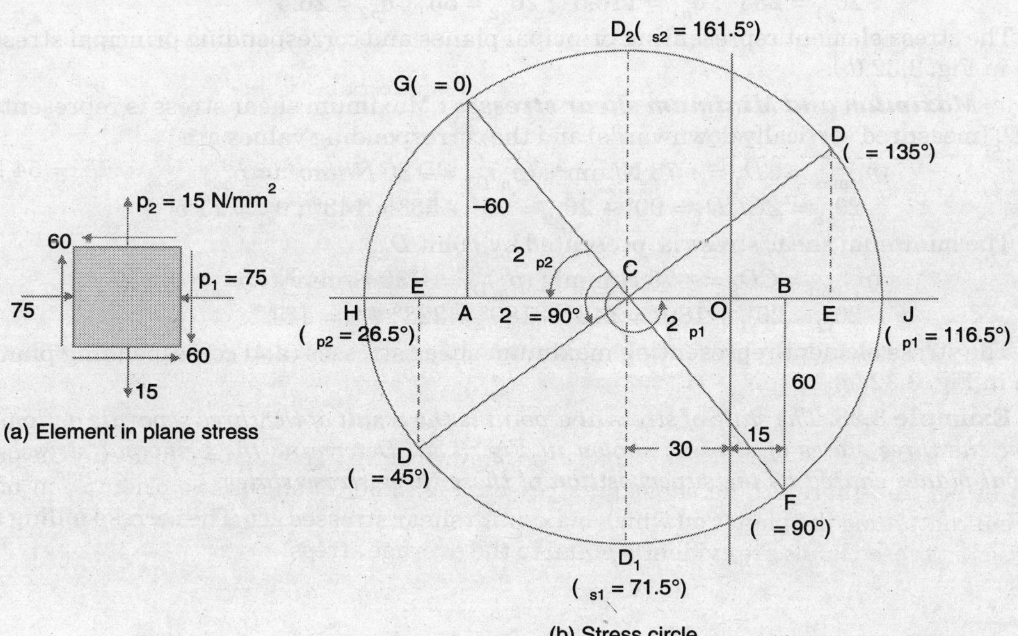

(a) Element in plane stress (b) Stress circle

Fig. 3.31

(*a*) ***Stresses on element rotated through* θ = 45°**

Draw *CD* at 2θ = 2 × 45° = 90° with *CG* (*i.e.* *x* face) in the anticlockwise direction. Then, stresses acting on a plane at θ = 45° is represented by *D*. Similarly, the coordinates of point *D′* represent the stresses on the other face, (*i.e.* at θ = 90° + 45° = 135° with *x*-face).

From Mohr's circle, we get the following values :

$(p_n)_D = -90$ N/mm²; $(p_t)_D = +45$ N/mm²

$(p_n)_{D'} = +30$ N/mm²; $(p_t)_{D'} = -45$ N/mm².

The corresponding stress element is shown in Fig. 3.32 (*a*)

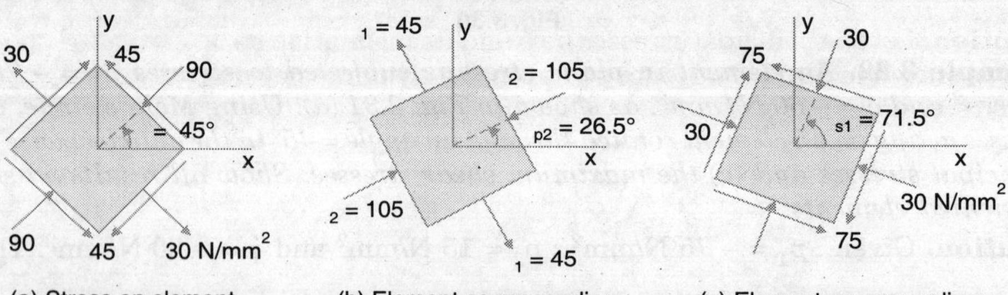

(a) Stress on element at = 45° (b) Element corresponding to principal stresses (c) Element corresponding to max. shear stress

Fig. 3.32

(*b*) ***Principal planes and principal stresses*** **:** The principal stresses are represented by point *H* and *J* on stress circle. The angle *GCH* is the angle $2\theta_{p2}$, from point *G*(θ = 0) to point *H*, representing the principal plane having the algebraically smaller principal stress. Hence from the stress circle, we obtain.

$$\sigma_1 = OJ = +45 \text{ N/mm}^2 ; \sigma_2 = OH = -105 \text{ N/mm}^2$$

$$2\theta_{p1} = 233° ; \theta_{p1} = 116.5° ; 2\theta_{p2} = 53° ; \theta_{p2} = 26.5°$$

The stress element representing principal planes and corresponding principal stresses is shown in Fig. 3.32 (*b*).

(*c*) ***Maximum and Minimum shear stresses*** **:** Maximum shear stress is represented by point D_1 (measured vertically downwards) and the corresponding values are

$$(p_t)_{max} = CD_1 = +75 \text{ N/mm}^2 ; (p_n)_{D1} = -30 \text{ N/mm}^2 (= p_{av})$$

$$2\theta_{s1} = \angle GCD_1 = 90° + 2\theta_{p2} = 90° + 53° = 143° ; \theta_{s1} = 71.5°$$

The minimum shear stress is presented by point D_2.

$$(p_t)_{min} = CD_2 = -75 \text{ N/mm}^2 ; (p_n)_{D2} = -30 \text{ N/mm}^2 (= p_{av})$$

$$2\theta_{s2} = 2\theta_{s1} + 180° = 143° + 180° = 323°; \theta_{s2} = 161.5°$$

The stress element representing maximum shear stresses (and corresponding planes) is shown in Fig. 3.32 (*c*).

Example 3.23. *The state of stress at a point is the result of the three separate actions that produce the three states of stresses shown in Fig. 3.33. Determine the principal stresses and principal planes caused by the superposition of these three stress states.*

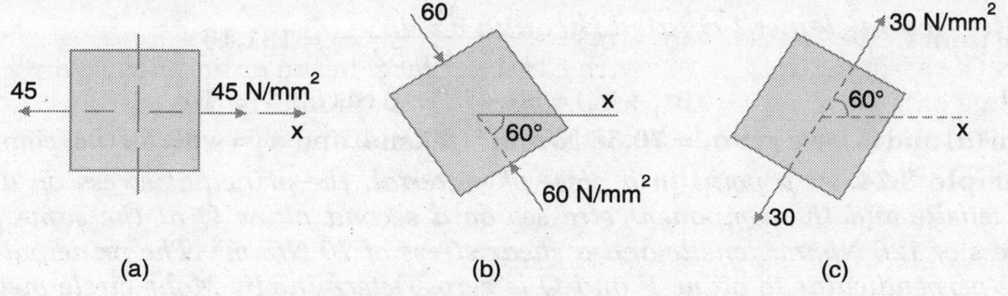

Fig. 3.33

Solution. Let the principal stresses be σ and its inclination be θ_p with x-axis as shown in Fig. 3.34. Hence stress on the element shown in Fig. 3.33 (a) is given by

$$p_{na} = \frac{1}{2}(\sigma_1 + \sigma_2) + \frac{1}{2}(\sigma_1 - \sigma_2)\cos 2\theta \qquad \text{...(3.6)}$$

Fig. 3.34

Here, $p_{na} = +45$, and $\theta = -\theta_p$ (since x-axis is rotated by θ_p in the clockwise direction with respect to x_p-axis.

$$\therefore \quad 45 = \frac{1}{2}(\sigma_1 + \sigma_2) + \frac{1}{2}(\sigma_1 - \sigma_2)\cos(-2\theta_p)$$

or $$(\sigma_1 + \sigma_2) + (\sigma_1 - \sigma_2)\cos 2\theta_p = 90 \qquad \text{...(1)}$$

Also, stress on the element shown in Fig. 3.33 (b) is given by Eq. 3.6 where

$$p_{na} = -60 \text{ and } \theta = -(60° + \theta_p)$$

$$\therefore \quad -60 = \frac{1}{2}(\sigma_1 + \sigma_2) + \frac{1}{2}(\sigma_1 - \sigma_2)\cos\{-2(60 + \theta_p)\}$$

or $$(\sigma_1 + \sigma_2) + (\sigma_1 - \sigma_2)\cos(120° + 2\theta_p) = -120$$

or $$(\sigma_1 + \sigma_2) + (\sigma_1 - \sigma_2)\{\cos 120° \cos 2\theta_p - \sin 120° \sin 2\theta_p\} = -120$$

or $$-(\sigma_1 + \sigma_2) + (\sigma_1 - \sigma_2)\{0.5\cos 2\theta_p + 0.866\sin 2\theta_p\} = 120 \qquad \text{...(2)}$$

Lastly, the stress on element shown in Fig. 3.33 (c) is given by Eq. 3.6, where

$$p_{na} = +30 \text{ and } \theta = (60° - \theta_p)$$

$$\therefore \quad 30 = \frac{1}{2}(\sigma_1 + \sigma_2) + \frac{1}{2}(\sigma_1 - \sigma_2)\cos 2(60° - \theta_p)$$

or $$(\sigma_1 + \sigma_2) + (\sigma_1 - \sigma_2)\cos(120° - 2\theta_p) = 60$$

or $$(\sigma_1 + \sigma_2) + (\sigma_1 - \sigma_2)\{\cos 120° \cos 2\theta_p + \sin 120° \sin 2\theta_p\} = 60$$

or $$(\sigma_1 + \sigma_2) + (\sigma_1 - \sigma_2)\{-0.5\cos 2\theta_p + 0.866\sin 2\theta_p\} = 60 \qquad \text{...(3)}$$

Thus we have three equations (*i.e.* Eqs. 1, 2 and 3) corresponding to the three unknowns σ_1, σ_2 and θ_p. Adding Eqs. (2) and (3), we get

$$(\sigma_1 - \sigma_2)\{1.732\sin 2\theta_p\} = 180 \qquad \text{...(I)}$$

Also, subtracting (3) from (1), we get

$$(\sigma_1 - \sigma_2)(1.5\cos 2\theta_p - 0.866\sin 2\theta_p) = 30 \qquad \text{...(II)}$$

Dividing (II) by (I), we get $$\frac{1.5\cos 2\theta_p - 0.866\sin 2\theta_p}{1.732\sin 2\theta_p} = \frac{30}{180} = \frac{1}{6}$$

or $$0.866\cot 2\theta_p - 0.5 = 0.1667$$

From which $\tan 2\theta_p = 1.2989$; $2\theta_p = 52.4°$; $\theta_p =$ **26.2°**

Also, from I, $(\sigma_1 - \sigma_2) = \dfrac{180}{1.732 \sin 52.4°} = 131.16$ ...(a)

and from (1), $(\sigma_1 + \sigma_2) = 90 - 131.16 \cos 52.4° \simeq 10$ ...(b)

From (a) and (b), we get σ_1 = **70.58 N/mm²** (tensile) and σ_2 = **– 60.58** (*i.e.* comp.).

Example 3.24. *At a point in a piece of material, the principal stress on a plane P is 35 N/mm² tensile and the component stresses on a second plane Q at the same point are a normal stress of 126 N/mm² tensile and a shear stress of 70 N/mm². The principal stress on a third plane perpendicular to plane P and Q is zero. Determine by Mohr circle method (a) the angle between planes P and Q (b) the other principal stress and location of its plane (c) maximum shear stress, and (d) the maximum obliquity of the resultant stress with the normal stress on any plane.*

Solution. Fig. 3.35 shows the position of planes and the corresponding stresses. Fig. 3.36 shows the Mohr's circle of stress which is drawn in the following steps.

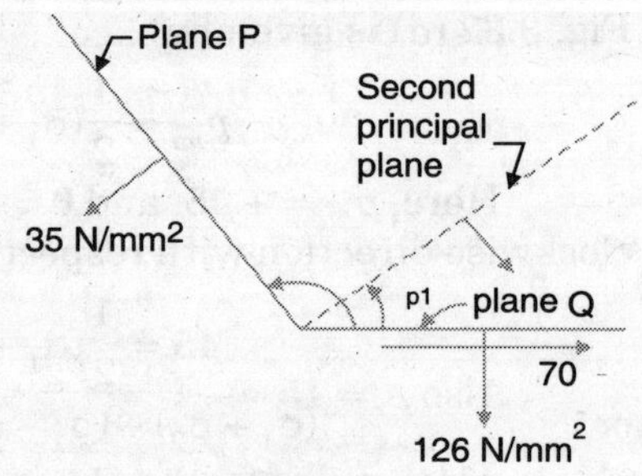

Fig. 3.35

1. Mark OA = 126 N/mm² in the positive direction.
2. Draw perpendicular AG in the downward direction (since q is positive on plane Q) and make it equal to 70 N/mm².
3. Mark OB = 35 N/mm² in the positive direction.
4. Since plane P is the principal plane (q being zero on it), the circle must pass through point B. Circle must also pass through point G. Hence join points B and G and draw its perpendicular bisector to cut the x-axis in point C, the centre of the circle.
5. With C as the centre and CB or CG as radius, draw the circle cutting the x-axis in point $B(J)$ and H.
6. Through C, draw CD perpendicular to x-axis. Point D corresponds to maximum shear stress.
7. To find the maximum obliquity of the resultant stress, draw line OD_1 tangential to the circle.

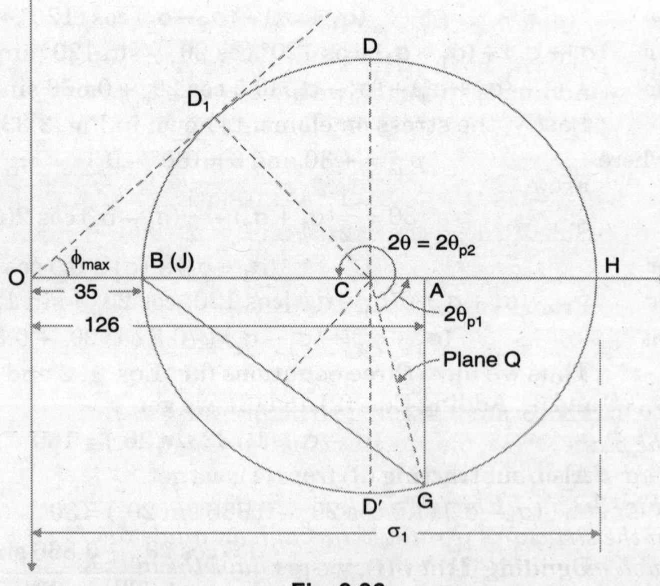

Fig. 3.36

From the above construction, we obtain the following :

(i) Angle between planes P and Q

$= \theta(\text{or } \theta_{p2}) = \dfrac{1}{2} \angle GCO$

$= \dfrac{1}{2}(256°) = \mathbf{128°}$

(ii) Other principal stress = $\sigma_1 = OH =$ **180 N/mm²**. Inclination of other principal plane with plane $Q = \theta_{p1} = \frac{1}{2}\angle GCH = \frac{1}{2} \times 76° =$ **38°**

(iii) Maximum shear stress = CD (or CD') = **71.7 N/mm²**

(iv) Maximum obliquity of the resultant stress $\varphi_{max} = \angle D_1OC =$ **41.5°**.

Example 3.25. *An element of plane stress is rotated through a known angle θ = 30° and the stresses on this rotated element are as shown in Fig. 3.37(a). Determine the normal and shear stresses (i.e. p_1, p_2 and q) on the original element (Fig. 3.37 b) whose sides are parallel to the x-y-axes.*

Solution. Let the stresses be p_1, p_2 and q.

Thus, there are three unknowns.

Here θ = + 30°. Hence

$$p_{n1} = -90 = \frac{p_1 + p_2}{2} + \frac{p_1 - p_2}{2} \cos 60° + q \sin 60°$$

or $(p_1 + p_2) + 0.5(p_1 - p_2) + 1.732\, q = -180$ ...(1)

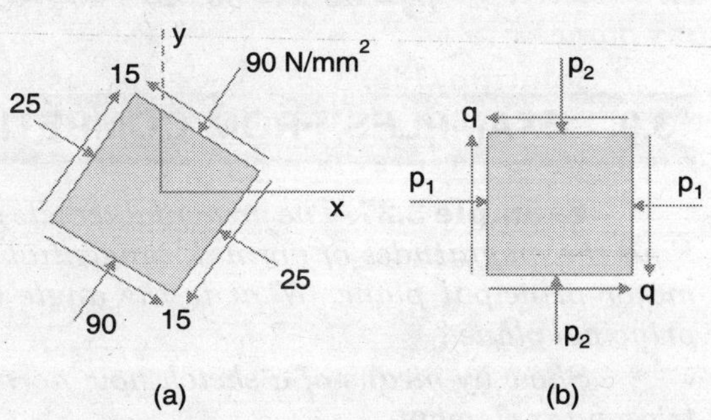

Fig. 3.37

Also $p_t = 15 = -\frac{p_1 - p_2}{2} \sin 60° + q \cos 60°$

or $-(p_1 - p_2)\, 0.866 + q = 30$ ...(2)

At θ = 30° + 90° = 120°, we have

$$p_{n2} = -25 = \frac{p_1 + p_2}{2} + \frac{p_1 - p_2}{2} \cos 240° + q \sin 240°$$

or $(p_1 + p_2) - 0.5(p_1 - p_2) - 1.732\, q = -50$ ...(3)

Thus, we have three equations corresponding to the three unknowns.

Adding (1) from (3), we have

$(p_1 + p_2) = -115$ ...(a)

From (2), $p_1 - p_2 = \frac{q - 30}{0.866} = 1.1547\,(q - 30)$ ...(b)

Substituting the values of $(p_1 + p_2)$ and $(p_1 - p_2)$ in (1), we get

$-115 + 0.5 \times 1.1547\,(q - 30) + 1.732\, q = -180$

From which $q = -20.65$ N/mm². Substituting the values of q in (b), we get

$p_1 - p_2 = -58.48$ ...(c)

From (a) and (c), we get $p_1 =$ **– 86.74 N/mm²** and $p_2 =$ **– 28.26 N/mm²**.

Example 3.26. *In a structure subjected to the plane stress, the stresses have the magnitudes and directions shown acting on a element A [Fig. 3.38(a)]. Element B located at the same point in the structure, is rotated through an angle θ_b of such magnitude that the stresses have the values shown in Fig. 3.38 (b). Calculate the normal stress p_b and the angle θ_b.*

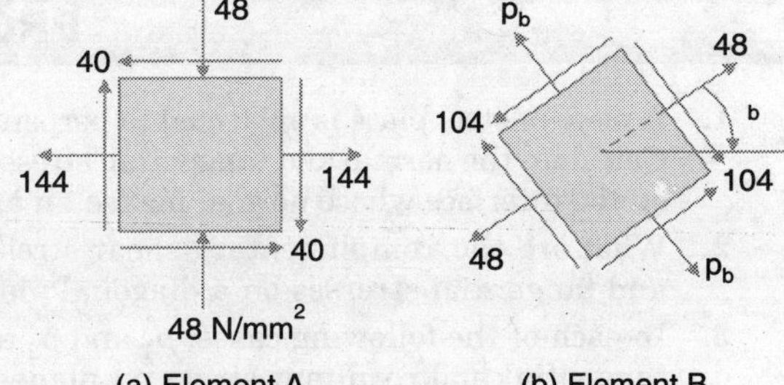

Fig. 3.38

Solution. Here $p_1 = +144$ N/mm², $p_2 = -48$ N/mm² and $q = -40$ N/mm²

Hence from Eq. 3.8, we get

$$p_{n1} = 48 = \frac{144-48}{2} + \frac{144+48}{2} \cos 2\theta_b - 40 \sin 2\theta_b$$

or $96 \cos 2\theta_b - 40 \sin 2\theta_b = 0$

From which $\tan 2\theta_b = 2.4$

or $2\theta_b = 67.38°$ or $\theta_b = \mathbf{33.69°}$.

Also, at $\theta = 90° + \theta_b = 90° + 33.69° = 123.69°$, we have

$$p_{n2} = p_b = \frac{144-48}{2} + \frac{144+48}{2} \cos (2 \times 123.69°) - 40 \sin (2 \times 123.69°)$$

or $p_b = 48 + (-36.92) - 40(-0.923) = \mathbf{48\ N/mm^2}$.

3.9. EXAMPLES FROM COMPETITIVE EXAMINATIONS

Example 3.27. *The principal tensile stresses at a point are $\sigma_x = 80$ MPa and $\sigma_y = 40$ MPa. Find the magnitudes of normal, tangential and the resultant stress on a plane at 20° with the major principal plane. What is the angle of obliquity of the resultant stress with the major principal plane ?*

Show by means of a sketch how normal stress (σ_n) and tangential stress (σ_t) act on the triangular element. (UPSC Engg. Services, IES, 2005)

Solution. Given : $\sigma_x = 80$ MPa ; $\sigma_y = 40$ MPa

$$p_n = p_x \cos^2 \theta + p_y \sin^2 \theta \quad ...(3.6)$$
$$= 80 \cos^2 20° + 40 \sin^2 20°$$
$$= \mathbf{75.32\ MPa}$$

$$p_t = \frac{p_x - p_y}{2} \sin 2\theta \quad ...(\text{Eq. } 3.7)$$
$$= \frac{80-40}{2} \sin 40° = \mathbf{12.856\ MPa}$$

$$p_r = \sqrt{p_n^2 + p_t^2}$$
$$= \sqrt{(75.32)^2 + (12.856)^2} = \mathbf{76.409\ MPa}$$

$$\alpha = \tan^{-1}\left(\frac{p_y}{p_x} \tan \theta\right) = \tan^{-1}\left(\frac{40}{80} \tan 20°\right) = \mathbf{10.31°}.$$

Fig. 3.39

The position of p_n, p_t and p_r acting on the triangular element are shown on Fig. 3.39.

PROBLEMS

1. A piece of steel plate is subjected to perpendicular stresses of 50 N/mm² and 30 N/mm², both tensile. Calculate the normal and tangential stresses and the magnitude and direction of the resultant stress on the interface whose normal makes an angle of 30° with the axis of the second stress.
2. What are the 'complimentary shear stresses' ? Using Mohr circle, derive expressions for normal and tangential stresses on a diagonal plane of a piece of material in pure shear.
3. In each of the following cases, p_1 and p_2 are perpendicular direct stresses. Calculate the normal, tangential and resultant stress on planes making angles of (*i*) 20° (*ii*) 40° and (*iii*) 80° with the

plane of p_1 and also for the plane having the maximum shear stress. Find also the obliquity of the resultant stress (*i.e.* its inclination to the normal).

(*a*) $p_1 = +40$ N/mm^2 ; $p_2 = +15$ N/mm^2 (*b*) $p_1 = +40$ N/mm^2 ; $p_2 = -15$ N/mm^2

(*c*) $p_1 = +20$ N/mm^2 ; $p_2 = -40$ N/mm^2.

Use Mohr circle method.

4. Calculate the principal stresses and the maximum shear stress for the following cases of two perpendicular direct stresses together with complementary shear stresses. Draw diagrams showing the positions of the principal planes and planes of maximum shear stress relative to the planes of the applied stresses.

(*a*) $p_1 = 60$ N/mm^2 ; $p_2 = 20$ N/mm^2 (tensile) and $q = 20$ N/mm^2

(*b*) $p_1 = 20$ N/mm^2 (tensile) ; $p_2 = 50$ N/mm^2 (comp.) and $q = 10$ N/mm^2

(*c*) $p_1 = 0$; $p_2 = 20$ N/mm^2 (comp.) ; $q = 40$ N/mm^2.

5. In a piece of material, a tensile stress p_1 and shearing stress q act on a given plane. Show that principal stresses are always of opposite sign.

If an additional tensile stress p_2 acts on a plane perpendicular to that on which p_1 acts and all the stresses are coplanar, find the condition that both principal stresses may be of the same sign.

6. Direct stresses of 80 N/mm^2 tension and 50 N/mm^2 compression are applied to an elastic material at a certain point, on planes at right angles to one another.

The greater principal stress in the material is limited to 100 N/mm^2 tension. To what shearing stress may the material be subjected on the given planes and what will then be the maximum shearing stress at the point ? Work from the first principles.

7. At a point in an elastic material under strain, the stresses on the three mutually perpendicular planes are as follows :

A normal tensile stress of 60 N/mm^2 and a shear stress of 40 N/mm^2 on one plane.

A normal compressive stress 40 N/mm^2 and a complementary shear stress of 40 N/mm^2 on the other plane.

No stress on the third plane. Find fully :

(*i*) the principal stresses and principal planes.

(*ii*) the maximum shear stress and its plane.

(*iii*) the position of the plane on which there is no normal stress.

8. The stresses at a particular point in a piece of material act upon the three planes whose relative angular positions are given by triangle *ABC*, in which *B* is a right angle and *C* is 30°. The normal stresses on these planes are 70 N/mm^2 tension on *AB*, 30 N/mm^2 compression on *BC* and 40 N/mm^2 tension on *AC*. Determine the magnitude and direction of the shearing stresses on the given planes and the magnitude of the greatest direct stress and the greatest shearing stress at the point.

9. Direct stresses of 120 N/mm^2 tension and 90 N/mm^2 compression are applied to an elastic material at a certain point, on planes at right angles. The greater principal stress is limited to 150 N/mm^2. What shearing stress may be applied to the given planes and what will be the maximum shearing stress at the point ? Work from the first principles.

10. The principal stresses at a point in a member subjected to two-dimensional stress are 120 N/mm^2 and 50 N/mm^2 tensile. Find the plane on which the resultant stress has maximum obliquity and the magnitude of this obliquity.

11. A right angled triangle *ABC* with the right angle at *C* represents planes in an elastic material. There are shearing stresses of 50 N/mm^2 acting along the planes *AC* and *CB* towards *C*, and normal tensile stresses on *AC* and *CB* of 80 N/mm^2 and 60 N/mm^2 respectively. There is no stress on the plane perpendicular to planes *AC* and *CB*.

Determine the positions of the plane *AB* when the resultant stress on *AB* has :

(*a*) the greatest magnitude (*b*) the least magnitude

(*c*) the greatest component normal to *AB* (*d*) the greatest tangential component along *AB*
(*e*) the least inclination to *AB*.

Analytical or graphical methods may be used. State for each plane found, its angular position relative to *AC* and the magnitude of the stress referred to. (*U.L.*)

12. At a certain point in a piece of material, the directions of principal stresses σ_1 and σ_2 are known. Measurements by strain gauges show that there is a tensile strain of 0.00065 in the direction of σ_1 and a compressive strain of 0.00045 in the direction of σ_2. Find the magnitudes of σ_1 and σ_2, stating whether compressive or tensile, and the value of maximum shear stress. Take $m = 4$ and $E = 200$ kN/mm².

ANSWERS

1. $p_n = 35$ N/mm² ; $p_t = 8.66$ N/mm² ; $p_r = 36.1$ N/mm² ; $\phi = 13°\ 54'$ N/mm².

2. $p_n = \pm q$; $p_t = 0$.

3.

		p_n	p_t	p_r	ϕ	
(*a*)	(*i*)	+ 37	+ 8.1	37.8	12°	$p_{t(\max)} = 12.5$
	(*ii*)	+ 29.6	+ 12.3	32.0	22.5°	
	(*iii*)	+ 15.8	+ 4.3	16.3	15.5°	
(*b*)	(*i*)	+ 33.5	+ 17.7	37.9	28°	
	(*ii*)	+ 17.3	+ 27.1	32.1	57.25°	$p_{t(\max)} = 27.5$
	(*iii*)	– 13.3	+ 9.4	16.3	144.5°	
(*c*)	(*i*)	+ 13.0	+ 19.3	23.3	66°	
	(*ii*)	– 4.78	+ 29.5	29.8	99°12°	
	(*iii*)	– 38.2	+ 10.3	39.5	164°52′	

$p_{t(\max)} = 30.0$

4.

	σ_1	σ_2	θ_1	θ_2	$p_{t(\max)}$
(*a*)	+ 68.3	+ 11.7	+ 22.25°	+ 112.25°	28.3
	+ 21.4	– 51.4	+ 8°	+ 98°	36.4
	+ 31.2	– 51.2	+ 38°	+ 128°	41.2

5. $p_1 p_2 > q^2$.

6. $q = 56.5$ N/mm² ; $p_{t(\max)} = 90$ N/mm².

7. (*i*) $\sigma_1 = +\ 74$ N/mm² ; $\sigma_2 = -\ 54$ N/mm² ; $\theta_1 = 19°$, $\theta_2 = 109°$, (*ii*) $p_{t(\max)} = 64$ N/mm², (*iii*) Plane of zero normal stress ; 69° with that of p_1.

8. Shear on *AB* = – 52.0 N/mm² ; Shear on *AC* = + 69.3 N/mm², $\sigma_1 = 92.1$ N/mm² (tension) ; $p_{t(\max)}$ = 72.1 N/mm².

9. $q_{\max} = 135$ N/mm².

10. Inclination of plane = 57° 9′ ; obliquity = 24° 18′.

11. (*a*) 121 N/mm² (tensile) at 39° 21′ to *AC* (*b*) 19 N/mm² (tensile) at 129° 21′ to *AC*
(*c*) Same as (*a*) (*d*) 51 N/mm² at 84° 21′ to *AC*
(*e*) 107° 45′.

12. $\sigma_1 = 114.7$ N/mm² (tensile) ; $\sigma_2 = 61.3$ N/mm² (comp.).

4

Analysis of Strain

4.1. PRINCIPAL STRAINS

Let σ_1, σ_2 and σ_3 be the three principal stresses acting on three mutually perpendicular planes at a point in an isotropic material. If all the three stresses are alike, the principal strains ε_1, ε_2 and ε_3 are obtained from the generalized Hooke's law :

$$\varepsilon_1 = \frac{\sigma_1}{E} - \frac{\sigma_2 + \sigma_3}{mE} \qquad \ldots(1)$$

$$\varepsilon_2 = \frac{\sigma_2}{E} - \frac{\sigma_3 + \sigma_1}{mE} \qquad \ldots(2)$$

$$\varepsilon_3 = \frac{\sigma_3}{E} - \frac{\sigma_1 + \sigma_2}{mE} \qquad \ldots(3)$$

If σ_1 is the greatest principal stress and σ_2 is the least principal stress, the greatest difference of principal strains is

$$\varepsilon_1 - \varepsilon_2 = \frac{\sigma_1 - \sigma_2}{E} + \frac{\sigma_1 - \sigma_2}{mE} = \frac{\sigma_1 - \sigma_2}{E}\left(1 + \frac{1}{m}\right) = \left(\frac{m+1}{m}\right)\left(\frac{\sigma_1 - \sigma_2}{E}\right) \qquad \ldots(4)$$

As proved earlier, $q_{max} = \dfrac{\sigma_1 - \sigma_2}{2}$ and $\dfrac{mE}{m+1} = 2N$

Hence $$\varepsilon_1 - \varepsilon_2 = \frac{1}{2N}(\sigma_1 - \sigma_2) = \frac{q_{max}}{N} \qquad \ldots(4.1)$$

$$\therefore \qquad \phi_{max} = \frac{q_{max}}{N} = \varepsilon_1 - \varepsilon_2 \qquad \ldots(4.2)$$

Thus the greatest shear strain is equal to the greatest difference of principal strains.

Example 4.1. *A circle of 300 mm in diameter is scribed on a mild steel plate before it is stressed as shown in Fig. 4.1. After stressing, the circle deforms to an ellipse. Calculate the lengths of the major and minor axes of the ellipse and also find their directions. Given $1/m = 0.28$; $E = 210$ kN/mm².*

Solution. The principal stresses are given by

$$\sigma_1 = \frac{p_1 + p_2}{2} \pm \sqrt{\left(\frac{p_1 - p_2}{2}\right)^2 + q^2}$$

$$= \frac{65 + 13}{2} \pm \sqrt{\left(\frac{65 - 13}{2}\right)^2 + (20)^2}$$

$$= 39 \pm 32.8$$

$\therefore$ $\sigma_1 = +\ 71.8\ \text{N/mm}^2$

and $\sigma_2 = +\ 6.2\ \text{N/mm}^2$

$$\tan 2\theta_p = \frac{2q}{p_1 - p_2} = \frac{2 \times 20}{65 - 13} = \frac{40}{52}$$

$$= 0.77\ ; \quad \text{or} \quad 2\theta_p = 37°\ 36'$$

$\therefore\ \theta_{p_1} = 18°\ 48'$ and $\theta_{p_2} = 108°\ 48'$

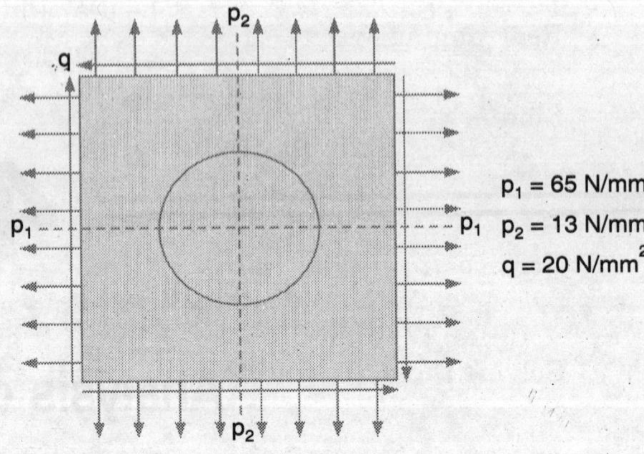

Fig. 4.1

The major principal strain $\varepsilon_1 = \dfrac{\sigma_1}{E} - \dfrac{\sigma_2}{mE} = (71.8 - 6.2 \times 0.28)\dfrac{1}{210 \times 1000} = 3.33 \times 10^{-4}$ (increase)

The minor principal strain $= \varepsilon_2 = \dfrac{\sigma_2}{E} - \dfrac{\sigma_1}{mE} = (6.2 - 71.8 \times 0.28)\dfrac{1}{210 \times 1000}$

$$= -\ 6.62 \times 10^{-5}\ \text{(increase)} = 6.62 \times 10^{-5}\ \text{(decrease)}$$

$\therefore$ Length of the diameter along the direction of σ_1

$$= d + \varepsilon_1 d = 300\ (1 + \varepsilon_1) = 300(1 + 3.33 \times 10^{-4}) = \mathbf{300.099\ mm}$$

Length of the diameter along the direction of σ_2

$$= d - \varepsilon_2 d = d(1 - \varepsilon_2) = 300(1 - 6.62 \times 10^{-5}) = \mathbf{299.98\ mm.}$$

Hence, the circle will be converted into an ellipse having major axis = 300.099 mm at 18° 48′ and the minor axis = 299.98 mm at 108° 48′ with the vertical centre line (measured in the clockwise direction).

4.2. COMPUTATION OF PRINCIPAL STRESSES FROM PRINCIPAL STRAINS

Many times, the principal strains are known and it is required to compute principal stresses. Let ε_1, ε_2 and ε_3 be the principal strains, and σ_1, σ_2 and σ_3 be the corresponding principal stresses. If μ (= $1/m$) is the Poisson's ratio, we have

$$E\varepsilon_1 = \sigma_1 - \mu\sigma_2 - \mu\sigma_3 \qquad ...(i)$$

$$E\varepsilon_2 = \sigma_2 - \mu\sigma_3 - \mu\sigma_1 \qquad ...(ii)$$

and

$$E\varepsilon_3 = \sigma_3 - \mu\sigma_1 - \mu\sigma_2 \qquad ...(iii)$$

Subtracting (*ii*) from (*i*), we get $E(\varepsilon_1 - \varepsilon_2) = (\sigma_1 - \sigma_2)\ (1 + \mu)$ ..(*iv*)

From (*i*) and (*iii*), eliminating σ_3, $E(\varepsilon_1 + \mu\varepsilon_3) = \sigma_1(1 - \mu^2) - \sigma_2(1 + \mu)\mu$ ...(*v*)

Multiplying (*iv*) by μ and subtracting from (*v*), we get

$$E[(1 - \mu)\varepsilon_1 + \mu(\varepsilon_2 + \varepsilon_3)] = \sigma_1(1 - \mu - 2\mu^2) = \sigma_1(1 + \mu)\ (1 - 2\mu)$$

Rearranging, we get $\sigma_1 = \dfrac{E(1-\mu)\varepsilon_1 + \mu(\varepsilon_2 + \varepsilon_3)]}{(1+\mu)(1-2\mu)}$ ...[4.3 (*a*)]

Similarly, $\sigma_2 = \dfrac{E[(1-\mu)\varepsilon_2 + \mu(\varepsilon_3 + \varepsilon_1)]}{(1+\mu)(1-2\mu)}$ ...[4.3 (*b*)]

and $\sigma_3 = \dfrac{E[(1-\mu)\varepsilon_3 + \mu(\varepsilon_1 + \varepsilon_2)]}{(1+\mu)(1-2\mu)}$ ...[4.3 (*c*)]

For *two dimensional stress system,* $\sigma_3 = 0$

$\therefore$ $E\,.\,\varepsilon_1 = \sigma_1 - \mu\sigma_2$ and $E\,.\,\varepsilon_2 = \sigma_2 - \mu\sigma_1$

Solving these, we get $\sigma_1 = \dfrac{E(\varepsilon_1 + \mu\varepsilon_2)}{1-\mu^2}$ and $\sigma_2 = \dfrac{E(\mu\varepsilon_1 + \varepsilon_2)}{1-\mu^2}$ ...(4.4)

4.3. STRAINS IN AN INCLINED DIRECTION

In § 3.4, we have done the analysis of stress, *i.e.* we have found expressions for p_n and p_t on any inclined plane (θ), if p_1 and p_2 in two mutually perpendicular directions are known along with shearing stress q. Similarly, we can easily find direct and shearing strains in any direction if the direct and shearing strains in two mutually perpendicular directions are given.

Consider a rectangular element $OABC$ in x-y plane (Fig. 4.2).

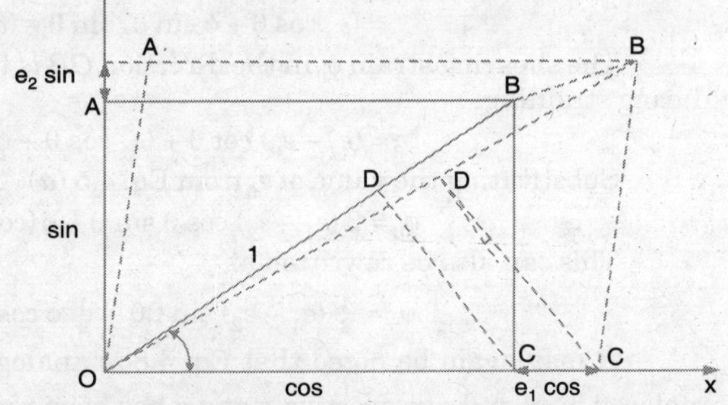

Fig. 4.2. Strains in inclined direction.

Let e_1 = Strain in x-direction

e_2 = Strain in y-direction

ϕ = Shearing strain relative to Ox and Oy.

It is required to find direct strain (e_θ) and shearing strain ϕ_θ along any direction OB, inclined at θ with x-direction.

Consider all the strains to be very small. If the diagonal OB is taken to be of unit length, side OA will be of length sin θ, and OC will be of length cos θ. In the strained condition, OA extends by small amount e_2 sin θ and OC extends by small amount e_1 cos θ. Also due to shearing strain, OA rotates by small angle ϕ. The rectangle $OABC$ thus distorts to $OA'B'C'$, in which point B moves to B'.

The movement of B parallel to x-axis is $= e_1 \cos\theta + \phi \sin\theta$ ...(*i*)

Similarly, movement of B parallel to y-axis is $= e_2 \sin\theta$ ...(*ii*)

Hence the movement of B parallel to OB is

$$= (e_1 \cos\theta + \phi \sin\theta)\cos\theta + (e_2 \sin\theta)\sin\theta \quad ...(iii)$$

This movement of B will be approximately equal to the extension of OB in the strained condition. However, since OB has been taken to be of unit length, the extension of OB is also equal to direct strain e_θ in the direction OB.

$$\therefore \quad e_\theta = (e_1 \cos\theta + \phi \sin\theta)\cos\theta + (e_2 \sin\theta)\sin\theta = e_1 \cos^2\theta + e_2 \sin^2\theta + \phi \sin\theta\cos\theta \qquad \text{...[4.5 (a)]}$$

or
$$e_\theta = e_1 \cos^2\theta + e_2 \sin^2\theta + \frac{\phi}{2}\sin 2\theta \qquad \text{...[4.5(b)]}$$

This can also be written in the form

$$e_\theta = \tfrac{1}{2}(e_1 + e_2) + \tfrac{1}{2}(e_1 - e_2)\cos 2\theta + \tfrac{1}{2}\phi \sin 2\theta \qquad \text{...(4.6)}$$

It should be noted that Eq. 4.5(*b*) is analogous to Eq. 3.8(*b*), where p_1 and p_2 have been replaced by e_1 and e_2 respectively and q has been replaced by $\frac{1}{2}\phi$.

Again, in order to find the shearing strain ϕ_θ, draw *CD* perpendicular to *OB* (Fig. 4.2). Point *D′* is the strained position of *D*. When *OB* extends to e_θ, *OD* extends by an amount $= e_\theta \, . \, OD = e_\theta \cos^2\theta$, (Since $OD = OC\cos\theta = \cos\theta \, . \, \cos\theta = \cos^2\theta$)

During straining, CD rotates in anti-clockwise direction through a small angle

$$= \frac{e_1 \cos^2\theta - e_\theta \cos^2\theta}{\cos\theta \sin\theta} = (e_1 - e_\theta)\cot\theta$$

At the same time, *OB* rotates in the clockwise direction through a small angle

$$= (e_1 \cos\theta + \phi\sin\theta)\sin\theta - (e_2 \sin\theta)\cos\theta$$

The shearing strain ϕ_θ in the direction *OB* is the amount by which angle *ODC* diminishes during straining.

$$\phi_\theta = (e_1 - e_\theta)\cot\theta + (e_1\cos\theta + \phi\sin\theta)\sin\theta - (e_2\sin\theta)\cos\theta$$

Substituting the value of e_θ from Eq. 4.5 (*a*)

$$\phi_\theta = 2(e_1 - e_2)\cos\theta\sin\theta - \phi(\cos^2\theta - \sin^2\theta) \qquad \text{...(4.7)}$$

This can also be rewritten as

$$\tfrac{1}{2}\phi_\theta = \tfrac{1}{2}(e_1 - e_2)\sin 2\theta - \tfrac{1}{2}\phi\cos 2\theta \qquad \text{...(4.8)}$$

It may again be noted that Eq. 4.8 is analogous to Eq. 3.9, where p_1 and p_2 have been replaced by e_1 and e_2 respectively and q has been replaced by $\frac{1}{2}\phi$.

Principal Strains. When the position of plane *OB* is such that the shearing strain ϕ_θ is zero, we get the maximum and minimum values of strain e_θ. These values occur at values of $\theta\,(= \theta_{p\epsilon})$ obtained by equating $\dfrac{de_\theta}{d\theta}$ to zero.

$$\therefore \quad \tan 2\theta_{p\epsilon} = \tan 2\theta = \frac{\phi}{e_1 - e_2} \qquad \text{...(4.9)}$$

The principal strains ε_1 and ε_2 are then given by

$$\varepsilon = \tfrac{1}{2}(e_1 + e_2) \pm \sqrt{\tfrac{1}{4}(e_1 - e_2)^2 + \phi^2} \qquad \text{...(4.10)}$$

or
$$\varepsilon_1 = \tfrac{1}{2}(e_1 + e_2) + \sqrt{\tfrac{1}{4}(e_1 - e_2)^2 + \phi^2} \qquad \text{...[4.10(a)]}$$

and
$$\varepsilon_2 = \tfrac{1}{2}(e_1 + e_2) - \sqrt{\tfrac{1}{4}(e_1 - e_2)^2 + \phi^2} \qquad \text{...[4.10(b)]}$$

4.4. MOHR CIRCLE OF STRAIN

It is observed that expression for linear strain e_θ and shearing strain ϕ_θ on any plane are analogous to the expressions for p_n and p_t. This suggests that the strains in any direction can be represented graphically by Mohr circle, in a similar way to the stress system.

Rewriting Eqs. 4.6 and 4.8

$$e_\theta - \frac{1}{2}(e_1 + e_2) = \frac{1}{2}(e_1 - e_2)\cos 2\theta + \frac{1}{2}\,\phi \sin 2\theta$$

and
$$\frac{1}{2}\,\phi_\theta = \frac{1}{2}(e_1 - e_2)\sin 2\theta - \frac{1}{2}\,\phi\cos 2\theta$$

Squaring each equation and adding, we get

$$[e_\theta - \tfrac{1}{2}(e_1 + e_2)]^2 + [\tfrac{1}{2}\,\phi_\theta]^2 = [\tfrac{1}{2}(e_1 - e_2)]^2 + [\tfrac{1}{2}\,\phi]^2$$

Thus all values of e_θ and $\frac{1}{2}\phi_\theta$ lie on a circle of radius $\sqrt{\frac{1}{2}(e_1 - e_2)^2 - [\frac{1}{2}\phi]^2}$ with its centre at point $[\frac{1}{2}(e_1 + e_2), 0]$.

Fig. 4.3 shows the Mohr circle of strains, in which points A and B correspond to strains e_1 and e_2 respectively and $AG = FB = \frac{1}{2}\,\phi$. The coordinates of point D represent e_θ and $\frac{1}{2}\,\phi_\theta$ (as marked) on a plane making angle θ with the direction of e_1.

Again, points H and J correspond to principal strains ε_1 and ε_2 respectively, since the shearing strain $\frac{1}{2}\,\phi_\theta$ at these points are each zero.

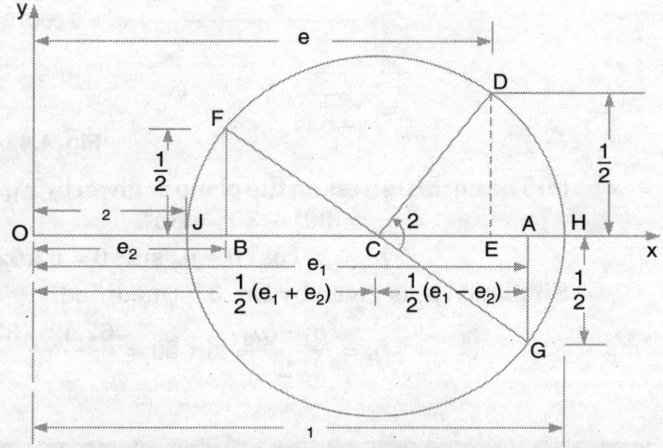

Fig. 4.3. Mohr circle of strains.

Example 4.2. *A flat brass plate was stretched by tensile forces acting in directions x and y at right angles. Strain gauges show that strain in x-direction was 0.00072 and that in the y-direction was 0.00016. Find (a) stresses acting in x- and y-directions, and (b) direct and shearing strains on a plane inclined at 30° to the x-direction (c) normal and shearing stresses on that plane. Take $E = 0.8 \times 10^5$ N/mm² and $\mu = 0.3$.*

Solution. (*a*) For two dimensional stress system, we have from Eqs. 4.4 (*a*), (*b*).

$$p_1 = \frac{E(e_1 + \mu e_2)}{1 - \mu^2} = \frac{0.8 \times 10^5\ (0.00072 + 0.3 \times 0.00016)}{1 - (0.3)^2} = \mathbf{67.52\ N/mm^2}$$

$$p_2 = \frac{E(\mu e_1 + e_2)}{1 - \mu^2} = \frac{0.8 \times 10^5\ (0.3 \times 0.00072 + 0.00016)}{1 - (0.3)^2} = \mathbf{33.05\ N/mm^2}.$$

(*b*) From Eq. 4.5, noting that $\phi = 0$, we get

$$e_\theta = e_1 \cos^2\theta + e_2 \sin^2\theta = 0.00072 \cos^2 30° + 0.00016 \sin^2 30° = \mathbf{0.00058}$$

Also, from Eq. 4.8

$$\frac{1}{2}\phi_\theta = \frac{1}{2}(e_1 - e_2)\sin 2\theta = \frac{1}{2}(0.00072 - 0.00016)\sin 60° = 0.000242$$

$$\therefore \qquad \phi_\theta = \mathbf{0.000484}$$

These values of e_θ and ϕ_θ can also be obtained from the construction of Mohr-circle of strains shown in Fig. 4.4.

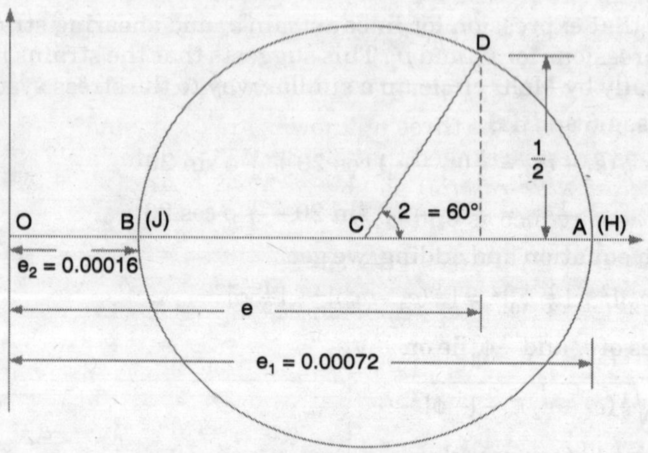

Fig. 4.4

(*c*) The normal stress on the plane is given by Eq. 3.6, noting that θ = the angle with *y*-direction = 90° – 30° = 60°.

$$\therefore \quad p_n = p_1 \cos^2 \theta + p_2 \sin^2 \theta = 67.52 \cos^2 60° + 33.05 \sin^2 60° = \mathbf{41.67\ N/mm^2}$$

Similarly, p_t is given by Eq. 3.7 (modified)

$$p_t = \frac{p_1 - p_2}{2} \sin 2\theta = \frac{67.52 - 33.05}{2} \sin 120° = \mathbf{14.93\ N/mm^2}.$$

4.5. STRAIN 'ROSETTES'

In order to determine actual stresses developed in a strained material, strains are measured by means of small gauges. Several types of gauges have been developed, but most convenient types are the electrical *strain gauges.* Electrical strain gauge consists of a small length of fine wire which is glued to the surface of the material. When the material is strained, the length of the strain gauge wire also changes, resulting in the change of its resistance which can be measured on a Wheat Stone Bridge.

In a two dimensional stress system, there are direct strains e_1 and e_2, and a shearing strain ϕ. However, the shear strain cannot be directly measured. Hence direct strains e_a, e_b and e_c are measured in any three convenient directions. From the measured values of these direct strains, principal strains ε_1 and ε_2 can be computed with reference to Fig. 4.5 as under.

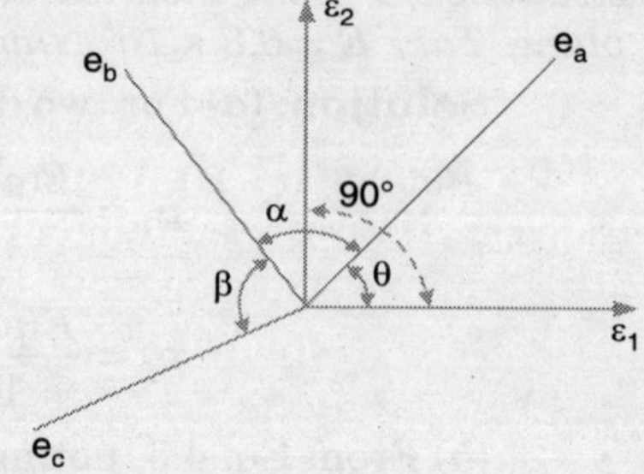

Fig. 4.5

Let e_a, e_b and e_c be inclined at θ, (θ + α) and (θ + α + β) with the directions of major principal strain ε_1.

From Eq. 4.6, we have

$$e_a = \tfrac{1}{2}(\varepsilon_1 + \varepsilon_2) + \tfrac{1}{2}(\varepsilon_1 - \varepsilon_2)\cos 2\theta \quad ...(i)$$

$$e_b = \tfrac{1}{2}(\varepsilon_1 + \varepsilon_2) + \tfrac{1}{2}(\varepsilon_1 - \varepsilon_2)\cos 2(\theta + \alpha) \quad ...(ii)$$

and
$$e_c = \tfrac{1}{2}(\varepsilon_1 + \varepsilon_2) + \tfrac{1}{2}(\varepsilon_1 - \varepsilon_2)\cos 2(\theta + \alpha + \beta) \quad ...(iii)$$

In the above equations, the three unknowns are : ε_1, ε_2 and θ. If the strain gauges are so arranged that $\alpha = \beta = 45°$, forming a 45° *rosette* (or a *rectangular rosette*) as shown in Fig. 4.6, the above pairs of equations reduce as under :

$$e_a = \tfrac{1}{2}(\varepsilon_1 + \varepsilon_2) + \tfrac{1}{2}(\varepsilon_1 - \varepsilon_2) \times \cos 2\theta \quad ...(a)$$

$$e_b = \tfrac{1}{2}(\varepsilon_1 + \varepsilon_2) - \tfrac{1}{2}(\varepsilon_1 - \varepsilon_2) \times \sin 2\theta \quad ...(b)$$

$$e_c = \tfrac{1}{2}(\varepsilon_1 + \varepsilon_2) - \tfrac{1}{2}(\varepsilon_1 - \varepsilon_2) \times \cos 2\theta \quad ...(c)$$

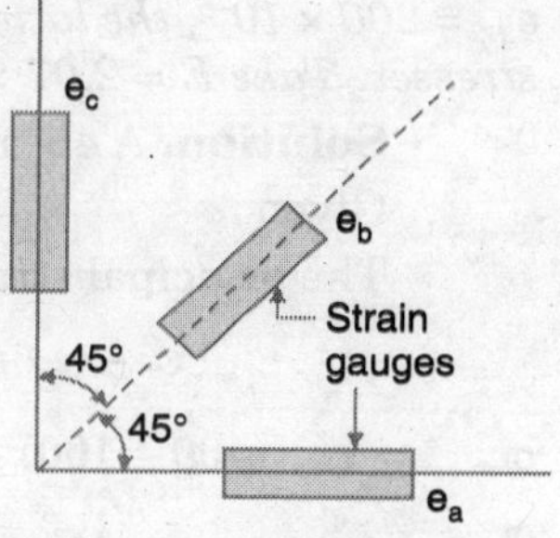

Fig. 4.6. A 45° strain rosette.

Eliminating θ, we get

$$\varepsilon^2 - (e_a + e_c)\varepsilon + [e_a e_c - \tfrac{1}{4}(2e_b - e_a - e_c)^2] = 0 \quad ...(4.11)$$

The roots of the above quadratic equation give the values of principal strains ε_1 and ε_2.

The directions of ε_1 and ε_2, relative to e_a, e_b and e_c are given by the roots of the equation

$$\tan 2\theta = \frac{2e_b - e_a - e_c}{e_c - e_a} \quad ...(4.12)$$

where θ is the angle measured between ε_1 and e_a, measured *clockwise* from the direction of ε_1.

Another form of *strain-rosettes* is a 120° rosette. Two forms of 120° rosettes are shown in Fig. 4.7 in which $\alpha = \beta = 120°$.

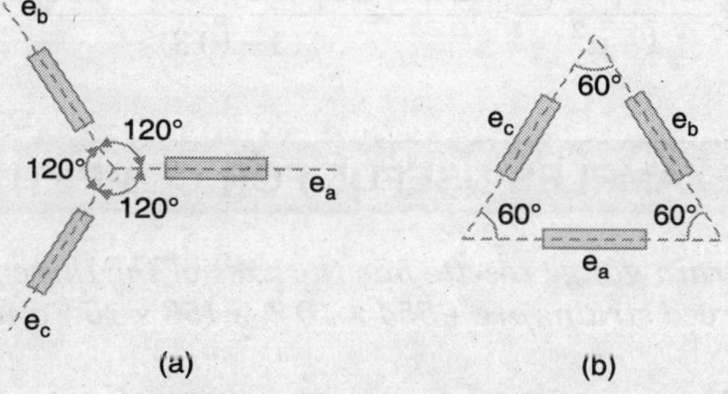

Fig. 4.7. 120° Rosettes

Putting $\alpha = \beta = 120°$ in Eqs. (*i*), (*ii*) and (*iii*), we get

$$e_a = \tfrac{1}{2}(\varepsilon_1 + \varepsilon_2) + \tfrac{1}{2}(\varepsilon_1 - \varepsilon_2)\cos 2\theta \quad ...(1)$$

$$e_b = \tfrac{1}{2}(\varepsilon_1 + \varepsilon_2) - \tfrac{1}{2}(\varepsilon_1 - \varepsilon_2)\left(\tfrac{1}{2}\cos 2\theta - \frac{\sqrt{3}}{2}\sin 2\theta\right) \quad ...(2)$$

and
$$e_c = \tfrac{1}{2}(\varepsilon_1 + \varepsilon_2) - \tfrac{1}{2}(\varepsilon_1 - \varepsilon_2)\left(\tfrac{1}{2}\cos 2\theta + \frac{\sqrt{3}}{2}\sin 2\theta\right) \quad ...(3)$$

The principal strains ε_1 and ε_2 are given by the roots of the equation

$$3\varepsilon^2 - 2(e_a + e_b + e_c)\varepsilon - (e_a^2 + e_b^2 + e_c^2 - 2e_a e_b - 2e_b e_c - 2e_c e_a) = 0 \quad ...(4.13)$$

The directions of the principal strains are given by

$$\tan 2\theta = \frac{\sqrt{3}\,(e_b - e_c)}{2e_a - e_b - e_c} \qquad \text{...(4.14)}$$

Example 4.3. *A rectangular rosette strain gauge records the following values for the linear strain at a point in a two dimensional stress system :* $e_x = 400 \times 10^{-6}$, $e_y = -100 \times 10^{-6}$ *and* $e_{45} = 200 \times 10^{-6}$, *the latter being at 45° to the X and Y axes. Determine the principal strains and stresses. Take* $E = 2.07 \times 10^5$ *N/mm² and* $\mu = 0.3$.

Solution. A 45° rosette is also known as rectangular rosette (See Fig. 4.6).

Given : $e_a = e_x = 400 \times 10^{-6}$; $e_b = e_{45} = 200 \times 10^{-6}$ and $e_c = e_y = -100 \times 10^{-6}$

The principal strains are given by roots of Eq. 4.11 :

$$\varepsilon^2 - (e_a + e_c)\varepsilon + [e_a e_c - \tfrac{1}{4}(2e_b - e_a - e_c)^2] = 0$$

or $$\varepsilon^2 - (400 - 100)\,10^{-6}\,\varepsilon + [-400 \times 100 - \tfrac{1}{4}(2 \times 200 - 400 + 100)^2]\,10^{-12} = 0$$

or $$\varepsilon^2 - 300 \times 10^{-6}\,\varepsilon - 42500 \times 10^{-12} = 0$$

From which $\varepsilon_1 = \mathbf{405 \times 10^{-6}}$ and $\varepsilon_2 = \mathbf{-105 \times 10^{-6}}$

$$\tan 2\theta = \frac{2e_b - e_a - e_c}{e_c - e_a} = \frac{2 \times 200 \times 10^{-6} - 400 \times 10^{-6} + 100 \times 10^{-6}}{-100 \times 10^{-6} - 400 \times 10^{-6}} \qquad \text{...(Eq. 4.12)}$$

$$\therefore \quad 2\theta = 11°\,18.6' \quad \text{or} \quad \theta = 5°\,39'\,3.$$

Knowing principal strains ε_1 and ε_2, principal stresses are given by Eq. 4.4.

$$\sigma_1 = \frac{E(\varepsilon_1 + \mu\varepsilon_2)}{1 - \mu^2} = \frac{2.07 \times 10^5\ (405 - 0.3 \times 105)\ 10^{-6}}{1 - (0.3)^2} \simeq \mathbf{84.96\ N/mm^2}$$

$$\sigma_2 = \frac{E(\mu\varepsilon_1 + \varepsilon_2)}{1 - \mu^2} = \frac{2.07 \times 10^5\ (0.3 \times 405 - 105)\ 10^{-6}}{1 - (0.3)^2} \simeq \mathbf{3.75\ N/mm^2}.$$

4.6. ADDITIONAL EXAMPLES USEFUL FOR COMPETITIVE EXAMINATIONS

Example 4.4. *A strain gauge rosette has the axes of the three gauges OA, OB and OC at 120° each other. The observed strains are* $+554 \times 10^{-6}$, -456×10^{-6} *and* $+64 \times 10^{-6}$ *along OA, OB and OC respectively.*

Determine the inclinations of the principal planes at O relative to OA and the magnitudes of the principal stresses. Determine also the strain at right angles to OA. Take $E = 2 \times 10^5$ *N/mm² and* $\mu = 0.3$

Solution. Refer Fig. 4.7 (*a*).

Given : $e_a = +554 \times 10^{-6}$; $e_b = -456 \times 10^{-6}$; $e_c = +64 \times 10^{-6}$

The principal strains ε_1 and ε_2 are given by the roots of Eq. 4.13 :

$$3\varepsilon^2 - 2(e_a + e_b + e_c)\varepsilon - (e_a^2 + e_b^2 + e_c^2 - 2e_a e_b - 2e_b e_c - 2e_c e_a) = 0$$

$$\therefore \quad 3\varepsilon^2 - 2(554 - 456 + 64)\,10^{-6}\,\varepsilon - [(554)^2 + (456)^2 + (64)^2 + 2(554)(456) + 2(456)(64) - 2(64)(554)]\,10^{-12} = 0$$

$$3\varepsilon^2 - 324 \times 10^{-6}\varepsilon - 1011652 \times 10^{-12} = 0$$

From which, we get $\varepsilon_1 = +637.3 \times 10^{-6}$ and $\varepsilon_2 = -529.3 \times 10^{-6}$

Inclination of principal planes is given by

$$\tan 2\theta = \frac{\sqrt{3}(e_b - e_c)}{2e_a - e_b - e_c} = \frac{\sqrt{3}(-456 - 64)}{2 \times 554 + 456 - 64} = -0.6$$

$$\therefore \quad 2\theta = -30° 58' \quad \text{or} \quad \theta = \mathbf{-15° 29'} \text{ and } \mathbf{+74° 31'}$$

Knowing the principal strains, principal stresses are given by Eq. 4.4.

$$\sigma_1 = \frac{E(\varepsilon_1 + \mu\varepsilon_2)}{1 - \mu^2} = \frac{2 \times 10^5 \,(637.3 - 0.3 \times 529.3)\, 10^{-6}}{1 - (0.3)^2} = \mathbf{105.2 \ N/mm^2}$$

and

$$\sigma_2 = \frac{E(\mu\varepsilon_1 + \varepsilon_2)}{1 - \mu^2} = \frac{2 \times 10^5 \,(0.3 \times 637.3 - 529.3)\, 10^{-6}}{1 - (0.3)^2} = \mathbf{-74.3 \ N/mm^2}$$

For a plane at 90° to OA, θ with the principal plane = 90° + 15° 29′ = 105° 29′.

Hence linear strain on this plane is computed from Eq. 4.6

$$e_\theta = \tfrac{1}{2}(\varepsilon_1 + \varepsilon_2) + \tfrac{1}{2}(\varepsilon_1 - \varepsilon_2) \cos 2\theta$$

or

$$e_\theta = \tfrac{1}{2}(637.3 - 529.3)\, 10^{-6} + \tfrac{1}{2}(637.3 + 529.3)\, 10^{-6} \times \cos 2\,(105° 29') = \mathbf{-446 \times 10^{-6}}.$$

Example 4.5. *The measured strains in three directions inclined at 60° to one another are 500×10^{-6}, -120×10^{-6} and 200×10^{-6}, as shown in Fig. 4.8 (a). Compute the magnitude and direction of principal strains in this plane. If there is no shear stress perpendicular to the given plane, determine the principal stresses at the point. Take $E = 2 \times 10^5$ N/mm² and $\mu = 0.3$.*

Solution.

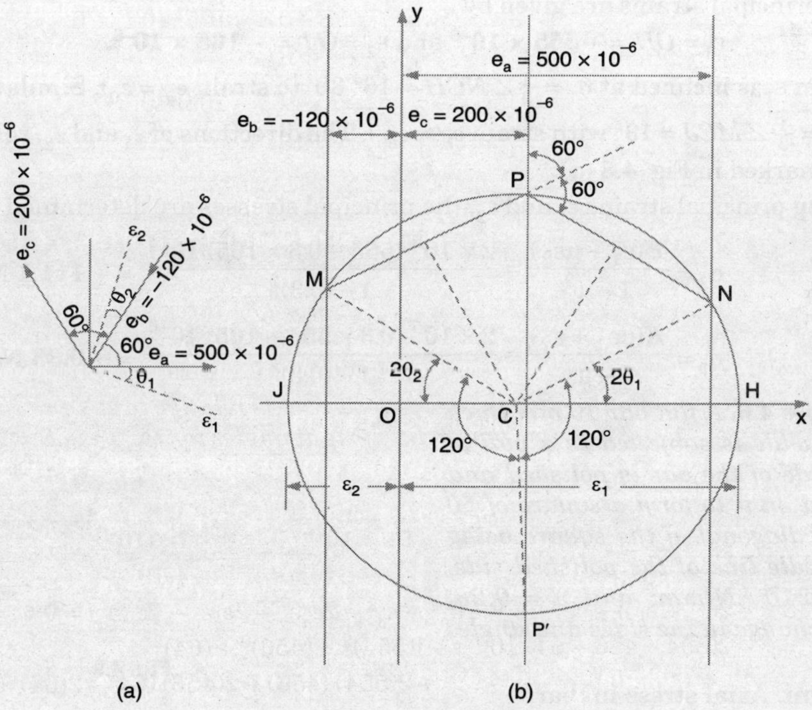

Fig. 4.8. Mohr circle of strains.

Fig. 4.8 (*a*) shows the magnitudes and directions of the three strains measured. Taking the strain $e_a = 500 \times 10^{-6}$ in the *x*-direction, we get

$$e_a = e_0 = +\ 500 \times 10^{-6}\ ;\ e_b = e_{60} = -\ 120 \times 10^{-6} \quad \text{and} \quad e_c = e_{120} = +\ 200 \times 10^{-6}$$

The Mohr circle of strains is shown in Fig. 4.8 (*b*), which is constructed in the following steps :

1. Draw a vertical line representing *y*-axis, and set off three vertical line at $+\ 500 \times 10^{-6}$, $+\ 200 \times 10^{-6}$ and $-\ 120 \times 10^{-6}$ distances (to some scale) from the *y*-axis.

2. On the intermediate strain (e_c) line, select any point *P*.

3. Through *P*, draw a line *PM* at 60° angle, intersecting the vertical line representing e_b (= e_{60}) at *M*. Similarly, draw another line *PN* at 120° angle, intersecting the vertical line representing e_a(= e_0) at *N*.

4. Thus, we have three points *P, M* and *N*, on the three vertical lines corresponding to the three strains. The Mohr circle will pass through *M, P* and *N*. To locate its centre, draw perpendicular bisectors of *MP* and *PN*, intersecting at *C*.

5. Through *C*, draw horizontal line, representing *x*-axis, intersecting the *y-axis* in *O*. Thus, we get origin *O* and the centre *C* of the Mohr circle. Draw the Mohr circle with radius *CM* (or *CN*). The circle will pass through *M, P* and *N*.

6. The strain conditions on the three given planes are represented by points *M, N* and *P′*, where *P′* is on the vertical through *P*. The justification of the construction lies from the fact that angles between the three radii vectors *CM, CN* and *CP′* are 120°, which is twice the angle between each pair of given direct strains.

7. The principal strains are given by

$$\varepsilon_1 = OH = +\ \mathbf{555 \times 10^{-6}} \text{ and } \varepsilon_2 = OJ = -\ \mathbf{165 \times 10^{-6}}.$$

8. Strain ε_1 is inclined at $\theta_1 = \frac{1}{2}\angle NCH = 16°\ 30'$ to strain e_a(= e_0). Similarly, strain ε_2 is inclined at $\theta_2 = \frac{1}{2}\ \angle MCJ = 13°$ with strain e_b(= e_{60}). The directions of ε_1 and ε_2, relative to e_a and e_b have been marked in Fig. 4.8 (*a*).

Knowing principal strains ε_1 and ε_2, the principal stresses are determined from Eq. 4.4 :

$$\sigma_1 = \frac{E(\varepsilon_1 + \mu\varepsilon_2)}{1-\mu^2} = \frac{2 \times 10^5\ (555 - 0.3 \times 165)\ 10^{-6}}{1-(0.3)^2} = +\ \mathbf{111.1\ N/mm^2}$$

$$\sigma_2 = \frac{E(\mu\varepsilon_1 + \varepsilon_2)}{1-\mu^2} = \frac{2 \times 10^5\ (0.3 \times 555 - 165)\ 10^{-6}}{1-(0.3)^2} = +\ \mathbf{0.33\ N/mm^2}.$$

Example 4.6. *A flat bar 10 mm thick and 100 mm wide is subjected to a pull of 96 kN. One side of the bar is polished and lines are ruled on it to form a square of 50 mm side, one diagonal of the square being along the middle line of the polished side. Taking E = 210 kN/mm² and μ = 0.28, calculate the change in the sides and angles of the square.*

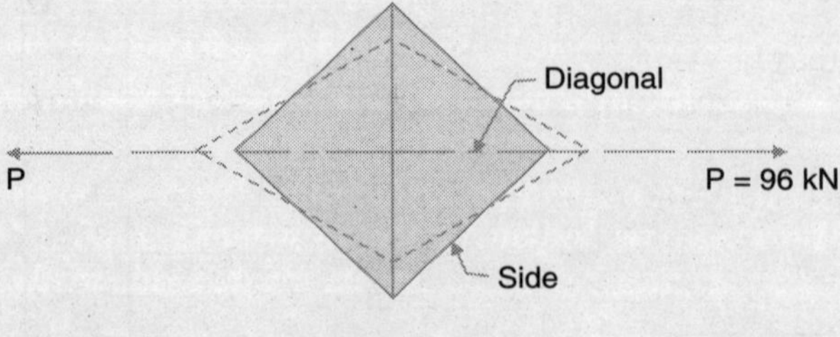

Fig. 4.9

Solution. Axial stress in bar,

$$p_1 = \frac{96 \times 1000}{100 \times 10} = 96 \text{ N/mm}^2 \text{ (tensile)}$$

Strain in one diagonal of square along the axis of the bar is

$$e_1 = \frac{p_1}{E} = \frac{96}{210 \times 10^3} = 4.571 \times 10^{-4} \text{ (tensile)}$$

$\therefore$ Strain in other diagonal $= e_2 = -\mu e_1 = -0.28 \times 4.571 \times 10^{-4}$

$= -1.28 \times 10^{-4}$ (compressive)

$\therefore$ Strain in the sides of square is given by

$$e_\theta = e_1 \cos^2 45° + e_2 \sin^2 45°$$

$$= 4.571 \times 10^{-4} \times 0.5 - 1.28 \times 10^{-4} \times 0.5 = 1.645 \times 10^{-4} \text{ (tensile)}$$

Hence change in length of each side of square $= (1.645 \times 10^{-4}) \times 50$

$=$ **0.008225 mm** (increase)

Total shear strain or change in the angles of the square $= \varphi_\theta$

$= (e_1 - e_2) \sin 2\theta$

$= (4.571 + 1.280) \times 10^{-4} \sin 90° =$ **5.851 × 10⁻⁴** radian

Thus the two angles of the square which are along the line of pull will be decreased and the other two will be increased by the above amount.

Example 4.7. *A brass plate 3 mm thick, is stretched in the plane of the plate in two directions at right angles. An extensometer, arranged in the X-direction gave an extension of 0.0413 mm on a gauge length of 50 mm, while another, arranged in Y-direction gave an extension of 0.0187 mm on a 100 mm gauge length. If the normal stress on a plane making angle θ with the Y-axis is 66 N/mm², find angle θ. Find also the decrease in the thickness of the plate. Take E for brass = 80 kN/mm² and μ = 0.3.*

Solution. Let σ_x and σ_y be the principal stresses in two directions (say x-and y-directions respectively) and ε_1 and ε_2 be the corresponding principal strains.

Then $\varepsilon_x = \dfrac{0.0413}{50} = 8.26 \times 10^{-4}$ and $\varepsilon_y = \dfrac{0.0187}{100} = 1.87 \times 10^{-4}$

Also, $\varepsilon_x = \dfrac{\sigma_x}{E} - \dfrac{\mu \sigma_y}{E}$ and $\varepsilon_y = \dfrac{\sigma_y}{E} - \dfrac{\mu \sigma_x}{E}$

$\therefore$ $8.26 \times 10^{-4} = \dfrac{1}{80 \times 10^3} [\sigma_x - 0.3\,\sigma_y]$ or $\sigma_x - 0.3\,\sigma_y = 66.08$ ...(*i*)

and $1.87 \times 10^{-4} = \dfrac{1}{80 \times 10^3} [\sigma_y - 0.3\,\sigma_x]$ or $\sigma_y - 0.3\,\sigma_x = 14.96$ ...(*ii*)

Solving (*i*) and (*ii*), we get $\sigma_x = 75.65$ N/mm² and $\sigma_y = 31.91$ N/mm²

Inclination of plane with y-axis $= \theta$. Hence inclination of plane with x-axis $= 90° - \theta = \theta_1$ (say).

The direct stress at θ_1 with x-axis is given by

$$p_n = \frac{1}{2}(\sigma_x + \sigma_y) + \frac{1}{2}(\sigma_x - \sigma_y)\cos 2\theta_1$$

or $66 = \dfrac{1}{2}(75.65 + 31.91) + \dfrac{1}{2}(75.65 - 31.91)\cos 2\theta_1$

from which $\cos 2\theta_1 = 0.5587$ or $2\theta_1 = 56°$ or $\theta_1 = 28°$

$\therefore$ $\theta = 90° - \theta_1 =$ **62°**

Strain in the direction of thickness

$$= -\frac{\mu\sigma_x}{E} - \mu\frac{\sigma_y}{E} = -\frac{0.3}{80000}(75.65 - 31.91) = -1.64 \times 10^{-4}$$

or $$\frac{\Delta t}{t} = -1.64 \times 10^{-4}$$

$\therefore$ $$\Delta t = -1.64 \times 10^{-4} \times 3 = -\mathbf{4.92 \times 10^{-4}\ mm}\ (i.e.\ \text{decrease})$$

4.7. EXAMPLES FROM COMPETITIVE EXAMINATIONS

Example 4.8. *In a stressed member, two strain gauges are fixed such that they are inclined at 30° to the known directions of the principal stresses. The strains measured in these two gauges are + 445 × 10^6 and – 32 × 10^{-6} respectively. E = 2.1 × 10^5 N/sq. mm and Poisson's ratio (μ) = 0.3. Determine the magnitude of principal stresses.*

(UPSC Engg. Services Exam. 2001)

Solution. Fig. 4.10 shows the Mohr circle of strain in which $OA = e_1 = +445 \times 10^{-6}$ and $OB = e_2 = -32 \times 10^{-6}$. Point C, the centre of Mohr circle, is midway between points A and B. Through C, draw a line inclined at 60° with x-axis.

Through A and B, draw perpendiculars to intersect the inclined line at points, D and D' respectively, giving $CD = CD'$ = radius of the Mohr circle. Draw the Mohr circle, which intersects the x-axis at points H and J. Then $OH = \varepsilon_1$ = major principal strain and $OJ = \varepsilon_2$ = minor principal strain.

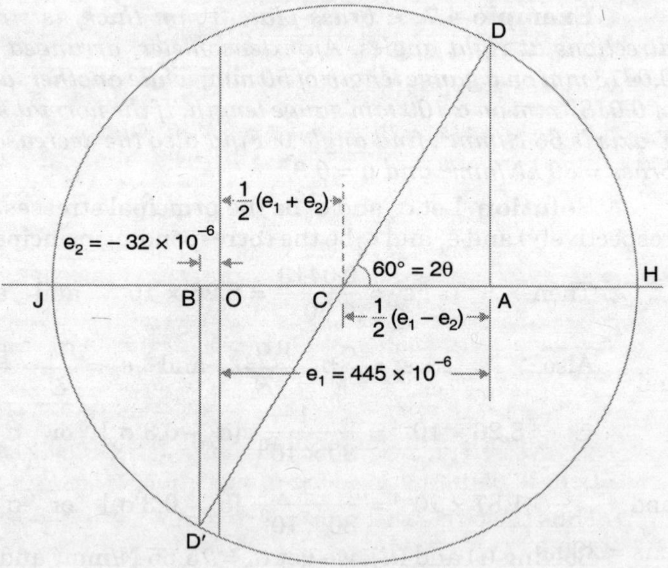

Fig. 4.10

Now, $CA = \frac{1}{2}(e_1 - e_2)$

$= \frac{1}{2}[445 - (-32)]\ 10^{-6}$

$= 238.5 \times 10^{-6}$

Radius $CD = CA \sec 60°$

$= 238.5 \times 10^{-6} \times 2$

$= 477 \times 10^{-6} = CH = CJ$

$$OC = \tfrac{1}{2}(e_1 + e_2) = \tfrac{1}{2}(445 - 32)\ 10^{-6} = 206.5 \times 10^{-6}$$

$\therefore$ Major principal stress, $\varepsilon_1 = OH = OC + CH = 206.5 \times 10^{-6} + 477 \times 10^{-6} = 683.5 \times 10^{-6}$

Minor principal stress, $\varepsilon_2 = -OJ = -(CJ - OC) = -(477 \times 10^{-6} - 206.5 \times 10^{-6})$

$= -270.5 \times 10^{-6}$

Now for a two dimensional stress system, from Eq. 4.4, we have

$$\sigma_1 = \frac{E(\varepsilon_1 + \mu\varepsilon_2)}{1-\mu^2} = \frac{2.1\times10^5\ (683.5 - 0.3\times270.5)\ 10^{-6}}{1-(0.3)^2} = \mathbf{139.0\ N/mm^2}$$

and
$$\sigma_2 = \frac{E(\mu\varepsilon_1 + \varepsilon_2)}{1-\mu^2} = \frac{2.1\times10^5\ (0.3\times683.5 - 270.5)\ 10^{-6}}{1-(0.3)^2} = \mathbf{-15.10\ N/mm^2.}$$

Example 4.9. *A mild steel plate is stressed as shown in the figure (Fig. 4.11). Before stressing, a circle of 300 mm diameter is drawn on the plate. Determine the lengths and directions of the major and minor axes of the ellipse into which the circle deforms after stressing. Poisson's ratio = 0.3 ; modulus of elasticity = 200 kN/mm².* (UPSC Engg. Services Exam. 1993)

Solution. *This problem was set from Example 4.1 of this book*. However, the values of $1/m$ and E have been slightly changed

p_2 = 13 N/mm²
20
p_1 = 65 N/mm²
p_1
20 N/mm²
p_2

Fig. 4.11

$$\sigma = \frac{p_1 + p_2}{2} \pm \sqrt{\left(\frac{p_1 - p_2}{2}\right)^2 + q^2}$$

$$\therefore \quad \sigma = \frac{65+13}{2} \pm \sqrt{\left(\frac{65-13}{2}\right)^2 + (20)^2}$$

$$= 39 \pm 32.8$$

$$\therefore \quad \sigma_1 = 39 + 32.8 = +\ 71.8\ \text{N/mm}^2$$

and
$$\sigma_2 = 39 - 32.8 = +\ 6.2\ \text{N/mm}^2.$$

$$2\theta_p = \tan^{-1}\frac{2q}{p_1 - p_2} = \tan^{-1}\frac{2\times20}{65-13} = 37°\ 36';\ \theta_{p1} = 18°\ 48' \text{ and } \theta_{p2} = 108°\ 48'$$

$$\varepsilon_1 = \frac{\sigma_1}{E} - \frac{\sigma_2}{mE} = \frac{71.8 - 6.2\times0.3}{2\times10^5} = 3.497\times10^{-4}\ \text{(increase)}$$

and
$$\varepsilon_2 = \frac{\sigma_2}{E} - \frac{\sigma_1}{mE} = \frac{6.2 - 71.8\times0.3}{2\times10^5} = -7.67\times10^5\ \text{(increase)} = 7.5\times10^{-5}\ \text{(decrease)}$$

∴ Length of diameter along σ_1 direction $= d + \varepsilon_1 d = 300(1 + 3.497\times10^{-4}) =$ **300.105 mm**

Length of diameter along σ_2 direction $= d + \varepsilon_2 d = 300(1 - 7.67\times10^{-5}) =$ **299.977 mm.**

Thus the circle will be converted into an ellipse with major axis = 300.105 mm and minor axis = 299.977 mm.

Example 4.10. *The strain measurements from a rectangular strain rosette were $e_0 = 600\times10^{-6}$, $e_{45} = 500\times10^{-6}$ and $e_{90} = 200\times10^{-6}$. Find the magnitude and direction of principal strains. If $E = 2\times10^5$ N/mm² and $\mu = 0.3$, find the principal stresses.*

(UPSC Engg. Services Exam. 1995)

Solution. Refer Fig. 4.6 and Example 4.3.

Given : $e_a = e_0 = 600\times10^{-6}$; $e_b = e_{45} = 500\times10^{-6}$ and $e_c = e_{90} = 200\times10^{-6}$

The principal stresses are given by the roots of Eq. 4.11 :

$$\varepsilon^2 - (e_a + e_c)\,\varepsilon + [e_a e_c - \tfrac{1}{4}(2e_b - e_a - e_c)^2] = 0$$

or $$\varepsilon^2 - (600 + 200)\, 10^{-6}\, \varepsilon + [600 \times 200 - \tfrac{1}{4}(2 \times 500 - 600 - 200)^2]\, 10^{-12} = 0$$

or $$\varepsilon^2 - 800 \times 10^{-6}\, \varepsilon + 110000 \times 10^{-12} = 0$$

From which $\varepsilon_1 = \mathbf{623.6 \times 10^{-6}}$ and $\varepsilon_2 = \mathbf{176.4 \times 10^{-6}}$

$$2\theta = \tan^{-1} \frac{2e_b - e_a - e_c}{e_c - e_a} = \tan^{-1} \frac{(2 \times 500 - 600 - 200)\, 10^{-6}}{(200 - 600)\, 10^{-6}} = -26°.56$$

or $\theta = \mathbf{-13°.28}$

The corresponding principal stresses are :

$$\sigma_1 = \frac{E(\varepsilon_1 + \mu\varepsilon_2)}{1 - \mu^2} = \frac{2 \times 10^5\, (623.6 + 0.3 \times 176.4)\, 10^{-6}}{1 - (0.3)^2} = \mathbf{148.69\ N/mm^2}$$

$$\sigma_2 = \frac{E(\mu\varepsilon_1 + \varepsilon_2)}{1 - \mu^2} = \frac{2 \times 10^5\, (0.3 \times 623.6 + 176.4)\, 10^{-6}}{1 - (0.3)^2} = \mathbf{79.89\ N/mm^2.}$$

PROBLEMS

1. At a certain point in a piece of material the directions of principal stresses σ_1 and σ_2 are known. Measurements by strain gauges show that there is a tensile strain of 0.00065 in the direction of σ_1 and a compressive strain of 0.00045 in the direction of σ_2. Find the magnitudes of σ_1 and σ_2, stating whether compressive or tensile, and the value of maximum shear stress. Take $m = 4$ and $E = 200$ kN/mm².
2. A thin sheet of material is subjected to a tensile stress of 80 MN/m², in a certain direction. One surface of the sheet is polished, and on this surface five lines are ruled to form a square of side 500 mm, one diagonal of the square being parallel to the direction of the tensile stresses. If $E = 200$ GN/m², and $\mu = 0.3$, estimate the alteration in the length of the sides of square, and the change in the angles at the corners of the square.
3. The measured strains with an electric rectangular rosette indicate that $\varepsilon_0 = -0.00050$, $\varepsilon_{45} = +0.002$ and $\varepsilon_{90} = +0.0003$. Taking $E = 2 \times 10^5$ N/mm² and $\mu = 0.3$, find the principal stresses at the point.

ANSWERS

1. $\sigma_1 = 114.7$ N/mm² (tensile) ; $\sigma_2 = 61.3$ N/mm² (comp.).
2. 7×10^{-4} mm ; 52×10^{-3} radians (0.0405°).
3. + 48.3 N/mm² ; – 105 N/mm².

5

Strain Energy, Resilience and Impact Loading

5.1. INTRODUCTION

When an axial load (tensile or compressive) is applied to a body, it deforms ; the amount of deformation will depend on the *manner* in which the load has been applied. If the applied load is not gradual, but is sudden, the maximum instantaneous stress and elongation will be much greater and can be easily calculated by considerations of the principle of work. If the elastic limit is not exceeded, the work done in straining the material is stored in it in the form of *strain energy*. As soon as the load is removed, the strain energy will be given up, the material behaving like a perfect spring. However, if the elastic limit is exceeded, the work done during non-elastic strain is spent in overcoming the cohesion of the particles of the material, causing them to slide one over another and the energy expended appears as heat in the material strained.

5.2. RESISTANCE-DEFORMATION DIAGRAM : STRAIN ENERGY

Let a member of length L be subjected to an axial load P. Let A be the area of section. Due to the application of the load, the member deforms and a corresponding stress or resistance is set up.

Let Δ = Total deformation

p = Stress corresponding to the above deformation

R = Final resistance = pA.

When the loading commences, deformation is zero and hence P and R are zero. When the deformation is Δ_x, the strain $e_x = \Delta_x / L$

$$\therefore \qquad p_x = \frac{E\Delta_x}{L} \quad \text{or} \quad R_x = E \cdot \frac{\Delta_x}{L} \cdot A \qquad \text{...(5.1)}$$

From Eq. 5.1, it is clear that R_x increases linearly with Δ_x. Fig. 5.1 shows the resistance-deformation diagram, which is a straight line within elastic limits.

The work done in straining of the material is equal to the shaded area of the resistance-deformation diagram and is stored in the body.

Energy stored $= \frac{1}{2} R\Delta = \frac{1}{2} p.A.e.L = \frac{1}{2} p.e.A.L$

or $U = \frac{1}{2}$ stress × strain × volume ...(5.2)

where U = strain energy stored and strain $e = p/E$

$$\therefore \quad U = \frac{1}{2}\frac{p^2}{E}.AL = \frac{p^2}{2E} \times \text{volume} \quad ...(5.3)$$

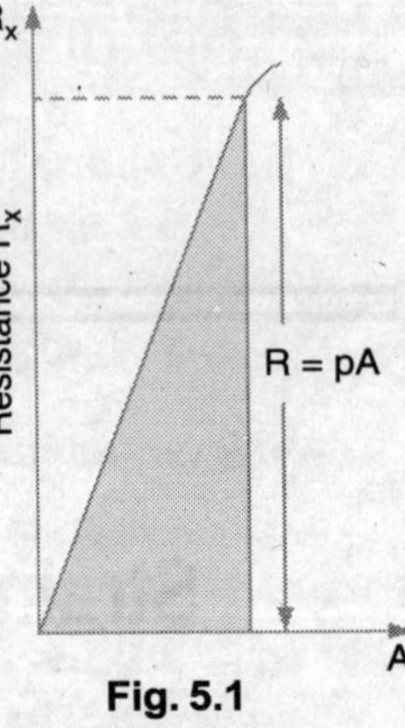

Fig. 5.1

5.3. PROOF RESILIENCE

The strain energy stored in a material is also known as *resilience* and is equal to $\frac{p^2}{2E} \times$ volume. If p is equal to the stress f at the elastic limit (generally known as "proof stress"), the corresponding strain energy is known as "Proof Resilience" U_p.

$$\text{Thus} \quad U_p = \frac{f^2}{2E} \times \text{volume} \quad ...[5.3\ (a)]$$

The quantity $f^2/2E$ is the proof resilience per unit volume and is known as *Modulus of Resilience* and may be looked upon as the property of the material.

Example 5.1. *Calculate the strain energy in a bar 3 m long and 40 mm in diameter when it is subjected to a tensile load of 100 kN. Take E = 2.05 × 10⁵ N/mm². What will then be the modulus of resilience of the material of the bar ?*

Solution. $U = \frac{p^2}{2E} \times \text{volume} = \left(\frac{P}{A}\right)^2 \frac{1}{2E} \times V$

Here, $A = \frac{\pi}{4}(40)^2 = 1256.6 \text{ mm}^2$; $V = 1256.6 \times 3000 \simeq 3670 \times 10^3 \text{ mm}^3$

$$U = \left(\frac{100 \times 1000}{1256.6}\right)^2 \times \frac{1}{2 \times 2.05 \times 10^5} \times 3670 \times 10^3$$

$= 5.67 \times 10^4$ mm-N = **0.0567 m-kN** = 56.7 m-N = **56.7 J.**

Also, modulus of resilience

$$= \left(\frac{P}{A}\right)^2 \frac{1}{2E} = \frac{5.67 \times 10^4}{3670 \times 10^3} = \mathbf{1.54 \times 10^{-2}\ mm\text{-}N/mm^3}$$

$= \mathbf{15.4\ m\text{-}kN/m^3}$. **Note.** (1 m-N = 1 J).

5.4. GRADUAL, SUDDEN AND IMPACT LOADINGS

(a) Gradual Loading

Let the axial load P be applied gradually to a bar. Let Δ be the corresponding deformation and R be the total resistance set up by the body.

Work done by the external load $= \frac{1}{2}P\Delta$

Work stored in the body $= \frac{1}{2}R\Delta = \frac{1}{2}pA\Delta$

Equating the work stored to the work done, we get

$$\tfrac{1}{2}pA\Delta = \tfrac{1}{2}P\Delta \quad \text{or} \quad p = P/A \qquad \text{...(5.4)}$$

Thus, the maximum stress set up is equal to the load divided by area of section of the bar.

(b) Sudden Loading

If the load P is applied suddenly, the value of load is P throughout the deformation. The deformation increases from zero value to its final value Δ.

Hence the work done on the bar $= P\Delta$.

However, the resistance set up in the body is zero when deformation is zero and is equal to R when the deformation is Δ. At any stage, the resistance varies linearly with deformation as given by equation 5.1. The work stored is therefore equal to $\frac{1}{2}R\Delta$ or equal to $\frac{1}{2}pA\Delta$.

Equating the work stored to the work supplied, we have

$$\tfrac{1}{2}pA\Delta = P\Delta \quad \text{or} \quad p = 2P/A \qquad \text{...(5.5)}$$

Thus, the maximum stress set up in the body in this case is twice that in the case of gradual loading.

(c) Impact Loading

Let the load P drop through a height h before it commences to stretch the body as shown in Fig. 5.2.

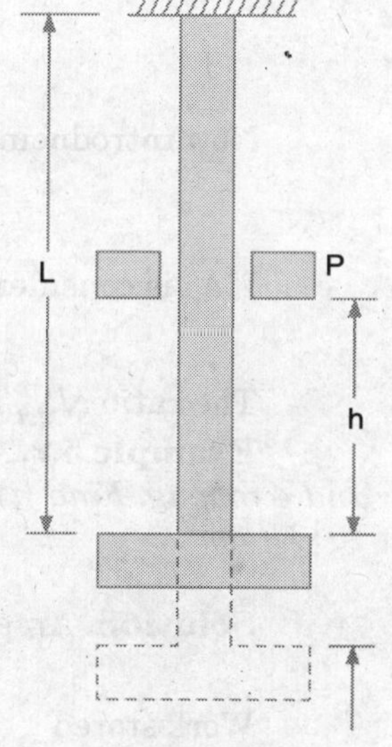

Fig. 5.2

When the weight falls on the collar at the lower end of the vertical bar, a small oscillation is set up and, provided the limit of proportionality is not exceeded, the collar will take up the same final position as in the case of a gradually applied load. However, the maximum instantaneous extension and stress are much greater than the steady value.

Assuming that the collar, the weight and the supports of the bar are infinitely rigid, the whole of the energy in the falling weight is expended in stretching the bar by an amount Δ.

Work done by the falling weight P is

$$= P(h + \Delta) = P\left(h + \frac{pL}{E}\right)$$

Work stored in the bar $= \frac{1}{2}R\Delta = \frac{1}{2}p.A \,.\, \dfrac{pL}{E} = \dfrac{p^2}{2E}AL$

Equating the two, we get

$$P\left(h + \frac{pL}{E}\right) = \frac{p^2}{2E}AL \quad \text{or} \quad \left(\frac{AL}{2E}\right)p^2 - \left(\frac{PL}{E}\right)p - Ph = 0$$

Multiplying throughout by $\frac{E}{AL}$,

$$\frac{p^2}{2} - \left(\frac{P}{A}\right)p - \frac{PEh}{AL} = 0$$

This is a quadratic in p and its solution is

$$p = \frac{P}{A} \pm \sqrt{\left(\frac{P}{A}\right)^2 + \left(4 \times \frac{1}{2}\right)\left(\frac{PEh}{AL}\right)} = \frac{P}{A}\left(1 + \sqrt{1 + \frac{2\,AEh}{PL}}\right) \quad ...(5.6)$$

the positive square root giving the maximum stress.

However, if Δ is negligible compared to h, then we have

$$Ph = \frac{p^2}{2E}\,.\,AL \quad \text{or} \quad p = \sqrt{\frac{2EPh}{AL}} \quad ...[5.6\,(a)]$$

Let us express the above equations in terms of deformation Δ.

Work done $= P(h + \Delta)$ and strain energy absorbed $= \frac{1}{2}P\Delta$

$$\therefore \quad P(h + \Delta) = \tfrac{1}{2}P\Delta \quad \text{But} \quad \Delta = \frac{PL}{AE} \text{ (Hooke's law) or } P = \frac{\Delta EA}{L}$$

$$\therefore \quad P(h + \Delta) = \frac{EA\Delta^2}{2L}$$

Multiplying all the sides by $2L/EA$, we get

$$\Delta^2 - \frac{2PL}{AE}\Delta - \frac{2PLh}{AE} = 0, \text{ solving which we get}$$

$$\Delta = \frac{PL}{AE} + \sqrt{\left(\frac{PL}{AE}\right)^2 + \frac{2PLh}{AE}}$$

Now introducing $\Delta_{st} = \frac{PL}{AE}$ = static deflection of bar due to P, we have

$$\Delta = \Delta_{st} + \sqrt{\Delta_{st}^2 + 2h\,\Delta_{st}} \quad ...(5.7)$$

If Δ_{st} is considered very small in comparison to h, we get

$$\Delta = \sqrt{2h\,\Delta_{st}} \quad ...[5.7\,(a)]$$

The ratio Δ/Δ_{st} is called the *impact factor*.

Example 5.2. *An axial pull of 60 kN is suddenly applied to a steel rod 50 mm in diameter and 4 m long. Find (i) the work done, (ii) maximum instantaneous elongation. Take E = 2.06 × 10⁵ N/mm².*

Solution. Area of section, $A = \frac{\pi}{4}(50)^2 = 1960 \text{ mm}^2$

Work stored $= \frac{1}{2}R\Delta = \frac{1}{2}pA\Delta$; Work done $= P.\Delta$

$$\therefore \quad \tfrac{1}{2}pA\Delta = P.\Delta \quad \text{or} \quad p = \frac{2P}{A} = \frac{2 \times 60000}{1960} = \mathbf{61.2\ N/mm^2}$$

and $$\Delta = \frac{pL}{E} = \frac{61.2 \times 4000}{2.06 \times 10^5} = \mathbf{1.19\ mm}$$

Work done on the rod and stored by it = $P\Delta$

$$= 60000 \times 1.19 = \mathbf{71400\ mm\text{-}N = 71.4\ m\text{-}N = 71.4\ J.}$$

(**Note.** 1 m-N = 1 J).

Example 5.3. *A bar 4 m long and 6 cm diameter hangs vertically and has a collar securely attached at the lower end. Find the maximum stress induced when*

(i) a weight of 300 kg falls 10 cm on the collar.

(ii) a weight of 3000 kg falls 1 cm on the collar.

Take E = 2.05 × 10⁶ kg/cm².

Solution. (*i*) Due to the weight of 300 kg, the extension will be extremely small in comparison to *h*.

Hence $$p = \sqrt{\frac{2EPh}{AL}} \text{ where } A = \frac{\pi}{4}(6)^2 = 28.2 \text{ cm}^2$$

$$\therefore \quad p = \sqrt{\frac{2 \times 2.05 \times 10^6 \times 300 \times 10}{28.2 \times 400}} = \sqrt{1.09 \times 10^6} = \mathbf{1045\ kg/cm^2.}$$

(*ii*) Due to weight of 3000 kg, the extension Δ will not be negligible in comparison to *h*.

$$\therefore \quad p = \frac{P}{A}\left(1 + \sqrt{1 + \frac{2AEh}{PL}}\right) = \frac{3000}{28.2}\left(1 + \sqrt{1 + \frac{2 \times 28.2 \times 2.06 \times 10^6 \times 1.0}{3000 \times 400}}\right)$$

$$= 106.5\,(1 + \sqrt{1 + 97}) = 106.5 \times 10.9 = \mathbf{1160\ kg/cm^2.}$$

Example 5.4. *A weight of 1.2 kN is dropped on a collar attached at the lower end of a vertical bar 4 m long and 30 mm in diameter. Calculate the height of drop if the maximum instantaneous stress is not to exceed 120 N/mm². Also, calculate the instantaneous elongation. Take E = 0.205 × 10⁶ N/mm².*

Solution. $$A = \frac{\pi}{4}(30)^2 = 705 \text{ mm}^2$$

Maximum instantaneous elongation $$= \frac{pL}{E} = \frac{120 \times 4000}{0.205 \times 10^6} = \mathbf{2.34\ mm}$$

Work supplied by the falling weight = $p(h + \Delta) = 1200\,(h + 2.34)$

Work stored by the bar $$= \frac{p^2}{2E} \cdot AL = \frac{120 \times 120}{2 \times 0.205 \times 10^6} \times 705 \times 4000 = 9.90 \times 10^4 \text{ mm-N}$$

Equating the work supplied to work stored

$$1200\,(h + 2.34) = 9.90 \times 10^4 \quad \text{or} \quad h + 2.34 = 82.5$$

$$\therefore \quad h = \mathbf{80.16\ mm.}$$

Example 5.5. *A steel rope lowers a load of 15 kN at a rate of 1 m/sec. When the length of rope unwound is 10 m, it suddenly gets jammed. Estimate the instantaneous stress induced in it due to the sudden stoppage and the maximum instantaneous elongation if the diameter of the rope is 30 mm. Take E = 2.05 × 10⁵ N/mm² and g = 981 m/sec².*

Solution. $$A = \frac{\pi}{4}(30)^2 = 706.8 \text{ mm}^2$$

K.E. of load $= \frac{WV^2}{2g} = \frac{15 \times 1000 \times (100)^2}{2 \times 981} = 7.645 \times 10^4$ cm-N $= 7.645 \times 10^5$ mm-N

This is to be absorbed by the rope at a stress p.

$$\therefore \quad U = \frac{p^2}{2E} \cdot A \cdot L = p^2 \frac{706.8 \times 10 \times 1000}{2 \times 2.05 \times 10^5} = 17.24\, p^2 \text{ mm-N}$$

Equating the K.E. to the energy to be absorbed,

$$17.24\, p^2 = 7.645 \times 10^5 \quad \text{or} \quad p^2 = \frac{7.645 \times 10^5}{17.24} = 4.43 \times 10^4$$

$$\therefore \quad p = \mathbf{210.6\ N/mm^2}$$

Instantaneous maximum elongation

$$\Delta = \frac{pL}{E} = \frac{210.6 \times 10000}{2.05 \times 10^5} = \mathbf{10.3\ mm.}$$

Example 5.6. *Two bars are subjected to equal gradually applied tensile loads. One bar is of 4 cm diameter throughout and the other, which has the same length, is turned down to diameter of 2 cm over the middle third of its length, the remainder having a diameter of 4 cm. Compare the strain energies of the two bars assuming that they are of the same material.*

Compare also the amount of energy which the two bars can absorb in simple tension without exceeding a given stress within the limits of proportionality.

Also compare the amount of energy per unit volume if the maximum stress is to be the same in both the cases.

Solution. The two bars have been shown in Fig. 5.3. Let suffix 1 stand for uniform bar and 2 for the other.

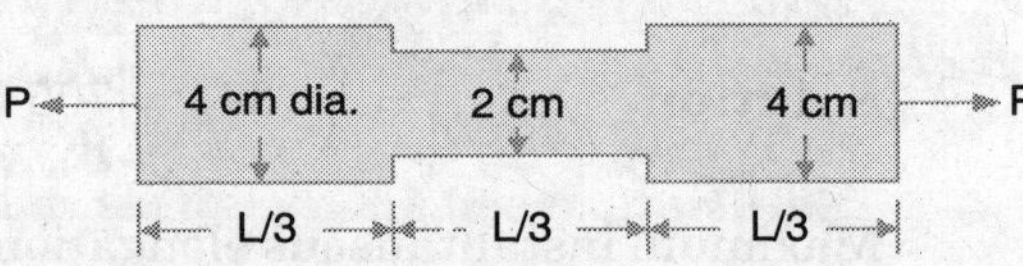

Fig. 5.3

For the uniform bar, the strain energy stored is

$$U_1 = \frac{f^2}{2E} \times V = \left(\frac{P}{\frac{\pi}{4} \times 16} \right)^2 \frac{1}{2E} \left(\frac{\pi}{4} 16 \right) L$$

$$= \frac{LP^2}{8\pi E} \qquad \text{...(1)}$$

For the non-uniform bar, the strain energy stored is

$$U_2 = \left\{ \left(\frac{P}{\frac{\pi}{4} \times 16} \right)^2 \frac{1}{2E} \left(\frac{\pi}{4} \times 16 \right) \frac{2}{3} L \right\} + \left\{ \left(\frac{P}{\frac{\pi}{4} \times 4} \right)^2 \frac{1}{2E} \left(\frac{\pi}{4} \times 4 \right) \frac{L}{3} \right\}$$

$$= \frac{LP^2}{8\pi E} \cdot \frac{2}{3} + \frac{LP^2}{2\pi E} \cdot \frac{1}{3} = \frac{LP^2}{12\pi E} + \frac{LP^2}{6\pi E} = \frac{LP^2}{4\pi E} \qquad \text{...(2)}$$

$$\because \quad \frac{U_1}{U_2} = \frac{LP^2}{8\pi E} \times \frac{4\pi E}{LP^2} = \mathbf{\frac{1}{2}}$$

Suppose P is the load applied to the non-uniform bar for a given stress. The maximum given stress occurs in the 2 cm diameter portion. If the same stress is to be attained in the uniform bar, the load required will be $(4/2)^2 P = 4P$, since the load is proportional to the square of the diameter. Hence replacing P by $4P$ in (1), we get

$$U_1 = \frac{L}{8\pi E}(4P)^2 = \frac{2LP^2}{\pi E} \qquad ...(3)$$

Also from (2), $U_2 = \dfrac{LP^2}{4\pi E}$

$$\therefore \quad \frac{U_1}{U_2} = \frac{2LP^2}{\pi E} \times \frac{4\pi E}{LP^2} = \mathbf{8}$$

Again, $V_1 = \dfrac{\pi}{4}\,16L = 4\pi L$ and $V_2 = \left(\dfrac{\pi}{4} \times 16\,\dfrac{2}{3}\,L\right) + \left(\dfrac{\pi}{4} \times 4 \times \dfrac{L}{3}\right) = \dfrac{8}{3}\,\pi L + \dfrac{\pi}{3}\,L = 3\pi L$

Strain energy per unit volume of uniform bar

$$= U_1 = \frac{2LP^2}{\pi E} \times \frac{1}{4\pi L} = \frac{P^2}{2\pi^2 E}$$

Strain energy per unit volume of non-uniform bar at the same stress

$$= U_2 = \frac{LP^2}{4\pi E} \times \frac{1}{3\pi L} = \frac{P^2}{12\pi^2 E}$$

$$\therefore \quad \frac{U_1}{U_2} = \frac{P^2}{2\pi^2 E} \times \frac{12\pi^2 E}{P^2} = \mathbf{6.}$$

Example 5.7. *Two bars A and B are each 20 cm long and are of the same material. Bar A is 2 cm in diameter for a length of 5 cm and 4 cm in diameter for the remaining 15 cm length. Bar B is 2 cm in diameter for a length of 15 cm and 4 cm in diameter for the remaining 5 cm length. An axial blow given to A produces a maximum instantaneous stress of 1200 kg/cm². Calculate the maximum instantaneous stress produced by the same blow on B.*

If each bar is stressed upto the elastic limit, calculate the ratio of energy stored by A and B at proof stress.

Also, calculate the ratio of strain energy per unit volume of each bar at the same stress.

Solution. When f is the stress in the 2 cm portion, the corresponding stress in the 4 cm diameter portion will be $f \times \dfrac{4}{16} = \dfrac{f}{4}$.

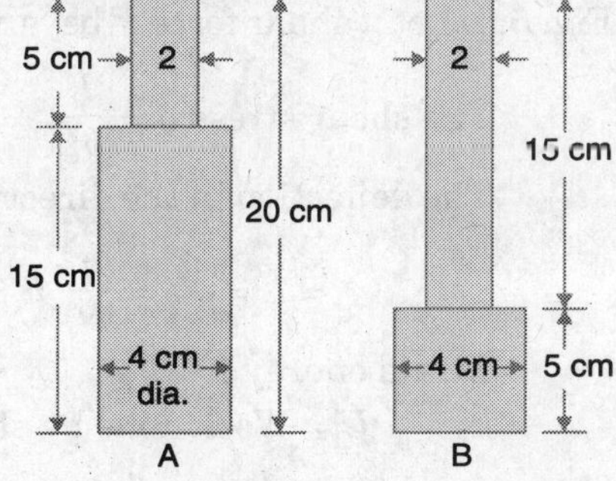

Fig. 5.4

The strain energy stored in bar A is given by

$$U_A = \frac{f_A^2}{2E}\left(\frac{\pi}{4} \times 4 \times 5\right) + \frac{(\frac{1}{4}f_A)^2}{2E}\left(\frac{\pi}{4} \times 16 \times 15\right)$$

$$= \frac{f_A^2}{2E}\,5\pi + \frac{f_A^2}{32E}\,60\pi = \frac{f_A^2\pi}{32E}(80 + 60)$$

$$= \frac{140\, f_A^2\pi}{32E} \qquad ...(1)$$

Similarly,

$$U_B = \frac{f_B^2}{2E}\left(\frac{\pi}{4} \times 4 \times 15\right) + \frac{(\frac{1}{4}f_B)^2}{2E}\left(\frac{\pi}{4} \times 16 \times 5\right)$$

$$= \frac{15\, f_B^2\pi}{2E} + \frac{f_B^2 5\pi}{8E} = \frac{f_B^2\pi}{8E}(60 + 5) = \frac{65}{8}\,\frac{f_B^2\pi}{E}$$

When an axial blow is delivered, the amount of energy stored will be the same.

$$\therefore \qquad U_A = U_B \quad \text{or} \quad \frac{65}{8} f_B^{\,2} \frac{\pi}{E} = \frac{140\, f_A^{\,2}\pi}{32E}$$

or $$f_B^{\,2} = f_A^{\,2}\, \frac{140}{32} \times \frac{8}{65} = (1200)^2 \times 0.538$$

or $$f_B = 1200 \times 0.734 = \mathbf{880.6\ kg/cm^2.}$$

If both the bars are allowed to reach the proof stress f, we have

$$f_A = f_B = f$$

$$\therefore \qquad U_A = \frac{140}{32} \times \frac{f^2\pi}{E} \text{ and } U_B = \frac{65}{8} \times \frac{f^2\pi}{E}$$

$$\frac{U_B}{U_A} = \frac{65}{8} \times \frac{32}{140} = \mathbf{1.855}$$

Also, $$V_A = \left(\frac{\pi}{4} \times 4 \times 5\right) + \left(\frac{\pi}{4} \times 16 \times 15\right) = 5\pi + 60\pi = 65\pi$$

$$V_B = \left(\frac{\pi}{4} \times 4 \times 15\right) + \left(\frac{\pi}{4} \times 16 \times 5\right) = 15\pi + 20\pi = 35\pi$$

$$U_B \text{ per unit volume} = \frac{65}{8} \frac{f^2\pi}{E} \times \frac{1}{35\pi} = \frac{65}{35 \times 8} \frac{f^2}{E}$$

$$U_A \text{ per unit volume} = \frac{140}{32} \cdot \frac{f^2\pi}{E} \times \frac{1}{65\pi} = \frac{140}{32 \times 65} \frac{f^2}{E}$$

$$\frac{U_B}{U_A} = \frac{65}{35 \times 8} \times \frac{32 \times 65}{140} = \mathbf{3.45.}$$

5.5. SHEAR RESILIENCE

Consider a block of length L, width b and height h, rigidly fixed at the base as shown in Fig. 5.5. Let a shear force F be applied gradually along the top surface as shown.

The shear stress $q = \dfrac{F}{bL}$

The deflection in the direction of force

$$= \Delta = \phi h = \frac{q}{N}\, h$$

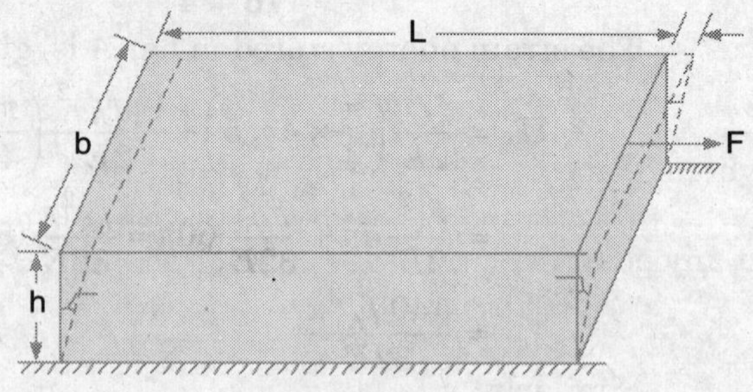

Fig. 5.5

Strain energy

U = Work done by shear force

$$= \frac{1}{2} F\Delta = \frac{1}{2} (qbL)\left(\frac{q}{N}\, h\right)$$

$$= \frac{1}{2} \frac{q^2}{N} (bLh) = \frac{q^2}{2N} \times V \qquad \text{...(5.8)}$$

$\therefore$ Shear strain energy per unit volume $= \dfrac{q^2}{2N}$

If q_p = proof shear stress, the strain energy per unit volume is equal to $\dfrac{q_p^{\,2}}{2N}$ and is known as *modulus of shear resilience*.

Example 5.8. *Calculate the local strain energy at a point in a material where shear stress is 80 N/mm². Take N = 0.84 × 10⁵ N/mm².*

Solution. Local strain energy per unit volume

$$= \frac{q^2}{2N} = \frac{80 \times 80}{2 \times 0.84 \times 10^5} = \mathbf{0.038\ mm\text{-}N/mm^3 = 3.8 \times 10^4\ m\text{-}N/m^3.}$$

5.6. STRAIN ENERGY IN TERMS OF PRINCIPAL STRESSES

(*a*) Let σ_1 and σ_2 be the principal stresses at any point in a strained material. The corresponding principal strains are given by

$$e_1 = \frac{\sigma_1}{E} - \frac{\sigma_2}{mE} \quad \text{and} \quad e_2 = \frac{\sigma_2}{E} - \frac{\sigma_1}{mE}$$

The strain energy per unit volume due to σ_1 will be

$$U_1 = \frac{1}{2} \times \text{stress} \times \text{strain} = \frac{1}{2}\sigma_1 e_1 \qquad \text{...(Eq. 5.2)}$$

$$= \frac{1}{2}\sigma_1\left(\frac{\sigma_1}{E} - \frac{\sigma_2}{mE}\right) = \frac{\sigma_1^{\,2}}{2E} - \frac{\sigma_1\sigma_2}{2mE}$$

Similarly, the strain energy per unit volume due to σ_2 will be

$$U_2 = \frac{1}{2}\sigma_2 e_2 = \frac{1}{2}\sigma_2\left(\frac{\sigma_2}{E} - \frac{\sigma_1}{mE}\right) = \frac{\sigma_2^{\,2}}{2E} - \frac{\sigma_1\sigma_2}{2mE}$$

Hence, the total strain energy per unit volume will be

$$U = U_1 + U_2 = \frac{\sigma_1^{\,2}}{2E} + \frac{\sigma_2^{\,2}}{2E} - \frac{\sigma_1\sigma_2}{mE}$$

$$= \frac{1}{2E}\left(\sigma_1^{\,2} + \sigma_2^{\,2} - \frac{2\sigma_1\sigma_2}{m}\right) \qquad \text{...(5.9)}$$

where σ_1 and σ_2 are of the same sign.

(*b*) Let us now consider a more general case in which the principal stresses are σ_1, σ_2 and σ_3 at any point in a strained material. The three principal strains will be

$$e_1 = \frac{\sigma_1}{E} - \frac{\sigma_2 + \sigma_3}{mE}\ ;\ e_2 = \frac{\sigma_2}{E} - \frac{\sigma_3 + \sigma_1}{mE}\ ;\ \sigma_3 = \frac{\sigma_3}{E} - \frac{\sigma_1 + \sigma_2}{mE}$$

Total strain energy per unit volume is

$$U = \frac{1}{2}\sigma_1 e_1 + \frac{1}{2}\sigma_2 e_2 + \frac{1}{2}\sigma_3 e_3$$

$$= \frac{\sigma_1}{2}\left(\frac{\sigma_1}{E} - \frac{\sigma_2 + \sigma_3}{mE}\right) + \frac{\sigma_2}{2}\left(\frac{\sigma_2}{E} - \frac{\sigma_3 + \sigma_1}{mE}\right) + \frac{\sigma_3}{2}\left(\frac{\sigma_3}{E} - \frac{\sigma_1 + \sigma_2}{mE}\right)$$

$$= \frac{1}{2E}\left(\sigma_1{}^2 + \sigma_2{}^2 + \sigma_3{}^2 - \frac{2\sigma_1\sigma_2 + 2\sigma_2\sigma_3 + 2\sigma_3\sigma_1}{m}\right) \quad ...(5.10)$$

where σ_1, σ_2 and σ_3 are of the same sign.

Example 5.9. *Find the strain energy stored in a steel bar 250 mm long and of cross-section 25 mm × 6 mm, when it is subjected simultaneously to an axial pull of 20 kN and a compressive stress of 80 N/mm² on its narrow edge. For steel, E = 2.04 × 10⁵ N/mm² and 1/m = 0.28.*

Solution. Total strain energy will be $U = \frac{1}{2E}\left[\sigma_1{}^2 + \sigma_2{}^2 - \frac{2\sigma_1\sigma_2}{m}\right] \times V$

$$\sigma_1 \text{ (corresponding to 20 kN load)} = \frac{20000}{25 \times 6} = 133.3 \text{ N/mm}^2 \text{ (tensile)}$$

$\sigma_2 = -\ 80$ N/mm² (*i.e.*, compressive)

$$\therefore \quad U = \frac{1}{2 \times 2.05 \times 10^5}[(133.3)^2 + (80)^2 + (2 \times 133.3 \times 80 \times 0.28)] \times (250 \times 25 \times 6)$$

= 2758 mm-N = **2.758 m-N = 2.758 J.**

5.7. RELATION BETWEEN ELASTIC MODULI FROM STRAIN ENERGY CONSIDERATION

In chapter 2, we have already derived the relation between the elastic *moduli*. We shall be deriving the same results from consideration of strain energy.

(*a*) **Relation between E and N**

From § 5.5, the strain energy per unit volume in the case of a state of simple shear = $q^2/2N$.

The state of simple shear is equivalent to two principal stresses σ_1 and σ_2 across the diagonal planes, such that $\sigma_1 = +\,q$ and $\sigma_2 = -\,q$.

The strain energy per unit volume, due to principal stresses is

$$U = \frac{1}{2E}\left(\sigma_1{}^2 + \sigma_2{}^2 - \frac{2\sigma_1\sigma_2}{m}\right) \quad ...(\text{Eq. } 5.9)$$

Substituting $\sigma_1 = +\,q$ and $\sigma_2 = -\,q$

We get $$U = \frac{1}{2E}\left(q^2 + q^2 + \frac{2q^2}{m}\right) = \frac{q^2}{E}\left(1 + \frac{1}{m}\right)$$

Equating the strain energy due to simple shear to that due to principle stresses,

$$\frac{q^2}{2N} = \frac{q^2}{E}\left(1 + \frac{1}{m}\right) \quad \text{or} \quad E = 2N\left(1 + \frac{1}{m}\right) \quad ...(1)$$

(*b*) **Relation between E and K**

Let a material be subjected to three equal principal stresses $\sigma_1 = \sigma_2 = \sigma_3 = \sigma$, acting on three mutually perpendicular planes.

The strain energy due to three principal stresses is given by

$$U = \frac{1}{2E}\left\{\sigma_1^2 + \sigma_2^2 + \sigma_3^2 - \frac{2(\sigma_1\sigma_2 + \sigma_2\sigma_3 + \sigma_3\sigma_1)}{m}\right\} \times V = \frac{3\sigma^2}{2E}\left(1 - \frac{2}{m}\right)V$$

Now $$K = \frac{\sigma}{e_v} = \frac{\sigma V}{\delta V} \quad \text{or} \quad \delta V = \frac{\sigma}{K}V$$

If δV is the change in volume due to uniform intensity of pressure σ applied gradually, the work done

$$= \frac{1}{2}\sigma\,\delta V = \frac{1}{2}\sigma\frac{\sigma}{K}V = \frac{\sigma^2}{2K}V$$

Equating the two strain energies,

$$\frac{3\sigma^2}{2E}\left(1 - \frac{2}{m}\right)V = \frac{\sigma^2}{2K}V \quad \text{or} \quad E = 3K\left(1 - \frac{2}{m}\right) \qquad ...(2)$$

Hence, we have $E = 2N\left(1 + \frac{1}{m}\right) = 3K\left(1 - \frac{2}{m}\right)$, as before.

5.8. IMPACT LOADING ON BEAMS

The maximum instantaneous deflection of a beam struck by a falling weight W can be found if it is assumed that the work done by the falling weight or the kinetic energy of the moving body can be stored in the resisting member (*i.e.*, beam) in the form of strain energy. It is further assumed that the mass of the beam may be neglected in comparison with the mass of the falling weight and that the beam is not stressed beyond elastic limit.

Fig. 5.6 (*a*) shows a cantilever, subjected to an impact loading by a falling weight W. The variation of end deflection of the beam with respect to time is shown in Fig. 5.6 (*b*), in which δ is the maximum deflection resulting from the falling weight transferring part of its kinetic energy to the beam while δ_s is the static deflection resulting when the beam and weight finally reach equilibrium. Fig. 5.6 (*c*) shows the cantilever loaded with an *equivalent static*

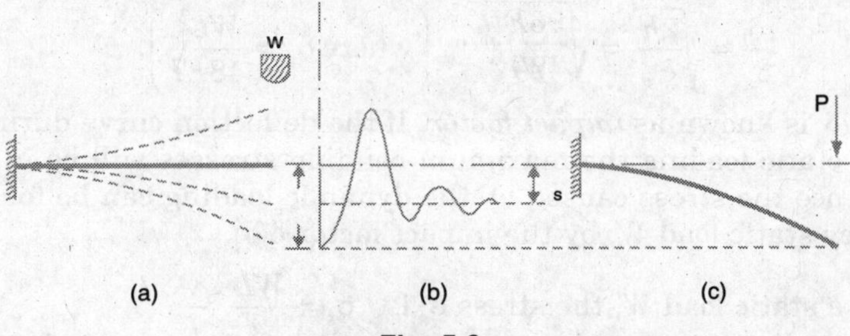

Fig. 5.6

load P causing the same maximum deflection δ. If it is assumed that the material behaves identically under the two loadings, the dynamic load W can be replaced by an equivalent static load P causing the same maximum deflection and same flexural stress.

Let us take the case of a simply supported beam of span L, subjected to the impact loading of a weight W falling through a height h. Let δ be the maximum instantaneous deflection and P be the equivalent static load causing the same maximum deflection δ.

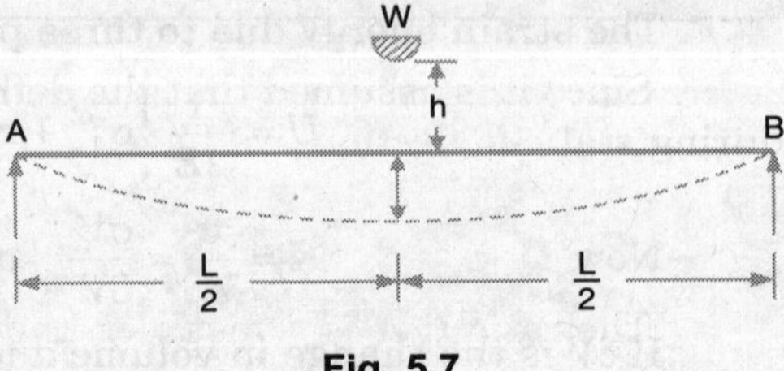

Fig. 5.7

For a simply supported beam with point load at mid-span,

$$\delta = \frac{PL^3}{48EI} \quad \text{or} \quad P = \delta \frac{48EI}{L^3} \qquad ...(i)$$

The total strain energy stored in the beam is equal to the work done by the force P

$$U = \frac{1}{2} P\delta = \delta^2 \frac{24EI}{L^3} \qquad ...(ii)$$

Work supplied by the falling weight = $W(h + \delta)$

$$\therefore \quad W(h + \delta) = \delta^2 \frac{24EI}{L^3} \quad \text{or} \quad \delta^2 - \frac{WL^3}{24EI}\delta - \frac{WL^3}{24EI}h = 0$$

Putting $\dfrac{WL^3}{48EI} = \delta_s$, we get $\delta^2 - 2\delta.\delta_s - 2h\delta_s = 0$

From which $\quad \delta = \delta_s + \sqrt{\delta_s^{\,2} + 2h\delta_s} \qquad ...(5.11)$

Putting $h = \dfrac{v^2}{2g}$, where v is the velocity of falling body, Eq. 5.11 is sometime expressed in the following form :

$$\delta = \delta_s + \sqrt{\delta_s^{\,2} + \frac{\delta_s}{g}v^2} \qquad ...[5.11\ (a)]$$

After having found δ, P can be found from (*i*).

If h is large in comparison to δ_s, Eq. 5.11 reduces to

$$\delta = \sqrt{2h\delta_s} = \sqrt{\frac{\delta_s}{g}.v^2} \qquad ...(5.12)$$

$$\therefore \quad \frac{\delta}{\delta_s} = \sqrt{\frac{2h}{\delta_s}} = \sqrt{\frac{96EIh}{WL^3}} \quad \left(\text{where } \delta_s = \frac{WL^3}{48EI}\right) \qquad ...(5.13)$$

The factor δ/δ_s is known as *impact factor*. If the deflection curve during impact is of the same shape as for static loading the maximum bending stresses will be in the same ratio as the deflections. Hence the stress caused by the dynamic loading can be found by multiplying the stress caused by static load W, by the impact factor δ/δ_s.

Thus under a static load W, the stress σ_s is $\quad \sigma_s = \dfrac{WL}{4}.\dfrac{y}{I}$

If σ is the stress due to impact loading,

$$\sigma = \frac{\delta}{\delta_s}.\sigma_s = \sqrt{\frac{6WhEy^2}{LI}} \qquad ...(5.14)$$

where y is the distance of the extreme fibre from the N.A.

Since it is assumed that the deflection curve during impact has the same shape as that during static deflection, we have

$$\frac{P}{W} = \frac{\delta}{\delta_s} \qquad \text{...(5.15)}$$

The ratio P/W is sometimes known as *load factor* and is equal to the impact factor.

Example 5.10. *A rectangular beam 200 mm × 100 mm is freely supported over a span of 2 m. A load of kN is dropped on to the middle of the beam from a height of 40 mm. Find (i) maximum instantaneous deflection, (ii) impact factor, and (iii) the stress induced in the beam. Take E = 1.1 × 10⁴ N/mm².*

Solution. $I = \frac{1}{12}bd^3 = \frac{1}{12} \times 100(200)^3 = 66.67 \times 10^5 \text{ mm}^4$

$$\delta_s = \frac{WL^3}{48EI} = \frac{10000(2000)^3}{48 \times 1.1 \times 10^4 \times 66.67 \times 10^6} = 2.27 \text{ mm}$$

$h = 40$ mm

Substituting in Eq. 5.11,

$$\delta = \delta_s + \sqrt{\delta_s^2 + 2h\delta_s} = 2.27 + \sqrt{(2.27)^2 + (2 \times 40 \times 2.27)} = \mathbf{15.77 \text{ mm}}$$

Impact factor $= \frac{\delta}{\delta_s} = \frac{15.77}{2.27} = \mathbf{6.93.}$ But $\frac{P}{W} = \frac{\delta}{\delta_s} = 6.93$

$\therefore \qquad P = 6.93\ W = 6.93 \times 10{,}000 = 69{,}300$ N

$$M_{max} = \frac{PL}{4} = \frac{69300 \times 2000}{4}$$

$$\therefore \qquad \sigma = \frac{M_{max}}{I}y = \frac{69300 \times 2000}{4 \times 66.67 \times 10^6} \times 100 = \mathbf{52 \text{ N/mm}^2}.$$

Example 5.11. *A simply supported aluminium alloy beam 4 cm wide and 3 cm deep is supported on a span of 4 m. A helical spring of modulus of 20 kg/cm supports the beam at its centre, as shown in Fig. 5.8 (a). The spring is initially unstressed and is in contact with the beam. When a 20 kg block drops 4 cm into the top of the beam, 90 percent of the energy is absorbed elastically by the system consisting of the beam A and the spring B. Determine (i) the maximum flexural stress in the beam for this loading, (ii) the load factor, (iii) the proportion of the effective energy absorbed by each of the components A and B. Take E = 0.8 × 10⁶ kg/cm².*

Solution. For the beam, $I = \frac{1}{12}(4)(3)^3 = 9 \text{ cm}^4$

Let $\delta_A = \delta_B = \delta$ be the instantaneous deflection of the beam and the spring [Fig. 5.8 (*b*)].

P = Equivalent static load acting on the beam in the downward direction.

Q = Equivalent static load on the spring causing the deflection δ_B.

$\therefore$ Net equivalent load on the beam = $(P - Q)$

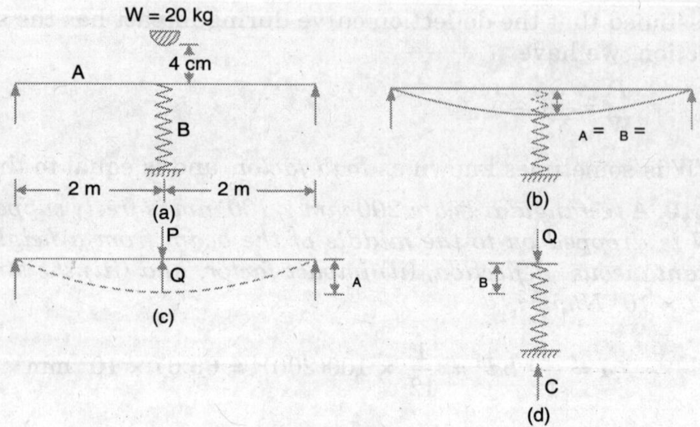

Fig. 5.8

$$\therefore \qquad \delta_A = \frac{(P-Q)L^3}{48EI} = (P-Q)\frac{(400)^3}{48 \times 0.8 \times 10^6 \times 9} = \frac{P-Q}{5.4}\ ;\ \delta_B = \frac{Q}{20}$$

Since $\delta_A = \delta_B$, we have $\frac{P-Q}{5.4} = \frac{Q}{20}$ From which $Q = 0.787\ P$ ...(1)

And $\delta = Q/20 = 0.03935\ P$ ...(2)

Again, applying the energy equation,

$$20(4+\delta) \times 0.9 = \frac{1}{2} P\delta \quad \text{or} \quad 20(4 + 0.03935P) \times 0.9 = \frac{1}{2} P^2(0.03935)$$

or $P^2 - 36P - 3660 = 0$, from which $P = 81.12$ kg

$\therefore$ $Q = 0.787 \times 81.12 = 63.84$

$$\therefore \quad \text{Load factor} = \frac{P}{W} = \frac{81.12}{20} = \mathbf{4.06}$$

Now, for the beam, reaction at each end $= \frac{1}{2}(P-Q) = \frac{1}{2}(81.12 - 63.84) = 8.64$ kg

$\therefore$ $M_{max} = 8.64 \times 200 = 1728$ kg-cm

$$\therefore \quad \sigma = \frac{M}{I} y = \frac{1728}{9} \times 1.5 = \mathbf{288\ kg/cm^2}$$

Energy absorbed by the beam $= \frac{1}{2}(P-Q)\ \delta = 8.64$

Energy absorbed by the spring $= \frac{1}{2}Q\delta = 31.92\ \delta$

Total energy absorbed by the system $= \frac{1}{2}P\delta = 40.56$

$\therefore$ Proportions absorbed by each component are :

$$\frac{8.64}{40.56} \times 100 \simeq \mathbf{21.3\%} \text{ for } A \text{ and } \frac{31.928}{40.56} \times 100 \simeq \mathbf{78.7\%} \text{ for } B.$$

5.9. STRAIN ENERGY OF UNIFORMLY TAPERING BAR

Consider an elementary strip of length dx at distance x from end B.

Diameter at $X = d_x = d_1 + \frac{d_2 - d_1}{L} x$

Area of cross-section = $\dfrac{\pi}{4}\left(d_1 + \dfrac{d_2 - d_1}{L}x\right)^2$

Force = P ; length of section = dx

$$\therefore \quad U = \int_0^L \frac{P^2 dx}{2A_x E}$$

or

$$U = \int_0^L \frac{P^2 dx}{2\dfrac{\pi}{4}\left\{d_1 + \dfrac{d_2 - d_1}{L}x\right\}^2 E}$$

$$= \int_0^L \frac{2P^2 dx}{\pi(d_1 + kx)^2 E}, \text{ where } k = \frac{d_2 - d_1}{L}.$$

$$\therefore \quad U = -\frac{2P^2}{\pi E}\cdot\frac{1}{k}\left[\frac{1}{d_1 + kx}\right]_0^L = -\frac{2P^2 L}{\pi E\,(d_2 - d_1)}\left[\frac{1}{d_1 + d_2 - d_1} - \frac{1}{d_1}\right]$$

or

$$U = \frac{2P^2 L}{\pi E\,(d_2 - d_1)}\left[\frac{1}{d_1} - \frac{1}{d_2}\right] = \frac{2P^2 L}{\pi E\, d_1 d_2} \qquad ...(5.16)$$

Fig. 5.9

5.10. STRAIN ENERGY OF UNIFORMLY TAPERING PLATE OF CONSTANT THICKNESS

Let the constant thickness of the plate be t.

Consider an elementary section dx at a distance x from B_1 (Fig. 5.10).

Width $\quad B_x = B_1 + \dfrac{B_2 - B_1}{L}x = B_1 + kx$, where $k = \dfrac{B_2 - B_1}{L}$

Area $\quad A_x = (B_1 + kx)\,t$

$$\therefore \quad U_x = \frac{P^2 (dx)}{2A_x E} = \frac{P^2 dx}{2(B_1 + kx)\,tE}$$

Total

$$U = \int_0^L U_x = \int_0^L \frac{P^2 dx}{2(B_1 + kx)\,tE}$$

$$= \frac{P^2}{2tE}\int_0^L \frac{dx}{B_1 + kx}$$

$$= \frac{P^2}{2tEk}\Big[\log_e (B_1 + kx)\Big]_0^L = \frac{P^2}{2ktE}\log_e \frac{B_1 + kL}{B_1}$$

or

$$U = \frac{P^2}{2ktE}\log_e \frac{B_2}{B_1} = \frac{P^2 L}{2tE\,(B_2 - B_1)}\log_e \frac{B_2}{B_1} \qquad ...(5.17)$$

Fig. 5.10

5.11. ADDITIONAL EXAMPLES USEFUL FOR COMPETITIVE EXAMINATIONS

Example 5.12. *An unknown weight falls through 8 mm on a collar rigidly attached to the lower end of a vertical bar, 4 m long and 800 mm² in section. If the maximum instantaneous extension is known to be 3 mm, what is the corresponding stress and the value of unknown weight ? Take E = 2 × 10⁵ N/mm².*

Solution. Let p be the stress.

Now, $$\Delta = \frac{pL}{E} \quad \therefore \quad p = \frac{\Delta E}{L} = \frac{3 \times 2 \times 10^5}{(4 \times 1000)} = \mathbf{150\ N/mm^2}$$

Let P be the value of unknown weight. From Eq. 5.6,

$$p = \frac{P}{A}\left[1 + \sqrt{1 + \frac{2AEh}{PL}}\right]$$

$$\therefore \quad 150 = \frac{P}{800}\left[1 + \sqrt{1 + \frac{2 \times 800 \times 2 \times 10^5 \times 8}{P \times 4000}}\right] \quad \text{or} \quad \frac{120000}{P} - 1 = \sqrt{1 + \frac{64 \times 10^4}{P}}$$

or $$\frac{144 \times 10^8}{P^2} + 1 - \frac{240000}{P} = 1 + \frac{64 \times 10^4}{P} \quad \text{or} \quad \frac{144 \times 10^8}{P} = 880000$$

From which P = **16363.6 N.**

Example 5.13. *Compare the amounts of strain energy in three circular bars shown in Fig. 5.11, assuming a uniform distribution of stresses over cross-section of the bars. Comment on the results, taking a large value of n.*

Solution. For the prismatic bar [Fig. 5.11 (*a*)] strain energy is

$$U = \frac{P^2 L}{2AE} \qquad ...(1)$$

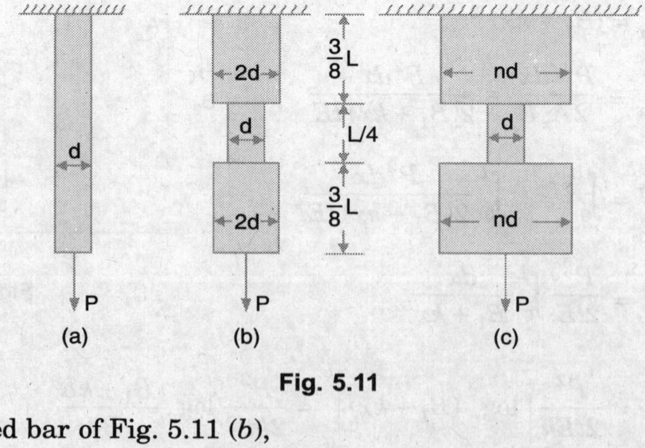

Fig. 5.11

For the grooved bar of Fig. 5.11 (*b*),

$$U_1 = \frac{P^2(L/4)}{2AE} + \frac{P^2(3L/4)}{2(4A)E} = \frac{7}{16}\frac{P^2 L}{2AE} \qquad ...(2)$$

For the grooved bar of Fig. 5.9 (*c*), area of grooved portion = A

Area of other portion = $\dfrac{\pi}{4}\,(nd)^2 = n^2 A$

$$\therefore \quad U_2 = \frac{P^2(L/4)}{2AE} + \frac{P^2(3L/4)}{2(n^2A)E} = \frac{P^2L}{2AE}\left(\frac{1}{4} + \frac{3}{4n^2}\right) = \frac{1}{4} \cdot \frac{P^2L}{2AE}\left(1 + \frac{3}{n^2}\right) \quad ...(3)$$

Hence $\frac{U_1}{U} = \frac{7}{16}$ ...(*a*) and $\frac{U_2}{U} = \frac{1}{4}\left(1 + \frac{3}{n^2}\right)$ ...(*b*)

From (*a*) and (*b*), it is seen that for a given maximum stress, the quantity of strain energy stored in a grooved bar is less than that of a bar of uniform thickness. Taking $n = 4$, for a bar having large outer diameter and a very narrow groove, we have

$$\frac{U_2}{U} = \frac{1}{4}\left(1 + \frac{3}{16}\right) = \frac{19}{64} = \frac{1}{3.368}$$

Hence, for such a case, it takes a very small amount of work to bring the tensile stress to a *dangerous limit.*

Example 5.14. *A bar ABC, 3 m long, has lower half portion BC of diameter d = 20 mm and upper half portion AB of diameter 2d = 40 mm. A sliding collar, having a mass of 25 kg, falls from a height h = 60 mm. Compute the maximum deflection Δ and the maximum tensile stress p caused by the impact. Take E = 2 × 10⁵ N/mm².*

Compare the above results with the case if the entire bar ABC were of uniform diameter d = 20 mm.

Solution.

(*a*) ***Non uniform bar.*** $A_{BC} = \frac{\pi}{4}(d)^2 = A(\text{say})$; $A_{AB} = \frac{\pi}{4}(2d)^2 = 4A$

If a static load P_1 acts on the free end of the bar, the elongation is

$$\Delta_1 = \frac{P_1(L/2)}{AE} + \frac{P_1(L/2)}{(4A)E} = \frac{5P_1L}{8AE}$$

$\therefore$ Equivalent stiffness

$$k = \frac{P_1}{\Delta_1} = \frac{8AE}{5L} = \frac{8\left(\frac{\pi}{4} \times 400\right) \times 2 \times 10^5}{5 \times 3000} = 33510 \text{ N/mm}$$

Now static deflection $\Delta_{st} = \frac{W}{k} = \frac{Mg}{k} = \frac{25 \times 9.81}{33510} = 0.00732$ mm

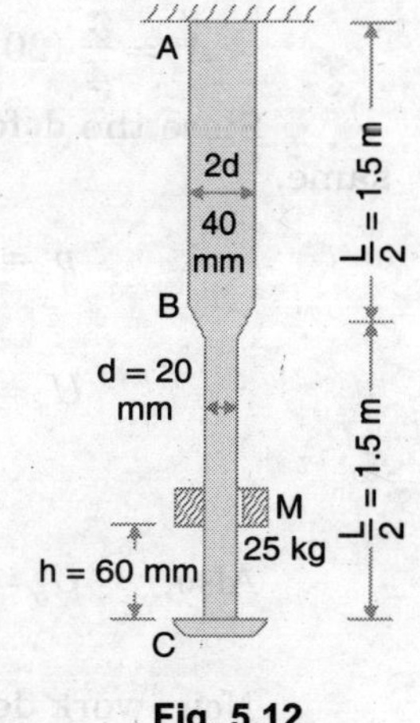

Fig. 5.12

The dynamic deflection is given by Eq. 5.7

$$\Delta = \Delta_{st} + \sqrt{(\Delta_{st})^2 + 2h\,\Delta_{st}}$$

or

$$\Delta = 0.00732 + \sqrt{(0.00732)^2 + 2 \times 60 \times 0.00732}$$

$$= 0.00732 + 0.93726 \approx \mathbf{0.9446 \text{ mm}}$$

Impact factor $= \frac{\Delta}{\Delta_{st}} = \frac{0.9446}{0.00732} = \mathbf{129}$

The static load that will cause the same deflection is

$$P = k\Delta = 33510 \times 0.9446 = 31654 \text{ N}$$

$$\therefore \quad p_{max} = \frac{P}{A} = \frac{31654}{\frac{\pi}{4}(400)} = \mathbf{100.76 \text{ N/mm}^2}$$

This occurs in the lower half.

(*b*) ***Uniform bar.*** If the bar were uniform, of diameter d = 20 mm,

$A = \frac{\pi}{4}$ (400) = 314.16 mm^2.

$$\Delta_{st} = \frac{WL}{AE} = \frac{MgL}{AE} = \frac{25 \times 9.81 \times 3000}{314.16 \times 2 \times 10^5} = 0.0117 \text{ mm}$$

$$\Delta = \Delta_{st} + \sqrt{(\Delta_{st})^2 + 2h\,\Delta_{st}} = 0.0117 + \sqrt{(0.0117)^2 + 2 \times 60 \times 0.0117}$$

$$= 0.0117 + 1.1855 = \mathbf{1.1967\ mm}$$

which is more than that for non-prismatic bar.

Alternatively, since Δ_{st} is very small in comparison to h,

$$\Delta \approx \sqrt{2h\,\Delta_{st}} = \sqrt{2 \times 60 \times 0.0117} = 1.1854$$

Impact factor $= \frac{\Delta}{\Delta_{st}} = \frac{1.1967}{0.0117} = 102.3$ which is less than that for non-prismatic bar.

$$p_{max} = \frac{E\Delta}{L} = \frac{2 \times 10^5 \times 1.1967}{3000} = \mathbf{79.78\ N/mm^2}$$

Also $p_{st} = \frac{W}{A} = \frac{Mg}{A} = \frac{25 \times 9.81}{314.16} = 0.78$ N/mm^2

$\therefore$ Ratio $\frac{p_{max}}{p_{st}} = \frac{79.78}{0.78} = \mathbf{102.3}$

which is the same as the corresponding ratio for deflections.

Example 5.15. *A vertical tie rod rigidly fixed at the top end consists of a steel rod 2 m long and 20 mm diameter encased throughout in a brass tube 20 mm internal diameter and 30 mm external diameter. The rod and the casing are fixed together at the ends. The compound bar is suddenly loaded in tension by a mass of 1530 kg falling freely through 3 mm before being arrested by the tie. Compute the maximum stresses in steel and brass. Take* $E_s = 2 \times 10^5$ *N/mm*^{2} *and* $E_b = 1 \times 10^5$ *N/mm*^{2}.

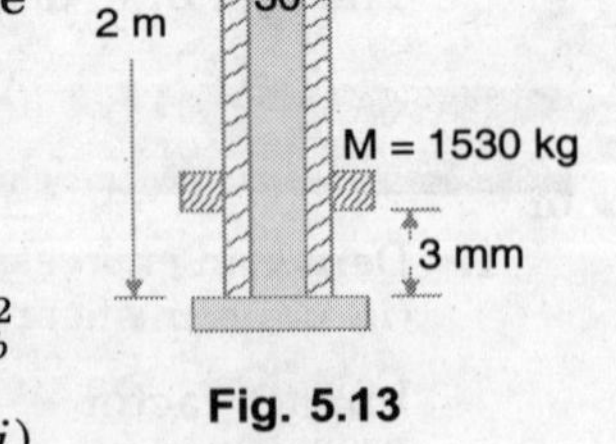

Fig. 5.13

Solution. Mass = 1530 kg ; hence weight $P = 1530 \times 9.81 = 15009$ N

$$A_s = \frac{\pi}{4}(20)^2 = 314.16 \text{ mm}^2 \; ; \; A_b = \frac{\pi}{4}(30^2 - 20^2) = 392.7 \text{ mm}^2$$

Since the deformation of both the rod and the tube is to be the same,

$$p_s = \frac{E_s}{E_b} \times p_b = \frac{2}{1} p_b = 2p_b$$

$$\therefore \quad U_s = \frac{p_s^2}{2E_s} V_s = \frac{(2p_b)^2}{2 \times 2 \times 10^5} (2000 \times 314.16) = 6.2832\, p_b^2 \quad ...(i)$$

Also, $U_b = \frac{p_b^2}{2E_b} V_b = \frac{p_b^2}{2 \times 1 \times 10^5} (2000 \times 392.7) = 3.927\, p_b^2$ ...(*ii*)

Now, work done by falling weight $= P(h + \Delta) = 15009 \left(3 + \frac{p_b + 2000}{1 \times 10^5}\right)$

$$= 15009\,(3 + 0.02\,p_b) = 45027 + 300.2\,p_b \quad ...(iii)$$

From the principle of conservation of energy

Work done by falling weight = $U_s + U_b$

or $$45027 + 300.2\, p_b = 6.2832\, p_b^2 + 3.927\, p_b^2 \quad \text{or} \quad p_b^2 - 29.4\, p_b - 4410 = 0$$

From which p_b = **82.72 N/mm²** ;

Hence $p_s = 2p_b$ = **165.43 N/mm²**.

Example 5.16. *A sliding mass of 500 kg is dropped down a vertical rod which is suspended from the top and is provided with a collar at the bottom end. The length of the rod is 2 m and its diameter is 12 mm. In order to reduce the shock, a buffer spring is placed on the collar which compresses 10 mm by a dead load of 2.5 kN. Taking into account the work done in compressing the spring and stretching the rod, find the height measured from the top of the spring from where the weight must be dropped in order to provide a momentary stress of 150 N/mm² in the rod. Take E = 2 × 10⁵ N/mm².*

Solution. $A = \dfrac{\pi}{4}(12)^2 = 113.1$ mm² ; Sliding mass M = 500 kg

∴ Sliding weight $W = 500 \times 9.81 = 4905$ N

Equivalent dead load, required to produce a stress of 150 N/mm² is

$$P_e = 150 \times 113.1 = 16965 \text{ N}$$

Compression of spring due to this dead load = x

$$= \frac{10}{2.5 \times 1000} \times 16965 = 67.86 \text{ mm}$$

Momentary extension of bar at a stress of 150 N/mm² is

$$\Delta = \frac{p}{E} L = \frac{150}{2 \times 10^5} \times 2000 = 1.5 \text{ mm}$$

Let h = height of fall, measured for the top of spring.

$$\therefore \quad W(h + x + \Delta) = \frac{1}{2} P_e x + \frac{p^2}{2E} AL$$

or $$4905\,(h + 67.86 + 1.5) = \frac{1}{2} \times 16965 \times 67.86 + \frac{(150)^2}{2 \times 2 \times 10^5} \times 113.1 \times 2000$$

or $$h + 69.36 = 119.95$$ From which h = **50.6 mm.**

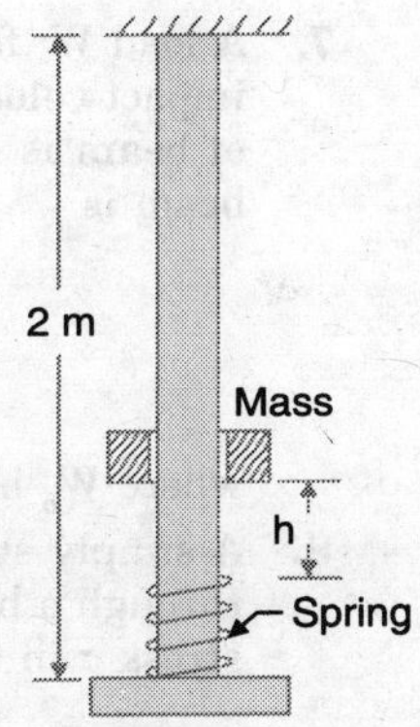

Fig. 5.14

PROBLEMS

1. Derive an expression for the strain energy in a material subjected to (*i*) a uniform tensile stress, (*ii*) uniform shear stress.

   Calculate the modulus of resilience in tension and in shear for steel having the following properties :

   Limits of proportionality in tension 300 N/mm² ; in shear 230 N/mm² ; $E = 2.05 \times 10^5$ N/mm² ; $N = 0.83 \times 10^5$ N/mm².

2. A wagon weighing 30 kN is attached to a wire rope and is moving down an incline at a speed of 70 cm per second. When the length of the rope unwound is 30 m, it gets jammed and the wagon is suddenly brought to rest. Calculate the maximum instantaneous stress and maximum instantaneous elongation produced if the diameter of the rope is 4 cm. $E = 2.05 \times 10^5$ N/mm² and g = 981 cm/sec².

3. A weight W falls a distance h before beginning to stretch a bar of length L and cross-sectional area A. Derive expressions for the maximum stress induced in the bar when

(*a*) the maximum extension is negligible compared with h.

(*b*) the maximum extension is of the same order as h.

A weight of 300 N is dropped from a height of 8 cm on to a collar at the end of a vertical steel bar 2 m long and 2 cm in diameter. Calculate the maximum instantaneous extension and stress produced in the section of the bar. Take $E = 2.05 \times 10^5$ N/mm².

4. Explain what do you understand by term "Strain energy" and derive from first principles an expression giving the strain energy in a tension bar of length L and cross-sectional area A when under a load F.

5. If σ_1 and σ_2 are the principal stresses at a point in a material in a two dimensional stress system, derive an expression for the strain energy per unit volume in terms of σ_1, σ_2, Poisson's ratio and Young's modulus.

Find the strain energy stored in a steel bar 50 cm long and of cross-section 5 cm × 1 cm when it is subjected to axial pull of 50 kN and to a compressive stress of 120 N/mm² on its narrow edges. Take $E = 2.05 \times 10^5$ N/mm² and $1/m = 0.29$.

6. An unknown weight falls 1 cm on to a collar rigidly attached to the lower end of a vertical bar, 4 m long and 3 cm in diameter. If the maximum instantaneous extension is known to be 4 mm, what is the corresponding stress and the value of the unknown weight ? Take $E = 2.05 \times 10^5$ N/mm².

7. A load W falls so as to strike the centre of a freely supported beam of uniform section with an impact velocity v. If the loss of energy at impact may be calculated by assuming 17/35 of the mass of beam is concentrated at the centre of the beam, prove that the maximum deflection of the beam is

$$\delta_s + \sqrt{\delta_s^{\,2} + \delta_s \frac{v^2}{g} \frac{1}{1 + \dfrac{17}{35}\dfrac{W_b}{W}}}$$

where W_b is the weight of the beam and δ_s is the static deflection produced by a load W.

8. A simply supported rectangular beam of span L is struck in the middle by a weight W falling through a height h. Show that the relation between the cross-sectional area A and the maximum stress σ in the beam is given by

$$A = \frac{18WhE}{L\sigma^2}.$$

ANSWERS

1. $U = \dfrac{f^2}{2E} \times V$; 0.219 mm-N/mm³ ; $U = \dfrac{q^2}{2N} \times V$; 0.319 mm-N/mm³.

2. 90 N/mm² ; 1.32 cm. **3.** 126 N/mm² ; 0.120 cm.

5. 19100 mm-N. **6.** 205 N/mm² ; 20700 N.

6

Theories of Failure

6.1. INTRODUCTION

Several hypothesis have been advanced as to the conditions which determine elastic failure of metals subjected to steady constant stresses. No great uniformity of opinion has been reached so far. This is due to the complex nature of failure. From the physical point of view, failure of a material represents either direct separation of particles from each other (so called *brittle failure*) or slipping of particles (so-called *ductile failure*) accompanied by considerable plastic deformations. Failure resulting from local crippling or elastic instability is not considered here.

According to the chief theories, perfect elasticity ceases when a certain limiting value is reached by one of the following :

1. The maximum principal stress.
2. The maximum principal strain.
3. The maximum shear stress.
4. The maximum strain energy.
5. The maximum shear strain energy.

6.2. MAXIMUM PRINCIPAL STRESS THEORY

This is the simplest and the oldest theory of failure, and is sometimes called *Rankine's theory*. According to this theory, permanent set takes place under a state of complex stress when the value of the maximum principal stress is equal to that of yield point stress as found in a simple tensile test, or the minimum principal stress (*i.e.*, the compressive stress) is equal to that of yield point stress in simple compression. Thus the conditions of yielding are :

$$\sigma_1 = f_y \qquad \text{...[6.1 }(a)]$$

or

$$|\sigma_3| = f_y' \qquad \text{...[6.1 }(b)]$$

where f_y is the yield point stress in simple tension and f_y' is the yield point stress in simple compression.

If maximum principal stress is the design criterion, the maximum principal stress must not exceed the working stress f for the material.

Hence $\sigma_1 \leq f$ ...(6.2)

This theory disregards the effect of other principal stresses and of the shearing stresses on other planes through the element. For brittle materials which do not fail by yielding but fail by brittle fracture, the maximum principal stress theory is considered to be reasonably satisfactory.

6.3. MAXIMUM PRINCIPAL STRAIN THEORY

This theory is credited to Saint Venant. According to this theory, a ductile material begins to yield when the maximum principal strain reaches the strain at which yielding occurs in simple tension, or when the minimum principal strain (*i.e.*, the compressive strain) equals the yield point strain in simple compression. In a three dimensional stress system, the strains are as under :

$$e_1 = \frac{\sigma_1}{E} - \frac{1}{mE}(\sigma_2 + \sigma_3)\ ;\ e_2 = \frac{\sigma_2}{E} - \frac{1}{mE}(\sigma_3 + \sigma_1)\ ;\ e_3 = \frac{\sigma_3}{E} - \frac{1}{mE}(\sigma_1 + \sigma_2)$$

If e_y = yield point strain (tensile) = $\frac{f_y}{E}$ and e_y' = yield point strain (compressive) = $\frac{f_y'}{E}$

Then $$e_1 = e_y \quad \text{or} \quad \sigma_1 - \frac{1}{m}(\sigma_2 + \sigma_3) = f_y \qquad \text{...[6.3 (a)]}$$

and $$e_3 = e_y' \quad \text{or} \quad \left| \sigma_3 - \frac{1}{m}(\sigma_1 + \sigma_2) \right| = f_y' \qquad \text{...[6.3 (b)]}$$

In a two dimensional stress system, having σ_1 and σ_2 as the major and minor principal stresses, Eq. 6.3 (*a*) reduces to

$$\sigma_1 - \frac{\sigma_2}{m} = f_y \qquad \text{...[6.3 (c)]}$$

Thus, the stress which, acting alone, will produce same maximum strain e_1 is equal to $\left(\sigma_1 - \frac{\sigma_2}{m}\right)$. The design criterion according to this theory, will be

$$\left(\sigma_1 - \frac{\sigma_2}{m}\right) \leq f,$$ where f is the working stress for the material. ...(6.4)

There are many cases in which the above theory may be shown to be invalid. For example, if we take a case when $\sigma_1 = \sigma_2$ (*i.e.*, equal tension in two perpendicular directions), we get from Eq. [6.3 (*c*)],

$$\sigma_1 - \frac{\sigma_1}{m} = f_y \quad \text{or} \quad \left(\frac{m-1}{m}\right)\sigma_1 = f_y$$

This gives $\sigma_1 > f_y$

Thus this theory indicates that the tensile stresses at yielding will be higher than the stress in simple tension. This conclusion is not supported by experiments.

6.4. MAXIMUM SHEAR STRESS THEORY

This theory is credited to J.J. Guest, and therefore known as *Guest's Theory*. This theory assumes that yielding begins when the maximum shearing stress equals the maximum shearing stress developed at the yielding in simple tension. Since the maximum shear is equal to half the difference between the maximum and minimum principal stress, and since the maximum shear in simple tension is equal to half the tensile stress, we have

$$q_{max} = \frac{1}{2}(\sigma_1 - \sigma_3) = \frac{f_y}{2}$$

Hence the condition of yielding is $(\sigma_1 - \sigma_3) = f_y$ ...(6.5)

In the case of two dimensional stress system, where σ_3 is zero, if f is the allowable stress, the design criterion is

$$(\sigma_1 - \sigma_2) = f_y \text{ (limited to negative values of } \sigma_2 \text{ when } \sigma_2 < 0).$$

This theory is fairly well justified for ductile materials and for the states of stress encountered in most load resisting members and machine parts. However, the theory does not give accurate results for the state of stress of pure shear in which the maximum amount of shear is developed (*i.e.* torsion test). Similarly, the theory would not be applicable in the case where the state of stress consists of triaxial tensile stresses of nearly equal magnitude, reducing the shearing stress to a small magnitude, so that failure would be by brittle fracture rather than by yielding.

6.5. MAXIMUM STRAIN ENERGY THEORY

This theory, originally put forward by Beltrami, is generally known as *Haigh's Theory*. According to this theory, if a body is brought to a particular state by various methods then the work done by passing from the initial to final state will be *independent* of the method applied. Hence, when a material is caused to take permanent set by stress which increases gradually from zero, then the initial strain energy is independent of the nature of the stresses and is almost constant in value.

The theory states that inelastic action at any point in a body due to any state of stress begins only when the energy per unit volume absorbed at the point is equal to the energy absorbed per unit volume of the material when subjected to the elastic limit under a uniaxial state of stress, as occurs in a simple tensile test.

In a three dimensional stress system, the strain energy per volume is given by Eq. 5.10 :

$$U = \frac{1}{2E}\left\{\sigma_1^2 + \sigma_2^2 + \sigma_3^2 - \frac{2\sigma_1\sigma_2 + 2\sigma_2\sigma_3 + 2\sigma_3\sigma_1}{m}\right\}$$

(where σ_1, σ_2 and σ_3 are of the same sign).

Hence the yield criterion can be represented as

$$\sigma_1^2 + \sigma_2^2 + \sigma_3^2 - \frac{2\sigma_1\sigma_2 + 2\sigma_2\sigma_3 + 2\sigma_3\sigma_1}{m} = f_y^2 \quad \text{...(6.6)}$$

Taking a two dimensional case ($\sigma_3 = 0$), the above criterion reduces to

$$\sigma_1^2 + \sigma_2^2 - \frac{2\sigma_1\sigma_2}{m} = f_y^2 \quad \text{...(6.7)}$$

If f is the working stress in the material, the design criterion may be stated as follows :

$$\left(\sigma_1^2 + \sigma_2^2 - \frac{2\sigma_1\sigma_2}{m}\right) \le f^2 \qquad ...(6.8)$$

6.6. MAXIMUM SHEAR STRAIN ENERGY THEORY : MISES-HENKY THEORY

The maximum shear strain energy theory, also called as *energy of distortion theory*, is associated with the names of von Mises and Henky. *This theory states that inelastic action at any point in a body under any combination of stresses begins only when the strain energy of distortion per unit volume absorbed at the point is equal to the strain energy of distortion absorbed per unit volume at any point in a bar stressed to the elastic limit under a state of uniaxial stress as occurs in a simple tension (or compression) test.*

The energy of distortion can be obtained by subtracting the energy of volumetric change from total strain energy. Under a system of stresses acting externally on a body, the total external work done causes : (*a*) change of volume due to the operation of the direct stress and (*b*) distortion due to the shearing stresses which does not affect the volumetric change.

Thus total $U = U_v + U_s$

Now $$U_v = \frac{1}{2}\,(\text{average }\sigma)\,.\,\Delta V = \frac{\sigma_1 + \sigma_2 + \sigma_3}{6}\,\Delta V$$

But $$\Delta V = \frac{1}{E}\left\{\sigma_1 + \sigma_2 + \sigma_3 - \frac{2}{m}(\sigma_1 + \sigma_2 + \sigma_3)\right\} = \frac{1}{E}(\sigma_1 + \sigma_2 + \sigma_3)\left(\frac{m-2}{m}\right)$$

$\therefore$ $$U_v = \frac{m-2}{6mE}\{\sigma_1 + \sigma_2 + \sigma_3\}^2 = \frac{m-2}{6mE}[\sigma_1^2 + \sigma_2^2 + \sigma_3^2 + 2(\sigma_1\sigma_2 + \sigma_2\sigma_3 + \sigma_3\sigma_1)]$$

$$= \frac{m-2}{6mE}(X + 2Y) \qquad ...(1)$$

where $X = \sigma_1^2 + \sigma_2^2 + \sigma_3^2$ and $Y = \sigma_1\sigma_2 + \sigma_2\sigma_3 + \sigma_3\sigma_1$.

Also, $$U = \frac{1}{2E}\left[\sigma_1^2 + \sigma_2^2 + \sigma_3^2 - \frac{2}{m}(\sigma_1\sigma_2 + \sigma_2\sigma_3 + \sigma_3\sigma_1)\right] = \frac{1}{2E}\left(X - \frac{2}{m}Y\right) \qquad ...(2)$$

Hence $$U_s = U - U_v = \frac{1}{2E}\left[X - \frac{2}{m}Y\right] - \frac{m-2}{6mE}(X + 2Y)$$

$$= \frac{1}{6E}\left[3X - \frac{6Y}{m} - X - 2Y + \frac{2X}{m} + \frac{4Y}{m}\right] = \frac{1}{3E}\left[X\left(1 + \frac{1}{m}\right) - Y\left(1 + \frac{1}{m}\right)\right]$$

$$= \frac{m+1}{3mE}(X - Y) = \frac{m+1}{3mE}\{\sigma_1^2 + \sigma_2^2 + \sigma_3^2 - (\sigma_1\sigma_2 + \sigma_2\sigma_3 + \sigma_3\sigma_1)\}$$

or $$U_s = \frac{m+1}{6mE}\{(\sigma_1 - \sigma_2)^2 + (\sigma_2 - \sigma_3)^2 + (\sigma_3 - \sigma_1)^2\} \qquad ...(6.9)$$

This is thus the expression for *strain energy of distortion*. Under the action of uniaxial stress f_y the strain energy of distortion will evidently be equal to $\frac{m+1}{3mE} f_y^2$. Hence the failure criterion may be stated as follows :

$$\frac{m+1}{6mE}\{(\sigma_1-\sigma_2)^2+(\sigma_2-\sigma_3)^2+(\sigma_3-\sigma_1)^2\}=\frac{m+1}{3mE}f_y^{\,2}$$

or $$\tfrac{1}{2}\{(\sigma_1-\sigma_2)^2+(\sigma_2-\sigma_3)^2+(\sigma_3-\sigma_1)^2\}=f_y^{\,2} \qquad ...(6.10)$$

In a two dimensional case ($\sigma_3 = 0$), this reduces to

$$\sigma_1^{\,2}+\sigma_2^{\,2}-\sigma_1\sigma_2=f_y^{\,2} \qquad ...(6.11)$$

If f is the working stress in the material, the design criterion may be stated as follows :

$$\sigma_1^{\,2}+\sigma_2^{\,2}-\sigma_1\sigma_2\le f^2 \qquad ...(6.12)$$

6.7. GRAPHICAL REPRESENTATION OF THEORIES FOR TWO DIMENSIONAL STRESS SYSTEM

(1) Maximum Stress Theory

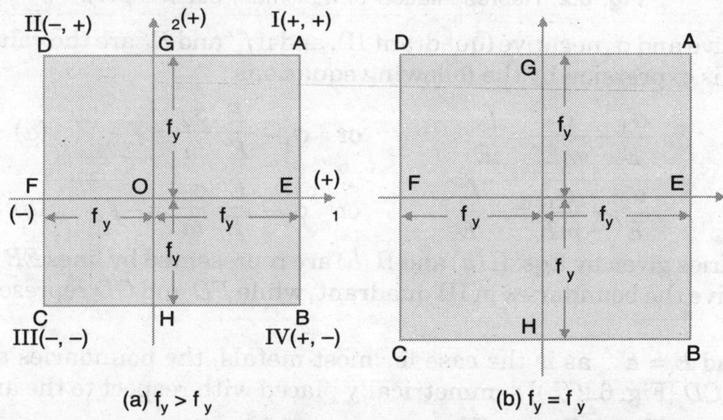

Fig. 6.1. Representation of maximum stress theory.

The failure criterion is given by Eq. 6.1 (*a*) and (*b*). Fig. 6.1 (*a*) shows the graphical representation of the maximum stress theory when *f* if greater than *f′*. The diagram is divided into four quadrants with the following signs of σ_1 and σ_2 : (*i*) I quadrant, both σ_1 and σ_2 positive, (*ii*) II quadrant, σ_1 negative (*i.e.*, compressive) and σ_2 positive (*i.e.*, tensile), (*iii*) III quadrant, both σ_1 and σ_2 negative and (*iv*) IV quadrant, σ_1 positive and σ_2 negative. The square *ABCD* represents the graphical representation of the maximum stress theory. The material will reach its elastic limit when the point (σ_1, σ_2) passes outside the square *ABCD*. If *f* = *f′*, then centre of the square will be at the origin, as shown in Fig. 6.1 (*b*).

(2) Maximum Strain Theory

The yield criterion is expressed as :

$$\sigma_1-\frac{\sigma_2}{m}=f_y \quad ...(a) \qquad \text{and} \qquad \sigma_2-\frac{\sigma_1}{m}=f_y' \quad ...(b) \qquad ...(I)$$

In the above case, both σ_1 and σ_2 are positive (*i.e.*, tensile) and the boundary representing elastic limit is shown by lines *AE* and *GA* representing Eqs. I (*a*) and I (*b*).

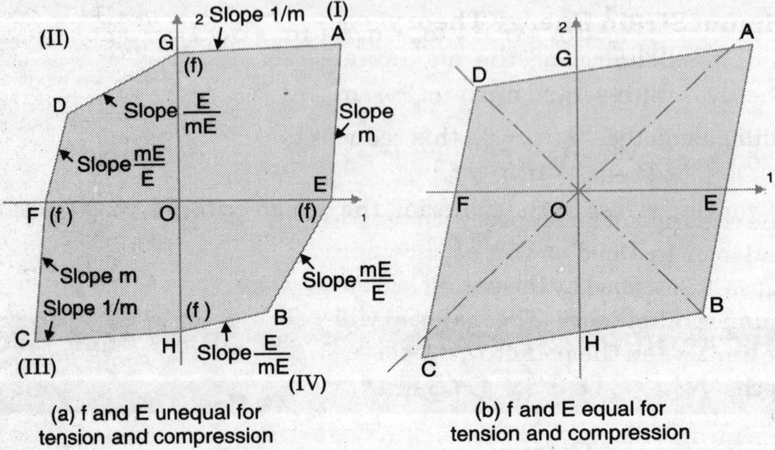

Fig. 6.2. Representation of maximum strain theory.

If σ_1 is positive and σ_2 negative (quadrant II), and if f_y' and E' are the values in compression, the yield criterion is expression by the following equations :

$$\left.\begin{aligned}\frac{\sigma_1}{E}-\frac{\sigma_2}{mE'}&=\frac{f_y}{E} \quad \text{or} \quad \sigma_1-\frac{E}{E'}\frac{\sigma_2}{m}=f_y \quad ...(a)\\ \frac{\sigma_2}{E'}-\frac{\sigma_1}{mE}&=-\frac{f_y'}{E'} \quad \text{or} \quad \sigma_2-\frac{E'}{E}\frac{\sigma_1}{m}=-f_y' \quad ...(b)\end{aligned}\right\} \quad ...(II)$$

The boundaries given by Eqs. II (*a*) and II (*b*) are represented by lines *EB* and *HB*. Similarly, lines *HC* and *FC* give the boundaries in III quadrant, while *FD* and *GD* represent the boundary in the II quadrant.

If $f_y = f_y'$ and $E = E'$, as is the case for most metals, the boundaries are represented by parallelogram *ABCD* [Fig. 6.2 (*b*)] symmetrically placed with respect to the axes.

(3) Maximum Shear Stress Theory

If σ_1 and σ_2 are of opposite sign, the greater shearing stress is given by $\frac{1}{2}(\sigma_1-\sigma_2)$ and the failure is represented by

$$(\sigma_1-\sigma_2)=f_y \quad ...(a) \quad \text{or by} \quad \sigma_2-\sigma_1=f_y \quad ...(b)$$

The boundaries of these equations are represented by parallel lines *FG* and *HE*.

If σ_1 and σ_2 are like (*i.e.* I or III quadrant), the maximum shear is equal to $\frac{\sigma_1}{2}$ or $\frac{\sigma_2}{2}$ whichever is more.

Hence the failure is represented by

$$\sigma_1=f_y \quad ...(a) \quad \text{or} \quad \sigma_2=f_y \quad ...(b)$$

The conditions are represented by lines *GA* and *AE* in the first quadrant and *FC* and *CH* in the third quadrant.

Hence according to maximum shear stress theory, material will reach its elastic limit when the point (σ_1, σ_2) passes outside the figure *AEHCFGA*.

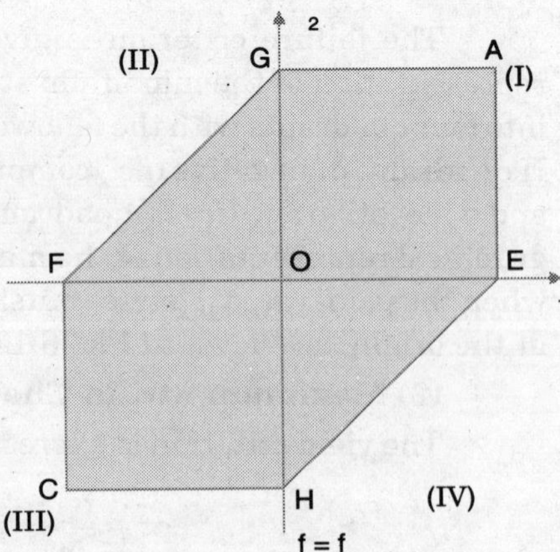

Fig. 6.3. Representation of maximum shear stress theory.

(4) Maximum Strain Energy Theory

For two dimensional case, the maximum strain energy theory is represented by the criterion

$$\sigma_1^2 + \sigma_2^2 - \frac{2}{m}\sigma_1\sigma_2 = f_y^2$$

This is the equation of an ellipse, with centre as the origin and axes inclined at 45°, as shown in Fig. 6.4. The ellipse is inscribed by the parallelogram given by maximum strain theory. The material will reach its elastic limit when the point (σ_1, σ_2) passes outside the ellipse.

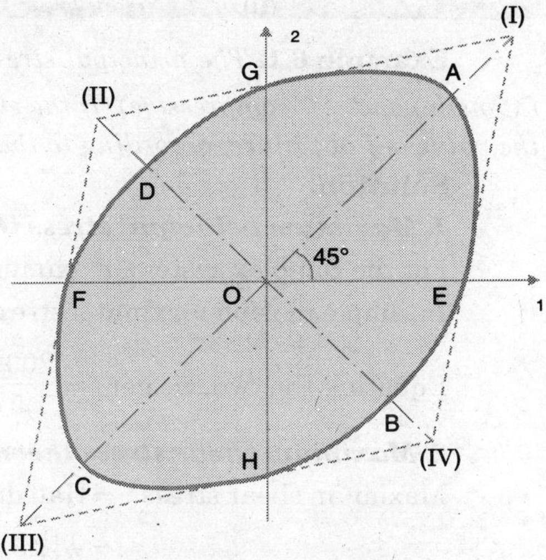

Fig. 6.4. Representation of maximum strain energy theory.

(5) Maximum Shear Strain Energy Theory

The yield criterion is given by

$$\sigma_1^2 + \sigma_2^2 - \sigma_1\sigma_2 = f_y^2$$

Thus this criterion is also represented by an ellipse, similar to Haigh's ellipse. Both the ellipse become identical when m has its smallest possible value of 2.

Fig. 6.5 shows the graphical representation of various theories on the same diagram, when $f = f'$ and $E = E'$.

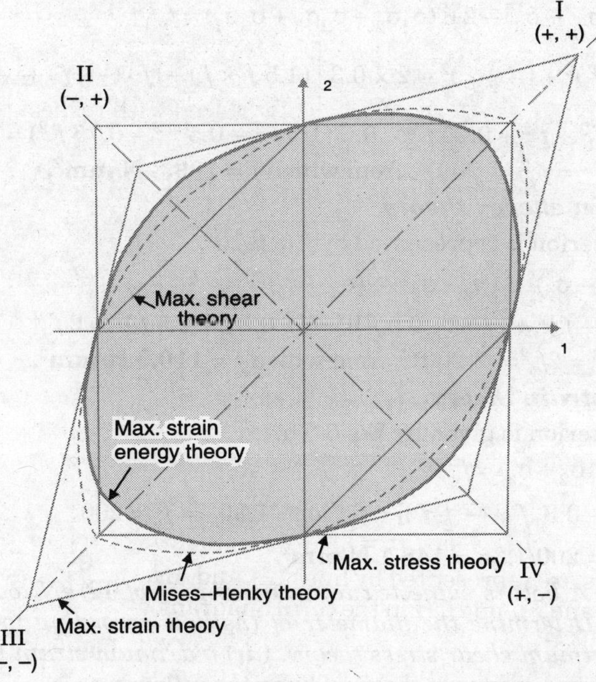

Fig. 6.5. Graphical representation of various theories.

6.8. ADDITIONAL EXAMPLES USEFUL FOR COMPETITIVE EXAMINATIONS

Example 6.1. *The principal stresses at a point in an elastic material are 1.5 f (tensile), f (tensile) and $\frac{1}{2}f$ (compressive). If the elastic limit in simple tension is 200 N/mm², determine the value of f at failure, according to the five different theories. Take μ = 0.3.*

Solution.

1. Maximum principal stress theory

For the complex system, maximum stress $= 1.5\,f$

In simple tension maximum stress $= 200$ N/mm².

Equating the two, we get $f = \frac{200}{1.5} =$ **133.3 N/mm².**

2. Maximum shear stress theory

Maximum shear stress = Half difference between principal stresses

$$= \tfrac{1}{2}[1.5\,f - (-\tfrac{1}{2}f)] = f$$

In simple tension, principal stresses are 200, 0, 0, and hence maximum shear stress

$$= \tfrac{1}{2}(200 - 0) = 100 \text{ N/mm}^2.$$

Equating the two, we get $f =$ **100 N/mm².**

3. Strain energy theory

The yield criterion is represented by Eq. 6.6

$$\sigma_1^2 + \sigma_2^2 + \sigma_3^2 - 2\,\mu\,(\sigma_1\sigma_2 + \sigma_2\sigma_3 + \sigma_3\sigma_1) = f_y^2$$

$$\therefore \quad (1.5\,f)^2 + (f)^2 + (-\tfrac{1}{2}\,f)^2 - 2 \times 0.3\left[(1.5\,f \times f) + (f)(-\tfrac{1}{2}f) + \left(-\frac{f}{2} \times 1.5\,f\right)\right] = f_y^2$$

or $$2.25\,f^2 + f^2 + 0.25\,f^2 - 0.6\,(1.5\,f^2 - 0.5\,f^2 - 0.75\,f^2) = f_y^2$$

or $3.35\,f^2 = f_y^2 = (200)^2$, from which $f =$ **109.3 N/mm².**

4. Shear strain energy theory

The failure criterion is represented by Eq. 6.10

$$\tfrac{1}{2}[(\sigma_1 - \sigma_2)^2 + (\sigma_2 - \sigma_3)^2 + (\sigma_3 - \sigma_1)^2] = f_y^2$$

$$\therefore \quad [1.5\,f - f]^2 + [f - (-0.5\,f)]^2 + [-0.5\,f - 1.5\,f]^2 = 2\,f_y^2$$

or $6.5\,f^2 = 2f_y^2 = 2(200)^2$, from which $f =$ **110.9 N/mm².**

5. Maximum strain theory

The failure criterion is given by Eq. 6.3 (*a*)

$$\sigma_1 - \mu(\sigma_2 + \sigma_3) = f_y$$

$$\therefore \quad 1.5\,f - 0.3\,[f + (-\tfrac{1}{2}\,f)] = f_y \quad \text{or} \quad 1.35\,f = f_y = 200$$

From which $f = 200/1.35 =$ **148.1 N/mm².**

Example 6.2. *A bolt is subjected to an axial pull of 12 kN together with a transverse shear force of 6 kN. Determine the diameter of the bolt according to (i) maximum principal stress theory, (ii) maximum shear stress theory, (iii) maximum strain theory, (iv) strain energy theory, and (v) shear strain energy theory. Given the following :*

Elastic limit in tension : 300 N/mm² ; Factor of safety : 3 ; Poisson's ratio : 0.3.

Solution. Permissible tensile stress = 300/3 = 100 N/mm².

Let d be the required diameter of the bolt, in mm.

Hence the applied stresses are

$$p = \frac{12000}{\pi d^2/4} = \frac{15278.9}{d^2} \text{ N/mm}^2 \text{ and } q = \frac{6000}{\pi d^2/4} = \frac{7639.4}{d^2} \text{ N/mm}^2.$$

(i) Maximum principal stress theory

The maximum principal stress in the bolt is given by

$$\sigma = \frac{p}{2} + \sqrt{\left(\frac{p}{2}\right)^2 + q^2} \quad (\text{where } p_1 = p \text{ and } p_2 = 0)$$

$$= \frac{1}{2}\frac{15278.9}{d^2} + \sqrt{\left(\frac{15278.9}{2d^2}\right)^2 + \left(\frac{7639.4}{d^2}\right)^2} = \frac{18443.2}{d^2}$$

Permissible simple stress in tension = 100 N/mm²

$$\therefore \quad \frac{18443.2}{d^2} = 100, \text{ from which } d = \mathbf{13.58\ mm.}$$

(ii) Maximum shear stress theory

$$\text{Max. shear stress} = \sqrt{\left(\frac{p}{2}\right)^2 + q^2} = \sqrt{\left(\frac{15278.9}{2d^2}\right)^2 + \left(\frac{7639.4}{d^2}\right)^2} = \frac{10803.8}{d^2}$$

Max. shear stress = 100/2 = 50 N/mm² (in simple tension)

$$\therefore \quad \frac{10803.8}{d^2} = 50, \text{ from which } d = \mathbf{14.7\ mm.}$$

(iii) Maximum strain theory

$$\sigma_1 = \frac{p}{2} + \sqrt{\left(\frac{p}{2}\right)^2 + q^2} = \frac{15278.9}{2d^2} + \frac{10803.8}{d^2}$$

$$= \frac{7639.4}{d^2} + \frac{10803.8}{d^2} = \frac{18443.2}{d^2}$$

$$\sigma_2 = \frac{p}{2} - \sqrt{\left(\frac{p}{2}\right)^2 + q^2} = \frac{7639.1}{d^2} - \frac{10803.8}{d^2} = -\frac{3161.4}{d^2}$$

$$\sigma_3 = 0$$

$$\therefore \quad \text{Maximum strain} = e_1 = \frac{\sigma_1}{E} - \frac{\sigma_2}{mE}$$

$$= \frac{1}{E}\left[\frac{18443.2}{d^2} - \frac{0.3\,(-3164.4)}{d^2}\right] = \frac{19392.5}{Ed^2}$$

$$\text{Maximum strain in simple tension} = \frac{f}{E} = \frac{100}{E}$$

Equating the two, we get $\frac{19392.5}{Ed^2} = \frac{100}{E}$, from which d = **13.93 mm.**

(iv) Maximum strain energy theory

The principal stresses are

$$\sigma_1 = \frac{18443.2}{d^2} \; ; \sigma_2 = -\frac{3164.4}{d^2} \text{ and } \sigma_3 = 0$$

$$\therefore \quad \text{Strain energy} = \frac{1}{2E}\left[\left(\frac{18443.2}{d^2}\right)^2 + \left(-\frac{3164.4}{d^2}\right)^2 - 2 \times 0.3\left(\frac{18443.2}{d^2}\right) \times \left(-\frac{3164.4}{d^2}\right)\right]$$

$$= \frac{1}{2E} \cdot \frac{3.8518 \times 10^8}{d^4}$$

$$\text{Strain energy in simple tension} = \frac{f^2}{2E} = \frac{(100)^2}{2E}$$

$$\therefore \quad \frac{1}{2E} \times \frac{3.8518 \times 10^8}{d^4} = \frac{(100)^2}{2E}, \text{ from which } d = \mathbf{14.01\ mm.}$$

(v) Shear strain energy theory

$$\text{Shear strain energy} = \frac{m+1}{6mE}\,[(\sigma_1 - \sigma_2)^2 + (\sigma_2 - \sigma_3)^2 + (\sigma_3 - \sigma_1)^2]$$

$$= \frac{m+1}{6mE} \times \frac{1}{2}\left[\left\{\frac{18443.2}{d^2} - \left(-\frac{3164.4}{d^2}\right)\right\}^2 + \left\{\left(-\frac{3164.4}{d^2}\right) - 0\right\}^2 + \left\{0 - \left(\frac{18443.2}{d^2}\right)\right\}^2\right]$$

$$= \frac{m+1}{3mE} \times \frac{1}{2}\left[\frac{4.669 \times 10^8}{d^4} + \frac{0.10 \times 10^8}{d^4} + \frac{3.402 \times 10^8}{d^4}\right] = \frac{m+1}{3mE} \times \frac{4.0855 \times 10^8}{d^4}$$

Shear strain energy under uniaxial stress f (= 100)

$$= \frac{m+1}{3mE} f^2 = \frac{m+1}{3mE} (100)^2$$

$$\therefore \quad \frac{4.0855 \times 10^8}{d^4} = (100)^2, \text{ from which } d = \mathbf{14.22\ mm.}$$

Example 6.3. *Two of the principal stresses at a point are 130 N/mm² and 90 N/mm². Determine the safe range of the third principal stress at the point by five different theories. Take $E = 2 \times 10^5$ N/mm², failure stress in tension test to be 210 N/mm², and $\mu = 0.25$. Failure stress in tension and compression is the same.*

Solution. The third principal stress (σ) may be the highest, the lowest or the intermediate one. Hence the range is $\sigma_3 \le \sigma \le \sigma_1$.

(i) Maximum principal stress theory

$$\sigma_1 = f_y = 210 \text{ N/mm}^2 \; ; \sigma_3 = f_y' = -\,210 \text{ N/mm}^2$$

Hence the range is $-210 \le \sigma \le 210$

(ii) Maximum strain theory

$$\sigma_1 - 0.25\,(130 + 90) = f_y = 210$$

or $$\sigma_1 = 210 + 55 = 265 \text{ N/mm}^2$$

Also, $$\sigma_3 = -\,0.25\,(130 + 90) = f_y' = -\,210$$

or $$\sigma_3 = -\,210 + 55 = -\,155 \text{ N/mm}^2$$

Hence the range is $-155 \le \sigma \le 265$ N/mm²

(iii) Maximum shear stress theory

$$\sigma_1 - 90 = f_y = 210$$

Hence $\quad \sigma_1 = 210 + 90 = 300 \text{ N/mm}^2$

Also, $\quad 130 - \sigma_3 = f_y = 210$

Hence $\quad \sigma_3 = 130 - 210 = -80 \text{ N/mm}^2$

Hence the range is $\quad -80 \le \sigma \le 300 \text{ N/mm}^2$

(iv) Maximum strain energy theory

$$\sigma^2 + (90)^2 + (130)^2 - 2 \times 0.25\,(130\,\sigma + 90\,\sigma + 90 \times 130) = (210)^2$$

or $\quad \sigma^2 - 110\,\sigma - 24950 = 0$

From which $\sigma_1 = 222.3 \text{ N/mm}^2$ and $\sigma_3 = -112.3 \text{ N/mm}^2$

Hence the range is $\quad -112.3 \le \sigma \le 222.3 \text{ N/mm}^2$

(v) Max. Distortion energy theory

$$(\sigma - 90)^2 + (\sigma - 130)^2 + (130 - 90)^2 = 2(210)^2$$

or $\quad \sigma^2 - 220\,\sigma - 30800 = 0$

From which $\sigma_1 = 317.1 \text{ N/mm}^2$ and $\sigma_3 = -97.1 \text{ N/mm}^2$

Hence the range is $\quad -97.1 \le \sigma \le 317.1 \text{ N/mm}^2$.

Example 6.4. *A body is under the action of two principal stresses of 40 N/mm² and – 70 N/mm², the third principal stress being zero. If the elastic limit in simple tension as well as compression is 200 N/mm², find the factor of safety, based on the elastic limit according to the five theories of failure. Take* $\mu = 0.3$.

Solution. Here $\sigma_1 = 40 \text{ N/mm}^2$, $\sigma_2 = 0$ and $\sigma_3 = -70 \text{ N/mm}^2$. Let the factor of safety be F

1. Maximum principal stress theory

Since σ_3 is numerically the largest stress, it will be the basis of failure.

$$\therefore \quad \sigma_3 = \frac{f_y}{F} \quad \text{or} \quad F = \frac{f_y}{\sigma_3} = \frac{200}{70} = \mathbf{2.86}$$

2. Maximum principal strain theory

Here, either $\quad \sigma_1 - \mu(\sigma_2 - \sigma_3) = \dfrac{f_y}{F} \quad$ or $\quad \sigma_3 - \mu(\sigma_1 - \sigma_2) = \dfrac{f_y}{F}$

Hence either $40 - 0.3\,(0 - 70) = \dfrac{200}{F} \quad$ or $\quad -70 - 0.3\,(40 + 0) = \dfrac{200}{F}$

Hence either $\quad F = \dfrac{200}{61} = 3.28 \quad$ or $\quad F = \dfrac{200}{82} = 2.44$

Hence actual $\quad F = \mathbf{2.44}$

3. Maximum shear stress theory

$$\sigma_1 - \sigma_3 = \frac{f_y}{F} \quad \text{or} \quad 40 - (-70) = \frac{200}{F}$$

From which $\quad F = \dfrac{200}{110} = \mathbf{1.818}$

4. Maximum strain energy theory

$$\sigma_1^2 + \sigma_2^2 + \sigma_3^2 - 2\,\mu\,(\sigma_1\sigma_2 + \sigma_2\sigma_3 + \sigma_3\sigma_1) = \left(\frac{f_y}{F}\right)^2$$

or $$(40)^2 + (0)^2 + (-70)^2 - 2 \times 0.3\ [40 \times 0 + 0(-70) + (-70)(40)] = \left(\frac{200}{F}\right)^2$$

or $$\left(\frac{200}{F}\right)^2 = 8180, \text{ from which } F = \mathbf{2.21}$$

5. Maximum shear strain energy theory

$$\sigma_1^2 + \sigma_2^2 + \sigma_3^2 - (\sigma_1\sigma_2 + \sigma_2\sigma_3 + \sigma_3\sigma_1) = \left(\frac{f_y}{F}\right)^2$$

or $$(40)^2 + (0)^2 + (-70)^2 - [40 \times 0 + 0(-70) + (-70)(40)] = \left(\frac{200}{F}\right)^2$$

or $$\left(\frac{200}{F}\right)^2 = 9300, \text{ from which } F = \mathbf{2.07.}$$

Example 6.5. *The B.M. M applied to a solid shaft carries a maximum direct stress f_y at elastic failure. Determine the numerical relationships between M and twisting moment T which, acting alone on the shaft, will produce elastic failure according to each of the following theories of failure. (a) Maximum principal stress (b) Maximum principal strain (c) Maximum shear stress, (d) Maximum strain energy and (e) shear strain energy. Take μ = 0.3.*

Solution. The elastic limit stress f_y caused by bending moment M is given by

$$f_y = \frac{32\,M}{\pi d^3}$$

Similarly, the shear stress τ produced by the torque T acting alone is

$$\tau = \frac{16\,T}{\pi d^3}$$

This shear stress is accompanied by complimentary shear stress, giving a state of shear stress, due to which the induced principal stress are

$$\sigma_1 = -\sigma_2 = \frac{16\,T}{\pi d^3}$$

Hence the principal stresses acting on the shaft are $\frac{16\,T}{\pi d^3}, 0, -\frac{16\,T}{\pi d^3}$

(a) Maximum principal stress theory

$$\frac{16\,T}{\pi d^3} = f_y = \frac{32\,M}{\pi d^3} \text{ at elastic failure}$$

$$\therefore \quad \frac{M}{T} = \mathbf{0.5}$$

(b) Maximum principal strain theory

$$\sigma_1 - \mu\sigma_2 = f_y$$

or $$\frac{16\,T}{\pi d^3} + \frac{0.3 \times 16\,T}{\pi d^3} = \frac{32\,M}{\pi d^3}$$

From which $$\frac{M}{T} = \frac{16 + 4.8}{32} = \mathbf{0.65}$$

(c) Maximum shear stress theory

$$\sigma_1 - \sigma_2 = f_y$$

or $$\frac{16\,T}{\pi d^3} + \frac{16\,T}{\pi d^3} = \frac{32\,M}{\pi d^3}$$

From which $$\frac{M}{T} = \mathbf{1}$$

(d) Maximum strain energy theory

$$\sigma_1^2 + \sigma_2^2 - 2\mu\sigma_1\sigma_2 = f_y^2$$

or $$\left(\frac{16\,T}{\pi d^3}\right)^2 + \left(\frac{16\,T}{\pi d^3}\right)^2 + 2 \times 0.3\left(\frac{16\,T}{\pi d^3}\right)^2 = \left(\frac{32\,M}{\pi d^3}\right)^2$$

or $$2.6\left(\frac{16\,T}{\pi d^3}\right)^2 = \left(\frac{32\,M}{\pi d^3}\right)^2$$

or $$1.612\left(\frac{16\,T}{\pi d^3}\right) = \frac{32\,M}{\pi d^3}$$

From which $$\frac{M}{T} = \frac{1.612}{2} = \mathbf{0.806}$$

(e) Maximum shear strain energy theory

$$\sigma_1^2 + \sigma_2^2 - \sigma_1\sigma_2 = f_y^2$$

or $$\left(\frac{16\,T}{\pi d^3}\right)^2 + \left(\frac{16\,T}{\pi d^3}\right)^2 + \left(\frac{16\,T}{\pi d^3}\right)^2 = \left(\frac{32\,M}{\pi d^3}\right)^2$$

or $$3\left(\frac{16\,T}{\pi d^3}\right)^2 = \left(\frac{32\,M}{\pi d^3}\right)^2$$

$\therefore$ $$\frac{M}{T} = \frac{\sqrt{3}}{2} = \mathbf{0.866.}$$

6.9. EXAMPLES FROM COMPETITIVE EXAMINATIONS

Example 6.6. *A bolt is subjected to a pull of 12 kN together with a transverse shear of 6 kN. Estimate the diameter of the bolt required if it is to be safe according to (i) maximum principal stress theory, (ii) maximum shear stress theory, and (iii) maximum strain energy theory. Elastic limit in simple tension = 300 N/mm². Factor of safety = 3. Poisson's ratio = 0.3.*

(UPSC. Engg. Services Exam. 1993)

Solution.

This problem was set exactly from Example 6.2 of this book. For solution, see Example 6.2.

Example 6.7. *The principal stresses at a point in an elastic material are 1.5 σ (tensile), σ (tensile) and 0.5 σ (compressive). The elastic limit in tension is 210 MPa and μ = 0.3. What would be the value of σ at failure when computed by the different theories of failure?*

(UPSC Engg. Services Exam. 1995)

Solution. *This problem was set from Example 6.1 of this book, exactly with the same data except that the value of elastic limit has been changed from 200 to 210 MPa.* Hence a short solution is given here. Note that 1 MPa = 1 N/mm².

1. Maximum principal stress theory

$$1.5\,\sigma = 210 \text{ MPa, from which } \sigma = 210/1.5 = \mathbf{140\ MPa}$$

2. Maximum shear stress theory

Max. shear stress $= \frac{1}{2}[1.5\,\sigma - (-0.5\,\sigma)] = \sigma$

In simple tension, the principal stresses are 210, 0, 0.

Hence Max. shear stress $= \frac{1}{2}(210 - 0) = 105$ MPa. Equating the two, we get

$$\sigma = \mathbf{105\ MPa}$$

3. Strain energy theory

From the yield criterion,

$$(1.5\,\sigma)^2 + (\sigma)^2 + (-0.5\,\sigma)^2 - 2 \times 0.3\,[(1.5\,\sigma \times \sigma) + (\sigma)(-0.5\,\sigma) + (-0.5\,\sigma \times 1.5\,\sigma)] = (210)^2$$

From which $3.35\,\sigma^2 = (210)^2$ or $\sigma = \mathbf{114.735\ MPa}$

4. Shear strain energy theory

$$[(1.5\,\sigma - \sigma)]^2 + [\sigma - (0.5\,\sigma)]^2 + [(-0.5\,\sigma - 1.5\,\sigma)]^2 = 2(210)^2$$

From which $6.5\,\sigma^2 = 2(210)^2$ or $\sigma = \mathbf{116.487\ MPa}$

5. Maximum strain theory

$$1.5\,\sigma - 0.3\,[\sigma + (-0.5\,\sigma)] = 210$$

From which $\sigma = 210/1.35 = \mathbf{155.556\ MPa.}$

PROBLEMS

1. Discuss in brief various prominent theories of failure.
2. Write a note on significance of theories of failure.
3. The principal stresses at a point in an elastic material are $1.5f$ (tensile), f (tensile) and $\frac{1}{2}f$ (compressive). If the elastic limit in simple tension is 200 N/mm^2, determine the value of f at failure, according to the five different theories. Take $\mu = 0.3$.
4. A bolt is subjected to an axial pull of 12 kN together with a transverse shear of 6 kN. Determine the diameter of the bolt according to (*i*) Maximum principal stress theory, (*ii*) Maximum shear stress theory, (*iii*) Maximum strain theory, (*iv*) Strain energy theory, and (*v*) Shear strain energy theory. Take elastic limit in tension = 300 N/mm^2, factor of safety = 3 and $\mu = 0.3$.
5. A circular shaft 100 mm diameter is subjected to combined bending and twisting of moments the B.M. being 3 times the twisting moment. If the direct tensile yield point of the material is 350 N/mm^2, and the factor of safety is 4, calculate the allowable twisting moment according to the following theories of elastic failure.

   (*a*) Maximum principal stress theory

   (*b*) Maximum shear stress theory

   (*c*) Shear strain energy theory if the simple shear is not to exceed 60 N/mm^2. (*Based on U.L.*)

ANSWERS

3. (*i*) 133.3 N/mm^2 (*ii*) 148.1 N/mm^2 (*iii*) 100 N/mm^2 (*iv*) 109.3 N/mm^2 (*v*) 110.9 N/mm^2
4. (*i*) 13.58 mm (*ii*) 13.93 mm (*iii*) 14.7 mm (*iv*) 14.01 mm (*v*) 14.22 mm
5. (*a*) 2.789 kN-m (*b*) 2.718 kN-m (*c*) 3.268 kN-m.

SECTION 2 BEAMS AND BENDING

7

Bending Moment and Shearing Force

7.1. BASIC CONCEPTS

Beam. Beam is a structural member which is acted upon by a system of external loads at right angle to its axis.

Bending. Bending implies deformation of a bar produced by loads perpendicular to its axis as well as force-couples acting in a plane passing through the axis of the bar.

Plane Bending. If the plane of loading passes through one of the principal centroidal axis of inertia of the cross-section of the beam, the bending is said to be *plane or direct*.

Oblique Bending. If the plane of loading does not pass through one of the principal centroidal axes of inertia of the cross-section of the beam, the bending is said to be *oblique*.

Movable Hinged Support. If an end of a beam is supported on a movable hinged support, it is free to rotate about the hinge and the translational displacement along the plane of rolling is possible. This permits the beam to change its length freely with variations in temperature (Fig. 7.1). The reaction arising at the support will be perpendicular to the plane of rolling and will be passing through the centre of the roller. Such a support is also known as roller support.

Immovable Hinged Support. If the ends of a beam are supported on an immovable hinged support, the ends of the beam are free to rotate but translational displacement is not possible in any direction (Fig. 7.2). The reaction arising at such support can be resolved in horizontal and vertical directions.

Simply Supported Beam. A beam is said to be simply supported when it is so supported at

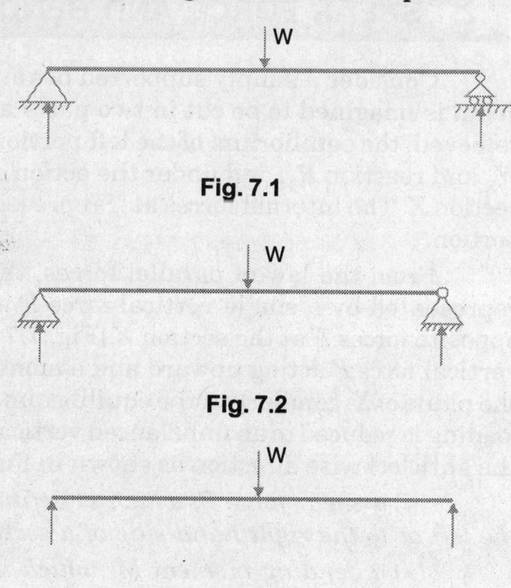

Fig. 7.1

Fig. 7.2

Fig. 7.3

its ends that the translational displacement is not possible, but rotation is possible (Fig. 7.3).

Fixed Support. If the end of a beam is fixed or clamped in position, the end is neither free to rotate nor free to have translational displacement. Such support can produce, in general, a reaction which can be resolved in the horizontal and vertical components and a fixing (or clamping) moment (Fig. 7.4).

Fixed Beam. Fixed beam is beam the ends of which are clamped in position so that they are not free to rotate (Fig. 7.4).

Cantilever Beam. A cantilever is a beam (Fig. 7.5) one end of which is clamped and the other end is free.

Continuous Beam. Continuous beam is a beam which is supported at several supports (Fig. 7.6).

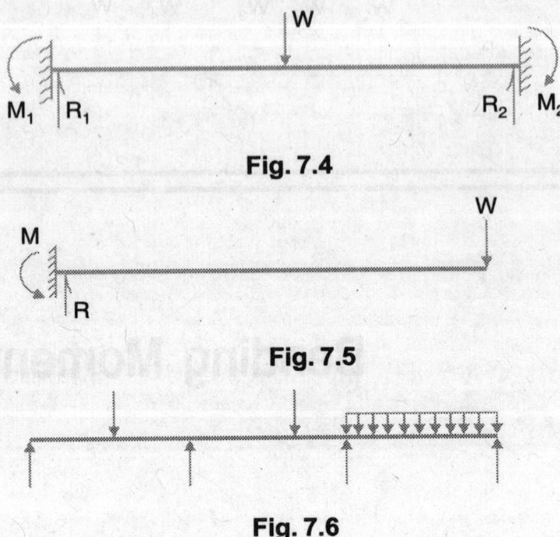

Fig. 7.4

Fig. 7.5

Fig. 7.6

Point Load. A *point load* or *concentrated load* is one which is considered to act at a point. In actual practice, the load has to be distributed over small area, because such small knife-edge contacts are generally neither possible no desirable.

Distributed Load. A *distributed load* is one which is distributed or spread in some manner over some length of the beam. If the spread is uniform (*i.e.*, at uniform rate, say w kg per metre run) it is said to be *uniformly distributed load* and is abbreviated as U.D.L.

If the spread is not at uniform rate, it is said to be non-uniformly distributed load. Triangular and trapezoidal distributed loads fall under this category.

7.2. SHEAR FORCE AND BENDING MOMENT

Consider a simply supported beam *AB* [Fig. 7.7 (*a*)] subjected to some point loads. If the beam is imagined to be cut in two parts at section X and the right hand portion of the beam is removed, the equilibrium of the left portion will be under the action of the external forces W_1, W_2, W_3 and reaction R_1, and under the action of internal forces which are distributed over the cross-section X. The internal forces at X represent the action of the right portion of the beam on the left portion.

From the law of parallel forces, the external forces W_1, W_2, W_3 and reaction R_1 can be represented by a single vertical force F acting at x' from the support. Introducing equal and opposite forces F at the section X [Fig. 7.7 (*b*)], the whole loading is equivalent to an unbalanced vertical force F acting upward and a moment M [Fig. 7.7 (*c*)] in the clockwise direction acting in the plane of X. Similarly if the equilibrium of the right hand portion of the beam is considered, the loading is reduced to an unbalanced vertical forces F, acting downwards and a moment M acting in the anticlockwise direction as shown in Fig. 7.7 (*d*). Hence the section X is subjected to :

(*i*) *a shear force F, which is defined as the algebraic sum of all vertical forces either to the left or to the right hand side of a section, and*

(*ii*) *a bending moment M, which is defined as the algebraic sum of the moments of all vertical forces either to the left or to the right of a section.*

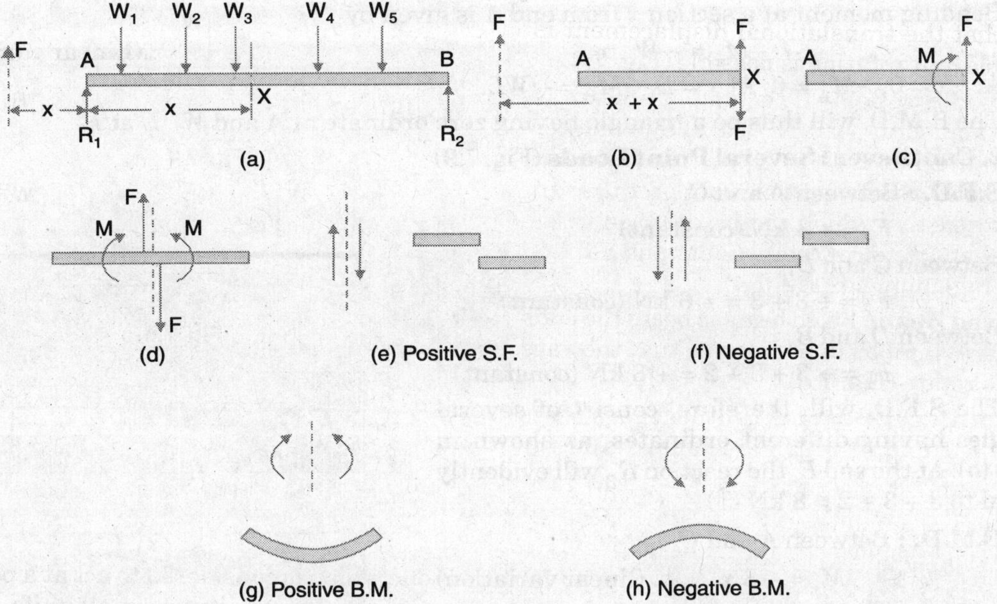

Fig. 7.7

Sign Conventions. For writing the general expressions for B.M. and S.F., we shall be adopting the following sign conventions :

(*a*) ***For Shear Force.*** Shear force having an upward direction to the left hand side of a section or downward direction to the right of the section will taken positive [Fig. 7.7(*e*)]. Similarly, a negative S.F. will be one that has a downward direction to the left of the section or upward direction to the right of the section [Fig. 7.7 (*f*)].

(*b*) ***For Bending Moment.*** A bending moment causing concavity upwards will be taken as positive and called as *saging B.M.* [Fig. 7.7 (*g*)]. Similarly, a bending moment causing convexity upwards will be taken as negative and called *hogging B.M.* [Fig. 7.7 (*h*)].

7.3. BENDING MOMENT AND SHEAR FORCE DIAGRAMS

A shear force diagram is the graphical representation of the variation of shear force along the length of the beam and is abbreviated as S.F.D.

A bending moment diagram is the graphical representation of the variation of the bending moment along the length of the beam and is abbreviated as B.M.D.

We will take different cases of beams and loadings for plotting S.F.D. and B.M.D.

1. Cantilever : Point Load at the End (Fig. 7.8)

At a section distant x from the end A, $F_x = +W$, and is constant for any position of the section. The S.F.D. will, therefore, be rectangle of height W. At the end B, the reaction R_B will evidently be equal to W (↑).

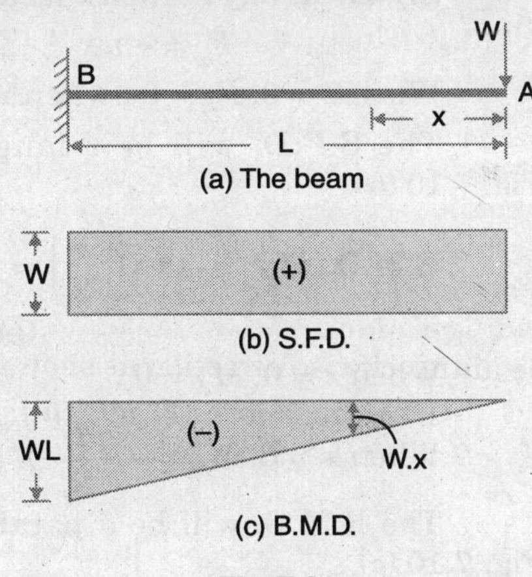

Fig. 7.8

Bending moment at a section x from end A is given by

$$M_x = -W \cdot x \qquad \text{...(Linear variation)}$$

At $x = 0$, $M_A = 0$; At $x = L$, $M_B = -WL$.

The B.M.D. will thus be a triangle having zero ordinate at A and $W \cdot L$ at B.

2. Cantilever : Several Point Loads (Fig. 7.9)

S.F.D. : Between A and C,

$$F_x = +3 \text{ kN (constant)}$$

Between C and D,

$$F_x = +3 + 3 = +6 \text{ kN (constant)}$$

Between D and B,

$$F_x = +3 + 3 + 2 = +8 \text{ kN (constant)}$$

The S.F.D. will, therefore, consist of several rectangles having different ordinates, as shown in Fig. 7.9 (*b*). At the end B, the reaction R_B will evidently be equal to $3 + 3 + 2 = 8$ kN (↑)

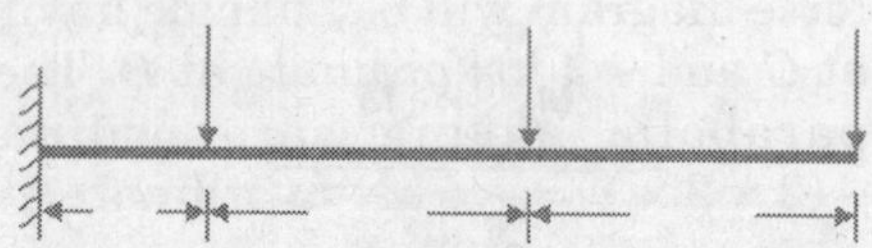

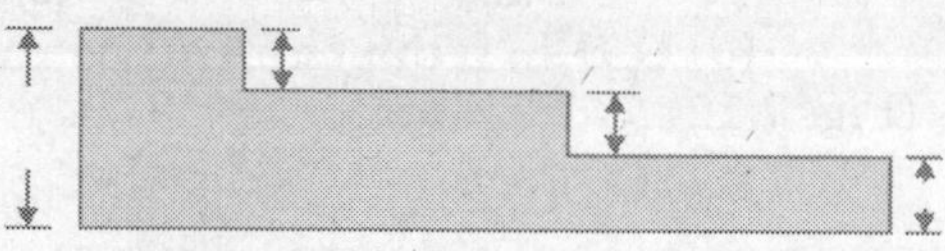

Fig. 7.9

B.M.D. : Between A and C,

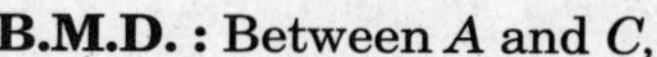

$$M_x = -3x \qquad \text{...(linear variation)}$$

When $x = 0$, $M_A = 0$.

When $x = 2$ m, $M_C = -3 \times 2 = -6$ kN-m

Between C and D, $M_x = -3x - 3(x - 2)$ ...(linear variation)

When $x = 2$ m, $M_C = -6$ kN-m (as before)

When $x = 4$ m, $M_D = -(3 \times 4) - (3 \times 2) = -18$ kN-m

Between D and B, $M_x = -3x - 3(x - 2) - 2(x - 4)$ ...(linear variation)

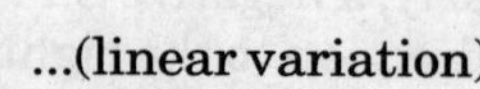

When $x = 4$ m, $M_D = -12 - 6 = -18$ kN-m (as before)

When $x = 5$ m, $M_B = -15 - 9 - 2 = -26$ kN-m.

The B.M.D. is shown in Fig. 7.9 (*c*).

3. Cantilever : U.D.L. Over Whole Span (Fig. 7.10).

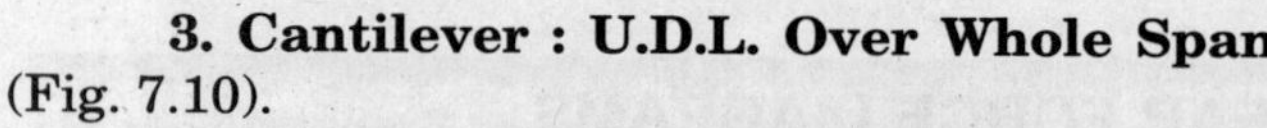

S.F.D. At any section x from A

$$F_x = +wx \qquad \text{...(linear variation)}$$

When $x = 0$, $F_A = 0$; when $x = L$, $F_B = +wL$.

The S.F.D. will be triangle, as shown in Fig. 7.10 (*b*).

B.M.D. $M_x = -(wx)\left(\dfrac{x}{2}\right) = -\dfrac{wx^2}{2}$ (parabolic variation)

When $x = 0$, $M_A = 0$;

When $x = L$, $M_B = -\dfrac{wL^2}{2}$.

The B.M.D. will be a parabola, as shown in Fig. 7.10 (*c*).

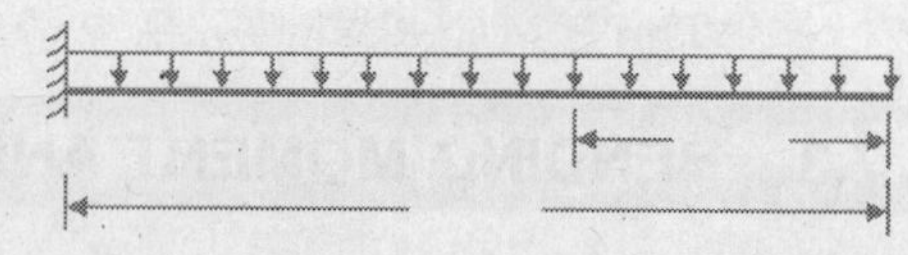

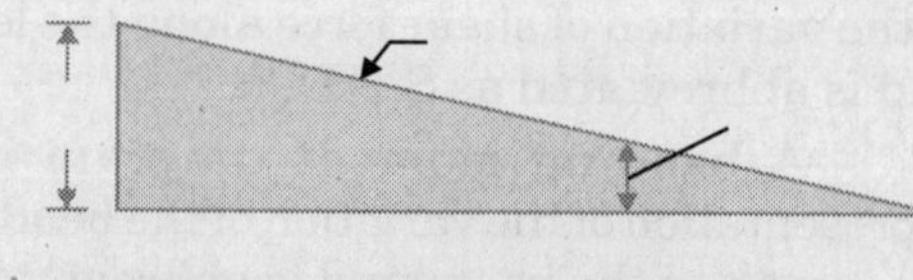

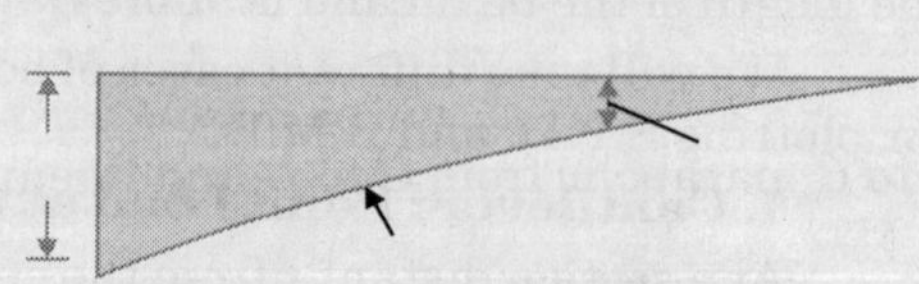

Fig. 7.10

4. Cantilever : U.D.L. not over the whole span

(*a*) U.D.L. from support to some distance (Fig. 7.11)

Since there is no loading from A to C, the shear force diagram will be triangle having zero ordinate at C and + 4 kN ordinate at B. The B.M.D. will be parabolic having zero ordinate at C and $-(2 \times 2 \times 1) = -4$ kN-m ordinate at B.

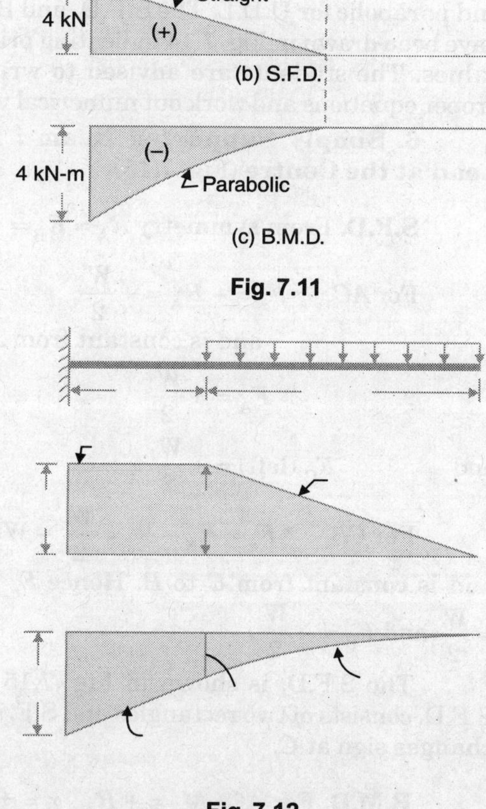

(a) The beam
(b) S.F.D.
(c) B.M.D.
Fig. 7.11

Fig. 7.12

(*b*) U.D.L. from the end to some distance (Fig. 7.12)

S.F.D. For AC, $F_x = +20x$ (linear variation)

At $x = 0$, $F_A = 0$;
At $x = 2$, $F_C = +20 \times 2 = +40$ kN
For CB, $F_x = +(20 \times 2) = +40$ kN (constant)

The S.F.D. is shown in Fig. 7.12 (*b*).

B.M.D. For AC, $M_x = -\dfrac{20x^2}{2}$ (parabolic)

At $x = 0$, $M_A = 0$;
At $x = 2$, $M_C = -40$ kN-m
For CB, $M_x = -40(x-1)$ (linear)
At $x = 2$ m, $M_C = -40$ kN-m (as before) ;
At $x = 3$ m, $M_B = -80$ kN-m
The B.M.D. is shown in Fig. 7.12 (*c*).

(*c*) U.D.L. Somewhere on the beam (Fig. 7.13)

The S.F.D. will be having zero ordinate from A to C, triangular from C to D and rectangular from D to B.

The B.M.D. will be having zero ordinate from A to C, parabolic from C to D and linear from D to B.

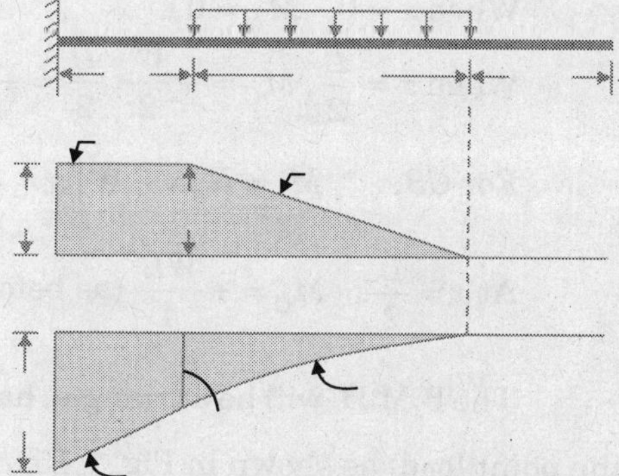
Fig. 7.13

5. Cantilever : Combination of Point Loads and U.D.L.

The same principle will be followed. The S.F.D. will be rectangular between point load to point load and triangular for U.D.L. Similarly, the B.M.D. will be triangular between point loads and parabolic for U.D.L. The S.F.D. and B.M.D. have been drawn in Fig. 7.14 indicating principal values. The students are advised to write the proper equations and work out numerical values.

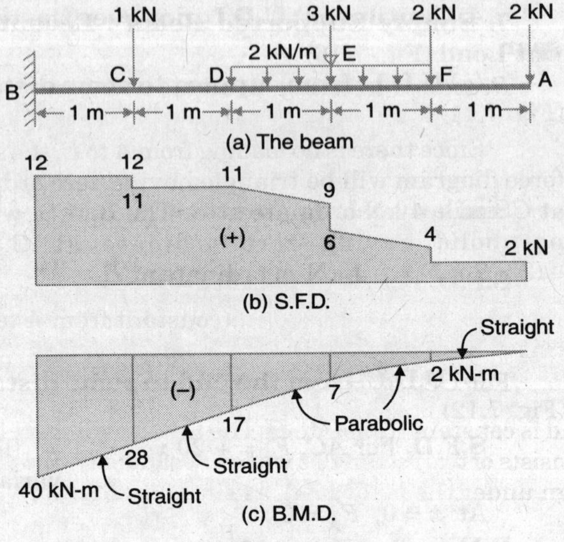

Fig. 7.14

6. Simply Supported Beam : Point Load at the Centre (Fig. 7.15)

S.F.D. From symmetry $R_A = R_B = \frac{W}{2}(\uparrow)$

For AC, $F_x = +R_A = +\frac{W}{2}$,

and is constant from A to C.

$\therefore \quad F_A = +\frac{W}{2}$

and $\quad F_C\,(\text{left}) = +\frac{W}{2}$

For CB, $\quad F_x = R_A - W = \frac{W}{2} - W = -\frac{W}{2}$

and is constant from C to B. Hence F_c (right) = $-\frac{W}{2}$ and $F_B = -\frac{W}{2}$.

The S.F.D. is shown in Fig. 7.15 (*a*). The S.F.D. consists of two rectangles and S.F. suddenly changes sign at C.

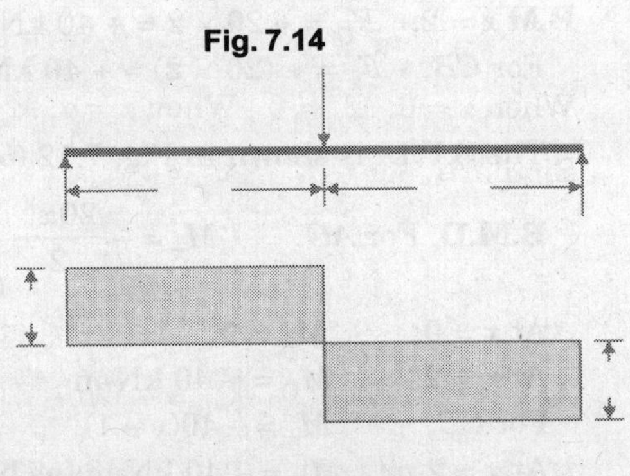

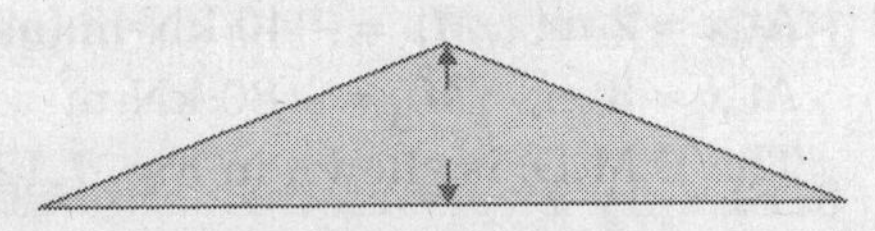

Fig. 7.15

B.M.D. For AC, $M_x = +R_A \cdot x = +\frac{W}{2}x$

(linear variation)

When $x = 0$, $M_A = 0$;

When $x = \frac{L}{2}$, $M_C = +\frac{W}{2} \cdot \frac{L}{2} = +\frac{WL}{4}$

For CB, $\quad M_x = R_A x - W\left(x - \frac{L}{2}\right) = \frac{W}{2}x - W\left(x - \frac{L}{2}\right)$ ...(linear)

At $x = \frac{L}{2}$, $\quad M_C = +\frac{WL}{4}$ (as before). At $x = L$, $M_B = \frac{WL}{2} - W\left(L - \frac{L}{2}\right) = 0$

The B.M.D. will be a triangle, having maximum ordinate of $+\frac{WL}{4}$ at the centre (*i.e.* under the point load) as shown in Fig. 7.15 (*c*).

7. Simply Supported Beam : Eccentric Point Load (Fig. 7.16)

Reactions : Taking moments at *B*,

$$R_A = \frac{Wb}{L}(\uparrow)$$

Similarly, taking moment at *A*, $R_B = \dfrac{Wa}{L}(\uparrow)$

S.F.D. For *AC*, $F_x = + R_A = + \dfrac{Wb}{L}$ ...(constant from *A* to *C*)

For *CB*, $F_x = \dfrac{Wb}{L} - W = \dfrac{W(b-L)}{L} = -\dfrac{Wa}{L}$,

and is constant from *C* to *B*. The S.F.D. will therefore consists of two rectangles with S.F. suddenly changing sign under the point load, as shown in Fig. 7.16 (*b*).

Fig. 7.16

B.M.D. For *AC*, $M_x = + R_A \,.\, x = + \dfrac{Wb}{L}x$ ...(linear variation)

When $x = 0$, $M_A = 0$; When $x = a$, $M_C = + \dfrac{Wab}{L}$

Similarly, for *CB*, $M_x = + R_A.\, x - W(x - a)$

or $M_x = + \dfrac{Wb}{L}x - W(x - a)$ ...(linear variation)

At $x = a$, $M_C = + \dfrac{Wab}{L}$. as before

At $x = L$, $M_B = + Wb - W(L - a) =$ zero.

Hence, the B.M.D. will be a triangle having maximum ordinate of $+\dfrac{Wab}{L}$ under the point load.

8. Simply Supported Beam : Several Point Loads (Fig. 7.17)

Taking moments about *B*,

$$R_A = \frac{1}{4}(10 \times 3 + 20 \times 2 + 10 \times 1) = 20 \text{ kN } (\uparrow)$$

Hence $R_B = (10 + 20 + 10) - 20 = 20$ kN (↑)

S.F.D. $F_A = + R_A = + 20$ kN

F_C (left) = + 20 kN

F_C (right) = + 20 – 10 = + 10 kN

F_D (left) = + 20 – 10 = + 10 kN

F_D (right) = + 20 – 10 – 20 = – 10 kN

F_E (left) = + 20 – 10 – 20 = – 10 kN

F_E (right) = + 20 – 10 – 20 – 10 = – 20 kN

$F_B = - R_B = - 20$ kN

The S.F.D. is shown in Fig. 7.17 (*b*)

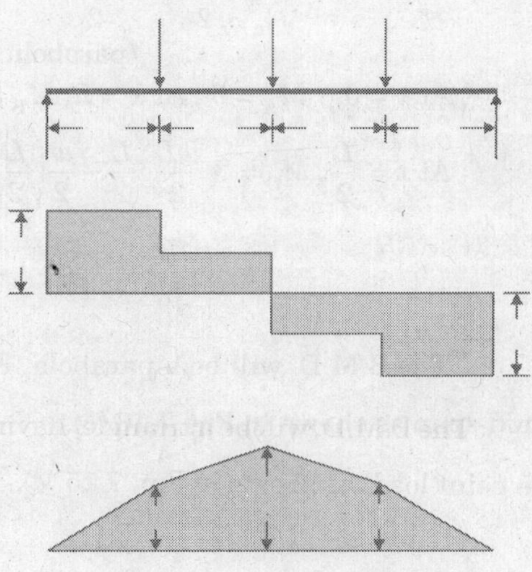

Fig. 7.17

B.M.D.

For AC, $M_x = + R_A . x = + 20x$ (linear)

$\therefore \quad M_A = 0 ; M_C = + 20 \times 1 = + 20$ kN-m

For CD, $M_x = + 20x - 10(x - 1)$ ...(linear)

$\therefore \quad M_C = + 20$ kN-m (as before)

$M_D = + 30$ kN-m

For DE, $M_x = + 20x - 10(x - 1) - 20(x - 2)$

$\therefore \quad M_D = + 30$ kN-m (as before)

and $\quad M_E = + 20$ kN-m

Lastly, $M_B = + 20 \times 4 - 10 \times 3 - 20 \times 2 - 10 \times 1 = 0$

The B.M.D. is shown in Fig. 7.17 (*c*).

9. Simply Supported Beam : U.D.L. Over the Whose Span (Fig. 7.18)

Due to symmetry, the reaction $R_A = R_B = \frac{wL}{2} (\uparrow)$

S.F.D. $F_x = + R_A - wx = \frac{wL}{2} - wx$ ...(linear variation)

At $x = 0$, $\quad F_A = + \frac{wL}{2}$;

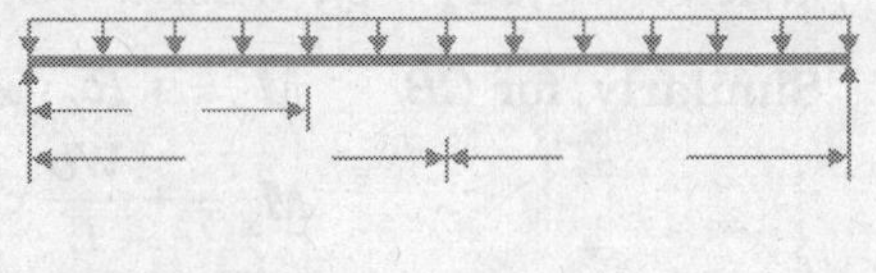

At $x = \frac{L}{2}$, $F_C = \frac{wL}{2} - \frac{wL}{2} = 0$

At $x = L$, $\quad F_B = \frac{wL}{2} - wL = - \frac{wL}{2}$

The S.F.D. is shown in Fig. 7.18 (*b*).

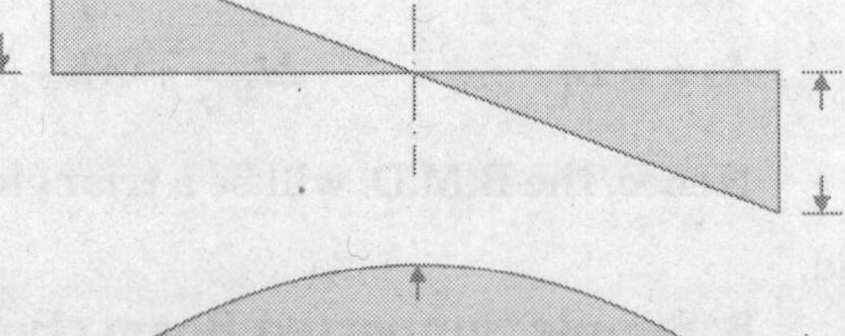

Fig. 7.18

B.M.D. $M_x = + R_A . x - wx . \frac{x}{2}$

$$= + \frac{wL}{2} x - \frac{wx^2}{2}.$$

(parabolic variation)

At $x = 0$, $\quad M_A = 0$; At $x = L$, $M_B = 0$

At $x = \frac{L}{2}$, $M_C = + \frac{wL}{2} . \frac{L}{2} - \frac{w}{2}\left(\frac{L}{2}\right)^2$

$$= + \frac{wL^2}{8}$$

The B.M.D. will be a parabola, Fig. 7.18 (*c*), with a maximum ordinate of $+ \frac{wL^2}{8}$ at the mid-span as shown in Fig. 7.18 (*c*).

10. Simply Supported Beam : Combination of Loads (Fig. 7.19)

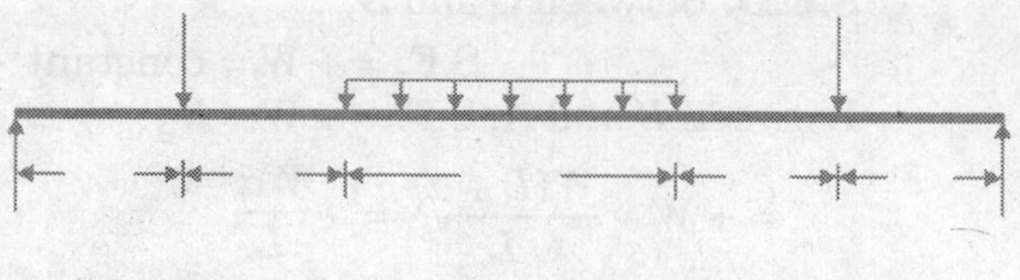

Reactions : Taking moments at *B*,

$6R_A = 1 \times 5 + (2 \times 2 \times 3) + 4 \times 1 = 21$

$R_A = 3.5$ kN ;

$R_B = (1 + 4 + 4) - 3.5 = 5.5$ kN

S.F.D. For *AC*, $F = + R_A = + 3.5$ kN

For *CD*, $F = 3.5 - 1 = 2.5$ kN

For *DE*, $F = + 3.5 - 1 - 2(x - 2)$

$= 6.5 - 2x$

(linear variation) ...(1)

which gives $F_D = + 2.5$ kN;

$F_E = 6.5 - 2 \times 4 = - 1.5$ kN

The S.F. is zero between *D* and *E* (Eq. 1), at $x = \dfrac{6.5}{2} = 3.25$ m from *A*.

For *EF*, $F = + 3.5 - 1 - 4$

$= - 1.5$ kN

For *FB*, $F = + 3.5 - 1 - 4 - 4$

$= - 5.5 \text{ kN} = - R_B$

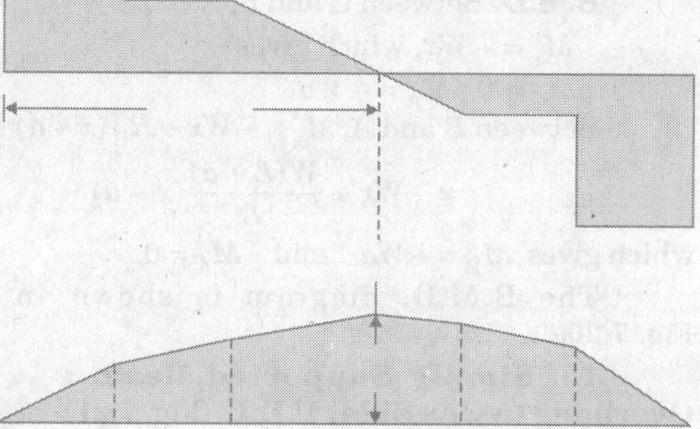

Fig. 7.19

B.M.D. For *AC*, $M_x = + R_A x = + 3.5\,x$ (linear), which gives

$M_A = 0$; $M_C = + 3.5$ kN-m.

For *CD*, $M_x = + 3.5\,x - 1(x - 1) = 2.5\,x + 1$, which gives

$M_C = 2.5 \times 1 + 1 = + 3.5$ kN-m and $M_D = 2.5 \times 2 + 1 = + 6$ kN-m

For *DE*, $M_x = + 3.5x - 1(x - 1) - \dfrac{2(x-2)^2}{2} = + 3.5x - x + 1 - (x^2 + 4 - 4x)$

$= 6.5\,x - x^2 - 3$ (parabolic)

This gives $M_D = 6.5 \times 2 - 2^2 - 3 = + 6$ kN-m (as before)

and $M_E = 6.5 \times 4 - 4^2 - 3 = 7$ kN-m

For *M* to be maximum in *DE*, $\dfrac{dM_x}{dx} = 0 = 6.5 - 2x$

which gives $x = 6.5/2 = 3.25$ m

Hence B.M. is maximum where S.F. is zero.

$M_{max} = 6.5(3.25) - (3.25)^2 - 3 = 7.5625$ kN-m

The S.F. and B.M. diagrams are shown in Fig. 7.19 (*b*) and (*c*) respectively.

11. Simply Supported Beam : Overhang to one Side : Point Load : (Fig. 7.20)

Taking moments about *A*, we get

$$R_B = \frac{W(L+a)}{L} (\uparrow)$$

$$\therefore \quad R_A = W - R_B = W - \frac{W(L+a)}{L} = - \frac{Wa}{L} \quad (\textit{i.e., acting downwards})$$

S.F.D. Between C and B,

$$\text{S.F.} = +W = \text{constant}$$

Between B and A, S.F. $= +W - R_B$

$$= +W - \frac{W(L+a)}{L} = -\frac{Wa}{L}$$

The S.F.D. is shown in Fig. 7.20 (*b*).

B.M.D. Between C and B,

$M_x = -Wx$, which gives

$M_C = 0$; $M_A = -Wa$.

Between B and A, $M_x = -Wx + R_B(x-a)$

$$= -Wx + \frac{W(L+a)}{L}(x-a)$$

which gives $M_B = -Wa$ and $M_A = 0$.

The B.M.D. diagram is shown in Fig. 7.20(*c*).

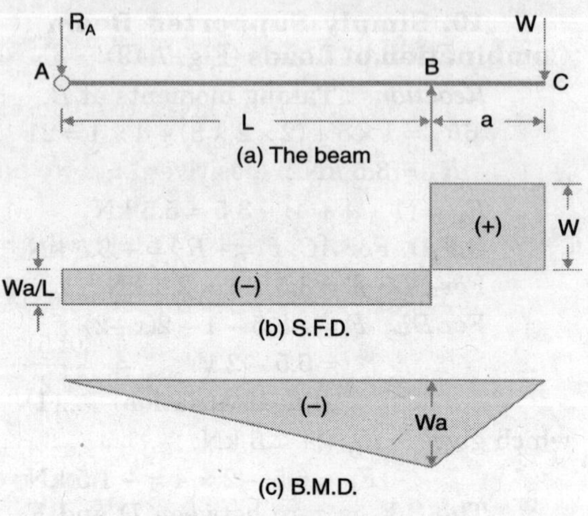

Fig. 7.20

12. Simply Supported Beam : Overhang to One Side : U.D.L. (Fig. 7.21)

Taking moments about A, we get

$$R_B = \frac{w(L+a)^2}{2L}$$

$$\therefore\ R_A = w(L+a) - \frac{w(L+a)^2}{2L}$$

$$= \frac{w(L+a)(L-a)}{2L}$$

If $L < a$, R_A will be acting downwards,

$L > a$, R_A will be acting upwards.

Let us take a numerical example, when $L = 4$ m, $a = 2$ m and $w = 1$ kN/m.

$$R_B = \frac{1(4+2)^2}{8} = \frac{36}{8} = 4.5 \text{ kN} ;$$

$$R_A = \frac{1 \times 6 \times 2}{8} = 1.5 \text{ kN.}$$

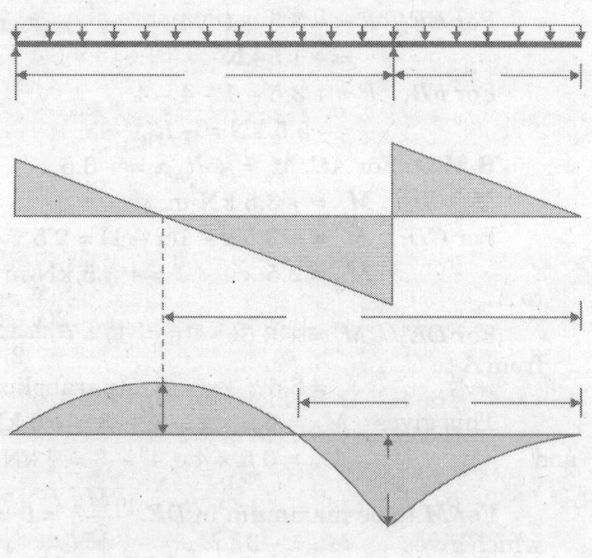

Fig. 7.21

S.F.D. For CB, $F_x = +x$ (x being measured from C)

$$F_C = 0 \text{ ; } F_B \text{ (right)} = +2 \text{ kN}$$

For BA, $F_x = +x - 4.5$, which gives

$$F_B \text{ (left)} = -2.5 \text{ kN} \text{ ; } F_A = +1.5 \text{ kN}$$

S.F. will be zero at $x = 4.5$ m. S.F.D. is shown in Fig. 7.21 (*b*).

B.M.D. For CB, $M_x = -\dfrac{wx^2}{2} = -\dfrac{x^2}{2}$, (parabolic variation)

which gives $M_C = 0$; $M_B = -2$ kN-m

For BA, $M_x = -\dfrac{x^2}{2} + 4.5(x-2)$, which gives

$$M_B = -2 \text{ kN-m (as before) and } M_A = 0.$$

For maximum positive B.M., we have $\dfrac{dM_x}{dx} = 0$

$\therefore \quad -\dfrac{2x}{2} + 4.5 = 0$, which gives $x = 4.5$ m

Thus, B.M. is maximum where S.F. is zero.

$$\therefore \quad M_{max} = -\frac{(4.5)^2}{2} + 4.5(2.5) = -\frac{20.25}{2} + 11.25 = +1.125 \text{ kN-m}$$

For M to be zero, we have $-\dfrac{x^2}{2} + 4.5(x-2) = 0$

or $\quad x^2 - 9x + 18 = 0$, which gives $x = 3$ m.

The B.M.D. is shown in Fig. 7.21 (*c*).

13. Simply Supported Beam : Overhangs to Both Sides : Point Loads (Fig. 7.22)

Reactions : Taking moments about A.

$$(3 \times 4) = (R_B \times 3) + (3 \times 2)$$

$$R_B = \frac{12-6}{3} = 2 \text{ kN};$$

$$R_A = 6 - 2 = 4 \text{ kN}$$

S.F.D. For DB,

$F = +3$ kN, constant from D to B.

For BA, $F = +3 - 2 = +1$ constant from B to A.

For AC, $F = +3 - 2 - 4 = -3$ kN, constant from A to C.

The S.F.D. is shown in Fig. 7.22 (*b*).

B.M.D. : For DB, $M_x = -3x$,

which gives $M_D = 0$; $M_B = -3$ kN-m

For BA, $M_x = -3x + 2(x-1) = -(x+2)$

which gives $\quad M_B = -3$; $M_A = -6$ kN-m

For AC, $M_x = -3x + 2(x-1) + 4(x-4)$

which gives $\quad M_B = -6$ kN-m ; $M_D = 0$.

The B.M.D. is shown in Fig. 7.22 (*c*).

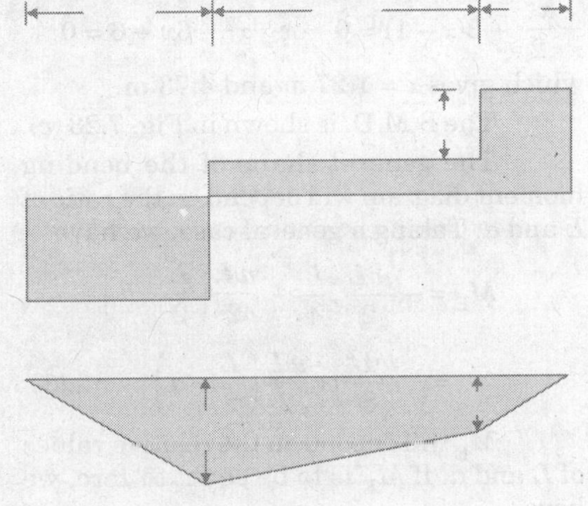

Fig. 7.22

14. Simply Supported Beam : Overhangs to Both Sides : U.D.L. (Fig. 7.23)

$$R_A = R_B = \frac{wL}{2} \text{ (due to symmetry)}$$

S.F.D. For DB, $\quad F_x = +wx$, which gives $F_D = 0$, F_B (right) $= +wa$

For BA, $\quad F_x = +wx - \dfrac{wL}{2}$

which gives $\quad F_B \text{ (left)} = +wa - \dfrac{wL}{2} = -w\left(\dfrac{L}{2} - a\right)$

Taking $a = 1$ m, $L = 6$ m, $w = 1$ kN/m ; $R_A = R_B = 3$ kN

$$F_B \text{ (right)} = +1 \text{ kN} ; F_B \text{ (left)} = -1\left(\frac{6}{2} - 1\right) = -2 \text{ kN}$$

$$F_A \text{ (right)} = +5 - 3 = +2 \text{ kN}$$

$$F_A \text{ (left)} = +5 - 3 - 3 = -1 \text{ kN} ; F_D = 0.$$

B.M.D. For DB, $\quad M_x = -\dfrac{x^2}{2}$ ...(parabolic variation)

which gives $M_D = 0$; $M_B = -0.5$ kN-m

For BA, $M_x = -\dfrac{x^2}{2} + 3(x-1)$, which gives $M_A = -12.5 + 12 = -0.5$.

Similarly, $\quad M_C = 0$.

Bending moment will be maximum at $x = 3$, where the S.F. is zero and its value there will be $= -\dfrac{9}{2} + 3(3-1) = +1.5$ kN-m.

Bending moment will be zero when

$$-\frac{x^2}{2} + 3(x-1) = 0 \quad \text{or} \quad x^2 - 6x + 6 = 0$$

which gives $x = 1.27$ m and 4.73 m.

The B.M.D. is shown in Fig. 7.23 (*c*).

The general shape of the bending moment diagram will depend on the ratio of L and a. Taking a general case, we have

$$M_E = -\frac{wL}{2}\cdot\frac{L}{4} + \frac{wL}{2}\left(\frac{L}{2} - a\right)$$

$$= -\frac{wL^2}{8} + \frac{wL}{2}\left(\frac{L}{2} - a\right) \quad \text{...(1)}$$

M_E will depend on the relative values of L and a. If M_E is to be equal to zero, we have

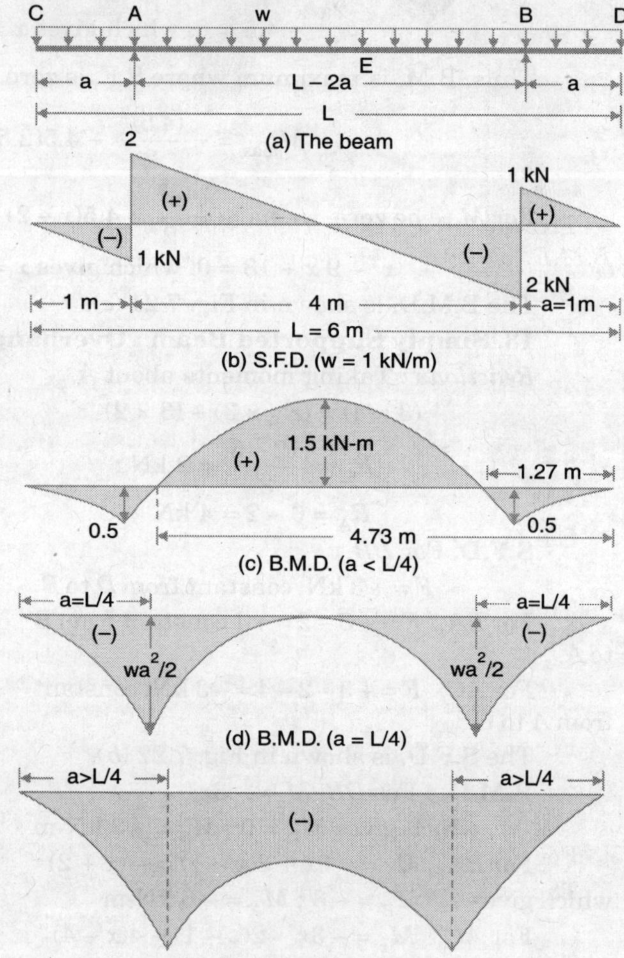

Fig. 7.23

$$-\frac{wL^2}{8} + \frac{wL}{2}\left(\frac{L}{2} - a\right) = 0$$

$$L = 4\left(\frac{L}{2} - a\right) = 2L - 4a$$

or $\quad a = \dfrac{L}{4}$

Thus, if $a = \dfrac{L}{4}$, the bending moment at E will be zero as shown in Fig. 7.23 (*d*).

Again, for M_E to be negative, we have

or $$\frac{wL^2}{8} > \frac{wL}{2}\left(\frac{L}{2} - a\right)$$

or $$L > 4\left(\frac{L}{2} - a\right) > (2L - 4a) \quad \text{or} \quad a > \frac{L}{4}$$

Hence if : $a = \frac{L}{4}$, $M_E = 0$ [See Fig. 7.23 (*d*)]

$a > \frac{L}{4}$, M_E is negative [See Fig. 7.23 (*e*)]

$a < \frac{L}{4}$, M_E is positive [See Fig. 7.23 (*c*)]

Example 7.1. *A simply supported beam of 9 m span is loaded as shown in Fig. 7.24. Draw the B.M.D. and S.F.D. indicating principal values.*

Solution.

Reactions :

$$9R_B = (40 \times 2) + (64 \times 7)$$
$$= 80 + 448 = 528$$

$\therefore$ $R_B = \frac{528}{9} = 58.7$ kN ;

$R_A = 104 - 58.7 = 45.3$ kN.

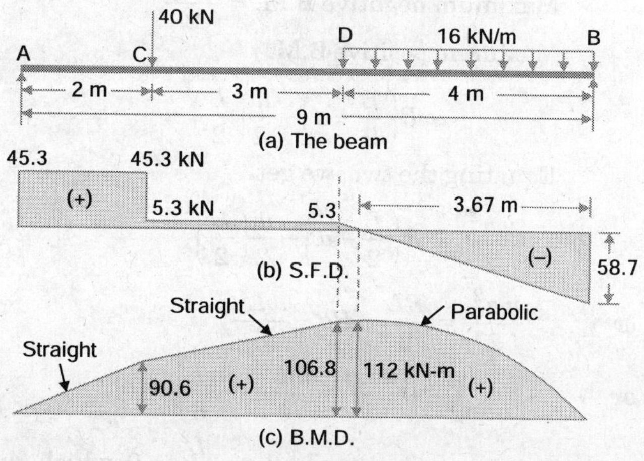

(a) The beam

(b) S.F.D.

(c) B.M.D.

Fig. 7.24

S.F.D. For *BD*, $F_x = -58.7 + 16x$ ...(linear variation)

This gives $F_B = -58.7$ kN

and $F_D = +5.3$ kN

Also, F is 0 at $x = \frac{58.7}{16} = 3.67$ m

For *DC*, $F_x = -58.7 + 64$
$= +5.3$ kN (constant)

For *CA*, $F_x = -58.7 + 64 + 40 = +45.3$ kN (constant)

The S.F.D. is shown in Fig. 7.24 (*b*).

B.M.D. For *BD*, $$M_x = +58.7x - \frac{16x^2}{2} = +58.7x - 8x^2 \quad \text{...(parabolic)}$$

which gives $M_B = 0$ and $M_D = +234.8 - 128 = +106.8$ kN-m.

For M to be maximum in *BD*, $\frac{dM_x}{dx} = 0$

$\therefore$ $+58.7 - 16x = 0$, which gives $x = 3.67$ m (*i.e.*, where S.F. is zero).

$$M_{max} = +58.7 \times 3.67 - 8(3.67)^2 = +216 - 104 = +112 \text{ kN-m}$$

For *DC*, $M_x = +58.7x - 64(x - 2)$ ...(linear)

which gives $M_D = +106.8$ kN-m (as before), $M_C = +90.6$ kN-m ; and $M_A = 0$.

Example 7.2. *Two slings are to be used in raising a newly cast reinforced concrete pile of span L and uniform cross-section, the pile remaining horizontal during the lift.*

Determine the most suitable position for the slings and sketch the shearing force and bending moment diagram. It may be assumed that the damage to the pile due to the shearing force is negligible and that failure would be by bending the pile under its own weight.

Solution. Let the self-weight of the pile = w/unit length.

The position of the slings will be most suitable if the maximum bending moment at any point is as minimum as possible. This can be there when the max. negative moment is equal to the max. positive moment. Let the position of each sling be at a from the corresponding ends.

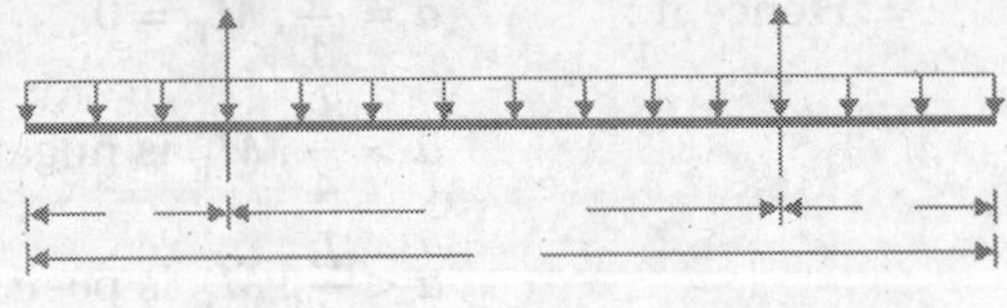

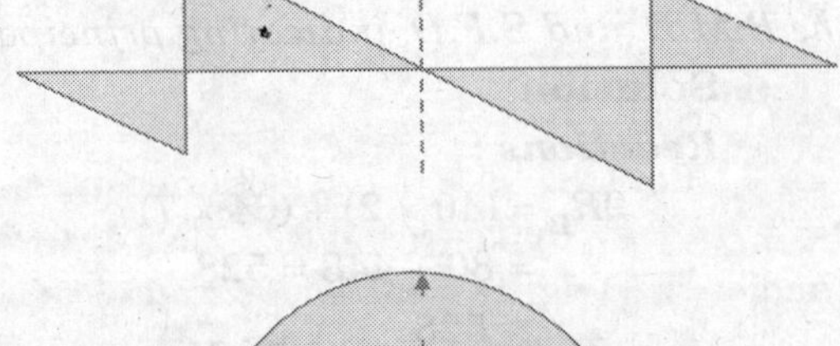

Fig. 7.25

$$R = \frac{wL}{2}$$

Maximum negative B.M. $= \dfrac{wa^2}{2}$

Maximum positive B.M.

$$= R\left(\frac{L}{2} - a\right) - \frac{w}{2}\left(\frac{L}{2}\right)^2$$

Equating the two, we get

$$\frac{wa^2}{2} = R\left(\frac{L}{2} - a\right) - \frac{w}{2}\left(\frac{L}{2}\right)^2$$

or $$\frac{wa^2}{2} = \frac{RL}{2} - Ra - \frac{wL^2}{8}$$

or $$\frac{wa^2}{2} + \frac{wLa}{2} - \frac{wL^2}{4} + \frac{wL^2}{8} = 0$$

or $$a^2 + La - \frac{L^2}{4} = 0 \text{ which gives } a = \mathbf{0.207\ L.}$$

$$\therefore \quad M_B = \frac{wa^2}{2} = \frac{w}{2}(0.207)^2 L^2 = 0.0215\ wL^2 = M_E$$

The B.M. diagram is shown in Fig. 7.25 (*c*).

S.F.D. Between D and C

$$F_x = + wx, \text{ (linear) which give } F_C \text{ (right)} = 0.207\ wL$$

For CB, $\quad F_x = + wx - 0.5\ wL,$

which gives $\quad F_C \text{ (left)} = + 0.207\ wL - 0.5\ wL = - 0.293\ wL$

S.F. will be zero at $x = \dfrac{L}{2}$. The S.F.D. is shown in Fig. 7.25 (*b*).

Example 7.3. *A beam ABCD 20 m long is loaded as shown in Fig. 7.26. The beam is supported at B and C and has an overhang of 2 m to the left of the support B and an overhang of K metres to the right of support C which is in the right hand half of the beam. Determine the value of K if the mid-point of the beam is a point of inflexion and for this arrangement, plot B.M. and S.F. diagrams indicating the principal numerical values.*

Solution.

Reactions : Since mid-point E is the point of inflexion, we have

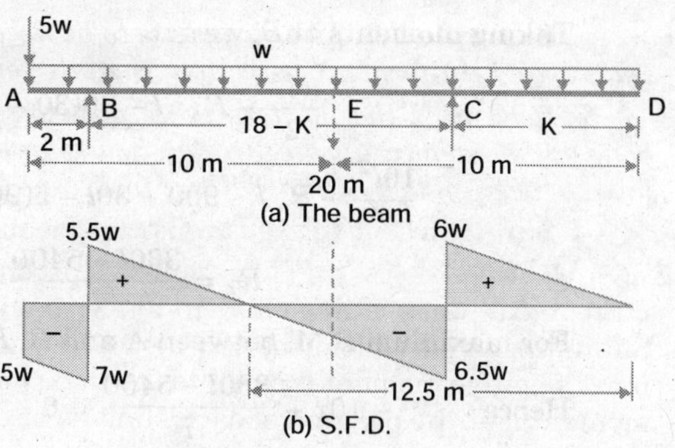

(a) The beam
(b) S.F.D.

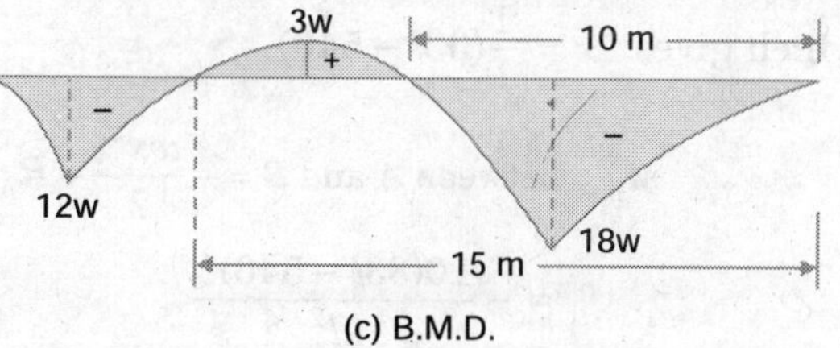

(c) B.M.D.

Fig. 7.26

$$M_E \text{ (left)} = -\frac{w(10)^2}{2} - 5w(10) + 8R_B = 0$$

$$\therefore \quad R_B = \frac{100}{8} w = 12.5\, w$$

and $\quad R_C = (20w + 5w) - 12.5w = 12.5w$

Again, $M_E \text{ (right)} = -\frac{w(10)^2}{2} + R_C(10 - K) = 0$

or $\quad -50w + 12.5w(10 - K) = 0,$

which gives $\quad$ **K = 6 m.**

S.F.D. For DC, $F = +\, wx$, from which

$F_C \text{ (right)} = +\, 6w$

For CB, $F = +\, wx - 12.5w$, from which

$F_C \text{ (left)} = -\, 6.5w$

and $\quad F_B \text{ (right)} = +\, 18w - 12.5w = +\, 5.5w$

S.F. is zero at $x = 12.5$ m.

For BA, $F_x = +\, wx - 12.5w - 12.5w$ from which $F_B \text{ (left)} = -\, 7w$

and $\quad F_A = -\, 7w + 2w = -\, 5w$

The S.F.D. is shown in Fig. 7.26 (*b*).

B.M.D. For CB, $\quad M_x = -\frac{wx^2}{2} + 12.5w(x - 6)$

which gives $\quad M_C = -\, 18w\, ;\; M_E = 0\, ;\; M_B = -\, 12w.$

For M to be zero, $-\frac{wx^2}{2} + 12.5w(x - 6) = 0$ which gives $x = 10$ and $x = 15$ m.

M will be maximum positive at $x = 12.5$ m, where S.F. is zero.

$$M_{max} = -\frac{w(12.5)^2}{2} + 12.5w(6.5) = 3w.$$

The B.M.D. is shown in Fig. 7.26 (*c*).

Example 7.4. *A horizontal beam, 30 m long, carries a uniformly distributed load of 10 kN/m over the whole length and a concentrated load of 30 kN at the right end. If the beam is freely supported at the left end, find the position of the second support so that the bending moment on the beam should be as small as possible.*

Draw the diagrams of shearing force and bending moment and insert the principal values.

Solution. *For the greatest B.M. to be as small as possible the* $+\, M_{max}$ *between A and B must be equal to* $-\, M_{max}$ *(or* M_B*).*

Let the support at B be at a distance l from A.

Taking moments at B, we get

$$\frac{wl^2}{2} - R_A \,.\, l - 30(30 - l) - \frac{w}{2}(30 - l)^2 = 0$$

or
$$\frac{10l^2}{2} - R_A\, l - 900 + 30l - 5(900 - 60l + l^2) = 0$$

$$\therefore \qquad R_A = \frac{330l - 5400}{l}$$

For maximum B.M. between A and B, $F_x = 0$

Hence
$$-10x + \frac{330l - 5400}{l} = 0$$

which gives $x = \frac{1}{l}(33l - 540)$.

$$\therefore \quad M_{max} \text{ between } A \text{ and } B = -\frac{wx^2}{2} + R_A \,.\, x$$

$$= -\frac{10(33l - 540)^2}{2l^2} + (330l - 5400)\frac{(33l - 540)}{l^2}$$

$$= +\frac{10(33l - 540)^2}{2l^2}$$

Thus the value of maximum positive B.M.

$$= \frac{10(33l - 540)^2}{2l^2} \qquad \text{...(1)}$$

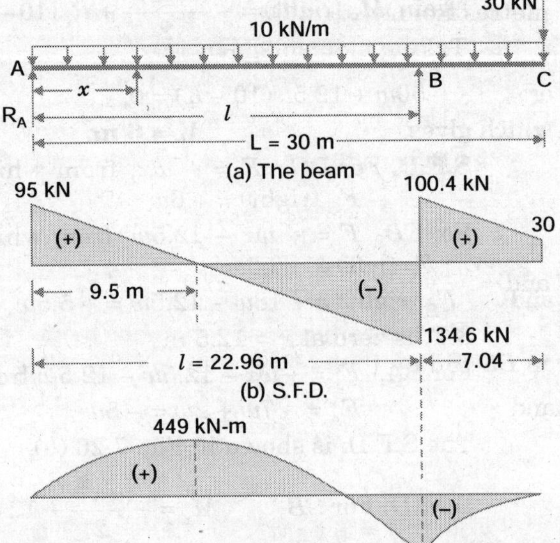

Fig. 7.27

Also, $M_B = -30(30 - l) - \frac{10}{2}(30 - l)^2$

$$= -\left(\frac{30 - l}{2}\right)(60 + 300 - 10l)$$

$$= -\frac{10}{2}(30 - l)(36 - l)$$

Thus the value of max. negative B.M. $= \frac{10}{2}(30 - l)(36 - l)$ ...(2)

Equating (1) and (2), we get

$$\frac{10(33l - 540)^2}{2l^2} = \frac{10}{2}(30 - l)(36 - l)$$

Simplifying and rearranging the above, we get

$$l^4 - 66l^3 - 9l^2 = 291600 - 35640\, l \quad \text{or} \quad y_1 = y_2$$

Plotting y_1 and y_2 between $l = 21$ to $l = 24$ (or solving by trial and error), we get $l =$ **22.96 m.**

For maximum B.M., $x = 33 - \frac{540}{22.96} = 9.5$ m

$$\therefore \qquad M_{max} = + \frac{10(33 \times 22.96 - 540)^2}{2(22.96)^2} = \mathbf{+\ 449\ kN\text{-}m}$$

The S.F. and B.M. diagrams are shown in Fig. 7.27 (*b*) and (*c*) respectively.

Example 7.5. *A girder 10 m long carrying a uniformly distributed load of w per metre run is to be supported on two piers 6 m apart so that the greatest bending moment on the girder shall be as small as possible. Find the distances of the piers from the ends of the girder and the maximum bending moment. Plot the B.M. and S.F. diagrams.*

Solution. (Fig. 7.28)

Let the distance of the pier *C* be at '*a*' metres from the end *D*.

Taking moments about *D*,

$$R_C \cdot a + R_B(6 + a) = 10w(5) \quad ...(i)$$

Also $$R_B + R_C = 10w \quad ...(ii)$$

From (*i*) and (*ii*), we get

$$R_C = \frac{5}{3} w(1 + a)$$

and $$R_B = \frac{5}{3} w(5 - a)$$

For maximum positive bending moment in *BC*, S.F. = 0

$$\therefore \quad F_x = + wx - R_C$$

$$= + wx - \frac{5}{3} w(1 + a) = 0$$

$$\therefore \quad 3x = 5(1 + a) \quad \text{or} \quad x = \frac{5}{3}(1 + a)$$

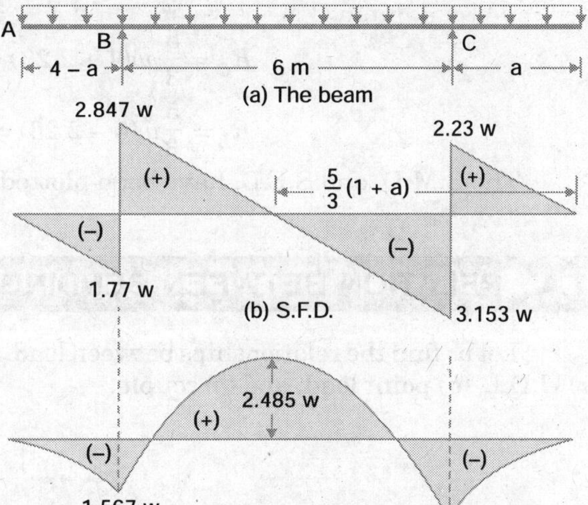

Fig. 7.28

Maximum negative moment $= M_C = \dfrac{wa^2}{2}$ ...(1)

Maximum positive moment $= R_C(x - a) - \dfrac{wx^2}{2}$

$$= \frac{5}{3}w(1+a)\left\{\frac{5}{3}(1+a) - a\right\} - \frac{w}{2}\left\{\frac{5}{3}(1+a)\right\}^2$$

$$= \frac{5}{9} w(1 + a)(5 + 2a) - \frac{25}{18} w(1 + a)^2$$

$$= \frac{25}{18} w(1+a)\left\{\frac{2}{5}(5 + 2a) - (1 + a)\right\}$$

$$= \frac{25}{18} w \frac{(1+a)}{5} \{10 + 4a - 5 - 5a\}$$

$$= \frac{5}{18} w(1 + a)(5 - a) = \frac{5}{18} w(5 + 4a - a^2) \quad ...(2)$$

Equating (1) and (2), we get

$$\frac{wa^2}{2} = \frac{5}{18} w(5 + 4a - a^2)$$

or $\quad 14a^2 - 20a - 25 = 0$, which gives $a \simeq 2.23$ m.

$$\therefore \quad M_C = M_{\max} = \frac{w}{2}(2.23)^2 = 2.485w$$

$$M_B = \frac{w}{2}(4 - a)^2 = \frac{w}{2}(4 - 2.23)^2 = 1.567w$$

$$R_C = \frac{5}{3} w(1 + 2.23) = 5.383w$$

$$R_B = \frac{5}{3} w(5 - 2.23) = \frac{5}{3} \times 2.77w = 4.617w$$

The B.M.D. and S.F.D. have been plotted in Fig. 7.28.

7.4. RELATION BETWEEN BENDING MOMENT AND SHEAR FORCE

Let us find the relationships between load, S.F. and B.M. for three different types of loading : (*a*) U.D.L. (*b*) point load, and (*c*) couple.

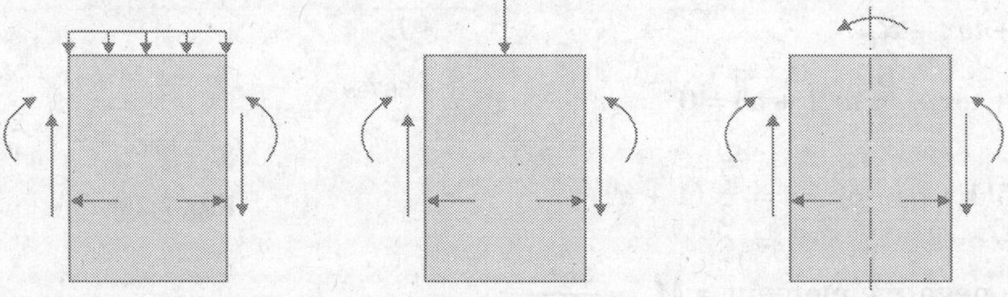

Fig. 7.29

(*a*) Uniformly distributed load (U.D.L.)

Fig. 7.29 (*a*) shows an element *ABCD* of a beam, cut out between two cross-section *dx* apart. Let the S.F. and B.M. on the face *AD* be *F* and *M* respectively, both acting in their positive directions. The corresponding stress resultants on the right hand section *BC* be $F + dF$ and $M + dM$ respectively where *dF* and *dM* be the increments in S.F. and B.M. in the length *dx* of the beam. Let the U.D.L. *w* be taken positive when acting downwards.

For the equilibrium of the forces in the vertical direction, we have

$$F - (F + dF) - w dx = 0$$

From which, we get

$$\frac{dF}{dx} = -w \quad \text{...(7.1)}$$

Thus the rate of change of S.F. (and hence the slope of the S.F.D.) is equal to $-w$.

Again consider the equilibrium of the element by taking the algebraic sum of moments about the left hand face *AD* of the element, considering counter clockwise moment as positive.

Thus $\quad -M-(F+dF)dx - wdx \cdot \dfrac{dx}{2} + (M+dM) = 0$

Neglecting the products and squares of small quantities, we get

$$F dx - dM = 0 \quad \text{or} \quad \frac{dM}{dx} = F \qquad \text{...(7.2)}$$

Thus the rate of change of B.M. along the length of the beam is equal to S.F.

Integrating equation 7.1 between two sections A and B on the beam, we get

$$\int_A^B dF = -\int_A^B wdx \qquad \text{...[7.1 }(a)]$$

or $$F_B - F_A = -\int_A^B wdx = -\left(A_w\right)_A^B \qquad \text{...[7.1 }(b)]$$

where $\left(A_w\right)_A^B$ is the area of load intensity diagram between A and B

Hence $$F_x = -\int_0^x wdx \qquad \text{...[7.1 }(c)]$$

Similarly, integrating equation 7.2 between sections A and B, we get

$$\int_A^B dM = \int_A^B Fdx \qquad \text{...[7.2 }(a)]$$

or $$M_B - M_A = \int_A^B Fdx = \left(A_F\right)_A^B \qquad \text{...[7.2 }(b)]$$

where $\left(A_F\right)_A^B$ is the area of S.F.D. between A and B.

Hence $$M_x = \int_0^x Fdx = -\int_0^x \int_0^x wdx \qquad \text{...(7.3)}$$

For B.M. to be maximum, $\dfrac{dM}{dx} = 0 = F$ (from equation 7.2)

Hence B.M. is maximum where S.F. is zero or changes sign.

(*b*) **Concentrated Load W**

Fig. 7.29 (*b*) shows a concentrated load acting on the beam element, load being considered positive when acting downward. Let the stress elements on the left hand face *AD*. be F and M, and the stress elements on the right hand face *BC* be $(F + F_1)$ and $(M + M_1)$ where F_1 and M_1 are the increments in S.F. and B.M. in the length dx, due to application of point load W.

From equilibrium of forces in vertical direction, we get

$$F - W - (F + F_1) = 0 \quad \text{or} \quad F_1 = -W$$

Thus there is an abrupt change F_1(equal to $-W$) due to application of point load as we pass from left to the right through the point of application of load.

Again, from summing up of the moments, about face *AD*, we get

$$-M - W\left(\frac{dx}{2}\right) - (F + F_1)dx + (M + M_1) = 0$$

or
$$M_1 = W\left(\frac{dx}{2}\right) + Fdx + F_1dx$$

Here, we observe that since the length dx is infinitesimally small, the increment M_1 is also infinitesimally small. Hence the B.M. does not change as we pass through the point of application of concentrated load. However, its derivative (dM/dx) undergoes abrupt change at the location of the concentrated load. At the left hand section AD, dM/dx is equal to F while at the right hand section BC, $dM/dx = F + F_1 = F - W$. Hence dM/dx decreases abruptly by an amount equal to W at the point of application of load.

(*c*) **Couple μ acting on the element** [Fig. 7.29 (*c*)]

Considering the equilibrium of the element in the vertical direction, we get

$$F - (F + F_1) = 0 \quad \text{or} \quad F_1 = 0$$

This shows that the S.F. does not change at the point of application of a couple.

Again, equilibrium of moments for the element yields

$$-M + \mu - (F + F_1)\,dx + (M + M_1) = 0$$

Neglecting the terms that contain differential dx, we get

$$M_1 = -\mu \qquad \text{...(7.4)}$$

Thus, there is an abrupt change in the B.M. at the point of application of a couple.

Example 7.6. *A simply supported beam of span L is loaded with a triangular load with intensity zero at one end to w per unit length at the other end. Plot the bending moment and shear force diagram indicating the principal values.*

Solution. *At any section distant x from A, the rate of loading* $w_x = \dfrac{wx}{L}$.

But
$$\frac{dF_x}{dx} = -w_x$$

or
$$F_x = -\int w_x dx$$
$$= -\int \frac{wx}{L}dx$$
$$= -\frac{w}{L}\frac{x^2}{2} + C_1$$

At $x = 0$, $F = +R_A$.

$\therefore$ $C_1 = +R_A$

$$F_x = -\frac{w}{2L}x^2 + R_A$$

(parabolic variation) ...(1)

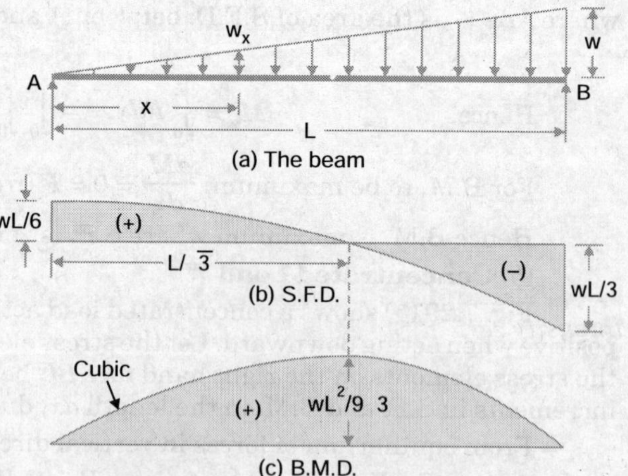

Fig. 7.30

Integrating further, $M_x = \int F_x\, dx = -\frac{w}{2L}\frac{x^3}{3} + R_A . x + C_2$

When $x = 0$, $M_A = 0$. $\therefore$ $C_2 = 0$

Hence $M_x = -\frac{w}{6L}x^3 + R_A . x$ (cubic variation) ...(2)

Equations (1) and (2) are the expressions for the S.F. and bending moment respectively at any point.

Now $R_A = \frac{(\frac{1}{2}wL)(\frac{1}{3}L)}{L} = \frac{1}{6}wL$; $R_B = \frac{wL}{2} - \frac{wL}{6} = \frac{wL}{3}$

$\therefore$ $F_x = -\frac{w}{2L}x^2 + \frac{1}{6}wL$

At $x = 0$, $F_A = +\frac{1}{6}wL$

At $x = L$, $F_B = -\frac{w}{2L}L^2 + \frac{1}{6}wL = -\frac{wL}{3} = -R_B$

F will be zero at $x = \frac{L}{\sqrt{3}}$

Again from (2), $M_x = -\frac{w}{6L}x^3 + \frac{wL}{6}x$

At $x = 0$, $M_A = 0$. At $x = L$, $M_B = -\frac{w}{6L}L^3 + \frac{wL^2}{6} = 0$

At $x = \frac{L}{\sqrt{3}}$, B.M. will be maximum.

$$\therefore \quad M_{max} = -\frac{w}{6L}\left(\frac{L}{\sqrt{3}}\right)^3 + \frac{wL}{6}.\frac{L}{\sqrt{3}} = -\frac{wL^2}{6\times 3\sqrt{3}} + \frac{wL^2}{6\sqrt{3}}$$

$$= +\frac{wL^2}{6\sqrt{3}}\left(\frac{2}{3}\right) = +\frac{wL^2}{9\sqrt{3}}.$$

The S.F.D. and B.M.D are shown in Figs. 7.30 (*b*) and (*c*) respectively.

Example 7.7. *A simply supported beam of span L is loaded with distributed load of intensity zero at the ends and w per unit length at the centre. Plot the B.M. and S.F. diagrams.*

Solution. The total load $= \frac{1}{2}wL$. Hence $R_A = R_B = \frac{wL}{4}$

At any section x, $w_x = \frac{2wx}{L}$

$$\therefore \quad F_x = -\int w_x dx = -\int \frac{2wx}{L}dx = -\frac{2w}{L}\frac{x^2}{2} + C_1 = -\frac{wx^2}{L} + C_1$$

At $x = 0$, $F_x = +R_A = +\dfrac{wL}{4} = C_1$

$\therefore \quad F_x = -\dfrac{wx^2}{L} + \dfrac{wL}{4}$

(parabolic variation) ...(1)

At $x = 0$, $F_A = +\dfrac{wL}{4}$;

At $x = \dfrac{L}{2}$, $F_C = -\dfrac{wL^2}{4L} + \dfrac{wL}{4} = 0$

Also, $M_x = \int F_x dx$

$= -\dfrac{wx^3}{3L} + \dfrac{wLx}{4} + 0$

(cubic variation) ...(2)

(Since $M_x = 0$ at $x = 0$)

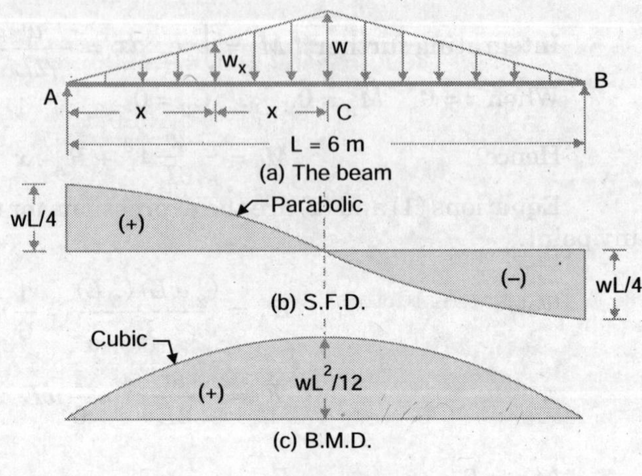

Fig. 7.31

From (2), at $x = \dfrac{L}{2}$, $M_{max} = -\dfrac{w}{3L}\left(\dfrac{L}{2}\right)^3 + \dfrac{wL}{4}\left(\dfrac{L}{2}\right) = +\dfrac{wL^2}{12}$.

The S.F.D. and B.M.D. are shown in Figs. 7.31 (*b*) and (*c*) respectively.

Example 7.8. *A horizontal simply supported beam of span 6 m carries a total load of 4000 N ; the load distribution varies parabolically from zero at each end to a maximum at mid-span. Calculate the values of the bending moment and S.F. at every one metre and plot the B.M. and S.F. diagrams.*

Solution. Let w = rate of loading at the centre. The total area of parabola $= \dfrac{2}{3} \times w \times 6 = 4w$ N

The total load given $= 4000$ N

$\therefore \quad 4w = 4000$ or $w = \dfrac{4000}{4} = 1000$ N/m

With D as origin, the equation of the parabola is $a = kx'^2$

When $x' = 3$ m, $\quad a = w = 1000$

$\therefore \quad k = \dfrac{1000}{9}$

Hence the equation to the parabola will be

$$a = \frac{1000}{9}x'^2 = \frac{1000}{9}(3-x)^2 \quad ...(1)$$

(when x is measured from support A).

Again, $w_x = 1000 - a$

$$= 1000 - \frac{1000}{9}(3-x)^2 \quad ...(2)$$

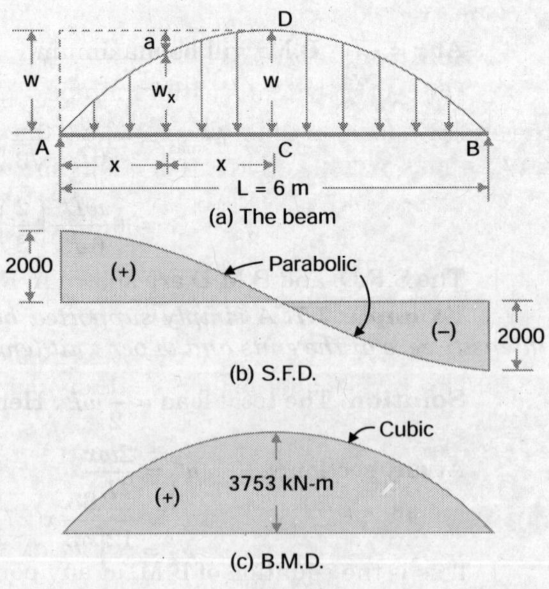

Fig. 7.32

$$F_x = -\int w_x dx = -1000x + \frac{1000}{9}\left(9x + \frac{x^3}{3} - \frac{6x^2}{2}\right) + C_1$$

$$= + \frac{1000}{27}x^3 - \frac{1000}{3}x^2 + C_1$$

At $x = 0$, $F_x = + R_A = + 2000$, $\therefore$ $C_1 = + 2000$

Hence $$F_x = + \frac{1000}{27}x^3 - \frac{1000}{3}x^2 + 2000 \quad ...(3)$$

Integrating further,

$$M_x = \int F_x dx = + \frac{1000}{27}\frac{x^4}{4} - \frac{1000}{9}x^3 + 2000\,x + C_2$$

At $x = 0$, $M_x = 0$ $\therefore$ $C_2 = 0$.

Hence, $$M_x = + \frac{1000}{108}x^4 - \frac{1000}{9}x^3 + 2000x \quad ...(4)$$

Equations (3) and (4), therefore, give the values of S.F. and B.M. The values of S.F. and B.M. for every 1 metre are tabulated below :

x (metres)	0	1	2	3	4	5	6
S.F. (N)	+ 2000	+ 1709	+ 963	0	– 963	– 1709	– 2000
B.M. (N-m)	0	+ 1898	+ 3260	+ 3753	+ 3260	+ 1898	0

The B.M. and S.F. diagrams are shown in Fig. 7.32.

Example 7.9. *A lintel of 3 metre span carries a stone wall 40 cm thick. Assuming that the effective load on the lintel is an equilateral triangle on the span as base, calculate the maximum B.M. Take unit weight of masonry as 22 kN/m³. Sketch the S.F. and B.M. diagrams.*

Solution. (Fig. 7.31)

The central height h of the triangle = 1.5 tan 60° ≃ 2.6 m. Volume of the wall transmitting the effective load on the wall = $\frac{1}{2} \times 3 \times 2.6 \times 0.40$ m³ = 1.56 m³. Total effective load on the lintel = W = 1.56 × 22 = 34.32 kN. If w = maximum intensity at the centre, we have

$$\tfrac{1}{2}w \times 3 = 34.32 \quad \text{or} \quad w = \frac{34.32}{1.5} = 22.88 \text{ kN/m}$$

From example 7.7, $$F_x = -\frac{wx^2}{L} + \frac{wL}{4} = \frac{22.88}{3}\left(\frac{9}{4} - x^2\right)$$

$$= 7.627\,(2.25 - x^2) \quad ...(1)$$

This is the equation of S.F. at any point.

Also, $$M_x = -\frac{wx^3}{3L} + \frac{wLx}{4} = \frac{wx}{12L}(3L^2 - 4x^2)$$

$$= \frac{22.88}{12 \times 3}x(27 - 4x^2) = 0.636x\,(27 - 4x^2). \quad ...(2)$$

This is the equation of B.M. at any point.

At $x = 1.5$ m, $M_{max} = 0.636 \times 1.5(27 - 4 \times 2.25) = $ **+ 17.17 kN-m**

The ordinates of S.F. and B.M. diagrams can be calculated giving different values to x.

Example 7.10. *A lintel of 4 metre span supports a concrete wall 15 cm thick. The height of the wall is 1 metre at one end and increases uniformly to 4 metre at the other end. Calculate the max. B.M. on the beam if the concrete weighs 23 kN/m³. Sketch the S.F. and B.M. diagrams.*

Solution. Intensity of load at A

$$= \frac{1 \times 1 \times 15}{100} \times 23$$

$= 3.45$ kN/m

Intensity of load at B

$$= \frac{4 \times 1 \times 15}{100} \times 23$$

$= 13.8$ kN/m

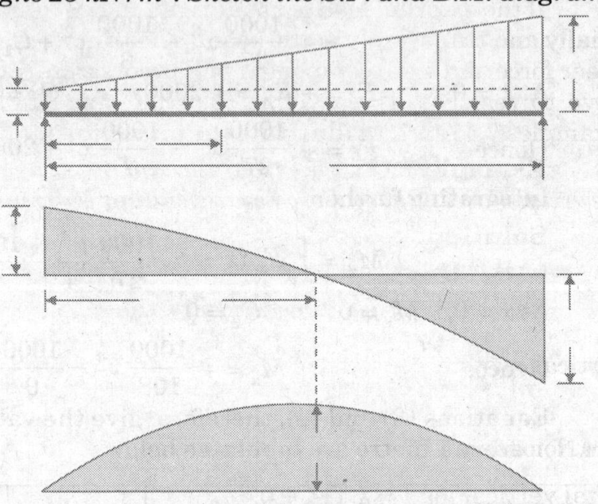
Fig. 7.33

The above loading condition can be divided into : (*i*) a uniformly distributed load of 3.45 kN per metre and (*ii*) a triangular load varying from 0 intensity at A to 10.35 kN metre at B.

Due to U.D.L., $R_A = 6.9$ kN

and $R_B = 6.9$ kN

Due to triangular load, $R_A = \frac{1}{3} \times \frac{1}{2} \times 4 \times 10.35 = 6.9$ kN

and $R_B = \frac{2}{3} \times \frac{1}{2} \times 4 \times 10.35 = 13.8$ kN

Total $R_A = 6.9 + 6.9 = 13.8$ kN ; $R_B = 6.9 + 13.8 = 20.7$ kN

At distance x from A, $w_x = 3.45 + \frac{10.35}{4} x$

$$F_x = -\int w_x \, dx = -\int \left(3.45 + \frac{10.35}{4} x\right) dx$$

$$= -3.45\,x - 1.2938\,x^2 + C_1$$

At $x = 0$, $F_x = +R_A = +13.8$; $\therefore$ $C_1 = +13.8$

Hence $F_x = -3.45\,x - 1.2938\,x^2 + 13.8$ ...(1)

Also, $M_x = \int F_x \, dx = -\frac{3.45\,x^2}{2} - \frac{1.2938\,x^3}{3} + 13.8\,x + C_2$

At $x = 0$, $M_A = 0$ $\therefore$ $C_2 = 0$

Hence $M_x = -1.725\,x^2 - 0.4313\,x^3 + 13.8\,x$ ...(2)

For position of zero S.F., we have from equation (1),

$$-3.45\,x - 1.2938\,x^2 + 13.8 = 0 \quad \text{or} \quad x^2 + 2.67\,x - 10.65 = 0$$

which gives $x = 2.19$ m

Hence the B.M. will be maximum at $x = 2.19$ m from A.

$$M_{max} = -1.725\,(2.19)^2 - 0.4313\,(2.19)^3 + 13.8\,(2.19)$$

$$= \mathbf{+17.42\ kN\text{-}m.}$$

The B.M. and S.F. diagrams have been shown in Fig. 7.33 (*b*) and (*c*) respectively.

7.5. INCLINED LOADING

If the external load is not at right angles to the axis of the beam, the loading can be resolved axially and transversly to the beam. The transverse component of the load will produce B.M. and shear force and axial component will produce pull or push. In such cases, therefore, the beam will have to resist (*i*) Shear force (*ii*) B.M. and (*iii*) Direct force, which may be pull or push. Examples 7.11 to 7.13 illustrate the procedure for analysis.

Example 7.11. *A horizontal beam 6 m long is loaded as shown in Fig. 7.34 (a). Construct the Axial Force, Shear Force and Bending Moment diagrams.*

Solution.

(*i*) **S.F.D.**

Resolving the system vertically, the vertical system consists of $\sqrt{2}, \dfrac{1}{\sqrt{2}}$ and $\dfrac{3}{2}$ kN loads at *C*, *D* and *E* respectively.

Total vertical load = 1.414 + 0.707 + 1.5 = 3.621 kN.

For vertical reaction V_A,

$$V_A \times 6 = (1.414 \times 5) + (0.707 \times 3.5) + (1.5 \times 2)$$

$$\therefore \quad V_A = 2.09 \text{ kN}$$

and $\quad V_B = 3.621 - 2.09 = 1.531$ kN

The S.F.D. has been shown in Fig 7.34 (*c*).

(*ii*) **Horizontal Thrust Diagram**

Horizontal system of loads consists of $\sqrt{2}, \dfrac{1}{\sqrt{2}}$ and $\dfrac{3\sqrt{3}}{2}$ tonnes at *C*, *D* and *E* respectively. The force at *C* is opposite in direction to those of *D* and *E*.

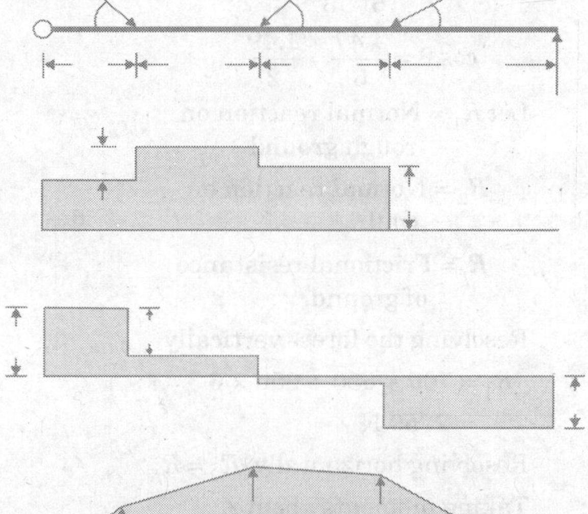

Fig. 7.34

$$\text{Horizontal reaction at hinge} \quad A = \frac{3\sqrt{3}}{2} + \frac{1}{\sqrt{2}} - \sqrt{2} = 1.893 \text{ t } (\rightarrow)$$

Horizontal reaction at *B* = 0, since *B* is a free support.

The horizontal thrust diagram (H.T.D.) is shown in Fig. 7.34 (*b*).

(*iii*) **B.M.D.**

Bending moment at any point will be due to only vertical component of the load.

B.M. for portion BE : $M_x = + V_B \cdot x = + 1.531\, x$

When $x = 0$, $M_B = 0$; At $x = 2$ m, $M_E = + 1.531 \times 2 = + 3.062$ kN-m

B.M. for portion ED : $M_x = + 1.531\, x - 1.5\,(x - 2)$

When $x = 2$ m, $M_E = + 3.062$ kN-m ; $x = 3.5$, $M_D = + 1.531 \times 3.5 - 1.5 \times 1.5 = 3.108$ kN-m

B.M. for portion DC : $M_x = + 1.531\, x - 1.5\,(x - 2) - 0.707\,(x - 3.5)$

When $x = 5$ m, $M_C = + 7.655 - 4.5 - 1.06 = + 2.095$ kN-m

$M_A = 0$

The B.M. diagram has been shown in Fig. 7.34(*d*).

Example 7.12. *A ladder 6 m long weighing 250 N/metre run, rests against a smooth vertical wall with the bottom resting on rough ground, the vertical distance between the ends of the ladder being 4 m. A man weighing 650 N is standing on the ladder at 1.5 metre from the bottom end and another man weighing 700 N on the ladder at 3 metre from the bottom. Construct the axial thrust, shear force and B.M. diagrams for the ladder and calculate the reactions at the ground and at the wall.*

Solution. Let the ladder be inclined at θ angle to horizontal.

$$\sin\theta = \frac{4}{6} = \frac{2}{3};$$

$$\cos\theta = \frac{4.47}{6} = \frac{1.49}{2}$$

Let R_1 = Normal reaction on rough ground.

R_2 = Normal reaction on wall.

R_f = Frictional resistance of ground.

Resolving the forces vertically

$$R_1 = 700 + 650 + 250 \times 6 = 2850 \text{ N}$$

Resolving horizontally, $R_2 = R_f$

Taking moments about A

$$R_2 \times 4 = 700 \times 3\cos\theta + 650 \times 1.5\cos\theta + 250 \times 6 \times \frac{4.47}{2}$$

$$= \left(2100 \times \frac{1.49}{2}\right) + \left(975 \times \frac{1.49}{2}\right) + \left(\frac{1500 \times 4.47}{2}\right)$$

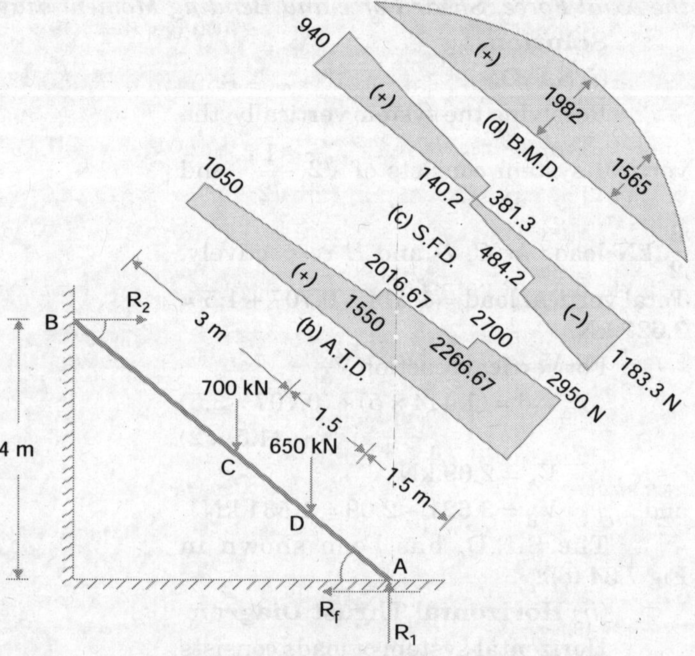

Fig. 7.35

$$\therefore \quad R_2 = \frac{1.49}{8}(2100 + 975 + 1500 \times 3) = \frac{1.49}{8} \times 7575 = 1410 \text{ N} = R_f$$

Resolving the system axially and transversely to AB :

At B, Axial component $P_B = R_2\cos\theta = 1410 \times \dfrac{1.49}{2} = 1050$ N

Transverse component $F_B = R_2\sin\theta = 1410 \times \dfrac{2}{3} = 940$ N

At A, Axial component $P_A = R_1\sin\theta + R_f\cos\theta$

$$= \left(2850 \times \frac{2}{3}\right) + \left(1410 \times \frac{1.49}{8}\right) = 2950 \text{ N}$$

Transverse component, $F_A = R_1 \cos\theta - R_f \sin\theta$

$$= \left(2850 \times \frac{1.49}{2}\right) - \left(1410 \times \frac{2}{3}\right) = 1183.3 \text{ N}$$

For ladder, axial load $= 250 \sin\theta = 250 \times \frac{2}{3} = 166.67$ N/m

Transverse load $= 250 \cos\theta = 250 \times \frac{1.49}{2} = 186.25$ N/m

At C, axial load $= 700 \sin\theta = 700 \times \frac{2}{3} = 466.67$ N

Transverse load $= 700 \cos\theta = 700 \times \frac{1.49}{2} = 521.5$ N

At D, axial load $= 650 \sin\theta = 650 \times \frac{2}{3} = 433.3$ N

Transverse load $= 650 \cos\theta = 650 \times \frac{1.49}{2} = 484.2$ N

The axial thrust diagram (A.T.D.) and S.F.D. are shown in Fig. 7.35 (*b*) and (*c*) respectively.

Max. B.M. occurs where S.F. changes its sign.

B.M. at point $C = M_C = +940 \times 3 - \left(186.25 \times 3 \times \frac{3}{2}\right) = +2820 - 838 = +1982$ N-m

B.M. at point $D = M_D = (+1183.3 \times 1.5) - \left(186.25 \times 1.5 \times \frac{1.5}{2}\right) = +1775 - 210 = +1565$ N-m.

The B.M.D. is shown in Fig. 7.35 (*d*).

Example 7.13. *A continuous member is bent in one plane and loaded in the same plane, as shown in Fig. 7.36 (a). Plot bending moment and shearing force diagrams. Assume the joints B and C to be rigid.*

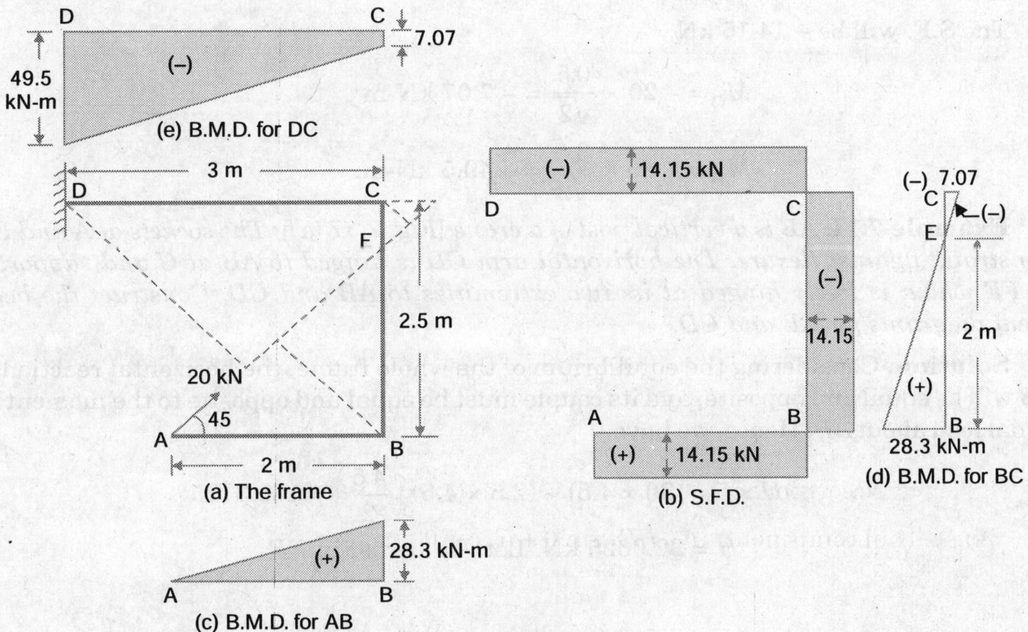

Fig. 7.36

Solution. To draw the shear force diagram, resolve the 20 kN force at A into two components along and perpendicular to AB.

$$\text{The perpendicular component} = \frac{20}{\sqrt{2}} = 14.15 \text{ kN}$$

$$\text{The horizontal or axial component} = \frac{20}{\sqrt{2}} = 14.15 \text{ kN}$$

***For AB* :**

For member AB there will be an axial force 14.15 kN and a transverse force of 14.15 kN. The transverse force will cause S.F. and bending moment in AB. The value of S.F. will, therefore, be + 14.15 kN.

The B.M. at A will be zero.

The B.M. at $B = +\,14.15 \times 2 = +28.30$ kN-m.

***For BC* :**

The axial force of 14.15 kN in AB will form the transverse load on BC. Thus the shear force in $BC = -\,14.15$. The member BC will also have an axial force of 14.15 kN.

The bending moment at any section is the product of the force of 20 kN and the perpendicular distance of that section from the line of action of the force.

Thus B.M. at $B = +\,28.30$ kN-m.

B.M. at E at 2 m from $B = 0$, since the line of action of 20 kN force passes through point E

$$\text{B.M. at } C = -\,20 \times \frac{0.5}{\sqrt{2}} = -\,7.07 \text{ kN-m}$$

***For CD* :**

The S.F. will be – 14.15 kN

$$M_C = -\,20 \times \frac{0.5}{\sqrt{2}} = -\,7.07 \text{ kN-m};$$

$$M_D = -\,20 \times \frac{3.5}{\sqrt{2}} = -\,49.5 \text{ kN-m.}$$

Example 7.14. *AB is a vertical post of a crane [Fig. 7.37 (a)]. The sockets at A and B offer no constraint against flexure. The horizontal arm CD is hinged to AB at C and supported by strut FE which is freely hinged at its two extremities to AB and CD. Construct the bending moment diagrams for AB and CD.*

Solution. Considering the equilibrium of the whole frame, the horizontal reactions at A and B will be equal and opposite, and its couple must be equal and opposite to the moment of the external load about AB. Hence, we have

$$H \times 5 = (30 \times 4.5) + (2.5 \times 4.5)\left(\frac{4.5}{2}\right) = 160.3125$$

$$\therefore \quad H = 32.0625 \text{ kN. Let } T = \text{thrust in } EF.$$

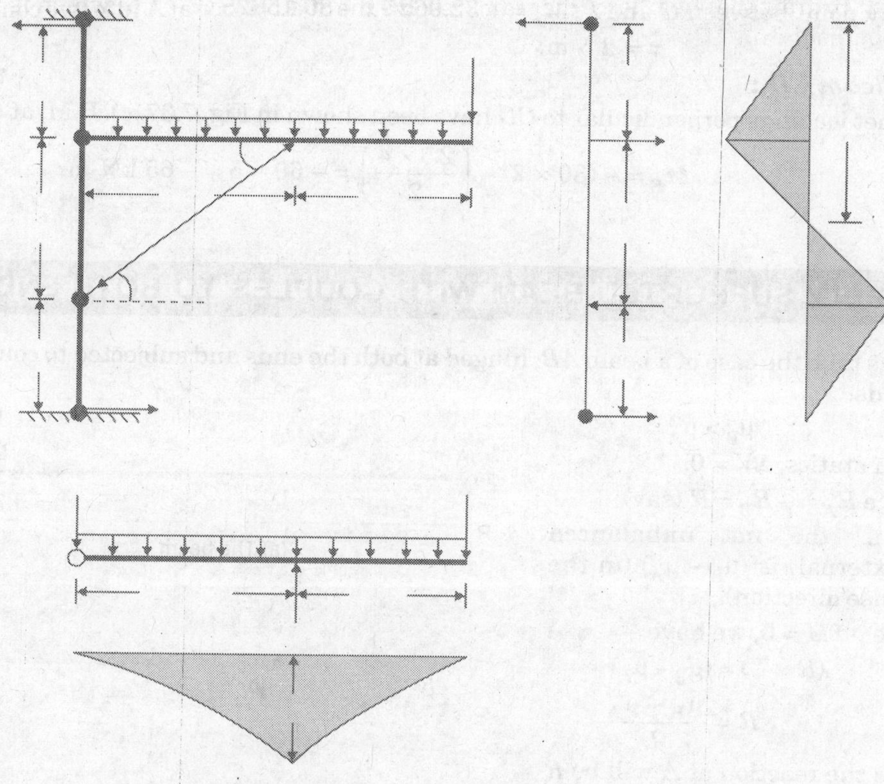

Fig. 7.37

For the rod *CD*, take moments about *C*.

$$M_C = 0 = T \sin \theta\,(2.5) - \frac{2.5 \times 4.5 \times 4.5}{2} - (30 \times 4.5)$$

$$\therefore \quad T = \frac{135 + 25.3125}{2.5\left(\dfrac{2}{3.2}\right)} = 102.6 \text{ kN}$$

The vertical reaction at $E = T \sin \theta = 102.6 \times \dfrac{2}{3.2} = 64.125$

Let P = pull in *CE*. Resolving horizontally at *E*, we get

$$P = T \cos \theta = 102.6 \times \frac{2.5}{3.2} = 80.15625 \text{ kN}$$

Horizontal load at $F = T \cos \theta = 80.15625$

For post AB :

B.M. at *A* and $B = 0$

$$M_C = +32.0625 \times 1.5 \simeq 48.094 \text{ kN-m}$$

$$M_F = (+32.0625 \times 3.5) - (80.15625 \times 2) = -48.094 \text{ kN-m}$$

At any point between C and F, $M_x = +\,32.0625\,x - 80.15625\,(x - 1.5)$ which is zero at $x = 2.5$ m.

For beam CD :

The net loadings perpendicular to CD have been shown in Fig. 7.37 (*c*). B.M. at C and $D = 0$

$$M_E = -\,(30 \times 2) - \left(\frac{2.5 \times 4}{2}\right) = -\,60 - 5 = -\,65 \text{ kN-m.}$$

7.6. FREELY SUPPORTED BEAM WITH COUPLES TO BOTH ENDS

Let us take the case of a beam AB, hinged at both the ends and subjected to couples μ_A and μ_B at the ends.

Let $\mu_B > \mu_A$

From statics, $\Sigma V = 0$.

Hence $R_A = -\,R_B = R$ (say)

Again, the net unbalanced moment (external) is $(\mu_B - \mu_A)$ (in the anti-clockwise direction).

Since $\Sigma M = 0$, we have

$$(R \times L) = (\mu_B - \mu_A)$$

$$\therefore \quad R = \frac{\mu_B - \mu_A}{L}$$

Thus the reaction at A will be R acting upwards and that at B will also be R, acting downwards.

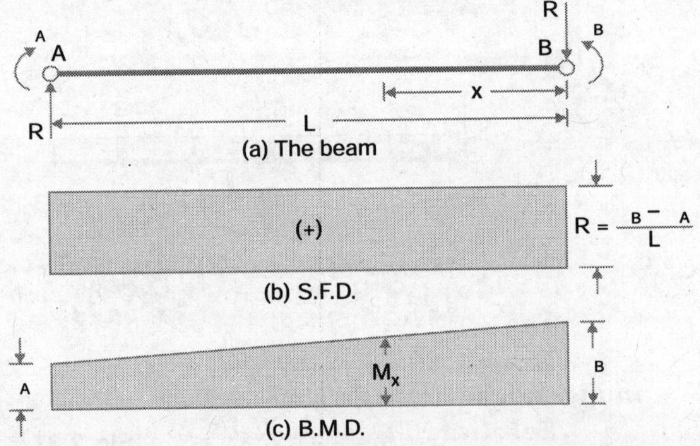

(a) The beam

(b) S.F.D.

(c) B.M.D.

Fig 7.38

S.F.D. : For any section x from B, we have

$F_x = +\,R = \dfrac{\mu_B - \mu_A}{L}$ which is constant from B to A. Hence the S.F.D. will be a rectangle, shown in Fig. 7.38 (*b*).

B.M.D. : $M_x = +\,\mu_B - R.x = \mu_B - (\mu_B - \mu_A)\dfrac{x}{L}$

At $x = 0$, $M_B = +\,\mu_B$, which decreases linearly.

At $x = L$, $M_A = +\,\mu_B - (\mu_B + \mu_A)\dfrac{L}{L} = +\,\mu_A$

Thus the B.M.D. will be a trapezium, as shown in Fig. 7.38 (*c*).

7.7. BEAM WITH AN INTERMEDIATE COUPLE

Consider a beam AB, hinged at A and B, subjected to couple μ at a distance a from A and b from B.

Since $\Sigma V = 0$, we get $R_A = -\,R_B = R$ (say).

The external couple is in the clockwise direction. Hence, the reaction at A will be acting downwards and that at B acting upwards, such that

$$R \times L = \mu \quad \text{or} \quad R = \frac{\mu}{L}$$

S.F.D. : At any distance x from B

$$F_x = -R = -\frac{\mu}{L},$$

which is constant from B to A. Hence, the S.F.D. will be a rectangle, as shown in Fig. 7.39 (b).

B.M.D. : For portion BC

$$M_x = +R.x = +\frac{\mu}{L}x$$

$$\therefore \quad M_C \text{ (right)} = +\frac{\mu b}{L}$$

For CA, $$M_x = +R.x - \mu = +\frac{\mu}{L}x - \mu \qquad \text{(linear variation)}$$

$$\therefore \quad M_C \text{ (left)} = +\frac{\mu b}{L} - \mu = -\frac{\mu a}{L}; \; M_A = 0.$$

(c) B.M.D.

Fig. 7.39

The B.M.D. and S.F.D. have been shown in Fig. 7.39 (b) and (c) respectively.

Example 7.15. *A beam AB of length 4 m is acted upon by forces and moments as shown in Fig. 7.40 (a). Plot the B.M. and S.F. diagrams.*

Solution. Taking moments of all the forces about B, we get

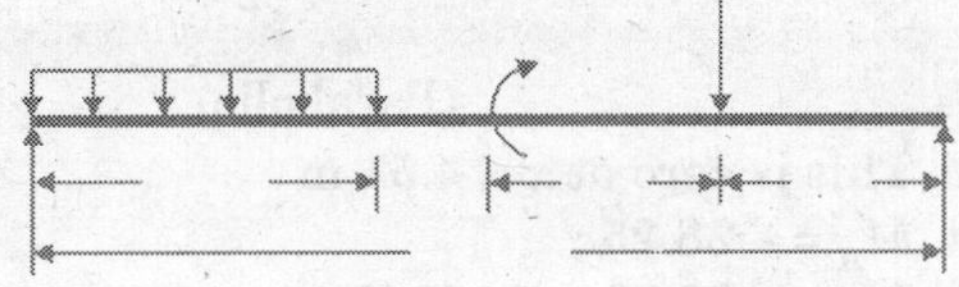

$$R_A = \frac{(2 \times 1) + (2 \times 1.5 \times 3.25) - 3}{4}$$

$$= 2.188 \text{ kN} \uparrow$$

$$R_B = (2 \times 1.5 + 2) - 2.188$$

$$= 2.812 \text{ kN} \uparrow$$

The S.F. diagram is shown in Fig. 7.40 (b). For S.F. to be zero between A and C, we have

$$F_x = +2.188 - 2x = 0$$

or $\quad x = 1.094$ m measured from A.

For any section in AC, we have

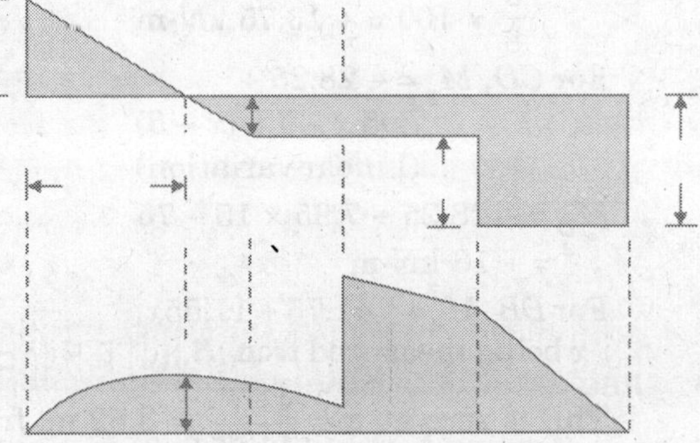

$$M_x = +2.188\,x - \frac{2x^2}{2}, \text{ which is maxi-}$$

mum at $x = 1.094$ m

Fig. 7.40

$\therefore$ $M_{max} = +2.188\,(1.094) - (1.094)^2 = +1.198$ kN-m

$M_C = +(2.188 \times 1.5) - (1.5)^2 = +1.032$ kN-m

For CD, $M_x = +2.188\,x - 3(x - 0.75)$ (Linear)

$\therefore$ M_D (left) $= +2.188\,(2) - (3 \times 1.25) = +0.624$ kN-m

M_D (right) $= +0.624 + 3 = +3.624$ kN-m

$M_E = +R_B \times 1 = +2.812 \times 1 = +2.812$ kN-m

$M_B = 0.$

The B.M. diagram has been shown in Fig. 7.40 (*c*).

Example 7.16. *A beam of 20 m span is hinged at each end. It carries a U.D.L. of $\frac{3}{4}$ kN/m on the left hand half of the beam together with a 12 kN load at 15 m from left hand end. In addition to this it is subjected to couples of 28.25 kN-m in anti-clockwise direction at left hand support and 41.75 kN-m in the clockwise direction at the right hand support. Find the reactions at the ends and draw the B.M. and S.F. diagrams inserting principal values.*

Solution. (Fig. 7.41)

Taking moments about B, we get

$$M_B = 0 = 28.25 + \left(10 \times \frac{3}{4} \times 15\right) + (12 \times 5) - R_A\,20 - 41.75$$

$$\therefore \quad R_A = \frac{28.25 + 112.5 + 60 - 41.75}{20} = 7.95 \text{ kN}$$

$$\therefore \quad R_B = 7.5 + 12 - 7.95 = 11.55 \text{ kN.}$$

The S.F. diagram has been shown in Fig. 7.41 (*b*).

For AC, at distance x from A,

$$M_x = -28.25 + 7.95\,x - \frac{3}{4}\frac{x^2}{2}$$

(Parabolic)

This is zero at $x = 4.51$ m

$M_A = -28.25$;

$M_C = -28.25 + 7.95 \times 10 - \frac{3}{8} \times 100 = +13.75$ kN-m

For CD, $M_x = -28.25 + 7.95\,x - 7.5\,(x - 5)$ ...(Linear variation)

$M_D = -28.25 + 7.95 \times 15 - 75 = +16$ kN-m

For DB, $M_x = -41.75 + 11.55x$, x being measured from B.

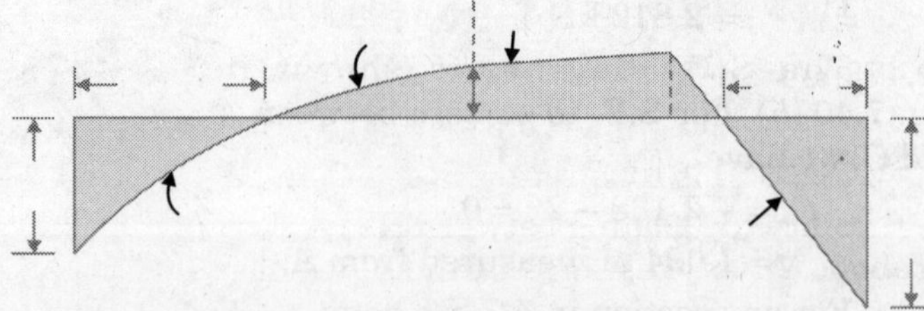

Fig. 7.41

This is zero at $x = \dfrac{41.75}{11.55} = 3.62$ m. from B.

$M_B = -41.75$; $M_D = -41.75 + 11.55 \times 5 = +16$ kN-m (as before)

Example 7.17. *A beam AB, 6.5 m long and supported at A, has a simple support of 1 m length between C and D. Assuming a uniformly distributed reaction between C and D, draw the B.M. and S.F. diagrams for the loading shown in Fig. 7.42 (a).*

Solution. (*a*) **Reactions :** Since the reaction at *CD* is uniformly distributed over a length of 1 m, it can be replaced by its resultant R_2 acting at the centre of *CD*. Taking moments about *A*, we get

$$(R_2 \times 4.5) = (\tfrac{1}{2} \times 3 \times \tfrac{1}{2})(\tfrac{2}{3} \times 3) + (\tfrac{1}{2} \times 6.5)$$ From which $R_2 = 1.056$ kN ↑

Taking moments about R_2, we get

$$(R_1 \times 4.5) + (\tfrac{1}{2} \times 2) = (\tfrac{1}{2} \times 3 \times \tfrac{1}{2}) \times (1.5 + \tfrac{1}{3} \times 3)$$

From which $R_1 = 0.194$ kN ↑

(*Check* : $R_1 + R_2 = 1.056 + 0.194 = 1.25$ = total load)

∴ Uniformly distributed reaction between *C* and *D*

$$r = \frac{R_2}{CD} = \frac{1.056}{1} = 1.056 \text{ kN/m.}$$

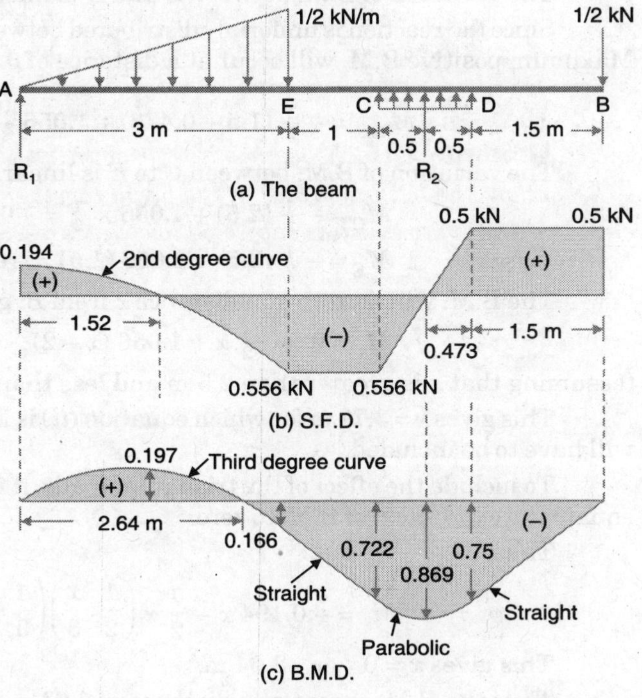

Fig. 7.42

(*b*) **S.F.D. :**

$F_B = +\tfrac{1}{2}$ kN ; $F_D = +\tfrac{1}{2}$ kN

∴ The S.F. is constant between B and D.

$$F_C = +\tfrac{1}{2} - (1 \times r) = +\tfrac{1}{2} - 1.056 = -0.556 \text{ kN}$$

The variation of S.F. between *D* and *C* is linear. The S.F. is zero at a distance of $\dfrac{0.5}{1.056}$ = 0.473 m from D.

$$F_E = 0.5 - 1.056 = -0.556 \text{ kN}$$

The S.F. is constant between *C* and *E*.

Since the loading between *E* to *A* is triangular, the variation of S.F. between *E* to *A* will be parabolic. At any distance *x* from *A*, we have

$$F_x = +R_1 - \frac{1}{2}x\left(\frac{1}{2}\frac{x}{3}\right) = +R_1 - \frac{x^2}{12} \quad \text{...(i)}$$

At $x = 0$, $F_A = +R_1 = +0.194$ kN

At $x = 3$ m, $F_E = +0.194 - \dfrac{3^2}{12} = -0.556$ kN

To get the value of *x* for zero S.F. equate (*i*) to zero.

∴ $x^2 = 12\,R_1 = 12 \times 0.194$ From which $x = 1.52$ m.

The complete S.F.D. is shown in Fig. 7.42 (*b*).

(*c*) **B.M.D. :**

$$M_B = 0 \ ; M_D = -\tfrac{1}{2} \times 1.5 = -0.75 \text{ kN-m}$$

The variation of B.M. between *B* and *D* is linear.

Since the reaction is uniformly distributed between *D* and *C*, the B.M. will vary parabolically. Maximum positive B.M. will occur at a distance of 0.473 m from *D* (where S.F. is zero).

$$\therefore \quad M_{max} = -\tfrac{1}{2}(1.5 + 0.473) + 1.056 \frac{(0.473)^2}{2} = -0.869 \text{ kN-m}$$

The variation of B.M. between *C* to *E* is linear.

$$M_C = -\tfrac{1}{2}(2.5) + 1.056 \times \tfrac{1}{2} = -0.722 \text{ kN-m}$$

$$M_E = -\tfrac{1}{2}(3.5) + 1.056\,(1.5) = -0.166 \text{ kN-m}$$

The B.M. will be zero at a distance *x* from *B*, given by

$$M_x = 0 = -\tfrac{1}{2}\,x + 1.056\,(x - 2) \qquad \text{...(ii)}$$

(assuming that *x* is greater than 2.5 m and less than 3.5 m)

This gives $x = 3.79$ m for which equation (*ii*) is invalid since the effect of the triangular load will have to be included.

To include the effect of the triangular loads, it is more convenient to measure *x* from *A* and equate the expression of B.M. to zero.

Thus,

$$M_x = +0.194\,x - \frac{1}{2}\,x\left(\frac{1}{2}\cdot\frac{x}{3}\right)\left(\frac{1}{3}\cdot x\right) = +0.194\,x - \frac{x^3}{36} \qquad \text{(cubic variation)}$$

This gives $x = 0$ or 2.64 m.

Thus the B.M. is zero at a distance of 2.64 m from *A*, or 3.86 m from *B*. The variation of B.M. between *E* and *A* will be a third degree curve. Maximum negative B.M. will occur at distance 1.52 m from *A*, where S.F. is zero.

$$\therefore \quad M_{max} = +R_1\,(1.52) - \frac{1}{2}\,(1.52)\left(\frac{1}{2}\times\frac{1.52}{3}\right)\left(\frac{1}{3}\times 1.52\right)$$

$$= +0.194 \times 1.52 - 0.098 = +0.197 \text{ kN-m}$$

The B.M.D. is shown in Fig. 7.42 (*c*).

7.8. LOADING AND B.M.D. FROM S.F.D.

If the S.F.D. for beam is given (or the B.M.D. is given), the loading diagram for the beam can very easily be drawn, if the following points are remembered :

(*i*) The S.F.D. will consist of rectangle (or series of rectangles) if the beam is loaded with point loads.

(*ii*) The S.F.D. will consist of inclined lines for the portion on which U.D.L. is acting.

(*iii*) The S.F.D. will consist of parabolic lines for the portion over which triangular or trapezium load distribution is acting.

(*iv*) The S.F.D. will consist of 'cubics' for the portion over which parabolic load distribution is acting.

Similarly

(*v*) The B.M.D. will consist of inclined lines, if the beam is loaded with point loads.

(*vi*) The B.M.D. will consist of parabolic lines for the portion over which U.D.L. is acting.

(*vii*) The B.M.D. will consist of 'cubic' or third degree polynomials if the load distribution is triangular.

(*viii*) The B.M.D. will consist of fourth degree polynomial if the load distribution is parabolic.

Examples 7.18, 7.20, 7.22, 7.23, 7.24 and 7.25 illustrate the procedure.

Example 7.18. *Fig. 7.43 (a) is the shearing force diagram for a beam which rests on two supports, one being at the left hand end. Deduce from S.F.D. the loading on the beam and draw the B.M. diagram giving principal values.*

Solution. By the inspection of the diagram, the other support will be at C, where the S.F. changes sign.

Since the S.F. is constant from D to C, there is a point load of 2 kN at D.

The value of $R_C = (8 + 2) = 10$ kN

Since the negative S.F. decreases from 8 kN at C, to 2 kN at B linearly, there is a U.D.L. from C to B of an intensity $\dfrac{8-2}{3}$ = 2 kN/m.

Since there is sudden increase of S.F., there is a point load at $B = 2 + 3.5 = 5.5$ kN

Again, since the S.F. varies from 3.5 kN at B to 7.5 kN at A linearly, there is U.D.L on AB of intensity $\dfrac{7.5-3.5}{3}$ = 2 kN/m.

The reaction at A is evidently 7.5 kN. The loading has been drawn on the beam in Fig. 7.43 (*b*).

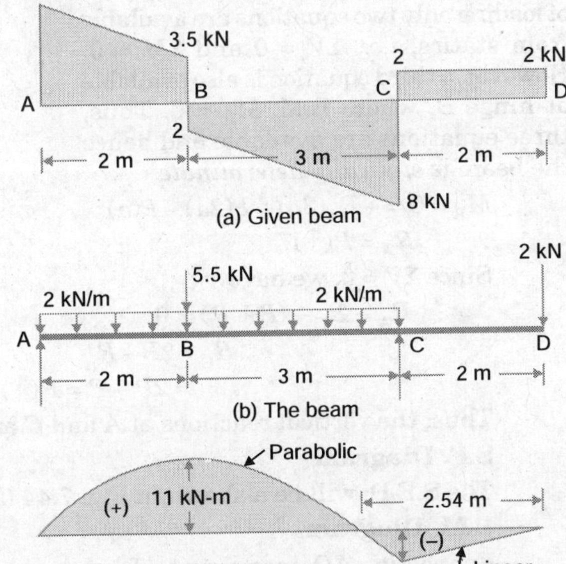

Fig. 7.43

B.M.D. :

Between D and C, $M_x = -2x$, from which $M_D = 0$ and $M_C = -2 \times 2 = -4$ kN-m

Between C and B, $M_x = -2x + 10(x-2) - \dfrac{2(x-2)^2}{2}$

$= -2x + 10(x-2) - (x-2)^2$. ...(parabolic)

B.M. will be maximum at $x = 5$ m, where S.F. changes sign.

$\therefore \quad M_{max} = M_B = -(2 \times 5) + 10(5-2) - (5-2)^2 = +11$ kN-m

For B.M. to be zero, $-2x + 10(x-2) - (x-2)^2 = 0$

or $\quad x^2 - 12x + 24 = 0$, which gives $x = 2.54$ m

The B.M.D. has been shown in Fig. 7.43 (*c*).

7.9. ADDITIONAL EXAMPLES USEFUL FOR COMPETITIVE EXAMINATIONS

Example 7.19. *For the beam shown in Fig. 7.44 (a), draw the shear force and bending moment diagrams.*

Solution. The beam is fixed at C and has a roller support at A. For vertical system of loading there will be two reaction components at *support C* : a vertical reaction $R_C(\uparrow)$ and a moment $M_C(\curvearrowright)$. Also, at roller support A, there will be a vertical reaction R_A at A. Thus, there are three reaction components for the beam : R_A, R_C and M_C. Ordinarily, for vertical system of loading only two equations are available from statics, *i.e.*, $\Sigma V = 0$ and $\Sigma M = 0$. However, a third equation is also available at hinge B, where B.M. $M_B = 0$. Thus, three equations are available and hence the beam is *statically determinate.*

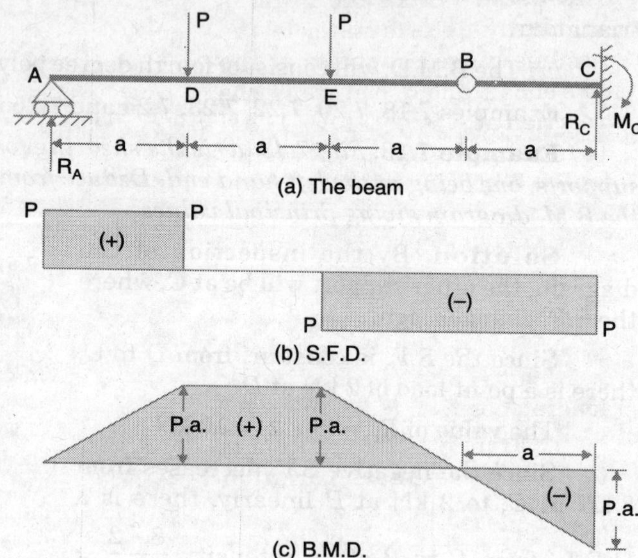

Fig. 7.44

$$M_B = 0 = +R_A(3a) - P(2a) - P(a)$$

$$\therefore \quad R_A = P(\uparrow)$$

Since $\Sigma V = 0$, we have

$$R_A + R_C - (P + P) = 0$$

$$\therefore \quad R_C = 2P - R_A = 2P - P = P(\uparrow)$$

Thus, the vertical reactions at A and C are known.

S.F. Diagram

The S.F.D. will be a shown in Fig. 7.44 (*b*).

B.M. Diagram

For section AD, measuring x from A,

$$M_x = +R_A x = +Px \quad \text{(linear)}$$

This gives $M_A = 0$ and $M_D = +P \cdot a$

For DE, measuring x from A,

$$M_x = +R_A \cdot x - P(x - a) = +P.x - Px + Pa = +Pa \text{ (constant)}$$

This gives $M_D = +Pa$ (as before) and $M_E = +Pa$

For EB, measuring x from A,

$$M_x = +R_A x - P(x - a) - P(x - 2a) = +Pa - Px + P.2a = -Px + 3Pa$$

This gives $M_E = +Pa$ (as before) and M_B = zero

For BC, measuring x from A, we get

$$M_x = +R_A x - P(x - a) - P(x - 2a) = +Pa - Px + P.2a = -Px + 3Pa$$

This gives M_B = zero and $M_C = -4Pa + 3Pa = -Pa$

The B.M.D. is shown in Fig. 7.44 (*c*).

Example 7.20. *The shear force diagram of a statically determinate beam is shown in Fig. 7.45 (a). Determine the loading on the beam and draw the bending moment diagram. Determine the magnitude of the maximum positive bending moment. Where does it occur ?*

Solution. Loading on the beam : Since the beam is statically determinate, let us assume hinged supports at the two ends *A* and *B*. Naturally, the reaction at *A* = S.F. at *A* = 11 kN (↓) and reaction at *B* = S.F. at *B* = 5 kN (↑).

At *F*, S.F. changes from – 5 kN to + 17 kN. Hence there exists a point load *P* = 5 + 17 = 22 kN (↓) at *F*.

Since S.F. is constant between *F* and *E*, there is no loading between *F* and *E*.

From *E* to *D*, and from *D* to *C*, there is linear variation (decreasing) in S.F. Hence there is U.D.L. from *E* to *C*. The rate $w_1 = \frac{17 - 13}{2}$ = 2 kN/m from *E* to *D*, and rate $w_2 = \frac{3+5}{4}$ = 2 kN/m from *D* to *C*. Hence $w_1 = w_2$ = 2 kN/m, acting upwards. Also, there is a point load P_1 = 10 kN (↑) at *D* and a point load P_2 = 6 kN (↑) at *C*. The complete loading on the beam is shown in Fig. 7.45 (*b*).

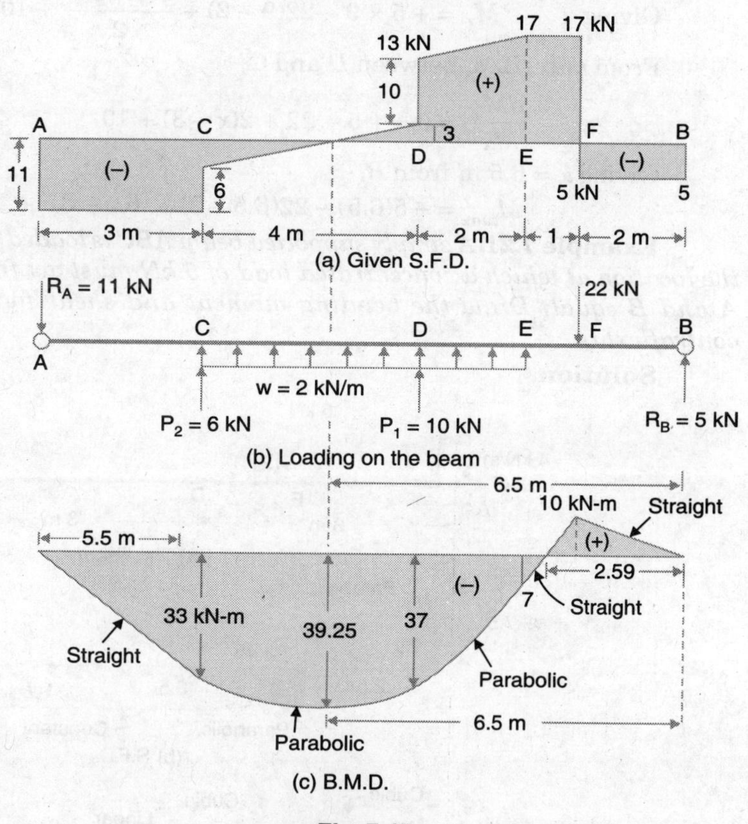

Fig. 7.45

B.M. Diagram :

The B.M.D. is shown in Fig. 7.45 (*c*).

$M_B = 0$; $M_F = + 5 \times 2 = + 10$ kN-m. There is linear variation of B.M. between *B* and *F*.

For *FE*, $M_x = + 5x - 22\,(x - 2)$ ...(Linear variation)

At $x = 3$, $M_E = + 5 \times 3 - 22(3 - 2) = - 7$ kN-m.

Thus, the B.M. changes sign between *F* and *E*, having zero value at $x = 44/17 = 2.59$ m from *B*.

For *ED*, $M_x = + 5x - 22(x - 2) + \frac{2(x-3)^2}{2}$ ...(Parabolic)

Giving $M_E = + 5 \times 3 - 22(3 - 2) = - 7$ kN-m (as before)

$$M_D = + 5 \times 5 - 22(5 - 2) + \frac{2(5-3)^2}{2} = - 37 \text{ kN-m.}$$

For DC, $$M_x = +5x - 22(x-2) + \frac{2(x-3)^2}{2} + 10(x-5).$$

Giving $$M_C = +5\times 9 - 22(9-2) + \frac{2(9-3)^2}{2} + 10(9-5) = -33 \text{ kN-m}.$$

From max. B.M. between D and C,

$$\frac{dM_x}{dx} = 0 = +5 - 22 + 2(x-3) + 10$$

Giving $x = 6.5$ m from B.

$$\therefore \quad M_{max} = +5(6.5) - 22(6.5-2) + (6.5-3)^2 + 10(6.5-5) = \mathbf{-39.25 \text{ kN-m.}}$$

Example 7.21. *A simply supported beam ABC is loaded as shown in Fig. 7.46 (a). Determine the location at which a concentrated load of 5 kN must act from end A to make the reactions at A and B equal. Draw the bending moment and shear force diagrams. Locate the point of contraflexure.*

Solution.

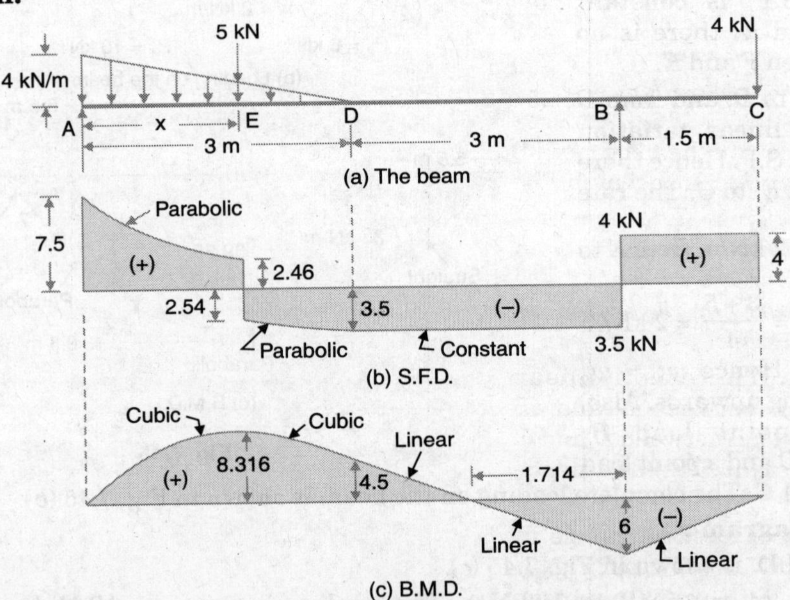

Fig. 7.46

Let the 5 kN load act at x from A. Taking moments about B,

$$6R_A = 5(6-x) + \tfrac{1}{2}\times 4\times 3(3 + \tfrac{2}{3}\times 3) - 4\times 1.5 \text{ or } R_A = \tfrac{1}{6}(54-5x) \quad ...(a)$$

Also, $$6R_B = (4\times 7.5) + \tfrac{1}{2}\times 4\times 3(\tfrac{1}{3}\times 3) + 5x \text{ or } R_B = \frac{1}{6}(36+5x)$$

For the two reactions to be equal,

$$\tfrac{1}{6}(54-5x) = \tfrac{1}{6}(36+5x) \text{ which gives } x = \mathbf{1.8 \text{ m.}}$$

$$\therefore \quad R_A = \tfrac{1}{6}(54 - 5\times 1.8) = 7.5 \text{ kN and } R_B = \tfrac{1}{6}(36 + 5\times 1.8) = 7.5 \text{ kN.}$$

S.F. Diagram : The S.F.D. is shown in Fig. 7.46 (*b*).

$$F_C = +4 \text{ kN} ; F_B \text{ (right)} = +4 \text{ kN} ; \; F_B \text{ (left)} = +4 - 7.5 = -3.5 \text{ kN}$$

$F_D = -3.5$ (constant S.F. between B and D, since there is no loading.)

At the location of 5 kN point load, rate of loading = $\frac{4}{3}(3 - 1.8) = 1.6$ kN/m

$$F_E \text{ (right)} = +4 - 7.5 + \tfrac{1}{2}(1.6 \times 1.2) = -2.54$$

$$F_E \text{ (left)} = -2.54 + 5 = +2.46 \text{ kN} ; \quad F_A = +R_A = +7.5 \text{ kN.}$$

B.M. Diagram :

$$M_C = 0 ; M_B = -4 \times 1.5 = -6 \text{ kN-m}$$

For *BD,* measuring *x* from *B,*

$$M_x = -4(x + 1.5) + 7.5x = 3.5x - 6 \qquad \text{(linear)}$$

which gives $M_B = -6$ kN-m and $M_D = 3.5 \times 3 - 6 = +4.5$ kN-m

Hence B.M. changes sign, having zero value at $x = 6/3.5 = 1.714$ m from B.

For DE, measuring *x* from *D*, we get

$$M_x = -4(4.5 + x) + 7.5(x + 3) - \left(\frac{4}{3}x\right)\frac{x}{2}\left(\frac{1}{3}x\right) \qquad \text{...(cubic)}$$

or

$$M_x = +4.5 + 3.5\,x - \frac{2}{9}\,x^3$$

At *E*, $x = 1.2$ m $\quad\therefore\quad M_E = +4.5 + 3.5 \times 1.2 - \frac{2}{9}(1.2)^3 = +8.316$ kN-m

This is evidently the maximum B.M. between *D* and *A*, since S.F. changes sign at *E*. The complete B.M. diagram is shown in Fig. 7.46 (*c*).

Example 7.22. *The bending moment diagram of a beam AD, hinged at A and D is shown in Fig. 7.47. Determine the loading on the beam and draw the S.F. diagram.*

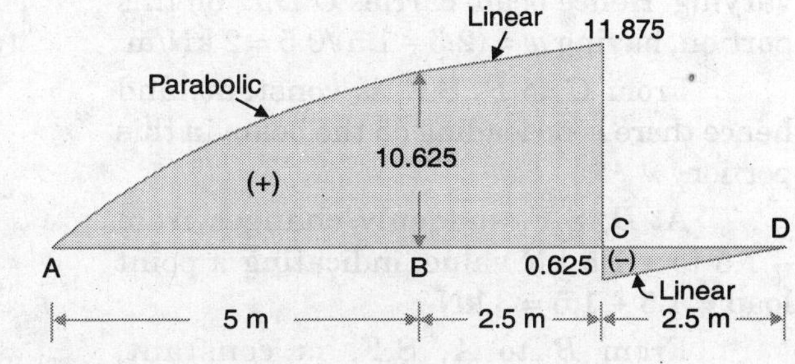

Fig. 7.47

Solution. There is a linear variation of B.M. between *D* and *C*. Hence there is no load between *D* and *C* (resulting in constant S.F.) and hence $R_D = 0.625/2.5 = 0.25$ kN (↓).

At *C*, there is sudden change in B.M. from – 0.625 kN-m to + 11.875 kN-m. Hence there is a anticlockwise moment = 0.625 + 11.875 = 12.5 kN-m at *C*, causing concavity upwards.

From *C* to *B*, there is linear variation of B.M., resulting in a constant

$$\text{Shear Force} = \frac{dM}{dx}$$

$$= \frac{\text{Initial B.M.} - \text{Final B.M.}}{2.5}$$

$$= \frac{+11.875 - (+10.625)}{2.5}$$

$$= +0.5 \text{ kN.}$$

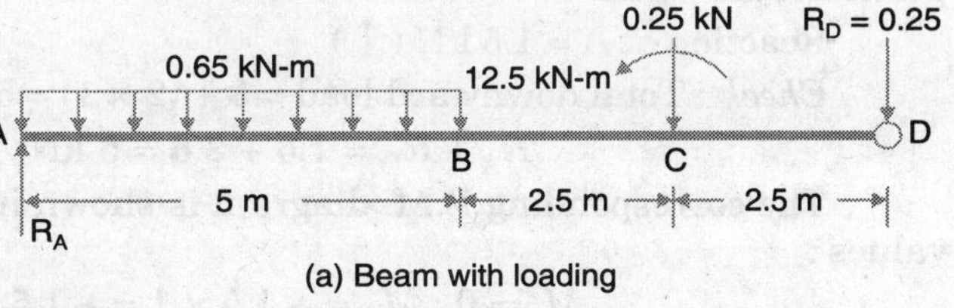

(a) Beam with loading

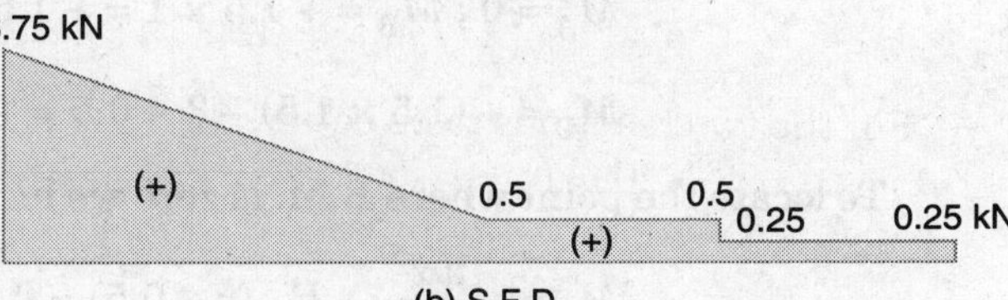

(b) S.F.D.

Fig. 7.48

Since $R_D = 0.25$ kN only, there exists a point load = 0.5 – 0.25 = 0.25 kN (↓) at *C*.

From *B* to *A*, there is parabolic variation of B.M. Hence there will be linear variation in S.F., indicating U.D.L. between *B* and *A*.

Taking moment about *A*, we have

$$M_A = 0 = -(0.25 \times 10) - (0.25 \times 7.5) + 12.5 - \frac{w(5)^2}{2}$$

This gives $w = 0.65$ kN/m

$\therefore$ $R_A = (0.65 \times 5) + 0.25 + 0.25 = 3.75$ kN (↑)

The loading diagram and S.F. diagram are shown in Fig. 7.48 (*a*) and (*b*) respectively.

Example 7.23. *The shear force diagram for a beam ADE, supported at A and D and having an overhang DE is shown in Fig. 7.49 (a). Determine the loading on the beam and draw the B.M. diagram.*

Solution. Since the S.F. is linearly varying from *E* to *D*, there is U.D.L. on the overhang portion of the beam having w = 1 kN/0.5 m = 2 kN/m.

At *D*, S.F. changes sign from + 1 kN to – 2.5 kN. Hence $R_D = 1 + 2.5 = 3.5$ kN (↑).

Again from *D* to *C*, S.F. is linearly varying. Hence beam carries U.D.L. on this portion, having $w = (2.5 - 1.5)/0.5 = 2$ kN/m.

From *C* to *B*, S.F. is constant, and hence there is no loading on the beam in this portion.

At *B*, S.F. suddenly changes from – 1.5 to + 1.5 kN value, indicating a point load = 1.5 + 1.5 = 3 kN.

From *B* to *A*, S.F. is constant, indicating that there is no loading in this portion of the beam.

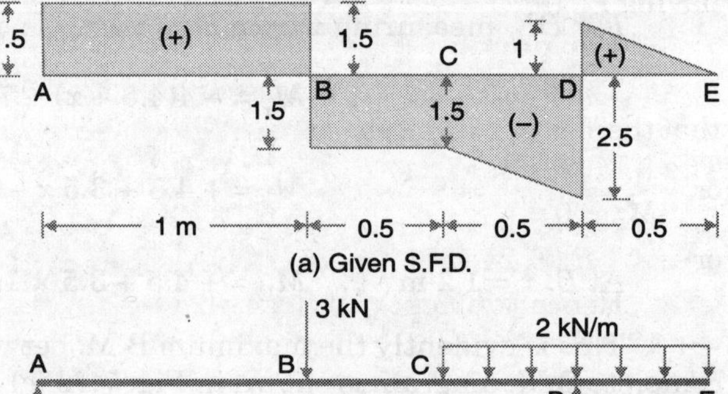

(a) Given S.F.D.

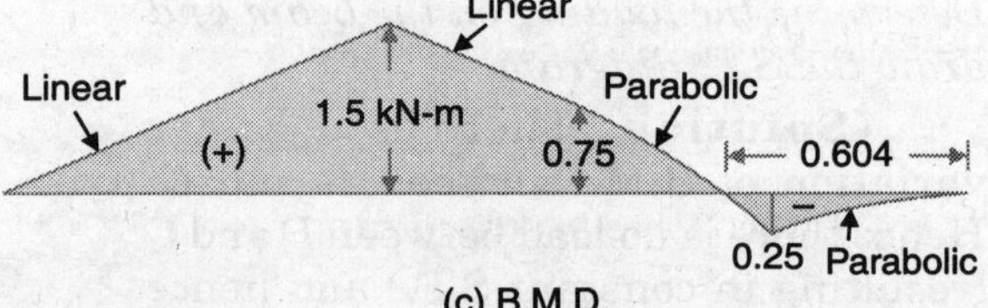

(b) Beam with loading

Linear
Linear
Parabolic
1.5 kN-m
(+)
0.75
0.604
–
0.25
Parabolic

(c) B.M.D.

Fig. 7.49

Reaction at *A* = 1.5 kN (↑)

Check : Total downward load = 3 + (2 × 1) = 5 kN

$$R_A + R_D = 1.5 + 3.5 = 5 \text{ kN}$$

The corresponding B.M. diagram is shown in Fig. 7.49 (*c*), having the following salient values :

$$M_A = 0 \; ; M_B = +1.5 \times 1 = +1.5 \text{ kN-m}$$

$$M_C = +(1.5 \times 1.5) - 3 \times 0.5 = +0.75 \text{ kN-m} \; ; M_D = -\frac{2(0.5)^2}{2} = -0.25 \text{ kN-m.}$$

To locate the point where B.M. is zero we have

$$M_x = -\frac{wx^2}{2} + R_D (x - 0.5) = 0$$

$\therefore$ $x^2 - 3.5(x - 0.5) = 0$, which gives $x = 0.604$ m (measured from *E*). The B.M.D. is shown in Fig. 7.49 (*c*).

Example 7.24. *A beam ABCD is supported at B and C and has overhangs AB and CD. The B.M. diagram for this beam is shown in Fig. 7.50 (a). Draw the loading on the beam and the S.F. diagram.*

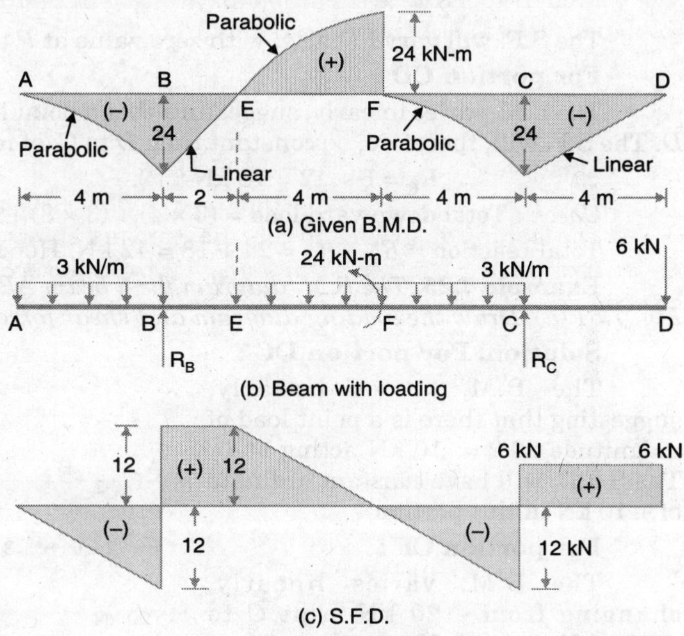

Fig. 7.50

Solution. For portion AB :

The B.M.D. is a second degree curve. Hence U.D.L. acts on *AB*, having

$$w = (24)\frac{2}{x^2} = \frac{24 \times 2}{4 \times 4}$$

$= 3$ kN/m.

The S.F. will vary linearly, having F_B (left) $= -3 \times 4 = -12$ kN.

For portion BE :

The B.M.D. is linear, suggesting that there is no load on the portion *BE*. The B.M. at *E* is zero.

$$M_E = 0 = -3 \times 4(2 + 2) + R_B \times 2$$

or $\quad R_B = 24$ kN (↑).

Hence F_B (right) $= -3 \times 4 + 24 = +12$ kN.

Also, $F_E = -3 \times 4 + 24 = +12$ kN

For portion EF :

The B.M.D. is a second degree curve, increasing from zero value at *E* to + 24 kN-m at *F*. Hence U.D.L. acts between *E* and *F*. Its magnitude (*w* kN/m) is given by writing expression for M_F (left) and equating it to + 24 kN-m.

$$\therefore \quad M_F \text{ (left)} = +24 = -3 \times 4(2 + 6) + 24 \times 6 - \frac{w(4)^2}{2} \text{ which gives } w = 3 \text{ kN/m.}$$

Alternatively, considering portion *EF* alone,

$$\frac{w(4)^2}{2} = 24 \quad \text{or} \quad w = 3 \text{ kN/m. The S.F. for } EF \text{ will vary linearly.}$$

At, *F*, $\quad S_F = -(3 \times 4) + 24 - 3 \times 4 = \text{zero.}$

At *F*, the B.M. suddenly decreases from + 24 kN-m to zero value suggesting that there is localised moment of 24 kN-m in the anticlockwise direction, as marked, causing concavity for the left portion of the beam and convexity to the right portion of the beam.

For portion FC :

The B.M. diagram is a second degree curve, suggesting that U.D.L. is acting on this portion. To find the value of *w*, write expression for M_C, for the left portion of the beam.

$$\therefore \quad M_C = -24 = -3 \times 4(2 + 10) + 24 \times 10 - 3 \times 4(2 + 4) - 24 - \frac{w(4)^2}{2}$$

This gives $w = 3$ kN/m.

(Note that the localised moment of 24 kN-m at *F* will cause convexity at *C*, and hence has been assigned negative sign in the above equation).

Alternatively, considering portion FC alone,

$$\frac{w(4)^2}{2} = 24 \quad \text{or} \quad w = 3 \text{ kN/m.}$$

The S.F. will vary linearly, with zero value at F to $-(3 \times 4) = -12$ kN at C.

For portion CD :

The B.M. varies linearly, suggesting that a point load of magnitude 24/4 = 6 kN (↓) acts at D. The S.F. will, therefore, be constant from D to C, of magnitude + 6 kN.

$$\therefore \quad R_C = 6 + 12 = 18 \text{ kN } (\uparrow).$$

Check : Total downward load = (3 × 4) + (3 × 8) + 6 = 42 kN

Total reaction = $R_B + R_C$ = 24 + 18 = 42 kN. Hence O.K.

Example 7.25. *The B.M. diagram for a beam ABCD, supported at B and C, is shown in Fig. 7.51 (a). Draw the loading diagram and shear force diagram for the beam.*

Solution. For portion DC :

The B.M. varies linearly, suggesting that there is a point load of magnitude 20/2 = 10 kN acting at D. The S.F.D. will have constant ordinate of + 10 kN in this portion.

For portion CF :

The B.M. varies linearly, changing from − 20 kN-m at C to + 130 kN-m at F. Hence there is no external loading in the portion CF. For finding reaction at C, write equation for M_F and equate it to + 130 kN-m.

$$M_F = +130 = -10 \times 7 + R_C \times 5$$

$$\therefore \quad R_C = 40 \text{ kN } (\uparrow).$$

The shear force diagram between C and F will have a constant ordinate = + 10 − 40 = − 30 kN.

For portion FE :

The B.M. varies linearly, decreasing from + 130 kN-m at F to + 110 kN-m at E. This suggests that there is a constant S.F. between F and E, resulting in a point load at F, the magnitude of which is given by

$$M_E = +110 = -(10 \times 9) + (40 \times 7) - W(2) \quad \text{or} \quad W = 40 \text{ kN } (\downarrow)$$

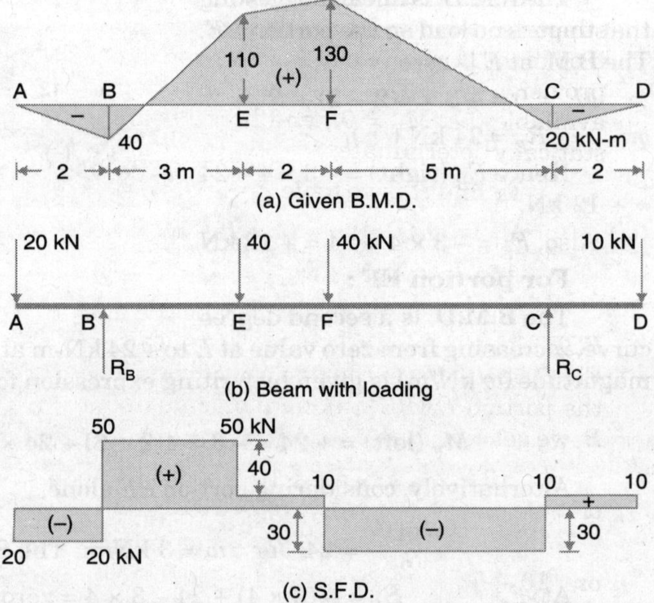

(a) Given B.M.D.

(b) Beam with loading

(c) S.F.D.

Fig. 7.51

Alternatively, S.F. between F and $E = \dfrac{dM}{dx} = \dfrac{+130 - (+110)}{2} = +10$ kN

$$\therefore \quad \text{Point load at } F = 30 + 10 = 40 \text{ kN } (\downarrow)$$

For portion EB :

The B.M. varies linearly, suggesting that the S.F. in this portion is *constant.* Its magnitude is given by

S.F. between E and $B = \dfrac{dM}{dx} = \dfrac{+110-(-40)}{3} = +50$ kN.

Hence magnitude of point load at E = 50 – 10 = 40 kN (↓).

For portion AB :

The B.M. varies linearly, suggesting that there is a point load at A, of magnitude = 40/2 = 20 kN. Hence shear force between A and B will have constant magnitude of – 20 kN.

$\therefore$ Reaction $R_B = 20 + 50 = 70$ kN (↑).

Check. Total downward load = 20 + 40 + 40 + 10 = 110 kN

Total reaction = 70 + 40 = 110 kN.

Example 7.26. *For a beam loaded as shown in Fig. 7.52 (a), draw the S.F. and B.M. diagram.*

Solution. Since the beam has three supports, it will be statically indeterminate if no hinge is provided at E. However, due to provision of hinge, one more equation is available (*i.e.* $M_E = 0$), and hence beam is statically determinate.

To find R_C, we have

M_E (right) $= 0 = -(4 \times 2) + R_C \times 1$

$\therefore$ $R_C = 8$ kN (↑)

The free body diagram for the portions *FCE* and *EBAD* of the beam are shown in Fig. 7.52 (*b*). From the free body diagram of the portion *DABE*, taking moments about *B*, we get

$$(3 \times 2 \times 1) - R_A \times 1 + 4 \times 1 = 0$$

or $R_A = 10$ kN (↑)

$$R_A + R_B + R_C = \Sigma W$$

or $10 + R_B + 8 = (3 \times 2) + 4 = 10$

$\therefore$ $R_B = 10 - 18 = -8$

or $R_B = 8$ kN (↓)

The S.F.D. and B.M.D. are shown in Fig. 7.52 (*c*) and (*d*) respectively.

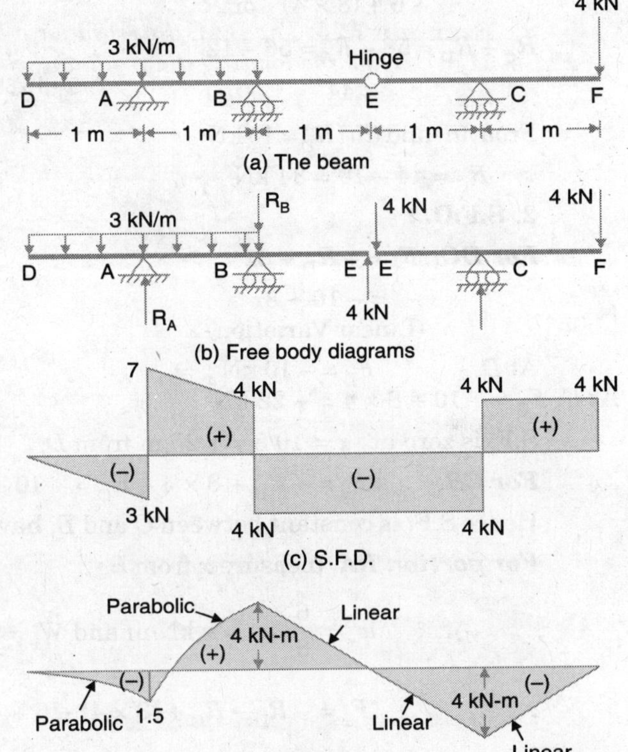

Fig. 7.52

Example 7.27. *A beam ABCD is supported at A, C and D, has an internal hinge at B, and is loaded as shown in Fig. 7.53 (a). Draw the bending moment and shear force diagrams, showing clearly the position and magnitude of maximum bending moment in the beam.*

Solution.

1. Determination of reactions

Due to hinge at B, the B.M. (M_B) at B is zero. Hence

$$M_B = +R_A.6 - \tfrac{1}{2} \times 6 \times 6\left(\tfrac{2}{3}6\right) = 0.$$

From which $R_A = 12$ kN ...(*i*)

Similarly, considering all the forces to the right hand side of *B*,

$$M_B = +R_D.6 + R_C.2 - 8 \times 4 \times 4 = 0$$

$$\therefore \quad 3R_D + R_C = 64 \quad ...(ii)$$

Also, for the whole beam,

$$R_A + R_C + R_D = (\tfrac{1}{2} \times 6 \times 6) + 6 + (8 \times 4) \quad \text{or}$$

$$R_C + R_D = 56 - R_A = 56 - 12 = 44 \quad ...(iii)$$

From (*ii*) and (*iii*) $R_D = 10$ kN

$\therefore \quad R_C = 44 - 10 = 34$ kN

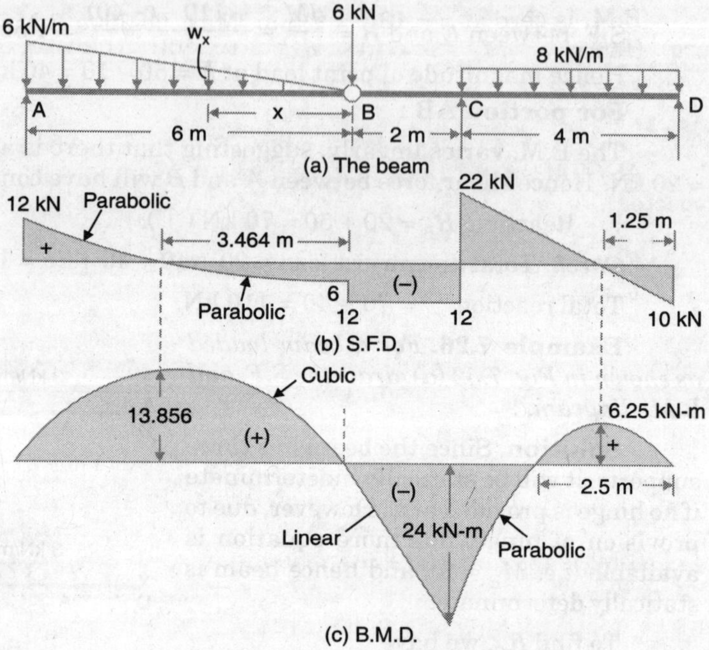

2. **S.F.D. :**

For DC, $F_x = -R_D + 8x = -10 + 8x$ (Linear Variation)

At *D*, $F_D = -10$ kN ;

At *C*, $F_C = -10 + 8 \times 4 = +22$ kN

Fig. 7.53

S.F. is zero at $x = 10/8 = 1.25$ m from *D*.

For CB, $F_x = -R_D + 8 \times 4 - R_C = -10 + 32 - 34 = -12$ kN

Hence S.F. is constant between *C* and *B*, having a value of – 12 kN

For portion BA, measure *x* from *B*.

$$w_x = \frac{6}{6} \times x = x \text{ kN/m and } W_x = \frac{1}{2} w_x \,.\, x = \frac{x^2}{2}$$

$$\therefore \quad F_x = -R_D - R_C + (8 \times 4) + 6 + \frac{x^2}{2} = -10 - 34 + 32 + 6 + \frac{x^2}{2} = -6 + \frac{x^2}{2}$$

...(Parabolic Variation)

At *B*, F_B (left) $= -6 + 0 = -6$ kN ; At *A*, $F_A = -6 + \frac{(6)^2}{2} = +12$ kN

S.F. is zero at *x* given by $F_x = 0 = -6 + \frac{x^2}{2}$. From which $x = \sqrt{12} = 3.464$ m from *B*.

The complete S.F.D. is shown in Fig. 7.53(*b*).

3. **B.M.D. :**

For DC, $M_x = +R_D\, x - \frac{8}{2} x^2 = +10x - 4x^2$ (Parabolic)

$M_D = 0$; $M_C = +10 \times 4 - 4(4)^2 = -24$ kN-m.

B.M. is zero at $x = 10/4 = 2.5$ m. Max. positive B.M. occurs at $x = 1.25$ m (where S.F. is zero). Hence $M_{max} = +10 \times 1.25 - 4(1.25)^2 = +6.25$ kN-m.

For CB, $M_x = +R_D\, x + R_C(x-4) - 8 \times 4(x-2)$

$= +10x + 34(x-4) - 32(x-2) = +12x - 72$ ...(Linear)

At C, $M_C = +12 \times 4 - 72 = -24$ kN-m (as before) ; At B, $M_B = +12 \times 6 - 72 = 0$ (as expected).

For BA, Measuring x from B, $w_x = \frac{6}{6} \times x = x$ kN/m

$$\therefore \quad M_x = +R_D(x+6) + R_C(x+2) - 8 \times 4(x+2+2) - 6x - \tfrac{1}{2}x^2\left(\tfrac{1}{3}x\right)$$

or
$$M_x = +10(x+6) + 34(x+2) - 32(x+4) - 6x - \frac{x^3}{6} = +6x - \frac{x^3}{6}$$

(Cubic variation)

At B, $M_B = 0$; At A, $M_A = +6 \times 6 - \frac{(6)^3}{6} =$ zero.

B.M. is max. at $x = 3.464$ m, where S.F. is zero

$$\text{Max.} = +6(3.464) - \frac{(3.464)^3}{6} = +13.856 \text{ kN-m.}$$

The complete B.M.D. is shown in Fig. 7.53(c).

Example 7.28. *A beam ABCD is loaded as shown in Fig. 7.54(a). Determine the reactions and draw the B.M. and S.F. diagrams.*

Solution.

1. **Determination of Reactions:**

Because of internal hinge at B, $M_B = 0 = -20 + R_C \times 2$.

From which $R_C = 10$ kN (↓)

Now $\Sigma V = 0$. Hence $R_A + R_C = 10$

$\therefore \quad R_A = 10 - R_C = 10 - 10 = 0.$

2. **S.F.D.:**

For DC, $F_x = 0$

For CB, $F_x = -R_C = -10$ kN (constant)

$\therefore \quad F_C = -10$ kN ; $F_B = -10$ kN ;

F_E (right) $= -10$ kN

For EA, $F_x = +R_C - 10 = 10 - 10 = 0$

$\therefore \quad F_E$ (left) $= -R_C + 10$

$= -10 + 10 = 0$; $F_A = 0$.

The complete S.F.D. is shown in Fig. 7.54(b).

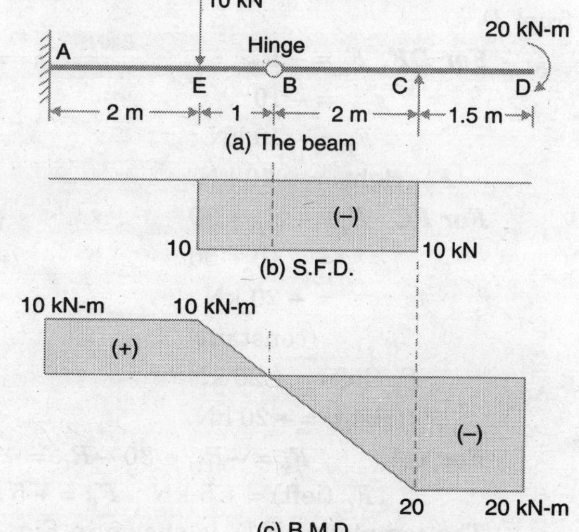

Fig. 7.54

3. **B.M.D.:** Measure x from D.

For DC, $M_x = -20$ kN-m (constant)

$\therefore \quad M_D = -20$ kN-m ; $M_C = -20$ kN-m.

For CE, $M_x = -20 + R_C(x - 1.5) = -20 + 10(x - 1.5) = -35 + 10x$ (linear)

At C, $M_C = -35 + 10 \times 1.5 = -20$ kN-m (as before)

At B, $M_B = -35 + 10 \times 3.5 =$ zero (as expected)

At E, $M_E = -35 + 10 \times 4.5 = +10$ kN-m

For EA, $M_x = -20 + R_C(x - 1.5) - 10(x - 4.5) = -20 + 10(x - 1.5) - 10(x - 4.5)$

or $M_x = +10$ kN-m (constant)

$\therefore$ At E, $M_E = +10$ kN-m ; At A, $M_A = +10$ kN-m

The complete B.M.D. is shown in Fig. 7.54(c).

Example 7.29. *A beam ABCD, has an internal hinge at B and is loaded as shown in Fig. 7.55(a). Determine reactions at A, C and D, and plot the shear force and B.M. diagrams, indicating principal values.*

Solution.

1. **Determination of reactions:**

Due to hinge at B, bending moment at B is zero. Hence considering L.H.S.,

$$M_B = 0 = +R_A \times 4 - 20 \text{ ; From which } R_A = \frac{20}{4} = 5 \text{ kN } (\uparrow) \quad ...(i)$$

Considering R.H.S., $M_B = 0 = -30 \times 3 + R_C \times 2 + R_D \times 6$ or $R_C + 3R_D = 45$ ...(ii)

Also, for the whole beam, $R_A + R_C + R_D = 30$

$$\therefore \quad R_C + R_D = 30 - R_A = 30 - 5 = 25 \quad ...(iii)$$

From (ii) and (iii),

$R_D = 10$ kN ($\uparrow$)

$\therefore$ $R_C = 25 - R_D$

$= 25 - 10$

$= 15$ kN ($\uparrow$).

2. **S.F.D.:** Measure x from D.

For DF, $F_x = -R_D$

$= -10$

$\therefore$ $F_D = -10$ kN ;

F_F (right) $= -10$ kN.

For FC, $F_x = -R_D + 30$

$= -10 + 30$

$= +20$ kN

(constant)

$\therefore$ F_F (left) $= +20$ kN

F_C (right) $= +20$ kN.

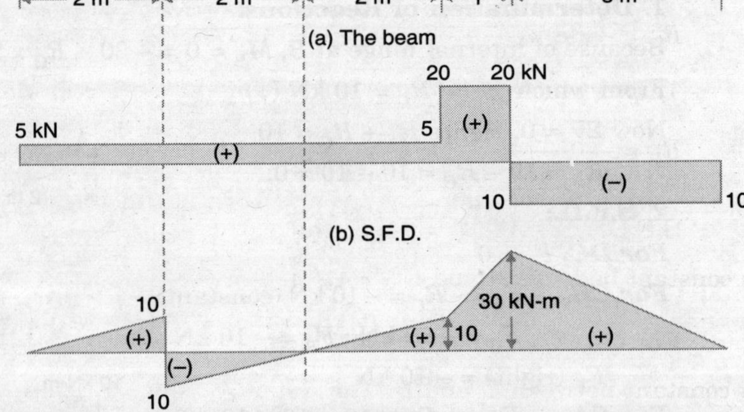

Fig. 7.55

For CA, $F_x = -R_D + 30 - R_C = -10 + 30 - 15 = +5$ (constant)

$\therefore$ F_C (left) $= +5$ kN ; $F_B = +5$ kN ; $F_A = +5$ kN.

The complete S.F.D. is shown in Fig. 7.55(a).

3. **B.M.D.** Measure x from D.

For DF, $M_x = +R_D\, x = +10\, x$ (linear)

$M_D = 0$; $M_F = + 10 \times 3 = + 30$ kN-m.

For FC, $M_x = + R_D\, x - 30(x - 3) = + 10\, x - 30(x - 3) = - 20\, x + 90$ (linear)

At F, $M_F = - 20 \times 3 + 90 = + 30$ kN-m (as before)

At C, $M_C = - 20 \times 4 + 90 = + 10$ kN-m.

For CE, $M_x = + R_D\, x - 30(x - 3) + R_C(x - 4) = + 10x - (30x - 90) + 15(x - 4)$

or $M_x = - 5x + 30$ (linear)

At C, $M_C = - 5 \times 4 + 30 = + 10$ kN-m (as before)

At B, $M_B = - 5 \times 6 + 30 = 0$ (as expected)

At E, M_E (right) $= - 5 \times 8 + 30 = - 10$ kN-m.

For EA, $M_x = + R_D\, x - 30(x - 3) + R_C(x - 4) + 20$

or $M_x = + 10x - (30x - 90) + 15(x - 4) + 20 = - 5x + 50$ (linear)

At E, M_E (left) $= - (5 \times 8) + 50 = + 10$ kN-m

At A, $M_x = - (5 \times 10) + 50 = 0$ (as expected)

The complete B.M.D. is shown in Fig. 7.55(*c*).

Example 7.30. *A simply supported beam ACB supports a vertical load P by means of a bracket CDE, as shown in Fig. 7.56(a). Draw the S.F. and B.M. diagrams for the beam.*

Solution. Apply equal and opposite forces P at point C [Fig. 7.56(*b*)]. Then the given system of forces is equivalent to (*i*) a downward force P at C and (*ii*) a moment $\mu = PL/4$ at C, as shown in Fig. 7.56(*c*).

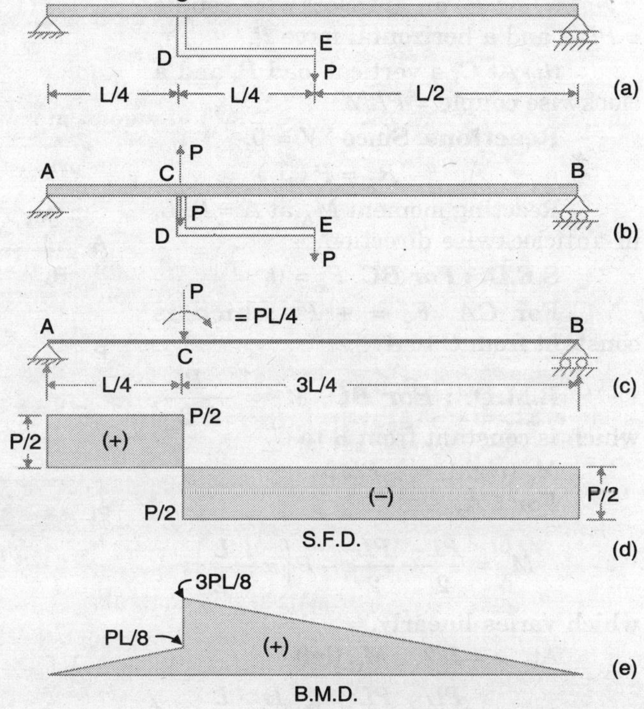

Fig. 7.56

Reactions. Taking moments about B,

$$R_A = \frac{1}{L}\left[P \times \frac{3L}{4} - \frac{PL}{4}\right] = \frac{P}{2}\ (\uparrow)$$

Taking moments about A,

$$R_B = \frac{1}{L}\left[\frac{PL}{4} + \frac{PL}{4}\right] = \frac{P}{2}\ (\uparrow)$$

S.F.D. : ***For AC***, $F_x = + \dfrac{P}{2}$, which is constant between A and C.

For CB, $F_x = \dfrac{P}{2} - P = - \dfrac{P}{2}$, which is constant between C and B. The S.F. diagram is shown in Fig. 7.56(*d*).

B.M.D. : ***For AC***, $M_x = \dfrac{P}{2} x$, which varies linearly. At $x = 0$, $M_A = 0$

At $x = \dfrac{L}{4}$, $M_C\text{ (left)} = \dfrac{P}{2} \times \dfrac{L}{4} = \dfrac{PL}{8}$

For CB, $M_x = \dfrac{P}{2}x + \dfrac{PL}{4} - P\left(x - \dfrac{L}{4}\right)$, which varies linearly.

At $x = \dfrac{L}{4}$, $M_C\text{ (right)} = \dfrac{P}{2} \times \dfrac{L}{4} + \dfrac{PL}{4} - P\left(\dfrac{L}{4} - \dfrac{L}{4}\right) = \dfrac{3PL}{8}$

At $x = L$, $M_B = \dfrac{PL}{2} + \dfrac{PL}{4} - P\left(L - \dfrac{L}{4}\right) = 0$, as expected.

The B.M.D. is shown in Fig. 7.56 (*e*).

Example 7.31. *Draw the S.F. and B.M. diagrams for the cantilever ACB shown in Fig. 7.57(a).*

Solution. Due to a load P applied to the rope, (*i*) a vertical force P and a horizontal force P are transmitted at D, and (*ii*) a horizontal force P is transmitted at E. These are shown in Fig. 7.57(*b*). Due to these, the following forces are transmitted to the cantilever, as shown in Fig. 7.57(*c*) :

(*i*) At B, an anticlockwise couple $= PL/2$ and a horizontal force P.

(*ii*) At C, a vertical load P, and a clockwise couple $= PL/2$.

Reactions. Since $\Sigma V = 0$,

$$R_A = P\ (\uparrow)$$

Reacting moment M_A at $A = P.\ L$, in anticlockwise direction.

S.F.D. : *For BC*, $F_x = 0$

For CA, $F_x = +P$, which is constant from C to A.

B.M.D. : *For BC*, $M_x = +\dfrac{PL}{2}$, which is constant from B to C.

$M_C\text{ (right)} = +PL/2$

For CA,

$$M_x = \frac{PL}{2} - \frac{PL}{2} - P\left(x - \frac{L}{2}\right),$$

which varies linearly.

At $x = L/2$, M_C (left)

$$= \frac{PL}{2} - \frac{PL}{2} - P\left(\frac{L}{2} - \frac{L}{2}\right) = 0$$

At $x = 3L/2$,

$$M_A = \frac{PL}{2} - \frac{PL}{2} - P\left(\frac{3L}{2} - \frac{L}{2}\right) = -PL$$

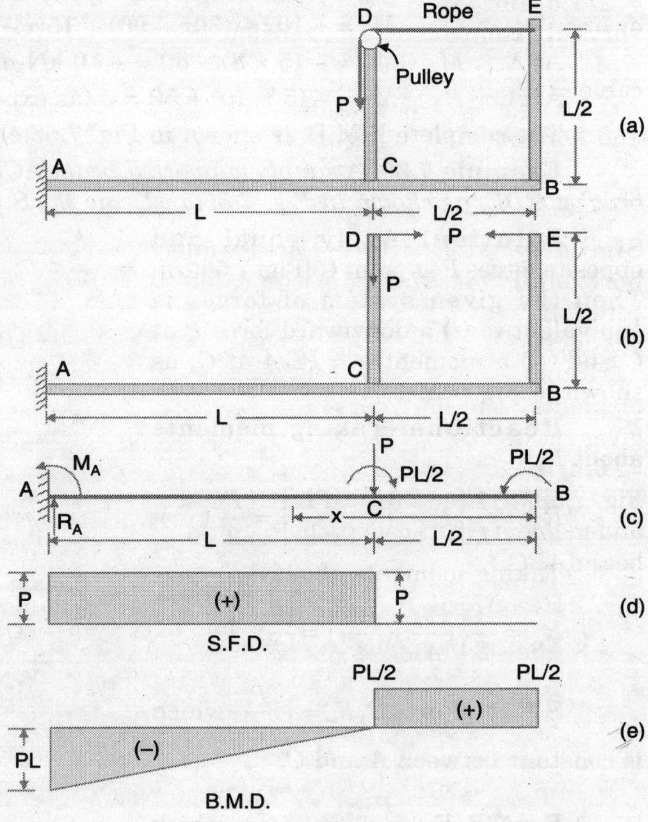

Fig. 7.57

The S.F. and B.M. diagrams are shown in Fig. 7.57 (*d*) and (*e*) respectively.

Example 7.32. *A beam ABCD is loaded by a force W = 8 kN by the arrangement shown in Fig. 7.58. The cable passes over a small frictionless pulley at B and is attached at E to the vertical arm. Compute the shear force and bending moment at the section, which is just to the left of the vertical arm. Also, sketch the S.F. and B.M. diagrams for the beam.*

Solution. Tension in the cable will be equal to the force W. Fig. 7.58 (*b*) shows the free body diagram. From geometry, we have $BE = \sqrt{(2)^2 + (1.5)^2} = 2.5$ m. Hence $\cos\theta = 2/2.5$ and $\sin\theta = 1.5/2.5$. At B, vertical component of tension in

$$\text{cable} = W \sin\theta = 8 \times \frac{1.5}{2.5} = 4.8 \text{ kN}$$

and horizontal component

$$= W \cos\theta = 8 \times \frac{2}{2.5} = 6.4 \text{ kN}.$$

Similarly, at E, the vertical and horizontal components of the cable tension are 4.8 kN and 6.4 kN respectively. Fig. 7.58 (*c*) shows various forces marked on the elements. The horizontal force at E gives rise to an anticlockwise moment = 6.4 × 1.5 = 9.6 kN-m at C. Fig. 7.58 (*d*) shows the vertical forces and moments acting on the horizontal beam $ABCD$.

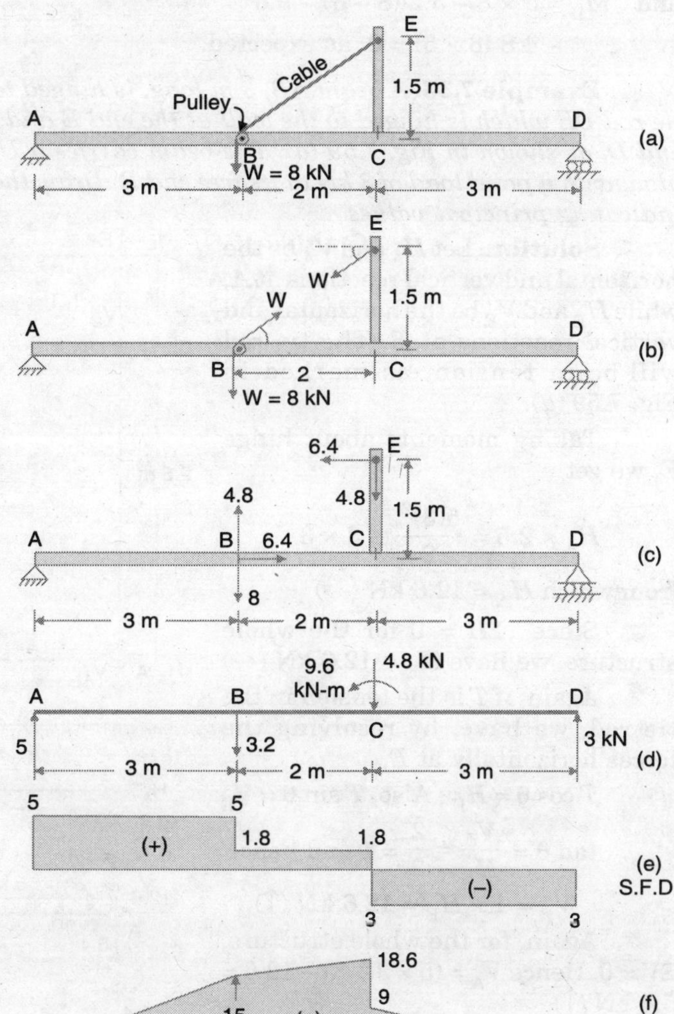

Fig. 7.58

Reactions :

Taking moments at D,

$$R_A = \tfrac{1}{8}(3.2 \times 5 + 4.8 \times 3 + 9.6)$$

$$= 5 \text{ kN } (\uparrow)$$

$$\therefore \quad R_D = 3.2 + 4.8 - 5 = 3 \text{ kN } (\uparrow)$$

S.F.D. : Fig. 7.58 (*e*) shows the S.F. diagram, which is self explanatory.

Hence S.F. just to the left of C = **1.8 kN**

B.M.D. :

For AB, $M_x = 5\,x$, which gives $M_B = 5 \times 3 = 15$ kN-m

For BC, $M_x = 5\,x - 3.2\,(x - 3)$, which gives M_C (left)

$$= 5 \times 5 - 3.2\,(5 - 3) = \mathbf{18.6 \text{ kN-m}}$$

For CD,

$$M_x = 5x - 3.2(x-3) - 9.6 - 4.8(x-5),$$

which gives M_C (right) = 9 kN-m

and $M_D = 5 \times 8 - 3.2(8-3) - 9.6$
$- 4.8(8-5) = 0$, as expected.

Example 7.33. *A beam AB, 3 m long, is hinged to a wall at end A and is supported by a tie rod DE which is hinged to the wall at the end E and also hinged to a vertical bracket CD at end D, as shown in Fig. 7.59 (a). The beam carries a U.D.L. of 5 kN/m throughout its length alongwith a point load of 3 kN at its free end B. Draw the S.F. and B.M. diagrams for the beam, indicating principal values.*

Solution. Let H_A and V_A be the horizontal and vertical reactions at *A*, while H_E and V_E be the horizontal and vertical reactions at *E*. The tie rod will be in tension, as marked in Fig. 7.59 (*a*).

Taking moments about hinge *E*, we get

$$H_A \times 2.5 = \frac{5(3)^2}{2} + 3 \times 3$$

from which $H_A = 12.6$ kN ($\rightarrow$)

Since $\Sigma H = 0$ for the whole structure, we have $H_E = 12.6$ kN ($\leftarrow$)

Again, if *T* is the tension in the tie rod, we have, by resolving the forces horizontally at *E*,

$T \cos\theta = H_E$; Also, $T \sin\theta = V_E$

$$\tan\theta = \frac{V_E}{H_E} = \frac{2}{2} = 1 \text{ and hence}$$

$$V_E = 1 \times H_E = 12.6 \text{ kN } (\uparrow)$$

Again, for the whole structure, $\Sigma V = 0$. Hence $V_A = (5 \times 3 + 3) - 12.6 = 5.4$ kN ($\uparrow$)

Thus, the reactions at *A* and *E* are completely known.

Again, resolving the tension in tie rod at end *D*, we have

H_D = Horizontal component of tension in tie rod

$= T\cos\theta = H_E = 12.6$ kN ($\leftarrow$)

V_D = Vertical component of tension in tie rod

$= T\sin\theta = V_E = 12.6$ kN ($\uparrow$)

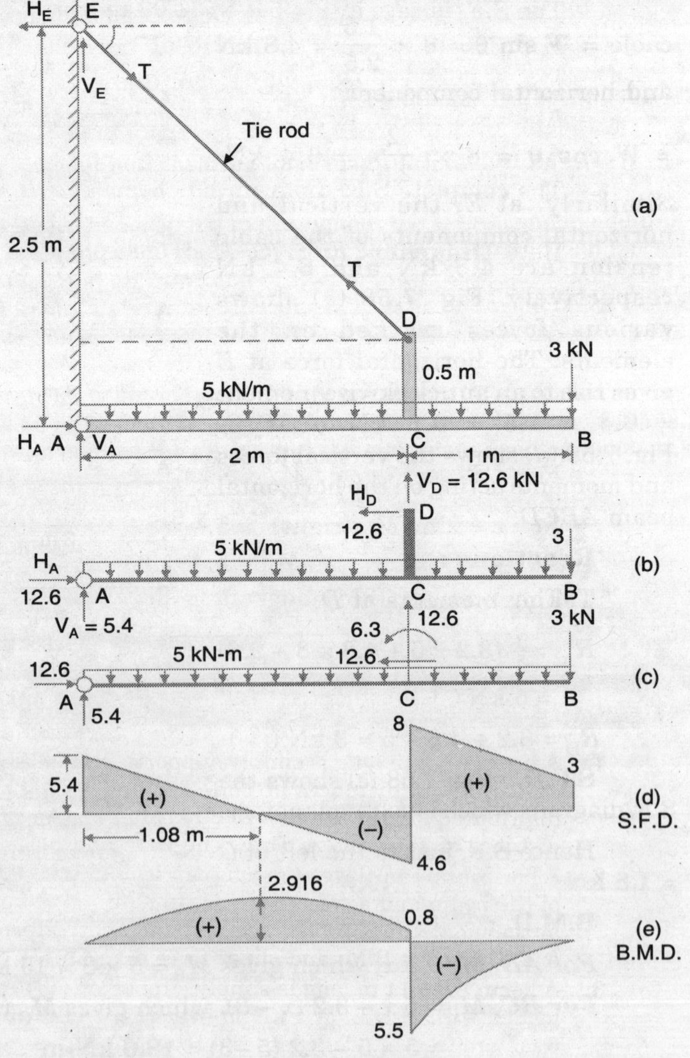

Fig. 7.59

Hence the beam *ACB* is subjected to the following loads and reactions :

(*i*) External loading

(*ii*) Reactions H_A and V_A at end *A*.

(*iii*) Horizontal force H_D and vertical force V_D at *D*.

These forces are shown marked in Fig. 7.59 (*b*). However, the horizontal load H_D (= 12.6 kN) acting at a lever arm of 0.5 m, gives rise to an anticlockwise moment = 12.6 × 0.5 = 6.3 kN-m, and a horizontal force of 12.6 kN (←), both acting at *C*. These final forces/reactions on beam *AB* have been marked in Fig. 7.59 (*c*).

S.F. Diagram :

For AC, $F_x = 5.4 - 5x$, which gives a linear variation.

At $x = 0$, $F_A = 5.4$ kN ; At $x = 2$ m, F_C (left) $= 5.4 - 5 \times 2 = -4.6$ kN.

The S.F. is zero at $x = 5.4/5 = 1.08$ m from *A*.

For CB,

$F_x = 5.4 - 5x + 12.6$, which again gives linear variation.

At $x = 2$ m, we get F_C (right) $= 5.4 - 5 \times 2 + 12.6 = +8$ kN.

At $x = 3$ m, we get $F_B = 5.4 - 5 \times 3 + 12.6 = +3$ kN.

The complete S.F. diagram has been shown in Fig. 7.59 (*d*).

B.M. Diagram : ***For AC,*** $M_x = 5.4x - \dfrac{5x^2}{2}$, which is a parabolic variation.

At $x = 2$ m, M_C (left) $= 5.4 \times 2 - \dfrac{5(2)^2}{2} = 0.8$ kN-m. B.M. is maximum at $x = 1.08$ m.

$$M_{max} = 5.4\,(1.08) - \frac{5}{2}\,(1.08)^2 = 2.916 \text{ kN-m}$$

For CB $M_x = 5.4x - \dfrac{5x^2}{2} - 6.3 + 12.6\,(x - 2)$

At $x = 2$ m, M_C (right) $= 5.4 \times 2 - 2.5\,(2)^2 - 6.3 + 12.6\,(2 - 2) = -5.5$ kN-m

At $x = 3$ m, $M_B = 5.4 \times 3 - 2.5\,(3)^2 - 6.3 + 12.6\,(3 - 2) = 0$, as expected.

The complete B.M. diagram is shown in Fig. 7.59 (*e*).

PROBLEMS

1. A beam *AB* 5 m long is simply supported at *A* and *B*. It is loaded with point loads of 20 kN, 30 kN and 20 kN at distances of 1 m, 3 m and 4 m respectively from the support *A*, and a uniformly distributed load at the rate of 20 kN/m over length of 2 m, the beginning of the U.D.L. being at a distance of 2 m from *A*. Draw the B.M. and S.F. diagrams, indicating the principal values.
2. A horizontal cantilever 4 m long carries a U.D.L. of 2 kN/m which extends over the middle 1.5 metre of the beam and act downwards. In addition to this, a concentrated load of 5 kN acts downward at the point 1 m from the support and another concentrated load of 3 kN acts upward at the free end. Draw the B.M. and shear force diagrams.
3. A beam *ABC* 11 m long is simply supported at *A* and *B* and has an overhang *BC* of 3 metres. It is loaded with U.D.L. of 4 kN-m over the whole length. A concentrated load of 30 kN is acting at 6 meters from *A*. Calculate the reactions. Find the position and value of maximum bending moment and plot the B.M. and S.F. diagrams.

4. A beam *ABCD* 10 metre long is supported at *B* and *C*. The overhangs *AB* and *CD* are 2 m and 3 m respectively. The overhang *AB* carries U.D.L. of 20 kN/m and *CD* carries U.D.L. of 10 kN/m. In addition, there are point loads of 20 kN, 40 kN and 20 kN, at distances of 1.5 m, 3 m and 8 metres from *A* respectively. Find the reactions and plot B.M. and S.F. diagrams, giving principal values.
5. A beam of length *L* carries an evenly distributed load and rests on two supports. How far from the ends must the support be placed if the greatest bending moment to which the beam is subjected is to be as small as possible ? Where are the points of contraflexure ?
6. A beam *ABCD*, 10 m long, is simply supported at *B* and *C* which are 4 m apart, and overhangs the support *B* by 3 m. The overhanging part *AB* carries U.D.L. of 20 kN/m and the part *BD* carries U.D.L. of 10 kN-m. Calculate the position and magnitude of the least value of the bending moment between the supports. Plot B.M. and S.F. diagrams.
7. A horizontal beam of length *L* carries a U.D.L. of *w* per unit length over the whole length. If the beam is freely supported at the left end, find the position of the second support so that the greatest bending moment on the beam shall be as small as possible.
Draw the diagrams of B.M. and S.F. and insert the principal value.
8. A horizontal beam 24 m long carries a uniformly distributed load, including its own weight of 6 kN-m run and in addition concentrated load of 26 kN at the left hand end. The beam is supported at two points 16.33 m apart, so chosen that each support carries half the total load.
Plot, on squared paper, the bending moment and shearing force diagrams, indicating clearly the principal value on each.
9. A beam *ABCD*, 16 m long is loaded with U.D.L. of 20 kN-m throughout the length, and in addition a concentrated load of 80 kN at the left hand end. The beam is supported at *B* and *C* and has an overhang of 1.5 m to the left support *B* and an overhang of *K* metres to the right of support *C* which is in the right hand half of the beam. Determine the value of *K* if the mid-point of the beam is the point of inflexion and for this arrangement plot the B.M. and S.F. diagrams, indicating the principal numerical values.
10. A girder 30 m long carrying a uniformly distributed load of 20 kN per m is to be supported on two piers 18 m apart so that the greatest bending moment on the girder shall be as small as possible. Find the distances of the piers from the ends of the girder and the maximum bending moment.
11. Fig. 7.60 shows the S.F. diagram for a beam *ABCD* having overhangs at both the supports. From the shear force diagram deduce the loading diagram. Draw the B.M.D. stating the principal values.
12. A beam with overhanging ends rests freely on two supports *A* and *B* and is loaded as shown in Fig. 7.61. What must be the intensity of loading in kN/m, on the beam between *C* and *B* if the shearing force is to be zero at a cross-section 2 m to the left of support *B* ? Draw the S.F. and B.M. diagrams and find out the point of contraflexure.

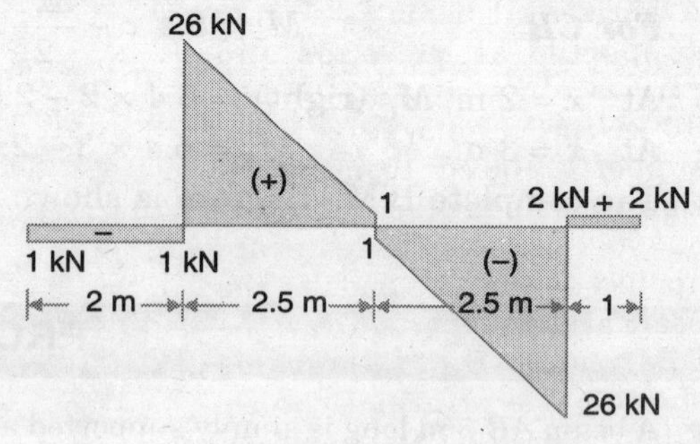

Fig. 7.60

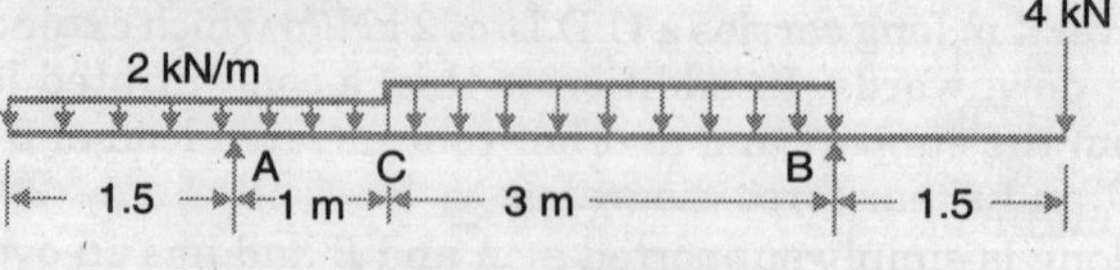

Fig. 7.61

13. A beam *AB* of 6 m span carries a distributed load with an intensity increasing from zero at *A* to 30 kN/m at *B*. Find the reactions and the location and value of maximum bending moment. Plot the B.M. and S.F. diagrams.

14. A horizontal beam simply supported on a span of 10 m carries a total load of 20 kN ; the load distribution varies parabolically from zero at each end to maximum at mid-span. Calculate the values of bending moment at intervals of 1 m and plot on squared paper, the B.M.D. for this loading.

15. A beam *ABC* is loaded with a couple of 12 kN-m shown in Fig. 7.62. Plot the B.M. and S.F. diagrams giving the main values.

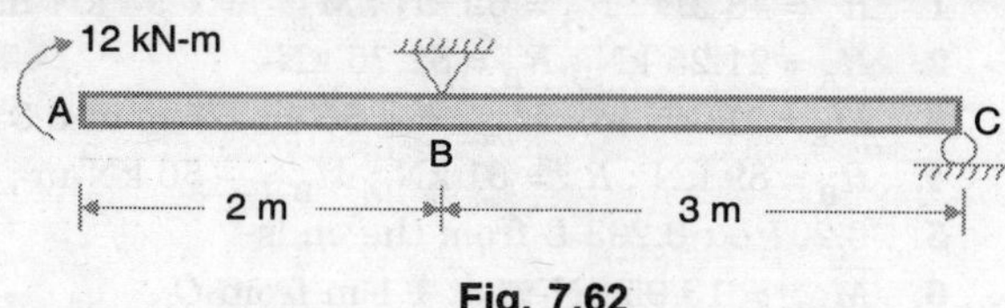

Fig. 7.62

16. A beam *ABCDE* is loaded as shown in Fig. 7.63. Plot on squared paper the B.M. and S.F. diagrams giving the principal values.

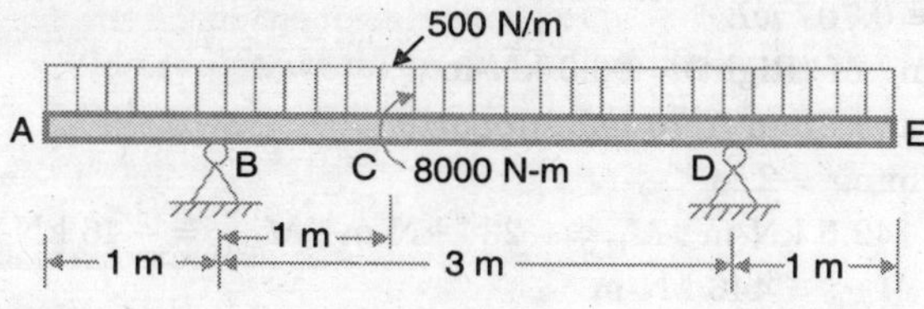

Fig. 7.63

17. A beam *AB* is loaded as shown in Fig. 7.64. Plot the B.M. and S.F. diagrams giving principal values.

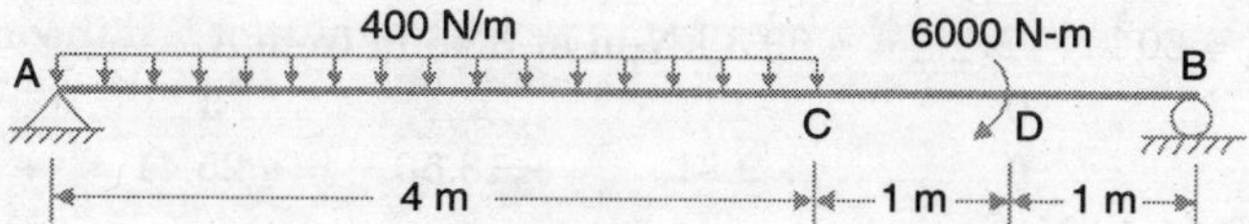

Fig. 7.64

18. A horizontal beam is simply supported at the ends and carries a distributed load of 30 kN/m between the supports placed 10 m apart. Counter-clockwise moments of 300 kN-m and 200 kN-m are applied to the two ends of the beam at the supports. Draw the B.M. and S.F. diagrams showing the principal values. Find the position and values of max. B.M.

19. Draw shearing force and bending moment diagrams for the beam shown in Fig. 7.65 (*a*). Indicate on the diagrams values of shearing force and bending moments (with proper units) at significant points. Also, show the location and magnitude of maximum bending moment.

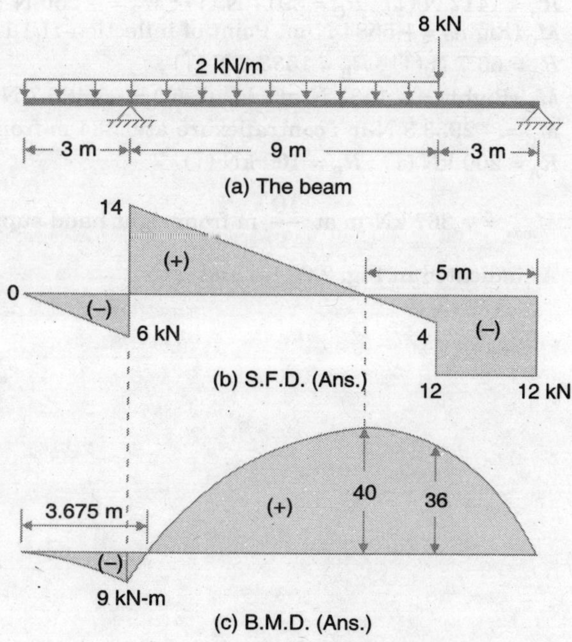

Fig. 7.65

ANSWERS

1. $R_A = 48$ kN ; $R_B = 62$ kN ; $M_{max} = +94$ kN-m
2. $R_A = 21.25$ kN ; $R_B = 52.75$ kN
3. $M_B = 18$ kN-m ; $M_{max} = +56.45$ kN-m at 5.31 m from A
4. $R_B = 89$ kN ; $R_C = 61$ kN ; $M_B = -50$ kN-m ; $M_C = -65$ kN-m
5. 0.207 L ; 0.293 L from the ends
6. $M_{min} = 13.95$ kN-m at 1.1 m from C
7. $l = \dfrac{L}{\sqrt{2}}$; $M = 0.043\, wL^2$

   $R_A = 0.293\, wL$; $R_B = 0.707\, wL$
8. M (Left) = – 64 kN-m ; M (Right) = 96.3 kN-m ;

   $M_{max} = +119$ kN-m at 7.83 m from left support.

   Distance of left overhang = 2 m
9. $K = 4.84$ m ; $M_B = -142.5$ kN-m ; $M_C = -234$ kN-m ; $M_{max} = +46$ kN-m
10. 6.68 m and 5.32 m ; $M_{max} = 446$ kN-m
11. Loading ; 1 kN at A ; U.D.L. of 10 kN/m from B and C.

    2 kN at E ; 2 kN at D.

    $M_B = -2$ kN-m, ; $M_C = -2$ kN-m ; $M_E = +31.75$ kN-m
12. 9.50 kN/m ; 0.20 m to the right of A ; 0.35 m to the left of B.
13. $R_A = 30$ kN ; $R_B = 60$ kN ; $M_{max} = +69.3$ kN-m at 3.46 m from A.
14.

Dist. from end	0	1	2	3	4	5
B.M. kN-m	0	+ 9.81	+ 18.56	+ 25.41	+ 29.76	+ 31.25

15. $R_B = 4$ kN (↓) ; $R_C = 4$ kN (↑)

    $M_A = M_B = +12$ kN-m ; $M_C = 0$
16. $R_B = 1417$ N (↓) ; $R_C = 3917$ N (↑) ; $M_B = -250$ N-m $= M_D$; M_C (Left) = – 2416 N-m ; M_C (Right) = + 5584 N-m, Point of inflection : (*i*) 1.09 m from E (*ii*) at C
17. $R_A = 66.7$ N (↑) ; $R_B = 1533.3$ N (↑) ;

    M_D (Right) = + 1533 N-m ; M_D (Left) = – 4466.7 N-m

    $M_C = -2933.3$ N-m : contraflexure at 0.334 m from A, also at D.
18. $R_A = 200$ kN (↑) ; $R_B = 100$ kN (↑)

    $M_{max} = +367$ kN-m at $\dfrac{10}{3}$ m from right hand support.
19. As indicated in Fig. 7.65 (*b*) and (*c*).

8

Centroids and Moments of Inertia of Plane Areas

8.1. INTRODUCTION : CENTRE OF GRAVITY AND CENTROID

The *centre of gravity* of a body is that point through which the resultant of system of parallel forces formed by the weights of all the particles of the body passes, for all positions of the body.

The *plane areas*, like triangle, rectangle, quadrilateral, circle etc., have only *areas* and no mass. The *centre of area* of such figures is known as *centroid*. In common terminology, we use the term *area* to mean *plane surface*. Strictly speaking area is a measure of size of the surface. The method of finding out the centroid of a figure is the same as finding out the centre of gravity of a body. Hence many Authors write *c.g.,* for *centroid* and *vice-versa*. The location of centroid of a plane area is an important geometrical property of the area.

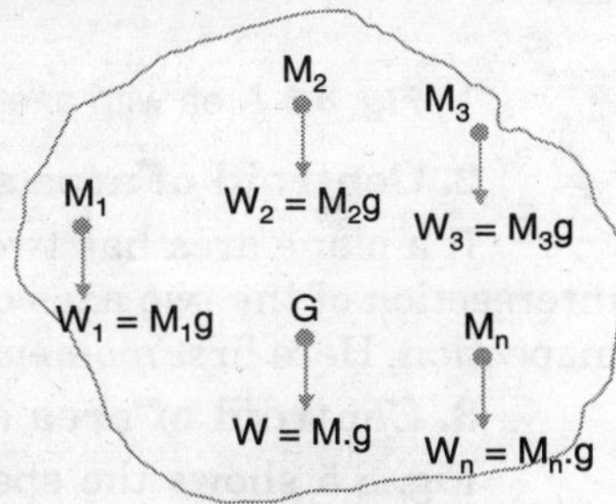

Fig. 8.1

8.2. CENTROID OF PLANE AREAS

Fig. 8.2 shows a plane area A, with its centroid G having its coordinates $\overline{x}, \overline{y}$, the values of which are required to be determined. Consider a small area dA, having co-ordinates x and y.

The total area $A = \int dA$ (by definition) ...(8.1)

The *first moments* of the area about x and y axes respectively are

$$Q_x = \int y \, . \, dA \quad \text{and} \quad Q_y = \int x . \, dA \quad ...(8.2)$$

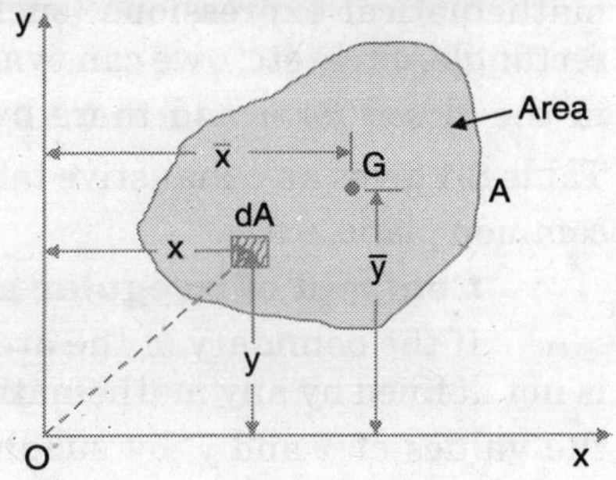

Fig. 8.2 Centroid of plane area

The coordinates $\bar{x}$ and $\bar{y}$ of the centroid G are evidently equal to the first moments divided by the area itself.

Hence $$\bar{x} = \frac{\int x\, dA}{\int dA} = \frac{Q_y}{A} \quad \text{and} \quad \bar{y} = \frac{\int y\, dA}{\int dA} = \frac{Q_x}{A} \qquad ...(8.3)$$

1. Centroid of area symmetric about an axis

If a plane area has an axis of symmetry, the centroid will lie on that axis of symmetry. This is so because the first moment about an axis of symmetry is zero. Fig. 8.3 shows a section having x-axis as the axis of symmetry ; hence the centroid G will lie on that axis. In such a case, only one distance need be calculated.

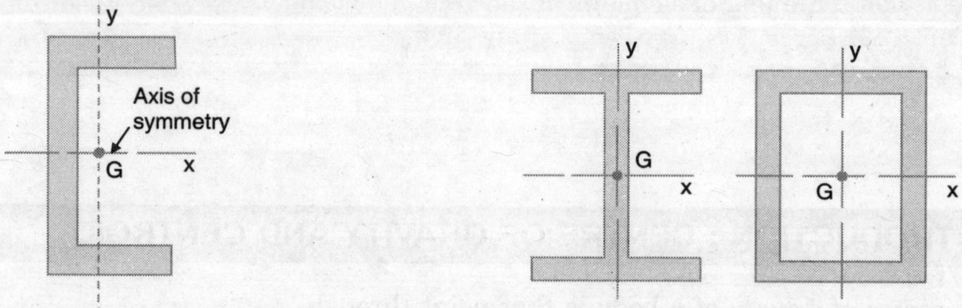

Fig. 8.3 Area with one axis of symmetry **Fig. 8.4** Area with two axes of symmetry

2. Centroid of area symmetric about two axes

If a plane area has two axes of symmetry, the centroid of such an area will be lie at the intersection of the two axes of symmetry. Thus, the centroid of such an area can be located by inspection. Here first moment of area about both the axes will be zero (Fig. 8.4).

3. Centroid of area symmetric about a point

Fig. 8.5 shows the special case of an area which is *symmetric about a point*. Such a point is called the *centre of symmetry* and is so located that every line in the area drawn through that point is symmetric about that point. The centroid of such an area will lie at that point, which can be located by inspection.

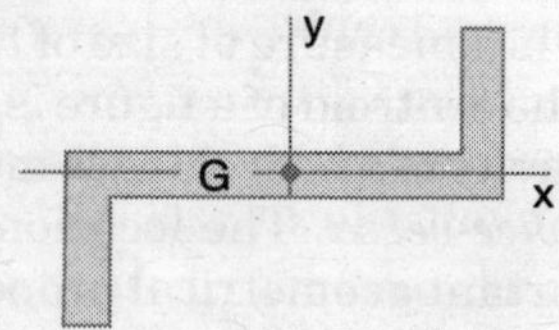

Fig. 8.5 Area symmetric about a point

Centroid of regular areas

If the boundaries of the areas are defined by simple mathematical expressions (such as in case of a triangle, rectangle, circle etc.), we can evaluate the integral of Eq. 8.3 in the *closed form* and there by get formulae for $\bar{x}$ and $\bar{y}$. Table 8.1 gives an exhaustive table for values of $\bar{x}$ and $\bar{y}$ for common plane areas.

Centroid of irregular areas

If the boundary of the area is *irregular curve*, which is not defined by any mathematical expression, we can find the values of $\bar{x}$ and $\bar{y}$ by subdividing the area into small elements of area ΔA.

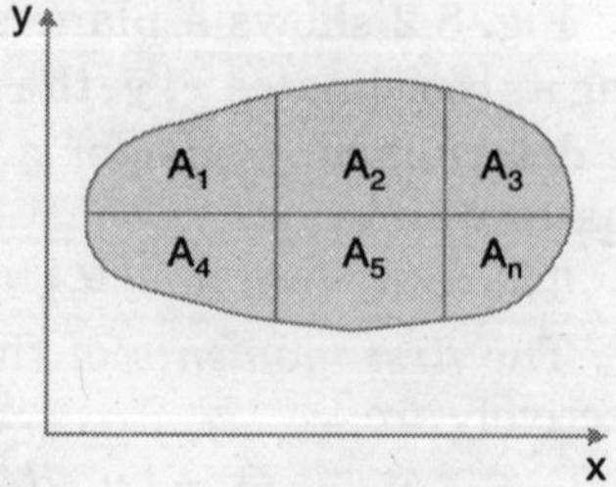

Fig. 8.6 Centroid of irregular area

Then $$A = \sum_{i=1}^{n} \Delta A_i \qquad \text{...[8.4(a)]}$$

$$Q_x = \sum_{i=1}^{n} y_i \, \Delta A_i ; \quad Q_y = \sum_{i=1}^{n} x_i \, \Delta A_i \qquad \text{...[8.4(b)]}$$

Hence $$\bar{x} = \frac{Q_y}{A} = \frac{\sum_{i=1}^{n} x_i \, \Delta A_i}{\sum_{i=1}^{n} \Delta A_i} \quad \text{and} \quad \bar{y} = \frac{Q_x}{A} = \frac{\sum_{i=1}^{n} y_i \, \Delta A_i}{\sum_{i=1}^{n} \Delta A_i} \qquad \text{...(8.5)}$$

Here n = total number of elements in the area and x_i and y_i are x and y coordinates of the centroid of area ΔA_i.

8.3. CENTROID OF COMPOSITE AREAS

Most of the sections used in engineering works consist of areas composed of two, three or more parts. Each part generally consists of a rectangular area, such as that shown in Figs. 8.7(*a*) and 8.7(*b*). Alternatively, a composite section, such as the one shown in Fig. 8.7(*c*), may be made up of regular section such as (*i*) a rectangular plate, (*ii*) an I-section and (*iii*) a channel section, and the properties (*i.e.*, area and location of the centroid) of each of these regular sections are known from available *section tables*.

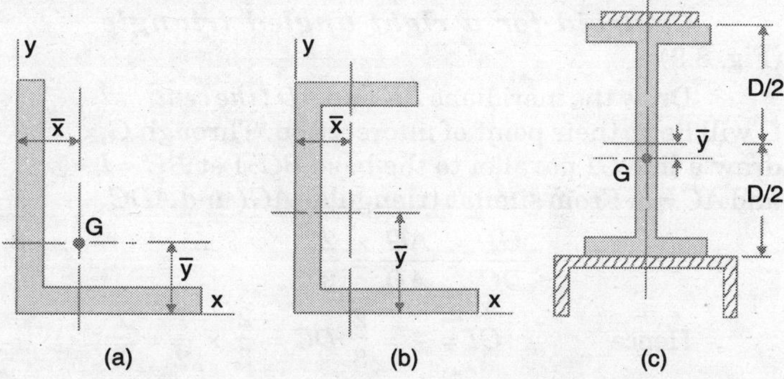

Fig. 8.7 Centroid of composite area

The area and first moment of a composite area may be calculated by summing the corresponding properties of its parts using Eq. 8.4. Thus :

$$A = \sum_{i=1}^{n} \Delta A_i = A_1 + A_2 + \ldots\ldots A_n$$

$$Q_x = \sum_{i=1}^{n} y_i \, \Delta A_i = y_1 A_1 + y_2 A_2 + \ldots\ldots y_n A_n \,; \; Q_y = \sum_{i=1}^{n} x_i \, \Delta A_i = x_1 A_1 + x_2 A_2 + \ldots\ldots x_n A_n$$

Hence $$\bar{x} = \frac{Q_y}{A} = \frac{x_1 A_1 + x_2 A_2 + \ldots\ldots x_n A_n}{A_1 + A_2 + \ldots\ldots A_n} \quad \text{and} \quad \bar{y} = \frac{Q_x}{A} = \frac{y_1 A_1 + y_2 A_2 + \ldots\ldots y_n A_n}{A_1 + A_2 + \ldots\ldots A_n} \qquad \text{...(8.6)}$$

8.4 SOME CASES OF CENTROID OF COMMON AREAS

1. Centroid of a triangular lamina

Fig. 8.8 shows a triangular lamina ABC, and we are required to locate its centroid G. Draw a narrow strip bc, parallel to the side BC. The centroid of the narrow strip bc will lie at the middle point m of the strip. Similarly, the centroid of a narrow strip at BC will lie at the middle point D. Hence the centroid of any narrow strip drawn parallel to BC will lie on the meridian AD. Similarly, the centroid of any narrow strip, drawn parallel to AC, will lie of the meridian BE, where E is the mid-point of AC. Lastly, the centroid of any narrow strip, drawn parallel to the side AB, will lie on the meridian CF. *Hence the centroid G of the whole triangular lamina will lie at the point of intersection of the three meridians.*

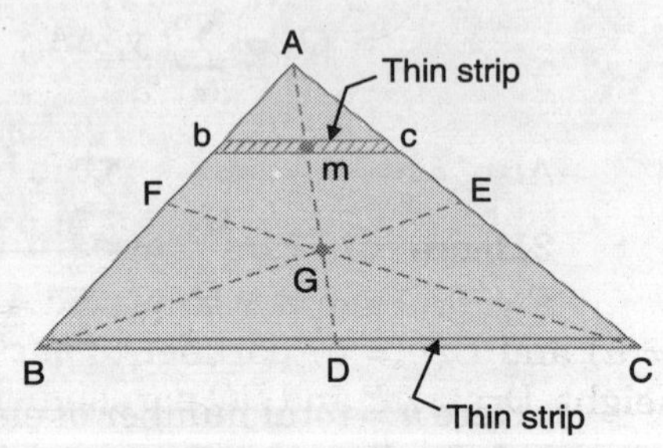

Fig. 8.8 Centroid of triangular Lamina

It can be shown that $DG = \frac{1}{3} DA$; $EG = \frac{1}{3} EB$ and $FG = \frac{1}{3} FC$

Centroid for a right angled triangle (Fig. 8.9)

Draw the meridians BE and AD ; *the centroid G will lie at their point of intersection.* Through G, draw a line GI parallel to the base BC. Let $BC = l$ and $AC = h$. From similar triangular AGI and ADC,

$$\frac{GI}{DC} = \frac{AG}{AD} = \frac{2}{3}$$

Hence $$GI = \bar{x} = \frac{2}{3} DC = \frac{2}{3} \times \frac{l}{2} = \frac{l}{3}$$

Similarly, it can be shown that $GJ = \bar{y} = \frac{h}{3}$

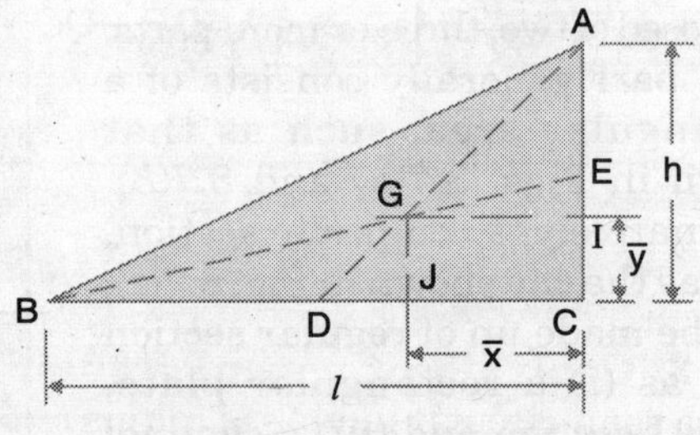

Fig. 8.9 Centroid of right angled triangle

Centroid of any triangle **(Fig. 8.10)**

Let l = length of the base BC

h = height of the triangle = AP

where AP is perpendicular to BC.

Let $BP = a$

and $PC = b$, such that $a + b = l$

Let the centroid G be at a distance $\bar{x}$ from B.

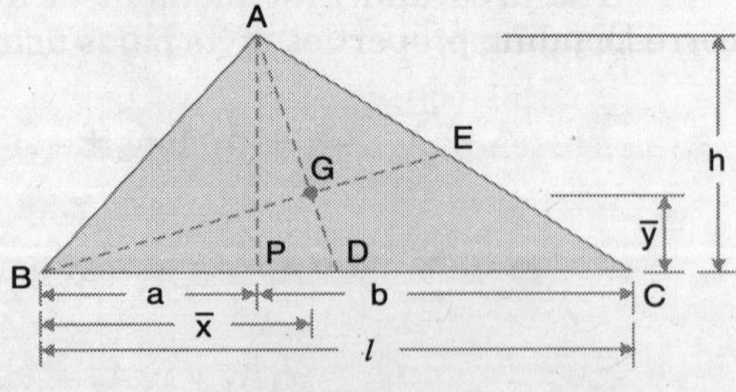

Fig. 8.10

Now sum of moments of areas ABP and APC about B = moment of total area about B.

$$\therefore \quad \frac{1}{2} a\, h \left(\frac{2}{3} a\right) + \frac{1}{2} b\, h \left(a + \frac{b}{3}\right) = \frac{1}{2}\, l\, h \bar{x}$$

or $$\frac{2}{3} a^2 + ab + \frac{b^2}{3} = l.\bar{x} \quad \text{or} \quad \frac{2a^2 + 3\,ab + b^2}{3} = \frac{(2a + b)\,(a + b)}{3} = l.\bar{x}$$

$$\therefore \quad \bar{x} = \frac{(2a+b)(a+b)}{3l} = \frac{(2a+b)\,l}{3l} = \frac{2a+b}{3} = \frac{a+(a+b)}{3} = \frac{a+l}{3} \quad ...(8.7)$$

If, however, $\bar{x}$ is measured from right hand end C, then

$$\bar{x} = \frac{b+l}{3} \quad ...(8.8)$$

Also, $\bar{y} = \dfrac{h}{3}$, where h is the height of the triangle.

2. Centroid of a trapezium

Fig. 8.11 shows a trapezoid $ABCD$, having side AB (= a) and CD (= b) parallel to each other, and h is its height. Draw AP and BQ perpendicular to DC.

Let the distance of centroid G be $\bar{y}$ from the base DC.

Fig. 8.11

Total area $\quad A = \dfrac{(a+b)}{2}\,h$

Taking moments of the areas about base DC, we have

$$\left(\frac{a+b}{2}\,h\right)\bar{y} = \frac{1}{2}\,h\,(DP+QC)\,\frac{h}{3} + ah\,.\,\frac{h}{2}$$

or

$$\frac{a+b}{2}\,h\bar{y} = \frac{h^2}{6}\,(b-a) + \frac{a\,h^2}{2} = \frac{h^2}{6}\,(b-a+3a)$$

or

$$\bar{y} = \frac{h}{3}\,\frac{b+2a}{b+a} \quad ...(8.9)$$

3. Centroid of a semicircular lamina : Fig. 8.12 shows a semi-circular lamina of radius r. Consider an elemental radial area CPQ, subtending an angle $d\theta$ at the centre C.

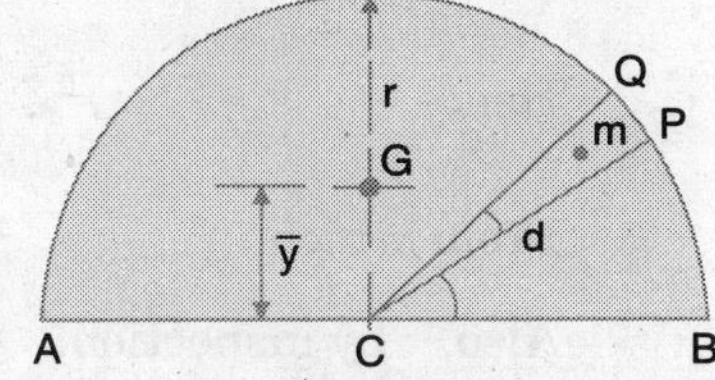

Fig. 8.12

Arc $\quad PQ = r\,.d\theta$

Elemental radial area $\quad = \dfrac{1}{2}\,r\,(r\,d\theta) = \dfrac{1}{2}\,r^2\,d\theta$

Distance of its centroid m from $C = \dfrac{2}{3}\,r$

$\therefore$ Height of the centroid of the elemental area from the base $= \left(\dfrac{2}{3}\,r\right)\sin\theta$

$\therefore$ Moment of elemental area about the base $= \left(\dfrac{1}{2}\,r^2\,d\theta\right)\left(\dfrac{2}{3}\,r\sin\theta\right) = \dfrac{r^3}{3}\sin\theta\,d\theta$

$\therefore$ Moment of the whole area about the base $= \displaystyle\int_0^{\pi}\frac{r^3}{3}\sin\theta\,d\theta = \frac{2r^3}{3}\int_0^{\frac{\pi}{2}}\sin\theta\,d\theta = \frac{2}{3}\,r^3$

Area of semicircle $\quad = \dfrac{\pi\,r^2}{2}$

Hence $\quad \dfrac{\pi\,r^2}{2}\,\bar{y} = \dfrac{2}{3}\,r^3$, from which $\quad \bar{y} = \dfrac{4r}{3\pi} \quad ...(8.10)$

4. Centroid of a quarter-circular lamina (Fig. 8.13)

From the previous discussions, we get

$$\bar{x} = \bar{y} = \frac{4r}{3\pi} \qquad \text{...(8.11)}$$

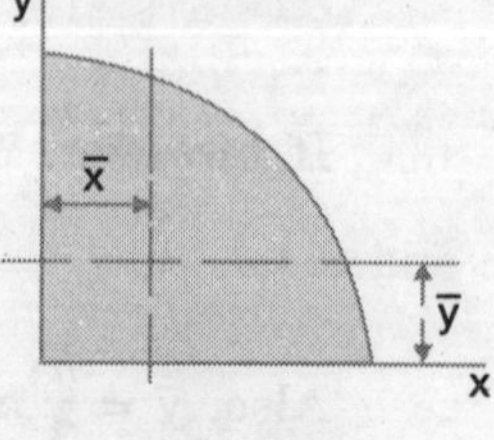

Fig. 8.13

5. Centroid of a circular sector (Fig. 8.14)

Consider a small angular sector CPQ, subtending an angle $d\theta$ at centre C. It's elemental area $= \frac{1}{2} r\,(r d\theta) = \frac{1}{2} r^2\, d\theta$. Distance of its centroid m from $C = \frac{2}{3} r$

$\therefore$ Height of centroid of the elemental area

$$\text{from the base} = \frac{2}{3} r \sin\theta$$

$\therefore$ Moment of elemental area about the base

$$= \left(\frac{1}{2} r^2\, d\theta\right)\left(\frac{2}{3} r \sin\theta\right) = \frac{r^3}{3} \sin\theta\, d\theta$$

$\therefore$ Moment of the whole area about the base

$$= \int_{\frac{\pi}{2}-\alpha}^{\frac{\pi}{2}+\alpha} \frac{r^3}{3} \sin\theta\, d\theta = \frac{2r^3}{3} \int_{\frac{\pi}{2}-\alpha}^{\frac{\pi}{2}} \sin\theta\, d\theta$$

$$= \frac{2r^3}{3} \left[-\cos\theta\right]_{\frac{\pi}{2}-\alpha}^{\pi/2} = \frac{2r^3}{3} \sin\alpha$$

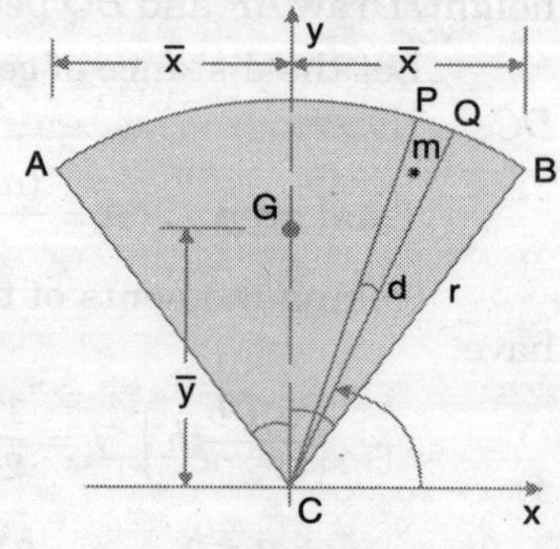

Fig. 8.14

Area of the sector $= \alpha\, r^2$

Hence $\qquad \alpha\, r^2 \,.\, \bar{y} = \dfrac{2r^3}{3} \sin\alpha$

From which $\qquad \bar{y} = \dfrac{2r \sin\alpha}{3\alpha} \qquad \text{...(8.12)}$

Also, by inspection, $\bar{x} = r \sin\alpha \qquad \text{..(8.13)}$

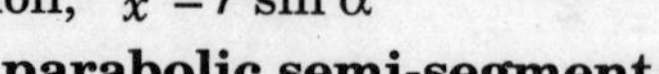

6. Centroid of a parabolic semi-segment

Fig. 8.15 shows a parabolic semi-segment CAB, the equation of the curve being

$$y = f(x) = h\left(1 - \frac{x^2}{b^2}\right)$$

Let us take a strip of thickness dx and height y, located at a distance x from the origin C.

$$\therefore \quad dA = y\,dx = h\left(1 - \frac{x^2}{b^2}\right) dx$$

$$\therefore \quad A = \int dA = \int_0^b h\left(1 - \frac{x^2}{b^2}\right) dx$$

$$= h\left[x - \frac{x^3}{3b^2}\right]_0^b = \frac{2bh}{3} \qquad \text{...(8.14)}$$

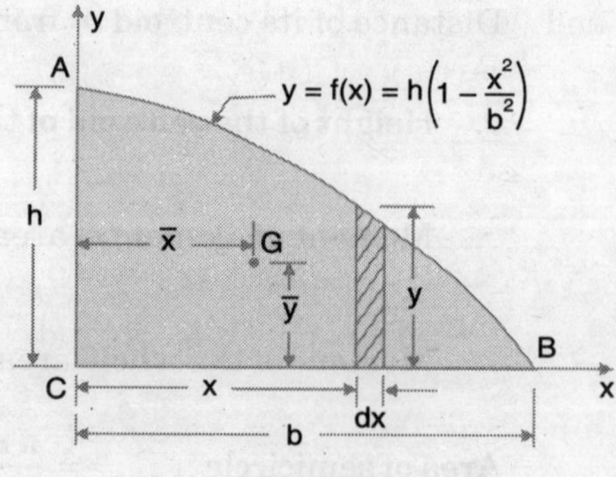

Fig. 8.15

$$Q_x = \int \frac{y}{2}\,dA = \int_0^b \frac{h^2}{2}\left(1-\frac{x^2}{b^2}\right)^2 dx = \frac{4bh^2}{15}$$

and
$$Q_y = \int x\,dA = \int_0^b hx\left(1-\frac{x^2}{b^2}\right) dx = \frac{b^2h}{4}$$

$$\therefore \quad \bar{x} = \frac{Q_y}{A} = \frac{b^2h}{4} \times \frac{3}{2bh} = \frac{3}{8}b \quad \text{and} \quad \bar{y} = \frac{Q_x}{A} = \frac{4bh^2}{15} \times \frac{3}{2bh} = \frac{2h}{5} \qquad ...(8.15)$$

Example 8.1. *Find the centroid of a 120 mm × 150 mm × 20 mm T section.*

Solution. Let the centroid be G at $\bar{y}$ above GH.

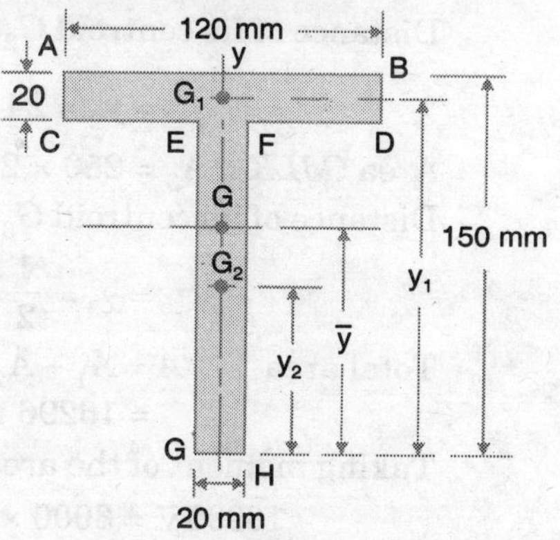

Fig. 8.16

Area $ABDC = A_1 = 120 \times 20 = 2400$ mm^2.

Distance of its centroid $G_1 = y_1 = 150 - 10 = 140$ mm

Area $EFHG = A_2 = (150 - 20) \times 20 = 2600$ mm^2

Distance of its centroid $G_2 = y_2 = \frac{1}{2}(150 - 20) = 65$ mm

Total area $A = A_1 + A_2 = 2400 + 2600 = 5000$ mm^2

Taking moments of the area along GH, we get

$$A \cdot \bar{y} = A_1 \cdot y_1 + A_2 \cdot y_2$$

$$\therefore \quad \bar{y} = \frac{A_1y_1 + A_2y_2}{A_1 + A_2} = \frac{2400 \times 140 + 2600 \times 65}{5000}$$

$= \mathbf{101\ mm}$ (measured from the bottom)

or $\bar{y} = 150 - 101 = 49$ mm, measured from the top.

Example 8.2. *Find the centroid of an unsymmetrical channel section showing Fig. 8.17.*

Solution. Let G be the centroid, distant $\bar{x}$ from face AI and $\bar{y}$ from face IJ.

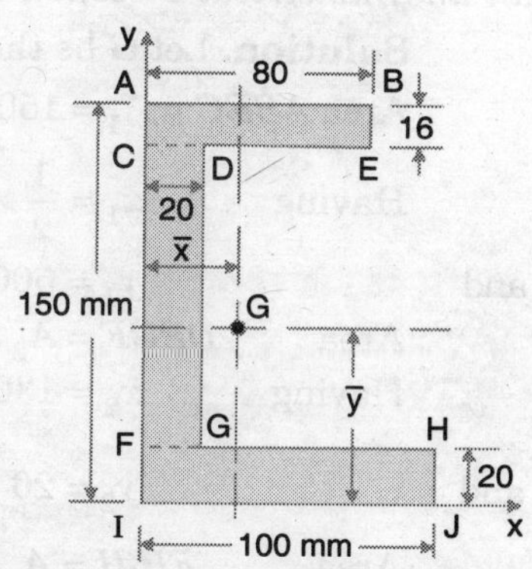

Fig. 8.17

Area $ABEC = A_1 = 80 \times 16 = 1280$ mm^2

Having $x_1 = 40$ mm (from AI)

and $y_1 = 150 - 8 = 142$ mm (from IJ)

Area $CDGF = A_2 = (150 - 16 - 20) \times 20 = 2280$ mm^2

Having $x_2 = 10$ mm (from AI) and

$$y_2 = 20 + \frac{1}{2}(150 - 16 - 20) = 77 \text{ mm}$$

Area $FHJI = A_3 = 100 \times 20 = 2000$ mm^2

Having $x_3 = 50$ mm (from AI) and $y_3 = 10$ mm (from IJ)

Total area $= A_1 + A_2 + A_3 = 1280 + 2280 + 2000$
$= 5560$ mm^2

Taking moments of areas about AI, we get

$$\bar{x} = \frac{A_1x_1 + A_2x_2 + A_3x_3}{A_1 + A_2 + A_3} = \frac{1280 \times 40 + 2280 \times 10 + 2000 \times 50}{5560} = \mathbf{31.29\ mm}$$

Similarly, taking moments of areas about IJ, we get

$$\bar{y} = \frac{A_1y_1 + A_2y_2 + A_3y_3}{A} = \frac{1280 \times 142 + 2280 \times 77 + 2000 \times 10}{5560} = \mathbf{67.86\ mm}$$

Example 8.3. *For the I-section shown in Fig. 8.18, determine the position of the centroid.*

Solution. Let G be the centroid of the section, distant $\bar{y}$ from bottom KL.

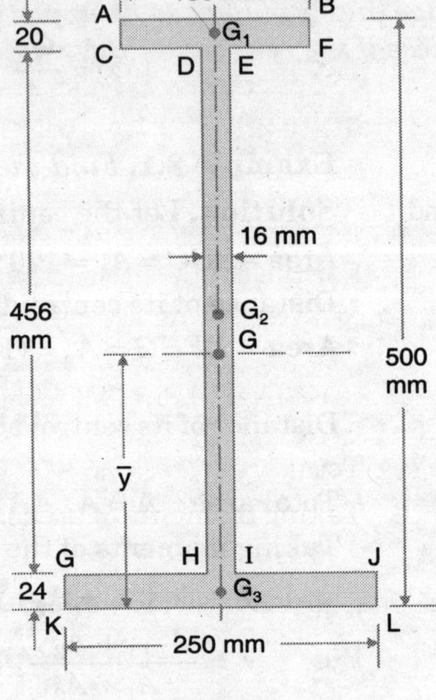

Fig. 8.18

Area of $ABFC = A_1 = 150 \times 20 = 3000\ mm^2$

Distance of its centroid G_1 from KL

$= y_1 = 500 - 10 = 490$ mm

Area $DEIH = A_2 = 456 \times 16 = 7296\ mm^2$

Distance of its centroid G_2 from KL

$= y_2 = 24 + \frac{1}{2} \times 456 = 252$ mm

Area $GJLK = A_3 = 250 \times 24 = 6000\ mm^2$

Distance of its centroid G_3 from KL

$= y_3 = \frac{1}{2} \times 24 = 12$ mm

Total area $\quad A = A_1 + A_2 + A_3 = 3000 + 7296 + 6000$

$= 16296\ mm^2$

Taking moment of the areas about KL, we get

$\therefore \quad 16296\bar{y} = 3000 \times 490 + 7296 \times 252 + 6000 \times 12$

From which $\quad \bar{y} = \mathbf{207.4\ mm}$

Example 8.4. *Determine the position of the centroid of an unsymmetrical Z-section shown in Fig. 8.19.*

Solution. Let G be the centroid, distant $\bar{x}$ from ACK and $\bar{y}$ from KIJ.

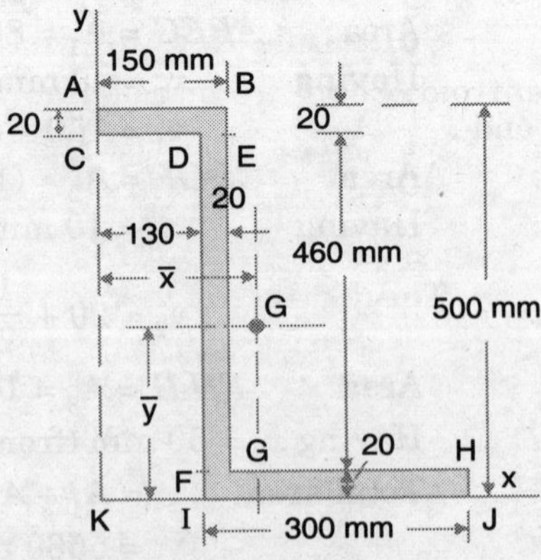

Fig. 8.19

Area $ABEC = A_1 = 150 \times 20 = 3000\ mm^2$

Having $\quad x_1 = \frac{1}{2} \times 150 = 75$ mm from ACK

and $\quad y_1 = 500 - 10 = 490$ mm from KIJ

Area $\quad DEGF = A_2 = 460 \times 20 = 9200\ mm^2$

Having $\quad x_2 = 130 + 10 = 140$ mm from ACK

and $\quad y_2 = 20 + \frac{1}{2} \times 460 = 250$ mm from KIJ

Area $\quad FHJI = A_3 = 300 \times 20 = 6000\ mm^2$

Having $\quad x_3 = 130 + \frac{1}{2} \times 300$

$= 280$ mm from ACK

and $\quad y_3 = 10$ mm from KIJ.

Total $\quad A = 3000 + 9200 + 6000 = 18200\ mm^2$

Taking moments of areas about ACK, we get

$\therefore \quad 18200\ \bar{x} = 3000 \times 75 + 9200 \times 140 + 6000 \times 280.$ From which $\quad \bar{x} = \mathbf{175.44\ mm}$

Similarly, taking moments of areas about *KIJ*, we get

$$18200\,\bar{y} = 3000 \times 490 + 9200 \times 250 + 6000 \times 10, \text{ from which } \quad \bar{y} = \mathbf{210.44\ mm}$$

Example 8.5. *Find the position of centroid of a uniform plate which is in the form of uniform trapezium whose parallel sides are 2 m and 4 m and height is 3 m.*

If it has a rectangular extension of the same weight per square meter attached to the 2 m edge and 2 m long so as to just fit that edge, find what should be the height of the rectangular piece if the centroid of the whole assembly is to be at a height of 4 m above the base.

Solution. For the trapezoidal plate, $A_1 = \dfrac{a+b}{2} \times H = \dfrac{4+2}{2} \times 3 = 9 \text{ m}^2$

and $\quad \bar{y}_1 = \dfrac{a+2b}{a+b} \times \dfrac{H}{3} = \dfrac{4+2\times 2}{4+2} \times \dfrac{3}{3} = 1.333 \text{ m}$

For the composite section, let A_2 be the area of the rectangular extension.

$\therefore \quad A_2 = 2 \times h = 2h \text{ m}^2$, and

$y_2 \text{ (from } AB) = 3 + \dfrac{h}{2} = (3 + 0.5\,h) \text{ m}$

Total area $\quad A = A_1 + A_2 = 9 + 2h \text{ m}^2$

Taking moments of areas about AB, we get

$$(9 + 2h)\,\bar{y} = 9 \times 1.333 + 2h\,(3 + 0.5h)$$

But $\quad \bar{y} = 4 \text{ m (given)}$

$\therefore \quad (9 + 2h)\,4 = 12 + 2h\,(3 + 0.5h)$

or $\quad h^2 - 2h - 24 = 0$

From which $\quad h = \mathbf{6\ m}$

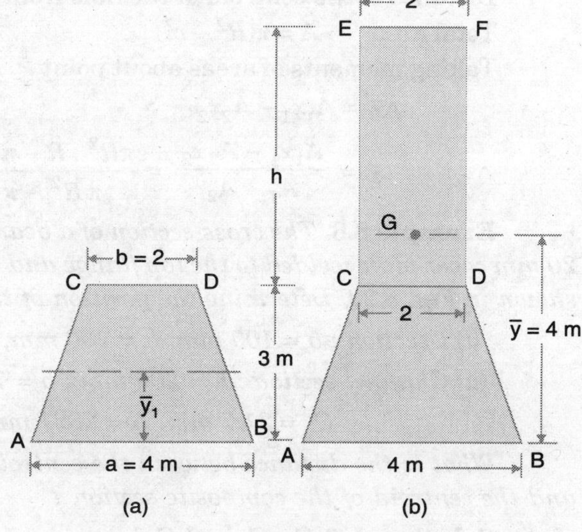

Fig. 8.20

Example 8.6. *Determine the position of the centroid of the plane figure shown in Fig. 8.21.*

Solution. Let the distance of the centroid G be $\bar{y}$ from the base.

Area of the trapezium $\quad = A_1 = \dfrac{120 + 240}{2} \times 120$

$\quad = 21600 \text{ mm}^2$

The distance of its centroid from the base

$$= y_1 = \frac{a+2b}{a+b} \times \frac{h}{3} = \frac{240 + 2\times 120}{240 + 120} \times \frac{120}{3} = 53.33 \text{ mm}$$

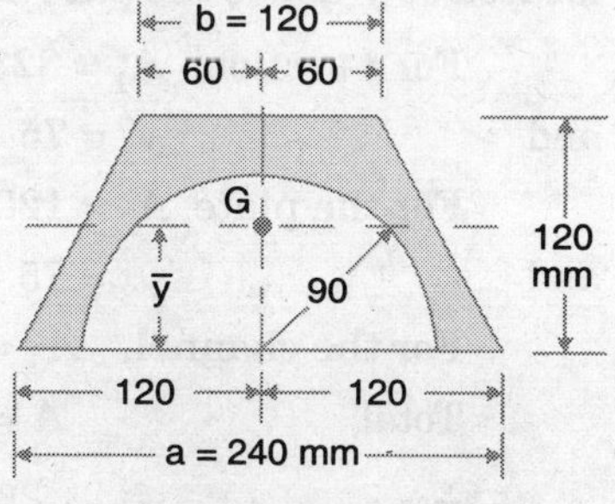

Fig. 8.21

Area of semi-circle $= A_2 = \dfrac{\pi r^2}{2} = \dfrac{\pi\,(90)^2}{2} = 12723.5 \text{ mm}^2$

The distance of its centroid from the base

$$= y_2 = \frac{4r}{3\pi} = \frac{4 \times 90}{3\pi} \approx 38.2 \text{ mm}$$

Total area $= A = A_1 - A_2 = 21600 - 12723.5 = 8876.5 \text{ mm}^2$

Now taking moments of the areas about the base,

$$A \cdot \bar{y} = A_1 \cdot y_1 - A_2 \cdot y_2$$

or $$\bar{y} = \frac{21600 \times 53.33 - 12723.5 \times 38.2}{8876.5} = \mathbf{75.02 \text{ mm}}$$

Example 8.7. *A circular sheet of metal has a radius R. If a hole of radius r is made as shown in Fig. 8.22, determine the position of the centroid of the remaining part.*

Solution. Let G be the position of the centroid, distance $\bar{x}$ from P.

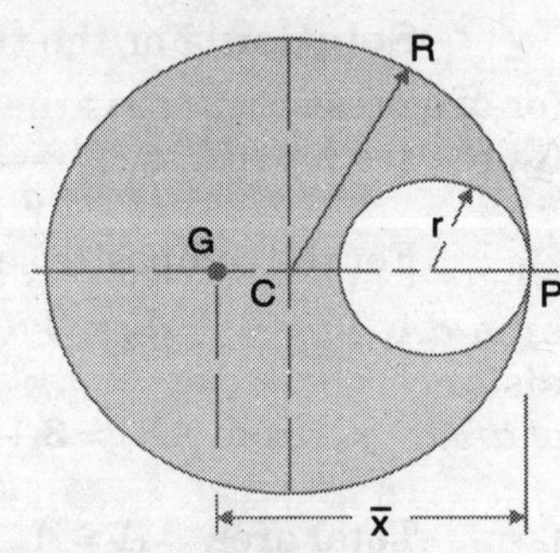

Fig. 8.22

Area of circular sheet $= A_1 = \pi R^2$

Distance of its centroid from $P = x_1 = R$

Area of hole $= A_2 = \pi r^2$

Distance of the centroid of the hole from $P = x_2 = r$

Total area $A = \pi(R^2 - r^2)$

Taking moments of areas about point P, we get

$$A\bar{x} = A_1 x_1 - A_2 x_2$$

$$\therefore \quad \bar{x} = \frac{A_1 x_1 - A_2 x_2}{A_1 - A_2} = \frac{\pi R^2 \cdot R - \pi r^2 \cdot r}{\pi R^2 - \pi r^2} = \frac{R^3 - r^3}{R^2 - r^2}$$

Example 8.8. *The cross-section of a beam is constructed of an I-section with a 120 mm × 20 mm cover plate welded to the top flange and a channel section welded to the bottom flange as shown in Fig. 8.23. Determine the position of the centroid, given the following :*

(i) *I-section : b = 100 mm, h = 200 mm, A = 3233 mm²*

(ii) *Channel section : h = 200 mm ; b = 75 mm,*

C_y = 21.7 mm, A = 2821 mm².

What is the distance between the centroid of the I-section and the centroid of the composite section ?

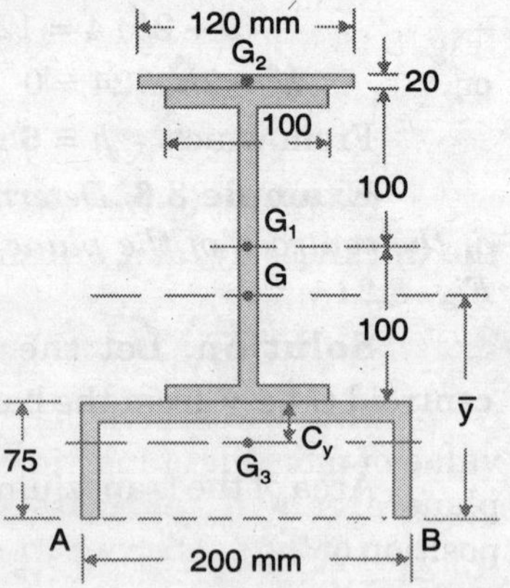

Fig. 8.23

Solution. Let G_1, G_2 and G_3 be the centroids of the three individual parts, and let G be the centroid of the composite section, located at $\bar{y}$ above the base line AB.

For I-section, $A_1 = 3233 \text{ mm}^2$

and $y_1 = 75 + 100 = 175$ mm (above AB)

For the plate, $A_2 = 120 \times 20 = 2400 \text{ mm}^2$

and $y_2 = 75 + 200 + 10 = 285$ mm

For the channel, $A_3 = 2821 \text{ mm}^2$ and $y_3 = 75 - 21.7 = 53.3$ mm

Total $A = A_1 + A_2 + A_3 = 3233 + 2400 + 2821 = 8454 \text{ mm}^2$

Taking moments about AB, we get $\bar{y} = \dfrac{A_1 y_1 + A_2 y_2 + A_3 y_3}{A_1 + A_2 + A_3}$

$$= \frac{(3233 \times 175) + (2400 \times 285) + (2821 \times 53.3)}{8454} = \mathbf{165.6\ mm}$$

Hence distance between G_1 and $G = y_1 - \bar{y} = 175 - 165.6 = \mathbf{9.4\ mm}$

8.5. MOMENT OF INERTIA

The axial (or equatorial) moment of inertia of a plane area (or of a cross-section) is the geometrical characteristic of the area (or the section) defined by the integrals :

$$I_{xx} = \int y^2 \,.\, dA \quad \text{and} \quad I_{yy} = \int x^2 dA \qquad ...(8.16)$$

where x and y are the co-ordinates of the differential element of area dA. Since this area dA is multiplied by the square of the distance, the *moment of inertia* is also called the *second moment of area*.

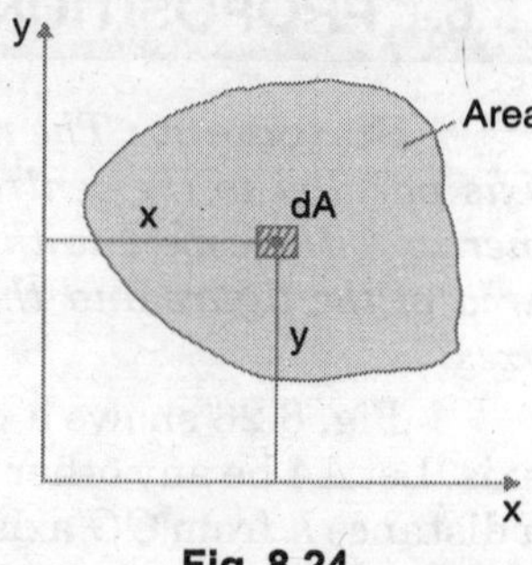

Fig. 8.24

To illustrate the method of finding the moment of inertia, let us find the moments in Fig. 8.25(*a*). Consider is thin strip of thickness dy, at a distance y from x-axis.

Then $\qquad dA = b\, dy$

Hence $\qquad I_{xx} = \int_{-h/2}^{+h/2} y^2\, b\, dy = \frac{b\, d^3}{12} \qquad$ ...[8.17 (*a*)]

Similarly, if we consider a thin strip of thickness dx, at a distance x from the y-y axis [Fig. 8.25 (*b*)], we have $dA = d \,.\, dx$

$\therefore \qquad I_{yy} = \int_{-b/2}^{b/2} x^2 d \,.\, dx = \frac{d\, b^3}{12} \qquad$ ...[8.17 (*b*)]

Lastly, let us find the moment of inertia about the base AA. Consider a strip of width b, and the thickness dy, at a distance y from AB [Fig. 8.25 (*c*)]. Evidently, $dA = b \,.\, dy$

$\therefore \qquad I_{AA} = \int_0^d y^2 b \,.\, dy = \frac{b\, d^3}{3} \qquad$ ...[8.17 (*c*)]

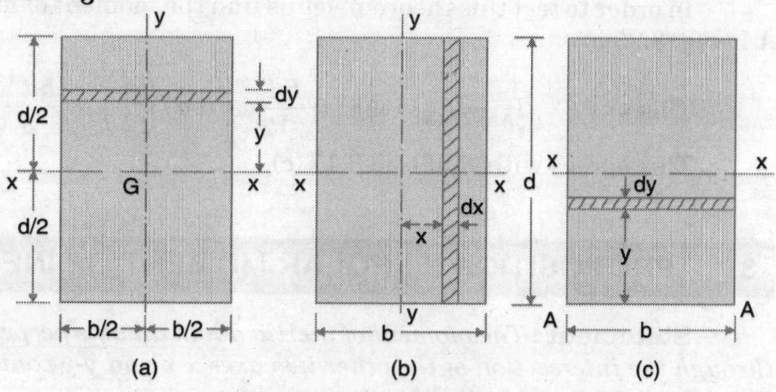

Fig. 8.25

Thus we observe that the value of moment of inertia of a plane area will vary with the position of the axis about which we wish to find the moment of inertia. Also, we observe that the moment of inertia is larger with respect to axis AA then with respect to the centroidal x-axis. Hence the moment of inertia increases as the reference axis is moved parallel to itself farther from the centroid. Also, the moments of inertia are always positive quantities, regardless of the positions of axes selected since coordinates x and y are squared in Eq. 8.16.

Before finding the expressions for moments of inertia for different sections we shall consider the following three *propositions* in the next three articles.

Proposition 1 : *Parallel axis theorem.*

Proposition 2 : *Polar moment of inertia.*

Proposition 3 : *Moment of inertia of composite section.*

8.6. PROPOSITION 1 : PARALLEL AXIS THEOREM

Statement : *The moment of inertia with respect to any axis parallel to the centroidal axis is equal to the moment of inertia with respect to the centroidal axis plus the product of area of the figure and the square of the distance between the axes.*

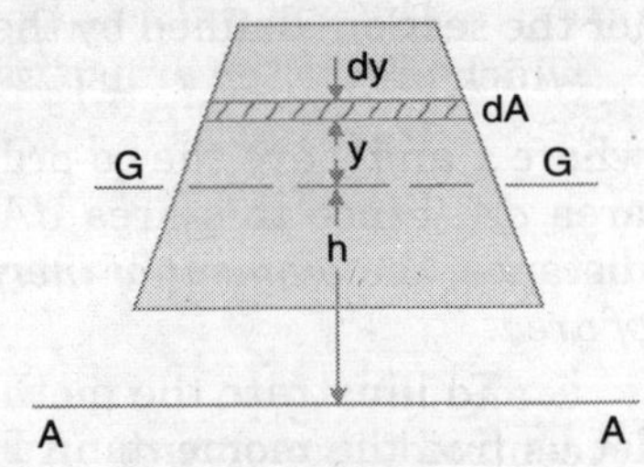

Fig. 8.26

Fig. 8.26 shows a plane area with G-G as its centroidal axis. Let AA be any other axis, parallel to the centroidal axis at a distance h from GG axis. Let us consider an elementary area dA at a distance y from the centroidal axis.

The $\quad I_{GG} = \Sigma\, y^2 dA$

Also, $\quad I_{AA} = \Sigma\, (h \pm y)^2 \,.\, dA$

(in which minus sign stands for the area dA below GG axis)

$$\therefore \quad I_{AA} = \Sigma\, (h^2 + y^2 \pm 2\, hy)\, dA = \Sigma\, h^2\, dA + \Sigma\, y^2 dA \pm 2\, h \Sigma\, y dA$$

where $\quad \Sigma\, y^2 dA = I_{GG}\,;\ \Sigma\, y\, dA = 0$

and $\quad \Sigma\, h^2 dA = h^2\, \Sigma\, dA = Ah^2$, where A is the area of cross-section

Hence $\quad I_{AA} = I_{GG} + Ah^2 \quad$ ...(8.18)

which proves the proposition.

From the parallel axis theorem, we conclude that the moment of inertia increases as the axis is moved parallel to itself farther from the centroid. Hence for a given direction of axes, *the moment of inertia about a centroidal axis is the least moment of inertia of the area.*

In order to test this theorem, let us find the moment of inertia of a rectangle about its base AA (Fig. 8.25 *c*)

This $\quad I_{AA} = I_{GG} + Ah^2 = \dfrac{bd^3}{12} + (bd)\left(\dfrac{d}{2}\right)^2 = \dfrac{bd^3}{3}$

This agrees with equation 8.17 (*c*).

8.7. PROPOSITION 2 : POLAR MOMENT OF INERTIA

Statement : *The moment of inertia about an axis perpendicular to the plane and passing through the intersection of the other two axes x-x and y-y contained by the plane is equal to the sum of moments of inertia about x-x and y-y.*

Fig. 8.27 shows an area A, which contains axes x-x and y-y in its plane. Let z-z axis be perpendicular to this plane, passing through the point O, the point of intersection of the x and y axes.

Let the elementary area dA have co-ordinates x, y and ρ with respect to x, y and z axes respectively, where ρ is the *polar distance*, given by the relation $\rho^2 = x^2 + y^2$.

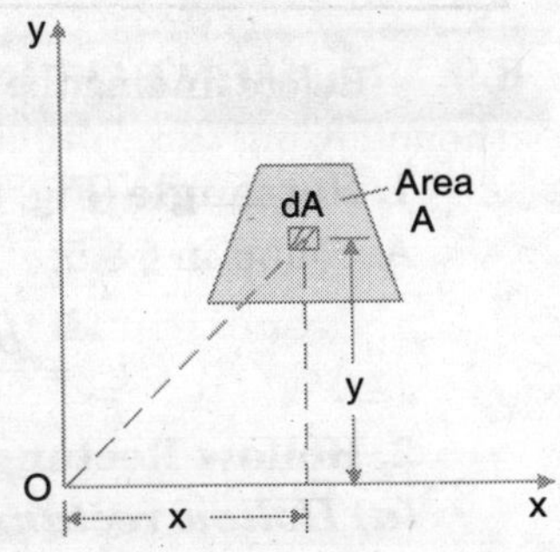

Fig. 8.27

By definition, the *polar moment of inertia* (I_p or I_{zz}) is given by

$$I_p \text{ (or } I_{zz}) = \int \rho^2 \, dA \quad ...(8.19)$$

Hence $I_p = \int \rho^2 \, dA = \int (x^2 + y^2) \, dA$

$= \int x^2 \, dA + \int y^2 \, dA$

But $\int y^2 \, dA = I_{xx}$ and $\int x^2 \, dA = I_{yy}$

Hence I_p (or $I_{zz}) = I_{xx} + I_{yy}$ ...(8.20)

which proves the proposition.

Parallel axis theorem for polar moments of inertia

Let I_{po} = Polar moment of inertia about the origin O.

I_{pg} = Polar moment of inertia about the centroid G.

Let x_g, y_g be the centroidal axes passing through the centroid G and let x_0 and y_0 be the two axes, parallel to the corresponding centroidal axes, passing through O.

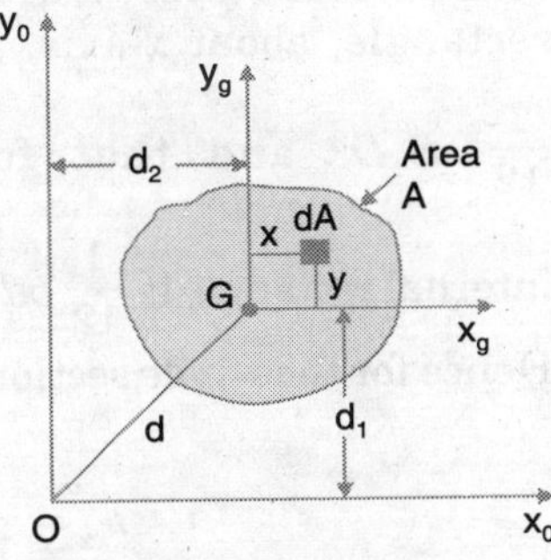

Fig. 8.28

Now $I_{pg} = I_{xg} + I_{yg}$ and $I_{po} = I_{xo} + I_{yo}$ (Eq. 8.20) ...(*a*)

Also, from the parallel axis theorem,

$I_{xo} = I_{xg} + Ad_1^2$ and $I_{yo} = I_{yg} + A\,d_2^2$ ...(*b*)

Adding these two, $I_{xo} + I_{yo} = I_{xg} + I_{yg} + A\,(d_1^2 + d_2^2)$ ...(*c*)

But $d_1^2 + d_2^2 = d^2$ (From Fig. 8.28)

$I_{xo} + I_{yo} = I_{po}$ and $I_{xg} + I_{yg} = I_{pg}$

Hence from (*c*), we get $I_{po} = I_{pg} + Ad^2$ ...(8.21)

This is the parallel axis theorem for polar moments of inertia, which can be stated as follows :

"The polar moment of inertia of an area with respect to any point O in its plane is equal to the polar moment of inertia with respect to the centroid G plus the product of the area and the square of the distance between the points O and G."

8.8. PROPOSITION 3 : MOMENT OF INERTIA OF COMPOSITE SECTION

Statement : *The moment of inertia of a composite section is equal to the sum of the moments of inertia of its separate parts.*

Fig. 8.29 shows a section consisting of three regular areas A_1, A_2 and A_3. The moment of inertia of the whole section about any axis AA is given by

$$I_{AA} = \int y^2 dA = \int y_1^2 dA_1 + \int y_2^2 dA_2 + \int y_3^2 dA_3$$

where dA_1, dA_2 and dA_3 refer to the individual parts 1, 2, 3.

$$\therefore \quad I_{AA} = I_{AA1} + I_{AA2} + I_{AA3} \quad ...(8.22)$$

which proves the proposition.

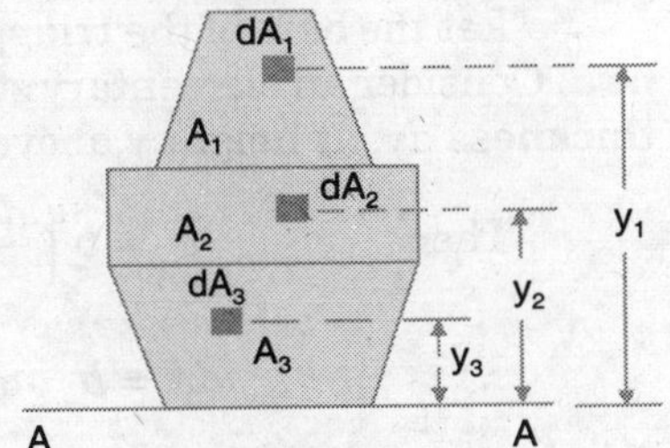

Fig. 8.29

8.9. MOMENT OF INERTIA OF SIMPLE COMMON SECTIONS

1. Rectangle (Fig. 8.25)

As found in § 8.5.

$$I_{xx} = \frac{bd^3}{12}\,;\, I_{yy} = \frac{db^3}{12} \quad \text{and} \quad I_{AA} = \frac{bd^3}{3} \qquad \text{...(Eq.8.17)}$$

2. Hollow Rectangular Section : I-section : Channel section

(a) Hollow rectangular section

Fig. 8.30 (a) shows a hollow rectangular section, having external dimensions $B \times D$, and internal dimensions $b \times d$. The moment of inertia of external rectangle, about x-axis, is $\frac{1}{12}$ $B.D^3$ and that for internal rectangle is $\frac{1}{12}\,bd^3$. Hence for the whole section.

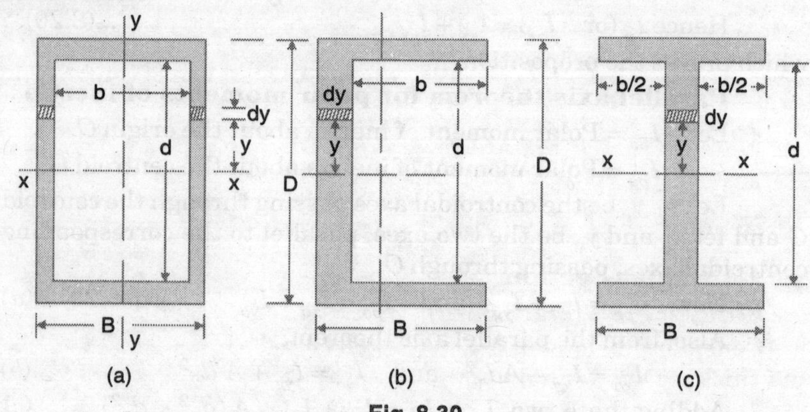

(a) (b) (c)

Fig. 8.30

$$I_{xx} = \frac{1}{12}\,BD^3 - \frac{1}{12}\,bd^3 = \frac{1}{12}\,(BD^3 - bd^3) \qquad \text{...[8.23 (a)]}$$

(b) Channel section **[Fig. 8.30 (b)] : A channel section can be formed by shifting one side-web of the hollow rectangular section to the other end. By doing so, quantity y^2dA is not changed. Hence for the channel section also,**

$$I_{xx} = \frac{1}{12}\,(BD^3 - bd^3) \qquad \text{...[8.23 (b)]}$$

(c) I-section **[Fig. 8.30 (c)] : An I-section can also be formed by shifting both the side webs of the hollow rectangular section to the centre. Here also, the quantity $y^2\,dA$ is not changed by such shifting.**

Hence $$I_{xx} = \frac{1}{12}\,(BD^3 - bd^3) \qquad \text{...[8.23 (c)]}$$

3. Triangular Section

Let the base of the triangle be b and height be d. Consider an elementary strip of width b_y and thickness dy, at height y above the base AB.

Then $$b_y = b\left(\frac{d-y}{d}\right)$$

$\therefore$ $$dA = b_y \cdot dy = \frac{b}{d}\,(d-y)\,dy$$

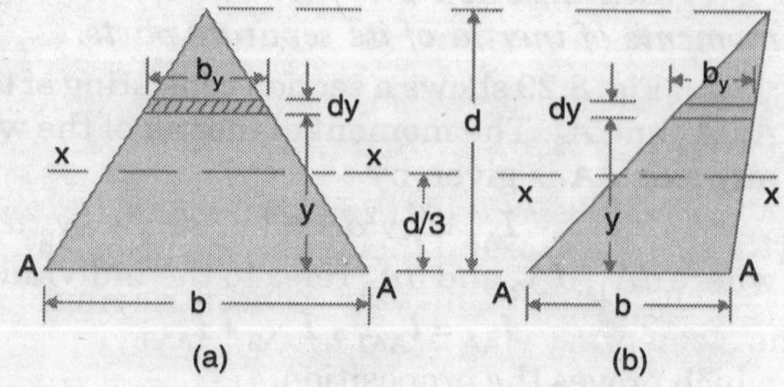

(a) (b)

Fig. 8.31

Hence $$I_{AA} = \Sigma y^2\, dA = \int_0^d y^2 \left(\frac{b}{d}\right)(d-y)\, dy = \frac{b}{d}\int_0^d (y^2 d - y^3)\, dy$$

or $$I_{AA} = \frac{b}{d}\left[\frac{y^3 d}{3} - \frac{y^4}{4}\right]_0^d = \frac{b}{d}\left[\frac{d^4}{3} - \frac{d^4}{4}\right] = \frac{b}{d}\cdot\frac{1}{12}\, d^4.$$

Hence $$I_{AA} = \frac{1}{12}\, bd^3 \qquad \text{...(8.24)}$$

Let x-axis be the centroidal axis. Then, by parallel axis theorem,

$$I_{AA} = I_{xx} + A\bar{x}^2$$

or $$I_{xx} = I_{AA} - A\bar{x}^2, \text{ when } \bar{x} = \text{distance of centroidal axis from } AA = \frac{d}{3}$$

$$\therefore \quad I_{xx} = \frac{1}{12}\, bd^3 - \left(\frac{1}{2}\, bd\right)\left(\frac{d}{3}\right)^2 = \frac{1}{12}\, bd^3 - \frac{bd^3}{18}$$

Hence $$I_{xx} = \frac{1}{36}\, bd^3 \qquad \text{...(8.25)}$$

4. Circular Section

First Method [Fig. 8.32 (*a*)]

Consider an elementary strip of width b_y and thickness dy, at a distance y from the x-axis. If r is the radius of the circle, then we have

$$y = r \sin \theta$$

$$\therefore \quad dy = r \cos \theta\, d\theta$$

Also, width of strip, $b_y = 2\, r \cos \theta$

$$\therefore \quad dA = b_y \,.\, dy$$
$$= 2\, r \cos \theta \,.\, r \cos \theta .\, d\theta$$
$$= 2r^2 \cos^2 \theta\, d\theta$$

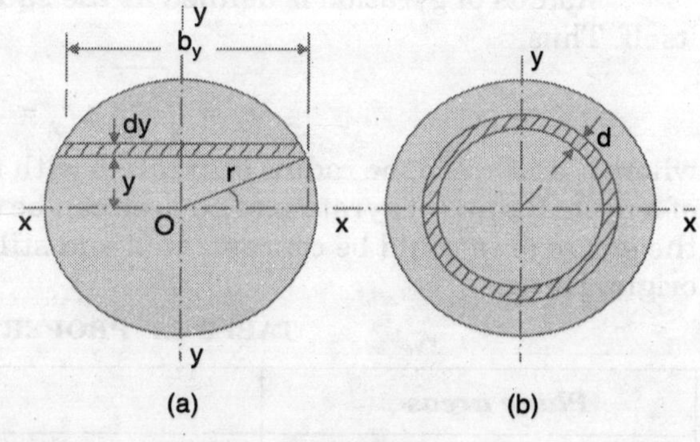

Fig. 8.32

Hence $$I_{xx} = \int y^2\, dA = \int_{-\pi/2}^{\pi/2} (r^2 \sin^2 \theta)(2r^2 \cos^2 \theta)\, d\theta$$

or $$I_{xx} = 4r^4 \int_0^{\pi/2} \sin^2 \theta\, (1 - \sin^2 \theta)\, d\theta = 4\, r^4 \left[\int_0^{\pi/2} \sin^2 \theta\, d\theta - \int_0^{\pi/2} \sin^4 \theta\, d\theta\right]$$

or $$I_{xx} = 4r^4 \left[\frac{\pi}{4} - \frac{3\,\pi}{16}\right] = \frac{\pi\, r^4}{4} = \frac{\pi}{64}\, d^4 \qquad \text{...(8.26)}$$

where d is the diameter of the circle.

Second method **(8.32 *b*)**

The moment of inertia about x-axis can be easily obtained by first finding the moment of inertia about z-axis and then applying proposition 2 for the polar moment of inertia.

Consider an elementary concentric ring of thickness $d\rho$, at a distance ρ from the centre. dA of the elementary ring $= 2\pi\rho d\rho$

$$\therefore \qquad I_p = I_{zz} = \Sigma\,\rho^2 dA = \int_0^r \rho^2 \,.\, 2\pi\rho \,.\, d\rho = \frac{2\pi r^4}{4} = \frac{\pi}{32} d^4 \qquad \text{...(8.27)}$$

But $\qquad I_{zz} = I_{xx} + I_{yy} = 2\,I_{xx}$, due to symmetry.

$$\therefore \qquad I_{xx} = \frac{I_{zz}}{2} = \frac{\pi}{64} d^4 \qquad \text{...(8.28)}$$

5. Hollow circular section

Consider a hollow circular section of external diameter D and internal diameter d.

Then,
$$I_{xx} = \frac{\pi}{64}\,(D^4 - d^4) \qquad \text{...(8.29)}$$

The moment of inertia for other sections are given in Table 8.1.

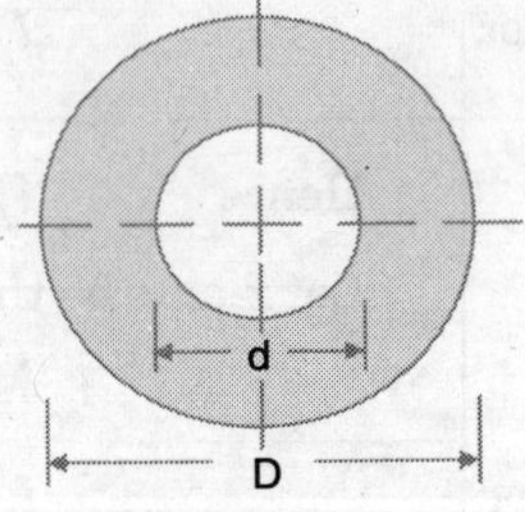

Fig. 8.33

8.10. RADIUS OF GYRATION

Radius of gyration is defined as the square root of moment of inertia divided by the area itself. Thus,

$$r_x = \sqrt{\frac{I_{xx}}{A}}\ ;\ r_y = \sqrt{\frac{I_{yy}}{A}} \qquad \text{...(8.30)}$$

where r_x and r_y are the radius of gyration with respect to x and y axes respectively. It has the unit of length. Radius of gyration of an area can be considered to be a distance from the axis at which the entire area could be concentrated and still have the same moment of inertia as that of the original area.

TABLE 8.1. PROPERTIES OF PLANE AREAS

Plane areas	*Properties*
(Rectangle: y, y, $\bar{x}$, d, G, x, $\bar{y}$, x, b)	**Rectangle** : $A = b\,.\,d\ ;\ \bar{x} = b/2\ ;\ \bar{y} = d/2$ $I_x = \frac{1}{12} bd^3\ ;\ I_y = \frac{1}{12} db^3\ ;\ I_{xy} = 0$ $I_x{}' = \frac{1}{3} bd^3\ ;\ I_y{}' = \frac{1}{3} db^3\ ;\ I_{x'y'} = \frac{b^2d^2}{4}$
(Triangle: y, c, d, $\bar{x}$, x, G, $\bar{y}$, x, b)	**Triangle** : $A = \frac{1}{2} bd\ ;\ \bar{x} = \frac{(b+c)}{3}\ ;\ \bar{y} = \frac{d}{3}$ $I_x = \frac{1}{36} bd^3\ ;\ I_y = \frac{bd}{36} (b^2 - bc + c^2)\ ;\ I_{xy} = \frac{bd^3}{72} (b - 2c)$ $I_x{}' = \frac{1}{12} bd^3\ ;\ I_y{}' = \frac{bd}{12} (3b^2 - 3\,bc + c^2)\ ;\ I_{x'y'} = \frac{bd^2}{24} (3b - 2c)$
(Trapezoid: y, a, d, G, x, $\bar{y}$, b)	**Trapezoid** : $A = d\,\frac{(a+b)}{2}\ ;\ \bar{y} = \frac{d}{3}\left(\frac{2a+b}{a+b}\right)$ $I_x = \frac{d^3\,(a^2 + 4\,ab + b^2)}{36\,(a+b)}\ ;\ I_x{}' = \frac{d^3\,(3a+b)}{12}$

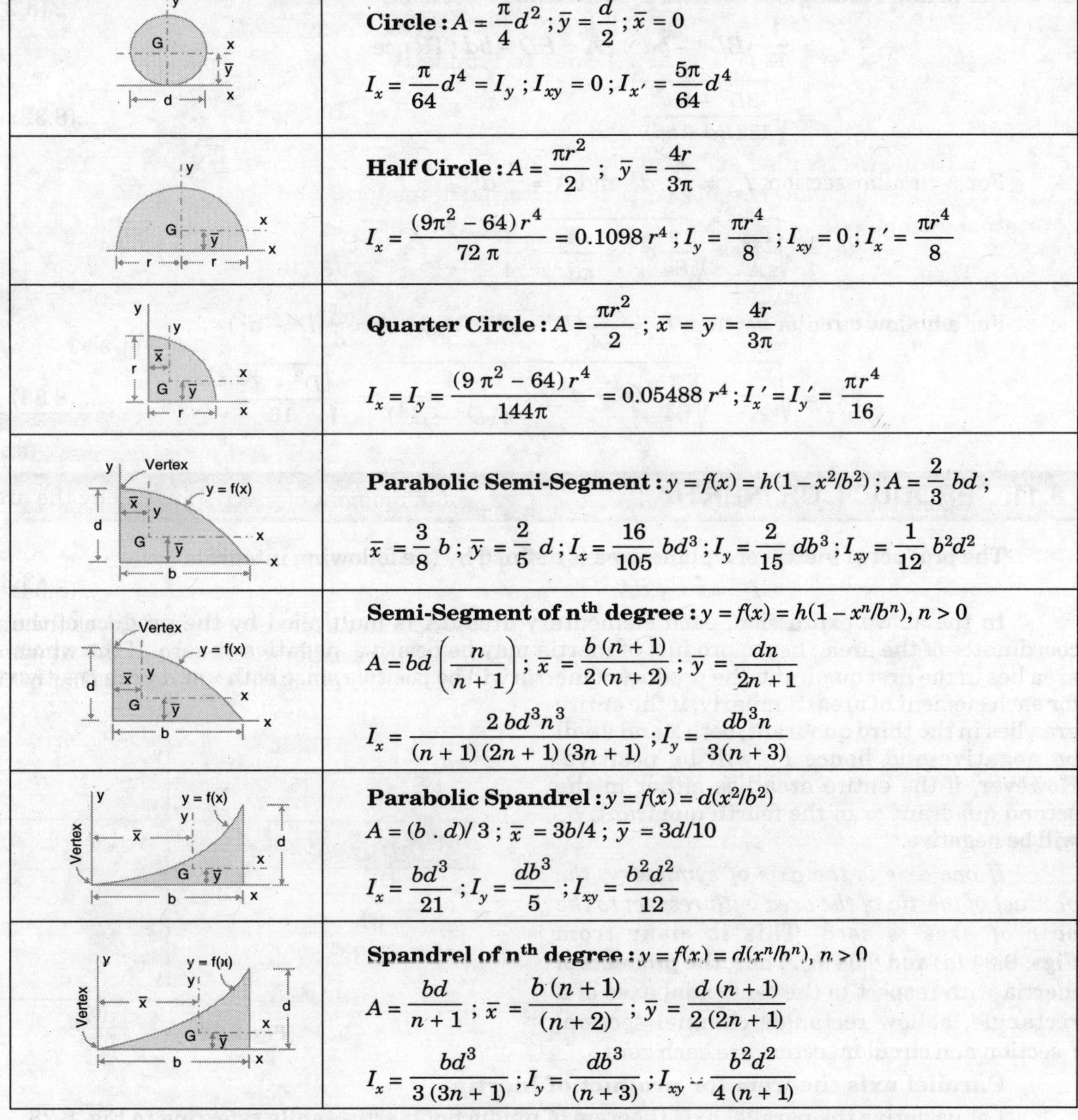

	Circle : $A = \frac{\pi}{4} d^2 \; ; \bar{y} = \frac{d}{2} \; ; \bar{x} = 0$ $I_x = \frac{\pi}{64} d^4 = I_y \; ; I_{xy} = 0 \; ; I_{x'} = \frac{5\pi}{64} d^4$
	Half Circle : $A = \frac{\pi r^2}{2} \; ; \; \bar{y} = \frac{4r}{3\pi}$ $I_x = \frac{(9\pi^2 - 64)\, r^4}{72\,\pi} = 0.1098\, r^4 \; ; I_y = \frac{\pi r^4}{8} \; ; I_{xy} = 0 \; ; I_x' = \frac{\pi r^4}{8}$
	Quarter Circle : $A = \frac{\pi r^2}{2} \; ; \; \bar{x} = \bar{y} = \frac{4r}{3\pi}$ $I_x = I_y = \frac{(9\,\pi^2 - 64)\, r^4}{144\pi} = 0.05488\, r^4 \; ; I_x' = I_y' = \frac{\pi r^4}{16}$
	Parabolic Semi-Segment : $y = f(x) = h(1 - x^2/b^2) \; ; A = \frac{2}{3}\, bd \; ;$ $\bar{x} = \frac{3}{8}\, b \; ; \bar{y} = \frac{2}{5}\, d \; ; I_x = \frac{16}{105}\, bd^3 \; ; I_y = \frac{2}{15}\, db^3 \; ; I_{xy} = \frac{1}{12}\, b^2 d^2$
	Semi-Segment of nth degree : $y = f(x) = h(1 - x^n/b^n),\ n > 0$ $A = bd\left(\frac{n}{n+1}\right) ; \bar{x} = \frac{b\,(n+1)}{2\,(n+2)} \; ; \bar{y} = \frac{dn}{2n+1}$ $I_x = \frac{2\, bd^3 n^3}{(n+1)\,(2n+1)\,(3n+1)} \; ; I_y = \frac{db^3 n}{3(n+3)}$
	Parabolic Spandrel : $y = f(x) = d(x^2/b^2)$ $A = (b\,.\,d)/\,3 \; ; \; \bar{x} = 3b/4 \; ; \; \bar{y} = 3d/10$ $I_x = \frac{bd^3}{21} \; ; I_y = \frac{db^3}{5} \; ; I_{xy} = \frac{b^2 d^2}{12}$
	Spandrel of nth degree : $y = f(x) = d(x^n/b^n),\ n > 0$ $A = \frac{bd}{n+1} \; ; \bar{x} = \frac{b\,(n+1)}{(n+2)} \; ; \bar{y} = \frac{d\,(n+1)}{2\,(2n+1)}$ $I_x = \frac{bd^3}{3\,(3n+1)} \; ; I_y = \frac{db^3}{(n+3)} \; ; I_{xy} = \frac{b^2 d^2}{4\,(n+1)}$

For a rectangular section, $I_{xx} = \frac{1}{12}\, bd^3$ and $I_{yy} = \frac{1}{12}\, db^3$

$$\therefore \quad k_x = \sqrt{\frac{I_{xx}}{A}} = \sqrt{\frac{1}{12}\frac{bd^3}{b\,d}} = \frac{d}{\sqrt{12}} = \frac{d}{2\sqrt{3}}$$

and
$$k_y = \sqrt{\frac{I_{yy}}{A}} = \sqrt{\frac{1}{12}\frac{db^3}{bd}} = \frac{b}{\sqrt{12}} = \frac{b}{2\sqrt{3}} \qquad ...(8.31)$$

For hollow rectangular section, *C*-section and *I*-section,

$$I_{xx} = \frac{1}{12}(BD^3 - bd^3) \ ; A = BD - bd \ ; \text{Hence}$$

$$k_x = \sqrt{\frac{BD^3 - bd^3}{12\,(Bd - bd)}} \quad ...(8.32)$$

For a circular section, $I_{xx} = \frac{\pi}{64} d^4$ and $A = \frac{\pi}{4} d^2$

$$\therefore \quad k_x = \sqrt{\frac{I_{xx}}{A}} = \sqrt{\frac{\pi}{64} d^4 \times \frac{4}{\pi d^2}} = \frac{d}{4}$$

For a hollow circular section, $I_{xx} = \frac{\pi}{64}(D^4 - d^4)$ and $A = \frac{\pi}{4}(D^2 - d^2)$

$$\therefore \quad k_x = \sqrt{\frac{I_{xx}}{A}} = \sqrt{\frac{\pi}{64}(D^4 - d^4) \times \frac{4}{\pi (D^2 - d^2)}} = \sqrt{\frac{D^2 + d^2}{16}} \quad ...(8.33)$$

8.11. PRODUCT OF INERTIA

The product of inertia of a plane area is defined by the following integral :

$$I_{xy} = \int x\, y\, dA \quad ...(8.34)$$

In the above expression, each elementary area dA is multiplied by the product of the coordinates of the area ; hence product of inertia may be positive, negative or zero. If the whole area lies in the first quadrant, the product of inertia will be positive since both x and y are positive for each element of area. Similarly, if the entire area lies in the third quadrant, both x and y will be negative and hence I_{xy} will be positive. However, if the entire area lies either in the second quadrant or in the fourth quadrant, I_{xy} will be negative.

If one axis is the axis of symmetry, the product of inertia of the area with respect to the pair of axes is zero. This is clear from Figs. 8.34 (*a*) and 8.34 (*b*). Thus, the products of inertia with respect to the centroidal axes of a rectangle, hollow rectangle channel section, *I*-section and circular section are each zero.

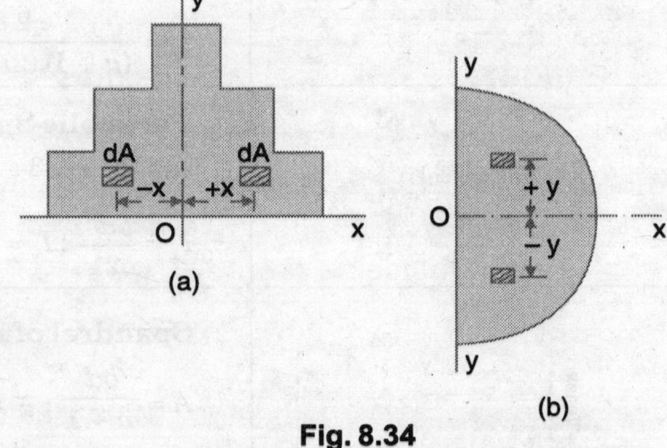

Fig. 8.34

Parallel axis theorem for product of inertia

Let us derive the parallel axis theorem of product of inertia, while referring to Fig. 8.28.

We get $I_{x0\,y0} = \int (x + d_2)(y + d_1)\, dA$ or $I_{x0\,y0} = \int xy\, dA + d_1 \int x\, dA + d_2 \int y\, dA + d_1 d_2 \int dA$

(where d_1 and d_2 are the coordinates of the centroid G)

In the above expression, $\int xy\, dA = I_{xg\,yg}$

$$\int x\, dA = 0 \ ; \int y\, dA = 0 \quad \text{and} \quad \int dA = A$$

Hence $$I_{x0y0} = I_{xgyg} + A \,.\, d_1 d_2 \quad ...(8.35)$$

Hence the product of inertia with respect to any pair of axes in its plane is equal to the product of inertia with respect to the parallel centroidal axes plus the product of the area and the coordinates of the centroid with respect to the pair of axes.

Example 8.9. *Find the moments of inertia I_{xx} and I_{yy} for a T-section shown in Fig. 8.35.*

Solution. $A = (150 \times 10) + (190 \times 10) = 3400 \text{ mm}^2$

To find the value of $\bar{y}$, take moments of the individual areas about base *AB*.

Thus $A\bar{y} = A_1 . y_1 + A_2 . y_2$

or $3400\,\bar{y} = (150 \times 10 \times 75) + (190 \times 10 \times 5)$

From which $\bar{y} = 35.88$ mm

Now $$I_{AA} = \frac{1}{3} \times 10 \times 150^3 + \frac{1}{3} \times 190 \times 10^3$$
$$= 1131.3 \times 10^4 \text{ mm}^4$$

Now $$I_{xx} = I_{AA} - A\bar{y}^2$$
$$= 1131.3 \times 10^4 - 3400\,(35.88)^2$$
$$= \mathbf{693.6 \times 10^4 \text{ mm}^4}$$

Also, $$I_{yy} = \frac{1}{12} \times 10(200)^3 + \frac{1}{12} \times 140\,(10^3)$$
$$= \mathbf{667.8 \times 10^4 \text{ mm}^4}$$

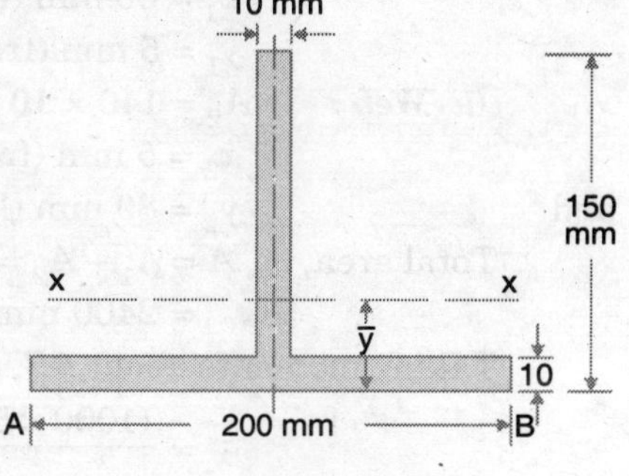

Fig. 8.35

Example 8.10. *Find the moments of inertia I_{xx} and I_{yy} for a channel section shown in Fig. 8.36.*

Solution. The section is symmetrical about the axis *xx*. Let the axis *yy* be at a distance $\bar{x}$ from the face *AA*. The whole area can be splitted into three parts.

(*i*) ***Top flange :*** $A_1 = 100 \times 12 = 1200 \text{ mm}^2$

$x_1 = 100/2 = 50$ mm, from face *AA*.

(*ii*) ***Bottom flange :*** $A_2 = 100 \times 12 = 1200 \text{ mm}^2$

$x_2 = 100/2 = 50$ mm, from face *AA*.

(*iii*) ***Web :*** $A_3 = (200 - 2 \times 12) \times 8 = 1408 \text{ mm}^2$

$x_3 = 4$ mm, from face *AA*.

∴ Total area, $A = A_1 + A_2 + A_3 = 1200 + 1200 + 1408$
$= 3808 \text{ mm}^2$

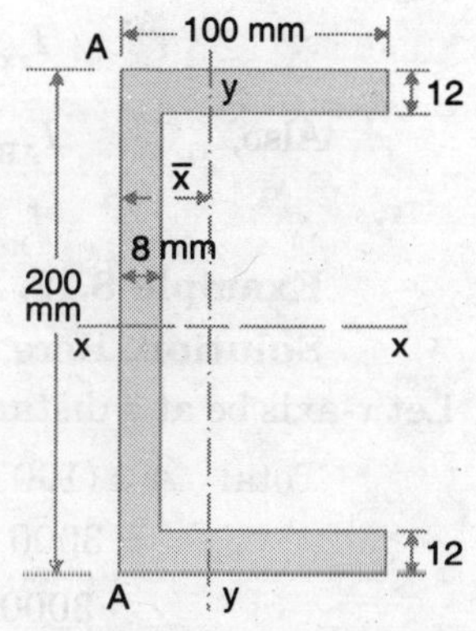

Fig. 8.36

Taking moments about *AB*, we get

$$\bar{x} = \frac{A_1x_1 + A_2x_2 + A_3x_3}{A_1 + A_2 + A_3}$$
$$= \frac{1200 \times 50 + 1200 \times 50 + 1408 \times 4}{3808}$$
$$= 32.99 \text{ mm}$$

Now $$I_{xx} = \frac{1}{12}\,(BD^3 - bd^3) = \frac{1}{12}\,(100 \times 200^3 - 92 \times 176^3) = \mathbf{2487 \times 10^4 \text{ mm}^4}$$

Also $$I_{AA} = 2\left(\frac{1}{3} \times 12 \times 100^3\right) + \frac{1}{3} \times 176 \times 8^3 = 803 \times 10^4 \text{ mm}^4$$

∴ $$I_{yy} = I_{AA} - A\bar{x}^2 = 803 \times 10^4 - 3808\,(32.99)^2 \simeq \mathbf{389 \times 10^4 \text{ mm}^4}$$

Example 8.11. *Find I_{xx} and I_{yy} for an unequal angle section shown in Fig. 8.37.*

Solution. Let the *x-x* axis be at a distance $\bar{y}$ from face *AC* and the *y-y* axis be at a distance $\bar{x}$ from face *AB*. The whole area can be splitted into two parts :

(*i*) ***Flange :*** $A_1 = 100 \times 10 = 1000 \text{ mm}^2$

$x_1 = 50$ mm (from face *AB*)

$y_1 = 5$ mm (from face *AC*)

(*ii*) ***Web :*** $A_2 = 140 \times 10 = 1400 \text{ mm}^2$

$x_2 = 5$ mm (from *AB*)

and $y_2 = 80$ mm (from *AC*)

Total area, $A = A_1 + A_2 = 1000 + 1400 = 2400 \text{ mm}^2$

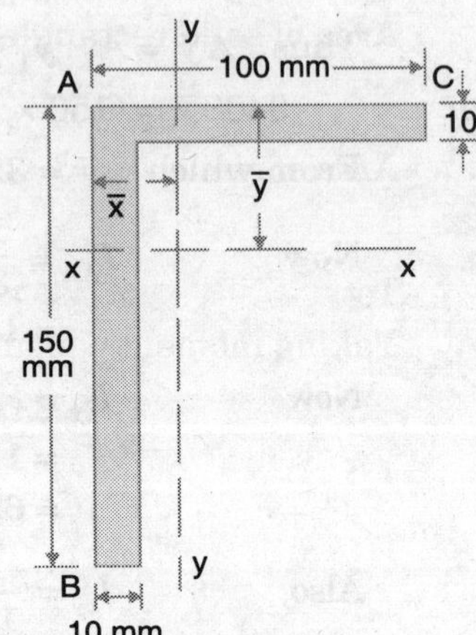

Fig. 8.37

Taking moments about *AB*,

$$\bar{x} = \frac{(1000 \times 50) + (1400 \times 5)}{2400} = 23.75 \text{ mm}$$

Taking moments about *AC*,

$$\bar{y} = \frac{(1000 \times 5) + (1400 \times 80)}{2400} = 48.75 \text{ mm}$$

Now $$I_{AC} = \frac{1}{3} \times 10 \times 150^3 + \frac{1}{3} \times 90 \times 10^3 = 1128 \times 10^4 \text{ mm}^4$$

$\therefore$ $$I_{xx} = I_{AC} - A\bar{y}^2 = 1128 \times 10^4 - 2400\,(48.75)^2 = \mathbf{557.6 \times 10^4\ mm^4}$$

Also, $$I_{AB} = \frac{1}{3} \times 10\,(100)^3 + \frac{1}{3} \times 140(10)^3 = 338 \times 10^4 \text{ mm}^4$$

$\therefore$ $$I_{yy} = I_{AB} - A\bar{x}^2 = 338 \times 10^4 - 2400\,(23.75)^2 = \mathbf{202.6 \times 10^6\ mm^4}$$

Example 8.12. *Find the moment of inertia for I-section shown in Fig. 8.38.*

Solution. Here, *y*-axis will be symmetrical, as shown.

Let *x*-axis be at a distance $\bar{y}$ above the base *AA*.

Total $A = (100 \times 30) + (250 \times 30) + (300 \times 50)$
$= 3000 + 7500 + 15000 = 25500 \text{ mm}^2$

$$\bar{y} = \frac{3000 \times 315 + 7500 \times 175 + 15000 \times 25}{25500} = 103.24 \text{ mm}$$

$$I_{AA} = \left(\frac{1}{3} \times 30 \times 300^3\right) + \left(\frac{1}{3} \times 270 \times 50^3\right) + \left[\left(\frac{1}{12} \times 100 \times 30^3\right) + (3000 \times 315^2)\right]$$

$= 57915 \times 10^4 \text{ mm}^4$

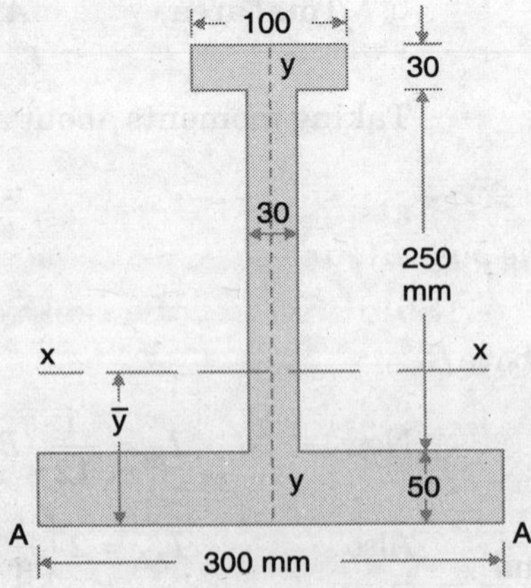

Fig. 8.38

Now $I_{xx} = I_{AA} - A\bar{y}^2 = 57915 \times 10^4 - 25500\,(103.24)^2$
$= \mathbf{30736 \times 10^4\ mm^4}$

Example 8.13. *Find the moment of inertia of a section shown in Fig. 8.39, about the centroidal axis parallel to the base.*

Solution. Let us first locate the centroidal axis.

Area of solid rectangle $= A_1 = 2a \times 3a = 6a^2$

$$y_1 = 1.5\,a \quad \text{(from } AA\text{)}$$

Area of circle $= \dfrac{\pi}{4}(1.5\,a)^2 = 1.767\,a^2$

$$y_2 = 2a \quad \text{(from } AA\text{)}$$

Total $A = A_1 - A_2 = 6a^2 - 1.767\,a^2 = 4.233\,a^2$

Taking moments about AA,

$$\bar{y} = \frac{6a^2\,(1.5\,a) - 1.767\,a^2\,(2a)}{(4.233\,a^2)} = 1.2913\,a$$

Fig. 8.39

$$I_{AA} = \frac{1}{3}(2\,a)(3\,a)^3 - \left[\frac{\pi}{64}(1.5\,a)^4 + \frac{\pi}{4}(1.5\,a)^2\,(2\,a)^2\right] = 10.683\,a^4$$

$$\therefore \quad I_{xx} = I_{AA} - A\bar{y}^2 = 10.683\,a^4 - 4.233\,a^2\,(1.2913\,a)^2 = \mathbf{3.6279\,a^4}$$

Example 8.14. *Find the moment of inertia of the shaded area shown in Fig. 8.40, about the axis AA.*

Solution. Moment of inertia of the triangle, about its base AA

$$= I_1 = \frac{1}{12}(2\,b)(2\,b)^3 = 1.3333\,b^4$$

Moment of inertia of outer half circle about its diameter

$$I_2 = \frac{1}{2}\,\frac{\pi}{64}\,(2\,b)^4 = 0.3927\,b^4$$

Moment of inertia of the inner full circle, about AA

$$I_3 = \frac{\pi}{64}(b)^4 = 0.0491\,b^4$$

Fig. 8.40

$\therefore$ Moment of inertia of whole section,

$$I = I_1 + I_2 - I_3$$
$$= 1.3333\,b^4 + 0.3927\,b^4 - 0.0491\,b^4$$
$$= \mathbf{1.6769\,b^4}$$

Example 8.15. *Find the moment of inertia of the shaded area shown in Fig. 8.41, about the base BC and also about the centroidal axis parallel to the base.*

Solution. Let the centroidal axis be at a distance $\bar{y}$ above BC

Area of triangle $= A_1 = \dfrac{1}{2}(5a)(4.5\,a) = 11.25\,a^2$

and $\quad y_1 = \dfrac{1}{3}(4.5a) = 1.5a$ from BC

Area of rectangular hole $= A_2 = a \times 1.5\,a = 1.5a^2$

and $\quad y_2 = 1.5\,a + 0.75\,a = 2.25\,a$ from BC

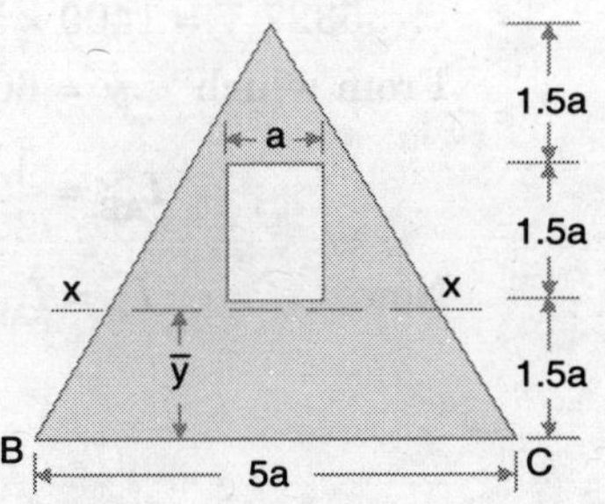

Fig. 8.41

$\therefore$ Total $A = A_1 - A_2 = 11.25\,a^2 - 1.5\,a^2 = 9.75\,a^2$

$$\therefore \quad \bar{y} = \frac{11.25\,a^2 \times 1.5\,a - 1.5\,a^2 \times 2.25\,a}{9.75\,a^2} = 1.3846\,a$$

$$\text{Now} \quad I_{BC} = \frac{1}{12}(5\,a)(4.5\,a)^3 - \left[\frac{1}{12}(a)(1.5\,a)^3 + 1.5\,a^2\,(2.25\,a)^2\right] = \mathbf{30.094\,a^4}$$

$$\text{Hence} \quad I_{xx} = I_{BC} - A\bar{y}^2 = 30.094\,a^4 - 9.75\,a^2\,(1.3846\,a)^2 = \mathbf{11.402\,a^4}$$

Example 8.16. *Fig. 8.42 shows the section of a beam. Find I_{xx} and I_{yy}.*

Solution. $I_{xx} = \frac{1}{12}\,90\,(120)^3 - \frac{\pi}{64}(80)^4$

$= 1296 \times 10^4 - 201 \times 10^4 = \mathbf{1095 \times 10^4\ mm^4}$

$I_{yy} = \frac{1}{12} \times 120\,(90)^3 - 2I = 729 \times 10^4 - 2I$

where I = moment of inertia of each semi-circle about yy-axis

For each semi-circle : $I_{vv} = \frac{1}{2}\frac{\pi}{64}(80)^4 = 100.5 \times 10^4\ \text{mm}^4$

$A = \frac{\pi}{2}(40)^2 = 2513\ \text{mm}^2\ ;\ \bar{h} = \frac{4r}{3\pi} = \frac{4 \times 40}{3\pi} = 17\ \text{mm}$

$\therefore \quad I_{cc} = I_{vv} - A\,h^2 = 100.5 \times 10^4 - 2513\,(17)^2 \simeq 28 \times 10^4\ \text{mm}^4$

Hence $\quad I = I_{cc} + A\bar{x}^2 = 28 \times 10^4 + 2513\,(45 - 17)^2 = 225 \times 10^4\ \text{mm}^4$

$\therefore \quad I_{yy} = 729 \times 10^4 - 2I = 729 \times 10^4 - 2 \times 225 \times 10^4 = \mathbf{279 \times 10^4\ mm^4}$

Fig. 8.42

Example 8.17. *Fig. 8.43 shows the section of a steel beam in the shape of an inverted semi-circular channel with flanges. Determine the position of the centroid and the second moment of the area of the section about the centroidal axis parallel to the base.*

Solution. Total area A

$= 2 \times 70 \times 10 + \frac{\pi}{2}(130^2 - 120^2)$

$= 1400 + 26546 - 22619$

$= 5327\ \text{mm}^2$

Let the centroidal axis x-x be at a distance $\bar{y}$ above AB. Distance of centroid of a semi-circle from its base $= \frac{4r}{3\pi} = 0.4244\,r$

Fig. 8.43

$\therefore \quad 5327\,\bar{y} = 1400 \times 5 + 26546\,(0.4244 \times 130) - 22619\,(0.4244 \times 120)$

From which $\quad \bar{y} = 60$ mm

$$I_{AB} = \frac{1}{3} \times 140 \times 10^3 + \frac{1}{2}\frac{\pi}{4}(130^4 - 120^4) = 3077 \times 10^4\ \text{mm}^4$$

$$\text{Now} \quad I_{xx} = I_{AB} - A\bar{y}^2 = 3077 \times 10^4 - 5327\,(60)^2 = \mathbf{1160 \times 10^4\ mm^4}$$

Example 8.18. *A compound section is built up of two 250 × 125 mm × 27.9 kg R.S.J. beams placed side by side with two 10 mm thick plates 300 mm wide welded to each top and bottom flange. Calculate the I_{xx} and I_{yy} of the section. For each 250 × 125 mm × 27.9 kg joist, area of section A = 35.53 cm², I_{xx} = 3718 cm⁴ and I_{yy} = 193.4 cm⁴.*

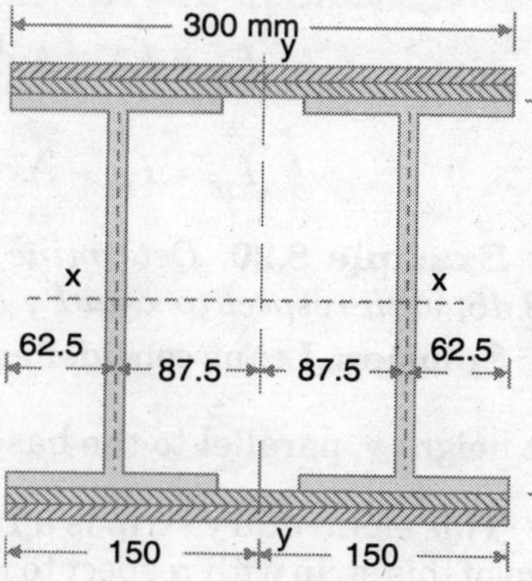

Fig. 8.44

Solution. The I_{xx} of the built up section will be equal to I_{xx} for the two joists, plus the I_{xx} of the rectangular plates.

Now I_{xx} of each 300 mm × 20 mm rectangular area of two plates will be

$$= \frac{1}{12} \times 300\,(20)^3 + 300 \times 20\,(125 + 10)^2$$

$$= 10955 \times 10^4 \text{ mm}^4$$

∴ For the whole built up section,

$$I_{xx} = 2 \times 3718 \times 10^4 + 2 \times 10955 \times 10^4 = \mathbf{29346 \times 10^4\ mm^4}$$

Now I_{yy} of the plates $= \frac{1}{12} \times 40(300)^3 = 9000 \times 10^4$ mm⁴

I_{yy} of the two joists = 2 [193.4 × 10⁴ + 3553 (87.5)²] = 5827 × 10⁴ mm⁴

∴ For the whole built up section, I_{yy} = 9000 × 10⁴ + 5827 × 10⁴ = **14827 × 10⁴ mm⁴**

Example 8.19. *For a parabolic semi-segment shown in Fig. 8.45, determine (a) moments of inertia about x_0, y_0 axes, and (b) moments of inertia about the centroidal axes x_g, y_g.*

Solution. As found in § 8.4 $\quad A = \dfrac{2bh}{3}, \bar{x} = \dfrac{3}{8}b$ and $\bar{y} = \dfrac{2h}{5}$

$$I_{y0} = \int_0^b x^2 \cdot y dx = \int_0^b x^2 h\left(1 - \frac{x^2}{b^2}\right) dx = \frac{2}{15}\, h\, b^3 \quad \textbf{(Ans.)}$$

In order to find I_{x0}, consider a strip of thickness dy at distance y from the x-axis. The width of the strip

$$= x = b\left(1 - \frac{y}{h}\right)^{1/2}$$

$$\therefore \quad I_{x0} = \int_0^h y^2 . x\, dy$$

$$= \int_0^h y^2 . b\left(1 - \frac{y}{h}\right)^{1/2} dy$$

$$= \frac{16\, b h^3}{105} \quad \textbf{(Ans.)}$$

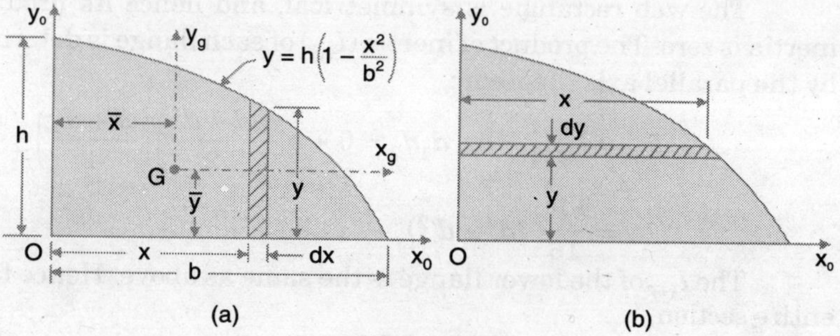

Fig. 8.45

Now, from the parallel axis theorem,

$$I_{xg} = I_{xo} - A\bar{y}^2 = \frac{16\,bh^3}{105} - \frac{2\,bh}{3}\left(\frac{2h}{5}\right)^2 = \frac{\mathbf{8}}{\mathbf{175}}\boldsymbol{bh^3} \quad \textbf{(Ans.)}$$

and

$$I_{yg} = I_{yo} - A\bar{x}^2 = \frac{2}{15}hb^3 - \frac{2\,bh}{3}\left(\frac{3}{8}b\right)^2 = \frac{\mathbf{19}}{\mathbf{480}}\boldsymbol{hb^3} \quad \textbf{(Ans.)}$$

Example 8.20. *Determine the product of inertia of a right angled triangle shown in Fig. 8.46, with respect to x and y axes, and (b) centroid axes.*

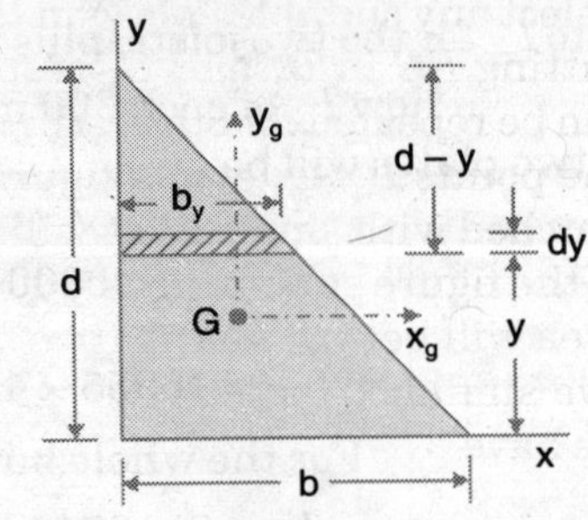

Fig. 8.46

Solution. Let us consider an elementary strip of thickness dy, at height y, parallel to the base. Width of strip, $b_y = \dfrac{d-y}{d}\,b.$

This elementary strip is a rectangle, and hence product of inertia of this strip with respect to its own centroidal axes, parallel to x-y axes is zero because of symmetry. Hence from the parallel axis theorem

$$d\,I_{xy} = (dA)\,d_1 d_2 = \left[\frac{(d-y)}{d}\,b\,dy\right](y)\left[\frac{(d-y)}{2d}\,b\right]$$

$$= \frac{b^2}{2d^2}\,(d-y)^2\,y\,dy$$

Hence for the whole triangle,

$$I_{xy} = \int d\,I_{xy} = \frac{b^2}{2d^2}\int_0^d (d-y)^2\;y\,dy = \frac{b^2 d^2}{24} \qquad \text{...(8.36)}$$

In order to find the product of inertia with respect to the centroidal axes, let us again apply the parallel axis theorem.

$$I_{xg\,yg} = I_{xy} - A\,.\,d_1 d_2 = \frac{b^2 d^2}{24} - \frac{bd}{2}\left(\frac{d}{3}\right)\left(\frac{b}{3}\right) = -\frac{b^2 d^2}{72} \qquad \text{...(8.37)}$$

Example 8.21. *Determine the product of inertia of the Z-section about its centroidal axes.*

Solution. The *Z*-section consists of (*i*) one web rectangle of width $(b - b')$ and height d, and (*ii*) two flange rectangles, each of width b' and height $(d - d')/2$.

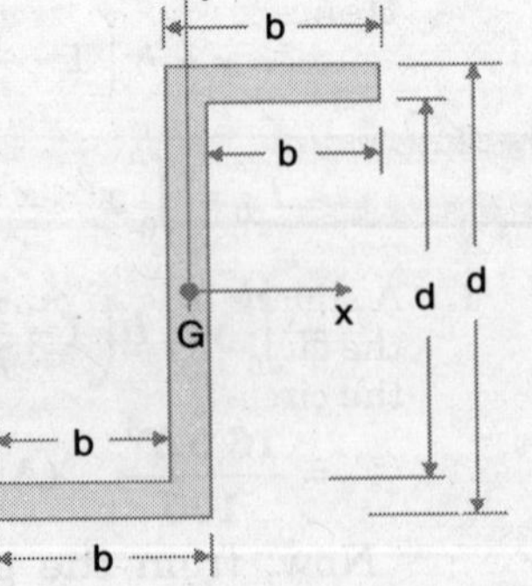

Fig. 8.47

The web rectangle is symmetrical, and hence its product of inertia is zero. The product of inertia (I_{xyf}) of each flange is determined by the parallel axis theorem :

$$I_{xyf} = I_{xg\,yg} + A\,.\,d_1 d_2 = 0 + b'\left(\frac{d-d'}{2}\right)\left(\frac{d+d'}{4}\right)\left(\frac{b}{2}\right)$$

$$= \frac{bb'}{16}\,(d^2 - d'^2)$$

The I_{xyf} of the lower flange is the same as above. Hence for the entire section,

$$I_{xy} = 2\,I_{xyf} = \frac{bb'}{8}\,(d^2 - d'^2) \qquad \text{...(8.38)}$$

8.12. GRAPHICAL METHOD FOR FIRST AND SECOND MOMENTS OF AREA

Let A be the area, the moment of which is to be found about axis PQ.

Draw a line $P'Q'$ parallel to PQ at a distance h, and to the other side of the area. Take any strip BC of thickness dy, distant y from the axis PQ. Project B and C on line $P'Q'$, getting B' and C' correspondingly. Select any suitable pole O in PQ and join $B'O$ and $C'O$, cutting BC in B_1 and C_1 respectively. This procedure can be repeated for other strips of the area. After joining the points B_1, C_1 etc., a figure, such as the one shown hatched with single lines will be obtained. Let the area of the figure so obtained by A_1. The first moment of the area will be $A_1 \times h$. To prove this proposition, consider two similar triangles in Fig. 8.48. Thus, for strip BC, we have

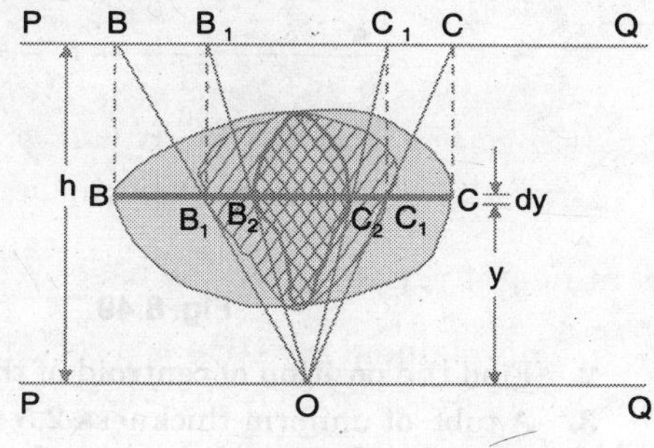

Fig. 8.48

$$\frac{B_1C_1}{y} = \frac{B'C'}{h} \quad \text{or} \quad h.B_1C_1 = y.B'C' = y.BC$$

or $$h \,.\, B_1C_1 \,.\, dy = y \,.\, BC \,.\, dy \qquad ...(1)$$

Now for the whole area,

$$A\bar{y} = \Sigma\, ydA = \Sigma\, h.dA_1 = h\, \Sigma\, dA_1 = h\, A_1 \qquad ...(2)$$

This proves the proposition.

The figure so obtained is also known as the *modulus figure*.

To find the second moment of area (*i.e.*, moment of inertia), find the *first moment of area* of the modulus figure obtained above. The same procedure will be followed and an area such as shown by the double-hatched lines will be obtained. The second moment of the original area will be equal to $h^2 . A_2$, where A_2 is the area of the figure so obtained.

We have already proved that

$$A\bar{y} = A_1 \,.\, h$$

Now $$A\bar{y}^2 = A\bar{y} \,.\, \bar{y} = (A_1 h)\, \bar{y} \qquad ...(3)$$

But $$A_1\bar{y} = h\, A_2 \quad \text{or} \quad A_1 h\, \bar{y} = h^2 A_2$$

Hence from (3), we get $A\bar{y}^2 = h^2 A_2$, which proves the proposition.

PROBLEMS

1. A square hole is punched out of a circular lamina, the diagonal of the square being the radius of the circle, as shown in Fig. 8.49. Find the position of the centroid of the remainder, if the radius of the circle is r.

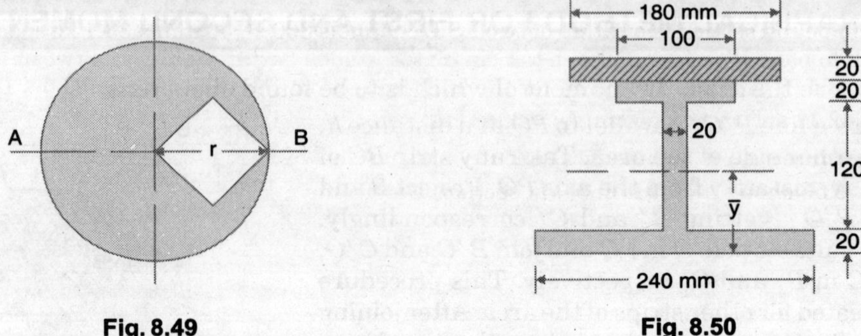

Fig. 8.49

Fig. 8.50

2. Find the position of centroid of the section shown in Fig. 8.50.
3. A tube of uniform thickness 2.5 mm has section in the shape of a regular hexagon as shown in Fig. 8.51, the outside hexagon having a side of 4 cm. Find the value of I_{xx} of the hollow hexagon.

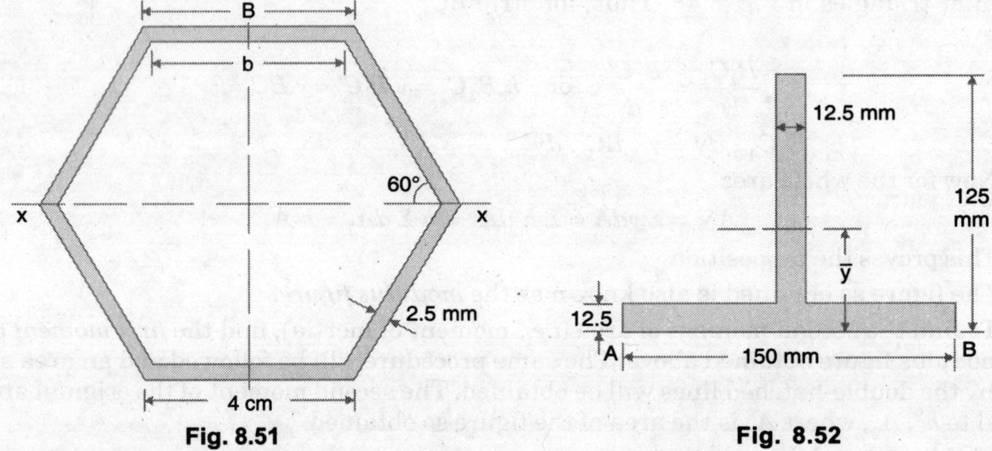

Fig. 8.51

Fig. 8.52

4. Find the position of the centroid and calculate the I_{xx} and I_{yy} for a 150 mm × 12.5 mm × 125 mm section shown in Fig. 8.52.
5. The thickness of flanges and web of a 150 mm × 75 mm I.S. channel section are 9 mm and 6 mm respectively (Fig. 8.53). Find the position of the centroid and its I_{xx} and I_{yy}.

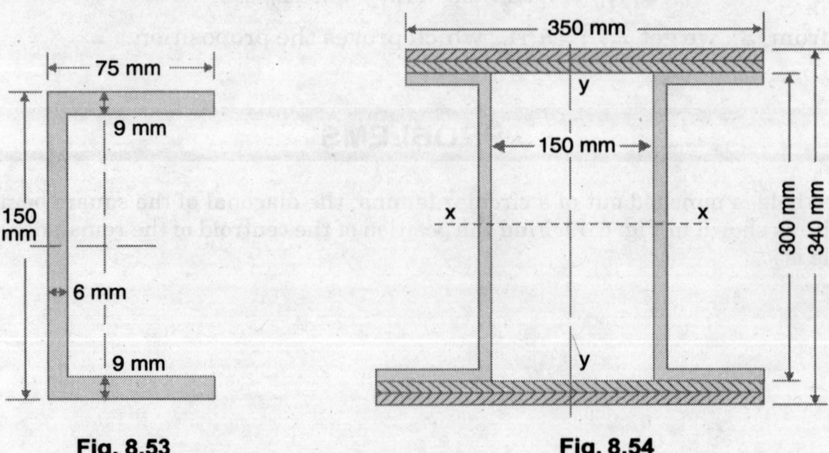

Fig. 8.53

Fig. 8.54

6. A compound section is built up of two I.S.L channels 300 × 100 mm × 33.1 kg spaced 150 mm apart back to back with four 350 mm × 1 cm plates welded two at the top and two at the bottom flange (Fig. 8.54). Find the I_{xx} and I_{yy} for the built up section. For each channel 300 × 100 mm × 32.1 kg, $A = 42.11$ cm^2, $\bar{y} = 2.55$ cm, $I_{xx} = 6048$ cm^4 and $I_{yy} = 346.0$ cm^4.

7. A beam section is symmetrical about y-axis, as shown in Fig. 8.55. Determine I_{xx} and I_{AB}.

8. Fig. 8.56 shows a cast iron beam section. Determine I_{xx}.

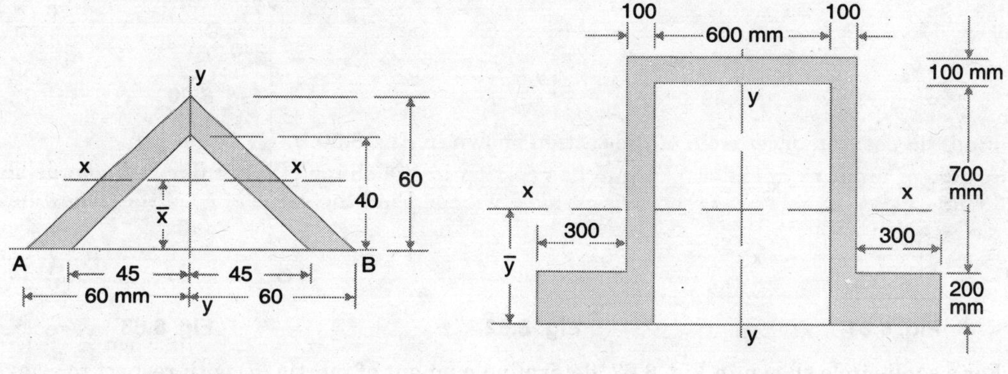

Fig. 8.55 Fig. 8.56

9. One quarter of a square of side is removed, as shown in Fig. 8.57. What are the coordinates of the centroid of the remaining part ?

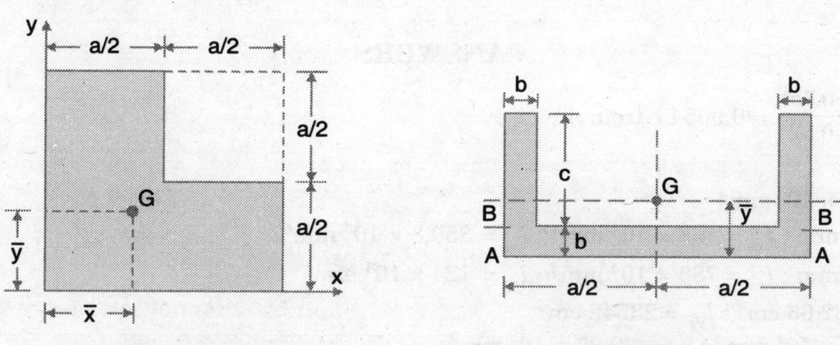

Fig. 8.57 Fig. 8.58

10. For a channel section shown in Fig. 8.58, determine the relationship between the dimensions a, b and c in order that the centroid G will be on BB.

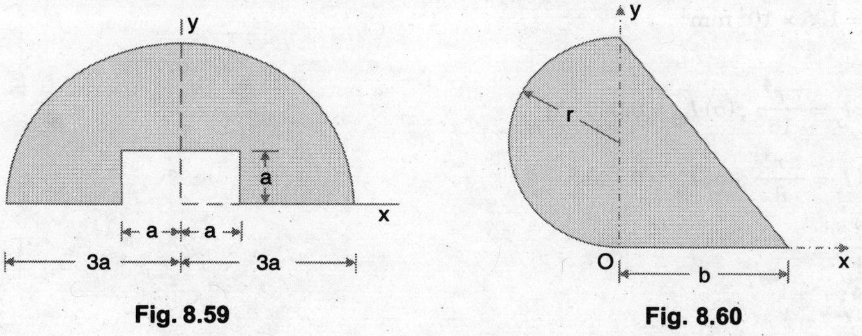

Fig. 8.59 Fig. 8.60

11. A semi-circular area of radius $3a$ has a rectangular cutout of dimension $2a \times a$ as shown in Fig. 8.59. Find the values of I_{xx} and I_{yy}.

12. Find the relationship between the radius r and the distance b for the composite area shown in Fig. 8.60 in order that the product of inertia I_{xy} will be zero.

13. For a quarter circle shown in Fig. 8.61, determine moment of inertia (a) with respect of xy axes, and (b) with respect to centroidal x-axis.

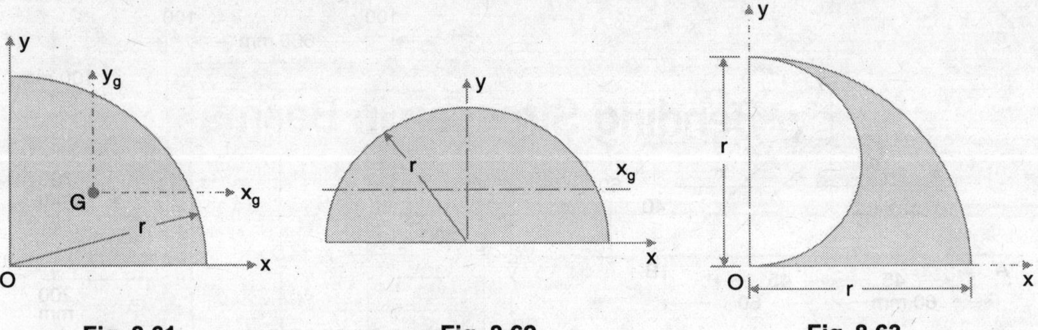

Fig. 8.61 **Fig. 8.62** **Fig. 8.63**

14. For a semi-circle shown in Fig. 8.62, determine moment of inertia (a) with respect to xy axes, and (b) with respect to the centroidal axis parallel to the diameter.

15. For the semi-circle shown in Fig. 8.61, determine the product of inertia with respect of xy axes.

16. Determine the product of inertia of the shaded area shown in Fig. 8.63, with respect to x- and y-axes.

ANSWERS

1. $\bar{x} = \dfrac{r(\pi - 0.75)}{(\pi - 0.5)} = 0.9054\, r$ from A ;
2. 90 mm
3. $I_{xx} = 35.5 \times 10^4\ \text{mm}^4$
4. $\bar{x} = 33.1$ mm ; $I_{xx} = 462 \times 10^4\ \text{mm}^4$; $I_{yy} = 350.8 \times 10^4\ \text{mm}^4$
5. $\bar{y} = 24.7$ mm ; $I_{xx} = 783 \times 10^4\ \text{mm}^4$; $I_{yy} = 124 \times 10^4\ \text{mm}^4$
6. $I_{xx} = 47982.66\ \text{cm}^4$; $I_{yy} = 23542\ \text{cm}^4$
7. $I_{AB} = 168 \times 10^4\ \text{mm}^4$; $I_{xx} = 39.97 \times 10^4\ \text{mm}^4$
8. $I_{xx} = 4.731 \times 10^4\ \text{mm}^4$
9. $\bar{x} = \bar{y} = 5a/12$
10. $2c^2 = ab$
11. $I_x = I_y = 195 \times 10^4\ \text{mm}^4$;
12. $b = 2r$
13. (a) $I_x = I_y = \dfrac{\pi r^4}{16}$; (b) $I_{xg} = 0.055\, r^4$;
14. (a) $I_x = I_y = \dfrac{\pi r^4}{8}$; (b) $I_{xg} = 0.11\, r^4$
15. $I_{xy} = \dfrac{r^4}{8}$;
16. $I_{xy} = \dfrac{r^4}{12}$.

9

Bending Stresses in Beams

9.1. BENDING AND SHEARING STRESSES IN A BEAM SECTION

We have already seen that due to the external loads and moments, bending moments and shearing forces are set up at all the sections of a beam and the beam deforms. The beam section has to, therefore, resist the action of bending moment and shearing force so induced. The process of bending will stop when it has set up full resistance to the bending moment and the shearing force. The stress produced at the section to resist the bending moment is known as the *bending stress* or the *longitudinal stress* and that to resist the action of shearing force is known as *shearing stress.* The shearing stress will be dealt with in chapter 10.

9.2. THEORY OF SIMPLE BENDING

A section resists the bending moment by setting up bending stresses. The bending moment, therefore, represents the resultant moment (called the moment of resistance) of internal forces distributed over the section. *The equations of statics are not sufficient to establish the law of the distribution and the magnitude of the bending stress occurring at the section.* The conditions of deformation of the beam are, therefore, used.

Consider two sections *mn* and m_1n_1 at distance δx apart, and subjected to bending moment *M* as shown in Fig. 9.1(*a*). If a specimen of a beam is taken and a network of lines are marked as shown in Fig. 9.1(*a*), the following points will be discovered :

(*i*) The section *mn* and m_1n_1 which were vertical and parallel before the application of the moment will rotate through an angle θ after deformation and will remain *straight*.

(*ii*) The fibre *ab* at the concave side of the beam shortens. Similarly the fibre *cd* on the concave side shortens; however, the fibre nn_1 on the convex side elongates. Also, there exists one fibre such as *ef* the length of which remains unchanged and indicates that this undergoes neither extension nor compression. The layer *ef* is called the *neutral layer*. The line of intersection of the neutral layer with the plane of cross-section of the beam is called the *neutral axis.*

(*iii*) From the experimental results considered, it follows that the amount of deformation of a layer is greater for farther layers and smaller for those which are nearer to the neutral axis.

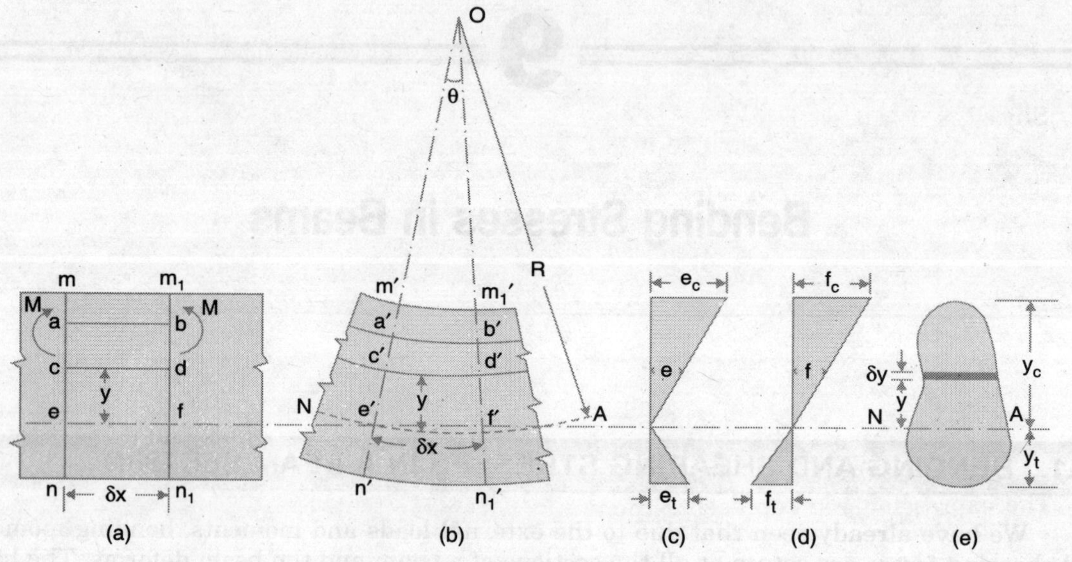

Fig. 9.1

Let us consider a layer such as *cd* at a distance *y* from the neutral axis. Fig. 9.1 (*b*) shows the deformed shape of the layers. The strain *e* of the layer *c′d′* is, therefore,

$$e = \frac{cd - c'd'}{cd}, \text{ compressive.}$$

Since the distance between the two sections *mn* and m_1n_1 is very small, it can be assumed that all the fibres are bent into a circular arc. Let the radius of the neutral fibre be *R*.

$$\theta = \frac{e'f'}{R} = \frac{ef}{R} = \frac{\delta x}{R}$$

$$\therefore \quad \delta x = R\theta = cd \quad \text{and} \quad c'd' = (R - y)\theta$$

Hence
$$e = \frac{cd - c'd'}{cd} = \frac{R\theta - (R - y)\theta}{R\theta} = \frac{y}{R} \qquad \text{...(9.1)}$$

Thus, it is proved that the strain of a fibre is proportional to its distance from neutral axis. The variation of strain is shown in Fig. 9.1(*c*).

Now, Stress = Strain × Elasticity

$$f = e.E = \frac{y}{R}E$$

Hence
$$\frac{f}{y} = \frac{E}{R} \qquad \text{...(9.2)}$$

After having established the law of stress distribution let us now determine the magnitude of stress from the equation of equilibrium. Consider the equilibrium of the section *mn* (say) subjected to the external moment *M* and internal stresses shown in Fig. 9.1 (*d*). Since there is no external force, the sum of all the compressive forces above the neutral layer must be equal to the sum of tensile forces below the neutral layer.

Let us take a strip δy thick at a distance y from N.A., and let its area be δa [Fig. 9.1(e)]. Since the stress is equal to f there, the force exerted by this small area will be $f\delta a$. The total compressive force

$$C = \sum_{y=0}^{y=y_c} f\delta a \text{. Similarly, the total tensile force } T = \sum_{y=0}^{y=y_t} f\delta a$$

Since $C + T = 0$, we have

$$\therefore \quad \sum_{y=y_t}^{y=y_c} f\delta a = 0$$

or
$$\sum_{y=y_t}^{y=y_c} \frac{E}{R} y.\delta a = 0$$

or
$$\sum_{y_t}^{y_c} y\,\delta a = 0 \left(\text{Since } \frac{E}{R} \text{ is not equal to zero.}\right) \quad ...(9.3)$$

The expression $y\delta a$ represents the statical moment of the elementary area δa about N.A. and $\sum_{y_t}^{y_c} y\delta a$ represents the sum of the moments of all such elementary areas of whole beam. Hence we have $A\bar{y} = 0$, where A is the total area of cross-section and $\bar{y}$ is the distance of centre of gravity from N.A. But $A \neq 0$, hence $\bar{y} = 0$. *Hence the neutral axis passes through the centroid of the section.*

Again, since the section mn under consideration is subjected to a bending moment M, it must be resisted by an internal moment. Such internal moment is evidently set up by the unbalanced forces C and T acting at some distance. This internal moment is known as *moment of resistance* (M_r).

In order to find the value of moment of resistance M_r, let us again consider the elementary area δa. The force set up by the area $= f.\delta a$ and the moment of the force about N.A. is thus $\delta M = f\,\delta a.y$.

Thus
$$\delta M = f.\delta a.y = \frac{E}{R}\delta a.y^2$$

Hence the total moment of resistance set up by such elementary areas will be

$$M_r = \sum_{y_t}^{y_c} \frac{E}{R}\delta a \,.\, y^2 = \frac{E}{R}\int_{y_t}^{y_c} y^2 \delta a \quad ...(9.4)$$

The expression in the integral is known as the *second moment of the area* or the *moment of inertia* about the N.A. and is represented by I.

Hence
$$M_r = \frac{E}{R} I \quad ...(9.5)$$

But for the equilibrium of the section, the moment of resistance and the bending moment M must be equal.

Hence $$M = M_r = \frac{E}{R} I \quad \text{or} \quad \frac{M}{I} = \frac{E}{R}$$

But from Eq. 9.2, $$\frac{E}{R} = \frac{f}{y}$$

Hence, we have $$\frac{M}{I} = \frac{f}{y} = \frac{E}{R} \qquad ...(9.6)$$

The term EI is known as *flexural rigidity* of the section. Since $\frac{1}{R} = \frac{M}{EI}$, it follows that the *curvature of the axis of beam in bending is directly proportional to bending moment and inversely proportional to the flexural rigidity.*

Assumptions Made in Simple Theory of Bending

The assumptions made in the simple theory of bending are summarised as under :

(1) The plane transverse sections remain plane and normal after the bending.

(2) The material is homogeneous, isotropic and obeys Hooke's law and the limits of elasticity are not exceeded.

(3) Every layer of material is free to expand or contract longitudinally and laterally under stress and does not exert pressure upon other.

(4) The modulus of direct elasticity E has the same value for compression as for tension.

Ordinary Bending

The case of *simple bending* discussed above is applicable only when the beam is subjected to pure moments, and the shearing force is absent. In ordinary bending that an engineer has to deal with, a beam cross-section has to resist both bending moment as well as shearing force. In the presence of shearing force, assumption (1), that the plane sections remain plane after bending will not be true. The initial plane cross-section in such a case will, therefore, be distored. However, since in many cases of continuous loading the greatest bending moment occurs where the shearing force is zero, the conditions correspond with those for *simple bending*, and the theory and equations discussed above are quite sufficient and gives results which enable the engineer to design beams and structures and calculate their stresses and strains with a reasonable degree of accuracy. Also, it generally happens that shearing force is maximum where the bending moment is small and the discussions made in chapter 10 for the shearing stresses are also reasonable.

Example 9.1. *A steel flat 10 cm wide and 2 cm thick is bent into a circular arc of 50 metres radius. Find the maximum intensity of stress induced. Take E = 2.05 × 10⁵ N/mm².*

Solution. Given : $R = 50$ m and $y_{max} = 20/2 = 10$ mm

Now, $$\frac{f}{y} = \frac{E}{R}$$

or $$f_{max} = \frac{E}{R}\, y_{max} = \frac{2.05 \times 10^5}{50 \times 1000} \times 10 = \mathbf{41\ N/mm^2}.$$

Example 9.2. *A beam 30 cm deep of symmetrical section has I = 8000 cm⁴ and is simply supported over a span of 8 metres. Calculate (a) the uniformly supported load it may carry, (b) the concentrated load it may carry at the centre, if the maximum bending stress is not to exceed 110 N/mm².*

Solution. We have, $\frac{M_r}{I} = \frac{f}{y}$

where $f = 110$ N/mm² ; $I = 8000 \times (10)^4$ mm⁴ $= 8 \times 10^7$ mm⁴ and $y = D/2 = 150$ mm

$$\therefore \quad M_r = \frac{f}{y} I = \frac{110}{150} \times 8 \times 10^7 = 5.87 \times 10^7 \text{ N-mm}$$

(*a*) Let the uniformly distributed load be w N/m

$$\therefore \quad M = \frac{wl^2}{8} = \frac{w \times 8 \times 8}{8} = 8w \text{ N-m} = 8000\, w \text{ N-mm}$$

Equating this to the moment of resistance,

$$8000\, w = 5.87 \times 10^7$$

$$\therefore \quad w = \frac{5.87 \times 10^7}{8000} = 7340 \text{ N/m} = \mathbf{7.34 \text{ kN/m.}}$$

(*b*) Let W N be the central load

$$M = \frac{WL}{4} = \frac{W \times 8}{4} = 2\, W \text{ N-m} = 2000\, W \text{ N-mm}$$

Equating this to the moment of resistance,

$$2000\, W = 5.87 \times 10^7$$

or
$$W = \frac{5.87 \times 10^7}{2000} = 29350 \text{ N} = \mathbf{29.35 \text{ kN.}}$$

9.3. DESIGN CRITERION AND SECTION MODULUS

To ensure strength in a beam, it is necessary that the greatest tensile and greatest compressive bending stress at the critical section *i.e.*, where B.M. is maximum, should not exceed the corresponding allowable stress for the material. With the notations of Fig. 9.1, we have

$$f_{t\,max} = \frac{M}{I} y_t \text{ and } f_{c\,max} = \frac{M}{I} y_c \qquad \text{...(9.7)}$$

Hence the condition of strength or of design is

$$f_{t\,max} \le f_t \text{ allowable} \quad \text{and} \quad f_{c\,max} \le f_c \text{ allowable.}$$

For a given value of allowable stress, the moment of resistance depends on the geometrical properties of the section.

We have $f_t = \frac{M}{I} y_t$

or in general, $f = \frac{M}{I} y$, where f is the maximum stress (tensile or compressive) and y the corresponding extreme fibre distance.

$$\therefore \quad M = f\frac{I}{y} = f.Z \quad ...(9.8)$$

where $Z = I/y$ and is known as *section modulus*. The section modulus, therefore, represents the strength of section. Greater the value of Z, stronger will be the section.

Section Modulus of a Rectangular Section

For a rectangular section of width b and depth d, moment of inertia $I = \frac{1}{12}bd^3$

Also, $y_{max} = y_t = y_c = D/2.$

$$\therefore \text{ Section modulus } \quad Z = \frac{I}{y_{max}} = \frac{1}{12}bd^3 \div \frac{d}{2} = \frac{1}{6}bd^2 \quad ...(9.9)$$

Section Modulus of a Circular Section

For a circular section of dia. D, $I = \frac{\pi}{64}D^4$

Also, $y_{max} = y_t = y_c = D/2$

$$\therefore \text{ Section modulus } \quad Z = \frac{I}{y_{max}} = \frac{\pi}{64}D^4 \div \frac{D}{2} = \frac{\pi}{32}D^3 \quad ...(9.10)$$

Section Modulus of Hollow Circular Section

For a hollow circular section of external diameter D and internal diameter d,

Moment of inertia $I = \frac{\pi}{64}(D^4 - d^4)$

$$y_{max} = y_t = y_c = D/2$$

$$\therefore \text{ Section modulus } Z_{xx} = \frac{I}{y_{max}} = \frac{\pi}{32}\frac{D^4 - d^4}{D} \quad ...(9.11)$$

Section Modulus of Hollow box section, I-section or C-section (Fig. 8.30)

$$I_{xx} = \frac{1}{12}(BD^3 - bd^3)$$

$$y_{max} = y_t = y_c = D/2$$

$$\therefore \quad Z_{xx} = \frac{I_{xx}}{y_{max}} = \frac{1}{6}\frac{BD^3 - bd^3}{D} \quad ...(9.12)$$

Example 9.3. *A floor has to carry a load of 8 kN per sq metre and is supported by joists 10 cm wide and 30 cm deep over a span of 5 metres. Calculate the spacing centre to centre of the joists if the maximum permissible bending stress is 9.5 N/mm².*

Solution. Section modulus of each joist $= \frac{1}{6}bd^2 = \frac{1}{6} \times 10 \times 30^2 = 1500 \text{ cm}^3$

$$= 15 \times 10^5 \text{ mm}^3.$$

$$M_r = f.Z = 9.5 \times 15 \times 10^5 = 142.5 \times 10^5 \text{ N-mm}$$

Let the spacing of the joists be x metres.

Rate of loading on the joist $= w = 8x \times 1$ kN/m

$$\therefore \quad \text{Max. B.M.} = \frac{wl^2}{8} = \frac{(8x)25}{8} = 25x \text{ kN-m} = 25x \times 10^6 \text{ N-mm}$$

Equating both, $25\,x \times 10^6 = 142.5 \times 10^5$

or $$x = \frac{142.5 \times 10^5}{25 \times 10^6} = \mathbf{0.57\ m.}$$

Example 9.4. *A timber joist of 6 metre span has to carry a load of 15 kN/metre. Find the dimensions of the joist if the maximum permissible stress is limited to 8 N/mm². The depth of the joist has to be twice the width.*

Solution. $$\text{B.M.} = \frac{wl^2}{8} = \frac{15 \times 36}{8} = 67.5 \text{ kN-m} = 67.5 \times 10^6 \text{ N-mm}$$

$\therefore$ $$M_r = f\,.\,Z = 67.5 \times 10^6 \text{ N-mm}$$

or $$Z = \frac{67.5 \times 10^6}{8} = 8.44 \times 10^6 \text{ mm}^3$$

But $$Z = \frac{1}{6}bd^2 = \frac{1}{6}b(2b)^2 = \frac{2}{3}b^3$$

$\therefore$ $$\frac{2}{3}b^3 = 8.44 \times 10^6$$

or $$b^3 = \frac{8.44 \times 10^6 \times 3}{2} = 12.66 \times 10^6$$

$\therefore$ b = **233 mm** and d = **466 mm.**

However, provide a section of 235 mm × 470 mm.

Example 9.5. *A 30 cm × 16 cm rolled steel joist of I-section has flanges 11 mm thick and web 8 mm thick. Find the safe uniformly distributed load that this section will carry over a span of 5 metres if the permissible skin stress is limited to 120 N/mm².*

Solution. $$I_{xx} = \frac{1}{12}(BD^3 - bd^3) = \frac{1}{12}(16 \times 30^3 - 15.2 \times 27.8^3)$$

$$= 8800 \text{ cm}^4 = 88 \times 10^6 \text{ mm}^4$$

$$Z_{xx} = I_{xx}/y_{max} = \frac{88 \times 10^6}{150} = 58.6 \times 10^4 \text{ mm}^3$$

Let w = safe uniformly distributed load in kN/m

$\therefore$ $$M_{max} = \frac{wl^2}{8} = \frac{w \times 25}{8} \text{ kN-m}$$

$$= \frac{w \times 25}{8} \times 10^6 \text{ N-mm} = 3.125 \times 10^6\, w \text{ N-mm}$$

$\therefore$ $$f_{max} = \frac{M}{Z} = \frac{3.125 \times 10^6\, w}{58.6 \times 10^4} = \frac{312.5}{58.6}\, w \text{ N/mm}^2$$

But this is equal to 120 N/mm²

$\therefore$ $$\frac{312.5}{58.6}w = 120 \quad \text{or} \quad w = \frac{120 \times 58.6}{312.5} = \mathbf{22.5\ kN/m.}$$

Example 9.6. *A cast iron water pipe has 50 cm bore and 2 cm metal thickness and is supported at two points 10 metres apart. Find the maximum stress in the metal when it is running full. Take unit weight of cast iron as 72 kN/m³ and that of water 10 kN/m³.*

Solution. Internal area of pipe $= \frac{\pi}{4}(50)^2 = 1960 \text{ cm}^2 = 0.1960 \text{ m}^2$

Area of section $= A = \frac{\pi}{4}(54^2 - 50^2) = \frac{\pi}{4}(2916 - 2500) = 327 \text{ cm}^2 = 0.0327 \text{ m}^2$

$$I_{xx} = \frac{\pi}{64}(54^4 - 50^4) = \frac{\pi}{64}(8.5 - 6.25) \times 10^6 = 1.105 \times 10^5 \text{ cm}^4 = 1.105 \times 10^9 \text{ mm}^4$$

$$Z_{xx} = \frac{1.105 \times 10^9}{270} = 4.09 \times 10^6 \text{ mm}^3$$

Weight of pipe per metre run = 0.0327 × 1 × 72 = 2.354 kN

Weight of water in pipe per metre run = (0.196 × 1 × 10) kN = 1.96 kN

Total weight per metre run = w = 2.354 + 1.96 = 4.314 kN

$$\therefore \quad \text{B.M.} = \frac{4.314 \times 100}{8} = 53.925 \text{ kN-m} = 53.925 \times 10^6 \text{ N-mm}$$

$$\therefore \quad f = \frac{M}{Z} = \frac{53.925 \times 10^6}{4.09 \times 10^6} = \mathbf{13.18\ N/mm^2.}$$

Example 9.7. *A tube of uniform thickness 2.5 mm has section in the shape of a regular hexagon as shown in Fig. 9.2, the outside haxagon having a side of 4 cm. Find the value of I_{xx} for the hollow hexagon. A length of the above tube rests in a horizontal position on two supports 2 metres apart. The tube carries two loads each of 2 kN situated between the supports and at equal distance from the ends of the tube. Find the least distance between the loads if the stresses due to bending is not to exceed 110 N/mm². Neglect the weight of the tube.*

Solution. For the outer hexagon,

$$h = B \sin 60° = \frac{B\sqrt{3}}{2} = \frac{4}{2}\sqrt{3} = 2\sqrt{3} \text{ cm}$$

$$I_{xx} = 2\left\{\frac{1}{3}Bh^3 + \frac{1}{12}Bh^3\right\} = \frac{5}{6}Bh^3$$

$$= \frac{5}{6}B\left(\frac{B\sqrt{3}}{2}\right)^3 = \frac{5}{16}\sqrt{3}\,B^4 = 0.541\,B^4$$

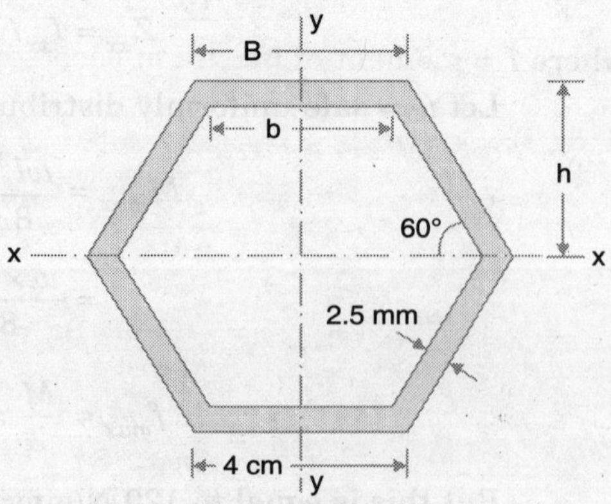

Fig. 9.2

Similarly, for the inner hexagon

$I_{xx} = 0.541\, b^4$

Net I_{xx} for the hollow hexagon

$= 0.541\,(B^4 - b^4)$

$= 0.541\,(B^2 + b^2)(B^2 - b^2)$

Now $b = B - 2t \tan 30° = 4 - \dfrac{0.5}{\sqrt{3}} = 3.711$ cm ; $b^2 = 13.8$

$B = 4, \quad B^2 = 16$

$\therefore \quad I_{xx} = 0.541(16 + 13.8)(16 - 13.8) = 35.5 \text{ cm}^4 = 35.5 \times 10^4 \text{ mm}^4$

and $$Z_{xx} = \frac{I_{xx}}{h} = \frac{35.5}{2\sqrt{3}} \text{ cm}^3 = \frac{35.5}{2\sqrt{3}} \times 10^3 \text{ mm}^3$$

Fig. 9.3 shows the loading on the beam along with the B.M. diagram, from which Max. B.M. = $2a$ kN-m, where a is the distance of the loads from the corresponding supports, in metres.

$$M_{max} = 2a \text{ kN-m} = 2 \times 10^6\, a \text{ N-mm.}$$

But $$M = f \cdot Z = 110 \times \frac{35.5}{2\sqrt{3}} \times 10^3$$

$$\therefore \quad 2 \times 10^6 a = \frac{110 \times 35.5}{2\sqrt{3}} \times 10^3$$

$$\therefore \quad a = \frac{110 \times 35.5 \times 10^3}{2\sqrt{3} \times 2 \times 10^6} = 0.564 \text{ m}$$

Fig. 9.3

$\therefore$ Least distance between the loads = 2 – (2 × 0.564) = **0.872 m.**

Example 9.8. *Fig. 9.4 shows the section of a beam. Find the ratio of its moment of resistance to bending in the plane YY to that for bending in the plane XX if the maximum stress due to bending is the same in both cases. For a semi-circle of radius r, the centroid is at a distance of 4r/3π from the centre.*

Solution. $$I_{xx} = \frac{1}{12} 9(12)^2 - \frac{\pi}{64}(8)^4 = 1296 - 201 = 1095 \text{ cm}^4$$

$$Z_{xx} = \frac{1095}{6} = 182.5 \text{ cm}^3$$

$$I_{yy} = \frac{1}{12} \times 12 \times 9^3 - 2I = 729 - 2I$$

where I = moment of inertia of each semi-circle about the YY-axis.

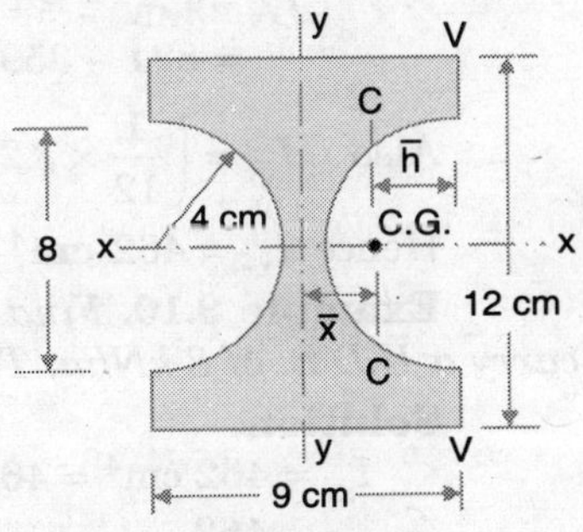

Fig. 9.4

For each semi-circle

$$I_{vv} = \frac{1}{2}\frac{\pi}{64}(8)^4 = 100.5 \text{ cm}^4$$

$$A = \frac{\pi}{2}(4)^2 = 8\pi = 25.1 \text{ cm}^2$$

$$\bar{h} = \frac{4r}{3\pi} = \frac{4 \times 4}{3\pi} = 1.7 \text{ cm}$$

$$I_{cc} = I_{vv} - Ah^2 = 100.5 - 25.1 \times 1.7^2 = 100.5 - 72.5 = 28 \text{ cm}^4.$$

again, $$I = I_{cc} + A\bar{x}^2 = 28 + 25.1(4.5 - 1.7)^2 = 28 + 197 = 225 \text{ cm}^4$$

Hence $I_{yy} = 729 - 2I = 729 - 2 \times 225 = 279 \text{ cm}^4$

and $Z_{yy} = \dfrac{279}{4.5} = 62.0 \text{ cm}^3$

Now $M = f \cdot Z$

$$\therefore \quad \frac{M_{yy}}{M_{xx}} = \frac{Z_{yy}}{Z_{xx}} = \frac{62.0}{182.5} = \mathbf{0.34.}$$

9.4. MOMENT OF INERTIA OF UNSYMMETRICAL SECTIONS

When the sections are unsymmetrical, the first step will be to find the position of the centroid or the centre of gravity through which the axis will pass. The section can then be divided into different rectangles and triangles (as the case may be) and the moment of inertia about an axis passing through the extreme edge of the section can be found out. The parallel axis theorem can then the used to find the moment of inertia about the centroidal axis. The procedure is illustrated in example 9.9, 9.10 and 9.11.

Example 9.9. *Find the position of the C.G. and calculate the I_{xx} and I_{yy} for a 15 cm × 1.25 cm × 12.5 cm Tee section shown in Fig. 9.5.*

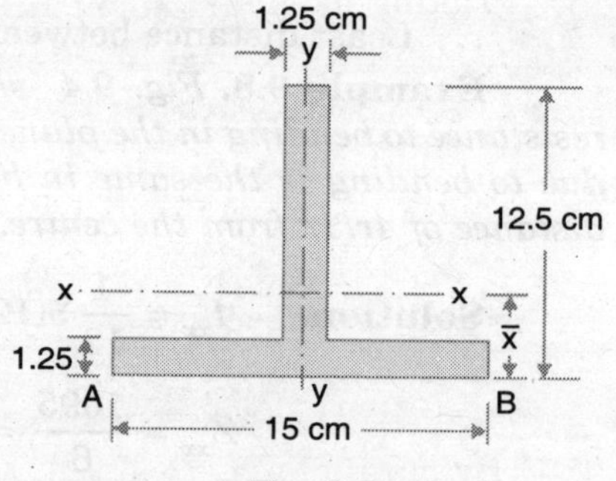

Fig. 9.5

Solution. Total area = (12.5 × 1.25) + (13.75 × 1.25)

$= 32.8 \text{ cm}^2$

To find the value of $\bar{x}$, take the moments of the areas about *AB*.

$$32.8\bar{x} = (12.5 \times 1.25 \times 6.25) + (1.25 \times 13.75)\, 0.625$$

or $$\bar{x} = \frac{97.8 + 10.75}{32.8} = 3.31 \text{ cm}$$

Now $I_{AB} = (\tfrac{1}{3} \times 1.25 \times 12.5^3) + (\tfrac{1}{3} \times 13.75 \times 1.25^3)$

$= 812 + 9 = 821 \text{ cm}^4$

$\therefore \quad I_{xx} = I_{AB} - A\bar{x}^2 = 821 - 32.8 \times 3.31^2$

$= 821 - 359 = 462 \text{ cm}^4$

Also $$I_{yy} = \left(\frac{1}{12} \times 1.25 \times 15^3\right) + \left(\frac{1}{12} \times 11.25 \times 1.25^3\right) = 349 + 1.8 = 350.8$$

Hence I_{xx} = **462 cm⁴** ; I_{yy} = **350.8 cm⁴**.

Example 9.10. *Find the span of the beam of section given in example 9.9 if it has to carry a U.D.L. of 8 kN/m. Take the allowable bending stress at 120 N/mm².*

Solution.

$I_{xx} = 462 \text{ cm}^4 = 462 \times 10^4 \text{ mm}^4$; $y_t = 3.31$ cm ; $y_c = 9.19$ cm

$$Z_t = \frac{462}{3.31} = 139.5 \text{ cm}^3 = 139.5 \times 10^3 \text{ mm}^3 \; ; \; Z_c = \frac{462}{9.19} = 50.4 \text{ cm}^3 = 50.4 \times 10^3 \text{ cm}^3$$

Since the section modulus represents the strength of the section, the moment of resistance that the section will have before failure is given by

$$M_r = f \cdot Z_c = 120 \times 50.4 \times 10^3 = 6.048 \times 10^6 \text{ N-mm} = 6.048 \text{ kN-m}$$

Now $$\text{B.M.} = \frac{wl^2}{8} = \frac{8l^2}{8} = l^2 \text{ kN-m}$$

Equating this to M_r, $l^2 = 6.048$ or l = **2.46 m.**

Example 9.11. *The thickness of flanges and web of a 15 cm × 7.5 cm I.S. channel-section are 9 mm and 6 mm respectively. Find the position of the c.g. of the section and its I_{xx} and I_{yy}.*

Solution. (Fig. 9.6)

The section is symmetrical about the axis *xx*. Let the axis *yy* be at a distance $\bar{y}$ from the face *AB*. The whole area can be splitted in three parts.

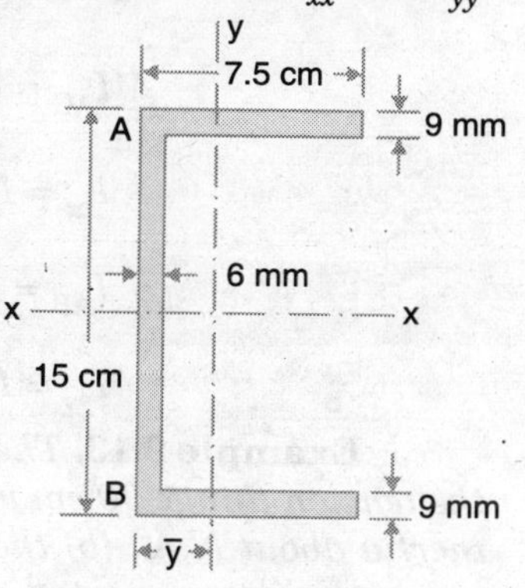

Fig. 9.6

(*i*) Top flange $A_1 = 7.5 \times 0.9 = 6.75\ \text{cm}^2$;

$$\bar{y}_1 = \frac{7.5}{2} = 3.75 \text{ cm from face } AB.$$

(*ii*) Bottom flange $A_2 = 7.5 \times 0.9 = 6.75\ \text{cm}^2$;

$$\bar{y}_2 = \frac{7.5}{2} = 3.75 \text{ cm from face } AB.$$

(*iii*) Web : $A_3 = (15 - 1.8) \times 0.6 = 13.2 \times 0.6 = 7.92\ \text{cm}^3$

$\bar{y}_3 = 0.3$ cm from face *AB*.

Total area $= A_1 + A_2 + A_3 = 6.75 + 6.75 + 7.92$

$= \mathbf{21.42\ cm^2}$.

Taking moments about *AB*, we get

$$\bar{y} = \frac{A_1\bar{y}_1 + A_2\bar{y}_2 + A_3\bar{y}_3}{A_1 + A_2 + A_3}$$

or

$$\bar{y} = \frac{(6.75 \times 3.75) + (6.75 \times 3.75) + (7.92 \times 0.3)}{21.42} = \mathbf{2.47\ cm}$$

$$I_{xx} = \frac{1}{12}(BD^3 - bd^3) = \frac{1}{12}[7.5 \times 15^3 - 6.9 \times 13.2^3] = \mathbf{783\ cm^4}$$

$$I_{AB} = 2\left(\frac{1}{3} \times 0.9 \times 7.5^3\right) + \frac{1}{3} \times 13.2 \times 0.6^3 = \mathbf{255\ cm^4}$$

By parallel axis theorem, $I_{AB} = I_{yy} + A\bar{y}^2$

$$I_{yy} = I_{AB} - A\bar{y}^2 = 255.0 - 21.42 \times 2.47^2 = \mathbf{124\ cm^4}.$$

Example 9.12. *Find the position of the c.g. and I_{xx} and I_{yy} for unequal angle section 12 cm × 10 cm × 1.25 cm.*

Solution. The whole area can be splitted in two parts :

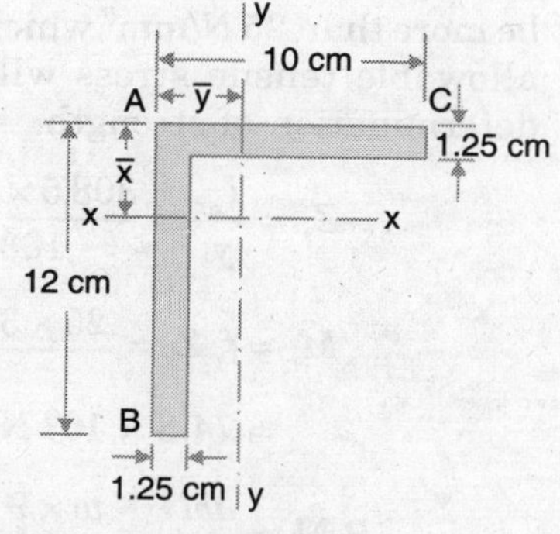

Fig. 9.7

(*a*) *Flange* : $A_1 = 10 \times 1.25 = 12.5\ \text{cm}^2$

$$\bar{x}_1 = \frac{1.25}{2} = 0.625 \text{ cm from face } AC.$$

$$\bar{y}_1 = \frac{10}{2} = 5 \text{ cm from face } AB.$$

(*b*) *Web* : $A_2 = 10.75 \times 1.25 = 13.45\ \text{cm}^2$

$$\bar{x}_2 = \frac{10.75}{2} + 1.25 = 6.625 \text{ cm}$$

$$\bar{y}_2 = \frac{1.25}{2} = 0.625 \text{ cm}$$

Total area $A = A_1 + A_2 = 12.5 + 13.45 = 25.95\ \text{cm}^2$

Taking moments about *AC*, $\bar{x} = \dfrac{A_1\bar{x}_1 + A_2\bar{x}_2}{A_1 + A_2} = \dfrac{(12.5 \times 0.625) + (13.45 \times 6.625)}{25.95} = 3.73$ cm.

Taking moments about AB, $\bar{y} = \dfrac{A_1\bar{y}_1 + A_2\bar{y}_2}{A_1 + A_2} = \dfrac{(12.5 \times 5) + (13.45 \times 0.625)}{25.95} = 2.74$ cm.

$$I_{AC} = \left(\frac{1}{3} \times 8.75 \times 1.25^3\right) + \left(\frac{1}{3} \times 1.25 \times 12^3\right) = 725.7 \text{ cm}^4.$$

$$\therefore \quad I_{xx} = I_{AC} - A\bar{x}^2 = 725.7 - 25.95\,(3.73)^2 = \mathbf{364.7\ cm^4}.$$

$$I_{AB} = \frac{1}{3} \times 10.75 \times 1.25^3 + \frac{1}{3} \times 1.25 \times 10^3 = 424 \text{ cm}^3$$

$$\therefore \quad I_{yy} = I_{AB} - A\bar{y}^2 = 424 - 25.95(2.74)^2 = \mathbf{229\ cm^4}.$$

Example 9.13. *The compression flange of a cast iron girder is 10 cm wide and 3 cm deep, the tension flange 30 cm wide and 5 cm deep and the web 25 cm × 3.00 cm. Find (a) moment of inertia about N.A., (b) the load per metre run which may be carried over a 3 metre span by a beam simply supported at its ends if the maximum permissible stresses are 95 N/mm² in compression and 25 N/mm² in tension (Fig. 9.8).*

Solution. (*a*) Total area= (10 × 3) + (25 × 3) + (30 × 5) = 30 + 75 + 150 = 255 cm².

$$\bar{x} = \frac{(30 \times 31.5) + (75 \times 17.5) + (150 \times 2.5)}{255} = 10.3 \text{ cm}$$

$$I_{AB} = \left(\frac{1}{3} \times 27 \times 5^3\right) + \left(\frac{1}{3} \times 3 \times 30^3\right) + \left\{\left(\frac{1}{12} \times 10 \times 3^3\right) + (30 \times 31.5^2)\right\}$$

$$= 57850 \text{ cm}^4$$

$$\therefore \quad I_{xx} = 57850 - 255 \times (10.3)^2 = 30850 \text{ cm}^4 = 308.5 \times 10^6 \text{ mm}^4.$$

(*b*) Now $y_c = 33 - 10.3 = 22.7$ cm = 227 mm ;

$y_t = 10.3$ cm = 103 mm

$$\frac{f_c}{y_c} = \frac{f_t}{y_t} \quad \text{or} \quad f_c = \frac{f_t}{y_t} y_c = f_t \frac{227}{103} = 2.2\, f_t$$

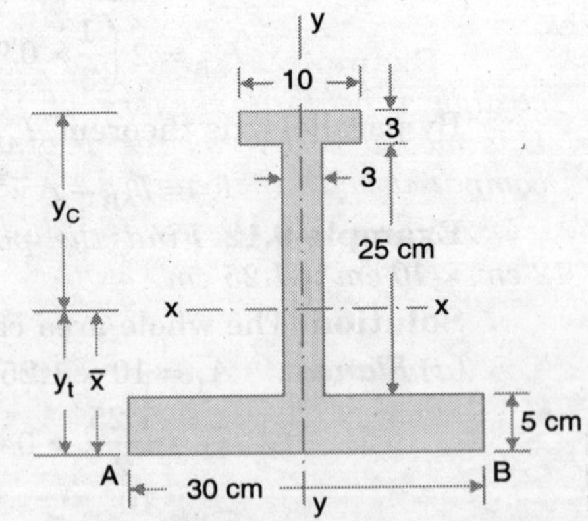

Fig. 9.8

When f_t reaches 25 N/mm², f_c = 25 × 2.2 = 55 N/mm², which is lower than the permissible limit. Should f_c be allowed to go to 95 N/mm², f_t will evidently be more than 25 N/mm² which is not permissible. Hence allowable tensile stress will be the criterion for the determination of strength.

$$\therefore \quad Z_t = \frac{I_{xx}}{y_t} = \frac{308.5 \times 10^6}{103} \text{ mm}^3$$

$$\therefore \quad M_r = f_t Z_t = \frac{25 \times 308.5 \times 10^6}{103}$$

$$= 74.8 \times 10^6 \text{ N-mm}$$

$$\text{B.M.} = \frac{wl^2}{8} = \frac{w \times 9}{8} \text{ (where } w \text{ is in kN/m)} = 1.125\, w \text{ kN-m} = 1.125\, w \times 10^6 \text{ N-mm}$$

Hence $\quad 1.125\, w \times 10^6 = 74.8 \times 10^6$

From which $\quad w = \dfrac{74.8 \times 10^6}{1.125 \times 10^6} = \mathbf{66.4\ kN/m.}$

Example 9.14. *Fig. 9.9 shows the section of steel beam in the shape of an inverted semi-circular channel with flanges. Determine the position of the neutral axis and the second moment of area of the section about xx.*

Calculate the moment of resistance of the beam if the maximum stress due to bending is 125 N/mm². For semi-circular area of radius r, the distance of the centroid from the diameter is 0.4244 r.

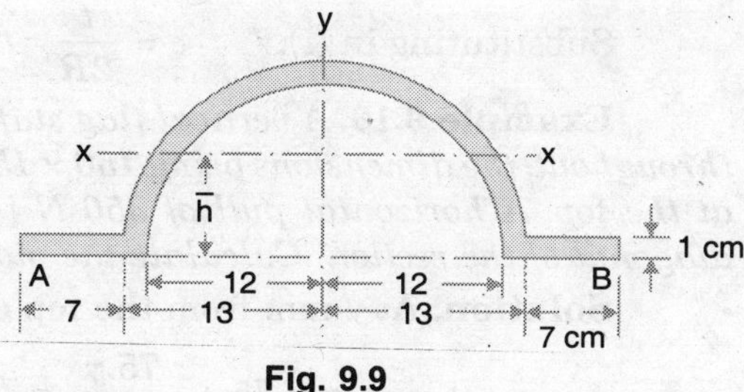

Fig. 9.9

Solution.

$$A = (2 \times 1 \times 7) + \frac{\pi}{2}(13^2 - 12^2) = 14 + 265 - 226 = 53 \text{ cm}^2$$

For finding the distance $\bar{h}$ of the C.G. from *AB*, take moments of areas about *AB*.

$$53\bar{h} = 2(7 \times 1)\frac{1}{2} + 0.4244 \times 13\left(\frac{\pi}{2} \times 13^2\right) - 0.4244 \times 12 \times \left(\frac{\pi}{2} \times 12^2\right) = 322 \text{ cm}^2$$

$$\bar{h} = \frac{322}{53} = 6.07 \text{ cm}$$

$$I_{AB} = \frac{1}{3} \times 14 \times 1^3 + \frac{1}{2}\frac{\pi}{4}(13^4 - 12^4) = 3105 \text{ cm}^4$$

$$I_{xx} = 3105 - 53(6.07)^2 = 3105 - 1960 = 1145 \text{ cm}^4$$

$$Z_{min} = \frac{1145}{13 - 6.07} = \frac{1145}{6.93} = 165.2 \text{ cm}^3 = 165.2 \times 10^3 \text{ mm}^3$$

$$\therefore \quad M_r = f.Z = 125 \times 165.2 \times 10^3 = 20.65 \times 10^6 \text{ N-mm} = \mathbf{20.65 \text{ kN-m.}}$$

Example 9.15. *A long rod of uniform rectangular section and thickness t, originally straight is bent into the form of a circular arc and the displacement d of the mid-point of length L is measured by means of dial gauge. The displacement d may be regarded as small as compared with the length L.*

Show that the longitudinal surface strain e in the rod is given by

$$e = \frac{4td}{L^2}. \qquad \text{(U.L.)}$$

Solution. At any height y from N.A.

$$\frac{f}{y} = \frac{E}{R} = \frac{M}{I}$$

$$\therefore \quad \frac{f}{E} = e = \frac{y}{R} = \frac{t}{2R} \qquad ...(i)$$

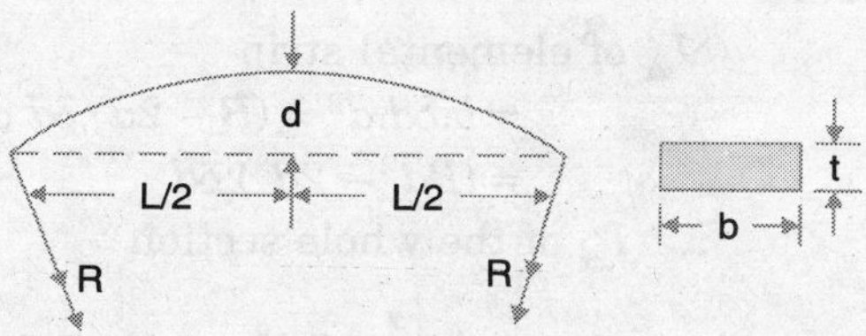

Fig. 9.10

From chord properties of a circle,

$$d(2R - d) = \frac{L^2}{4} \quad \text{or} \quad 2Rd - d^2 = \frac{L^2}{4}$$

Neglecting d^2 in comparison to $2Rd$, we get

$$2Rd = \frac{L^2}{4} \quad \text{or} \quad 2R = \frac{L^2}{4d} \qquad ...(ii)$$

Substituting in (*i*), $e = \frac{t}{2R} = t \div \frac{L^2}{4d} = \frac{4dt}{L^2}$ (proved)

Example 9.16. *A vertical flag staff standing 10 m above the ground, is of square section throughout, the dimensions being 150 × 150 mm at the ground, tapering uniformly to 75 × 75 mm at the top. A horizontal pull of 350 N is applied at the top, the direction of loading being a diagonal of the section. Calculate the maximum stress due to bending.*

Solution. At y mm from the top of flag staff, side x of the section is

$$x = 75 + \frac{75\,y}{10000} = 75 + \frac{y}{133.3} \text{ mm}$$

$$I = \frac{x^4}{12};\ Z = \frac{I\sqrt{2}}{x} = \frac{x^3\sqrt{2}}{12}$$

$$\pm f = \frac{M}{Z} = \frac{350y \times 12}{x^3\sqrt{2}}, \text{ in which } y = (x - 75)\,133.3$$

or $$\pm f = \frac{350 \times 12}{x^3\sqrt{2}} \times 133.3\,(x - 75) = 395940\,(x^{-2} - 75\,x^{-3})$$

For f to be maximum, $\frac{df}{dx} = 0$

$\therefore$ $$\frac{df}{dx} = 395940\,(-\,2x^{-3} + 225x^{-4}) = 0$$

$\therefore$ $2x - 225 = 0$ from which $x = 112.5$ mm

$\therefore$ $y = (x - 75)\,133.3 = (112.5 - 75)\,133.3 \simeq 5000 \text{ mm} = 5 \text{ m}.$

$\therefore$ $$\pm f = \frac{350 \times 12}{\sqrt{2}} \times \frac{y}{x^3} = \frac{350 \times 12}{\sqrt{2}} \times \frac{5000}{(112.5)^3} = \mathbf{10.43\ N/mm^2}.$$

Example 9.17. *Prove that the moment of resistance of a beam of square section with the diagonal in the plane of bending is increased by flattening the top and bottom corners as shown in Fig. 9.11 and that the moment of resistance is maximum when* $y = \frac{8}{9}y_1$. (U.L.)

Solution. Consider an elementary strip of width b and thickness δd at a distance d from N.A.

δI_{xx} of elemental strip

$$= b.\delta d.d^2 = (B - 2d).\delta d\ d^2$$

$$= (Bd^2 - 2d^3)\,\delta d$$

$\therefore$ I_{xx} of the whole section

$$= 2\int_0^y (Bd^2 - 2d^3)\,\delta d$$

$$= \frac{y^3}{3}(2B - 3y) = I_1 \text{ (say)} \quad ...(1)$$

When $y = \frac{B}{2}, I_{xx} = \frac{B^3}{24}\left(2B - \frac{3B}{2}\right)$

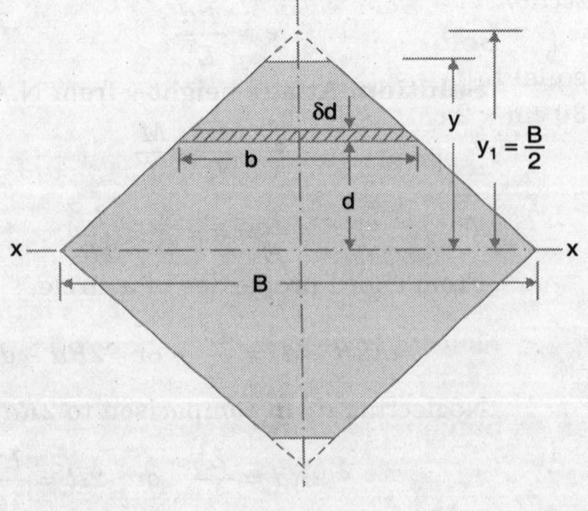

Fig. 9.11

$$= \frac{B^4}{48} = I_2 \text{ (say)} \qquad \text{...(2)}$$

Now $\quad M_1 = f_1 Z_1 \quad$ and $\quad M_2 = f_2 Z_2$

$\therefore \quad M_1 > M_2$, if $Z_1 > Z_2$ for the same value of f.

$$Z_1 = \frac{I_1}{y} = \frac{y^2}{3}(2B - 3y) = \frac{2}{3}By^2 - y^3$$

$$Z_2 = \frac{I_2}{B/2} = \frac{B^3}{24}$$

For Z_1 to be maximum, $\dfrac{dZ_1}{dy} = \dfrac{4By}{3} - 3y^2 = 0$,

from which,
$$y = \frac{4}{9}B = \frac{8}{9}y_1 \qquad \text{(proved)}$$

$$Z_1 = \frac{2}{3}B \cdot \frac{16B^2}{81} - \frac{64B^3}{729} = \frac{32B^3}{729}$$

$$Z_2 = \frac{B^3}{24}$$

$\therefore \quad Z_1$ is greater than Z_2. Hence $M_1 > M_2$.

The reason for this increase in the moment of resistance is quite obvious ; the removal of the small corner areas decreases the extreme fibre distance in greater proportion than it reduces the moment of inertia of the section.

9.5. MOMENT OF INERTIA OF BUILT-UP SECTIONS : ILLUSTRATIVE EXAMPLES

Example 9.18. *A compound section is built up of two 250 × 125 mm × 27.9 kg I.S.J. beams placed side by side with two 1 cm plates 300 mm wide riveted to each top and bottom flange. Calculate the I_{xx} and I_{yy} of the section. For each 250 × 125 mm × 27.9 kg joist area of section, $A = 35.53$ cm^2 ; $I_{xx} = 3718$ cm^4 ; $I_{yy} = 193.4$ cm^4.*

Solution. The I_{xx} of the built up section will be equal to I_{xx} for the joists plus I_{xx} of two rectangular area 30 cm × 2 cm.

I_{xx} of each 30 × 2 cm rectangular area will be

$$= \frac{1}{12} \times 30.0 \times 2^3 + 30.0 \times 2(12.5 + 1)^2$$

$$= 10970 \text{ cm}^4$$

Moment of inertia of the built-up section is

$$I_{xx} = (2 \times 3718) + (2 \times 10970) = \mathbf{29376\ cm^4}.$$

Similarly, I_{yy} for two joists is

$$= 2[193.4 + 35.53 \times 8.75^2] = 5826 \text{ cm}^4$$

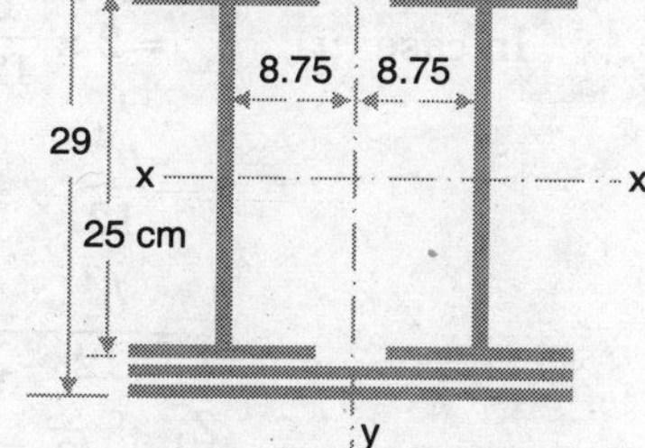

Fig. 9.12

I_{yy} for the plates $= \dfrac{1}{12} \times 4 \times 30^3 = 9000 \text{ cm}^4$

Moment of inertia of the built-up section

$$I_{yy} = 5826 + 9000 = \mathbf{14826\ cm^4}.$$

Example 9.19. *A compound section is built up of two I.S.L. channels 30 × 10 cm × 33.1 kg spaced 15 cm apart back to back with four 35 cm × 1 cm plates riveted, two at the top and two at the bottom flange. Find the I_{xx} and I_{yy} of the built-up section. For each channel 300 × 100 mm × 33.1 kg, A = 42.11 cm²; $\bar{y}$ = 2.55 cm; I_{xx} = 6048 cm⁴; I_{yy} = 346.0 cm⁴.*

Solution. The I_{xx} of each 35 cm × 2 cm rectangular area is

$$= \frac{1}{12} \times 35 \times 2^3 + 35 \times 2 \times 16^2$$
$$= 17943.33 \text{ cm}^4$$

Moment of inertia of built-up section is

$$I_{xx} = (2 \times 6048) + (2 \times 17943.33)$$
$$= \mathbf{47982.66\ cm^4}.$$

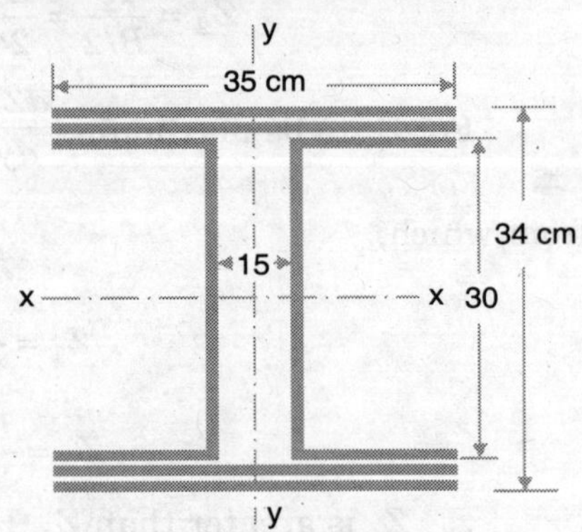

Fig. 9.13

Similarly, for the built-up section,

$$I_{yy} = 2[346 + 42.11(7.5 + 2.55)^2] + \frac{1}{12} \times 4 \times 35^3$$
$$= 9198 + 14292 = \mathbf{23490\ cm^4}.$$

9.6. STRENGTH OF SECTION

We have already seen that for a given value of maximum allowable stress, the moment of resistance depends upon the section modulus Z. *Thus the section modulus represents the strength of a given section. The strength of a section does not, therefore, depend on the sectional area but depends on the disposition of this area with respect to the centroidal axis.*

Example 9.20. *For a given stress, compare the moments of resistance of a beam of square section placed (i) with two sides horizontal, (ii) with a diagonal horizontal.*

Solution. Let b be the side of square.

In case (*i*), $Z_1 = \dfrac{1}{6} bd^2 = \dfrac{1}{6} b^3$

In case (*ii*), $I_{xx} = 2 \times \dfrac{1}{12} b\sqrt{2}\left(\dfrac{1}{2} b\sqrt{2}\right)^3$

$$= \frac{b^4}{12};\ y = \frac{b}{\sqrt{2}}$$

$$Z_2 = \frac{b^4}{12} \div \frac{b}{\sqrt{2}} = \frac{b^3}{6\sqrt{2}}$$

$$\frac{Z_1}{Z_2} = \sqrt{2} = 1.414$$

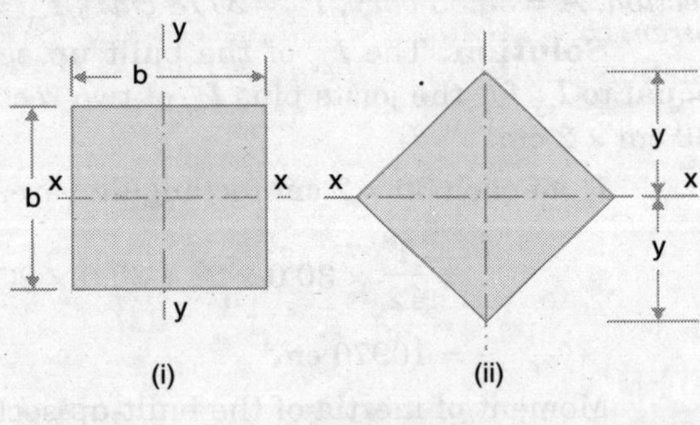

Fig. 9.14

Hence arrangement (*i*) is 41.4% more strong than (*ii*).

Example 9.21. *Compare the flexural strength of the following three beams of equal weight :*

(i) I-section 30 cm × 15 cm, having 2 cm thick flanges and 1.25 cm thick web.

(ii) Rectangular section having depth equal to twice the width.

(iii) Solid circular section.

Solution. For the *I*-section

$$A_1 = (2 \times 15 \times 2) + (1.25 \times 26) = 60 + 32.5 = 92.5 \text{ cm}^2$$

$$I = \frac{1}{12}(15 \times 30^3 - 13.75 \times 26^3) = 13.61 \times 10^3 \text{ cm}^4$$

or $$Z_1 = \frac{13.61 \times 10^3}{15} = 907.4 \text{ cm}^3.$$

For the rectangular section

$$A_2 = A_1 = 92.5 \text{ cm}^2$$

$\therefore$ $$bd = 92.5 \quad \text{or} \quad 2b^2 = 92.5$$

$\therefore$ $$b = 6.8 \text{ cm} \quad \text{and} \quad d = 13.6$$

$\therefore$ $$Z_2 = \frac{1}{6}bd^2 = \frac{1}{6} \times 6.8 \times 13.6^2$$

$$= 209.6 \text{ cm}^3.$$

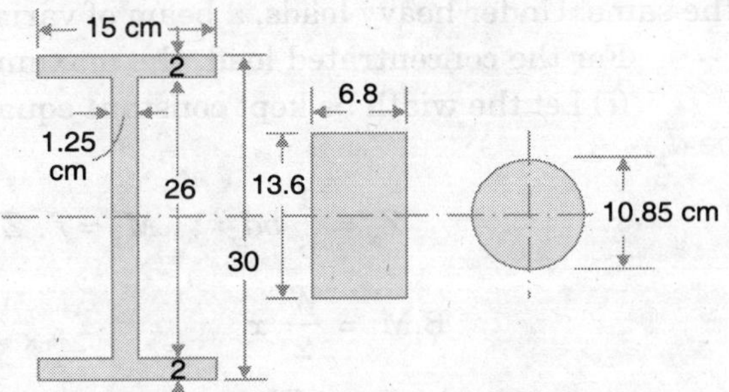

Fig. 9.15

For the circular section

$$A_3 = A_2 = A_1 = 92.5 \text{ cm}^2$$

$\therefore$ $$\frac{\pi}{4}d^2 = 92.5 \quad \text{or} \quad d = 10.85 \text{ cm}$$

$\therefore$ $$Z_3 = \frac{\pi}{32}d^3 = \frac{\pi}{32}(10.85)^3 = 125.5 \text{ cm}^3$$

$\therefore$ $$\frac{Z_1}{Z_2} = \frac{907.4}{209.6} = \mathbf{4.33} \text{ and } \frac{Z_1}{Z_3} = \frac{907.4}{125.5} = \mathbf{7.23}$$

The disposition of sectional area with respect to the axis of bending is shown in Fig. 9.15 for the three cases. It is clear that circular section is very uneconomical for a beam.

Example 9.22. *Compare the weights of two beams of the same material and of equal strength, one being of circular section and solid, and the other being of hollow circular section, the internal diameter being 7/8 of the external.*

Solution. Let d = diameter of solid section and D = external diameter of hollow section.

Z for the solid section $= \dfrac{\pi}{32}d^3$

Z for the hollow section $= \dfrac{\pi}{32D}\left\{D^4 - \left(\dfrac{7}{8}D\right)^4\right\}$

$$= \frac{\pi}{32}D^3(1 - 0.586) = 0.414 \times \frac{\pi}{32}D^3$$

The resistance to bending being proportional to the modulus of the section, we have

$$\frac{\pi}{32}d^3 = 0.414\frac{\pi}{32}D^3$$

or $\left(\frac{D}{d}\right)^3 = \frac{1}{0.414} = 2.42$ or $\left(\frac{D}{d}\right) = 1.342.$

$$\text{Now} \quad \frac{\text{weight of solid one}}{\text{weight of hollow one}} = \frac{d^2}{D^2 - \left(\frac{7}{8}D\right)^2} = \left(\frac{d}{D}\right)^2 \frac{1}{1-0.765} = \left(\frac{1}{1.342}\right)^2 \frac{1}{0.235}$$

$$= \mathbf{2.369.}$$

Example 9.23. ***Beam of uniform strength** : A beam of span L is subjected to a concentrated load W at the centre. Design the beam of uniform strength by (i) keeping width constant and varying the depth, (ii) keeping the depth constant and varying the width.*

Solution. In general practice, a beam of constant cross-section throughout length is provided. This is not economical as the moment of resistance required at all the sections is not the same. Under heavy loads, a beam of variable section is designed to have uniform strength.

For the concentrated load, the maximum B.M. will be at the centre.

(*i*) Let the width be kept constant equal to b. The depth at a distance x from the support be d_x.

$$\therefore \quad Z_x = \frac{1}{6} bd_x^2 ; \quad M_x = f . Z_x = f \frac{1}{6} bd_x^2$$

$$\text{B.M.} = \frac{W}{2} x$$

$$\therefore \quad f \frac{1}{6} bd_x^2 = \frac{W}{2} x$$

$$\text{or} \quad d_x^2 = \frac{3Wx}{fb}$$

$$\text{or} \quad d_x = \sqrt{\frac{3W}{fb}} \sqrt{x} = C\sqrt{x}$$

W
x
d_x
L
b
Plan
(a)
W
d
L
x
b
b_x
Plan
(b)

Fig. 9.16

Thus the depth will be proportional to $\sqrt{x}$ and the variation of depth will be parabolic, as shown in Fig 9.16(*a*).

At the centre, when $x = \frac{L}{2}$, $d = \sqrt{\frac{3WL}{2fb}}$.

(*ii*) Let the depth d be constant throughout and b be varied.

At any section x from the support, if b_x is the width, $Z_x = \frac{1}{6} b_x d^2$

$$\therefore \quad M_x = f Z_x = f . \frac{1}{6} b_x d^2$$

$$\text{Hence} \quad f . \frac{1}{6} b_x d^2 = \frac{W}{2} x \quad \text{or} \quad b_x = \frac{3W}{fd^2} x = C'x$$

Thus the width b_x is proportional to the distance. The variation of b is linear, having a value of zero at the supports and $\frac{3WL}{2fd^2}$ at the centre, as shown in Fig. 9.16 (*b*).

9.7. THE MODULUS FIGURE

The modulus figure is the first moment of area about some chosen axis. We have already seen in chapter 8 that every strip of the modulus figure is reduced by a ratio y/h. This reduction is very much useful if we want to find the shape of the section which will have uniform bending stress throughout the depth of the section.

We have seen that at any distance y from the N.A., the stress f is equal to $f_c \dfrac{y}{y_c}$, and therefore, varies with the position of the fibre. The whole compression (or tension) area having varying bending stress can be converted into a *modified area* having uniform stress f_c (or f_t).

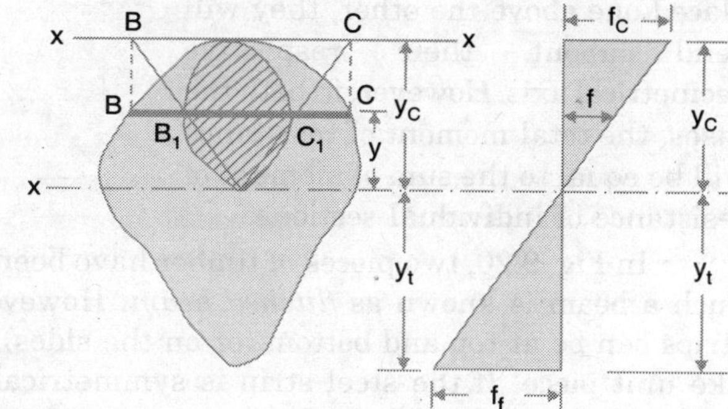

Fig. 9.17

To find the *modified area* draw the modulus figure of compression area about axis xx, as described in § 8.12 and illustrated in Fig. 9.17. It can be easily shown that the modulus figure (shown hatched) so obtained will be modified area having constant stress f_c.

Elementary area of strip $BC = \delta A = BC.\delta y$.

Elementary thrust δC of this strip $= f\delta A = f_c \dfrac{y}{y_c} \delta A$

But $\dfrac{y}{y_c} \delta A = \delta A_1$, where δA_1 = area of B_1C_1 of modulus figure.

$$\therefore \quad \delta C = f_c \cdot \delta A_1$$

Hence $\quad C = f_c \, \Sigma\delta A_1 = f_c \cdot A_1$

This proves the proposition.

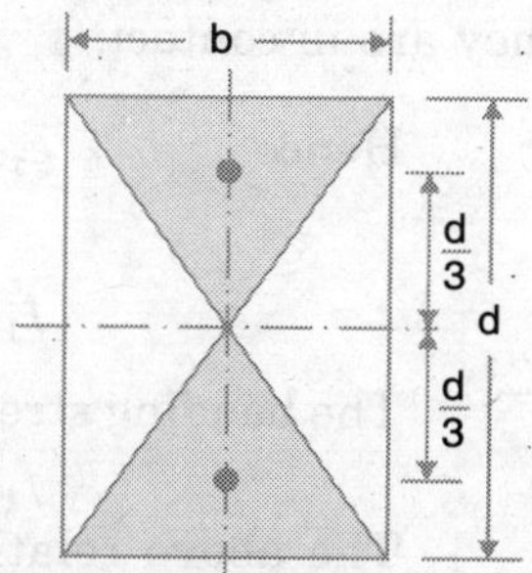

Fig. 9.18

The modulus figure for rectangular section has been shown in Fig. 9.18 and that for I-section in Fig. 9.19.

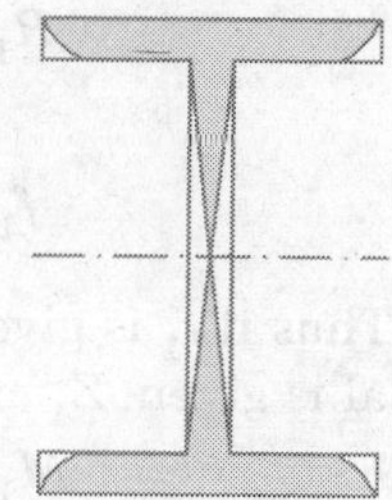

Fig. 9.19

For the rectangular section (Fig. 9.19), the total thrust

$$C = f_c \times \frac{1}{2} b \cdot \frac{d}{2} = \frac{bd}{4} f_c$$

$$\therefore \quad M = C \times \frac{2d}{3} = \left(\frac{bd}{4} f_c\right) \frac{2d}{3}$$

$$= \frac{1}{6} bd^2 f_c = Zf_c$$

which is the same as proved earlier.

9.8. BEAMS OF COMPOSITE SECTION : FLITCHED BEAM

Many times, the section of a beam of a particular material is strengthened by other material, *e.g.*, presence of steel strips in wooden beams or steel bars in concrete beams. Such structures are statically indeterminate.

If both the materials have been rigidly joined together, they will behave like a unit piece and the bending will take place about the *combined axis.* On the other hand, if both the materials have been simply placed one above the other, they will bend about their respective geometrical axis. However, in both the cases, the total moment of resistance will be equal to the sum of moment of resistance of individual sections.

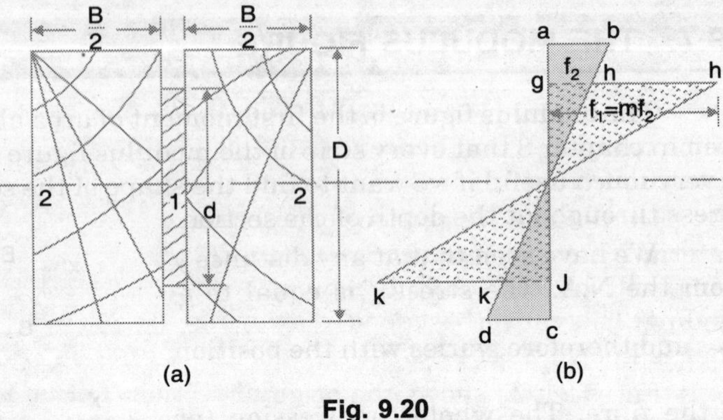

Fig. 9.20

In Fig. 9.20, two pieces of timber have been strengthened by a steel strip between them. Such a beam is known as *flitched beam*. However, the alternative arrangement of the steel strips can be at top and bottom (or on the sides). In each case, the whole section will behave like unit piece. If the steel strip is symmetrically added, the common axis of bending will remain the same as that of the timber.

From statics $M = M_1 + M_2$ ...(1)

where suffix 1 stands for steel and 2 for timber.

At any distance y from the N.A., the strain in both the materials will be the same, since they are in contact.

Hence $$e_1 = e_2 \quad \text{or} \quad \frac{f_1}{E_1} = \frac{f_2}{E_2}$$

$$\therefore \quad f_1 = \frac{E_1}{E_2} f_2 = mf_2, \quad \text{where} \quad m = \text{modular ratio.} \qquad ...(2)$$

The bending stress diagram has also been drawn in Fig. 9.20 from which it is clear that

$$f_1 = gh' = m.gh = mf_2.$$

The above relation can also be obtained by considering the fact that the radius of curvature at any level will be the same for both the materials. Thus,

$$R_1 = R_2 \quad \text{or} \quad \frac{E_1 y}{f_1} = \frac{E_2 y}{f_2}$$

$$\therefore \quad f_1 = f_2 \frac{E_1}{E_2} = mf_2.$$

Thus if f_1 is given, f_2 can be found and *vice versa.* If the geometry of the section of the material is given, Z_1 and Z_2 can be calculated. Hence

$$M_1 = f_1 Z_1 \quad \text{and} \quad M_2 = f_2 Z_2$$

and $$M = M_1 + M_2 = f_1 Z_1 + f_2 Z_2$$

Thus, the total moment of resistance can be calculated.

Also, if it is required to find the *stresses* in the section for a given bending moment M, the following equation will be useful :

$$R_1 = R_2 \quad \text{or} \quad \frac{E_1 I_1}{M_1} = \frac{E_2 I_2}{M_2}$$

$$\therefore \quad \frac{M_1}{M_2} = \frac{E_1 I_1}{E_2 I_2} \qquad ...(3) \qquad ...(9.13)$$

Combining (1) and (3), M_1 and M_2 can be calculated and hence f_1 and f_2 can be calculated.

However, if the sections of both the materials are not symmetrical, the first thing will be to find the new axis of bending. This can be best done by drawing what is known as *equivalent section*.

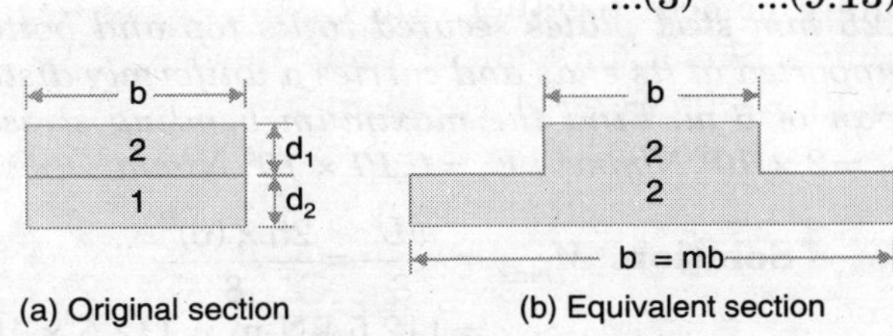

Fig. 9.21

Fig. 9.21 (*a*) shows the original section. Let us find the equivalent section of plate 1 in terms of plate 2. For the original section of plate 1, we have $M_1 = f_1 Z_1$.

If M_2' and Z_2' are moment of resistance and section modulus of equivalent section of plate 1, we have $M_2' = f_2 Z_2'$.

If $M_1 = M_2'$, we have $f_1 Z_1 = f_2 Z_2'$ or $Z_2' = \frac{f_1}{f_2} Z_1 = mZ_1$

Hence $\qquad b' = mb \qquad ...(9.14)$

(Since Z_2' and Z_1 are proportional to b).

Example 9.24. *A flitched timber beam consists of two joists 100 mm wide 300 mm deep with a steel plate 200 mm deep and 15 mm thick placed symmetrically between and clamped to them. Calculate the total moment to resistance of the section if the allowable stress in joist is 7.5 N/mm². Take $E_s = 20\ E_w$.*

Solution. Fig. 9.22 shows the arrangement and also the stress distribution in timber. From the stress diagram, the stress f_w in wood at the level of steel

$$= 7.5 \times \frac{100}{150} = 5 \text{ N/mm}^2.$$

Fig. 9.22

Since the strain in timber = strain in wood, we have

maximum stress in steel $= m \times 5 = 20 \times 5 = 100$ N/mm².

Now Z_w for wood $= \frac{1}{6} bd^2 = \frac{1}{6} \times 200 \times 300^2 = 3 \times 10^6$ mm³

Z_s for steel $= \frac{1}{6} bd^2 = \frac{1}{6} \times 15 \times 200^2 = 0.1 \times 10^6$ mm²

Moment of resistance of timber section

$$= f_w Z_w = 7.5 \times 3 \times 10^6 = 22.5 \times 10^6 \text{ N-mm}$$

Moment of resistance of steel section

$$= f_s Z_s = 100 \times 0.1 \times 10^6 = 10 \times 10^6 \text{ N-mm}$$

Total moment of resistance offered by the composite section

$$= (22.5 + 10)10^6 = 32.5 \times 10^6 \text{ N-mm} = \mathbf{32.5\ kN\text{-}m.}$$

Example 9.25. *A 300 × 300 mm timber is strengthened by the addition of 300 mm × 6.25 mm steel plates secured to its top and bottom surfaces. The composite beam is simply supported at its ends and carries a uniformly distributed load of 25 kN/m run over an effective span of 6 m. Find the maximum bending stress in the steel and timber at the mid-span.* $E_s = 2 \times 10^5$ *N/mm²* ; $E_t = 0.10 \times 10^5$ *N/mm²*.

Solution. $M_{max} = \dfrac{wL^2}{8} = \dfrac{25 \times (6)^2}{8}$

$= 112.5$ kN-m $= 112.5 \times 10^6$ N-mm

Since the section in composite,

Strain for steel = strain for timber, at the same distance from N.A.

$$e_s = e_t$$

or $$\frac{f_s}{E_s} = \frac{f_t}{E_t} \quad \text{or} \quad \frac{M_s \cdot y}{I_s E_s} = \frac{M_t \cdot y}{I_t E_t}$$

Hence $$\frac{M_t}{M_s} = \frac{I_t E_t}{I_s E_s}\ ; \quad \text{Also} \quad M_r = M_t + M_s$$

Now $$I_s = \frac{1}{12} \times 300(312.5^3 - 300^3) = 87.9 \times 10^6 \text{ mm}^4$$

$$I_t = \frac{1}{12} \times 300(300)^3 = 675 \times 10^6 \text{ mm}^4$$

Steel plate
6.25
300 mm
6.25
300 mm

Fig. 9.23

Hence $$\frac{M_t}{M_s} = \frac{675}{87.9} \times \frac{0.1}{2} = 0.384 \quad \text{and} \quad M_t + M_s = 112.5 \times 10^6$$

or $0.384\, M_s + M_s = 112.5 \times 10^6$ From which $M_s = 81.3 \times 10^6$ N-mm ; $M_t = 31.2 \times 10^6$ mm

$$\therefore \quad f_s = \frac{M_s}{Z_s} = \frac{81.3 \times 10^6}{87.9 \times 10^6} \times 156.25 = \mathbf{144.5\ N/mm^2}$$

$$f_t = \frac{M_t}{Z_t} = \frac{31.2 \times 10^6}{675 \times 10^6} \times 150 = \mathbf{6.93\ N/mm^2.}$$

Example 9.26. *Two rectangular bars, one of brass and the other of steel, each 40 mm × 10 mm are placed together to form a beam 4 cm wide and 20 mm deep on two supports 1 metre apart, the brass bar being on the top of steel. Determine the maximum central load which can be applied to the beam if the bars are (a) separate and can bend independently, (b) firmly secured to each other throughout their length. Given* $E_s = 2.05 \times 10^5$ *N/mm²* ; f_{max} *for brass* = 75 *N/mm²* ; f_{max} *for steel = 112.5 N/mm²* ; $E_b = 0.82 \times 10^5$ *N/mm²*.

Solution. (*a*) When the bars are separate, each one will bend about its own axis and the total moment of resistance will be equal to the sum of individual moments of resistance about their own axis. The radius at the junction will approximately be the same.

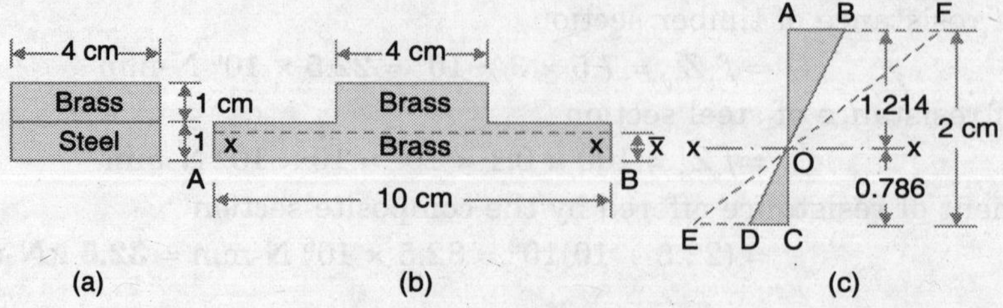

Fig. 9.24

Hence, we have $M_r = M_b + M_s$ for equilibrium ...(1)

and $R_b = R_s$, from compatibility

or $$\frac{M_b}{E_b I_b} = \frac{M_s}{E_s I_s} \quad ...(2)$$

Now, from (2), $\dfrac{M_s}{M_b} = \dfrac{E_s I_s}{E_b I_b} = \dfrac{2.05}{0.82} \times 1 = 2.5$, since $I_s = I_b$

Substituting in (1), we get

$$M_r = M_b + 2.5\, M_b = 3.5\, M_b \quad ...(3)$$

Now $M = \dfrac{W \times 1}{4}$ N-m (where W is in N) $= \dfrac{W}{4} \times 1000 = 250\, W$ N-mm

Hence $$M_b = \frac{M_r}{3.5} = \frac{M}{3.5} = \frac{250W}{3.5} = 71.43\, W \quad ...(4)$$

Again, $$\frac{f_b}{f_s} = \frac{M_b}{M_s} \cdot \frac{Z_s}{Z_b} = \frac{M_b}{M_s} = \frac{1}{2.5}$$

Taking $f_s = 112.5$ N/mm², $f_b = \dfrac{112.5}{2.5} = 45$ N/mm²

Hence when the stress in steel has reached 112.5, the stress in brass will be 45 N/mm². Should the stress in brass be allowed to go upto 75 N/mm², the corresponding stress in steel will be above the permissible unit. Hence $f_b = 45$ N/mm²

$$Z_b = \frac{1}{6} \times 40\,(10)^2 = 666.7 \text{ mm}^3\ ;\ M_b = f_b Z_b = 45 \times 666.7 = 30000 \text{ N-mm}$$

But $M_b = 71.43\, W$, from (4)

$\therefore$ $71.43\, W = 30000$ From which $W \simeq$ **420 N.**

(*b*) When the bars are composite they cannot bend about their own axis, but will bend about the common axis. In order to find the common axis, the whole section will have to be converted into an equivalent section (of brass or steel) as shown in Fig. 9.24 (*b*).

In order to convert the whole section into the equivalent of brass, the width at the top of brass will be = 4 cm (original width of brass) and the width at the bottom

$$= 4 \times \frac{E_s}{E_b} = 4 \times \frac{2.05}{0.85} = 10 \text{ cm}$$

The equivalent section of brass is shown in Fig. 9.24 (*b*).

In order to find the C.G. of the section, take the moment of the area about *AB*.

$\therefore$ $\{(4 \times 1) + (10 \times 1)\}\bar{x} = (4 \times 1 \times 1.5) + (10 \times 1 \times 0.5) = 6 + 5 = 11.$

From which $\bar{x} = 0.786$ cm

$\therefore$ I_{xx} of the equivalent section

$$= \frac{1}{3}\{(10 \times 0.786^3) + (4 \times 1.214^3) + (10 - 4)(0.214)^3\} = 4.02 \text{ cm}^4$$

$$= 4.02 \times 10^4 \text{ mm}^4$$

Now $f_{s,\,max} = 112.5$ N/mm²

Corresponding $f_b = 112.5/2.5 = 45$ N/mm², which is lesser than the allowable value in brass.

Hence in Fig. 9.24 (*c*), $EC = 112.5$ N/mm² ; $CD = 45$ N/mm²

$$AB = \frac{45 \times 1.214}{0.786} = 69.5, \text{ which is less than } 75 \text{ N/mm}^2$$

M_r for the transformed section

$$= f_b Z = \frac{69.5 \times 4.02 \times 10^4}{1.214 \times 10} = 230140 \text{ N-mm}$$

$$\left(\text{or} \quad M_r = \frac{45 \times 4.02 \times 10^4}{0.786 \times 10} = 230140 \text{ N-mm} \right)$$

But $\quad M = 250\,W$ N-mm (as before)

$\therefore \quad 250W = 230140 \quad$ or $\quad$ **W = 920 N.**

9.9. STRAIN ENERGY DUE TO BENDING

We know that strain energy per unit volume $= \dfrac{f^2}{2E}$.

Consider two transverse sections at a small distance δx apart and let δA be a small transverse area over which the direct stress f is acting at a height h from N.A.

We have $\quad f = \dfrac{M_x y}{I}$

Strain energy of the elementary piece of length δx and δA are

$$= \frac{f^2}{2E} \times \delta x \times \delta A = \frac{M_x{}^2 y^2}{2I^2 E} \; \delta x \,.\, \delta A$$

Strain energy of the whole piece of length δx between the transverse sections

$$= \delta U = \Sigma \frac{M_x{}^2 y^2}{2I^2 E} \; \delta x \,.\, \delta A = \frac{M_x{}^2 \delta x}{2I^2 E} \; \Sigma \delta A . y^2 = \frac{M_x{}^2 \delta x}{2I^2 E} \times I \; (\text{Since } \Sigma\delta A \,.\, y^2 = I).$$

$$\therefore \quad \delta U = \frac{M_x{}^2 \delta x}{2EI}$$

The above expression can also be derived, referring to Fig. 9.1

$$\delta U = \text{Work done by B.M.} = \text{average B.M.} \times \text{change in angle}$$

$$= \tfrac{1}{2} M_x \delta\theta = \tfrac{1}{2} M_x \frac{\delta x}{R} = \tfrac{1}{2} M_x \delta x \frac{M_x}{EI} = \frac{M_x{}^2 \delta x}{2EI}$$

Hence total strain energy of the beam is

$$U = \frac{1}{2EI} \int_0^L M_x{}^2 dx \qquad \text{...(9.15)}$$

where L is the span of the beam and I is constant.

(a) For a simply supported beam with central concentrated load

$$M_x = \frac{W}{2} x$$

$$\therefore \quad U = \frac{1}{2EI} \times 2 \int_0^{L/2} \left(\frac{W}{2} x \right)^2 dx = \frac{1}{EI} \frac{W^2 L^3}{96} = \frac{W^2 L^3}{96 EI} \qquad \text{...(9.16)}$$

(*b*) **For a cantilever with U.D.L. *w* :**

At any section at distance x from the free end,

$$M_x = \frac{wx^2}{2}$$

$$\therefore \quad U = \frac{1}{2EI}\int_0^L \frac{w^2x^4}{4}\,dx = \frac{w^2L^5}{2EI \times 20} = \frac{w^2L^5}{40EI} \quad ...(9.17)$$

(*c*) **Simply supported beam with constant *M* throughout**

$$M_x = M$$

$$\therefore \quad U = \frac{1}{2EI}\int_0^L M^2\,dx = \frac{M^2L}{2EI} \quad ...(9.18)$$

Example 9.27. *Compare the total strain energies of two equal uniform beams (simply supported at their ends), one carrying a concentrated central load and the other carrying a uniformly distributed load when (a) the total load on each beam is the same.*

(b) the maximum stress due to bending in each beam is the same.

Solution. For simply supported beam loaded with a central point load W_1

$$U_1 = \frac{W_1^2L^3}{96\,EI} \quad ...(1)$$

For a simply supported beam loaded with U.D.L., w per unit length,

$$U_2 = \int_0^L \frac{M^2}{2EI}\,dx = \frac{1}{2EI}\int_0^L \left(\frac{wLx}{2} - \frac{wx^2}{2}\right)^2 dx$$

$$= \frac{1}{2EI}\int_0^L \left[\frac{w}{2}(Lx - x^2)\right]^2 dx = \frac{w^2}{8EI}\int_0^L (L^2x^2 - 2Lx^3 + x^4)\,dx$$

$$= \frac{w^2L^5}{240\,EI} = \frac{W_2^2L^3}{240\,EI}, \text{ where } W_2 = wL \quad ...(2)$$

(*a*) If the total load for both the beams is the same,

$$W_1 = W_2 = W$$

$$\therefore \quad \frac{U_1}{U_2} = \frac{W^2L^3}{96\,EI} \times \frac{240\,EI}{W^2L^3} = \frac{240}{96} = \mathbf{2.5.}$$

(*b*) If $f_2 = f_1$, we must have $M_2 = M_1$

or $$\frac{W_2L}{8} = \frac{W_1L}{4} \quad \text{or} \quad W_2 = 2W_1$$

Hence $$\frac{U_1}{U_2} = \frac{W_1^2L^3}{96\,EI} \times \frac{240\,EI}{(2W_1)^2L^3} = \frac{240}{96 \times 4} = \mathbf{0.625.}$$

Example 9.28. *A flat ribbon of steel, 3 mm wide and 0.5 mm thick is wound round a cylinder 500 mm in diameter. Find the maximum stress in the steel and the energy stored per metre length of the ribbon. The modulus of elasticity E = 2 × 10⁵ N/mm².*

Solution.

$$f = \frac{E}{R}y = \frac{2 \times 10^5 \times 0.25}{250} = 200 \text{ N/mm}^2.$$

When the ribbon is wound round the drum, it is wound to a constant radius of curvature and hence, it will be subjected to constant bending moment M, where $M = EI/R$.

Hence $$U = \frac{M^2L}{2EI} = \left(\frac{EI}{R}\right)^2 \frac{L}{2EI} = \frac{EIL}{2R^2} = \frac{2 \times 10^5 \times 3(0.5)^3 \times \pi \times 500}{12 \times 2 \times 250 \times 250} = 78.5 \text{ mm-N}$$

$\therefore$ Strain energy stored per metre length $= \dfrac{78.5 \times 1000}{\pi \times 500} = $ **50 mm-N.**

9.10. FORCE ON A PARTIAL AREA OF A BEAM SECTION AND ITS MOMENT ABOUT N.A.

Sometimes, it is required to find the magnitude of force on a partial area of a beam section subjected to bending, and the moment of this force about the N.A. Fig. 9.25 shows a beam section subjected to bending moment, due to which maximum bending stress f_{max} is induced at the extreme fibre situated at y_{max} distance from the N.A. It is now required to find the force P acting on a partial area A shown shaded.

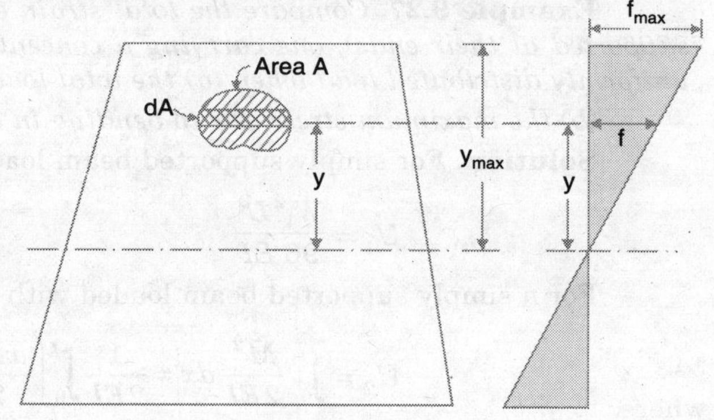

Fig. 9.25

Consider an elementary area dA of the shaded area, at distance y from N.A., where the stress f is given by

$$f = \frac{f_{max}}{y_{max}} . y$$

$\therefore$ Force δP on elementary area $= f\,dA = \dfrac{f_{max}}{y_{max}} . y\,dA$

$\therefore$ Total force P_A on shaded area $= \dfrac{f_{max}}{y_{max}} . \Sigma y\,dA$

$$\therefore \quad P_A = \frac{f_{max}}{y_{max}} . A\bar{y} \qquad \text{...(9.19)}$$

where $\bar{y}$ is the distance of centroid of the shaded area from the N.A.

Again, moment of elementary force δP about the about N.A. is

$$\delta M = \delta P . y = \frac{f_{max}}{y_{max}} . y^2 . dA$$

$\therefore$ Total moment of the force on the shaded area, about N.A. is

$$M_A = \frac{f_{max}}{y_{max}} . \Sigma y^2 . dA = \frac{f_{max}}{y_{max}} . I_A \qquad \text{...(9.20)}$$

where I_A is the moment of inertia of the shaded area about the neutral axis.

Example 9.29. *A beam having a solid rectangular cross-section with the dimensions shown in Fig. 9.26 (a) is subjected to a sagging B.M. of 80 kN-m acting around the horizontal axis. (a) Find the compressive force, and its moment about N.A., acting on the shaded area of the cross-section, (b) Find the tensile force, and its moment about N.A., acting of the cross-hatched area of the section.*

Solution.

$$I = \frac{1}{12} \times 120\,(180)^3$$
$$= 5832 \times 10^4 \text{ mm}^4$$

$$\therefore \quad f_{max} = \frac{M}{I}\, y_{max} = \frac{80 \times 10^6}{5832 \times 10^4} \times 90$$
$$= 123.46 \text{ N/mm}^2.$$

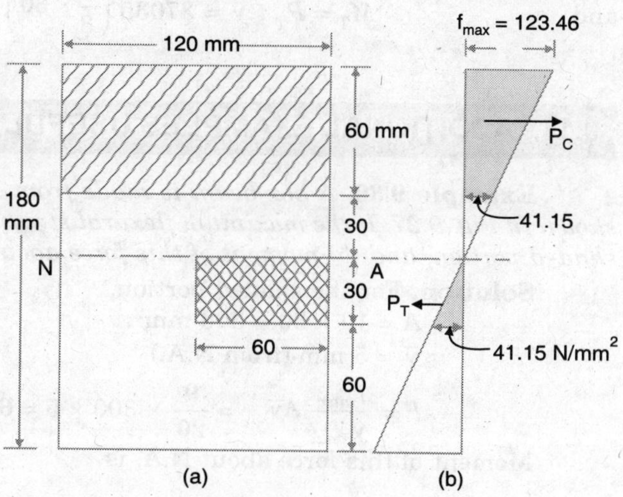

Fig. 9.26

(*a*) Compressive force on the shaded area :

$$P_C = \frac{f_{max}}{y_{max}} \,.\, A\bar{y}$$
$$= \frac{123.46}{90}\,(120 \times 60)(30 + 30)$$
$$= 592593 \text{ N} \simeq \mathbf{592.59\ kN}$$

Moment of this force about N.A. is

$$M_C = \frac{f_{max}}{y_{max}} \,.\, I_A$$

where I_A = moment of inertia of the shaded area about N.A.

$$= \frac{1}{12} \times 120(60)^3 + (120 \times 60)60^2 = 2808 \times 10^4 \text{ mm}^4$$

$$\therefore \quad M_C = \frac{123.46}{90} \times 2808 \times 10^4 = 38.519 \times 10^6 \text{ N-mm} = \mathbf{38.519\ kN\text{-}m.}$$

(*b*) Tensile force on the cross-hatched area :

$$P_T = \frac{f_{max}}{y_{max}} \,.\, A\bar{y} = \frac{123.46}{90}\,(30 \times 60)15 = 37038 \text{ N} = \mathbf{37.04\ kN}$$

$$I_A = \tfrac{1}{3} \times 60(30)^3 = 54 \times 10^4 \text{ mm}^4$$

$$\therefore \quad M_T = \frac{f_{max}}{y_{max}}\, I_A = \frac{123.46}{90} \times 54 \times 10^4 = 740700 \text{ N-mm} = \mathbf{0.741\ kN\text{-}m.}$$

Alternative Solution

The stress diagram is shown in Fig. 9.26 (*b*), with stresses at various levels marked.

(*a*) ***Compressive force on shaded area***

$$P_C = \text{average stress} \times \text{area} = \frac{123.46 + 41.15}{2}\,(120 \times 60)$$
$$= \mathbf{592593\ N.}$$

Centroid of this force from top fibre is

$$\bar{y} = \frac{60}{3}\left[\frac{2 \times 41.15 + 123.46}{41.15 + 123.46}\right] = 25 \text{ mm}$$

$$\therefore \quad M_C = P_C\,(90 - \bar{y}) = 592593\,(90 - 25) = 38.519 \times 10^6 \text{ N-mm} = \mathbf{38.519\ kN\text{-}m.}$$

(*b*) ***Tensile force on cross-hatched area***

$$P_T = \text{average stress} \times \text{area} = \frac{0 + 41.15}{2}(30 \times 60) = \mathbf{37035\ N}$$

and $$M_T = P_T \cdot \bar{y} = 37035\left(\frac{2}{3} \times 30\right) = 740700 \text{ N-mm} \simeq \mathbf{0.741\ kN\text{-}m.}$$

9.11. ADDITIONAL EXAMPLES USEFUL FOR COMPETITIVE EXAMINATIONS

Example 9.30. *A box beam is made from 10 mm × 30 mm pieces screwed together as shown in Fig. 9.27. If the maximum flexural stress is 9 N/mm², compute (a) force acting on the shaded portion, and (b) moment of this force about the N.A.*

Solution. For the shaded portion,

$$A = 10 \times 30 = 300 \text{ mm}^2.$$

$$\bar{y} = 5 \text{ mm (from N.A.)}$$

$$P = \frac{f_{max}}{y_{max}} A\bar{y} = \frac{9}{20} \times 300 \times 5 = \mathbf{675\ N}$$

Moment of this force about N.A. is

$$M = \frac{f_{max}}{y_{max}} \cdot I_A$$

where I_A = moment of inertia of shaded area about N.A.

$$= \frac{1}{12} \times 10(30)^3 + 10 \times 30(5)^2 = 30000 \text{ mm}^4$$

$$\therefore \quad M = \frac{9}{20} \times 30000 = \mathbf{13500\ N\text{-}mm.}$$

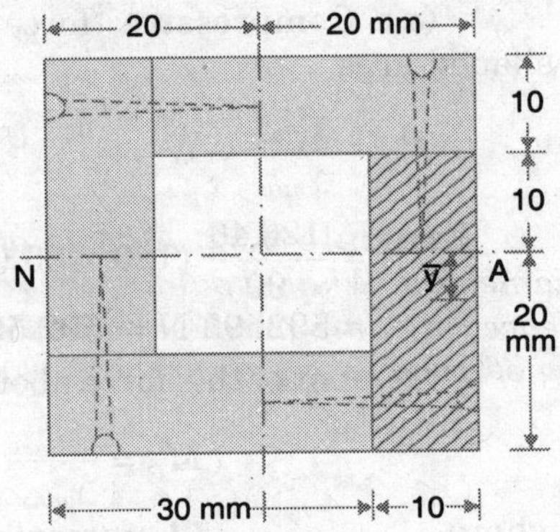

Fig. 9.27

Example 9.31. *A 150 mm × 300 mm rectangular section is subjected to a sagging B.M. of 400 kN-m around the 'strong axis'. The material of the beam is non-isotropic and is such that the modulus of elasticity in tension is 1.5 times as great as in compression. If the stresses do not exceed the elastic limit, find the maximum tensile and compressive stresses in the beam.*

Solution.

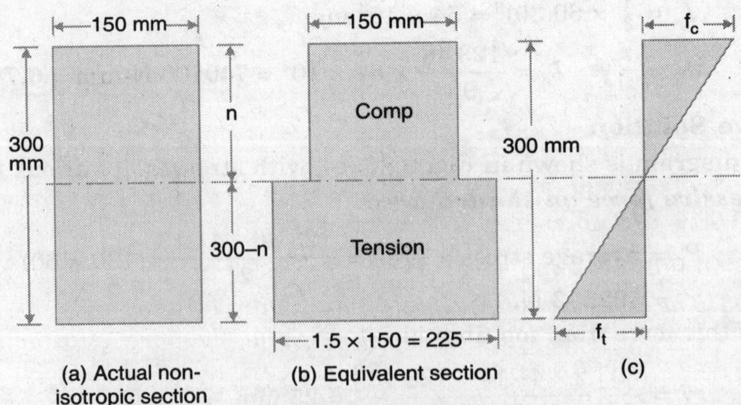

Fig. 9.28

Fig. 9.28 (a) shows the actual non-isotropic section, with the depth of N.A. n below the top compression fibre. Modular ratio = E_t/E_c = 1.5.

Fig. 9.28 (b) shows the equivalent isotropic section in which the width of tension portion = 1.5 × 150 = 225 mm.

Now $\sum M_{NA} = 0$

$$\therefore \quad 225(300 - n)\left(\frac{300 - n}{2}\right) = 150 \times n \times \frac{n}{2}$$

or $$(300 - n)^2 = 0.6667\, n^2 \quad \text{or} \quad 300 - n = 0.8165\, n \quad \text{or } n = 165.15 \text{ mm}$$

$$I_{NA} = \tfrac{1}{3} \times 150(165.15)^3 + \tfrac{1}{3} \times 225(300 - 165.15)^3 = 409.15 \times 10^6 \text{ mm}^4$$

$$f_c = \frac{M}{I} y_c = \frac{400 \times 10^6}{409.15 \times 10^6} \times 165.15 = \mathbf{161.5\ N/mm^2}$$

and $$f_t = \frac{M}{I} y_t = \frac{400 \times 10^6}{409.15 \times 10^6} (300 - 165.15) = \mathbf{131.8\ N/mm^2}$$

Example 9.32. *A beam of rectangular cross-section is to be cut from a circular log of diameter d. What should be the ratio of depth to width for maximum strength in pure bending ? Hence calculate the depth and width of the strongest beam that can be cut out of a cylindrical log of wood whose diameter is 600 mm.*

Solution. (Fig. 9.29). At the outer most fibre of rectangular section,

$$f_{max} = \frac{M}{bh^3/12} \times \frac{h}{2} = \frac{6M}{bh^2}$$

$$\therefore \quad M = f_{max} \frac{bh^2}{6}$$

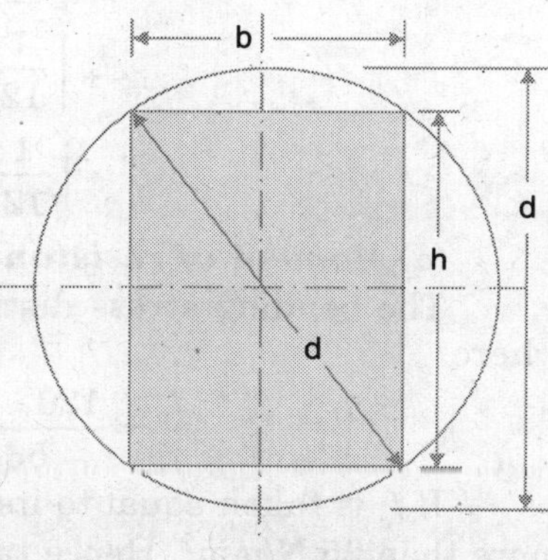

Fig. 9.29

For maximum strength, *i.e.*, maximum value of M, the product bh^2 must be made maximum.

Now $h^2 = d^2 - b^2$

$$\therefore \quad bh^2 = b(d^2 - b^2) = bd^2 - b^3$$

For max. value of bh^2, $\frac{d}{db}(bh^2) = 0$

$$\therefore \quad d^2 - 3b^2 = 0 \quad \text{or} \quad b^2 + h^2 - 3b^2 = 0$$

or $$h^2 = 2b^2 \quad \text{or} \quad \frac{h}{b} = \sqrt{2}. \textbf{ Ans.}$$

For given value of d = 600 mm, we have

$$b^2 + h^2 = d^2 = (600)^2 \quad \text{or} \quad b^2 + 2b^2 = 360000$$

From which b = **346.4 mm** and $h = \sqrt{2}\, b$ = **489.9 mm.**

Example 9.33. *The cross-section of a cast iron machine element used as a beam is shown in Fig. 9.30 (a). The beam resists bending moment about the horizontal neutral axis. The permissible stresses in tension and compression are 25 N/mm² and 30 N/mm² respectively. Determine the moment of resistance of the section about the horizontal N.A., for both positive and negative B.M.*

Solution.

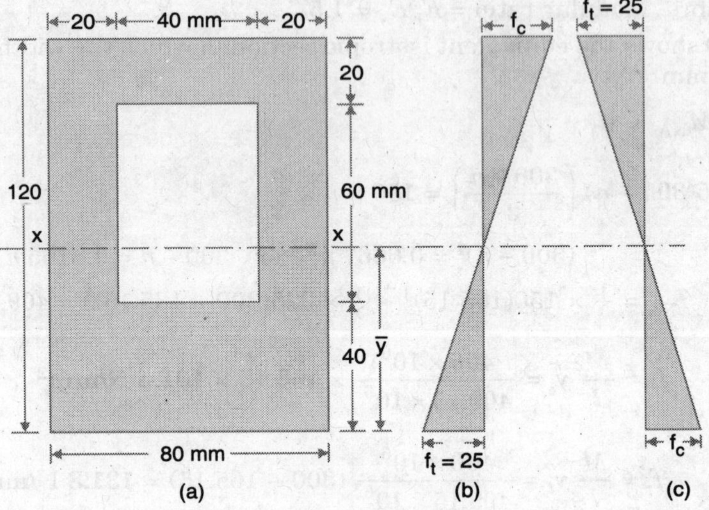

Fig. 9.30

$$A = (2 \times 20 \times 120) + (40 \times 20) + (40 \times 40) = 7200 \text{ mm}^2$$

$$\bar{y} = \frac{(2 \times 20 \times 120 \times 60) + (40 \times 20 \times 110) + (40 \times 40 \times 20)}{7200} = 56.67 \text{ mm}$$

$$I_{xx} = 2\left[\frac{1}{12} \times 20 \times 120^3 + 20 \times 120(60 - 56.67)^2\right]$$

$$+ \left[\frac{1}{12} \times 40 \times 20^3 + 40 \times 20(120 - 56.67 - 10)^2\right]$$

$$+ \left[\frac{1}{12} \times 40 \times 40^3 + 40 \times 40(56.67 - 20)^2\right] = 1048 \times 10^4 \text{ mm}^4.$$

(*a*) ***Moment of resistance for maximum sagging B.M.***

The bending stress distribution diagram for sagging B.M. is shown in Fig. 9.30 (*b*), where

$$\frac{f_c}{f_t} = \frac{120 - 56.67}{56.67} = 1.1175$$

If f_c is taken equal to its maximum value = 30 N/mm², f_t = 30/1.1175 = 26.84 which is more than 25 N/mm². Hence take f_t = 25 N/mm², for which f_c = 1.1175 × 25 = 27.94 N/mm².

$$\therefore \qquad M = \frac{f_t}{y_t} . I = \frac{25}{56.67} \times 1048 \times 10^4 = 4.623 \times 10^6 \text{ N-mm} = \mathbf{4.623 \text{ kN-m.}}$$

(*b*) ***Moment of resistance of maximum hogging B.M.***

The bending stress diagram for hogging B.M. is shown in Fig. 9.30 (*c*), from which

$$\frac{f_t}{f_c} = \frac{120 - 56.67}{56.67} = 1.1175.$$

If f_c is taken equal to its maximum value of 30 N/mm², f_t = 30 × 1.1175 = 33.53 N/mm² which is more than its permissible value of 25 N/mm². Hence take f_t = 25 N/mm², for which

$$f_c = \frac{25}{1.1175} = 22.37 \text{ N/mm}^2.$$

Now $$M = \frac{f_t}{y_t} . I = \frac{25}{120 - 56.67} \times 1048 \times 10^4 = 4.137 \times 10^6 \text{ N-mm} = \mathbf{4.137 \text{ kN-m.}}$$

Example 9.34. *A beam, simply supported over a span of 2 m is subjected to a central point load W. The beam section is symmetrical about y-axis, as shown in Fig. 9.31. Determine the magnitude of W if the maximum compressive and tensile stresses are not to exceed 100 N/mm² and 90 N/mm² respectively.*

Solution.

$$A = (\tfrac{1}{2} \times 120 \times 60) - (\tfrac{1}{2} \times 90 \times 40) = 1800 \text{ mm}^2$$

$$\bar{y} = \frac{\frac{1}{2} \times 120 \times 60 \times \frac{60}{3} - \frac{1}{2} \times 90 \times 40 \times \frac{40}{3}}{1800} = 26.67 \text{ mm}$$

$$I_{AB} = \frac{1}{12}\,[120(60)^3 - 90(40)^3] = 168 \times 10^4 \text{ mm}^4$$

$$\therefore \quad I_{xx} = I_{AB} - A\bar{y}^2 = 168 \times 10^4 - 1800(26.67)^2 = 39.97 \times 10^4 \text{ mm}^4.$$

Now, from the stress diagram,

$$\frac{f_c}{f_t} = \frac{60 - 26.67}{26.67} = 1.2497.$$

If $f_t = 90$, $f_c = 1.2497 \times 90 = 112.47$ N/mm², which more than the permissible value of 100 N/mm².

Again, if $f_c = 100$ N/mm², $f_t = 100/1.2497 = 80.02$ N/mm², which is less than permissible value of 90 N/mm².

Hence $f_c = 100$ N/mm² and $f_t = 80.02$ N/mm².

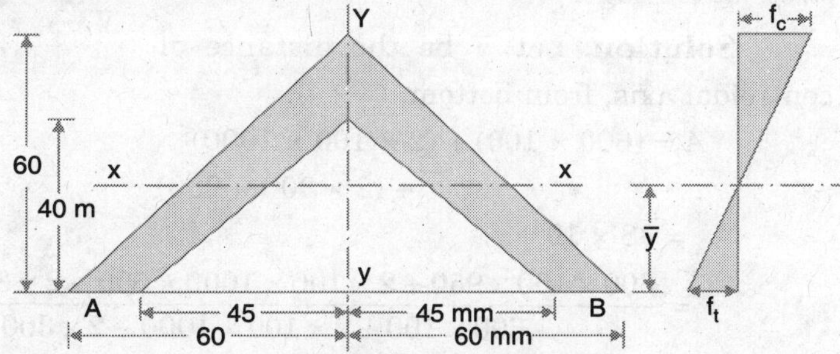

Fig. 9.31

$$\therefore \quad M = \frac{f_c}{y_c}\,.\,I = \frac{100}{60 - 26.67} \times 39.97 \times 10^4 = 1.199 \times 10^6 \text{ N-mm} = 1199 \text{ N-m}$$

Now, $\frac{WL}{4} = M = 1199$ N-m or $W = 1199 \times \frac{4}{2} =$ **2398 N.**

Example 9.35. *A prismatic beam of trapezoidal section, shown in Fig. 9.32 is subjected to sagging B.M. Calculate the ratio of b_1 to b_2 for maximum economy if the allowable stresses in tension and compression are 30 N/mm² and 40 N/mm² respectively.*

Solution. Let the depth of N.A. be n below the top fibre.

Now $\quad \frac{M}{I}\,.\,n = f_c = 40$ and $\frac{M}{I}\,(d - n) = f_t = 30$

$$\therefore \quad \frac{n}{d-n} = \frac{40}{30} = \frac{4}{3} \quad \text{or} \quad \frac{d-n}{n} = \frac{3}{4}$$

$$\therefore \quad \frac{d}{n} - 1 = \frac{3}{4} \quad \text{or} \quad n = \frac{4}{7}\,d \qquad \text{...}(i)$$

Again, from geometry,

$$n = \frac{d}{3}\left[\frac{b_1 + 2b_2}{b_1 + b_2}\right] \quad ...(ii)$$

$$\therefore \quad \frac{d}{3}\left[\frac{b_1 + 2b_2}{b_1 + b_2}\right] = \frac{4}{7}d \quad \text{or} \quad \frac{b_1 + 2b_2}{b_1 + b_2} = \frac{12}{7}$$

or $$\frac{b_2}{b_1 + b_2} = \frac{5}{7} \quad \text{or} \quad \frac{b_1 + b_2}{b_2} = \frac{7}{5}$$

$$\therefore \quad \frac{b_1}{b_2} = \frac{7}{5} - 1 = \mathbf{\frac{2}{5}}.$$

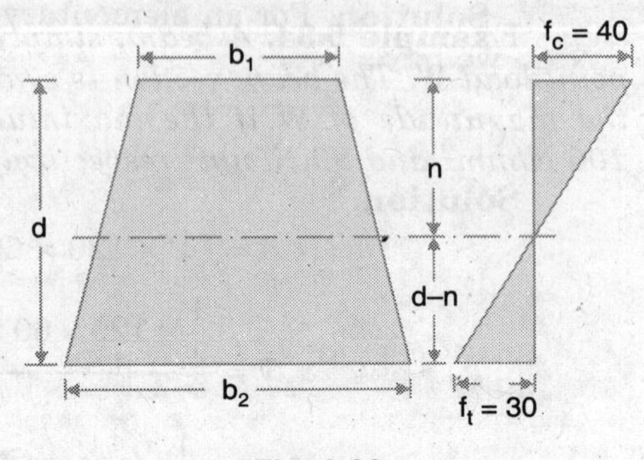

Fig. 9.32

Example 9.36. *Fig. 9.33 (a) shows a cast iron beam section which is subjected to a sagging B.M. causing tensile stress of 30 N/mm² at the bottom edge. Determine (a) magnitude of B.M., and (b) the compressive stress at top edge.*

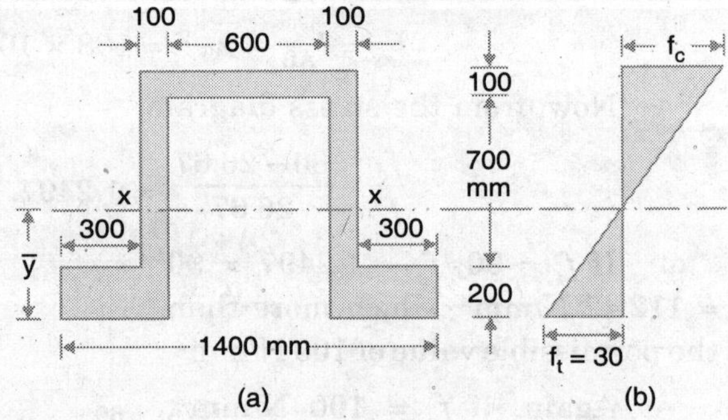

Fig. 9.33

Solution. Let $\overline{y}$ be the distance of centroidal axis, from bottom,

$$A = (600 \times 100) + (2 \times 100 \times 1000) + (2 \times 300 \times 200)$$

$$= 38 \times 10^4$$

$$\overline{y} = \frac{600 \times 100 \times 950 + 2 \times 100 \times 1000 \times 500 + 2 \times 300 \times 200 \times 100}{600 \times 100 + 2 \times 100 \times 1000 + 2 \times 300 \times 200} = 444.74 \text{ mm}$$

$$I_{xx} = \left[\frac{1}{12} \times 600(100)^3 + 600 \times 100(1000 - 444.74 - 50)^2\right]$$

$$+ 2\left[\frac{1}{12} \times 100 \times 1000^3 + 100 \times 1000(500 - 444.74)^2\right]$$

$$+ 2\left[\frac{1}{12} \times 300(200)^3 + 300 \times 200(444.74 - 100)^2\right] = 4.731 \times 10^{10} \text{ mm}^4$$

$$\therefore \quad M = \frac{f}{y}I = \frac{30}{444.74} \times 4.731 \times 10^{10} = 3.19 \times 10^9 \text{ N-mm} = \mathbf{3190 \text{ kN-m}}.$$

Again, stress at the top edge is given by

$$\frac{f_c}{1000 - 444.74} = \frac{30}{444.74} \quad \text{or} \quad f_c = \frac{30}{444.74}(1000 - 444.74) = \mathbf{37.46 \text{ N/mm}^2}.$$

Example 9.37 *Show that the maximum bending stress in a beam of rectangular section of depth d is given by*

$$f_{max} = \frac{M.d}{6I.n}(2n + 1),$$

if instead of Hooke's law the stress strain relation is $f^n = E.\,\varepsilon$*, where n is a number dependent on the properties of the material.*

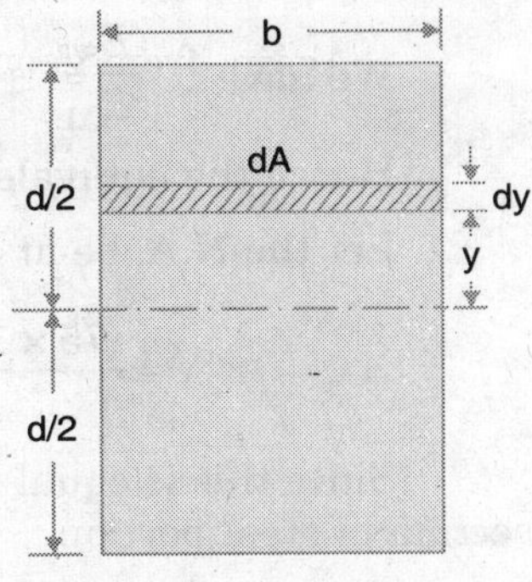

Fig. 9.34

Solution. For an elementary area dA, located at y from N.A., we have

$$dM = f_y \, . \, dA \, . \, y = (E\varepsilon)^{1/n} \, . \, y \, . \, bdy.$$

$$\therefore \quad M = 2\int_0^{d/2} (E\varepsilon)^{1/n} \, by \, dy$$

But $$\varepsilon = \frac{\varepsilon_{max}}{d/2} \, . \, y = \frac{f_{max}^n}{E} \, . \, \frac{2y}{d}$$

$$\therefore \quad M = 2\int_0^{d/2} \left(f_{max}^n \, . \, \frac{2y}{d} \right)^{1/n} \times by \, dy$$

or $$M = 2\left(\frac{d}{2}\right)^{-1/n} . \, b \, f_{max} \left[\frac{y^{2+\frac{1}{n}}}{2+\frac{1}{n}} \right]_0^{d/2}$$

$$= f_{max} \, . \, b \, . \, 2\left(\frac{d}{2}\right)^2 \frac{n}{2n+1}$$

or $$f_{max} = \frac{2n+1}{n} \, . \, \frac{M}{2b\left(\frac{d}{2}\right)^2} = \frac{2n+1}{n} \, . \, \frac{M}{\frac{bd^2}{2}}$$

But $$I = \frac{1}{12} bd^3 \quad \text{or} \quad \frac{bd^2}{2} = \frac{6I}{d}$$

$$\therefore \quad f_{max} = \frac{2n+1}{n} \, . \, \frac{M}{6I/d} = \frac{M \, . \, d}{6In} (2n+1).$$

Example 9.38. *A composite beam having the cross-sectional dimensions shown in Fig. 9.35(a) is subjected to a bending moment of 400 kN-m. Materials are fastened so that the beams act as a unit. Determine the maximum bending stress in each material. Take* $E_{ST} = 2 \times 10^5 \; N/mm^2$, $E_{AL} = 0.667 \times 10^5 \; N/mm^2$ *and* $E_{CI} = 1 \times 10^5 \; N/mm^2$.

Solution.

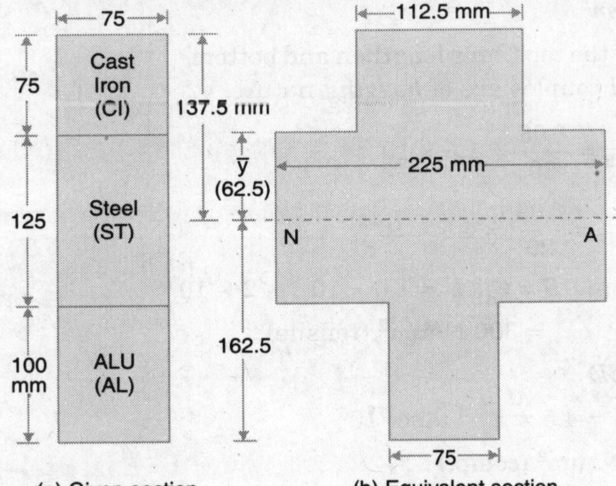

Fig. 9.35

We have $\dfrac{E_{ST}}{E_{AL}} = 3$ and $\dfrac{E_{CI}}{E_{AL}} = 1.5$

Hence the equivalent section, in terms of aluminium will be as shown in Fig. 9.35 (*b*).

Let the N.A. be at a distance of $\bar{y}$ from the junction of cast iron and steel sections.

$$\therefore \quad \bar{y} = \frac{(75 \times 100 \times 175) + (225 \times 125 \times 125/2) - (112.5 \times 75 \times 75/2)}{(75 \times 100) + (225 \times 125) + (112.5 \times 75)} = 62.5 \text{ mm.}$$

Since this is equal to half the depth of steel section, the N.A. passes through mid-depth section of steel portion.

$$\therefore \quad I = \left[\frac{112.5 \times 75^3}{12} + 112.5 \times 75\left(\frac{75}{2} + 62.5\right)^2\right] + \left[\frac{1}{12} \times 225\,(125)^3\right]$$

$$+ \left[\frac{1}{12} \times 75(100)^3 + 75 \times 100(50 + 62.5)^2\right] = 2.261 \times 10^8 \text{ mm}^4$$

Now $$f_{AL} = \frac{M}{I}(y_{max})_{AL} = \frac{400 \times 10^6 \times 162.5}{2.261 \times 10^8} = \mathbf{287.5\ N/mm^2}$$

$$f_{CI} = 1.5\,\frac{M}{I}\,(y_{max})_{CI} = \frac{1.5 \times 400 \times 10^6}{2.261 \times 10^8} \times 137.5 = \mathbf{364.9\ N/mm^2}$$

$$f_{ST} = 3\,\frac{M}{I}\,(y_{max})_{ST} = \frac{3 \times 400 \times 10^6}{2.261 \times 10^8} \times 62.5 = \mathbf{331.7\ N/mm^2}.$$

Example 9.39. *In a laboratory test of a beam loaded by end couples, the fibres at layer AB (Fig. 9.36) are found to lengthen by 0.03 mm and the fibres at CD shorten by 0.09 mm in the 20 mm gauge length. Locate the neutral axis and determine the flexural stresses at top and bottom fibres. What is the nature of end couples ? Take E = 2 × 10⁵ N/mm².*

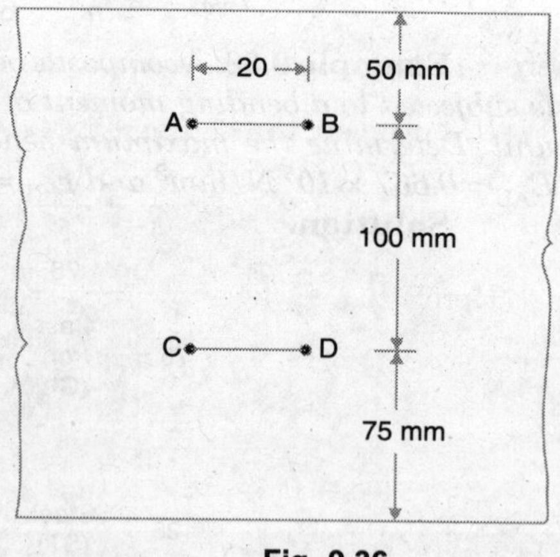

Fig. 9.36

Solution. Since the top fibres lengthen and bottom fibres shorten, the end couples are of *hogging* nature.

For fibre *AB*, $\varepsilon_{AB} = \dfrac{0.03}{20} = 1.5 \times 10^{-3}$

For fibre *CD*, $\varepsilon_{CD} = \dfrac{0.09}{20} = 4.5 \times 10^{-3}$

$\therefore$ Stress in fibre $AB = \varepsilon_{AB}.E = 1.5 \times 10^{-3} \times 2 \times 10^5$

$= 300$ N/mm² (tensile)

Stress in fibre *CD*

$= \varepsilon_{CD}.E = 4.5 \times 10^{-3} \times 2 \times 10^5$

$= 900$ N/mm² (comp.)

The stress distribution diagram is shown in Fig. 9.37.

Let the depth of N.A. be n mm from top fibre.

From geometry, $\dfrac{300+900}{100} = \dfrac{300}{n-50}$

From which n = **75 mm.**

$\therefore$ Stress at top fibre $= f_T = \dfrac{1200}{100} \times 75$

= **900 N/mm²** (tensile)

Stress at bottom fibre $= f_B = \dfrac{1200}{100}$ (225 – 75)

= **1800 N/mm²** (comp.)

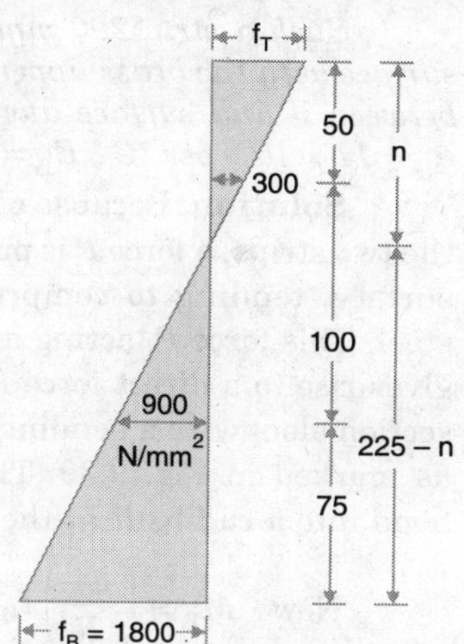

Fig. 9.37

Example 9.40. *A simple beam weighs W and supports a concentrated load P at its mid-span. If the error in calculating the bending stress is 5% when the weight W is neglected, what is the ratio P/W ?*

Solution. If the self weight W of the beam is neglected,

$$M = \frac{PL}{4}$$

$$\therefore \quad f_1 = \frac{M}{I} y_{max} = \frac{PL}{4} \cdot \frac{y_{max}}{I} = kP \quad ...(1)$$

where $k = \dfrac{L\, y_{max}}{4I}$

P
W
L

Fig. 9.38

If, however, the self weight is considered,

$$f_2 = \left(\frac{PL}{4} + \frac{WL}{8}\right)\frac{y_{max}}{I} = \left(P + \frac{W}{2}\right)\frac{L\, y_{max}}{4I} = k\left(P + \frac{W}{2}\right) \quad ...(2)$$

Hence % error $= \dfrac{f_2 - f_1}{f_2} \times 100$

$$\therefore \quad \frac{k\left(P + \frac{W}{2}\right) - kP}{\left(P + \frac{W}{2}\right)k} \times 100 = 5 \quad \text{or} \quad \frac{\frac{W}{2}}{P + \frac{W}{2}} = \frac{1}{20} \quad \text{or} \quad \frac{W}{2P + W} = \frac{1}{20}$$

or $\dfrac{2P + W}{W} = 20$ or $\dfrac{2P}{W} = 19$ or $\dfrac{P}{W}$ = **9.5.**

Example 9.41. *A straight bimetallic strip of brass of rectangular section of width b and thickness t is joined along its length by a strip of steel of the same dimensions, thus forming a composite bar of width b and thickness 2t. If the bar is uniformly heated and is quite free to bend, show that it bend to a radius.*

$$R = \frac{E_B{}^2 + E_S{}^2 + 14E_B E_S}{12E_B E_S(a_B - a_S)} \cdot \frac{t}{T}$$

where α_B and α_S are the coefficients of linear expansion and T is the rise in temperature.

Such a strip 200 mm long with the steel and brass each 1.5 mm thick rests on a level surface with the brass upper most. If the strip is initially straight, find the maximum clearance between it and surface due to a rise in temperature of 100°C. Take $\alpha_B = 19 \times 10^{-6}$ *per °C,* $\alpha_S = 11 \times 10^{-6}$ *per °C ;* $E_B = 95000$ *N/mm²* *and* $E_S = 205000$ *N/mm².* (U.L.)

Solution. Because of unequal expansion of the two strips, a force P is produced at the common surface, tending to compress brass and extend steel. This force P, acting at the common surface, gives rise to a direct force P at the centre of each section alongwith a bending moment in each strip, as marked in Fig. 9.39. The combined unit will bend into a radius R at the common surface.

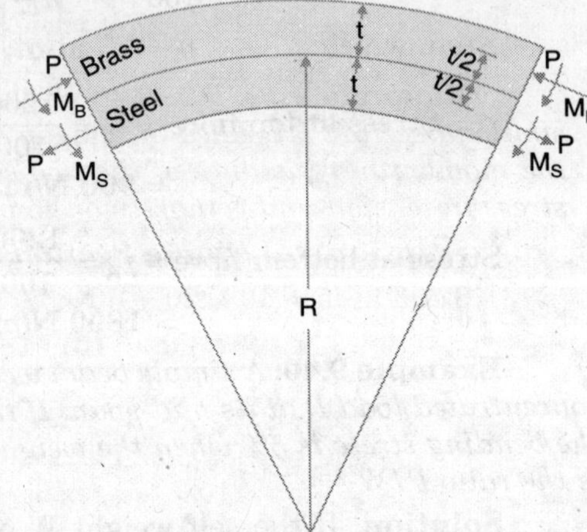

Fig. 9.39

Now $$M_B = \left(\frac{E_B}{R}\right) I_B = \frac{bt^3}{12} \cdot \frac{E_B}{R} \quad ...(i)$$

and $$M_S = \left(\frac{E_S}{R}\right) I_S = \frac{bt^3}{12} \cdot \frac{E_S}{R} \quad ...(ii)$$

Also, for equilibrium of the section,

$$M_B + M_S = P.\,t$$

Hence $$P.t = \frac{bt^3}{12R}\,(E_B + E_S) \quad ...(a)$$

The difference in linear strains at the central axis of each strip.

$$= \frac{t/2}{R} - \left(-\frac{t/2}{R}\right) = \frac{t}{R} \quad \text{(See Eq. 9.1)}$$

Hence allowing for force P and temperature, the compatibility equation is

$$\frac{t}{R} = \left(\alpha_B T - \frac{P}{bt\,E_B}\right) - \left(\alpha_S T + \frac{P}{bt\,E_S}\right)$$

or $$(\alpha_B - \alpha_S)\,T = \frac{P}{bt}\left(\frac{1}{E_B} + \frac{1}{E_S}\right) + \frac{t}{R} \quad ...(b)$$

Substituting the value of $\frac{P}{bt}$ from (a), we get

$$(\alpha_B - \alpha_S)\,T = \frac{t}{12R} \cdot \frac{(E_B + E_S)^2}{E_B E_S} + \frac{t}{R} = \frac{t}{12R} \times \frac{E_B{}^2 + E_S{}^2 + 14E_B E_S}{E_B E_S}$$

$$\therefore \quad R = \frac{E_B{}^2 + E_S{}^2 + 14E_B E_S}{12E_B E_S\,(\alpha_B - \alpha_S)} \cdot \frac{t}{T} \quad \textbf{(Proved)}$$

Substituting the numerical values,

$$R = \frac{95000^2 + 205000^2 + 14 \times 95000 \times 205000}{12 \times 95000 \times 205000\,(8 \times 10^{-6})} \times \frac{1.5}{100} = 2740 \text{ mm}$$

The clearance h is given by the property of a circle

$$(2R-h)\,h = \frac{L}{2}\times\frac{L}{2}.\ \text{Taking}\quad (2R-h) \simeq 2R,$$

$$2Rh = \left(\frac{L}{2}\right)^2 = (100)^2 \quad \therefore \quad h = \frac{(100)^2}{2R} = \frac{10000}{2\times 2740} = \mathbf{1.83\ mm.}$$

Example 9.42. *Fig. 9.40 shows a rolled steel beam of an unsymmetrical I-section. If a similar I-section is welded on the top of it to form a symmetrical section, determine the ratio of the moment of resistance of the two sections to that of the single section. Assume permissible stress in tension and compression to be the same.*

Solution. (*a*) *Single Section :* (Fig. 9.40)

Total area = 40 × 20 + 80 × 20 + 80 × 20 = 4000 mm²

$$\bar{y} = \frac{(40\times 20\times 110)+(80\times 20\times 60)+(80\times 20\times 10)}{4000} = 50 \text{ mm}$$

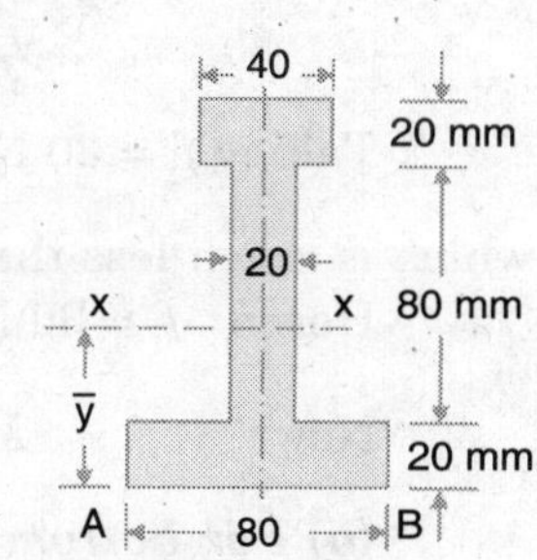

Fig. 9.40

$$I_{AB} = \left(\frac{1}{3}\times 60\times 20^3\right)+\left(\frac{1}{3}\times 20\times 100^3\right)$$

$$+\left\{\frac{1}{12}\times 40\times 20^3 + 40\times 20\,(110)^2\right\}$$

$$= 1653 \times 10^4 \text{ mm}^4$$

$$\therefore \quad I_{xx} = 1653\times 10^4 - 4000\,(50)^2 = 653\times 10^4 \text{ mm}^4$$

$$y_t = \bar{y} = 50 \text{ mm}\ ;\ y_c = 120-50 = 70 \text{ mm}$$

Now $M_r = f.I/y$, in general. Using the larger value of y to get minimum moment of resistance, we have

$$M_{r1} = \frac{f}{y_c}I = \frac{f}{70}\times 653\times 10^4 = 93286\,f \quad ...(1)$$

(*b*) *Double Section* (Fig. 9.41)

The centroid of the section will be at the centre of the figure.

$$\therefore \quad I = 2\,[653\times 10^4 + 4000\,(70)^2] = 5226\times 10^4 \text{ mm}^4$$

$$y = 120 \text{ mm}$$

$$\therefore \quad M_{r2} = \frac{f}{y}\times I = \frac{f}{120}\times 5226\times 10^4 = 435500\,f$$

$$\therefore \quad \frac{M_{r2}}{M_{r1}} = \frac{435500}{93286} = \mathbf{4.668}.$$

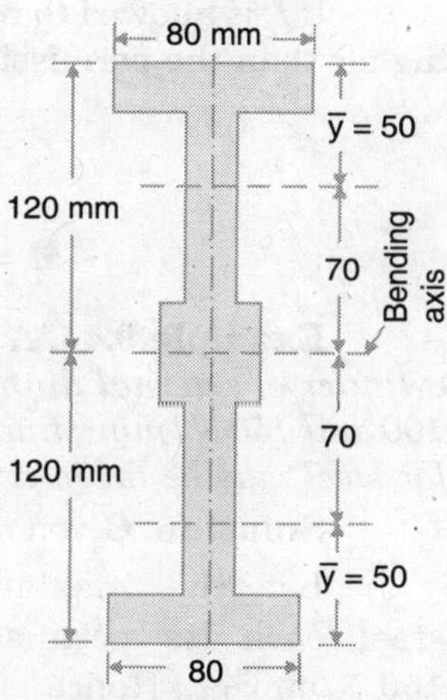

Fig. 9.41

Example 9.43. *The cross-section of a conveyer beam is shown in Fig. 9.42. The beam is subjected to a B.M. in the vertical plane yy. Determine the maximum permissible bending moment (a) for the bottom flange to be in tension (b) for the bottom flange in compression. The safe bending stress in tension and compression are 30 N/mm² and 120 N/mm² respectively.*

Solution.

$A = 40 \times 80 + 80 \times 160 + 20 \times 120$

$= 3200 + 12800 + 2400 = 18400 \text{ mm}^2$

$$\bar{y} = \frac{3200 \times 20 + 12800 \times 80 + 2400 \times 180}{18400} = 82.60 \text{ mm}$$

$\therefore \quad y_1 = \bar{y} = 82.60 \text{ mm} \; ; \; y_2 = 240 - 82.6 = 157.4 \text{ mm}$

$$I_{AB} = \frac{1}{3} \times 80\,(40)^3 + \left\{\frac{1}{12} \times 160 \times 80^3 + 12800\,(80)^2\right\} + \left\{\frac{1}{12} \times 20\,(120)^3 + 2400\,(180)^2\right\}$$

$= 17109 \times 10^4 \text{ mm}^4$

Fig. 9.42

$\therefore \quad I_{xx} = 17109 \times 10^4 - 18400\,(82.60)^2 = 4555 \times 10^4 \text{ mm}^4$

(a) For bottom face to be in tension

$$y_t = y_1 = 82.60 \text{ mm} \; ; \; y_c = y_2 = 157.4 \text{ mm}$$

Taking $f_t = 30 \text{ N/mm}^2$, corresponding $f_c = \frac{f_t}{y_t} \,.\, y_c = \frac{30}{82.6} \times 157.4 = 57.17 \text{ N/mm}^2$ which is much less than the permissible value of 120 N/mm².

Hence $f_t = 30 \text{ N/mm}^2$ is the governing stress.

Now $\quad M = f_t \,.\, \frac{I}{y_t} = 30 \times \frac{4555 \times 10^4}{82.6} = 16.55 \times 10^6$ N-mm = **16.55 kN-m**

(b) For bottom face to be in compression

Here $\quad y_c = y_1 = 82.60 \text{ mm} \; ; \; y_t = y_2 = 157.4 \text{ mm}$

If f_c is allowed to reach the permissible value of 120 N/mm², corresponding f_t will be much higher than the permissible value of 30 N/mm². Hence max. f_t should be taken as 30 N/mm².

$\therefore \quad f_c = \frac{f_t}{y_t} \,.\, y_c = \frac{30}{157.4} \times 82.6 = 15.74 \text{ N/mm}^2$ only

$\therefore \quad M = f_t \,.\, \frac{I}{y_t} = 30 \times \frac{4555 \times 10^4}{157.4} = 8.68 \times 10^6$ N-m = **8.68 kN-m.**

Example 9.44. *A solid rod of bronze 20 mm in diameter is surrounded by a fitting steel cylinder of external diameter 28 mm. If the permissible bending stress in bronze and steel are 100 and 150 N/mm², find the moment of resistance of the composite section. The Young's modulus for steel may be taken as 1.75 times that of bronze.*

Solution. Given $m = E_s/E_b = 1.75$

Let the maximum stress in steel reach its permissible value of 150 N/mm². Hence corresponding maximum stress reached in bronze.

$$= 150 \times \frac{10}{14} \times \frac{1}{m} = \frac{150 \times 10}{14 \times 1.75}$$

$= 61.22 \text{ N/mm}^2$, which is less than the permissible value of 100 N/mm².

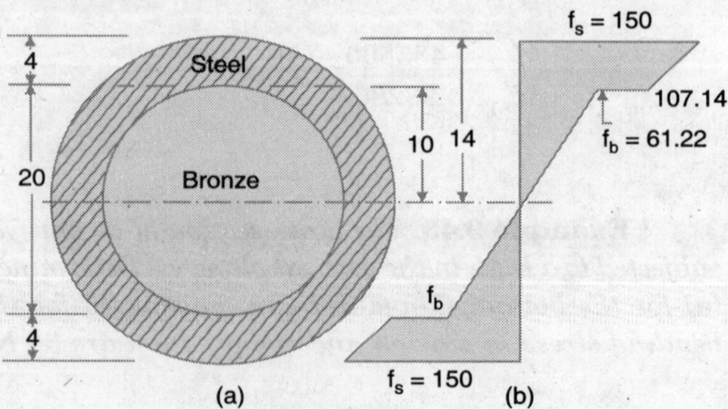

Fig. 9.43

For bronze component,

$$Z_b = \frac{\pi}{32}\,(20)^3 = 785.4 \text{ mm}^3$$

For steel component, $Z_s = \dfrac{\pi}{32}\dfrac{(28^4 - 20^4)}{28} = 1594.1 \text{ mm}^3$

$$\therefore \quad M_{rb} = f_b\,Z_b = 61.22 \times 785.4 = 48082 \text{ N-mm}$$

$$M_{rs} = f_s Z_s = 150 \times 1594.1 = 239120 \text{ N-mm}$$

$\therefore$ Total $M_r = 48082 + 239120 = 287202$ N-m $\approx$ **0.287 kN-m.**

Example 9.45. *Two planks of thicknesses t_1 and t_2 rest just one above another forming a beam as shown in Fig. 9.44, and support a uniformly distributed load of w Newton per metre run. Find the ratio of the maximum stress in the two beams.*

Solution. Since the two planks are not rigidly jointed, they will bend separately about their individual bending axes. However, since the thicknesses are small, one may assume that the radii of curvature of neutral layer of each plank is the same.

Fig. 9.44

Let maximum B.M. for the first layer of thickness $t_1 = M_1$

Let maximum B.M. for the second layer of thickness $t_2 = M_2$

Radius of curvature for each plank $= R$

$$\therefore \quad \frac{1}{R} = \frac{M_1}{EI_1} = \frac{M_2}{EI_2} \text{ or } \frac{M_1}{M_2} = \frac{I_1}{I_2} = \frac{bt_1^3/12}{bt_2^3/12} = \frac{t_1^3}{t_2^3} \quad \text{...(1)}$$

If f_1 and f_2 are the maximum bending stresses in the two planks at mid span,

$$M_1 = f_1 Z_1 = f_1 \cdot \frac{1}{6}\, b\, t_1^{\,2} \text{ and } M_2 = f_2 Z_2 = f_2 \cdot \frac{1}{6}\, b\, t_2^{\,2}$$

$$\therefore \quad \frac{M_1}{M_2} = \frac{f_1}{f_2} \cdot \frac{t_1^2}{t_2^2} \quad \text{...(2)}$$

Hence from (1) and (2), we get $\dfrac{t_1^3}{t_2^3} = \dfrac{f_1}{f_2}\dfrac{t_1^2}{t_2^2}$. From which $\boldsymbol{\dfrac{f_1}{f_2} = \dfrac{t_1}{t_2}}$

Example 9.46. *Find the condition at which any further decrease of depth H of the beam section shown in Fig. 9.45 will result in an increase in section modulus.*

Solution.

$$I_{xx} = \frac{1}{12}\, b\, d^3 + \frac{1}{12}\, a\, H^3$$

$$\therefore \quad Z = \frac{I_{xx}}{H/2} = \frac{2I_{xx}}{H} = \frac{bd^3}{6H} + \frac{aH^2}{6}$$

Rate of change of Z with respect to H is given by

$$\frac{dZ}{dH} = -\frac{bd^3}{6H^2} + \frac{aH}{3}$$

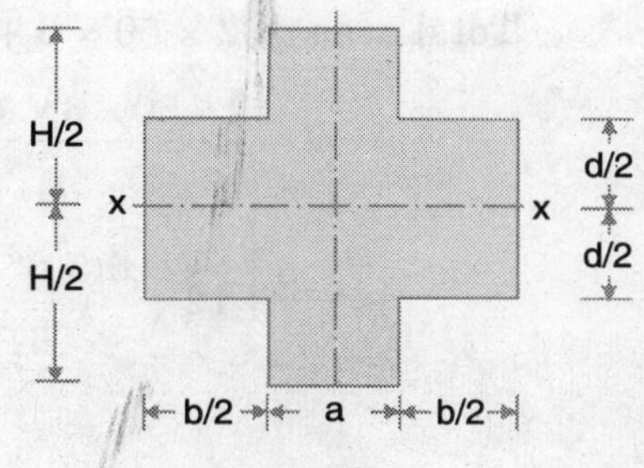

Fig. 9.45

The condition for Z to increase when H decreases is

$$\frac{bd^3}{6H^2} > \frac{aH}{3} \text{ or } \frac{\mathbf{b}}{\mathbf{2a}} > \frac{\mathbf{H^3}}{\mathbf{d^3}}$$

Example 9.47. *A uniformly tapering vertical post of height H having diameter D at the base and a diameter d at the top is fixed at the base. A horizontal force P is applied at the top of the post. Determine the maximum bending stresses for the post and state where it occurs.*

Solution. Let us take a horizontal section at a distance x from the top. Its diameter d_x is given by

$$d_x = d + \frac{D-d}{H}x = d + kx, \text{ where } k = \frac{D-d}{H}$$

$$\therefore \quad Z = \frac{\pi d_x^{\ 3}}{32} \text{ and } M = P \,.\, x$$

$$\text{Stress } f = \frac{M}{Z} = \frac{P\,.\,x}{\pi d_x^3/32} = \frac{32P}{\pi}\cdot\frac{x}{d_x^3} = \frac{32P}{\pi}\cdot\frac{x}{(d+kx)^3}$$

$$\text{For } f_{max},\ \frac{df}{dx} = 0 = \frac{32P}{\pi}\,\frac{(d+kx)^3 - x(3)(d+kx)^2\,.\,k}{(d+kx)^3}$$

or $\quad (d + kx)^2(d + kx - 3kx) = 0 \quad$ or $\quad d - 2kx = 0$

$$\therefore \quad x = \frac{d}{2k} = \frac{d}{2(D-d)}H,$$

which gives the location of maximum stress.

$$\text{Again,} \quad f_{max} = \frac{32P}{\pi}\cdot\frac{d/2k}{\left(d+\dfrac{d}{2}\right)^3} = \frac{32P}{\pi}\times\frac{8}{27}\cdot\frac{d}{2kd^3}$$

$$= \frac{32\times 8P}{27\times 2\pi}\cdot\frac{1}{d^2}\cdot\frac{H}{D-d} = \mathbf{\frac{128}{27\,\pi}\,\frac{PH}{(D-d)d^2}} \quad \textbf{(Answer)}$$

Fig. 9.46

Example 9.48. *A composite beam is made by bolting on a steel channel 50 mm × 150 mm × 6 mm to a 150 mm × 150 mm timber beam as shown in Fig. 9.47 (a). The composite beam is freely supported over a span of 4 m. Find the maximum uniformly distributed load that the beam can carry safely, if the allowable bending stresses in timber and steel are 8 N/mm² and 160 N/mm² respectively. Take E for steel = 2 × 10⁵ N/mm² and E for timber = 0.1 × 10⁵ N/mm².*

Solution. Modular ratio $\quad m = 2/0.1 = 20$

$\therefore$ Width of steel component equivalent to timber section = 150/20 = 7.5 mm

The equivalent section is shown in Fig. 9.47 (*b*).

Total area $= 2 \times 50 \times 6 + 138 \times 6 + 150 \times 7.5 = 2553 \text{ mm}^2$

$$\bar{y} = \frac{2\times 50\times 6\times 175 + 138\times 6\times 153 + 150\times 7.5\times 75}{2553} = 123.8 \text{ mm}$$

$$I_{xx} = \left[\frac{2}{12}\times 6(50)^3 + 2\times 50\times 6\,(26.2+25)^2\right] + \left[\frac{1}{12}\times 138(6)^3 + 138\times 6(26.2+3)^2\right]$$

$$+ \left[\frac{1}{12}\times 7.5\,(150)^3 + 150\times 7.5\,(123.8-75)^2\right] = 719.5\times 10^4 \text{ mm}^4$$

Let us permit the stress in steel to reach a value of 160 N/mm². Hence stress in steel at 76.2 mm from N.A.

$$= 160 \text{ N/mm}^2$$

∴ Stress in timber at 123.8 mm from N.A.

$$= \frac{123.8}{76.2} \times \frac{160}{20}$$

$$= 13 \text{ N/mm}^2$$

This is more than the permissible value of 8 N/mm² in timber.

Hence maximum stress in timber = 8 N/mm².

Corresponding stress in steel

$$= \frac{76.2}{123.8} \times 8 \times 20$$

$$= 98.48 \text{ N/mm}^2$$

Fig. 9.47

$$\therefore \quad M_{max} = \frac{wL^2}{8} = \frac{w(4)^2}{8} = 2w = 2w \times 1000 \text{ N-mm}$$

Now $\frac{M}{I} = \frac{f}{y}$ or $\frac{2000\,w}{719.5 \times 10^4} = \frac{98.48}{76.2}$

From which $w = 4649$ N/m = 4.649 kN/m.

PROBLEMS

1. A beam of symmetrical section 30 cm deep and I = 12000 cm⁴, carries U.D.L. of 16 kN/m. Calculate the maximum span of the beam if the maximum bending stress is not to exceed 160 N/mm². With this span, calculate the maximum central load if the bending stress is not to exceed the limit given above.
2. A timber joist of 4 metre span has to carry a total load of 80 kN uniformly distributed over the length. Calculate the dimensions of the joist if the maximum permissible stress is limited to 8 N/mm². The ratio of depth to width is to be 1.5.
3. A 20 × 10 cm rolled steel joist of I-section has flanges 0.75 cm thick and web 0.55 cm thick. Find the safe uniformly distributed load that this section will carry over a span of 4 metres if the permissible skin stress is limited to 140 N/mm².
4. A cast iron water pipe has 30 cm bore and 1.5 cm metal thickness and is supported at two points 8 metres apart. Find the maximum stress in the metal when it is running full. Take the unit weight of cast iron as 72 kN/m³ and that of water as 10 kN/m³.
5. Compare the flexural strength of the following three beams of equal weight
   (*i*) *I*-section 20 cm × 15 cm, having 2 cm as flange thickness and 1 cm as web thickness.
   (*ii*) Rectangular section having depth equal to 1.75 width.
   (*iii*) Solid circular section.
6. Compare the weight of two beams of the same material and of equal strength, one being of circular section and solid, and the other being of hollow circular section, the internal diameter being 3/4th of the external.
7. A beam *ABC* is loaded as shown in Fig. 9.48. The beam is made of 12.5 cm × 10 cm × 1.25 cm inverted Tee section. Find the weight *W* if the maximum bending stress is not to exceed 120 N/mm².

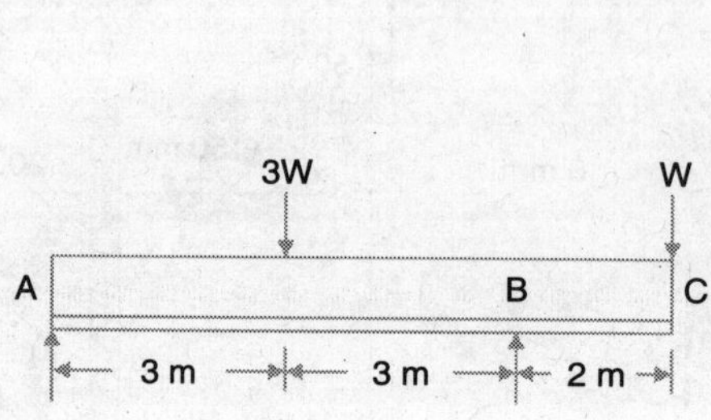

Fig. 9.48

Fig. 9.49

8. A cast iron beam has the cross-section shown in Fig. 9.49. The section is subjected to bending moment in the plane *xy*. Determine the permissible value of this bending moment if the safe fibre stress intensity due to bending is 154.3 N/mm^2 in tension and 123.4 N/mm^2 in compression for the following cases :

(*a*) bottom fibre in tension (*b*) top fibre in tension.

9. A flitched beam is made up of two timber joists each 10 cm wide and 25 cm deep, with a 2 cm steel plate 15 cm deep placed symmetrically between them and firmly attached to both. The plate is recessed into grooves cut in the inner faces of joists so that the overall dimensions of the built-up section may be taken as 20 cm × 25 cm. Calculate the moment of resistance of the combined section when the maximum bending stress in the timber is 8 N/mm^2. What is then the maximum stress in steel ?

Take E for steel = 2.1 × 10^5 N/mm^2; E for timber = 0.105 × 10^5 N/mm^2.

10. A compound beam is formed by joining two bars rigidly together, one of steel and the other of brass, each bar being 4 cm wide. The bars are of thickness t_1 and t_2 so that the total depth is $(t_1 + t_2)$.

If Young's Modulus for steel is twice that for brass, find the ratio of t_1 to t_2 so that the neutral axis of the section is at the dividing line of the two bars.

If the total depth of the section is to be 3 cm, and the stresses in the steel and brass are not to exceed 112.5 N/mm^2 and 42 N/mm^2 respectively, determine the maximum moment of resistance of the beam.

11. A timber beam 7.5 cm wide by 15 cm deep is to be reinforced by bending strips of aluminium alloy 7.5 cm wide on the top and bottom faces of the timber over the whole length of the beam. If the moment of resistance of the composite beam is to be 4 times that of timber alone and the same value of maximum bending stress in timber is reached in both cases, determine the thickness of alloy strip and the ratio of the maximum bending stresses in the alloy strip and timber. Take E for alloy strip = 7.15 E for timber.

ANSWERS

1. 6.93 m ; 55.4 kN **2.** 24 cm × 36 cm **3.** 11.8 kN/m

4. 12.7 N/mm^2 **5.** $\frac{Z_1}{Z_2} = 3.56$; $\frac{Z_1}{Z_3} = 5.57$ **6.** 1.77

7. 1.07 kN **8.** (*a*) 9.31 kN-m ; (*b*) 4.47 kN-m

9. 23.44 kN-m ; 96 N/mm^2 **10.** $\frac{t_1}{t_2} = 0.707$; $M_r = 295$ N-m

11. Thickness of each alloy strip = 9.3 mm $\frac{f_{max} \text{ for alloy}}{f_{max} \text{ for timber}} = 8.03$.

10

Shearing Stresses in Beams

10.1. GENERAL

We have already seen the existence of bending or longitudinal stress and its variation along the cross-section of the beam. While discussing the simple theory of bending, we have neglected the presence of the shear, and the distortion of the plane section due to it. We stated that the result could be applied with sufficient accuracy to the cases where shear is present. This does not imply that the shear stresses can be neglected altogether. *The effect of shear on the bending stresses is not of any practical importance, but the shearing stresses must be considered for their own importance.* We shall, in this chapter, discuss the variation of shear stress along the section and find an expression for the same. In the elementary treatment, we shall consider only the shear stresses in transverse planes parallel to the shearing force and the complimentary shear stresses in longitudinal planes parallel to the axis of the beam. The presence of other shear stresses is neglected.

10.2. DISTRIBUTION OF SHEARING STRESSES

Let us consider two normal sections mn and m_1n_1 of a beam at an infinitely small distance δx (Fig. 10.1). Let the bending moment at the section mn be M and that at m_1n_1 be $(M + \delta M)$. Let us investigate the amount of horizontal shear stress at a plane AB, distant y_1 from the N.A.

Due to the bending moment M at the section mn, there will be bending stresses across the section and the value of the stress will depend on the position of the layer with respect of N.A. For any layer at a distance y from N.A., the intensity of bending stress is $f = \dfrac{M}{I}\, y$. Hence the compressive thrust on the strip of δa area at this height will be equal to $f\,\delta a$ or $= \dfrac{M}{I}\, y\,\delta a$.

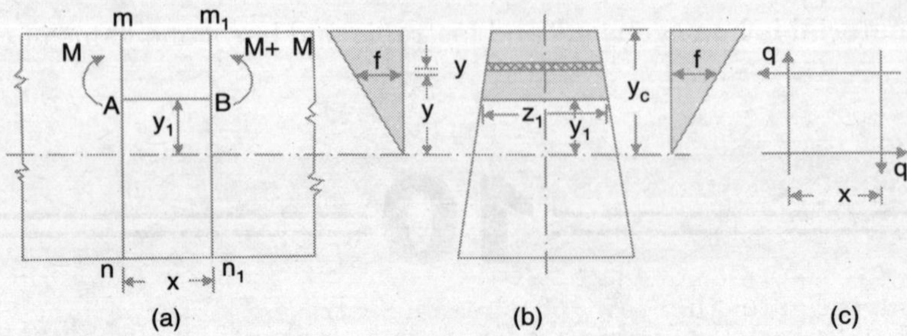

Fig. 10.1

Similarly, the elementary thrust on the strip of δa area at the height y from N.A. at the section m_1n_1 will equal to $\left(\frac{M + \delta M}{I} y\delta a\right)$, assuming that the section of the beam is constant from mn to m_1n_1.

Hence elementary unbalanced horizontal thrust, between mn and m_1n_1

$$= \frac{M + \delta M}{I} y\delta a - \frac{M}{I} y\delta a = \frac{\delta M}{I} y\delta a.$$

Total unbalanced horizontal thrust for the layers between $y = y_1$ to $y = y_c$ will be equal to $H = \sum_{y=y_1}^{y=y_c} \frac{\delta M}{I} y \,.\, \delta a$. For the equilibrium of the portion ABm_1m, a horizontal shear force at AB must act to counter-balance this, as no other horizontal force is acting. Let the shear intensity (balancing shear) at AB be q_1. Assuming that this is uniform over the width of the section, the horizontal shear force at $AB = q_1\delta x.z_1$, where z_1 is the width of the beam at the height y_1.

Equating this to the unbalanced horizontal thrust, we get

$$q_1\delta x.z_1 = \sum_{y=y_1}^{y=y_c} \frac{\delta M}{I} y\delta a \quad \text{or} \quad q_1 = \frac{\delta M}{\delta x.I.z_1} \sum_{y=y_1}^{y=y_c} y\delta a$$

But $\delta M/\delta x = F$ and $\Sigma y\delta a$ is the moment of the area above AB (*i.e.*, the shaded area) about N.A.

Hence $$q_1 = \frac{F}{I\,.\,z_1} (A\bar{y})$$

or dropping the suffix, we get

$$q = \frac{F}{I\,.\,z} (A\bar{y}) \quad \text{...(10.1)}$$

Assumptions Made

It is important to know the assumptions made. They are summarized below :

(*i*) For all values of y, q is uniform across the width of a section of whatever shape. This is not strictly correct because the tangential value must be zero at the boundaries of the section.

Hence it is understood that q used is average value across the section.

(*ii*) $F = dM/dx$ is derived from the assumption that the bending stress varies linearly across the section and is zero at the centroid. Actually, the stress curve is not a straight line passing through the centroid of the section.

(*iii*) The material is homogeneous and isotropic and has same value of E for tension and compression.

10.3. SHEARING STRESS DISTRIBUTION OVER BEAM OF RECTANGULAR SECTION

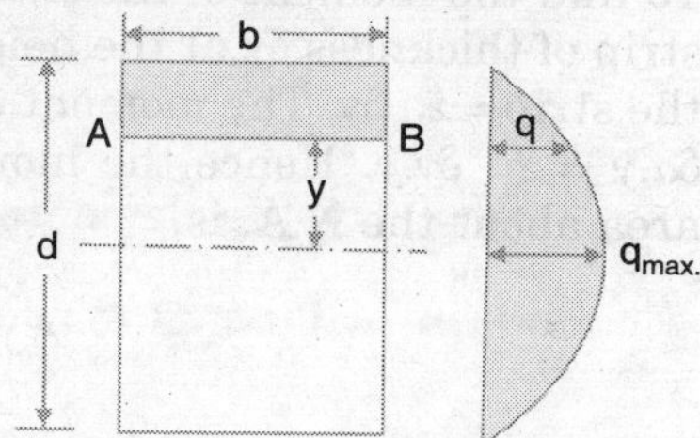

Fig. 10.2

Let us investigate the law of shearing stress distribution over the section of a beam of rectangular cross-section (Fig. 10.2).

On any layer AB, at height y from N.A., the intensity of shear stress q, from Eq. 10.1, is given by

$$q = \frac{F}{I\,.\,z}\,(A\overline{y})$$

where I = moment of inertia of whole section

z = width of the section at $AB = b$

$A\overline{y}$ = moment of the area above AB = moment of the shaded area about N.A.

Substituting the values, we get

$$q = \frac{F}{Ib}\times\left\{b\left(\frac{d}{2}-y\right)\right\}\left\{\frac{1}{2}\left(\frac{d}{2}-y\right)+y\right\}$$

$$= \frac{F}{Ib}\left\{b\left(\frac{d}{2}-y\right)\right\}\left\{\frac{1}{2}\left(\frac{d}{2}+y\right)\right\}$$

$$= \frac{F}{Ib}\times\frac{b}{2}\left(\frac{d^2}{4}-y^2\right) = \frac{F}{2I}\left(\frac{d^2}{4}-y^2\right) \quad \text{...(10.2)}$$

This is the equation of a parabola and hence the variation of q is parabolic.

At $\quad y = \pm\frac{d}{2},\quad q = 0$

At N.A., $\quad y = 0,\quad$ hence $q = q_{max} = \frac{F}{2I}\frac{d^2}{4} = \frac{Fd^2}{8I} = \frac{Fd^2}{8\times\frac{1}{12}bd^3} = \frac{3}{2}\frac{F}{bd}$

The mean shear stress on the section $= \frac{F}{\text{area}} = \frac{F}{bd}$

Hence $\quad \frac{q_{max}}{q_{mean}} = \frac{3}{2} = 1.5$

Thus, the maximum shear stress at the N.A. is 50% more than the mean value. The shear stress distribution has been shown in Fig. 10.2

10.4. SHEARING STRESS DISTRIBUTION OVER BEAM OF SOLID CIRCULAR SECTION

Consider a beam of solid circular cross-section, of diameter d (radius r).

The intensity of shear stress at level AB, at height y from N.A. is given by

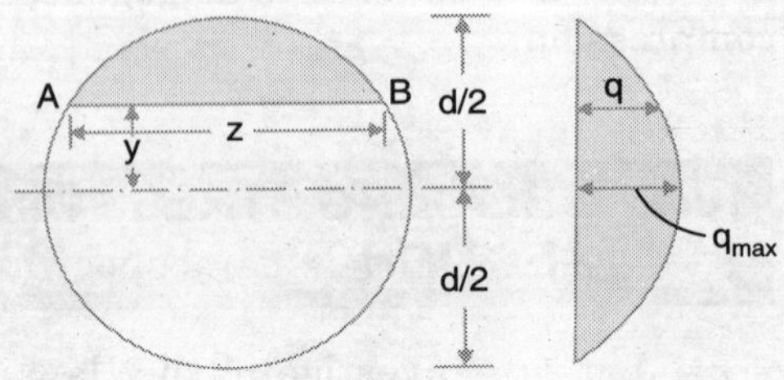

Fig. 10.3

$$q = \frac{F}{I \, . \, z} (A\bar{y})$$

where z = width of the section at the height

$y = 2(r^2 - y^2)^{1/2}$.

$A\bar{y}$ = moment of the shaded area, about N.A.

To find the moment of the shaded area, consider a strip of thickness δy at the height y. The area δa of the strip $= z \, . \, \delta y$. The moment of elementary area $= \delta a . y = z \, . \, \delta y . y$. Hence the moment of the shaded area about the N.A. is,

$$A\bar{y} = \int_{y=y}^{y=r} yzdy$$

But $\quad z = 2\sqrt{r^2 - y^2} \quad$ or $\quad z^2 = 4(r^2 - y^2)$

$$\therefore \quad 2zdz = 4(-\,2y)dy = -\,8ydy \quad \text{or} \quad ydy = -\frac{1}{4}\,zdz$$

When $y = r, z = 0$; when $y = y, z = z$

$$\therefore \quad A\bar{y} = \int_z^0 \left(-\frac{z^2}{4}\right) dz = \frac{1}{4}\int_0^z z^2 dz = \frac{z^3}{12} \quad \text{...(10.3)}$$

$$\therefore \quad q = \frac{F}{Iz}\,A\bar{y} = \frac{F}{Iz} \cdot \frac{z^3}{12} = \frac{Fz^2}{12I} = \frac{F}{12I}\,\{4(r^2 - y^2)\} = \frac{F}{3I}\,(r^2 - y^2) \quad \text{...(10.4)}$$

The variation of q is, therefore, parabolic.

At $y = r$, $\quad q = 0$ and at $y = 0$, $q = q_{max} = \frac{F}{3I} r^2 = \frac{Fd^2}{12I}$ at N.A.

$$q_{max} = \frac{Fd^2}{12\,\frac{\pi d^4}{64}} = \frac{4}{3} \cdot \frac{F}{\frac{\pi d^2}{4}} = \frac{4}{3}\,\frac{F}{\text{area}} = \frac{4}{3}\,q_{mean}$$

Hence q_{max} is 33.33% greater than the mean.

10.5. SHEAR STRESS DISTRIBUTION OVER BEAM OF I-SECTION

Let us take the case of a beam of I-section with the following dimensions :

Flange width B ; Flange thickness t_1 ; Overall depth D

Depth of web $d = D - 2t_1$; Web thickness t

In the flange

In the flange, consider a plane at any height y from N.A. The shear intensity

$$q = \frac{F}{I \, . \, z} (A\bar{y}). \quad \text{Here,} \quad z = B$$

$A\bar{y}$ = moment (about N.A.) of the area above that plane

$$= \left\{B\left(\frac{D}{2} - y\right)\right\}\left\{\left(\frac{D}{2} + y\right)\frac{1}{2}\right\} = \frac{B}{2}\left(\frac{D^2}{4} - y^2\right)$$

$$\therefore \qquad q = \frac{F}{IB} \cdot \frac{B}{2}\left(\frac{D^2}{4} - y^2\right) = \frac{F}{2I}\left(\frac{D^2}{4} - y^2\right) \qquad ...(10.5)$$

The variation is parabolic, which gives

$$q = 0 \text{ at } y = \frac{D}{2}$$

At the junction with the web,

$$y = \frac{d}{2} \text{ and } q = q_f$$

where $q_f = \dfrac{F}{2I}\left(\dfrac{D^2}{4} - \dfrac{d^2}{4}\right) = \dfrac{F}{8I}(D^2 - d^2)$ ...(10.6)

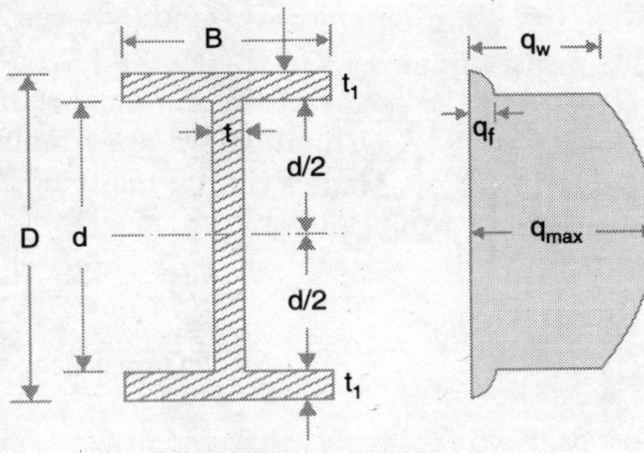

Fig. 10.4

In the web

The shear intensity q at height y is given by

$$q = \frac{F}{I \cdot z} \times (A\bar{y}) \text{ where } z = \text{width of the web} = t$$

$$A\bar{y} = B\left(\frac{D}{2} - \frac{d}{2}\right)\left(\frac{D}{2} + \frac{d}{2}\right)\frac{1}{2} + t\left(\frac{d}{2} - y\right)\left(\frac{d}{2} + y\right)\frac{1}{2}$$

$$= \frac{B}{2}\left(\frac{D^2}{4} - \frac{d^2}{4}\right) + \frac{t}{2}\left(\frac{d^2}{4} - y^2\right) = \frac{B}{8}(D^2 - d^2) + \frac{t}{8}(d^2 - 4y^2)$$

$$\therefore \qquad q = \frac{F}{It}\left\{\frac{B}{8}(D^2 - d^2) + \frac{t}{8}(d^2 - 4y^2)\right\}$$

$$= \frac{F}{8I}\frac{B}{t}(D^2 - d^2) + \frac{F}{8I}(d^2 - 4y^2) \qquad ...(10.7)$$

At the junction with the flange, $y = d/2$. Hence

$$q = q_w = \frac{F}{8I}(D^2 - d^2)\frac{B}{t} = \frac{B}{t}q_f. \qquad ...(10.8)$$

Thus, at the junction, the shear stress in the web suddenly increases and becomes B/t times that in the flange.

At $y = 0$, $\qquad q = q_{max} = \dfrac{F}{8I}\dfrac{B}{t}(D^2 - d^2) + \dfrac{Fd^2}{8I}$

or $\qquad q_{max} = \dfrac{F}{8I}\left[\dfrac{B}{t}(D^2 - d^2) + d^2\right]$ ...(10.9)

The shear stress distribution has been shown in Fig. 10.4.

10.6. SHEAR STRESS DISTRIBUTION FOR OTHER SECTIONS

The shear stress distribution for some of the typical sections are shown in Fig. 10.5.

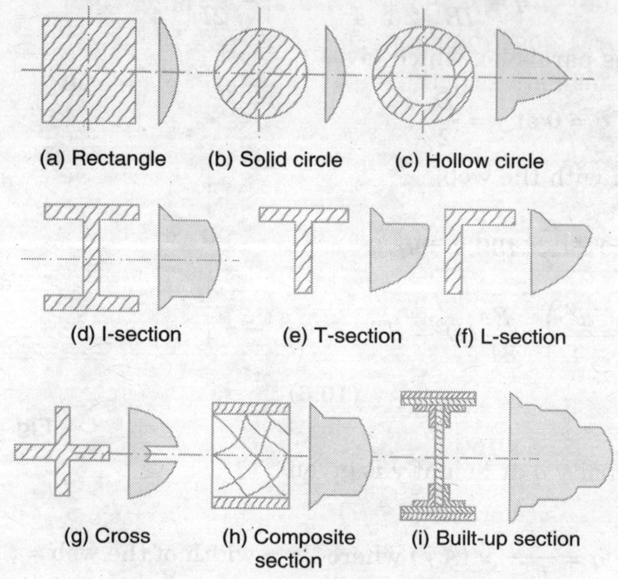

Fig. 10.5

Example 10.1. *A 400 mm × 150 mm I-girder has 20 mm thick flanges and 13 mm thick web. Calculate the maximum intensity of shear stress and sketch the distribution of shear stress across the section, the S.F. at the cross-section being 160 kN.*

Solution.

$$I_{xx} = \frac{1}{12} \times 150 \times 400^3 - \frac{1}{12} \times 137 \times 360^3 = 26734 \times 10^4 \text{ mm}^4$$

In the flange, at any distance y from N.A.,

$$q = \frac{F}{2I}\left(\frac{D^2}{4} - y^2\right)$$

At the junction with web, $q_f = \frac{F}{8I}(D^2 - d^2) = \frac{F}{8I}(400^2 - 360^2)$

$$= \frac{3800\,F}{I} = \frac{3800 \times 160000}{26734 \times 10^4} = 2.27 \text{ N/mm}^2$$

In the web, at the junction, this suddenly increases to 2.27 × 150/13 = 26.24 N/mm²

$$q_{max} \text{ at the N.A.} = \frac{B}{t}\frac{F}{8I}(D^2 - d^2) + \frac{Fd^2}{8I} = 26.24 + \frac{160000\,(360)^2}{8 \times 26734 \times 10^4} = \mathbf{35.93\ N/mm^2}$$

Refer Fig. 10.4 for shear stress distribution.

Example 10.2. *The cross-section of a joist is a tee-section 150 mm × 100 mm × 13 mm with 150 mm side horizontal. Find the maximum intensity of shear stress and sketch the distribution of stress across the section if it has to resist a shear force of 80 kN.*

Solution.

$$\bar{x} = \frac{(100 \times 13 \times 50) + (137 \times 13 \times 6.5)}{(100 \times 13) + (137 \times 13)} = \frac{76576.5}{3081} = 24.85 \text{ mm}$$

$$I_{AB} = \tfrac{1}{3} \times 13(100)^3 + \tfrac{1}{3} \times 137(13)^3 = 4.434 \times 10^6 \text{ mm}^4$$

$$I_{xx} = 4.434 \times 10^6 - 3081\,(24.85)^2 = 2.53 \times 10^6 \text{ mm}^4$$

The intensity of shear stress in the flange at the junction

$$q = \frac{F}{I.z}(A\bar{y})$$

$$= \frac{F}{I \times 150}\{150 \times 13(24.85 - 6.5)\}$$

$$= \frac{238.6\,F}{I} = \frac{238.6 \times 80000}{2.53 \times 10^6}$$

$$= 7.55 \text{ N/mm}^2$$

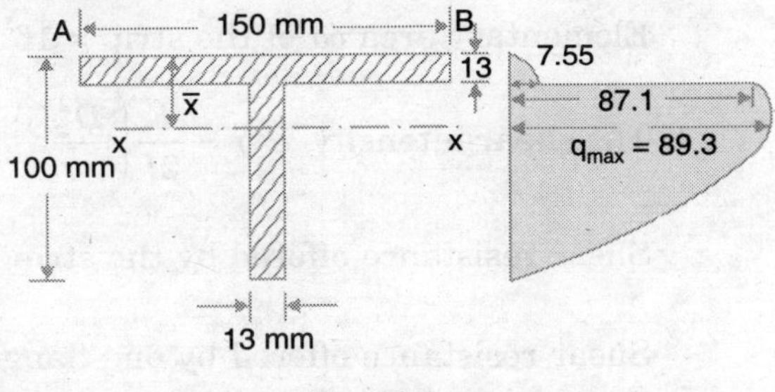

Fig. 10.6

Intensity of shear stress in the web at the junction

$$= \frac{7.55 \times 150}{13} = 87.1 \text{ N/mm}^2$$

The maximum intensity at the centroid

$$= \frac{F}{I \times 13}\left\{150 \times 13(24.85 - 6.5) + \frac{13(24.85 - 13)^2}{2}\right\}$$

$$= \frac{F}{I}\{2753.1 + 70.3\} = \frac{2823.4 \times 80000}{2.53 \times 10^6} = 89.3 \text{ N/mm}^2$$

The variation of shear stress has been shown in Fig. 10.6.

Example 10.3. *A rolled steel joist 20 cm deep × 16 cm wide has flanges 2.2 cm thick and web 1.2 cm thick. Calculate the proportion in which the flange and web resist* (i) *the B.M.,* (ii) *the shear force.*

Solution.

$$I = \frac{1}{12} \times 16 \times 20^3 - \frac{1}{12} \times 14.8 \times 15.6^3 = 10667 - 4682 = 5985 \text{ cm}^4$$

(i) Resistance to B.M. : Consider a section of δy thickness in the web at a distance y from N.A.

Stress $f = \dfrac{M}{I}\,y$ at the strip.

Elementary force offered by the strip $= f\delta a = \dfrac{M}{I}\,.\,y(1.2\delta y)$

Moment of elementary force, about N.A. $= \delta M = \dfrac{M}{I}\,1.2y^2\delta y$

Moment of resistance offered by the web $= M_w = \displaystyle\sum_{y=-7.8}^{y=+7.8} \delta M$

or

$$M_w = \int_{-7.8}^{+7.8} 1.2\frac{M}{I}\,y^2 dy = \frac{2.4M}{I}\int_0^{7.8} y^2 dy = \frac{2.4M}{I}\left(\frac{y^3}{3}\right)_0^{7.8}$$

or $$M_w = \frac{2.4M}{3I}(7.8)^3 = \frac{379.64M}{I} = \frac{379.64}{5985} M = 0.0634M$$

Moment of resistance offered by the flanges = $M - 0.0634M = 0.9366\ M$

Hence the flanges take up 93.66% of B.M. and web 6.34% only.

(ii) Resistance to S.F. : Consider an elementary strip of δy thickness of the flange, at a distance y from N.A.

Elementary area δa of the strip = $16\ \delta y$

The shear intensity $$q = \frac{F}{2I}\left(\frac{D^2}{4} - y^2\right) = \frac{F}{2I}(100 - y^2)$$

Shear resistance offered by the strip $$= q\delta a = \frac{F}{2I}(100 - y^2)\, 16\delta y = \frac{8F}{I}(100 - y^2)\,\delta y$$

Shear resistance offered by one flange $$= \int_{7.8}^{10} \frac{8F}{I}(100 - y^2)\,\delta y$$

Shear resistance offered by both flanges

$$= \frac{2 \times 8F}{I}\left(100y - \frac{y^3}{3}\right)_{7.8}^{10} = \frac{16F}{I}\left\{\left(1000 - \frac{1000}{3}\right) - \left(780 - \frac{7.8^3}{3}\right)\right\}$$

$$= \frac{16 \times 44.85F}{I} = \frac{16 \times 44.85}{5985} F = 0.1199\ F.$$

Shearing resistance by the web = $0.8801\ F$.

∴ Web takes 88.01% shear and flange take 11.99% shear only.

Example 10.4. *A beam of square section is used as beam with one diagonal horizontal. Find the magnitude and location of maximum shear stress in the beam. Sketch the shear stress distribution across the section.*

Solution. Let B be the length of the side.

Let b = length of the diagonal = $\sqrt{2}B$

$$I_{AB} = 2 \times \frac{1}{12}(b)\left(\frac{b}{2}\right)^3 = \frac{b^4}{48}$$

At a distance x from C,

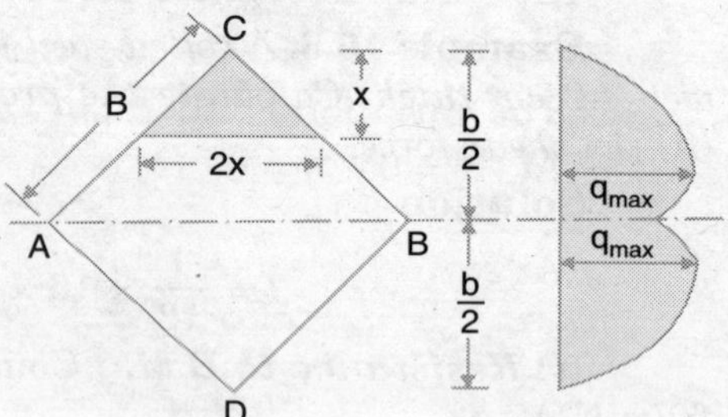

Fig. 10.7

$$q = \frac{F}{I(2x)} \times \text{moment of shaded area about N.I.}$$

$$= \frac{F}{I(2x)}\left\{x^2\left(\frac{b}{2} - \frac{2}{3}x\right)\right\}$$

$$= \frac{48Fx}{2b^4}\left\{\frac{3b - 4x}{6}\right\} = \frac{4Fx}{b^4}(3b - 4x) \qquad \text{...(10.10)}$$

The variation is parabolic.

In the above expression, the first term $4Fx/b^4$ increases with the increasing value of x, while the second term $(3b - 4x)$ decreases with the increasing value of x.

At $x = 0$, $q = 0$.

At $x = \frac{b}{2}$, $q_n = \frac{2F}{b^3}(3b - 2b) = \frac{2F}{b^2} = \frac{F}{\text{area}} = q_{mean}$

Thus the shear intensity at N.A. is the mean intensity. For maximum shear intensity,

$$\frac{dq}{dx} = 0. \text{ Hence } 3b - 8x = 0, \text{ which gives } x = \frac{3}{8}b$$

Hence the q_{max} will occur at $\frac{3}{8}b$ from C or at $\frac{b}{8}$ from N.A.

$$q_{max} = \frac{4F}{b^4}\left(\frac{3}{8}b\right)\left(3b - \frac{12}{8}b\right) = \frac{9F}{4b^2} = \frac{9}{8}\frac{F}{B^2} \quad ...(10.11)$$

$$q_{mean} = \frac{F}{\text{area}} = \frac{F}{B^2} = \frac{2F}{b^2}$$

$$\therefore \quad \frac{q_{max}}{q_{mean}} = \frac{9}{8} = \mathbf{1.125.}$$

Example 10.5. *A beam of triangular cross-section with base b and height h, is used with the base horizontal. Calculate the intensity of maximum shear stress and plot the variation of shear stress intensity along the section.*

Solution.

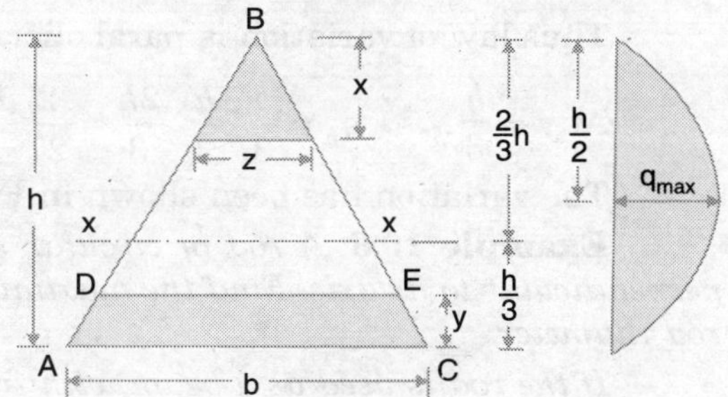

Fig. 10.8

$$I_{xx} = \frac{1}{36}bh^3$$

At any section x from the apex B,

$$q = \frac{F}{I.z}(A\bar{y}) \text{ where } z = b\frac{x}{h}$$

$$A\bar{y} = \left(\frac{1}{2}zx\right)\left(\frac{2}{3}h - \frac{2}{3}x\right) = \frac{1}{3}zx(h - x)$$

$$q = \frac{F}{Iz}\frac{1}{3}zx(h - x) = \frac{F}{3I}x(h - x) \quad ...(10.12)$$

The variation is, therefore, parabolic.

At $x = 0$, $q = 0$

At N.A., $$q_n = \frac{F}{3I}\left(\frac{2}{3}.h\right)\left(\frac{1}{3}h\right) = \frac{2}{27}\frac{F}{I}h^2 = \frac{2}{27}\frac{36F}{bh^3}h^2 = \frac{72}{27}\frac{F}{bh} = \frac{8}{3}\frac{F}{bh} \quad ...(10.13)$$

or $$q_n = \frac{4}{3}\frac{F}{\text{area}} = \frac{4}{3}q_{mean}$$

Thus the intensity at N.A. is 4/3 of mean.

For maximum intensity above N.A., $\frac{dq}{dx} = 0 = h - 2x$

$$\therefore \quad x = h/2$$

Hence maximum stress occurs at half the height.

$$q_{max} = \frac{F}{3I}\frac{h^2}{4} = \frac{F}{12I}h^2 = \frac{3F}{bh^3}h^2 = \frac{3F}{bh} = \frac{3F}{2\text{ area}} = \frac{3}{2}q_{mean}. \quad ...(10.14)$$

Similarly, to plot the shear distribution for the portion below N.A., take a layer DE at a distance y from the base. The intensity of shear stress q is given by,

$$q = \frac{F}{Iz'}(A\bar{y}) \text{ where } z' = DE = \frac{b}{h}(h-y)$$

$$A\bar{y} = \frac{1}{2}(b+z')y\left\{\frac{h}{3} - \frac{y}{3}\left(\frac{b+2z'}{b+z'}\right)\right\}$$

$$= \frac{y}{6}\left\{b + \frac{b}{h}(h-y)\right\}\left[h - y\left\{\frac{b + \frac{2b}{h}(h-y)}{b + \frac{b}{h}(h-y)}\right\}\right]$$

$$= \frac{by}{6h}\{(2h-y)\}\left[h - y\left\{\frac{3h-2y}{2h-y}\right\}\right] = \frac{by}{3h}(h-y)^2$$

$$\therefore \quad q = \frac{F}{I\frac{b}{h}(h-y)} \cdot \frac{by}{3h}(h-y)^2 = \frac{Fy}{3I}(h-y) \quad \text{...(10.15)}$$

Thus law of variation is parabolic and is similar to that in Eq. 10.12. At $y = 0$, $q = 0$.

At $y = \frac{h}{3}$, $\quad q_n = \frac{F}{3I}\frac{h}{3} \cdot \frac{2h}{3} = \frac{2}{27}\frac{Fh^2}{I}$, which is the same as found earlier.

The variation has been shown in Fig. 10.8.

Example 10.6. *A rod of circular section is subjected to a shearing force on a plane perpendicular to its axis. Find the maximum shearing stress in terms of the shearing force and rod diameter.*

If the rod is used as a beam with free ends and a central concentrated load, express the free length in terms of diameter for which the maximum shearing stress, due to the shearing force, is half the maximum direct stress. (U.L.)

Solution. As already proved, $\quad q_{max.} = \frac{16F}{3\pi D^2}$

For simply supported beam, $\quad M_{max.} = \frac{WL}{4}$

$$\therefore \quad \pm f_b = \frac{WL}{4Z} = \frac{32WL}{4\pi D^3} = \frac{8WL}{\pi D^3} \quad \text{...(1)}$$

$$q_{max.} = 16\,\frac{W}{2} \cdot \frac{1}{3\pi D^2} = \frac{8W}{3\pi D^2} \quad \text{...(2)}$$

But $\quad q_{max.} = \frac{f_b}{2}$

$$\therefore \quad \frac{8W}{3\pi D^2} = \frac{1}{2} \cdot \frac{8WL}{\pi D^3} \quad \text{or} \quad L = \frac{8W}{3\pi D^2} \times \frac{2\pi D^3}{8W} = \mathbf{\frac{2}{3}\,D.}$$

Example 10.7. *A steel bar rolled to the section shown in Fig. 10.9 is subjected to shearing force of 200 kN applied in the direction YY. Making the usual assumptions, determine the average shearing stress at sections A, B, C and D, and find the ratio of maximum to the mean shearing stress in the section. Draw a diagram to show the variation of the average shearing stress the section.*

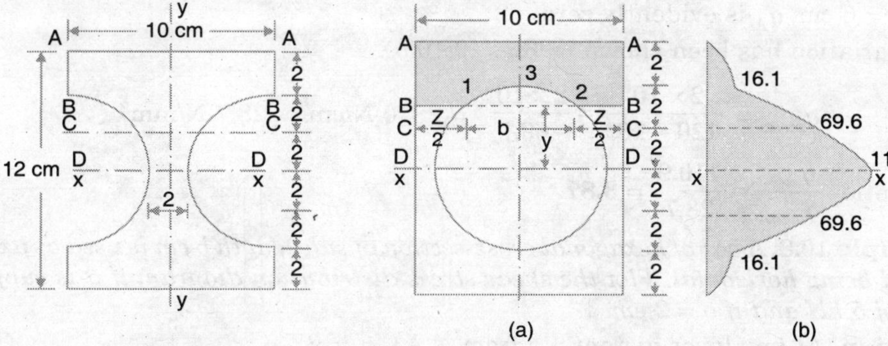

Fig. 10.9 **Fig. 10.10**

Solution. Fig. 10.10 (*a*) shows the equivalent section.

At any layer distant y from the N.A., $q = \dfrac{F}{I \, . \, z}(A\bar{y})$

Here $\quad I = \dfrac{1}{12} \times 10 \times 12^3 - \dfrac{\pi}{64} 8^4 = 1440 - 201 = 1239 \text{ cm}^2$

F = shearing force = 200 kN = 2×10^5 N

z = solid width of the resisting layer = $10 - b = 10 - 2(r^2 - y^2)^{1/2}$

where $\quad b$ = length $\overline{1-2} = 2(r^2 - y^2)^{1/2}$; r = radius of circle

$A\bar{y}$ = moment of area of the shaded portion, about N.A.

$= [10(6 - y)\,(6 + y)\tfrac{1}{2} - A\bar{y} \text{ of segment 1.3.2}] = 5(36 - y^2) - (A\bar{y} \text{ of segment 1.3.2})$

$A\bar{y}$ of segment 1.3.2 $= \dfrac{b^3}{12}$ (see Eq. 10.3) $= \dfrac{2}{3}\,(r^2 - y^2)^{3/2}$

$\therefore \quad A\bar{y}$ of shaded portion $= 5(36 - y^2) - \dfrac{2}{3}\,(16 - y^2)^{3/2}$

At $y = 0$, $\quad z = 10 - 4 \times 2 = 2$ cm.

$$A\bar{y} = 5(36 - 0) - \frac{2}{3}\,(16 - 0)^{3/2} = 180 - 42.6 = 137.4 \text{ cm}^3$$

$$\therefore \quad q_D = q_{max} = \frac{F}{I . z} A\bar{y} = \frac{2 \times 10^5}{1239 \times 2} \times 137.4 = 11090 \text{ N/cm}^2 = 110.9 \text{ N/mm}^2$$

At $y = 2$ cm, $\quad z = 10 - 2(16 - 4)^{1/2} = 10 - 6.93 = 3.07$ cm

$$A\bar{y} = 5(36 - 4) - \frac{2}{3}\,(16 - 4)^{3/2} = 160 - 27.6 = 132.4 \text{ cm}^3$$

$$\therefore \quad q_c = \frac{2 \times 10^5}{1239 \times 3.07} \times 132.4 = 6960 \text{ N/cm}^2 = 69.6 \text{ N/mm}^2$$

At $\; y = 4$ cm, $\; z = 10$ cm. Hence $A\bar{y} = 5(36 - 16) = 100$

$$\therefore \quad q_B = \frac{2 \times 10^5 \times 100}{1239 \times 10} = 1610 \text{ N/cm}^2 = 16.1 \text{ N/mm}^2$$

At $x = 6$ cm, q_A is evidently zero.

The variation has been shown in Fig. 10.10(*b*).

$$q_{mean} = \frac{2\times10^5}{120-16\pi} = \frac{2\times10^5}{69.7} = 2870 \text{ N/cm}^2 = 28.7 \text{ N/mm}^2$$

$$\therefore \quad \text{Ratio } \frac{q_{max}}{q_{mean}} = \frac{110.9}{28.7} = \mathbf{3.87.}$$

Example 10.8. *A bar of hexagonal cross-section of side length b cm is used as a cantilever, one diagonal being horizontal. Plot the shear stress distribution diagram if it is subjected to a shear force of 5 kN and if b = 2 cm.*

Solution. At any layer distant $\pm y$ from N.A.

$$q = \frac{F}{I.z}(A\bar{y})$$

Fig. 10.11

where $I = I_x$ of rectangle + I_x of triangles

$$= 2\left\{\frac{1}{3}bd^3 + \frac{1}{12}bd^3\right\} = \frac{5}{6}bd^3$$

$$= \frac{5}{6}b\,\frac{3\sqrt{3}}{8}\,b^3 = 0.54b^4$$

$$z = b + 2\left(\frac{b}{2}\right)\left(\frac{d-y}{d}\right) = b + \frac{2}{\sqrt{3}}\left(\frac{\sqrt{3}}{2}b - y\right) = 2b - \frac{2}{\sqrt{3}}y = \frac{2}{\sqrt{3}}(\sqrt{3}b - y)$$

At $y = h$, $\quad z' = \frac{2}{\sqrt{3}}(\sqrt{3}b - h)$

$A\bar{y}$ = moment of the shaded area about N.A.

$$= \sum_{h=y}^{h=d}\left\{\frac{2}{\sqrt{3}}(\sqrt{3}b - h)\,\delta h\right\}h = \frac{2}{\sqrt{3}}\left(\frac{\sqrt{3}bh^2}{2} - \frac{h^3}{3}\right)_y^d$$

$$= \frac{2}{\sqrt{3}}\left\{\left(\frac{\sqrt{3}bd^2}{2} - \frac{d^3}{3}\right) - \left(\frac{\sqrt{3}by^2}{2} - \frac{y^3}{3}\right)\right\}$$

$$= \frac{2}{\sqrt{3}}\left\{\left(\frac{\sqrt{3}}{2}\,\frac{3}{4}b^3 - \frac{1}{3}\,\frac{3\sqrt{3}}{8}b^3\right) - \left(\frac{\sqrt{3}by^2}{2} - \frac{y^3}{3}\right)\right\}$$

$$= \frac{2}{\sqrt{3}}\left\{\frac{\sqrt{3}}{4}b^3 - \left(\frac{\sqrt{3}}{2}by^2 - \frac{y^3}{3}\right)\right\}$$

$$\therefore \quad q = \frac{F}{Iz}(A\bar{y}) = \frac{F\sqrt{3}}{(0.54b^4)\,2(\sqrt{3}b - y)}\cdot\frac{2}{\sqrt{3}}\left\{\frac{\sqrt{3}}{4}b^3 - \left(\frac{\sqrt{3}}{2}by^2 - \frac{y^3}{3}\right)\right\}$$

or

$$q = \frac{F}{0.54b^4\,(1.73b - y)}\{0.433b^3 - y^2(0.865b - 0.333y)\}$$

which is the equation for shearing stress.

When $b = 2$ cm and $F = 5000$ kN, $d = \sqrt{3} = 1.73$ cm

$$q = \frac{5000}{0.54 \times 16(3.46 - y)} \{3.46 - y^2(1.730 - 0.333y)\}$$

$$= \frac{578.7}{3.46 - y} \{3.46 - y^2(1.730 - 0.333y)\} \quad ...(1)$$

$$\text{Area of hexagon } = (b \times 2d) + 4 \left\{\frac{1}{2} \cdot \frac{b}{2} d\right\}$$

$$= 2bd + bd = 3bd = 3b \frac{\sqrt{3}}{2} b = 3 \frac{\sqrt{3}}{2} b^2 = 6\sqrt{3} = 10.4 \text{ cm}^2$$

$$\therefore \quad q_{mean} = \frac{5000}{10.40} = 480 \text{ N/cm}^2 = \mathbf{4.8\ N/mm^2}$$

$$\text{At } y = 0, \quad q_n = \frac{578.7}{3.46}(3.46) = 578.7 \text{ N/cm}^2 \simeq \mathbf{5.79\ N/mm^2}$$

The shear stress diagram can now be plotted.

10.7. PRINCIPAL STRESSES AND PRINCIPAL PLANES AT A POINT IN A BEAM SECTION

We have seen that at a point in the cross-section of a beam subjected to B.M. and S.F., there is bending stress f and a shear stress q given by

$$f = \frac{M}{I} y \quad \text{and} \quad q = \frac{F}{I \cdot z}(A\bar{y})$$

From Eq. 3.12, the principal stresses are given by

$$\sigma = \frac{f}{2} \pm \sqrt{\frac{f^2}{4} + q^2}$$

The major principal stress σ_1 will be of the same sign as the bending stress f, *i.e.*, compressive above the N.A. and tensile below the N.A. The minor principal stress σ_2 will be of the opposite sign, *i.e.*, tensile above N.A. and compressive below N.A.

The position of the principal planes will be given by $2\theta = 2q/f$, where θ is the inclination of the principal plane with that of f.

At the top, when $q = 0$, $\theta_1 = 0$.

As q goes on increasing, θ_1 goes on increasing. At the N.A. when $f = 0$, $2\theta_1 = 90°$ and $\theta_1 = 45°$. Thus at N.A., there is a state of simple shear.

The maximum shear anywhere in the beam section will be equal to $\sqrt{\frac{f^2}{4} + q^2}$, its plane being at 45° to that of σ_1.

The variation of the principal stress along the section of the beam has been shown in examples 10.9 and 10.10.

Example 10.9. *A beam AB of 4 metre span is loaded with a central point load of 60 kN. The section of the beam is a rectangle 15 cm wide and 30 cm deep. At a cross-section distant 2 metre from the support, calculate the bending and shear stresses at points 15 cm, 10 cm, 4 cm and 0 cm from the neutral axis of the section. Find also the principal planes and principal stress at these points.*

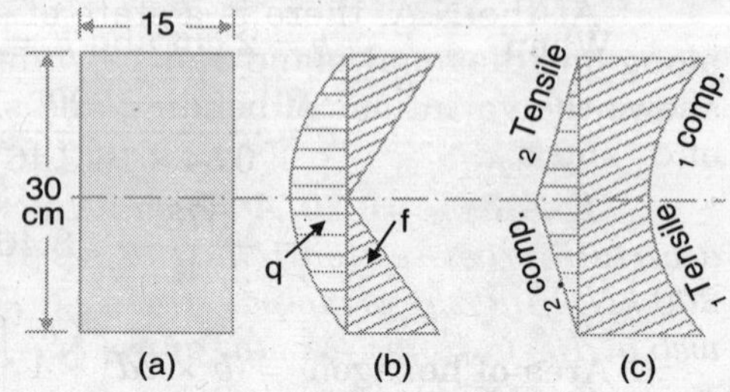

Fig. 10.12

Solution.

$$I = \frac{1}{12} \times 15 \times 30^3 = 33750 \text{ cm}^4 = 337.5 \times 10^6 \text{ mm}^4$$

$$M = 30 \times 2 = 60 \text{ kN-m} = 60 \times 10^6 \text{ N-mm}$$

$$F = 30 \text{ kN} = 30000 \text{ N}$$

(i) At a point 150 mm from N.A.

$$f = \frac{M}{I}y = \frac{60 \times 10^6}{337.5 \times 10^6} \times 150 = 26.7 \text{ N/mm}^2 \text{ (compressive)}$$

$$q = 0$$

$$\therefore \quad \sigma_1 = f = 26.7 \text{ N/mm}^2 \text{ (compressive)} ; \sigma_2 = 0 ; \theta_1 = 0.$$

(ii) At a point 100 mm from N.A.

$$f = \frac{60 \times 10^6 \times 100}{337.5 \times 10^6} = 17.8 \text{ N/mm}^2 \text{ (comp.)}$$

$$q = \frac{F}{2I}\left(\frac{d^2}{4} - y^2\right) = \frac{30000}{2 \times 337.5 \times 10^6}\left[\frac{(300)^2}{4} - (100)^2\right] = 0.56 \text{ N/mm}^2$$

$$\sigma = \frac{17.8}{2} \pm \sqrt{\frac{(17.8)^2}{4} + (0.56)^2} = 8.9 \pm 8.92$$

$$\sigma_1 = 17.82 \text{ N/mm}^2 \text{ (comp.) and } \sigma_2 = 0.02 \text{ N/mm}^2 \text{ (tensile)}$$

$$\tan 2\theta = \frac{2q}{f} = \frac{2 \times 0.56}{17.8} = 0.063 ; \theta_1 = 1°48' \text{ and } \theta_2 = 91°48'$$

(iii) At a point 40 mm from N.A.

$$f = \frac{60 \times 10^6 \times 40}{337.5 \times 10^6} = 7.11 \text{ N/mm}^2$$

$$q = \frac{3000}{2 \times 337.5 \times 10^6}\left[\frac{(300)^2}{4} - (40)^2\right] = 0.93 \text{ N/mm}^2$$

$$\therefore \quad \sigma_1 = 7.23 \text{ N/mm}^2 \text{ (comp.)} \quad \text{and} \quad \sigma_2 = 0.12 \text{ N/mm}^2 \text{ (tensile)}$$

$$\tan 2\theta = \frac{2q}{f} = \frac{2 \times 0.93}{7.11} = 0.261 \quad \therefore \quad \theta_1 = 7°18' \text{ and } \theta_2 = 97°18'.$$

(iv) At the N.A.

$$f = 0 ; q = \frac{30000\,(300)^2}{8 \times 337.5 \times 10^6} = 1 \text{ N/mm}^2 ; \theta_1 = \theta_2 = 45° ; \sigma_1 = \sigma_2 = 1 \text{ N/mm}^2.$$

At the N.A. there is a state of simple shear and principal stress are 1 N/mm^2 (comp.) along one diagonal plane and 1 N/mm^2 (tensile) along another diagonal plane. Fig. 10.12 (*b*) shows the variations of bending and shear stresses while Fig. 10.12 (*c*) shows the variations of σ_1 and σ_2.

Example 10.10. *A 30 cm × 15 cm R.S.J. has flanges 2 cm thick and web 1.3 cm thick. At a certain cross-section it has to resist a bending moment of 100 kN-m and a shear force of 200 kN. Find the principal stresses (a) at the top, (b) in the flange at 13 cm from N.A., (c) in the web at 13 cm from N.A., (d) at the N.A. Plot the variation of principal stresses along the section.*

Solution.

$$I = \frac{1}{12} \times 15 \times 30^3 - \frac{1}{12} \times 13.7 \times 26^3 = 13600 \text{ cm}^4 = 136 \times 10^6 \text{ mm}^4$$

(a) At the top

$$f = \frac{M}{I} y = \frac{100 \times 10^6}{136 \times 10^6} \times 150 = 110.3 \text{ N/mm}^2 \text{ ; } q = 0$$

$$\sigma_1 = 110.3 \text{ N/mm}^2 \text{ (comp.) ; } \sigma_2 = 0$$

(b) In the flange at 13 cm from N.A.

$$f = \frac{110.3 \times 130}{150} = 95.6 \text{ N/mm}^2$$

$$q = \frac{F}{8I}(D^2 - d^2) = \frac{200000}{8 \times 136 \times 10^6}(300^2 - 260^2) = 4.12 \text{ N/mm}^2$$

$$\therefore \quad \sigma = \frac{95.6}{2} \pm \sqrt{\left(\frac{95.6}{2}\right)^2 + (4.12)^2} = 47.8 \pm 48.0$$

$$\therefore \quad \sigma_1 = 95.8 \text{ N/mm}^2 \text{ (comp.) and } \sigma_2 = 0.2 \text{ N/mm}^2 \text{ (tensile)}$$

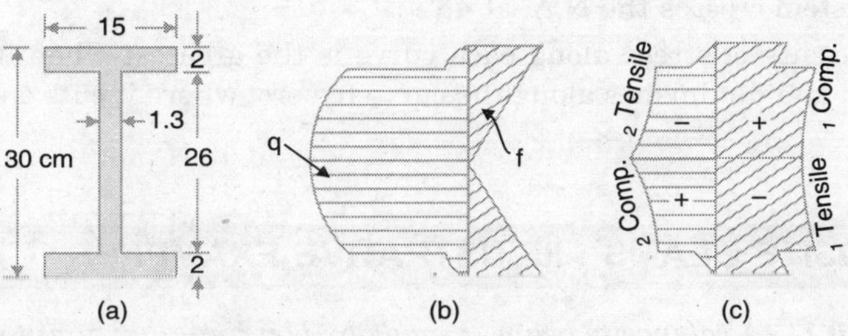

Fig. 10.13

(c) In the flange at 13 cm from N.A.

$$f = 95.6 \text{ N/mm}^2 \text{ ; } q = 4.12 \times \frac{150}{13} = 47.5 \text{ N/mm}^2$$

$$\sigma = \frac{95.6}{2} + \sqrt{\left(\frac{95.6}{2}\right)^2 + (47.5)^2} = 47.8 \pm 67.4$$

$$\therefore \quad \sigma_1 = 115.2 \text{ N/mm}^2 \text{ (comp.) and } \sigma_2 = 19.6 \text{ N/mm}^2 \text{ (tensile).}$$

Thus the intensity of principal stress (= 115.2) at the junction may exceed that at the extreme outside layers of the section when it is subjected to heavy B.M. and S.F. simultaneously.

(d) At the N.A.

$$f = 0$$

$$q = \frac{B}{t}\frac{F}{8I}(D^2 - d^2) + \frac{Fd^2}{8I} = 47.5 + \frac{200000}{8 \times 136 \times 10^6}(260)^2 = 59.9 \text{ N/mm}^2$$

$\therefore$ $\sigma_1 = 59.9 \text{ N/mm}^2$; $\sigma_2 = 59.9 \text{ N/mm}^2$

The variations of σ_1 and σ_2 have been plotted in Fig. 10.13 (*c*).

10.8. CURVES OF PRINCIPAL STRESSES

We have seen that the direction of the principal planes vary from layer to layer. If the directions for principal planes are determined at all points in a cross-section and for a number of sections from one end of the beam to the other, it is then possible to draw curves which will be "envelopes" of the principal planes as shown in Fig. 10.14. The curves are such that the tangent and normal at any point give the directions of the two principal stresses at that point.

As is clear in Fig. 10.14, there are two systems of curves. The following are the properties of the two systems of the curves :

(*i*) The tangent and normal at any point on a curve give the directions of two principal stresses at a point.

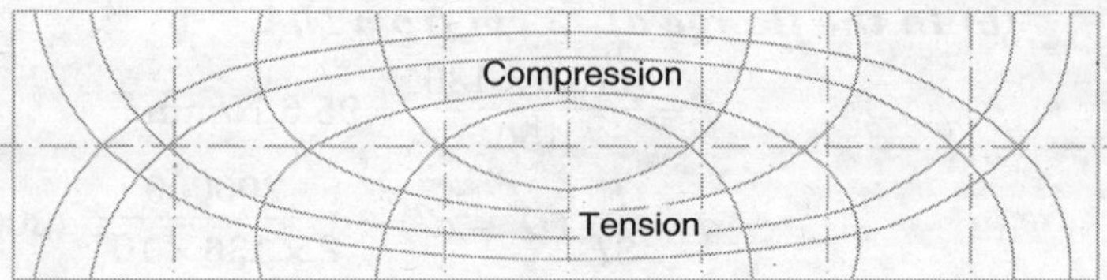

Fig. 10.14

(*ii*) The two systems cross each other at right angles.

(*iii*) Each system crosses the N.A. at 45°.

(*iv*) The intensity of stress along each curve is the greatest when it is parallel to the length of the beam and diminishes along the curve to zero, where it cuts a face of the beam at right angles.

10.9. COMPOSITE BEAMS : ILLUSTRATIVE EXAMPLES

Example 10.11. *A composite beam is made by bonding duralumin sheets to the upper and lower surface of a rectangular wooden spar. The wooden spar is 120 mm deep and 60 mm wide and duralumin sheets 60 mm by 1.5 mm. Calculate the greatest load which the beam can carry when loaded as a cantilever if the maximum shearing stress in the wood is limited to 5.5 N/mm². For duralumin E = 7.73 × 10⁴ N/mm² ; for wood, E = 7.03 × 10³ N/mm².*

Solution. Let suffix *D* stand for duralumin and *W* stand for wood.

We know that $\frac{\text{shear stress}}{\text{shear strain}}$ = Modulus of rigidity.

$$\therefore \quad q \propto N$$

At any point $\frac{q_D}{q_W} = \frac{N_D}{N_W} = \frac{E_D}{E_W} = \frac{7.73 \times 10^4}{7.03 \times 10^3} = 11$

Fig. 10.15 (*b*) shows the shear stress distribution diagram in which

bc = shear stress in wood at the junction.

bd = shear stress in duralumin at the junction.

fe = maximum shear stress in wood = 5.5 N/mm² (given)

The *first step* will be to determine the values of bc and bd.

The distribution of the shear stress in wood will be parabolic, as for the rectangular beam. Let the equation to the parabola be $y = kx^2$, where x is measured from N.A. towards the top.

and y = ordinate of the parabola as marked in Fig. 10.15 (*c*).

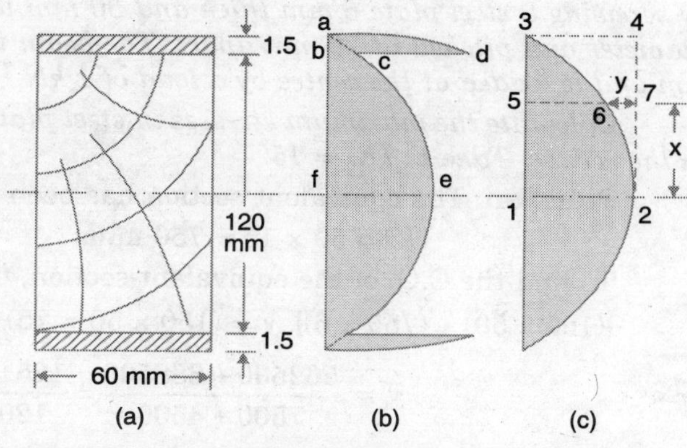

Fig. 10.15

$$\therefore \quad \overline{5-6} = \overline{3-4} - \overline{6-7}$$

$$= k(61.5)^2 - k(x)^2$$

$$\therefore \quad \frac{\overline{5-6}}{\overline{1-2}} = \frac{\overline{5-6}}{\overline{3-4}} = \frac{(61.5)^2 - x^2}{(61.5)^2}$$

or $$\overline{5-6} = \frac{(61.5)^2 - x^2}{(61.5)^2} \times \overline{(1-2)}$$

Thus, at any distance x from N.A., the ordinate is given by the equation given above.

For the parabolic shear stress diagram, we have, therefore

$$bc = \frac{(61.5)^2 - (60)^2}{(61.5)^2} \times fe$$

where $x = 60$ mm and $fe = q_{(w\ max.)} = 5.5$ N/mm²

$$\therefore \quad bc = \frac{3782 - 3600}{3782} \times 5.5 = 0.265 \text{ N/mm}^2$$

$$bd = q_D = 11q_W = 11 \times 0.265 = 2.915 \text{ N/mm}^2$$

Thus the stresses are completely known.

Now $$W = \text{total load taken} = W_D + W_W = F_D + F_W \qquad ...(i)$$

But $$q = \frac{F}{I\,.\,z}(A\bar{y}) \quad \text{or} \quad F = \frac{q\,.\,I\,.\,z}{A\bar{y}}, \text{ with usual symbols}$$

$$\therefore \quad F_D = \frac{q_D I_D\,.\,z_D}{(A\bar{y})_D} \text{ and } F_W = \frac{q_W I_W\,.\,z_W}{(A\bar{y})_W} \qquad ...(ii)$$

Now $$I_D = \frac{1}{12} \times 60(123^3 - 120^3) = 664335 \text{ mm}^3$$

$$(A\bar{y})_D = 60 \times 1.5 \times 60.75 = 5467.5$$

$$I_W = \frac{1}{12} \times 60 \times 120^3 = 8.64 \times 10^6 \text{ mm}^3 \ ; (A\bar{y})_W = 60 \times 60 \times 30 = 108000$$

$$z_D = z_W = 60 \text{ mm}$$

Substituting these values in (*ii*),

$$F_D = \frac{2.915 \times 664335 \times 60}{5467.5} = 21251 \text{ N}$$

$$F_W = \frac{5.5 \times 8.64 \times 10^6 \times 60}{108000} = 26400 \text{ N}$$

$$\therefore \quad W = F_D + F_W = 21251 + 26400 = 47651 \text{ N} = \mathbf{47.65 \text{ N.}}$$

Example 10.12. *A rectangular wooden beam 50 mm wide and 150 mm deep is reinforced by screwing a steel plate 6 mm thick and 50 mm wide on to the bottom. The screws are 5 mm diameter and pitched at 70 mm apart. The beam is simply supported at the ends over a span 3 m and is loaded at the centre by a load of 1 kN.*

Calculate the maximum stresses in steel plate and timber and maximum shearing stress in the screws. Take $E_S/E_T = 15$.

Solution. The equivalent section has been drawn in Fig. 10.16 (*b*) in which

$$CD = 50 \times 15 = 750 \text{ mm.}$$

To find the C.G. of the equivalent section, take moments about *AB*.

$$\{(150 \times 50) + (750 \times 6)\}\ \bar{y} = (150 \times 50 \times 75) + (750 \times 6 \times 153)$$

or

$$\bar{y} = \frac{562500 + 688500}{7500 + 4500} = \frac{1251000}{12000} = 104.25 \text{ mm}$$

$$I_{xx} = \tfrac{1}{3}[50(104.25)^3 + 750(51.75)^3 - 700(45.75)^3]$$

$$= \tfrac{1}{3}[56.65 + 103.94 - 67.03] \times 10^6 = 31.19 \times 10^6 \text{ mm}^4$$

$$M_{max} = \frac{WL}{4} = \frac{1000 \times 3000}{4} = 0.75 \times 10^6 \text{ N-mm} = \frac{f\,I}{y_{max}}$$

$$\therefore \quad f = \frac{0.75 \times 10^6 \times 104.25}{31.19 \times 10^6}$$

$$= 2.51 \text{ N/mm}^2$$

$\therefore$ Max. comp. stress in timber

$$= 2.51 \text{ N/mm}^2$$

Tensile stress in timber

$$= \frac{0.75 \times 10^6 \times 45.75}{31.19 \times 10^6} = \mathbf{1.1 \text{ N/mm}^2}$$

Maximum tensile stress in steel

$$= \left\{1.1 \times \frac{51.75}{45.75}\right\} \times 15 = \mathbf{18.66 \text{ N/mm}^2}$$

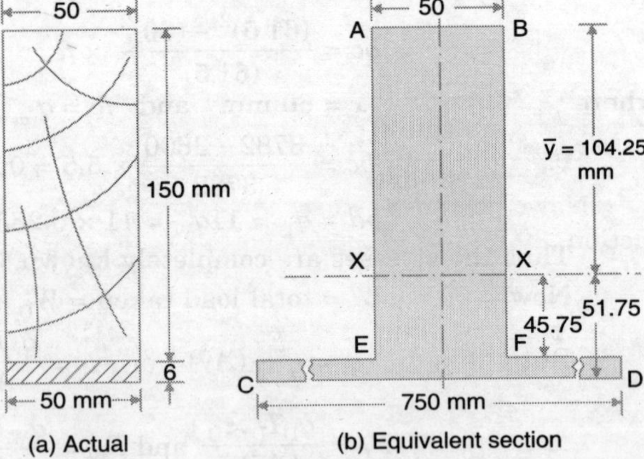

Fig. 10.16

Shear stress in screws

Area of screws $= \frac{\pi}{4}(5)^2 = 19.63 \text{ mm}^2$; Pitch = 70 mm

Let q = max. horizontal shear stress in screw.

Max. horizontal force F_{max} in each end screw

$$= q \times b \times \text{pitch} = 70\,bq = 70b\,\frac{F}{Ib}(A\bar{y}) = \frac{70F}{I}(A\bar{y})$$

$$F = \frac{1000}{2} = 500 \text{ N} \; ; \; I = 31.19 \times 10^6 \text{ mm}^4$$

$(A\bar{y})$ at $EF = (750 \times 6)\,[51.75 - 3] = 219375 \text{ mm}^2$

$$\therefore \quad F_{max} = \frac{70 \times 500 \times 219375}{31.19 \times 10^6} = 246.2 \text{ N}$$

$$\therefore \quad \text{Max. shear stress in screw} = \frac{246.2}{19.63} = \mathbf{12.5\ N/mm^2.}$$

10.10. ADDITIONAL EXAMPLES USEFUL FOR COMPETITIVE EXAMINATIONS

Example 10.13. *A beam having the cross-section shown in Fig. 10.17 carries a vertical load W at midspan over a simply supported span of 2 m. If the allowable working stress in bending is 8 N/mm², find the spacing of lag screws. Each screw can transmit a shear force of 3000 N.*

Solution. Let the spacing of the lag screws be *s*.

$$I = \frac{2 \times 50(240)^3}{12} + 2\left[\frac{1}{12} \times 300(30)^3 + 30 \times 300(135)^2\right] = 4.446 \times 10^8 \text{ mm}^4$$

$$Z = \frac{4.446 \times 10^8}{150} = 2.964 \times 10^6 \text{ mm}^3$$

$$M = f.Z = 8 \times 2.964 \times 10^6 = 23.712 \times 10^6 \text{ N-mm} = 23712 \text{ N-m}$$

$$\text{But} \quad M = \frac{WL}{4} \quad \text{or} \quad W = \frac{4M}{L} = \frac{4 \times 23712}{2} = 47424 \text{ N}$$

$$\therefore \quad F = \frac{W}{2} = \frac{47424}{2} = 23712 \text{ N}$$

Now, shear stress at the junction, 30 mm from top (or bottom)

$$q' = \frac{FA\bar{y}}{I.z}$$

$$= \frac{23712 \times 30 \times 300 \times 135}{4.446 \times 10^8 \times (2 \times 50)}$$

$$= 0.648 \text{ N/mm}^2$$

Since there are two screws in each row, total shear transmitted by each screw

$$= \tfrac{1}{2}(100 \times s \times 0.648) = 32.4s \text{ N}$$

Equating this to shear value of each screw, we have

$$32.4.s = 3000 \quad \text{or} \quad s = \mathbf{92.6\ mm.}$$

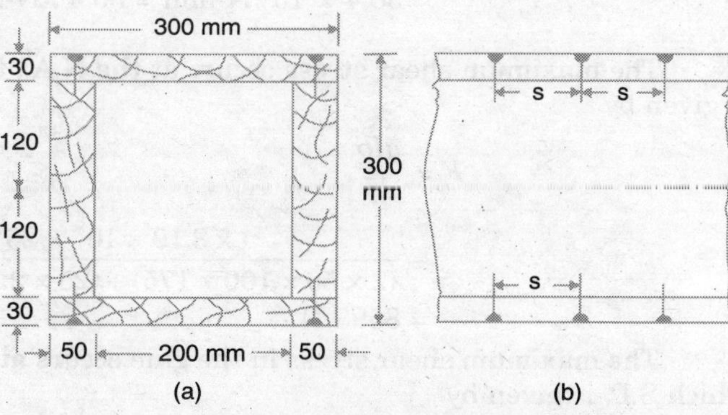

Fig. 10.17

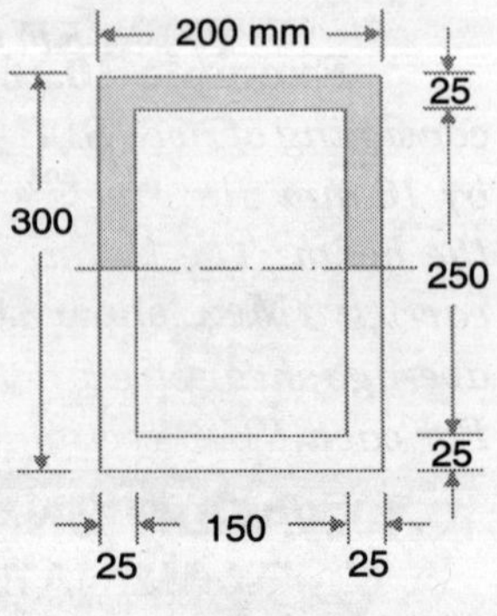

Fig. 10.18

Example 10.14. *A box section having the dimensions shown in Fig. 10.18 is used for a beam on simple supports. In a certain region along the beam, there is a constant, linear change in bending moment of Є kN-m per m along the axis of the beam. Determine the maximum shearing stress at a section in this region.*

Solution.

$$I = \frac{1}{12}\,[200 \times 300^3 - 150 \times 250^3] = 2.547 \times 10^8 \text{ mm}^4$$

$$F = \frac{\Delta M}{\Delta x} = 6 \text{ kN} = 6000 \text{ N}$$

Maximum shear stress evidently occurs at N.A.

$$q_{max} = \frac{FA\bar{y}}{Iz} = \frac{6000}{2.547 \times 10^8\,(2 \times 25)} \times [25 \times 200\,(137.5) + 2 \times 25 \times 125\,(62.5)] = \mathbf{0.51\ N/mm^2}.$$

Example 10.15. *A beam is made up of four 50 mm × 100 mm full sized Douglas fir pieces which are glued to a 25 mm × 450 mm Douglas fir plywood web, as shown in Fig. 10.19. Determine the maximum allowable shear force and maximum allowable bending moment which the section can carry if the allowable bending stress is 10 N/mm², the allowable shear stress in wood is 1 N/mm² and the allowable shearing stress in the glued joint is 0.2 N/mm².*

Solution.

$$I = \frac{1}{12}\,[125 \times 450^3 - 100 \times 250^3] = 8.19 \times 10^8 \text{ mm}^4$$

$$M = \frac{f.I}{y_{max}} = \frac{10 \times 8.19 \times 10^8}{225} = 36.4 \times 10^6 \text{ N-mm} = 36.4 \text{ kN-m} \quad \text{...}(i)$$

$$M = \frac{f.I}{y_{max}} = \frac{10 \times 8.19 \times 10^8}{225}$$

$$= 36.4 \times 10^6 \text{ N-mm} = 36.4 \text{ kN-m} \quad \text{...Eq. } (i)$$

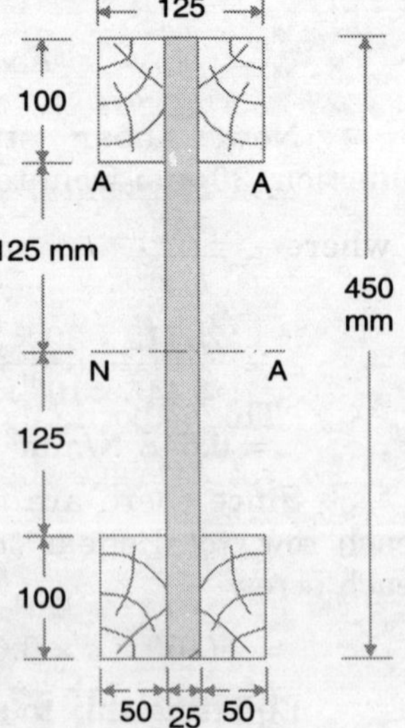

Fig. 10.19

The maximum shear stress occurs at the N.A., for which S.F. is given by

$$F = \frac{qIb}{A\bar{y}}$$

$$= \frac{1 \times 8.19 \times 10^8 \times 25}{(2 \times 50 \times 100 \times 175) + (25 \times 225 \times 112.5)}$$

$$= 8593 \text{ N}$$

The maximum shear stress in the glue occurs at level *A-A*, for which S.F. is given by

$$F = \frac{qIb}{A\bar{y}} = \frac{0.2 \times 8.19 \times 10^8\,(2 \times 50)}{2(50 \times 100 \times 175)} = 9360 \text{ N}$$

Hence Allowable B.M. = **36.4 kN-m.**

and Allowable S.F. = **8.593 kN.**

Example 10.16. *Fig. 10.20 shows a built-up beam section consisting of two ISLB 100 placed one above the other and connected by 16 mm dia. rivets spaced at 120 mm centres along the length of the beam. The beam, when uniformly loaded over a span of 2 m carries a maximum bending stress of 130 N/mm². Calculate the average shear stress developed in the rivets near the end of the beam. For each ISLB section,* $I_{xx} = 168 \times 10^4$ *mm⁴ and A = 1021 mm².*

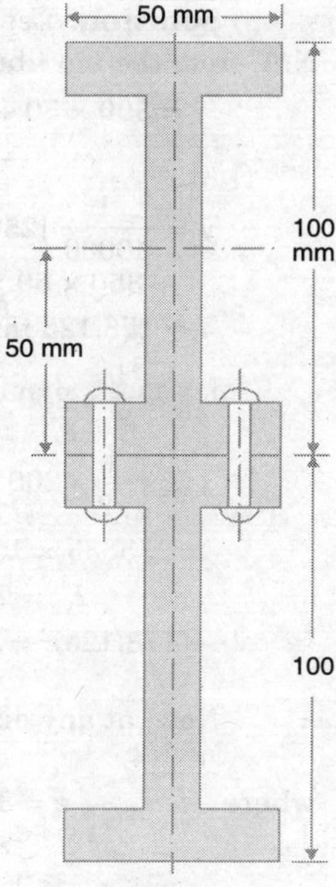

Fig. 10.20

Solution. Size of ISLB 100 : Flange width = 50 mm,

Depth of section = 100 mm

For the built-up section, I about N.A.

$$= 2[168 \times 10^4 + 1021\,(50)^2] = 846.5 \times 10^4 \text{ mm}^4$$

$$\therefore \quad Z = \frac{846.5 \times 10^4}{100} = 84650 \text{ mm}^3$$

$$\therefore \quad M = f\,.\,Z = 130 \times 84650 = 11.005 \times 10^6 \text{ N-mm}$$

$$= 11.005 \text{ kN-m}$$

$$\therefore \quad M = \frac{wL^2}{8} = 11.005$$

$$\therefore \quad F = \frac{wL}{2} = \frac{wL^2}{8}\cdot\frac{4}{L} = \frac{11.005 \times 4}{2} = 22.01 \text{ kN.}$$

Now, at the rivet level (*i.e.*, at N.A.)

$$q_{max} = \frac{F(A\bar{y})}{I\,.\,z} = \frac{(22.01 \times 10^3) \times (1021 \times 50)}{846.5 \times 10^4 \times 50}$$

$$= 2.65 \text{ N/mm}^2$$

$\therefore$ Horizontal shear $= q_{max} \times b \times s$

where b = width of section at N.A. = 50 mm

s = spacing of rivets = 120 mm.

$\therefore$ Horizontal shear force = 2.65 × 50 × 120 − 15900 N

This is borne by two rivets.

$\therefore$ Average shear stress in rivet cross-section

$$= \frac{15900}{2 \times \frac{\pi}{4}(16)^2} = \mathbf{158.2\ N/mm^2.}$$

Example 10.17. *A cast iron bracket, subjected to bending has a cross-section of I-shape with unequal flanges, as shown in Fig. 10.21 (a). If the section is subjected to a shear force of 1600 kN, draw the shear stress distribution over the depth of the section, indicating the principal values.*

Solution. Let $\bar{y}$ the distance of the N.A. from the top fibre. Total Area

$$= 300 \times 50 + 300 \times 50 + 200 \times 50 = 40000 \text{ mm}^2$$

$$\bar{y} = \frac{1}{40000} [250 \times 50 \times 25 + 350 \times 50 \times 175 + 200 \times 50 \times 375]$$

$$= 178.125 \text{ mm}$$

$$I_{AB} = \frac{1}{3} \times 250 \times 50^3 + \frac{1}{3} \times 50 \times 350^3 + \left\{\frac{1}{12} \times 200\,(50)^3 + 200 \times 50\,(375)^2\right\}$$

$$= 213333 \times 10^4 \text{ mm}^4$$

$$\therefore \quad I_{xx} = 213333 \times 10^4 - 40000\,(178.125)^2 = 86419 \times 10^4 \text{ mm}^4$$

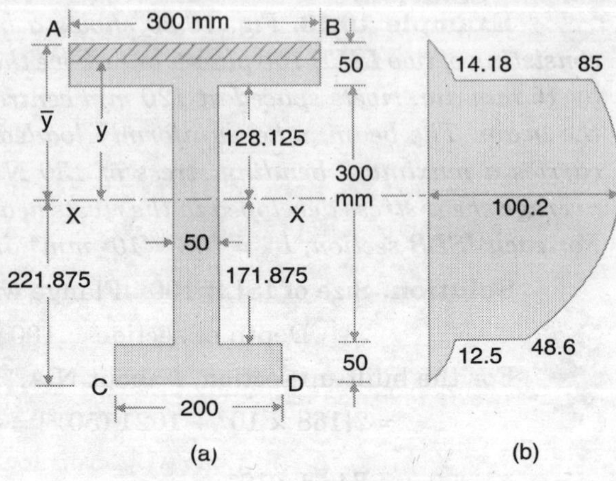

Fig. 10.21

Now, at any distance y from the N.A., $q = \dfrac{F}{I\,.\,z}(A\bar{y})$

where $\quad z = 300$ mm $\quad$ and $\quad A\bar{y} = 300\,(\bar{y} - y)\,(\bar{y} + y)\,\frac{1}{2} = 150\,(\bar{y}^2 - y^2)$

$$= 150\,(178.125^2 - y^2) = 150\,(31729 - y^2)$$

$$\therefore \quad q = \frac{F}{I.z}(A\bar{y}) = \frac{1600 \times 10^3}{86419 \times 10^4 \times 300} \times 150\,(31729 - y^2) = 9.257 \times 10^{-4}\,(31729 - y^2)$$

The variation is parabolic, giving $q = 0$ at $y = 178.125$ mm.

At the junction with the web, $q_f = 9.257 \times 10^{-4}\,(31729 - 128.125^2) =$ **14.18 N/mm²**

At the junction with the web, $q_w = 14.18 \times \dfrac{300}{50} = 85$ N/mm²

At the N.A.,

$$q_n = \frac{F}{I\,.\,z}(A\bar{y}) = \frac{1600 \times 10^3}{86419 \times 10^4 \times 50}\left[300 \times 50\,(178.125 - 25) + 50 \times 128.125 \times \frac{128.125}{2}\right]$$

$$= 85 + 15.2 = 100.2 \text{ N/mm}^2$$

Similarly, at the junction of web with lower flange,

$$q_f = \frac{1600 \times 10^3}{86419 \times 10^4 \times 300}\,[200 \times 50\,(221.875 - 25)] = 12.15 \text{ N/mm}^2$$

and $$q_w = 12.15 \times \frac{200}{50} = 48.6 \text{ N/mm}^2$$

The shear stress distribution is shown in Fig. 10.21 (*b*).

Example 10.18. *A laminated wooden beam 120 mm wide and 180 mm deep in made of three 120 mm × 60 mm planks glued together to resist longitudinal shear. The beam is simply supported over a span of 2.5 m. If the allowable shearing stress in the glued joint is 0.5 N/mm², find the safe U.D.L. the beam can carry.*

Solution.

$$I_{xx} = \frac{1}{12} \times 120\,(180)^3 = 5832 \times 10^4 \text{ mm}^4$$

$$F = \frac{wL}{2} = \frac{w \times 2.5}{2} = 1.25\, w \text{ N}$$

(where w is the safe U.D.L. in N/m)

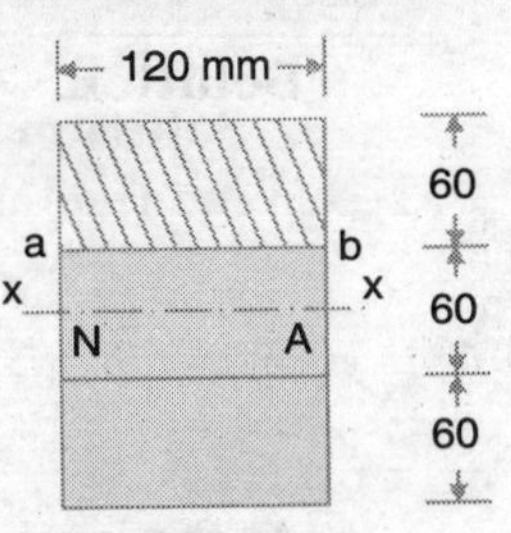

Fig. 10.22

At the glued joint ab,

$$q = \frac{F}{I \cdot z}(A\bar{y}) = \frac{1.25\, w}{5832 \times 10^4 \times 120}[120 \times 60 \times 60]$$

$$= 0.7716 \times 10^{-4}\, w \text{ N/mm}^2$$

But permissible value of shear stress is 0.5 N/mm²

$\therefore$ $0.7716 \times 10^{-4}\, w = 0.5$ from which $w =$ **6480 N/m.**

Example 10.19. *A steel beam of I-section, 200 mm deep and 160 mm wide has 16 mm thick flanges and 10 mm thick web. The beam is subjected to a shear force of 200 kN. Determine the shear stress distribution over the beam section if the web of the beam is kept horizontal.*

Solution.

$$I_{xx} = 2 \times \frac{1}{12} \times 16\,(160)^3 + \frac{1}{12} \times 168\,(10)^3$$

$$= 1094 \times 10^4 \text{ mm}^4$$

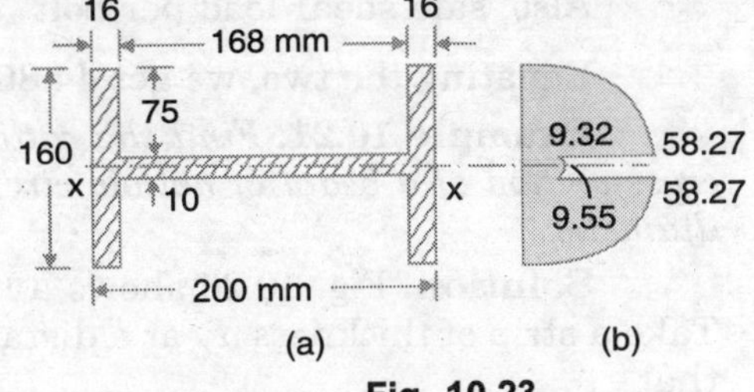

Fig. 10.23

Shear stress in the flange, just above the web

$$q_f = \frac{F}{Iz}(A\bar{y}) \text{ where } z = 2 \times 16 = 32 \text{ mm}$$

$$A\bar{y} = 2 \times 16 \times 75\,(37.5 + 5) = 102000 \text{ mm}^3$$

$$\therefore \quad q_f = \frac{200 \times 10^3}{1094 \times 10^4 \times 32} \times 102000 = \mathbf{58.27 \text{ mm}^2}$$

Shear stress in the web, just at the junction

$$q_w = q_f \times \frac{32}{200} = 58.27 \times \frac{32}{200} = \mathbf{9.32 \text{ N/mm}^2}$$

Shear stress at N.A.

$$q_n = \frac{F}{Iz}(A\bar{y}), \text{ where } z = 200 \text{ mm}$$

$$A\bar{y} = 2 \times 16 \times 75\,(37.5 + 5) + 200 \times 5 \times 2.5 = 104500 \text{ mm}^3$$

$$\therefore \quad q_n = \frac{200 \times 10^3}{1094 \times 10^4 \times 200} \times 104500 = \mathbf{9.55 \text{ N/mm}^2}$$

The shear stress distribution is shown in Fig. 10.23 (b).

Example 10.20. *Three planks, each of 150 mm × 50 mm are bolted together by 10 mm diameter bolts to form an I-section, as shown in Fig. 10.24. The beam carries a central point load of 20 kN. If the shear stress in the bolt is not to exceed 100 N/mm² find the pitch of the bolts.*

Solution.

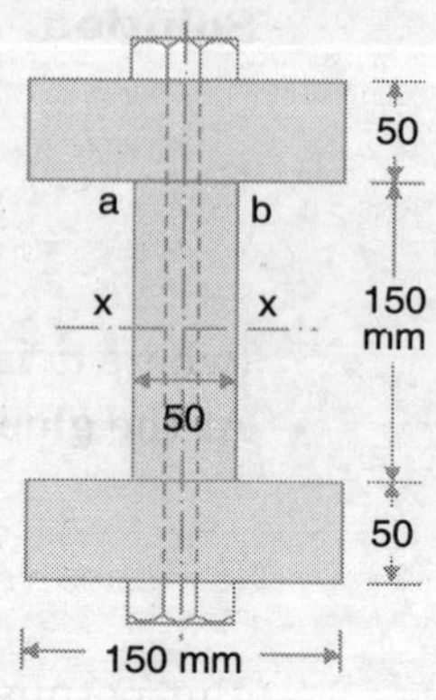

Fig. 10.24

$$I_{xx} = \frac{1}{12}\,[150\,(250)^3 - 100\,(150)^3]$$
$$= 16719 \times 10^4 \text{ mm}^4$$
$$F = \frac{20 \times 10^3}{2} = 10000 \text{ N}$$

At the level *ab*, the junction of the flange with the web

$$q = \frac{F}{Iz}\,(A\bar{y})$$

∴ Horizontal shear load per mm length of the beam is given by

$$Q = \frac{F}{I}\,(A\bar{y}) = \frac{10000}{16719 \times 10^4}\,(150 \times 50 \times 100) = 44.86 \text{ N/mm}$$

Let p be the pitch of the bolts, in mm.

∴ Horizontal shear load per pitch length $= 44.86\,p$ N ...(*i*)

Also, safe shear load per bolt $= f_s \dfrac{\pi}{4}\, d^2 = 100 \times \dfrac{\pi}{4}\,(10)^2 = 7854$ N ...(*ii*)

Equating the two, we get $44.86\,p = 7854$ from which $p =$ **175 mm.**

Example 10.21. *Find the ratio of maximum to mean intensity of vertical shear in the cross-section of a beam of hollow circular section, the outside diameter being twice the inside diameter.*

Solution. Fig. 10.25 shows a hollow circular section. Take a strip of thickness dy at a distance y from the N.A., so that

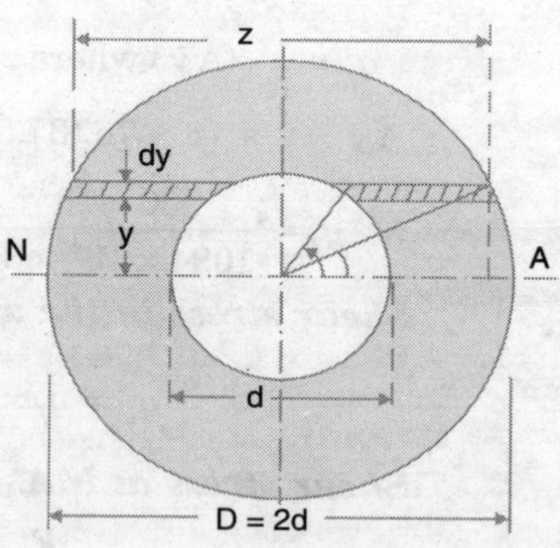

Fig. 10.25

$$y = \frac{D}{2}\sin\theta = \frac{d}{2}\sin\phi\,;$$

$$dy = \frac{D}{2}\cos\theta\, d\theta = \frac{d}{2}\cos\phi\, d\phi$$

Now $\quad q = \dfrac{F}{I\,.\,z}(A\bar{y})$

where $\quad z = 2\left[\dfrac{D}{2}\cos\theta - \dfrac{d}{2}\cos\varphi\right] = D\cos\theta - d\cos\varphi$

Area of strip $= (D\cos\theta - d\cos\varphi)\, dy$

$$= D\cos\theta\,.\,\frac{D}{2}\cos\theta\, d\theta - d\cos\varphi\,.\,\frac{d}{2}\cos\varphi\, d\varphi$$

$$= \frac{D^2}{2}\cos^2\theta\, d\theta - \frac{d^2}{2}\cos^2\varphi\, d\varphi.$$

Also, $A\bar{y} = \displaystyle\int_0^{\pi/2} \frac{D^2}{2}\cos^2\theta\, d\theta\,.\,y - \int_0^{\pi/2} \frac{d^2}{2}\cos^2\varphi\, d\varphi\,.\,y$

$$= \int_0^{\pi/2} \frac{D^2}{2}\cos^2\theta\, d\theta\,.\,\frac{D}{2}\sin\theta - \int_0^{\pi/2} \frac{d^2}{2}\cos^2\varphi\, d\varphi\,.\,\frac{d}{2}\sin\varphi$$

$$= \frac{D^3}{4}\int_0^{\pi/2} \cos^2\theta \sin\theta \, d\theta - \frac{d^3}{4}\int_0^{\pi/2} \cos^2\varphi \sin\varphi \, d\varphi$$

$$= \frac{D^3}{4} \cdot \frac{\cos^3\theta}{3} - \frac{d^3}{4} \cdot \frac{\cos^3\varphi}{3}$$

$$\therefore \qquad q = \frac{F \times \frac{1}{12}[D^3 \cos^3\theta - d^3 \cos^3\varphi]}{\frac{\pi}{64}(D^4 - d^4) \times (D\cos\theta - d\cos\varphi)}$$

Now shear stress will be maximum at the N.A. where $\theta = \varphi = 0$

$$\therefore \qquad q_{max} = \frac{\frac{F}{12}[D^3 - d^3]}{\frac{\pi}{64}(D^4 - d^4)(D-d)}. \text{ Also } q_{av} = \frac{F}{\frac{\pi}{4}(D^2 - d^2)}$$

$$\therefore \quad \frac{q_{max}}{q_{av}} = \frac{\frac{F}{12}(D^3 - d^3)}{\frac{\pi}{64}(D^4 - d^4)(D-d)} \times \frac{\frac{\pi}{4}(D^2 - d^2)}{F} = \frac{4(D^3 - d^3)}{3(D^2 + d^2)(D-d)}$$

For the present problem, $d = D/2$

$$\therefore \quad \frac{q_{max}}{q_{av}} = \frac{4\left[D^3 - \left(\frac{D}{2}\right)^3\right]}{3\left[D^2 + \left(\frac{D}{2}\right)^2\right]\left[D - \frac{D}{2}\right]} = \frac{4 \times \frac{7}{8}D^3}{3 \times \frac{5}{4}D^2 \times \frac{1}{2}D} = \frac{\mathbf{28}}{\mathbf{15}}.$$

PROBLEMS

1. A 61 cm × 19 cm I-joist has 2.7 cm thick flanges and 1.5 cm thick web. Calculate the maximum intensity of shear stress and sketch the distribution of shear stress across the section, the S.F. at the cross-section being 600 kN.
2. The cross-section of a joist is a tee-section 12.5 × 12.5 × 1.2 cm with 12.5 cm side horizontal. Find the maximum intensity of shear stress and sketch the distribution of shear across the section if it has to resist a shear force of 90 kN.
3. A beam of I-section is 25 cm deep and 20 cm wide. The flanges are 2.5 cm thick and the web 1.3 cm thick. Calculate the proportion in which the flanges and web resist (*i*) B.M. and (*ii*) shear force at a cross-section.
4. The section of a beam is an isosceles triangle with base 20 cm and side angles 30°. It is used with the base horizontal and carries a shear force of 60 kN at a particular section. Find the magnitude of maximum shear intensity and shear intensity at N.A.
5. A bar of regular hexagonal section on side a is subjected to a shearing force S which acts at right angles to one of the diagonals of the hexagon. Derive a formula for the shearing stress q at a distance x from the diagonal, and hence find the ratio of q to the mean shearing stress for the particular value $x = 0$ and $x = a/2$. Indicate by means of sketch how the shearing stress varies across the depth of the section. (*U.L.*)

6. At a section of a rectangular beam, the B.M. is 50 kN-m and shear force 150 kN. The section of the beam is a rectangle 10 cm wide and 20 cm deep. Calculate the bending and shearing stresses at 10 cm, 5 cm and 0 cm from the neutral axis and find the principal planes and principal stresses at these points.
7. A 25 cm × 15 cm R.S.J. has flanges 2 cm thick and web 1.2 cm thick. At a certain cross-section, it has to resist a bending moment of 80 kN-m and a shearing force of 160 kN. Find the principal planes and principal stresses (*a*) at the top, (*b*) in the web at 10.5 cm from N.A. and (*c*) at N.A.
8. Prove that in a thin circular tube, the maximum intensity of shear is twice the mean intensity over the whole section.
9. A cylindrical tube made of an alloy of density S has a diameter D at one end and $D/2$ at the other, with a straight taper in between. The wall thickness is every where constant and is every small compared with D. The tube is freely supported at the ends so that the axis, of length L, is horizontal. Calculate the maximum shear stress in tube due to its own weight.
10. A rectangular beam of height d, breadth b and length l is supported at the ends and carries a concentrated load P at the middle. Show that the principal stress at a point in the central cross-section and distance $\frac{d}{4}$ from the top are $\frac{3}{8}\frac{Pl}{bd^2}\left(1+\sqrt{1+\frac{9}{4}\frac{d^2}{l^2}}\right)$.

ANSWERS

1. 76.1 N/mm² 2. 85.2 N/mm²
3. M = 6.4% by web and 93.6% by flanges ; F = 10.8% by flanges and 89.2% by web.
4. 15.6 N/mm² ; 13.85 N/mm².
5. $$q = \frac{4S(9a^3 - 18ax^2 + 4x^2 + 4x^3\sqrt{3})}{15a^4\sqrt{3}(3a - x\sqrt{3})}$$

At $x = 0$, $q/q_{mean} = 1.20$; At $x = \frac{a}{2}$; $q/q_{mean} = 1.008$.

6. At $h = 10$ cm ; $\sigma_1 = 75$ N/mm² ; $\sigma_2 = 0$
At $h = 5$ cm ; $\sigma_1 = 39.3$ N/mm² (comp.) ; $\sigma_2 = 1.81$ N/mm² (tensile) ; $\theta_1 = 12° 6'$; $\theta_2 = 102° 6'$
At $h = 0$, $\sigma_1 = \sigma_2 = 11.25$ N/mm² ; $\theta_1 = 45°$, $\theta_2 = 135°$
7.

σ_1 (N/mm²) ;	σ_2 (N/mm²)	θ_1	θ_2
(*a*) 112.5 (comp.)	zero	0°	90°
(*b*) 117.8 (comp.)	22.9 (tension)	23° 48′	113° 48′
(*c*) 61.9 (comp.)	61.9 (tension)	45°	135°

9. $\frac{4SL}{3}$.

11

Deflection of Beams I : Double Integration Method

11.1. INTRODUCTION

The manner in which the bending moment and shearing force is resisted by the cross-section of beam has been explained in the previous two chapters. Obviously, the cross-section of a beam must be *strong* enough to resist the bending and shearing stresses. This is one criterion of beam design, generally known as *strength criterion*. There is another aspect of bending, which is equally important : the *deflection* of the beam. The maximum deflection of the beam must not exceed a given limit, *i.e.,* the beam must be stiff enough to resist against the deflection. Thus, the *stiffness* of the beam, which is inversely proportional to the deflection, is the second criterion of beam design.

The study of deflection is not only important as a design criterion but also as only means of determining B.M. and S.F. in cases of beams with restraints, *e.g.,* continuous beams, fixed beams etc. The end reactions and consequently B.M. and S.F. cannot be calculated in such cases without studying the deformation of the axis of the beam. In the present chapter, we shall discuss the stiffness aspect and find the slope and deflection of the centre line of the beam at any section.

11.2. DIFFERENTIAL EQUATION OF THE DEFLECTION CURVE OF BEAM

Fig. 11.1 shows the deflected shape or the deflection curve of a loaded beam. When a beam is subjected to pure bending couple M, it is bent into a circular arc the radius of curvature of which is given by Eq. 8.5

$$\frac{1}{R} = \frac{M}{EI} \qquad ...(i)$$

However, when the beam is not subjected to a pure bending couple, as is generally the case, its radius of curvature will not be constant, but will change from point to point. The first step will therefore be to find an expression for the radius of curvature of the deflection curve at any given point.

Let m and n be two points at a very small distance δs along the deflection curve. Let x and y be the coordinates of point m and $(x + \delta x)$ and $(y + \delta y)$ be the coordinates of the point n. We take the origin of the coordinates at the end, with x-axis directed to the right and the y-axis directed downwards. Thus xy plane is the *plane of bending*. The deflection y of the beam at any point m at a distance x from the origin is the translation (or displacement) of that point in the y-direction, measured from the x-axis to the deflection curve. We will consider *downward* deflection as *positive* and *upward* deflection as *negative*. The equation of the deflection curve is obtained when y is expressed as the function of x.

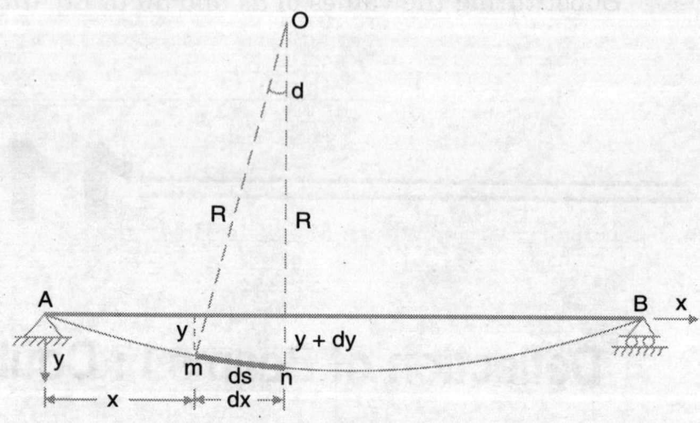

Fig. 11.1

In Fig. 11.2, let the tangent mT_1 to the curve at m be making an angle θ and the tangent nT_2 at n be making an angle $\theta + d\theta$ with the x-axis. Let R be the radius of the curve at m. Since δs is extremely small, we have

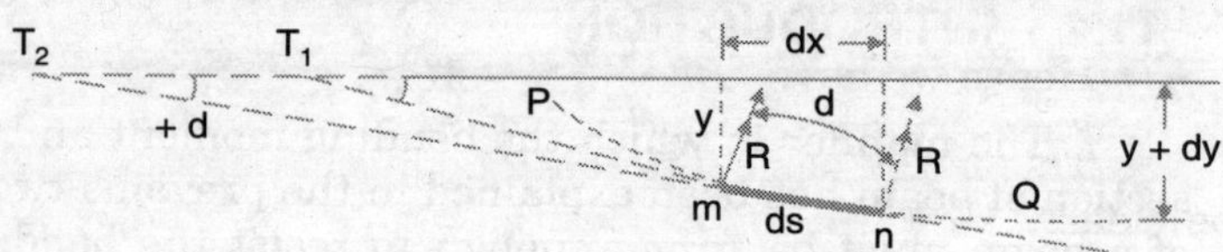

Fig. 11.2

$$ds = R\,d\theta \quad \text{and} \quad \frac{1}{R} = \left|\frac{d\theta}{ds}\right| \qquad \text{...}(ii)$$

The bars in the above expression indicate that we consider here only the numerical value of curvature. Regarding sign, it should be clearly borne is mind that B.M. is taken positive in Eq. (*i*) if it produces concavity. *Hence curvature is positive when the centre of curvature is above the curve*, as in Figs. 11.1 and 11.2. By close inspection of Fig. 11.2, we observe that for such a curvature, the angle θ decreases as the point m moves along the curve from point P to Q. Thus, we have proper sign for Eq. (*ii*), which must be written as

$$\frac{1}{R} = -\frac{d\theta}{ds} \qquad \text{...}(iii)$$

Also, $$ds = \sqrt{dx^2 + dy^2} \quad \text{or} \quad ds = dx\sqrt{1+\left(\frac{dy}{dx}\right)^2} \qquad \text{...}(iv)$$

Now $$\tan\theta = \frac{dy}{dx}$$

Differentiating, we get $\sec^2\theta\, d\theta = \dfrac{d^2y}{dx^2}dx$ or $(1 + \tan^2\theta)\, d\theta = \dfrac{d^2y}{dx^2}dx$

or $$d\theta = \frac{\dfrac{d^2y}{dx^2}dx}{1+\left(\dfrac{dy}{dx}\right)^2} \qquad \text{...}(v)$$

Substituting the values of ds and $d\theta$ in Eq. (*iii*), we have

$$\frac{1}{R} = -\frac{\frac{d^2y}{dx^2}dx}{1+\left(\frac{dy}{dx}\right)^2} \times \frac{1}{dx\sqrt{1+\left(\frac{dy}{dx}\right)^2}} = -\frac{\frac{d^2y}{dx^2}}{\left[1+\left(\frac{dy}{dx}\right)^2\right]^{3/2}} \quad \text{...(vi)}$$

But $\frac{1}{R} = \frac{M}{EI}$ where M is the B.M. at m.

$$\text{Hence} \quad \frac{M}{EI} = -\frac{\frac{d^2y}{dx^2}}{\left[1+\left(\frac{dy}{dx}\right)^2\right]^{3/2}} \quad \text{or} \quad EI\frac{d^2y}{dx^2} = -M\left[1+\left(\frac{dy}{dx}\right)^2\right]^{3/2} \quad \text{...(11.1)}$$

Since $\frac{dy}{dx}$ (*i.e.*, the slope of the tangent to the curve at any point m) is extremely small, $\left(\frac{dy}{dx}\right)^2$ will be still smaller and hence $\left[1+\left(\frac{dy}{dx}\right)^2\right]$ will be very nearly equal to unity.

$$\text{Hence we have} \quad EI\frac{d^2y}{dx^2} = -M \quad \text{...[11.1(a)]}$$

This is the basic differential equation for the deflected shape of the beam. This should be clearly borne in mind that the above equation takes into account the effect of bending moment only. The effect of shear on deflection is extremely small and may be neglected.

The basic differential equation can also be written as follows :

$$\frac{d^2y}{dx^2} = \frac{d}{dx}\left(\frac{dy}{dx}\right) = \frac{d\theta}{dx}\left(\text{or } \frac{di}{dx}\right) = -\frac{M}{EI} \quad \text{...[11.1(b)]}$$

This equation can be integrated in each particular case to find the angle of rotation θ or i (usually called the slope) or the deflection y, provided the bending moment M is known.

SIGN CONVENTIONS

We adopt the following sign conventions :

(*i*) x is positive when measured to the right

(*ii*) y is positive when measured downwards

(*iii*) θ (or i) is positive when the rotation is clockwise from the x-axis

(*iv*) M is positive when sagging, *i.e.*, when it produces compression in the upper part of the beam.

11.3. SLOPE AND DEFLECTION AT A POINT

There are several methods to find the slope and deflection of the centre line of a beam at a certain location along its length. However, only the following methods will be discussed in the present chapter and in subsequent chapters :

(*i*) Double integration method ;
(*ii*) Macaulay's method.
(*iii*) Area-moment method ;
(*iv*) Conjugate beam method.
(*v*) Strain energy method.

11.4. DOUBLE INTEGRATION METHOD

We have already derived the differential equation :

$$EI\frac{d^2y}{dx^2} = -M \text{ where } M \text{ is sagging bending moment.} \quad ...(i)$$

Integrating, $$EI\frac{dy}{dx} = EI \cdot i = -\int M \quad ...(ii)$$

and $$EIy = -\iint M \quad ...(iii)$$

Thus, integrating the original differential equation once, we get slope, and integrating it further, we can get value of deflection at any point. On integration, two constants C_1 and C_2 will be introduced. These constants can very easily be determined by using the end conditions.

We shall now take different cases.

(A) CANTILEVERS

(1) Cantilever with Uniformly Distributed load (Fig. 11.3).

At any distance x from A, we have

$$EI\frac{d^2y}{dx^2} = -M = -\left[-\frac{w(L-x)^2}{2}\right] = \frac{w(L-x)^2}{2}$$

Integrating, $$EI\frac{dy}{dx} = -\frac{w(L-x)^3}{6} + C_1$$

At $x = 0$, $\frac{dy}{dx} = 0$, $\therefore \quad C_1 = \frac{wL^3}{6}$

Hence $$EI\frac{dy}{dx} = -\frac{w(L-x)^3}{6} + \frac{wL^3}{6} \quad ...(1)$$

Fig. 11.3

Integrating further, $$EIy = +\frac{w(L-x)^4}{24} + \frac{wL^3}{6}x + C_2$$

At $x = 0, y = 0, \quad \therefore \quad C_2 = -\frac{wL^4}{24}$. Hence $$EIy = \frac{w(L-x)^4}{24} + \frac{wL^3}{6}x - \frac{wL^4}{24} \quad ...(2)$$

At $x = 0$, we get from (1), $EI\frac{dy}{dx} = EI\, i_A = -\frac{wL^3}{6} + \frac{wL^3}{6} = 0$, as expected.

At $x = L$, we have $EI\, i_B = +\frac{wL^3}{6}$, from (1) or $i_B = \frac{wL^3}{6EI}$ radians ....(11.2)

And $$EIy_B = \frac{wL^4}{6} - \frac{wL^4}{24} = \frac{wL^4}{8} \quad \text{or} \quad y_B = \frac{wL^4}{8EI} \quad ...(11.3)$$

(2) Cantilever Partially Loaded with w over a Length L_1 from the Fixed End (Fig. 11.4).

Let the U.D.L. w extend upto a length L_1 from A.

From equations 11.2 and 11.3, we have

$$i_C = \frac{wL_1^3}{6EI} \quad \text{and} \quad y_C = \frac{wL_1^4}{8EI}$$

Since there is no load on CB, the portion CB remains straight.

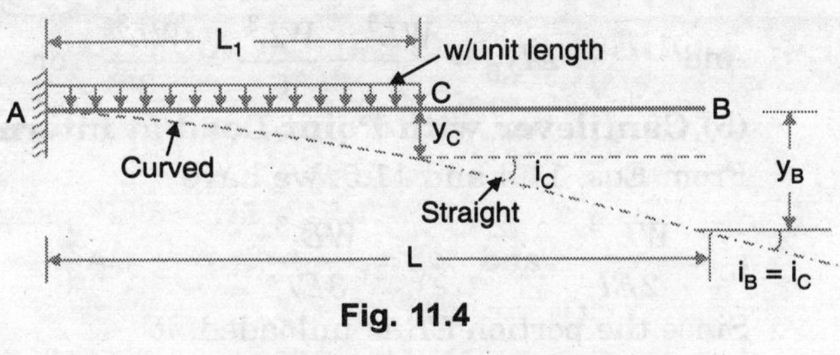

Fig. 11.4

Hence $i_B = i_C = \dfrac{wL_1^3}{6EI}$

And $y_B = y_C + i_C (L - L_1)$

$$= \frac{wL_1^4}{8EI} + \frac{wL_1^3}{6EI}(L - L_1).$$

(3) Cantilever Partially Loaded with w from the Free End (Fig. 11.5).

The slope and deflection may be looked upon as the difference between the slopes and deflections with (*a*) entire span loaded from A to B and (*b*) upward U.D.L. from A to C.

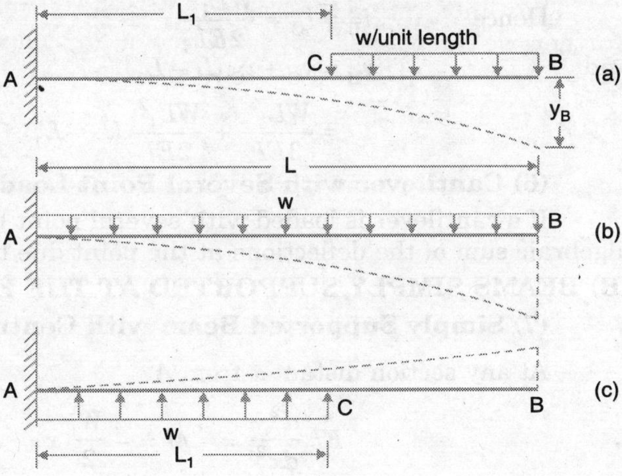

Fig. 11.5

$$\therefore\ i_B = \frac{wL^3}{6EI} - \frac{wL_1^3}{6EI} = \frac{w}{6EI}(L^3 - L_1^3)$$

and $$y_B = \frac{wL^4}{8EI} - \left\{\frac{wL_1^4}{8EI} + \frac{wL_1^3}{6EI}(L - L_1)\right\}$$

(4) Cantilever with Point Load at the Free End (Fig. 11.6).

At any point distance x from A, we have

Fig. 11.6

$$EI\frac{d^2y}{dx^2} = -M = -[-W(L-x)] = W(L-x)$$

Integrating, $EI\dfrac{dy}{dx} = -\dfrac{W(L-x)^2}{2} + C_1$

At the fixed end, $x = 0$,

and $\dfrac{dy}{dx} = 0, \quad \therefore\ C_1 = \dfrac{WL^2}{2}$

$$\therefore \qquad EI\frac{dy}{dx} = -\frac{W(L-x)^2}{2} + \frac{WL^2}{2} \qquad ...(1)$$

Integrating further $EIy = +\dfrac{W(L-x)^3}{6} + \dfrac{WL^2}{2}x + C_2$

At the fixed end, $x = 0$, and $y = 0$. $\therefore\ C_2 = -\dfrac{WL^3}{6}$

Hence $$EIy = \frac{W(L-x)^3}{6} + \frac{WL^2}{2}x - \frac{WL^3}{6} \qquad ...(2)$$

Eqs. (1) and (2) give slope and deflection respectively, at any point X.

At $x = L$, $EIi_B = \dfrac{WL^2}{2}$ or $i_B = \dfrac{WL^2}{2EI}$ ...(11.4)

And $\quad EIy_B = \dfrac{WL^3}{2} - \dfrac{WL^3}{6} = \dfrac{WL^3}{3}$ or $y_B = \dfrac{WL^3}{3EI}$ ...(11.5)

(5) Cantilever with Point Load at intermediate point (Fig. 11.7).

From Eqs. 11.4 and 11.5, we have

$$i_C = \frac{WL_1^{\ 2}}{2EI} \quad \text{and} \quad y_C = \frac{WL_1^{\ 3}}{3EI}$$

Since the portion CB is unloaded, it will remain straight.

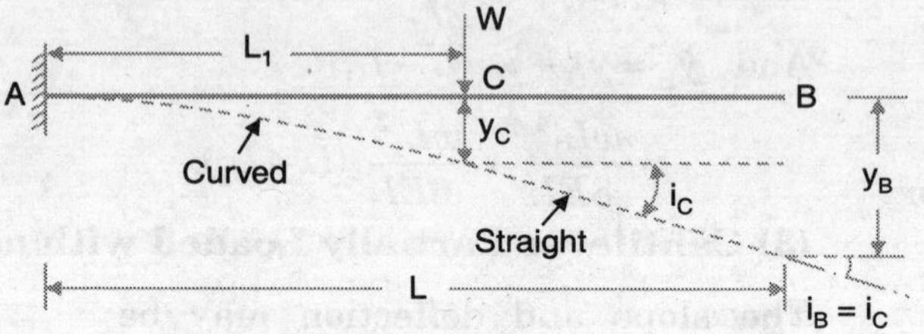

Hence $\quad i_B = i_C = \dfrac{WL_1^{\ 2}}{2EI}$

and $\quad y_B = y_C + i_C\,(L - L_1)$

$$= \frac{WL_1^{\ 3}}{3EI} + \frac{WL_1^{\ 2}}{2EI}\,(L - L_1). \qquad ...[11.5\,(a)]$$

Fig. 11.7

(6) Cantilever with Several Point Loads.

If a cantilever is loaded with several point loads, the deflection at any point will be the algebraic sum of the deflections at the point due to the point loads acting individually.

(B) BEAMS SIMPLY SUPPORTED AT THE ENDS

(7) Simply Supported Beam with Central Point Load W (Fig. 11.8).

At any section distant x from A

$$EI\frac{d^2y}{dx^2} = -M = -\frac{W}{2}x$$

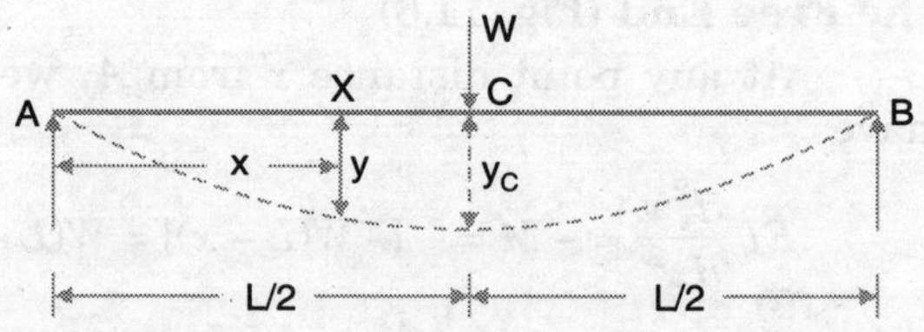

Integrating, $\quad EI\dfrac{dy}{dx} = -\dfrac{W}{2}\dfrac{x^2}{2} + C_1$

and $\quad EIy = -\dfrac{Wx^3}{12} + C_1x + C_2$

At $x = 0$, $y = 0$ $\therefore$ $C_2 = 0$.

Fig. 11.8

Also, due to symmetry, at $x = \dfrac{L}{2}$, $i_C = 0 = -\dfrac{W}{4}\left(\dfrac{L}{2}\right)^2 + C_1$ or $C_1 = \dfrac{WL^2}{16}$

Hence the equations for slope and deflection are

$$EI\frac{dy}{dx} = -\frac{W}{4}x^2 + \frac{WL^2}{16} \quad \text{and} \quad EIy = -\frac{Wx^3}{12} + \frac{WL^2}{16}x$$

At $x = 0$, $\quad i_A = \dfrac{1}{EI}\left(\dfrac{WL^2}{16}\right) = \dfrac{WL^2}{16EI}$; By symmetry $i_B = -\dfrac{WL^2}{16EI}$ ...(11.6)

Also, at $x = \dfrac{L}{2}$, $\quad y_{max} = \dfrac{1}{EI}\left(-\dfrac{W}{12}\dfrac{L^3}{8} + \dfrac{WL^2}{16}\cdot\dfrac{L}{2}\right) = \dfrac{wL^3}{48EI}$ ...(11.7)

(8) Simply Supported Beam with U.D.L. (Fig. 11.9).

At any section distant x from A

$$EI\frac{d^2y}{dx^2} = -M = -\left[\frac{wL}{2}x - \frac{wx^2}{2}\right] = -\frac{wL}{2}x + \frac{wx^2}{2}$$

$$\therefore \quad EI\frac{dy}{dx} = -\frac{wL}{2}\frac{x^2}{2} + \frac{wx^3}{6} + C_1$$

Due to symmetry, at $x = L/2$, $i_C = 0$

$$\therefore \quad 0 = -\frac{wL}{4}\left(\frac{L}{2}\right)^2 + \frac{w}{6}\left(\frac{L}{2}\right)^3 + C_1$$

or $$C_1 = \frac{wL^3}{24}$$

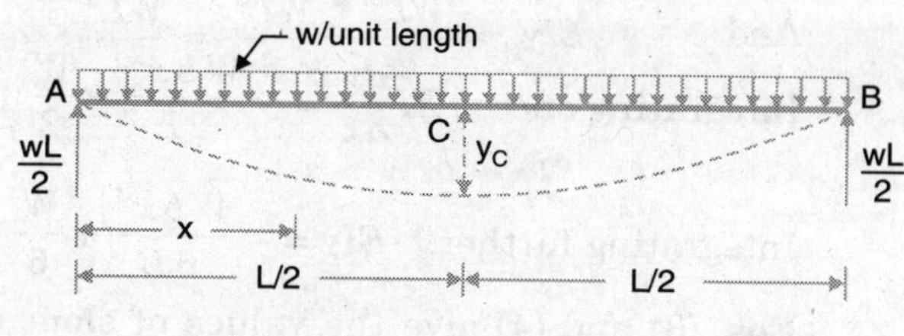

Fig. 11.9

Hence, $$EI\frac{dy}{dx} = -\frac{wL}{4}x^2 + \frac{wx^3}{6} + \frac{wL^3}{24} \quad ...(1)$$

Integrating further, $$EIy = -\frac{wL}{12}x^3 + \frac{wx^4}{24} + \frac{wL^3}{24}x + C_2$$

At $x = 0, y = 0 \quad \therefore \quad C_2 = 0$

Hence $$EIy = -\frac{wL}{12}x^3 + \frac{wx^4}{24} + \frac{wL^3}{24}x \quad ...(2)$$

Eqs. (1) and (2) give slope and deflection respectively, at any point

At A, $x = 0$, Hence $i_A = \dfrac{wL^3}{24EI}$ (from 1) ...(11.8)

Also, $i_B = -\dfrac{wL^3}{24EI}$ by symmetry

At $x = \dfrac{L}{2}$, $EIy_C = EIy_{max} = -\dfrac{wL}{12}\left(\dfrac{L}{2}\right)^3 + \dfrac{w}{24}\left(\dfrac{L}{2}\right)^4 + \dfrac{wL^3}{24}\left(\dfrac{L}{2}\right) = \dfrac{5}{384}wL^4$

$$\therefore \quad y_{max} = y_C = \frac{5}{384}\frac{wL^4}{EI} \quad ...(11.9)$$

(9) Simply Supported Beam with Eccentric Point Load

Fig. 11.10 shows a beam AB of length L, simply supported at the ends, and loaded with a point load W at point C, distant a from end A and b from end B,

Naturally, $R_A = \dfrac{Wb}{L}$ and $R_B = \dfrac{Wa}{L}$.

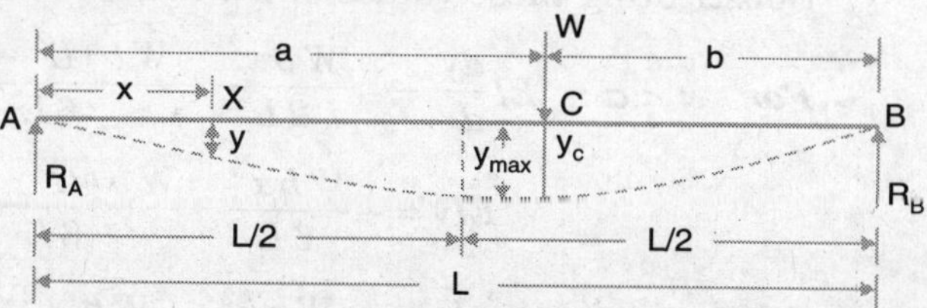

Fig. 11.10

For $x < a$:

$$EI\frac{d^2y}{dx^2} = -M = -\left[\frac{Wb}{L}x\right]$$

Integrating once $$EI\frac{dy}{dx} = -\frac{Wb}{L}\cdot\frac{x^2}{2} + C_1 \quad ...(1)$$

Integrating further $$EIy = -\frac{Wbx^3}{6L} + C_1x + C_2 \quad ....(2)$$

Eqs. (1) and (2) give the values of slope and deflection for the portion AC only.

For $x > a$: $$EI\frac{d^2y}{dx^2} = -\frac{Wb}{L}x + W(x - a)$$

Integrating once, $$EI\frac{dy}{dx} = -\frac{Wbx^2}{2L} + \frac{W}{2}(x-a)^2 + C_3 \quad ...(3)$$

Integrating further, $$EIy = -\frac{Wbx^3}{6L} + \frac{W}{6}(x-a)^3 + C_3x + C_4 \quad ...(4)$$

Eqs. (3) and (4) give the values of slope and deflection for the portion *CB* only. The constants of integration appearing in the above four equations can be found from the following conditions : (1) at $x = a$, the slope dy/dx $(= i_C)$ for the two parts of the beam should be equal ; (2) at $x = a$, the deflection $y(= y_C)$ for the two parts of the beam should be equal ; (3) at $x = 0$, the deflection y_A is zero and (4) at $x = L$, the deflection y_B is zero.

For the first condition, the slopes determined by Eqs. (1) and (3) are equal when $x = a$.

$$\therefore \quad -\frac{Wba^2}{2L} + C_1 = -\frac{Wba^2}{2L} + C_3 \text{ ; This gives } \quad C_1 = C_3$$

For the second condition, the deflections determined by Eqs. (2) and (4) should be equal when $x = a$.

$$\therefore \quad -\frac{Wba^3}{6L} + C_1a + C_2 = -\frac{Wba^3}{6L} + C_3a + C_4 \text{ ; This gives } \quad C_2 = C_4$$

Again, applying the third condition, *i.e.*, at $x = 0$, $y = 0$ we get from Eq. (2)

$$0 = C_2. \text{ Hence } C_4 = 0$$

Lastly, applying the fourth condition, *i.e.*, at $x = L$, $y = 0$, we get from (4)

$$0 = -\frac{WbL^3}{6L} + \frac{W}{6}(L-a)^3 + C_3L + 0 \text{ which gives}$$

$$C_3 = C_1 = \frac{Wb(L^2 - b^2)}{6L}$$

Substituting these values of C_1, C_2, C_3 and C_4 in Eqs. (1) to (4), we get

For $\boldsymbol{x < a}$**:** $$EI\frac{dy}{dx} = -\frac{Wbx^2}{2L} + \frac{Wb(L^2-b^2)}{6L} = \frac{Wb}{6L}(L^2 - b^2 - 3x^2) \quad ...(I) ...[11.10(a)]$$

$$EIy = -\frac{Wbx^3}{6L} + \frac{Wb(L^2-b^2)}{6L}x = \frac{Wb}{6L}x(L^2 - b^2 - x^2) \quad ...(II) ...[11.10(b)]$$

For $\boldsymbol{x > a}$**:** $$EI\frac{dy}{dx} = -\frac{Wbx^2}{2L} + \frac{Wb(L^2-b^2)}{6L} + \frac{W}{2}(x-a)^2$$

$$= \frac{Wb}{6L}(L^2 - b^2 - 3x^2) + \frac{W}{2}(x-a)^2 \quad ...(III) ...[11.10(c)]$$

and $$EIy = -\frac{Wbx^3}{6L} + \frac{Wb(L^2-b^2)}{6L}x + \frac{W}{6}(x-a)^3$$

$$= \frac{Wbx}{6L}(L^2 - b^2 - x^2) + \frac{W}{6}(x-a)^3 \quad ...(IV) ...[11.10(d)]$$

From the above four equations, the slope and deflection at any point can be found.

Slopes at the ends

At $x = 0$, we get from Eq. (I), $i_A = \dfrac{Wb}{6EIL}(L^2 - b^2) = \dfrac{Wab}{6EIL}(L + b)$ ...[11.11(*a*)]

At $x = L$, we get from Eq. (III), $i_B = -\dfrac{Wa}{6EIL}(L^2 - a^2) = -\dfrac{Wab}{6EIL}(L + a)$ ...[11.11(*b*)]

Maximum deflection

Eqs. (I) and (II) give slope and deflection for the portion *AC*. Maximum deflection will occur in *AC*, if $a > b$. For maximum deflection, $dy/dx = 0$

Hence from (I), $0 = \dfrac{Wb}{6L}(L^2 - b^2 - 3x^2)$. This gives $x = \sqrt{\dfrac{L^2 - b^2}{3}}$ ...(11.12)

Hence from (II), $y_{max} = \dfrac{Wb}{6EIL}\sqrt{\dfrac{L^2 - b^2}{3}}\left[L^2 - b^2 - \left(\dfrac{L^2 - b^2}{3}\right)\right]$

or $y_{max} = \dfrac{Wb}{3EIL}\left(\dfrac{L^2 - b^2}{3}\right)^{3/2} = \dfrac{Wb(L^2 - b^2)^{3/2}}{9\sqrt{3}\,EIL}$ ...(11.13)

Deflection under the load

From (II), at $x = a$, $y_C = \dfrac{Wab}{6EIL}(L^2 - b^2 - a^2) = \dfrac{Wa^2b^2}{3EIL}$ ...(11.14)

Central point load : If the load *W* is applied at the mid-span (*i.e.*, if $a = b = L/2$) the maximum deflection will evidently be at the mid-span and its magnitude is given by Eq. 11.14.

$$\Big[y\Big]_{x=L/2}^{a=b} = \frac{W}{3EIL}\left(\frac{L}{2}\right)^2\left(\frac{L}{2}\right)^2 = \frac{WL^3}{48EI} \quad ...(11.15)$$

Deflection at mid-span

From Eq. (II), $\Big[y\Big]_{x=L/2}^{a>b} = \dfrac{Wb}{6EIL}\cdot\dfrac{L}{2}\left(L^2 - b^2 - \dfrac{L^2}{4}\right) = \dfrac{Wb}{48EI}(3L^2 - 4b^2)$ ...(11.16)

Deflection at mid-span versus maximum deflection

From Eq. 11.12, we find that when $b = L/2$, $x = 0.5\,L$. In the most unfavourable case, when *b* is extremely small (*i.e.*, when *W* is very near to the support), we get $x \simeq L/\sqrt{3} = 0.577\,L$ from Eq. 11.12. Thus the point of maximum deflection is only $0.077\,L$ (or $L/13$) away from the centre. Also, the difference between Eqs. 11.13 and 11.16 in the most unfavourable case (when *b* approaches zero) is only 2.5% of the maximum deflection. *Hence we conclude that the deflection at the mid-span is a close approximation to the maximum deflection.*

CONSIDERATION OF UNITS

While computing the numerical values of slope and deflection, proper consideration of units must be made. Generally, span is expressed in metres while *E* and *I* are expressed in mm units. Various formulae developed in the previous article, for various standard cases, can't give correct results unless proper care is taken for units.

(*a*) Computation of deflection : We find from various equations for deflection that *L* appears in third power, *i.e.*, $y = k_1L^3/EI$. Hence for working out the numerical value of deflection in mm, the value of *L* must be substituted in metres and then the whole expression must be multiplied by $(1000)^3$. Thus, $y = k_1L^3/EI \times 10^9$ mm, when *L* is substituted in metres and *E* and *I* are expressed in mm units. However, *W* and *E* must have the same units of force (*i.e.*, N or kN).

(b) Computation of slope : We find from various equations for slope that L appears in second power, *i.e.*, $i = k_2L^2/EI$. Hence for working out the numerical value of slope in radians, L must be substituted in metres and then the whole expression must be multiplied by $(1000)^2$. Thus $i = k_2L^2/EI \times 10^6$ radians, when L is substituted in metres and E and I are expressed in mm units. However, W and E must have the same units of force.

Example 11.1. *A cantilever 2 metre long is loaded as shown in Fig. 11.11. Calculate the deflection at the free end if the section is rectangular, 100 mm × 200 mm. Take E = 0.105 × 10⁵ N/mm².*

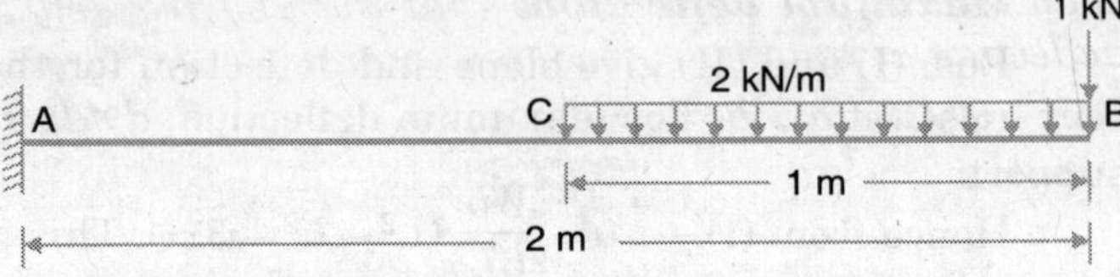

Fig. 11.11

Solution.

$$I = \frac{1}{12} \times 100(200)^3 = 66.67 \times 10^6 \text{ mm}^4\ ;\ E = 0.105 \times 10^5 \text{ N/mm}^2$$

Due to U.D.L.,

$$y_{B1} = \frac{wL^4}{8EI} - \left[\frac{wL_1^{\ 4}}{8EI} + \frac{wL_1^{\ 3}}{6EI}(L - L_1)\right] = \frac{2000}{EI}\left[\frac{(2)^4}{8} - \frac{(1)^4}{8} - \frac{(1)^3\ (2-1)}{6}\right](1000)^3$$

$$= \frac{3.417 \times 10^{12}}{EI} = \frac{3.417 \times 10^{12}}{0.105 \times 10^5 \times 66.67 \times 10^6} = 4.88 \text{ mm}$$

Due to point load,

$$y_{B2} = \frac{WL^3}{3EI} = \frac{1000(2)^3\ (1000)^3}{3 \times 0.105 \times 10^5 \times 66.67 \times 10^6} = 3.81 \text{ mm}$$

∴ Total deflection of $B = 4.88 + 3.81 =$ **8.69 mm.**

Example 11.2. *A girder of uniform section and constant depth of 500 mm is freely supported over a span of 8 metres. Calculate the deflection for a uniformly distributed load on it such that the maximum bending stress induced is 90 N/mm². Take E = 2.04 × 10⁵ N/mm².*

If for the above girder, the flanges are so proportioned that there is a uniform flange stress of 90 N/mm² throughout the beam, calculate the central deflection.

Solution. Let w be the U.D.L in N/m length.

$$\text{Central B.M.} \quad = \frac{wL^2}{8} = \frac{w(8)^2}{8} = 8w \text{ N-m} = 8000\ w \text{ N-mm}$$

$$\text{Maximum stress} \quad f = \frac{M}{Z} = \frac{M}{I}\ y.$$

$$\therefore \quad 90 = \frac{8000w}{I} \times 250 \quad \text{or} \quad \frac{w}{I} = \frac{90}{8000 \times 250} = 4.5 \times 10^{-5}$$

$$y_{\max} = y_C = \frac{5}{384}\frac{wL^4}{EI} = \frac{5}{384}\frac{w}{I}\frac{(8)^4\ (1000)^3}{2.04 \times 10^5} = 26.14 \times 10^4\ \frac{w}{I}$$

$$= (26.14 \times 10^4)\ (4.5 \times 10^{-5}) = \mathbf{11.76\ mm.}$$

If a girder is of constant depth and uniform strength, it will bend to a circular arc of radius R. The radius R is found from the relation

$$\frac{f}{y} = \frac{E}{R} \quad \text{or} \quad \frac{1}{R} = \frac{f}{yE} = \frac{90}{250 \times 2.04 \times 10^5}$$

For a circular arc, the central deflection is given by

$$y_C = \frac{L^2}{8R} = \frac{64(1000)^2}{8} \times \frac{90}{250 \times 2.04 \times 10^5} = \mathbf{14.12\ mm.}$$

Example 11.3. *A uniform beam of length L is simply and symmetrically supported on a span l. Find the ratio L/l so that the upward deflection at each end equals the downward deflection at mid-span, due to a central point load.*

Solution. We know that due to central point load, the slope at the supports of a simply supported beam = $Wl^2/16\,EI$ and the deflection at the centre = $Wl^3/48\,EI$.

Hence $\quad i_B = \dfrac{Wl^2}{16\,EI}$

Since the angle is very small,

$$y_A = i_B\ (AB) = \frac{Wl^2}{16\,EI}\ \frac{(L-l)}{2} \quad ...(i)$$

Also, deflection at the centre $= \dfrac{Wl^3}{48EI}$ ...(*ii*)

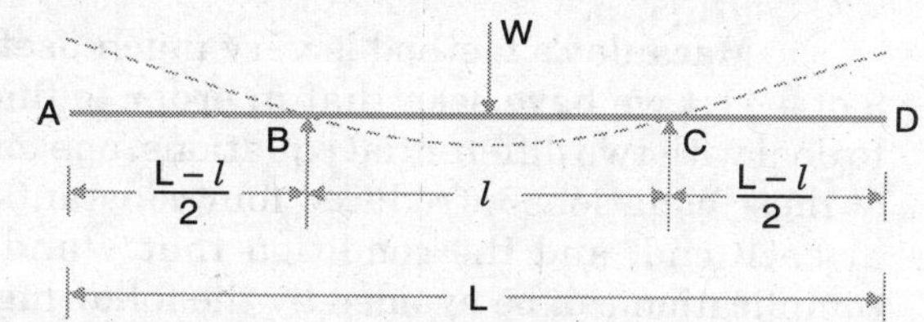

Fig. 11.12

Equating (*i*) and (*ii*) as per conditions given, we have

$$\frac{Wl^2}{16EI}\ \frac{(L-l)}{2} = \frac{Wl^3}{48EI} \quad \text{or} \quad (L-l) = \frac{2}{3}l \quad \text{or} \quad \frac{L}{l} = \frac{2}{3} + 1 = \mathbf{\frac{5}{3}}.$$

Example 11.4. *A uniform circular bar of length L extends by an amount Δ under a tensile pull P. Show that if the bar is used as a beam simply supported at its ends and carrying a central load W, the maximum deflection y is given by* $y = \dfrac{W\,\Delta.L^2}{3\,Pd^2}$ *where d = diameter of the bar. If L = 80 d and maximum bending stress due to W is equal to 0.9 times the tensile stress due to the pull P, find the ratio y/Δ.*

Solution. Due to the tensile load *P*, the extension Δ is given by

$$\Delta = \frac{PL}{\frac{\pi}{4}d^2E} = \frac{4PL}{\pi d^2 E} \quad \text{or} \quad E = \frac{4\,PL}{\pi d^2 \Delta} \quad ...(i)$$

Also, due to central load, the central deflection is given by

$$y = y_{\max} = \frac{WL^3}{48EI} = \frac{WL^3}{48E\frac{\pi}{64}d^4} = \frac{4WL^3}{3E\pi d^4} \quad ...(ii)$$

Substituting the value of *E* in (*ii*), we get

$$y = \frac{4WL^3}{3\pi d^4} \times \frac{\pi d^2 \Delta}{4\,PL} = \frac{W\,\Delta\,L^2}{3\,Pd^2} \quad \text{(Proved)} \quad ...(iii)$$

$$\text{Now max. bending stress} = f = \frac{M}{Z} = \frac{WL}{4} \times \frac{32}{\pi d^3} = \frac{8W}{\pi} \times \frac{80d}{d^3} = \frac{640W}{\pi d^2} \quad ...(iv)$$

$$\text{Max. tensile stress} = p = \frac{P}{\frac{\pi}{4}d^2} = \frac{4P}{\pi d^2} \quad ...(v)$$

Since $f = 0.9\, p$, we have, $\dfrac{640\,W}{\pi d^2} = 0.9\,\dfrac{4P}{\pi d^2}$ or $\dfrac{P}{W} = \dfrac{640}{3.6}$ ...(*vi*)

But, from (*iii*), $\dfrac{P}{W} = \dfrac{\Delta L^2}{3yd^2} = \dfrac{\Delta\,(80d)^2}{3yd^2} = \dfrac{6400\,\Delta}{3y}$ ...(*vii*)

Hence, equating (*vi*) and (*vii*), $\dfrac{6400\,\Delta}{3y} = \dfrac{640}{3.6}$ or $\dfrac{y}{\Delta} = \dfrac{6400}{3} \times \dfrac{3.6}{640} = \dfrac{36}{3} = \mathbf{12.}$

11.5. MACAULAY'S METHOD

Macaulay's method is very much useful when point loads are acting on a beam. In case 9 of § 11.4 we have seen that in order to find the expression for slope and deflection, we have to deal with two differential equations, one for AC ($x < a$) and other for CB ($x > a$). The integration of these equations introduced four constants which were determined from the conditions $y = 0$ at each end, and the condition that y and dy/dx should be continuous at C. However, the complication can be avoided by the following artifice given by Macaulay. The method is known as *Macaulay's method.*

Macaulay's method essentially consists in the *manner* of successive integration for the expression for the bending moment. Though law of bending moment varies from section to section, the constants of integration are valid for all the sections of the beam. For the case 9 of § 11.4 we can write the expression for B.M. as (Fig. 11.10) :

$$EI\,\frac{d^2y}{dx^2} = -M = -\frac{Wbx}{L}\;\Big|\; + W(x-a) \quad ...(1)$$

In the above expression, there is a separation line. The quantity to the left of this line is valid for the portion AC only, while both the quantities are valid for the portion CB.

Integrating the above expression, $EI\,\dfrac{dy}{dx} = -\dfrac{Wbx^2}{2L} + C_1 \;\Big|\; + \dfrac{W(x-a)^2}{2}$ ...(2)

It is to be noted that the integration of $(x - a)$ has been made *as a whole* and not for the individual terms. This *manner* of integration makes Macaulay's method effective, because the constant of integration C_1 is valid for the slopes at all sections of the beam and not for AC alone.

Integrating once more the above expression,

$$EIy = -\frac{Wbx^3}{6L} + C_1x + C_2 \;\Big|\; + \frac{W(x-a)^3}{6} \quad ...(3)$$

At $x = 0, y = 0 \quad \therefore \quad C_2 = 0$

At $x = L, y = 0$; Substituting this in (3) and including all the terms, we get

$$0 = -\frac{WbL^3}{6L} + C_1L + \frac{W(L-a)^3}{6}$$

$$\therefore \quad C_1L = \frac{WbL^2}{6} - \frac{Wb^3}{6} = \frac{Wb}{6}(L^2 - b^2) \quad \text{or} \quad C_1 = \frac{Wb}{6L}(L^2 - b^2)$$

Substituting the values in (2) and (3), we get

$$EI\,\frac{dy}{dx} = -\frac{Wbx^2}{2L} + \frac{Wb}{6L}(L^2 - b^2) \;\Big|\; + \frac{W(x-a)^2}{2}$$

or $$EI\frac{dy}{dx} = \frac{Wb}{6L}(L^2 - b^2 - 3x^2) \;\Big|\; + \frac{W(x-a)^2}{2} \quad ...(4)$$

and $$EIy = -\frac{Wbx^3}{6L} + \frac{Wb}{6L}(L^2 - b^2)\,x \;\Big|\; + \frac{W(x-a)^3}{6}$$

or $$EIy = \frac{Wbx}{6L}(L^2 - b^2 - x^2) \;\Big|\; + \frac{W(x-a)^3}{6} \quad ...(5)$$

Equations (4) and (5) are, therefore, standard equations for slope and deflection for the whole beam.

To write the equations for slope and deflection for the portion *AC* only, we have to neglect the terms to the right hand side of Macaulay's line. Thus, for the portion *AC*, we have

$$EI\frac{dy}{dx} = \frac{Wb}{6L}(L^2 - b^2 - 3x^2) \quad \text{and} \quad EIy = \frac{Wbx}{6L}(L^2 - b^2 - x^2)$$

which are exactly the same as equations 11.10 and 11.11. The rest of the procedure for finding the maximum deflection and its position is the same as described in double integration method of case 9 of § 11.4.

Example 11.5. *A horizontal beam of uniform section and length L rests on support at its ends. It carries a U.D.L. w per unit length which extends over a length l from the right hand support. Determine the value of l in order that the maximum deflection may occur at the left hand end of the load, and if the maximum deflection is wL^2/kEI, determine the value of k.*

Solution. Taking moments about *B*,

$$R_A = \frac{wl^2}{2L}$$

Fig. 11.13

At any section *x* from *A*, $EI\frac{d^2y}{dx^2} = -M$

$$= -R_A\,x \;\Big|\; + \frac{w}{2}\{x - (L-l)\}^2 \quad ...(i)$$

Integrating the above $$EI\frac{dy}{dx} = -\frac{wl^2}{2L}\frac{x^2}{2} + C_1 \;\Big|\; + \frac{w}{6}\{x - (L-l)\}^3 \quad ...(ii)$$

Integrating once more, $$EIy = -\frac{wl^2}{12L}x^3 + C_1x + C_2 \;\Big|\; + \frac{w}{24}\{x - (L-l)\}^4 \quad ...(iii)$$

At $x = 0$, $y = 0$ ∴ $C_2 = 0$

At $x = L$, $y = 0 = -\frac{wl^2}{12}L^2 + C_1L + \frac{w}{24}l^4$ or $C_1 = \frac{wl^2L}{12} - \frac{wl^4}{24L}$ ...(*iv*)

Again, if *y* is to be maximum at *C*, $\frac{dy}{dx} = 0$ when $x = (L - l)$.

Hence from (*ii*), $$0 = -\frac{wl^2}{4L}(L-l)^2 + \frac{wl^2L}{12} - \frac{wl^4}{24L}$$

or $7l^2 - 12Ll + 4L^2 = 0$ from which l = **0.453 *L*.**

Substituting in (*iv*), $C_1 = \dfrac{w(0.453L)^2 L}{12} - \dfrac{w(0.453L)^4}{24L} = 0.01535\, wL^3$

Substituting the value of $x = (L - l) = 0.547\, L$ and $C_1 = 0.01535\, wL^3$ in (*iii*)

we get $$y_{\max} = \frac{1}{EI}\left\{\frac{-w\,(0.453L)^2\,(0.547L)^3}{12L} + 0.01535\, wL^3\,(0.547L)\right\}$$

$$= \frac{wL^4}{EI}\{-0.0028 + 0.0084\} = 0.0056\,\frac{wL^4}{EI} = \frac{wL^4}{178.6EI}$$

Hence $\boldsymbol{k = 178.6.}$

Example 11.6. *A beam AB of length L is loaded as shown in Fig. 11.14. Deduce standard expressions for slope and deflection at any point.*

Solution. When the load does not reach the end support, the case can be treated in the same way as in Example 11.5 by supposing the load to extend from C to B and superimposing an upward load w from D to B, as shown in Fig. 11.15.

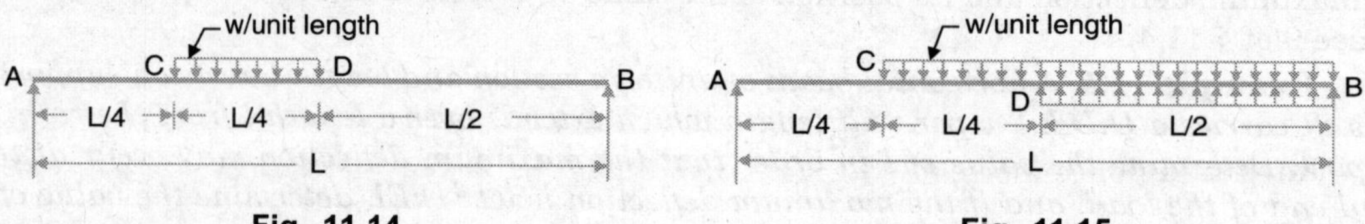

Fig. 11.14 **Fig. 11.15**

Hence at any point x from A, we have

$$EI\frac{d^2y}{dx^2} = -M = -R_A \cdot x \quad \Bigg| + \frac{w\left(x - \dfrac{L}{4}\right)^2}{2} \quad \Bigg| - \frac{w\left(x - \dfrac{L}{2}\right)^2}{2} \qquad \text{...}(i)$$

Taking moments about B, $R_A = \dfrac{1}{L}\left\{w\dfrac{L}{4}\left(\dfrac{L}{2} + \dfrac{L}{8}\right)\right\} = \dfrac{5}{32}\, wL$

Hence $$EI\frac{d^2y}{dx^2} = -\frac{5}{32}w\,Lx \quad \Bigg| + \frac{w}{2}\left(x - \frac{L}{4}\right)^2 \quad \Bigg| - \frac{w}{2}\left(x - \frac{L}{2}\right)^2$$

$\therefore$ $$EI\frac{dy}{dx} = -\frac{5}{64}\, wLx^2 + C_1 \quad \Bigg| + \frac{w}{6}\left(x - \frac{L}{4}\right)^3 \quad \Bigg| - \frac{w}{6}\left(x - \frac{L}{2}\right)^3 \qquad \text{...}(ii)$$

and $$EIy = -\frac{5}{192}\, wLx^3 + C_1x + C_2 \quad \Bigg| + \frac{w}{24}\left(x - \frac{L}{4}\right)^4 \quad \Bigg| - \frac{w}{24}\left(x - \frac{L}{2}\right)^4 \qquad \text{...}(iii)$$

At $x = 0, y = 0$ $\therefore$ $C_2 = 0$

At $x = L, y = 0.$ $$0 = -\frac{5}{192}\, wL^4 + C_1L + \frac{w}{24}\,\frac{81}{256}\, L^4 - \frac{wL^4}{24 \times 16}$$

or $$C_1 = \frac{95}{256 \times 24}\, wL^3 \qquad \text{...}(iv)$$

Substituting the values of C_1 and C_2 in (*ii*) and (*iii*), we get the standard equations for slope and deflection at any point.

Example 11.7. *A beam AB of length L is loaded with a couple applied at an intermediate point. Calculate the slopes at the ends and the deflection under the point of application of the couple.*

Solution. Since there is no external vertical load, the reactions at A and B will be equal and opposite.

For equilibrium of the beam

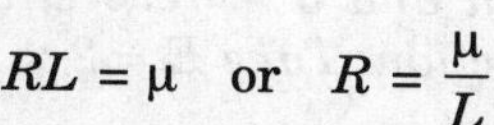

$$RL = \mu \quad \text{or} \quad R = \frac{\mu}{L}$$

Fig. 11.16

At any distance x form A, we have

$$EI\frac{d^2y}{dx^2} = -M = +\frac{\mu}{L}x \;\Big|\; - \mu\,(x-a)^0 \qquad \ldots(1)$$

Integrating,
$$EI\frac{dy}{dx} = +\frac{\mu}{2L}x^2 + C_1 \;\Big|\; - \mu(x-a) \qquad \ldots(2)$$

Integrating once more, $EIy = \dfrac{\mu}{6L}x^3 + C_1.x + C_2 \;\Big|\; - \dfrac{\mu\,(x-a)^2}{2}$ $\qquad \ldots(3)$

At $x = 0, y = 0 \quad \therefore \quad C_2 = 0$

At $x = L$, $\quad y = 0 = \dfrac{\mu}{6L}L^3 + C_1L - \dfrac{\mu b^2}{2} \quad$ or $\quad C_1 = \dfrac{\mu b^2}{2L} - \dfrac{\mu L}{6} = \dfrac{\mu}{6L}\,(3b^2 - L^2)$ $\qquad \ldots(4)$

Substituting the value of C_1 in (2) and (3), we get

$$EI\frac{dy}{dx} = \frac{\mu}{2L}x^2 + \frac{\mu}{6L}(3b^2 - L^2) \;\Big|\; - \mu\,(x-a) \qquad \ldots(a)$$

and
$$EIy = \frac{\mu}{6L}x^3 + \frac{\mu}{6L}(3b^2 - L^2)\,x \;\Big|\; - \frac{\mu}{2}\,(x-a)^2 \qquad \ldots(b)$$

To calculate the slope at the ends, substitute $x = 0$ and $x = L$ in (*a*). Thus, we have

$$i_A = \frac{\mu}{6EIL}\,(3b^2 - L^2)$$

and
$$i_B = \frac{1}{EI}\left(\frac{\mu}{2L}L^2 + \frac{\mu}{6L}(3b^2 - L^2) - \mu b\right) = \frac{\mu}{6EIL}\,(3L^2 + 3b^2 - L^2 - 6Lb)$$

or
$$i_B = \frac{\mu}{6EIL}\,(2L^2 + 3b^2 - 6Lb)$$

To calculate the deflection under the point of application of the couple, substitute $x = a$ in (*b*). Thus, we have

$$EIy_C = \frac{\mu}{6L}\,a^3 + \frac{\mu}{6L}\,(3b^2 - L^2)\,a = \frac{\mu a}{6L}\,(a^2 + 3b^2 - L^2)$$

$$\therefore \quad y_C = \frac{\mu}{6EIL}\,(a^2 + 3b^2 - L^2)$$

Special Case : When $a = b = L/2$,

$$i_A = \frac{\mu}{6EIL}\left(\frac{3L^2}{4} - L^2\right) = -\frac{\mu L}{24EI} \text{ and } i_B = \frac{\mu}{6\,EIL}\left(2L^2 + \frac{3L^2}{4} - \frac{6L^2}{2}\right) = -\frac{\mu L}{24EI}.$$

Hence both the slopes are equal, and are in anti-clockwise direction.

$$y_C = \frac{\mu L}{12EIL}\left(\frac{L^2}{4} + \frac{3L^2}{4} - L^2\right) = 0.$$

Example 11.8. *A beam, simply supported at ends A and B is loaded with two point loads of 60 kN and 50 kN at distance 1 m and 3 m respectively from end A. Determine the position and magnitude of maximum deflection. Take E = 2 × 10⁵ N/mm² and I = 8500 cm⁴.*

Solution. (Fig. 11.17).

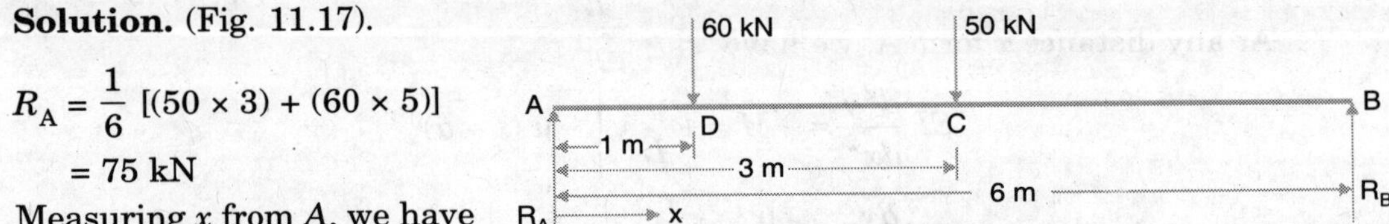

Fig. 11.17

$$R_A = \frac{1}{6}\,[(50 \times 3) + (60 \times 5)]$$

$$= 75 \text{ kN}$$

Measuring x from A, we have

$$EI\frac{d^2y}{dx^2} = -M_x$$

$$= -75\,x \mid + 60(x-1) \mid + 50(x-3)$$

Integrating,
$$EI\frac{dy}{dx} = -\frac{75x^2}{2} + C_1 \left| + \frac{60\,(x-1)^2}{2} \right| + \frac{50(x-3)^2}{2} \quad \text{...[1(a)]}$$

and
$$EIy = -75\,\frac{x^3}{6} + C_1x + C_2 \left| + \frac{30\,(x-1)^3}{3} \right| + \frac{25(x-3)^3}{3} \quad \text{...[2(a)]}$$

At $x = 0$, $y = 0$. Hence from 2(a), $C_2 = 0$.

Also, at $x = 6$ m, $y = 0$. Hence from Eq. 2(a), we get

$$0 = -\frac{75}{6}\,(6)^3 + C_1 \times 6 + 10(6-1)^3 + \frac{25}{3}\,(6-3)^3. \text{ From which } C_1 = 204.2$$

Hence the slope and deflection equations are :

$$EI\frac{dy}{dx} = -37.5\,x^2 + 204.2 \mid + 30(x-1)^2 \mid + 25\,(x-3)^2 \quad \text{...(1)}$$

and
$$EIy = -12.5\,x^3 + 204.2\,x \mid + 10(x-1)^3 \mid + 8.33\,(x-3)^3 \quad \text{...(2)}$$

Now, for maximum deflection, $\frac{dy}{dx} = 0$.

Also, maximum deflection always occurs near the mid-span. Hence, by inspection, it will be between D and C region where maximum loading is there. Therefore, including the corresponding terms of Eq. (1), we get

$$0 = -37.5\,x^2 + 204.2 + 30(x-1)^2$$

or $x^2 + 8x - 31.22 = 0$, which gives $x = 2.87$ m.

Substituting this value of x in Eq. (2), in the first two terms (*i.e.*, terms upto point C).

$$Ely_{max} = -12.5(2.87)^3 + 204.2(2.87) + 10(2.87-1)^3$$

or
$$Ely_{max} = 355.9 \text{ kN-m}^3 = 355.9 \times 10^3\,(1000)^3 \text{ N-mm}^3 = 355.9 \times 10^{12} \text{ N-mm}^3$$

or
$$y_{max} = \frac{355.9 \times 10^{12}}{EI} = \frac{355.9 \times 10^{12}}{(2 \times 10^5) \times (8500 \times 10^4)} = \mathbf{20.94\ mm.}$$

Example 11.9. *A beam AB of 8 m span is simply supported at the ends and is loaded as shown in Fig. 11.18. Determine (i) deflection at C, (ii) maximum deflection, and (iii) slope at end A. Take $E = 2 \times 10^5$ N/mm² and $I = 1000$ cm⁴.*

Solution.

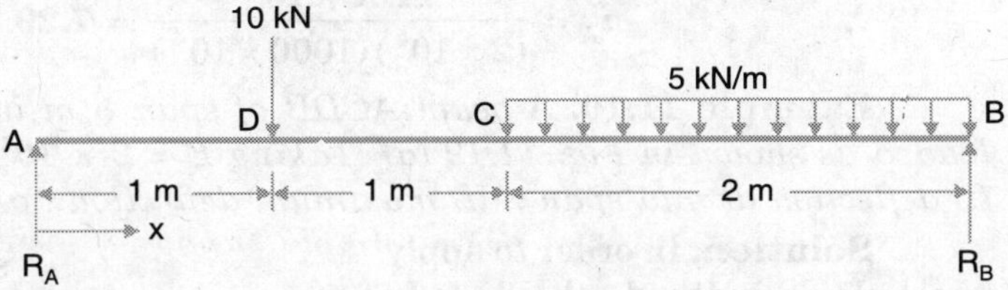

$$R_A = \frac{1}{4}[(10 \times 3) + (5 \times 2 \times 1)] = 10 \text{ kN}$$

Fig. 11.18

Measuring x from A, we have

$$EI\frac{d^2y}{dx^2} = -M_x = -10x \;\Big|\; +10(x-1) \;\Big|\; +\frac{5(x-2)^2}{2}$$

$$\therefore \quad EI\frac{dy}{dx} = -10\frac{x^2}{2} + C_1 \;\Big|\; +\frac{10(x-1)^2}{2} \;\Big|\; +\frac{5(x-2)^3}{6} \qquad \text{...[1(a)]}$$

and
$$EIy = -\frac{10}{6}x^3 + C_1x + C_2 \;\Big|\; +\frac{10(x-1)^3}{6} \;\Big|\; +\frac{5(x-2)^4}{24} \qquad \text{...[2(a)]}$$

At $x = 0$, $y = 0$. Hence $C_2 = 0$

At $x = 4$ m, $y = 0 = -\frac{10}{6}(4)^3 + C_1 \times 4 + \frac{10}{6}(4-1)^3 + \frac{5}{24}(4-2)^4$,

which gives $C_1 = 14.58$.

Hence, the slope and deflection equations are

$$EI\frac{dy}{dx} = -5x^2 + 14.58 \;\Big|\; +5(x-1)^2 \;\Big|\; +\frac{5}{6}(x-2)^3 \qquad \text{...(1)}$$

and
$$EIy = -\frac{5}{3}x^3 + 14.58\,x \;\Big|\; +\frac{5}{3}(x-1)^3 \;\Big|\; +\frac{5}{24}(x-2)^4 \qquad \text{...(2)}$$

For maximum deflection, $dy/dx = 0$. Also, the maximum deflection will be very near to the mid-point C. Let us assume that it occurs in sector DC. Hence from (1), including the terms upto point C, we get

$$0 = -5x^2 + 14.58 + 5(x-1)^2 \quad \text{or} \quad 10x = 19.58 \text{ which gives } x = 1.958 \text{ m.}$$

Hence from Eq. (2), $EIy_{max} = -\frac{5}{3}(1.958)^3 + 14.58(1.958) + \frac{5}{3}(1.958-1)^3$

or $$EIy_{max} = 17.5 \text{ kN-m}^3 \text{ units} = 17.5 \times 10^3 \times (1000)^3 = 17.5 \times 10^{12} \text{ N-mm}^3$$

$$\therefore \quad y_{max} = \frac{17.5 \times 10^{12}}{EI} = \frac{17.5 \times 10^{12}}{2 \times 10^5 (1000)(10)^4} = \mathbf{8.75 \text{ mm}}$$

For deflection at C, put $x = 2$ m in Eq. (2), in the terms upto point C

$$EIy_C = -\frac{5}{3}(2)^3 + 14.58(2) + \frac{5}{3}(2-1)^3 = 17.49 \text{ kN-m}^3 \text{ units}$$

or $$y_C = \frac{17.49(10^{12})}{(2 \times 10^5)(1000 \times 10^4)} = \mathbf{8.747 \text{ mm.}}$$

Thus, we note that deflection at the mid-span is very nearly equal to max. deflection.

Also, putting $x = 0$ in (1), we get $EI\ i_A = 14.58$ kN-m² $= 14.58 \times 10^3 \times (1000)^2$ N-mm²

$$\therefore \quad i_A = \frac{14.58 \times 10^9}{(2 \times 10^5)(1000 \times 10^4)} = \mathbf{7.29 \times 10^{-3}\ radian.}$$

Example 11.10. *A beam ACDB of span 8 m is supply supported at ends A and B is loaded as shown in Fig. 11.19 (a). Taking E = 2 × 10⁵ N/mm² and I = 8000 cm⁴ ; determine (i) deflection at mid-span ; (ii) maximum deflection ; and (iii) slope at end A.*

Solution. In order to apply Macaulay's method, the U.D.L. must be continuous upto the end of the beam. Hence extend the U.D.L. from *D* to *B* and apply a reverse U.D.L. from *B* to *D* as shown in Fig. 11.19 (*b*).

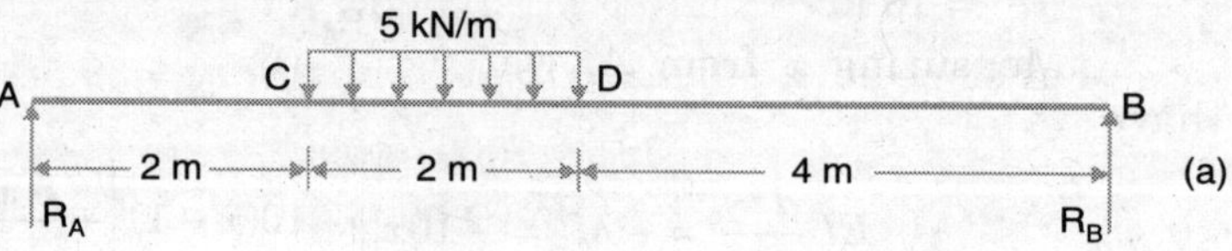

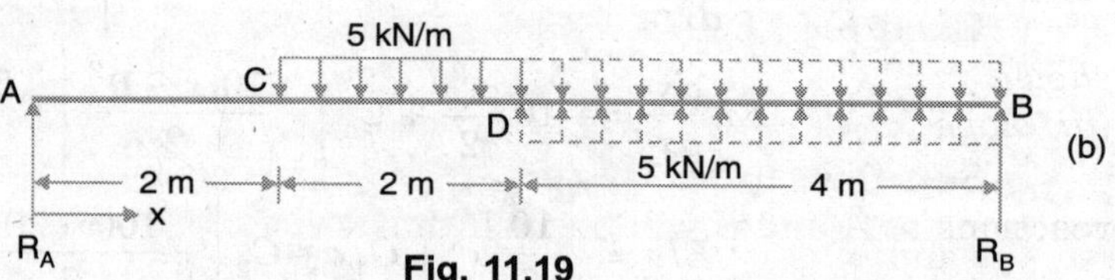

Fig. 11.19

$$R_A = \frac{1}{8}\ [5 \times 2(4 + 1)]$$

$$= 6.25 \text{ kN.}$$

Measuring x from A, we have

$$EI \frac{d^2y}{dx^2} = -M_x = -6.25\,x \; \Bigg| + \frac{5(x-2)^2}{2} \Bigg| - \frac{5(x-4)^2}{2}$$

$$\therefore \quad EI \frac{dy}{dx} = -6.25 \frac{x^2}{2} + C_1 \Bigg| + \frac{5}{6}(x-2)^3 \Bigg| - \frac{5}{6}(x-4)^3 \quad \text{...[1(a)]}$$

and

$$EIy = -\frac{6.25}{6} x^3 + C_1 x + C_2 \Bigg| + \frac{5}{24}(x-2)^4 \Bigg| - \frac{5}{24}(x-4)^4 \quad \text{...[2(a)]}$$

At $x = 0$, $\quad y = 0 \quad \therefore \quad C_2 = 0$

At $x = 8$ m, $\quad y = 0 = -\dfrac{6.25}{6}(8)^3 + C_1 \times 8 + \dfrac{5}{24}(8-2)^4 - \dfrac{5}{24}(8-4)^4$

which gives $C_1 = 39.58$. Hence the slope and deflection equations are :

$$EI \frac{dy}{dx} = -3.125\,x^2 + 39.58 \Bigg| + \frac{5}{6}(x-2)^3 \Bigg| - \frac{5}{6}(x-4)^3 \quad \text{...(1)}$$

and

$$EIy = -1.042\,x^3 + 39.58\,x \Bigg| + \frac{5}{24}(x-2)^4 \Bigg| - \frac{5}{24}(x-4)^3 \quad \text{...(2)}$$

For deflection at mid-span, put $x = 4$ m in Eq. (2) and include terms upto D only.

$$\therefore \quad EIy = -1.042\,(4)^3 + 39.58(4) + \frac{5}{24}(4-2)^4 = 94.97 \text{ kN-m}^3 \text{ units.}$$

$$\therefore \quad y = \frac{94.97 \times 1000 \times (1000)^3}{(2 \times 10^5)(8000 \times 10^4)} = \mathbf{5.94\ mm.}$$

For maximum deflection, put $dy/dx = 0$ in Eq. (1) and include terms upto D only since max. deflection will be very near to the mid-point D (between C and mid-point).

$$\therefore \qquad 0 = -3.125\,x^2 + 39.58 + \frac{5}{6}(x-2)^3$$

This is a cubic equation. Solving this by trial and error, for values of x between $\sqrt{39.58/3.125}$ (= 3.6 m) and 4 m, we get $x \simeq 3.75$ m.

Hence $\quad EIy_{max} = -1.042\,(3.75)^3 + 39.58\,(3.75) + \frac{5}{24}(3.75-2)^4 =$ **95.43 kN-m³.**

$$\therefore \qquad y_{max} = \frac{95.43\,(1000)\,(1000)^3}{(2\times10)^5\,(8000\times10^4)} = \mathbf{5.964\ mm}$$

For slope at A, put $x = 0$ in (1). Thus $EI\,i_A = 39.58$ kN-m²

$$\therefore \qquad i_A = \frac{39.58\,(1000)\,(1000)^2}{(2\times10^5)\,(8000\times10^4)} = \mathbf{2.47\times10^{-3}\ radian.}$$

Example 11.11. *A simply supported beam AB is loaded with a couple μ at the end B as shown in Fig. 11.20. Determine (i) equation for the elastic curve, (ii), maximum deflection, (iii) slopes at the ends. Takes EI constant along the beam.*

Solution. Since the external moment is in anticlockwise direction, the directions of reactions at A and B will be such that a clockwise moment is generated.

$$\therefore \quad R = \frac{\mu}{L} \uparrow \text{ at } A$$

and $\qquad R = \frac{\mu}{L} \downarrow$ at B.

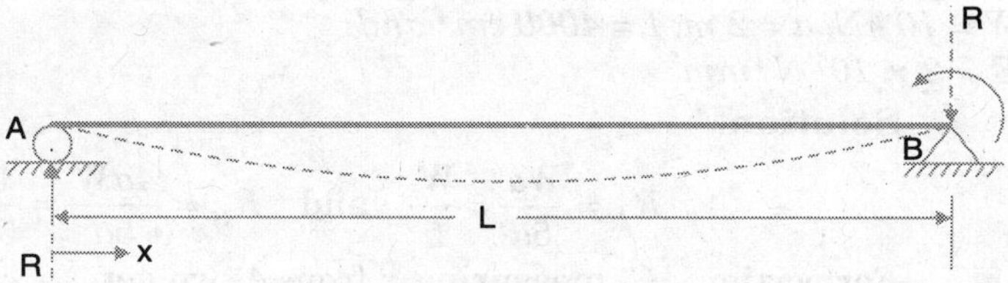

Fig. 11.20

At any distance x from A,

$$EI\frac{d^2y}{dx^2} = -M_x = -\frac{\mu}{L}x$$

Integrating successively, $EI\dfrac{dy}{dx} = -\dfrac{\mu}{L}\dfrac{x^2}{2} + C_1$ ...[1(*a*)]

and $\qquad EIy = -\dfrac{\mu}{L}\dfrac{x^3}{6} + C_1x + C_2$ ...[2(*a*)]

At $x = 0$, $y = 0$. Hence $C_2 = 0$. At $x = L$, $y = 0$. Hence from 2(*a*),

$$0 = -\frac{\mu}{L}\frac{L^3}{6} + C_1L \text{ which gives } C_1 = \frac{\mu L}{6}$$

Hence the slope and deflection equations are

$$EI\frac{dy}{dx} = -\frac{\mu}{2L}x^2 + \frac{\mu L}{6} \qquad \text{...(1)}$$

and

$$EIy = -\frac{\mu}{6L}x^3 + \frac{\mu L}{6}x \qquad \text{...(2)}$$

Hence the equation for elastic curve is

$$y = \frac{\mu}{6EI}\left(Lx - \frac{x^3}{L}\right)$$

For maximum deflection, $dy/dx = 0$; hence from (1),

$$\frac{\mu}{2L}x^2 = \frac{\mu L}{6}, \text{ from which } x = \frac{L}{\sqrt{3}}$$

Substituting in Eq. (2),

$$EIy_{\max} = -\frac{\mu}{6L}\left(\frac{L}{\sqrt{3}}\right)^3 + \frac{\mu L}{6} \times \frac{L}{\sqrt{3}} = \frac{\mu L^2}{9\sqrt{3}} \quad \text{Hence} \quad y_{\max} = \frac{\mu L^2}{9\sqrt{3}\ EI} \ \textbf{Ans.}$$

For slope at end A, put $x = 0$ in Eq. (1)

$$\therefore \quad EIi_A = \frac{\mu L}{6} \quad \text{or} \quad i_A = \frac{\mu L}{6EI} \ \textbf{Ans.} \qquad \text{...[11.17}(a)]$$

For slope at end B, put $x = L$ in Eq. (1)

$$\therefore \quad EIi_B = -\frac{\mu}{2L}(L^2) + \frac{\mu L}{6} = -\frac{1}{3}\mu L \quad \text{or} \quad i_B = -\frac{\mu L}{3EI} \ \textbf{Ans.} \qquad \text{...[11.17}(b)]$$

Example 11.12. *A beam ACB, simply supported at the ends, has moment of inertia 4I for the length AC and I for length CB, and is loaded with a point load W at C. Determine (i) slope at end A, (ii) deflection at mid-span, (iii) maximum deflection. Compute the numerical values if W = 10 kN, a = 2 m, I = 4000 cm⁴ and E = 2 × 10⁵ N/mm².*

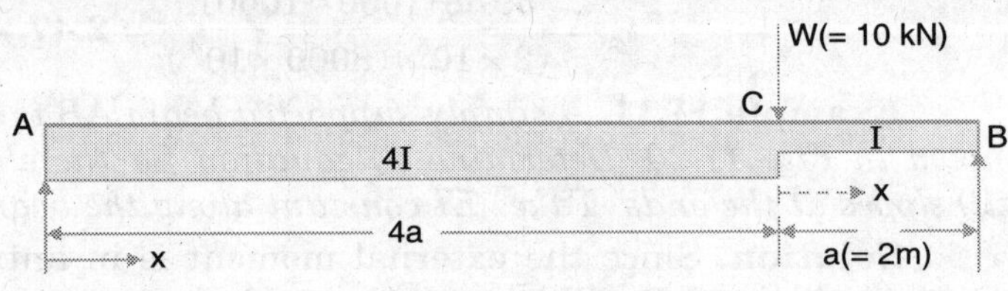

Fig. 11.21

Solution.

$$R_A = \frac{Wa}{5a} = \frac{W}{5} \quad \text{and} \quad R_B = \frac{4aW}{5a} = \frac{4}{5}W$$

For portion AC, measuring x from A, we get

$$E\frac{d^2y}{dx^2} = -\frac{M_x}{I_x} = \frac{1}{4I}\left[-\left(\frac{W}{5}x\right)\right]$$

Integrating,

$$E\frac{dy}{dx} = -\frac{W}{40I}x^2 + C_1 \qquad \text{...[1}(a)]$$

and

$$Ey = -\frac{W}{120I}x^3 + C_1x + C_2 \qquad \text{...[2}(a)]$$

At $x = 0, y = 0$. Hence $C_2 = 0$. Also, at $x = 4a$,

$$\left(E\frac{dy}{dx}\right)_C = Ei_C = -\frac{W}{40I}(16a^2) + C_1 \qquad \text{...[1}(b)]$$

and

$$Ey_C = -\frac{W}{120I}(4a)^3 + 4aC_1 \qquad \text{...[2}(b)]$$

These values of i_C and y_C become initial constants for portion CB. Measuring x from C, for the portion CB, we get

$$E\frac{d^2y}{dx^2} = -\frac{M_x}{I_x} = \frac{1}{I}\left[-\left\{\frac{W}{5}(4a + x) + Wx\right\}\right]$$

$$\therefore \quad E\frac{dy}{dx} = -\frac{4}{5}\frac{Wax}{I} - \frac{Wx^2}{10I} + \frac{Wx^2}{2I} + C_1' \qquad \text{...[1}(c)]$$

and

$$Ey = -\frac{2}{5}\frac{Wax^2}{I} - \frac{Wx^3}{30I} + \frac{Wx^3}{6I} + C_1'x + C_2' \qquad \text{...[2}(c)]$$

At $x = 0$, $\quad E\dfrac{dy}{dx} = Ei_C = -\dfrac{W}{40I}(16a^2) + C_1$, from 1(*b*).

Hence from 1(*c*), $\quad C_1' = -\dfrac{2Wa^2}{5I} + C_1$

Also, at $x = 0$, $\quad Ey_C = -\dfrac{W}{120I}(4a)^3 + 4aC_1$, from 2(*b*).

Hence from 2(*c*), $-\dfrac{W}{120I}(4a)^3 + 4aC_1 = C_2'$

Substituting the values of C_1' and C_2' in Eq. 2(*c*), we get

$$Ey = -\frac{2}{5}\frac{Wax^2}{I} - \frac{Wx^3}{30I} - \frac{Wx^3}{6I} + \left(C_1 - \frac{2}{5}\frac{Wa^2}{I}\right)x + \left(4aC_1 - \frac{8}{15}\frac{Wa^3}{I}\right)$$

This is the expression for deflection at any points in *CB*.

At $x = a$, $y = y_B = 0$.

$$0 = -\frac{2}{5}\frac{Wa^3}{I} - \frac{Wa^3}{30I} + \frac{Wa^3}{6I} + C_1 a - \frac{2}{5}\frac{Wa^3}{I} + 4aC_1 - \frac{8}{15}\frac{Wa^3}{I}$$

which gives $\quad C_1 = \dfrac{6}{25}\dfrac{Wa^2}{I}$.

Substituting this value of C_1 in Eq. 1(*a*) and 2(*a*), we get slope and deflection equation for portion *AC* as below :

$$E\frac{dy}{dx} = -\frac{W}{40I}x^2 + \frac{6}{25}\frac{Wa^2}{I} \quad ...(i)$$

and

$$Ey = -\frac{W}{120I}x^3 + \frac{6}{25}\frac{Wa^2}{I}x \quad ...(ii)$$

For max. deflection, which will evidently take place between *A* and *C*, put $dy/dx = 0$.

$\therefore \quad 0 = -\dfrac{W}{40I}x^2 + \dfrac{6}{25}\dfrac{Wa^2}{I}$ which gives $x = 3.098\ a$.

Substituting the value of *x* in (*ii*), we get

$$Ey_{max} = -\frac{Wa^3}{120I}(3.098)^3 + \frac{6}{25}\frac{Wa^2}{I}(3.098)$$

$$\therefore \quad y_{max} = \mathbf{0.4957\frac{Wa^3}{EI}} \quad ...(a)$$

Also, at $x = 2.5a$ (mid-span)

$$Ey_{centre} = -\frac{Wa^3}{120I}(2.5)^3 + \frac{6}{25}\frac{Wa^3}{I}(2.5)$$

or

$$y_{centre} = \mathbf{0.4698\frac{Wa^3}{EI}} \quad ...(b)$$

Thus, the deflection at the mid-span is very near to the max. deflection.

Also, at $x = 0$, we get from (*i*),

$$Ei_A = \frac{6}{25}\frac{Wa^2}{I} \quad \text{or} \quad i_A = \mathbf{\frac{6}{25}\frac{Wa^2}{EI}} \quad ...(c)$$

Substituting the numerical values in N-mm units, we have

$$\frac{Wa^2}{EI} = \frac{10000(2)^2}{(2\times10^5)(4000\times10^4)} \times (1000)^2 = 5 \times 10^{-3}$$

$$\frac{Wa^3}{EI} = 5 \times 10^{-3} \times (2 \times 1000) = 10$$

Hence $y_{max} = 0.4957\,\frac{Wa^3}{EI} = 0.4957 \times 10 \simeq \mathbf{4.96\ mm}$

$$y_{centre} = 0.4698\,\frac{Wa^3}{EI} = 0.4698 \times 10 \simeq \mathbf{4.7\ mm}$$

$$i_A = \frac{6}{25}\frac{Wa^2}{EI} = \frac{6}{25} \times 5 \times 10^{-3} = \mathbf{1.2 \times 10^{-3}\ radian.}$$

Note. The solution of the above problem by conjugate beam method is extremely short and easy. See example 12.3.

Example 11.13. *An over-hanging beam ABC is supported at A and B, and is loaded with point load W at end C. Determine (i) deflection at C, and (ii) maximum deflection between A and B. Compute the numerical values if L = 4 m, b = 2 m, W = 10 kN, I = 1200 cm⁴ and E = 2 × 10⁵ N/mm².*

Solution.

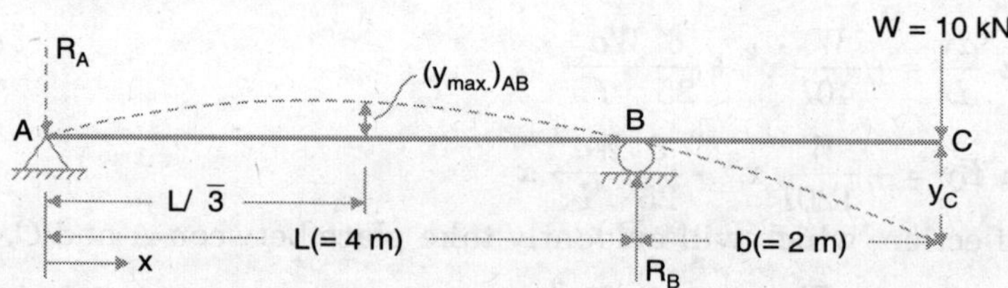

Fig. 11.22

$$R_A = \frac{Wb}{L}\downarrow. \text{ Also } R_B = \frac{W(b+L)}{L}\uparrow$$

Measuring x from A, we have

$$EI\frac{d^2y}{dx^2} = -M_x = +\frac{Wb}{L}x \;\Big|\; -\frac{W(b+L)}{L}(x-L)$$

$$\therefore \quad EI\frac{dy}{dx} = +\frac{Wbx^2}{2L} + C_1 \;\Big|\; -\frac{W(b+L)}{L}\cdot\frac{(x-L)^2}{2} \quad \text{...[1(a)]}$$

and

$$EIy = +\frac{Wbx^3}{6L} + C_1x + C_2 \;\Big|\; -\frac{W(b+L)}{L}\frac{(x-L)^3}{6} \quad \text{...[2(a)]}$$

At $x = 0$, $y = 0$. Hence $C_2 = 0$, from $2(a)$.

Also, at $x = L$, $y = 0$. Hence from $2(a)$, $0 = \frac{Wb}{6L}(L)^3 + C_1L$ From which $C_1 = -\frac{WbL}{6}$.

Hence the slope and deflection equations are

$$EI \frac{dy}{dx} = \frac{Wbx^2}{2L} - \frac{WbL}{6} \Bigg| - \frac{W(b+L)}{2L}(x-L)^2 \quad ...(1)$$

and
$$EIy = \frac{Wbx^3}{6L} - \frac{WbL}{6}x \Bigg| - \frac{W(b+L)}{6L}(x-L)^3 \quad ...(2)$$

For maximum deflection, between A and B, put $\frac{dy}{dx} = 0$ in (1), in the expressions upto B.

$$\therefore \quad 0 = \frac{Wbx^2}{2L} - \frac{WbL}{6}. \text{ This gives } x = \frac{L}{\sqrt{3}}$$

Hence $$(y_{max})_{AB} = \frac{1}{EI}\left[\frac{Wb}{6L} \times \left(\frac{L}{\sqrt{3}}\right)^3 - \frac{WbL}{6}\left(\frac{L}{\sqrt{3}}\right)\right] = -\mathbf{\frac{WbL^2}{9\sqrt{3}\,EI}}$$

Also, at $x = (L + b)$,

$$EIy_C = \frac{Wb}{6L}(L+b)^3 - \frac{WbL}{6}(L+b) - \frac{W(b+L)}{6L}(b+L-L)^3$$

or
$$y_C = \mathbf{\frac{Wb^2}{3\,EI}(L+b)}$$

Substituting the numerical values, we get

$$(y_{max})_{AB} = -\frac{10000 \times 2(4)^2}{9\sqrt{3}\,(2 \times 10^5)(1200 \times 10^4)} \times (1000)^3 = \mathbf{-8.55\ mm} = 8.55 \text{ mm } (\uparrow)$$

and
$$y_C = \frac{10000(2)^2\,(4+2)}{3(2 \times 10^5)(1200 \times 10^4)} \times (1000)^3 = \mathbf{33.3\ mm} \downarrow.$$

Example 11.14. *A beam ABC is loaded as shown in Fig. 11.23. Compute (i) deflection at free end C, (ii) maximum deflection and (iii) slope at end A if I = 3000 cm⁴ and E = 2 × 10⁵ N/mm².*

Solution. Taking moments about B, we get

$10R_A + 400 \times 3 \times 1.5 - 4000 \times 4 = 0 \therefore R_A = 1420$ N ↑.

Hence $R_B = 4000 + 400 \times 3 - 1420 = 3780$ N ↑

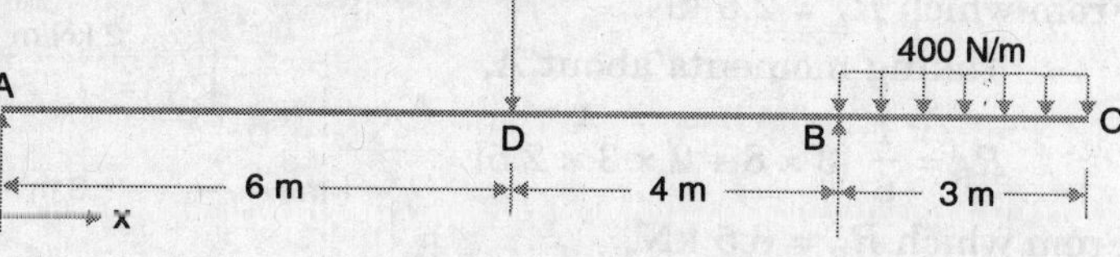

Fig. 11.23

$$\therefore \quad EI\frac{d^2y}{dx^2} = -M_x = -1420\,x \;\Big|\; + 4000(x-6) \;\Big|\; - 3780\,(x-10) + \frac{400\,(x-10)^2}{2}$$

$$\therefore \quad EI\frac{dy}{dx} = -\frac{1420x^2}{2} + C_1 \;\Bigg|\; + \frac{4000\,(x-6)^2}{2} \;\Bigg|\; - \frac{3780(x-10)^2}{2} + \frac{400}{6}(x-10)^3 \quad ...[1(a)]$$

and
$$EIy = -\frac{1420x^3}{6} + C_1x + C_2 \;\Bigg|\; + \frac{4000\,(x-6)^3}{6} \;\Bigg|\; - \frac{3780(x-10)^3}{6} + \frac{400}{24}(x-10)^4 \quad ...[1(b)]$$

At $x = 0$, $y = 0 \therefore C_2 = 0$; At $x = 10$ m, $y = 0$.

Hence from 1(*b*), taking terms upto B,

$$0 = -\frac{1420}{6}(10)^3 + 10C_1 + \frac{4000}{6}(10-6)^3. \text{ From which } C_1 = 19400.$$

Hence the slope and deflection equations are

$$EI\frac{dy}{dx} = -710x^2 + 19400 \Big| + 2000(x-6)^2 \Big| - 1890(x-10)^2 + 66.67(x-10)^3 \quad ...(1)$$

and
$$EIy = -236.7x^3 + 19400x \Big| + 666.7(x-6)^3 \Big| - 630(x-10)^3 + 16.67(x-10)^4 \quad ...(2)$$

For deflection at C, put $x = 13$ m in Eq. (2).

$$\therefore \quad EIy_C = -236.7(13)^3 + 19400(13) + 666.7(13-6)^3 - 630(13-10)^3 + 16.67(13-10)^4$$

or
$$EIy_C = -54812 \text{ N-m}^3 = -54812 \times (1000)^3 \text{ N-mm}^3$$

or
$$y_C = -\frac{54812 \times 10^9}{(2\times10^5)(3000\times10^4)} = -9.14 \text{ mm} = \mathbf{9.14\ mm}\ (\uparrow)$$

The max. deflection between A and B will evidently occur near the mid-span, in the portion AD. Putting dy/dx in Eq. (1) and including the terms applicable for point D, we get

$$0 = -710x^2 + 19400. \text{ From which } x = 5.23 \text{ m}$$

$$\therefore \quad EIy_{max} = -236.7(5.23)^3 + 19400(5.23) = 67600 \text{ N-m}^3 = 67600\,(1000)^3 \text{ N-mm}^3$$

$$\therefore \quad y_{max} = \frac{67600\times10^9}{(2\times10^5)(3000\times10^4)} = \mathbf{11.27\ mm} \downarrow$$

Again, for slope at end A, put $x = 0$ in Eq. (1)

$$\therefore \quad EIi_A = 19400 \text{ N-m}^2 = 19400 \times (1000)^2 \text{ N-mm}^2$$

or
$$i_A = \frac{19400\,(1000)^2}{(2\times10^5)(3000\times10^4)} = \mathbf{3.12 \times 10^{-3}\ radian.}$$

Example 11.15. *An overhanging beam ABC, supported at A and B is loaded as shown in Fig. 11.24(a). Determine (i) Deflection at free end C, and (ii) maximum deflection between A and B. Take I = 450 cm⁴ and E = 2 × 10⁵ N-mm².*

Solution. Taking moments about B,

$$R_A = \frac{1}{6}\,[2\times3\times3.5 - 3\times2]$$

From which $R_A = 2.5$ kN.

Taking moments about A,

$$R_B = \frac{1}{6}\,[3\times8 + 2\times3\times2.5]$$

From which $R_B = 6.5$ kN.

For continuity, extend U.D.L. from E to C, and apply U.D.L. in negative direction (*i.e.*, upwards) from C to E as shown in Fig. 11.24(*b*).

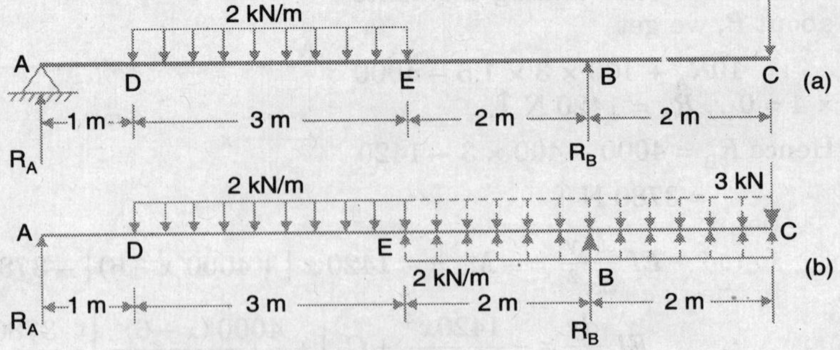

Fig. 11.24

Measuring x from A, we have

$$EI\frac{d^2y}{dx^2} = -M_x = -2.5x \Big| + \frac{2(x-1)^2}{2} \Big| - \frac{2(x-4)^2}{2} \Big| - 6.5(x-6)$$

$$\therefore \quad EI\frac{dy}{dx} = -\frac{2.5x^2}{2} + C_1 \Big| + \frac{(x-1)^3}{3} \Big| - \frac{(x-4)^3}{3} \Big| - \frac{6.5(x-6)^2}{2} \quad ...[1(a)]$$

and $$EIy = -\frac{2.5x^3}{6} + C_1x + C_2 \left| + \frac{(x-1)^4}{12} \right| - \frac{(x-4)^4}{12} \left| - \frac{6.5(x-6)^3}{6} \right. \quad ...[2(a)]$$

At $x = 0$, $y = 0$. Hence $C_2 = 0$.

At $x = 6$, $y = 0$. Hence from Eq. 2(*a*), including appropriate terms upto *B*, we get

$$0 = -\frac{2.5(6)^3}{6} + 6C_1 + \frac{(6-1)^4}{12} - \frac{(6-4)^4}{12}.$$ From which $C_1 = +6.542$

Hence the slope and deflection equations are

$$EI\frac{dy}{dx} = -1.25x^2 + 6.542 \left| + \frac{(x-1)^3}{3} \right| - \frac{(x-4)^3}{3} \left| - \frac{6.5(x-6)^3}{2} \right. \quad ...(1)$$

$$EIy = -0.417x^3 + 6.542x \left| + \frac{(x-1)^4}{12} \right| - \frac{(x-4)^4}{12} \left| - \frac{6.5(x-6)^3}{6} \right. \quad ...(2)$$

For deflection at *C*, put $x = 8$ m in Eq. (2)

$$\therefore \quad EIy_C = -0.417(8)^3 + 6.542 \times 8 + \frac{(8-1)^4}{12} - \frac{(8-4)^4}{12} - \frac{6.5(8-6)^3}{6}$$

or $$EIy_C = 8.915 \text{ kN-m}^3 = 8.915 \times 10^3 (1000)^3 \text{ N-mm}^3 = 8.915 \times 10^{12} \text{ N-mm}^3$$

$$\therefore \quad y_C = \frac{8.915 \times 10^{12}}{(2 \times 10^5)(450 \times 10^5)} \triangleq \mathbf{9.9\ mm} \downarrow$$

The maximum deflection in span *AB* will evidently occur in range *DE*. Putting $dy/dx = 0$ in Eq. (1), and including all the terms upto point *E*, we get

$$0 = -1.25x^2 + 6.542 + \frac{(x-1)^3}{3} \quad \text{or} \quad (x-1)^3 - 3.75x^2 + 19.63 = 0$$

Solving this by trial, we get $x = 2.46$ m. Substituting this value in Eq. (2),

$$EIy_{max} = -0.417(2.46)^3 + 6.542(2.46) + \frac{(2.46-1)^4}{12}$$

$$= 10.26 \text{ kN-m}^3 = 10.26 \times 10^3(1000)^3 \text{ N-mm}^3$$

$$\therefore \quad y_{max} = \frac{10.26 \times 10^{12}}{(2 \times 10^5)(450 \times 10^4)} = \mathbf{11.4\ mm} \downarrow.$$

11.6. PROPPED CANTILEVERS AND PROPPED BEAMS

If a beam or a cantilever is propped at some point, the reaction of the prop can very easily by found out from the consideration of the deflection of the point at which it is propped.

Let us first take the case of a cantilever *AB* propped at its free end *B* [Fig. 11.25 (*a*)]. Let the prop be rigid. If the prop were not there [Fig. 11.25(*b*)], point *B* would be deflected by an amount y_B due to the external loading. The effect of introducing the prop is to exert an upward force *P* to cancel the downward deflection due to the external loading. If y_B' is the upward deflection of the point *B* [Fig. 11.25(*c*)] due to the prop reaction *P* only, then we have

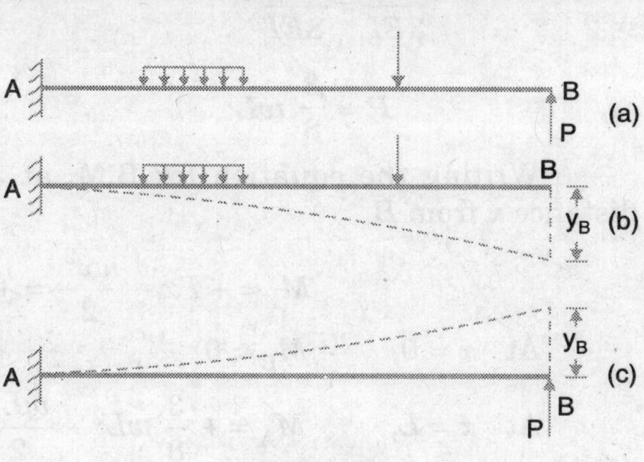

Fig. 11.25

$$y_B = y_B' \qquad ...[11.18(a)]$$

If the prop is not rigid, but yields by an amount δ we have the equation,

$$y_B = y_B' + \delta \qquad ...[11.18(b)]$$

Similarly, let a simply supported beam *AB* be propped at a point *C* by rigid prop (Fig. 11.26). The effect of introducing the prop is to exert an upward force *P* to cancel the downward deflection of *C* due to external loading.

Thus $y_C = y_C'$ ...[11.18(c)]

where y_C is the downward deflection of *C* due to external loading and y_C' is the upward deflection of *C* due to *P*.

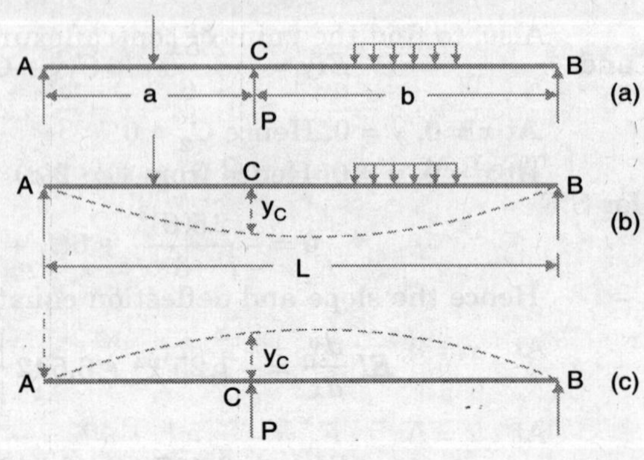

Fig. 11.26

If the prop is elastic and yields by an amount δ, we have

$$y_C = y_C' + \delta \qquad ...[11.18(d)]$$

Example 11.16. *A cantilever of length L is loaded with uniformly distributed load of w per unit length over the whole span. It is propped at the free end. Calculate the prop reaction if (a) the prop is rigid, (b) prop yields by an amount λ under unit load.*

Plot the B.M. and S.F. diagrams for the first case and derive general equations for slope and deflection.

Solution. (Fig. 11.27).

(*a*) The downward deflection of *B* due to the reaction *P* (in absence of the prop) is

$$y_B = \frac{wL^4}{8EI} \text{ (downwards)}$$

The upward deflection of *B* due to the reaction *P* (in absence of load) is

$$y_B' = \frac{PL^3}{3EI}$$

Equating the two,

$$\frac{PL^3}{3EI} = \frac{wL^4}{8EI}$$

or $$\boldsymbol{P = \frac{3}{8} wL} \qquad ...(i)$$

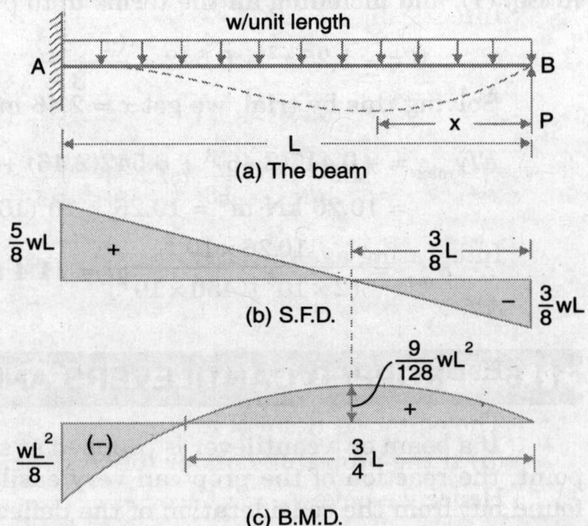

Fig. 11.27

Writing the equation for B.M. at any distance *x* from *B*

$$M_x = +Px - \frac{wx^2}{2} = +\frac{3}{8}wLx - \frac{wx^2}{2} \qquad ...(ii)$$

At $x = 0$, $M_B = 0$

At $x = L$, $M_A = +\frac{3}{8}wL^2 - \frac{wL^2}{2} = -\frac{wL^2}{8}$

Also, to find the point of contraflexure, equate (*ii*) to zero.

$$\frac{3}{8}wLx - \frac{wx^2}{2} = 0 \quad \text{or} \quad x = \frac{3}{8} \times 2L = \frac{3}{4}L$$

The B.M. diagram will be parabolic as shown in Fig. 11.27 (*c*). Also, writing the equation for S.F.

$$F_x = -P + wx = -\frac{3}{8}wL + wx \qquad \text{(Linear variation) ...(i)}$$

At $x = 0$, $\quad F_B = -\frac{3}{8}wL$

At $x = L$, $\quad F_A = -\frac{3}{8}wL + wL = +\frac{5}{8}wL$

Also, $\quad F_x = 0 = -\frac{3}{8}wL + wx \quad \therefore \quad x = \frac{3}{8}L$

Hence maximum + ve B.M. occurs at this location, its value being given by

$$M_{max} = \frac{3}{8}wL\left(\frac{3}{8}L\right) - \frac{w}{2}\left(\frac{3}{8}L\right)^2 = +\frac{9}{128}wL^2$$

Again, we have $EI\dfrac{d^2y}{dx^2} = -M_x = -\dfrac{3}{8}wLx + \dfrac{wx^2}{2}$

Integrating, $\quad EI\dfrac{dy}{dx} = -\dfrac{3}{8}\dfrac{wLx^2}{2} + \dfrac{wx^3}{6} + C_1$

At $x = L$, $\quad \dfrac{dy}{dx} = 0 \quad \therefore \quad C_1 = \dfrac{3}{16}wL^3 - \dfrac{wL^3}{6} = \dfrac{wL^3}{48}$

$$\therefore \quad EI\frac{dy}{dx} = -\frac{3}{16}wLx^2 + \frac{wx^3}{6} + \frac{wL^3}{48} \qquad \text{...(I)}$$

Integrating again, $\quad EIy = -\dfrac{3}{16}\dfrac{wLx^3}{3} + \dfrac{wx^4}{24} + \dfrac{wL^3}{48}x + C_2$

At $x = 0$, $y = 0$ $\quad \therefore \quad C_2 = 0$

$$\text{Hence} \quad EIy = -\frac{1}{16}wLx^3 + \frac{wx^4}{24} + \frac{wL^3}{48}x \qquad \text{...(II)}$$

The equation I and II are equations for slope and deflection at any point.

(*b*) If the prop sinks under the load, the sinking of the prop will be given by $\delta = P\lambda$. Hence we have $y_B = y_B' + \delta$

or $$\frac{wL^4}{8EI} = \frac{PL^3}{3EI} + P\,.\,\lambda = P\left(\frac{L^3}{3EI} + \lambda\right)$$

or $$P = \frac{wL^4}{8EI} \times \frac{3EI}{(L^3 + 3EI\lambda)} = \frac{3}{8} \times \frac{wL^4}{L^3 + 3EI\lambda}. \quad \textbf{Ans.}$$

Example 11.17. *A beam AB of length L is simply supported on two props at the ends and carries an U.D.L. of w per unit length. A prop is introduced in the middle of the beam. Calculate the beam reactions and plot the B.M.D. if the prop is rigid. Also, find the position and magnitude of maximum deflection.*

If all the props are elastic having σ as the stiffness, calculate the reaction of the central prop.

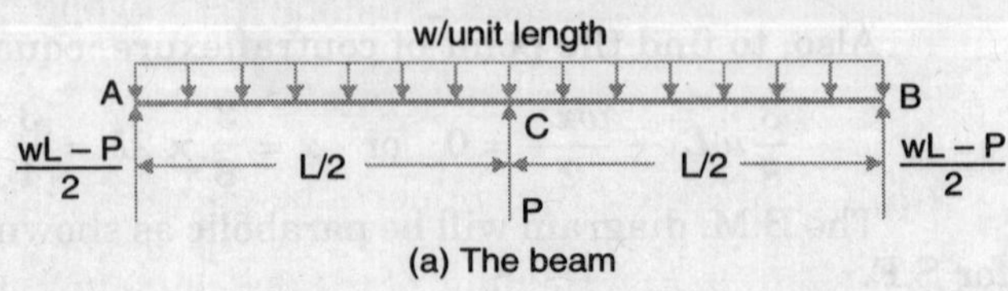

(a) The beam

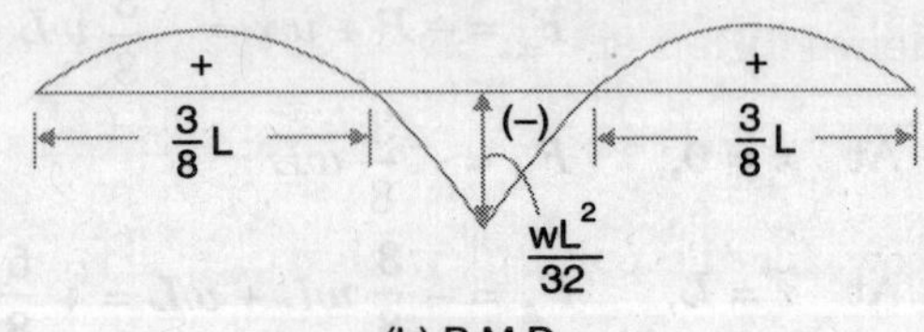

(b) B.M.D.

Fig. 11.28

Solution. (*a*) The reactions at the ends will be $\dfrac{wL-P}{2}$ each, where P is the reaction of the prop in the middle.

The downward deflection of C due to U.D.L. is

$$y_C = \frac{5}{384}\frac{wL^4}{EI}.$$

The upward deflection of C due to prop reaction is $y_C' = \dfrac{PL^3}{48EI}$.

Equating both, $\dfrac{PL^3}{48EI} = \dfrac{5}{384}\dfrac{wL^4}{EI}$ or $P = \dfrac{5}{8}wL.$

Also, at any distance x from A for the portion AC, we have

$$EI\frac{d^2y}{dx^2} = -M_x = -\left(\frac{wL-P}{2}\right)x + \frac{wx^2}{2} = -\frac{3}{16}wLx + \frac{wx^2}{2} \quad ...(i)$$

This is zero at $x = 0$ and at $x = \dfrac{3L}{8}$, the latter being the point of contraflexure. The B.M.D. will be parabolic, as shown in Fig. 11.28(*b*).

$$\text{At} \quad x = \frac{L}{2}, \quad M_C = +\frac{3}{16}wL \cdot \frac{L}{2} - \frac{w}{2}\left(\frac{L}{2}\right)^2 = +\frac{3}{32}wL^2 - \frac{wL^2}{8} = -\frac{wL^2}{32}$$

Integrating (*i*), we get $EI\dfrac{dy}{dx} = -\dfrac{3}{16}wL\dfrac{x^2}{2} + \dfrac{wx^3}{6} + C_1$ ...(*ii*)

and $EIy = -\dfrac{3}{16}wL\dfrac{x^3}{6} + \dfrac{wx^4}{24} + C_1x + C_2$ ...(*iii*)

At $x = 0, \quad y = 0 \quad \therefore \quad C_2 = 0$

At $x = \dfrac{L}{2}, \quad y = 0 = -\dfrac{3}{96}wL\dfrac{L^3}{8} + \dfrac{wL^4}{24 \times 16} + C_1\dfrac{L}{2}$

$$\therefore \quad C_1 = \frac{3wL^3}{96 \times 4} - \frac{wL^3}{24 \times 8} = \frac{wL^3}{384}$$

Again, for maximum deflection,

$$EI\frac{dy}{dx} = 0 = -\frac{3}{32} \cdot wLx^2 + \frac{wx^3}{6} + \frac{wL^3}{384}$$

or $\quad 64x^3 - 36Lx^2 + L^3 = 0.$ Solving, which we get $x = 0.27\,L$

$$EIy_{max} = wL^4\left\{-\frac{3}{96}(0.27)^3 + \frac{1}{24}(0.27)^4 + \frac{1}{384}(0.27)\right\}$$

$$= wL^4 \{-0.0006152 + 0.0002214 + 0.0007\}$$

$$y_{max} = \frac{0.0003062\, wL^4}{EI}.$$

(*b*) Let all the three supports be elastic. Since stiffness is defined as force required to depress it by unity, the yielding of the end supports $= \frac{wL - P}{2\sigma}$ and that of the central prop is $\frac{P}{\sigma}$.

Hence difference in levels of the central prop and the end props is given by

$$\delta = \frac{P}{\sigma} - \frac{wL - P}{2\sigma}$$

Now, for the point C, we have $y = y_C' + \delta$

or $$\frac{5}{384}\frac{wL^4}{EI} = \frac{PL^3}{48EI} + \frac{P}{\sigma} - \frac{wL - P}{2\sigma}$$

or $$P\left\{1 + \frac{48EI}{L^3\sigma} + \frac{48EI}{L^3 2\sigma}\right\} = \frac{5}{384}\frac{wL^4}{EI} \times \frac{48\,EI}{L^3} + \frac{wL}{2\sigma} \times \frac{48EI}{L^3}$$

$$\therefore \quad P\left(1 + \frac{72EI}{\sigma L^3}\right) = \frac{5}{8}wL + \frac{24wL\,EI}{\sigma L^3} \quad \text{or} \quad P = \frac{wL\left(\frac{5}{8} + \frac{24EI}{\sigma L^3}\right)}{\left(1 + \frac{72EI}{\sigma L^3}\right)}.$$

Example 11.18. *A bridge across a river has a span 2L and is constructed with beams resting on the banks and supported at the middle on a pontoon. When the bridge is unloaded the three supports are all at the same level and the pontoon is such that the vertical displacement is equal to the load on it multiplied by a constant λ. Show that the load on the pontoon, due to a concentrated load placed one quarter of the way along the bridge, is given by* $\frac{11W}{16\left(1 + \frac{6EI\lambda}{L^3}\right)}$, *where I is the moment of inertia of the sections of the beams.*

Solution. From Eq. 11.16, the deflections at C due to the point load is given by

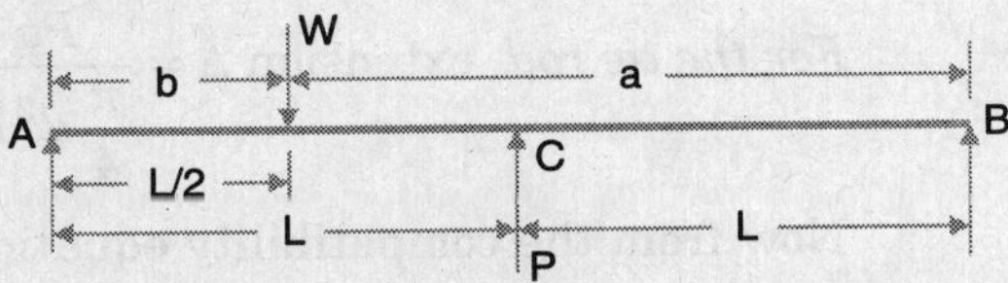

Fig. 11.29

$$y_C = \frac{wb}{48EI}(3l^2 - 4b^2), \text{ where } l = \text{span.}$$

For our case, l = span = $2L$; $b = L/2$ and $a > b$.

Hence $$y_C = \frac{W}{48EI}\frac{L}{2}\left\{3(2L)^2 - 4\left(\frac{1}{2}L\right)^2\right\} = \frac{wL}{96EI}(12L^2 - L^2) = \frac{11WL^3}{96EI} \text{ downwards}$$

Again, deflection of C due to the prop reaction is

$$y_C' = \frac{Pl^3}{48EI} = \frac{P}{48EI}(2L)^3 = \frac{PL^3}{6EI}$$

and downward displacement of the prop $= \delta = P\lambda$.

Hence $y_C = y_C' + P\lambda$

or $$\frac{11WL^3}{96EI} = \frac{PL^3}{6EI} + P\lambda$$

or $$P\left(1+\frac{6EI\lambda}{L^3}\right) = \frac{11WL^3}{96EI} \times \frac{6EI}{L^3} = \frac{11}{16}W$$

$\therefore$ $$P = \frac{11W}{16\left(1+\frac{6EI\lambda}{L^3}\right)}.$$ Hence proved.

Example 11.19. *Two equal steel beams are encastered at one end and connected by a steel rod as shown in Fig. 11.30. Show that the pull in tie rod is* $P = \dfrac{5WL^3}{32\left(\frac{6aI}{\pi d^2} + L^3\right)}$ *where d is the diameter of the rod and I is the moment of inertia of the section of each beam about its neutral axis.*

Solution. Let the tension in the tie rod be *P*.

For the cantilever AB,

$$y_B = \left\{\frac{W\left(\frac{L}{2}\right)^3}{3EI} + \frac{W}{2EI}\left(\frac{L}{2}\right)^2 \times \frac{L}{2}\right\} - \frac{PL^3}{3EI}$$

$$= \frac{WL^3}{24EI} + \frac{WL^3}{16EI} - \frac{PL^3}{3EI} = \frac{5}{48}\frac{WL^3}{EI} - \frac{PL^3}{3EI}$$ (downwards) ...(*i*)

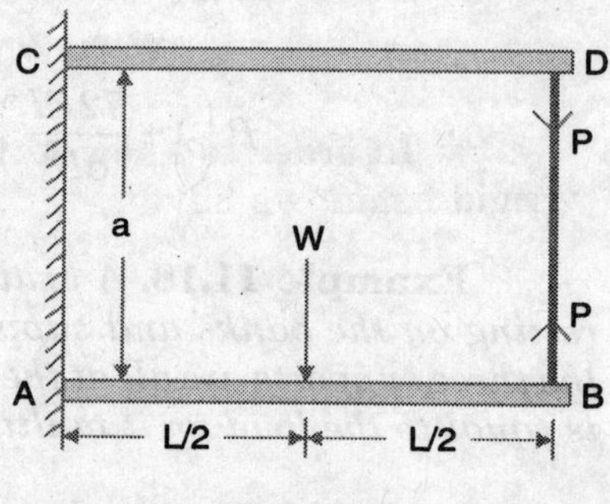

Fig. 11.30

For the cantilever CD : $y_D = \dfrac{PL^3}{3EI}$ (downwards) ...(*ii*)

For the tie rod, extension $\Delta = \dfrac{Pa}{\frac{\pi}{4}d^2E} = \dfrac{4Pa}{\pi d^2 E}$ ...(*iii*)

Now from the compatibility equation, $y_B = y_D + \Delta$

or $$\frac{5}{48}\frac{WL^3}{EI} - \frac{PL^3}{3EI} = \frac{4Pa}{\pi d^2 E} + \frac{PL^3}{3EI} \quad \text{or} \quad \frac{5WL^3}{48I} = P\left(\frac{2L^3}{3I} + \frac{4a}{\pi d^2}\right)$$

or $$P = \frac{5WL^3}{48I\left(\frac{2}{3}\frac{L^3}{I} + \frac{4a}{\pi d^2}\right)} = \frac{5WL^3}{48I\,\frac{2}{3I}\left(L^3 + \frac{4a}{\pi d^2} \times \frac{3I}{2}\right)}$$

$$= \frac{5WL^3}{32\left(L^3 + \frac{6aI}{\pi d^2}\right)}.$$ Hence proved.

Example 11.20. *A rigid beam ABC is pinned to a wall at C and supported by two springs at A and B as shown in Fig. 11.31. Determine the slope of the beam when it carries load W at a distance c from C. The spring at A and B stretch δ_1 and δ_2 respectively under unit load. The unloaded beam is horizontal.*

Fig. 11.31

Solution. Let P_1 = force in spring at A and

P_2 = force in spring at B.

Hence Δ_1 = stretch of spring at A

$$= P_1\delta_1 \text{ ; or } P_1 = \frac{\Delta_1}{\delta_1}$$

and Δ_2 = stretch of spring at B

$$= P_2\delta_2 \text{ ; or } P_2 = \frac{\Delta_2}{\delta_2}$$

Due to pinned joint, moment at $C = 0$

$$\therefore \quad W.c = P_1(a+b) + P_2 a = \frac{\Delta_1}{\delta_1}(a+b) + \frac{\Delta_2}{\delta_2}a \qquad \text{...}(i)$$

In order to know Δ_1 and Δ_2 another equation will be needed. From deformation of the rigid beam, we have

$$\frac{\Delta_1}{\Delta_2} = \frac{a+b}{a} \quad \text{or} \quad \Delta_1 = \frac{a+b}{a}\,.\,\Delta_2$$

Substituting in (*i*), we get

$$W.c = \left(\frac{a+b}{a}\right)\frac{\Delta_2}{\delta_1}(a+b) + \frac{\Delta_2}{\delta_2}a = \Delta_2\left\{\frac{(a+b)^2}{a\delta_1} + \frac{a}{\delta_2}\right\}$$

or

$$\Delta_2 = \frac{W.c\,a\,\delta_1\delta_2}{(a+b)^2\,\delta_2 + a^2\delta_1}$$

$$\therefore \quad \text{Slope of the beam} = \theta = \frac{\Delta_2}{a} = \frac{Wc\,\delta_1\delta_2}{(a+b)^2\,.\,\delta_2 + a^2\delta_1}.$$

Example 11.21. *An experimental beam of steel and of uniform section, I = 5000 cm⁴ is simply supported in a horizontal position by rigid supports 4 m apart. An elastic prop is placed under the beam at the mid-span with initial clearance between the top of the prop and the under-side of the beam. The beam is centrally loaded with a series of increasing loads. The observed central deflection is 8 mm for a concentrated load of 100 kN, and when the results are plotted, the slope of the load-deflection line is 27.3 kN per mm after the contact with the prop has occurred. Determine the stiffness of the prop and the initial clearance. Take E = 2.04 × 10⁵ N/mm².*

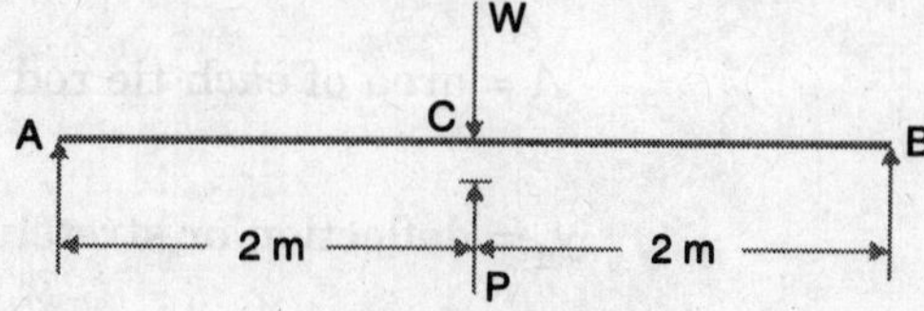

Fig. 11.32

Solution. The stiffness of the whole structure (*i.e.*, the beam and the prop) = slope of load-deflection curve.

Let σ_b = stiffness of the beam and σ_p = stiffness of the prop.

Then $\sigma_b + \sigma_p = 27.3$ (kN/mm)

But
$$\sigma_b = \frac{\text{Load}}{\text{Deflection}} = \frac{W}{y} = \frac{W}{(WL^3/48EI)} = \frac{48EI}{L^3}$$

$$= \frac{48 \times 2.04 \times 10^5 \times 50 \times 10^6}{(4)^3\ (1000)^3} = 7650 \text{ N/mm} = 7.65 \text{ kN/mm}$$

$\therefore$ $\sigma_p = 27.3 - 7.65 =$ **19.65 kN/mm**

Let δ = initial clearance (mm) $\therefore$ Deflection of prop = $8 - \delta$

Force on prop = (stiffness of prop) × (deflection of prop)

or $P = \sigma_p(8 - \delta) = 19.65 \times (8 - \delta)$ kN.

Again, deflection of beam = $\dfrac{\text{Net load on the beam}}{\text{Stiffness of the beam}}$

$$\therefore \quad 8 = \frac{100 - P}{\sigma_b} = \frac{100 - 19.65(8 - \delta)}{7.65}$$

Simplifying and rearranging, we get

$$19.65\ \delta = 61.2 - 100 + 157.2 = 117.4 \quad \text{or} \quad \delta = \mathbf{5.97\ mm.}$$

Example 11.22. *A horizontal steel beam I = 8820 cm⁴ carries a uniformly distributed load of 60 kN over its length of 4 m. The beam is supported by three vertical steel tie rods, each 2 m long, one at each end and one in middle, the diameter of each rod being 25 mm. Calculate the deflection of the centre of the beam below the end points and the stress in each of the rods. Take E = 2.04 × 10⁵ N/mm².*

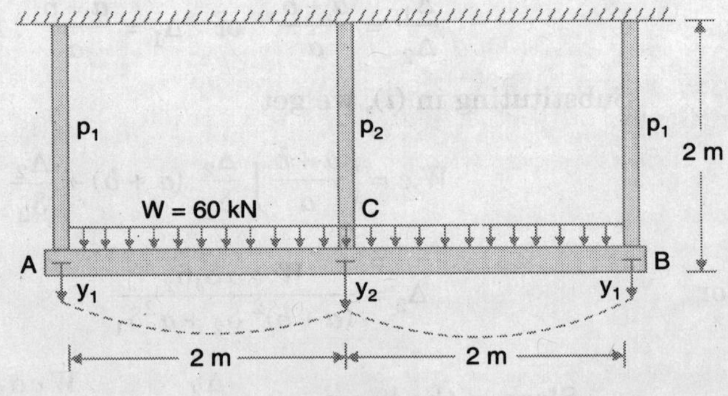

Fig. 11.33

Solution. Let p_1 be the stress in outer tie rods and p_2 be the stress in the middle tie rod.

$$A = \text{area of each tie rod} = \frac{\pi}{4}(25)^2 = 490.9 \text{ mm}^2.$$

$$y_1 = \text{deflection or stretch of each outer tie rod} = \frac{2000\ p_1}{E}$$

$$y_2 = \text{deflection of stretch of middle tie rod} = \frac{2000\ p_2}{E}$$

Now, for the beam AB, $(y_2 - y_1)$ = net downward deflection of C with respect to ends

$$\therefore \quad EI(y_2 - y_1) = \frac{5}{384}WL^3 - \frac{P_2L^3}{48} \quad \text{where } W = \text{total U.D.L.} \qquad ...(i)$$

Substituting the values, we get

$$EI\left(\frac{2000p_2}{E}-\frac{2000p_1}{E}\right)=\frac{5}{384}\times 60000\,(4000)^3-\frac{p_2\times 490.9\,(4000)^3}{48}$$

or $$2000\times 88.2\times 10^6\,(p_2-p_1)=\frac{5\times 60000\times 64\times 10^9}{384}-\frac{490.9\times 64\times 10^9\,p_2}{48}$$

or $$p_2-p_1=283.5-3.71\,p_2 \quad\text{or}\quad 4.71p_2-p_1=283.5 \qquad ...(i)$$

Also, from statics, $2p_1A+p_2A=60000$ ...(*ii*)

or $$p_1+0.5p_2=\frac{60000}{2\times 490.9}=61.1 \qquad ...(iii)$$

Adding (*i*) and (*iii*), we get $5.21p_2=344.6$

From which $p_2=\dfrac{344.6}{5.21}=66.1$ N/mm² and $p_1=61.1-33.05\simeq 28$ N/mm²

Hence stress in outer rods = **28 N/mm²** ; Stress in middle rod = **66.1 N/mm².**

Example 11.23. *A cantilever AB, loaded at the end B is supported by a short cantilever CD of the same cross-section as cantilever AB (Fig. 11.34). Prove that the pressure X between the two beams at C is given by*

$$X=\frac{3P}{4}\left(\frac{L}{L_1}-\frac{1}{3}\right).$$

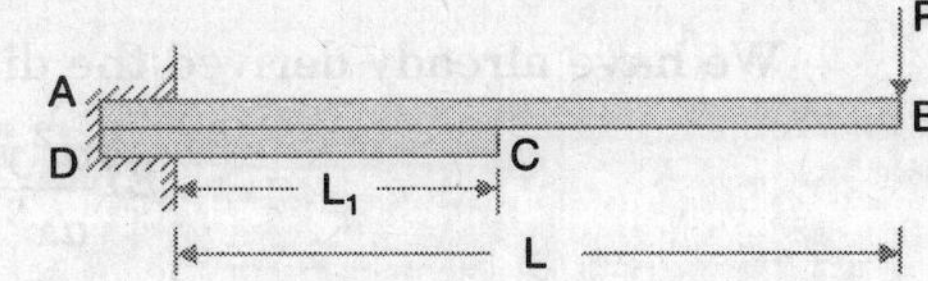

Fig. 11.34

Solution. Due to the bending of both the cantilevers, the upper cantilever will have larger angular deflection than the lower. Hence, there will be contact between the two cantilevers only at *D* and *C*. Let the pressure at *C* be *X*.

For the cantilever AB : At any distance *x* from fixed end,

$$EI\frac{d^2y}{dx^2}=-M=P\,(L-x)\ \text{(due to the load } P \text{ only)}$$

$$\therefore \quad EI\frac{dy}{dx}=P\left(Lx-\frac{x^2}{2}\right)+0 \qquad \left(\text{since } \frac{dy}{dx}=0 \text{ at } x=0\right)$$

and $$EIy=P\left(\frac{Lx^2}{2}-\frac{x^3}{6}\right)+0 \qquad (\text{since } y=0 \text{ at } x=0)$$

$$\therefore \quad EIy_{C1}=P\left(\frac{LL_1^2}{2}-\frac{L_1^3}{6}\right)\downarrow$$

where y_{C1} is the deflection of *C* due to load *P*. ...(*i*)

Deflection of *C* due to the upward pressure *X* will be

$$y_{C1}=\frac{XL_1^3}{3EI} \qquad \text{(upwards) ...}(ii)$$

Hence the net deflection of *C* is given by

$$y_C=y_{C_1}-y_{C_2}=\frac{P}{EI}\left(\frac{LL_1^2}{2}-\frac{L_1^3}{6}\right)-\frac{XL_1^3}{3EI}\ \text{(downwards)} \qquad ....(iii)$$

For the cantilever DC :

$$y_{C'} = \frac{XL_1^3}{3EI} \quad \text{(downwards)} \quad ...(iv)$$

Since both the beams are in contact at C, we have

$$y_C = y_C'$$

or
$$\frac{P}{EI}\left(\frac{LL_1^2}{2} - \frac{L_1^3}{6}\right) - \frac{XL_1^3}{3EI} = \frac{XL_1^3}{3EI} \quad \text{or} \quad \frac{2XL_1^3}{3EI} = \frac{PL_1^3}{2EI}\left(\frac{L}{L_1} - \frac{1}{3}\right)$$

or
$$X = \frac{3P}{4}\left(\frac{L}{L_1} - \frac{1}{3}\right)$$
which is the required expression.

11.7. VARIABLE LOADING ON A BEAM OF UNIFORM SECTION

We have already derived the differential equation

$$EI\frac{d^2y}{dx^2} = -M_x$$

Differentiating it, we get $EI\frac{d^3y}{dx^3} = -\frac{dM_x}{dx} = -F_x$ ...(11.19)

Differentiating further, $EI\frac{d^4y}{dx^4} = -\frac{dF_x}{dx} = w_x$ ...[11.19(*a*)]

where w_x is the rate of loading at a point distant x from the origin.

Thus, if a beam is loaded with non-uniformly distributed load, the rate of loading at any distance x from the origin can be written and equation 11.19(*a*) obtained. Integrating 11.19(*a*) successively, we can get expression for S.F., B.M., slope and deflection at any point. In doing so, some constants will be introduced which may be very easily determined by knowing the end conditions.

Example 11.24. *A cantilever of length L carries uniformly varying load of intensity zero at the free end to w/unit length at the fixed end. Derive an expression for deflection at any point. Find also the slope and deflection of the free end.*

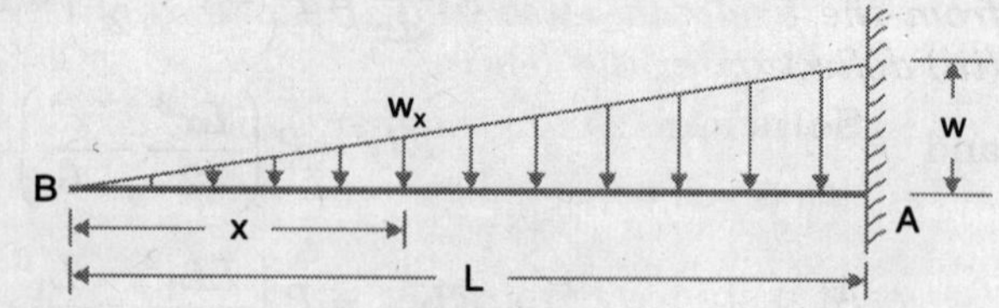

Fig. 11.35

Solution. At distance x from the free end

$$EI\frac{d^4y}{dx^4} = \text{rate of loading} = w_x = \frac{wx}{L}$$

$$\therefore \quad EI\frac{d^3y}{dx^3} = -F_x = \frac{wx^2}{2L} + C_1$$

At $x = 0$, $F_B = 0$ $\quad \therefore \quad C_1 = 0$

$$\therefore \quad EI\frac{d^3y}{dx^3} = -F_x = \frac{wx^2}{2L}$$

Integrating further, $EI\dfrac{d^2y}{dx^2} = -M_x = \dfrac{wx^3}{6L} + C_2$

At $x = 0$, $M_B = 0$ $\therefore$ $C_2 = 0$

Hence $EI\dfrac{d^2y}{dx^2} = \dfrac{wx^3}{6L}$

Integrating further, $EI\dfrac{dy}{dx} = \dfrac{wx^4}{24L} + C_3$

At $x = L$, $\dfrac{dy}{dx} = 0 = \dfrac{wL^4}{24L} + C_3$ or $C_3 = -\dfrac{wL^3}{24}$

Hence $EI\dfrac{dy}{dx} = \dfrac{wx^4}{24L} - \dfrac{wL^3}{24}$ ...(1)

Integrating further , $EIy = \dfrac{wx^5}{120L} - \dfrac{wL^3}{24}x + C_4$

At $x = L$, $y = 0$ $\therefore$ $\dfrac{wL^5}{120L} - \dfrac{wL^4}{24} + C_4 = 0$ or $C_4 = \dfrac{wL^4}{30}$

Hence $EIy = \dfrac{wx^5}{120L} - \dfrac{wL^3x}{24} + \dfrac{wL^4}{30}$ ...(2)

which is the required expression.

To obtain slope at the free end, substitute $x = 0$ in (1).

Thus $i_B = -\dfrac{wL^3}{24EI}$ (*i.e.*, anti-clockwise rotation)

To obtain deflection of the free end, substitute $x = 0$ in (2).

Thus $y_B = \dfrac{wL^4}{30EI}$.

Example 11.25. *A simply supported beam of length L carries uniformly varying load of intensity zero at one end to w at the other end. Derive an expression for deflection at a point x from the first end. Also find (i) slope at the ends, (ii) maximum deflection and its position, (iii) deflection at the centre.*

Solution. Taking moment about B,

$$R_A = \frac{wL}{2} \times \frac{L}{3} \times \frac{1}{L} = \frac{wL}{6}$$

At distance x from A,

$$EI\frac{d^4y}{dx^4} = w_x = \frac{wx}{L} \quad ...(i)$$

Integrating (*i*), we get

$$EI\frac{d^3y}{dx^3} = -F_x = \frac{wx^2}{2L} + C_1$$

At $x = 0$, $F_x = +R_A = +\dfrac{wL}{6}$, $\therefore$ $C_1 = -\dfrac{wL}{6}$

Hence $EI\dfrac{d^3y}{dx^3} = \dfrac{wx^2}{2L} - \dfrac{wL}{6}$ ...(*ii*)

Fig. 11.36

Integrating further, $$EI\frac{d^2y}{dx^2} = -M_x = \frac{wx^3}{6L} - \frac{wLx}{6} + 0 \text{ (since } M = 0 \text{ at } x = 0) \quad ...(iii)$$

Integrating further, $$EI\frac{dy}{dx} = \frac{wx^4}{24L} - \frac{wLx^2}{12} + C_3 \quad ...(iv)$$

and $$EIy = \frac{wx^5}{120L} - \frac{wLx^3}{36} + C_3x + C_4 \quad ...(v)$$

At $x = 0, y = 0 \quad \therefore \quad C_4 = 0$

At $x = L$, $y = 0 = \frac{wL^5}{120L} - \frac{wL.L^3}{36} + C_3L$ from which $C_3 = \frac{wL^3}{36} - \frac{wL^3}{120} = \frac{7wL^3}{360}$

Substituting in (*iv*) and (*v*), we get

$$EI\frac{dy}{dx} = \frac{wx^4}{24L} - \frac{wLx^2}{12} + \frac{7}{360}wL^3 \quad ...(I)$$

and $$EIy = \frac{wx^5}{120L} - \frac{wLx^3}{36} + \frac{7}{360}wL^3x \quad ...(II)$$

Substituting $x = 0$ in (I), we get $i_A = \frac{7wL^3}{360EI}$

Substituting $x = L$ in (I), we get $i_B = \frac{1}{EI}\left[\frac{wL^3}{24} - \frac{wL^3}{12} + \frac{7}{360}wL^3\right] = -\frac{wL^3}{45EI}$

Substituting $x = \frac{L}{2}$ in (II), we get

$$EIy_C = \frac{w}{120L}\left(\frac{L}{2}\right)^5 - \frac{wL}{36}\left(\frac{L}{2}\right)^3 + \frac{7}{360}wL^3\left(\frac{L}{2}\right)$$

or $$y_C = \frac{1}{EI}\left[\frac{wL^4}{120\times 32} - \frac{wL^4}{36\times 8} + \frac{7wL^4}{360\times 2}\right] = \frac{5}{768}\frac{wL^4}{EI} = 0.00651\frac{wL^4}{EI}$$

To find the position of maximum deflection, equate (I) to zero.

$$\therefore \quad \frac{wx^4}{24L} - \frac{wLx^2}{12} + \frac{7}{360}wL^3 = 0$$

Solving this for x, we get $x = \mathbf{0.519\ L}$

Substituting this value of x in (II), we get

$$EIy_{max} = \frac{w}{120L}(0.519\,L)^5 - \frac{wL}{36}(0.519L)^3 + \frac{7}{360}wL^3\,(0.519L)$$

or $$y_{max} = 0.00652\frac{wL^4}{EI}.$$

Again we see here that y_{max} and y_C are almost equal. The same observation was made while dealing with an eccentric point load. *Thus, in simply supported beam, whatever may be the type and location of loading, the central deflection will be approximately equal to the maximum deflection.*

11.8. BEAM OF VARIABLE CROSS-SECTION

If a beam has variable cross-section, and if the cross-section varies uniformly according to some law, the expression $EI\dfrac{d^2y}{dx^2} = -M_x$ can be rearranged in the form $E\dfrac{d^2y}{dx^2} = -\dfrac{M_x}{I_x}$ in which both M_x and I_x are functions of x. The above expression can be integrated to get slope and deflection.

If a beam has different moment of inertia for different segments of its length, the expression can be used separately for separate portions of the length, I being constant for a particular segment of the beam. This method consists of selecting an origin at one end of a beam and completing the integrations until i and y are obtained for first segment. The values of i and y at the *end* of the first segment become the *initial constants* for the next segment. This process is repeated until the far end of the beam is reached, where the unknown constants are determined from the boundary condition. A new origin of x is used at every juncture of the segments.

Example 11.26. *A cantilever of uniform length is to be turned from a mild steel bar 50 mm in diameter. A load of 5 kN is to be supported from the free end, and the maximum stress is to be limited to 70 N/mm². Determine the maximum permissible length of the cantilever and the deflection at the end when it is loaded. E = 2 × 10⁵ N/mm².*

Solution. (Fig. 11.37)

At any section x from fixed end, we have

$$M_x = W(L - x)$$

$$= f.Z_x = f\frac{\pi}{32}D_x^{\,3}$$

where D_x = diameter of the section there

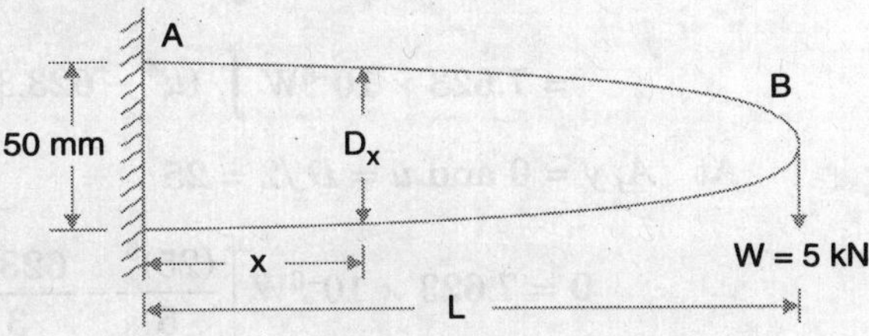

Fig. 11.37

$$\therefore \quad D_x^{\,3} = \frac{32W(L-x)}{\pi f}$$

$$= \frac{32 \times 5000\,(L-x)}{\pi \times 70}$$

$$= 728\,(L-x) \qquad \text{...(1)}$$

At A, $x = 0$, $D_A = 50$ mm. Hence $(50)^3 = 728L$ or $L = \dfrac{125000}{728} = \mathbf{171.8\ mm.}$

Again $$E\frac{d^2y}{dx^2} = -\frac{M_x}{I_x} = \frac{W(L-x)64}{\pi D_x^{\,4}} = \frac{64W(L-x)}{\pi \times 728\,(L-x)\,D_x}$$

or $$E\frac{d^2y}{dx^2} = 0.028\,WD_x^{-1} \qquad \text{...(2)}$$

Let $D_x = 2u$. Hence from (1), $(L - x) = \dfrac{D_x^{\,3}}{728} = \dfrac{8u^3}{728}$

$$\therefore \quad x = L - \frac{8u^3}{728} \quad \text{and} \quad \frac{dx}{du} = -\frac{24u^2}{728} = -0.033\,u^2 \qquad \text{...(3)}$$

Integrating (2), we get

$$E\frac{dy}{dx} = \int 0.028\, W D_x^{-1}\, dx = 0.028\, W \int \frac{1}{2u} dx = 0.014\, W \int \frac{1}{u} \cdot \frac{dx}{du}\, du$$

$$= 0.014W \int \frac{1}{u} (-0.033\, u^2)\, du = -462 \times 10^{-6}\, W \int u du$$

$$= -462 \times 10^{-6}\, W \left(\frac{u^2}{2}\right) + A$$

At the fixed end, $\frac{dy}{dx} = 0$ when $u = \frac{D_x}{2} = \frac{50}{2} = 25$ mm

$\therefore \quad 0 = -\,462 \times 10^{-6}\, W \left(\frac{25^2}{2}\right) + A \quad$ From which $A = 0.144\, W$

Hence $\quad E\frac{dy}{dx} = -\,231 \times 10^{-6} W\, u^2 + 0.144W$

Integrating further,

$$Ey = -\,231 \times 10^{-6}\, W \int (u^2 - 623.3) \frac{dx}{du}\, du$$

$$= -\,231 \times 10^{-6} W \int (u^2 - 623.3)(-\,0.033u^2)\, du$$

$$= 7.623 \times 10^{-6} W \int (u^4 - 623.3u^2)\, du = 7.623 \times 10^{-6} W \left(\frac{u^5}{5} - \frac{623.3u^3}{3}\right) + B$$

At A, $y = 0$ and $u = D_x/2 = 25$

$\therefore \quad 0 = 7.623 \times 10^{-6} W \left[\frac{(25)^5}{5} - \frac{623.3}{3}(25)^3\right] + B.$ From which $B = 9.86\ W$

Hence $\quad Ey = 7.623 \times 10^{-6} W \left(\frac{u^5}{5} - \frac{623.3u^3}{3}\right) + 9.86W$

At B, $\quad u = 0$ and $y = y_B$. Hence $Ey_B = 9.86\ W$

or $\quad y_B = \frac{9.86\ W}{E} = \frac{9.86 \times 5000}{2 \times 10^5} = $ **0.25 mm.**

Example 11.27. *A freely supported beam of span L carries a central load W. The sectional area of the beam is so designed that the moment of inertia of section increases uniformly from I at ends to 1.5I at the middle. Calculate the central deflection.*

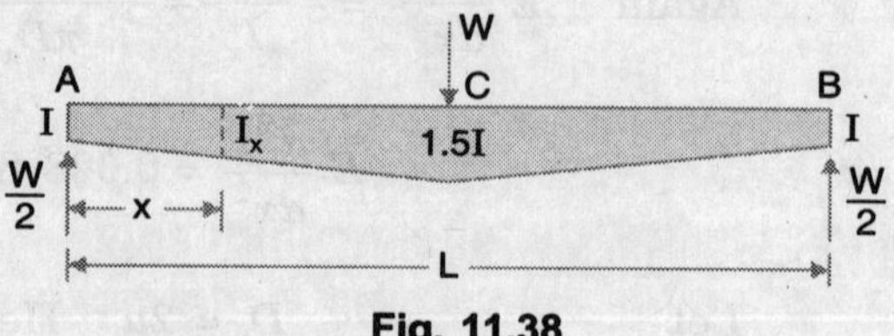

Fig. 11.38

Solution. At a section distant x from A, we have,

$$I_x = I + 0.5I\,\frac{2x}{L}$$

$$E\frac{d^2y}{dx^2} = -\frac{M_x}{I_x} = \frac{-\frac{W}{2}x}{I + 0.5I\,\frac{2x}{L}} = \frac{-\frac{W}{2}x}{I\left(1+\frac{x}{L}\right)} = -\frac{Wx}{2I\left(1+\frac{x}{L}\right)} = -\frac{W}{2I}\left\{\frac{L\left(1+\frac{x}{L}\right)-L}{1+\frac{x}{L}}\right\}$$

or $$E\frac{d^2y}{dx^2} = -\frac{WL}{2I} + \frac{WL}{2I\left(1+\frac{x}{L}\right)}$$

Integrating, $E\dfrac{dy}{dx} = -\dfrac{WLx}{2I} + \dfrac{WL^2}{2I}\log_e\left(1+\dfrac{x}{L}\right) + C_1$

At $x = \dfrac{L}{2}$, $\dfrac{dy}{dx} = 0 = -\dfrac{WL}{2I}\left(\dfrac{L}{2}\right) + \dfrac{WL^2}{2I}\log_e\left(1+\dfrac{1}{2}\right) + C_1$

From which $C_1 = \dfrac{WL^2}{4I} - \dfrac{WL^2}{2I}\log_e 1.5$

Substituting in (1), we get $E\dfrac{dy}{dx} = -\dfrac{WLx}{2I} + \dfrac{WL^2}{4I} + \dfrac{WL^2}{2I}\log_e\dfrac{2}{3}\left(1+\dfrac{x}{L}\right)$ ...(1)

Integrating, $Ey = -\dfrac{WLx^2}{4I} + \dfrac{WL^2x}{4I} + \dfrac{WL^2}{2I}\left\{x\log_e\dfrac{2}{3}\left(1+\dfrac{x}{L}\right) - \int\dfrac{\frac{2}{3}\frac{x}{L}}{\frac{2}{3}\left(1+\frac{x}{L}\right)}dx\right\} + C_2$

or $$Ey = -\frac{WLx^2}{4I} + \frac{WL^2x}{4I} + \frac{WL^2}{2I}\left\{x\log_e\frac{2}{3}\left(1+\frac{x}{L}\right) - \int\left[1-\frac{1}{\left(1+\frac{x}{L}\right)}\right]dx\right\} + C_2$$

or $$Ey = -\frac{WLx^2}{4I} + \frac{WL^2x}{4I} + \frac{WL^2}{2I}\left\{x\log_e\frac{2}{3}\left(1+\frac{x}{L}\right) - x + L\log_e\left(1+\frac{x}{L}\right)\right\} + C_2$$

At $x = 0$, $y = 0$. Hence $C_2 = 0$

$$\therefore \quad Ey = -\frac{WLx^2}{4I} + \frac{WL^2x}{4I} + \frac{WL^2}{2I}\left\{x\log_e\frac{2}{3}\left(1+\frac{x}{L}\right) - x + L\log_e\left(1+\frac{x}{L}\right)\right\} \quad ...(2)$$

Hence the central deflection at $x = L/2$ is given by

$$Ey_C = -\frac{WL}{4I}\left(\frac{L}{2}\right)^2 + \frac{WL^2}{4I}\left(\frac{L}{2}\right) + \frac{WL^2}{2I}\left\{\frac{L}{2}\log_e\frac{2}{3}\left(1+\frac{1}{2}\right) - \frac{L}{2} + L\log_e\left(1+\frac{1}{2}\right)\right\}$$

$$= -\frac{WL^3}{16I} + \frac{WL^3}{8I} - \frac{WL^3}{4I} + \frac{WL^3}{2I}\log_e\frac{3}{2}$$

$$= -\frac{3}{16}\frac{WL^3}{I} + 0.2027\frac{WL^3}{I} = 0.01523\frac{WL^3}{I}$$

Hence $y_C = 0.01523\,\dfrac{WL^3}{EI}$.

11.9. RELATION BETWEEN MAXIMUM STRESS AND MAXIMUM DEFLECTION

Example 11.28. *A simply supported beam of span L carries a central load W. Find the ratio of the maximum deflection to the maximum skin stress.*

Solution. Let f = maximum skin stress and δ = maximum deflection.

Now B.M. M at the centre = $WL/4$

$$\therefore \quad f = \frac{M}{Z} = \frac{WL}{4} \times \frac{d}{2I} = \frac{WLd}{8I}$$

Similarly, $\delta = \dfrac{WL^3}{48\,EI}$ Hence $\dfrac{\delta}{f} = \dfrac{WL^3}{48EI} \times \dfrac{8I}{WLd} = \boldsymbol{\dfrac{L^2}{6Ed}}$.

Example 11.29. *A beam of uniform section and span L, is subjected to a constant bending moment M. Find the ratio of the maximum deflection to the maximum skin stress.*

Solution. We have $\dfrac{M}{I} = \dfrac{E}{R}$ or $M = \dfrac{EI}{R}$

When M is constant, R is also constant. Hence a constant bending moment will bend the beam into a circle of radius R. Let δ = Central deflection (Fig. 11.39)

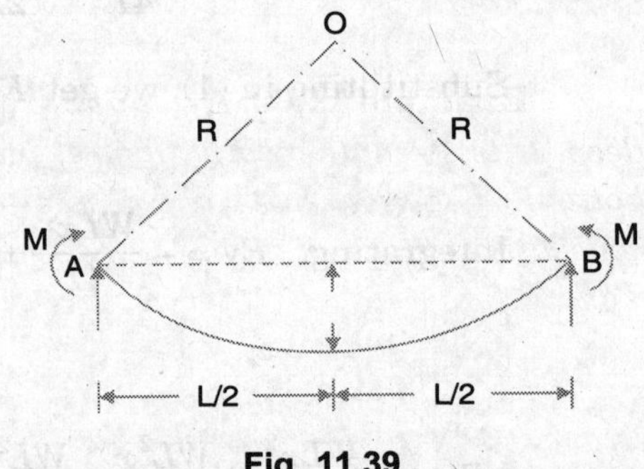

Fig. 11.39

From the property of the circle,

$$\delta.2R = \frac{L}{2} \cdot \frac{L}{2}; \quad \text{or} \quad \delta = \frac{L^2}{8R}$$

But $\dfrac{1}{R} = \dfrac{M}{EI}$. Hence $\delta = \dfrac{L^2}{8} \cdot \dfrac{M}{EI}$

Also, $f = \dfrac{M}{Z} = \dfrac{Md}{2I}$

$$\therefore \quad \frac{\delta}{f} = \frac{L^2 M}{8EI} \times \frac{2I}{Md} = \boldsymbol{\frac{L^2}{4Ed}}.$$

11.10. DEFLECTION OF COMPOSITE BEAMS

A composite beam is the one which is made up of two or more materials, joined together in such a manner that they behave as a single unit. For such a beam, the deflection of each component/material, at a given section of the beam, will be the same. The slope and deflection of such beams are found by the same expressions, as given for beams of one single material. However, the *flexural rigidity* (*EI*) for such a composite beam is found out by adding the flexural rigidities of each component.

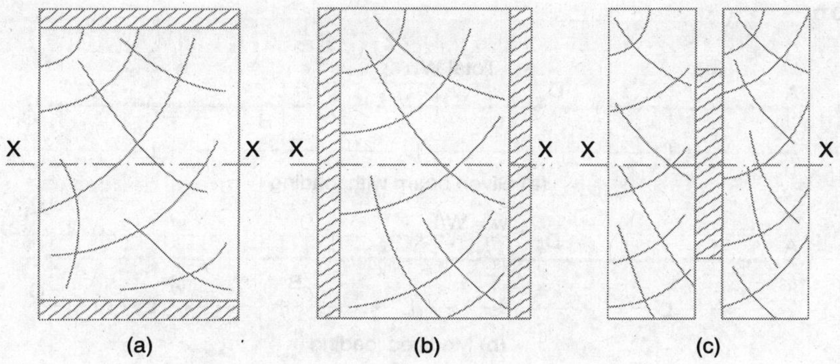

Fig. 11.40. Beams of composite section

Fig. 11.40 shows some composite beams (flitched beams) made up of two materials. For each one of these beams,

$$\Sigma EI = E_1 I_1 + E_2 I_2 \qquad \text{...(11.20)}$$

Example 11.30. *A composite beam consists of a timber section 100 mm wide and 200 mm deep. It is strengthened with steel plates 100 mm wide and 20 mm thick, fixed at the top and bottom. The composite beam is simply supported over a span of 4 metres and carries a concentrated load of 80 kN at the mid-span. Compute the deflection of the beam under the load. Take E for steel and timber as 2×10^5 N/mm² and 0.1×10^5 N/mm² respectively.*

Solution. (Fig. 11.41) For a beam with central point load,

$$y_C = WL^3/48\Sigma EI$$

$$(EI)_{\text{timber}} = 0.1 \times 10^5 \left[\frac{1}{12} \times 100(200)^3\right]$$

$$= 66.7 \times 10^{10} \text{ N-mm}^2$$

$$(EI)_{\text{steel}} = 2 \times 10^5 \times 2 \left[\frac{1}{12} \times 100(12)^3 + 100 \times 12\,(106)^2\right]$$

$$= 539.9 \times 10^{10} \text{ N-mm}^2$$

$$\therefore \quad \Sigma EI = 66.7 \times 10^{10} + 539.9 \times 10^{10} = 606.6 \times 10^{10} \text{ N-mm}^2$$

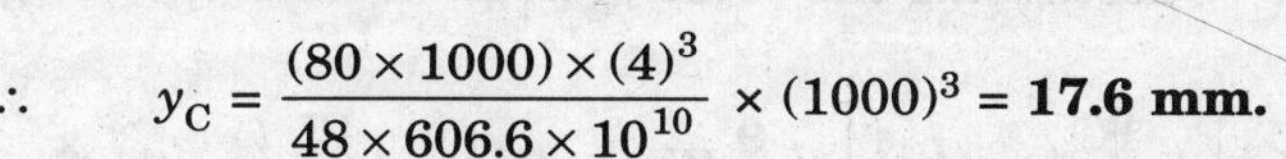

$$\therefore \quad y_C = \frac{(80 \times 1000) \times (4)^3}{48 \times 606.6 \times 10^{10}} \times (1000)^3 = \mathbf{17.6 \text{ mm.}}$$

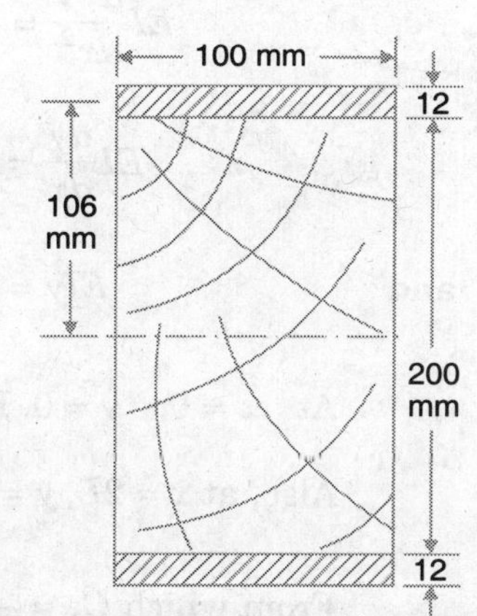

Fig. 11.41

11.11. ADDITIONAL EXAMPLES USEFUL FOR COMPETITIVE EXAMINATIONS

Example 11.31. *A overhanging beam ABC is loaded as shown in Fig. 11.42(a). Determine the slope and deflection of the free end C of the beam.*

Solution.

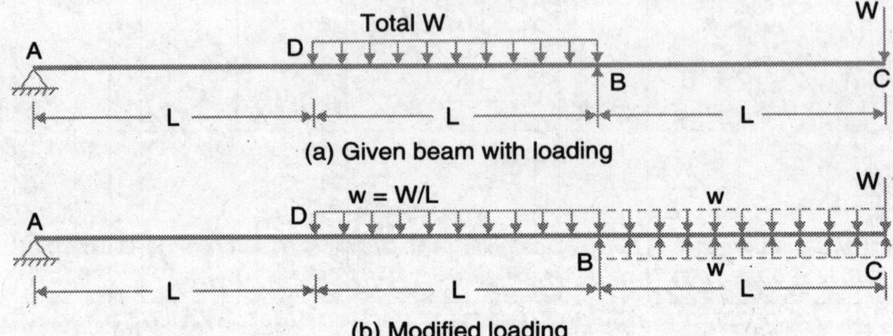

(a) Given beam with loading

(b) Modified loading

Fig. 11.42

$$R_B = \frac{1}{2L}\left[W \times 3L + W\left(L + \frac{L}{2}\right)\right] = \frac{9}{4}W \uparrow$$

$$\therefore \quad R_A = (W + W) - \frac{9}{4}W = -\frac{W}{4} \;(i.e., \downarrow)$$

Total U.D.L. = W $\quad\therefore\quad w = W/L$.

In order to effectively apply Macaulay's method, extend the U.D.L. from B to C, and apply upward load, from B to C, as shown in Fig. 11.42(b).

Measuring x from A, we have

$$EI\frac{d^2y}{dx^2} = -M_x = \frac{W}{4}x \;\Big|\; + \frac{W}{L}\frac{(x-L)^2}{2} \;\Big|\; - \frac{9}{4}W(x-2L) - \frac{W}{L}\frac{(x-2L)^2}{2} \quad \text{...(i)}$$

$$\therefore \quad EI\frac{dy}{dx} = \frac{W}{8}x^2 + C_1 \;\Big|\; + \frac{W}{6L}(x-L)^3 \;\Big|\; - \frac{9}{8}W(x-2L)^2 - \frac{W}{6L}(x-2L)^3 \quad \text{...(ii)}$$

and

$$EIy = \frac{Wx^3}{24} + C_1x + C_2 \;\Big|\; + \frac{W}{24L}(x-L)^4 \;\Big|\; - \frac{9}{24}W(x-2L)^3 - \frac{W}{24L}(x-2L)^4 \quad \text{...(iii)}$$

At $x = 0$, $y = 0$. Hence $C_2 = 0$.

Also, at $x = 2L$, $y = 0 = \frac{W}{24}(2L)^3 + C_1(2L) + \frac{W}{24L}(2L-L)^4$

From which $C_1 = -\frac{3}{16}WL^2$. Substituting the value of C_1 in (ii) and (iii), we get

$$EI\frac{dy}{dx} = \frac{W}{8}x^2 - \frac{3}{16}WL^2 \;\Big|\; + \frac{W}{6L}(x-L)^3 \;\Big|\; - \frac{9}{8}W(x-2L)^2 - \frac{W}{6L}(x-2L)^3 \quad \text{...(a)}$$

and

$$EIy = \frac{W}{24}x^3 - \frac{3}{16}WL^2x \;\Big|\; + \frac{W}{24L}(x-L)^4 \;\Big|\; - \frac{9}{24}W(x-2L)^3 - \frac{W}{24L}(x-2L)^4 \quad \text{...(b)}$$

At end C, $x = 3L$. Hence from (a),

$$EIi_C = \frac{W}{8}(3L)^2 - \frac{3}{16}WL^2 + \frac{W}{6L}(3L-L)^3 - \frac{9}{8}W(3L-2L)^2 - \frac{W}{6L}(3L-2L)^3$$

or $$i_C = -\frac{47}{48}\frac{WL^2}{EI}.$$

Also, from (*b*)

$$EIy_C = \frac{W}{24}(3L)^3 - \frac{3}{16}WL^2(3L) + \frac{W}{24L}(3L-L)^4 - \frac{9}{24}W(3L-2L)^3 - \frac{W}{24L}(3L-2L)^4$$

From which $y_C = \dfrac{13}{16}\dfrac{WL^3}{EI}$.

Example 11.32. *A beam AB of span 5L is simply supported at A and B. A cantilever DC of length 3L, which is fixed at D, meets the beam AB at the mid-point C, thereby forming a rigid joint at C. Find out the reactions at A and B, if a vertical load W is applied at the end C of the cantilever. Both the beam and the cantilever have the same moment of inertia, and are made of the same material. What will be the values of reactions at A and B, if the cantilever and beam have equal span.*

Solution. (*a*) Let the load shared by the cantilever be *P*. Load shared by the beam $= W - P$

For the cantilever, $$y_C = \frac{P(3L)^3}{3EI} = 9\frac{PL^3}{EI} \quad ...(i)$$

For the beam $$y_C = \frac{(W-P)(5L)^3}{48EI} = \frac{125}{48}\frac{(W-P)L^3}{EI} \quad ...(ii)$$

Equating the two, we get $$\frac{9PL^3}{EI} = \frac{125}{48}\frac{(W-P)L^3}{EI}$$

$$\therefore \quad \frac{9\times 48}{125}P = (W-P).$$

From which, $P = \dfrac{125}{557}W$

$\therefore$ Load shared by beam

$$= W - P = W - \frac{125}{557}W = \frac{432}{557}W$$

Hence $R_A = R_B = \dfrac{1}{2}\times\dfrac{432}{557}W$

$$= \frac{216}{557}W = \mathbf{0.388\ W}.$$

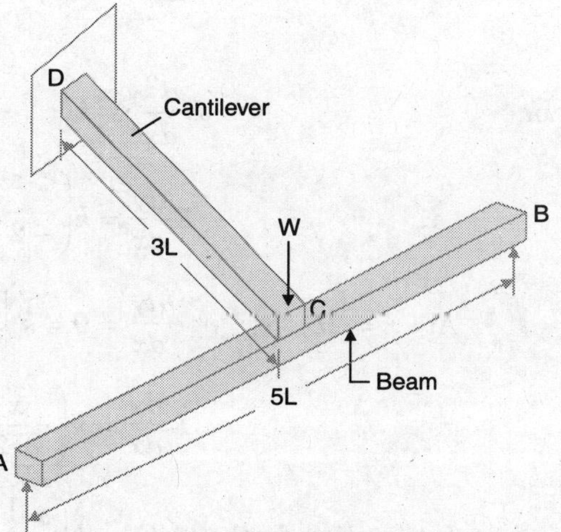

Fig. 11.43

(*b*) Let the cantilever and the beam have equal span $= l$

$\therefore$ For the cantilever, $$y_C = \frac{Pl^3}{3EI} \quad ...(a)$$

For the beam, $$y_C = \frac{(W-P)l^3}{48EI} \quad ...(b)$$

$$\therefore \quad \frac{Pl^3}{3EI} = \frac{(W-P)l^3}{48EI} \quad \text{or} \quad 16P = (W-P) \text{ From which, } P = \frac{W}{17}$$

Load shared by the beam $= W - P = W - \frac{W}{17} = \frac{16}{17} W$

$$\therefore \quad R_A = R_B = \frac{1}{2} \times \frac{16}{17} W = \frac{8}{17} W = \mathbf{0.471\ W.}$$

Example 11.33. *A cantilever of a circular section tapers uniformly from a diameter (d) at the free end to 2d at the fixed end. It carries a concentrated load W at the free end. Calculate the diameter of a cantilever of uniform cross-section, which would have the same deflection.*

Solution.

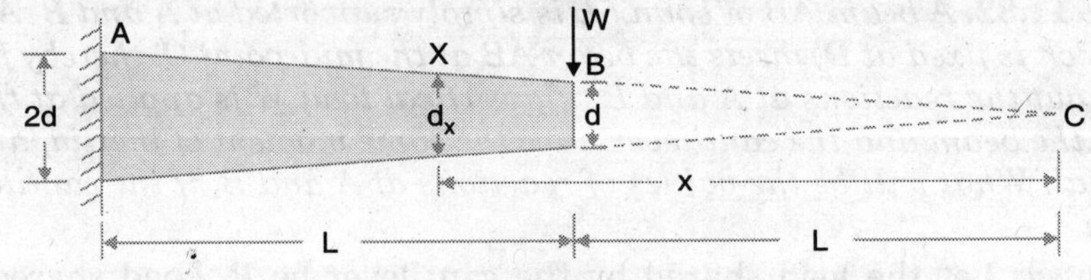

Fig. 11.44

Extend the upper and lower faces of the cantilever to meet at C, as shown in Fig. 11.44. Evidently, $BC = AB = L$.

At any section x from C, the diameter is given by $d_x = \frac{x}{L} d$ and $I_x = \frac{\pi}{64}\left(\frac{x^4 d^4}{L^4}\right)$

Now $\quad M_x = -W(x - L)$

Now $$E\frac{d^2y}{dx^2} = -\frac{M_x}{I_x} = \frac{W(x-L)}{\frac{\pi}{64}\frac{x^4 d^4}{L^4}} = \frac{64}{\pi d^4} WL^4 \left(\frac{x-L}{x^4}\right)$$

or $$E\frac{d^2y}{dx^2} = k(x^{-3} - Lx^{-4}) \text{ where } k = \frac{64\,WL^4}{\pi d^4} \quad ...(i)$$

$$\therefore \quad E\frac{dy}{dx} = k\left(\frac{x^{-2}}{-2} - \frac{Lx^{-3}}{-3}\right) + C_1 = k\left(-\frac{x^{-2}}{2} + \frac{Lx^{-3}}{3}\right) + C_1$$

At $x = 2L$, $$\frac{dy}{dx} = 0 = k\left[-\frac{1}{2}(2L)^{-2} + \frac{L}{3}(2L)^{-3}\right] + C_1 \text{ From which, } C_1 = \frac{kL^{-2}}{12}$$

$$\therefore \quad E\frac{dy}{dx} = k\left(-\frac{x^{-2}}{2} + \frac{Lx^{-3}}{3}\right) + \frac{kL^{-2}}{12}$$

Integrating further, $$Ey = k\left(\frac{x^{-1}}{2} - \frac{Lx^{-2}}{6} + \frac{xL^{-2}}{12}\right) + C_2$$

At $x = 2L$, $$y = 0 = k\left[\frac{(2L)^{-1}}{2} - \frac{L}{6}(2L)^{-2} + \frac{L^{-2}}{12}(2L)\right] + C_2 \text{ From which, } C_2 = -\frac{3}{8}\frac{k}{L}$$

$$\therefore \quad Ey = k\left[\frac{1}{2x} - \frac{L}{6x^2} + \frac{x}{12L^2} - \frac{3}{8L}\right]$$

For the free end, $x = L$

$$\therefore \quad y_B = \frac{k}{E}\left[\frac{1}{2L} - \frac{L}{6L^2} + \frac{L}{12L^2} - \frac{3}{8L}\right] = \frac{k}{24\,EL}$$

Substituting the value of k, $$y_B = \frac{1}{24EL}\left(\frac{64WL^4}{\pi d^4}\right) = \frac{64}{24\pi}\frac{WL^3}{Ed^4} \quad ...(a)$$

For a cantilever of uniform diameter D, $I = \frac{\pi}{64}D^4$,

$$\therefore \quad y_B = \frac{WL^3}{3EI} = \frac{WL^3}{3E\left(\frac{\pi}{64}D^4\right)} = \frac{64}{3\pi}\frac{WL^3}{ED^4} \quad ...(b)$$

Equating the two, $$\frac{64}{24\pi}\frac{WL^3}{Ed^4} = \frac{64}{3\pi}\frac{WL^3}{ED^4}$$

From which, $D^4 = 8d^4$ or $D = \mathbf{1.6818\ d.}$

Example 11.34. *A beam PQRS is initially horizontal. A load W is then applied at R. In order to bring the position of the beam at Q, back to the original level PQRS, a vertical force W_1 is placed at Q. Find W_1.*

Solution. Initially, when a load W is placed at R, the *downward* deflection of Q is given by Eq. 11.11.

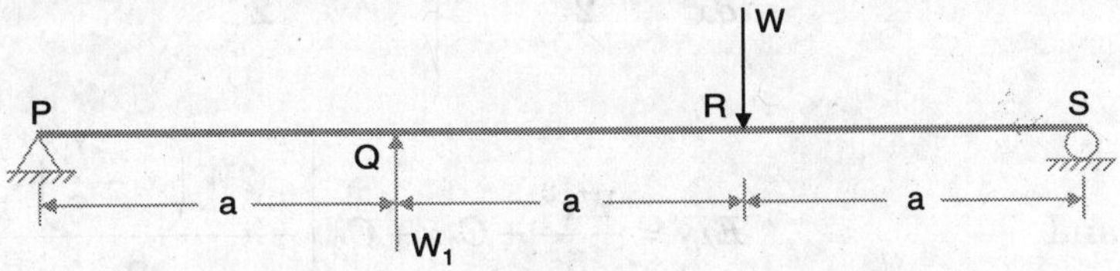

$$y = \frac{Wbx}{6EIL}(L^2 - b^2 - x^2),$$

with usual notations.

Fig. 11.45

Here, $x = a$, $b = a$ and $L = 3a$.

$$\therefore \quad y_Q = \frac{W}{6EI}\frac{a\times a}{3a}[(3a)^2 - (a)^2 - (a)^2] = \frac{7}{18}\frac{Wa^3}{EI}\ (\downarrow) \quad ...(a)$$

In general, the deflection under a point load P is given by $y = \dfrac{Pa^2b^2}{3EIL}$

When the upward load W_1 is applied at Q, we have $P = W_1$; $a = a$; $b = 2a$; $L = 3a$

$$\therefore \quad y_Q = \frac{W_1}{3EI}\frac{(a)^2\,(2a)^2}{3a} = \frac{4}{9}\frac{W_1a^3}{EI}\uparrow \quad ...(b)$$

Since the final deflection at Q is zero, we have

$$-\frac{4}{9}\frac{W_1a^3}{EI} + \frac{7}{18}\frac{Wa^3}{EI} = 0 \quad \text{or} \quad W_1 = \frac{7}{18}\times\frac{9}{4}W = \mathbf{\frac{7}{8}W.}$$

Example 11.35. *A beam 3 m long, simply supported at its ends, is carrying a point load W at its mid-span. If the slope at the ends of the beam is not to exceed 1°, find the deflection at the mid-span.*

Solution. For a simply supported beam with central point load W, we have

$$i_A = \frac{WL^2}{16EI} \quad \text{and} \quad y_C = \frac{WL^3}{48EI}$$

Here $\quad i_A = 1° = \dfrac{1\times\pi}{180} = 0.01745$ radian

Now $\quad y_C = \dfrac{WL^3}{48EI} = \dfrac{WL^2}{16EI}\times\dfrac{L}{3} = i_A \times \dfrac{L}{3} = 0.01745\left(\dfrac{3000}{3}\right) = \mathbf{17.45\ mm.}$

Example 11.36. *A beam ABCD has equal over hangs to both the sides and is loaded as shown in Fig. 11.46. Find (a) slopes at the free ends and at the supports (b) deflections at the free ends and at the mid-span.*

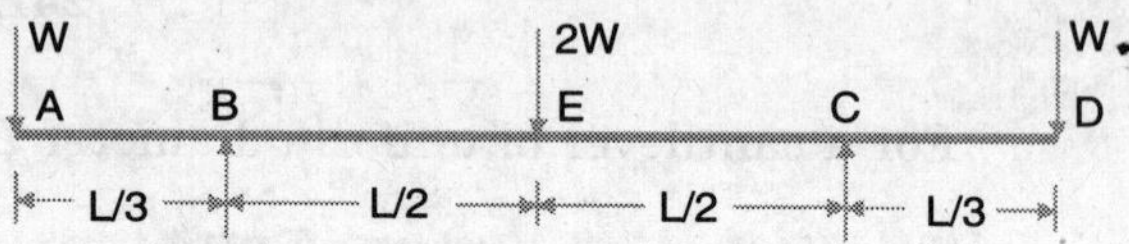

Fig. 11.46

Solution.

$$R_B = R_C = 4W/2 = 2\ W$$

Measuring x from A, we have

$$EI\frac{d^2y}{dx^2} = -M_x = W.x \;\Big|\; -2W\left(x-\frac{L}{3}\right), \text{ for the portion upto point } E.$$

$$\therefore \quad EI\frac{dy}{dx} = \frac{Wx^2}{2} + C_1 \;\Big|\; -\frac{2W\left(x-\frac{L}{3}\right)^2}{2} \quad ...(1)$$

and

$$EIy = \frac{Wx^3}{6} + C_1x + C_2 \;\Big|\; -\frac{2W\left(x-\frac{L}{3}\right)^3}{6} \quad ...(2)$$

Due to symmetry, at $x = \left(\dfrac{L}{3}+\dfrac{L}{2}\right) = \dfrac{5}{6}L, \quad \dfrac{dy}{dx} = 0$

$$\therefore \quad 0 = \frac{W}{2}\left(\frac{5}{6}L\right)^2 + C_1 - W\left(\frac{5}{6}L - \frac{L}{3}\right)^2, \text{ which gives } C_1 = -\frac{7}{72}WL^2$$

Also, at $x = \dfrac{L}{3}$, $y = 0$. Hence from (2)

$$0 = \frac{W}{6}\left(\frac{L}{3}\right)^3 - \frac{7}{72}WL^2\left(\frac{L}{3}\right) + C_2, \text{ from which } C_2 = +\frac{17}{648}WL^3$$

Hence the slope and deflection equations are :

$$EI\frac{dy}{dx} = \frac{Wx^2}{2} - \frac{7}{72}WL^2 \;\Big|\; -W\left(x-\frac{L}{3}\right)^2 \quad ...(I)$$

and

$$EIy = \frac{Wx^3}{6} - \frac{7}{72}WL^2x + \frac{17}{648}WL^3 \;\Big|\; -\frac{W}{3}\left(x-\frac{L}{3}\right)^3 \quad ...(II)$$

Slope at free ends : Substituting $x = 0$, in (I), we get

$$i_A = \frac{1}{EI}\left(-\frac{7}{72}WL^2\right) = -\frac{\mathbf{7WL^2}}{\mathbf{72EI}}$$

Slopes at supports : Substituting $x = L/3$, in (I), we get

$$i_B = \frac{1}{EI}\left[\frac{W}{2}\left(\frac{L}{3}\right)^2 - \frac{7}{72}WL^2\right] = -\frac{\mathbf{WL^2}}{\mathbf{24EI}}$$

Deflection at free end : Substituting $x = 0$ in (II), we get

$$y_A = \frac{1}{EI}\left(+\frac{17}{648}WL^3\right) = +\frac{\mathbf{17}}{\mathbf{648}}\frac{\mathbf{WL^3}}{\mathbf{EI}}$$

Deflection at mid-span : Substituting $x = \frac{5}{6}L$ in (II), we get

$$y_E = \frac{1}{EI}\left[\frac{W}{6}\left(\frac{5}{6}L\right)^3 - \frac{7}{72}WL^2\left(\frac{5}{6}L\right) + \frac{17}{648}WL^3 - \frac{W}{3}\left(\frac{5}{6}L - \frac{L}{3}\right)^3\right] = 0.$$

Example 11.37. *A propped cantilever of span L is subjected to a couple μ at a section distant L/4 from the propped end. Find the reaction at the prop and the bending moment at the fixed end.*

Solution. Let us remove the prop, so that the beam is subjected to only a couple at *C*. Measuring *x* from *A*, we have

$$EI\frac{d^2y}{dx^2} = -M_x = \mu$$

$$EI\frac{dy}{dx} = \mu x + C_1$$

Fig. 11.47

At $x = 0$, $\frac{dy}{dx} = 0$. Hence $C_1 = 0$. Hence $EI\frac{dy}{dx} = \mu x$ ...(*i*)

Integrating further, $EIy = \frac{\mu x^2}{2} + C_2$

At $x = 0$, $y = 0$. Hence $C_2 = 0$. Hence $EIy = \frac{\mu x^2}{2}$ ...(*ii*)

Hence deflection at *B* due to μ acting at *C* is given by

$$y_B = \mu\left(\frac{3}{4}L\right)^2\frac{1}{2EI} + \mu\left(\frac{3}{4}L\right)\frac{1}{EI} \times \frac{L}{4} = \frac{15}{32}\mu L^2 \quad ...(a)$$

Now remove μ and put an upward force *P* at *B*. The upward deflection at *B* is given by,

$$y_B' = \frac{PL^3}{3EI} \quad ...(b)$$

Equating the two, we get $\frac{PL^3}{3EI} = \frac{15}{32}\mu L^2$. From which $P = \frac{\mathbf{45}}{\mathbf{32}}\frac{\boldsymbol{\mu}}{\mathbf{L}}$.

B.M. at $A = M_A = PL - \mu = \frac{45}{32}\frac{\mu}{L}\cdot L - \mu = \frac{\mathbf{13}}{\mathbf{32}}\boldsymbol{\mu}$.

Example 11.38. *Fig. 11.48 shows two cantilevers, the end of one being vertically above the other, and its connected to it by a spring AB.*

Initially, the system is unstrained. A weight W, placed at A, causes a vertical deflection at A of δ_1 and a vertical deflection at B of δ_2. When the spring is removed, the weight W at A causes a deflection at A of δ_3. Find the extension of the spring, when it is refixed and the weight W is transferred to B.

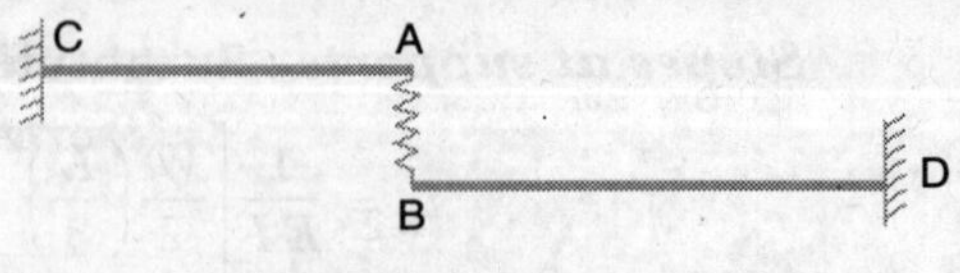

Fig. 11.48

Solution.

Let L_1 = length of cantilever CA ; L_2 = length of cantilever DB,

δ_4 = deflection of A with load at B ; δ_5 = deflection of B with load at B.

Given : δ_1 = deflection of A with load at A, with spring in position.

δ_2 = deflection of B with load at A, with spring in position

δ_3 = deflection of A with load at A, without spring.

When the load is at A, let P_{SA} be the force in the spring. When the load is at B, let P_{SB} be the force in the spring.

$$\therefore \quad \delta_1 = \frac{(W - P_{SA})\, L_1^3}{3EI} \qquad \text{...(i)}$$

$$\text{and} \quad \delta_2 = \frac{P_{SA} L_2^3}{3EI} \quad \text{or} \quad \frac{\delta_2}{P_{SA}} = \frac{L_2^3}{3EI} \qquad \text{...(ii)}$$

Compression of spring $AB = \delta_1 - \delta_2$

$$\therefore \quad \text{Stiffness of spring} = s = \frac{\text{force}}{\text{compression}} = \frac{P_{SA}}{\delta_1 - \delta_2} \qquad \text{...(iii)}$$

Again, when the spring is removed, and load W is placed at A,

$$\text{Deflection of } A = \delta_3 = \frac{W L_1^3}{3EI} \quad \text{or} \quad \frac{\delta_3}{W} = \frac{L_1^3}{3EI} \qquad \text{...(iv)}$$

Dividing (*i*) and (*iv*), we get $\dfrac{\delta_1}{\delta_3} = \dfrac{W - P_{SA}}{W} = 1 - \dfrac{P_{SA}}{W}$

$$\text{or} \quad \frac{P_{SA}}{W} = 1 - \frac{\delta_1}{\delta_3} = \frac{\delta_3 - \delta_1}{\delta_3} \quad \text{or} \quad P_{SA} = \frac{W(\delta_3 - \delta_1)}{\delta_3} \qquad \text{...(v)}$$

Substituting this value of P_{SA} in Eq. (*iii*), we get

$$s = \frac{W(\delta_3 - \delta_1)}{\delta_3\,(\delta_1 - \delta_2)} \qquad \text{...(vi)}$$

When the cantilever BD is loaded with W at B, the deflection of B is given by

$$\delta_5 = \frac{(W - P_{SB})\, L_2^3}{3EI}$$

and deflection of A is given by $\delta_4 = \dfrac{P_{SB} L_1^3}{3EI} = \dfrac{P_{SB}\,\delta_3}{W}$ (from *iv*).

Substituting the values of $\dfrac{L_2^3}{3EI}$ (from *ii*), and P_{SA} (from *v*), we get

$$\delta_5 = \frac{(W - P_{SB})\,\delta_2}{P_{SA}} = \frac{(W - P_{SB})\,\delta_2\delta_3}{W\,(\delta_3 - \delta_1)} \qquad \text{...(vii)}$$

Now stiffness $\quad s = \dfrac{\text{force}}{\text{extension}} = \dfrac{P_{SB}}{\delta_5 - \delta_4}$

$\therefore \quad \delta_5 - \delta_4 = \dfrac{P_{SB}}{s} = \dfrac{P_{SB} \cdot \delta_3 (\delta_1 - \delta_2)}{W(\delta_3 - \delta_1)}$ (from *vi*) ...(I)

But $\quad \delta_5 - \delta_4 = \dfrac{(W - P_{SB})\delta_2\delta_3}{W(\delta_3 - \delta_1)} - \dfrac{P_{SB}\,\delta_3}{W} = \dfrac{\delta_3}{W}\left[\dfrac{(W - P_{SB})\delta_2}{\delta_3 - \delta_1} - P_{SB}\right]$ ...(II)

Equating (I) and (II), we get

$$\frac{P_{SB} \cdot \delta_3 (\delta_1 - \delta_2)}{W(\delta_3 - \delta_1)} = \frac{\delta_3}{W}\left[\frac{(W - P_{SB})\delta_2}{\delta_3 - \delta_1} - P_{SB}\right]$$

or $\quad \dfrac{P_{SB}(\delta_1 - \delta_2)}{\delta_3 - \delta_1} = \dfrac{(W - P_{SB})\delta_2 - P_{SB}(\delta_3 - \delta_1)}{\delta_3 - \delta_1}$

or $\quad P_{SB}\delta_1 - P_{SB}\delta_2 = W\delta_2 - P_{SB}\,\delta_2 - P_{SB}\,\delta_3 + P_{SB}\delta_1$

or $\quad W\,\delta_2 = P_{SB}\,.\,\delta_3$ from which $P_{SB} = \dfrac{W\,\delta_2}{\delta_3}$.

Substituting the value of P_{SB} in (I), we get

$$\delta_5 - \delta_4 = \frac{\dfrac{W\,\delta_2}{\delta_3} . \delta_3 (\delta_1 - \delta_2)}{W(\delta_3 - \delta_1)} = \frac{\delta_2(\delta_1 - \delta_2)}{\delta_3 - \delta_1} \quad \textbf{(Ans.)}$$

Example 11.39. *A cantilever of length L carries a point load W at the free end. If the moment of inertia of the section increases uniformly from I at the free end to 2 I at the fixed end, calculate the deflection of the free end.*

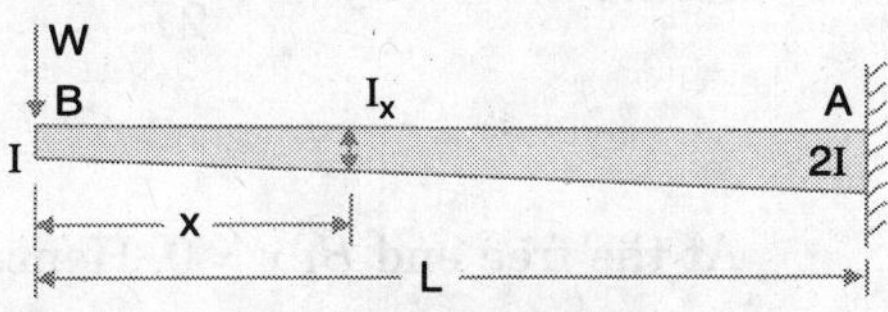

Fig. 11.49

Solution. (Fig. 11.49)

At any distance x from the free end, $I_x = I\left(1 + \dfrac{x}{L}\right)$

Now $\quad E\dfrac{d^2y}{dx^2} = -\dfrac{M_x}{I_x} = \dfrac{W\,.\,x}{I\left(1 + \dfrac{x}{L}\right)} = \dfrac{W\,L}{I}\,\dfrac{\dfrac{x}{L}}{\left(1 + \dfrac{x}{L}\right)}$

or $\quad E\dfrac{d^2y}{dx^2} = \dfrac{W\,L}{I}\left[1 - \dfrac{1}{1 + \dfrac{x}{L}}\right] = \dfrac{WL}{I} - \dfrac{WL}{I}\,\dfrac{1}{1 + \dfrac{x}{L}}$

Integrating, $\quad E\dfrac{dy}{dx} = \dfrac{W\,L}{I}\,x - \dfrac{W\,L^2}{I}\log_e\left(1 + \dfrac{x}{L}\right) + C_1$

At $x = L$, $\quad E\dfrac{dy}{dx} = 0 = \dfrac{W\,L}{I}\,.\,L - \dfrac{W\,L^2}{I}\log_e 2 + C_1$ or $C_1 = -\dfrac{W\,L^2}{I}(1 - \log_e 2)$

Hence $\quad E\dfrac{dy}{dx} = \dfrac{W\,L}{I}\,x - \dfrac{W\,L^2}{I}\log_e\left(1 + \dfrac{x}{L}\right) - \dfrac{W\,L^2}{I}(1 - \log_e 2)$

$$\therefore \quad Ey = \frac{WLx^2}{2I} - \frac{WL^2}{I}\left[x \log_e\left(1+\frac{x}{L}\right) - \int \frac{x}{L}\frac{1}{1+\frac{x}{L}}\,dx\right] - \frac{WL^2}{I}(1-\log_e 2)\,x + C_2$$

or
$$Ey = \frac{WLx^2}{2I} - \frac{WL^2}{I}\left[x \log_e\left(1+\frac{x}{L}\right) - \int \left\{1 - \frac{1}{1+\frac{x}{L}}\right\} dx\right] - \frac{WL^2}{I}(1-\log_e 2)x + C_2$$

or
$$Ey = \frac{WLx^2}{2I} - \frac{WL^2}{I}\left[x \log_e\left(1+\frac{x}{L}\right) - x + L\log_e\left(1+\frac{x}{L}\right)\right] - \frac{WL^2}{I}(1-\log_e 2)\,x + C_2$$

At the fixed end (*i.e.*, $x = L$), deflection $y = 0$. Hence

$$0 = \frac{WL^3}{2I} - \frac{WL^2}{I}[L\log_e 2 - L + L\log_e 2] - \frac{WL^3}{I}(1-\log_e 2) + C_2$$

From which $C_2 = -\dfrac{WL^3}{2I} + \dfrac{WL^2}{I}[L\log_e 2 - L + L\log_e 2 + L - L\log_e 2]$

or
$$C_2 = -\frac{WL^3}{2I} + \frac{WL^3}{I}\log_e 2 = \frac{WL^3}{I}(\log_e 2 - 0.5)$$

Hence
$$-Ey = \frac{WLx^2}{2I} - \frac{WL^2}{I}\left[x\log_e\left(1+\frac{x}{L}\right) - x + L\log_e\left(1+\frac{x}{L}\right)\right]$$
$$-\frac{WL^2}{I}(1-\log_e 2)x + \frac{WL^3}{I}(\log_e 2 - 0.5)$$

At the free end B, $x = 0$. Hence deflection of end B is given by

$$y_B = \frac{1}{E}\left[\frac{WL^3}{I}(\log_e 2 - 0.5)\right] = \frac{WL^3}{EI}(\log_e 2 - 0.5) = \mathbf{0.1931}\,\frac{\boldsymbol{WL^3}}{\boldsymbol{EI}}.$$

Example 11.40. *Two uniform steel bars of the same rectangular section are built-in together with the longer bar immediately over the shorter bar to form a cantilever as shown in Fig. 11.50. Before loading, the bars are in contact from A to B, but after loading it may be assumed that separation takes place with contact occurring only in the built-in part at A, and at the tip of the lower bar at B. Determine (a) the reaction at B (b) the deflection at C and (c) the maximum gap between the bars.*

Solution. Let the reaction between the two bars, at B, be P.

Deflection of bar AC. Bar AC is subjected a downward load W at end C and an upward load P at point B. Measuring x from A, we have, for the portion AB [Fig. 11.50(*b*)]

$$EI\frac{d^2y}{dx^2} = -M_x = W(L-x) - P\left(\frac{3}{4}L - x\right)$$

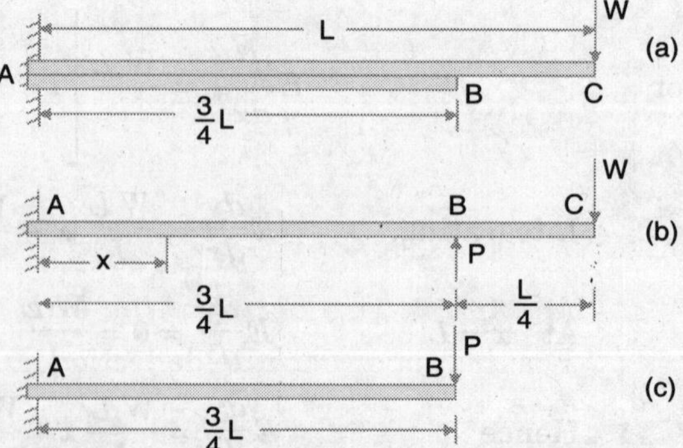

Fig. 11.50

Integrating,

$$EI\frac{dy}{dx} = W\left(Lx - \frac{x^2}{2}\right) - P\left(\frac{3}{4}Lx - \frac{x^2}{2}\right) + C_1$$

At $x = 0$, $\dfrac{dy}{dx} = 0$. Hence $C_1 = 0$

$$\therefore \quad EI\frac{dy}{dx} = W\left(Lx - \frac{x^2}{2}\right) - P\left(\frac{3}{4}Lx - \frac{x^2}{2}\right) \quad ...(i)$$

Integrating, $E\,I\,y = W\left(\dfrac{Lx^2}{2} - \dfrac{x^3}{6}\right) - P\left(\dfrac{3}{4}\dfrac{Lx^2}{2} - \dfrac{x^3}{6}\right) + C_2$ ...(ii)

At $x = 0$, $y = 0$. Hence $C_2 = 0$

$$\therefore \quad E\,I\,y = W\left(\frac{Lx^2}{2} - \frac{x^3}{6}\right) - P\left(\frac{3}{8}Lx^2 - \frac{x^3}{6}\right) \quad ...(iii)$$

For point B, $x = 3L/4$ and $y = y_B$

$$\therefore \quad E\,I\,y_B = W\left[\frac{L}{2}\left(\frac{3}{4}L\right)^2 - \frac{1}{6}\left(\frac{3}{4}L\right)^3\right] - P\left[\frac{3}{8}L\left(\frac{3}{4}L\right)^2 - \frac{1}{6}\left(\frac{3}{4}L\right)^3\right]$$

or $$y_B = \frac{9L^3}{128\,EI}[3W - 2P] \quad ...(a)$$

Deflection of bar AB : Bar AB is subjected to a downward load P at B [Fig. 11.50 (c)]

Hence $$y_B{}' = \frac{P\left(\frac{3}{4}L\right)^3}{3EI} = \frac{9PL^3}{64\,EI} \quad ...(b)$$

Equating (a) and (b), we have $\dfrac{9PL^3}{64\,EI} = \dfrac{9L^3}{128\,EI}[3W - 2P]$ or $P = \dfrac{3}{4}W$

Deflection at C :

$$y_C = \frac{WL^3}{3EI} - \left[\frac{P\left(\frac{3}{4}L\right)^3}{3EI} + \frac{P\left(\frac{3}{4}L\right)^2}{2EI}\cdot\frac{L}{4}\right] = \frac{WL^3}{3EI} - \frac{81}{384}\frac{PL^3}{EI}$$

$$= \frac{WL^3}{3EI} - \frac{81}{384}\left(\frac{3}{4}W\right)\frac{L^3}{EI} = \frac{\mathbf{269\,WL^3}}{\mathbf{1536EI}}.$$

Maximum gap between the bars :

For the upper bar, the deflection curve is given by Eq. (iii), for portion AB :

$$E\,I\,y_u = W\left(\frac{Lx^2}{2} - \frac{x^3}{6}\right) - P\left(\frac{3}{8}Lx^2 - \frac{x^3}{6}\right) = W\left(\frac{Lx^2}{2} - \frac{x^3}{6}\right) - \frac{3}{4}W\left(\frac{3}{8}Lx^2 - \frac{x^3}{6}\right)$$

or $$E\,I\,y_u = \frac{7}{32}\,W\,L\,x^2 - \frac{W\,x^3}{24} \qquad \text{...(I)}$$

The differential equation for the lower cantilever is

$$E\,I\frac{d^2y}{dx^2} = \frac{3}{4}\,W\left(\frac{3}{4}L - x\right) = \frac{9}{16}\,WL - \frac{3}{4}\,Wx$$

$$E\,I\frac{dy}{dx} = \frac{9}{16}\,W\,L\,x - \frac{3}{8}\,W\,x^2 \text{ (constant of integration being zero)}$$

and $$E\,I\,y_l = \frac{9}{32}\,W\,L\,x^2 - \frac{3}{24}\,W\,x^3 = \frac{9}{32}\,W\,L\,x^2 - \frac{1}{8}\,W\,x^3 \qquad \text{...(II)}$$

(The constant of integration being zero)

Hence, at any distance x from A, the gap δ between the two beams is given by

$$E\,I\,\delta = y_l - y_u = \left(\frac{9}{32}\,W\,L x^2 - \frac{1}{8}\,W\,x^3\right) - \left(\frac{7}{32}\,W\,L x^2 - \frac{W\,x^3}{24}\right)$$

or $$E\,I\,\delta = \frac{1}{16}\,W\,L\,x^2 - \frac{1}{12}\,W\,L\,x^3$$

For maximum gap, $\dfrac{d\delta}{dx} = 0 = \dfrac{1}{8}\,W\,L\,x - \dfrac{1}{4}\,W\,x^2$ which gives $x = \dfrac{L}{2}$

$\therefore$ $$E\,I\,\delta_{\max} = W\,L^3\left(\frac{1}{64} - \frac{1}{96}\right) \quad \text{or} \quad \delta_{\max} = \frac{\boldsymbol{W\,L^3}}{\boldsymbol{192\,E\,I}} \quad \textbf{(Ans.)}$$

Example 11.41. *A cantilever of length L carries a point load W at its free end. The member is circular in section, having diameter D for a distance L/2 from the fixed end and a diameter D/2 for the remaining length. Find the deflection at the free end.*

Solution. For the portion AC, $I_{AC} = \dfrac{\pi\,D^4}{64}$

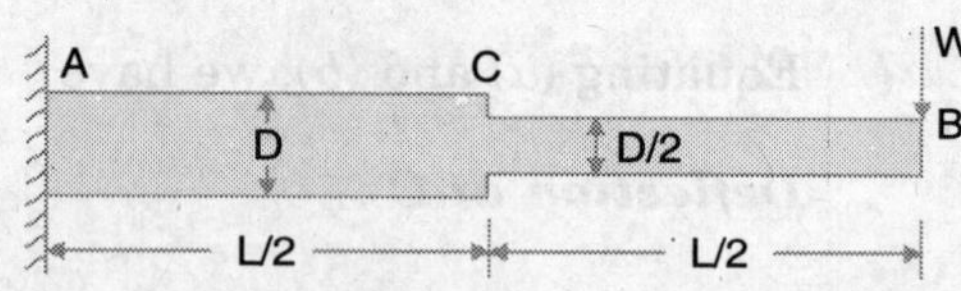

Fig. 11.51

For the portion CB, $I_{CB} = \dfrac{\pi}{64}\left(\dfrac{D}{2}\right)^4 = \dfrac{\pi\,D^4}{64 \times 16}$

Hence if $I_{CB} = I$, $I_{AC} = 16\,I$

For portion AC : Measuring x from A, we get

$$E\frac{d^2y}{dx^2} = -\frac{M_x}{I_{AC}} = \frac{W\,(L-x)}{16\,I}$$

Integrating, $$E\frac{dy}{dx} = \frac{W}{16\,I}\left(Lx - \frac{x^2}{2}\right) + C_1 \text{ At } x = 0, \ \frac{dy}{dx} = 0. \text{ Hence } C_1 = 0$$

$\therefore$ $$E\frac{dy}{dx} = \frac{W}{16\,I}\left(Lx - \frac{x^2}{2}\right) \qquad \text{...(I)}$$

Integrating further, $Ey = \frac{W}{16\,I}\left(\frac{Lx^2}{2} - \frac{x^3}{6}\right) + C_2$ **At** $x = 0,\ y = 0$. **Hence** $C_2 = 0$

$$\therefore \quad Ey = \frac{W}{16\,I}\left(\frac{Lx^2}{2} - \frac{x^3}{6}\right) \qquad \text{...(II)}$$

Eqs. (I) and (II) are the equations of slope and deflection for portion AC. **At** $x = L/2$, **the slope and deflection at** C **are as under :**

$$i_C = \frac{W}{16EI}\left[L.\frac{L}{2} - \frac{1}{2}\left(\frac{L}{2}\right)^2\right] = \frac{3WL^2}{128EI} \qquad \text{...}(a)$$

and

$$y_C = \frac{W}{16EI}\left[\frac{L}{2}\left(\frac{L}{2}\right)^2 - \frac{1}{6}\left(\frac{L}{2}\right)^3\right] = \frac{5WL^3}{16 \times 48\,EI} \qquad \text{...}(b)$$

For portion CB : **Measuring** x **from** C, **we have**

$$E\frac{d^2y}{dx^2} = -\frac{M_x}{I_{CB}} = \frac{W\left(\frac{L}{2} - x\right)}{I}$$

Integrating

$$E\frac{dy}{dx} = \frac{W}{I}\left(\frac{L}{2}x - \frac{x^2}{2}\right) + C_3 \qquad \text{...}(i)$$

At $x = 0$, $\frac{dy}{dx} = i_C = \frac{3WL^2}{128\,EI}$. **Hence from** (i), **we get** $C_3 = \frac{3WL^2}{128\,I}$

$$\therefore \quad E\frac{dy}{dx} = \frac{W}{I}\left(\frac{L}{2}x - \frac{x^2}{2}\right) + \frac{3WL^2}{128\,I} \qquad \text{..(III)}$$

Integrating further, $Ey = \frac{W}{I}\left(\frac{Lx^2}{4} - \frac{x^3}{6}\right) + \frac{3WL^2}{128\,I}x + C_4$ ...(ii)

At $x = 0$, $y = y_C = \frac{5WL^2}{16 \times 48EI}$. **Hence from** (ii), **we get** $C_4 = \frac{5WL^3}{16 \times 48\,I}$

$$\therefore \quad Ey = \frac{W}{I}\left(\frac{Lx^2}{4} - \frac{x^3}{6}\right) + \frac{3WL^2}{128\,I}x + \frac{5WL^3}{16 \times 48\,I} \qquad \text{...(IV)}$$

Eqs. (III) and (IV) are the equations for slope and deflection for the portion CB.

At the free end, $x = L/2$. **Hence from IV, we get**

$$Ey_B = \frac{W}{I}\left[\frac{L}{4}\left(\frac{L}{2}\right)^2 - \frac{1}{6}\left(\frac{L}{2}\right)^3\right] + \frac{3WL^2}{128\,I}\left(\frac{L}{2}\right) + \frac{5WL^3}{16 \times 48\,I}$$

From which we get $y_B = \frac{\mathbf{23}}{\mathbf{384}}\frac{\boldsymbol{WL^3}}{\boldsymbol{EI}}$.

The solution of this problem (or similar problem) by area moment method, is relatively easier. See examples 12.12. Also see example 13.11 for solution by conjugate beam method.

Example 11.42. *A horizontal cantilever of uniform section and length L carries a uniformly distributed load w per unit length throughout its length. The cantilever is supported by a rigid prop at a distance kL from the fixed end, the level of the beam at prop being the same as that at the fixed end. Determine the value of k which will make the bending moment at the prop the same as that at the fixed end. Also, sketch the S.F. and B.M. diagrams for the propped cantilever.*

Solution. Let the reaction at the prop be P.

$$M_C = -\frac{w(1-k)^2 L^2}{2} \quad ...(i)$$

Also, $M_A = P.kL - \dfrac{wL^2}{2}$ ...(ii)

Equating the two, we get

$$PkL - \frac{wL^2}{2} = -\frac{w(1-k)^2 L^2}{2}$$

or $$Pk = \frac{wL}{2} - \frac{wL}{2}(1-k)^2$$

$$= \frac{wL}{2}[1 - 1 + 2k - k^2]$$

$$= \frac{wL}{2}k(2-k)$$

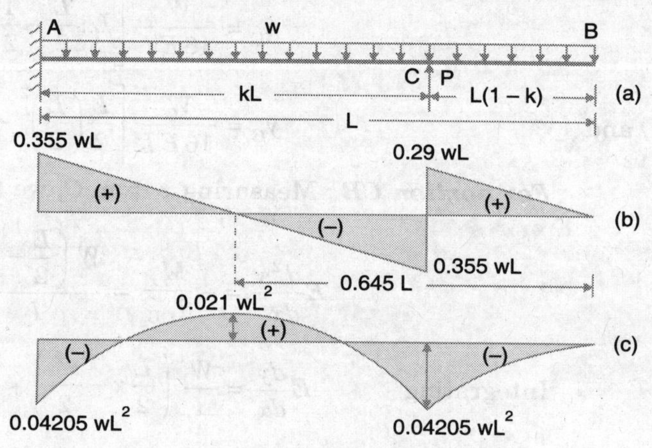

Fig. 11.52

$$\therefore \qquad P = \frac{wL}{2}(2-k) \quad ...(iii)$$

Thus, we know the value of prop reaction P in terms of k. In order to find the value of k, we use the compatibility conditions at C. Since the deflection at C is zero, the downward deflection due to load w, in absence of the prop, should be equal to the upward deflection (y_c') of C due to upward force P alone, applied at C.

Hence $\qquad y_C = y_C'$

In order to find the y_C due to U.D.L., in absence of the prop, consider a section distant x from A. Then we have

$$EI\frac{d^2y}{dx^2} = -M_x = \frac{w}{2}(L-x)^2 = \frac{w}{2}(L^2 + x^2 - 2Lx)$$

Integrating, $$EI\frac{dy}{dx} = \frac{w}{2}\left(L^2x + \frac{x^3}{3} - Lx^2\right) + C_1$$ At $x = 0$, $\dfrac{dy}{dx} = 0$. Hence $C_1 = 0$

$$\therefore \quad EI\frac{dy}{dx} = \frac{w}{2}\left[L^2x + \frac{x^3}{3} - Lx^2\right]$$

Integrating further, $EIy = \frac{w}{2}\left[\frac{L^2x^2}{2} + \frac{x^4}{12} - \frac{Lx^3}{3}\right] + C_2$

At $x = 0$, $y = 0$. Hence $C_2 = 0$

$$\therefore \quad EIy = \frac{w}{2}\left[\frac{L^2x^2}{2} + \frac{x^4}{12} - \frac{Lx^3}{3}\right]$$

At $x = kL$, $y = y_C$

$$\therefore \quad EIy_C = \frac{w}{2}\left[\frac{L^2}{2}(kL)^2 + \frac{1}{12}(kL)^4 - \frac{L}{3}(kL)^3\right]$$

From which $y_C = \frac{w}{24\,EI}k^2(k^2 - 4k + 6)$ ...(*a*)

Now upward deflection of C due to P alone, is

$$y_C' = \frac{P(kL)^3}{3EI} = \frac{wL}{2}(2-k)\frac{k^3L^3}{3EI} = \frac{wL^4(2-k)k^3}{6EI} \quad \text{...(b)}$$

Equating the two, we get $\frac{w}{24\,EI}k^2(k^2 - 4k + 6) = \frac{wL^4(2-k)k^3}{6EI}$

Simplifying, we get $k^2 - 4k + 6 = 4k(2-k)$ or $5k^2 - 12k + 6 = 0$ From which $k = \mathbf{0.71}$

Hence $P = \frac{wL}{2}(2-k) = \frac{wL}{2}(2 - 0.71) = 0.645\,wL$

and $R_A = wL - P = wL - 0.645\,wL = 0.355\,wL$

Length $AC = kL = 0.71\,L$; Length $CB = L(1-k) = 0.29\,L$

S.F. Diagram : $F_B = 0$; $F_C\,(right) = +wL(1-k) = +0.29\,wL$

$F_C\,(left) = 0.29\,wL - P = 0.29\,wL - 0.645\,wL = -0.355\,wL$

$F_A = +R_A = +0.355\,wL$

Evidently, S.F. is zero at $\frac{kL}{2} = \frac{0.71\,L}{2} = 0.355\,L$ from A.

The S.F. diagram is shown in Fig. 11.52 (*b*).

B.M. Diagram : $M_C = -\frac{w(0.29\,L)^2}{2} = -0.04205\,wL^2$

For CA, measuring x from B,

$$M_x = -\frac{wx^2}{2} + P(x - 0.29\,L) = -\frac{wx^2}{2} + 0.645\,wL(x - 0.29\,L) \quad \text{...(I)}$$

Maximum positive B.M. occurs at $kL/2$ from C, where S.F. is zero.

At this location, $x = 0.29\,L + 0.71\,\dfrac{L}{2} = 0.645\,L$

$$\therefore \quad M_{max} = -\frac{w}{2}(0.645\,L)^2 + 0.645\,w\,L\,(0.645\,L - 0.29\,L) = 0.021\,w\,L^2$$

Also, putting $x = L$ in (I), we get

$$M_A = -\frac{w\,L^2}{2} + 0.645\,w\,L\,(L - 0.29\,L) = -\,0.04205\,w\,L^2$$

The complete B.M. diagram is shown in Fig. 11.52 (*c*).

PROBLEMS

1. A beam of uniform section and length $2l$ is simply supported at its ends and carries two point loads each of magnitude W situated at equal distances a on either side of mid-span. Show that the maximum deflection Δ is given by the formula.

$$\Delta = \frac{W(2l^3 - 3la^2 + a^3)}{6EI} \qquad (U.L.)$$

2. A uniform beam of length $3L$ is simply and symmetrically supported over a span of L. It carries a load of W_1 at each end and a load of W_2 at mid-span. Find the ratio of W_1 to W_2 if the deflection at the mid-span is equal to that at each end.

3. A uniform beam of length $4L$ is simply and symmetrically supported over a span of $2L$. It carries a load W_1 at each end and a total uniformly distributed load of W_2 on the span between the supports. Find the ratio of W_1 to W_2 if the deflection at the mid-span is equal to that at each end.

4. A horizontal cantilever of effective length $3a$ carries two concentrated loads : W at a distance a from the fixed end and W' at distance a from the free end. Obtain a formula for the maximum deflection due to this loading.

   If the cantilever is made of 25 cm × 15 cm R.S.J. 3 metre long having moment of inertia $I = 8240$ cm^4, determine W and W' to give a maximum deflection of 8 mm when the maximum stress due to bending is 100 N/mm^2. Take $E = 2.05 \times 10^5$ N/mm^2.

5. A uniform section beam of length L is simply supported at its ends and carries a single concentrated load W at a distance of $L/3$ from one end.

   Working from fundamental beam theory, derive formula for the deflection (*a*) under the load, (*b*) at the centre, (*c*) at the point of maximum deflection.

6. Fig. 11.53 shows a horizontal beam freely supported at its ends on two similar cantilevers. If I, the second moment of area, of the cantilever section is twice that for the beam, derive an expression is terms of W, L, E and I for the deflection at the centre of the beam.

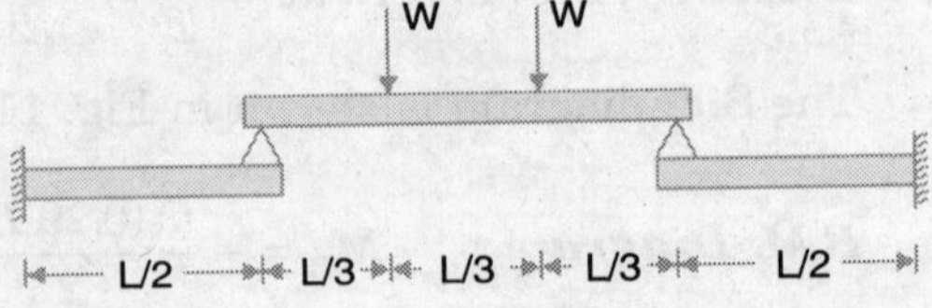

Fig. 11.53

7. A cantilever of length L and constant flexural rigidity EI carries a uniformly distributed load of intensity w/unit length on the middle half on its length. Determine the ratio of the slope and the ratio of the deflection at the centre and free end of the cantilever. (*U.L.*)

8. A horizontal cantilever of uniform section and length L carries a load W at the free end. Derive from the first principles the slope and deflection at the free end in terms of W, L, E.

If the load W is placed at $L/4$ from the free end, use the expression found above to obtain (*a*) the deflection of the free end, (*b*) the load on a prop at the free end if the level there is that of the fixed end.

Sketch the bending moment diagram for this case, stating on it the chief values. (*U.L.*)

9. A beam of uniform section and of length $2L$ is freely supported by rigid supports at its ends and by an elastic prop at its centre. If the prop deflects by an amount which is α times the load which it carries, and if the beam carries total distributed load of W show that the load carried by the prop is $\dfrac{5W}{8\left(1+\dfrac{6EI\alpha}{L^3}\right)}$.

If $L = 3$ m, $I = 500$ cm^4, $W = 50$ kN and $\alpha = 2 \times 10^{-3}$ mm/N, find the position and value of max. bending moment in the beam. Sketch approximately to scale, the shearing force and bending moment diagrams. Take $E = 2.04 \times 10^5$ N/mm^2.

10. A horizontal cantilever L long has its free end attached to a vertical tie rod $1.5L$ long, which is initially unstrained. If the moment of inertia of the section of the cantilever is I and the area of the cross-section of the tie rod is a, determine the load taken by the tie rod and the deflection of the cantilever when a distributed load of w unit/length is placed on the whole length of the cantilever. E for both the cantilever and rod is the same.

11. A horizontal propped cantilever of length L is securely fixed at one end, freely supported at the other, and is subjected to a bending couple M in a vertical plane containing the longitudinal axis of the beam. If the couple is applied about an axis $0.75L$ from the fixed end of the cantilever, determine the end fixing moment and the reaction at the freely supported end.

Sketch the shape of the bending moment diagram. (*U.L.*)

12. A beam of length L is supported at each end with a couple applied at an intermediate point. Deduce an expression for the deflection and hence calculate the deflection at the point of application of the moment.

13. A beam of length $2L$ carries a uniformly distributed load w/unit length and rests on three supports, two at the ends and one in the middle. Find how much the middle support be lower than the end ones in order that the pressures on the three supports shall be equal .

14. A simply supported beam of span L has a second moment of area of $2I$ for the left hand half of the span and I for the right hand half. The beam carries a load of intensity $2w$ uniformly distributed over the left hand half of the span and of intensity w uniformly distributed over the right hand half. Obtain an expression for deflection at the centre of the span. If a prop is now placed at the centre of the span to restore the beam to its original level at this point, find the force on the prop.

(*U.L.*)

15. A freely supported beam of span L carries a central load W. The section of the beam is so designed that the moment of inertia of the section increases from I at the ends to $2I$ at the middle. Calculate the central deflection.

16. A cantilever balcony projecting L from the wall is supported by two propped cantilever beams each of length L and at a distance $L/4$ between them as shown in Fig. 11.54. Both the beams are supported at the free ends by a cross girder of the length $L/4$. The cross girder is supported at

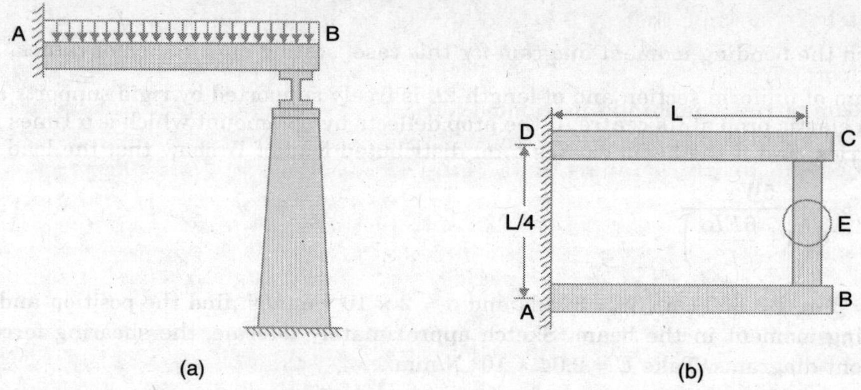

Fig. 11.54

its centre by a rigid pillar. Calculate the load on the pillar if the total distributed load on the balcony is W.

17. A cantilever of length L carries a load W at the free end. It is propped at the same level as the fixed end at a distance l from the fixed end. Show that the prop reaction P is given by

$$P = \frac{W(3L - l)}{2l}.$$

Show also that the deflection under the load is given by

$$y = \frac{W(L - l)^2\,(4L - l)}{12EI}.$$

18. A load W acts at the free end of a cantilever of constant depth h but of variable width b as shown in Fig. 11.55. Find the deflection of the free end.

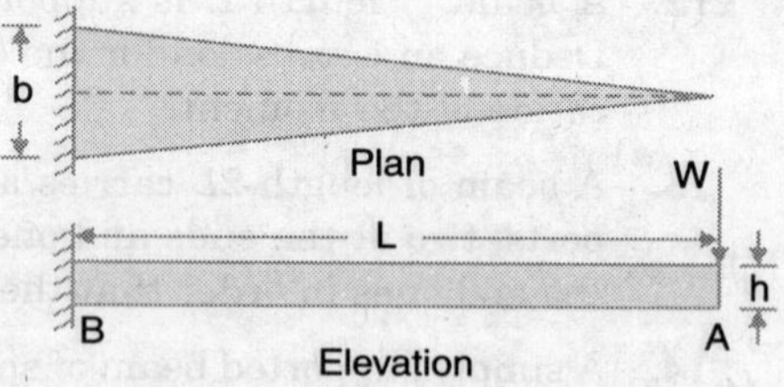

Fig. 11.55

19. A cantilever of length L is subjected to a clockwise couple of moment μ at the free end. Calculate the slope and deflection at the free end.

20. A simple supported beam of length L is subjected to a counter-clockwise couple of moment μ at the end as shown in Fig. 11.56. Find (*i*) the slopes at the ends, (*ii*) the central deflection and (*iii*) the maximum deflection and its location.

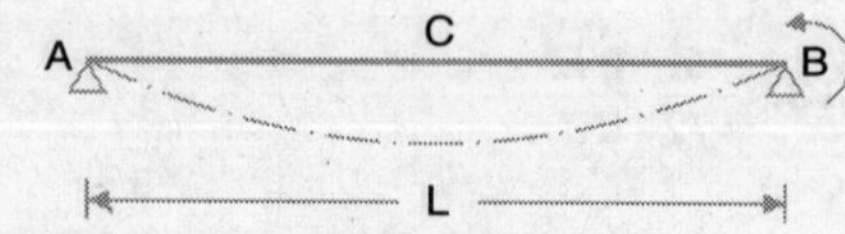

Fig. 11.56

21. A simply supported beam of length L carries a distributed load, the intensity of which varies as shown in Fig. 11.57. Derive an equation of elastic line.

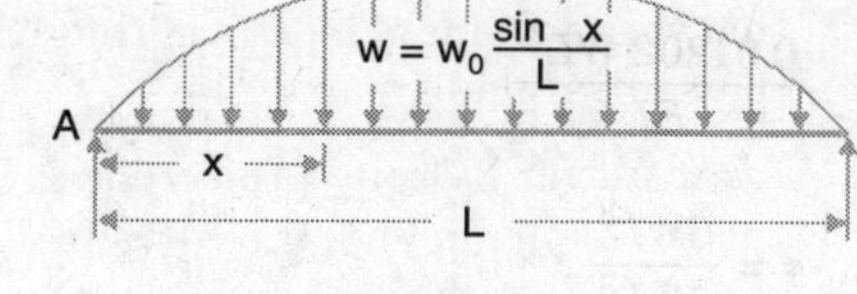

Fig. 11.57

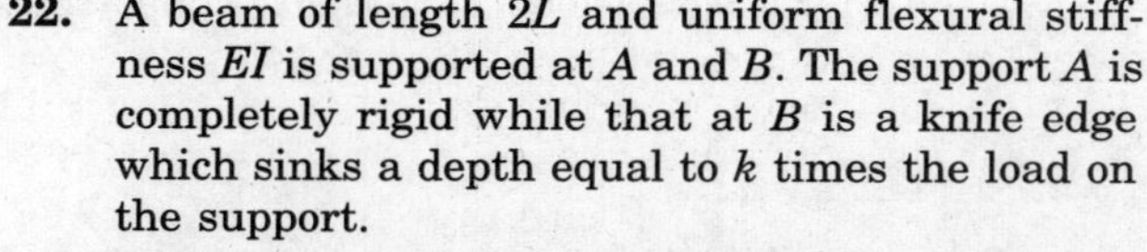

22. A beam of length $2L$ and uniform flexural stiffness EI is supported at A and B. The support A is completely rigid while that at B is a knife edge which sinks a depth equal to k times the load on the support.

If the beam carries a uniformly distributed vertical load of intensity w over the whole length, prove that the load on the support B is equal to

$$\frac{17wL^4}{8(L^3 + 3EIk)}.$$

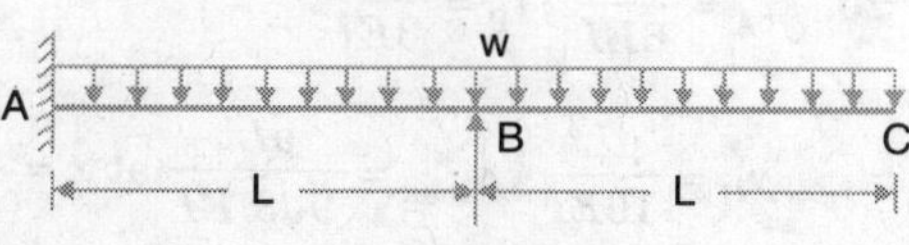

Fig. 11.58

ANSWERS

2. $\frac{W_1}{W_2} = \frac{2}{23}$.

3. $\frac{W_1}{W_2} = \frac{13}{88}$.

4. $y_A = \frac{2a^3}{3EI}(2W + 7W'), W = 19 \text{ kN} ; W' = 23.5 \text{ kN}$.

5. (a) $\frac{0.0164WL^3}{EI}$; (b) $\frac{0.01775WL^3}{EI}$; (c) $\frac{0.01792\,WL^3}{EI}$.

6. $\frac{73\,WL^3}{648\,EI}$.

7. Ratio of slopes = $\frac{25}{26}$; Ratio of deflection = $\frac{43}{112}$.

8. $i = \frac{WL^2}{2EI}$; $y = \frac{WL^3}{3EI}$.

(a) $y = \frac{27\,WL^3}{128\,EI}$;

(b) $P_2 = \frac{81\,W}{128}$.

9. $M_{max} = 12.25$ kN-m at 1.71 m from either end support.

10. $P = \frac{3awL^3}{4(2aL^2 + 9I)}$; $\delta = \frac{9wL^4}{8E(2aL^2 + 9I)}$.

11. $\frac{45}{32}\frac{W}{L}$.

12. $y = \frac{Wa}{3EI\,l}(l - a)(l - 2a)$, under the load

13. $\frac{7}{72}\frac{WL^4}{EI}$.

14. $y = \frac{11}{768}\frac{WL^4}{EI}$; $P = \frac{11}{12}WL$.

15. $\dfrac{0.01202\, WL^3}{EI}$.

16. $\dfrac{192}{513}\, W$.

18. $y = \dfrac{6WL^3}{Ebh^3}$.

19. $i = \dfrac{\mu L}{EI}$; $y = \dfrac{\mu L^2}{2EI}$.

20. $i_A = \dfrac{\mu L}{6EI}$; $i_B = \dfrac{\mu L}{3EI}$.

$y_C = \dfrac{\mu L^2}{16EI}$; $y_{max} = \dfrac{\mu L^2}{9\sqrt{3}\, EI}$ at $x = \dfrac{L}{\sqrt{3}}$.

21. $y = \dfrac{w_0 L^4}{\pi^4 EI} \sin \dfrac{\pi x}{L}$.

12

Deflection of Beams II : Area Moment Method

12.1. INTRODUCTION

Area moment method, also known as *moment area method,* is a very useful and simple method for finding slopes and deflections of the beams. The method utilises the properties of the area of the bending moment diagram and also the moment of that area. Depending as it does on the geometry of the elastic curve, the area moment method emphasizes the physical significance of slope and deflection. The method is especially suited when the deflection and angle of rotation at only one point of the beam is required, because it is possible to find these quantities directly without first finding the complete equation for the deflection curve. The method is also known as *Mohr's method* because it is based on two theorems given by Mohr and discussed in § 12.3. The area moment method is subject to the same limitations as the double integration method.

12.2. DEVIATION, DEFLECTANCE AND DEFLECTION

Fig. 12.1 shows the elastic curve *ACB* of a beam *AB*, subject to external loading. AB_2 is the tangent to the elastic curve at point *A*. Point *C* is deflected position of any point on the beam, while C_1 is the original position of the point. Line $A\,C_1B$ is the original position of the centre line of the beam, before loading. The interval C_1C is therefore the

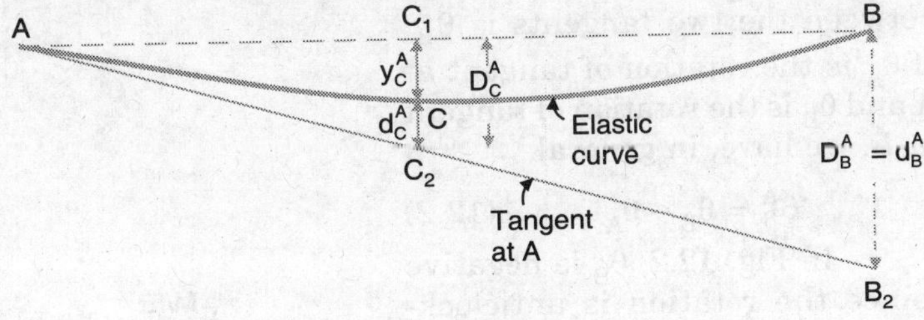

Fig. 12.1. Basic Definitions.

deflection of the point and is abbreviated as y_C^A in the area moment method. Here, the lower suffix to *y* denote the deflection of the *point* under consideration and the upper suffix to *y* will mean the point with respect to which the deflection is being calculated. The interval CC_2 is called the *deviation* of the point *C* from the tangent at *A* and is abbreviated as d_C^A. The lower suffix to d will mean the deviation of the point under consideration and the upper suffix to d will mean the point with respect to which deviation is being calculated. Thus, the *deviation* of point B with respect to the tangent at A is abbreviated as d_B^A. Naturally, the deviation of a point on the elastic curve may have infinite number of values depending upon the reference point chosen for the tangent. The total interval C_1C_2 is called the *deflectance* of point C and is abbreviated as D_C^A. Here again, the lower suffix to D will mean the point under consideration and the upper suffix to D will mean the point with respect to which the *deflectance* is being calculated.

Study of Fig. 12.1 shows that the relation between the deflection, deviation and deflectance of a point can be expressed by the equation :

$$y_C^A = D_C^A - d_C^A \quad ...(12.1)$$

Evidently, if the horizontal line through the *point of reference passes through the point where deflection is required (such as point B),* we have

$$D_B^A = d_B^A \quad ...[12.1(a)]$$

Hence $\quad y_B^A = D_B^A - d_B^A = 0$

In order to determine the deflection of a point, it is first required to find the *deviation,* which in turn can be found from the Mohr's theorems discussed in the next article.

12.3. MOHR'S THEOREMS : AREA MOMENT EQUATIONS

(*a*) **Rotation.** Fig. 12.2(*a*) shows the elastic curve *AmnB* of a beam. *AB′* is the tangent to the curve at *A* and *BB″* is the tangent to the curve at *B*. Evidently, *BB′* is the *deviation* (d_B^A) of *B* with respect to tangent at *A*. The angle between the two tangents is θ_B^A. If θ_A is the rotation of tangent at *A* and θ_B is the rotation of tangent at *B*, we have, in general,

$$\theta_B^A = \theta_B - \theta_A \quad ...(12.2)$$

In Fig. 12.2, θ_B is negative (since the rotation is anticlockwise).

Hence $\theta_B^A = (-\theta_B) - \theta_A$

$$= -(\theta_B + \theta_A) \quad ...[12.2(a)]$$

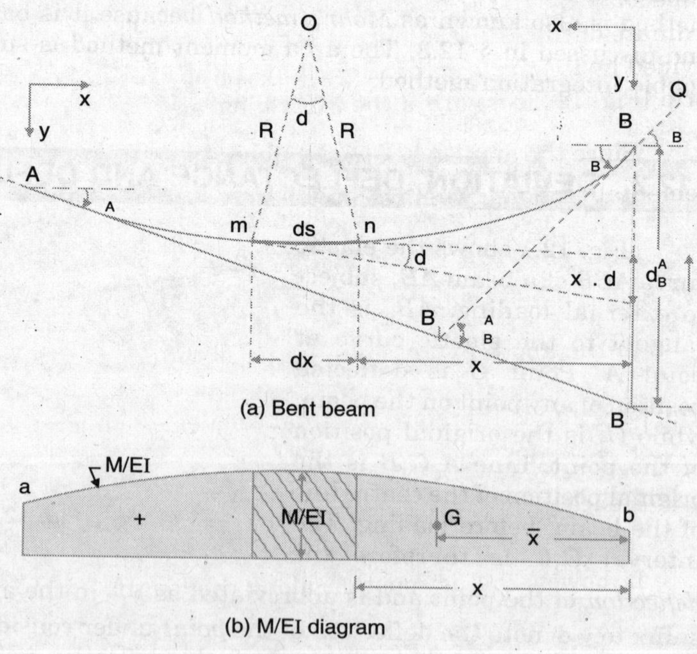

(a) Bent beam

(b) M/EI diagram

Fig. 12.2

In order to find the values of θ_B^A and d_B^A, let us consider two sections m and n at a very small distance ds along the curve. After bending, the normal sections at m and n will subtend an angle $d\theta$ at the centre of curvature. Hence $ds = R \cdot d\theta$

and $$\frac{1}{R} = \left|\frac{d\theta}{ds}\right| \quad ...(i)$$

The bars in the above expression indicate that we consider here only the numerical value of curvature. Regarding sign, B.M. is taken positive if it produces concavity. *Hence curvature is positive when the centre of curvature is above the curve,* as in Fig. 12.2. By close inspection of Fig. 11.2 we observe that for such a curvature, the angle θ decreases as the point m moves along the curve from point P to Q. Thus we have proper sign for Eq. (*i*) which must be written as

$$\frac{1}{R} = -\frac{d\theta}{ds} \quad \text{But} \quad \frac{1}{R} = \frac{M}{EI}$$

Hence $$d\theta = -\frac{M}{EI}\,ds$$

Since the curvature is very small, we can take ds equal to dx.

Hence $$d\theta = -\frac{M}{EI}\,dx \quad ...(12.3)$$

The term $(M/EI)\,dx$ has a simple geometric interpretation. Directly below the deflection curve is shown the M/EI diagram. It should be noted that the M/EI diagram will have the same shape as the M-diagram, only if EI is constant along the entire length of the beam. The ordinate of the (M/EI) diagram is equal to the ordinate of the B.M. diagram (*i.e.*, M-diagram) divided by the *flexural* rigidity EI at that point. The term $(M/EI)\,dx$ is the area of the shaded strip [Fig. 12.2(*b*)] within the M/EI diagram.

Since the angle θ_B^A between the two tangents at A and B consists of such small angles of elemental sections, we have :

$$\theta_B^A = \sum_B^A d\theta = -\int_B^A \frac{M\,dx}{EI} \quad ...(12.4)$$

In the above expression, $\frac{M}{EI}\,dx$ is the area of the $\frac{M}{EI}$ diagram between m and n. The term $\int_B^A \frac{M}{EI}\,dx$ is therefore the total area of $\frac{M}{EI}$ diagram between B and A, say $\Sigma A_{M/EI}$.

Hence $$\theta_B^A = -\int_B^A \frac{M\,dx}{EI} = -\sum_B^A A_{M/EI} \quad ...(12.5)$$

If, however EI is constant along the length of the beam, we have

$$EI\,\theta_B^A = -\sum_B^A A_M \quad ...[12.5(a)]$$

where A_M is the area of the B.M. diagram between B and A.

The use of bending moment diagram for calculating the slope and deflection of beam was developed by Mohr. Hence Eq. 12.5 can be stated by the following *first theorem* by Mohr :

Mohr's First Theorem : *The angle* θ_B^A *between the tangents to the deflection curve at two points A and B is equal to the negative of the area of M/EI diagram between the points.*

(*b*) **Deviation :** Let us now determine the deviation d_B^A of B from the tangent at A. Multiplying Eq. 12.3 by x', we have

$$x'\, d\theta = -\frac{M \cdot x'}{EI}\, dx$$

Since the deviation d_B^A consists of summation of such $x' \cdot d\theta$ of elementary sections, we have

$$d_B^A = \sum_B^A x'\, d\theta = -\int_B^A \frac{M . x'}{EI}\, dx \text{ (where } x' \text{, is zero at } B) \quad \text{...(12.6)}$$

In the above expression, $\frac{M}{EI}\, dx.x'$ is the moment of the elementary area of the $\frac{M}{EI}$ diagram about the point B.

Hence $$d_B^A = -\sum_B^A A_{M/EI} \cdot \bar{x} \text{ (where } x' = 0 \text{ at } B) \quad \text{...(12.7)}$$

It must be remembered that the variable x′ has its origin at the point at which deviation is being computed. Eq. 12.7 can be stated by Mohr's second theorem :

Mohr's Second Theorem : *The deviation of B from tangent at A is equal to the negative of the statical moment (or the first moment) with respect to B, of the M/EI diagram area between A and B.*

If however, EI is constant along the length of the beam, we have

$$EI d_B^A = -\sum_B^A A_M \bar{x} \quad \text{...[12.7(a)]}$$

where A_M is the area of B.M. diagram between B and A.

Mohr's second theorem is useful is finding the deflection because it relates the position of a point of the beam to the tangent at some other point.

12.4. SIGN CONVENTIONS

The following sign conventions are followed in the area-moment method :

1. B.M. is taken *positive* when it causes *concavity* at the upper side (*i.e.,* if it causes compression in the upper fibres).

2. Area of *M/EI* diagram is taken *positive* if *M* is *positive.*

3. The rotation θ at any point is taken *positive* if the rotation is *clockwise*.

4. The deflection y at any point is *positive* if it is *downwards* with respect to the original axis of the beam.

5. The angle θ_B^A is *positive* if the tangent at B rotates *clockwise* with respect to tangent at A. Thus θ_B^A in Fig. 12.2 (a) is negative since the rotation of tangent at B is anticlockwise with respect to tangent AB' at A.

6. The deviation d_B^A is taken *positive* when it is measured *downward* from the tangent at A. Thus is Fig. 12.2 (a), d_B^A is *negative* since it is measured *upward* from the tangent at A.

7. $\bar{x}$ is taken positive when it is measured from point B towards point A.

In the above sign conventions, point B must be to the right of A.

If the area of B.M. diagram is *positive*, the first moment is also positive; hence the *deviation* comes out to be negative from Eq. 12.6 and the point B is above the tangent at A. This situation is illustrated in Fig. 12.2. If, however, as we move from A to B in the x direction, the area of B.M. diagram is negative, then the first moment of the area is also negative; then the deviation comes out to be *positive* from Eq. 12.6 which means that point B is below the tangent at A. This situation is illustrated in Fig. 12.3.

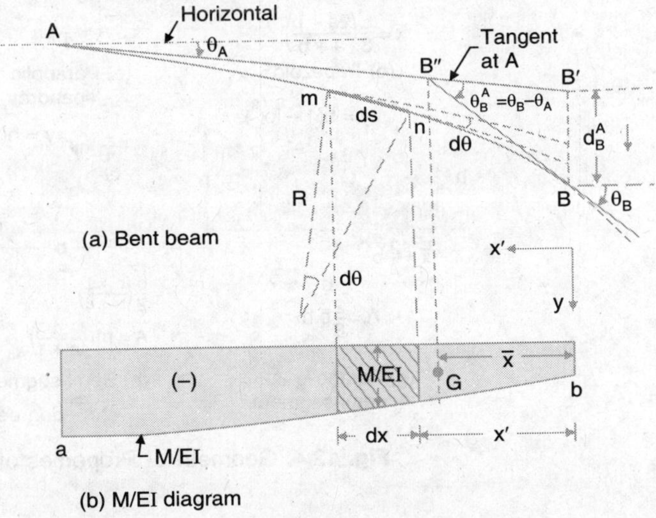

Fig. 12.3

Note. The above sign conventions may seem complicated at the first instance. In many practical cases, one can find by inspection itself whether the beam deflects upwards or downwards and whether the angle of rotation is clockwise or anticlockwise. Hence, for such obvious cases, it may not be necessary to follow the above sign conventions. Instead, we can calculate the absolute values and determine the direction by inspection.

12.5. GEOMETRICAL PROPERTIES OF M/EI DIAGRAMS

The *first moment of area* $\left(\dfrac{M \,.\, x'}{EI}\right)$ *of the M/EI diagram can be found either by integration or by taking the product of the area of the diagram and the distance* $\bar{x}$ *from point* B to the centroid G of the area. The latter method is more convenient because the *M/EI* diagram usually consists of common geometrical figures such as rectangles, triangles and parabolic segments. Fig. 12.4 shows expressions for area A and centroidal distance $\bar{x}$ for some common and familiar B.M. diagrams.

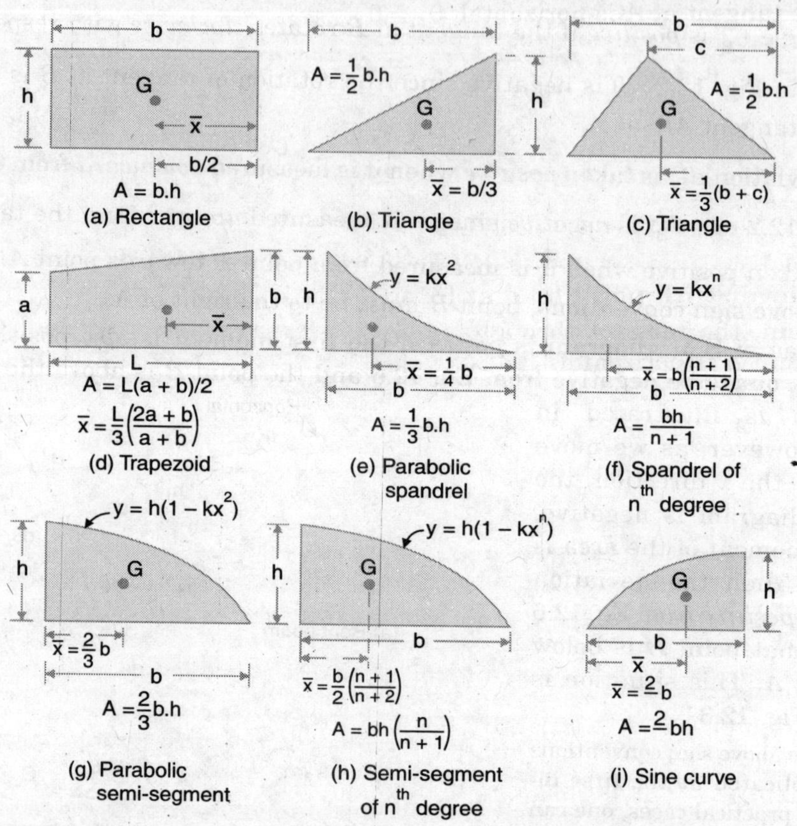

Fig. 12.4. Geometrical Properties of Familiar Segments

12.6. SLOPE AND DEFLECTION FROM MOHR'S THEOREMS

1. Cantilever with point load at the free end. Fig. 12.5 shows a cantilever *AB* of span *L*, with a point load *W* at the free end *B*. The B.M. diagram is shown in Fig. 12.5 (*b*). Let us take fixed end *A* as the reference point. Hence the tangent at the fixed end and the horizontal line through it will coincide.

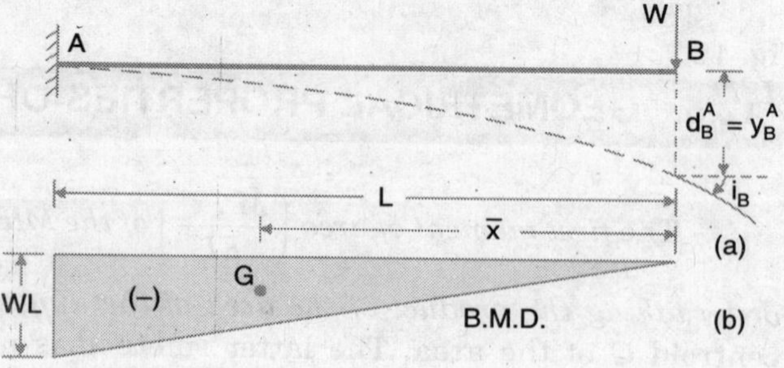

Fig. 12.5

$$\sum_{B}^{A} A_{M/EI} = \frac{1}{2}(L)(-WL)\left(\frac{1}{EI}\right)$$

$$= -\frac{WL^2}{2EI}$$

From Eq. 12.5, $\qquad \theta_B^A = \theta_B - \theta_A = -\sum_{B}^{A} A_{M/EI} = +\frac{WL^2}{2EI}$

Since the tangent at A is horizontal, $\theta_A = 0$

Hence $\qquad \theta_B \text{ (or } i_B) = \dfrac{WL^2}{2EI}$

Also,
$$d_B^A = -\sum_B^A A_{M/EI} \cdot \bar{x} = -\left[\frac{1}{2}(L)(-WL)\left(\frac{2}{3}L\right)\frac{1}{EI}\right] = \frac{WL^3}{3EI}$$

$$\therefore \quad y_B^A = d_B^A = \frac{WL^3}{3EI} \qquad \text{...(12.8)}$$

2. Cantilever with point load at intermediate point

Here again, the tangent through the reference point A is horizontal.

Hence $\theta_A = 0$.

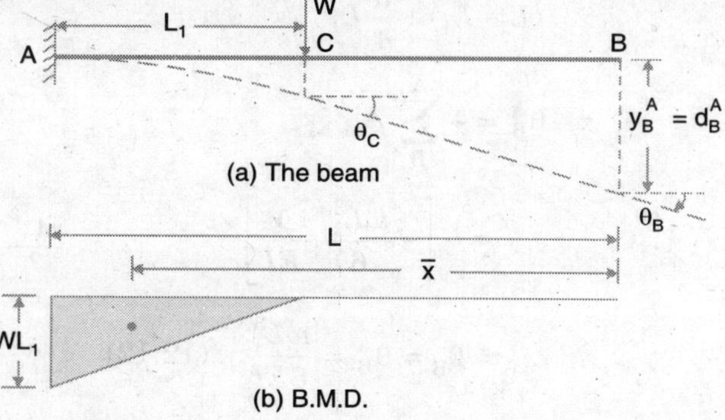

(a) The beam

(b) B.M.D.

Fig. 12.6

$$\therefore \quad \theta_B^A = \theta_B - \theta_A = -\sum_B^A A_{M/EI}$$

$$= -\left[\frac{1}{2}(L_1)(-WL_1)\frac{1}{EI}\right]$$

$$\therefore \quad i_B = \theta_B = \frac{WL_1{}^2}{2EI}$$

Also, by inspection $y_B^A = d_B^A$

$$\therefore \quad y_B^A = d_B^A = -\sum_B^A A_{M/EI} \cdot \bar{x}$$

$$= -\left[\frac{1}{2}(L_1)(-WL_1)\left(L - L_1 + \frac{2}{3}L_1\right)\frac{1}{EI}\right] = \frac{WL_1^2}{2EI}(L - L_1) + \frac{WL_1^3}{3EI}$$

Hence $\qquad y_B^A = \dfrac{WL_1^3}{3EI} + \dfrac{WL_1^2}{2EI}(L - L_1) \qquad \text{...(12.9)}$

3. Cantilever with uniformly distributed load over the whole span

The B.M. diagram, shown in Fig. 12.7 has a maximum ordinate of $wL^2/2$.

(a) The beam

(b) B.M.D.

Fig. 12.7

$$A_M = \frac{1}{3}(L)\left(-\frac{wL^2}{2}\right) = -\frac{wL^3}{6}$$

$$\bar{x} = \frac{3}{4}L$$

$$\therefore \quad \theta_B^A = -\sum_B^A A_{M/EI}$$

$$i_B = \theta_B^A = -\left[-\frac{wL^3}{6}\left(\frac{1}{EI}\right)\right] = \frac{wL^3}{6EI} \qquad \text{...(12.10)}$$

Also, $y_B^A = d_B^A = -\sum_B^A A_{M/EI} \cdot \bar{x} = -\left[-\frac{wL^3}{6}\left(\frac{3}{4}L\right)\left(\frac{1}{EI}\right)\right]$

$\therefore \quad y_B = y_B^A = \frac{wL^4}{8\,EI}$ ...(12.11)

4. Cantilever, partially loaded with w over a length L_1 from fixed end

$A_M = \frac{1}{3}(L_1)\left(-\frac{wL_1^2}{2}\right) = -\frac{wL_1^3}{6}$

$\bar{x} = (L - L_1) + \frac{3}{4}L_1$

$\therefore \quad \theta_B^A = -\sum_B^A A_{M/EI}$

$= -\left[-\frac{wL_1^3}{6} \cdot \frac{1}{EI}\right]$

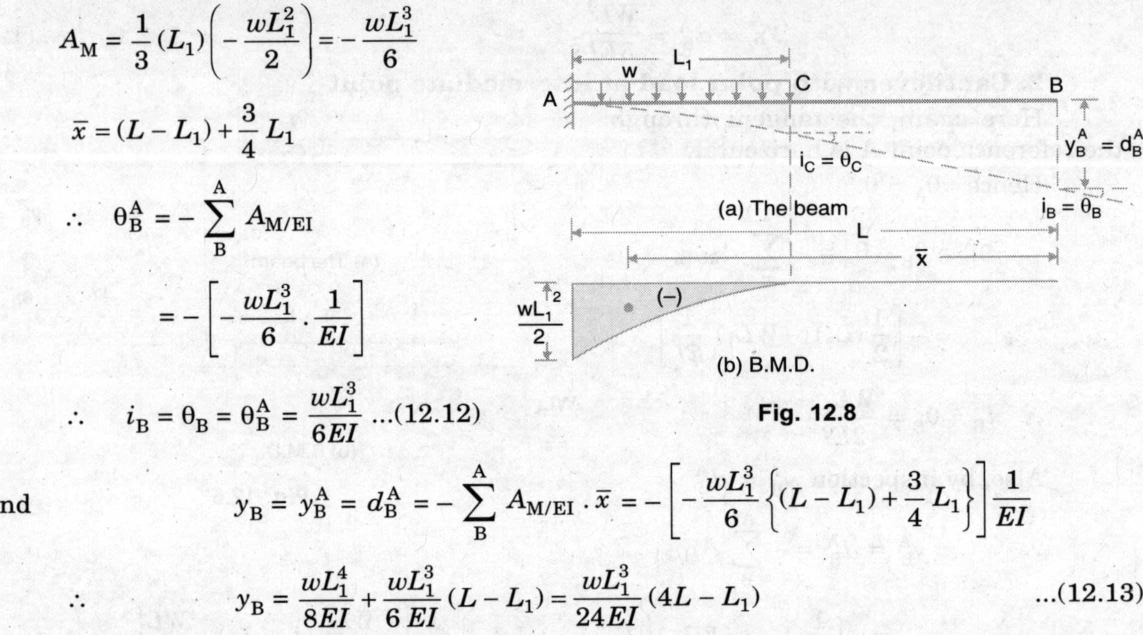

Fig. 12.8

$\therefore \quad i_B = \theta_B = \theta_B^A = \frac{wL_1^3}{6EI}$ ...(12.12)

and $\quad y_B = y_B^A = d_B^A = -\sum_B^A A_{M/EI} \cdot \bar{x} = -\left[-\frac{wL_1^3}{6}\left\{(L - L_1) + \frac{3}{4}L_1\right\}\right]\frac{1}{EI}$

$\therefore \quad y_B = \frac{wL_1^4}{8EI} + \frac{wL_1^3}{6\,EI}(L - L_1) = \frac{wL_1^3}{24EI}(4L - L_1)$ ...(12.13)

when $L_1 = L$, we get $y_B = \frac{wL^4}{8\,EI}$, which is the same as Eq. 12.11.

5. Cantilever partially loaded with w from the free end

The B.M. diagram is shown in Fig. 12.9 (*b*). The variation of B.M. is parabolic from B to C, and linear from C to A.

Ordinate $\quad cc_1 = \frac{w(L - L_1)^2}{2} = a\,a_1$

Ordinate $\quad a\,a_2 = w\,(L - L_1)\left(L_1 + \frac{L - L_1}{2}\right) = \frac{w}{2}(L - L_1)(L + L_1)$

$\therefore \quad a_1 a_2 = \frac{w}{2}(L - L_1)(L + L_1) - \frac{w}{2}(L - L_1)^2 = wL_1(L - L_1)$

The total area can be splitted into three portions : A_{M1}, A_{M2} and A_{M3}

Area $\quad A_{M1} = -\frac{1}{3}(L - L_1)\frac{w}{2}(L - L_1)^2 = -\frac{w}{6}(L - L_1)^3$

$\bar{x}_1 = \frac{3}{4}(L - L_1)$, from b

Area A_{M2}

$$= -L_1 \cdot \frac{w}{2}(L - L_1)^2$$

$$= -\frac{wL_1}{2}(L - L_1)^2$$

$$\bar{x}_2 = \frac{L_1}{2} + (L - L_1)$$

$$= L - \frac{L_1}{2}, \text{ from } b$$

Area A_{M3}

$$= -\frac{1}{2} \cdot L_1 \cdot w L_1 (L - L_1)$$

$$= -\frac{w{L_1}^2}{2}(L - L_1) \quad \text{and} \quad \bar{x}_3 = (L - L_1) + \frac{2}{3}L_1 = L - \frac{L_1}{3} \text{ from } b$$

Fig. 12.9

Now $\quad i_B = \theta_B^A = -\sum_B^A A_M \cdot \frac{1}{EI} = \frac{1}{EI}\left[\frac{w}{6}(L - L_1)^3 + \frac{wL_1}{2}(L - L_1)^2 + \frac{wL_1^2}{2}(L - L_1)\right]$

or $$i_B = \frac{w}{6EI}(L - L_1)(L^2 + L_1^2 + LL_1) \qquad \text{...(12.14)}$$

When $L_1 = \frac{L}{2}$, $i_B = \frac{w}{6EI}\left(L - \frac{L}{2}\right)\left(L^2 + \frac{L^2}{4} + \frac{L^2}{2}\right) = \frac{7wL^3}{48\,EI}$ ...[12.14(*a*)]

Also, $\quad y_B = d_B^A = -\sum_B^A A_M \cdot \bar{x} \cdot \frac{1}{EI} = \frac{1}{EI}\left[\frac{w}{6}(L - L_1)^3 \times \frac{3}{4}(L - L_1) + \frac{wL_1}{2}(L - L_1)^2 \times \left\{L - \frac{L_1}{2}\right\} + \frac{w{L_1}^2}{2}(L - L_1)\left(L - \frac{L_1}{3}\right)\right]$

or $$y_B = \frac{1}{EI}\left[\frac{w}{8}(L - L_1)^4 + \frac{wL_1}{2}(L - L_1)^2\left(L - \frac{L_1}{2}\right) + \frac{w}{2}{L_1}^2(L - L_1)\left(L - \frac{L_1}{3}\right)\right]$$

Expanding, rearranging and simplifying, we get

$$y_B = \frac{wL^4}{8\,EI} - \left\{\frac{wL_1^4}{8\,EI} + \frac{wL_1^3}{6\,EI}(L - L_1)\right\} \qquad \text{...(12.15)}$$

When $L_1 = \frac{L}{2}$, $y_B = \frac{wL^4}{8\,EI} - \left\{\frac{w}{8\,EI}\left(\frac{L}{2}\right)^4 + \frac{w}{6EI}\left(\frac{L}{2}\right)^3\left(L - \frac{L}{2}\right)\right\} = \frac{11}{384}\frac{wL^4}{EI}$ ...[12.15(*a*)]

6. Simply supported beam with U.D.L.

Because of symmetrical loading, the tangent at C will be horizontal, as shown in Fig. 12.10 (*a*). Hence the central deflection y_C^A is given by

$y_C^A = -y_A^C$

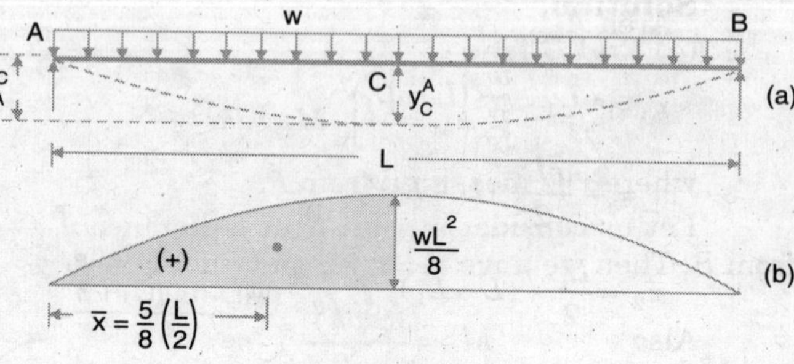

Fig. 12.10

Since the tangent at C coincides with the horizontal line through C, D_A^C will be zero.

$$\therefore \quad y_A^C = d_A^C = -\sum_A^C \frac{A_M}{EI} \cdot \bar{x}$$

$$\text{Now } A_M = \frac{2}{3} \times \frac{L}{2} \times \frac{wL^2}{8} = \frac{wL^3}{24}$$

and $$\bar{x} = \frac{5}{8}\left(\frac{L}{2}\right) = \frac{5}{16} L$$

$$\therefore \quad y_A^C = -\left[\frac{wL^3}{24} \times \frac{5}{16} L \times \frac{1}{EI}\right] = -\frac{5}{384}\frac{wL^4}{EI}$$

$$\text{Hence} \quad y_C^A = -y_A^C = \frac{5}{384}\frac{wL^4}{EI} \qquad \text{...(12.16)}$$

7. Simply supported beam with central point load

Here also, because of symmetrical loading, the tangent at C will be horizontal, as shown in Fig. 12.11 (*a*). Hence the central deflection is given by

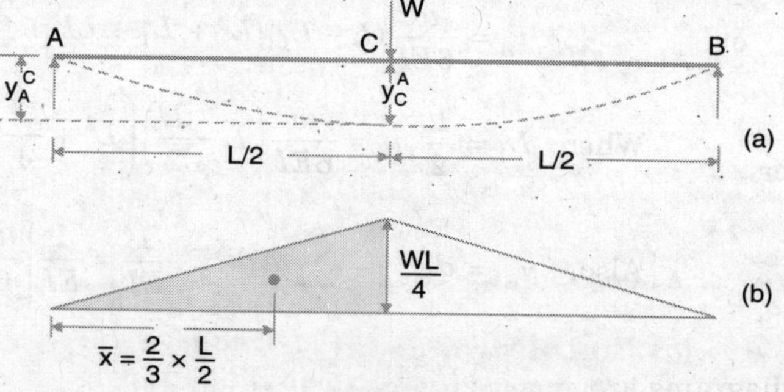

Fig. 12.11

$y_C^A = -y_A^C$

Since the tangent at C coincides with the horizontal line through C, D_A^C will be zero.

$$\text{Hence} \quad y_A^C = d_A^C = -\sum_B^C \frac{A_M}{EI} \cdot \bar{x}$$

Here A_M = shaded area between A and C

$$= \frac{1}{2} \times \frac{L}{2} \times \frac{WL}{4} = \frac{WL^2}{16}$$

$\bar{x}$ = distance of the C.G. of the shaded area from $A = \frac{2}{3} \times \frac{L}{2} = \frac{L}{3}$

$$y_A^C = -\left[\frac{WL^2}{16} \times \frac{L}{3} \times \frac{1}{EI}\right] = -\frac{WL^3}{48\,EI} \quad \text{and} \quad y_C = y_C^A = -y_A^C = +\frac{WL^3}{48\,EI} \qquad \text{...(12.17)}$$

8. Simply supported beam with general loading : See § 12.8

Example 12.1. (*a*) *A cantilever AB, of length L is loaded with U.D.L. throughout its length. Find the slope and deflection at any point P, distant c from the free end.*

(*b*) *If the end B is propped, find* (*i*) *reaction at the prop, and* (*ii*) *slope at the propped end.*

Solution.

(*a*) **End *B* free**

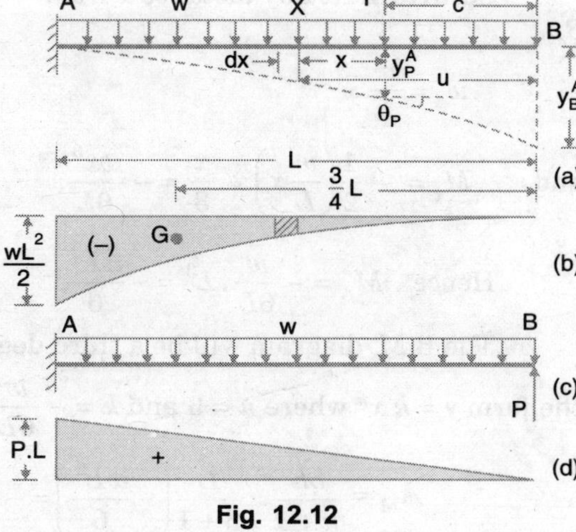

Fig. 12.12

$$EI\,\theta_P^A = -\int_P^A A_M = -\int_P^A M_x \,.\, dx$$

where x is measured from P.

Let us consider a point X at a distance u from B. Then we have $u = x + c$ and $du = dx$.

Also $\qquad M_x = -\dfrac{wu^2}{2}$

Hence $\qquad EI\,\theta_P^A = -\displaystyle\int_P^A \left(-\frac{wu^2}{2}\right) du$

$$= -\int_C^L \left(-\frac{wu^2}{2}\right) du$$

$$\therefore \qquad \theta_P^A = \frac{1}{EI}\left[\frac{wu^3}{6}\right]_C^L = \frac{w}{6EI}[L^3 - c^3]$$

When $c = 0$, we have $\theta_B^A = \dfrac{wL^3}{6\,EI}$, which is the same as Eq. 12.10.

$$\text{Also, } EIy_P^A = -\int_P^A x\,M_x\,dx = -\int_P^A (u-c)\,M_x\,du = \int_C^L (u-c)\frac{wu^2}{2}du$$

or $$y_P^A = \frac{w}{2EI}\left[\frac{u^4}{4} - \frac{cu^3}{3}\right]_C^L = \frac{W}{2EI}\left[\left(\frac{L^4}{4} - \frac{cL^3}{3}\right) - \left(\frac{c^4}{4} - \frac{c^4}{3}\right)\right] = \frac{w}{24\,EI}(3L^4 - 4cL^3 + c^4)$$

When $c = 0$, we have $y_B^A = \dfrac{wL^4}{8\,EI}$ which is the same as Eq. 12.11.

(*b*) **End *B* propped.** Let P be the prop reaction. The component bending moment diagrams are shown in Fig. 12.12 (*b*) and (*d*) respectively.

Now $$y_B^A = \sum_B^A (A_1\bar{x}_1)_{M/EI} + \sum_B^A (A_2\bar{x}_2)_{M/EI} = 0$$

where A_1 is the area of B.M. diagram due to external load and A_2 is the B.M. diagram due to prop reaction.

$$\therefore \quad -\frac{1}{3}\frac{w\,L^2}{2}L\,.\,\frac{3}{4}L + \frac{1}{2}PL\,.\,L\,.\,\frac{2}{3}L = 0 \quad \text{or} \quad \frac{wL^4}{8} = \frac{PL^3}{3} \text{ from which } P = \frac{3}{8}wL$$

Also, $$EI\,\theta_B^A = -\sum_B^A (A_1 + A_2)_M = -\left[-\frac{1}{3}.\frac{wL^2}{2}.L + \frac{3}{8}wL\,.\,\frac{L^2}{2}\right] = \frac{wL^3}{48}$$

Hence $\qquad \theta_B^A = \dfrac{wL^3}{48\,EI}.$

Example 12.2. *A cantilever beam AB, supporting a triangularly distributed load of maximum intensity w is shown in Fig. 12.13. Find the slope and deflection at the free end.*

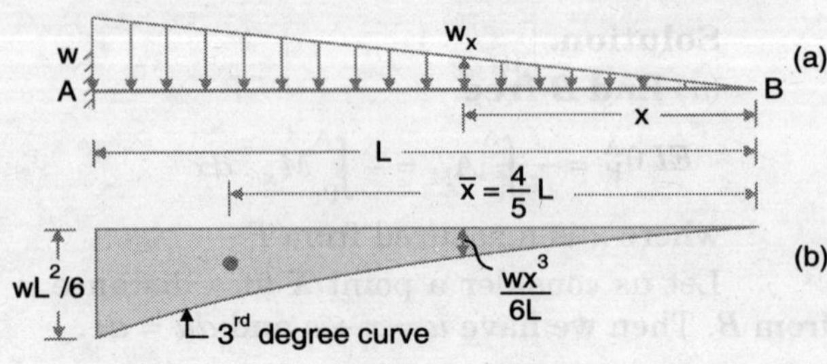

Fig. 12.13

Solution. At any distance x from B,

$$w_x = \frac{w}{L}x$$

and $$M_x = -\frac{1}{2}\left(\frac{w}{L}x\right)x \cdot \frac{x}{3} = -\frac{wx^3}{6L}$$

Hence $$M_A = -\frac{w}{6L} \cdot L^3 = -\frac{wL^2}{6}$$

The B.M. diagram will be a third degree curve, as shown in Fig. 12.13 (*b*), and will be of the form $y = k\,x^n$ where $n = 3$ and $k = -\dfrac{w}{6L}$.

$$A_M = \frac{bh}{n+1} = \frac{L}{3+1}\left(-\frac{wL^2}{6}\right) = -\frac{wL^3}{24};\ \bar{x} = b\left(\frac{n+1}{n+2}\right) = L\left(\frac{3+1}{3+2}\right) = \frac{4}{5}L$$

Now, $$\theta_B^A = -\sum_B^A \frac{A_M}{EI} = \frac{wL^3}{24\,EI}$$

Also, $$y_B^A = -\sum_B^A \frac{A_M}{EI} \cdot \bar{x} = \frac{1}{EI}\left(\frac{wL^3}{24} \times \frac{4}{5}L\right) = \frac{wL^4}{30\,EI}.$$

Example 12.3. *A cantilever AB of length L, is subjected to a couple μ in the clockwise direction at the free end. Determine the slope and deflection of the free end.*

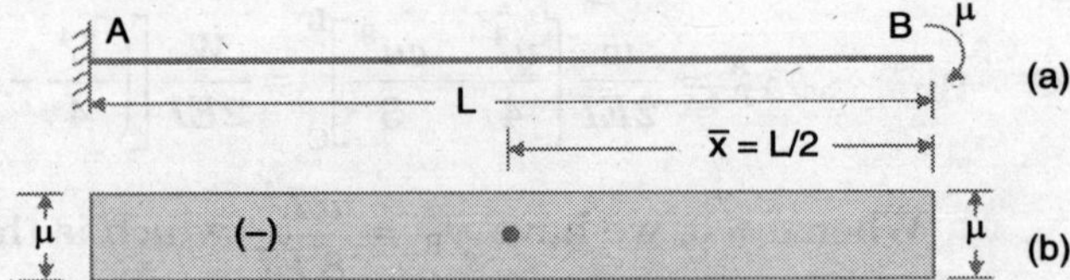

Fig. 12.14

Solution. The B.M. diagram is shown in Fig. 12.14 (*b*). Evidently, it will be a rectangle of height equal to μ, since B.M. at any section is – μ, which is constant all along the length of the cantilever.

Now $$\theta_B^A = -\sum_B^A A_{M/EI}$$

$$= -\left[-\frac{\mu}{EI} \cdot L\right] = \frac{\mu L}{EI}$$

Also, $$y_B^A = -\sum_B^A A_{M/EI} \cdot \bar{x} = -\left[-\frac{\mu L}{EI} \cdot \frac{L}{2}\right] = \frac{\mu L^2}{2EI}.$$

Example 12.4. *A cantilever AB of length L is subjected to a concentrated load P and a couple μ acting at the free end, as shown in Fig. 12.15. Determine the slope and deflection at the free end.*

Solution. The component B.M. diagrams are shown in Fig. 12.15 (*b*) and (*c*) respectively.

Now $$\theta_B^A = -\sum_B^A A_{M/EI}$$

$$= -\left[\mu L - \frac{1}{2}PL\,.\,L\right]\frac{1}{EI}$$

$$= \frac{L}{2EI}[PL - 2\mu]$$

and $$y_B^A = -\sum_B^A A_{M/EI}\,.\,\bar{x}$$

$$= -\frac{1}{EI}\left[\mu L\,.\,\frac{L}{2} - \frac{1}{2}\,.\,PL\,.\,L\,.\,\frac{2L}{3}\right]$$

or $$y_B^A = \frac{PL^3}{3\,EI} - \frac{\mu\,L^2}{2\,EI} = \frac{L^2}{6\,EI}\,[2PL - 3\mu].$$

Fig. 12.15

Example 12.5. *A beam of length L is simply supported at the ends and carries a load W at a distance a from the left support. Calculate the deflection at the centre.*

Solution. In order to produce symmetry, so that slope at C becomes zero, add a load W at a distance a from B. Evidently, the deflection at C will be *doubled.* The corresponding bending moment diagram is shown in Fig. 12.16 (*b*).

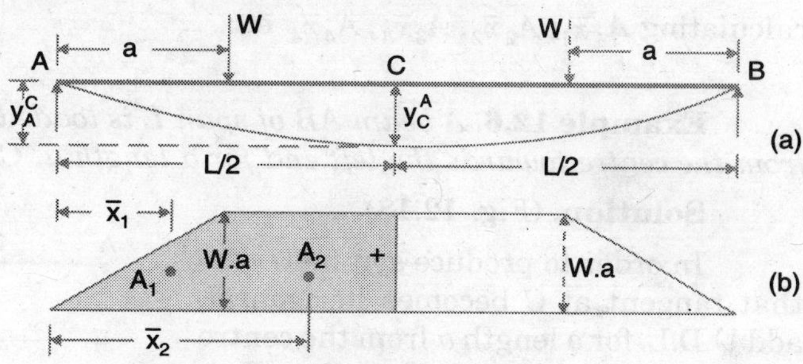

Fig. 12.16

Now, from Fig. 12.16 (*a*),

$$y_C^A = -\,y_A^C = -\,d_A^C$$

But $$y_A^C = d_A^C = -\sum_A^C A_{M/EI}\,.\,\bar{x}$$

$\therefore$ $$y_C^A = \sum_A^C A_{M/EI}\,.\,\bar{x} = \frac{1}{EI}\left[\frac{1}{2}\times a \times Wa \times \frac{2}{3}a + \left(\frac{L}{2} - a\right).\,Wa\,.\,\frac{1}{2}\left(a + \frac{L}{2}\right)\right]$$

or $$y_C^A = \frac{1}{EI}\left[\frac{Wa^3}{3} + \frac{Wa}{8}(L - 2a)\times(L + 2a)\right]$$

or $$y_C^A = \frac{Wa}{24\,EI}[8a^2 + 3L^2 - 12a^2].\ \text{Hence}\ y_C = \frac{1}{2}y_C^A = \frac{Wa}{48\,EI}[3L^2 - 4a^2].$$

12.7. USE OF MOMENT DIAGRAMS BY PARTS

When a beam is subjected to several loads of various types, it is more convenient to use the moment diagrams by parts, sometimes also known as the *component bending moment diagram* to compute $\int M.x.dx$. This is done by plotting the graph of each bending moment component corresponding to each load separately, *i.e.,* separate graph for each term of the bending moment equation.

For example, let us take the simply supported beam AB, loaded as shown in Fig. 12.17 (a). The equation of B.M. at any point P, distant x from A is

$$M_x = +R_A . x - W_1 x_1 - W_2 x_2 - \frac{w x_2^2}{2}$$

$$= M_1 + M_2 + M_3 + M_4$$

Fig. 12.17 (b) shows the usual B.M. diagram. In Fig. 12.17 (c), the terms M_1, M_2, M_3 and M_4 have been plotted *separately*. It is very clear that Fig. 12.17 (c) is more convenient for calculating $A_1\bar{x}_1, A_2\bar{x}_2, A_3\bar{x}_3, A_4\bar{x}_4$ etc.

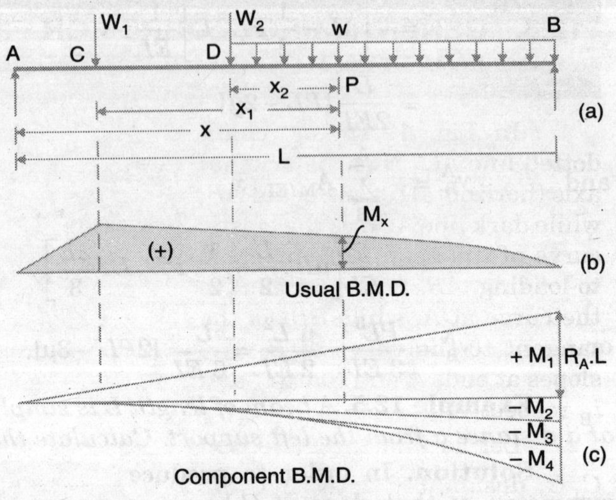

Fig. 12.17

Example 12.6. *A beam AB of span L is loaded with U.D.L. w per unit length, extending from the centre towards the left end for a length a. Compute the central deflection.*

Solution. (Fig. 12.18).

In order to produce symmetry, so that tangent at C becomes horizontal, add U.D.L. for a length a from the centre towards right hand end. Then $R_A = R_B = w\ a$. The deflection at C will then be doubled.

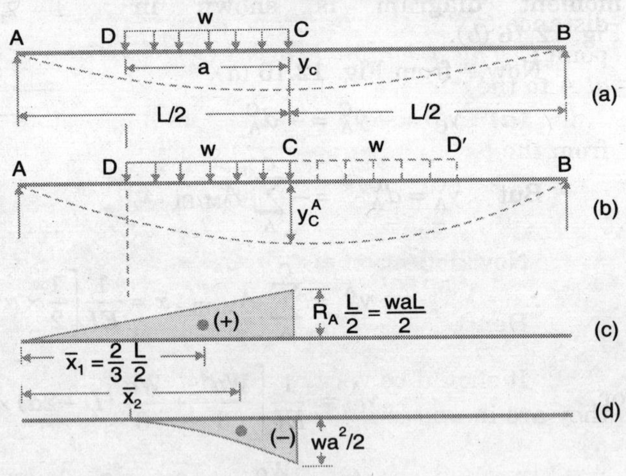

Fig. 12.18

Also, $y_C^A = -y_A^C = -d_A^C$

But $d_A^C = -\sum_A^C \frac{A_M}{EI} . \bar{x}$

Hence $y_C^A = \sum_A^C \frac{A_M}{EI} \bar{x}$

The component B.M. diagrams have been shown in Fig. 12.18 (c) and (d), from which

$$y_C^A = \frac{1}{EI}\left[\frac{1}{2}.\frac{waL}{2} \times \frac{L}{2} \times \frac{2}{3}\frac{L}{2} - \frac{1}{3} a \times \frac{wa^2}{2}\left(\frac{L}{2} - a + \frac{3}{4} a\right)\right].$$

or $$y_C^A = \frac{wa}{24\ EI}\ [L^3 - a^2\ (2L - a)] = \frac{wa}{24\ EI}\ [L^3 - 2La^2 + a^3]$$

Hence actual $$y_C = \frac{1}{2}\ y_C^A = \frac{wa}{48\ EI}\ [L^3 - 2La^2 + a^3].$$

12.8. GENERAL PROCEDURE FOR FINDING SLOPE AND DEFLECTION AT ANY POINT OF SIMPLY SUPPORTED BEAM

In Fig. 12.19 (*a*), chain dotted line AC_1B is the original axis (horizontal) of the beam AB, while dark line ACB is the elastic curve of the beam AB subjected to loading. AB_2 is the tangent to the curve at A, while BB_1 is the tangent to the curve at B. The slopes at ends A and B are i_A and i_B respectively.

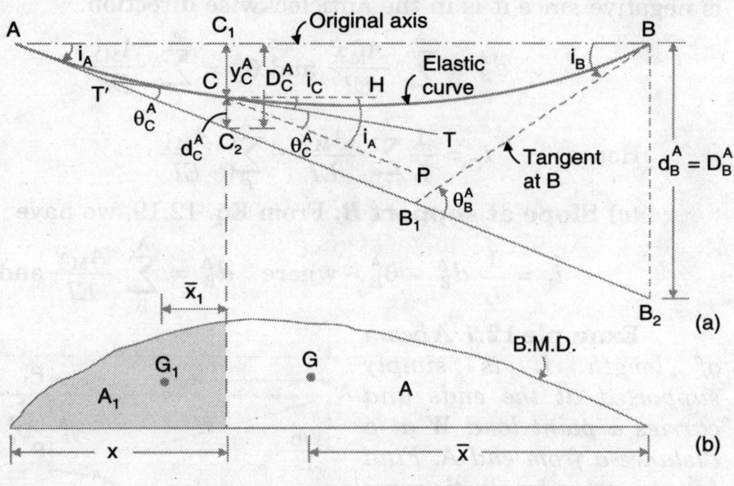

Fig. 12.19

Let us find the deflection C_1C, designated as y_C^A, of the point C. Note that this deflection is positive, since it is in the *downward* direction from the original horizontal axis. The distance C_2C is the *deviation* of point C. *This deviation is negative since it is in the upward direction measured from the tangent at A to the curve.*

(*a*) Deflection at C. Let us find deflection C_1C of the point C located at a distance x from the left hand support A. From the geometry of Fig. 12.19 (*a*),

$$C_1C_2 = \frac{x}{L}\,BB_2. \quad \text{But } BB_2 = D_B^A = d_B^A. \quad \text{Hence} \quad C_1C_2 = \frac{x}{L}\,d_B^A$$

Now deflection of $\quad C = y_C = C_1C = C_1C_2 - CC_2 = \dfrac{x}{L}\,d_B^A - d_C^A$

Hence
$$y_C = \frac{x}{L}\,d_B^A - d_C^A \qquad \text{...(12.18)}$$

It should be very clearly noted that both d_B^A $(= B_2B)$ and d_C^A $(= C_2C)$ are *negative* since they are in *upward direction* when measured from the tangent at A to the elastic curve.

$$\therefore \qquad d_B^A = \sum_B^A \frac{A_M\bar{x}}{EI} \quad \text{and} \quad d_C^A = \sum_C^A \frac{A_{M1}\bar{x}_1}{EI}$$

Hence
$$y_C = \frac{x}{L}\sum_B^A \frac{A_M\bar{x}}{EI} - \sum_C^A \frac{A_{M1}\bar{x}_1}{EI} \qquad \text{...[12.18(}a\text{)]}$$

(*b*) Slope at C. Through C, draw horizontal line CH, tangent line CT and line CP parallel to the tangent at A, Evidently, we have.

$$\angle HCT = i_C = \text{slope at } C$$

$$\angle CT'B_1 = \angle TCP = \theta_C^A = \text{rotation of tangent at } C \text{ with respect to tangent at } A$$

$$\angle HCP = i_A = \text{slope at } A.$$

Now slope at $C = i_C = i_A - \theta_C^A$ [From Fig. 12.19(*a*)]. But $i_A \approx \tan i_A = \dfrac{1}{L}\,d_B^A$

Hence $\quad i_C = \frac{1}{L}\, d_B^A - \theta_C^A$ ...(12.19)

Here again d_B^A is negative since it is measured upwards from the tangent at *A*. Also, θ_C^A is negative since it is in the anticlockwise direction.

$$\therefore \quad d_B^A = \sum_B^A \frac{A_M \bar{x}}{EI} \text{ and } \theta_C^A = \sum_C^A \frac{A_{M1}}{EI}$$

Hence $\quad i_C = \frac{1}{L}\sum_B^A \frac{A_M\bar{x}}{EI} - \sum_C^A \frac{A_{M1}}{EI}$ ...[12.19(*a*)]

(*c*) Slope at support *B*. From Eq. 12.19, we have

$$i_B = \frac{1}{L}\, d_B^A - \theta_B^A, \text{ where } d_B^A = \sum_B^A \frac{A_M\bar{x}}{EI} \text{ and } \theta_B^A = \sum_B^A \frac{A_M}{EI}. \quad \text{...[12.19(b)]}$$

Example 12.7. *A beam of length L is simply supported at the ends and carries a point load W at a distance a from end A. Find (a) expression for elastic curve of the beam (b) central deflection (c) maximum deflection (d) deflection under the load and (e) slopes at the ends.*

Solution. Consider a point *P* at a distance *x* from *A*. The component bending moment diagram for the length *AB* are shown in Fig. 12.20 (*b*) and (*c*), while that of the length *AP* is shown in Fig. 12.20 (*d*). The deviation d_B^A is *negative* since it is in the *upward* direction from the tangent at *A*.

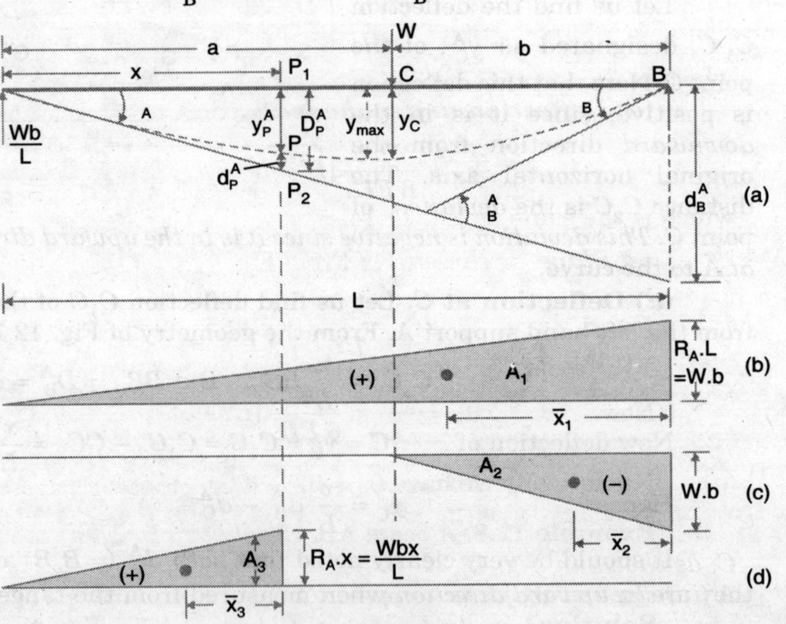

Fig. 12.20

Hence $\quad d_B^A = \frac{1}{EI}\sum_B^A A_M\,\bar{x} = \frac{1}{EI}\sum_B^A (A_1\bar{x}_1 + A_2\bar{x}_2)$

$$= \frac{1}{EI}\left[\frac{1}{2}Wb\,.\,L\,.\,\frac{L}{3} - \frac{1}{2}Wb\,.\,b.\,\frac{b}{3}\right] = \frac{Wb}{6\,EI}[L^2 - b^2] \quad \text{...(1)}$$

Similarly $\quad d_P^A = \frac{1}{EI}\sum_P^A A_M\,\bar{x} = \frac{1}{EI}\left[\frac{1}{2}.\,x\,.\,\frac{Wbx}{L}.\,\frac{x}{3}\right] = \frac{Wbx^3}{6EIL}$ ...(2)

(a) Expression for the elastic curve :

From Fig. 12.20 (*a*), $\quad D_P^A = P_1P_2 = d_B^A\,.\,\frac{x}{L} = \frac{Wbx}{6\,EIL}\,[L^2 - b^2]$

Hence $\quad y_P^A = D_P^A - d_P^A = \frac{Wbx}{6\,EIL}[L^2 - b^2] - \frac{Wbx^3}{6\,EIL} = \frac{Wbx}{6\,EIL}[L^2 - b^2 - x^2]$ ...(3)

which is the required expression for the elastic curve [same as Eq. 11.10(*a*)].

(*b*) ***Central deflection :***

At $x = \frac{L}{2}$, central deflection $= \frac{Wb}{6\,EIL}\cdot\frac{L}{2}\left[L^2 - b^2 - \frac{L^2}{4}\right] = \frac{Wb}{48\,EI}[3L^2 - 4b^2]$

This is the same as Eq. 11.16

(*c*) ***Maximum deflection :***

For maximum deflection, we obtain from Eq. 3

$$\frac{dy_P^A}{dx} = 0 = L^2 - b^2 - 3x^2, \text{ from which } x = \sqrt{\frac{L^2 - b^2}{3}}$$

$$\therefore \quad y_{max} = \frac{Wb}{6\,EIL}\left(\frac{L^2 - b^2}{3}\right)^{1/2}\left[L^2 - b^2 - \left(\frac{L^2 - b^2}{3}\right)\right] = \frac{Wb\,(L^2 - b^2)^{3/2}}{9\sqrt{3}\,EIL}$$

which is the same as Eq. 11.13.

(*d*) *Deflection under the load :* Putting $x = a$ in Eq. 3

$$y_C = \frac{Wb\,a}{6\,EIL}[L^2 - b^2 - a^2] = \frac{Wab}{6\,EIL}\cdot 2ab = \frac{Wa^2b^2}{3\,EIL}$$

which is the same as Eq. 11.14.

(*e*) ***Slope at the ends :*** From Fig. 12.20, θ_B^A is negative. Hence

$$\theta_B^A = \sum_B^A \frac{A_M}{EI} = \left[\frac{1}{2}Wb.L - \frac{1}{2}Wb.b\right]\frac{1}{EI} = \frac{Wb}{2\,EI}(L - b) = \frac{Wab}{2\,EI}$$

Now $\quad \theta_A = \frac{d_B^A}{L} = \frac{Wb}{6\,EIL}[L^2 - b^2] = \frac{Wab}{6\,EIL}(L + b)$

Numerically we have $\theta_B = \theta_B^A - \theta_A = \frac{Wab}{2\,EI} - \frac{Wab}{6\,EIL}[L + b] = \frac{W\,ab}{6\,EIL}[L + a]$.

Example 12.8. *A beam AB, simply supported at the ends is subjected to a point load at C, as shown in Fig. 12.21. Using area moment method, compute (i) deflection at C (ii) slope at A (iii) slope at B and (iv) slope at C. Take I = 6 × 10⁸ mm⁴ and E = 2 × 10⁵ N/mm².*

Solution.

$M_C = \frac{Wab}{L} = \frac{20 \times 6 \times 2}{8}$

$= 30$ kN-m

$= 30 \times 10^6$ N-mm

$EI\,d_B^A = \sum_B^A A_M\,\bar{x}$

$= \frac{1}{2} \times 8000 \times 30 \times 10^6$

$\times \frac{1}{3} \times (8000 + 2000)$

$= 4 \times 10^{14}$

(The values have been substituted in N-mm units)

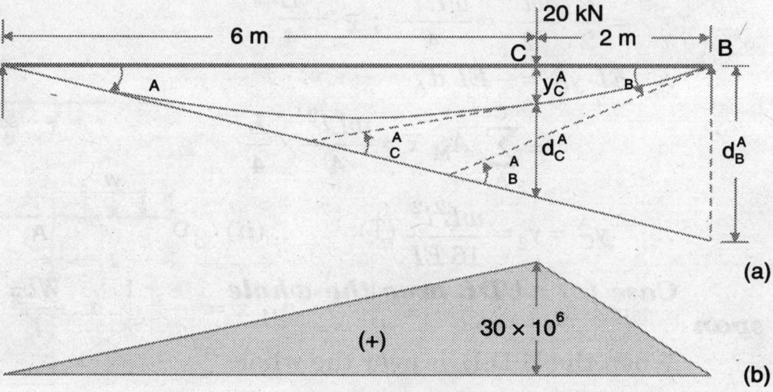

Fig. 12.21

$$EI\, d_C^A = \sum_C^A A_M\,\bar{x} = \frac{1}{2} \times 6000 \times 30 \times 10^6 \times \frac{1}{3} \times 6000 = 1.8 \times 10^{14}$$

Now, $$y_C^A = \frac{x}{L} d_B^A - d_C^A = \frac{1}{EI}\left[4\times10^{14}\times\frac{6}{8} - 1.8\times10^{14}\right] = \frac{1.2}{EI}\times10^{14}$$

$$= \frac{1.2\times10^{14}}{2\times10^5\times6\times10^8} = \mathbf{1\ mm}$$

Also, $$\theta_A = \frac{d_B^A}{L} = \frac{4\times10^{14}}{8000\,EI} = \frac{4\times10^{14}}{8000\times2\times10^5\times6\times10^8} = \mathbf{0.4167\times10^{-3}\ radian}$$

$$\theta_B^A = \sum_B^A \frac{A_M}{EI} = \frac{\frac{1}{2}\times8000\times30\times10^6}{2\times10^5\times6\times10^8} = \mathbf{1\times10^{-3}\ radian}$$

$\therefore$ θ_B (numerically) $= \theta_B^A - \theta_A = 1\times10^{-3} - 0.4167\times10^{-3} = \mathbf{0.5833\times10^{-3}\ radian.}$

Also, $$\theta_C^A = \sum_C^A \frac{A_M}{EI} = \frac{\frac{1}{2}\times6000\times30\times10^6}{2\times10^5\times6\times10^8} = \mathbf{0.75\times10^{-3}\ radian}$$

$\therefore$ θ_C (numerically) $= \theta_C^A - \theta_A = (0.75 - 0.4167)\times10^{-3} = \mathbf{0.3333\times10^{-3}\ radian.}$

Example 12.9. *A beam of uniform section and length (L + 2l) is simply supported over a span L with two equal over hanging lengths l. Derive expression for the deflection at mid-span due to a uniformly distributed load covering (a) the length L between the supports (b) the two overhanging lengths. If the beam carries uniformly distributed load on the whole length, find the ratio l/L so that the beam at the mid-span is just level with the supports.*

Solution. ***Case (a) : UDL on mid-span [Fig. 12.22 (a), (b)]***

$$EI\, y_C^A = -EI\, y_A^C = -EI d_A^C$$

$$\therefore\quad EI\, y_C^A = +\sum_A^C A_M\,\bar{x} = \frac{2}{3}\left(\frac{wL^2}{8}\right)\left(\frac{L}{2}\right)\left(\frac{5}{8}\frac{L}{2}\right) \quad\text{or}\quad y_C^A = y_1 = \frac{5}{384}\frac{wL^4}{EI}\ (\downarrow) \qquad ...(i)$$

Case (b) : UDL on overhangs [Fig. 12.22 (c), (d)]

When the load is on the overhangs only,

$$A_M = \text{shaded area}$$

$$= \frac{L}{2}\cdot\frac{wl^2}{2} = \frac{wL\,l^2}{4}\ ;\ \bar{x} = \frac{L}{4}$$

$$\therefore\quad EI\, y_C^A = -EI\, d_A^C$$

$$= \sum_A^C A_M\,\bar{x} = \frac{wLl^2}{4}\times\frac{L}{4}$$

$$\therefore\quad y_C^A = y_2 = \frac{wL^2l^2}{16\,EI}\ (\uparrow) \qquad ...(ii)$$

Case (c) : UDL over the whole span

When the U.D.L. is over the whole span

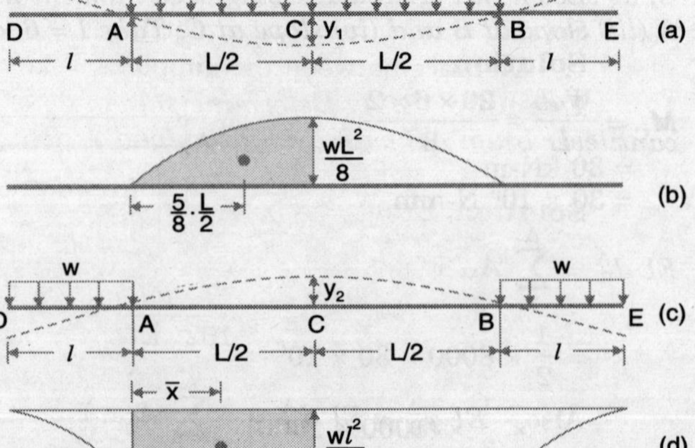

Fig. 12.22

$$y_C^A = y_1 + y_2 = \frac{5}{384}\frac{wL^4}{EI} - \frac{wL^2l^2}{16\,EI}. \quad \text{But} \quad y_C^A = 0 \text{ (given condition)}$$

$$\therefore \quad \frac{5}{384}\frac{wL^4}{EI} = \frac{wL^2l^2}{16\,EI} \quad \text{or} \quad \frac{5}{24}L^2 = l^2. \text{ From which } \frac{l}{L} = \sqrt{\frac{5}{24}} = \mathbf{0.4564.}$$

Example 12.10. *A uniform section straight edge of length L is loaded by its own weight only and is freely supported two points. Find the distance between the two supports :*

(a) so that, with the supports at the same level, the two ends of the beam remain horizontal.

(b) so that the deviation, from the straight line is as small as possible.

Solution. *(a) Required Condition :*

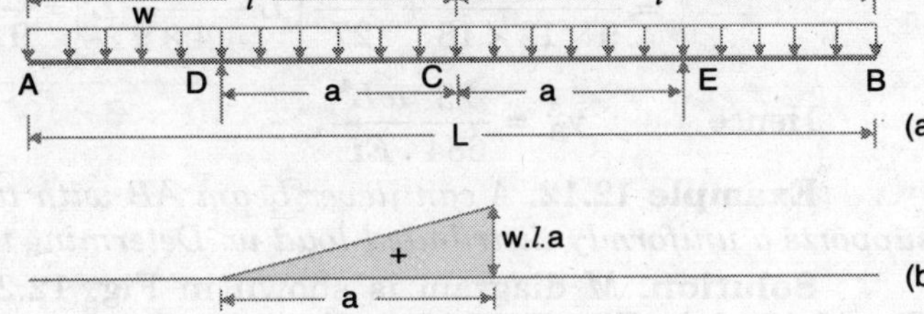

Fig. 12.23

$$i_A = i_B = 0$$

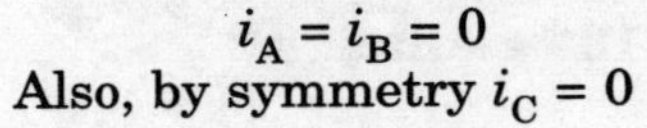

Also, by symmetry $i_C = 0$

$$\therefore \quad EI\,\theta_C^A = -\sum_C^A A_M = 0$$

$$\therefore \quad \frac{1}{2}a\,.\,w\,l\,a - \frac{1}{3}\,.\,l\,\frac{wl^2}{2} = 0$$

From which $a^2 = \dfrac{l^2}{3}$.

or $\quad a = \dfrac{l}{\sqrt{3}} = 0.5774\,l$

$\therefore$ Distance between the supports $= 2a = 0.5774\,(2l) = \mathbf{0.5774\,L}$

(b) For the deviation to be as small as possible, $y_A = y_B = y_C$

$$\therefore \quad EI\,y_A^C = \sum_A^C A_M\,\bar{x} = 0$$

$$\text{Hence} \quad \frac{1}{2}a\,.\,w\,l\,a\left(l - \frac{a}{3}\right) - \frac{1}{3}\times l\times\frac{wl^2}{2}\times\frac{3}{4}l = 0 \quad \text{or} \quad \frac{w}{2}\left(a^2l - \frac{a^3}{3}\right) - \frac{wl^3}{8} = 0$$

$$\text{or} \quad 4\,a^2l - \frac{4a^3}{3} - l^3 = 0 \quad \text{or} \quad 4a^3 - 12a^2l + 3l^3 = 0.$$

Let $a = nl$. Then $4n^3 - 12n^2 + 3 = 0$

Solving this by trial or by plotting, we get $n = 0.554$ and $a = 0.554\,l$

$\therefore$ Distance between the supports $= 2a = 0.554\,(2l) = 0.554\,L$.

Example 12.11. *Determine the angle of rotation and deflection at the free end of a cantilever beam AB with a uniform load w acting over the middle third of the length.*

$$\textbf{Solution.}\ EI\,\theta_B^A = -\sum_B^A A_M = \left[\frac{1}{3}\times\frac{wL^2}{18}\times\frac{L}{3} + \frac{1}{2}\left(\frac{wL^2}{18} + \frac{wL^2}{6}\right)\frac{L}{3}\right] = \frac{7}{162}wL^3$$

$$\therefore \quad i_B = \theta_B^A = \frac{7}{162}\frac{wL^3}{EI}$$

$$\text{Also,} \quad EI\,y_B^A = EI\,d_B^A = -\sum_B^A A_M\,.\,\bar{x}$$

Fig. 12.24

$$= \left[\frac{1}{3}\times\frac{wL^2}{18}\times\frac{L}{3}\left(\frac{L}{3}+\frac{3}{4}\cdot\frac{L}{3}\right)\right]+\left[\frac{1}{2}\left(\frac{wL^2}{18}+\frac{wL^2}{6}\right)\frac{L}{3}\left\{L-\frac{\frac{2wL^2}{18}+\frac{wL^2}{6}}{\frac{wL^2}{18}+\frac{wL^2}{6}}\times\frac{1}{3}\frac{L}{3}\right\}\right]$$

or
$$EI\ y_B^A = \frac{wL^3}{9\times 18}\times\frac{7}{12}L+\frac{L}{6}\times\frac{2}{9}wL^2\times\left\{L-\frac{\frac{5}{18}wL^2}{\frac{4}{18}wL^2}\times\frac{L}{9}\right\}$$

$$= \frac{7wL^4}{9\times 12\times 18}+\frac{wL^3}{27}\left[L-\frac{5}{4\times 9}L\right]=\frac{7}{9\times 12\times 18}wL^4+\frac{31}{27\times 36}wL^4$$

Hence $\quad y_B^A = \frac{23}{684}\frac{wL^4}{EI}.$

Example 12.12. *A cantilever beam AB with two different moments of inertia I and 2I supports a uniformly distributed load w. Determine the slope and deflection at the free end.*

Solution. M diagram is shown in Fig. 12.25 (b) while M/EI diagram is shown in Fig. 12.25 (c). In Fig. 12.25 (c), ordinate

at C (right) $= \frac{wL^2}{8}\times\frac{1}{EI}=\frac{wL^4}{8EI}$, ordinate

at C (left) $= \frac{wL^2}{8}\times\frac{1}{2EI}=\frac{wL^2}{16EI}$ and

ordinate at A $= \frac{wL^2}{2}\times\frac{1}{2\,EI}=\frac{wL^2}{4\,EI}$.

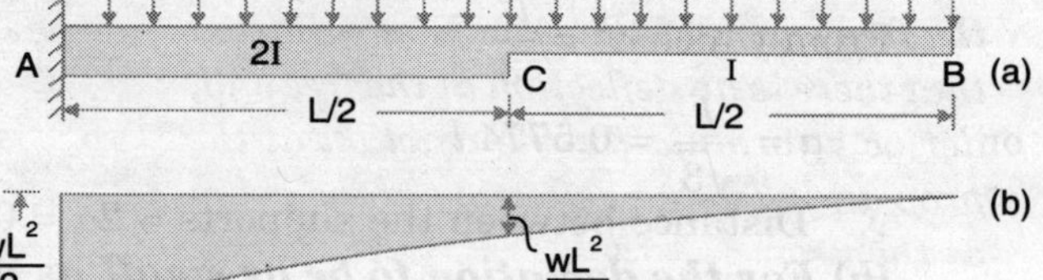

The complete M/EI diagram can be splitted into two parts A_1 and A_2.

(i) Area A₁ : Parabolic variation

$$A_1 = \frac{1}{3}\frac{wL^2}{8EI}\cdot\frac{L}{2}=\frac{wL^3}{48\ EI}$$

$$\bar{x}_1 = \frac{3}{4}\left(\frac{L}{2}\right)=\frac{3}{8}L \text{ (from } B)$$

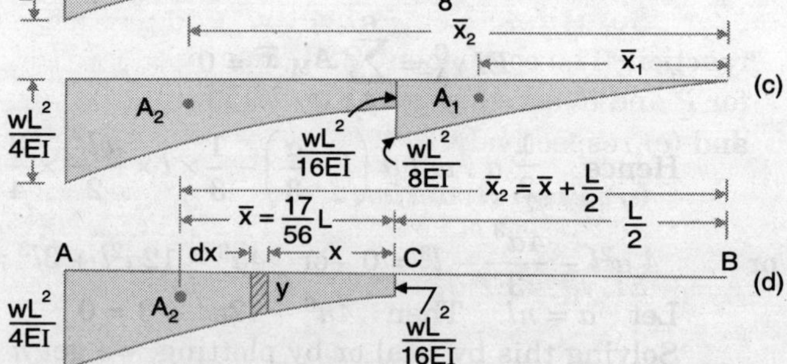

Fig. 12.25

(ii) Area A₂ : This is *parabolic sector*. Fig. 12.25 (d) shows the details of the sector. Measuring x from C, towards A, the equation of the curve is

$$y = \frac{M_x}{EI_x}=\frac{w}{2}\left(\frac{L}{2}+x\right)^2\times\frac{1}{2EI}=\frac{w}{4EI}\left(\frac{L}{2}+x\right)^2;\quad dA = y\ .\ dx$$

$$\therefore\quad A_2 = \int_0^{L/2} dA = \int_0^{L/2} y\ .\ dx = \int_0^{L/2}\frac{w}{4\ EI}\left(\frac{L}{2}+x\right)^2 dx$$

$$= \frac{w}{4\ EI}\int_0^{L/2}\left(\frac{L^2}{4}+x^2+Lx\right)dx = \frac{w}{4EI}\left[\frac{L^2}{4}x+\frac{x^3}{3}+\frac{Lx^2}{2}\right]_0^{L/2}=\frac{7\ wL^3}{96\ EI}$$

Also $\quad Q = \int_0^{L/2} xdA = \frac{w}{4\ EI}\int_0^{L/2}\left(\frac{L^2}{4}x+x^3+Lx^2\right)dx = \frac{w}{4\ EI}\left[\frac{L^2}{4}\cdot\frac{x^2}{2}+\frac{x^4}{4}+\frac{Lx^3}{3}\right]_0^{L/2}$

$$= \frac{17\,wL^4}{4 \times 192\,EI}$$

$$\therefore \quad \bar{x} = \frac{Q}{A_2} = \frac{17\,wL^4}{4 \times 192\,EI} \times \frac{96\,EI}{7\,wL^3} = \frac{17}{56}L$$

Now $\quad \theta_B^A = -\sum_B^A A_{M/EI} = \frac{wL^3}{48\,EI} + \frac{7\,wL^3}{96\,EI} = \frac{9\,wL^3}{96\,EI} = \mathbf{\frac{3wL^3}{32\,EI}}$

Also, $\quad y_B^A = -\sum_B^A A_{M/EI} \,.\, \bar{x} = \frac{wL^3}{48\,EI} \times \frac{3}{8}L + \frac{7\,wL^3}{96\,EI}\left(\frac{17}{56}L + \frac{L}{2}\right) = \mathbf{\frac{17}{256}\frac{wL^4}{EI}}$.

12.9. ADDITIONAL EXAMPLES USEFUL FOR COMPETITIVE EXAMINATIONS

Example 12.13. *A cantilever of effective length L carries a total load wL uniformly distributed throughout the length. If the cantilever is propped at a point L/4 from the free end and the prop is so adjusted that there is no deflection at the free end, derive a formula for the reaction at the prop and also the deflection of the beam at the prop.*

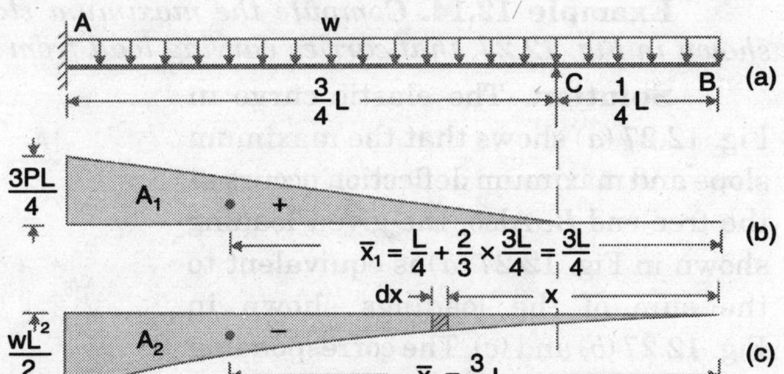

Fig. 12.26

Solution. Let P be the prop reaction. The component B.M. diagram for P and w are shown in Fig. 12.26 (*b*) and (*c*) respectively.

(a) Prop Reaction :

$$EI\ y_B^A = EI\ d_B^A = -\sum_B^A A\bar{x} = 0 \text{ (Since the deflection at } A \text{ is zero)}$$

$$\therefore \quad -\frac{1}{2}\frac{3L}{4}\left(\frac{3PL}{4}\right)\left(\frac{3L}{4}\right) + \frac{1}{3}L\,.\,\frac{wL^2}{2}\,.\,\frac{3}{4}L - 0$$

or $\quad \frac{27\,PL^3}{128} = \frac{wL^4}{8}$ from which $\mathbf{P = \frac{16}{27}\,wL}$

(b) Deflection of the beam at the prop :

$$EI\ y_C^A = EI\ d_C^A = -\sum_C^A A\,\bar{x} \qquad \text{...(i)}$$

(i) $A\bar{x}$ *due to P :*

$$\left(A\bar{x}\right)_C^A \text{ due to } P = \frac{1}{2} \times \frac{3}{4}L \times \frac{3PL}{4}\left(\frac{2}{3}\,.\,\frac{3}{4}L\right) = \frac{9PL^3}{64} = \frac{9}{64}\left(\frac{16}{27}wL\right)L^3 = \frac{wL^4}{12}$$

(ii) $A\bar{x}$ due to w:

To find $(A\bar{x})_C^A$ due to w, consider an elementary strip dx at a distance x from A.

Ordinate $y = \dfrac{wx^2}{2}$. Hence $dA = ydx = \dfrac{wx^2}{2}\,dx$; $\bar{x}$ of this strip, from $C = x - L/4$.

$$\delta(A\bar{x}) = \frac{wx^2}{2}\,dx\,(x - L/4)$$

$$\therefore \quad \sum_C^A (A\bar{x})_w = \int_{L/4}^{L} \frac{wx^2}{2}\left(x - \frac{L}{4}\right)dx = \frac{w}{2}\int_{L/4}^{L}\left(x^3 - \frac{Lx^2}{4}\right)dx$$

$$= \frac{w}{2}\left[\frac{x^4}{4} - \frac{Lx^3}{12}\right]_{L/4}^{L} = \frac{171\,wL^4}{2048}$$

Substituting the values in (*i*), we get

$$EI\,y_C^A = -\left[\frac{wL^4}{12} - \frac{171\,wL^4}{2048}\right] = \mathbf{\frac{wL^4}{6144\,EI}}.$$

Example 12.14. *Compute the maximum slope and deflection for a cantilever beam, shown in Fig. 12.27, that carries varying load from zero at the wall to w_0 at the free end.*

Solution. The elastic curve in Fig. 12.27 (*a*) shows that the maximum slope and maximum deflection occurs at the free end *B*. Also, the given loading shown in Fig. 12.27 (*a*) is equivalent to the sum of the loadings shown in Fig. 12.27 (*b*) and (*c*). The corresponding B.M. diagrams for the modified loadings are shown in Fig. 12.27 (*d*).

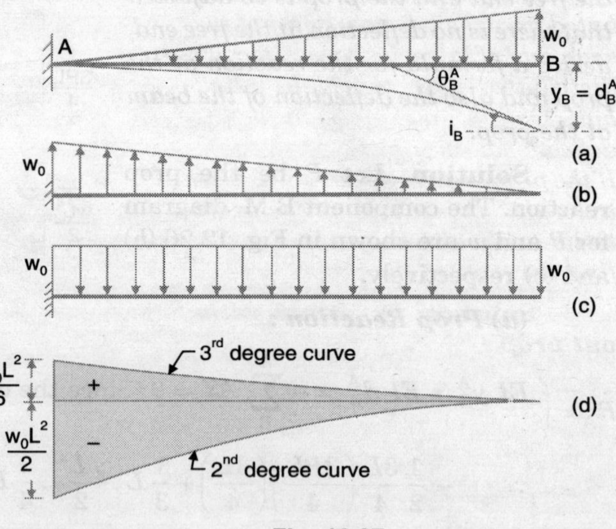

Fig. 12.27

Now $i_B = \theta_B^A = -\dfrac{1}{EI}\sum_B^A A_M$

$$= -\frac{1}{EI}\left[\frac{1}{4}\times\frac{w_0L^2}{6}\times L - \frac{1}{3}\,\frac{w_0L^2}{2}\times L\right]$$

$$= \mathbf{\frac{w_0L^2}{8\,EI}}.$$

Also, $\qquad y_B = d_B^A = -\dfrac{1}{EI}\sum_B^A A_M\,.\,\bar{x}$

$$= -\frac{1}{EI}\left[\frac{1}{4}\times\frac{w_0L^2}{6}\times L\times\frac{4}{5}\,L - \frac{1}{3}\,.\,\frac{w_0L^2}{2}\,.\,L\times\frac{3}{4}\,L\right] = \mathbf{\frac{11}{120}\,\frac{w_0L^4}{EI}}.$$

Example 12.15. *A cantilever is acted upon by a uniformly distributed moment of intensity m per unit distance along the axis of the beam. Find the slope and deflection at the free end.*

Solution.

$$M_B = 0\ ;\ M_A = mL$$

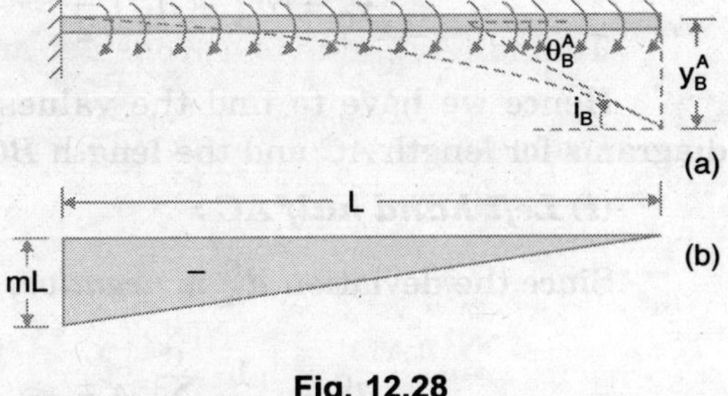

Fig. 12.28

The B.M. diagram is shown in Fig. 12.28 (*b*), the variation being linear.

$$\text{Now} \quad i_B = \theta_B^A = -\sum_B^A \frac{A_M}{EI}$$

$$= -\frac{1}{EI}\left[-\frac{1}{2} \times mL \times L\right] = \frac{mL^2}{2EI}$$

$$\text{Also,} \quad y_B = d_B^A = -\sum_B^A \frac{A_M}{EI} \cdot \bar{x} = -\frac{1}{EI}\left[-\frac{1}{2} \times mL \times L \times \frac{2}{3} L\right] = \frac{mL^3}{3EI}.$$

Example 12.16. *A simply supported beam of span 2 L has a second moment of area of 3 I for the left half of the span and 2 I for the right half. The beam carries a load of intensity 3 w uniformly distributed over the left hand half of the span and of intensity 2 w uniformly distributed over the right hand half. Obtain an expression for the deflection at the centre of the span. If the prop is now placed at the centre of the span to restore the beam to its original level at this point, find the force in the prop.*

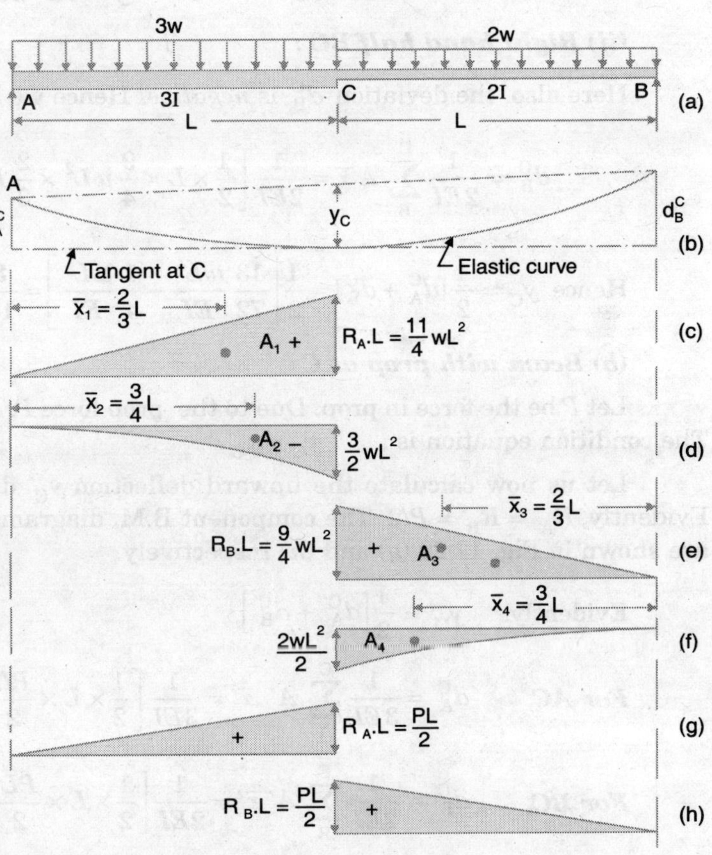

Fig. 12.29

Solution. (*a*) *Beam without prop :*

$$R_A = \left(3wL \times \frac{3}{2} L + 2wL \times \frac{L}{2}\right) \frac{1}{2L}$$

$$= \frac{11}{4} wL\ ;$$

$$R_B = (3wL + 2wL) - \frac{11}{4} wL$$

$$= \frac{9}{4} wL$$

The deflection curve is shown in Fig. 12.29 (*b*). If a tangent is drawn at *C*, to the elastic curve, it is evident that

$$y_C = (d_A^C + d_B^C)\,\frac{1}{2}$$

Hence we have to find the values of d_A^C and d_B^C. The component bending moment diagrams for length AC and the length BC have been drawn separately.

(i) Left hand half AC :

Since the deviation d_A^C is *negative*, we have

$$d_A^C = \frac{1}{3EI}\sum_A^C A\,\bar{x} = \frac{1}{3EI}\left[\frac{1}{2}\times L\times\frac{11}{4}wL^2\times\frac{2}{3}L - \frac{1}{3}\times L\times\frac{3}{2}wL^2\times\frac{3}{4}L\right]$$

$$= \frac{1}{3EI}\left[\frac{11}{12}wL^4 - \frac{3}{8}wL^4\right] = \frac{13}{72}\frac{wL^4}{EI} \qquad ...(i)$$

(ii) Right hand half BC :

Here also, the deviation d_B^C is *negative*. Hence we have

$$d_B^C = \frac{1}{2EI}\sum_B^C A\,\bar{x} = \frac{1}{2EI}\left[\frac{1}{2}\times L\times\frac{9}{4}wL^2\times\frac{2}{3}L - \frac{1}{3}L\,.\,wL^2\times\frac{3}{4}L\right] = \frac{wL^4}{4EI} \qquad ...(ii)$$

Hence $y_C = \frac{1}{2}[d_A^C + d_B^C] = \frac{1}{2}\left[\frac{13}{72}\frac{wL^4}{EI} + \frac{1}{4}\frac{wL^4}{EI}\right] = \frac{\mathbf{31}}{\mathbf{144}}\frac{\boldsymbol{wL^4}}{\boldsymbol{EI}}.$

(b) Beam with prop at C

Let P be the force in prop. Due to the prop force P, let y_C' be the upward deflection of C. The condition equation is : $y_C = y_C'$.

Let us now calculate the upward deflection y_C' due to prop force P acting upwards. Evidently, $R_A' = R_B' = P/2$. The component B.M. diagrams for AC and BC due to this loading are shown in Fig. 12.29 (g) and (h) respectively.

Evidently $\quad y_C' = \frac{1}{2}\left[d_A^{C'} + d_B^{C'}\right]$

For AC, $\quad d_A^{C'} = \frac{1}{3EI}\sum_A^C A'\,\bar{x}' = \frac{1}{3EI}\left[\frac{1}{2}\times L\times\frac{PL}{2}\times\frac{2}{3}L\right] = \frac{PL^3}{18\,EI} \qquad ...(iii)$

For BC, $\quad d_B^{C'} = \frac{1}{2EI}\sum_B^C A'\,\bar{x}' = \frac{1}{2EI}\left[\frac{1}{2}\times L\times\frac{PL}{2}\times\frac{2}{3}L\right] = \frac{PL^3}{12\,EI} \qquad ...(iv)$

$\therefore \quad y_C' = \frac{1}{2}\left[d_A^{C'} + d_B^{C'}\right] = \frac{1}{2}\left[\frac{PL^3}{18\,EI} + \frac{PL^3}{12\,EI}\right] = \frac{5}{72}\frac{PL^3}{EI}$

Now $\quad y_C' = y_C$. Hence $\quad \frac{5}{72}\frac{PL^3}{EI} = \frac{31}{144}\frac{wL^4}{EI}$. From which $\quad P = \frac{\mathbf{31}}{\mathbf{10}}\,\boldsymbol{wL}.$

Example 12.17. *A beam with variable moment of inertia is loaded as shown in Fig. 12.30. Determine the deflection at the mid-span.*

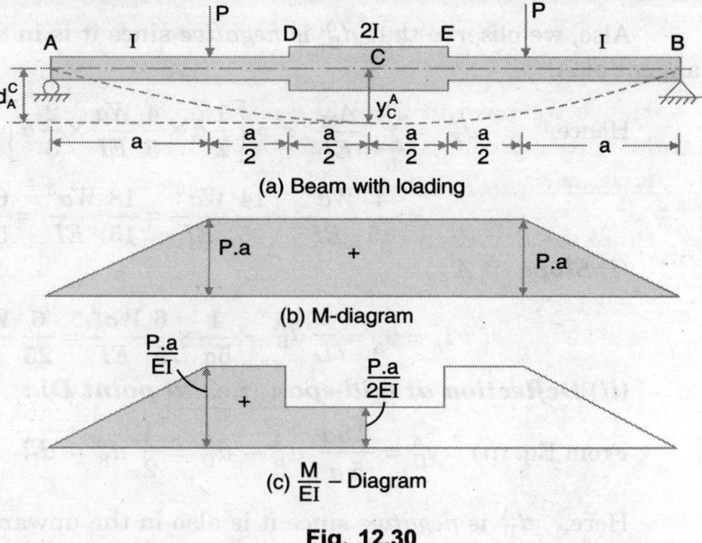

Fig. 12.30

Solution. Due to symmetry, both with regard to geometry as well as loading, the tangent at *C* will be horizontal.

Hence $y_C^A = -\, d_C^A$

$$= \sum_A^C \frac{A_M \bar{x}}{EI}$$

Fig. 12.30 (*b*) shows *M*-diagram. Fig. 12.30 (*c*) shows *M/EI* diagram, which can be splitted into three parts.

$$\sum_A^C \frac{A_M \bar{x}}{EI} = \left[\frac{1}{2} \times a \times \frac{Pa}{EI} \times \frac{2}{3}a\right] + \left[\frac{Pa}{EI} \times \frac{a}{2}\left(a + \frac{a}{4}\right)\right] + \left[\frac{Pa}{2EI} \times \frac{a}{2}\left(a + \frac{a}{2} + \frac{a}{4}\right)\right]$$

$$= \frac{Pa^3}{3\,EI} + \frac{5Pa^3}{8\,EI} + \frac{7Pa^3}{16\,EI} = \frac{67}{48}\frac{Pa^3}{EI}$$

Hence $\quad y_C^A = \mathbf{\frac{67}{48}\frac{Pa^3}{EI}}$ **(Ans.)**

Example 12.18. *A beam ACB, simply supported at the ends, has moment of inertia 4 I for the length AC and I for the length CB, and is loaded with point load W at C. Determine (i) slope at end A (ii) Deflection at mid-span and (iii) Maximum deflection.*

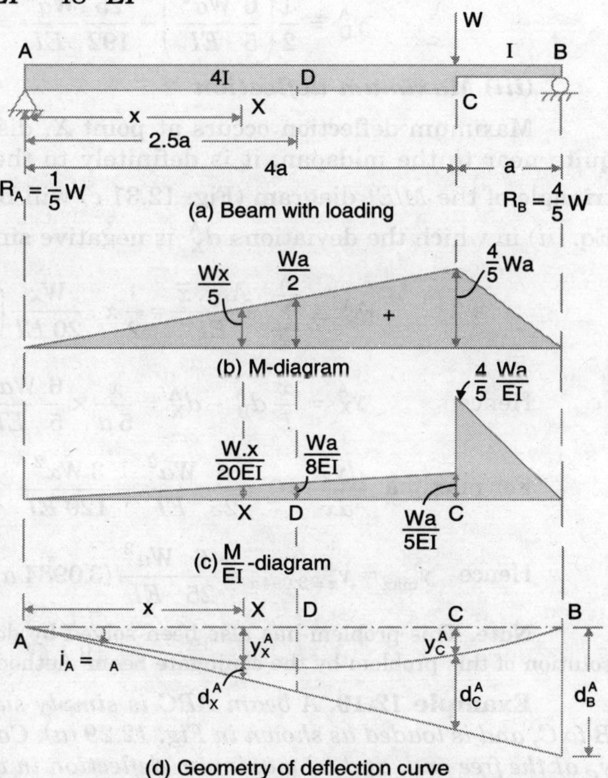

Fig. 12.31

Solution. Fig. 12.31 (*b*) shows the *M* diagram, while Fig. 12.31 (*c*) shows *M/EI* diagram which is self explanatory.

Fig. 12.31 (*d*) shows the geometry of the deflection curve from which we find that

$$i_A = \theta_A = \frac{1}{L}\, d_B^A \qquad \ldots(i)$$

Also, for any point *X*, distant *x* from *A*, we have

$$y_X^A = \frac{x}{L}\, d_B^A - d_X^A \qquad \ldots(ii)$$

Also, we observe that d_B^A is *negative* since it is in *upward* direction with respect to the tangent at *A*.

$$\text{Hence}\qquad d_B^A = \sum_B^A \frac{A_M}{EI}.\bar{x} = \left[\frac{1}{2}a \times \frac{4}{5}\frac{Wa}{EI} \times \frac{2}{3}a\right] + \left[\frac{1}{2} \times 4a \times \frac{Wa}{5EI}\left(a + \frac{1}{3} \times 4a\right)\right]$$

$$= \frac{4}{15}\frac{Wa^3}{EI} + \frac{14}{15}\frac{Wa^3}{EI} = \frac{18}{15}\frac{Wa^3}{EI} = \frac{6}{5}\frac{Wa^3}{EI} \qquad \text{...(iii)}$$

(i) Slope at A :

$$i_A = \theta_A = \frac{1}{L}d_B^A = \frac{1}{5a} \times \frac{6}{5}\frac{Wa^3}{EI} = \mathbf{\frac{6}{25}\frac{Wa^2}{EI}}$$

(ii) Deflection at mid-span (i.e., at point D) :

$$\text{From Eq. } (ii)\quad y_D^A = \frac{2.5a}{5\,a}d_B^A - d_D^A = \frac{1}{2}d_B^A - d_D^A \qquad \text{...(iv)}$$

Here, d_D^A is *negative* since it is also in the upward direction.

$$\therefore\qquad d_D^A = \sum_D^A \frac{A_M.\bar{x}}{EI} = \frac{1}{2} \times 2.5\,a \times \frac{Wa}{8\,EI}\left(\frac{1}{3} \times 2.5\,a\right) = \frac{25}{192}\frac{W\,a^3}{EI}$$

Substituting the values of d_B^A and d_D^A in (*iv*), we get

$$y_D^A = \frac{1}{2}\left(\frac{6}{5}\frac{Wa^3}{EI}\right) - \frac{25}{192}\frac{Wa^3}{EI} = \frac{902}{1920}\frac{Wa^3}{EI} = \mathbf{0.4698\,\frac{Wa^3}{EI}}$$

(iii) Maximum deflection

Maximum deflection occurs at point *X*, distant *x* from *A*. Since its location is *always* quite near to the midspan, it is definitely to the left of point *C*. Hence only the left hand triangle of the ***M/EI*** diagram (Fig. 12.31 *c*) will be considered. The deflection y_X^A is given by Eq. (*ii*) in which the deviations d_X^A is negative since it is in the upward direction. Hence

$$d_X^A = \sum_X^A \frac{A_M.\bar{x}}{EI} = \frac{1}{2}x.\frac{Wx}{20\,EI}\left(\frac{x}{3}\right) = \frac{W\,x^3}{120\,EI}$$

$$\text{Hence}\qquad y_X^A = \frac{x}{L}d_B^A - d_X^A = \frac{x}{5\,a} \times \frac{6}{5}\frac{Wa^3}{EI} - \frac{Wx^3}{120\,EI} = \frac{6}{25}\frac{Wx\,a^2}{EI} - \frac{Wx^3}{120\,EI}$$

For maxima, $\dfrac{dy_X^A}{dx} = 0 = \dfrac{6}{25}\dfrac{Wa^2}{EI} - \dfrac{3\,Wx^2}{120\,EI}$ From which, $x = 3.0984\,a$.

Hence $y_{max} = y_{x=3.094a}^A = \dfrac{6}{25}\dfrac{Wa^2}{EI}(3.0984\,a) - \dfrac{W}{120\,EI}(3.0984\,a)^3$ or $y_{max} = \mathbf{0.4957\,\dfrac{Wa^3}{EI}}$.

Note. This problem has also been solved by double integration method at example 11.12. For solution of this problem by the conjugate beam method, see example 13.13.

Example 12.19. *A beam ABC is simply supported at A and B, and it overhangs from B to C, and is loaded as shown in Fig. 12.29 (a). Calculate (a) slopes at A, B and C (b) deflection y_C at the free end, and (c) maximum deflection in the span AB. Take E = 2 × 10⁵ N/mm² and I = 1.5 × 10⁹ mm⁴.*

Solution. Fig. 12.32 (*b*) shows the deflection curve, in which $A\,B_1$ is the tangent to the curve at A and A_1BC_1 is the tangent to the curve at support B. As marked in Fig. 12.32 (*b*), angles i_A, i_B and i_C are all positive. The B.M. diagram is shown in Fig. 12.32 (*c*). For CB, the B.M. varies parabolically, with

$$M_B = -\frac{6(3)^2}{2} = -27 \text{ kN-m.}$$

$R_A = \frac{1}{7.5}(48 \times 4.5 - 6 \times 3 \times 1.5) = 25.2$ kN (↑), $R_B = 48 + 18 - 25.2 = 40.8$ kN (↑)

Hence $M_D = 25.2 \times 3 = +75.6$ kN-m. The variation of B.M. in AD is linear. For B.M. in DB, measuring x from B, we have

$$M_x = -6 \times 3\,(x + 1.5) + 40.8\,x = 22.8\,x - 27.$$

This variation is linear, which gives $M_D = 22.8 \times 4.5 - 27 = 75.6$ kN-m, as found earlier. The point of contraflexure occurs at $x = 27/22.8 = 1.1842$ m from B. This point of inflexion has been marked by a dark dot on the deflection curve. The component bending moment diagrams for span AB and span CB are shown in Fig. 12.32 (*d*).

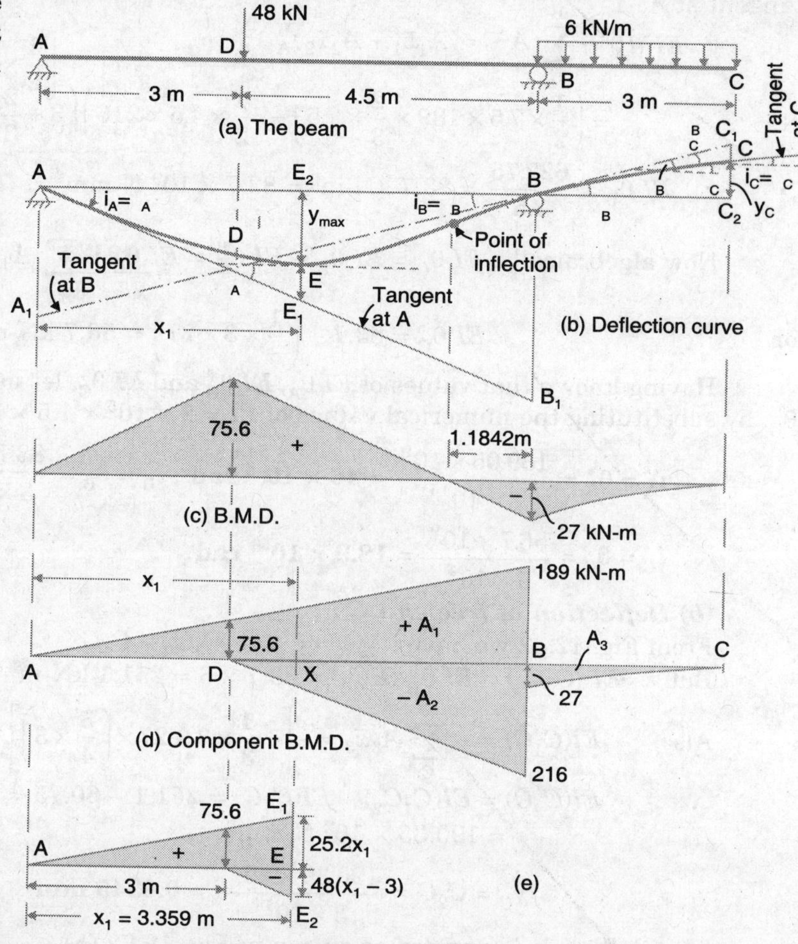

Fig. 12.32

(a) Rotations at A, B and C :

Now $$i_A = \theta_A = \frac{d_B^A}{AB} = \frac{d_B^A}{7.5} \qquad \ldots(i)$$

Note that $d_B^A = B_1B$ is negative since it is in upward direction from the tangent $A\,B_1$.

$$\therefore \quad EI\,d_B^A = \sum_B^A A\,\bar{x} = (A_1\bar{x}_1 + A_2\bar{x}_2)_B^A$$

$$= \left[\frac{1}{2} \times 7.5 \times 189 \times \frac{1}{3} \times 7.5 - \frac{1}{2} \times 4.5 \times 216 \times \frac{1}{3} \times 4.5\right] = 1042.9 \text{ kN-m}^3.$$

$$\therefore \quad EI\,\theta_A = \frac{EId_B^A}{7.5} = \frac{1042.9}{7.5} = 139.05 \text{ kN-m}^2 = 139.05 \times 10^9 \text{ N-mm}^2 \qquad \ldots(ii)$$

Similarly, $$i_B = \theta_B = \frac{d_A^B}{BA} = \frac{d_A^B}{7.5} \qquad \ldots(iii)$$

Here also, d_A^B is negative since it is in upward direction from the tangent BA_1.

$$\therefore\ EI\, d_A^B = \sum_A^B A\,\bar{x} = (A_1\bar{x}_1 + A_2\bar{x}_2)_A^B$$

$$= \left(\frac{1}{2}\times 7.5\times 189\times\frac{2}{3}\times 7.5\right) - \left(\frac{1}{2}\times 4.5\times 216\right)\left(3+\frac{2}{3}\times 4.5\right) = 627.75 \text{ kN-m}^3$$

$$\therefore\quad EI\,\theta_B = \frac{627.75}{7.5} = 83.7 \text{ kN-m}^2 = 83.7\times 10^9 \text{ N-mm}^2$$

Now algebraically, $EI\,\theta_C = EI\,\theta_B + EI\,\theta_C^B = EI\,\theta_B + \sum_C^B A_3$

or
$$EI\,\theta_C = 83.7 - \left(\frac{1}{3}\times 3\times 27\right) = 56.7 \text{ kN-m}^2 = 56.7\times 10^9 \text{ N-mm}^2$$

Having known that values of $EI\,\theta_A$, $EI\,\theta_B$ and $EI\,\theta_C$, let us find the values of θ_A, θ_B and θ_C, by substituting the numerical values of $EI = 2\times 10^5\times 1.5\times 10^9 = 3\times 10^{14}$ N-mm².

$$\therefore\quad i_A = \theta_A = \frac{139.05\times 10^9}{3\times 10^{14}} = \mathbf{46\times 10^{-5}\ rad}\ ;\ i_B = \theta_B = \frac{83.7\times 10^9}{3\times 10^{14}} = \mathbf{27.9\times 10^{-5}\ rad.}$$

$$i_C = \theta_C = \frac{56.7\times 10^9}{3\times 10^4} = \mathbf{18.9\times 10^{-5}\ rad.}$$

(b) Deflection of free end C :

From Fig. 12.32 we have : $y_C = C_2C = C_1C_2 - C_1C$

But $\quad EI\,(C_1C_2) = EI\,\theta_B\,(BC_2) = 83.7\times 3 = 251.1$ kN-m³

$$\text{Also,}\quad EI(C_1C) = -\sum_C^B A_3\bar{x}_3 = -\left[-\frac{1}{3}\times 3\times 27\times\left(\frac{3}{4}\times 3\right)\right] = 60.75 \text{ kN-m}^3$$

$$\therefore\quad EI(C_2C) = EI(C_1C_2) - EI(C_1C) = 251.1 - 60.75 = 190.35 \text{ kN-m}^3$$

$$= 190.35\times 10^2 \text{ N-mm}^2$$

$$\therefore\quad y_C = C_2C = \frac{190.35\times 10^{12}}{3\times 10^{14}} = \mathbf{0.6345\ mm}$$

This deflection is upward as shown in Fig. 12.32 (*b*).

(c) Maximum deflection in AB :

The maximum downward deflection in *AB* occurs at a point *E*, the location of which is to be found. Let us assume that point *E* is between *D* and *B*, and is at a distance x_1 from *A*. If this is not, the calculations will indicate so, and then we will assume that *E* is between *A* and *D*.

Note that at point *E*, the point of maximum deflection, the tangent will be horizontal. Hence the area of *M/E I* diagram will be equal to the angle of rotation θ_A.

$$\text{Hence}\quad EI\,\theta_A = \sum_A^E A \qquad \ldots(iv)$$

The component *M*-diagram between *A* and *E* is shown separately in Fig. 12.32 (*e*). In this diagram, ordinate $EE_1 = R_A.x_1 = +\,25.2\,x_1$ kN-m while ordinate $EE_2 = -\,48\,(x_1 - 3)$.

$$\sum_A^E A = \frac{1}{2}x_1\,(25.2\,x_1) - \frac{1}{2}(x_1 - 3)\,48\,(x_1 - 3)$$

$$= 12.6\,x_1^2 - 24x_1^2 + 144x_1 - 216 = -\,11.4x_1^2 + 144x_1 - 216$$

In the above expression x_1 has units of metres. Hence substituting the value of $EI\theta_A$ in kN-m² in (*iv*), we get

$$139.05 = -11.4\, x_1^2 + 144\, x_1 - 216$$

or $$11.4\, x_1^2 - 144\, x_1 + 355.05 = 0, \text{ from which we get } x_1 = 3.359 \text{ m.}$$

The maximum deflection y_{max} is numerically equal to the offset of point A from the horizontal tangent at E.

Hence $$EIy_{max} = \sum_A^E A\bar{x} = \left[\frac{1}{2} x_1 \times 25.2\, x_1 \left(\frac{2}{3} x_1\right)\right] - \left[\frac{1}{2}(x_1 - 3)\, 48\, (x_1 - 3)\left\{3 + \frac{2}{3}(x_1 - 3\right\}\right]$$

Substituting $x_1 = 3.359$ m, we get

$$EI\, y_{max} = 308.26 \text{ kN-m}^3 = 308.26 \times 10^{12} \text{ N-mm}^3$$

$$\therefore \quad y_{max} = \frac{308.26 \times 10^{12}}{3 \times 10^4} = \mathbf{1.028 \text{ mm.}}$$

PROBLEMS

1. A cantilever beam AB is subjected to a concentrated load W and a couple μ acting at the free end (Fig. 12.33). Determine the slope and the deflection at the free end.
2. Determine the deflection at points B and C, of the cantilever beam shown in Fig. 12.34.

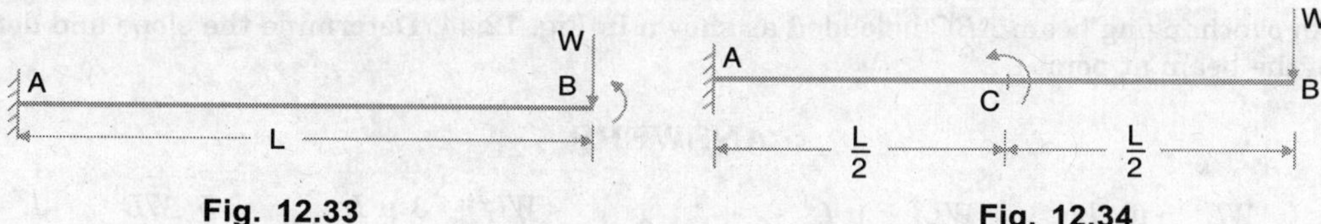

Fig. 12.33 Fig. 12.34

3. Determine the deflections at points B and C of the cantilever beam shown in Fig. 12.35.

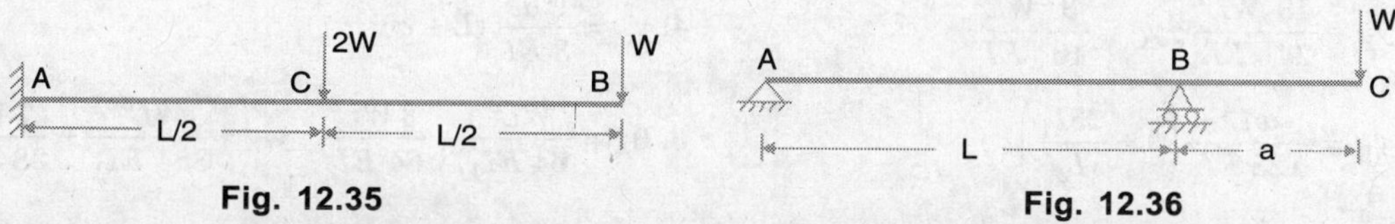

Fig. 12.35 Fig. 12.36

4. Determine the deflection at the end of the overhang for the beam, loaded as shown in Fig. 12.36.
5. Determine the deflection at the free end B of the cantilever beam AB shown in Fig. 12.37.

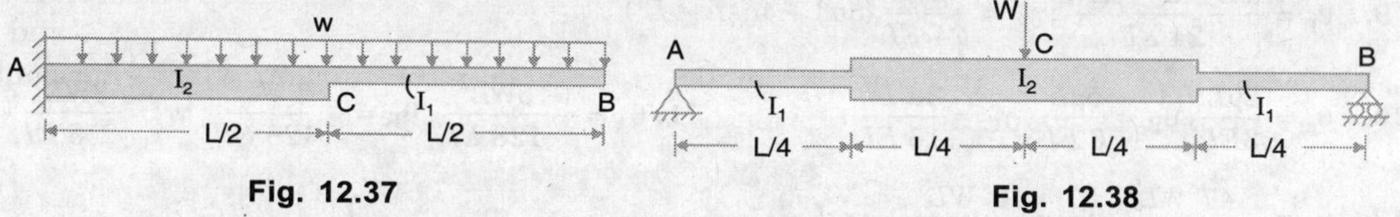

Fig. 12.37 Fig. 12.38

6. A simple beam AB, shown in Fig. 12.38 has two different moments of inertia I_1 and I_2. Determine (*a*) slope at end A, and (*b*) deflection at the mid-span.
7. Determine the deflection at the mid-span of an overhanging beam, loaded as shown in Fig. 12.39.

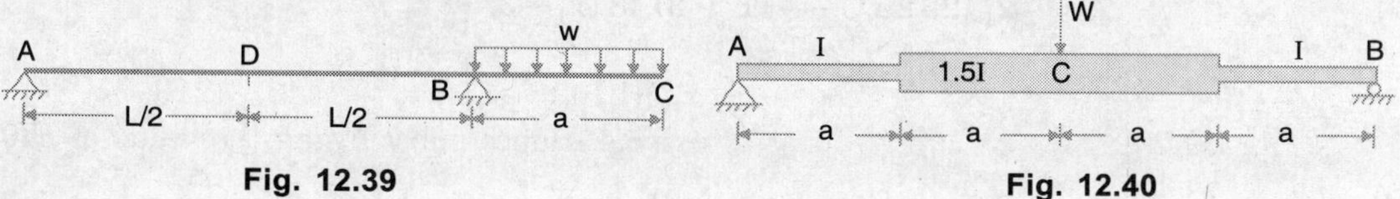

Fig. 12.39 Fig. 12.40

8. The middle half of the beam shown in Fig. 12.40 has a moment of inertia 1.5 times that of the rest of the beam. Find the mid-span deflection.

9. A simple beam with overhang is loaded as shown in Fig. 12.41. Find (*i*) the slope at *B*, and (*ii*) the deflection at *C*.

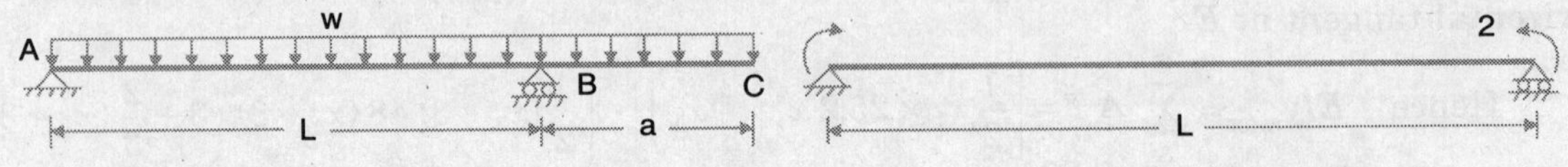

Fig. 12.41

Fig. 12.42

10. A simple beam *AB* is acted upon by couples μ and 2μ at the ends, as shown in Fig. 12.42. Determine (*i*) slopes at *A* and *B*, and (*ii*) the deflections at the mid-span.

11. Find the slopes at the ends and central deflection for a simply supported beam shown in Fig. 12.43.

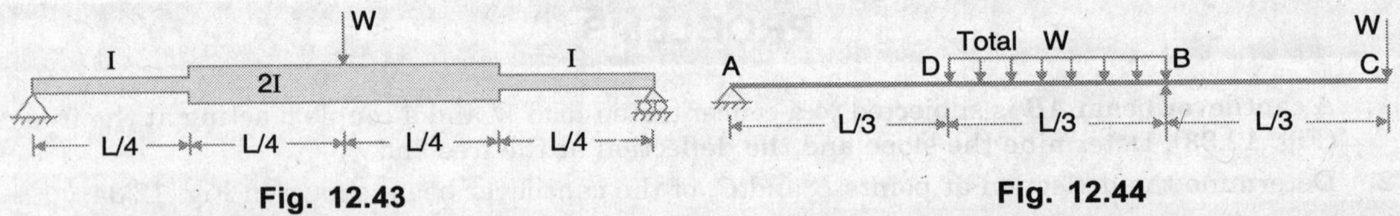

Fig. 12.43

Fig. 12.44

12. An overhanging beam *ABC* is loaded as shown in Fig. 12.44. Determine the slope and deflection of the beam at point *C*.

ANSWERS

1. $\theta_B = \dfrac{WL^2}{2\,EI} - \dfrac{\mu_0\,L}{EI}$; $y_B = \dfrac{WL^3}{3\,EI} - \dfrac{\mu\,L^2}{2\,EI}$

2. $y_B = \dfrac{WL^3}{3\,EI} - \dfrac{3}{8}\dfrac{\mu\,L^2}{EI}$; $y_C = \dfrac{5}{48}\dfrac{WL^3}{EI} - \dfrac{\mu L^2}{8\,EI}$

3. $y_B = \dfrac{13}{24}\dfrac{WL^3}{EI}$; $y_C = \dfrac{3}{16}\dfrac{WL^3}{EI}$

4. $y_C = \dfrac{Wa^2}{3\,EI}(L + a)$

5. $y_B = \dfrac{wL^4}{128\,EI_1}\left(1 + \dfrac{15I_1}{I_2}\right)$

6. $\theta_A = \dfrac{WL^2}{64\,EI_1} + \dfrac{3\,WL^2}{64\,EI_2}$; $y_C = \dfrac{WL^3}{384\,EI_1} + \dfrac{7\,WL^3}{384\,EI_2}$

7. $y_D = \dfrac{wa^2L^2}{32\,EI}$

8. $y_C = \dfrac{17\,wa^3}{18\,EI}$

9. $\theta_B = \dfrac{wL\,(4a^2 - L^2)}{24\,EI}$; $y_C = \dfrac{wa}{24\,EI}(3a^3 + 4a^2L - L^3)$

10. $\theta_A = \dfrac{2\mu L}{3EI}$; $\theta_B = \dfrac{5\mu L}{6\,EI}$; $y_C = \dfrac{3\mu\,L^2}{16\,EI}$

11. $\theta_A = +\dfrac{5WL^2}{128\,EI}$; $\theta_B = -\dfrac{5\,WL^2}{128\,EI}$; $y_C = \dfrac{3WL^3}{256\,EI}$

12. $\theta_C = -\dfrac{47}{48}\dfrac{WL^2}{EI}$; $y_C = \dfrac{13}{16}\dfrac{WL^3}{EI}$.

13

Deflection of Beams III : Conjugate Beam Method

13.1. INTRODUCTION

In chapter 11 and 12, we have found the slope and deflection at any point in a beam using double integration method (including Macaulay's method) and the area moment method respectively. These methods of determining slope and deflection are suitable only for simple cases, where the beam is having uniform flexural rigidity. If, however, the flexural rigidity, *EI,* is not uniform, the methods become quite laborious. In such cases conjugate beam method presents an easy approach. It is an ingenious method and the student will find its application very interesting. A *conjugate beam is an imaginary secondary beam,* which when loaded with the *M/EI* diagram of the real beam, yield directly the slope and deflection of the real beam in the form of shear force and bending moment of the conjugate beam. Evidently, a *conjugate beam* is a mirror image of the real beam, but with support conditions changed so as to meet the boundary conditions. *Conjugate beam method* is also known as the *method of elastic weight.*

13.2. THE CONJUGATE BEAM

We start with the deflection equation, the *successive differentiation* of which give the following relations :

$$EI\,y = \text{Deflection} \quad \text{...(1)}$$

$$EI\frac{dy}{dx} = \text{slope} \quad \text{...(2)}$$

$$EI\frac{d^2y}{dx^2} = -M \text{ (i.e., bending moment)} \quad \text{...(3)}$$

$$EI\frac{d^3y}{dx^3} = -\frac{dM}{dx} = -F \quad \text{(i.e., shear force)} \quad \text{...(4)}$$

$$EI\frac{d^4y}{dx^4} = -\frac{d^2M}{dx^2} = -\frac{dF}{dx} = w \quad \text{(i.e., rate of loading)} \quad \text{...(5)}$$

From the above equations, we observe that there is similarity of relations among load, shear and moment (Eqs. 5, 4 and 3) and among moment, slope and deflection (Eqs. 3, 2 and 1). This similarity directly suggests us a method of calculating slope and deflection using the usual method of calculating shear and bending moment in a beam.

The method, therefore, uses a *fictitious secondary beam,* known as *conjugate beam*, which is loaded with not the actual loads, but with the *elastic weight* M/EI corresponding to the actual loads. Then, the *shear force at any point* on the *conjugate beam* will give the *slope* at the *corresponding point* of the *real beam.* Similarly, the *B.M.* at any point on the *conjugate beam* will give the *deflection* at the *corresponding point* of the *real beam*. In order to show this, we will consider two *propositions.*

Proposition 1. *Similarity between S.F. of conjugate beam and slope of real beam*

We have: *load on conjugate beam* = w' = *M/EI* of real beam

$$\therefore \quad \text{Shear force } F_x' = -\int_0^x w'\,dx = -\int_0^x \frac{M}{EI}\,dx \text{ ; But } -\frac{M}{EI} = \frac{d^2y}{dx^2}, \text{ from (3)}$$

$$\therefore \qquad F_x' = \int \frac{d^2y}{dx^2}\,dx = \frac{dy}{dx} = \text{slope} \qquad \text{...(13.1)}$$

This shows that the slope at any point of a real beam, relative to the original axis of the beam, is equal to the shear force at the corresponding point of the conjugate beam.

Proposition 2. *Similarity between B.M. of conjugate beam and deflection of real beam*

$$\text{Shear force} \quad F_x' = -\int_0^x w'\,dx = -\int_0^x \frac{M}{EI}\,dx = \int_0^x \frac{d^2y}{dx^2}\,dx$$

$$\therefore \quad M_x' = \int_0^x F_x'.\,dx = -\int_0^x\int_0^x w'\,dx = -\int_0^x\int_0^x \frac{M}{EI}\,dx = \iint \frac{d^2y}{dx^2}\,dx = \int_0^x \frac{dy}{dx}\,dx = y \quad \text{...(13.2)}$$

This shows that the deflection at any point of a real beam, relative to the original axis of the beam, is equal to the bending moment at the corresponding point of the conjugate beam.

Before we apply this method to simply supported beams, cantilevers and propped cantilevers, we shall consider the sign conventions.

13.3. SIGN CONVENTIONS

The following sign conventions will be adopted.

1. Loading on the conjugate beam. The loading on the conjugate beam is equal to the *M/EI* diagram of the real beam. This loading is also known as the *elastic weight.* The loading or the elastic weight acts *downwards* if the *M/EI* diagram is *positive* and *upwards* if *M/EI* diagram is *negative*. If however, *M/EI* diagram is partly positive and partly negative, the elastic weight or the loading will act downward for the positive portion of the diagram and upward for the negative portion of the *M/EI* diagram.

2. Slope of the real beam. The slope of the real beam will be *clockwise* if the S.F. is positive and anticlockwise is the S.F. is negative.

3. Deflection of the real beam. The deflection of the *real beam* will be *downward* (or positive) if the B.M. at the corresponding point of the conjugate beam is *positive* and *upward* if the B.M. is *negative.*

13.4. SIMPLY SUPPORTED BEAMS

Fig. 13.1 (*a*) shows a simply supported beam *AB*, called the *real beam*, with loading, along with the B.M. diagram. Fig. 13.1 (*b*) shows the *conjugate beam A′ B′* with the loading equal to the *M/EI* diagram of the real beam. Corresponding to a simply supported real beam, the conjugate beam will also be simply supported. This is so because the bending moment at the ends of the conjugate beam have to be zero so as to give zero deflection at the ends of the real beam. *Hence the conjugate beam of a simply supported real beam will also be simply supported at the ends.* The application of the conjugate beam method to simply supported beam is illustrated in Example 13.1 to 13.4.

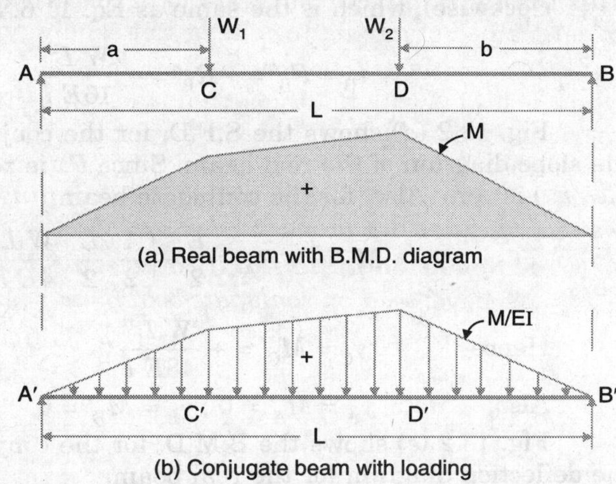

Fig. 13.1

Example 13.1. *Determine the slopes at the ends and the deflection at the midspan of a simply supported beam carrying a central point load W. Also, plot the slope and deflection diagrams for the beam.*

Solution. Fig. 13.2 (*a*) shows the real beam with the loading, while Fig. 13.2 (*b*) shows the B.M.D. having a maximum ordinate of *W L*/4 under *C*. Fig. 13.2 (*c*) shows the conjugate beam *A′C′B′* with loading (*i.e.,* elastic weight) having maximum ordinate of *WL*/4 *EI* at *C′*. Ends *A′* and *B′* of the conjugate beam will be simply supported. Since *M/EI* is *positive*, the loading on the conjugate beam acts downwards.

$$\text{Reaction } R_A' = \frac{1}{2}\left[\frac{1}{2} L \frac{W L}{4 E I}\right]$$

$$= \frac{W L^2}{16 E I} (\uparrow) = R_B' ;$$

$\therefore$ S.F. at $A' = F_A'$

$$= + R_A' = + \frac{W L^2}{16 E I}$$

But shear at any point of the conjugate beam is equal to slope of the real beam at the corresponding point.

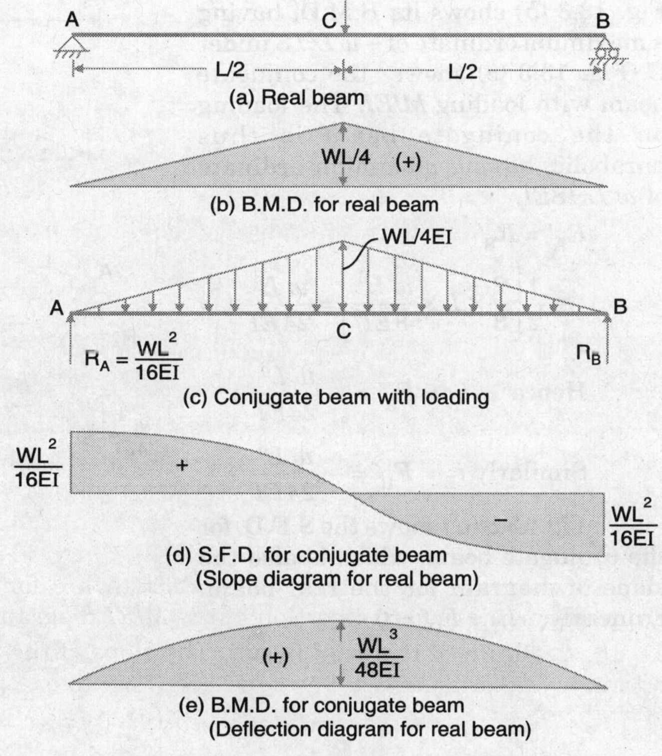

Fig. 13.2

$$\therefore \qquad i_A = F_A' = +\frac{W L^2}{16E I}$$

(*i.e.,* clockwise), which is the same as Eq. 11.6. Similarly,

$$i_B = F_B' = -R_B' = -\frac{W L^2}{16E I} \qquad (\text{i.e., anticlockwise})$$

Fig. 13.2 (*d*) shows the S.F.D. for the conjugate beam ; obviously this also represents the slope diagram of the real beam. Since F_C is zero, the slope at the centre of the real beam (*i.e.,* i_C) is zero. Also, for the conjugate beam,

$$M_C' = +R_A' \cdot \frac{L}{2} - \left(\frac{1}{2}\cdot\frac{L}{2}\cdot\frac{W L}{4E I}\right)\left(\frac{1}{3}\frac{L}{2}\right) = \frac{W L^2}{16E I}\cdot\frac{L}{2} - \frac{W L^3}{96E I} = +\frac{W L^3}{48E I}$$

Hence $\qquad y_C = M_C' = +\dfrac{W L^3}{48E I}$ $\qquad$ (*i.e.,* downward)

Also, $\qquad y_A = M_A' = 0 \; ; y_B = M_B' = 0.$

Fig. 13.2 (*e*) shows the B.M.D. for the conjugate beam; obviously, this also represents the deflection diagram for the real beam.

Example 13.2. *Determine the slopes at the ends and the deflection at the mid-span of a simply supported beam carrying U.D.L. w throughout its length. Also, plot the slope and deflection diagrams for the beam.*

Solution. Fig. 13.3 (*a*) shows the real beam *ACB* with loading, while Fig. 13.3 (*b*) shows its B.M.D. having a maximum ordinate of $+ w L^2/8$ under *C*. Fig. 13.3 (*c*) shows the conjugate beam with loading *M/EI*. The loading on the conjugate beam is thus parabolic, having maximum ordinate of $w L^2/8EI$.

$$R_A' = R_B'$$

$$= \frac{1}{2}\left[\frac{2}{3} \times L \times \frac{w L^2}{8EI}\right] = \frac{w L^3}{24EI} \; (\uparrow)$$

Hence $\qquad i_A = F_A' = +\dfrac{w L^3}{24EI}$;

Similarly $i_B = F_B' = -\dfrac{w L^3}{24EI}$

Fig. 13.3 (*d*) shows the S.F.D. for the conjugate beam, which is also the slope of diagram for the real beam. Evidently, $\; i_C = F_C' = 0$

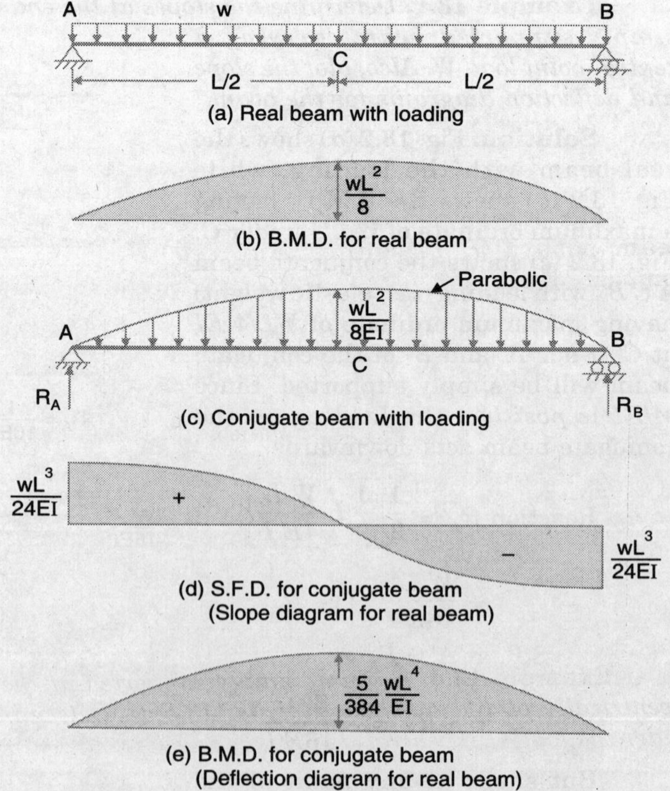

Fig. 13.3

Again $$M_C' = R_A' \cdot \frac{L}{2} - \left[\frac{2}{3} \times \frac{L}{2} \times \frac{wL^2}{8EI}\left(\frac{3}{8}\cdot\frac{L}{2}\right)\right] = \frac{wL^3}{24EI}\frac{L}{2} - \frac{wL^4}{128} = +\frac{5}{384}\frac{wL^4}{EI}$$

$$\therefore \quad y_C = M_C' = +\frac{5}{384}\frac{wL^4}{EI} \quad \textit{(i.e., downwards)}$$

Fig. 13.3 (*e*) shows the B.M.D. for the conjugate beam ; evidently, this is also the deflection diagram for the real beam. Also $y_A = M_A' = 0$

and $y_B = M_B' = 0$

Example 13.3. *Calculate the slope at the ends and deflection at the centre of a beam loaded with a couple μ at the centre.*

Solution. Fig. 13.4 (*a*) shows the real beam with the loading, while Fig. 13.4 (*b*) shows its B.M.D.

Evidently, $R_A = \mu/L\ (\downarrow)$

and $R_B = \mu/L\ (\uparrow)$

$\therefore \quad M_A = 0$;

$$M_C\ \text{(left)} = -\frac{\mu}{L} \times \frac{L}{2} = -\frac{\mu}{2};$$

$$M_C\ \text{(right)} = -\frac{\mu}{L} \times \frac{L}{2} + \mu = +\frac{\mu}{2}$$

$$M_B = 0$$

Fig. 13.4 (*c*) shows the conjugate beam with *M*/*E I* loading. For the conjugate beam

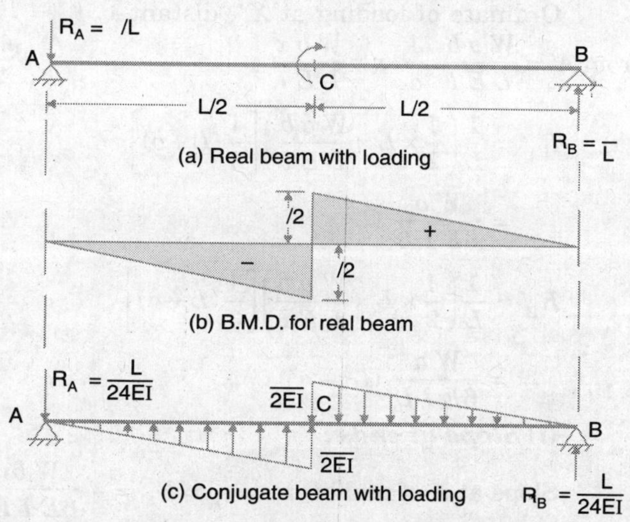

Fig. 13.4

$$R_A' = \frac{1}{L}\left[\frac{1}{2} \times \frac{L}{2} \times \frac{\mu}{2EI}\left(\frac{L}{2} + \frac{1}{3}\frac{L}{2}\right) - \frac{1}{2} \times \frac{L}{2} \times \frac{\mu}{2EI}\left(\frac{2}{3}\cdot\frac{L}{2}\right)\right] = \frac{\mu L}{24EI}\ (\downarrow)$$

$$R_B' = \frac{\mu L}{24EI}\ (\uparrow)$$

$i_A = F_A' = -\dfrac{\mu L}{24EI}$ (*i.e.*, anticlockwise wise); $i_B = F_B' = -\dfrac{\mu L}{24EI}$ (*i.e.*, anticlockwise wise)

Also, $$M_C' = -R_A' \cdot \frac{L}{2} + \left(\frac{1}{2} \times \frac{L}{2} \times \frac{\mu}{2EI} \times \frac{1}{3}\frac{L}{2}\right) = -\frac{\mu L}{24EI}\cdot\frac{L}{2} + \frac{\mu L^2}{48EI} = 0$$

$\therefore \quad y_C = M_C' = 0.$

Example 13.4. *A beam, simply supported at the ends, is subjected to a point load W, eccentrically placed as shown in Fig. 13.5(a). Determine (i) slopes at the ends (ii) deflection under the load (iii) central deflection, and (iv) maximum deflection and its location.*

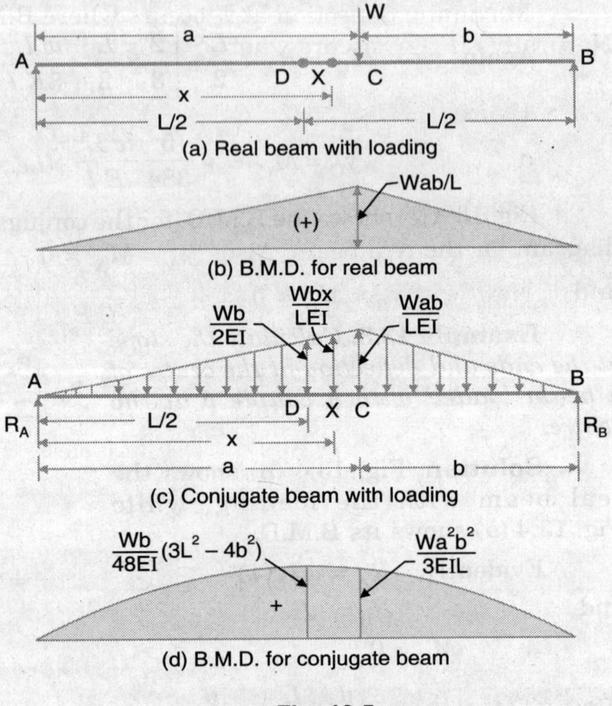

(a) Real beam with loading
(b) B.M.D. for real beam
(c) Conjugate beam with loading
(d) B.M.D. for conjugate beam

Fig. 13.5

Solution. Fig. 13.5 (*a*) shows the real beam with loading while Fig. 13.5 (*b*) shows its B.M.D. having a maximum ordinate of *Wab/L* under the point load. Fig. 13.5 (*c*) shows the conjugate beam with *M/EI* loading.

Ordinate of loading at $C' = W\,a\,b/LEI$.

Ordinate of loading at D'

$$= \frac{W\,a\,b}{L\,E\,I} \cdot \frac{1}{a} \cdot \frac{L}{2} = \frac{W\,b}{2E\,I}$$

Ordinate of loading at X', distant x from $A' = \dfrac{W\,a\,b}{L\,E\,I} \cdot \dfrac{1}{a} \cdot x = \dfrac{W\,b\,x}{L\,E\,I}$

$$R_A' = \frac{1}{L}\left(\frac{1}{2} \times L \times \frac{W\,a\,b}{L\,E\,I}\right)\left[\frac{1}{3}(L+b)\right]$$

$$= \frac{W\,b}{6E\,I\,L}\,(L^2 - b^2)$$

$$R_B' = \frac{1}{L}\left(\frac{1}{2} \times L \times \frac{W\,a\,b}{L\,E\,I}\right)\left[\frac{1}{3}(L+a)\right]$$

$$= \frac{W\,a}{6E\,I\,L}\,(L^2 - a^2)$$

(i) Slope at ends :

Slope at A of real beam $= i_A = F_A' = +\dfrac{W\,b}{6E\,I\,L}\,(L^2 - b^2)$ (*i.e.*, clockwise)

Slope at B of real beam $= i_B = F_B' = -\dfrac{W\,a}{6E\,I\,L}\,(L^2 - a^2)$ (*i.e.*, anticlockwise)

(ii) Deflection under the load : We have $y_C = M_C'$

Now, $$M_C' = R_A' \,.\, a - \frac{1}{2}\,a \times \frac{W\,ab}{L\,E\,I}\left(\frac{1}{3}\,a\right) = \frac{W\,a\,b}{6E\,I\,L}\,(L^2 - b^2) - \frac{W\,a\,b}{6E\,I\,L}\,(a^2)$$

$$= \frac{W\,a\,b}{6E\,I\,L}\,(L^2 - b^2 - a^2) = \frac{W\,a^2\,b^2}{3E\,I\,L}$$

$$\therefore \quad y_C = M_C' = \frac{W\,a^2\,b^2}{3E\,I\,L}$$

(iii) Central deflection : We have $y_D = M_D'$

Now $$M_D' = R_A' \,.\, \frac{L}{2} - \frac{1}{2} \cdot \frac{L}{2} \cdot \frac{W\,b}{2E\,I}\left(\frac{1}{3} \cdot \frac{L}{2}\right) = \frac{W\,b}{6E\,I\,L}\,(L^2 - b^2) \,.\, \frac{L}{2} - \frac{W\,b\,L^2}{48\,E\,I}$$

$$= \frac{W\,b}{6E\,I\,L} \cdot \frac{L}{2}\left[L^2 - b^2 - \frac{L^2}{4}\right] = \frac{W\,b}{48E\,I}\,(3L^2 - 4b^2)$$

(iv) Maximum deflection :

Let maximum deflection occur at point X, between D and C, distant x from A.

Maximum deflection at X occurs where B.M. is maximum at X' of the conjugate beam. Naturally $(M_x')_{max}$ occurs where S.F. is zero, *i.e.,* where $F_x' = 0$.

Hence $$F_x' = 0 = R_A' - \frac{1}{2}x\left(\frac{W\,b\,x}{L\,E\,I}\right) = \frac{W\,b}{6E\,I\,L}(L^2 - b^2) - \frac{W\,b\,x^2}{2E\,I\,L}$$

This gives $$x = \left(\frac{L^2 - b^2}{3}\right)^{1/2}$$

Now $$M_x' = R_A' \,.\, x - \frac{1}{2}x\left(\frac{W\,b\,x}{L\,E\,I}\right)\frac{x}{3} = \frac{W\,b\,x}{6E\,I\,L}(L^2 - b^2) - \frac{W\,b}{6E\,I\,L}\,x^3$$

$$\therefore \quad (M_x')_{max} = \frac{W\,b}{6E\,I\,L}(L^2 - b^2)\left(\frac{L^2 - b^2}{3}\right)^{1/2} - \frac{W\,b}{6E\,I\,L}\left(\frac{L^2 - b^2}{3}\right)^{3/2}$$

$$= \frac{W\,b}{6E\,I\,L}\left(\frac{L^2 - b^2}{3}\right)^{1/2}\left[L^2 - b^2 - \left(\frac{L^2 - b^2}{3}\right)\right]$$

$$\therefore \quad y_{max} = (M_x')_{max} = \frac{W\,b}{3E\,I\,L}\left(\frac{L^2 - b^2}{3}\right)^{3/2} = \frac{W\,b\,(L^2 - b^2)^{3/2}}{9\sqrt{3}\,E\,I\,L}$$

13.5. CANTILEVERS

In the case of cantilevers, the B.M. diagram is negative for the usual downward loading, as shown in Fig. 13.6 (*b*). Hence the loading on the conjugate beam will be in the upward direction, as shown in Fig. 13.6 (*c*). Also, since the S.F. at any point in the conjugate beam gives the slope of the corresponding point of the real beam, the conjugate beam will have free end at A', corresponding to the fixed end A of the real beam, so as to give $y_A = M_A' = 0$. Again, the end B of the real beam deflects downward by some finite amount. To satisfy this condition, end B' of the conjugate beam should have some bending moment. This is possible only if end B' of the conjugate beam is *fixed* instead of being free. *Thus in the case of a cantilever beam, the conjugate beam is also a cantilever, but with fixed support on the opposite end.*

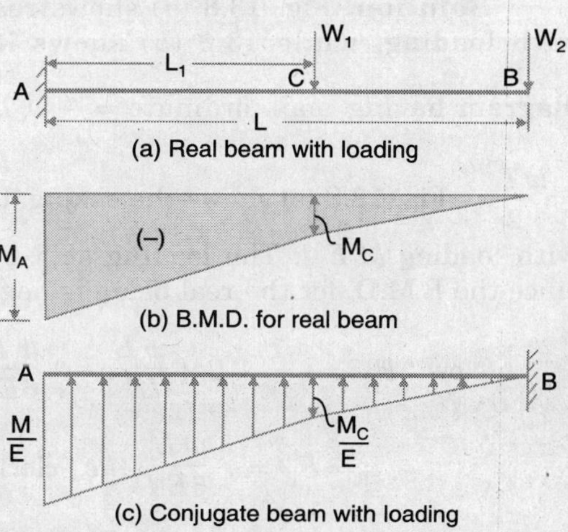

Fig. 13.6

Example 13.5 to 13.8 illustrate the method of application.

Example 13.5. *A cantilever AB is subjected to point load W at the free end. Determine the slope and deflection of the free end.*

Solution. Fig. 13.7 (*a*) shows the real beam with loading while Fig. 13.7 (*b*) shows its B.M. diagram. Fig. 13.7 (*c*) shows the conjugate beam with loading *M/E I*. Since B.M.D. is negative, the loading on the conjugate beam will act upward. The conjugate beam will have end *A′* free (corresponding to the fixed end *A*) while end *B′* will be fixed, corresponding to the free end *A* of the real beam.

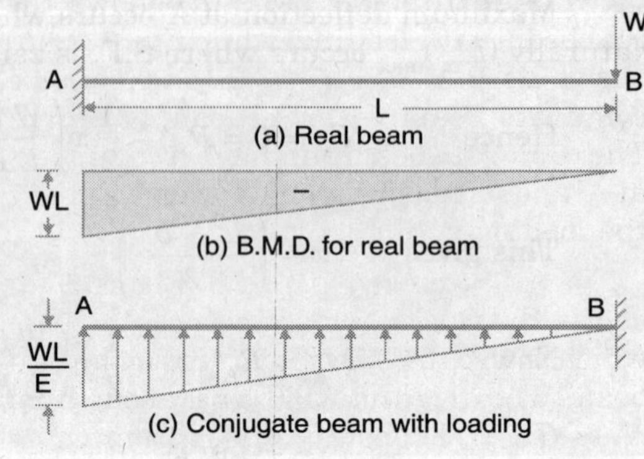

Fig. 13.7

Now $F_B' = +\frac{1}{2} \times L \times \frac{W L}{E I} = +\frac{W L^2}{2E I}$

$$i_B = F_B' = +\frac{W L^2}{2E I} \text{ (i.e., clockwise)}$$

Again $M_B' = +\frac{1}{2} \times L \times \frac{W L}{E I}\left(\frac{2}{3} L\right) = +\frac{W L^3}{3E I}$

$$\therefore \quad y_B = M_B' = +\frac{W L^3}{3E I} \text{ (i.e., downwards).}$$

Example 13.6. *A cantilever AB is subjected to uniformly distributed load w throughout its length. Determine the slope and deflection at the free end.*

Solution. Fig. 13.8 (*a*) shows real beam with loading, while 13.8 (*b*) shows its B.M. diagram having max. ordinate $= -\, w\, L \,.\, \frac{L}{2} = -\frac{w L^2}{2}$. Fig. 13.8 (*c*) shows the conjugate beam, with loading *M/E I*. The loading acts upwards, since the B.M.D. for the real beam is negative.

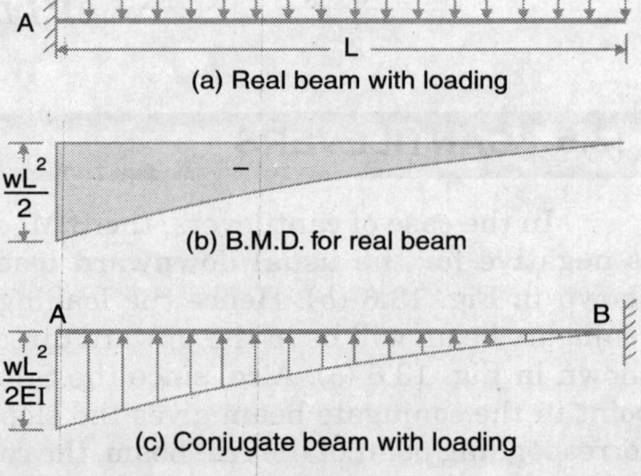

Fig. 13.8

Now $F_B' = +\frac{1}{3} \times L \times \frac{w L^2}{2E I} = +\frac{w L^3}{6E I}$

$$\therefore \quad i_B = F_B' = +\frac{w L^3}{6E I} \text{ (i.e., clockwise)}$$

Also, $M_B' = +\frac{1}{3} \times L \times \frac{w L^2}{2E I}\left(\frac{3}{4} L\right) = +\frac{w L^4}{8E I} \quad \therefore \quad y_B = M_B' = +\frac{w L^4}{8E I}$ (*i.e.*, downwards)

Example 13.7. *A cantilever AB is loaded as shown in Fig. 13.9 (a). Draw the slope and deflection diagram for the cantilever. Hence find the slope and deflection of (a) point under the loading, and (b) free end.*

Solution. Fig. 13.9 (*a*) shows the real beam with loading, while Fig. 13.9 (*b*) shows its B.M.D. Fig. 13.9 (*c*) shows the conjugate beam with the loading *M/E I*. The conjugate beam will have free support at *A′* and fixed support at *B′* and will have the loading acting upwards.

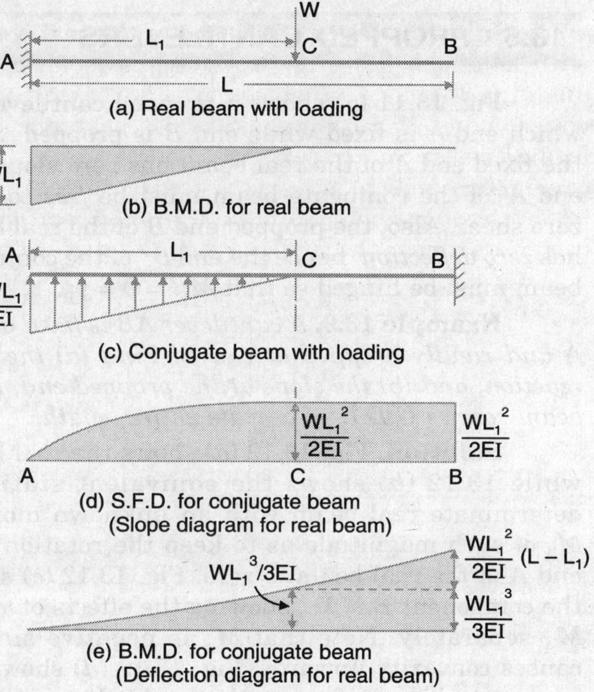

(a) Real beam with loading
(b) B.M.D. for real beam
(c) Conjugate beam with loading
(d) S.F.D. for conjugate beam (Slope diagram for real beam)
(e) B.M.D. for conjugate beam (Deflection diagram for real beam)

Fig. 13.9

(a) Slopes at C and B. Fig. 13.9 (*d*) shows the S.F.D. for the conjugate beam which is also the slope diagram for the real beam. The diagram will be parabolic from *A′* to *C′*, and constant shear from *C′* to *B′*. This shows that $i_C = i_B$.

Now $F_C' = F_B'$

$$= \frac{1}{2} \times L_1 \times \frac{W L_1}{E I} = \frac{W L_1^2}{2E I}$$

$$\therefore \quad i_C = i_B = F_C' = F_B'$$

$$= \frac{W L_1^2}{2 E I} \text{ (i.e., clockwise)}$$

(b) Deflection at C and B. Fig. 13.9 (*e*) shows the B.M.D. for the conjugate beam, which is also the deflection diagram for the real beam.

$$y_C = M_C' = + \frac{1}{2} \cdot \frac{W L_1}{E I} . L_1 . \frac{2}{3} L_1 = + \frac{W L_1^3}{3E I} \text{ (i.e. downwards)}$$

$$y_B = M_B' = + \frac{1}{2} . \frac{W L_1}{E I} L_1 \left[L - L_1 + \frac{2L_1}{3} \right] = \frac{W L_1^3}{3E I} + \frac{W L_1^2}{2E I} (L - L_1) \text{ (i.e., downwards)}$$

Example 13.8. *A cantilever AB is subjected to a moment μ at free end. Find slope and deflection of the free end.*

Solution. Fig. 13.10 (*a*) shows the real beam with loading, while Fig. 13.10 (*b*) shows its B.M.D. which is positive having a constant ordinate of μ. Fig. 13.10 (*c*) shows the conjugate beam, with the loading equal to *M/E I*. The conjugate beam will have end *A′* free and end *B′* fixed.

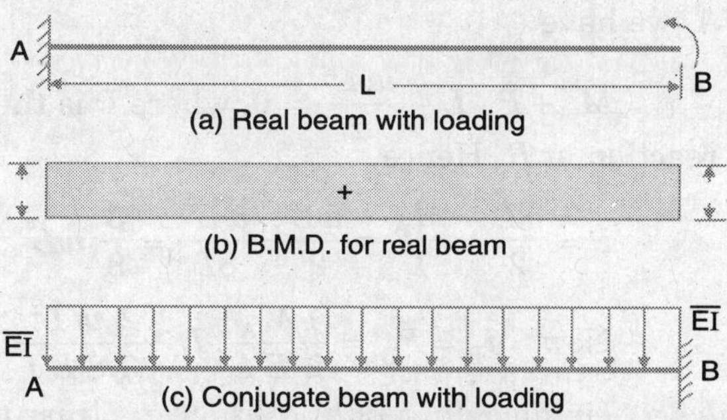

(a) Real beam with loading
(b) B.M.D. for real beam
(c) Conjugate beam with loading

Fig. 13.10

Now $i_B = F_B' = - \frac{\mu}{E I} . L = - \frac{\mu L}{E I}$

(*i.e.*, anticlockwise)

$$y_B = M_B' = - \frac{\mu}{E I} . L . \frac{L}{2} = - \frac{\mu L^2}{2E I}$$

(*i.e.*, upwards).

13.6. PROPPED CANTILEVERS

Fig. 13.11 (*a*) shows a propped cantilever, in which end *A* is fixed while end *B* is propped. Since the fixed end *A* of the real beam has zero slope, the end *A*′ of the conjugate beam must be *free* to have zero shear. Also, the propped end *B* of the real beam has *zero deflection,* hence the end *B* ′ of the conjugate beam must be hinged so that $M_B{}' = 0 = y_B$.

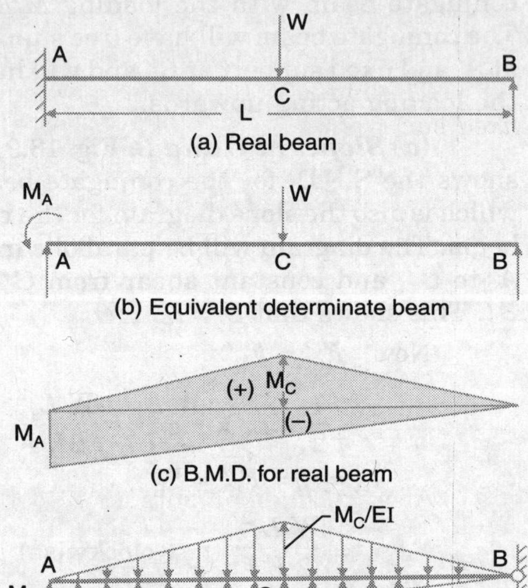

Fig. 13.11

Example 13.9. *A cantilever AB is fixed at end A and rigidly propped at end B. Find (a) the prop reaction, and (b) the slope at the propped end, if the beam carries U.D.L. w over its entire length.*

Solution. Fig. 13.12 (*a*) shows the real beam while 13.12 (*b*) shows the equivalent statically determinate real beam with an unknown moment M_A of such magnitude as to keep the rotation i_A of end *A* of the real beam to zero. Fig. 13.12 (*c*) shows the component B.M.D., showing the effects of *w* and M_A separately. Note that M_A is negative since it causes convexity upwards. Fig. 13.12 (*d*) shows the conjugate beam along with the loading *M*/*EI* in appropriate directions. The end *B* of the real beam does not deflect ; hence end *B* ′ of the conjugate beam should have zero B.M., which is possible by providing at hinge at end *B* ′ of the conjugate beam. For the conjugate beam,

$$M_B{}' = 0 = -\frac{2}{3}\cdot\frac{wL^2}{8EI}\times L\times\frac{L}{2}+\frac{1}{2}\frac{M_A\,.\,L}{EI}\cdot\frac{2}{3}L$$

This gives $M_A = \dfrac{wL^2}{8}$

From Fig. 13.12 (*b*), taking moments about *A*, we have

$M_A + P\,.\,L - \dfrac{wL^2}{2} = 0$, where *P* is the prop reaction at *B*. Hence,

$$P = \frac{wL}{2}-\frac{M_A}{L}=\frac{wL}{2}-\frac{wL^2}{8L}=\mathbf{\frac{3}{8}\,wL}$$

Now $\quad i_B = F_B{}' = \dfrac{1}{2}\dfrac{M_A}{EI}\,.\,L - \dfrac{2}{3}\dfrac{wL^2}{8EI}L$

$$= \frac{wL^3}{16EI}-\frac{wL^3}{12EI}=-\mathbf{\frac{wL^3}{48EI}}$$

(*i.e.,* anticlockwise)

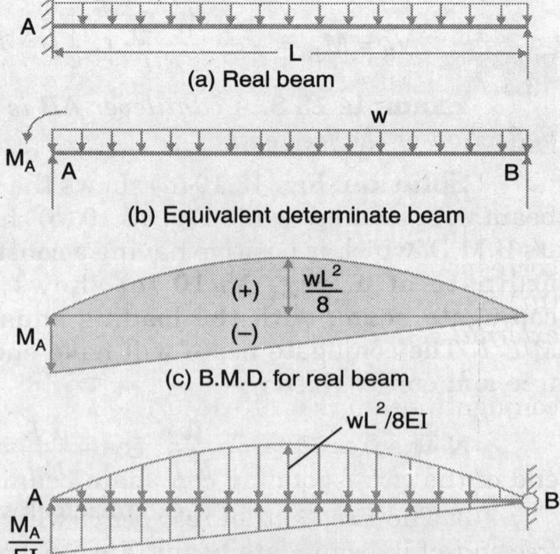

Fig. 13.12

Example 13.10. *If the prop at B [Fig. 13.13(a)] sinks by λ under unit load, determine the prop reaction.*

Solution. Let P = prop reaction.

The condition equation is $y_B = P \cdot \lambda$

The end B' of the conjugate beam is *partially fixed* such that

$$M_B' = y_B = P\lambda \quad \ldots(i)$$

But
$$M_B' = \frac{1}{2}\frac{M_A}{EI} L \cdot \frac{2}{3} L - \frac{2}{3}\frac{wL^2}{8EI} \cdot L \cdot \frac{L}{2}$$

$$= \frac{M_A L^2}{3EI} - \frac{wL^4}{24EI} \quad \ldots(ii)$$

(a) Real beam

(b) Equivalent determinate beam

(c) Conjugate beam with loading

Fig. 13.13

Substituting in (i), we get

$$\frac{M_A L^2}{3EI} - \frac{wL^4}{24EI} = P\lambda \quad \ldots(a)$$

Also, taking moments about end A of the real beam [Fig. 13.13 (b)]

$$M_A + PL - \frac{wL^2}{2} = 0 \quad \text{or} \quad M_A = \frac{wL^2}{2} - PL \quad \ldots(b)$$

Substituting in (ii), the value of M_A, we get

$$\left(\frac{wL^2}{2} - PL\right)\frac{L^2}{3EI} - \frac{wL^4}{24EI} = P\lambda \quad \text{or} \quad P\left(\lambda + \frac{L^3}{3EI}\right) = \frac{wL^4}{6EI} - \frac{wL^4}{24EI} = \frac{wL^4}{8EI}$$

$$\therefore \quad P = \frac{3}{8}\frac{\mathbf{wL^4}}{\mathbf{L^3 + 3EI\lambda}}$$

13.7. END CONDITIONS AND INTERNAL CONDITIONS OF A CONJUGATE BEAM

(a) External Conditions

In the previous three articles, we have considered the end conditions of conjugate beam corresponding to three types of ends of a real beam : (i) simple or hinged end (ii) free end, and (iii) fixed end. In that process, we have drawn the following conclusions with regard to the *external conditions :*

1. The *hinged end* of a real beam, where B.M. is zero, will result in a *hinged end* of the conjugate beam [Fig. 13.14 *(a)*].

2. The *free end* of a real beam where both S.F. and B.M. are zero will result in the *fixed* end of the corresponding conjugate beam [Fig. 13.14 *(b)*].

3. The *fixed end* of *real beam* where both slope and deflection are zero will give rise to *free end* of its conjugate beam [Fig. 13.14 *(c)*].

4. A *simple support* (or knife) edge support or roller support of real beam, where B.M. is zero, will result in a roller support (or simple support) of the corresponding conjugate beam [Fig. 13.14 *(d)*].

(b) Internal Conditions

Let us now take the internal conditions of a real beam.

1. *Internal hinge.* Consider an internal hinge at intermediate point C of the real beam, where B.M. is zero [Fig. 13.14 (*e*)]. Also, the deflection to the left of the hinge is equal to the deflection to the right of the hinge. In addition to this, the shear to the left of the hinge is equal to the shear to the right of the hinge. These conditions are represented by the following equations for the real beam :

$$y_L = y_R \quad ...(i)$$

$$F_L = F_R \quad ...(ii)$$

and $$M_L = M_R = 0 \quad ...(iii)$$

These conditions are satisfied by providing a *roller support* or *interior pin* at C' of the conjugate beam, where the following equations are applicable.

$$M_L' = M_R' \quad ...(i)$$

$$i_L' = i_R' \quad ...(ii)$$

and $$y_L' = y_R' = 0 \quad ...(iii)$$

2. *Interior pin or knife edge* [Fig. 13.14(*f*)]. For such an internal condition of a real beam, (*i*) the deflection is zero, (*ii*) the slope at the left is equal to the slope at the right and (*iii*) the moment at the left is equal to the moment at the right. These are represented by the following equations for the *real beam.*

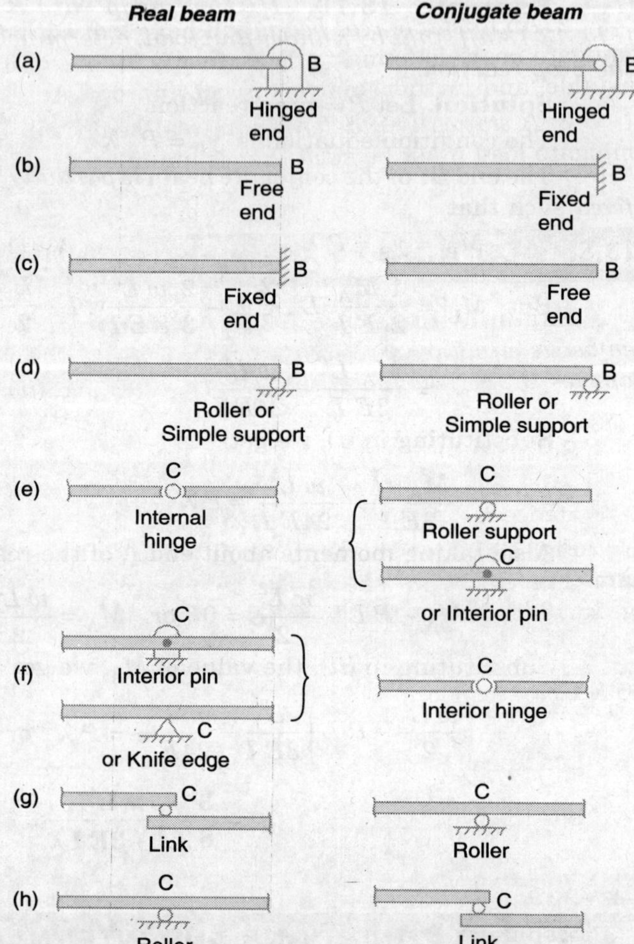

Fig. 13.14. External and Internal Conditions

$$y_L = y_R = 0 \quad ...(i)$$

$$i_L = i_R \quad ...(ii)$$

$$M_L = M_R \quad ...(iii)$$

The conjugate beam must have the following conditions for the corresponding point C'.

$$M_L' = M_R' = 0 \quad ...(i) \; ; \; F_L' = F_R' \quad ...(ii) \quad \text{and} \quad y_L' = y_R' \quad ...(iii)$$

These conditions are satisfied by providing an *interior hinge at* C'.

3. *Internal Link* [Fig. 13.14 (*g*)]. For an internal link at point C in a real beam, there will be an roller at the corresponding point C' of the conjugate beam.

4. *Internal roller* [Fig. 13.14 (*h*)]. For an internal roller at point C in a real beam, there will be an internal link at corresponding point C' of the conjugate beam.

Important points. In addition to the above support conditions and internal conditions, the following points should be carefully noted.

1. A *stable* and *statically determinate real beam* will have a conjugate beam which is also stable and statically determinate.

2. An *unstable real beam* will have *statically indeterminate* conjugate beam. Hence if a conjugate beam is found to be statically indeterminate, it is concluded that the real beam is unstable, and further analysis is not appropriate.

3. A *statically indeterminate* real beam will have *unstable* conjugate beam. Hence its conjugate load must be such that is maintains equilibrium.

13.8. ADDITIONAL EXAMPLES USEFUL FOR COMPETITIVE EXAMINATIONS

Example 13.11. *A cantilever of length L carries a point load W at its free end. The member is circular in section having diameter D for a distance L/2 from the fixed end and a diameter D/2 for the remaining length. Find the slope and deflection at point C and B.*

Solution. For the portion AC, $I_{AC} = \frac{\pi D^4}{64}$. Similarly, $I_{CB} = \frac{\pi}{64}\left(\frac{D}{2}\right)^4 = \frac{\pi D^4}{64 \times 16}$

Hence if $I_{CB} = I$, $I_{AC} = 16\,I$

Fig. 13.15 (a) shows the real beam, while Fig. 13.15 (b) shows the BMD for the real beam. Fig. 13.15 (c) shows the conjugate beam, with the end A' free and end B' fixed; Fig. 13.15 (d) shows loading M/EI for the conjugate beam. Thus, loading at end $A' = (WL)\left(\frac{1}{16EI}\right) = \frac{WL}{16EI}$;

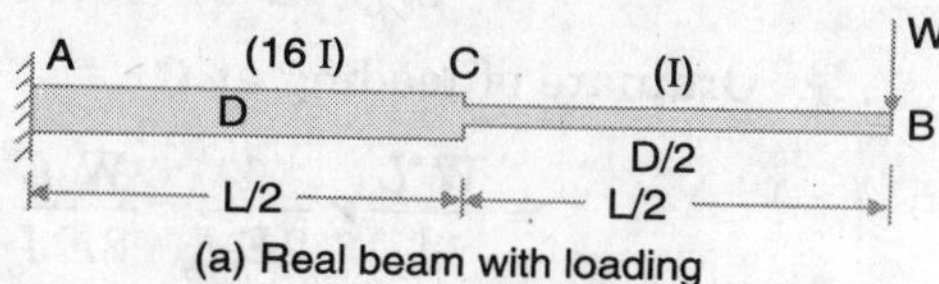

(a) Real beam with loading

WL (–) WL/2

(b) B.M.D. for real beam

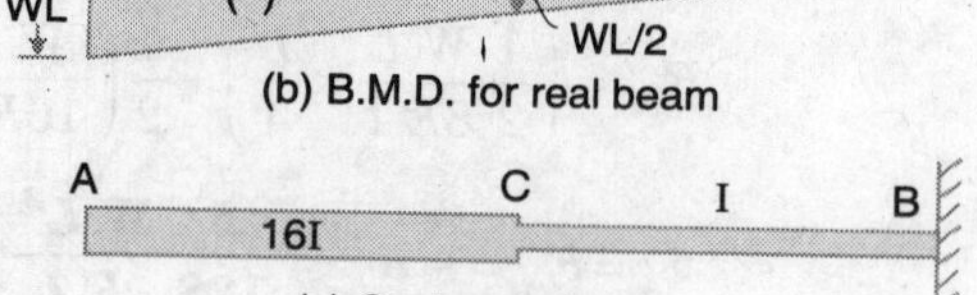

(c) Conjugate beam

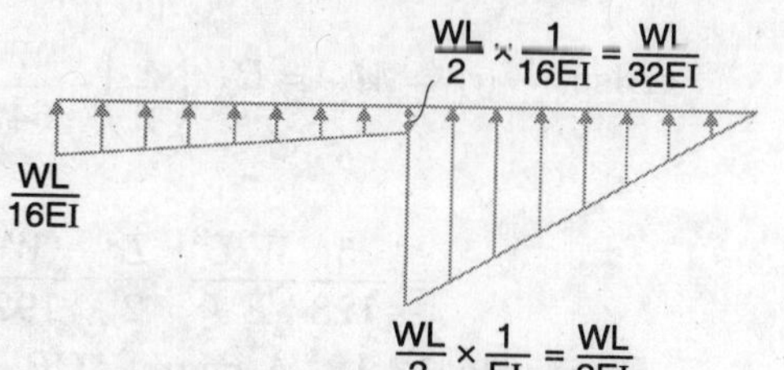

(d) Loading on conjugate beam

Fig. 13.15

$$\text{Loading at } C' \text{(left)} = \frac{WL}{2}\left(\frac{1}{16EI}\right) = \frac{WL}{32EI}$$

Loading at C' (right)

$$= \frac{WL}{2}\left(\frac{1}{EI}\right) = \frac{WL}{2EI}$$

Loading at $B' = 0$

(a) Slopes at C and B

$$i_C = F_C' = +\frac{1}{2}\left(\frac{WL}{16EI} + \frac{WL}{32EI}\right)\frac{L}{2}$$

$$= +\frac{\mathbf{3\,WL^2}}{\mathbf{128\,EI}} \quad (i.e., \text{ clockwise})$$

$$i_B = F_B'$$

$$= +\frac{1}{2}\left(\frac{WL}{16EI} + \frac{WL}{32EI}\right)\frac{L}{2} + \frac{1}{2}\left(\frac{WL}{2EI}\right) \times \frac{L}{2}$$

$$= \frac{3WL^2}{128EI} + \frac{WL^2}{8EI} = \frac{\mathbf{19}}{\mathbf{128}}\,\frac{\mathbf{WL^2}}{\mathbf{EI}} \quad \textbf{(clockwise)}$$

(b) Deflections at C and B

$$y_C = M_C' = \frac{1}{2}\left(\frac{WL}{16EI} + \frac{WL}{32EI}\right)\frac{L}{2}\left[\frac{1}{3}\,\frac{L}{2}\,\frac{\frac{2WL}{16EI} + \frac{WL}{32EI}}{\frac{WL}{16EI} + \frac{WL}{32EI}}\right] = \frac{\mathbf{5}}{\mathbf{768}}\,\frac{\mathbf{WL^3}}{\mathbf{EI}}$$

$$y_B = M_B{}' = \frac{3}{128}\frac{W\,L^2}{E\,I}\left[\frac{L}{2}+\frac{5L}{18}\right]+\frac{1}{2}\left(\frac{W\,L}{2\,E\,I}\right)\frac{L}{2}\left(\frac{2}{3}\frac{L}{2}\right) = \mathbf{\frac{23}{384}\frac{W\,L^3}{E\,I}}.$$

Note. The above problem has also been solved by double integration method in Example 11.41 and by area moment method in example 12.12.

Example 13.12. *Using conjugate beam method, find slopes at the ends and central deflection for a simply supported beam shown in Fig. 13.16 (a).*

Solution. For the real beam,

$$M_C = \frac{W\,L}{4};\ M_D = \frac{W}{2}\cdot\frac{L}{2} = \frac{W\,L}{8} = M_E.\ \text{Also, } M_A = M_B = 0$$

Fig. 13.16 (*b*) shows the conjugate beam with its loading, *M/EI*.

Ordinate of loading at D' (left)

$$= \frac{W\,L}{8}\cdot\frac{1}{E\,I} = \frac{W\,L}{8E\,I}$$

Ordinate of loading at D' (right)

$$= \frac{W\,L}{8}\cdot\frac{1}{2E\,I} = \frac{W\,L}{16E\,I}$$

Ordinate of loading at C'

$$= \frac{W\,L}{4}\times\frac{1}{2E\,I} = \frac{W\,L}{8E\,I}$$

(a) Real beam with loading

(b) Conjugate beam with loading

Fig. 13.16

For the conjugate beam,

$$R_A{}' = \left(\frac{1}{2}\frac{W\,L}{8E\,I}\times\frac{L}{4}\right)+\frac{1}{2}\left(\frac{W\,L}{16E\,I}+\frac{W\,L}{8E\,I}\right)\frac{L}{4} = \frac{W\,L^2}{64E\,I}+\frac{3W\,L^2}{128E\,I} = \frac{5}{128}\frac{W\,L^2}{E\,I}$$

$$i_A = F_A{}' = +R_A{}' = \mathbf{\frac{5}{128}\frac{W\,L^2}{E\,I}}\ \text{(clockwise)};\ i_B = F_B{}' = -\mathbf{\frac{5}{128}\frac{W\,L^2}{E\,I}}\ \text{(\textit{i.e.}, anticlockwise)}$$

Also,
$$y_C = M_C{}' = R_A{}'\left(\frac{L}{2}\right)-\frac{W\,L^2}{64E\,I}\left(\frac{L}{4}+\frac{1}{3}\frac{L}{4}\right)-\frac{3}{128}\frac{W\,L^2}{E\,I}\left[\frac{1}{3}\frac{L}{4}\,\frac{\frac{2W\,L}{16E\,I}+\frac{W\,I}{8E\,I}}{\frac{W\,L}{16E\,I}+\frac{W\,L}{8E\,I}}\right]$$

$$= \frac{5}{128}\frac{W\,L^2}{E\,I}\cdot\frac{L}{2}-\frac{W\,L^3}{192E\,I}-\frac{1}{128\times 3}\frac{W\,L^3}{E\,I} = \mathbf{\frac{3}{256}\frac{W\,L^3}{E\,I}}.$$

Example 13.13. *A beam ACB, simply supported at the ends, has moment of inertia 4I for the length AC and I for the length CB, and is loaded with point load W at C. Determine (i) slope at end A, (ii) deflection at mid-span, and (iii) maximum deflection. Compute the numerical values taking W = 8 kN, a = 2.5 m, I = 5000 cm⁴ and E = 2 × 10⁵ N/mm².*

Solution. Fig. 13.17 (*a*) shows the real beam *ABC*, while Fig. 13.17 (*b*) shows the B.M.D. for the real beam, with salient values. Fig. 13.17 (*c*) shows the conjugate beam $A'\,C'\,B'$ with the loading diagram (*M/E I* diagram). For the real beam,

$$R_A = \frac{Wa}{5a} = \frac{W}{5}\ (\uparrow);\quad R_B = \frac{4Wa}{5a} = \frac{4}{5}W\ (\uparrow),\quad M_C = \frac{W(4a)(a)}{5a} = \frac{4}{5}Wa$$

$\therefore$ Ordinate of M/EI diagram just to the left of $C' = \frac{4}{5}Wa\left(\frac{1}{4EI}\right) = \frac{Wa}{5EI}$

Ordinate of M/EI diagram just to the right of $C' = \frac{4}{5}Wa\left(\frac{1}{EI}\right) = \frac{4}{5}\frac{Wa}{EI}$

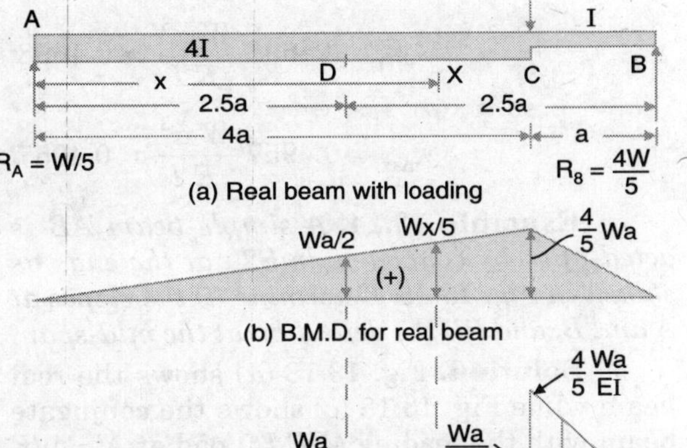

Fig. 13.17

Also, $M_D = R_A\,(2.5\,a)$

$$= \frac{W}{5} \times 2.5\,a = \frac{Wa}{2}$$

$\therefore$ Ordinate of M/EI diagram at D'

$$= \frac{Wa}{2} \times \frac{1}{4EI} = \frac{Wa}{8EI}$$

Similarly, at X, distant x from A,

$$M_x = \frac{Wx}{5}.$$

$\therefore$ Ordinate of M/EI diagram at X'

$$= \frac{Wx}{5} \times \frac{1}{4EI} = \frac{Wx}{20EI}$$

$$R_A' = \frac{1}{5a}\left[\left\{\frac{1}{2}\times a \times \frac{4Wa}{5EI}\left(\frac{2}{3}a\right)\right\} + \left\{\frac{1}{2}\times 4a \times \frac{Wa}{5EI}\left(a + \frac{1}{3}4a\right)\right\}\right] = \frac{6}{25}\frac{Wa^2}{EI}$$

$$R_B' = \frac{1}{5a}\left[\left\{\frac{1}{2}\times 4a \times \frac{Wa}{5EI}\left(\frac{2}{3}4a\right)\right\} + \left\{\frac{1}{2}\times a \times \frac{4Wa}{5EI} \times \left(4a + \frac{1}{3}a\right)\right\}\right] = \frac{14}{25}\frac{Wa^2}{EI}$$

$$\therefore \quad i_A = F_A' = +R_A' = + \mathbf{\frac{6Wa^2}{25EI}}$$

$$M_D' = \frac{6}{25}\frac{Wa^2}{EI} \times 2.5\,a - \left\{\frac{1}{2}\times 2.5a \times \frac{Wa}{8EI}\left(\frac{1}{3}\times 2.5\,a\right)\right\} = 0.4698\frac{Wa^3}{EI}$$

$$\therefore \quad y_D = M_D' = \mathbf{0.4698\frac{Wa^3}{EI}}.$$

Also, the maximum B.M. in the conjugate beam occurs at x', some where between D and C, where S.F. is zero.

$$\therefore \quad F_x' = 0 = R_A' - \frac{1}{2}\frac{Wx}{20EI}\,.\,x \quad \text{or} \quad \frac{Wx^2}{40EI} = R_A' = \frac{6}{25}\frac{Wa^2}{EI},$$

which gives $\quad x = \mathbf{3.098\,a}$

$$\therefore \quad (M_X')_{max} = \frac{6}{25}\frac{Wa^2}{EI}(3.098a) - \left[\frac{1}{2}\frac{W}{20EI}(3.098\,a)(3.098\,a)\left(\frac{1}{3}\times 3.098\,a\right)\right]$$

$$= 0.7436\frac{Wa^3}{EI} - 0.2479\frac{Wa^3}{EI} = 0.4957\frac{Wa^3}{EI}$$

$$\therefore \quad y_{max} = (M_X')_{max} = \mathbf{0.4957 \frac{W a^3}{E I}}$$

Substituting the *numerical values,* we get

$$i_A = \frac{6}{25}\frac{W a^2}{E I} = \frac{6}{25}\frac{8000\,(2.5)^2}{(2\times10^5)\,(5000\times10^4)}\,(1000)^2 = \mathbf{1.2 \times 10^{-3}\ radian}$$

$$y_D = 0.4698\,\frac{W a^3}{E I} = 0.4698 \times \frac{8000\,(2.5)^3}{(2\times10^5)\,(5000\times10^4)}\,(1000)^3 = \mathbf{5.87\ mm}\ (\downarrow)$$

$$y_{max} = 0.4957\,\frac{W a^3}{E I} = 0.4957 \times \frac{8000\,(2.5)^3}{(2\times10^5)\,(5000\times10^4)}\,(1000)^3 = \mathbf{6.20\ mm}\ (\downarrow)$$

Example 13.14. *A simple beam AB is acted upon by couples μ and 2μ at the ends as shown in Fig. 13.18. Determine (i) the slopes at A and B, and (ii) the deflection at the mid-span.*

Solution. Fig. 13.18 (*a*) shows the real beam while Fig. 13.18 (*c*) shows the conjugate beam with the loading *M/E I*. Load at *A′* = μ/*E I* while the load at *B′* = 2 μ/*E I*.

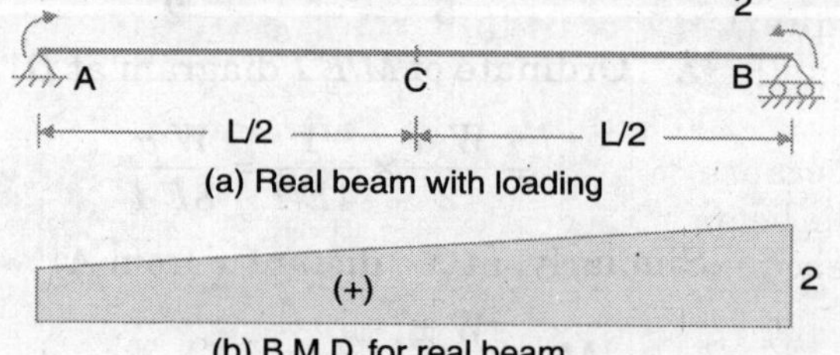

(a) Real beam with loading

(b) B.M.D. for real beam

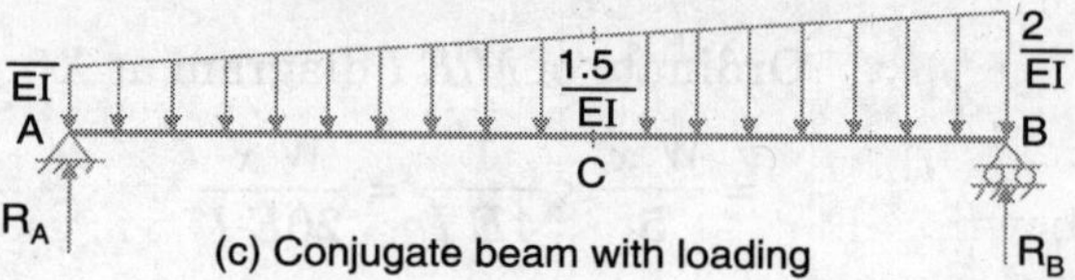

(c) Conjugate beam with loading

Fig. 13.18

For the conjugate beam,

$$\text{Reaction } R_A' = \frac{1}{L}\left[\frac{1}{2}\left(\frac{\mu}{EI}+\frac{2\mu}{EI}\right)\times L\left\{\frac{L}{3}\,\frac{\frac{2\mu}{EI}+\frac{2\mu}{EI}}{\frac{\mu}{EI}+\frac{2\mu}{EI}}\right\}\right] = \frac{1}{L}\left[\frac{3\mu L}{2EI}\times\frac{4L}{9}\right] = \frac{2\mu L}{3EI}$$

$$R_B' = \frac{1}{L}\left[\frac{1}{2}\left(\frac{\mu}{EI}+\frac{2\mu}{EI}\right)L\left\{\frac{L}{3}\,\frac{\frac{2\times2\mu}{EI}+\frac{\mu}{EI}}{\frac{2\mu}{EI}+\frac{\mu}{EI}}\right\}\right] = \frac{1}{L}\left[\frac{3\mu L}{2EI}\times\frac{5}{9}L\right] = \frac{5\mu L}{6EI}$$

Now $\quad i_A = F_A' = +R_A' = +\mathbf{\frac{2\mu L}{3EI}}$ and $i_B = F_B' = -R_B' = -\mathbf{\frac{5\mu L}{6EI}}$

$$y_C = M_C' = R_A' \times \frac{L}{2} - \left[\frac{1}{2}\left(\frac{\mu}{EI}+\frac{1.5\mu}{EI}\right)\frac{L}{2}\left\{\frac{1}{3}\cdot\frac{L}{2}\,\frac{\frac{2\mu}{EI}+\frac{1.5\mu}{EI}}{\frac{\mu}{EI}+\frac{1.5\mu}{EI}}\right\}\right]$$

$$= \frac{2\mu L}{3EI}\cdot\frac{L}{2} - \left[\frac{5}{8}\frac{\mu L}{EI}\times\frac{7}{30}L\right] = \frac{\mu L^2}{3EI} - \frac{7}{48}\frac{\mu L^2}{EI} = \mathbf{\frac{3}{16}\frac{\mu L^2}{EI}}.$$

Example 13.15. *A beam ABC is fixed at end A, and has a roller support at end C. It is also provided with an internal hinge at B. Determine the slope and deflection at the hinge B, when loaded with a point load W as shown in Fig. 13.19 (a).*

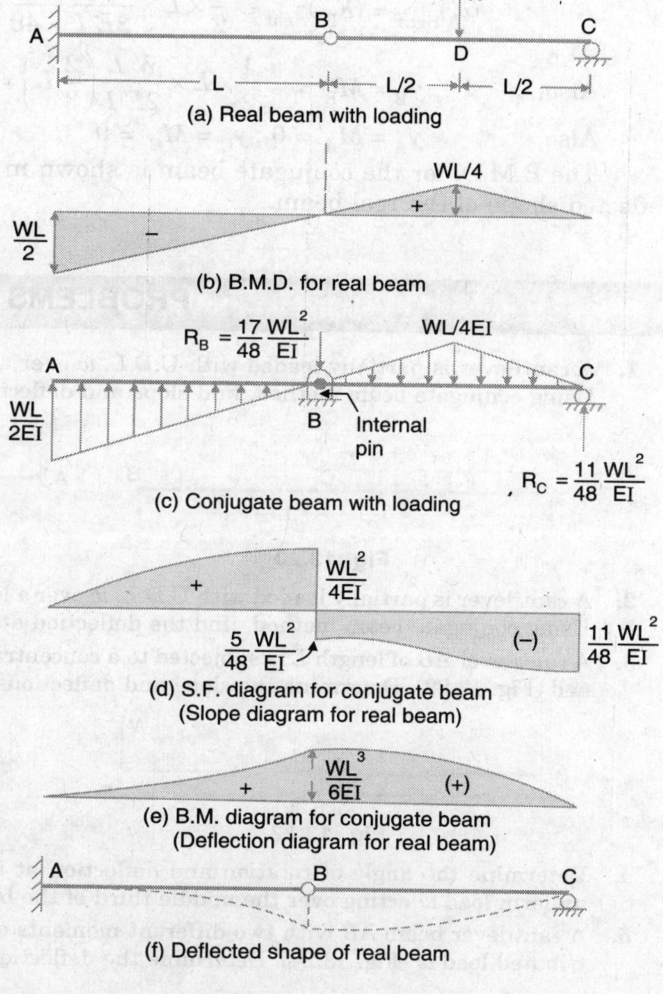

Fig. 13.19

Solution. Fig. 13.19 (*a*) shows the real beam *ABC*. There are three reaction components : R_A, M_A and R_C. Against this, there are three condition equations available from statical equilibrium :

$\Sigma V = 0$, $M_C = 0$ and $M_B = 0$ (because of internal hinge at *B*). Hence the beam is statically determinate. For reaction at *C*, take moment about *B*.

$$\therefore \quad M_B = 0 = W \cdot \frac{L}{2} - R_C L$$

This gives $R_C = \dfrac{W}{2}$.

The B.M.D. for the real beam is shown in Fig. 13.19 (*b*), which is self explanatory. The conjugate beam *A′B′C′*, will have *free* end at *A′*, roller support at *C′* and will have as *interior pin support* at *B′*. The conjugate beam with the elastic loading equal to *M/E I* is shown in Fig. 13.19 (*c*). For reaction $R_B{}'$ at *B′* of the conjugate beam, take moment about *C′*.

$$R_B{}' \times L + \left(\frac{WL}{4EI} \times \frac{L}{2}\right)\left(\frac{L}{2}\right) - \left(\frac{1}{2} \times L \times \frac{WL}{2EI}\right)\left(L + \frac{2}{3}L\right) = 0, \text{ which gives}$$

$$R_B{}' = \frac{1}{L}\left[\frac{WL^3}{EI} \times \frac{5}{12} - \frac{WL^3}{16EI}\right] = \frac{17}{48}\frac{WL^2}{EI} \ (\downarrow)$$

Hence
$$R_C{}' = \frac{1}{2} \times L \times \frac{WL}{4EI} + \frac{17}{48}\frac{WL^2}{EI} - \frac{1}{2} \times L \times \frac{WL}{2EI} = \frac{11}{48}\frac{WL^2}{EI} \ (\uparrow)$$

The S.F. diagram for the conjugate beam is shown in Fig. 13.19 (*d*).

Now
$$(i_B)_{\text{left}} = (F_B{}')_{\text{left}} = +\frac{1}{2} \times L \times \frac{WL}{2EI} = +\frac{\mathbf{WL^2}}{\mathbf{4EI}} \ (i.e., \text{ clockwise})$$

$$(i_B)_{right} = (F_B')_{right} = \frac{1}{2} \times L \times \frac{W\,L}{2E\,I} - \frac{17}{48}\frac{W\,L^2}{E\,I} = -\frac{5}{48}\frac{W\,L^2}{E\,I} \text{ (i.e., anticlockwise)}$$

Also, $$y_B = M_B' = +\frac{1}{2} \times L \times \frac{W\,L}{2E\,I}\left(\frac{2}{3}L\right) = +\frac{W\,L^3}{6E\,I} \text{ (i.e., } \downarrow)$$

Also, $$y_A = M_A' = 0\,;\, y_C = M_C' = 0$$

The B.M.D. for the conjugate beam is shown in Fig. 13.19 (*e*). Fig. 13.19 (*f*) shows the deflected shape of the real beam.

PROBLEMS

1. A cantilever is partially loaded with U.D.L. w over a length L_1, from the fixed end (Fig. 13.20). Using conjugate beam method, find slope and deflection at the end.

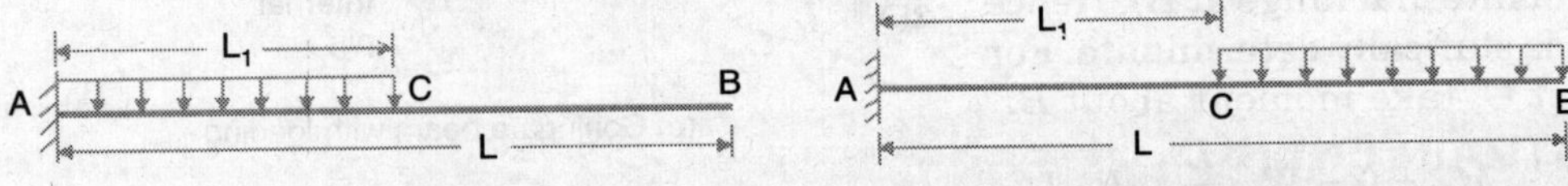

Fig. 13.20 Fig. 13.21

2. A cantilever is partially loaded with U.D.L. w over a length $(L_1 - L)$ from the free end (Fig. 13.21). Using conjugate beam method, find the deflection at the free end.

3. A cantilever AB of length L is subjected to a concentrated load W and a couple μ acting at the free end (Fig. 13.22). Determine the slope and deflections at the free end.

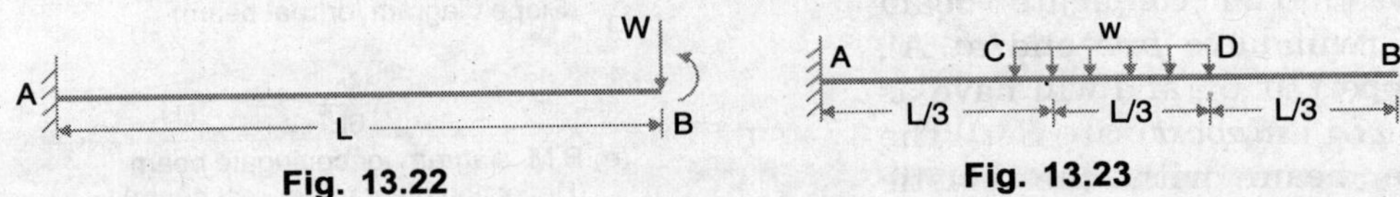

Fig. 13.22 Fig. 13.23

4. Determine the angle of rotation and deflection at the free end of a cantilever beam AB with uniform load w acting over the middle third of the beam (Fig. 13.23).

5. A cantilever beam AB with two different moments of inertia I and $2I$ supports a uniformly distributed load w (Fig. 13.24). Determine the deflection at the free end.

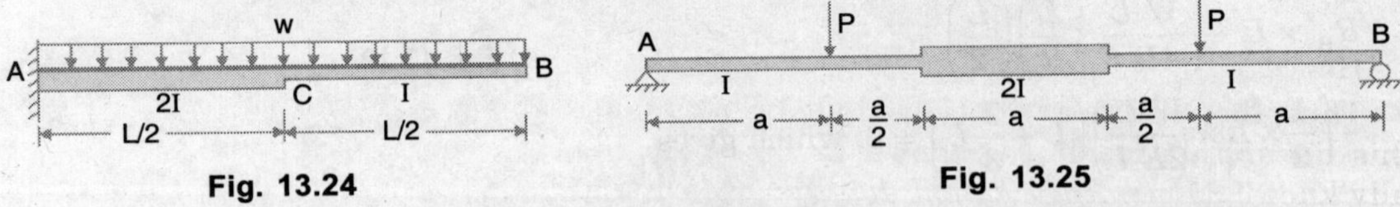

Fig. 13.24 Fig. 13.25

6. A beam with variable moment of inertia is loaded as shown in Fig. 13.25. Determine the deflection at the mid-span.

ANSWERS

1. $i_B = \dfrac{w\,L_1^3}{6E\,I}\,;\, y_B = \dfrac{w\,L_1^4}{8E\,I} + \dfrac{w\,L_1^3}{6E\,I}(L - L_1)$

2. $\dfrac{w\,L^4}{8E\,I} - \left\{\dfrac{w\,L_1^4}{8E\,I} + \dfrac{w\,L_1^3}{6E\,I}(L - L_1)\right\}$

3. $\dfrac{W\,L^2}{2E\,I} - \dfrac{\mu\,L}{E\,I}\,;\, \dfrac{W\,L^3}{3E\,I} - \dfrac{\mu\,L^2}{2E\,I}$

4. $i_B = \dfrac{7\,w\,L^3}{162E\,I}\,;\, y_B = \dfrac{23\,w\,L^4}{648E\,I}$

5. $y_B = \dfrac{17\,w\,L^4}{256E\,I}$

6. $y_C = \dfrac{67}{48}\dfrac{P\,a^3}{E\,I}$.

14

Deflection of Beams IV : Strain Energy Method

14.1. STRAIN ENERGY OF BENDING

In chapter 5, the concept of strain energy was discussed with respect to the axially loaded members. We have seen there that the strain energy per unit volume is equal to $f^2/2E$, where f is the axial stress. We shall now apply the concept of strain energy of the bending of beams. To start with, consider a beam of span L, subjected to couple M at the ends (*i.e.*, pure bending). The resulting deflection curve will be a circular arc of constant curvature the radius of curvature R being given by

$$\frac{1}{R} = \frac{M}{EI} \qquad ...(i)$$

If θ is the angle substended at the centre,

$$\theta = L/R \qquad ...(ii)$$

Hence $$\theta = ML/EI \qquad ...(14.1)$$

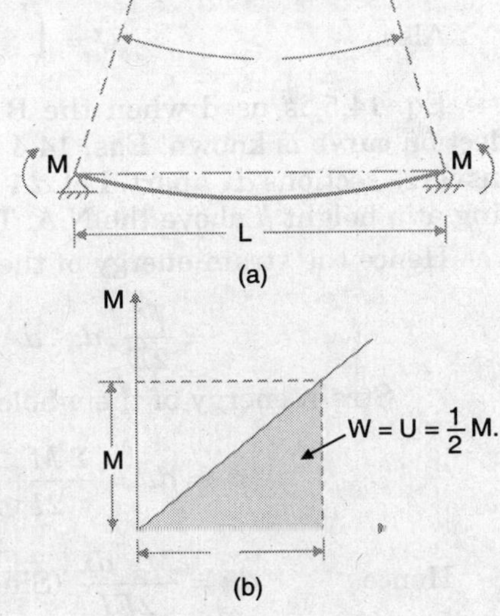

Fig. 14.1

Thus, there is linear relationship between M and θ. This linear relationship is shown in Fig. 14.1 (*b*), usually known as *moment rotation diagram*. Assuming that moment M is applied gradually in magnitude from zero to its maximum value M, the work (W) done in the process is represented by the shaded area. The work is stored in the form of strain energy in the beam given by

$$U = W = M\theta/2 \qquad ...(14.2)$$

Hence from Eqs. 14.1 and 14.2, $$U = \frac{M^2 L}{2\,EI} \qquad ...[14.3\ (a)]$$

Also, $$U = \frac{EI\,\theta^2}{2L}$$ ...[14.3 (b)]

Let us now take the usual case of beam bending in which the bending moment varies along the length of the beam. Let the bending moment at any section X, distant x from the origin, be M_x. Consider another section distant dx from the previous one. Such an element, of length dx, of the beam is shown in Fig. 14.2.

Let $d\theta$ be the angle between the sides of the element. Then the angle $d\theta$ is given by

$$d\theta \approx \frac{d^2y}{dx^2}\,dx = \frac{M_x\,dx}{EI}$$

Hence strain energy stored in the element is

$$du = \frac{1}{2}M_x \,.\, d\theta = \frac{M_x^{\,2}dx}{2EI} \quad ...(14.4)$$

Also $$du = \frac{EI}{2}\left(\frac{d^2y}{dx^2}\right)^2 dx \quad ...[14.4(a)]$$

Hence total strain energy stored in the whole beam is given by

$$U = \int_0^L \frac{M_x^{\,2}dx}{2EI} \quad ...(14.5)$$

Also, $$U = \int_0^L \frac{EI}{2}\left(\frac{d^2y}{dx^2}\right)^2 dx \quad ...[14.5(a)]$$

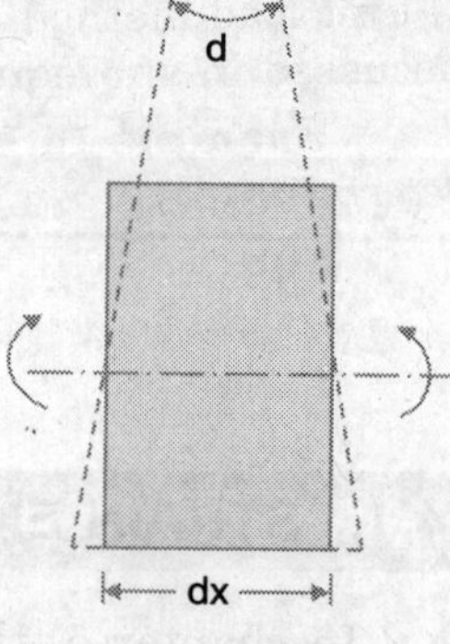

Fig. 14.2

Eq. 14.5 is used when the B.M. is known while Eq. 14.5 (*a*) can be used when the deflection curve is known. Eqs. 14.4 and 14.5 can be derived *alternatively* by considering two transverse sections dx apart. Let dA be small transverse area over which the direct stress f is acting at a height h above the N.A. Then $f = M_x \cdot y/I$

Hence the strain energy of the elementary volume of length dx and area dA is

$$= \frac{f^2}{2E} \,.\, dx \,.\, dA = \frac{M_x^{\,2}y^2}{2I^2E} \,.\, dx \,.\, dA$$

∴ Strain energy of the whole piece of length dx between the transverse-sections

$$= du = \frac{\Sigma\, M_x^{\,2}\, y^2}{2I^2\,E} \,.\, dx\; dA = \frac{M_x^{\,2}dx}{2I^2\,E}\,\Sigma\, dA \,.\, y^2 = \frac{M_x^{\,2}.dx}{2I^2E}\,(I)$$

Hence $$du = \frac{M_x^{\,2}dx}{2EI} \quad (\text{Since } \Sigma\, dA \,.\, y^2 = I) \quad ...(\text{Eq. } 14.4)$$

and $$U = \int_0^L \frac{M_x^{\,2}dx}{2EI} \quad ...(\text{Eq. } 14.5)$$

14.2. DEFLECTION OF BEAMS FROM STRAIN ENERGY CONSIDERATION

When a beam is subjected to a point load, the deflection under the point load can be found by equating the total strain energy of the beam, using Eq. 14.5, to the work (W) done by load P, which is equal to $\frac{1}{2} P \cdot y$, where y is the required deflection. This is illustrated in Example 14.1. It is important to observe that the only deflection we can find is the *deflection under the load*, and not the deflection else where, by this *direct method*. Similarly, when a beam is subjected to a couple μ, the rotation θ at the point of application can be found by equating the total strain energy of the beam (using Eq. 14.5) to the work done by couple μ, which is equal to $\frac{1}{2} \mu \theta$. This is illustrated in Example 14.2. Here also, we can find the rotation only at the point where the couple is applied, and not else where. Thus, the *direct method* has limited use. However we can find the deflection at *any point* of the beam by two other methods : (*i*) unit load method, and (*ii*) fictitious load method or Castigliano's method. These have been discussed in subsequent articles.

Example 14.1. *A cantilever AB of length L is loaded with a point load W at its free end. Determine (a) strain energy of the beam and (b) the deflection of the free end.*

Solution. At any section X, distant x from B,

$$M_x = -W \cdot x$$

$$\therefore \quad U = \int \frac{M_x^2 dx}{2EI} = \int_0^L \frac{(-W \cdot x)^2 \, dx}{2\,EI}$$

$$= \frac{W^2}{2\,EI}\left(\frac{x^3}{3}\right)_0^L = \frac{W^2 L^3}{6EI} \qquad \text{...(1)}$$

Fig. 14.3

It is to be noted that strain energy is always a *positive* quantity, and that the load appears in the second power.

$$\text{Work done by load} = \frac{1}{2} W y_B \qquad \text{...(2)}$$

Equating the two, $\frac{1}{2} W y_B = \frac{W^2 L^3}{6EI}$, From which $y_B = \frac{\mathbf{WL^3}}{\mathbf{3\,EI}}$. **(Ans.)**

Example 14.2. *A cantilever AB of length L is subjected to a couple μ acting at the free end. Determine (a) the strain energy of the beam, and (b) rotation of the free end.*

Solution. At any section X, distant x from B, we have

$$M_x = -\mu$$

$$\therefore \quad U = \int \frac{M_x^2 dx}{2EI} = \int_0^L \frac{(-\mu)^2 \, dx}{2\,EI}$$

$$= \frac{\mu^2}{2\,EI}\Big[x\Big]_0^L = \frac{\mu^2 \mathbf{L}}{\mathbf{2\,EI}}$$

Fig. 14.4

If i_B is the rotation at the free end, work done by during loading of the beam is $\frac{1}{2} \mu \, i_B$

Hence $\frac{1}{2} \mu \, i_B = \frac{\mu^2 L}{2\,EI}$, From which $i_B = \frac{\mu \mathbf{L}}{\mathbf{EI}}$ (Clockwise)

Example 14.3. *A cantilever AB of length L is subjected simultaneously to both a concentrated load W and a couple μ, as shown in Fig. 14.5. Determine the strain energy of the beam. Is the method of super position applicable ? Can you find the slope and deflection of the beam ?*

Solution. At any distance x from the free end,

$$M_x = -W \cdot x - \mu$$

$$\therefore \quad U = \int \frac{M_x^2 dx}{2EI} = \frac{1}{2EI}\int_0^L (-W \cdot x - \mu)^2 \, dx$$

Fig. 14.5

$$= \frac{1}{2EI}\left[\frac{W^2x^3}{3} + \mu^2 x + \frac{2W\mu x^2}{2}\right]_0^L = \frac{1}{2EI}\left[\frac{W^2L^3}{3} + W\mu L^2 + \mu^2 L\right]$$

$$= \frac{W^2L^3}{6EI} + \frac{W\mu L^2}{2EI} + \frac{\mu^2 L}{2EI} \qquad \text{...(1)} \quad \textbf{(Ans.)}$$

The above expression for U contains three terms. The first term gives the strain energy due to W *acting alone* while the last term gives the value of strain energy due to μ acting alone. However, when both W and μ acts together, the term in the middle also appears. *This shows that the strain energy due to two or more loads cannot be found by method of super position.* This is because the strain energy U is a quadratic function of loads and not a linear function.

$$\text{Again work done} = \frac{W y_B}{2} + \frac{\mu \, i_B}{2} \qquad \text{...(2)}$$

When (1) and (2) are equated, there appear two unknowns y_B and i_B. Hence slope and deflection cannot be found by this method.

14.3. DEFLECTION BY UNIT LOAD METHOD

In the previous article, we have seen that we can find the deformation of the beam only at the point of application of the load/ moment, by the direct method. For finding the deflection of the beam at any other point, we will use the *unit load method.* Fig. 14.6 (*a*) shows a simply supported beam with external loading. It is required to find the deflection at any point P, distant x_P from the L.H. support, due to the external loading. Consider a section X, distant x from A, and another section distant dx from the previous one. Let M be the B.M. there. Imagine an elementary point load dW at point P where the deflection is desired. Due to the addition of this load, the work supplied

$$= \frac{1}{2} dW \cdot y_P \qquad \text{...(1)}$$

where y_P is the desired deflection at P.

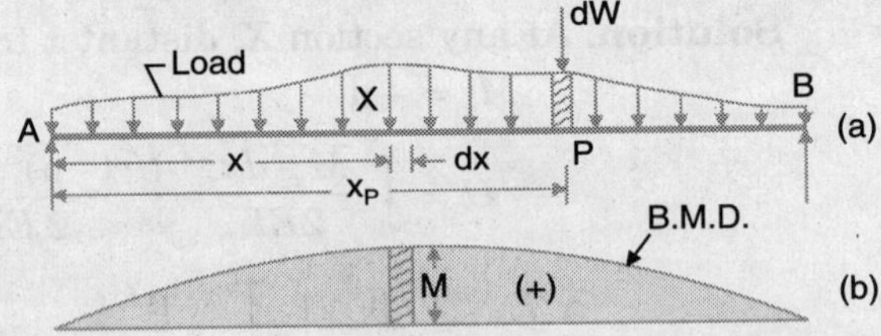

Fig. 14.6

Again, due to the addition of dW at P, the B.M. at X will be increased to $(M + dM)$ where dM is the increment due to dW. If m is the B.M. at X due to unit load at P, the incremental B.M. $dM = m \, dW$. The change in slope of the tangent at X, due to dW at P, is given by

$$EI \frac{di}{dx} = dM = m dW$$

$$\therefore \quad di = \frac{m \, dW \cdot dx}{EI} \qquad \text{...(2)}$$

Work stored in the strip of length $dx = \frac{1}{2}(M + m\,dW).\,di = \frac{1}{2}(M + mdW).\,\frac{m\,dW\,.\,dx}{EI}$

$\therefore$ Work stored in the whole beam due to dW at P

$$= \frac{1}{2\,EI}\int_0^L M\,.\,m\,.\,dW\,.\,dx \qquad ...(3)$$

(neglecting the small quantities of higher order). Equating (1) and (3), we get

$$\frac{1}{2}\,dW\,.\,y_P = \frac{1}{2\,EI}\int_0^L M\,m\,dW\,.\,dx$$

or

$$y_P = \int_0^L \frac{M\,.\,m\,.\,dx}{EI} \qquad ...(14.6)$$

Procedure. In order to find the deflection at any point P, proceed as follows.

1. Find the B.M. at any section X, distant x from the support due to external loading. Thus M is known in terms of x.

2. Put unit load at point P under consideration and find the B.M. m at the section X. Thus m is also known in terms if x.

3. Integrate Eq. 14.6 between limits $x = 0$ to $x = L$ and get the value of deflection y_P at the point under consideration.

Example 14.4. *A cantilever AB of length L is loaded with a point load W at distance L_1 from the support. Calculate the deflection at the free end.*

Solution. Apply unit load at B. Measuring x from A, we have

$$y_B = \int_0^L \frac{M\,m\,dx}{EI} = \int_0^{L_1} \frac{M\,m\,dx}{EI} + \int_{L_1}^{L} \frac{M\,m\,dx}{EI}$$

For AC : $M = -\,W(L_1 - x)$ and $m = -\,1(L - x)$

$$\therefore \int_0^{L_1} \frac{Mmdx}{EI} \int_0^{L_1} \frac{W(L_1 - x)(L - x)}{EI}\,dx$$

$$= \frac{W}{EI}\int_0^{L_1} L\,L_1 - x(L + L_1) + x^2$$

$$= \frac{W}{EI}\left[LL_1x - \frac{x^2}{2}(L + L_1) + \frac{x^3}{3}\right]_0^{L_1}$$

$$= \frac{W}{EI}\left[LL_1^{\,2} - \frac{L_1^{\,2}}{2}(L + L_1) + \frac{L_1^{\,3}}{3}\right]$$

$$= \frac{W}{EI}\left[\frac{LL_1^{\,2}}{2} - \frac{L_1^{\,3}}{6}\right] = \frac{W}{EI}\left[\frac{3LL_1^{\,2} - 3L_1^{\,3} + 2L_1^{\,3}}{6}\right] = \frac{WL_1^{\,3}}{3\,EI} + \frac{WL_1^{\,2}}{2\,EI}(L - L_1) \qquad ...(1)$$

A, x, C, B, W, 1, L_1, L

Fig. 14.7

For CB : $M = 0$ and $m = -\,1\,(L - x)$

Hence $$\int_{L_1}^{L} \frac{M\,m\,dx}{EI} = 0$$

$$\therefore \quad y_B = \int_{L_0}^{L} \frac{M\,m\,dx}{EI} = \frac{WL_1^{\,3}}{3EI} + \frac{WL_1^{\,2}}{2\,EI}(L - L_1)$$

This is the same as Eq. 11.5 (*a*).

Example 14.5. *A cantilever is loaded with uniformly distributed load throughout its length. Calculate (a) the deflection at the free end (b) deflection of a point C distant c from the free end.*

Solution. (*a*) ***Deflection at B :*** Apply a unit load at *B*, as shown in Fig. 14.8 (*b*).

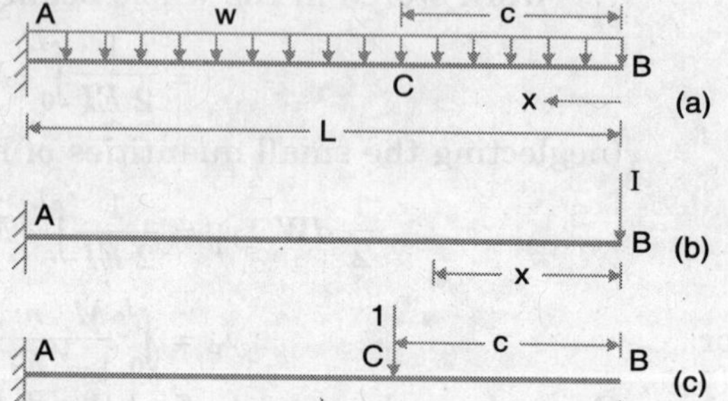

Fig. 14.8

Then $y_B = \int_0^L \frac{M\,m\,dx}{EI}$

Here $M = -\frac{w\,x^2}{2}$ and $m = -1\,x = -x$

$$y_B = \frac{1}{EI}\int_0^L -\frac{wx^2}{2}(-x)\,dx$$

$$= \frac{1}{EI}\int_0^L \frac{wx^3}{2} = \frac{\boldsymbol{wL^4}}{\boldsymbol{8\,EI}} \quad \textbf{(Ans.)}$$

(*b*) ***Deflection at C :*** Apply the unit load at *C*, as shown in Fig. 14.8 (*c*).

Then $y_C = \int_0^L \frac{M\,m\,dx}{EI} = \int_0^c \frac{M\,m\,dx}{EI} + \int_c^L \frac{M\,m\,dx}{EI}$

For BC $\quad M = -\frac{wx^2}{2}$; $m = 0 \quad \therefore \quad \int_0^c \frac{M\,m\,dx}{EI} = \text{zero}$

For CA $\quad M = -\frac{wx^2}{2}$; $m = -1(x - c)$

$$\therefore \quad \int_c^L \frac{M\,m\,dx}{EI} = \frac{1}{EI}\int_c^L \frac{wx^2}{2}(x-c)\,dx = \frac{1}{EI}\left[\frac{wx^4}{8} - \frac{w\,cx^3}{6}\right]_c^L$$

$$= \frac{w}{EI}\left[\left\{\frac{L^4}{8} - \frac{cL^3}{6}\right\} - \left\{\frac{c^4}{8} - \frac{c^4}{6}\right\}\right] = \frac{w}{24EI}[3L^4 - 4cL^3 + c^4]$$

Hence $\quad y_C = \int_0^L \frac{M\,m\,dx}{EI} = \frac{w}{24EI}[3L^4 - 4cL^3 + c^4]$ **Ans.**

Example 14.6. *Find the central deflection of a simply supported beam loaded with a point load W distant a from A and b from B.*

Solution. Apply additional load *W* at a distance *b* from *A*. Due to this application, the deflection at *C* will be *doubled*. Now apply unit load at *C*, where deflection is desired.

Fig. 14.9

Then $\quad y_C = \int_0^L \frac{M\,m\,dx}{EI}$

$$= 2\int_0^{L/2} \frac{M\,m\,dx}{EI}, \text{ due to symmetry.}$$

For AE, $\quad M = W\,.\,x$ and $m = \frac{1}{2}\,.\,x$

$$\therefore \quad \int_0^b \frac{M\,m\,dx}{EI} = \frac{1}{EI}\int_0^b \frac{Wx^2}{2}dx = \frac{Wb^3}{6\,EI} \qquad \text{...}(i)$$

For EC, $M = Wx - W(x - b)$ and $m = \dfrac{1}{2}x$

$$\therefore \int_b^{L/2} \frac{M\,m\,dx}{EI} = \int_b^{L/2} \frac{1}{EI}\left[\frac{Wx^2}{2} - \frac{Wx}{2}(x-b)\right]dx = \frac{1}{EI}\left[\frac{Wx^3}{6} - \frac{Wx^3}{6} + \frac{Wbx^2}{4}\right]_b^{L/2}$$

$$= \frac{Wb}{4\,EI}\left[\frac{L^2}{4} - b^2\right] = \frac{Wb}{16\,EI}(L^2 - 4b^2) \quad \text{...(ii)}$$

Now $$y_C = 2\int_0^{L/2} \frac{M\,m\,dx}{EI} = 2\left[\int_0^b \frac{M\,m\,dx}{EI} + \int_b^{L/2} \frac{M\,m\,dx}{EI}\right]$$

$$= \frac{2}{EI}\left[\frac{Wb^3}{6} + \frac{Wb}{16}(L^2 - 4b^2)\right] = \frac{2Wb}{48EI}[8b^2 + 3L^2 - 12b^2]$$

$$= \frac{Wb}{24EI}(3L^2 - 4b^2)$$

Hence the central deflection due to single point load = $\dfrac{\boldsymbol{Wb}}{\boldsymbol{48EI}}$ $\boldsymbol{(3L^2 - 4b^2)}$.

14.4. DEFLECTION BY FICTITIOUS LOAD METHOD USING CASTIGLIANO'S FIRST THEOREM

The concept of elastic strain energy can be very useful in the study of deflections of various points of a structure under load. Instead of directly equating the *external work* to the internal strain energy, considerable simplification is obtained by *Castigliano's first theorem,* which states that deflection caused by any external force is equal to the partial derivative of the strain energy with respect of that force. A generalised statement of the theorem* is as follows :

"If there is any elastic system in equilibrium under the action of a set of forces W_1, W_2, W_3, ... W_n and corresponding displacements δ_1, δ_2, δ_3, ... δ_n, and a set of moments M_1, M_2, M_3, ... M_n and the corresponding rotations φ_1, φ_2, φ_3, ... φ_n, then the partial derivative of the total strain energy U with respect to any one of the forces or moments, taken individually would yield its corresponding displacement in its direction of action".

Expressed mathematically,

$$\frac{\partial U}{\partial W_1} = \delta_1 \text{ and } \frac{\partial U}{\partial M_1} = \varphi_1 \quad \text{...(14.7)}$$

The above theorem of Castigliano can be used for computing the deflection and slope at any point of a beam. The procedure is discussed below. Consider a beam subjected to loads W_1, W_2, ... W_n. It is desired to find deflection under load W_1.

The strain energy of bending is given by

$$U = \int_0^L \frac{M^2\,dx}{2\,EI} \text{ (where } M \text{ is a function of } W_1) \quad \text{...(Eq. 14.5)}$$

*For the proof of the theorem, see Author's book THEORY OF STRUCTURES

$$\therefore \qquad \delta_1 = \frac{\partial U}{\partial W_1} = \int_0^L M\left(\frac{\partial M}{\partial W_1}\right)\frac{dx}{EI} \qquad \text{...(14.8)}$$

In the above expression, $\partial M/\partial W_1$ is evaluated by differentiating *inside the integral sign, before integrating.* If no load is acting at a point where the deflection is desired, a *fictitious load W* is applied at the point, in the direction the deflection is required. Then, *after differentiating,* but *before integrating,* the fictitious load is set to zero. The method is also sometimes known as the *fictitious load method.* If, however, the rotation φ_1 is required in the direction of M_1, Eq. 14.8 is modified as follows :

$$\varphi_1 = \frac{\partial U}{\partial M_1} = \int_0^L M\left(\frac{\partial M}{\partial M_1}\right)\frac{dx}{EI} \qquad \text{...(14.9)}$$

where M is a function of M_1.

Example 14.7. *Calculate the central deflection, and the slope at ends of a simply supported beam carrying a U.D.L. w per unit length over the whole span.*

Solution. (*a*) ***Central deflection :***

Since no point load is acting at the centre where the deflection is required, apply a *fictitious load W* there, as shown in Fig. 14.10 (*a*). The reactions at A and B will be $\left(\frac{wL}{2}+\frac{W}{2}\right)(\uparrow)$ each.

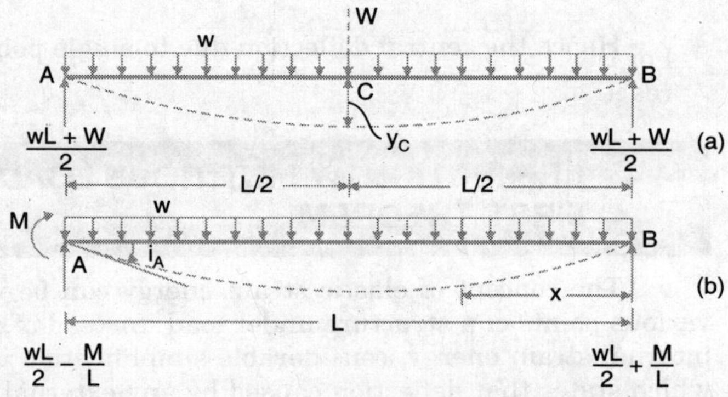

Fig. 14.10

$$\therefore\ y_C = \frac{\partial U}{\partial W} = \frac{1}{EI}\int_0^L M_x \frac{\partial M_x}{\partial W}dx \qquad \text{...(1)}$$

where M_x is the B.M. at any section distant x from A.

$$M_x = \left(\frac{wL}{2}+\frac{W}{2}\right)x - \frac{wx^2}{2}; \quad \frac{\partial M_x}{\partial W} = \frac{x}{2}$$

Hence $\quad y_C = \dfrac{2}{EI}\displaystyle\int_0^{L/2}\left[\left(\frac{wL}{2}+\frac{W}{2}\right)x - \frac{wx^2}{2}\right]\frac{x}{2}\,dx$

Putting $W = 0$, we get

$$y_C = \frac{2}{EI}\int_0^{L/2}\left(\frac{wL}{2}x - \frac{wx^2}{2}\right)\frac{x}{2}\,dx = \frac{2}{EI}\left[\frac{wLx^3}{12} - \frac{wx^4}{16}\right]_0^{L/2} = \mathbf{\frac{5}{384}\frac{wL^4}{EI}}$$

(*b*) ***Slope at ends :***

To obtain slope at end A, apply *fictitious moment M* at A, as shown in Fig. 14.10 (*b*). The reactions at A and B will be respectively $\left(\frac{wL}{2}-\frac{M}{L}\right)$ and $\left(\frac{wL}{2}+\frac{M}{L}\right)$.

Measuring x from B, we have, $\quad i_A = \dfrac{\partial U}{\partial M} = \dfrac{1}{EI}\displaystyle\int_0^L M_x \cdot \frac{\partial M_x}{\partial M}\cdot dx \qquad$ ...(2)

where M_x is the B.M. at any point distant x from B, and is a function of M.

$$M_x = \left(\frac{wL}{2} + \frac{M}{L}\right)x - \frac{wx^2}{2} \,;\, \frac{\partial M_x}{\partial M} = \frac{x}{L}$$

Substituting in Eq. (2), we get

$$i_A = \frac{1}{EI}\int_0^L \left\{\left(\frac{wL}{2} + \frac{M}{L}\right)x - \frac{wx^2}{2}\right\}\frac{x}{L}\,dx$$

Putting $M = 0$, we get

$$i_A = \frac{1}{EI}\int_0^L \left(\frac{wL}{2}x - \frac{wx^2}{2}\right)\frac{x}{L}\,dx = \frac{1}{EI}\left[\frac{wx^3}{6} - \frac{wx^4}{8L}\right]_0^L$$

$$= + \frac{\boldsymbol{wL^3}}{\boldsymbol{24\,EI}} \quad (i.e.\text{ clockwise}).$$

Example 14.8. *Using Castigliano's first theorem, determine the deflection and rotation of the over hanging end A of the beam loaded as shown in Fig. 14.11 (a).*

Solution. (*a*) ***Slope at A.***

For the loading of Fig. 14.11 (*a*) the reaction

$$R_B = \frac{M}{L}\ (\downarrow) \text{ and } R_C = \frac{M}{L}\ (\uparrow)$$

Fig. 14.11

$$i_A = \frac{\partial U}{\partial M}$$

$$= \frac{1}{EI}\int_A^B M_x \cdot \frac{\partial M_x}{\partial M}\cdot dx + \frac{1}{EI}\int_C^B M_x \cdot \frac{\partial M_x}{\partial M}\,dx \quad ...(i)$$

For AB : Measuring x from A.

$$M_x = M \,;\, \frac{\partial M_x}{\partial M} = 1$$

$$\therefore \quad \frac{1}{EI}\int_A^B M_x \frac{\partial M_x}{\partial M}\,dx = \frac{1}{EI}\int_0^{L/3} (M \times 1)\,dx = \frac{M}{EI}\Big[x\Big]_0^{L/3} = \frac{ML}{3\,EI} \quad ...(ii)$$

For CB : Measuring x from C, $M_x = \frac{M}{L}x \,;\, \frac{\partial M_x}{\partial M} = \frac{x}{L}$

$$\therefore \quad \frac{1}{EI}\int_C^B M_x \frac{\partial M_x}{\partial M}\,dx = \frac{1}{EI}\int_0^L \left(\frac{M}{L}x\right)\left(\frac{x}{L}\right)dx = \frac{M}{EIL^2}\left[\frac{x^3}{3}\right]_0^L = \frac{ML}{3\,EI} \quad ...(iii)$$

Hence from (*i*), $\quad i_A = \frac{ML}{3\,EI} + \frac{ML}{3\,EI} = \frac{\boldsymbol{2ML}}{\boldsymbol{3EI}}$ (clockwise)

(*b*) ***Deflection at A.*** To find the deflection at A, apply a fictitious load W at A, in upward direction, as shown in Fig. 14.11 (*b*).

$$R_B = \left(M + \frac{4}{3}WL\right)\frac{1}{L}\ (\downarrow) \quad \text{and} \quad R_C = \left(M + \frac{1}{3}WL\right)\frac{1}{L}\ (\uparrow)$$

Now $$y_A = \frac{\partial U}{\partial W} = \frac{1}{EI}\int_A^B M_x \frac{\partial M_x}{\partial W}\,dx + \frac{1}{EI}\int_C^B M_x \cdot \frac{\partial M_x}{\partial W}\,.\,dx \qquad ...(iv)$$

For AB, Measuring x from A,

$$M_x = M + Wx\ ;\ \frac{\partial M_x}{\partial W} = x$$

For CB, Measuring x from C,

$$M_x = \left(M + \frac{1}{3}WL\right)\frac{1}{L}x\ ;\ \frac{\partial M_x}{\partial W} = \frac{x}{3}$$

Substituting the values in (*iv*), we get

$$y_A = \frac{1}{EI}\int_0^{L/3}(M + Wx)\,x\,dx + \frac{1}{EI}\int_0^L\left(M + \frac{1}{3}WL\right)\frac{x}{L}\cdot\frac{x}{3}\,dx$$

Putting $W = 0$, $$y_A = \frac{1}{EI}\int_0^{L/3} M\,x\,dx + \frac{1}{EI}\int_0^L \frac{Mx^2}{3L}\,dx = \frac{M}{EI}\left[\frac{x^2}{2}\right]_0^{L/3} + \frac{M}{3\,EIL}\left[\frac{x^3}{3}\right]_0^L$$

$$= \frac{ML^2}{18\,EI} + \frac{ML^2}{9\,EI} = \mathbf{\frac{ML^2}{6\,EI}}\ (\uparrow).$$

Example 14.9. *A vertical load W is applied to the rigid cantilever frame shown in Fig. 14.12. Assuming EI to be constant throughout the frame, determine the horizontal and vertical displacement of the point C. Neglect axial deformations.*

Solution. (*a*) ***Vertical deflection of C***

$$y_{CV} = \frac{\partial U}{\partial W} = \frac{1}{EI}\int M_x \frac{\partial M_x}{\partial W}\,dx \qquad ...(1)$$

For CB : Measuring x from C,

$$M_x = -\,Wx\ ;\ \frac{\partial M_x}{\partial W} = -\,x$$

For BA : Measuring x from B,

$$M_x = -\frac{WL}{2}\ ;\ \frac{\partial M_x}{\partial W} = -\,L/2$$

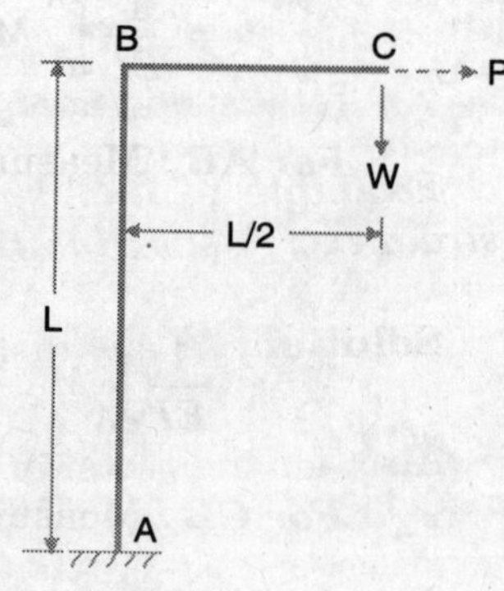

Fig. 14.12

Substituting in (1), we get

$$y_{CV} = \frac{1}{EI}\left[\int_0^{L/2} Wx\,.\,x\,dx + \int_0^L \frac{WL}{2}\cdot\frac{L}{2}\,dx\right]$$

$$= \frac{1}{EI}\left[\frac{Wx^3}{3}\right]_0^{L/2} + \frac{1}{EI}\left[\frac{WL^2}{4}x\right]_0^L$$

$$= \frac{1}{EI}\left[\frac{WL^3}{24} + \frac{WL^3}{4}\right] = \mathbf{\frac{7WL^3}{24\,EI}}\ (\downarrow)$$

(*b*) ***Horizontal deflection of C***. Apply a fictitious horizontal load P at C. Then

$$y_{CH} = \frac{\partial U}{\partial P} = \frac{1}{EI}\int M_x \frac{\partial M_x}{\partial P}\,dx \qquad ...(2)$$

For CB : Measuring x from C, $M_x = -\,Wx\ ;\ \partial M_x / \partial P = 0$

For BA : Measuring x from B, $M_x = -\dfrac{WL}{2} - P\,.\,x\ ;\ \partial M_x / \partial P = -\,x$

Substituting the values in (2),

$$y_{CH} = \frac{1}{EI}\left[\int_0^{L/2} (-Wx)(0)\,dx + \int_0^L \left(\frac{WL}{2} + Px\right) x\,dx\right]$$

Putting $P = 0$, $\quad y_{CH} = \frac{1}{EI}\int_0^L \frac{WL}{2} x\,dx = \frac{1}{EI} \cdot \frac{WL}{2}\left[\frac{x^2}{2}\right]_0^L = \frac{WL^3}{4EI}$ $(\rightarrow)$.

14.5. STRAIN ENERGY DUE TO SHEAR

Let us now determine the strain energy due to shear at any section of a beam subjected to external loading. Taking an elementary strip of length dx of a beam, distant x from the support, the shear stress at the layer distant y from the neutral axis of the section is given by

$$q = \frac{F}{Iz}(A\,\bar{y})$$

where I = moment of inertia of the whole section

z = width of the section at that layer

$A\,\bar{y}$ = moment of the area *above* that layer, about N.A.

The volume V of the strip of thickness $dy = dx \,.\, z \,.\, dy$

Strain energy stored in the elementary strip $= dU = \frac{q^2}{2N} \times V = \frac{q^2}{2N} \,.\, z \,.\, dx \,.\, dy$

Hence total energy stored in section of length $dx = \int_{y_t}^{y_c} \frac{q^2}{2N} z\,dx \,.\, dy$

$\therefore$ Total strain energy stored in the whole beam $= U_S = \int_0^L \int_{y_t}^{y_c} \frac{q^2}{2N} z \,.\, dx \,.\, dy$ ...(14.10)

Example 14.10. *A simply supported beam of span L carries a central load W. Calculate the strain energy stored in the beam due to shear, if the beam is a rectangular section, b × d.*

Solution. At any distance x from the support, $\quad F_x = \frac{W}{2}$

Also, for the rectangular beam section, $\quad I = \frac{1}{12} b\,d^3$

$$\therefore \quad q = \frac{F}{2I}\left(\frac{d^2}{4} - y^2\right) = \frac{3W}{bd^3}\left(\frac{d^2}{4} - y^2\right)$$

or

$$q^2 = \frac{9\,W^2}{b^2 d^6}\left(\frac{d^4}{16} + y^4 - \frac{d^2 y^2}{2}\right)$$

The total strain energy in the beam, due to shear, is given by Eq. 14.10.

$$U_S = 2\int_0^{L/2}\int_{y_t}^{y_c} \frac{q^2}{2N} z\,dx\,dy = 2\int_0^{L/2}\int_{-\frac{d}{2}}^{+\frac{d}{2}} \frac{b}{2N} \cdot \frac{9W^2}{b^2 d^6}\left(\frac{d^4}{16} + y^4 - \frac{d^2 y^2}{2}\right) dx\,dy$$

$$= \frac{9W^2}{Nbd^6} \int_0^{L/2} \left(\frac{d^4}{16} y + \frac{y^5}{5} - \frac{d^2 y^3}{6} \right)_{-\frac{d}{2}}^{+\frac{d}{2}} dx = \frac{9W^2}{Nbd^6} \int_0^{L/2} \left(\frac{d^4}{16} d + \frac{1}{5} \frac{d^5}{16} - \frac{d^2}{6} \cdot \frac{d^3}{4} \right) dx$$

$$= \frac{9W^2}{Nbd^6} \cdot \frac{d^5}{30} \int_0^{L/2} dx = \frac{9W^2}{30\,Nbd} \cdot \frac{L}{2} = \frac{\mathbf{3}}{\mathbf{20}} \frac{\mathbf{W^2 L}}{\mathbf{N\,bd}} . \qquad ...(14.11)$$

14.6. DEFLECTION DUE TO SHEAR

Upto this stage (*i.e.*, in Chapters 11, 12 and 13), only the effects of bending deformations were considered in finding deflections. *Shear* deformations produce additional deflection. The shear deformations, for a beam of rectangular cross-section, cause an element of the beam of length dx to be deformed as shown in Fig. 14.13 (a). Since the shear stresses vary over the depth of the beam, the cross-sections become curved surfaces. The line ab represents the original axis of the beam, while the line ac shows the position of the line after the shear deformations have occurred. Let angle γ_c between the lines ab and ac be the value of shear strain at the N.A. The value of γ_c is maximum at the N.A., and it decreases in the outer layers. At the outer most layers, γ_c becomes zero ; hence the sides of this layer are at right angles.

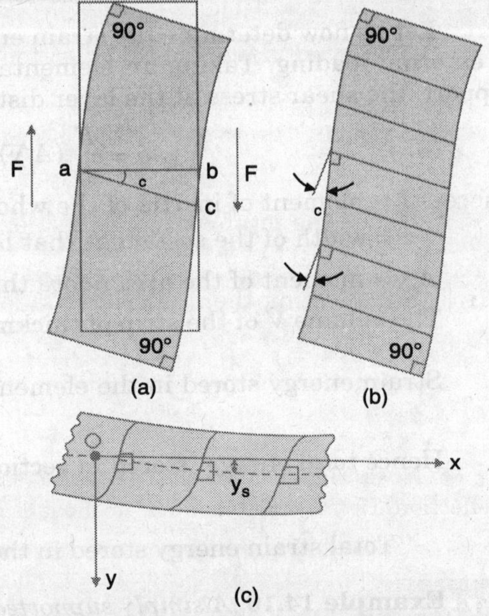

Fig. 14.13. Shear deformations

Fig. 14.13 (c) shows the curvature of the beam axis produced by shear deformation *alone*. The plane sections do not remain plane after bending, but are warped due to shearing strain. Small rectangular elements on the central axis become rhombus, the vertical edges of which remain vertical during deformations. The exact method of calculating the deflection curve due to shear in based on the fact that the slope of the deflection curve at any cross-section is simply equal to the shear strain at the N.A. However, we will find an approximate formula for deflection y_S due to shear. In the approximate method, the strain energy stored due to shear in the beam is equated to the work done by the load in producing shear.

Now, strain energy U of the whole beam due to shear is

$$U = \int_0^L \int_{y_t'}^{y_c} \frac{q^2}{2N} z \,.\, dy \,.\, dx \qquad ...(i)$$

Also, work done by the load $= \frac{1}{2} P \,.\, y_S$. (where y_S is the deflection due to shear)

Equating the two, $\frac{1}{2} P \cdot y_s = \int_0^L \int_{y_t}^{y_c} \frac{q^2}{2N} z \, dy \, dx$ ...(14.12)

Thus, Eq. 14.12 gives the *additional deflection* y_s due to shear.

Example 14.11. *Calculate the deflection due to shear and the total deflection for the beam of example 14.10. Take W = 50 kN, L = 4 m, b = 100 mm, d = 200 mm, E = 2 × 10⁵ N/mm² and N = 0.82 × 10⁵ N/mm².*

Solution. As computed in example 14.10, strain energy due to shear is given by

$$U = \frac{3}{20} \frac{W^2 L}{N bd} \quad \text{...(Eq.14.11)}$$

Also, work done due to deflection due to shear = $\frac{1}{2} W \cdot y_s$

Equating the two, $\frac{1}{2} W y_s = \frac{3}{20} \frac{W^2 L}{N bd} \quad \therefore \quad y_s = \frac{3}{10} \frac{WL}{N bd}$ ...(14.13)

Hence the total central deflection is $y = \frac{WL^3}{48 EI} + \frac{3}{10} \frac{WL}{N bd}$ **(Ans.)**

Substituting the numerical values, noting that $I = \frac{1}{12} (100)(200)^3 = 6.67 \times 10^7$

$$y = \frac{50000\,(4)^3}{48 \times 2 \times 10^5 \times 6.67 \times 10^7} (1000)^3 + \frac{3}{10} \frac{50000\,(4 \times 1000)}{0.82 \times 10^5 \times 200 \times 400}$$

$= 4.997 + 0.009 = \mathbf{5.006.}$

Note. Here we find that deflection due to shear is only 0.009 mm which is only 0.18% of the total deflection. Hence deflection due to shear is usually neglected.

14.7. IMPACT LOADING ON BEAMS

So far, we have assumed that the load W was applied *gradually* on the beam. If, on the other hand, the load W is dropped through a height h, causing impact on the beam, the beam will have *maximum instantaneous deflection* δ. The beam will vibrate for some time and finally it will come to rest. Let P be the *equivalent static load* which cause the same deflection δ. *The dynamic deflection* δ caused by impact load may be determined under certain simplified conditions (or assumptions) by equating the work done by the load with the strain energy stored in the beam.

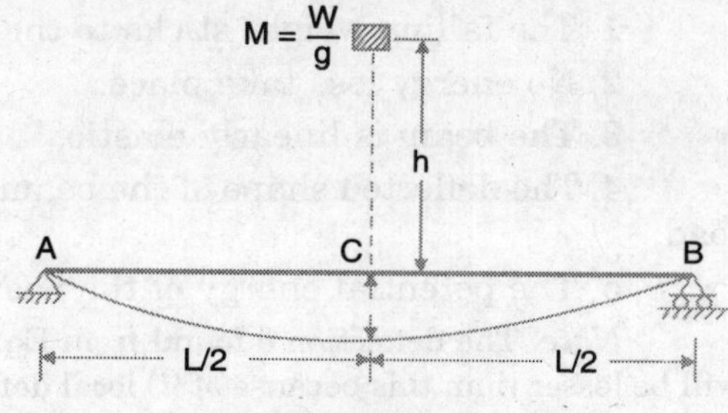

Fig. 14.14

Work done = $W(h + \delta)$

Strain energy of the beam = $\frac{1}{2} P \cdot \delta$

Equating the two, we get $W(h + \delta) = \frac{1}{2} P \delta$ ...(1)

Since P is the equivalent static load exerted on the beam when its deflection is maximum (*i.e.,* δ), we have

$$\delta = \frac{PL^3}{48\,EI} \quad \text{or} \quad P = \frac{48\,EI\,\delta}{L^3} \qquad \text{...(2)}$$

From (1) and (2), the values of P and δ can be easily found.

Substituting the value of P in (1), we get

$$W(h + \delta) = \frac{24\,EI\,\delta^2}{L^3} \qquad \text{...(3)}$$

This is a quadratic equation in δ, and can be solved for its positive root :

$$\delta = \frac{WL^3}{48\,EI} + \left[\left(\frac{WL^3}{48\,EI}\right)^2 + 2h\left(\frac{WL^3}{48\,EI}\right)\right]^{1/2} \qquad \text{...[14.14 (a)]}$$

Denotating the *static deflection* of the beam due to W as δ_{st} :

$\delta_{st} = \frac{WL^3}{48\,EI}$, Eq. 14.14 (*a*) takes the form

$$\delta = \delta_{st} + (\delta_{st}^{\,2} + 2\,h\,\delta_{st})^{1/2} \qquad \text{...[14.14(b)]}$$

From Eq. 14.14(*b*), we find that the dynamic deflection δ is always much higher than the static deflection.

Case (i) : h = 0 : If $h = 0$, *i.e.,* when the load is *applied suddenly* but without any free fall, we get from Eq. 14.14(*b*).

$$\delta = \delta_{st} + \delta_{st} = 2\delta_{st}$$

Hence dynamic deflection is twice the static deflection.

Case (ii) : h very large : When h is very large compared to δ, then the term containing h predominates and hence Eq. 14.14 (*b*) can be written as

$$\delta = \sqrt{2h\,\delta_{st}} \qquad \text{...(14.15)}$$

Assumptions. The above analysis is based on the following assumptions :

1. The falling weight sticks to the beam and moves/vibrates with it.
2. No energy loss take place.
3. The beam is linearly elastic.
4. The deflected shape of the beam is the same under a dynamic load as under the static load.
5. The potential energy of the beam due to change in the position is disregarded.

Note. The deflection δ found from Eq. 14.14(*b*), represents its upper limit. The actual deflection will be lesser than this because of (*i*) local deformation of contact surfaces (*ii*) tendency of falling mass to bounce upward, and (*iii*) mass of the beam itself. All these tend to cause loss of energy, due to which the actual deformations will be lesser than the one given by Eq. 14.14(*b*).

Example 14.12. *A rectangular beam 200 mm × 300 mm is freely supported over a span of 4 m. A load of 50 kN is dropped on to the middle of the beam from a height of 20 mm. Find the maximum instantaneous deflection and the stress induced in the beam. Take E = 2 × 10^5 N/mm^2.*

Solution. $I = \frac{1}{12}(200)(300)^3 = 45 \times 10^7 \text{ mm}^4$

Let P = equivalent static load in N ; $W = 50000$ N

Now $W(h + \delta) = \frac{1}{2} P\,\delta$ or $P\delta = 2 \times 50000\,(20 + \delta)$ ...(1)

where $\delta = \frac{PL^3}{48\,EI} = \frac{P(4)^3\,(1000)^3}{48 \times 2 \times 10^5 \times 45 \times 10^7} = 1.481 \times 10^{-5}\,P$ mm ...(2)

Substituting δ in (1), we have

$$P\,(1.481 \times 10^{-5}\,P) = 100000\,(20 + 1.481 \times 10^{-5}\,P)$$

or
$$P^2 = 13.5 \times 10^{10} + 100000\,P$$

This is a quadratic equation in P, solving which, we get $P = 4.208 \times 10^5$ N = 420.8 kN

$\therefore$ $\delta = 1.481 \times 10^{-5}\,P = 1.48 \times 10^{-5}\,(4.208 \times 10^5) = 6.231$ mm

Alternatively, $\delta_{st} = \frac{WL^3}{48\,EI} = \frac{50000\,(4)^3\,(1000)^3}{48 \times 2 \times 10^5 \times 45 \times 10^7} = 0.74$ mm

$\therefore$ $\delta = \delta_{st} + (\delta_{st}^{\,2} + 2\,h\,\delta_{st})^{1/2} = 0.74 + [(0.74)^2 + 2 \times 20 \times 0.74]^{1/2} = 6.231$ mm

$$M_{max} = \frac{PL}{4} = \frac{4.208 \times 10^5 \times 4000}{4} = 1.052 \times 10^8 \text{ N-mm}$$

$\therefore$ $f_{max} = \frac{M}{I}\,y = \frac{1.052 \times 10^8}{45 \times 10^7} \times 150 =$ **35.07 N/mm²**.

14.8. ADDITIONAL EXAMPLES USEFUL FOR COMPETITIVE EXAMINATIONS

Example 14.13. *A simply supported beam of span L has an overhang of length a on the left. The vertical load W is applied at the end of the overhang. Calculate the deflection of the point of application of the load by (i) Unit load method (ii) Castigliano's first theorem.*

Solution. (*a*) **Unit load method:**

Apply unit load at C. Then

Fig. 14.15

$$y_C = \int_{x=0}^{a+L} \frac{M\,m\,dx}{EI}$$

$$= \int_0^a \frac{M\,m\,dx}{EI} + \int_a^{a+L} \frac{M\,m\,dx}{EI}$$

$$= \int_0^a \frac{M\,m\,dx}{EI} + \int_0^L \frac{M\,m\,du}{EI}$$ (where u is measured from B)

For CA, $M = -\,W\,.\,x$ and $m = -\,1.x$

$\therefore$ $\int_0^a \frac{M\,m\,dx}{EI} = \int_0^a \frac{Wx^2}{EI}\,.\,dx = \frac{Wa^3}{3\,EI}$

For BA : $R_B = \frac{Wa}{L}\,(\downarrow)$ due to point load and $R'_B = \frac{a}{L}\,(\downarrow)$ due to unit load

$$\therefore \qquad M = -\frac{Wa}{L}u \quad \text{and} \quad m = -\frac{a}{L}u$$

$$\therefore \qquad \int_0^L \frac{M\,m\,du}{EI} = \frac{1}{EI}\int_0^L \frac{Wa}{L}u\left(\frac{a}{L}u\right)du = \frac{Wa^2}{L^2EI}\int_0^L u^2\,du = \frac{Wa^2L}{3EI}$$

Hence $\qquad y_C = \frac{Wa^3}{3EI} + \frac{Wa^2L}{3EI} = \frac{\boldsymbol{Wa^2}}{\boldsymbol{3EI}}$ **(*a* + *L*) Ans.**

(*b*) **Castigliano's first theorem :**

$$y_C = \frac{\partial U}{\partial W} = \frac{1}{EI}\int M_x \frac{\partial M_x}{\partial W}dx = \frac{1}{EI}\left[\int_0^a M_x \frac{\partial M_x}{\partial W}dx + \int_0^L M_u \frac{\partial M_u}{\partial W}du\right]$$

For CA, $\qquad M_x = -W.x \quad \text{and} \quad \frac{\partial M_x}{\partial W} = -x$

$$\therefore \qquad \int_0^a M_x \frac{\partial M_x}{\partial W}dx = \int_0^a (Wx)\,x\,dx = \frac{Wa^3}{3}$$

For BA, $\qquad M_u = -\frac{Wa}{L}u \quad \text{and} \quad \frac{\partial M_u}{\partial W} = -\frac{a}{L}u$

$$\therefore \qquad \int_0^L M_u \frac{\partial M_u}{\partial W}du = \int_0^L \frac{Wau}{L}\left(\frac{a}{L}u\right)du = \frac{Wa^2}{L^2}\left(\frac{u^3}{3}\right)_0^L = \frac{Wa^2L}{3}$$

Hence $\qquad y_C = \frac{1}{EI}\left[\frac{Wa^3}{3} + \frac{Wa^2L}{3}\right] = \frac{\boldsymbol{Wa^2}}{\boldsymbol{3EI}}$ **(*a* + *L*).**

Example 14.14. *A cantilever of length L carries a point load W at its free end. The member is circular in section having diameter D for a distance L/2 from the fixed end and a diameter D/2 for the remaining length. Find the deflection at the free end.*

Solution. Let us solve this problem by the direct method.

$$U = \int_0^L \frac{{M_x}^2\,dx}{2\,EI_x} \quad \text{where } x \text{ is measured from } B.$$

For the portion $\qquad BC = I_{BC} = \frac{\pi}{64}\left(\frac{D}{2}\right)^4 = \frac{\pi D^4}{64 \times 16}$

For the portion *CA*, $\quad I_{CA} = \frac{\pi}{64}D^4$

Hence if $\qquad I_{BC} = I, \quad I_{CA} = 16\,I$

Fig. 14.16

$$\therefore \qquad U = \int_0^{L/2} \frac{{M_x}^2\,dx}{2\,EI} + \int_{L/2}^L \frac{{M_x}^2\,dx}{2\,E(16\,I)} \qquad \text{...(1)}$$

For *BC*, $\qquad U_{BC} = \int_0^{L/2} \frac{(Wx)^2}{2\,EI}dx = \frac{W^2}{2\,EI}\left[\frac{x^3}{3}\right]_0^{L/2} = \frac{W^2L^3}{48\,EI}$

For *CA*, $\qquad U_{CA} = \int_{L/2}^L \frac{(Wx)^2}{2\,E(16\,I)} = \frac{W^2}{32\,EI}\left[\frac{x^3}{3}\right]_{L/2}^L = \frac{7W^2L^3}{96 \times 8\,EI}$

$$\therefore \qquad U = U_{BC} + U_{CA} = \frac{W^2L^3}{48\,EI} + \frac{7\,W^2L^3}{96 \times 8EI} = \frac{23}{768}\,\frac{W^2L^3}{EI}$$

Now, work supplied $= \frac{1}{2}\,W \,.\, y_B$

$$\therefore \qquad \frac{1}{2}\,W \,.\, y_B = \frac{23}{768}\,\frac{W^2L^3}{EI} \text{ from which } y_B = \frac{\mathbf{23}}{\mathbf{384}}\,\frac{\boldsymbol{WL^3}}{\boldsymbol{EI}}.$$

Example 14.15. *A freely supported beam of span L carries a central load W. The sectional area of the beam is so designed that the moment of inertia of the section increases uniformly from I at the ends to 1.5 I in the middle. Calculate the central deflection.*

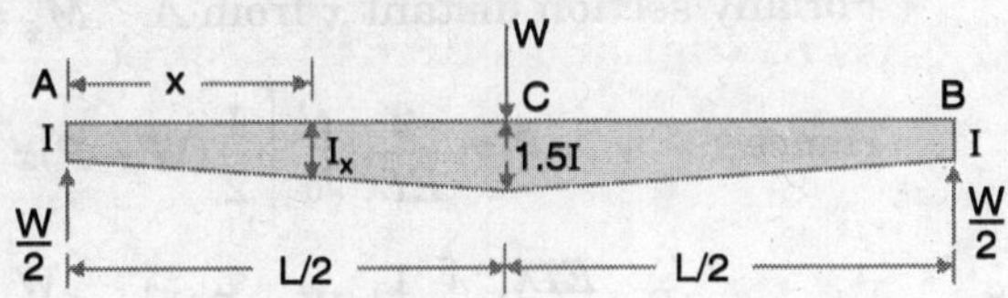

Fig. 14.17

Solution. The central deflection is given by

$$y_C = \frac{\partial U}{\partial W} = \frac{1}{E}\int_0^L \frac{M_x}{I_x}\,\frac{\partial M_x}{\partial W}\,dx = \frac{2}{E}\int_0^{L/2} \frac{M_x}{I_x}\,.\,\frac{\partial M_x}{\partial W}\,.\,dx, \text{ due to symmetry.}$$

At any distance x from A,

$$I_x = I + \frac{I}{2}\,.\,\frac{2}{L}\,x = I\left(1+\frac{x}{L}\right); \quad M_x = \frac{W}{2}\,x \text{ and } \frac{\partial M_x}{\partial W} = \frac{x}{2}$$

$$y_C = \frac{2}{E}\int_0^{L/2} \frac{1}{I\left(1+\frac{x}{L}\right)}\,\frac{W}{2}\,x\,.\,\frac{x}{2}\,dx = \frac{W}{2\,EI}\int_0^{L/2} \frac{x^2}{\left(1+\frac{x}{L}\right)}\,dx$$

Putting $x + L = t$ in the above integral and simplifying, we get

$$y_C = \frac{WL}{2\,EI}\int_L^{3L/2}\left(t - 2L + \frac{L^2}{t}\right)dt = \frac{WL}{2\,EI}\left[\frac{t^2}{2} - 2Lt + L^2 \log_e t\right]_L^{3L/2} = \mathbf{0.01523}\,\frac{\boldsymbol{WL^3}}{\boldsymbol{EI}}$$

[**Note.** The problem was solved by double integration method in Example 11.27]

Example 14.16. *A beam of uniform section and of length 2L is freely supported by rigid supports at its ends and by an elastic prop at its centre. If the prop deflects by an amount λ times the load it carries, and if the beam carries a total distributed load of W, show that the load carried by the prop is* $\dfrac{5W}{8\left(1+\dfrac{6EI\lambda}{L^3}\right)}$.

Solution. Let the prop reaction $= P$

Total load on the beam $= W$

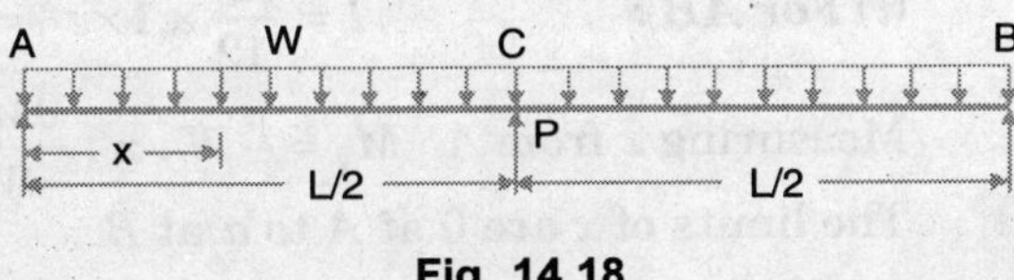

Fig. 14.18

$$\therefore \qquad \text{U.D.L.} \quad w = \frac{W}{2L}$$

Reactions at ends $= R = \frac{1}{2}\,(W - P)$

The deflection at the prop is given by

$$y_C = \frac{\partial U}{\partial P} = \frac{1}{EI}\int M_x\,\frac{\partial M_x}{\partial P}\,.\,dx$$

Since the prop deflects by an amount λ times the load it carries, we have

$$y_C = -P\lambda$$

The minus sign has been used since the deflection is in a direction opposite to the line of action of P.

Hence $\quad -P.\lambda = \dfrac{2}{EI}\displaystyle\int_0^L M_x \dfrac{\partial M_x}{\partial P}.dx \quad$ or $\quad P = -\dfrac{2}{EI\lambda}\displaystyle\int_0^L M_x \dfrac{\partial M_x}{\partial P}.dx \qquad$ ...(1)

For any section distant x from A, $\quad M_x = \dfrac{1}{2}(W-P)x - \dfrac{W}{2L}\dfrac{x^2}{2}; \dfrac{\partial M_x}{\partial P} = -\dfrac{x}{2}$

Hence $\quad P = \dfrac{2}{EI\lambda}\displaystyle\int_0^L \left[\dfrac{1}{2}(W-P)x - \dfrac{W}{2L}.\dfrac{x^2}{2}\right]\dfrac{x}{2}\,dx$

or $\quad P.\dfrac{EI\lambda}{2} = \left[\dfrac{1}{12}(W-P)x^3 - \dfrac{W}{32L}x^4\right]_0^L = \dfrac{WL^3}{12} - \dfrac{PL^3}{12} - \dfrac{WL^3}{32} = \dfrac{5WL^3}{96} - \dfrac{PL^3}{12}$

or $\quad P\left(\dfrac{6EI\lambda}{L^3}+1\right) = \dfrac{5W}{8}$ From which $\quad P = \dfrac{5W}{8\left(1+\dfrac{6EI\lambda}{L^3}\right)}$.

Example 14.17. *Obtain an expression for vertical displacement of point A in the bent cantilever shown in Fig. 14.19 (a).*

Solution. The vertical displacement of A is given by

$$y_A = \frac{\partial U}{\partial W}$$

$$= \frac{1}{E}\int \frac{M_x}{I_x}\frac{\partial M_x}{\partial W}dx \qquad \text{...(1)}$$

where the integration is carried over the whole frame. A bending moment is designated negative if it causes convexity to the *inside* of the frame. The B.M.D. for the whole frame is shown in Fig. 14.19 (*b*). Assume width of the frame as unity.

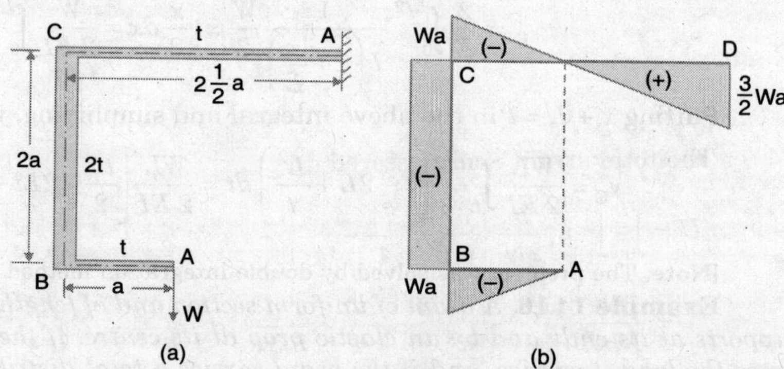

Fig. 14.19

(*i*) ***For AB :*** $\quad I = \dfrac{1}{12}\times 1\times t^3 = \dfrac{t^3}{12}$.

Measuring x from A, $M_x = -W.x;\ \dfrac{\partial M_x}{\partial W} = -x$

The limits of x are 0 at A to a at B.

(*ii*) ***For BC :*** $\quad I = \dfrac{1}{12}\times 1(2t)^3 = \dfrac{2t^3}{3};\ M_x = -Wa;\ \dfrac{\partial M_x}{\partial W} = -a$

The limits of x are 0 at B to $2a$ at C.

(*iii*) ***For CD :*** $\quad I = \dfrac{1}{12}\times 1(t)^3 = \dfrac{t^3}{12}$.

Measuring x from C, $M_x = -W(a-x)$; $\dfrac{\partial M_x}{\partial W} = -(a-x)$

The limits of x are 0 at C to $5a/2$ at D. Substituting the values in (1), we get

$$y_A = \frac{1}{E}\left[\int_0^a Wx(x)\frac{12}{t^3}dx + \int_0^{2a} Wa(a)\frac{3}{2t^3}dx + \int_0^{5a/2} W(a-x)(a-x)\frac{12}{t^3}dx\right]$$

$$= \frac{1}{Et^3}\left[\left(4Wx^3\right)_0^a + \left(\frac{3}{2}Wa^2x\right)_0^{2a} + 12W\left(a^2x + \frac{x^3}{3} - \frac{2ax^2}{2}\right)_0^{5a/2}\right]$$

$$= \frac{1}{Et^3}\left[4Wa^3 + 3Wa^3 + 12W\left(\frac{5}{2}a^3 + \frac{125}{24}a^3 - \frac{25}{4}a^3\right)\right] = \frac{\mathbf{24.5\,Wa^3}}{\mathbf{Et^3}}.$$

Example 14.18. *For a given cantilever of rectangular cross-section, length L, depth d and carrying a concentrated load at the free end, show that*

$$\frac{\textit{Deflection due to shearing strain}}{\textit{Bending deflection}} = \textit{constant} \times \left(\frac{d}{L}\right)^2$$

and find the value of the constant for a steel cantilever. Hence find the least value of (L/d) if the deflection due to shearing strain is not to exceed 1% of the total deflection.

Take $E = 2 \times 10^5$ *N/mm²* *and* $N = 0.765 \times 10^5$ *N/mm².*

Solution. At any layer distant y from the N.A., shear stress is given by

$$q = \frac{F}{2I}\left(\frac{d^2}{4} - y^2\right), \text{ where } F = W \text{ and } I = \frac{1}{12}bd^3$$

$$\therefore \quad q = \frac{6W}{bd^3}\left(\frac{d^2}{4} - y^2\right)$$

The total strain energy stored due to shear is given by Eq. 14.10, from which

$$U = \int_0^L \int_{-\frac{d}{2}}^{+\frac{d}{2}} \frac{b}{2N}\left\{\frac{6W}{bd^3}\left(\frac{d^2}{4} - y^2\right)\right\}^2 dx\,dy = \frac{b}{2N}\int_0^L \int_{-\frac{d}{2}}^{+\frac{d}{2}} \frac{36W^2}{b^2d^6}\left(\frac{d^4}{16} - \frac{d^2y^2}{2} + y^4\right)dx\,dy$$

$$= \frac{18W^2}{Nb\,d^6}\int_0^L \left(\frac{d^5}{16} - \frac{d^2}{6}\cdot\frac{d^3}{4} + \frac{1}{5}\frac{d^5}{16}\right)dx = \frac{18W^2}{Nbd^6}\cdot\frac{d^5}{30}\cdot L = \frac{3}{5}\frac{W^2L}{Nbd} \quad \text{...(1)}$$

Also, work done $= \dfrac{1}{2}Wy_S$ ...(2)

Equating the two, we get $\dfrac{1}{2}Wy_S = \dfrac{3W^2L}{5Nbd}$ From which $y_S = \dfrac{6WL}{5Nbd}$ ...(3)

Deflection due to bending $= y_b = \dfrac{WL^3}{3EI} = \dfrac{4WL^3}{Ebd^3}$ ...(4)

$$\therefore \quad \frac{y_S}{y_b} = \frac{6WL}{5Nbd} \times \frac{bd^3E}{4WL^3} = \frac{3}{10}\frac{E}{N}\left(\frac{d}{L}\right)^2 = k\left(\frac{d}{L}\right)^2$$

Hence $\quad k = \dfrac{\mathbf{3}}{\mathbf{10}}\dfrac{\mathbf{E}}{\mathbf{N}}$ **(Ans.)**

Substituting numerical values, $k = \dfrac{3}{10} \times \dfrac{2 \times 10^5}{0.765 \times 10^5} = 0.7843$

Again $\quad y_S = 0.01\,(y_S + y_b)$ ...(Given) or $\dfrac{y_S}{y_b} = \dfrac{1}{99} = \left(\dfrac{d}{L}\right)^2 k$

$$\therefore \quad \left(\frac{d}{L}\right)^2 = \frac{1}{99k} = \frac{1}{99 \times 0.7843} \quad \text{From which} \quad \frac{L}{d} = (99 \times 0.7843)^{1/2} = \mathbf{8.81}$$

PROBLEMS

1. A cantilever AB of length L is loaded with U.D.L. w over a length L_1 from the fixed end (Fig. 14.20). Using the method of strain energy, calculate the deflection at the free end.

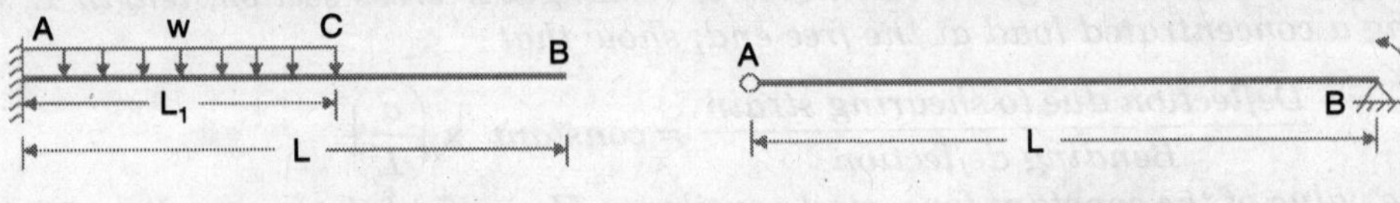

Fig. 14.20 Fig. 14.21

2. A simply supported beam AB is loaded with a couple μ at the end B (Fig. 14.21). Determine the slopes at the ends, using strain energy method.

3. A beam AB of length L is loaded with U.D.L. w/unit length extending from the centre towards the left for a length a (Fig. 14.22). Calculate the central deflection.

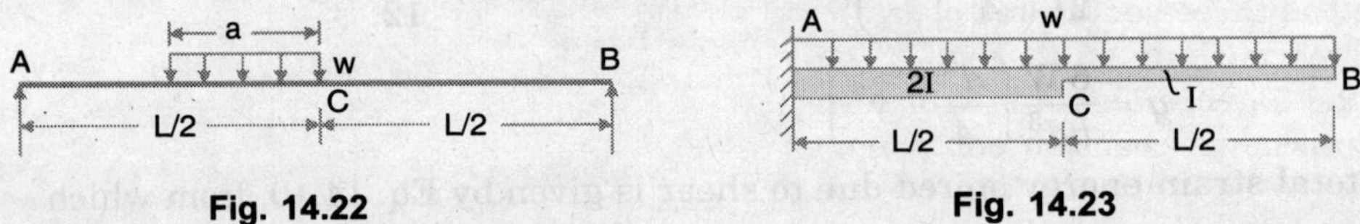

Fig. 14.22 Fig. 14.23

4. A cantilever beam AB with two different moments of inertia I and $2I$ supports a uniformly distributed load W (Fig. 14.23). Determine the deflection at the free end, using strain energy method.

5. A simply beam with overhang is loaded as shown in Fig. 14.24. Find the deflection at the free end.

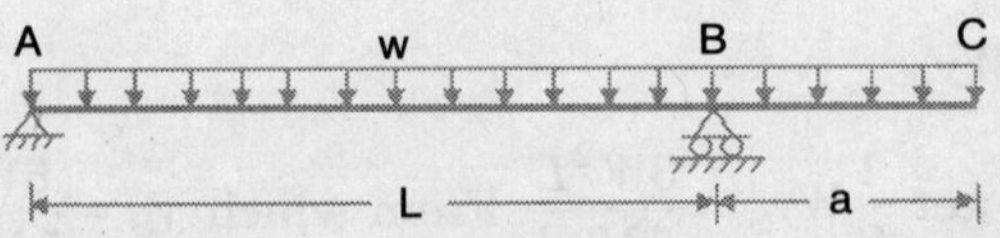

Fig. 14.24

ANSWERS

1. $y_B = \dfrac{wL_1^4}{8\,EI} + \dfrac{wL_1^3}{6\,EI}(L - L_1)$; **2.** $i_A = \dfrac{\mu L}{6\,EI}$; $i_B = -\dfrac{\mu L}{3\,EI}$

3. $y_C = \dfrac{wa}{24\,EI}\left\{L^3 - 2La^2 + a^3\right\}$; **4.** $y_B = \dfrac{17\,wL^4}{256\,EI}$

5. $y_C = \dfrac{wa}{24\,EI}(3a^3 + 4a^2L - L^3)$.

15

Fixed Beams

15.1. STATICALLY INDETERMINATE BEAMS

A beam is said to be statically determinate if all the reaction components can be calculated by applying the three conditions of static equilibrium, *i.e.,* $\Sigma V = 0$, $\Sigma H = 0$ and $\Sigma M = 0$, where ΣV is the algebraic sum of all vertical forces, ΣH is the algebraic sum of all horizontal forces and ΣM is the algebraic sum of moment of all the forces about a point. However, when the number of unknown reaction components exceeds the static conditions of equilibrium, the beam is said to be *statically indeterminate.* To determine the reaction components at the supports of such beams, we will have to consider the equations of deformations or displacements, in addition to the equations of statics. For vertical system of loading, there are two conditions of statical equilibrium available (*i.e.,* $\Sigma V = 0$ and $\Sigma M = 0$). For such loading, Fig. 15.1(*a*), 15.1(*b*), 15.1(*c*), 15.1(*d*) and 15.1(*e*) are the examples of statically determinate beams while Fig. 15.1(*f*), 15.1(*g*) and 15.1(*h*) are the examples of statically indeterminate beams.

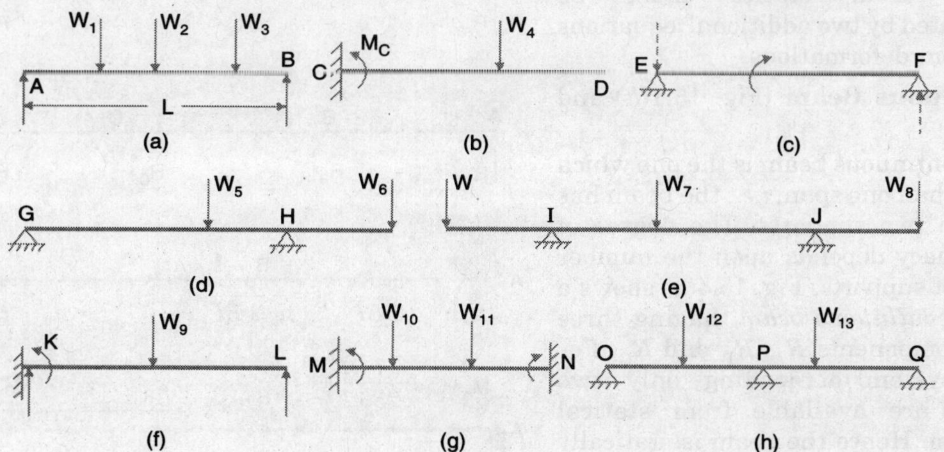

Fig. 15.1. Statically determinate and Statically indeterminate Beams

Although there are a wide variety of statically indeterminate beams, we shall be considering here three basic types of statically indeterminate beams : (*i*) propped cantilever (*ii*) fixed beam, and (*iii*) continuous beam.

1. Propped Cantilever [Fig. 15.1(*f*) and 15.2]

A propped cantilever is a cantilever which is propped either at the free end or at some intermediate point. There are *three* reaction components : R_A, R_B and M_A. For vertical system of loading, only two equations of statics are available, *i.e.*, $\Sigma V = 0$ and $\Sigma M = 0$. *Hence the beam is statically indeterminate to first degree.* The two equations of statics must be supplemented by the third equation expressing the conditions of compatibility (or continuity) of deformations. Since end *B* (*i.e.*, the propped end) cannot yield, we have

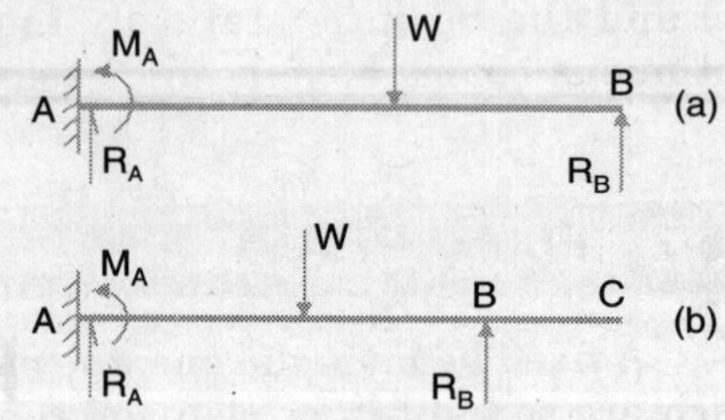

Fig. 15.2. Propped Cantilever

$$y_B \text{ due to } W = y_B' \text{ due to } R_B$$

Such cases have been discussed fully in the previous four chapters.

2. Fixed Beam [Fig. 15.1(*g*) and 15.3]

In the case of a fixed beam, the ends are *constrained* or *built-in* so that the tangents there remain in horizontal position. A fixed beam is also known as a *built-in* or *encastre* beam. In order to satisfy the condition of zero slope at each end, a fixidity moment will be induced at each end. Due to this, there will be *four* reaction components at the ends : R_A , M_A, R_B, M_B. For vertical system of loading, only two equations are available from statical equilibrium (*i.e.*, $\Sigma V = 0$ and $\Sigma M = 0$), *Hence a fixed beam is a statically indeterminate beam of second order.* To analyse such a beam, the two equations from statics must be supplemented by two additional equations arising from deformations.

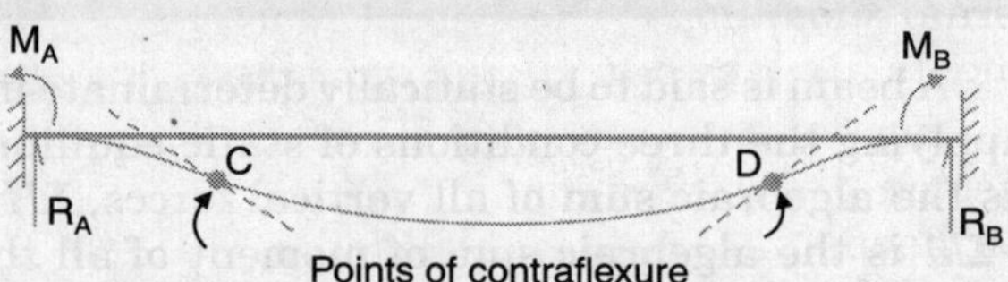

Fig. 15.3. Fixed Beam

3. Continuous Beam [Fig. 15.1(*h*) and 15.4].

A continuous beam is the one which has more than one span, *i.e.*, the beam has more than two supports. The degree of indeterminacy depends upon the number and type of supports. Fig. 15.4(*a*) shows a *two span continuous beam,* having three reaction components R_A, R_B and R_C. For vertical system of loading only two equations are available from statical equilibrium. Hence the beam is statically indeterminate to the first degree. Similarly, Fig. 15.4(*b*) shows a *three span*

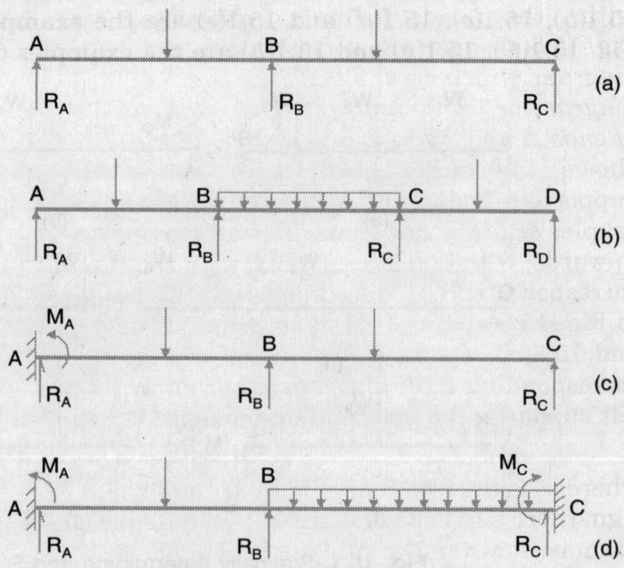

Fig. 15.4. Continuous Beams

continuous beam having four reaction components R_A, R_B, R_C and R_D; the beam is therefore statically indeterminate to *second degree.* The two span continuous beam of Fig. 15.4(*c*) has one end *A* fixed. Hence there are four reaction components (*i.e.*, R_A, R_B, R_C and M_A) and the beam is statically indeterminate to second degree. Similarly, the two span continuous beam of Fig. 15.4(*d*) has ends *A* and *C* fixed; hence there are five reaction components (R_A, R_B, R_C, M_A, M_C) and the beam is statically indeterminate to *third degree.* The methods of analyzing continuous beams have been discussed in Chapter 16.

15.2. FIXED BEAM

A fixed beam is the one, the ends of which are so built-in or clamped that the tangent at each end remain horizontal when the beam is subjected to external loading. Hence a fixed beam may be looked upon as a simply supported beam subjected to end moments of such magnitude that the slope at each end is zero when the beam is subjected to external loading. The end reactive moments are known as *fixing moments.*

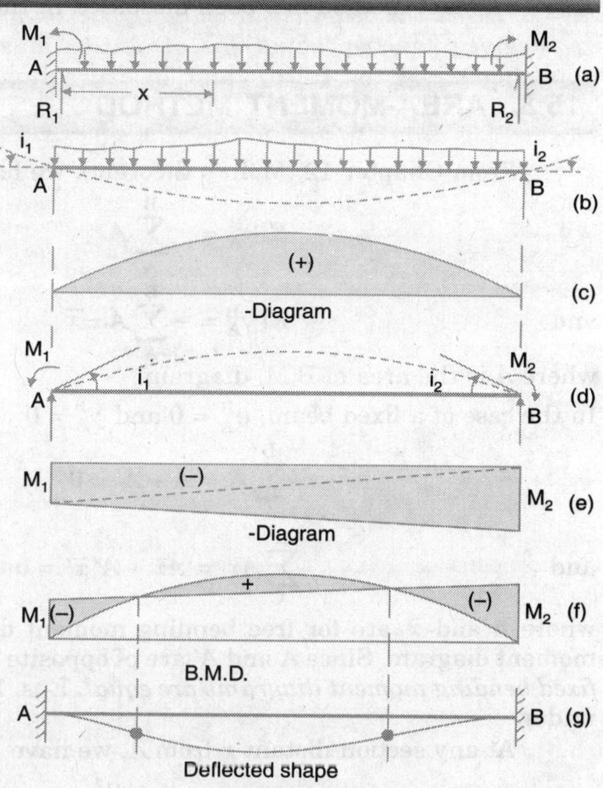

Fig. 15.5. Effects of Fixidity

Fig. 15.5(*a*) shows a fixed beam subjected to external loading. When the beam bends under external loading, the ends will try to rotate; this is prevented by the reactive couples M_1 and M_2, *induced* at the two ends so as to reduce slope at each end to zero value. Fig. 15.5(*b*) shows the same beam, with the same loading, but with the ends simply supported. Fig. 15.5(*c*) shows the corresponding B.M. diagram, which we shall call the *free B.M. diagram* or the *μ-diagram. The rotations at ends A* and *B* are i_1 and i_2. Fig. 15.5(*d*) shows the same beam with simply supported ends and subjected to end couples M_1, M_2; the beam will bend convex upwards, with end slopes i_1 and i_2 in the *reverse directions* to those shown in Fig. 15.5(*b*). The corresponding B.M. diagram, known as the *fixing moment diagram* or the *μ′-diagram* is shown in Fig. 15.5(*e*). The B.M. is negative throughout. By super-imposing the effects of Fig. 15.5(*b*) and 15.5(*d*), we get a fixed beam of Fig. 15.5(*a*), with slopes at each end equal to zero; the corresponding B.M. diagram is shown in Fig. 15.5(*f*). Thus, at any section distant *x* from the left supports, the net bending moment M_x is given by,

$$M_x = \mu_x + \mu_x' \qquad ...(15.1)$$

where μ_x is the free B.M. and μ_x' is the fixed B.M. at the section. Since μ_x and μ_x' are of opposite sign, there will be two points of contraflexure, as shown in Fig. 15.5(*f, g*). The bent shape of the beam is shown in Fig. 15.5(*g*). It is important and interesting to note that the bending moment in a fixed beam [Fig. 15.5(*f*)] is much lesser that the B.M. at the corresponding point in the simply supported beam [Fig. 15.5(*c*)]. *The effect of fixidity of the beam at the ends is thus to reduce the bending moment.* However, the B.M. M_x at any point cannot be evaluated unless M_1 and M_2 are evaluated first. There are several methods of analysing a fixed beam. But we shall discuss here the following methods:

(*i*) Area moment method

(*ii*) Double integration method

(*iii*) Method of super-position (flexibility method)

(*iv*) Method of three moment theorem.

The last method has been discussed in the next Chapter.

15.3. AREA-MOMENT METHOD

From Chapter 12 (Mohr's theorem), we have

$$EI\theta_A^B = -\sum_A^B A \quad \text{...(1)}$$

and

$$EIy_A^B = -\sum_A^B A \cdot \bar{x} \quad \text{...(2)}$$

where A is the area of B.M. diagram.

In the case of a fixed beam, $\theta_A^B = 0$ and $y_A^B = 0$

$$\therefore \quad \sum_A^B A = A + A' = 0 \quad \text{...(15.2)}$$

and

$$\sum_A^B A\bar{x} = A\bar{x} + A'\bar{x}' = 0 \quad \text{...(15.3)}$$

where A and $\bar{x}$ are for free bending moment diagram and A' and $\bar{x}'$ are for fixed bending moment diagram. Since A and A' are of opposite sign, it is concluded that the *areas of free and fixed bending moment diagrams are equal.* Eqs. 15.2 and 15.3 can be derived independently as under.

At any section distant x from A, we have

$$M_x = -EI\frac{d^2y}{dx^2} = \mu_x + \mu_x' \quad \text{...(3)}$$

Integrating over the whole length,

$$-\left[EI\frac{dy}{dx}\right]_0^L = \int_0^L \mu_x dx + \int_0^L \mu_x' dx$$

or

$$-EI(\theta_B - \theta_A) = A + A'$$

But $\theta_B = 0$ and θ_A is also zero.

Hence $A + A' = 0$ ...(Eq. 15.2)

Also, multiplying both sides of Eq. 3 by x and integrating over the whole length

$$-\int_0^L EIx\frac{d^2y}{dx^2} = \int_0^L x\mu_x dx + \int_0^L x\mu_x' dx$$

or $$-EI\left[x\frac{dy}{dx}-y\right]_0^L = A\bar{x} + A'\bar{x}'$$

or $$-EI\,[(L\theta_B - y_B) - (0.\theta_A - y_A)] = A\bar{x} + A'\bar{x}'$$

But θ_B, y_B, θ_A and y_A are each zero.

$$\therefore \quad A\bar{x} + A'\bar{x}' = 0 \qquad \text{...(Eq. 15.3)}$$

Values of A, A', $\bar{x}$ and $\bar{x}'$

The values of A and $\bar{x}$ will depend upon the type of external loading. However, the area of fixing B.M. diagram is given by

$$A' = -\frac{M_1 + M_2}{2}\,.\,L \qquad \text{...(15.4)}$$

The value of $A'\bar{x}'$ of the fixing moment diagram can be found by splitting the diagram into two triangles by a dotted line a shown in Fig. 15.5(e).

Thus $$A'\bar{x}' = -\left[\left(\frac{1}{2}M_1L\,.\,\frac{L}{3}\right)+\left(\frac{1}{2}M_2L\,.\,\frac{2}{3}L\right)\right]$$

or $$A'\bar{x}' = -\left[\frac{M_1L^2}{6}+\frac{M_2L^2}{3}\right] = -\frac{L^2}{6}(M_1 + 2M_2)$$

where $\bar{x}'$ is measured from end A.

We shall now consider different cases of loading on a fixed beam.

15.4. FIXED BEAM WITH U.D.L.

Fig. 15.6(a) shows a fixed beam with U.D.L. Fig. 15.6(b) shows the μ-diagram. Due to symmetry, we have $M_1 = M_2 = M$. Hence μ′ diagram will be a rectangle.

Hence $A' = -M\,.\,L$

The μ-diagram will be parabola with maximum ordinate equal to $\frac{wL^2}{8}$.

$$\therefore \quad A = \left(\frac{2}{3}L\right)\left(\frac{wL^2}{8}\right) = \frac{wL^3}{12} \qquad \text{...(15.5)}$$

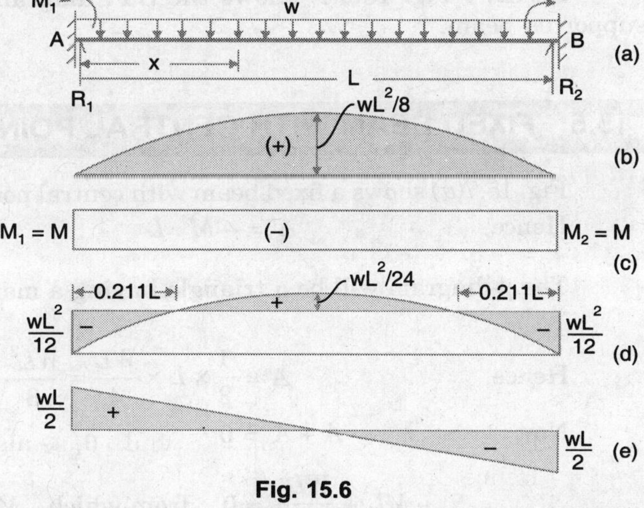

Fig. 15.6

Now $A + A' = 0$

$$\therefore \quad \frac{wL^3}{12} + (-ML) = 0$$

From which $$M = \frac{wL^2}{12} \qquad \text{...(15.6)}$$

B.M. Diagram : Fig. 15.6(d) shows the B.M. diagram obtained by super-position of μ_x and μ_x'-diagrams. The B.M. at any point distant x from the end A is given by

$$M_x = \mu_x + \mu_x' = \left(\frac{wL}{2}x - \frac{wx^2}{2}\right) - \frac{wL^2}{12} \quad ...(1)$$

For points of contraflexure, equate M_x to zero.

Thus, $\frac{wL}{2}x - \frac{wx^2}{2} - \frac{wL^2}{12} = 0$ or $x^2 - Lx + \frac{L^2}{6} = 0$

or $\left(x - \frac{L}{2}\right)^2 = \frac{L^2}{12}$ which gives $x = \frac{L}{2} \pm \frac{L}{2\sqrt{3}} = 0.5\,L \pm 0.289\,L = 0.211\,L$ or $0.789L$

At the centre,
$$M_C = \frac{wL}{2}\cdot\frac{L}{2} - \frac{w}{2}\left(\frac{L}{2}\right)^2 - \frac{wL^2}{12} = +\frac{wL^2}{24} \quad ...(15.7)$$

Thus, the central B.M. is 1/3 of the B.M. for a freely supported beam. The slope and deflection at any point can be found by Mohr's theorem.

Maximum deflection : By inspection, maximum deflection occurs at the centre. Hence, we have, from Mohr's theorem :

$$EI\,y_C^A = -\sum_C^A A\bar{x} = -\left[\left(\frac{2}{3}\times\frac{L}{2}\times\frac{wL^2}{8}\right)\left(\frac{3}{8}\frac{L}{2}\right) - \left(\frac{wL^2}{12}\times\frac{L}{2}\times\frac{L}{4}\right)\right]$$

$$= -\left[\frac{wL^4}{128} - \frac{wL^4}{96}\right] = \frac{wL^4}{384}$$

$$\therefore \quad y_C^A = \frac{wL^4}{384\,EI} \quad ...(15.8)$$

This is 1/5 of the deflection of a freely supported beam with U.D.L.

S.F.D. : Fig. 15.6(e) shows the S.F. diagram, which is similar to that for a simply supported beam.

15.5 FIXED BEAM WITH CENTRAL POINT LOAD

Fig. 15.7(a) shows a fixed beam with central point load W. Due to symmetry, $M_1 = M_2 = M$.

Hence, $A' = -M\,.\,L$

The μ-diagram will be a triangle having a maximum ordinate of $+\frac{WL}{4}$ at the centre.

Hence, $A = \frac{1}{2}\times L\times\frac{WL}{4} = \frac{WL^2}{8}$

Now $A + A' = 0$

$$\therefore \quad -ML + \frac{WL^2}{8} = 0, \text{ from which } M = \frac{WL}{8} \quad ...(15.9)$$

Thus, the fixing moments are known.

(a) ***B.M. diagram :***

Fig. 15.7(*d*) shows the B.M. diagram obtained by super-imposing μ and μ′-diagrams. At any point distant x from A.

$$M_x = \mu_x + \mu_x' = \frac{W}{2}x - \frac{WL}{8}$$

$$\text{At} \quad x = \frac{L}{2}, M_C = \frac{W}{2}.\frac{L}{2} - \frac{WL}{8} = + \frac{WL}{8} \qquad \text{...[15.9(a)]}$$

Thus, the central B.M. is half of the B.M. for a freely supported beam. For point of contraflexure,

$$M_x = 0 = \frac{W}{2}x - \frac{WL}{8}$$

This gives $x = \frac{L}{4}$.

(b) ***Maximum deflection :***

The slope and deflection at any point can be found using Mohr's theorems. The maximum deflection will evidently occur at C, for which

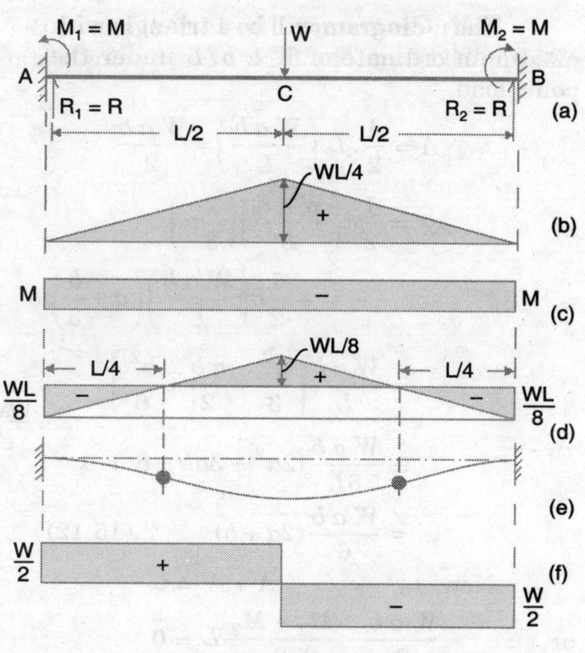

Fig. 15.7

$$E\,I\,y_C^A = -\sum_C^A A\bar{x} = -\left[\left(\frac{1}{2}\times\frac{L}{2}\times\frac{WL}{4}\times\frac{1}{3}.\frac{L}{2}\right) - \left(M.\frac{L}{2}.\frac{L}{4}\right)\right] = -\left[\frac{WL^3}{96} - \frac{WL^3}{64}\right] = \frac{WL^3}{192}$$

$$\therefore \quad y_C^A = \frac{WL^3}{192\,EI} \qquad \text{...(15.10)}$$

This is 1/4 of the corresponding deflection for a simply supported beam. The deflected shape of the beam is shown in Fig. 15.7(*e*), in which the beam has two points of contraflexure at $L/4$ from each end.

(c) ***S.F.D. :*** The S.F.D. shows in Fig. 15.7(*f*) is similar to the one for the simply supported beam.

15.6. FIXED BEAM WITH ECCENTRIC POINT LOAD

Fig. 15.8(*a*) shows a fixed beam subjected to a point load W placed at a distance a from A and b from B. Due to eccentric loading, M_1 and M_2 will not be equal. Let $a > b$. In that case, $M_2 > M_1$ and the μ′-diagram will be a trapezium.

$$\therefore \quad A' = -\frac{M_1 + M_2}{2}L$$

and

$$A'\bar{x} = -\left[\frac{M_1 + M_2}{2}L\left\{\frac{L}{3}\frac{M_1 + 2M_2}{M_1 + M_2}\right\}\right] = -\frac{L^2}{6}(M_1 + 2M_2) \qquad \text{...(15.11)}$$

(with A as the origin)

The μ-diagram will be a triangle with maximum ordinate of $W\,a\,b/L$ under the point load.

$$\therefore \quad A = \frac{1}{2}.L.\left(\frac{W\,a\,b}{L}\right) = \frac{W\,a\,b}{2}$$

$$A\bar{x} = \frac{1}{2}a\left(\frac{W\,a\,b}{L}\right)\left(\frac{2}{3}a\right) + \frac{1}{2}b\left(\frac{W\,a\,b}{L}\right)\left(a+\frac{b}{3}\right)$$

$$= \frac{W\,a\,b}{L}\left(\frac{a^2}{3}+\frac{a\,b}{2}+\frac{b^2}{6}\right)$$

$$= \frac{W\,a\,b}{6L}(2a^2+3ab+b^2)$$

$$= \frac{W\,a\,b}{6}(2a+b) \quad \text{...(15.12)}$$

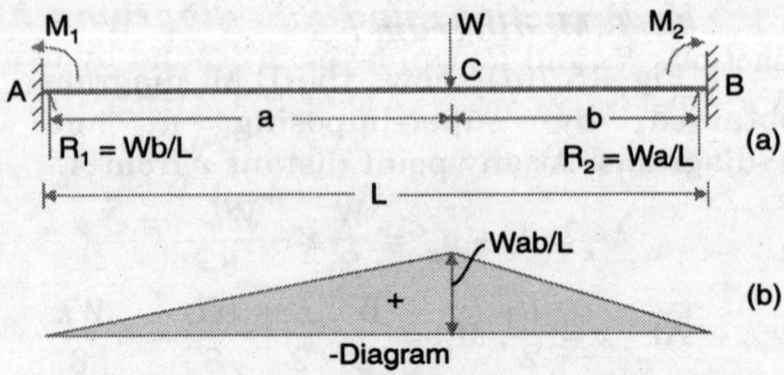

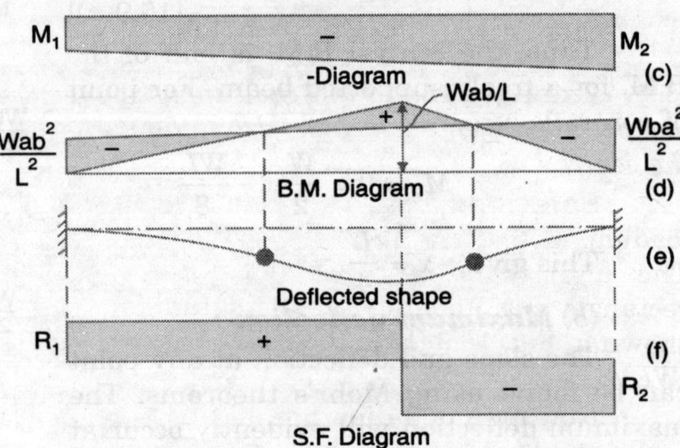

Fig. 15.8

Now $\quad A + A' = 0$

or $\quad \dfrac{W\,a\,b}{2} - \dfrac{M_1+M_2}{2}L = 0$

From which $\quad M_1 + M_2 = \dfrac{W\,a\,b}{L} \quad$ ...(1)

Also, $\quad A\bar{x} + A'\bar{x}' = 0$

or $\quad \dfrac{W\,a\,b}{6}(2a+b) - \dfrac{L^2}{6}(M_1+2M_2) = 0$

or $\quad M_1 + 2M_2 = \dfrac{W\,a\,b}{L^2}(2a+b) \quad$ ...(2)

Solving Eqs. (1) and (2), we get $\quad M_1 = \dfrac{W\,a\,b^2}{L^2} \quad$ and $\quad M_2 = \dfrac{W\,ba^2}{L^2} \quad$ ...(15.13)

Thus the fixing moments M_1 and M_2 are known.

B.M. diagram : Fig. 15.8(*d*) shows the B.M.D. obtained by superimposing μ and μ′ diagrams. The B.M. at any point in *AC*, distant *x* from *A* is given by

$$M_x = \mu_x + \mu_x' = \frac{Wb}{L}x - \left[M_1 + (M_2 - M_1)\frac{x}{L}\right] = \frac{W\,b\,x}{L} - \frac{W\,a\,b^2}{L^2} - \frac{W\,a\,b\,(a-b)}{L^3}x$$

For point of contraflexure in *AC*, equate M_x to zero.

$$0 = \frac{W\,b\,x}{L} - \frac{W\,a\,b^2}{L^2} - \frac{W\,a\,b\,(a-b)}{L^3}x$$

which gives $\quad x = \dfrac{a\,L}{3a+b} \quad$ ...(15.14)

End reactions and S.F. diagram : After having known the end moments, the end reactions R_1 and R_2 can be easily found by using the two equations of statics.

Taking moments about B, we get $R_1L - M_1 - Wb + M_2 = 0$

or $$R_1L = M_1 - M_2 + Wb = \frac{Wab^2}{L^2} - \frac{Wba^2}{L^2} + Wb = \frac{Wb^2}{L^2}\left(a - \frac{a^2}{b} + \frac{L^2}{b}\right)$$

From which $$R_1 = \frac{Wb^2}{L^3}(3a + b) \quad ...[15.15(a)]$$

Similarly, $$R_2 = \frac{Wa^2}{L^3}(a + 3b) \quad ...[15.15(b)]$$

Example 15.1. *Find the support moments of a built-in beam loaded at third points by two point loads W each. Also, draw the B.M. and S.F. diagrams and compute the deflection at the centre.*

Solution. Due to symmetrical loading, $M_1 = M_2 = M$ and $R_1 = R_2 = R$. The μ'-diagram will therefore be a rectangle [Fig. 15.9(*c*)]. The μ diagram is shown in Fig. 15.9 (*b*) in which $\mu_D = \mu_E = (WL)/3$.

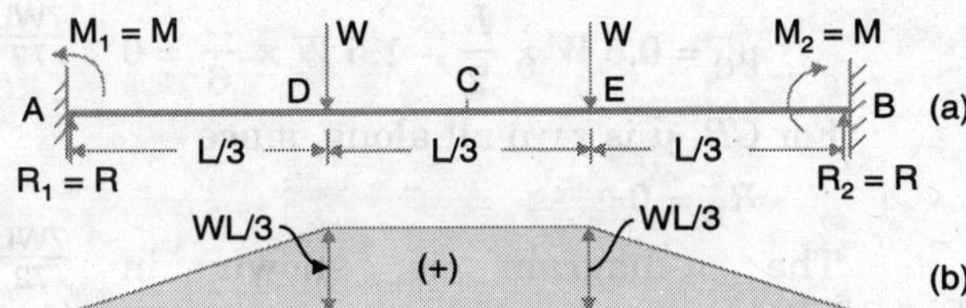

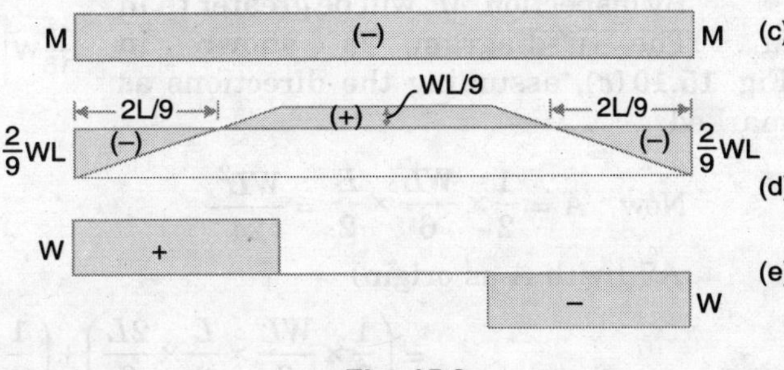

Fig. 15.9

Now area of μ-diagram.

$$A = \left(2 \times \frac{1}{2} \times \frac{L}{3} \times \frac{WL}{3}\right) + \left[\frac{WL}{3} \times \frac{L}{3}\right]$$

$$= \frac{2WL^2}{9}$$

Area of μ' diagram $= A' = -M \cdot L$

Now $A + A' = 0$ or $A = -A'$

$$\therefore \quad ML = \frac{2WL^2}{9} \quad \text{or} \quad M = \mathbf{\frac{2}{9}WL}$$

Also, due to symmetry,

$$R_1 = R_2 = R = W$$

Now, at any distance x from A, the B.M. in AD is given by

$$M_x = Wx - \frac{2}{9}WL \quad ...(1)$$

This will be zero at $\quad x = \frac{2}{9}L$

At the mid-span, $\quad M_C = \mu_C + \mu_C' = \frac{WL}{3} - \frac{2}{9}WL = +\frac{WL}{9}$

The B.M. and S.F. diagrams are shown in Fig. 15.9 (*d*) and (*e*).

Again $$EI\,y_C^A = -\sum_C^A A\bar{x} = \left[\frac{2}{9}WL \cdot \frac{L}{2} \cdot \frac{L}{4}\right] - \left[\left(\frac{WL}{3} \times \frac{L}{6} \times \frac{L}{12}\right) + \left(\frac{1}{2}\frac{WL}{3} \times \frac{L}{3}\right) \times \left(\frac{L}{6} + \frac{L}{9}\right)\right]$$

$$= \frac{WL^3}{36} - \left[\frac{WL^3}{216} + \frac{5WL^3}{18 \times 18}\right] = \frac{5}{648}\frac{WL^3}{EI}.$$

Example 15.2. *A fixed beam is loaded as shown. Determine the end reactions and fixing couples and draw the B.M. and S.F. diagrams for the beam.*

Solution. Let us first plot the μ-diagram for which

$$R_A = \frac{1}{L}\left(1.5\,W \times \frac{2L}{3} - \frac{WL}{2}\right)$$

$= 0.5\,W\,(\uparrow)$

Similarly,

$$R_B = \frac{1}{L}\left(1.5\,W \times \frac{L}{3} - \frac{WL}{2}\right) = 0$$

$\therefore \quad \mu_A = 0, \mu_D = 0.5\,W \times \dfrac{L}{3} = \dfrac{WL}{6}$

$$\mu_C = 0.5\,W \times \frac{L}{2} - 1.5\,W \times \frac{L}{6} = 0$$

For CB, μ is zero all along, since

$R_B = 0$

The μ-diagram is shown in Fig. 15.10 (*b*).

By inspection, M_1 will be greater than M_2. The μ′-diagram is shown in Fig. 15.10 (*c*), assuming the directions as marked.

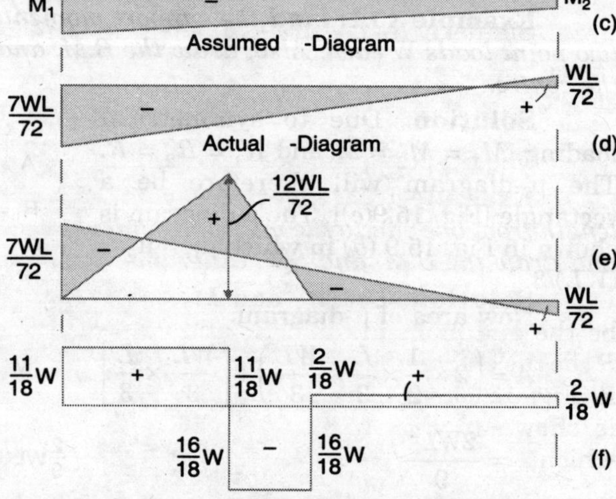

Fig. 15.10

Now $A = \dfrac{1}{2} \times \dfrac{WL}{6} \times \dfrac{L}{2} = \dfrac{WL^2}{24}$

$A\bar{x}$ (with A as origin)

$$= \left(\frac{1}{2} \times \frac{WL}{6} \times \frac{L}{3} \times \frac{2L}{9}\right) + \left(\frac{1}{2} \times \frac{WL}{6} \times \frac{L}{6}\right)\left(\frac{L}{3} + \frac{L}{18}\right) = \frac{5}{432}\,WL^3$$

Also, $\quad A' = -\dfrac{M_1 + M_2}{2} \times L \quad$ and $\quad A'\bar{x}' = -(M_1 + 2M_2)\,\dfrac{L^2}{6}$

Now, $\quad A + A' = 0 \quad$ or $\quad \dfrac{WL^2}{24} - \dfrac{M_1 + M_2}{2} L = 0$

or $$M_1 + M_2 = \frac{WL}{12} \qquad \text{...(1)}$$

and $\quad A\bar{x} + A'\bar{x}' = 0, \quad$ or $\quad \dfrac{5}{432}\,WL^3 - (M_1 + 2M_2)\,\dfrac{L^2}{6} = 0$

or $$M_1 + 2M_2 = \frac{5}{72}\,WL \qquad \text{...(2)}$$

Solving (1) and (2), we get $\quad M_1 = +\dfrac{7\,WL}{72} \quad$ and $\quad M_2 = -\dfrac{WL}{72}$

The plus sign with M_1 shows that M_1 is in the *same direction* as marked in Fig. 15.10 (*a*), while the – sign with M_2 shows that the direction of M_2 is reverse to the assumed one, *i.e.*, M_2 is in the anticlockwise direction. The actual μ′-diagram is shown in Fig. 15.10 (*d*), while Fig. 15.10 (*e*) shows the resultant B.M. diagram.

For reaction R_1 of the fixed beam, take moments about *B*. Thus.

$$R_1 L + W\frac{L}{2} - 1.5\, W \times \frac{2L}{3} - M_2 - M_1 = 0$$

$$\therefore \quad R_1 = \frac{1}{L}\left[-\frac{WL}{2} + WL + \frac{7\,WL}{72} + \frac{WL}{72}\right] = \frac{11}{18}\, W\,(\uparrow)$$

Similarly $R_2 L + M_1 + M_2 + \dfrac{WL}{2} - 1.5W\dfrac{L}{3} = 0$

$$\therefore \quad R_2 = \frac{1}{L}\left[-\frac{7}{72}WL - \frac{1}{72}WL - \frac{WL}{2} + 0.5\,WL\right] = -\frac{1}{9}W = \frac{1}{9}W\,(\downarrow)$$

The S.F. diagram is shown in Fig. 15.10.

Example 15.3. *A beam of span L has its ends fixed and carries a U.D.L. of w per unit length from one end to the midspan. Calculate the fixing moments and reactions at the supports and draw the B.M. and S.F. diagrams.*

Solution. Let M_1 and M_2 be the end moments and R_1 and R_2 be the end reactions, as marked in Fig. 15.11 (*a*). The μ′-diagram is shown in Fig. 15.11 (*c*) from which

$$A' = -(M_1 + M_2)\frac{L}{2}$$

and $\quad A'\bar{x}' = -(M_1 + 2M_2)\dfrac{L^2}{6}$

(with *A* as origin).

The μ-diagram is shown in Fig. 15.11 (*b*). Area of μ-diagram

$$= A = \int_0^{L/2} \mu_x dx + \int_{L/2}^{L} \mu_x dx$$

M1 w M2 A C B (a) R1 L/2 L/2 R2 wL²/16 + (b) – M2 (c) M1 + 11/192 wL² – – 5/192 wL² (d) 13/32 wL + – 3/32 wL (e)

Fig. 15.11

For a simply supported beam,

$$V_1 = \frac{3}{8}wL \text{ and } V_2 = \frac{wL}{8}$$

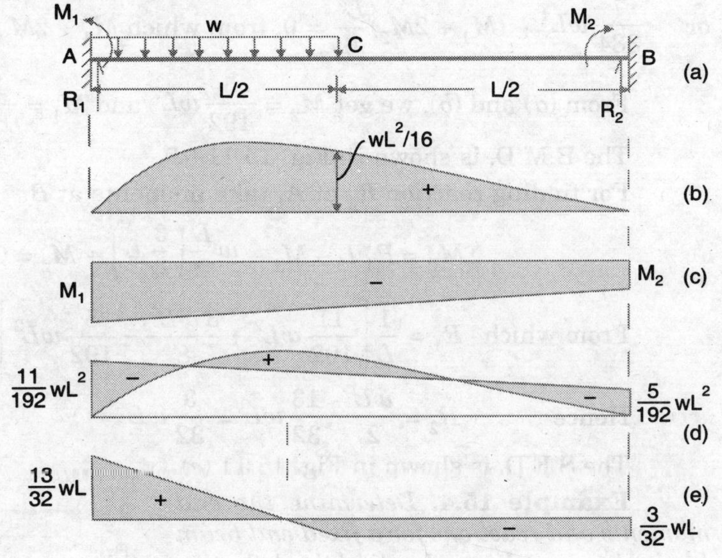

For *AC* $\quad \mu_x = \dfrac{3}{8}wL\,x - \dfrac{wx^2}{2}$

At $x = \dfrac{L}{2}$, $\quad \mu_C = \dfrac{3}{8}wL\,.\,\dfrac{L}{2} - \dfrac{w}{2}\left(\dfrac{L}{2}\right)^2 = \dfrac{wL^2}{16}$

Hence $\quad A = \left[\displaystyle\int_0^{\frac{L}{2}} \left(\frac{3}{8}wL\,x - \frac{wx^2}{2}\right) dx\right] + \left[\frac{1}{2} \times \frac{L}{2} \times \frac{wL^2}{16}\right]$

$$= \left[\frac{3}{8}wL\frac{x^2}{2} - \frac{wx^3}{6}\right]_0^{\frac{L}{2}} + \frac{wL^3}{64} = \frac{3}{8}\frac{wL}{2}\left(\frac{L}{2}\right)^2 - \frac{w}{6}\left(\frac{L}{2}\right)^3 + \frac{wL^3}{64} = \frac{wL^3}{24}$$

and $$A\bar{x} = \left[\int_0^{\frac{L}{2}}\left(\frac{3}{8}wL\,x - \frac{wx^2}{2}\right)x\,dx\right] + \left[\frac{1}{2}\times\frac{L}{2}\times\frac{wL^2}{16}\left(\frac{L}{2}+\frac{L}{6}\right)\right]$$

$$= \left[\frac{3}{8}wL\frac{x^3}{3} - \frac{wx^4}{8}\right]_0^{\frac{L}{2}} + \frac{wL^3}{64}\times\frac{2L}{3}$$

$$= \left[\frac{3}{24}wL\left(\frac{L}{2}\right)^3 - \frac{w}{8}\left(\frac{L}{2}\right)^4\right] + \frac{wL^4}{96} = \frac{7}{384}wL^4$$

Now, from area moment equations, $A + A' = 0$

or $$\frac{wL^3}{24} - (M_1 + M_2)\frac{L}{2} = 0, \text{ from which } M_1 + M_2 = \frac{wL^2}{12} \quad ...(a)$$

and $$A\bar{x} + A'\bar{x}' = 0$$

or $$\frac{7}{384}wL^4 - (M_1 + 2M_2)\frac{L^2}{6} = 0, \text{ from which } M_1 + 2M_2 = \frac{7}{64}wL^2 \quad ...(b)$$

From (*a*) and (*b*), we get $M_2 = \frac{5}{192}wL^2$ and $M_1 = \frac{11}{192}wL^2$

The B.M.D. is shown in Fig. 15.11 (*d*).

For finding reaction R_1 at A, take moments at B.

$$\Sigma M_B = R_1 L - M_1 - w\frac{L}{2}\left(\frac{3}{4}L\right) + M_2 = 0$$

From which $$R_1 = \frac{1}{L}\left[\frac{11}{192}wL^2 + \frac{3\,wL^2}{8} - \frac{5}{192}wL^2\right] = \frac{13}{32}wL$$

Hence $$R_2 = \frac{wL}{2} - \frac{13}{32}wL = \frac{3}{32}wL$$

The S.F.D. is shown in Fig. 15.11 (*e*).

Example 15.4. *Determine the end moments and reactions for a fixed end beam loaded by a couple μ_0. Also, find the deflection at the point C where the couple acts.*

Solution. Let M_1 and R_1 be the reaction components at A. Release end A apply M_1 and R_1 at A. The basic determinate structure (or the released structure) is shown in Fig. 15.12 (*b*), which is a cantilever.

The component B.M. diagram showing the effects of R_1, M_1 and μ_0 separately is shown in Fig. 15.12 (*c*). There

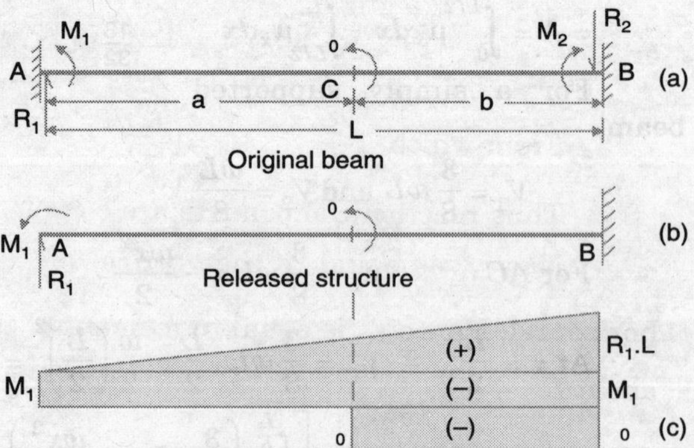

Fig. 15.12

are two unknowns : M_1 and R_1, and hence two condition equations are required. As a first condition, since both the ends of the beam have zero slopes, the change of slope between A and B is zero. Hence from area moment theorem

$$\sum_A^B A_M = 0 \text{ (Since } EI \text{ is constant throughout)}$$

$$\therefore \quad \frac{1}{2}R_1L\,.\,L - M_1\,.\,L - \mu_0 b = 0 \quad \text{or } R_1 - \frac{2M_1}{L} = \frac{2\mu_0 b}{L^2} \qquad ...(1)$$

The second condition is obtained from the fact that tangent to the deflection curve at A passes through point B. This means that the moment of the area of B.M. diagram between A and B, taken about B, is zero.

$$\therefore \quad \sum_A^B A_M\bar{x} = 0 \qquad \text{(Since } EI \text{ is constant throughout)}$$

$$\therefore \quad \left(\frac{1}{2}R_1L\,.\,L\right)\left(\frac{L}{3}\right) - M_1L\,.\,\frac{L}{2} - b\,\mu_0\,.\,\frac{b}{2} = 0$$

or $$R_1 - \frac{3M_1}{L} = \frac{3\mu_0 b^2}{L^3} \qquad ...(2)$$

Solving (1) and (2), we get

$$M_1 = \frac{\mu_0 b}{L^2}(2L - 3b) = \frac{\mu_0 b}{L^2}(2a - b) \quad \text{and} \quad R_1 = \frac{6\mu_0\,ab}{L^3}\ (\uparrow)$$

Since $\Sigma V = 0$ for the whole beam, we get $R_2 = -R_1 = -\dfrac{6\mu_0\,ab}{L^3}$ (*i.e.*, $\downarrow$)

Taking moments about A and assuming M_2 to be clockwise, we get

$$M_2 - M_1 - \mu_0 + \frac{6\mu_0\,ab}{L^2} = 0$$

or $$M_2 = \frac{\mu_0\,b}{L^2}(2a - b) + \mu_0 - \frac{6\mu_0\,ab}{L^2}$$

From which $$M_2 = \frac{\mu_0\,a}{L^2}(a - 2b)$$

Thus reaction components are completely known.

Now y_C is equal to the vertical offset from the tangent at A, since tangent at A is horizontal. Hence y_C is equal to moment of the area of $\dfrac{M}{EI}$ diagram between A and C, taken about C.

$$\therefore \quad y_C = \sum_C^A \frac{A_M}{EI}\bar{x} = \left[\frac{1}{2}a\,.\,R_1a\left(\frac{a}{3}\right) - M_1\,.\,a\,.\,\frac{a}{2}\right]\frac{1}{EI}$$

$$= \left(\frac{R_1 a^3}{6} - \frac{M_1 a^2}{2}\right)\frac{1}{EI} = \frac{1}{EI}\left[\frac{6\mu_0\, a^4 b}{6L^3} - \frac{\mu_0\, b}{2L^2}(2a-b)\,a^2\right] = \frac{\mu_0\, a^2 b^2\,(a-b)}{2L^3\, EI}$$

Special Case. If the couple acts at the mid-point $\left(i.e., a = b = \frac{L}{2}\right)$

$$\therefore \qquad M_1 = -M_2 = \frac{\mu_0}{4} \quad \text{and} \quad R_1 = -R_2 = \frac{3}{2}\frac{\mu_0}{L} \quad \text{and} \quad y_C = 0.$$

15.7. DOUBLE INTEGRATION METHOD

We shall now apply the double integration method (including Macaulay's method) to various standard cases of loadings on a fixed beam.

1. U.D.L. w over the whole span :

The loading is symmetrical.

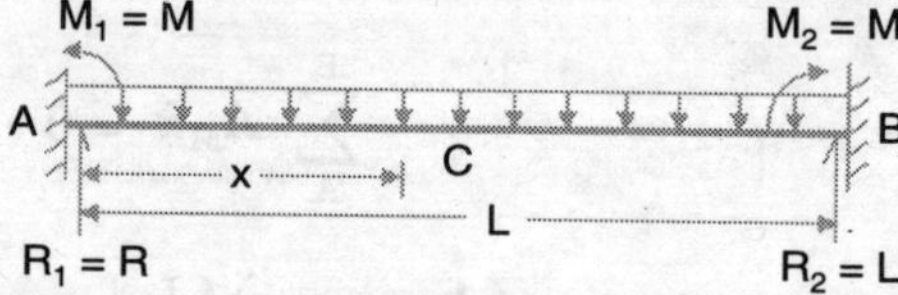

Fig. 15.13

Hence $M_1 = M_2 = M$ and $R_1 = R_2 = R = \frac{wL}{2}$.

At any distance x from end A,

$$EI\frac{d^2y}{dx^2} = -M_x = -\left[-M + \frac{wL}{2}x - \frac{wx^2}{2}\right] = M - \frac{wL}{2}x + \frac{wx^2}{2} \qquad \text{...(1)}$$

$$\text{Integrating,} \qquad EI\frac{dy}{dx} = Mx - \frac{wL}{4}x^2 + \frac{w}{6}x^3 + 0 \left(\text{Since } \frac{dy}{dx} = 0 \text{ at } x = 0\right) \qquad \text{...(2)}$$

$$\text{Integrating further, } EIy = \frac{Mx^2}{2} - \frac{wL}{12}x^3 + \frac{w}{24}x^4 + 0 \text{ (Since } y = 0 \text{ at } x = 0) \qquad \text{...(3)}$$

Now at $x = L$, $\frac{dy}{dx} = 0$. Hence from (2), we get $0 = ML - \frac{wL}{4}(L^2) + \frac{w}{6}(L)^3$

From which $M = \frac{wL^2}{4} - \frac{wL^2}{6} = \frac{wL^2}{12}$, which is the same as Eq. 15.6.

The maximum deflection at $x = \frac{L}{2}$ is given by

$$EIy_C = \frac{wL^2}{24}\left(\frac{L}{2}\right)^2 - \frac{wL}{12}\left(\frac{L}{2}\right)^3 + \frac{w}{24}\left(\frac{L}{2}\right)^4 = \frac{wL^4}{96} - \frac{wL^4}{96} + \frac{wL^4}{384}$$

$$\therefore \qquad y_C = \frac{wL^4}{384\, EI}, \text{ as before.}$$

2. Point load at mid-span :

Here also, due to symmetry,

$$M_1 = M_2 = M \text{ and } R_1 = R_2 = R = \frac{W}{2}.$$

Fig. 15.14

At any distance x from A,

$$EI\frac{d^2y}{dx^2} = -M_x = M - \frac{W}{2}x \,\Big|\, + W\left(x - \frac{L}{2}\right)$$

$$\therefore \quad EI\frac{dy}{dx} = Mx - \frac{Wx^2}{4} + 0 \,\Big|\, + \frac{W}{2}\left(x - \frac{L}{2}\right)^2 \quad ...(1) \quad \left(\text{Since } \frac{dy}{dx} = 0 \text{ at } x = 0\right)$$

and
$$EIy = \frac{Mx^2}{2} - \frac{Wx^3}{12} + 0 \,\Big|\, + \frac{W}{6}\left(x - \frac{L}{2}\right)^3 \quad ...(2) \quad (\text{Since } y = 0 \text{ at } x = 0)$$

At $x = L$, $\quad \dfrac{dy}{dx} = 0 = ML - \dfrac{WL^2}{4} + \dfrac{W}{2}\left(L - \dfrac{L}{2}\right)^2$

From which $\quad M = \dfrac{WL}{4} - \dfrac{WL}{8} = \dfrac{WL}{8}$ which is the same as found earlier.

The central deflection is found by substituting $x = L/2$ in (2).

$$EIy_C = \frac{M}{2}\left(\frac{L}{2}\right)^2 - \frac{W}{12}\left(\frac{L}{2}\right)^3 = \frac{WL}{8 \times 2} \times \frac{L^2}{4} - \frac{WL^3}{96} = \frac{WL^3}{192}$$

$\therefore \quad y_C = \dfrac{WL^3}{192\,EI}$ which is the same as found earlier.

3. Eccentric point load. Let M_1 be the reactive moment and R_1 be the vertical reaction at A, both of which are presently unknown. At any distance x from A,

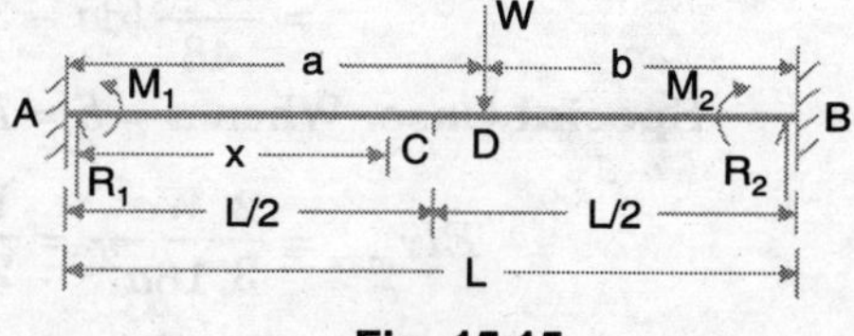

Fig. 15.15

$$EI\frac{d^2y}{dx^2} = -M_x = M_1 - R_1x \mid + W(x - a) \quad ...(1)$$

$$\therefore \quad EI\frac{dy}{dx} = M_1x - R_1\frac{x^2}{2} + 0 \,\Big|\, + \frac{W(x-a)^2}{2} \quad ...(2)$$

and
$$EIy = M_1\frac{x^2}{2} - R_1\frac{x^3}{6} + 0 \,\Big|\, + \frac{W(x-a)^3}{6} \quad ...(3)$$

At $x = L$, $\quad \dfrac{dy}{dx} = 0$ and $y = 0$. Hence from (2) and (3), we obtain

$$0 = M_1L - R_1\frac{L^2}{2} - \frac{W}{2}b^2 \quad ...(a)$$

and
$$0 = M_1\frac{L^2}{2} - R_1\frac{L^3}{6} + \frac{W}{6}b^3 \quad ...(b)$$

Solving (*a*) and (*b*), we get $\quad M_1 = \dfrac{Wab^2}{L^2} \quad$ and $\quad R_1 = \dfrac{Wb^2}{L^3}(3a + b)$

Substituting the values of M_1 and R_1 and $x = L$ in (1), we get

$$EI\frac{d^2y}{dx^2}\ (\text{at } x = L) = M_2 = \frac{Wab^2}{L^2} - \frac{Wb^2}{L^3}(3a + b)L + Wb = \frac{Wba^2}{L^2}$$

For maximum deflection, $\frac{dy}{dx} = 0$. Assuming that this occurs in *AD*, we get

$$0 = \frac{Wab^2}{L^2}x - \frac{Wb^2}{L^3}(3a+b)\frac{x^2}{2}, \text{ from (2). This gives } x = \frac{2aL}{3a+b}$$

$$\therefore \quad EIy_{max} = \frac{Wab^2}{2L^2}\left[\frac{2aL}{3a+b}\right]^2 - \frac{Wb^2}{6L^3}(3a+b)\left[\frac{2aL}{3a+b}\right]^3$$

$$= \frac{Wa^3b^2}{(3a+b)^2}\left[2 - \frac{4}{3}\right] = \frac{2}{3}\frac{Wa^3b^2}{(3a+b)^2} \qquad \text{...(15.16)}$$

The deflection under the load is obtained by substituting $x = a$ in (3)

$$EIy_D = \frac{Wab^2}{2L^2}(a^2) - \frac{Wb^2}{6L^3}(3a+b)(a^3) = \frac{Wa^3b^2}{6L^3}(3L - 3a - b)$$

$$= \frac{Wa^3b^2}{6L^3}(3a + 3b - 3a - b) = \frac{Wa^3b^3}{3L^3} \qquad \text{...(15.17)}$$

The deflection at the midspan is obtained by substituting at $x = L/2$ in (3)

$$EIy_C = \frac{Wab^2}{2L^2}\left(\frac{L}{2}\right)^2 - \frac{Wb^2}{6L^3}(3a+b)\left(\frac{L}{2}\right)^3 = \frac{Wab^2}{8} - \frac{Wb^2(3a+b)}{48}$$

$$= \frac{Wb^2}{48}[6a - 3a - b] = \frac{Wb^2}{48}(3a - b) = \frac{Wb^2}{48}(3L - 4b) \qquad \text{...(15.18)}$$

Special Case. When $a = b = L/2$, we get from Eq. 15.16,

$$EIy_{max} = \frac{2}{3}\frac{Wa^5}{16a^2} = \frac{W}{24}a^3 = \frac{W}{24}\left(\frac{L}{2}\right)^3 = \frac{WL^3}{192}$$

Also, $$EIy_D = \frac{Wa^6}{3L^3} = \frac{W}{3L^3}\left(\frac{L}{2}\right)^6 = \frac{WL^3}{192} = EIy_{max}$$

and $$EIy_C = \frac{W}{48}\left(\frac{L}{2}\right)^2\left(3L - \frac{4L}{2}\right) = \frac{WL^3}{192} = EIy_{max}$$

Hence, $y_C = y_D = y_{max}$, as expected for a central point load.

4. Moment at intermediate point: Let M_1 be the reactive moment and R_1 be the vertical reaction, in the directions marked in Fig. 15.16. At any distance x from *A*

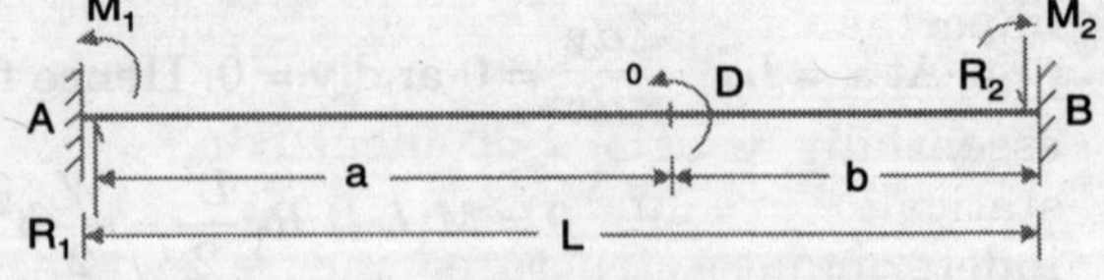

Fig. 15.16

$$EI\frac{d^2y}{dx^2} = -M_x$$

$$= -[-M_1 + R_1x \mid - \mu_0(x-a)^0]$$

or $$EI\frac{d^2y}{dx^2} = M_1 - R_1x \mid + \mu_0(x-a)^0 \qquad \text{...(1)}$$

$$\therefore \quad EI\frac{dy}{dx} = M_1x - R_1\frac{x^2}{2} + 0 \mid + \mu_0(x-a) \qquad \text{...(2)}$$

and $$EIy = M_1\frac{x^2}{2} - \frac{R_1x^3}{6} + 0 \Big| + \frac{\mu_0(x-a)^2}{2} \qquad \text{...(3)}$$

At $x = L$, $\dfrac{dy}{dx} = 0$ and $y = 0$. Hence from (2) and (3), we get

$$0 = M_1 L - \frac{R_1}{2} L^2 + \mu_0 b \qquad \text{...}(a)$$

and
$$0 = M_1 \frac{L^2}{2} - \frac{R_1}{6} L^3 + \frac{\mu_0\, b^2}{2} \qquad \text{...}(b)$$

Solving (a) and (b), we get $M_1 = \dfrac{\mu_0\, b}{L^2}\,(2a - b)$ ...[15.19(a)]

and
$$R_1 = \frac{6\mu_0\, ab}{L^3} \qquad \text{...[15.20}(a)]$$

At $x = L$,
$$E\,I\,\frac{d^2y}{dx^2} = M_2 = M_1 - R_1 L + \mu_0 = \frac{\mu_0\, b}{L^2}\,(2a - b) - \frac{6\mu_0\, ab}{L^2} + \mu_0$$

$$= \frac{\mu_0}{L^2}\,[b(2a - b) - 6ab + L^2] = \frac{\mu_0\, a}{L^2}\,(a - 2b) \qquad \text{...[15.19}(b)]$$

Also,
$$R_2 = -R_1 = \frac{6\mu_0\, ab}{L^3}\ (\downarrow) \qquad \text{...[15.20}(b)]$$

For the deflection under the load, put $x = a$ in Eq. (3)

$$\therefore \quad E\,I\,y_D = M_1 \frac{a^2}{2} - \frac{R_1}{6} a^3 = \frac{\mu_0\, b}{2L^2}\, a^2\,(2a - b) - \frac{6\mu_0\, ab}{6L^3}\, a^3$$

$$= \frac{\mu_0\, ba^2}{L^2}\left[\frac{2a - b}{2} - \frac{a^2}{L}\right] = \frac{\mu_0\, a^2 b^2}{2L^3}\,(a - b) \qquad \text{...(15.21)}$$

Special Case. If $a = b = \dfrac{L}{2}$: $M_1 = -M_2 = \dfrac{\mu_0}{4}$ and $R_1 = -R_2 = \dfrac{3}{2}\dfrac{\mu_0}{L}$.

15.8. METHOD OF SUPERPOSITION (FLEXIBILITY METHOD OR FORCE METHOD)

The method of superposition is an important tool in analysing statically indeterminate structures. The method essentially consists of identifying the statical redundants from the indeterminate structure, and removing the redundant reactions, thus making the structure statically determinate. Such a structure is known as the *released structure* or the *basic determinate structure*. The displacements of the released structure are now found. Next, the redundant reactions themselves are visualised as loads acting on the released structure, and the corresponding displacements are calculated. The

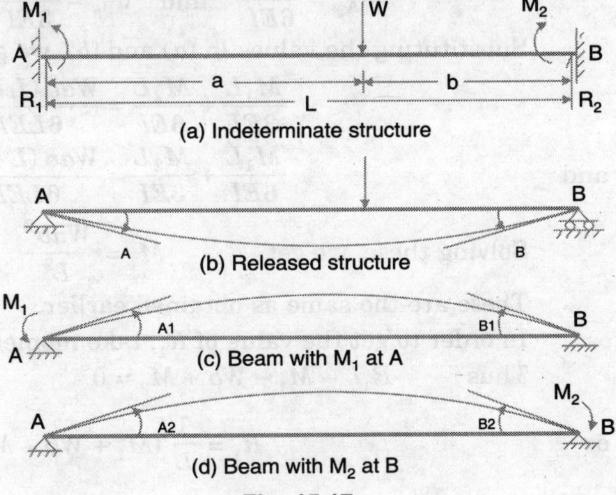

Fig. 15.17

principle of superposition is then applied, according to which the *final displacements* due to both the actual loads and the redundant reactions must be equal to the sum of those displacements as calculated separately. In the case of redundant constraints, these displacements are either zero (in case of non-yielding supports) or have known magnitudes (in case of yielding supports). Thus, equations of superposition expressing these relationships are obtained, the solution of which give the redundant reactions. The method is known as the *method of consistent deformation* or the *flexibility method* or the *force method.* We shall now illustrate the method for analysing a fixed beam subjected to an eccentric point load.

Fig. 15.17 (*a*) shows the fixed beam having four reaction components : M_1, M_2, R_1 and R_2. Since only two equations are available from statical equilibrium, the beam is statically indeterminate to second degree, *i.e.,* there are two redundant reactions. Let us choose M_1 and M_2 as the redundants. Remove these redundants, thus obtaining the *basic determinate structure* or the *released structure*, as shown in Fig. 15.17 (*b*). Let θ_A and θ_B be the slopes at the ends A and B, of the released structure, due to external load. Now remove the external load, and apply redundant moment M_1 at A, as shown in Fig. 15.17 (*c*), the corresponding rotations at A and B being θ_{A_1} and θ_{B_1}. Now remove M_1, and apply redundant moment M_2 at B, as shown in Fig. 15.17 (*d*), the corresponding rotations being θ_{A_2} and θ_{B_2} at A and B. Since the angles of rotations at both ends in the original beam (*i.e.,* fixed beam) are zero, we have the following two equations of compatibility:

$$\Sigma\theta_A = \theta_A - \theta_{A_1} - \theta_{A_2} = 0 \qquad ...(a)$$

and

$$\Sigma\theta_B = \theta_B - \theta_{B_1} - \theta_{B_2} = 0 \qquad ...(b)$$

Now consider Fig. 15.17(*b*). The angles of rotations at the ends, due to load W are:

$$\theta_A = \frac{Wab\,(L+b)}{6LEI} \quad \text{and} \quad \theta_B = \frac{Wab\,(L+a)}{6LEI} \qquad ...(11.11)$$

Again, consider Fig. 15.17 (*c*). The angles of rotations at the ends, due to M_1 at A are:

$$\theta_{A_1} = \frac{M_1 L}{3EI} \quad \text{and} \quad \theta_{B_1} = \frac{M_1 L}{6EI} \qquad ...[11.17(a)]$$

Finally, consider Fig. 15.17 (*d*). The angles of rotations at ends, due to M_2 at B are:

$$\theta_{A_2} = \frac{M_2 L}{6EI} \quad \text{and} \quad \theta_{B_2} = \frac{M_2 L}{3EI} \qquad ...[11.17(b)]$$

Substituting the values in (*a*) and (*b*), we get

$$\frac{M_1 L}{3EI} + \frac{M_2 L}{6EI} = \frac{Wab\,(L+b)}{6LEI} \qquad ...\text{(I)}$$

and

$$\frac{M_1 L}{6EI} + \frac{M_2 L}{3EI} = \frac{Wab\,(L+a)}{6LEI} \qquad ...\text{(II)}$$

Solving these, we get $\quad M_1 = \dfrac{Wab^2}{L^2} \quad$ and $\quad M_2 = \dfrac{Wba^2}{L^2}$

These are the same as obtained earlier.

In order to get the value of R_1, take moments about B.

Thus $\quad R_1 L - M_1 - Wb + M_2 = 0$

or

$$R_1 = \frac{1}{L}\,(M_1 + Wb - M_2) = \frac{1}{L}\left[\frac{Wab^2}{L^2} + Wb - \frac{Wba^2}{L^2}\right]$$

From which $R_1 = \dfrac{Wb^2}{L^3}(L + 2a) = \dfrac{Wb^2}{L^3}(3a + b)$

Similarly $R_2 = \dfrac{Wa^2}{L^3}(L + 2b) = \dfrac{Wa^2}{L^3}(a + 3b).$

15.9. FIXED BEAM WITH UDL OVER A PART LENGTH FROM ONE END

Consider a section X distant x from A (Fig. 15.18). Load acting on an elemental distance $dx = w\ dx$. Due to this elemental point load $w\ dx$, the fixed end moments at A and B are:

$$dM_1 = \frac{(w\,dx)\,x\,(L-x)^2}{L^2}$$

and
$$dM_2 = \frac{(w\,dx)\,(L-x)\,(x^2)}{L^2}$$

Fig. 15.18

$$\therefore \quad \text{Total } M_1 \text{ at } A = \int_0^a \frac{wx\,(L-x)^2}{L^2}\,dx = \frac{w}{L^2}\int_0^a (L^2x + x^3 - 2Lx^2)dx$$

$$= \frac{w}{L^2}\left(\frac{L^2x^2}{2} + \frac{x^4}{4} - \frac{2Lx^3}{3}\right)_0^a$$

$$= \frac{w}{L^2}\left[\frac{L^2a^2}{2} + \frac{a^4}{4} - \frac{2La^3}{3}\right] = \frac{wa^2}{12L^2}\,[6L^2 + 3a^2 - 8La] \qquad \text{...[15.22(a)]}$$

$$\text{Similarly, total } M_2 \text{ at } B = \int_0^a \frac{w\,dx\,(L-x)\,x^2\,dx}{L^2} = \frac{w}{L^2}\int_0^a (Lx^2 - x^3)\,dx$$

$$= \frac{w}{L^2}\left[\frac{Lx^3}{3} - \frac{x^4}{4}\right]_0^a = \frac{wa^3}{12L^2}\,[4L - 3a] \qquad \text{...[15.22(b)]}$$

For reaction R_1, take moments about B.

Thus $\Sigma M_B = 0 = R_1L + M_2 - M_1 - wa\left(L - a + \dfrac{a}{2}\right)$

$$\therefore \quad R_1 = \frac{1}{L}\left[M_1 - M_2 + wa\left(L - \frac{a}{2}\right)\right]$$

$$= \frac{1}{L}\left[\frac{wa^2}{12L^2}(6L^2 + 3a^2 - 8La) - \frac{wa^3}{12L^2}(4L - 3a) + \frac{wa\,(2L-a)}{2}\right]$$

$$= \frac{wa}{2L^3}\,[a^3 - 2La^2 + 2L^3] \qquad \text{...[15.23(a)]}$$

Similarly, taking moments about A,

$$R_2L + M_1 - M_2 - \frac{wa^2}{2} = 0$$

From which $R_2 = \dfrac{1}{L}\left[M_2 - M_1 + \dfrac{wa^2}{2}\right]$

$$= \frac{1}{L}\left[\frac{wa^3}{12L^2}(4L-3a) - \frac{wa^2}{12L^2}(6L^2+3a^2-8La) + \frac{wa^2}{2}\right]$$

$$= \frac{wa^3}{2L^3}(2L-a) \qquad \text{...[15.23(b)]}$$

Special Case (1). ***When a = L (i.e., U.D.L. over the whole span)***

$$M_1 = \frac{w}{12L^2}(L)^2\ [6L^2 + 3L^2 - 8L\,.\,L] = \frac{wL^2}{12}$$

$$M_2 = \frac{w}{12L^2}(L)^3\ [4L - 3L] = \frac{wL^2}{12}$$

$$R_1 = \frac{w}{2L^3}\,.\,L\ [L^3 - 2L\,.\,L^2 + 2L^3] = \frac{wL}{2}$$

and $$R_2 = \frac{w}{2L^3}(L)^3\ (2L - L) = \frac{wL}{2}$$

These results match with the results of the standard case.

(2) ***When*** $a = \frac{L}{2}$ ***(i.e., U.D.L. over left half span)***

$$M_1 = \frac{w}{12L^2}\left(\frac{L}{2}\right)^2\left[6L^2 + 3\left(\frac{L}{2}\right)^2 - 8L\left(\frac{L}{2}\right)\right] = \frac{11}{192}\,wL^2$$

$$M_2 = \frac{w}{12L^2}\left(\frac{L}{2}\right)^3\left[4L - 3\frac{L}{2}\right] = \frac{5}{192}\,wL^2$$

$$R_1 = \frac{w}{2L^3}\left(\frac{L}{2}\right)\left[\left(\frac{L}{2}\right)^3 - 2L\left(\frac{L}{2}\right)^2 + 2L^3\right] = \frac{13}{32}\,wL$$

$$R_2 = \frac{w}{2L^3}\left(\frac{L}{2}\right)^3\left(2L - \frac{L}{2}\right) = \frac{3}{32}\,wL$$

These results match with the results of example 15.3 solved by the area moment method.

15.10. FIXED BEAM WITH TRIANGULAR LOAD HAVING ZERO INTENSITY AT ONE END AND *w* AT THE OTHER END

Consider a section X, distant x from A (Fig. 15.19). Intensity of loading at $X = \frac{w\,.\,x}{L}$. Hence point load dW acting on an element at distance dx is $dW = \frac{w}{L}\,x\,dx$.

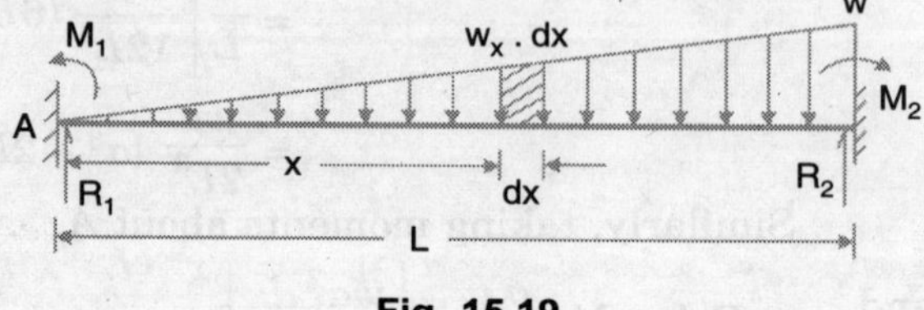

Fig. 15.19

Fixed end moments due to this elemental point load are given by

$$dM_1 = \left(\frac{w}{L}x\,dx\right)\frac{x(L-x)^2}{L^2} = \frac{wx^2\,(L-x)^2\,dx}{L^3}$$

and $$dM_2 = \left(\frac{w}{L}x\,dx\right)\frac{x^2(L-x)}{L^2} = \frac{wx^3}{L^3}\,(L-x)\,dx$$

∴ Total $$M_1 = \int_0^L \frac{wx^2}{L^3}(L-x)^2\,dx = \frac{w}{L^3}\int_0^L (L^2x^2 + x^4 - 2Lx^3)\,dx$$

$$= \frac{w}{L^3}\left[L^2\frac{L^3}{3} + \frac{L^5}{5} - \frac{2L}{4}(L)^4\right] = \frac{wL^2}{30} \qquad \text{...[15.24(a)]}$$

and Total $$M_2 = \int_0^L \frac{wx^3}{L^3}(L-x)\,dx = \frac{w}{L^3}\int_0^L (Lx^3 - x^4)\,dx = \frac{w}{L^3}\left[L.\frac{L^4}{4} - \frac{L^5}{5}\right] = \frac{wL^2}{20}$$

...[15.24(b)]

For R_1, take moments about B.

Thus $$\Sigma M_B = 0 = R_1L + M_2 - M_1 - \frac{wL^2}{6}$$

∴ $$R_1 = \frac{1}{L}\left[\frac{wL^2}{30} - \frac{wL^2}{20} + \frac{wL^2}{6}\right] = \frac{3}{20}wL \qquad \text{...[15.25(a)]}$$

Similarly, $$\Sigma M_A = 0 = R_2L + M_1 - M_2 - \frac{wL^2}{3}$$

From which $$R_2 = \frac{1}{L}\left[M_2 - M_1 + \frac{wL^2}{3}\right] = \frac{1}{L}\left[\frac{wL^2}{20} - \frac{wL^2}{30} + \frac{wL^2}{3}\right] = \frac{7}{20}wL$$

...[15.25(b)]

15.11. FIXED BEAM WITH TRIANGULAR LOAD FOR A GIVEN DISTANCE FROM ONE END

Consider a section X distant x from A (Fig. 15.20). The intensity of loading at X is $w_x = \frac{w}{a}x$. Hence load dW acting on an elemental distance $dx = \frac{w}{a}\,.\,xdx$.

Hence fixed end moments due to dW load are given by

$$dM_1 = \left(\frac{w}{a}x\,dx\right)\frac{x}{L^2}(L-x)^2$$

$$= \frac{wx^2}{aL^2}(L-x)^2\,dx$$

and $$dM_2 = \left(\frac{w}{a}x\,dx\right)\frac{x^2\,(L-x)}{L^2}$$

$$= \frac{wx^3}{aL^2}(L-x)\,dx$$

Fig. 15.20

$\therefore$ Total $\quad M_1 = \int_0^a \frac{wx^2}{aL^2}(L-x)^2\,dx = \frac{w}{aL^2}\int_0^a (x^2L^2 + x^4 - 2Lx^3)\,dx$

$$= \frac{w}{aL^2}\left[L^2\frac{a^3}{3} + \frac{a^5}{5} - 2L\frac{a^4}{4}\right] = \frac{wa^2}{30L^2}[10L^2 + 6a^2 - 15La] \qquad \text{...[15.26(a)]}$$

and $\quad$ Total $M_2 = \int_0^a \frac{wx^3}{aL^2}(L-x)\,dx = \frac{w}{aL^2}\int_0^a (Lx^3 - x^4)\,dx$

$$= \frac{w}{aL^2}\left[L\frac{a^4}{4} - \frac{a^5}{5}\right] = \frac{wa^3}{20L^2}[5L - 4a] \qquad \text{...[15.26(b)]}$$

For reaction R_1, take moments about B.

Hence $\quad R_1L + M_2 - M_1 - \frac{1}{2}aw\left(L - \frac{2}{3}a\right) = 0$

$$\therefore \quad R_1 = \frac{1}{L}\left[M_1 - M_2 + \frac{aw(3L-2a)}{6}\right]$$

$$= \frac{1}{L}\left[\frac{wa^2}{30L^2}(10L^2 + 6a^2 - 15La) - \frac{wa^3}{20L^2}(5L-4a) + \frac{aw}{6}(3L-2a)\right]$$

$$= \frac{wa^2}{60L^3}(20L^2 + 24a^2 - 45La) + \frac{aw}{6L}(3L-2a) \qquad \text{...[15.27(a)]}$$

Similarly, $\quad R_2L + M_1 - M_2 - \frac{1}{2}aw \,.\, \frac{2}{3}a = 0$

$$R_2 = \frac{1}{L}\left(M_2 - M_1 + \frac{wa^2}{3}\right)$$

$$= \frac{1}{L}\left[\frac{wa^3}{20L^2}(5L-4a) - \frac{wa^2}{30L^2}(10L^2 + 6a^2 - 15La) + \frac{wa^2}{3}\right]$$

$$= \frac{wa^2}{60L^3}(45La - 20L^2 - 24a^2) + \frac{wa^2}{3L} \qquad \text{...[15.27(b)]}$$

Special Case. When $a = L$ (*i.e.,* triangular load over the whole span)

$$M_1 = \frac{wL^2}{30L^2}[10L^2 + 6L^2 - 15L\,.\,L] = \frac{wL^2}{30}$$

$$M_2 = \frac{wL^3}{20L^2}(5L - 4L) = \frac{wL^2}{20}$$

$$R_1 = \frac{wL^2}{60L^3}(20L^2 + 24L^2 - 45L\,.\,L) + \frac{Lw}{6L}(3L - 2L) = \frac{3}{20}wL$$

and $\quad R_2 = \frac{wL^2}{60L^3}(45L\,.\,L - 20L^2 - 24L^2) + \frac{wL^2}{3L} = \frac{7}{20}wL$

These values are the same as found in the previous article.

15.12. FIXED BEAM WITH TRIANGULAR LOAD HAVING MAXIMUM INTENSITY AT MID-SPAN AND ZERO AT ENDS

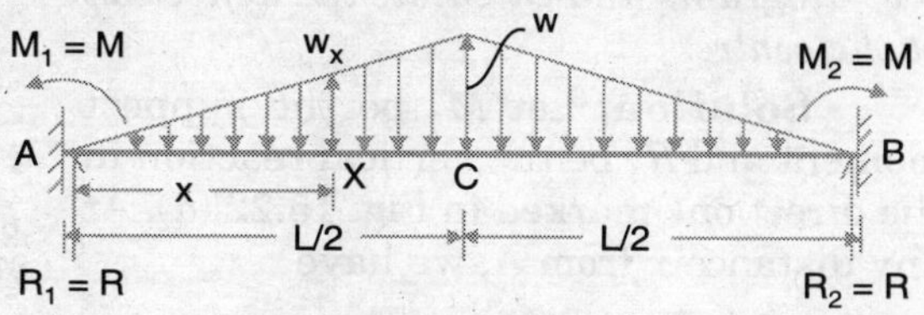

Fig. 15.21 shows the loading. Because of symmetry,

$$M_1 = M_2 = M$$

and $R_1 = R_2 = R = \frac{1}{2}\left(\frac{1}{2} Lw\right) = \frac{wL}{4}$

Fig. 15.21

At any section X distant x from A,

$$w_x = \frac{w}{L/2} \cdot x = \frac{2w}{L} x$$

For portion AC of the beam, we have

$$EI \frac{d^2y}{dx^2} = -M_x = M - \frac{wL}{4} x + \frac{1}{2}\left(\frac{2w}{L} x\right) x \cdot \frac{x}{3}$$

$$= M - \frac{wLx}{4} + \frac{wx^3}{3L} \qquad ...(1)$$

$$\therefore \quad EI \frac{dy}{dx} = Mx - \frac{wLx^2}{8} + \frac{wx^4}{12L} + 0 \qquad ...(2)$$

and

$$EIy = \frac{Mx^2}{2} - \frac{wLx^3}{24} + \frac{wx^5}{60L} + 0 \qquad ...(3)$$

Due to symmetry, $\frac{dy}{dx} = 0$ at $x = \frac{L}{2}$

$$\therefore \quad 0 = M\left(\frac{L}{2}\right) - \frac{wL}{8}\left(\frac{L}{2}\right)^2 + \frac{w}{12L}\left(\frac{L}{2}\right)^4$$

This gives

$$M = \frac{5}{96} wL^2 \qquad ...(15.28)$$

Thus the reaction components at the ends are completely known. For deflection at mid-span, put $x = L/2$ in (3), alongwith the value of M.

$$\therefore \quad EIy_C = \frac{5}{96} wL^2 \cdot \frac{1}{2}\left(\frac{L}{2}\right)^2 - \frac{wL}{24}\left(\frac{L}{2}\right)^3 + \frac{w}{60L}\left(\frac{L}{2}\right)^5 = \frac{7}{3840} wL^4$$

$$\therefore \quad y_C = \frac{7}{3840} \frac{wL^4}{EI} \qquad ...(15.29)$$

Example 15.5. *Find the support moments of a fixed beam loaded at third points by two point loads of W each as shown in Fig. 15.22 (a). Draw the B.M. and S.F. diagrams and calculate the deflection at the centre.*

(a) Beam with loading

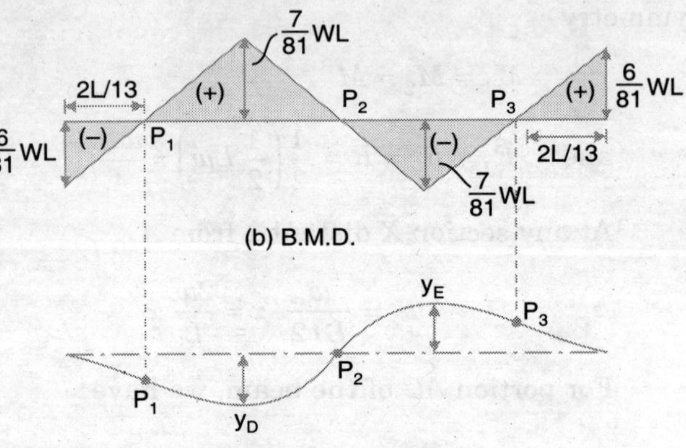

(b) B.M.D.

(c) Deflected shape

(d) S.F.D.

Fig. 15.22

Solution. Let M_1 be the support moment and R_1 be the vertical reaction in the directions marked in Fig. 15.22 (*a*). At any distance x from A, we have

$$EI\frac{d^2y}{dx^2} = -M_x$$

$$= M_1 - R_1 x \,\Bigg|\, + W\left(x - \frac{L}{3}\right)\Bigg|$$

$$- W\left(x - \frac{2L}{3}\right) \quad ...(1)$$

Integrating successively,

$$EI\frac{dy}{dx} = M_1 x - R_1\frac{x^2}{2} + 0 \,\Bigg|$$

$$+ \frac{W}{2}\left(x - \frac{L}{3}\right)^2 \Bigg| - \frac{W}{2}\left(x - \frac{2L}{3}\right)^2 \quad ...(2)$$

and $$EIy = \frac{M_1x^2}{2} - \frac{R_1x^3}{6} + 0 \,\Bigg|\, + \frac{W}{6}\left(x - \frac{L}{3}\right)^3 \Bigg| - \frac{W}{6}\left(L - \frac{2L}{3}\right)^3 \quad ...(3)$$

At $x = L$, $$\frac{dy}{dx} = 0 = M_1L - \frac{R_1L^2}{2} + \frac{W}{2}\left(L - \frac{L}{3}\right)^2 - \frac{W}{2}\left(L - \frac{2L}{3}\right)^2$$

or $$18M_1 - 9R_1L + 3WL = 0 \quad ...(a)$$

Also, at $x = L$, $$y = 0 = \frac{M_1}{2}L^2 - \frac{R_1}{6}L^3 + \frac{W}{6}\left(L - \frac{L}{3}\right)^3 - \frac{W}{6}\left(L - \frac{2L}{3}\right)^3$$

or $$81M_1 - 27R_1L + 7WL = 0 \quad ...(b)$$

Solving Eqs. (*a*) and (*b*), we get $M_1 = +\mathbf{\frac{2}{27}}\,\boldsymbol{WL}$ and $R_1 = +\mathbf{\frac{13}{27}}\,\boldsymbol{W}$

The plus sign with M_1 and R_1 values shows that the *assumed directions* are correct.

Again, substituting $x = L$, and values of R_1 and M_1 in Eq. (1), we get

$$EI\frac{d^2y}{dx^2} = -M_x = M_1 - R_1L + W\left(L - \frac{L}{3}\right) - W\left(L - \frac{2L}{3}\right)$$

But $$M_x = M_2$$

$\therefore$ $$-M_2 = M_1 - R_1L + \frac{1}{3}WL$$

or $$M_2 = -M_1 + R_1L - \frac{1}{3}WL = -\frac{2}{27}WL + \frac{13}{27}WL - \frac{1}{3}WL = +\frac{2}{27}WL$$

Thus $M_2 = +\frac{2}{27}WL$; this means that M_2 is of sagging nature, and is therefore in the anticlockwise direction as marked in Fig. 15.22 (*a*).

B.M.D. : *For AD*, we get the equation for M_x from (1) :

$$M_x = -M_1 + R_1x = -\frac{2}{27}WL + \frac{13}{27}Wx$$

From which, $$M_A = -\frac{2}{27}WL = -\frac{6}{81}WL \text{ (i.e., hogging)}$$

$$M_D = -\frac{2}{27}WL + \frac{13}{27}W\left(\frac{L}{3}\right) = +\frac{7}{81}WL \text{ (i.e., sagging)}$$

B.M. is zero at $x = \frac{2}{13}L$

For DE, $$M_x = -\frac{2}{27}WL + \frac{13}{27}Wx - W\left(x - \frac{L}{3}\right)$$

From which $M_E = -\frac{2}{27}WL + \frac{13}{27}W\left(\frac{2}{3}L\right) - W\left(\frac{2L}{3} - \frac{L}{3}\right) = -\frac{7}{81}WL$ (*i.e.*, hogging)

B.M. is zero at $x = L/2$. Hence $M_C = 0$.

For EB, $$M_x = -\frac{2}{27}WL + \frac{13}{27}Wx - W\left(x - \frac{L}{3}\right) + W\left(x - \frac{2L}{3}\right)$$

From which $M_B = -\frac{2}{27}WL + \frac{13}{27}WL - W\left(L - \frac{L}{3}\right) + W\left(L - \frac{2L}{3}\right) = +\frac{2}{27}WL$

B.M. is zero at $x = \frac{11}{13}L$ from A $\left(\text{or } \frac{2}{13}L \text{ from } B\right)$

The complete B.M.D. is shown in Fig. 15.22 (*b*), including the positions of the three points of contraflexure P_1, P_2, P_3.

Deflection. From Eq. (3) deflection at various points is given by

$$EIy = \frac{1}{27}WLx^2 - \frac{13}{27 \times 6}Wx^3 \left| + \frac{W}{6}\left(x - \frac{L}{3}\right)^3 \right| - \frac{W}{6}\left(L - \frac{2L}{3}\right)^3 \quad ...(4)$$

This gives $$EIy_D = \frac{1}{27}WL\left(\frac{L}{3}\right)^2 - \frac{13}{27 \times 6}W\left(\frac{L}{3}\right)^3 = +\frac{5}{4374}WL^3 \text{ (i.e., } \downarrow)$$

$$EIy_C = \frac{1}{27}WL\left(\frac{L}{2}\right)^2 - \frac{13W}{27 \times 6}\left(\frac{L}{2}\right)^3 + \frac{W}{6}\left(\frac{L}{2} - \frac{L}{3}\right)^3 = 0$$

$$EIy_E = \frac{1}{27}WL\left(\frac{2L}{3}\right)^2 - \frac{13}{27 \times 6}W\left(\frac{2L}{3}\right)^3 + \frac{W}{6}\left(\frac{2L}{3} - \frac{L}{3}\right)^3 = -\frac{5}{4374}WL^3 \text{ (i.e., } \uparrow)$$

Fig. 15.22 (*c*) shows the deflected shape of the beam, including the positions of three points of contraflexure P_1, P_2, P_3.

S.F.D. : From statics, since $\Sigma V = 0$, we get $R_2 = -R_1 = -\frac{13}{27}W \left(i.e., \ \frac{13}{27}W\downarrow\right)$. The S.F.D. is shown in Fig. 15.22 (*d*).

Example 15.6. *A fixed beam of span 4 m carries a distributed load whose intensity varies from 30 kN/m at the left end to 60 kN/m at the right end. Find the fixing moments and reactions at the ends.*

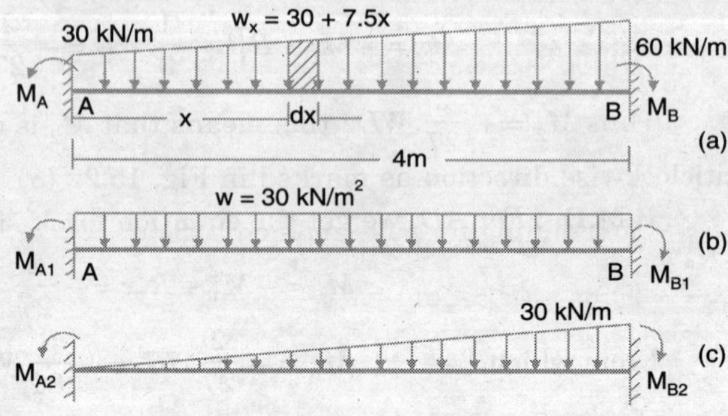

Fig. 15.23

Solution. The given loading is equivalent to (*a*) U.D.L. of intensity 30 kN/m throughout, plus (*b*) a triangularly varying load having zero intensity at *A* to 30 kN/m at *B*. These two effects have been shown separately in Fig. 15.23 (*b*) and (*c*) respectively.

For case (a) of loading : *w* throughout the length

$$M_{A_1} = M_{B_1} = \frac{wL^2}{12} = \frac{30(4)^2}{12} = 40 \text{ kN-m}$$

For case (b) of loading : Triangularly varying load

$$M_{A_2} = \frac{wL^2}{30} = \frac{30(4)^2}{30} = 16 \text{ kN-m} \quad \text{...[Eq. 15.24}(a)]$$

$$M_{B_2} = \frac{wL^2}{20} = \frac{30(4)^2}{20} = 24 \text{ kN-m} \quad \text{...[Eq. 15.24}(b)]$$

$\therefore$ Total $M_A = M_{A_1} + M_{A_2} = 40 + 16 = \mathbf{56\ kN\text{-}m}$

Total $M_B = M_{B_1} + M_{B_2} = 40 + 24 = \mathbf{64\ kN\text{-}m.}$

$$R_1 = \frac{w_1 L}{2} + \frac{3}{20} w_2 L = \frac{1}{2} \times 30 \times 4 + \frac{3}{20} \times 30 \times 4 = \mathbf{78\ kN} \quad \text{...[Eq. 15.25}(a)]$$

$$R_2 = \frac{w_1 L}{2} + \frac{7}{20} w_2 L = \frac{1}{2} \times 30 \times 4 + \frac{7}{20} \times 30 \times 4 = \mathbf{102\ kN} \quad \text{...[Eq. 15.25}(b)]$$

Alternative independent solution [Fig. 15.23(a)]

At any distance *x* from *A*, rate of loading $w_x = 30 + \frac{60-30}{4} x = 30 + 7.5x$

$$\therefore \quad dM_A = \{(30 + 7.5x)\, dx\} \frac{x(4-x)^2}{16}$$

and
$$dM_B = \{(30 + 7.5x)\, dx\} \frac{x^2(4-x)}{16}$$

$$\therefore \text{ Total} \quad M_A = \int_0^4 \frac{(30+7.5x)x\,(4-x)^2}{16}\, dx = \frac{1}{16}\int_0^4 (7.5x^4 - 30x^3 - 120x^2 + 480x)\, dx$$

$$= \frac{1}{16}\left[\frac{7.5x^5}{5} - 30\left(\frac{x^4}{4}\right) - 120\left(\frac{x^3}{3}\right) + 480\left(\frac{x^2}{2}\right)\right]_0^4 = \mathbf{56\ kN\text{-}m}$$

Similarly, Total $M_B = \int_0^4 (30 + 7.5x)\frac{x^2(4-x)}{16} = \frac{1}{16}\int_0^4 (120x^2 - 7.5x^4)\,dx$

$$= \frac{1}{16}\left[120\left(\frac{x^3}{3}\right) - 7.5\left(\frac{x^5}{5}\right)\right]_0^4 = \mathbf{64\ kN\text{-}m}$$

For finding R_1, take moments about B.

Thus $4R_1 + M_B - M_A - 30 \times 4 \times 2 - \frac{1}{2} \times 30 \times 4 \times \frac{4}{3} = 0$

$\therefore$ $R_1 = \frac{1}{4}[-64 + 56 + 240 + 80] = \mathbf{78\ kN}$ and $R_2 = \frac{1}{2}(30 + 60)\,4 - 78 = \mathbf{102\ kN.}$

Example 15.7. *A fixed beam AB of span 4.5 m is subjected to a concentrated couple of 400 kN-m applied at a section 3 m from end A. Find the end moments from the first principles and draw the B.M. and S.F. diagrams.*

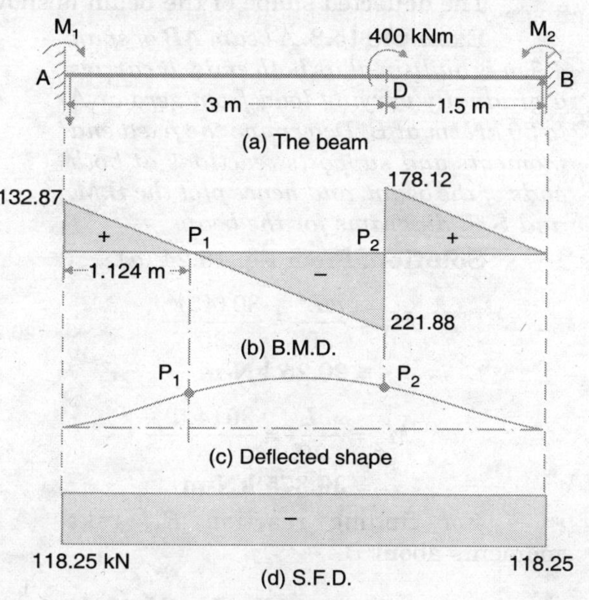

Fig. 15.24

Solution. Let M_1 and R_1 be the reaction components at A in the directions marked. At any section distant x from A.

$$EI\frac{d^2y}{dx^2} = -M_x$$

$$= R_1x - M_1 \mid -400\,(x-3)^0 \quad ...(1)$$

$$\therefore \quad EI\frac{dy}{dx} = R_1\frac{x^2}{2} - M_1x + 0 \mid -400\,(x-3) \quad ...(2)$$

$$EIy = R_1\frac{x^3}{6} - \frac{M_1x^2}{2} + 0 \mid -\frac{400(x-3)^2}{2} \quad ...(3)$$

At $x = 4.5$ m, $\frac{dy}{dx} = 0 = \frac{R_1}{2}(4.5)^2 - M_1(4.5) - 400\,(4.5 - 3)$

or $R_1 - 0.444\,M_1 = 59.259$ ...(*a*)

Also, at $x = 4.5$ m, $y = 0 = \frac{R_1}{6}(4.5)^3 - \frac{M_1}{2}(4.5)^2 - 200\,(4.5 - 3)^2$

or $R_1 - 0.667\,M_1 = 29.63$ ...(*b*)

Solving (*a*) and (*b*), we get M_1 = **132.87 kN-m** and R_1 = **118.25 kN.** *Plus signs with the values of M_1 and R_1 show that the assumed directions of M_1 and R_1 are correct.*

From statics, $R_2 = -R_1 = -118.25$ (*i.e.*, **118.25** ↑)

For the whole beam, $\Sigma M = 0 \quad \therefore \quad 132.87 + 400 - 118.25\,(4.5) = M_2$

From which $M_2 = \mathbf{0}$.

The B.M. is zero at $x = 132.87/118.25 = 1.124$ m from A. The complete B.M.D. is shown in Fig. 15.24 (b) including the positions of two points of inflexion, P_1 and P_2. The S.F. diagram is shown in Fig. 15.24 (d). The S.F. will be **– 118.25 kN** throughout.

Deflection : The deflection at any point is given by

$$EIy = \frac{118.25}{6}x^3 - \frac{132.87}{2}x^2 \left| - \frac{400}{2}(x-3)^2 \right.$$

$$= 19.708\,x^3 - 66.435\,x^2 \,|\, - 200\,(x-3)^2$$

$$\therefore \quad EIy_{P_1} = 19.708\,(1.124)^3 - 66.435\,(1.124)^2 = -\,55.95$$

and $$EIy_D = 19.708\,(3)^3 - 66.435(3)^2 - 200(3-3)^2 = -\,65.8$$

The deflected shape of the beam is shown in Fig. 15.24 (c).

Example 15.8. *A beam AB of span 4.5 m is built-in at its both ends. It carries a gradually varying load from zero at A to 30 kN/m at B. Determine the fixed end moments and supports reactions at both ends of the beam, and hence plot the B.M. and S.F. diagrams for the beam.*

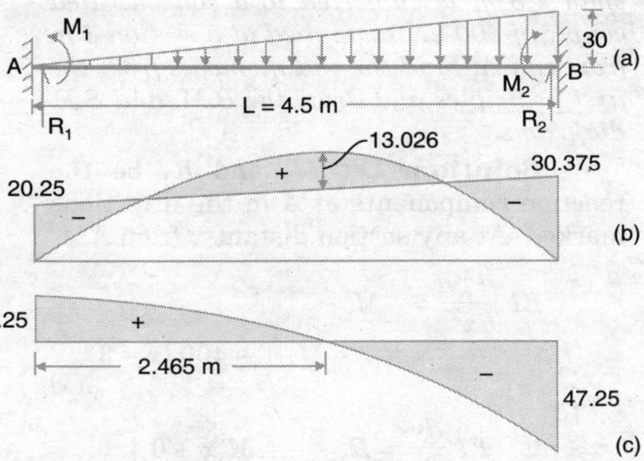

Fig. 15.25

Solution. From Eq. 15.24 (a),

$$M_1 = \frac{wL^2}{30} = \frac{30\,(45)^2}{30}$$

$$= \mathbf{20.25\ kN\text{-}m}$$

$$M_2 = \frac{wL^2}{20} = \frac{30\,(4.5)^2}{20}$$

$$= \mathbf{30.375\ kN\text{-}m}$$

For finding reaction R_1, take moments about B.

or $$R_1 \,.\, L + M_2 - M_1 - \frac{1}{2} \times L \times w \,.\, \frac{L}{3} = 0$$

$$R_1 = \frac{M_1 - M_2}{L} + \frac{wL}{6} = \frac{20.25 - 30.375}{4.5} + \frac{30 \times 4.5}{6}$$

$$= -\,2.25 + 22.5 = \mathbf{20.25\ kN}\ (\uparrow)$$

Similarly, for finding reaction R_2, take moments about A.

Thus $$R_2 \,.\, L + M_1 - M_2 - \frac{1}{2}Lw \,.\, \frac{2L}{3} = 0$$

or $$R_2 = \frac{M_2 - M_1}{L} + \frac{wL}{3} = \frac{30.375 - 20.25}{4.5} + \frac{30 \times 4.5}{3} = 2.25 + 45 = \mathbf{47.25\ kN}\ (\uparrow)$$

Check. $R_1 + R_2 = 20.25 + 47.25 = 67.5$ kN ; Total load $= \frac{1}{2} \times 4.5 \times 30 = 67.5$ kN.

At any distance x from A, $w_x = \dfrac{30}{4.5}\,x = 6.667x$

$$F_x = R_1 - \tfrac{1}{2}\,(6.667x)x = 20.25 - 3.3333x^2$$

The variation is parabolic. S.F. is zero at $x = \sqrt{(20.25/3.3333)} = 2.465$ m from A. The S.F.D. is shown in Fig. 15.25 (c).

Again, at any distance x from A, $M_x = R_1 x - M_1 - \dfrac{1}{2} w_x \,.\, x \,.\, \dfrac{x}{3} = 20.25x - 20.25 - 1.111x^3$

This is a cubic variation. At $x = 2.465$ m, $M_x = +\ 13.026$ kN-m $= M_{max}$

The B.M. diagram is shown in Fig. 15.25 (b).

Example 15.9. *A beam AB is fixed at A and B and carries a load which varies uniformly from w at A to 2w at B. The moment of inertia of the beam also varies uniformly having its value I at A and 2I at B. Determine (a) the fixing moments at A and B, and (b) the maximum deflection.*

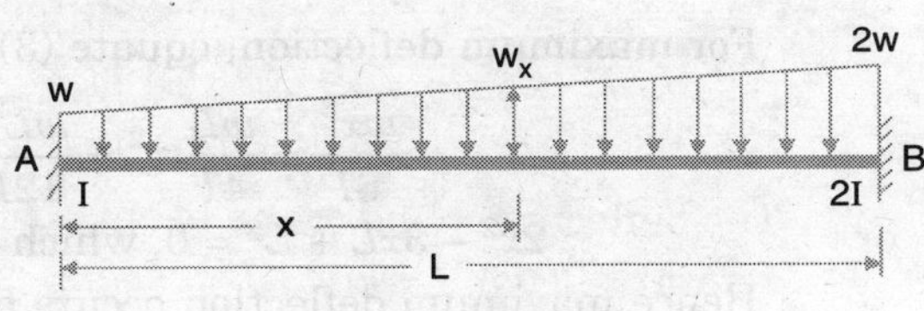

Fig. 15.26

Solution. At any distance x from A.

$$E\frac{d^4y}{dx^4} = \frac{w_x}{I_s} = \frac{w + w\dfrac{x}{L}}{I + I\dfrac{x}{L}} = \frac{w\,(L+x)}{I\,(L+x)} = \frac{w}{I}$$

$$\therefore \qquad E\frac{d^3y}{dx^3} = \frac{w}{I}x + C_1 \qquad ...(1)$$

$$E\frac{d^2y}{dx^2} = \frac{w}{2I}x^2 + C_1x + C_2 \qquad ...(2)$$

$$E\frac{dy}{dx} = \frac{w}{6I}x^3 + \frac{C_1}{2}x^2 + C_2x + 0, \quad \text{since } \frac{dy}{dx} = 0 \text{ at } x = 0 \qquad ...(3)$$

and
$$Ey = \frac{w}{24I}x^4 + \frac{C_1}{6}x^3 + \frac{C_2}{2}x^2 + 0, \quad \text{since } y = 0 \text{ at } x = 0 \qquad ...(4)$$

At $x = L$, $\dfrac{dy}{dx} = 0$ and $y = 0$. Hence substituting in (3) and (4), we get

$$\frac{wL^2}{I} + 3C_1L + 6C_2 = 0 \qquad ...(a)$$

and
$$\frac{wL^2}{I} + 4C_1L + 12C_2 = 0 \qquad ...(b)$$

Solving (a) and (b), we get $C_1 = -\dfrac{wL}{2I}$ and $C_2 = \dfrac{wL^2}{12I}$

Substituting the values of C_1 and C_2 in (2), we get

$$E\frac{d^2y}{dx^2} = \frac{w}{2I}x^2 - \frac{wL}{2I}x + \frac{wL^2}{12I} \qquad ...(I)$$

This is the equation for B.M. at any point

At $x = 0$, $\qquad E\dfrac{d^2y}{dx^2} = \dfrac{wL^2}{12I}$

Hence $$M_A = -EI_A \frac{d^2y}{dx^2} = -\frac{\boldsymbol{wL^2}}{\boldsymbol{12}} \text{ (Since } I_A = I).$$

Also, at $x = L$, $$\frac{Ed^2y}{dx^2} = \frac{w}{2I}(L)^2 - \frac{wL}{2I}(L) + \frac{wL^2}{12I} = \frac{wL^2}{12I}$$

Hence $$M_B = -EI_B \frac{d^2y}{dx^2} = \left(-E\frac{d^2y}{dx^2}\right)I_B = \left(-\frac{wL^2}{12I}\right)(2I) = -\frac{\boldsymbol{wL^2}}{\boldsymbol{6}} \textbf{ (Ans.)}$$

For maximum deflection, equate (3) to zero

$$\therefore \quad \frac{wx^3}{6I} - \frac{wL}{4I}x^2 + \frac{wL^2}{12I}x = 0$$

or $$2x^2 - 3xL + L^2 = 0, \text{ which gives } x = L/2$$

Hence maximum deflection occurs at mid-span. Hence from (4)

$$\therefore \quad y_{max} = \frac{wL^4}{EI}\left[\frac{1}{384} - \frac{1}{96} + \frac{1}{96}\right] = \frac{\boldsymbol{wL^4}}{\boldsymbol{384EI}} \textbf{ (Ans.)}$$

Example 15.10. *A fixed beam of length 4 m is loaded as shown in Fig. 15.27 (a). Find the end moments and reactions and plot the B.M. and S.F. diagrams.*

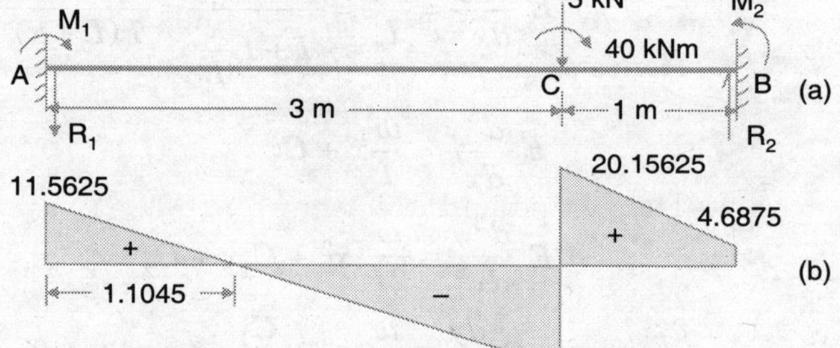

Fig. 15.27

Solution. Let us assume that M_1 acts in the clockwise direction and R_1 acts downwards. Measuring x from A, we have

$$EI\frac{d^2y}{dx^2} = -M_x = -M_1 + R_1x \,|\, + 5(x-3) - 40(x-3)^0 \quad ...(1)$$

$$\therefore \quad EI\frac{dy}{dx} = -M_1x + \frac{R_1x^2}{2} + 0 \,\bigg|\, + \frac{5(x-3)^2}{2} - 40(x-3) \quad ...(2)$$

and $$EIy = -\frac{M_1x^2}{2} + \frac{R_1x^3}{6} + 0 \,\bigg|\, + \frac{5(x-3)^3}{6} - \frac{40(x-3)^2}{2} \quad ...(3)$$

At $x = 4$ m, $$\frac{dy}{dx} = 0 = -M_1(4) + \frac{R_1}{2}(4)^2 + 2.5(4-3)^2 - 40(4-3)$$

From which $R_1 = \dfrac{M_1}{2} + 4.6875$. Substituting in (3), we get

$$EIy = -\frac{M_1}{2}x^2 + \frac{1}{6}\left(\frac{M_1}{2} + 4.6875\right)x^3 \,\bigg|\, + \frac{5}{6}(x-3)^3 - \frac{40(x-3)^2}{2} \quad ...(4)$$

At $x = L = 4$ m, $y = 0$. Hence from (4), we get

$$0 = -\frac{M_1}{2}(4)^2 + \frac{1}{6}\left(\frac{M_1}{2} + 4.6875\right)(4)^3 + \frac{5}{6}(4-3)^3 - 20\,(4-3)^2$$

From which $M_1 = \mathbf{+\ 11.5625\ kN\text{-}m}$

The plus sign shows that the moment at A acts in the same direction as assumed.

Also, $$R_1 = \frac{11.5625}{2} + 4.6875 = \mathbf{+\ 10.46875\ kN}$$

Here also, plus sign shows that R_1 acts in the same direction as assumed *i.e.,* it acts in the downward direction.

Again, substituting $x = L$ in (1), we get

$$(-M_x)_{x=L} = -M_2 = EI\frac{d^2y}{dx^2} = -11.5625 + 10.46875\,(4) + 5\,(4-3) - 40\,(4-3)$$

or $-M_2 = -4.6875$ or $M_2 = \mathbf{4.6875}$ kN-m

(*i.e.,* M_2 is of sagging nature, as marked.)

B.M. diagram $M_1 = +\ 11.5625$; $M_2 = +\ 4.6875$

For *AC*, $M_x = 11.5625 - 10.46875\,x$

Hence M_x is zero at $x = 11.5625/10.46875 = 1.1045$ m

$$M_C\text{ (left)} = 11.5625 - 10.46875 \times 3 = -\ 19.84375\text{ kN-m}$$

$$M_C\text{ (right)} = 11.5625 - 10.46875 \times 3 + 40 = +\ 20.15625$$

The B.M. diagram is shown in Fig. 15.27 (*b*)

S.F. diagram $R_2 = (R_1 + 5)\,(\uparrow) = (10.46875 + 5) = 15.46875$ kN $(\uparrow)$

The S.F. diagram is shown in Fig. 15.27.

15.13. EFFECT OF SINKING OF A SUPPORT

Fig. 15.28 (*a*) shows a fixed beam *AB*, the right hand end *B* of which settles downwards by an amount δ. Due to this settlement or sinking of support *B*, moments and vertical reactions will be induced at both the ends. In order to know their directions, consider a simply supported beam *AB* [Fig. 15.28(*b*)], the right hand end *B* of which settles downwards by amount δ. Due to this settlement, the beam will rotate in the clockwise direction, and the slope at each end will change from 0 to θ in the clockwise direction. Since ends *A* and *B* are built-in, the fixidity will not permit this rotation θ of the tangents, and consequently, fixing couples (*M*)

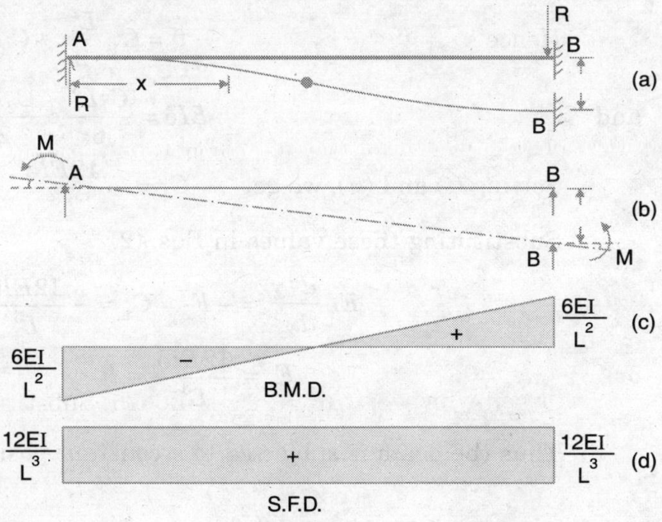

Fig. 15.28

will be induced at each end in such directions so as to make the tangents there horizontal. Hence the fixing moment M will be induced in the anticlockwise direction at each end. If however, the left hand A sinks downwards relative to the right hand one, the beam will have a tendency to rotate in the anticlockwise direction and hence the fixing moment will be induced in the clockwise direction at each end.

For the present case, shown in Fig. 15.28, fixing moment at each end will be M (*i.e.*, equal) and will be in the anticlockwise direction since the beam has the tendency to rotate in the clockwise direction due to sinking of the support B. The vertical reactions at A and B will be *equal* and *opposite* (since $\Sigma V = 0$) and will be of such magnitude so as to provide a total couple equal to the total moments induced, since ΣM has to be zero.

Hence we have, from statics $R \times L = 2M$ or $R = \dfrac{2M}{L}$

At any distance x from the support A,

$$EI\frac{d^4y}{dx^4} = \text{rate of loading } w = 0 \quad \text{...(1)}$$

Integrating successively, we get $EI\dfrac{d^3y}{dx^3} = C_1 = -F_x$ ...(2)

$$EI\frac{d^2y}{dx^2} = C_1x + C_2 = -M_x \quad \text{...(3)}$$

$$EI\frac{dy}{dx} = C_1\frac{x^2}{2} + C_2x + 0, \text{ since } \frac{dy}{dx} = 0 \text{ at } x = 0 \quad \text{...(4)}$$

and $$EIy = \frac{C_1x^3}{6} + \frac{C_2x^2}{2} + 0, \text{ since } y = 0 \text{ at } x = 0 \quad \text{...(5)}$$

In the above five equations, there are only two unknowns, C_1 and C_2, which can be evaluated by using the boundary conditions that at $x = L$, $\dfrac{dy}{dx} = 0$ and $y = +\delta$.

Hence $$0 = C_1\frac{L^2}{2} + C_2L \quad \text{...(}i\text{)}$$

and $$EI\delta = \frac{C_1L^3}{6} + \frac{C_2L^2}{2} \quad \text{...(}ii\text{)}$$

Solving (*i*) and (*ii*), we get $C_1 = -\dfrac{12EI\delta}{L^3}$ and $C_2 = \dfrac{6EI\delta}{L^2}$

Substituting these values in Eqs. (2)

$$EI\frac{d^3y}{dx^3} = -F_x = C_1 = -\frac{12EI\delta}{L^3} \quad \text{...(}a\text{)}$$

or $$F_x = \frac{12EI\delta}{L^3} = R \quad \text{...(15.30)}$$

Thus the beam is subjected to a constant positive S.F. of magnitude $\dfrac{12EI\delta}{L^3}$

Also, from Eq. (3), $EI\dfrac{d^2y}{dx^2} = -M_x = C_1x + C_2 = \dfrac{-12EI\delta}{L^3}x + \dfrac{6EI\delta}{L^2}$

At $x = 0$, we get $\quad -M_x = \dfrac{6EI\delta}{L^2} \quad$ or $\quad M_{x=0} = M_1 = -\dfrac{6EI\delta}{L^2}$ ...[15.31(*a*)]

Also, at $x = L$, $\quad EI\dfrac{d^2y}{dx^2} = -M_x = -\dfrac{12EI\delta}{L^3}L + \dfrac{6EI\delta}{L^2} = -\dfrac{6EI\delta}{L^2}$

$$\therefore \qquad (M_x)_{x=L} = M_2 = +\frac{6EI\delta}{L^2} \qquad \text{...[15.31(b)]}$$

The B.M. and S.F. diagrams are shown in Fig. 15.28 (*c*) and (*d*) respectively. We observe that there is a point of contraflexure at the midspan. If the beam is loaded with U.D.L., and the right hand end also sinks by δ, the fixing moment at the higher end *A* will be $-\left(\dfrac{wL^2}{12} + \dfrac{6EI\delta}{L^2}\right)$ and that at the lower end will be $\dfrac{6EI\delta}{L^2} - \dfrac{wL^2}{12}$.

15.14. EFFECT OF ROTATION OF A SUPPORT : PARTIALLY FIXED BEAM

Let us now take the case of a beam which is perfectly fixed at end *A* (so that angle of rotation is zero there) and *partially fixed* at *B* so that its tangent rotates through an angle θ_B in the *anticlockwise* direction.

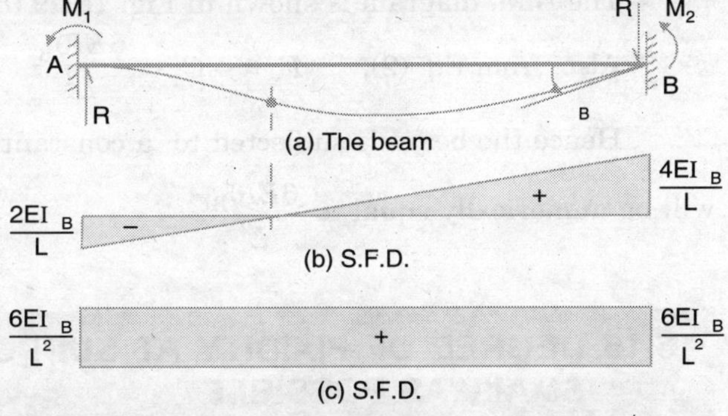

Fig. 15.29

Let M_1 and M_2 be the end moments at *A* and *B* respectively in the directions marked. Since there is no external load on the beam, the vertical reactions R_1 and R_2 will be equal and opposite such that

$$R \times L = M_1 + M_2$$

or $$R = \frac{M_1 + M_2}{L}$$

Now at any distance *x* from *A*,

$$EI\frac{d^4y}{dx^4} = \text{rate of loading} = 0 \qquad \text{...(1)}$$

Integrating successively,

$$EI\frac{d^3y}{dx^3} = C_1 = -F_x \qquad \text{...(2)}$$

$$EI\frac{d^2y}{dx^2} = C_1x + C_2 = -M_x \qquad \text{...(3)}$$

$$EI\frac{dy}{dx} = \frac{C_1x^2}{2} + C_2x + 0 \left(\text{since } \frac{dy}{dx} = 0 \text{ at } x = 0\right) \qquad \text{...(4)}$$

and $$EIy = \frac{C_1x^3}{6} + \frac{C_2x^2}{2} + 0 \text{ (since } y = 0 \text{ at } x = 0) \qquad \text{...(5)}$$

In the above equations, there are only two unknowns : C_1 and C_2

At $x = L$, $\quad \dfrac{dy}{dx} = -\theta_B$. Hence from Eq. (4) : $-EI\theta_B = \dfrac{C_1L^2}{6} + C_2L$ ...(*i*)

Also, at $x = L$, $y = 0 = \frac{C_1L^3}{6} + \frac{C_2L^2}{2}$ ...(ii)

Solving (i) and (ii), we get

$$C_1 = -\frac{6EI\theta_B}{L^2} \text{ and } C_2 = +\frac{2EI\theta_B}{L}$$

Substituting in Eq. (3), we get

$$EI\frac{d^2y}{dx^2} = -M_x = C_1x + C_2 = -\frac{6EI\theta_B}{L^2}x + \frac{2EI\theta_B}{L} \quad ...(15.32)$$

At $x = 0$, $M_x = M_1$

$$\therefore \quad -M_1 = \frac{2EI\theta_B}{L} \text{ or } M_1 = -\frac{2EI\theta_B}{L} \quad ...[15.33(a)]$$

Also, at $x = L$, $M_x = M_2$

$$\therefore \quad -M_2 = -\frac{6EI\theta_B}{L^2}.L + \frac{2EI\theta_B}{L} = -\frac{4EI\theta_B}{L}$$

or

$$M_2 = +\frac{4EI\theta_B}{L} \quad ...[15.33(b)]$$

The B.M. diagram is shown in Fig. 15.29 (*b*).

Also, from Eq. (2), $F_x = -C_1 = \frac{6EI\theta_B}{L^2}$ ...(15.34)

Hence the beam is subjected to a constant S.F. of $\frac{6EI\theta_B}{L^2}$. The reaction R at each end will be numerically equal to $\frac{6EI\theta_B}{L^2}$.

15.15. DEGREE OF FIXIDITY AT SUPPORTS TO HAVE MAXIMUM B.M. AS SMALL AS POSSIBLE

We have seen that when a simply supported beam is subjected to U.D.L., the B.M. at the mid-span is $wL^2/8$ while B.M. at the ends is zero. When the ends of the same beam are fixed, the net B.M. at the mid-span is equal to $wL^2/24$ (see Fig. 15.6), while the B.M. at each end is $wL^2/12$. Thus for a fixed beam, the maximum hogging B.M. is *twice* the maximum sagging B.M. If the degree of fixidity at each end is *reduced*, the hogging B.M. at each end will decrease and consequently the sagging B.M. at the mid-span will increase. Hence for the maximum B.M. to be as small as possible, the maximum sagging moment should be equal to maximum hogging moment. Let the B.M. be M_0. Now for a simply supported beam, central B.M. = $wL^2/8$.

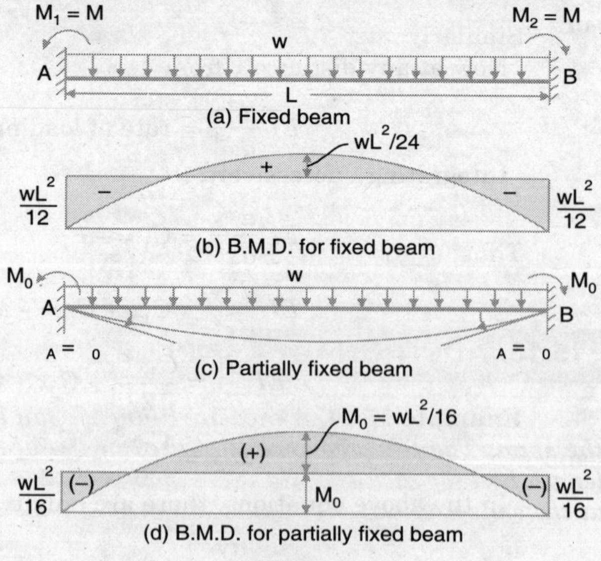

Fig. 15.30

Hence we have the condition

$$M_0 = \frac{wL^2}{8} - M_0 \text{ from which } M_0 = \frac{wL^2}{16}.$$

For such a *partially fixed beam,* let θ_0 to be the slope at each end (Note that $\theta = 0$ for a fixed beam).

Due to symmetry $\quad R_1 = R_2 = R = \dfrac{wL}{2}$

Hence at any section distant x from A,

$$EI\frac{d^2y}{dx^2} = -M_x = -\left[\frac{wL}{2}x - M_0 - \frac{wx^2}{2}\right] = -\frac{wL}{2}x + \frac{wL^2}{16} + \frac{wx^2}{2}$$

Integrating, $\quad EI\dfrac{dy}{dx} = -\dfrac{wL}{4}x^2 + \dfrac{wL^2}{16}x + \dfrac{wx^3}{6} + C_1$

Due to symmetry, at $\quad x = \dfrac{L}{2},\ \dfrac{dy}{dx} = 0 = -\dfrac{wL}{4}\left(\dfrac{L}{2}\right)^2 + \dfrac{wL^2}{16}\left(\dfrac{L}{2}\right) + \dfrac{w}{6}\left(\dfrac{L}{2}\right)^3 + C_1$

From which $\quad C_1 = \dfrac{wL^3}{96}$

Hence $\quad EI\dfrac{dy}{dx} = -\dfrac{wL}{4}x^2 + \dfrac{wL^2}{16}x + \dfrac{wx^3}{6} + \dfrac{wL^3}{96}$

This is the equation for the slope. At $x = 0$, $\dfrac{dy}{dx} = \theta_A$

$$\therefore \quad EI\theta_A = \frac{wL^3}{96} \quad \text{or} \quad \theta_A = +\frac{wL^3}{96EI} \qquad \text{...[15.35(a)]}$$

Similarly, at $x = L$, $\dfrac{dy}{dx} = \theta_B$

$$\therefore \quad EI\theta_B = -\frac{wL}{4}(L^2) + \frac{wL^2}{16}(L) + \frac{w}{6}(L)^3 + \frac{wL^3}{96} = -\frac{wL^3}{96}$$

$$\text{Hence} \quad \theta_B = -\frac{wL^3}{96EI} \qquad \text{...[15.35(b)]}$$

Thus θ_A and θ_B are numerically equal.

15.16. ADDITIONAL EXAMPLES USEFUL FOR COMPETITIVE EXAMINATIONS

Example 15.11. *A encastre beam of span L carries a load wL uniformly distributed over the span. The second moment of area of the beam section is not the same throughout ; for a length L/4 for each end, its value is 2I and for a middle length L/2, it is I. Determine the B.M. at the ends and sketch the B.M. diagram.*

Solution. Fig. 15.31 (*b*) shows the component B.M.D. for U.D.L., for half of the length of the beam. Fig. 15.31 (*c*) shows the component B.M.D. for reaction R_1 or half the length of the beam while Fig. 15.31 (*d*) shows the component B.M.D. for fixing moment M_1 (= M_2 = M) for half the length. From Mohrs theorem,

$$E\Sigma\theta_A^C = -\sum_A^C \frac{A}{I \text{ or } 2I} = 0$$

Since the tangent at C is horizontal due to symmetry.

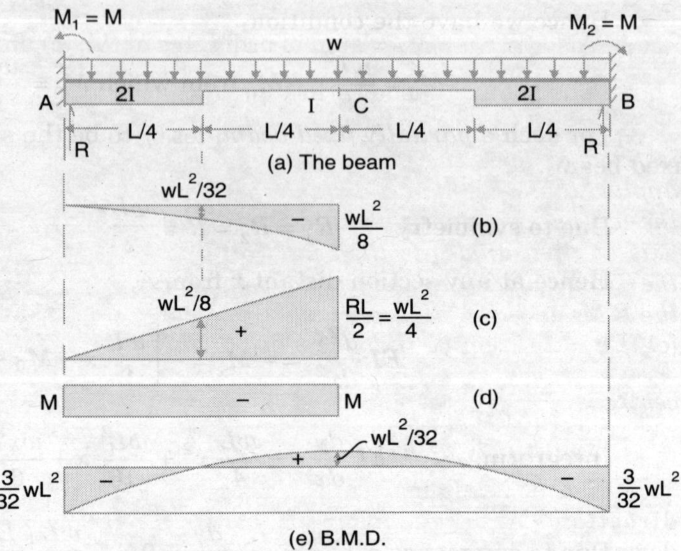

Fig. 15.31

Now for B.M.D. of Fig. 15.31 (b)

$$\sum_A^C \frac{A}{I \text{ or } 2I} = -\left(\frac{1}{3} \times \frac{wL^2}{32} \times \frac{L}{4} \times \frac{1}{2I}\right) - \left\{\left(\frac{1}{3} \times \frac{wL^2}{8} \times \frac{L}{2}\right) - \left(\frac{1}{3} \times \frac{wL^2}{32} \times \frac{L}{4}\right)\right\}\frac{1}{I}$$

$$= -\frac{wL^3}{768I} - \frac{wL^3}{48I} + \frac{wL^3}{384I} = -\frac{15}{768}\frac{wL^3}{I} \quad \text{...(1)}$$

For B.M.D. of Fig. 15.31 (c)

$$\sum_A^C \frac{A}{I \text{ or } 2I} = \left(\frac{1}{2} \cdot \frac{wL^2}{8} \times \frac{L}{4} \times \frac{1}{2I}\right) + \left\{\frac{1}{2}\left(\frac{wL^2}{8} + \frac{wL^2}{4}\right)\frac{L}{4} \cdot \frac{1}{I}\right\} \quad \text{...(2)}$$

$$= \frac{wL^3}{128I} + \frac{3wL^3}{64I} = \frac{7}{128}\frac{wL^3}{I}$$

For B.M.D. of Fig. 15.31 (d)

$$\sum_A^C \frac{A}{I \text{ or } 2I} = \left(-M_1 \frac{L}{4} \cdot \frac{1}{2I}\right) + \left(-M_1 \times \frac{L}{4} \times \frac{1}{I}\right) = -\frac{3M_1L}{8I} \quad \text{...(3)}$$

$$\therefore \text{ Total } \quad \sum_A^C \frac{A}{I \text{ or } 2I} = -\frac{15}{768}\frac{wL^3}{I} + \frac{7}{128}\frac{wL^3}{I} - \frac{3}{8}\frac{M_1L}{I} = 0$$

or
$$\frac{3}{8}\frac{M_1L}{I} = \frac{7}{128}\frac{wL^3}{I} - \frac{15}{768}\frac{wL^3}{I} = \frac{27}{768}\frac{wL^3}{I} \text{ from which}$$

$$M_1 = \mathbf{\frac{3}{32}\, wL^2} = M_2$$

Also, $\quad M_C$ = sum of ordinates of (*b*), (*c*) and (*d*) $= -\dfrac{wL^2}{8} + \dfrac{wL^2}{4} - \dfrac{3}{32}wL^2 = +\dfrac{wL^2}{32}$

The B.M.D. is shown in Fig. 15.31 (*e*).

Example 15.12. *A beam of span L is fixed at both the ends. A couple μ is applied to the beam at a distance nL from the left hand support, about a horizontal axis at right angles to the beam. (a) find the fixing couples and the reactions, (b) plot the B.M. and S.F. diagrams when n = 2/3 (c) if n = 1/2, prove that the slope at the centre is* $\dfrac{\mu L}{16EI}$.

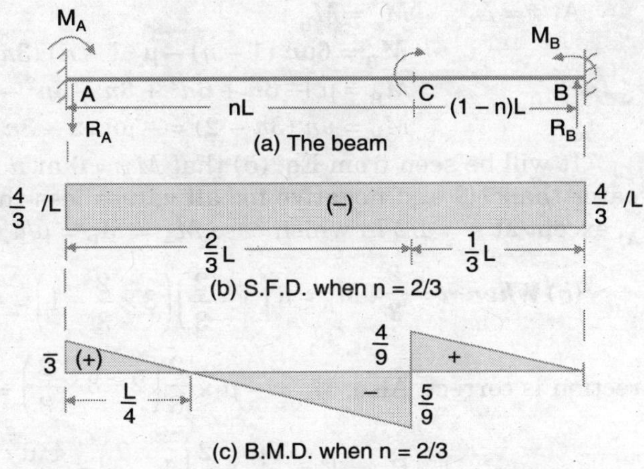

Fig. 15.32

Solution. (*a*) $R_A = -R_B = R$

Let us assume M_A in the clockwise direction and reaction R acting downwards. Measuring x from end A, we have

$$EI\frac{d^2y}{dx^2} = -M_x = Rx - M_A \mid -\mu(x-nL)^0 \qquad \text{...(1)}$$

$$\therefore \quad EI\frac{dy}{dx} = \frac{Rx^2}{2} - M_A \,.\, x + 0 \mid -\mu(x-nL) \qquad \text{...(2)}$$

$$EIy = \frac{Rx^3}{6} - M_A\frac{x^2}{2} + 0 \left| -\frac{\mu(x-nL)^2}{2} \right. \qquad \text{...(3)}$$

At $x = L$, $\quad \dfrac{dy}{dx} = 0 = \dfrac{R}{2}L^2 - M_A \,.\, L - \mu(L - nL)$

or $$R = \frac{1}{L}[2M_A + 2\mu(1-n)] = \frac{2}{L}[M_A + \mu(1-n)] \qquad \text{...(4)}$$

Again, at $x = L$, $y = 0$ and R as in (4). Hence from (3)

$$0 = \frac{2}{L}[M_A + \mu(1-n)]\frac{L^3}{6} - M_A\frac{L^2}{2} - \frac{\mu(1-n)^2L^2}{2}$$

or $$2M_A + 2\mu(1-n) - 3M_A - 3\mu(1-n)^2 = 0$$

or $$M_A = 2\mu(1-n) - 3\mu(1-n)^2 = \mu(2 - 2n - 3 - 3n^2 + 6n)$$

or $$M_A = \mu(-3n^2 + 4n - 1) = \mu(1-n)(3n-1) \qquad \text{...(5)}$$

Hence $$R = \frac{2}{L}[\mu(1-n)(3n-1) + \mu(1-n)]$$

$$= \frac{2}{L}[\mu(4n - 3n^2 - 1 + 1 - n)] = \frac{6\mu n(1-n)}{L} \qquad \text{...(6)}$$

Substituting the values of M_A and L in the moment equation (*i.e.*, Eq. 1)

$$-M_x = EI\frac{d^2y}{dx^2} = Rx - M_A \Big| -\mu = \frac{6\mu n(1-n)}{L}x - \mu(1-n)(3n-1) \Big| -\mu$$

At $x = L$, $M_x = M_B$;

$$\therefore \quad -M_B = 6\mu n(1-n) - \mu(1-n)(3n-1) - \mu$$

or $$M_B = \mu[-6n + 6n^2 + 3n - 3n^2 - 1 + n + 1] = \mu[3n^2 - 2n]$$

or $$M_B = \mu n(3n-2) = -\mu n(2-3n) \quad ...(7)$$

It will be seen from Eq. (5) that $M_A = 0$ at $n = 1/3$. It will be positive for all values of n greater than 1/3 and negative for all values less than 1/3. The sign of M_B will be opposite to M_A, except at $n = 1/2$ in which case $M_A = M_B = \mu/4$, *in the same direction as* μ.

(*b*) ***When n*** $= \frac{2}{3}$: $M_A = \mu\left(1-\frac{2}{3}\right)\left(3\times\frac{2}{3}-1\right) = +\frac{\mu}{3}$. The plus sign shows that the assumed direction is correct. Also, $M_B = -\mu\times\frac{2}{3}\left(2-3\times\frac{2}{3}\right) = 0$.

$$R_A = R = \frac{6\mu}{L}\times\frac{2}{3}\left(1-\frac{2}{3}\right) = \frac{4}{3}\frac{\mu}{L}, \text{ acting downwards}$$

$$R_B = -R = \frac{4}{3}\frac{\mu}{L}, \text{ acting upwards.}$$

The S.F. diagram will be a rectangle as shown in Fig. 15.32 (*b*).

At any distance x from A,

$$M_x = \frac{\mu}{3} - \frac{4}{3}\frac{\mu}{L}x \Big| + \mu.$$

At $x = 0$, $M_A = +\frac{\mu}{3}$; At $x = \frac{2}{3}L$,

$$M_{C\,(\text{left})} = \frac{\mu}{3} - \frac{4}{3}\frac{\mu}{L}\times\frac{2}{3}L = -\frac{5}{9}\mu\,;\; M_{C\,(\text{right})} = -\frac{5}{9}\mu + \mu = +\frac{4}{9}\mu.$$

At $x = L$, $M_B = \frac{\mu}{3} - \frac{4}{3}\frac{\mu}{L}\,.\,L + \mu = 0$; Also, M_x is zero at $x = \frac{L}{4}$.

(*c*) ***When n*** $= \frac{1}{2}$: $M_A = \mu\left(1-\frac{1}{2}\right)\left(\frac{3}{2}-1\right) = \frac{\mu}{4}$, in the same direction as μ.

Also, $$R = \frac{6\mu}{L}\cdot\frac{1}{2}\left(1-\frac{1}{2}\right) = \frac{3}{2}\frac{\mu}{L}\,(\downarrow)$$

$$\therefore \quad \frac{EId^2y}{dx^2} = -M_x = \frac{3}{2}\frac{\mu}{L}x - \frac{\mu}{4} \Big| -\mu(x-nL)^0$$

$$\therefore \quad EI\frac{dy}{dx} = \frac{3}{2}\frac{\mu}{L}\frac{x^2}{2} - \frac{\mu}{4}x + 0 \Big| -\mu(x-nL).$$

At $x = \frac{L}{2}$, $$\left(\frac{dy}{dx}\right)_{\text{centre}} = \frac{1}{EI}\left[\frac{3}{4}\frac{\mu}{L}\cdot\frac{L^2}{4} - \frac{\mu}{4}\frac{L}{2}\right] = \frac{\mu L}{16EI} \quad \textbf{(Proved)}$$

Example 15.13. *A horizontal shaft of length L is subjected at its centre to a vertical load W. The shaft fits in bearings at its ends and when the slope of the shaft at the end is θ, the bearing exerts a bending moment on the shaft of magnitude k θ. Find (a) the B.M. at the ends and (b) deflection of the shaft at mid-span.*

Solution. The shaft is partially fixed at the ends. Let the support moment, exerted by the bearings be μ. Then $\mu = k\,\theta$.

or $$\theta = \frac{\mu}{k} = \frac{dy}{dx} \text{ at the end.}$$

Also, due to symmetry, $R_1 = R_2 = R = \frac{W}{2}$, at each end.

At any point distant x from the left support

$$EI\frac{d^2y}{dx^2} = -M_x = \mu - \frac{W}{2}x \;\Big|\; + W\left(x - \frac{L}{2}\right) \quad \text{...(1)}$$

$$\therefore \quad EI\frac{dy}{dx} = \mu x - \frac{W}{2}\frac{x^2}{2} + C_1 \;\Big|\; + \frac{W}{2}\left(x - \frac{L}{2}\right)^2 \quad \text{...(2)}$$

At $x = 0$, $$\frac{dy}{dx} = \theta = \frac{\mu}{k} \text{ (given)}$$

$$\therefore \quad C_1 = EI\left(\frac{dy}{dx}\right)_{x=0} = EI\cdot\frac{\mu}{k}$$

Hence from (2), $$EI\frac{dy}{dx} = \mu x - \frac{W}{4}x^2 + EI\frac{\mu}{k} \;\Big|\; + \frac{W}{2}\left(x - \frac{L}{2}\right)^2 \quad \text{...(3)}$$

Also, at $x = \frac{L}{2}$, $\frac{dy}{dx} = 0$, due to symmetry.

$$\therefore \quad 0 = \frac{\mu L}{2} - \frac{W}{4}\left(\frac{L}{2}\right)^2 + EI\frac{\mu}{k} \quad \text{or} \quad \mu\left(\frac{L}{2} + \frac{EI}{k}\right) = \frac{WL^2}{16}$$

From which $$\mu = \frac{WL^2k}{8(Lk + 2EI)} \text{ (Ans.)}$$

Substituting this value of μ in (3), we get, for half the shaft,

$$EI\frac{dy}{dx} = \frac{WL^2k}{8(Lk + 2EI)}\cdot x - \frac{W}{4}x^2 + EI\frac{\mu}{k}$$

Integrating, $$EIy = \frac{WL^2k}{8(Lk + 2EI)}\cdot\frac{x^2}{2} - \frac{W}{4}\cdot\frac{x^3}{3} + EI\frac{\mu}{k}x + 0 \text{ (Since } y = 0 \text{ at } x = 0)$$

At $x = \frac{L}{2}$, we get the maximum deflection.

$$\therefore \quad EIy_{max} = \frac{WL^2k}{16(Lk + 2EI)}\left(\frac{L}{2}\right)^2 - \frac{W}{12}\left(\frac{L}{2}\right)^3 + EI\frac{\mu}{k}\left(\frac{L}{2}\right)$$

or $$EIy_{\max} = \frac{WL^4k}{64(Lk+2EI)} - \frac{WL^3}{96} - \frac{EIL}{2k}\left[\frac{WL^2k}{8(Lk+2EI)}\right]$$

or $$y_{\max} = \frac{WL^3}{192EI}\left[\frac{3kL}{kL+2EI} - 2 + \frac{12EI}{(kL+2EI)}\right]$$

$$= \frac{WL^3}{192EI}\left[\frac{3kL-2kL-4EI+12EI}{kL+2EI}\right] = \frac{WL^3}{192EI}\left[\frac{kL+8EI}{kL+2EI}\right] \quad \textbf{Ans.}$$

Example 15.14. *A beam AB of constant section and span L is fixed at end A and carries a central load W. During loading B sinks by an amount δ and rotates through an angle δ/L. Determine the fixing moment at A if the rotation at B is (a) anticlockwise, (b) clockwise.*

Solution. Let M_A and R_A be the reaction components at end A. Measuring x from A,

$$EI\frac{d^2y}{dx^2} = -M_x = M_A - R_A x \Big| + W\left(x - \frac{L}{2}\right) \quad ...(1)$$

$$\therefore \quad EI\frac{dy}{dx} = M_A x - R_A x^2 + 0 \Big| + \frac{W}{2}\left(x-\frac{L}{2}\right)^2 \quad ...(2)$$

and $$EIy = \frac{M_A x^2}{2} - \frac{R_A x^3}{6} + 0 \Big| + \frac{W}{6}\left(x-\frac{L}{2}\right)^3 \quad ...(3)$$

At $x = L$, $y = \delta$. Hence from (3), $EI\delta = M_A\dfrac{L^2}{2} - \dfrac{R_A L^3}{6} + \dfrac{W}{6}\cdot\dfrac{L^3}{8}$

$$\text{or} \quad 24M_A - 8R_A L + WL = \frac{48EI\delta}{L^2} \quad ...(I)$$

(a) For anticlockwise rotation

At $x = L$, $\dfrac{dy}{dx} = -\dfrac{\delta}{L}$. Hence from (2), $-EI\dfrac{\delta}{L} = M_A \cdot L - R_A\dfrac{L^2}{2} + \dfrac{WL^2}{8}$

or $$8M_A - 4R_A \cdot L + WL = -\frac{8EI\delta}{L^2} \quad ...(II)$$

From (I) and (II), $\boldsymbol{M_A = \dfrac{WL}{8} + \dfrac{8EI\delta}{L^2}}$

(b) For clockwise rotation

At $x = L$, $\dfrac{dy}{dx} = \dfrac{\delta}{L}$. Hence from (2), $EI\dfrac{\delta}{L} = M_A L - R_A\dfrac{L^2}{2} + \dfrac{WL^2}{8}$

or $$8M_A - 4R_A L + WL = \frac{8EI\delta}{L^2} \quad ...(III)$$

Hence from (I) and (III), $\boldsymbol{M_A = \dfrac{WL}{8} + \dfrac{4EI\delta}{L^2}}$

Example 15.15. *A beam AB of uniform section and 4 m span is built-in at the ends, and is loaded as shown in Fig. 15.33 (a). Determine the fixing moments and reactions at the ends and plot the B.M. and S.F. diagrams. Also, find magnitude and position of maximum deflection.*

Take E = 2 × 10^5 N/mm² and I = 5 × 10^8 mm⁴.

Solution. Let M_A and R_A be the reaction components at A, in the directions marked. Measuring x from A, we have

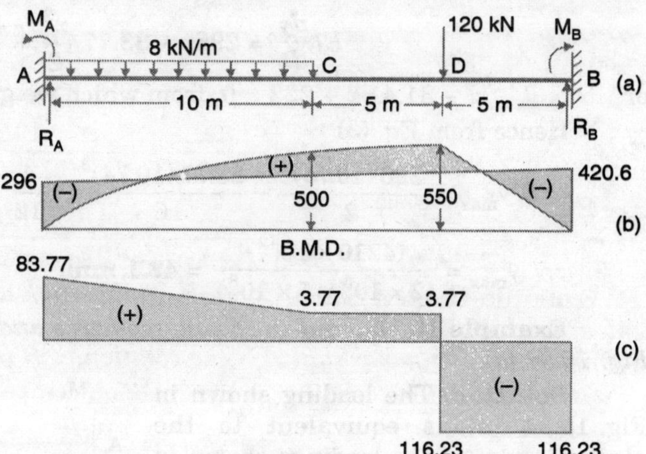

Fig. 15.33

$$EI\frac{d^2y}{dx^2} = -M_x$$
$$= M_A - R_A x + \frac{8x^2}{2} \Bigg| - \frac{8(x-10)^2}{2} \Bigg| + 120\,(x-15) \quad ...(1)$$

$$\therefore \quad EI\frac{dy}{dx} = M_A x - R_A\frac{x^2}{2} + \frac{4x^3}{3} + 0 \Bigg| - \frac{8(x-10)^3}{6} \Bigg| + \frac{120\,(x-15)^2}{2} \quad ...(2)$$

and $$EIy = M_A\frac{x^2}{2} - \frac{R_A x^3}{6} + \frac{4x^4}{12} + 0 \Bigg| - \frac{8\,(x-10)^4}{24} \Bigg| + \frac{120\,(x-15)^3}{6} \quad ...(3)$$

At $x = 20$ m, $\frac{dy}{dx} = 0$. Hence from (2)

$$0 = M_A\,(20) - \frac{R_A}{2}\,(20)^2 + \frac{4}{3}\,(20)^3 - \frac{8\,(20-10)^3}{6} + \frac{120}{2}\,(20-15)^2$$

or $$M_A - 10R_A = -\,541.7 \quad ...(I)$$

Also, at $x = 20$ m, $y = 0$. Hence from (3)

$$0 = \frac{M_A}{2}\,(20)^2 - \frac{R_A}{6}\,(20)^3 + \frac{4}{12}\,(20)^4 - \frac{8}{24}\,(20-10)^4 + \frac{120}{6}\,(20-15)^3$$

or $$M_A - 6.667\,R_A = -\,262.5. \quad ...(II)$$

Solving (I) and (II), we get $R_A = +$ **83.77 kN** and $M_A = +$ **296 kN.**

The plus signs with R_A and M_A show that the assumed directions of R_A and M_A are correct.

Again $R_B = [(8 \times 10) + 120] - 83.77 =$ **116.23 kN** (↑)

Again putting $x = 20$ m in Eq. (1), we get

$$(-M_x)_{x=20} = 296 - 83.77\,(20) + 4(20)^2 - 4(20-10)^2 + 120(20-15) = 420.6$$
$$(M_x)_{x=20} = M_B = -\,420.6 \text{ kN-m}$$

The B.M. and S.F. diagrams are shown in Fig. 15.33 (*b*) and (*c*) respectively.

Since the concentrated load is greater than the total U.D.L. and acts at an equal distance from the nearest end, zero slope will evidently occur between $x = 10$ m to $x = 15$ m. Hence from Eq. (2), noting that $x \simeq 10$ m, we get

$$EI\frac{dy}{dx} = 296x - 83.77\frac{x^2}{2} + \frac{4x^3}{3} - \frac{8}{6}(x-10)^3 = 0$$

or $\quad x^2 - 31.41\,x + 222 = 0$, from which we get $x = 10.74$ m

Hence from Eq. (3)

$$EIy_{max} = \frac{296\,(10.74)^2}{2} - \frac{83.77\,(10.74)^3}{6} + \frac{4}{12}(10.74)^4 - \frac{8\,(10.74-10)^4}{24} = 4210 \text{ kN-m}^3$$

$$\therefore \quad y_{max} = \frac{(4210\times10^{12})}{(2\times10^5\times5\times10^8)} = \mathbf{42.1\ mm.}$$

Example 15.16. *Find the fixing moments and end reactions for a fixed beam shown in Fig. 15.34 (a).*

Solution. The loading shown in Fig. 15.34 (*a*) is equivalent to the algebraic sum of the loadings shown in Fig. 15.34 (*b*) and (*c*).

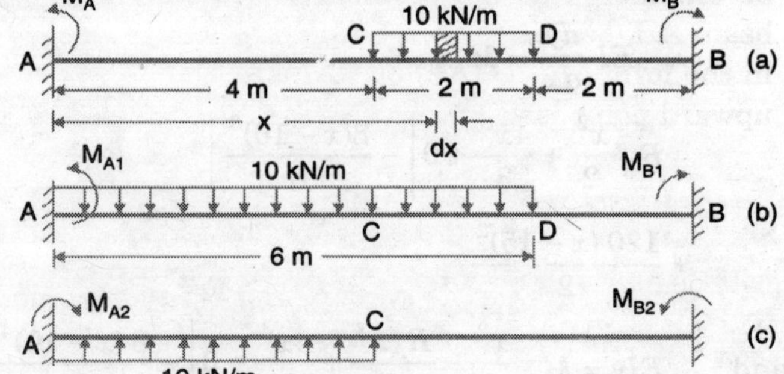

Fig. 15.34

For loading of Fig. 15.34 (*b*)

$$M_{A_1} = \frac{wa^2}{12L^2}(6L^2 - 8La + 3a^2) \qquad ...[15.22(a)]$$

$$= \frac{10(6)^2}{12(8)^2}[6(8)^2 - 8\times8\times6 + 3(6)^2]$$

$= 50.625$ kN-m (hogging)

$$M_{B_1} = \frac{wa^3}{12L^2}(4L - 3a) \qquad ...[15.22(b)]$$

$$= \frac{10(6)^3}{12(8)^2}[4\times8 - 3\times6] = 39.375 \text{ kN-m (hogging)}$$

For the loading of Fig. 15.34 (*c*),

$$M_{A_2} = \frac{10(4)^2}{12(8)^2}[6(8)^2 - 8\times8\times4 + 3(4)^2] = 36.667 \text{ kN-m (sagging)}$$

$$M_{B_2} = \frac{10(4)^3}{12(8)^2}[4\times8 - 3\times4] = 16.667 \text{ kN-m (sagging)}$$

$\therefore$ Total $\quad M_A = M_{A_1} - M_{A_2} = 50.625 - 36.667 = \mathbf{13.958\ kN\text{-}m\ (hogging)}$

Total $\quad M_B = M_{B_1} - M_{B_2} = 39.375 - 16.667 = \mathbf{22.708\ kN\text{-}m\ (hogging)}$

Alternative Independant Solution [Refer Fig. 15.34(*a*)]

At any distance *x* from *A*, consider elementary load $w\,dx = 10\,dx$, due to which

$$dM_A = \frac{(10\,dx)\,x\,(8-x)^2}{64} \quad \text{and} \quad dM_B = \frac{(10\,dx)\,(8-x)\,x^2}{64}$$

$$\therefore \quad M_A = \int_4^6 \frac{10x\,(8-x)^2}{64}dx = \frac{5}{32}\left[\int_4^6 (x^3 - 16x^2 + 64x)\,dx\right] = 13.958 \text{ kN-m}$$

and $$M_B = \int_4^6 \frac{10}{64}(8-x)\,x^2\,dx = \frac{5}{32}\left[\int_4^6 (8x^2 - x^3)\,dx\right] = 22.708 \text{ kN}$$

End Reactions. Taking moments about B, $R_A = \frac{1}{8}$ [13.958 – 22.708 + 60] = **6.406 kN (↑)**

Similarly, $R_B = \frac{1}{8}$ [22.708 – 13.958 + 20 × 5] = **13.594 kN (↑)**

Example 15.17. *A beam AB fixed at A and B and 7 m long, carries U.D.L. of 30 kN/m over the whole span and is propped at C, distant 4 m from A. The prop at C sinks slightly such that the pressure on the prop is 98 kN. Find the support moments and reactions at A and B.*

Solution. The above problem may be considered as the problem of a fixed beam AB, subjected to a downward load in the form of U.D.L. of 30 kN/m and an upward point load P = 98 kN

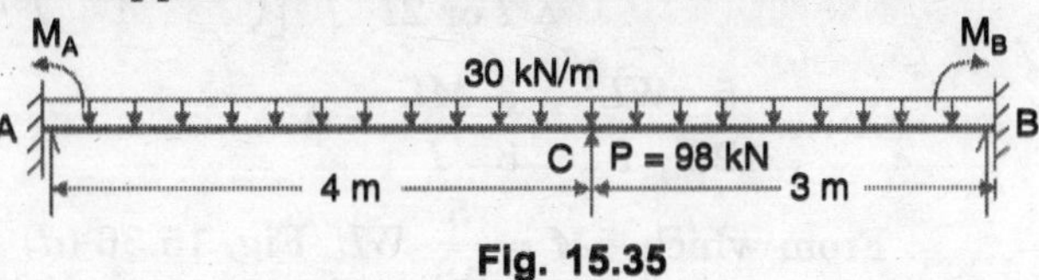

Fig. 15.35

$$\therefore \quad M_A = \frac{wL^2}{12} - \frac{Pab^2}{L^2} = \frac{30(7)^2}{12} - \frac{98 \times 4(3)^2}{(7)^2} = \textbf{50.5 kN-m (hogging)}$$

$$M_B = \frac{wL^2}{12} - \frac{Pba^2}{L^2} = \frac{30(7)^2}{12} - \frac{98 \times 3(4)^2}{(7)^2} = \textbf{26.5 kN-m (hogging)}$$

For finding reaction at A, take moments at B.

$\therefore \quad R_A L + M_B - M_A - 30 \times 7 \times 3.5 + 98 \times 3 = 0$

From which R_A = **66.429 kN**

Similarly, $R_B L + M_A - M_B + 98 \times 4 - 30 \times 7 \times 3.5 = 0$

From which R_B = **45.571 kN.**

Example 15.18. *A fixed beam AB of length L carries a concentrated load W at the mid-span. The moment of inertia of the beam from either end to a distance of L/4 is I and 2I for the remaining length, as shown in Fig. 15.36 (a). Determine the end moments and plot the B.M.D.*

Solution. From the geometrical point of view as well as loading point of view, the beam is symmetrical. Hence $M_1 = M_2 = M$ and $R_1 = R_2 = R = \frac{W}{2}$. The component B.M. diagrams due to R and M are shown in Fig. 15.36 (b) and (c) respectively. From Mohr's theorems,

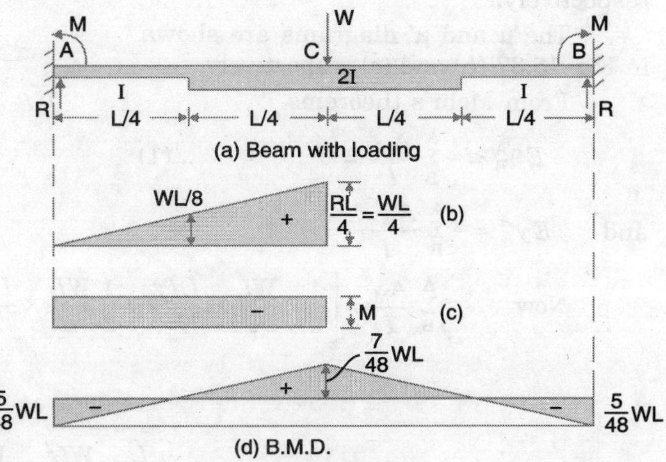

Fig. 15.36

$$E\theta_A^C = -\sum_A^C \frac{A_M}{I \text{ or } 2I} = 0$$

(since the tangent at C is horizontal).

For B.M.D. of Fig. 15.36 (b) :

$$\sum_A^C \frac{A_M}{I \text{ or } 2I} = \left(\frac{1}{2}\times\frac{L}{4}\times\frac{WL}{8}\right)\frac{1}{I}+\frac{1}{2}\left(\frac{WL}{8}+\frac{WL}{4}\right)\times\frac{L}{4}\times\frac{1}{2I} = \frac{5}{128}\frac{WL^2}{I}$$

For B.M.D. of Fig. 15.36 (c)

$$\sum_A^C \frac{A_M}{I \text{ or } 2I} = -\left[\left(M\times\frac{L}{4}\times\frac{1}{I}\right)+\left(M\times\frac{L}{4}\times\frac{1}{2I}\right)\right] = -\frac{3}{8}\frac{ML}{I}$$

$$\therefore \quad \frac{5}{128}\frac{WL^2}{I}-\frac{3}{8}\frac{ML}{I} = 0$$

From which $M = \dfrac{5}{48}WL$. Fig. 15.36 (*d*) shows the B.M. diagram.

Note. If the beam were of uniform I, the support moment due to central point load would be $\dfrac{1}{8}WL\left(=\dfrac{6}{48}WL\right)$.

Example 15.19. *A fixed beam ACB has variable moment of inertia as shown in Fig. 15.37 (a). Determine the fixed end moments at A and B due to a concentrated load W applied at C. Hence find the fixed support moments if (a) $I_2 = I_1$ and (b) $I_2 = 2I_1$.*

Solution. Let M_1 and M_2 ($> M_1$) be the fixed end moments at A and B respectively.

The μ and μ′ diagrams are shown in Fig. 15.37 (*b*) and (*c*) respectively.

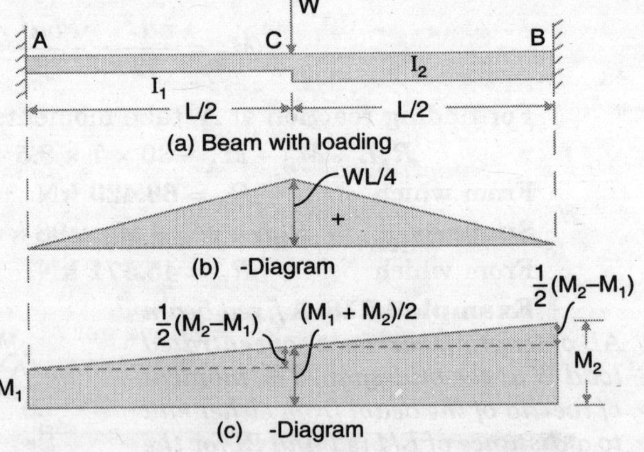

Fig. 15.37

From Mohr's theorems,

$$E\theta_B^A = -\sum_B^A \frac{A_M}{I} = 0 \quad \text{...(1)}$$

and

$$Ey_B^A = -\sum_B^A \frac{A_M\,\bar{x}}{I} = 0 \quad \text{...(2)}$$

Now

$$\sum_B^A \frac{A_M}{I} = \left[\frac{1}{2}\times\frac{WL}{4}\times\frac{L/2}{I_1}+\frac{1}{2}\frac{WL}{4}\times\frac{L/2}{I_2}\right]-\left[\frac{1}{2}\left(M_1+\frac{M_1+M_2}{2}\right)\frac{L/2}{I_1}\right.$$

$$\left.+\frac{1}{2}\left(\frac{M_1+M_2}{2}+M_2\right)\frac{L/2}{I_2}\right] = 0$$

or

$$(3M_1+M_2)\frac{L}{I_1}+(3M_2+M_1)\frac{L}{I_2} = \frac{WL^2}{2I_1}+\frac{WL^2}{2I_2}$$

or

$$M_1\left(\frac{3}{I_1}+\frac{1}{I_2}\right)+M_2\left(\frac{3}{I_2}+\frac{1}{I_1}\right) = \frac{WL}{2}\left(\frac{1}{I_1}+\frac{1}{I_2}\right)$$

or $$M_1 + M_2\left(\frac{3I_1 + I_2}{3I_2 + I_1}\right) = \frac{WL}{2}\left(\frac{I_1 + I_2}{I_1 + 3I_2}\right) \qquad ...(a)$$

Also $$\sum_B^A \frac{A_M\,\bar{x}}{I} = \left[\left\{\frac{1}{2}\frac{WL}{4} + \frac{L/2}{I_1}\left(\frac{L}{2} + \frac{L}{6}\right)\right\} + \left\{\frac{1}{2} \times \frac{WL}{4} \times \frac{L/2}{I_2}\left(\frac{L}{3}\right)\right\}\right]$$

$$- \left[\left\{\frac{M_1}{I_1}\frac{L}{2}\left(\frac{L}{2} + \frac{L}{4}\right)\right\} + \left\{\frac{1}{2}.\frac{L}{2}\left(\frac{M_2 - M_1}{2I_1}\right)\left(\frac{L}{2} + \frac{L}{6}\right)\right\}\right.$$

$$\left. + \left\{\frac{M_1 + M_2}{2I_2} \times \frac{L}{2} \times \frac{L}{4}\right\} + \left\{\frac{1}{2} \times \frac{L}{2}\left(\frac{M_2 - M_1}{2I_2}\right) \times \frac{L}{6}\right\}\right] = 0$$

or $$\frac{3}{8}\frac{M_1 L^2}{I_1} + \frac{(M_2 - M_1)\,L^2}{12I_1} + \frac{(M_1 + M_2)\,L^2}{16I_2} + \frac{(M_2 - M_1)\,L^2}{48I_2} = \frac{WL^3}{24I_1} + \frac{WL^3}{48I_2}$$

or $$M_1\left(\frac{3}{8I_1} - \frac{1}{12I_1} + \frac{1}{16I_2} - \frac{1}{48I_2}\right) + M_2\left(\frac{1}{12I_1} + \frac{1}{16I_2} + \frac{1}{48I_2}\right) = \frac{WL}{24}\left(\frac{1}{I_1} + \frac{1}{2I_2}\right)$$

or $$M_1 + M_2\frac{2(I_1 + I_2)}{I_1 + 7I_2} = \frac{WL}{2}\left(\frac{I_1 + 2I_2}{I_1 + 7I_2}\right) \qquad ...(b)$$

Solving (*a*) and (*b*), we get

$$M_1 = \frac{WL}{2}\frac{I_1\,(I_1 + 3I_2)}{I_1^2 + I_2^2 + 14I_1I_2} \text{ and } M_2 = \frac{WL}{2}\frac{I_2\,(I_2 + 3I_1)}{I_1^2 + I_2^2 + 14I_1I_2}$$

Special Cases

(*a*) When $I_2 = I_1 = I$, we get $M_1 = M_2 = \dfrac{WL}{8}$

(*b*) When $I_2 = 2I_1 = 2I$, we get $M_1 = \dfrac{7}{66}WL$ and $M_2 = \dfrac{10}{66}WL$

PROBLEMS

1. A beam *AB*, fixed at the ends is loaded with three points loads as shown in Fig. 15.38. Determine the fixing moments and plot the B.M. and S.F. diagrams.

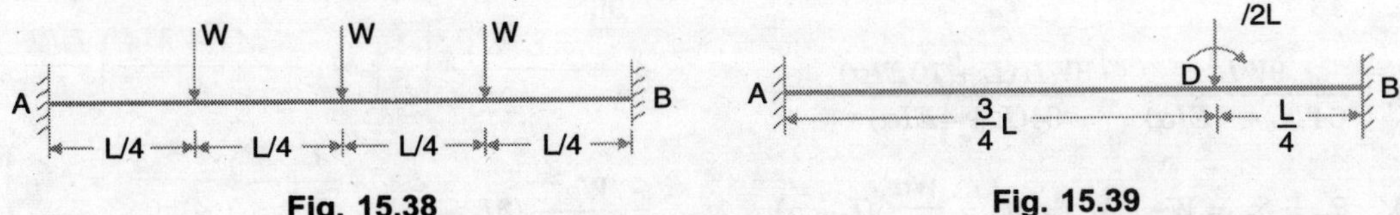

Fig. 15.38 Fig. 15.39

2. A steel beam *AB* of span 6 m is fixed at the ends, and is 60 mm wide and 100 mm deep. If the support *B* sinks by 6 mm, determine the fixing moments. Take $E = 2 \times 10^5$ N/mm^2.

3. A fixed beam of length L is loaded as shown in Fig. 15.39. Find fixing moments and end reactions and plot the B.M. and S.F. diagrams.
4. A fixed beam of variable moment of inertia is loaded as shown in Fig. 15.40. Determine the fixing moments at the ends.

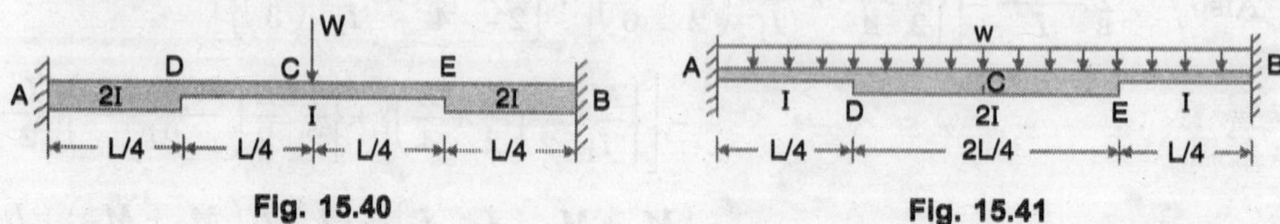

Fig. 15.40

Fig. 15.41

5. A fixed beam AB of length L carries U.D.L. over the whole span. The moment of inertia of the beam from the either end to a distance of $L/4$ is I and for the middle half, it is $2I$. Determine the support moments (Fig. 15.41).
6. A beam of span L is fixed at both the ends. A couple μ is applied to the beam at the middle about a horizontal axis at right angles to the beam. Prove that the fixing couple at each support is $\mu/4$ in the same direction and that the slope at the centre is $(\mu L) / (16EI)$. *(Cambridge University)*
7. A beam AB of flexural rigidity EI and span L carries U.D.L. of intensity w per unit length. It is encastre at A and B, but support B settled during the application of the load by an amount δ.

   Show that if $\delta = \dfrac{wL^4}{72EI}$, there is no fixing moment at B. *(U.L.)*
8. A beam AB, of span L and constant flexural rigidity EI is encastre at B and supported at A so that it rotates through an angle αM_A, where M_A is the moment at A. Calculate the value of the moments at both supports when it carries a load W at a distance of $L/4$ from A.
9. Find the fixed end moments and end reactions for a fixed-end beam AB loaded as shown in Fig. 15.42. Also, determine the maximum deflection.

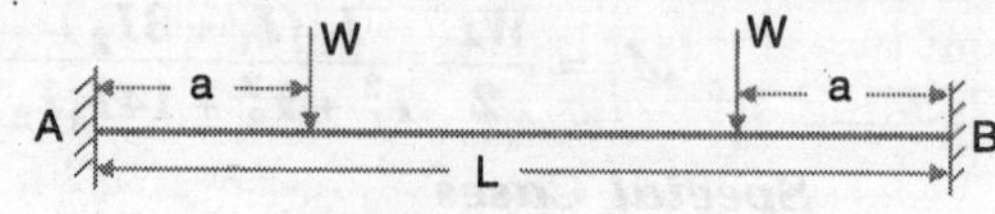

Fig. 15.42

ANSWERS

1. $M_A = M_B = \dfrac{5}{16} WL$

2. 1 kN-m

3. $M_A = \dfrac{37}{128}\mu$ (↻); $M_B = \dfrac{15}{128}\mu$ (↺); $R_A = \dfrac{67}{64}\dfrac{\mu}{L}$ (↓); $R_B = \dfrac{99}{64}\dfrac{\mu}{2}$ (↑)

4. $\dfrac{7}{48} WL$

5. $\dfrac{7}{96} wL^2$

8. $\dfrac{9WL^2}{64(L + 4EI\alpha)}$; $\dfrac{3WL(L + 10EI\alpha)}{64(L + 4EI\alpha)}$

9. $R_A = R_B = W$; $M_A = M_B = \dfrac{Wa}{L}(L - a)$; $y_{max} = \dfrac{Wa^2}{24EI}(3L - 4a)$.

16

Continuous Beams

16.1. INTRODUCTION

A continuous beam is the one which has more than one span, and more than two supports. It is thus *statically indeterminate beam* since the number of reaction components are more than two while only two equations are available from statical equilibrium, for vertical system of loading. However, the *degree of indeterminacy* depends upon the number and type of supports.

The continuous beam, shown in Fig. 16.1 (*a*) has three supports, the beam being simply supported at the ends ; the beam is, therefore, statically indeterminate to single degree since it has three reaction components (*i.e.*, R_A, R_B and R_C). On the other hand, the same beam, with one end fixed [Fig. 16.1 (*b*)] and other simply supported has four reaction components (*i.e.*, M_A, R_A, R_B and R_C) ; the beam is therefore statically indeterminate to *second degree*. If, however, both the ends of the two span beam are fixed [Fig. 16.1 (*c*)], there are five reaction components (*i.e.*, M_A, R_A, R_B, R_C and M_C) and the beam is statically indeterminate to *third degree*. Lastly, the beam shown in Fig. 16.1 (*d*) has six reaction components, while *three* equations are available from statical equilibrium, the beam is therefore statically indeterminate to *third degree*.

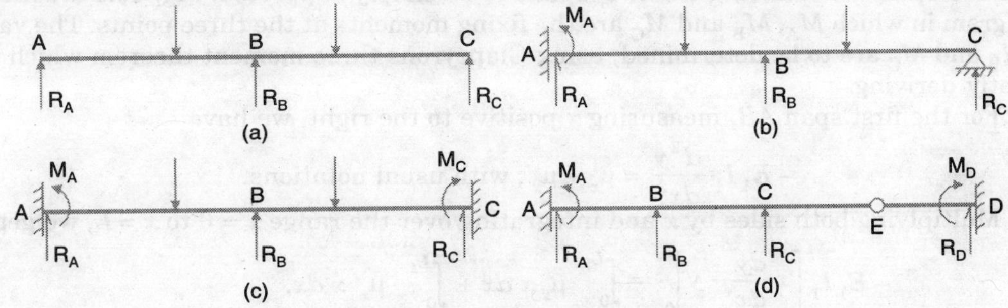

Fig. 16.1. Continuous beams

Following are some of the important methods for the analysis of a continuous beam:

1. Three moment theorem method.
2. Method of consistent deformation.
3. Slope deflection method and
4. Moment distribution method.

However, we shall discuss in this Chapter, only the first method. The remaining methods have been dealt with in Author's book 'Theory of Structures'.

16.2. CLAPEYRON'S THEOREM OF THREE MOMENTS : DERIVATION

Let us take two consecutive spans AB (= L_1) and BC (= L_2) of a continuous beam subjected to a general system of loading, as shown in Fig. 16.2 (*a*). We will use suffix 1 (*i.e.*, L_1, E_1, I_1) for the first span AB and suffix 2 (*i.e.*, L_2, E_2, I_2 etc.) for the second span BC. Fig. 16.2 (*b*) shows the deflected position of the two spans, after the supports A, B and C have settled to positions A', B' and C', by amounts δ_A, δ_B and δ_C respectively below the original centre line.

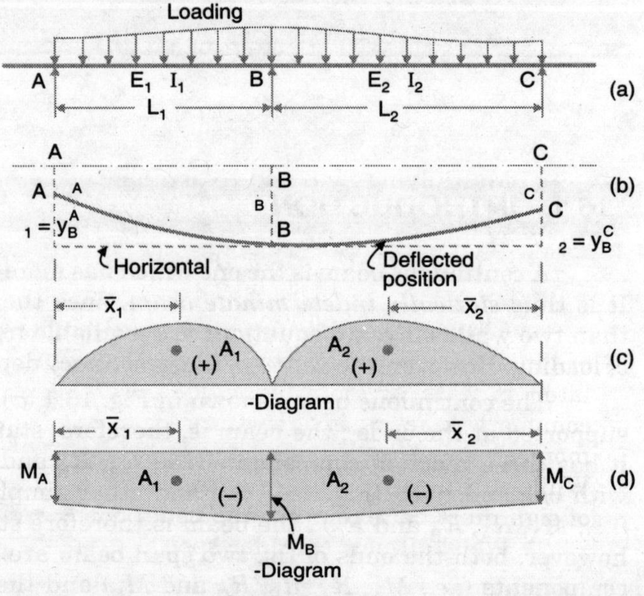

Fig. 16.2

From the deflected position of two spans, we have

$$y_B^A = \delta_B - \delta_A = \delta_1 \text{ (say)}$$

where y_B^A is the deflection of B with respect to A. Similarly

$$y_B^C = \delta_B - \delta_C = \delta_2 \text{ (say)}.$$

where y_B^C is the deflection of B with respect to C.

The μ and μ′ diagrams can be constructed as usual. Fig. 16.1 (*c*) shows the μ-diagrams for the two spans, considering the two spans to be simply supported. Fig. 16.1 (*d*) shows the μ′–diagram in which M_A, M_B and M_C are the fixing moments at the three points. The values of M_A, M_B and M_C are to be determined, using Clapeyrons three moment theorem which we are presently deriving.

For the first span AB, measuring x positive to the right, we have

$$-E_1 I_1 \frac{d^2 y}{dx^2} = \mu_x + \mu_x', \text{ with usual notations.}$$

Multiplying both sides by x and integrating over the range $x = 0$ to $x = L$, we get

$$-E_1 I_1 \left[x\frac{dy}{dx} - y \right]_0^{L_1} = \int_0^{L_1} \mu_x . x\, dx + \int_0^{L_1} \mu_x'\, x\, dx.$$

At $x = L_1$, $\quad \dfrac{dy}{dx} = i_B$ and $y_B^A = \delta_1$

$$\therefore \quad -E_1 I_1 [L_1 i_B - \delta_1] = A_1 \bar{x}_1 + A_1' \bar{x}_1'$$

From which $$i_B = -\frac{1}{E_1 I_1 L_1}[A_1\bar{x}_1 + A_1'\bar{x}_1'] + \frac{\delta_1}{L_1} \quad ...(1)$$

Similarly, considering span BC, taking C as the origin, and measuring x positive to the left, we can obtain:

$$i_B' = -\frac{1}{E_1 I_2 L_2}(A_2\bar{x}_2 + A_2'\bar{x}_2') + \frac{\delta_2}{L_2} \quad ...(2)$$

Due to continuity of the span, $i_B = -i_B'$ (or $i_B + i_B' = 0$). Hence

$$\frac{1}{E_1 I_1 L_1}(A_1\bar{x}_1 + A_1'\bar{x}_1') - \frac{\delta_1}{L_1} + \frac{1}{E_2 I_2 L_2}(A_2\bar{x}_2 + A_2'\bar{x}_2') - \frac{\delta_2}{L_2} = 0$$

Now, for a trapezoidal μ' diagram, we know that

$$A_1'\bar{x}_1' = (M_A + 2M_B)\frac{L_1^2}{6} \quad \text{and} \quad A_2'\bar{x}_2' = (M_C + 2M_B)\frac{L_2^2}{6}.$$

$$\therefore \quad \frac{A_1\bar{x}_1}{E_1 I_1 L_1} + \frac{L_1}{6E_1 I_1}(M_A + 2M_B) - \frac{\delta_1}{L_1} + \frac{A_2\bar{x}_2}{E_2 I_2 L_2} + \frac{L_2}{6E_2 I_2}(M_C + 2M_B) - \frac{\delta_2}{L_2} = 0$$

Multiplying by 6 and re-arranging, we get

$$M_A \cdot \frac{L_1}{E_1 I_1} + 2M_B\left(\frac{L_1}{E_1 I_1} + \frac{L_2}{E_2 I_2}\right) + M_C\frac{L_2}{E_2 I_2} + \left(\frac{6A_1\bar{x}_1}{E_1 I_1 L_1} + \frac{6A_2\bar{x}_2}{E_2 I_2 L_2}\right) = 6\left(\frac{\delta_1}{L_1} + \frac{\delta_2}{L_2}\right) \quad ...(16.1)$$

This is the *generalised theorem of three moments*. The three moments equation was originally developed by French engineers B.P.E. Clapeyron and H. Bertot in 1857 and modified later (in 1860) by Mohr to include the effects of uneven settlements. However, the equation is commonly known as Clapeyron's equation of three moments. This equation is known as three moment equation because it relates the three consecutive bending moments in the beam. While substituting the *numerical values of* A_1 and A_2 for a given system of loading, proper care of sign must be taken. For the usual downward loading, A_1 and A_2 will be positive.

We shall now consider special cases.

16.3. CASE I : EI CONSTANT : GENERAL LOADING

This is a very common case. Taking $E_1 I_1 = E_2 I_2 = E I$, we get from Eq. 16.1,

$$M_A \cdot L_1 + 2M_B(L_1 + L_2) + M_C \cdot L_2 + \frac{6A_1\bar{x}_1}{L_1} + \frac{6A_2\bar{x}_2}{L_2} = 6EI\left(\frac{\delta_1}{L_1} + \frac{\delta_2}{L_2}\right) \quad ...(16.2)$$

16.4. CASE II : EI CONSTANT : NO SETTLEMENT : GENERAL LOADING

Given $E_1 I_1 = E_2 I_2 = E I$ and $\delta_1 = 0$; $\delta_2 = 0$

Hence from Eq. 16.2, we have

$$M_A L_1 + 2M_B(L_1 + L_2) + M_C L_2 + \frac{6A_1\bar{x}_1}{L_1} + \frac{6A_2\bar{x}_2}{L_2} = 0 \quad ...(16.3)$$

16.5. CASE III : EI CONSTANT : NO SETTLEMENT : U.D.L. THROUGHOUT

Given $E_1 I_1 = E_2 I_2 = E I$; $\delta_1 = \delta_2 = 0$

Let U.D.L. on span AB be w_1 and U.D. L. of span BC be w_2.

Then $$A_1 = \frac{2}{3}\left(\frac{w_1 L_1^2}{8}\right) L_1 = \frac{w_1 L_1^3}{12}$$

$$A_1 \bar{x}_1 = \frac{w_1 L_1^3}{12} \times \frac{L_1}{2} = \frac{w_1 L_1^4}{24}$$

Similarly, $$A_2 \bar{x}_2 = \frac{w_2 L_2^4}{24}$$

Hence from Eq. 16.3 : $M_A L_1 + 2M_B (L_1 + L_2) + M_C L_2 + \frac{6}{L_1}\left(\frac{w_1 L_1^4}{24}\right) + \frac{6}{L_2}\left(\frac{w_2 L_2^4}{24}\right) = 0$

or $$M_A L_1 + 2M_B (L_1 + L_2) + M_C L_2 = -\frac{w_1 L_1^3}{4} - \frac{w_2 L_2^3}{4} \quad ...(16.4)$$

16.6. CASE IV : EI CONSTANT : SUPPORTS SETTLE : U.D.L.

From Eqs. 16.2 and 16.4

$$M_A L_1 + 2M_B (L_1 + L_2) + M_C L_2 = -\left(\frac{w_1 L_1^3}{4} - \frac{w_2 L_3^3}{4}\right) + 6E I\left(\frac{\delta_1}{L_1} + \frac{\delta_2}{L_2}\right) \quad ...(16.5)$$

16.7. VALUES OF FACTOR $\frac{6A\bar{x}}{L}$

The usefulness of the three moment equation for finding unknowns M_A, M_B and M_C depends upon the ease with which the expressions $(6A . \bar{x})/L$ can be found. We shall now consider few cases of loading.

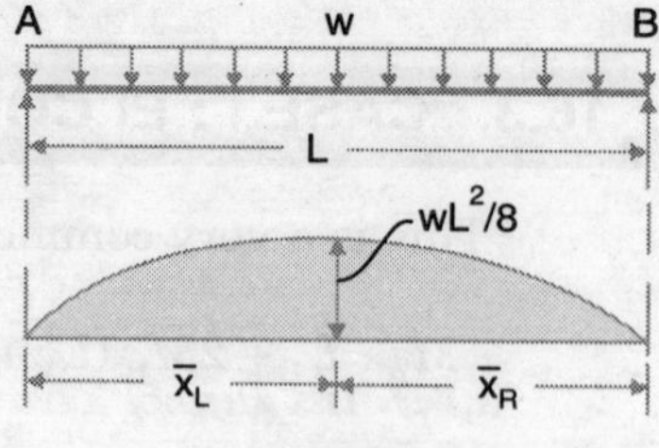

Fig. 16.3

(a) U.D.L.: (Fig. 16.3) $A = \frac{2}{3} \times L \times \frac{w L^2}{8} = \frac{w L^3}{12}$

$$\bar{x}_L = L/2 \; ; \; \bar{x}_R = L/2$$

$$\therefore \quad \frac{6A\bar{x}_L}{L} = \frac{6A\bar{x}_R}{L} = \frac{w L^3}{4} \quad ...(16.6)$$

(b) Triangularly distributed load :

Fig. 16.4 (a), (b) and (c) show three cases of triangularly distributed loads.

Triangularly distributed load of Fig. 16.4 (a).

The component B.M.D. is shown in Fig. 16.4 (a iii)

$$A\bar{x}_L = \frac{1}{2} \times L \times \frac{w L^2}{6} \times \frac{2}{3} L - \frac{1}{4} \times L \times \frac{w L^2}{6} \times \frac{4}{5} L = \frac{1}{45} w L^4$$

$$A\bar{x}_R = \frac{1}{2} \times L \times \frac{w L^2}{6} \times \frac{1}{3} L - \frac{1}{4} L \times \frac{w L^2}{6} \times \frac{1}{5} L = \frac{7}{360} w L^4$$

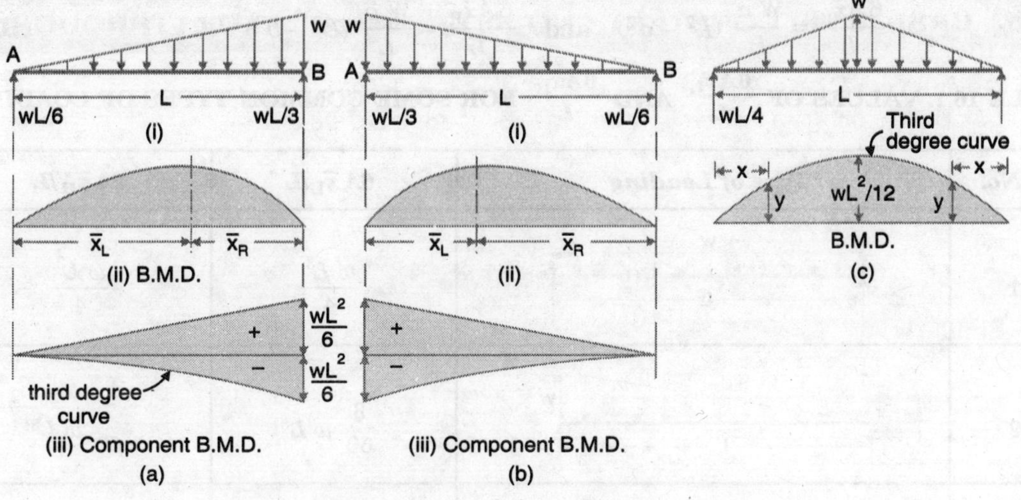

Fig. 16.4. Factor $\frac{6A\bar{x}}{L}$ for triangularly distributed loads.

$$\therefore \quad \frac{6A\bar{x}_L}{L} = \frac{6}{L} \times \frac{1}{45} w L^4 = \frac{2}{15} w L^3 = \frac{8}{60} w L^3 \qquad \text{...[16.6 (a)]}$$

$$\frac{6A\bar{x}_R}{L} = \frac{6}{L} \times \frac{7}{360} w L^4 = \frac{7}{60} w L^3 \qquad \text{...[16.6 (b)]}$$

Triangularly distributed load of Fig. 16.4 (b)

Evidently,
$$\frac{6A\bar{x}_L}{L} = \frac{7}{60} w L^3 \quad \text{and} \quad \frac{6A\bar{x}_R}{L} = \frac{8}{60} w L^3 \qquad \text{...(16.7)}$$

Triangularly distributed load of Fig. 16.4 (c)

From Example 7.7,

ordinates
$$y = \frac{w L x}{4} - \frac{w x^3}{3L} \text{ and } y' = \frac{w L x'}{4} - \frac{w x'^3}{3L}$$

$$\therefore \quad A\bar{x}_L = \int_0^{L/2} \left(\frac{w L x}{4} - \frac{w x^3}{3 L} \right) x \, dx + \int_0^{L/2} \left(\frac{w L x'}{4} - \frac{w x'^3}{3 L} \right) (L - x') \, dx'$$

$$= \frac{w L^4}{120} + \frac{17}{960} w L^4 = \frac{5}{192} w L^4$$

$$\frac{6A\bar{x}_L}{L} = \frac{6A\bar{x}_R}{L} = \frac{6}{L} \times \frac{5}{192} w L^4 = \frac{5}{32} w L^3 \qquad \text{...(16.8)}$$

(c) *Point load :* The component B.M.D. is shown in Fig. 16.5 (*c*).

$$A\bar{x}_L = \frac{1}{2} \times L \times W b \times \frac{2 L}{3} - \frac{1}{2} \times b \times W b \left(L - \frac{b}{3} \right)$$

$$= \frac{W b L^2}{3} - \frac{W b^2}{6} (3L - b) = \frac{W a}{6} (L^2 - a^2)$$

$$A\bar{x}_R = \frac{1}{2} \times L \times W b \times \frac{L}{3} - \frac{1}{2} \times b \times W b \times \frac{b}{3}$$

$$= \frac{W b L^2}{6} - \frac{W b^3}{6} = \frac{W b}{6} (L^2 - b^2)$$

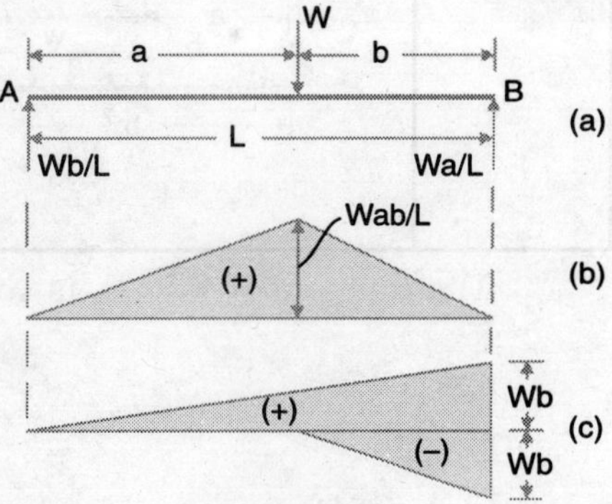

Fig. 16.5

$$\therefore \quad \frac{6A\bar{x}_L}{L} = \frac{W\,a}{L}\,(L^2 - a^2) \quad \text{and} \quad \frac{6A\bar{x}_R}{L} = \frac{W\,b}{L}\,(L^2 - b^2) \qquad \text{...(16.9)}$$

TABLE 16.1. VALUES OF $\frac{6A\bar{x}_L}{L}$ AND $\frac{6A\bar{x}_R}{L}$ FOR SOME COMMON TYPES OF LOADINGS

S.No.	*Type of Loading*	$6A\bar{x}_L/L$	$6A\bar{x}_R/L$
1.	w; L	$\frac{wL^3}{4}$	$\frac{wL^3}{4}$
2.	w; L	$\frac{8}{60}\,wL^3$	$\frac{7}{60}\,wL^3$
3.	w; L	$\frac{7}{60}\,wL^3$	$\frac{8}{60}\,wL^3$
4.	w; L/2; L/2; L	$\frac{5}{32}\,wL^3$	$\frac{5}{32}\,wL^3$
5.	W; a; b; L	$\frac{W\,a}{L}\,(L^2 - a^2)$	$\frac{W\,b}{L}\,(L^2 - b^2)$
6.	a; b; L	$-\frac{\mu}{L}\,(3a^2 - L^2)$	$+\frac{\mu}{L}\,(3b^2 - L^2)$
7.	a; d; w; b; c	$\frac{w}{4L}\,[b^2(2L^2 - b^2)$ $- a^2(2L^2 - a^2)]$	$\frac{w}{4L}\,[d^2(2L^2 - d^2)$ $- c^2(2L^2 - c^2)]$

(d) Couple at intermediate point :

The component B.M.D. is shown in Fig. 16.6 (c).

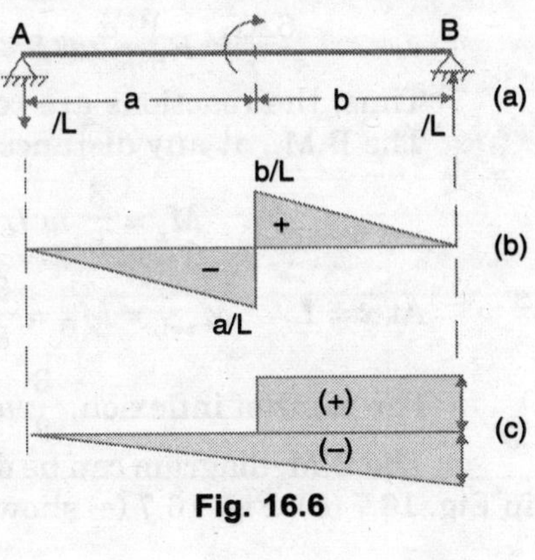

Fig. 16.6

$$A\bar{x}_L = \mu\, b\left(L - \frac{b}{2}\right) - \frac{1}{2} \times L \times \mu\left(\frac{2}{3}L\right)$$

$$= \frac{\mu\, b\,(2L - b)}{2} - \frac{\mu\, L^2}{3} = -\frac{\mu}{6}\,(3a^2 - L^2)$$

$$A\bar{x}_R = \mu\, b\,.\,\frac{b}{2} - \frac{1}{2}L\,\mu \times \frac{L}{3}$$

$$= \frac{\mu\, b^2}{2} - \frac{\mu\, L^2}{6} = \frac{\mu}{6}\,(3b^2 - L^2)$$

$$\therefore \quad \frac{6A\bar{x}_L}{L} = -\frac{\mu}{L}\,(3a^2 - L^2)$$

and
$$\frac{6A\bar{x}_R}{L} = +\frac{\mu}{L}\,(3b^2 - L^2) \quad \text{...(16.10)}$$

Table 16.1 gives the values of $\frac{6A\bar{x}_L}{L}$ and $\frac{6A\bar{x}_R}{L}$ for some common types of loadings.

Example 16.1. *A beam ABC of length 2L rests on three supports equally spaced and is loaded with U.D.L. w / unit length throughout the length of the beam. Draw the S.F. and B.M. diagrams for the beam.*

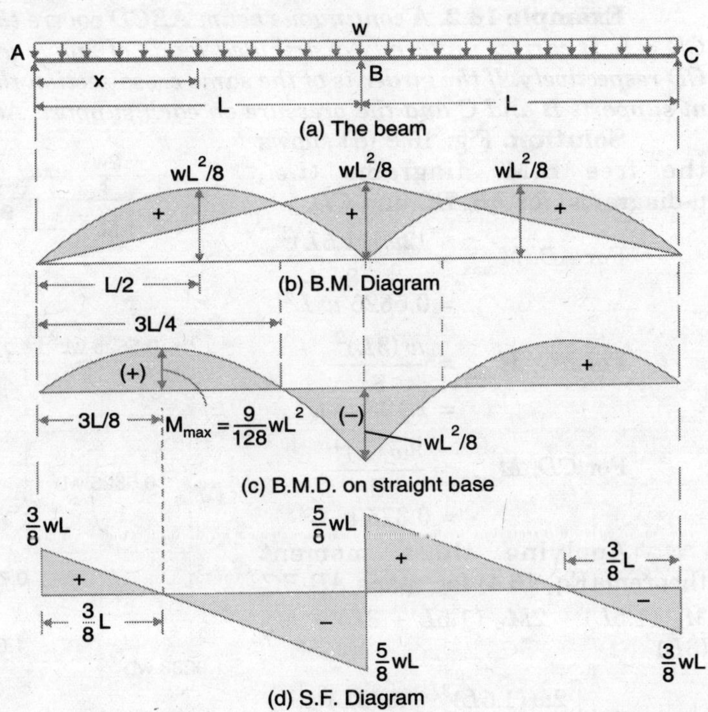

Fig. 16.7

Solution. Applying three moment theorem for U.D.L., we get, from Eq. 16.4

$$M_A\, L + 2M_B\,(L + L) + M_C\, L = -\left(\frac{w\,L^3}{4} + \frac{w\,L^3}{4}\right)$$

But $M_A = 0$ and $M_C = 0$

$$4M_B = -\frac{w\,L^2}{2}$$

or
$$M_B = -\frac{w\,L^2}{8}$$

Thus the fixing moments are known.

For R_A, write equation for B.M. at B. Thus,

$$R_A\,.\,L - w\,L\,.\,\frac{L}{2} = M_B = -\frac{w\,L^2}{8}$$

$$\therefore \quad R_A = \frac{w\,L}{2} - \frac{w\,L}{8} = \frac{3}{8}\,w\,L\,(\uparrow) \quad \text{and} \quad R_C = \frac{3}{8}\,w\,L\,(\uparrow), \text{ by symmetry.}$$

$\therefore \qquad R_B = 2w\,L - (R_A + R_C) = 2w\,L - \dfrac{3}{4}\,w\,L = \dfrac{5}{4}\,w\,L\,(\uparrow)$

Thus, the reactions are completely known.

The B.M., at any distance x from A, for the span AB, is given by

$$M_x = \frac{3}{8}\,w\,L\,x - \frac{w\,x^2}{2} \text{ (Parabolic variation)}$$

At $x = L$, $\quad M_{x=L} = M_B = \dfrac{3}{8}\,w\,L\,.\,L - \dfrac{w\,L^2}{2} = -\dfrac{w\,L^2}{8}$

For point of inflexion, $\dfrac{3}{8}\,w\,L\,x - \dfrac{w\,x^2}{2} = 0$, which gives $x = \dfrac{3}{4}\,L.$

The B.M. diagram can be drawn by super-imposing μ-diagram over μ′-diagram, as shown in Fig. 16.7 (*b*). Fig. 16.7 (*c*) shows the B.M. diagram on a straight base. Fig. 16.7 (*d*) shows the S.F. diagram. B.M. is maximum where S.F. is zero. This is obtained at $x = \dfrac{3}{8}\,L.$

$$M_{max} = \frac{3}{8}\,w\,L\left(\frac{3}{8}\,L\right) - \frac{w}{2}\left(\frac{3}{8}\,L\right)^2 = \frac{9}{128}\,w\,L^2.$$

Example 16.2. *A continuous beam ABCD covers three spans, AB = 1.5 L, BC = 3 L and CD = L. It carries uniformly distributed loads of 2w, w and 3w per metre run on AB, BC and CD respectively. If the girder is of the same cross-section throughout, find the bending moments at supports B and C and the pressure on each support. Also plot the B.M. and S.F. diagrams.*

Solution. Fig. 16.8 (*a*) shows the free B.M. diagrams (*i.e.*, μ-diagrams) for *AB*, *BC* and *CD*.

For AB, $M_{max} = \dfrac{(2w)\,(1.5L)^2}{8}$

$= 0.5625\;w\;L^2$

For BC, $M_{max} = \dfrac{w\,(3L)^2}{8}$

$= 1.125\;w\;L^2$

For CD, $M_{max} = \dfrac{3w\,(L)^2}{8}$

$= 0.375\;w\;L^2$

Applying three moment theorem (Eq. 16.4) for spans *AB-BC*

$M_A\,(1.5L) + 2M_B\,(1.5L + 3L) + M_C\,(3L)$

$$= -\left[\frac{2w\,(1.5L)^3}{4} + \frac{w\,(3L)^3}{4}\right]$$

where $M_A = 0$

$\therefore \quad M_B + 0.3333\;M_C$

$= -\,0.9375\;w\;L^2 \quad ...(1)$

Similarly, applying three moments theorem for *BC-CD*, we get

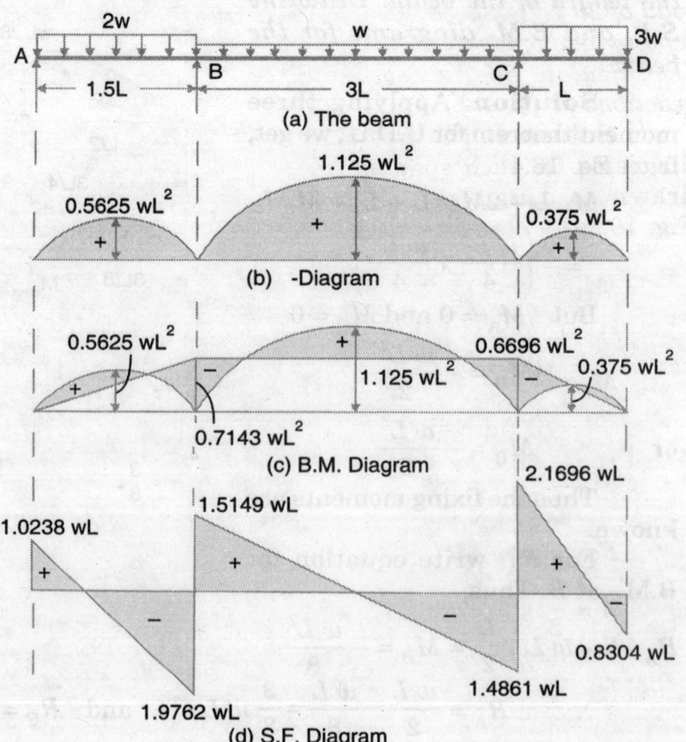

Fig. 16.8

$$M_B(3L) + 2M_C(3L+L) + M_D(L) = -\left[\frac{w(3L)^3}{4} + \frac{3w(L)^3}{4}\right], \text{ where } M_D = 0$$

$\therefore$ $$M_B + 2.6667\,M_C = -2.5\,w\,L^2 \quad \ldots(2)$$

From (1) and (2), we get $M_C = -\mathbf{0.6696\,w\,L^2}$ and $M_B = -\mathbf{0.7143\,w\,L^2}$

Reactions. For reaction at A write the expression for B.M. at B.

$$\therefore \quad R_A(1.5L) - 2w(1.5L)\left(\frac{1.5L}{2}\right) = M_B = -0.7143\,w\,L^2$$

From which $R_A = \mathbf{1.0238\,w\,L}$

For R_B, write the expression for B.M. at C.

$$\therefore \quad R_A(4.5\,L) + R_B(3L) - 2w(1.5L)(3.75L) - w(3L)(1.5L) = M_C = -0.6696\,w\,L^2$$

or $$1.0238\,w\,L(4.5L) + 3R_B L - 15.75\,w\,L^2 = -0.6696\,w\,L^2$$

From which $R_B = \mathbf{3.4911\,w\,L}$

Similarly, for R_D write the equation for B.M. at C, considering right hand side,

$$\therefore \quad R_D(L) - 3w\left(\frac{L^2}{2}\right) = M_C = -0.6696\,w\,L^2 \quad \therefore \quad R_D = \mathbf{0.8304\,w\,L}$$

Hence $$R_C = \{2w(1.5L) + w(3L) + 3w(L)\} - \{1.0238\,w\,L + 3.4911\,w\,L + 0.8304\,w\,L\}$$
$$= \mathbf{3.6547\,w\,L}$$

The B.M. and S.F. diagrams are shown in Fig. 16.8 (*c*) and (*d*) respectively.

Example 16.3. *A beam ABCD, 16 m long is continuous over three spans : AB = 6 m, BC = 5 m and CD = 5 m, the supports being at the same level. There is a uniformly distributed load of 20 kN/m over BC. On AB, there is a point load of 80 kN at 2 m from A. On CD, there is a point load of 60 kN at 3 m from D. Calculate the moments and reactions at the supports.*

Solution. The free B.M. diagrams for each span can be drawn as usual, as shown in Fig. 16.9 (*b*). Here, we have

$$M_E = \frac{80 \times 2 \times 4}{6}$$
$$= 106.67 \text{ kN-m}$$

$$M_F = \frac{20(5)^2}{8}$$
$$= 62.5 \text{ kN-m}$$

$$M_G = \frac{60 \times 2 \times 3}{5}$$
$$= 72 \text{ kN-m}$$

Fig. 16.9

For span AB

$$A = \frac{1}{2} \times 6 \times 106.67$$
$$= 320.01$$

$$\bar{x} = \frac{1}{3}(6+2) = \frac{8}{3}$$

$$\therefore \quad A\bar{x} = 320.01 \times \frac{8}{3}$$
$$= 853.36$$

with A as origin.

For span BC $A = \frac{2}{3} \times 5 \times 62.5 = 208.33$

With C as origin, $A\bar{x} = 208.33 \times 2.5 = 520.83$;

With B as origin, $A\bar{x} = 208.33 \times 2.5 = 520.83$.

For span CD $A = \frac{1}{2} \times 5 \times 72 = 180$; $A\bar{x} = 180 \times \frac{1}{3}(5 + 3) = 480$, with D as origin.

Applying three moment theorem equation for spans *AB-BC*

$$6M_A + 2M_B(6+5) + 5M_C + \frac{6A_1\bar{x}_1}{6} + \frac{6A_2\bar{x}_2}{5} = 0$$

or $$6 \times 0 + 22M_B + 5M_C = -\left[\frac{6}{6} \times 853.36 + \frac{6}{5} \times 520.83\right] = -1478.4 \quad ...(1)$$

Similarly, applying three moment theorem equation for spans *BC-CD*

$$5M_B + 2M_C(5+5) + 5M_D + \frac{6A_1\bar{x}_1}{5} + \frac{6A_2\bar{x}_2}{5} = 0$$

or $$5M_B + 20M_C + 5 \times 0 = -\left[\frac{6}{5} \times 520.83 + \frac{6}{5} \times 480\right] = -1200.1 \quad ...(2)$$

Solving (1) and (2), we get $M_B = -$ **56.79 kN-m** and $M_C = -$ **45.81 kN-m**

For finding reaction R_A, write equation for B.M. at B. Thus

$R_A \times 6 - 80 \times 4 = M_B = -56.79$, from which $R_A =$ **43.87 kN**

For finding reaction R_B, write equation for B.M. at C. Thus

$R_A \times 11 - 80 \times 9 + R_B \times 5 - 20 \times 5 \times 2.5 = M_C = -45.81$

or $43.87 \times 11 - 720 + 5R_B - 250 = -45.81$, from which $R_B =$ **88.32 kN**

For reaction R_D, write equation for B.M. at C, considering R.H.S.

$\therefore$ $R_D \times 5 - 60 \times 2 = M_C = -45.81$, from which $R_D =$ **14.84 kN**

Hence $R_C = (80 + 20 \times 5 + 60) - (43.87 + 88.32 + 14.84) =$ **92.97 kN**

The B.M. and S.F. diagrams are shown in Fig. 16.9 (*c*) and (*d*) respectively.

Example 16.4. *Solve Example 16.3 if the support B sinks by 5 mm. I for the section is 9300 cm⁴ and E = 2.10 × 10⁵ N/mm².*

Solution. Applying three moments theorem for spans *AB-BC*, we get

$$22M_B + 5M_C + 1478.4 = 6EI\left(\frac{\delta_1}{L_1} + \frac{\delta_2}{L_2}\right)$$

In which $$\frac{6A_1\bar{x}_1}{L_1} + \frac{6A_2\bar{x}_2}{L_2} = 1478.4 \text{ from Example 16.3}$$

Here $E = 2.1 \times 10^5$ N/mm² $= 210$ kN/mm² ; $I = 9300 \times 10^4$ mm⁴

$\therefore$ $EI = 2.1 \times 10^5 \times 9300 \times 10^4$ kN/mm² $= 19530 \times 10^9$ kN/mm² $= 19530$ kN-m²

$$\delta_1 = 5 \text{ mm} = \frac{5}{1000} \text{ m} = \frac{1}{200} \text{ m} = \delta_2$$

Thus, substituting all the values in kN and m units, we get

$$22M_B + 5M_C + 1478.4 = 6 \times 19530\left(\frac{1}{200 \times 6} + \frac{1}{200 \times 5}\right)$$

or $$M_B + 0.2273\,M_C = -57.435 \quad ...(1)$$

Similarly, *for span BC-CD* :

$$\delta_1 = \text{movement of } C \text{ with respect to } B = -5 \text{ mm} = -\frac{1}{200} \text{ m}$$

(The movement of C being upwards with respect to B)

$$\delta_2 = \text{movement of } C \text{ with respect to } D = 0$$

$$\therefore \quad 5M_B + 20M_C + 1200.1 = 6EI\left(\frac{\delta_1}{L_1} + \frac{\delta_2}{L_2}\right) = 6 \times 19530\left(-\frac{1}{200 \times 5} + 0\right)$$

or $$M_B + 4M_C = -216.58 \quad \text{...(2)}$$

Solving (1) and (2), we get $M_B = -45.03$ kN-m ; $M_C = -54.61$ kN-m

For finding R_A, write equation for B.M. at B.

Thus, $R_A \times 6 - 80 \times 4 = M_B = -45.03$, from which $\boldsymbol{R_A = 45.83}$ **kN.**

For finding R_B, write equation for B.M. at C.

Thus $45.83 \times 11 \times R_B \times 5 - 80 \times 9 - 20 \times 5 \times 2.5 = M_C = -54.61$

From which $\boldsymbol{R_B = 82.26}$ **kN**

For finding R_D, write equation for B.M. at C, considering R.H.S. Thus,

$R_D \times 5 - 60 \times 2 = M_C = -54.61$, from which $\boldsymbol{R_D = 13.08}$ **kN**

For finding R_C, write equation for B.M. at B, considering R.H.S. Thus

$13.08 \times 10 + R_C \times 5 - 60 \times 7 - 20 \times 5 \times 2.5 = M_B = -45.03$

From which $\boldsymbol{R_C = 98.83}$ **kN**

Check. Total load = 80 + 60 + 20 × 5 = 240 kN

Total reaction = 45.83 + 82.26 + 98.83 + 13.08 = 240 kN.

16.8. APPLICATION OF CLAPEYRON'S THEOREM FOR FIXED BEAM

A fixed beam can also be analyzed by the application of three moment theorem, by imagining a zero span to the left of A and a zero span to the right of B, as shown in Fig. 16.10 (a). Applying three moment theorem for the span $A'A - AB$, we have

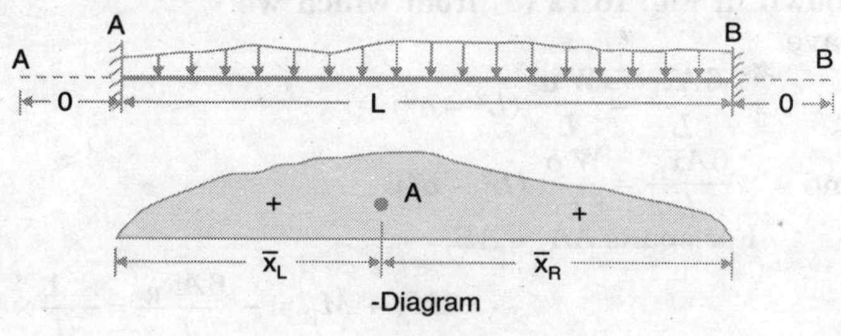

Fig. 16.10

$$0 + 2M_A(0 + L) + M_B \cdot L + 0 + \frac{6A\bar{x}_R}{L} = 0$$

or $$2M_A L + M_B \cdot L + \frac{6A\bar{x}_R}{L} = 0$$

or $$2M_A + M_B + \frac{6A\bar{x}_R}{L^2} = 0 \quad \text{...(1)}$$

Similarly, applying the theorem for the span $AB - BB'$, we get

$$M_A \cdot L + 2M_B(L + 0) + 0 + \frac{6A\bar{x}_L}{L} = 0 \quad \text{or} \quad M_A + 2M_B + \frac{6A\bar{x}_L}{L^2} = 0 \quad \text{...(2)}$$

Solving Eqs. (1) and (2), M_A and M_B can be found.

Example 16.5. *Find the fixing moments for a fixed beam subjected to U.D.L. throughout the span.*

Solution. Imagine a zero span AA' to the left of A and zero span BB' to the right of B, as shown in Fig. 16.11 (a). The μ-diagram is shown in Fig. 16.11 (b), from which we have

$$A = \frac{2}{3} \times L \times \frac{wL^2}{8} = \frac{wL^3}{12}$$

$$\bar{x}_L = \frac{L}{2} = \bar{x}_R$$

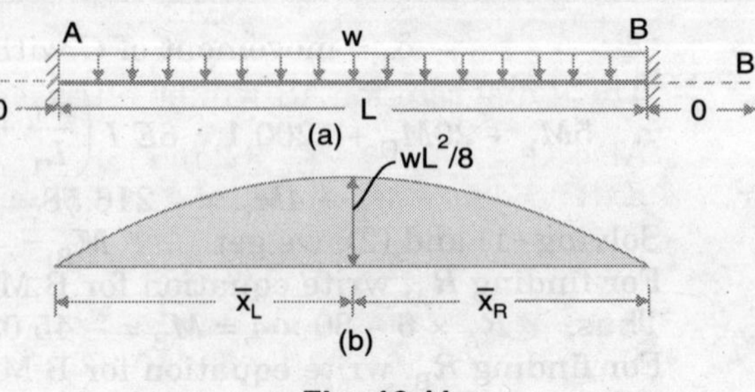

Fig. 16.11

Applying three moment equation for span $AA' - AB$, we get

$$2M_A + M_B + \frac{6A\bar{x}_R}{L^2} = 0$$

or $$2M_A + M_B + \frac{6}{L^2}\left(\frac{w\,L^3}{12}\right)\left(\frac{L}{2}\right) = 0$$

or $$2M_A + M_B = -\frac{wL^2}{4} \quad ...(1)$$

Similarly, applying three moment equation for span $AB - BB'$, we get

$$M_A + 2M_B + \frac{6A\bar{x}_R}{L^2} = 0 \quad \text{or} \quad M_A + 2M_B = -\frac{w\,L^2}{4} \quad ...(2)$$

From (1) and (2), we get $\qquad \boldsymbol{M_A = -\frac{w\,L^2}{12}}$ and $\boldsymbol{M_B = -\frac{w\,L^2}{12}}$.

Example 16.6. *Find the fixing moments for a fixed beam, subjected to an eccentric point load W.*

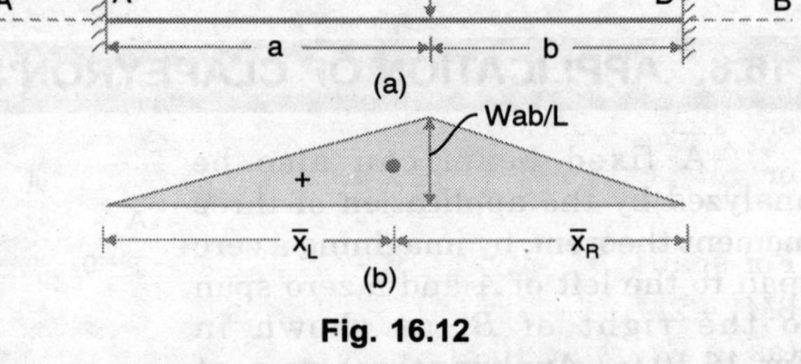

Fig. 16.12

Solution. Imagine span AA' and BB' of zero length. The μ-diagram is shown in Fig. 16.12 (*b*) from which we have

$$\frac{6A\bar{x}_L}{L} = \frac{W\,a}{L}\,(L^2 - a^2)$$

and $$\frac{6A\bar{x}_R}{L} = \frac{W\,b}{L}\,(L^2 - b^2)$$

For spans $AA' - AB$,

$$2M_A + M_B = -\frac{6A\bar{x}_R}{L} = -\frac{1}{L} \times \frac{W\,b}{L}\,(L^2 - b^2) = -\frac{W\,b}{L^2}\,(L^2 - b^2) \quad ...(1)$$

and for spans $AB - BB'$, $$M_A + 2M_B = -\frac{6A\bar{x}_L}{L^2} = -\frac{1}{L} \times \frac{W\,a}{L}\,(L^2 - a^2) = -\frac{W\,a}{L^2}\,(L^2 - a^2) \quad ...(2)$$

Solving (1) and (2), we get $\boldsymbol{M_A = -\frac{W\,a\,b^2}{L^2}}$ and $\boldsymbol{M_B = -\frac{W\,b\,a^2}{L^2}}$.

Example 16.7. *Find the prop reaction and fixing moment for a propped cantilever loaded as shown in Fig. 16.13.*

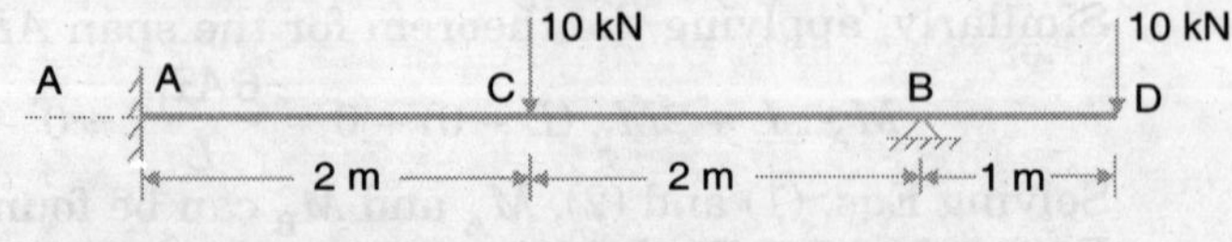

Fig. 16.13

Solution. Considering cantilever portion BD, we get

$$M_B = -10 \times 1 = -10 \text{ kN-m}$$

Also, for span AB,

$$\mu_C = \frac{10 \times 4}{4} = 10 \text{ kN}$$

The μ-diagram for AB will be a triangle, having $A = \frac{1}{2} \times 4 \times 10 = 20$.

Also $\bar{x}_L = \bar{x}_R = 2$ m

Extend A to A' such that $AA' = 0$. Then for $A'A - AC$, we have

$$0 + 2M_A (4) + M_B (4) + \frac{6}{4} (20) (2) = 0 \quad \text{or} \quad M_A + 0.5\, M_B = -7.5 \qquad ...(1)$$

But $M_B = -10$ kN-m. Hence $M_A = -7.5 + 0.5 \times 10 =$ **$-$ 2.5 kN-m**

Now, $R_B \times 4 - 10 \times 5 - 10 \times 2 = M_A = -2.5$, from which R_B = **16.875 kN** (↑)

Hence $R_A = 10 + 10 - 16.875 =$ **3.125 kN** (↑).

Example 16.8. *Solve Example 16.3 if end A is fixed.*

Solution. Imagine a point A' to the left of A, such that $AA' = 0$.

Then for the spans $A'A - AB$,

$$0 + 2\, M_A\, (0 + 6) + M_B \times 6 = -\left[0 + \frac{6}{6}\left(\frac{1}{2} \times 6 \times 106.67\right) \times \left(\frac{6+4}{3}\right)\right]$$

or $\quad 12M_A + 6M_B = -1066.7$

or $\quad M_A + 0.5\, M_B = -88.89 \qquad ...(1)$

For span $AB - BC$,

$6M_A + 2M_B (6 + 5) + 5M_C = -1478.4$

(as in Example 16.3)

or $\quad M_A + 3.667\, M_B + 0.833\, M_C = -246.4 \qquad ...(2)$

For span $BC - CD$, $5M_B + 20M_C + 5M_D = -1200.1$ (as in Example 16.3)

or $\quad M_B + 4M_C + M_D = -240$ where $M_D = 0$

$\therefore \quad M_B + 4M_C = -240 \qquad ...(3)$

From (1) and (2), we get $\quad M_B + 0.263\, M_C = -49.735 \qquad ...(4)$

From (3) and (4), we get $\quad M_C$ = **$-$ 50.914 kN-m**

Hence from (4), we get $\quad M_B$ = **$-$ 36.345 kN-m**

Lastly from (1), we get $\quad M_A$ = **$-$ 70.718 kN-m**

For reaction at D, write equation for B.M. at C.

$\therefore \quad R_D \times 5 - 60 \times 2 = M_C = -50.914$, from which R_D = **13.817 kN**

For reaction at C, write equation for B.M. at B

$\therefore\; 13.817 \times 10 + R_C \times 5 - 60 \times 7 - 20 \times 5 \times \frac{5}{2} = M_B = -36.345$, from which R_C = **99.097 kN**

For reaction at B, write equation for B.M. at A.

$\therefore \quad 13.817 \times 16 + 99.097 \times 11 + R_B \times 6 - 60 \times 13 - 100 \times 8.5 - 80 \times 2 = M_A = -70.718$

From which, we get $\quad R_B$ = **68.024 kN**

For reaction at A, write equation for B.M. at B, considering L.H.S.

$R_A \times 6 + M_A - 80 \times 4 = M_B$

or $\quad 6R_A - 70.718 - 320 = -36.345$ from which R_A = **59.062 kN**

Fig. 16.14

Check. Total load = 80 + 100 + 60 = 240 kN

Total reaction = 59.062 + 68.024 + 99.097 + 13.817 = 240 kN

The B.M. and S.F. diagrams are shown in Fig. 16.14 (*b*) and (*c*) respectively.

16.9. ADDITIONAL EXAMPLES USEFUL FOR COMPETITIVE EXAMINATIONS

Example 16.9. *A straight elastic beam of uniform section rests on four similar elastic supports which are placed L apart. The supports are such that they are compressed by d for each unit of load upon them. Show that when a uniformly distributed load of total amount W comes on them, the reaction at central supports are each*

$$\frac{W\left(\frac{11}{6}+\frac{3EId}{L^3}\right)}{\left(5+\frac{12EId}{L^3}\right)} \quad \textit{(Cambridge)}$$

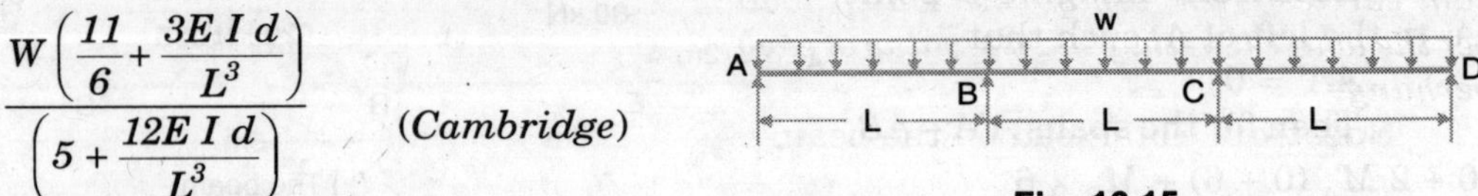

Fig. 16.15

Solution. Total load = W ;

$\therefore$ U.D.L., $w = W/3L$

Let R_A, R_B, R_C and R_D be the reactions at A, B, C and D respectively. By symmetry,

$$R_A = R_D \text{ and } R_B = R_C.$$

Sinking of support at $A = R_A \cdot d$; Sinking of support at $B = R_B \cdot d$.

Sinking of support at $C = R_C \cdot d$; Sinking of support at $D = R_D \cdot d$.

Also, $M_A = M_D = 0$ and $M_B = M_C$.

δ_1 = deflection of B with respect to $A = \delta_B - \delta_A = (R_B - R_A)\,d$

δ_2 = deflection of B with respect to $C = \delta_B - \delta_C = (R_B - R_C)\,d = 0$.

Now applying three moment theorem equation for spans $AB - BC$,

$$M_A \cdot L + 2M_B\,(L + L) + M_C\,L + \frac{6A\bar{x}_1}{L} + \frac{6A_2\bar{x}_2}{L} = 6EI\left(\frac{\delta_1}{L} + \frac{\delta_2}{L}\right)$$

or
$$4M_B \cdot L + M_B\,L + \frac{w\,L^3}{4} + \frac{w\,L^3}{4} = \frac{6EId}{L}\,(R_B - R_A)$$

or
$$5M_B \cdot L + \left(\frac{W}{3L}\right)\frac{L^3}{2} = \frac{6EId}{L}\,(R_B - R_A)$$

or
$$M_B = \frac{1}{5}\left[-\frac{WL}{6} + \frac{6EId}{L^2}(R_B - R_A)\right] = -\frac{1}{5}\left[\frac{WL}{6} - \frac{6EId}{L^2}(R_B - R_A)\right]$$

Now writing equation for B.M. at B, we have

$$R_A \cdot L - \frac{W}{3L}\left(\frac{L^2}{2}\right) = M_B = -\frac{1}{5}\left[\frac{WL}{6} - \frac{6EId}{L^2}(R_B - R_A)\right] \quad \text{...(1)}$$

Also, $$2R_B + 2R_A = W \quad \text{or} \quad R_A = \frac{W}{2} - R_B \quad \text{...(2)}$$

Substituting the value of R_A in (1), we get

$$\left(\frac{W}{2} - R_B\right)L - \frac{WL}{6} = -\frac{1}{5}\left[\frac{WL}{6} - \frac{6EId}{L^2}\left\{R_B - \left(\frac{W}{2} - R_B\right)\right\}\right]$$

$$\frac{W}{2}\,L - R_B\,L - \frac{WL}{6} = -\frac{WL}{30} + \frac{6EId}{5L^2}\left(2R_B - \frac{W}{2}\right)$$

or $$R_B\left[\frac{12EId}{5L^2}+L\right]=\frac{W}{2}L-\frac{WL}{6}+\frac{WL}{30}+\frac{3EIdW}{5L^2}$$

or $$R_B=\frac{\frac{11}{30}WL+\frac{3EIdW}{5L^2}}{\frac{12EId}{5L^2}+L}=\frac{\frac{WL}{5}\left[\frac{11}{6}+\frac{3EId}{L^3}\right]}{\frac{L}{5}\left[5+\frac{12EId}{L^3}\right]}=\frac{W\left(\frac{11}{6}+\frac{3EId}{L^3}\right)}{\left(5+\frac{12EId}{L^3}\right)}.$$ Hence proved.

Example 16.10. *A continuous beam ABCDE has three supports with two equal over hangs as shown in Fig. 16.16. The overhangs are so selected that the reaction of the three supports are equal when the beam carries a uniformly distributed load over the whole length. Find the value of each overhang.*

Fig. 16.16

Solution. Total load on the beam $= w\,(2L + 2a) = 2w\,(L + a)$

Since all the reactions are equal, we have $R_B = R_C = R_D = \frac{2}{3}\,w\,(L + a)$

Now $$M_B = -\frac{wa^2}{2} = M_D$$

Applying three moments theorem equation for spans $BC - CD$, we get

$$M_B\,.\,L + 2M_C\,(L + L) + M_D\,.\,L = -\frac{wL^3}{4} - \frac{wL^3}{4}$$

or $$-\frac{wa^2}{2}\,.\,L + 4M_C\,.\,L - \frac{wa^2}{2}\,.\,L = -\frac{wL^3}{2}$$

From which, we get $$M_C = \frac{1}{4L}\left[-\frac{wL^3}{2}+\frac{wa^2L}{2}+\frac{wa^2L}{2}\right]=\frac{w}{4}\left(a^2-\frac{L^2}{2}\right)$$

or $$M_C = -\frac{w}{8}\,(L^2 - 2a^2) \qquad ...(1)$$

But $$M_C = R_B\,.\,L - \frac{w(a+L)^2}{2} = \frac{2w}{3}\,(L + a)\,L - \frac{w}{2}\,(L + a)^2$$

$$= -\frac{w\,(L+a)}{6}\,[3(L + a) - 4L] = -\frac{w\,(L+a)}{6}\,(3a - L) \qquad ...(2)$$

Equating (1) and (2), we get

$$\frac{w}{8}\,(L^2 - 2a^2) = \frac{w}{6}\,(L + a)\,(3a - L) \quad \text{or} \quad 18a^2 + 8aL - 7L^2 = 0$$

From which, we get $a = \mathbf{0.4398\ L.}$

Example 16.11. *A bridge of uniform cross-section rests on rigid abutments at the ends and three equal pontoons as shown as Fig. 16.17 and has a concentrated load W at the middle. When bridge is unloaded, the pontoons just touch it without carrying any force. With the load W at the middle and the two end pontoons removed, the central deflection is one half what it would be with no pontoons. Find the reactions and draw the B.M. diagram for the bridge due to central load with three pontoons in position.*

Solution. Due to symmetry,

$R_A = R_E$ and $R_B = R_D$

$M_B = M_D$

Also, $M_A = M_E = 0$

Thus, there are five unknowns :

R_A, R_B, R_C, M_B and M_C.

If R is the reaction on any pontoon, its settlement will be equal to $R.k$ where k is a constant. The value of k can be found from the data given in the problem when there is only a central pontoon. In that case, if P is the reaction on the pontoon, we have

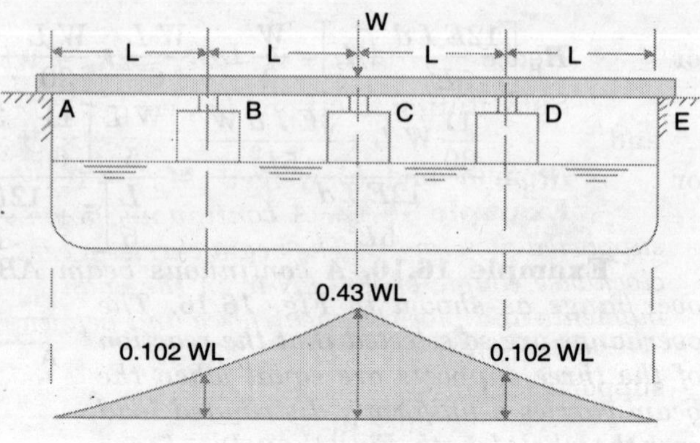

Fig. 16.17

$$\text{Central deflection} = \frac{(W-P)(4L)^3}{48EI} = \frac{4(W-P)L^3}{3EI}, \text{ with the pontoon.}$$

$$\text{Central deflection, with no pontoon} = \frac{W(4L)^3}{48EI} = \frac{4WL^3}{3EI}$$

As per the given condition, $\dfrac{4(W-P)L^3}{3EI} = \dfrac{1}{2}\left[\dfrac{4}{3}\dfrac{WL^3}{EI}\right]$ from which, we get $P = \dfrac{W}{2}$.

$\therefore$ Central deflection with the pontoon $= P \,.\, k = \dfrac{W}{2}\, k.$

But this is equal to $\dfrac{1}{2}\left[\dfrac{4}{3}\dfrac{WL^3}{EI}\right]$

$$\therefore \qquad \frac{W}{2}\, k = \frac{1}{2}\left[\frac{4}{3}\frac{WL^3}{EI}\right]$$

From which $\quad k = \dfrac{4L^3}{3EI} \quad$ Also, $\dfrac{6EIk}{L^2} = 8L$

$$\delta_1 = \delta_B - \delta_A = (R_B k - 0) = R_B \,.\, k \text{ (Since } \delta_A = 0)$$

$$\delta_2 = \delta_B - \delta_C = (R_B - R_C)\, k$$

Now applying three moment theorem equation for spans $AB - BC$, we get

$$M_A \,.\, L + 2M_B\,(L+L) + M_C \,.\, L + \frac{6A_1\bar{x}_1}{L} + \frac{6A_2\bar{x}_2}{L} = 6EI\left(\frac{\delta_1}{L} + \frac{\delta_2}{L}\right)$$

or $\qquad 0 + 4M_B + M_C + 0 = \dfrac{6EIk}{L^2}\,(2R_B - R_C) = 8L\,(2R_B - R_C) \qquad$ ...(1)

Similarly, for span $BC - CD$,

$M_B L + 2M_C\,(L+L) + M_D \,.\, L = \dfrac{6EIk}{L}\,(R_C - R_B + R_C - R_D)$ where $R_B = R_D$ and $M_B = M_D$

$\therefore \qquad M_B + 2M_C = \dfrac{6EIk}{L^2}\,(R_C - R_B) = 8L\,(R_C - R_B) \qquad$ ...(2)

Again $\qquad M_B = R_A \,.\, L \;$ or $\; R_A = M_B/L$

Also, $\qquad M_C = R_A \,.\, 2L + R_B \,.\, L, \;$ or $\; R_B = \dfrac{M_C}{L} - 2R_A = \dfrac{M_C}{L} - \dfrac{2M_B}{L} = \dfrac{M_C - 2M_B}{L}$

Hence $R_C = W - 2R_A - 2R_B = W - \dfrac{2M_B}{L} - \dfrac{2M_C - 4M_B}{L} = W - \dfrac{2(M_C - M_B)}{L}$

Substituting in (1) and (2), we get $52M_B - 31M_C = -8WL$ ...(a)

and $31M_B - 26M_C = -8W\,L$ ...(b)

From (a) and (b), we get $M_B = \mathbf{+\,0.102\,W\,L} = M_D$ and $M_C = \mathbf{+\,0.43\,W\,L.}$

Example 16.12. *A continuous beam ABC of constant moment of inertia is simply supported at A and C. The beam carries a central point load of 40 kN in span AB and a central clockwise couple of 300 kN-m in the span BC, as shown in Fig. 16.18 (a). Find the support moments and plot the shear force and bending moment diagrams.*

Solution. For simply supported span *AB*,

$$M_D = \frac{40 \times 10}{4} = 100 \text{ kN-m}$$

For simply supported span *BC*,

μ_E (left) = – 150 kN-m

μ_E (right) = + 150 kN-m.

$$A_1\bar{x}_1 = \frac{1}{2} \times 10 \times 100 \times 5 = 2500$$

$A_2 = 0$. Hence $A_2\bar{x}_2 = 0$

Applying three moment theorem for spans *AB* – *BC*, we get

$$M_A\,(10) + 2M_B\,(10 + 15) + M_C\,(15) + \frac{6A_1\bar{x}_1}{10} + \frac{6A_2\bar{x}_2}{15} = 0$$

But $M_A = 0\,;\, M_C = 0$

$$\therefore\quad 50M_B = -\frac{6}{10}\,(2500)$$

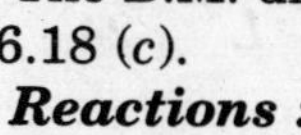

$$-\frac{6}{15}\,(0)$$

From which $M_B = \mathbf{-\,30\text{ kN-m}}$

The B.M. diagram is shown in Fig. 16.18 (*c*).

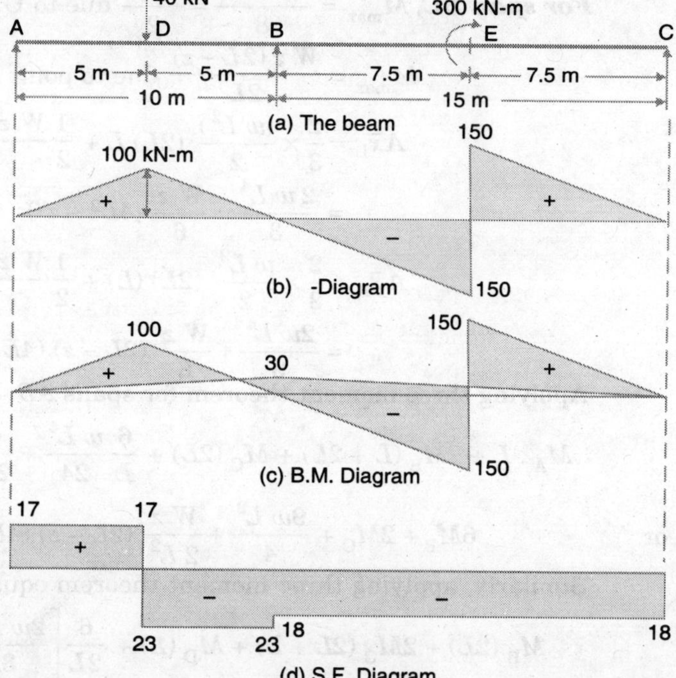

Fig. 16.18

Reactions :

For finding reaction R_A, write equation for B.M. at *B*, considering L.H.S.

Thus, $R_A \times 10 - 40 \times 5 = M_B = -30$, from which $R_A = \mathbf{17\text{ kN}}$

For finding reaction R_C, write equation for B.M. at *B*, considering R.H.S.

Thus $R_C \times 15 - 300 = M_B = -30$, from which $R_C = \mathbf{18\text{ kN}}$

For finding reaction R_B, write equation for B.M. at *A*.

Thus $18 \times 25 - 300 + 10\,R_B - 40 \times 5 = M_A = 0$

From which $R_B = \mathbf{5\text{ kN}}$

Check. Total vertical load = 40 kN (↓) ; Total reaction = 17 + 18 + 5 = 40 kN (↑)

The S.F. diagram is shown in Fig. 16.18 (*b*).

Example 16.13. *A continuous beam ABCD has a weight w per unit length, and rests on four knife edge supports A, B, C and D. The middle span carries a load W as shown in Fig. 16.19. Find the B.M at B and C and reactions at A and D.* (Cambridge)

Solution. Self weight = w

Point load = W on BC.

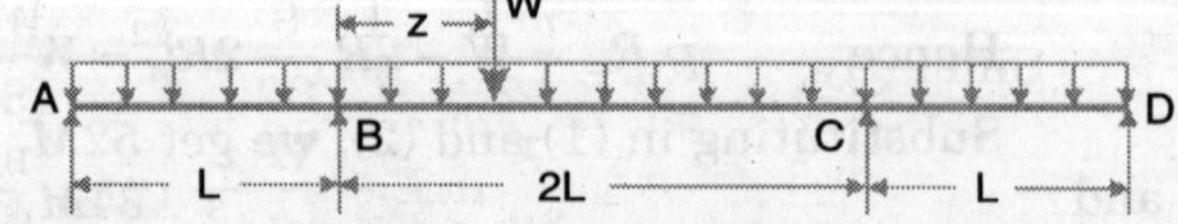

Fig. 16.19

For span AB or CD

$$M_{max} = \frac{wL^2}{8}$$

$$A\bar{x} = \frac{2}{3} \times \frac{wL^2}{8} \times L \times \frac{L}{2} = \frac{wL^4}{24}$$

For span BC, $M_{max} = \dfrac{w(2L)^2}{8} = \dfrac{wL^2}{2}$ due to U.D.L.

$$M_{max} = \frac{Wz(2L-z)}{2L} \text{ due to point load.}$$

$$A\bar{x}_L = \frac{2}{3} \times \frac{(wL^2)}{2}(2L)L + \frac{1}{2}\frac{Wz(2L-z)}{2L} \times 2L \times \frac{1}{3} \times (2L+z)$$

$$= \frac{2wL^4}{3} + \frac{Wz}{6}(4L^2 - z^2)$$

$$A\bar{x}_R = \frac{2}{3} \times \frac{wL^2}{2}(2L)(L) + \frac{1}{2}\frac{Wz(2L-z)}{2L} \times 2L \times \frac{1}{3}(2L + 2L - z)$$

$$= \frac{2wL^4}{3} + \frac{Wz}{6}(2L-z)(4L-z)$$

Applying three moment theorem for spans $AB - BC$, we get

$$M_A \cdot L + 2M_B(L + 2L) + M_C(2L) + \frac{6}{L}\frac{wL^4}{24} + \frac{6}{2L}\left[2\frac{wL^4}{3} + \frac{Wz}{6}(2L-z)(4L-z)\right] = 0$$

or $$6M_B + 2M_C + \frac{9wL^2}{4} + \frac{Wz}{2L^2}(2L-z)(4L-z) = 0 \quad \text{...(1)}$$

Similarly, applying three moment theorem equation for spans $BC - CD$

$$M_B(2L) + 2M_C(2L + L) + M_D(L) + \frac{6}{2L}\left[\frac{2wL^4}{3} + \frac{Wz}{6}(4L^2 - z^2)\right] + \frac{6}{L} \times \frac{wL^4}{24} = 0$$

or $$2M_B + 6M_C + \frac{9}{4}wL^2 + \frac{Wz}{2L^2}(2L-z)(2L+z) = 0 \quad \text{...(2)}$$

Solving (1) and (2), we get

$$M_B = -\left[\frac{9}{32}wL^2 + \frac{W.z}{16L^2}(2L-z)(5L-2z)\right] \textbf{ Ans.}$$

and $$M_C = -\left[\frac{9}{32}wL^2 + \frac{Wz}{16L^2}(2L-z)(L+2z)\right] \textbf{ Ans.}$$

For reaction at A, write down equation for B.M. at B. Thus

$$R_A \cdot L - \frac{wL^2}{2} = M_B = -\left[\frac{9}{32}wL^2 + \frac{Wz}{16L^2}(2L-z)(5L-2z)\right]$$

$$\therefore \quad R_A = \frac{wL}{2} - \frac{9}{32}wL - \frac{Wz}{16L^3}(2L-z)(5L-2z)$$

$$= \frac{7}{32}wL - \frac{Wz}{16L^3}(2L-z)(5L-2z)$$

Similarly, $R_B = \frac{wL}{2} - \frac{1}{L}\left[\frac{9}{32} wL^2 + \frac{Wz}{16L^2}(2L - z)(L + 2z)\right]$

$$= \frac{7}{32} wL - \frac{Wz}{16L^3}(2L - z)(L + 2z).$$

Example 16.14. *A three span beam ABCD has different moments of inertia and is loaded as shown in Fig. 16.20 (a). Find the reactions and support moments and draw the S.F. and B.M. diagrams.*

Solution. The μ-diagram is shown in Fig. 16.20 (*b*).

For span AB,

$$M_{max} = \frac{12 \times 3}{4} = 9 \text{ kN-m}$$

$$A\bar{x} = \frac{1}{2} \times 9 \times 3 \times \frac{3}{2} = 20.25$$

For span BC

B.M is zero throughout.

Hence $A\bar{x} = 0$

For span CD

$$M_{max} = \frac{4(3)^2}{8} = 4.5 \text{ kN-m}$$

$$A\bar{x} = \frac{2}{3} \times 3 \times 4.5 \times 1.5 = 13.5$$

Here, $M_A = 0$ and $M_D = 0$.

Writing three moment theorem equation for spans $AB - BC$, we get

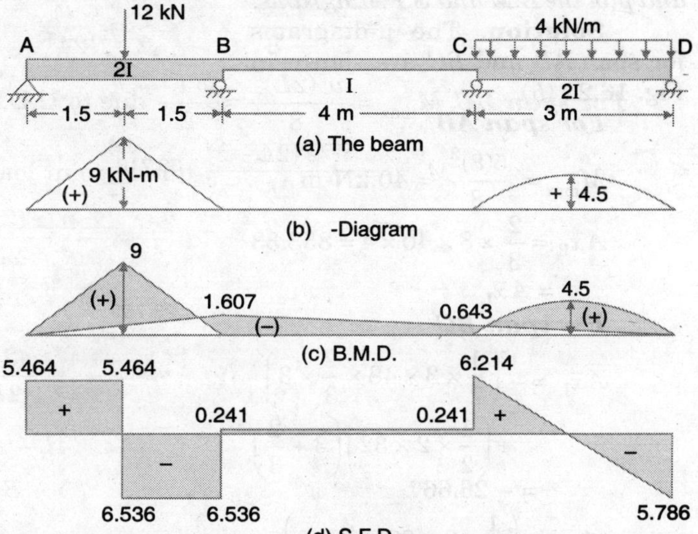

Fig. 16.20

$$M_A\left(\frac{L_1}{I_1}\right) + 2M_B\left(\frac{L_1}{I_1} + \frac{L_2}{I_2}\right) + M_C\left(\frac{L_2}{I_2}\right) + \frac{6A_1\bar{x}_1}{I_1L_1} + \frac{6A_2\bar{x}_2}{I_2L_2} = 0$$

$$0\left(\frac{3}{2I}\right) + 2M_B\left(\frac{3}{2I} + \frac{4}{I}\right) + M_C\left(\frac{4}{I}\right) + \frac{6}{2I \times 3}(20.25) + \frac{6}{I \times 4}(0) = 0$$

or $$11M_B + 4M_C + 20.25 = 0 \quad ...(1)$$

Similarly, writing three moment theorem equation for spans $BC - CD$, we get

$$M_B\left(\frac{4}{I}\right) + 2M_C\left(\frac{4}{I} + \frac{3}{2I}\right) + M_D\left(\frac{3}{2I}\right) + \frac{6}{4I}(0) + \frac{6}{2I \times 3}(13.5) = 0$$

or $$4M_B + 11M_C + 13.5 = 0 \quad ...(2)$$

Solving (1) and (2), we get $M_B = -1.607$ kN-m and $M_C = -0.643$ kN-m

The B.M. diagram is shown in Fig. 16.20.

Reactions : For finding R_A, write equation for B.M. at *B*.

Thus $R_A \times 3 - 12 \times 1.5 = M_B = -1.607$, from which R_A = **5.464** (↑).

For finding R_B, write down equation for B.M. at *C*, considering L.H.S.

∴ $5.464(7) - 12(5.5) + R_B \times 4 = M_C = -0.643$, from which R_B = **6.777 kN** (↑).

For finding R_D, write equation for B.M. at *C*, considering R.H.S.

$R_D(3) - 4 \times 3 \times 1.5 = M_C = -0.643$ from which R_D = **5.786** (↑)

Lastly, for finding R_C, write equation for B.M. at *B*, considering R.H.S.

∴ $5.786 \times 7 + R_C(4) - 4 \times 3 \times 5.5 = M_B = -1.607$, from which R_C = **5.973 kN** (↑)

Check. Total load = 12 + 4 × 3 = 24 kN

Total reaction = 5.464 + 6.777 + 5.786 + 5.973 = 24 kN

S.F. diagram : The S.F. diagram is shown in Fig. 16.20 (*d*).

Example 16.15. *For a two span beam shown in Fig. 16.21, find the support moments and reactions, and plot the B.M and S.F. diagrams.*

Solution. The μ-diagrams for span *AB* and *BC* are shown in Fig. 16.21 (*b*).

A A 5 kN/m B 80 kN-m C C
8 m 3 m 2 m

(a) The beam

32 + 40 + − 48

(b) -Diagram

36.718 40 32 − + 6.564 + − 6.482 48

(c) B.M.D.

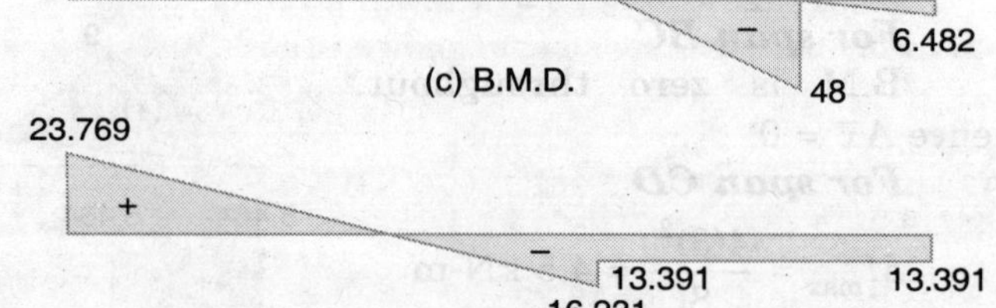

(d) S.F.D.

Fig. 16.21

For span AB

$$M_{\max} = \frac{5(8)^2}{8} = 40 \text{ kN-m}$$

$$A\bar{x}_R = \frac{2}{3} \times 8 \times 40 \times 4 = 853.33$$

$$= A\bar{x}_L$$

For span BC

$$A\bar{x}_L = -\left(\frac{1}{2} \times 3 \times 48 \times \frac{2}{3} \times 3\right) + \left(\frac{1}{2} \times 2 \times 32\right)\left(3 + \frac{2}{3}\right)$$

$$= -\ 26.667$$

$$A\bar{x}_R = \left(\frac{1}{2} \times 2 \times 32 \times \frac{2}{3} \times 2\right) - \left(\frac{1}{2} \times 3 \times 48\right)\left(2 + \frac{3}{3}\right) = -\ 173.333$$

Imagine a point *A′* to the left of *A* such that *AA′* = 0. Hence writing three moment theorem equation for spans *A′A* – *AB*, we get

$$0 + 2M_A (0 + 8) + M_B (8) + 0 + \frac{6}{8} (853.33) = 0$$

or $$2M_A + M_B + 80 = 0 \qquad ...(1)$$

Writing three moment theorem equation for spans *AB* – *BC*, we get

$$M_A (8) + 2M_B (8 + 5) + M_C (5) + \frac{6}{8} (853.33) + \frac{6}{5} (-\ 173.333) = 0$$

or $$8M_A + 26M_B + 5M_C + 432 = 0 \qquad ...(2)$$

Imagine a point *C′* to the right of *C* such that *CC′* = 0.

Hence for spans *BC* – *CC′*

$$M_B (5) + 2M_C (5 + 0) + 0 + \frac{6}{5} (-\ 26.667) + 0 = 0$$

or $$5M_B + 10M_C - 32 = 0 \qquad ...(3)$$

Now from (1) and (2), $$22M_B + 5M_C + 112 = 0 \qquad ...(4)$$

From (3) and (4), we get $M_B = -\ 6.564$ and $M_C = +\ 6.482$ kN-m (*i.e.*, (↻))

Hence from (1), $M_A = -$ **36.718**. The B.M. diagram is shown in Fig. 16.21 (*c*).

Reactions. For reaction at *A*, write equation for B.M. at *B*. Thus

$$R_A (8) + M_A - 5 \times 8 \times 4 = M_B$$

or $8R_A - 36.718 - 160 = -\ 6.564$ from which R_A = **23.769 kN** (↑)

For reaction at C, write equation for B.M. at B, considering R.H.S. Thus,

$$R_C(5) + M_C - 80 = M_B$$

or $\quad 5R_C + 6.482 - 80 = -6.564$, from which R_C = **13.391 kN** (↑)

For reaction at B, write equation for B.M. at A, considering R.H.S. Thus

$$M_C - 80 + R_B \times 8 + R_C \times 13 - 5 \times 8 \times 4 = M_A$$

or $\quad +6.482 - 80 + 8R_B + 13.391 \times 13 - 160 = -36.718$ from which R_B = **2.840** (↑)

Check. Total reaction $= 23.769 + 13.391 + 2.840 = 40.0$ kN

Total load $= 5 \times 8 = 40$ kN

The S.F. diagram is shown in Fig. 16.21 (*d*).

PROBLEMS

1. Determine the lengths of the overhangs for a continuous beam shown in Fig. 16.22 so that the moments over the supports will be equal.
2. For the continuous beam shown in Fig. 16.23 determine support moments at B and C.

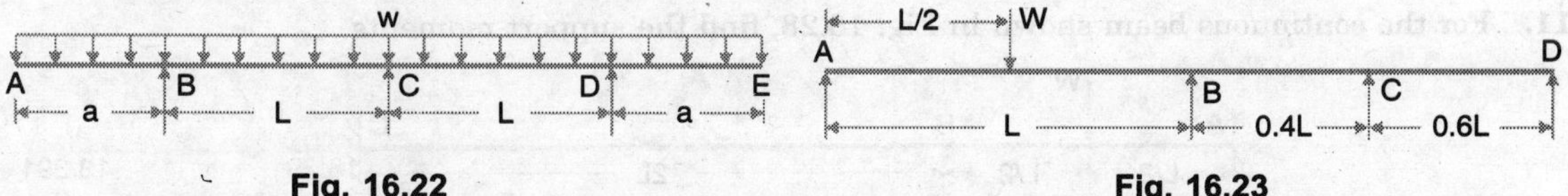

Fig. 16.22 Fig. 16.23

3. For the continuous beam shown in Fig. 16.24, determine the support moments at B and C.

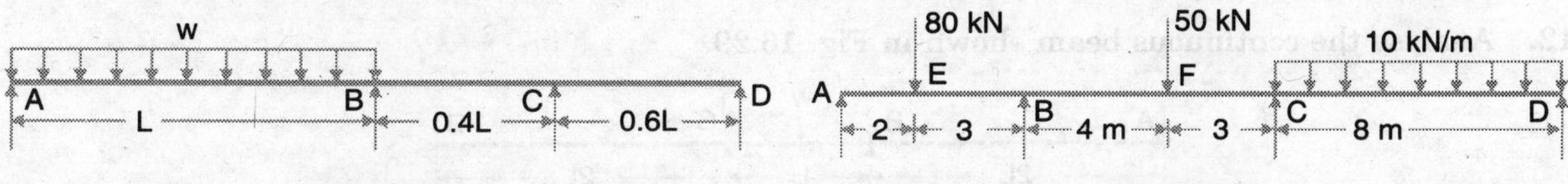

Fig. 16.24 Fig. 16.25

4. A continuous beam *ABCD* is loaded as shown in Fig. 16.25. Find the B.M. and reactions at the four supports and hence plot the B.M. and S.F. diagrams.
5. Solve problem 4 if the support B sinks by 10 mm below A and C. The moment of inertia for the whole beam $= 85 \times 10^6$ mm^4 and $E = 2.1 \times 10^5$ N/mm^2.
6. Solve problem 4 if the end A is fixed and end D is simply supported.
7. Analyse the continuous beam shown in Fig. 16.26.

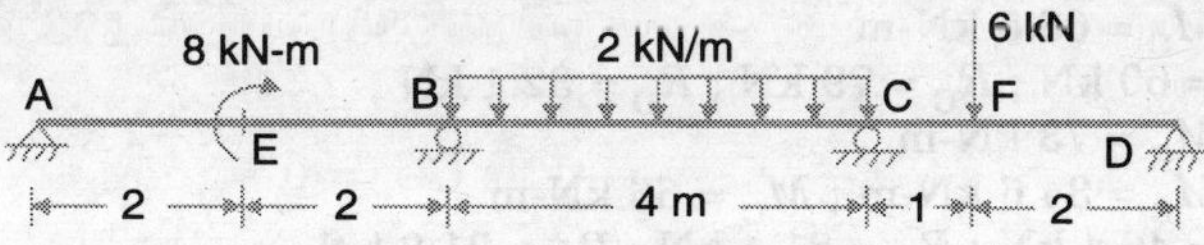

Fig. 16.26

8. *ABCD* is a straight uniform beam of length $4L$. It is freely supported at its ends A and D, and at two intermediate supports B and C distant L from either end. The supports A and D are rigid but those at B and C are such that they deflect by an amount λ for each unit of load which is placed upon them the beam carries a uniformly distributed load w per unit length along its entire length. Show that the reactions at the supports are

$$\frac{wL}{8}\left[\frac{7L^3 + 48EI\lambda}{4L^3 + 3EI\lambda}\right] \text{ and } \frac{3wL}{8}\left[\frac{19L^3}{4L^3 + 3EI\lambda}\right] \qquad \textit{(Cambridge)}$$

9. A uniform continuous girder *ABC* rests upon three similar floating supports, situated at each end and at the middle point *B*. The buoyancy of each float is such that every additional kN of load increases its immersion by *h*. Initially, all floats are equally immersed. If a load *W* is placed on the girder at *B*, show that the proportion carried by the central float is $\dfrac{W\left(1+\dfrac{3hEI}{a^3}\right)}{\left(1+\dfrac{9hEI}{a^3}\right)}$, where 2*a* is the length of the girder. *(Cambridge)*

10. For the continuous beam shown in Fig. 16.27, find the support moments.

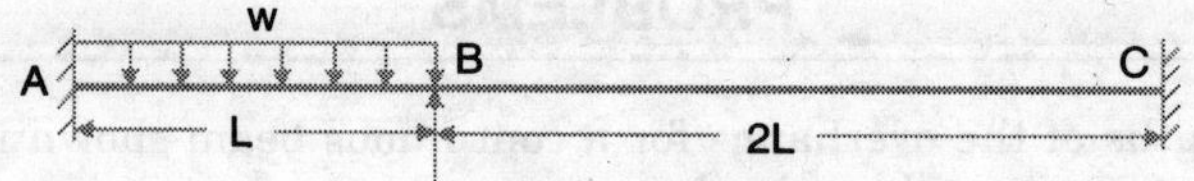

Fig. 16.27

11. For the continuous beam shown in Fig. 16.28, find the support moments.

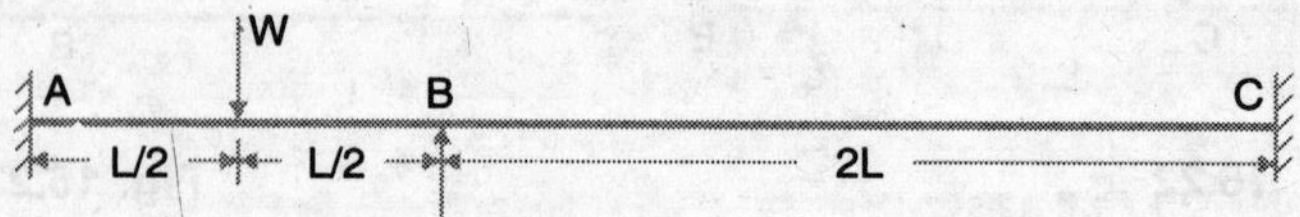

Fig. 16.28

12. Analyse the continuous beam shown in Fig. 16.29.

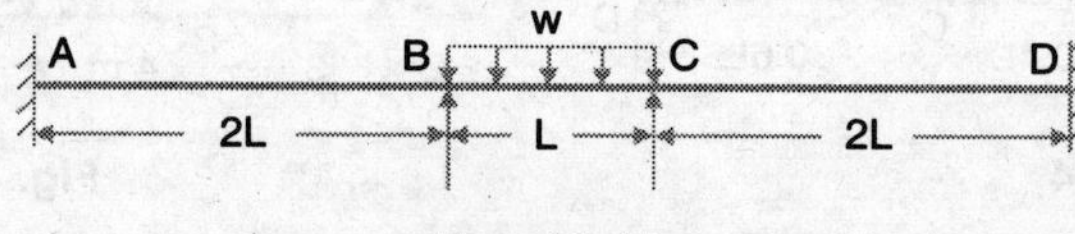

Fig. 16.29

ANSWERS

1. $a = L/\sqrt{6}$;

2. $M_B = 0.138\ WL$; $M_C = 0.028\ WL$

3. $M_B = 0.092\ wL^2$; $M_C = 0.018\ wL^2$

4. $M_B = 45.4$ kN-m ; $M_C = 63.5$ kN-m
$R_A = 38.9$ kN ; $R_B = 60$ kN ; $R_C = 79$ kN ; $R_D = 32.1$ kN

5. $M_B = 27.4$ kN-m ; $M_C = 73$ kN-m

6. $M_A = 46.7$ kN-m ; $M_B = 34.6$ kN-m ; $M_C = 66$ kN-m
$R_A = 50.4$ kN ; $R_B = 46.4$ kN ; $R_C = 81.4$ kN ; $R_D = 31.8$ kN

7. $M_B = 1690$ N-m ; $M_C = 3230$ N-m

10. $M_A = \dfrac{wL^2}{9}$; $M_B = \dfrac{wL^2}{36}$; $M_C = \dfrac{wL^2}{72}$

11. $M_A = \dfrac{WL}{6}$; $M_B = \dfrac{WL}{24}$; $M_C = \dfrac{WL}{48}$

12. $M_A = M_D = \dfrac{wL^2}{48}$; $M_B = M_C = \dfrac{wL^2}{24}$.

SECTION 3 COMPRESSION MEMBERS

CHAPTERS

17

Combined Direct and Bending Stresses

17.1. INTRODUCTION

It often happens that a structural member is subjected to bending stresses as well as direct stresses simultaneously. A beam, for example, may be subjected to axial load in addition to the transverse loads. Sometimes a compression member such as a column may be subjected to an axial load which may not pass through the geometrical axis of the member, thus causing bending as well as direct stresses. In all such cases, the position of neutral layer changes or altogether vanishes.

We know that longitudinal stresses (either tensile or compressive) are set up in a beam cross-section subjected to bending. If the beam is also subjected to a direct force causing direct stress (tensile or compressive), the resulting stress at any layer in the cross-section will be the algebraic sum of the two.

17.2. ECCENTRIC LOADING

Consider a short prismatic bar subjected to a direct load P, the line of action of which is parallel to the axis of the bar and intersects an axis of symmetry (*i.e.,* geometric axis) at a distance e from the centroid of the section, as shown in Fig. 17.1.

If two equal and opposite forces, each of value P, are introduced at the centroid of the section, the whole system reduces to (*i*) a direct compressive force P acting at the centroid and (*ii*) a couple $M = P.e$ causing bending in a vertical plane passing through the longitudinal axis of the bar and the original point of application of the load.

The axial force will cause a direct compressive stress given by

$$p_0 = \frac{P}{A}$$

The bending couple will cause longitudinal tensile and compressive stresses. The fibre stress p_b at any distance y from N.A. is given by

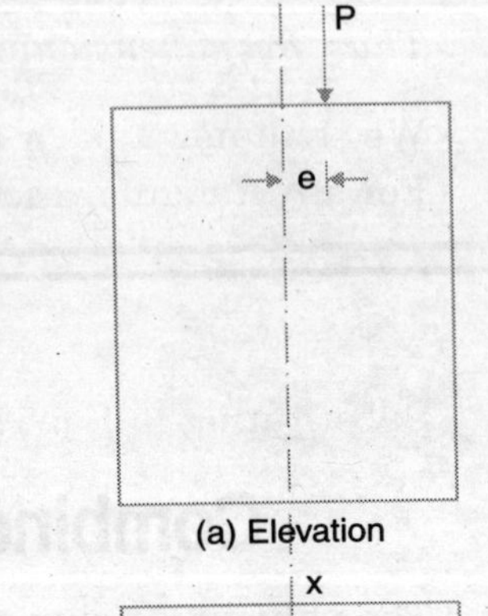

$$p_b = \frac{My}{I_x} = \frac{P\,e.y}{I_x} \text{ (tensile or compressive)}$$

Hence the total stress at any section is given by

$$f = \frac{P}{A} \pm \frac{Pey}{I_x} \quad ...(17.1)$$

The extreme fibre stress due to bending is given by

$$p_b = \frac{M}{Z_x} = \frac{Pe}{Z_x}$$

Hence the maximum total stresses are given by

$$f_{max} = p_0 + p_b = \frac{P}{A} + \frac{M}{Z_x} \quad ...(17.2)$$

and

$$f_{min} = p_0 - p_b = \frac{P}{A} - \frac{M}{Z_x} \quad ...[17.2(a)]$$

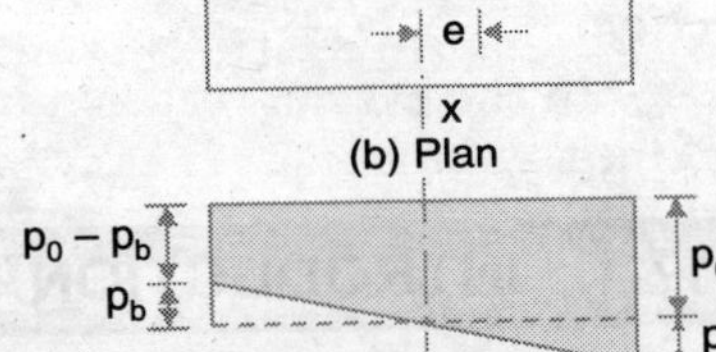

(c) Stress distribution

Fig. 17.1

If p_0 is greater than p_b, the stress throughout the section will be of the same sign. If, however, p_0 is less than p_b, the stress will change sign, being partly tensile and partly compressive across the section. Thus there can be three possible pressure distributions as indicated in Fig. 17.2. When $p_0 = p_b$, $f_{max.} = 2p_0$ and $f_{min.} = 0$.

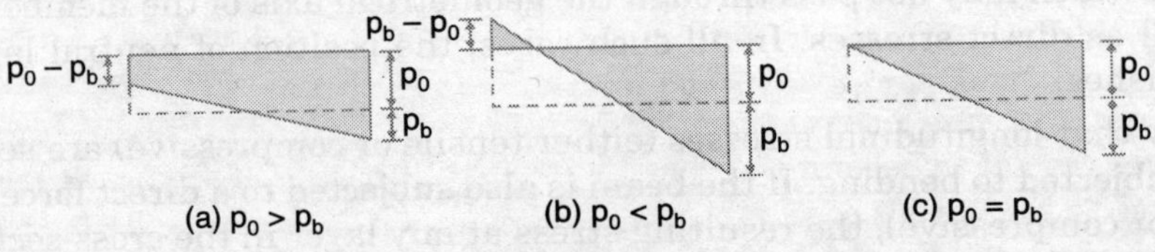

Fig. 17.2

17.3. MIDDLE THIRD RULE

In Fig. 17.2 (*b*), $p_0 < p_b$ and, therefore, stress changes sign, being partly tensile and partly compressive across the section. In masonry and concrete structures, the development of tensile stress in the section is not desirable, as they are weak in tension. This limits the eccentricity e to a certain value which will be investigated now for different sections.

In order that the stress may not change sign from compressive to tensile, we have

$$p_0 \geq p_b \geq \frac{M}{Z}$$

or

$$\frac{P}{A} \geq \frac{Ped}{2I} \geq \frac{Ped}{2Ak^2} \quad \text{or} \quad e \leq \frac{2k^2}{d} \quad ...(17.3)$$

where k = radius of gyration of the section with regard to N.A. and d is the depth of the section.

Thus, for no tension in the section, the eccentricity must not exceed $\dfrac{2k^2}{d}$.

We shall now take a rectangular section and find out the limiting value of e.

For a rectangular section of width b and depth d,

$$I = \frac{1}{12}bd^3;\ A = bd \quad \text{and} \quad k^2 = \frac{I}{A} = \frac{d^2}{12}.$$

Substituting in Eq. 17.3, we get $e \le \dfrac{2d^2}{d \times 12} \le \dfrac{d}{6}$

$$\therefore \qquad e_{max} = \frac{\mathbf{d}}{\mathbf{6}} \qquad \text{...(17.4)}$$

The value of eccentricity can be on either side of the geometrical axis. Thus, *the stress will be of the same sign throughout the section if the load line is within the middle third of the section.*

In the case of rectangular section, the maxium intensities of extreme stresses are given by

$$f = \frac{P}{A} \pm \frac{Pe}{Z_x} = \frac{P}{bd} \pm \frac{6Pe}{bd^2} = \frac{P}{bd}\left[1 \pm \frac{6e}{d}\right] \qquad \text{...(17.5)}$$

17.4. THE CORE OF A SECTION

If the line of action of the force is on neither of the centre lines of the section, the bending is unsymmetrical. However, there is certain area within which the line of action of the force P must cut the cross-section if the stress is not to become tensile. The area is known as *core* or *kernal* of the section. We shall now investigate the shape of core or kernal of different sections.

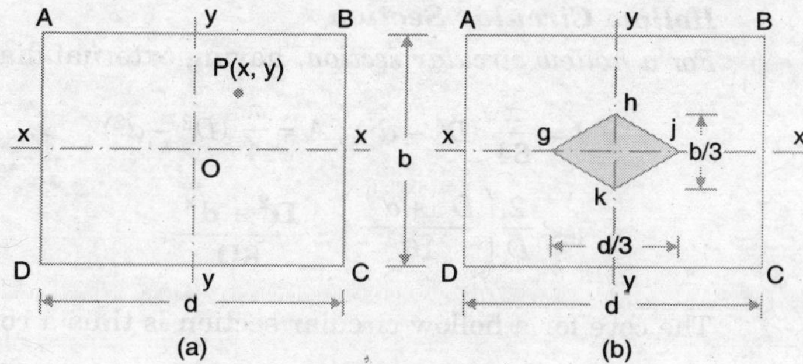

Fig. 17.3

Rectangular Section

Let the point of application of the load P have the co-ordinates (x, y), with reference to the axes shown in Fig. 17.3 (a) in which x is positive when measured to the right of O and y is positive when measured upwards.

The stress at any point having co-ordinates (x', y') will be

$$f = \frac{P}{bd} + \frac{Pyy'}{\frac{1}{12}db^3} + \frac{Pxx'}{\frac{1}{12}bd^3} = \frac{12P}{bd}\left(\frac{1}{12} + \frac{yy'}{b^2} + \frac{xx'}{d^2}\right) \qquad \text{...(1)}$$

...(17.6)

At D, $x' = -\dfrac{d}{2}$ and $y' = -\dfrac{b}{2}$ and, therefore, f will be minimum.

Thus, at D,

$$f_D = \frac{6P}{bd}\left(\frac{1}{6} - \frac{y}{b} - \frac{x}{d}\right) \qquad \text{...(2)}$$

The value of f reaches zero when

$$\frac{y}{b}+\frac{x}{d}=\frac{1}{6} \quad \text{or} \quad \frac{6y}{b}+\frac{6x}{d}=1 \qquad ...(17.7)$$

Thus, the deviation of the load line is governed by the straight line of equation 17.7, whose intercepts on the axes are respectively $\frac{b}{6}$ and $\frac{d}{6}$. This is true for the load line in the first quadrant. Similar limits will apply in other quadrants and the stress will be wholly compressive throughout the section, if the line of action of P fall within the rhombus *ghjk* [Fig. 17.3 (*b*)], the diagonals of which are of length $\frac{d}{3}$ and $\frac{b}{3}$ respectively. This rhombus is called the *core* of the rectangular section.

Circular Section

We have seen that for no tension, $e \le \frac{2k^2}{d}$

For a solid circular section

$$\therefore \qquad I=\frac{\pi}{64}d^4, A=\frac{\pi}{4}d^2; \quad k^2=\frac{I}{A}=\frac{d^2}{16}$$

$$\therefore \qquad e \le \frac{2}{d}\frac{d^2}{16} \le \frac{\mathbf{d}}{\mathbf{8}} \qquad ...(17.8)$$

d/4

D

Fig. 17.4

Thus, in order that tension is not developed, the load line must fall within *middle fourth* of the section. The '*core*' in this case is a circle with the same centre and diameter $\frac{d}{4}$.

Hollow Circular Section

For a hollow circular section, having external diameter D and internal diameter d,

$$I=\frac{\pi}{64}(D^4-d^4)\,;\; A=\frac{\pi}{4}(D^2-d^2)\,; \quad \therefore \quad k^2=\frac{I}{A}=\frac{D^4-d^4}{16(D^2-d^2)}=\frac{D^2+d^2}{16}$$

$$\therefore \qquad e \le \frac{2}{D}\left(\frac{D^2+d^2}{16}\right) \le \frac{\mathbf{D^2+d^2}}{\mathbf{8D}} \qquad ...(17.9)$$

The core for a hollow circular section is thus a concentric circle of diameter $\frac{D^2+d^2}{4D}$.

Other Sections

Eq. 17.6 can be re-written in the form

$$f=\frac{P}{A}\left(1+\frac{yy'}{k_x^{\,2}}+\frac{xx'}{k_y^{\,2}}\right)$$

where k_x and k_y are radii of gyration of the area of section about the axes of X and Y respectively.

For zero stress at the point, we must have

$$\frac{yy'}{k_x^{\,2}}+\frac{xx'}{k_y^{\,2}}=-1 \qquad ...(17.10)$$

For I-section, the four corners will be the limiting points, where $x'=\frac{b}{2}$ and $y'=\frac{d}{2}$.

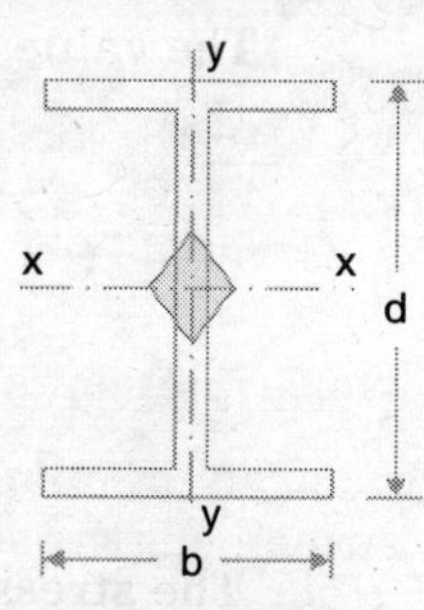

Fig. 17.5

Hence, we have, $\frac{yd}{2k_x^2} + \frac{xb}{2k_y^2} = -1$

$$\therefore \quad y = -\frac{k_x^2}{k_y^2} \cdot \frac{b}{d} x - \frac{2k_x^2}{d} \quad ...(17.11)$$

Equation 17.11 is the equation of the bounding line which limits the deviation of load from the centroid for no change in the sign of the stress, for I-section. The equations to the three other bounding lines will be similar, thus forming a rhombus having the principal axes of I-section as diagonals, as shown in Fig. 17.5.

Example 17.1. *Find the maximum and minimum stress intensities in a short C.I. column of hollow circular section attached with a projection bracket carrying a load of 60 kN. The external and internal diameter of the column are 300 mm and 250 mm respectively. The load line is off the axis of the column by 300 mm.*

Solution. Area of section $A = \frac{\pi}{4}(300^2 - 250^2) = 2.16 \times 10^4 \text{ mm}^2$

Modulus of section $Z = \frac{\pi}{32}\left(\frac{D^4 - d^4}{D}\right) = \frac{\pi}{32 \times 300}(300^4 - 250^4) = 137.24 \times 10^4 \text{ mm}^3$

$$\therefore \quad p_0 = \frac{P}{A} = \frac{60000}{2.16 \times 10^4} = 2.78 \text{ N/mm}^2 \text{ (comp.)}$$

$$p_b = \pm\frac{M}{Z} = \pm\frac{60000 \times 300}{137.24 \times 10^4} = \pm 13.12 \text{ N/mm}^2$$

$$\therefore \quad f_{max} = p_0 + p_b = 2.78 + 13.12 = \mathbf{15.9 \text{ N/mm}^2} \text{ (comp.)}$$

$$f_{min} = p_0 - p_b = 2.78 - 13.12 = -10.34 \text{ N/mm}^2$$

$$= \mathbf{10.34 \text{ N/mm}^2} \text{ (tensile).}$$

Example 17.2. *Determine the maximum and minimum stress intensities set up in steel flat section 150 mm wide and 20 mm thick subjected to a pull of 200 kN, which is off the geometrical axis by 4 mm in the plane which bisects the thickness.*

Solution. Area of section $A = 150 \times 20$

$= 3000 \text{ mm}^2$

Section modulus $Z = \frac{1}{6} \times 20 \times (150)^2$

$= 75000 \text{ mm}^3$

B.M. $M = P.e. = 200 \times 4$

$= 800 \text{ kN-mm}$

$= 8 \times 10^5 \text{ N-mm.}$

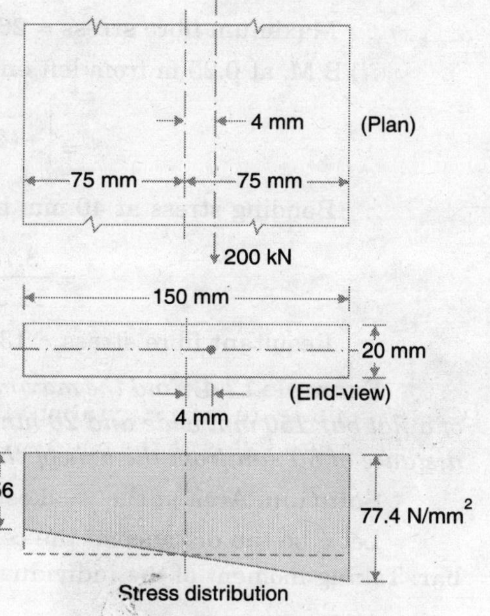

Fig. 17.6

$$p_0 = \frac{P}{A} = \frac{200 \times 1000}{3000} = 66.7 \text{ N/mm}^2;$$

$$p_b = \frac{M}{Z} = \frac{8 \times 10^5}{75000} = 10.7 \text{ N/mm}^2$$

$$\therefore \quad f_{max} = p_0 + p_b = 66.7 + 10.7 = \mathbf{77.4\ N/mm^2} \text{ (tensile)}$$

$$f_{min} = p_0 - p_b = 66.7 - 10.7 = \mathbf{56\ N/mm^2} \text{ (tensile)}$$

The stress distribution has been shown in Fig. 17.6.

Example 17.3. *A beam of rectangular section of 80 mm to 120 mm carries a uniformly distributed load of 40 kN / m over a span of 1 m and an axial compressive force of 5 kN. Calculate (i) maximum fibre stress, (ii) fibre stress at a point 0.25 m from the left end of the beam and 40 mm below the neutral axis.*

Solution. (*i*) Max. B.M. $= \frac{wL^2}{8} = \frac{40 \times 1^2}{8} = 5 \text{ kN-m} = 5 \times 10^6 \text{ N-mm}$

$$Z = \frac{1}{6} \times 80(120)^2 = 1.92 \times 10^5 \text{ mm}^3;$$

$$I = Z \times \frac{d}{2} = 1.92 \times 10^5 \times \frac{120}{2} = 11.52 \times 10^5 \text{ mm}^4$$

$$\therefore \quad p_b = \frac{M}{Z} = \frac{5 \times 10^6}{1.92 \times 10^5} = 26.04 \text{ N/mm}^2$$

Also,

$$p_0 = \frac{P}{A} = \frac{5 \times 1000}{80 \times 120} = 0.52 \text{ N/mm}^2$$

$\therefore$ Maximum fibre stress = 26.04 + 0.52 = **26.56 N/mm²**.

(*ii*) B.M. at 0.25 m from left end

$$= \left[+(20 \times 0.25) - \frac{40 \times 0.25^2}{2} \right] = 3.75 \times 10^6 \text{ N-mm}$$

$\therefore$ Bending stress at 40 mm below N.A.

$$= \frac{3.75 \times 10^6}{11.52 \times 10^6} \times 40 = 13.02 \text{ N/mm}^2 \text{ (tensile)}$$

$\therefore$ Resultant fibre stress = 13.02 – 0.52 = **12.5 N/mm²** (tensile).

Example 17.4. *Find the maximum and minimum stress intensities at the weakest section of a flat bar 150 mm wide and 20 mm thick carrying axial load of 150 kN if it is punched at a distance of 50 mm from the axis of the bar with 20 mm wide hole.*

Solution. Area at the weakest section = A = (150 – 20) 20 = 2600 mm²

Let e be the distance of the centre of gravity of the net section from the axis yy of the bar. Taking moment of the individual areas about yy, we get

$$e = \frac{(150 \times 20 \times 0) - 20 \times 20\,(-50)}{(150 \times 20) - (20 \times 20)} = \frac{20000}{2600}$$

$$\simeq 7.7 \text{ mm.}$$

$$I_{yy} = \frac{1}{12} \times 20 \times 150^3 - \left\{\frac{1}{12} \times 20\,(20)^3 + (20 \times 20)\,(50)^2\right\}$$

$$= 4.61 \times 10^6 \text{ mm}^4$$

$$I_{gg} = I_{yy} - (a_1 - a_2)\, e^2$$

$$= 4.61 \times 10^6 - (150 \times 20 - 20 \times 20) \times 7.7^2$$

$$= 4.456 \times 10^6 \text{ mm}^4$$

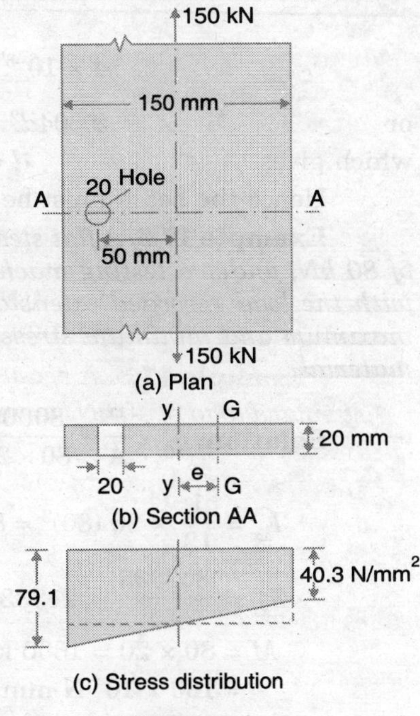

Fig. 17.7

$$p_0 = \frac{P}{A} = \frac{150 \times 1000}{2600} = 57.69 \text{ N/mm}^2 \text{ (tensile)}$$

$$M = P.e = (150 \times 1000) \times 7.7$$

$$= 1155000 \text{ N-mm}$$

$$y_c = \frac{150}{2} - 7.7 = 67.3 \text{ mm}$$

and $$y_t = \frac{150}{2} + 7.7 = 82.7 \text{ mm}$$

$$p_{bc} = \frac{M\,.\,y_c}{I_{gg}} = \frac{1155000 \times 67.3}{4.456 \times 10^6}$$

$$= 17.44 \text{ N/mm}^2 \text{ (comp.)}$$

$$p_{bt} = \frac{M\,.\,y_t}{I_{gg}} = \frac{1155000 \times 82.7}{4.456 \times 10^6} = 21.43 \text{ N/mm}^2 \text{ (tensile)}$$

$$f_{max} = p_0 + p_{bt} = 57.69 + 21.43 = \mathbf{79.12} \text{ N/mm}^2 \text{ (tensile)}$$

$$f_{min} = p_0 + p_{bc} = 57.69 - 17.44 = \mathbf{40.25} \text{ N/mm}^2 \text{ (tensile).}$$

Example 17.5. *A cantilever 160 mm broad is at the wall subjected to a shearing force of 160 kN and a bending moment of 100 kN-m. Assuming a uniformly varying pressure between the beam and its seating, find what length of the beam must be built into the wall in order that the pressure shall not exceed 4 N/mm².*

Solution. Let b = width of seating = 160 mm;

d = length of seating (or of beam built into wall), in mm

Seating pressure $$p_0 = \frac{R}{bd} = \frac{160}{160 \times d} = \frac{1}{d} \text{ kN/mm}^2$$

Total B.M. at the centre of seating

$$= \left(100 \times 10^3 + 160 \times \frac{d}{2}\right) \text{ kN-mm} = (10^5 + 80d) \text{ kN-mm}$$

$$\therefore \quad f_{max} = p_0 + p_b = \frac{1}{d} + \frac{6(10^5 + 80d)}{bd^2} = \frac{1}{d} + \frac{6 \times 10^5 + 480d}{160d^2}$$

But $$f_{max} = 4 \text{ N/mm}^2 = 4 \times 10^{-3} \text{ kN/mm}^2$$

$\therefore \quad 4 \times 10^{-3} = \frac{1}{d} + \frac{6 \times 10^5 + 480d}{160d^2}$

or $\quad 0.004d^2 = d + 3750 + 3d \quad$ or $\quad 0.004d^2 - 4d - 3750 = 0$

which gives $\quad d = 1590$ mm = **1.59** m.

Hence the beam must be built into the wall for a length of 1.59 m.

Example 17.6. *A flat steel bar 80 mm × 20 mm was subjected to an eccentric tensile load of 80 kN, under a testing machine as shown in Fig. 17.8 (a). An extensometer placed in line with the load recorded extension of 0.085 mm on a gauge length of 200 mm. Calculate the maximum and minimum stresses set up in the bar and also find out Young's Modulus of the material.*

Fig. 17.8

Solution. $p_0 = \frac{P}{A} = \frac{80000}{80 \times 20} = 50 \text{ N/mm}^2$

$I_{xx} = \frac{1}{12} \times 20\,(80)^3 = 853333 \text{ mm}^4$

$Z_{xx} = \frac{853333}{40} = 21333 \text{ mm}^3$

$M = 80 \times 20 = 1600$ kN-mm

$= 160 \times 10^4$ N-mm

$\therefore \quad p_b = \frac{160 \times 10^4}{21333} = 75 \text{ N/mm}^2$

Maximum tensile stress

$= p_0 + p_b = 50 + 75 = 125 \text{ N/mm}^2$

Maximum compressive stress

$= p_0 - p_b = 75 - 50 = 25 \text{ N/mm}^2$

The stress diagram has been plotted in Fig. 17.8 (*b*).

In order to locate the position of neutral layer, we have

$\frac{h}{25} = \frac{80 - h}{125}$ from which $h = 13.33$ mm

Let x = stress caused by the measured strain of 0.085/200

Thus, $\quad \frac{x}{125} = \frac{66.67 - 20}{66.67} \quad$ or $\quad x = \frac{46.67 \times 125}{66.67} = 87.5 \text{ N/mm}^2$

Now, $\quad E = \frac{\text{Stress}}{\text{Strain}} = \frac{87.5}{\frac{0.085}{200}} = \frac{87.5 \times 200}{0.085} = \mathbf{2.059 \times 10^5 \text{ N/mm}^2}$.

Example 17.7. *A short hollow cylindrical cast iron column having outside diameter 30 cm and inside diameter 20 cm was cast in a factory. On inspection it was found that the bore is eccentric in such a way that the thickness varies from 3 cm at one end to 7 cm at the other. Calculate the extreme intensities of stress induced in the section, if column carries a load of 800 kN along the axis of the bore.*

Solution. Net area of column $= \frac{\pi}{4}(30^2 - 20^2)$

$$= \frac{\pi}{4} \times 500 = 393 \text{ cm}^2 = 39300 \text{ mm}^2$$

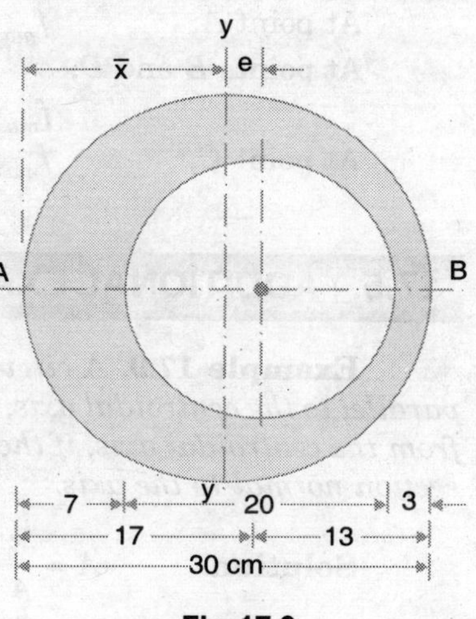

Fig 17.9

Let $\bar{x}$ be the distance of C.G. of section from end A, then

$$393\bar{x} = \frac{\pi}{4} \times 30^2 \times 15 - \frac{\pi}{4} \times 20^2 \times 17$$

$$= 10{,}600 - 5{,}340$$

$\therefore \quad \bar{x} = 13.38 \text{ cm} = 133.8 \text{ mm}$

$\therefore \quad$ Eccentricity, $e = 17 - 13.38 = 3.62 \text{ cm} = 36.2 \text{ mm}$

Moment, $\quad M = (800 \times 1000)\ 36.2$

$$= 289.6 \times 10^5 \text{ N-mm}$$

$$\therefore \quad p_{bt} = \frac{289.6 \times 10^5 \times \bar{x}}{I_{yy}} = \frac{289.6 \times 10^5 \times 133.8}{I_{yy}}$$

$$p_{bc} = \frac{289.6 \times 10^5\ (300 - \bar{x})}{I_{yy}} = \frac{289.6 \times 10^5 \times 166.2}{I_{yy}}$$

$$I_{yy} = \left[\frac{\pi}{64} \times 30^4 + \frac{\pi}{4} \times 30^2\ (15 - 13.38)^2\right] - \left[\frac{\pi}{64} \times 20^4 + \frac{\pi}{4} \times 20^2\ (17 - 13.38)^2\right]$$

$$= 29665 \text{ cm}^4 = 296.65 \times 10^6 \text{ mm}^4$$

$$\therefore \quad p_{bt} = \frac{289.6 \times 10^5 \times 133.8}{296.65 \times 10^6} = 13.05 \text{ N/mm}^2\ ;\ p_{bc} = \frac{289.6 \times 10^5 \times 166.2}{296.65 \times 10^6} = 16.23 \text{ N/mm}^2$$

Also, $\quad p_0 = \dfrac{800 \times 1000}{39300} = 20.36 \text{ N/mm}^2$

$\therefore \quad$ Max. stress $= p_0 + p_{bc} = 20.36 + 16.23 =$ **36.59 N/mm²** (comp.)

Min. stress $= p_0 - p_{bt} = 20.36 - 13.05 =$ **7.31 N/mm²** (comp.).

Example 17.8. *Calculate the normal stresses at the four outside corners of a horizontal section of short hollow pier 1.5 metre square outside and 1 metre square inside supporting a vertical point load of 140 kN on a diagonal and 0.8 metre from the vertical axis of the pier. Neglect the self weight of the pier.*

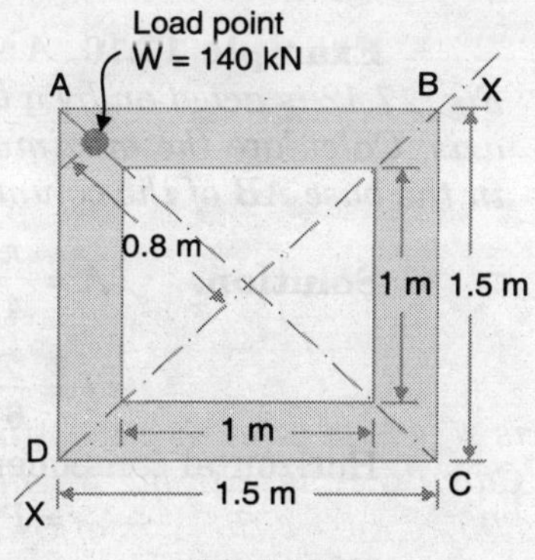

Fig 17.10

Solution. Uniform stress $p_0 = \dfrac{W}{A} = \dfrac{140}{2.25 - 1} = \dfrac{140}{1.25}$

$$= 112 \text{ kN/m}^2 \text{ (comp.)}$$

$$I_{xx} = \frac{1}{12}\ (1.5^4 - 1^4) = \frac{4.08}{12} = 0.34 \text{ m}^4\ ;$$

$$Z_{xx} = \frac{I_{xx}\sqrt{2}}{b} = \frac{0.34\sqrt{2}}{1.5} = 0.321 \text{ m}^3$$

Moment $\quad M = 140 \times 0.8 = 112$ kN-m

p_b at point $A = \dfrac{M}{Z} = \dfrac{112}{0.321} = 348.9 \text{ kN/m}^2$ (comp.)

p_b at points B and $D = 0$

and p_b at point $C = 348.9 \text{ kN/m}^2$ (tensile)

At point A, $f_{max} = p_0 + p_b = 112 + 348.9 = 460.9 \text{ kN/m}^2 = 0.461 \text{ N/mm}^2$

At points B and D,

$$f_{max} = p_0 = 112 \text{ kN/m}^2 = \mathbf{0.112 \ N/mm^2} \text{ (comp).}$$

At point C, $f_{max.} = p_0 - p_b = 112 - 348.9 = -236.9 \text{ kN/m}^2 \simeq \mathbf{-0.237 \ N/mm^2}$.

17.5. ADDITIONAL EXAMPLES USEFUL FOR COMPETITIVE EXAMINATIONS

Example 17.9. *A circular rod of 20 mm diameter carries a pull along a line which is parallel to the centroidal axis, but is displaced from it. Determine the distance of the line of pull from the centroidal axis, if the maximum stress is 20 percent greater than the mean stress on a section normal to the axis.*

Solution. $A = \frac{\pi}{4}(20)^2 = 314.16 \text{ mm}^2$;

$$Z = \frac{\pi}{32}(20)^3 = 785.4 \text{ mm}^3$$

Let P be the axial tension and e be its eccentricity.

$$\therefore \quad M = P.e$$

$$\text{Mean Stress, } p_{mean} = \frac{P}{A} = \frac{P}{314.16} \quad ...(1)$$

$$\text{Max. stress, } p_{max} = \frac{P}{A} + \frac{M}{Z} = \frac{P}{314.16} + \frac{P.e}{785.4} \quad ...(2)$$

$$\text{Now, } p_{max} = p_{mean} \times \frac{120}{100}$$

$$\therefore \quad \frac{P}{314.16} + \frac{Pe}{785.4} = \frac{P}{314.16} \times 1.2$$

$$\text{or} \quad \frac{1}{314.16} + \frac{e}{785.4} = \frac{1.2}{314.16}$$

$$\text{From which} \quad e = \left(\frac{1.2}{314.16} - \frac{1}{314.16}\right) 785.4 = \mathbf{0.5 \ mm}$$

Fig. 17.11

Example 17.10. *A short hollow cylindrical column shown in Fig. 17.12 is acted on by a load of 10 kN inclined at 30° to the column axis. Calculate the maximum tensile and compressive stresses set up in the base AB of the column.*

Solution. $A = \frac{\pi}{4}(100^2 - 75^2) = 3436 \text{ mm}^2$

$$I = \frac{\pi}{64}(100^4 - 75^4) = 335.6 \times 10^4 \text{ mm}^2$$

Horizontal component of the load

$$= P_h = P \sin 30° = 10 \sin 30° = 5 \text{ kN}$$

Vertical component of the load

$$= P_v = P \cos 30° = 10 \cos 30° = 8.66 \text{ kN}$$

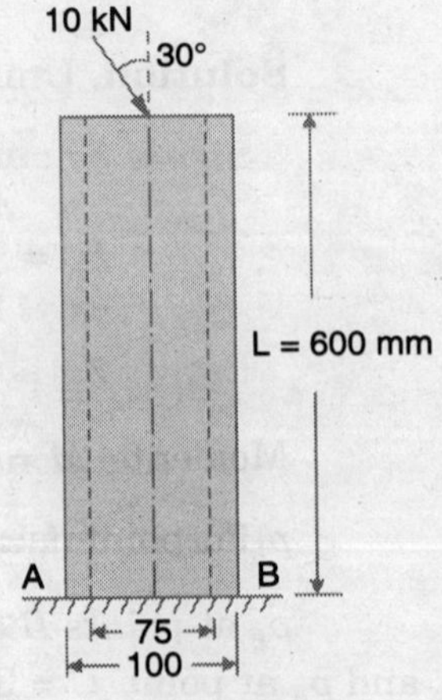

Fig. 17.12

$$M_{AB} = P_h \times L = 5 \times 10^3 \times 600 = 3 \times 10^6 \text{ N-mm}$$

$$p_0 = \frac{P_v}{A} = \frac{8.66 \times 10^3}{3436} = 2.52 \text{ N/mm}^2$$

$$p_b = \pm \frac{M}{I} y = \pm \frac{3 \times 10^6}{335.6 \times 10^4} \times 50 = \pm 44.700 \text{ N/mm}^2$$

$\therefore$ Stress at $B = p_0 + p_b = 2.52 + 44.70 =$ **47.22 N/mm²** (comp.)

Stress at $A = p_0 - p_b = 2.52 - 44.70 = -42.18$ N/mm² = **42.18 N/mm²** (tensile)

Example 17.11. *A masonry pier of 2 m × 3 m supports a vertical load of 50 kN as shown in Fig. 17.13. Find (a) stresses developed at each corner of the pier (b) what additional load should be placed at the centre of the pier, so that there is no tension any where in the pier section ? (c) what are the stresses at the corners with the additional load in the centre ?*

Solution. $B = 3$ m ; $D = 2$ m ; $A = B \times D = 3 \times 2 = 6$ m²

$$I_{xx} = \frac{1}{12} \times 3\,(2)^3 = 2 \text{ m}^4 \; ; \; I_{yy} = \frac{1}{12} \times 2\,(3)^3 = 4.5 \text{ m}^4$$

Eccentricity about x-x axis $= e_x = 0.4$ m

Eccentricity about y-y axis $= e_y = 0.8$ m

$\therefore$ Moment $M_{xx} = P \cdot e_x = 50 \times 0.4 = 20$ kN-m

Moment $M_{yy} = P \cdot e_y = 50 \times 0.8 = 40$ kN-m

Distance of A and B from y-axis $= x_A = x_B = 1.5$ m

Distance of A and D from x-axis $= y_A = y_D = 1$ m

Fig. 17.13

Stress at corner A: $p_A = \frac{P}{A} + \frac{M_{xx}}{I_{xx}} \cdot y_A - \frac{M_{yy}}{I_{yy}} \cdot x_A = \frac{50}{6} + \frac{20}{2} \times 1 - \frac{40}{4.5} \times 1.5$

$= 8.33 + 10 - 13.33 =$ **5 kN/m²**

Stress at corner B: $p_B = \frac{P}{A} + \frac{M_{xx}}{I_{xx}} y_B + \frac{M_{yy}}{I_{yy}} \cdot x_B = \frac{50}{6} + \frac{20}{2} \times 1 + \frac{40}{4.5} \times 1.5$

$= 8.33 + 10 + 13.33 =$ **31.66 kN/m²**

Stress at corner C: $p_C = \frac{P}{A} - \frac{M_{xx}}{I_{xx}} \cdot y_C + \frac{M_{yy}}{I_{yy}} \cdot x_C = \frac{50}{6} - \frac{20}{2} \times 1 + \frac{40}{4.5} \times 1.5$

$= 8.33 - 10 + 13.33 =$ **11.66 kN/m²**

Stress at corner D: $p_D = \frac{P}{A} - \frac{M_{xx}}{I_{xx}} \cdot y_D - \frac{M_{yy}}{I_{yy}} \cdot x_D = \frac{50}{6} - \frac{20}{2} \times 1 - \frac{40}{4.5} \times 1.5$

$= 8.33 - 10 - 13.33 =$ **– 15 kN/m²** (*i.e.* tension)

Additional load at the centre for no tension any where

Let W (kN) be the additional central load.

$\therefore$ Additional compressive stress $p_0 = \frac{W}{A} = \frac{W}{6}$ kN/m²

For no tension to develop at D, p_0 should be equal to p_D.

$$\therefore \quad \frac{W}{6} = 15, \text{ from which } W = \mathbf{90 \text{ kN}}$$

Final stresses at corners due to the additional load

$$W = 90 \text{ kN} ; \quad \therefore \quad p_0 = 90/6 = 15 \text{ kN/m}^2$$

$$\therefore \quad p_{AF} = p_A + p_0 = 5 + 15 = \mathbf{20 \text{ kN/m}^2} \text{ (comp.)}$$

$$p_{BF} = p_B + p_0 = 31.66 + 15 = \mathbf{46.66 \text{ kN/m}^2} \text{ (comp.)}$$

$$p_{CF} = p_C + p_0 = 11.66 + 15 = \mathbf{26.66 \text{ kN/m}^2} \text{ (comp.)}$$

$$p_{DF} = p_D + p_0 = -15 + 15 = \mathbf{zero}$$

Example 17.12. *A short C.I. column has a rectangular section 160 mm × 200 mm with a circular hole of 80 mm diameter as shown in Fig. 17.14. It carries an eccentric load of 100 kN, located as shown in figure. Determine the values of the stresses at the four corners of the section.*

Solution. The net section is symmetrical about *X-X* axis. Let the *Y-Y* axis be situated at a distance $\bar{x}$ from the face *AD*.

Let a_1 = area of solid rectangle
$= 160 \times 200 = 32000 \text{ mm}^2$

$\bar{x}_1$ = distance of c.g. of rectangle from *AD*
$= 80$ mm

Let a_2 = area of the circular hole

$$= \frac{\pi}{4} (80)^2 = 5026 \text{ mm}^2$$

$\bar{x}_2$ = distance of c.g. of hole from *AD* = 60 mm

Fig. 17.14

$$\text{Hence} \quad \bar{x} = \frac{a_1\bar{x}_1 - a_2\bar{x}_2}{a_1 - a_2} = \frac{32000 \times 80 - 5026 \times 60}{32000 - 5026}$$

$$= \frac{2258440}{26974} = 83.73 \text{ mm}$$

$$I_{XX} = \frac{1}{12} \times 160\,(200)^3 - \frac{\pi}{64}\,(80)^4 = 10466 \times 10^4 \text{ mm}^4$$

$$I_{YY} = \left[\frac{1}{12} \times 200\,(160)^3 + 200 \times 160\,(83.73 - 80)^2\right] - \left[\frac{\pi}{64}(80)^4 + \frac{\pi}{4}(80)^2\,(83.73 - 60)^2\right]$$

$$= 6871 \times 10^4 - 484 \times 10^4 = 6387 \times 10^4 \text{ mm}^4$$

The stress at any point, in general, is given by

$$p = \frac{P}{A} + \frac{M_{XX}}{I_{XX}} \cdot y + \frac{M_{YY}}{I_{YY}} \cdot x$$

where *x* and *y* are the coordinates of any point with respect to the coordinate axes,

$$M_{XX} = P \,.\, e_X = 100 \times 10^3 \times 50 = 5 \times 10^6 \text{ N-mm};$$

$$M_{YY} = P \,.\, e_Y = 100 \times 10^3 \times (83.73 - 60) = 2.373 \times 10^6 \text{ N-mm}$$

Stress at point A : By inspection

$$p_A = \frac{P}{A} + \frac{M_{XX}}{I_{XX}} \cdot y_A + \frac{M_{YY}}{I_{YY}} \cdot x_A$$

where $\quad y_A = 100 \text{ mm and } x_A = 83.73 \text{ mm}$

$$\therefore \quad p_A = \frac{100 \times 10^3}{26974} + \frac{5 \times 10^6}{10466 \times 10^4} \times 100 + \frac{2.373 \times 10^6}{6387 \times 10^4} \times 83.73 = \mathbf{11.6\ N/mm^2}$$

Stress at point B : By inspection

$$p_B = \frac{P}{A} + \frac{M_{XX}}{I_{XX}} \cdot y_B - \frac{M_{YY}}{I_{YY}} \cdot x_B$$

where $\quad y_B = 100$ mm and $x_B = 160 - 83.73 = 76.27$ mm

$$\therefore \quad p_B = \frac{100 \times 10^3}{26974} + \frac{5 \times 10^6}{10466 \times 10^4} \times 100 - \frac{2.373 \times 10^6}{6387 \times 10^4} \times 76.27$$

$$= \mathbf{5.65\ N/mm^2}$$

Stress at point C : By inspection

$$p_C = \frac{P}{A} - \frac{M_{XX}}{I_{XX}} y_C - \frac{M_{YY}}{I_{YY}} \cdot x_C$$

where $\quad y_C = 100$ mm and $x_C = 160 - 83.73 = 76.27$ mm

$$\therefore \quad p_C = \frac{100 \times 10^3}{26974} - \frac{5 \times 10^6}{10466 \times 10^4} \times 100 - \frac{2.373 \times 10^6}{6387 \times 10^4} \times 76.27$$

$$= -3.90 \text{ N/mm}^2 = 3.90 \text{ N/mm}^2 \quad \text{(tension)}$$

Stress at point D : By inspection

$$p_D = \frac{P}{A} - \frac{M_{XX}}{I_{XX}} y_D + \frac{M_{YY}}{I_{YY}} \cdot x_D$$

where $\quad y_D = 100$ mm and $x_D = \bar{x} = 83.23$

$$\therefore \quad p_D = \frac{100 \times 10^3}{26974} - \frac{5 \times 10^6}{10466 \times 10^4} \times 100 + \frac{2.373 \times 10^6}{6387 \times 10^4} \times 83.23 = \mathbf{2.02\ N/mm^2}$$

Example 17.13. *A bracket shown in Fig. 17.15 (a) is pulled with a force of P = 5 kN acting as shown. (a) Calculate the maximum and minimum normal stresses on the section XX; (b) Show the distribution of stress across the section by means of a diagram (c) How far from the upper edge of the XX plane is the horizontal plane of zero stress ?*

Solution. Horizontal component of $P = 5 \cos 45° = 3.536$ kN

Vertical component of $P = 5 \sin 45° = 3.536$ kN

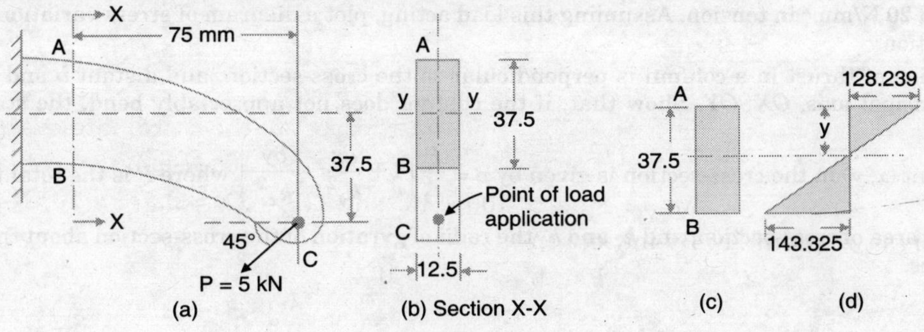

Fig. 17.15

The horizontal component of P, being normal to section XX, will induce direct stress p_0, while both the horizontal and vertical components of P will induce bending moments M_1 and M_2 about the section AB. Both the moments M_1 and M_2 will cause tension at the upper edge (A) of the section and compression at the lower edge (B).

Now M_1 due to horizontal component = 3.536 × 37.5 = 132.6 kN-mm = 132.6 × 10^3 N-mm

M_2 due to vertical component = 3.536 × 75 = 265.2 kN-m = 265.2 × 10^3 N-mm

Area of cross-section at $XX = A$ = 12.5 × 37.5 = 468.75 mm^2

$$I \text{ about } y\text{-}y \text{ axis} = \frac{bd^3}{12} = \frac{12.5}{12}(37.5)^3 = 54932 \text{ mm}^2$$

$$\text{Direct stress} \quad p_0 = \frac{P_H}{A} = \frac{3.536 \times 10^3}{468.75} = 7.543 \text{ N/mm}^2$$

$$\text{Bending stress} \quad p_b = \pm \frac{M}{I}\left(\frac{d}{2}\right) = \pm \frac{(132.6 + 265.2)\,10^3}{54932} \times \frac{37.5}{2} = \pm\, 135.782 \text{ N/mm}^2$$

$\therefore$ f_t (at the top edge) $= p_b - p_0$ = 135.782 – 7.543 = **128.239 N/mm²** (tension)

f_b (at the bottom edge) $= p_b + p_0$ = 135.782 + 7.543 = **143.325 N/mm²** (comp)

Fig. 17.15(d) shows the final stress distribution across the cross-section at XX. If y is the distance of zero stress from the top, we have

$$\frac{y}{37.5 - y} = \frac{128.239}{143.325}, \text{ from which } y = \frac{128.239 \times 37.5}{128.239 + 143.325} \approx \mathbf{17.71 \text{ mm.}}$$

PROBLEMS

1. A masonry column of circular cross-section is subjected to a thrust parallel to the axis but at a distance e from it. Show that if there is to be no tension produced in the column the value of e must not exceed $\frac{1}{8}$ of the diameter of the column.

   A bar of diameter 1 cm is subjected to a tensile test, but due to defective centering the resultant force cuts the cross-section at a point distant 0.3 mm from the centre. Show that for this cross-section, if the elastic limit is not exceeded, the ratio of maximum stress to mean stress is about $1\frac{1}{4}$.

2. A short hollow cast iron column having 20 cm external diameter and 12 cm internal diameter is subjected to vertical compressive load acting at an eccentricity of 6 cm from the axis of the column. Find the greatest allowable load if the maximum permissible stresses are 80 N/mm^2 in compression and 20 N/mm^2 in tension. Assuming this load acting, plot a diagram of stress variation across the section.

3. A line of thrust in a column is perpendicular to the cross-section, and distant b and a from the principal axis, OX, OY. Show that, if the column does not appreciably bend, the stress at any point (x, y) in the cross-section is given by $p = \frac{P}{A}\left(1 + \frac{ax}{k_y{}^2} + \frac{by}{k_x{}^2}\right)$ where P is the total thrust, A is the area of cross-section, and k_x and k_y the radii of gyration of the cross-section about the principal axes.

4. In drilling, a part of spindle *A* of the drilling machine shown in Fig. 17.16 is subjected to an axial load of 10 kN. Determine the diameter of the solid cast iron column *B*. The allowable tensile stress is 40 N/mm^2.

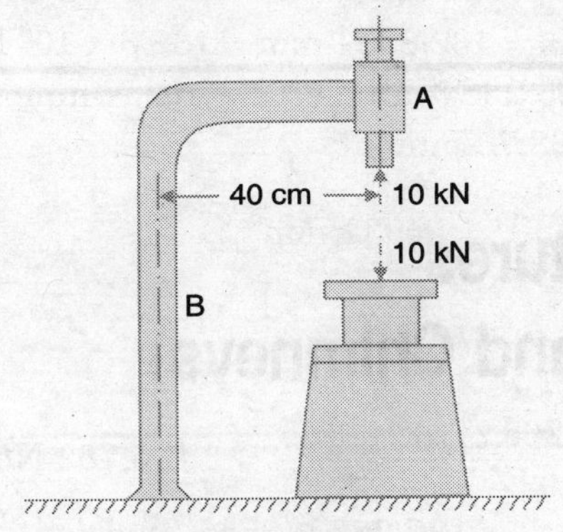

Fig. 17.16

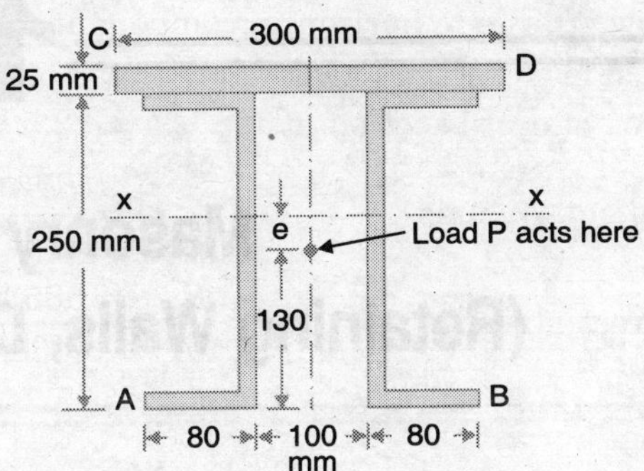

Fig. 17.17

5. A short horizontal strut of built-up section as shown in Fig. 17.17 is subjected to a horizontal compression load of 1100 kN acting along an axis which intersects the section midway between the two channels and 13 cm from the bottom of the channels.

Calculate the maximum and minimum intensities of stress in the material, if I_{xx} for one channel is 3816.8 cm^4 and area of one channel is 38.67 cm^2.

6. What depth of metal may be safely removed from one of the narrow sides of a tie bar 10 cm wide and 3 cm thick sustaining and axial load of 100 kN in order to limit the maximum stress over the reduced width to 100 N/mm^2.

ANSWERS

2. 527 kN ; f_{max} = 72.4 N/mm^2 (comp.), $f_{min.}$ = 20 N/mm^2 (tensile).

4. 10.17 cm.

5. $f_{max.}$ at *AB* = 162.9 N/mm^2 (comp.),

$f_{min.}$ at *CD* = 34 N/mm^2 (comp.).

6. 2.8 cm.

18

Masonry Structures (Retaining Walls, Dams and Chimneys)

18.1. INTRODUCTION

Masonry is one of the most common building medium used for construction. Common examples of masonry structures are : (*i*) masonry walls and masonry columns, (*ii*) masonry retaining walls, (*iii*) masonry dams, (*iv*) masonry chimneys and (*v*) masonry arches. *Masonry* may be defined as the construction of *building units* bonded together with *mortar*. The building units (commonly known as *masonry units*) may be stones, bricks or precast blocks of concrete. When stones are used as the building units or building blocks, we have *stone masonry*. Similarly, in *brick masonry*, bricks are used as the building units. A *composite masonry* is a construction in more than one type of building units. Masonry work is one of the major building crafts and one of the oldest.

Masonry walls are used almost in all the buildings. These may be either load bearing walls or partition walls. Unless otherwise specified, a *wall* may be defined as a vertical load-bearing member, the width (*i.e.,* length) of which exceeds four times the thickness. In contrast to this, a *column* is an isolated load-bearing member, the width of which does not exceed four times the thickness. Load bearing walls carry super-imposed loads, transferred through roof etc., in addition to their own weight. Load bearing walls are structurally efficient when the load is uniformly distributed and when the structure is so planned that eccentricity of loading on the wall is as small as possible. The strength of a wall is measured in terms of its resistance to the stresses set up in it by its own weight, by super-imposed loads and by lateral pressure such as wind etc., and its stability by its resistance to overturning by lateral forces and buckling caused by excessive slenderness.

An *arch* is a structure constructed of wedge-shaped units (stones or bricks) jointed together with mortar and spanning an opening to support the weight of the wall above it along with super-imposed loads. Due to wedge like form, the units support each other, the load tends to make them compact and enables them to transmit the pressure downwards to their supports. An arch transmits the super-imposed load to the side walls (or abutments) through friction between the surfaces of voussoirs and the cohesion of mortar. Every element of arch remains in compression. It has also to bear transverse shears. Masonry arches are commonly used over openings (*i.e.,* doors, windows etc.,) in buildings. They are also used in bridges.

A *masonry retaining wall* is the masonry structure constructed for maintaining the ground surface (or earthfill) at different elevations on either side of it. The material supported by the retaining wall is called *backfill*, the top surface of which may either be horizontal or inclined. Such retaining walls are also known as *gravity retaining walls* as they resist the lateral earth pressure by their own weight in contrast to the cantilever and counter-fort retaining walls in which pressure is resisted by bending action. Due to this reason, masonry retaining walls are thicker in section.

A *masonry dam* is a masonry structure which retains water on its one side, usually known as the upstream side. The upstream face of a masonry dam may be either straight and vertical or may be inclined, depending upon its height. The top width of a masonry dam depends both upon the roadway requirements as well as on economy. A masonry dam may be either a *low height dam* or a *high dam*. We shall discuss here only the low height dams. A masonry dam is subjected to a number of horizontal forces such as lateral water pressure (which is the major force), wave pressure, silt pressure and earthquake pressure, and vertical forces consisting of its own weight plus the component of weight of water supported by its upstream inclined face (if any). Due to these forces, the resultant force acts inclined on the plane lamina at any height, the point of application of which is away from the centre of lamina. The normal stress distribution is therefore not uniform. In order that no tension should develop any where on the base, the resultant should be within the middle third. Also, the dam should be safe against overturning.

Masonry chimneys are used to escape the flue gases to such a height that diffusion of gases takes place so that they do not contaminate the surroundings. The height of chimney is fixed to provide the required draft. Masonry chimneys are generally built in brick masonry, using either the common bricks or radial bricks. Though masonry chimneys may have circular, square, hexagonal or octagonal shape in plan, circular plan shape chimneys are usually preferred. Chimneys are lined from inside with fire bricks. Apart from self weight and the weight of lining, chimneys are subjected to lateral forces due to wind and/or earthquake.

18.2. GENERAL CONDITIONS OF STRENGTH AND STABILITY OF MASONRY STRUCTURES

Masonry structures, such as retaining walls, dams, chimneys etc., are subjected to net vertical force ΣV and net horizontal for ΣH, due to self weight, super-imposed loads and lateral loads. The resultant R of ΣV and ΣH strikes the base at some eccentricity with respect to the *c.g.* of the section. Due to this, the normal stress distribution at any section, including the base section, will not be uniform. If the eccentricity (e) is large, there may be a possibility of development of tensile normal stress at the heel of the section. Masonry units are jointed with the help of mortar which is supposed to be weak in tension. Also, the maximum normal compressive stress at the other end may exceed the permissible compressive stress for the material. In addition to these strength criteria, the structure may be unstable because of the overturning effect of lateral force and unstable against the sliding effect. Hence we have four *safety criteria*.

(a) Strength criteria

1. The masonry structure should be safe against development of *tension* any where in the section, *i.e.*, no tension should develop.
2. The masonry structure should be safe against *crushing*, *i.e.*, the maximum compressive stress should be equal to or less than the safe compressive stress for the material.

(b) Stability criteria

3. The structure should be safe against *overturning, i.e.,* the restoring moment should be greater than the overturning moment.
4. The structure should be safe against *sliding, i.e.,* the horizontal shear at any plane should be less than the available frictional force between the successive layers.

Strength criteria

Let us consider a plane lamina of width b, subjected to a net vertical force ΣV and a net horizontal force ΣH. Let the resultant force R strike the base at point D at some eccentricity e, as marked in Fig. 18.1(a). This resultant R can be again resolved into two components: (i) vertical component ΣV acting at an eccentricity e and, (ii) a horizontal component, or horizontal shear ΣH acting horizontally along the section as shown by dotted arrows in Fig. 18.1(a). If we introduce equal and opposite ΣV forces at the centre C of the lamina, we get the following system of forces:

(i) An axial force ΣV acting at C

(ii) A bending couple

$$M = e \, \Sigma V$$

and (iii) A horizontal force ΣH along the base AB.

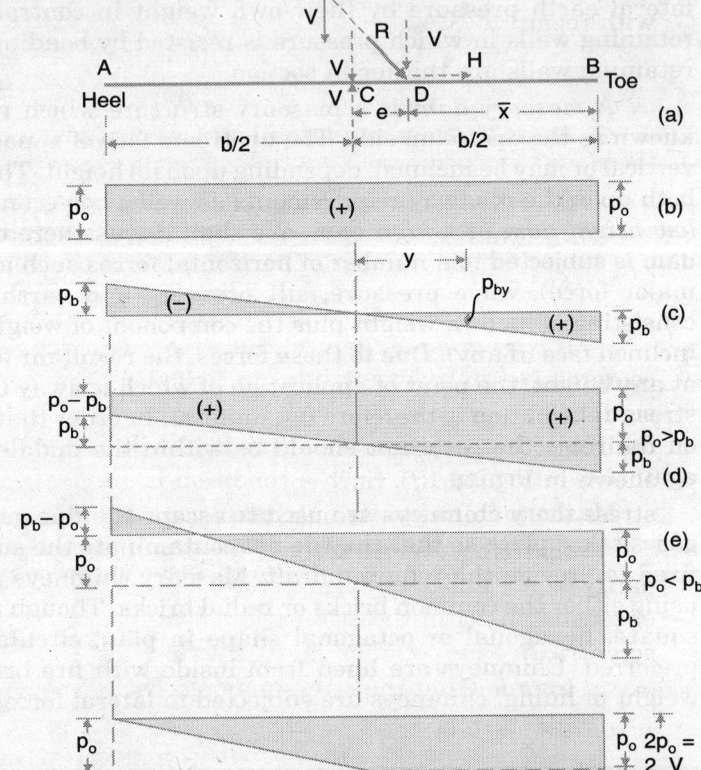

Fig. 18.1. Strength Criteria

The axial force ΣV will cause a direct compressive stress p_0 given by

$$p_0 = \frac{\Sigma V}{A} \qquad ...(a)$$

where A is the area of cross-section of the base. The distribution of p_0 is shown in Fig. 18.1(b).

The bending couple (or moment) will cause longitudinal tensile and compressive stresses. The fibre stress p_b at any distance y from N.A. is given by

$$p_{by} = \frac{M}{I_x} \cdot y = \frac{\Sigma V \, . \, e}{I_x} \, . \, y \qquad ...(b)$$

Hence the total stress at any section is given by

$$f = \frac{\Sigma V}{A} + \frac{\Sigma V \, . \, e \, . \, y}{I_x} \qquad ...(18.1)$$

The *extreme fibre stress* p_b due to bending is given by

$$p_b = \frac{M}{Z_x} = \frac{\Sigma V.e}{Z_x} = \frac{\Sigma V.e}{I_x}.\frac{b}{2} \qquad ...(18.2)$$

where b = depth of the section and, $I_x = Ak^2$, where k is the radius of gyration of the section with regard to N.A.

$$\therefore \qquad p_b = \frac{\Sigma V.e.b}{2\,Ak^2} \qquad ...[18.2(a)]$$

Again the maximum total stresses are given by the following :

At toe, $$f_{max} = p_0 + p_b = \frac{\Sigma V}{b} + \frac{\Sigma V.e}{Z_x} = \frac{\Sigma V}{b}\left(1+\frac{b.e}{Z_x}\right) \qquad ...[18.3(a)]$$

At heel, $$f_{min} = p_0 - p_b = \frac{\Sigma V}{b} - \frac{\Sigma V.e}{Z_x} = \frac{\Sigma V}{b}\left(1-\frac{b.e}{Z_x}\right) \qquad ...[18.3(b)]$$

If p_0 is greater than p_b, the stress throughout the section will be of the same sign (*i.e.,* compressive) as shown in Fig. 18.1(*d*). If however, p_0 is less than p_b, the stress will change sign, being partly tensile and partly compressive across the section. Such a condition is not desirable. If $p_0 = p_b$, stress at the heel will be zero, while the stress at the toe will be $2p_0$, as shown in Fig. 18.1(*f*). Such a condition is an ideal condition, wherein, we are in a position to stress the material to its maximum capacity without causing tension at the other end.

Limiting eccentricity : In Fig. 18.1(*e*), $p_0 < p_b$ and therefore stress changes sign, being partly tensile and partly compressive across the section. In masonry structures, the development of tensile stress in the section is not desirable as they are weak in tension. This limits the eccentricity e to a certain value which will now be investigated for different sections.

In order that the stress may not change the sign from compressive to tensile, we have

$$p_0 \geq p_b \geq \frac{M}{Z} \quad \text{or} \quad \frac{\Sigma V}{A} \geq \frac{\Sigma V.e.b}{2I} \geq \frac{\Sigma V.e.b}{2Ak^2}$$

or $$e \leq \frac{2\,k^2}{b} \qquad ...(18.4)$$

Thus for no tension in the section, the eccentricity must not exceed $\frac{2\,k^2}{b}$.

Rectangular section : For a rectangular section of *unit width* and depth b,

$$I = \frac{1\times 1}{12}b^3 = \frac{b^3}{12}\,; \qquad A = b \times 1 = b \qquad \text{and} \quad k^2 = \frac{I}{A} = \frac{b^2}{12}$$

Hence $$e \leq \frac{2}{b}.\frac{b^2}{12} \leq \frac{b}{6} \quad \text{or} \quad e_{max} = \frac{b}{6} \qquad ...(18.5)$$

This value of eccentricity can be on either side of the geometrical axis. Thus, the stress will be of the same sign throughout the section if the load line is within the *middle third* of the section. Thus, in case of rectangular section, the maximum intensities of extreme fibre stresses are given by

$$f = \frac{\Sigma V}{A} \pm \frac{\Sigma V.e}{Z_x} = \frac{\Sigma V}{b\times 1} \pm \frac{\Sigma V.e}{\frac{1}{6}\times 1\times b^2} = \frac{\Sigma V}{b}\left(1 \pm \frac{6e}{b}\right) \qquad ...(18.6)$$

when $$e = \frac{b}{6}, \quad f = \frac{\Sigma V}{b}(1 \pm 1)$$

or $$f_{max} = \frac{2 \Sigma V}{b} \text{ and } f_{min} = 0 \qquad \ldots(18.7)$$

This condition is shown in Fig. 18.1(*f*)

Solid circular section

For a solid circular section of diameter b,

$$I = \frac{\pi}{64} b^4 ; A = \frac{\pi}{4} b^2 ; \quad k^2 = \frac{I}{A} = \frac{b^2}{16}$$

$$\therefore \quad e \le \frac{2k^2}{b} \le \frac{2}{b} \times \frac{b^2}{16} \le \frac{b}{8} \qquad \ldots(18.8)$$

Thus, in order that tension is not developed, the load line must fall within the *middle fourth* of the section.

Hollow circular section

For a hollow circular section, having external diameter B and internal diameter b,

$$\mathrm{I} = \frac{\pi}{64}(B^4 - b^4) ; A = \frac{\pi}{4}(B^2 - b^2) ; \quad k^2 = \frac{I}{A} = \frac{B^4 - b^4}{16(B^2 - b^2)} = \frac{B^2 + b^2}{16}$$

$$\therefore \quad e \le \frac{2k^2}{B} \le \frac{2}{B}\left(\frac{B^2 + b^2}{16}\right) \le \frac{B^2 + b^2}{8B} \qquad \ldots[18.8(a)]$$

Stability criteria

Safety against overturning

The lateral forces (*i.e.*, forces due to earth pressure in retaining walls, water pressure in dams and wind forces in chimneys and tall walls) cause overturning moments (ΣM_O) in the masonry structure endangering its stability. This must be resisted entirely or partly by the body forces (*i.e.*, weight) of the structure, by producing the resisting moments (ΣM_R) about the toe. In fact, the resisting moments must be much more than the overturning moments. The *stability factor or factor of safety against overturning* is given by

$$\text{F.S.} = \frac{\Sigma M_R}{\Sigma M_O} \qquad \ldots[18.9(a)]$$

The minimum value of F.S. against overturning should be 1.5.

Safety against sliding. The masonry structure should be safe against sliding, under the action of lateral forces (ΣH), by the development of appropriate frictional forces. Let μ be the coefficient of friction. Mobilized frictional force = $\mu \Sigma V$.

Fig. 18.2

This should be much more than the lateral force (ΣH) acting at the base. The factor of safety against sliding is given by

$$\text{F.S.} = \frac{\mu \Sigma V}{\Sigma H} \qquad \ldots[18.9(b)]$$

This factor of safety should be greater than 1.5.

Example 18.1. *Find the maximum and minimum stress intensities induced on the base of a masonry wall 6 m high, 4 m wide and 1.5 m thick, subjected to a horizontal wind pressure of 1.5 kN/m² acting on the 4 m side. The unit weight of masonry may be taken as 22.4 kN/m³.*

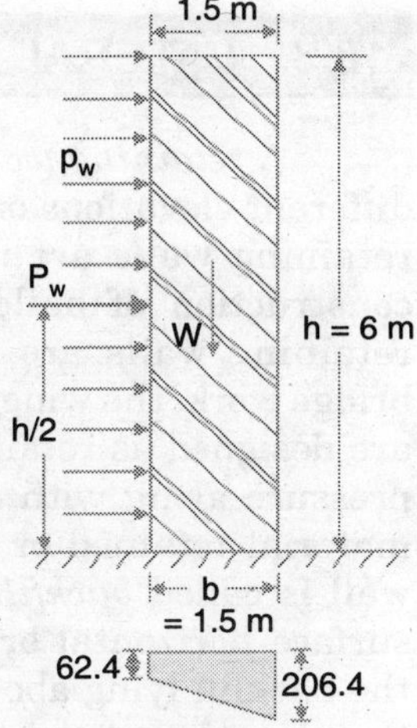

Fig. 18.3

Solution. The direct stress (or axial stress) due to self weight of masonry is given by:

$$p_0 = \gamma H = 22.4 \times 6 = 134.4 \text{ kN/m}^2$$

The wind pressure $P_w = p_w.L.\ h = 1.5 \times 4 \times 6 = 36$ kN

Moment due to wind, $M_w = P_w \cdot \dfrac{h}{2} = 36 \times \dfrac{6}{2} = 108$ kN-m

$$Z = \frac{1}{6} Lb^2 = \frac{1}{6} \times 4\,(1.5)^2 = 1.5 \text{ m}^3$$

$$\therefore \quad p_b = \frac{M_w}{Z} = \frac{108}{1.5} = 72 \text{ kN/m}^2.$$

$$\therefore \quad f_{max} = p_0 + p_b = 134.4 + 72 = \mathbf{206.4\ kN/m^2} \text{ (comp.)}$$

$$f_{min} = p_0 - p_b = 134.4 - 72 = \mathbf{62.4\ kN/m^2} \text{ (comp.)}$$

Example 18.2. *Calculate the normal stresses at the four outside corners of a horizontal section of short hollow pier 1.5 metre square outside and 1 metre square inside supporting a vertical point load of 60 kN on a diagonal and 0.8 metre from the vertical axis of the pier. The unit weight of pier masonry may be taken as 21 kN/m³. The height of pier is 3 m.*

Solution. Total self weight, $W_S = (1.5^2 - 1^2) \times 3 \times 21 = 78.75$ kN

$$\therefore \quad \text{Uniform stress,} \quad p_0 = \frac{W + W_S}{A} = \frac{60 + 78.75}{1.5^2 - 1^2} = 111 \text{ kN/m}^2 \text{ (compressive)}$$

$$I_{xx} = \frac{1}{12}(1.5^4 - 1^4) = 0.339 \text{ m}^4$$

$$Z_{xx} = \frac{I_{xx}\sqrt{2}}{b} = \frac{0.339\sqrt{2}}{1.5} = 0.319 \text{ m}^3$$

Moment $M = 60 \times 0.80 = 48$ kN-m

$$\therefore \quad p_b \text{ at point } A = \frac{M}{Z} = \frac{48}{0.319}$$

$$= 150.5 \text{ kN/m}^2 \text{ (comp.)}$$

p_b at point $C = 150.5$ kN/m² (tensile)

p_b at points B and $D = 0$

Hence f_{max} (at A) $= p_0 + p_b$

$= 111 + 150.5$

$= 261.5$ kN/m² (comp.)

f_{max} (at C) $= p_0 - p_b = 111 - 150.5$

$= -39.5$ kN/m² (*i.e.*, tensile)

f_{max} (at B and D) $= p_0 = 111$ kN/m² (comp.)

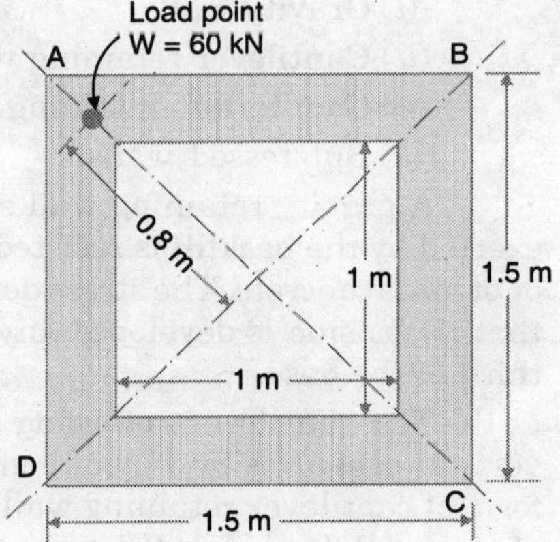

Fig. 18.4

18.3. LATERAL EARTH PRESSSURE ON RETAINING WALLS

A *retaining wall* or retaining structure is used for maintaining the ground surfaces at different elevations on either side of it. Whenever embankments are involved in construction, retaining walls are usually necessary. In the construction of buildings having basements, retaining walls are mandatory. Similarly in bridge work, the wing walls and abutments etc., are designed as retaining walls, to resist earth pressure along with super-imposed loads. The material retained or supported by a retaining wall is called *backfill* which may have its top surface horizontal or inclined. The position of the backfill lying above the horizontal plane at the elevation of the top of a wall is called the *surcharge*, and its inclination to the horizontal is called the *surcharge angle* β.

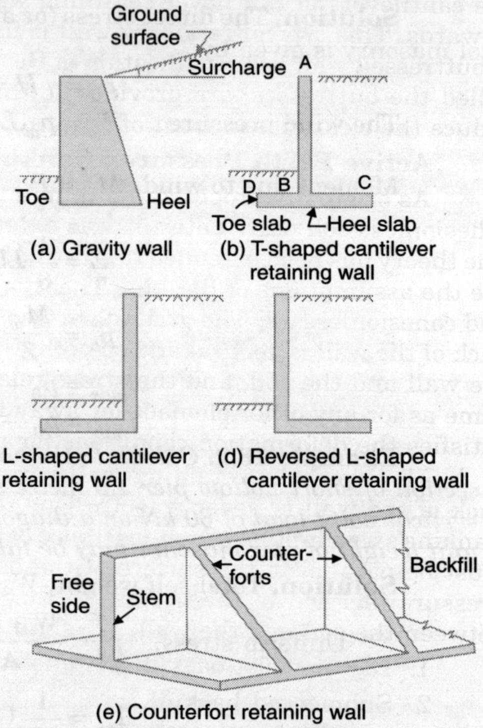

In the design of retaining walls or other retaining structures, it is necessary to compute the lateral earth pressure exerted by the retaining mass of soil. The equation of finding out the lateral earth pressure against retaining wall is one of the oldest in the Civil Engineering field. The *plastic state of stress,* when the failure is imminent, was investigated by Rankine in 1860. A lot of theoretical and experimental work has been done in this field and many theories and hypothesis have been proposed.*

Retaining walls may be classified according to their mode of resisting the earth pressure, and according to their shape. Following are some of common types of retaining walls (Fig. 18.5).

Fig. 18.5. Various Types of Retaining Walls

(*i*) Gravity walls

(*ii*) Cantilever retaining walls : (*a*) T-shaped (*b*) L-shaped

(*iii*) Counterfort retaining walls.

(*iv*) Buttressed walls.

A gravity retaining wall shown in Fig. 18.5 (*a*) is the one in which the earth pressure exerted by the backfill is resisted by dead weight of the wall, which is either made of masonry or of mass concrete. The stress developed in the wall is very low. These walls are so proportioned that no tension is developed anywhere, and the resultant of forces remain within the middle third of the base.

The cantilever retaining wall resists the horizontal earth pressure as well as other vertical pressures by way of bending of various components acting as cantilevers. A common form of cantilever retaining wall is the T-shaped wall shown in Fig. 18.5(*b*). The wall consists of stem *AB*, heel slab *BC* and toe slab *DB*. Each of these bend as cantilevers, about *B*. They

*For detailed account of various earth pressure theories, the reader may refer to the Author's book 'Soil Mechanics and Foundations'.

are, therefore, reinforced on the tension face. Another form of cantilever retaining walls are the L-shaped walls shown in Fig. 18.5(*c*) and (*d*). They also resist the soil pressures by bending.

A counterfort retaining wall is shown in Fig. 18.5(*e*). The vertical stem and the heel slab are strengthened by providing counterforts at some suitable interval. Because of provision of counterforts, the vertical stem as well as the heel slab act as *continuous slab*, in contrast to the cantilevers of cantilever retaining wall. The toe slab however acts as cantilever bending upwards. This type of retaining wall is used when backfill of greater height is to be retained. A buttressed wall is a modification of the counterfort retaining wall in which the counterforts, called the buttresses, are provided to the other side of the backfill. However the buttresses reduce the clearance in front of the wall, and therefore these walls are not commonly used.

Active Earth Pressure : Rankine's Theory

As originally proposed, Rankine's theory of lateral earth pressure is applied to uniform cohesionless soil only. Later, it was extended to include cohesive soils, by Resal and by Bell. The theory has been extended to stratified, partially submerged and submerged soils. Following are the assumptions of Rankine Theory : (1) The soil mass is semi-infinite, homogeneous, dry and cohesionless. (2) The ground surface is plane which may be horizontal or inclined. (3) The back of the wall is vertical and smooth. In other words, there are no shearing stresses between the wall and the soil, and the stress relationship for any element adjacent to the wall is the same as for any other element for away from the wall. (4) The wall yields about the base and satisfies the deformation conditions for plastic equilibrium.

However, the retaining walls are constructed of masonry or concrete, and hence the back of the wall is never smooth. Due to this, frictional forces develop. As a consequence of Rankine's assumption of non-existence of frictional forces at the wall face, the resultant pressure must be parallel to the surface of the backfill. The existence of friction makes the resultant pressure inclined to the normal to the wall at an angle that approaches the friction angle between the soil and the wall. We shall consider the following cases of cohesionless backfill :

1. Dry or moist backfill with no surcharge.
2. Submerged backfill.
3. Backfill with uniform surcharge.
4. Backfill with sloping surfaces.

1. Dry or moist backfill with no surcharge. Fig. 18.6 (*a*) shows a retaining wall of height *h*, having dry or moist backfill, with no surcharge. According to Rankine's theory, the intensity of *active earth pressure*, p_a, trying to move the wall away from the fill, is given by

$$p_a - K_a \gamma H \qquad \text{...(18.10)}$$

where K_a = co-efficient of active earth pressure,

$$= \frac{1-\sin\phi}{1+\sin\phi} \qquad \text{...(18.11)}$$

ϕ = angle of internal friction, for the backfill,

γ = unit weight of soil (backfill)

H = height of retaining wall.

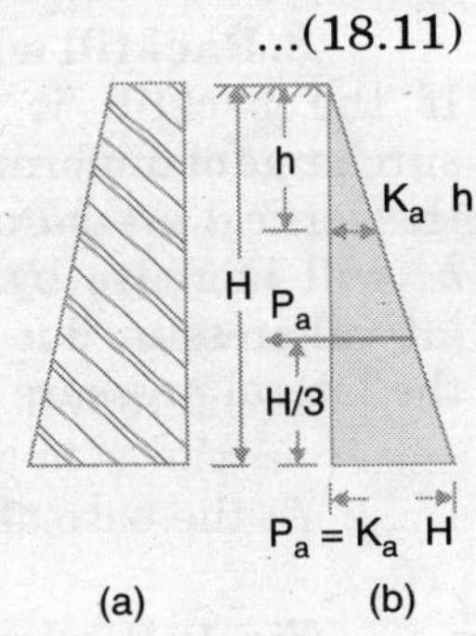

Fig. 18.6

At any height *h* below the horizontal ground surface, the intensity of active earth pressure is $K_a \gamma\, h$. Fig. 18.6 (*b*) shows the distribution of active earth pressure P_a over the retaining wall. The total active earth pressure P_a or the resultant pressure per unit length of the wall is given by

$$P_a = \frac{1}{2} K_a . \gamma . H^2 \qquad \text{...(18.12)}$$

acting at $H/3$ above the base of retaining wall. If the soil is dry, γ is the dry weight of the soil, and if wet, γ is the moist weight, to be substituted in Eq. 18.10 or 18.12.

2. Submerged backfill. In this case, the back fill behind the retaining wall is saturated with water. The lateral pressure is made up of two components.

(*i*) lateral pressure due to submerged weight γ' of soil, and

(*ii*) lateral pressure due to water.

Thus, at any depth h below the surface,

$$p_a = K_a \gamma' h + \gamma_w \cdot h \qquad \text{...(18.13)}$$

The pressure at the base of the retaining wall ($h = H$) is given by

$$p_a = K_a \gamma' H + \gamma_w H \qquad \text{...(18.14)}$$

If the free water stands to both the sides of the wall [Fig. 18.7(*b*)], the water pressure need not be considered, and the net lateral pressure given by

$$p_a = K_a \gamma' H \qquad \text{...(18.15)}$$

If the backfill is partly submerged, *i.e.*, the backfill is moist to a depth H_1 below the ground level, and then it is submerged, the lateral pressure intensity at the base of the wall is given by

$$p_a = K_a \cdot \gamma H_1 + K_a \gamma' H_2 + \gamma_w \cdot H_2 \qquad \text{...(18.16)}$$

The above expression is based on the assumption that the value of ϕ is the same for the moist as well as submerged soil. If it is different (say ϕ_1 and ϕ_2 respectively), the earth pressure coefficient K_{a1} and K_{a2} for both the portions will be different. As ϕ decreases, K_a increases. The lateral pressure intensity [Fig. 18.8 (*c*)] at the base of wall is

$$p_a = K_{a2} \gamma H_1 + K_{a2} \gamma' H_2 + \gamma_w H_2 \qquad \text{...(18.17)}$$

3. Backfill with uniform surcharge. If the backfill is horizontal and carries surcharge of uniform intensity q per unit area, the vertical pressure increment, at any depth h, will increase by q. The increase in the lateral pressure due to this will be $K_a q$. Hence the lateral pressure at any depth h is given by

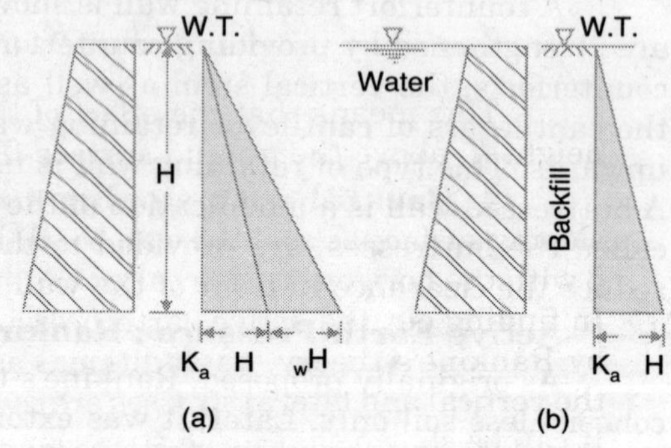

Fig. 18.7. Submerged Backfill

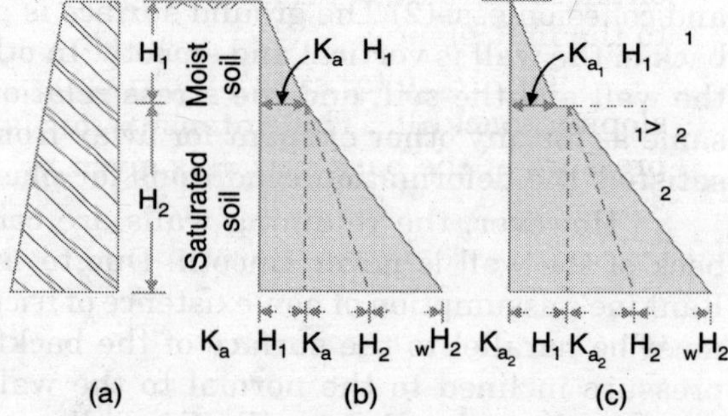

Fig. 18.8. Partly Submerged Backfill

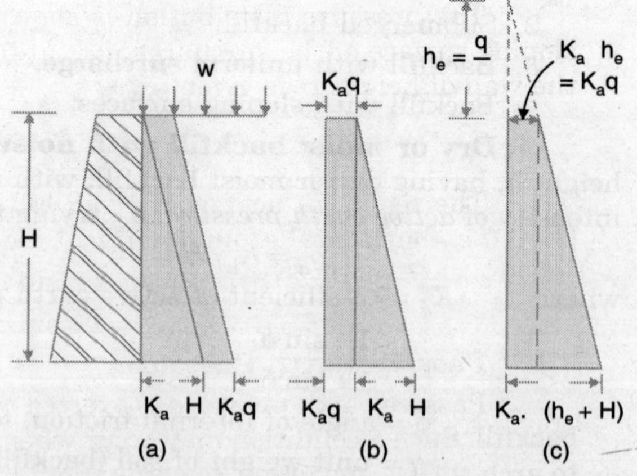

Fig. 18.9. Backfill with Uniform Surcharge

$$p_a = K_a \cdot \gamma \cdot h + K_a \cdot q \qquad \text{...(18.18)}$$

At the base of the wall, the pressure intensity is :

$$p_a = K_a \cdot \gamma \cdot H + K_a \cdot q \qquad \text{...[18.18 (a)]}$$

Fig. 18.9 (*a*) and (*b*) show two alternative methods of plotting the lateral pressure diagram for this case. The lateral pressure increment $K_a \cdot q$ due to the surcharge is the same at every point of the back of the wall, and does not vary with height h. The height of fill h_e equivalent to the uniform surcharge intensity is given by the relation :

$$K_a \cdot \gamma \cdot h_e = K_a q \quad \text{or} \quad h_e = \frac{q}{\gamma} \qquad \ldots(18.19)$$

This means that the effect of the surcharge of intensity q is the same as that of a fill of height h_e above the ground surface [Fig. 18.9 (*c*)].

4. Backfill with sloping surface. Let the sloping surface behind the wall be inclined at an angle β with the horizontal ; β is called the *surcharge angle*. In finding out the active earth pressure for this case by Rankine's theory, an additional assumption that the vertical and lateral stresses are *conjugate* is made. It can be shown that if the stress on a given plane at a given point is parallel to another plane, the stress on the latter plane at the same point must be parallel to the first plane.

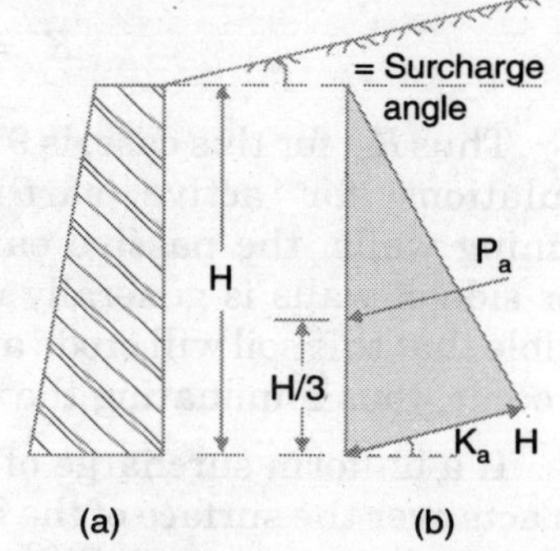

Fig. 18.10. Lateral Pressure Distribution for Sloping Surcharge

Fig. 18.10 (*a*) shows the retaining wall with a sloping backfill. The intensity of lateral earth pressure at the base of wall is given by

$$p_a = \gamma H \cdot \cos\beta \, \frac{\cos\beta - \sqrt{\cos^2\beta - \cos^2\phi}}{\cos\beta + \sqrt{\cos^2\beta - \cos^2\phi}}$$

or

$$p_a = K_a \gamma H \qquad \ldots[18.20\ (a)]$$

where

$$K_a = \cos\beta \, \frac{\cos\beta - \sqrt{\cos^2\beta - \cos^2\phi}}{\cos\beta + \sqrt{\cos^2\beta - \cos^2\phi}} \qquad \ldots[18.20\ (b)]$$

The pressure distribution is shown in Fig. 18.10 (*b*). The pressure acts *parallel* to the sloping surface of the surcharge, *i.e.*, at β with the horizontal. The total active pressure P_a for the wall of height H is given by

$$P_a = \frac{1}{2} K_a \gamma H^2.$$

The resultant acts at $H/3$ above the base, in direction parallel to the surcharge. When $\beta = 0$ (*i.e.*, horizontal ground surface), Eq. 18.20 (*b*) reduces to

$$K_a = \frac{1 - \sin\phi}{1 + \sin\phi} \text{ which is the same as Eq. 18.11.}$$

Passive Earth Pressure

Passive earth pressure is exerted on a wall when it has a tendency to move towards the backfill. Such a condition may occur when the retaining wall supports an arch, and is subjected to arch thrust, moving it towards the fill. Another condition of passive earth pressure may be when the wall supports soil of different heights on both the sides as shown in Fig. 18.11. Due to active earth pressure from the right hand side, the wall moves to the left. The soil to the left is thus compressed and in turn exert passive earth pressure, resisting such movement.

If h is the height of fill, the intensity of passive pressure at height h is given by

$$p_p = K_p \cdot \gamma \cdot h \qquad \ldots(18.21)$$

where K_p = coefficient of passive earth pressure $= \dfrac{1 + \sin\phi}{1 - \sin\phi} = \dfrac{1}{K_a}$ ...(18.22)

The passive pressure distribution will thus be a triangle. The total pressure is given by

$$P_p = K_p \cdot \frac{\gamma h^2}{2} \quad \text{...(18.23)}$$

acting at $h/3$ above base.

It should be noted that K_p is very much greater than K_a. For example when $\phi = 30°$,

$$K_a = \frac{1-\sin\phi}{1+\sin\phi} = \frac{1}{3} \quad \text{and} \quad K_p = \frac{1+\sin\phi}{1-\sin\phi} = 3.$$

Thus K_p, for this case, is 9 times K_a. In making calculations for active earth pressure behind retaining walls, the passive earth pressure due to other side of walls is generally neglected, since it is possible that this soil will erode away or that shearing will occur, thus eliminating the passive pressure.

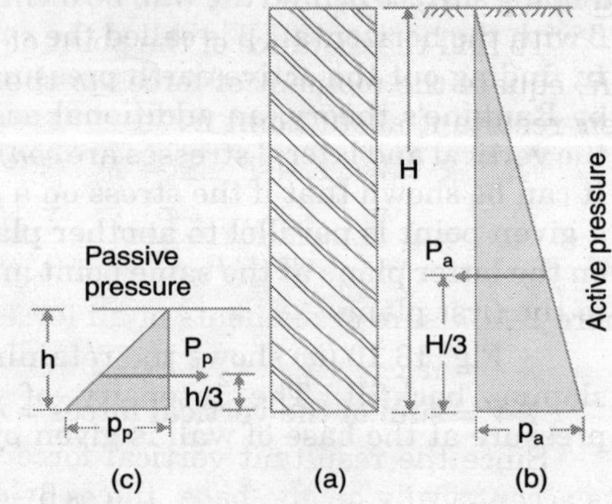

Fig. 18.11

If a uniform surcharge of intensity q per unit area acts over the surface of the backfill, the increase in the passive pressures will be equal to $K_p \cdot q$. The passive pressure intensity at depth h given by

$$p_p = K_p\,(\gamma h + q) \quad \text{...(18.24)}$$

If the backfill is having its top surface inclined at an angle β, the passive pressure is given by

$$p_p = \gamma h \cos\beta \, \frac{\cos\beta + \sqrt{\cos^2\beta - \cos^2\phi}}{\cos\beta - \sqrt{\cos^2\beta - \cos^2\phi}} \quad \text{...(18.25)}$$

or

$$p_p = K_p \cdot \gamma h \quad \text{...[18.26 (a)]}$$

where

$$K_p = \cos\beta \, \frac{\cos\beta + \sqrt{\cos^2\beta - \cos^2\phi}}{\cos\beta - \sqrt{\cos^2\beta - \cos^2\phi}} \quad \text{...[18.26 (b)]}$$

18.4. DESIGN OF GRAVITY RETAINING WALL

A gravity retaining wall is the one which resists the lateral earth pressure by its own weight, in contrast to the cantilever and counterfort retaining walls in which the pressure is resisted by bending action. A gravity retaining wall is therefore thicker in section. They are constructed of mass concrete, brick or stone masonry. The criterion of design of gravity retaining walls are as follows:

1. The base width of the wall must be such that the maximum pressure exerted on the foundation soil does not exceed the safe bearing capacity of the soil.
2. No tension should be developed anywhere in the wall.
3. The wall must be safe against sliding.
4. The wall must be safe against overturning.

Let P_a be the resultant active pressure acting on the wall face. P_h and P_v are its components in horizontal and vertical directions. If W is the weight of the wall acting through its centroid, the resultant R will pass through point of intersection of W and P_a. The resultant R striking the base at some point can be resolved into vertical and horizontal components R_v and R_h.

Resolving the forces into vertical and horizontal directions, we get

$$R_v = W + P_v \quad \text{...(18.27)}$$

and $$R_h = P_h \quad \text{...(18.28)}$$

To find the distance of the point of application of R, equate the moment of forces to the moment of their resultant about point B.

$$R_v \cdot \bar{x} = W \cdot x_1 + P_v \cdot x_2 + P_h \cdot \bar{z}$$

or $$\bar{x} = \frac{Wx_1 + P_v x_2 + P_h \cdot \bar{z}}{W + P_v} = \frac{\Sigma M}{\Sigma V} \quad \text{...(18.29)}$$

where ΣM = sum of moments of all forces about B

$$= W \cdot x_1 + P_v \cdot x_2 + P_h \cdot \bar{z}$$

ΣV = sum of the vertical forces = $R_v = W + P_v$

Since the resultant vertical force R_v (= ΣV) acts eccentrically on the base, the bearing pressure on the soil beneath the base are the combination of direct and bending stress. *Assuming* a linear distribution of pressure, the pressure intensities at the ends of the base are given by :

$$f_1 = \frac{R_v}{b}\left(1 + \frac{6e}{b}\right) \quad \text{...[18.30 (a)]}$$

and $$f_2 = \frac{R_v}{b}\left(1 - \frac{6e}{b}\right) \quad \text{...[18.30 (b)]}$$

where b = base width of the wall

e = eccentricity or the distance from the mid-point of the base to the point of application of = $\bar{x} - b/2$.

Fig. 18.12

Fig. 18.12 (*b*), (*c*), (*d*) shows three cases of pressure distribution below the base, When R strikes the base within the middle third (*i.e.*, $e < b/6$), f_1 and f_2 will be compresive, as shown in Fig. 18.12 (*b*). Fig. 18.12 (*c*) shows the special case when $e = b/6$, so that

$$f_1 = \frac{2\,R_v}{b} \quad \text{and} \quad f_2 = 0.$$

When $e > b/6$, (Fig. 18.12 *d*), tension is developed at B. Since soil is generally considered incapable to resist tension, the pressure will be redistributed along the intact base, of width $3b_1$, where b_1, is the distance of the point of application of R from end C. The intensity f_1 will then be given by

$$f_1 = \frac{2\,R_v}{3\,b_1} \quad \text{...[18.31 (a)]}$$

Substituting $\quad b_1 = \dfrac{b}{2} - e$, we get $\quad f_1 = \dfrac{4\,R_v}{3\,(b-2e)}$ ...[18.31 (b)]

The following are the criteria of design of gravity retaining walls:

1. The maximum pressure f_1 should not exceed the bearing capacity of soil.
2. For no tension to develop, $e < b/6$, or $\bar{x} \le 2b/3$
3. For no sliding to occur, $R_h < R_v \,.\, \mu$

where $\mu = \tan \delta'$ = coefficient of friction between the base of the wall and the soil. The factor of safety F against sliding is given by

$$F = \frac{R_v \,.\, \mu}{R_h} \qquad ...(18.32)$$

The minimum value of F should be 1.5.

4. For the wall to be stable against overturning, R must pass within the base width. However if the requirement of no tension is full-filled, complete safety against overturning is automatically assured.

Example 18.3. *Compute the intensities of active and passive earth pressure at depth of 8 metres in dry cohensionless sand with an angle of internal friction of 30° and the unit weight of 18 kN/m³. What will be the intensities of active and passive earth pressure if the water level rises to the ground level ? Take saturated unit weight of sand as 22 kN/m³.*

Solution (*a*) ***Dry soil :***

$$K_a = \frac{1-\sin\phi}{1+\sin\phi} = \frac{1-\sin 30°}{1+\sin 30°} = \frac{1}{3}\ ;\ K_p = \frac{1+\sin\phi}{1-\sin\phi} = \frac{1}{K_a} = 3.$$

$$p_a = K_a\,\gamma\,H = \frac{1}{3} \times 18 \times 8 = 48 \text{ kN/m}^2\ ;\quad p_p = K_p \,.\, \gamma\,H = 3 \times 18 \times 8 = 432 \text{ kN/m}^2.$$

(*b*) ***Submerged backfill***

$$\gamma' = \gamma_{sat} - \gamma_w = 22 - 9.81 = 12.19 \text{ kN/m}^3$$

$$p_a = K_a\,\gamma'\,H + \gamma_w H = \frac{1}{3} \times 12.19 \times 8 + 9.81 \times 8 \approx 111 \text{ kN/m}^2$$

$$p_p = K_p\,\gamma'\,H + \gamma_w\,H = 3 \times 12.19 \times 8 + 9.81 \times 8 = 371 \text{ kN/m}^2.$$

Example 18.4. *A retaining wall, 4 m high, has a smooth vertical back. The backfill has a horizontal surface in level with the top of the wall. There is uniformly distributed surcharge load of 36 kN/m² intensity over the backfill. The unit weight of the backfill is 18 kN/m³, its angle of shearing resistance is 30° and cohesion is zero. Determine the magnitude and the point of application of active pressure per metre length of the wall.*

Solution. $\quad K_a = \dfrac{1-\sin\phi}{1+\sin\phi} = \dfrac{1}{3}$

The lateral pressure intensity due to surcharge is given by

$$p_1 = K_a \,.\, q = \frac{1}{3} \times 36 = 12 \text{ kN/m}^2$$

The pressure intensity due to backfill at depth H = 4 m is given by

$$p_2 = K_a\,\gamma\,H = \frac{1}{3} \times 18 \times 4 = 24 \text{ kN/m}^2$$

The total pressure intensity at the base of the wall is given by

$$p_a = p_1 + p_2 = 12 + 24 = 36 \text{ kN/m}^2.$$

Fig. 18.13 shows the pressure distribution diagrams for p_1 and p_2. The resultant total pressure P_1 due to intensity p_1 is given by

$$P_1 = p_1 H = 12 \times 4$$
$$= 48 \text{ kN/m acting at } 4/2 = 2 \text{ m}$$

from the base.

The resultant total pressure, P_2 due to intensity p_2 is given by

$$P_2 = \frac{1}{2} p_2 H = \frac{1}{2} \times 24 \times 4 = 48 \text{ kN/m}$$

acting at $\frac{1}{3} \times 4 = 1.33$ m from the base.

$$\therefore \qquad P = P_1 + P_2 = 48 + 48 = 96 \text{ kN}$$

per metre length of the wall.

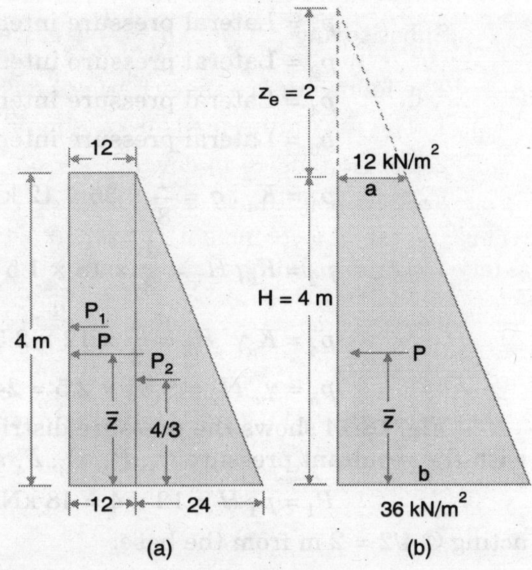

Fig. 18.13

The resultant P acts at a distance $\bar{z}$ above the base, given by taking the moments about the base:

$$\bar{z} = \frac{(48 \times 2) + \left(48 \times \frac{4}{3}\right)}{96} = 1.67 \text{ m}$$

Alternative solution : The equivalent height of surcharge is given by

$$z_e = \frac{q}{\gamma} = \frac{36}{18} = 2 \text{ m}.$$

Thus, the problem reduces to the calculation of earth pressure due to height $Z = H + z_e = 4 + 2 = 6$ m

The pressure intensity at the base of the wall is given by

$$p_a = \mathrm{K}_a \, . \, \gamma Z = \frac{1}{3} \times 18 \times 6 = 36 \text{ kN/m}^2$$

The pressure intensity at the top of the wall is evidently 12 kN/m². The pressure distribution is thus trapezoidal, having pressure intensities of 12 and 36 kN/m².

$$P = \frac{1}{2}(12 + 36) \times 4 = 96 \text{ kN/m}$$

The distance of point of application from the base is given by

$$\bar{z} = \frac{H}{3} \cdot \frac{2a + b}{a + b} = \frac{4}{3} \cdot \frac{2 \times 12 + 36}{12 + 36} = 1.67 \text{ m}.$$

Example 18.5. *In example 18.4, if the water table rises behind the wall to an elevation 1.5 m below the top, determine the total active pressure and its point of application. Take submerged weight of sand as 12 kN/m³. Assume that there is no change in the angle of shearing resistance due to submergence.*

Solution. Let

p_1 = Lateral pressure intensity due to surcharge
p_2 = Lateral pressure intensity due to dry soil
p_4 = Lateral pressure intensity due to submerged soil
p_5 = Lateral pressure intensity due to water.

$$\therefore \quad p_1 = K_a \cdot q = \frac{1}{3} \times 36 = 12 \text{ kN/m}^2$$

$$p_2 = K_a \gamma H_1 = \frac{1}{3} \times 18 \times 1.5 = 9 \text{ kN/m}^2$$

$$p_4 = K_a \gamma' H_2 = \frac{1}{3} \times 12 \times 2.5 = 10 \text{ kN/m}^2$$

$$p_5 = \gamma_w H_2 = 9.81 \times 2.5 = 24.53 \text{ kN/m}^2$$

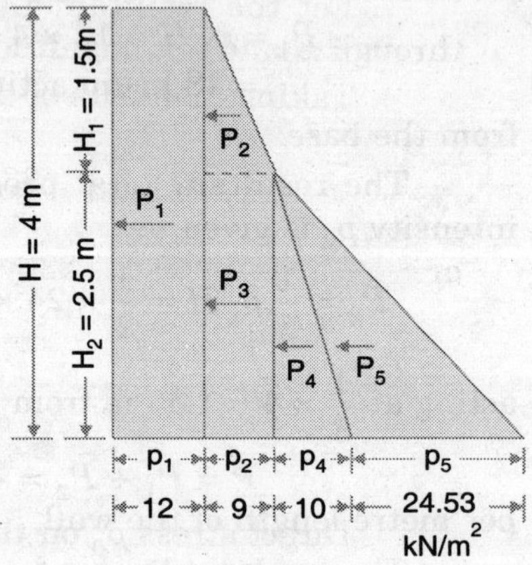

Fig. 18.14

Fig. 18.14 shows the pressure distribution diagram with the resultant pressure P_1, P_2, P_3, P_4 and P_5

$$P_1 = p_1 H = 12 \times 4 = 48 \text{ kN/m}$$

acting @ 4/2 = 2 m from the base.

$$P_2 = \frac{1}{2} p_2 H_1 = \frac{1}{2} \times 9 \times 1.5 = 6.75 \text{ kN/m}$$

acting @ $2.5 + \frac{1.5}{3} = 3$ m from the base

$$P_3 = p_2 H_2 = 9 \times 2.5 = 22.5 \text{ kN/m}$$

acting @ 1.25 m from the base.

$$P_4 = \frac{1}{2} p_4 H_2 = \frac{1}{2} \times 10 \times 2.5 = 12.5 \text{ kN/m acting @ 0.833 m from the base.}$$

$$P_5 = \frac{1}{2} p_5 H_2 = \frac{1}{2} \times 24.53 \times 2.5 = 30.66 \text{ kN/m acting @ 0.833 from the base.}$$

$$\therefore \quad \text{Total pressure } P = P_1 + P_2 + P_3 + P_4 + P_5$$

$$= 48 + 6.75 + 22.5 + 12.5 + 30.66 = \mathbf{120.41 \text{ kN/m}}$$

The distance $\bar{z}$ of the point of application of P above the base is obtained by taking moments about the base.

$$\bar{z} = \frac{1}{120.41} [(48 \times 2) + (6.75 \times 3) + (22.5 \times 1.25) + (12.5 \times 0.833) + (30.66 \times 0.833)] = \mathbf{1.50 \text{ m.}}$$

Example 18.6. *A masonry retaining wall, trapezoidal in section, with one face vertical is 1 m wide at top and 3 m at the base and 8 m high. The material retained on the vertical face exerts a lateral pressure varying from zero at top to 25 kN/m² at the base. If the unit weight of masonry is 21 kN/m³, calculate the maximum and minimum stress intensities induced in the base.*

Solution. Consider 1 m length of the wall (Fig. 18.15)

$$\text{Weight of masonry} \quad W = \left(\frac{3+1}{2}\right) 8 \times 21 = 336 \text{ kN}$$

$$\bar{x} = \frac{(1\times 8\times 0.5)+\left(\frac{1}{2}\times 2\times 8\right)\left(1+\frac{2}{3}\right)}{(1\times 8)+\left(\frac{1}{2}\times 2\times 8\right)} = 1.084 \text{ m}$$

Pressure per metre run = 25 × 1 = 25 kN/m

Total lateral thrust $P = \frac{1}{2} \times 25 \times 8 = 100$ kN

Let the resultant (R) of load W and thrust P pass through E and its magnitude be R. Let $AE = y$ m.

Taking moments of all the forces about E, we have

$$\left(100\times\frac{8}{3}\right) - 336\,(y-\bar{x}) = 0$$

or $$266.7 - 336y + 336\bar{x} = 0$$

From which $$y = 1.88 \text{ m}$$

∴ Eccentricity, $$e = y - \frac{b}{2} = 1.88 - 1.5 = 0.38 \text{ m}$$

Direct stress p_0 on the base $$= \frac{W}{A} = \frac{W}{b} = \frac{336}{3} = 112 \text{ kN/m}^2$$

Moment M on the base $= W.e = 336 \times 0.38 = 127.5$ kN-m

Modulus of section, $$Z = \frac{1}{6} \times 1 \times 3^2 = 1.5 \text{ m}^3$$

Bending stress, $$p_b = \frac{M}{Z} = \frac{127.5}{1.5} = 85 \text{ kN/m}^2$$

∴ $f_{max} = p_0 + p_b = 112 + 85 =$ **197 kN/m²** (comp.) at extreme point D

and $f_{min.} = p_0 - p_b = 112 - 85 =$ **27 kN/m²** (comp.) at extreme point A.

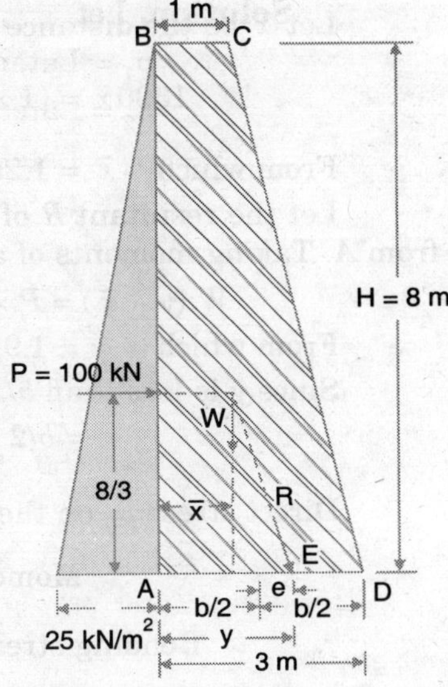

Fig. 18.15

Example 18.7. *A masonry wall, trapezoidal in section with one face vertical is 3 m long, 8 m high, 1 m wide at top and 4 m wide at bottom. In the middle of the length, an inclined thrust of 141.4 kN is transmitted at the top of angle of 45° to the horizontal. Calculate the extreme stress intensities at the base, if the masonry weighs 20 kN/m³.*

Solution. (Fig. 18.16)

Vertical component of inclined thrust

$$W_1 = 141.4 \sin 45° = 100 \text{ kN}$$

Horizontal component of inclined thrust

$$= P = 141.4 \cos 45° = 100 \text{ kN}$$

Weight of masonry

$$= W_2 = 3 \times \frac{1+4}{4} \times 8 \times 20 = 1200 \text{ kN}$$

Total vertical load $$W = W_1 + W_2 = 100 + 1200 = 1300 \text{ kN}$$

Total horizontal load, $P = 100$ kN

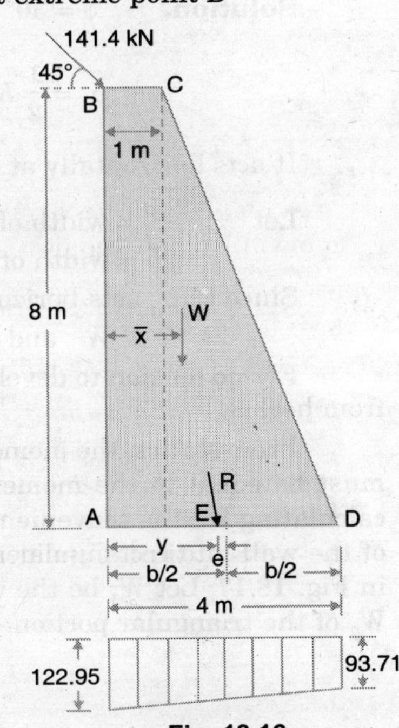

Fig. 18.16

Let $\bar{x}$ be the distance of the total vertical load from vertical face. Then

$$1300\bar{x} = (1 \times 8 \times 3 \times 20 \times 0.5) + \tfrac{1}{2} \times 3 \times 8 \times 3 \times 20 \left(1+\frac{3}{3}\right)$$

From which $\bar{x} = 1.292$ m

Let the resultant R of the two forces W and P meet the base at some point E distant y from A. Taking moments of all the forces about E, we have

$$W(y - \bar{x}) = P \times h \quad \text{or} \quad 1300\,(y - 1.292) = 100 \times 8$$

From which $y = 1.91$ m

Since y is less than $b/2$, the resultant falls to the left of the middle point.

$$\therefore \quad e = b/2 - y = 2 - 1.91 = 0.09 \text{ m}$$

$$\text{Direct stress } p_0 \text{ on the base} = \frac{W}{A} = \frac{1300}{4 \times 3} = 108.33 \text{ kN/m}^2$$

$$\text{Moment } M = W \times e = 1300 \times 0.09 = 117 \text{ kN-m}$$

$$\therefore \quad \text{Bending stress, } p_b = \frac{M}{Z} = \frac{117}{\frac{1}{6} \times 3 \times 4^2} = 14.62 \text{ kN/m}^2$$

$$\therefore \quad f_{max.} = 108.33 + 14.62 = 122.95 \text{ kN/m}^2 \text{ (comp.) at end } A.$$

$$f_{min.} = 108.33 - 14.62 = 93.71 \text{ kN/m}^2 \text{ (comp.) at end } D.$$

Example 18.8. Design of masonry retaining wall

Design a gravity retaining wall, 5 m high with vertical back to retain a dry cohesionless backfill of unit weight 18 kN/m³ and angle of shearing resistance 30°. Find also the factor of safety against sliding assuming the angle of friction between the base of the wall and the foundation soil as 30°. The wall is to be 1 m wide at the top, and to be constructed of brick masonry having unit weight 20 kN/m³. Use Rankine's theory.

Solution. $\phi = 30°$; $\therefore \; K_a = \dfrac{1}{3}$

$$p_a = \frac{1}{2} K_a \gamma H^2 = \frac{1}{2} \times \frac{1}{3} \times 18\,(5)^2 = 75 \text{ kN/m}$$

It acts horizontally at a height of $\frac{5}{3}$ m from the base.

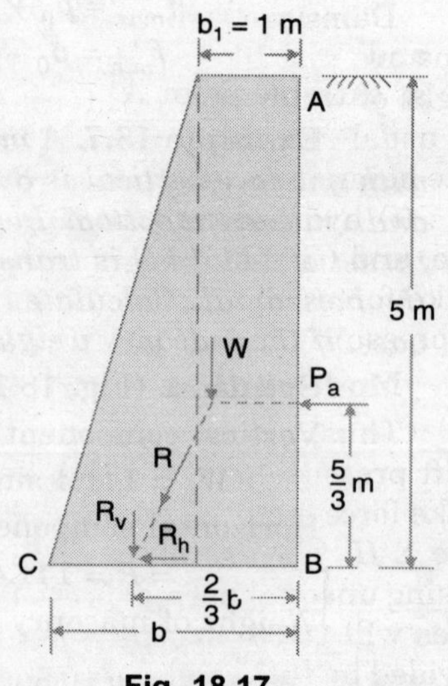

Fig. 18.17

Let b_1 = width of wall at its top

b = width of wall at its bottom

Since p_a acts horizontally, we have

$R_v = W$ and $R_h = p_a$.

For no tension to develop, R_v must act at a distance $2b/3$ from heel B.

From statics, the moment of the forces W and P_a about B must be equal to the moment of the resultant about B. For calculating W, it is convenient to divide the trapezoidal section of the wall into rectangular and triangular portions as shown in Fig. 18.17. Let W_1 be the weight of rectangular portion and W_2 of the triangular portion.

$$\therefore \quad R_v \times \frac{2b}{3} = W_1 \times \frac{b_1}{2} + W_2 \left(b_1 + \frac{b - b_1}{3} \right) + P_a \times \frac{H}{3} \quad ...(18.33)$$

The above equation is valid only for the case when P_a acts horizontally.

Now, $$R_v = W = \frac{b + b_1}{2} H\rho = \frac{b+1}{2} \times 5 \times 20 = 50\,(b+1)$$

$$W_1 = (b_1 \times H)\,\rho = 1 \times 5 \times 20 = 100 \text{ kN}$$

$$W_2 = \tfrac{1}{2}\,(b - b_1)\,H\rho = \tfrac{1}{2}\,(b-1)\,5 \times 20 = 50\,(b-1)$$

$$P_a = 75 \text{ kN.}$$

Substituting these values in Eq. 18.33, we get

$$50\,(b+1) \times \frac{2b}{3} = 100 \times \tfrac{1}{2} + 50\,(b-1)\left(1 + \frac{b-1}{3}\right) + \frac{75 \times 5}{3}$$

or $b^2 + b - 8.5 = 0$ **From which $b = 2.45$ m. Hence adopt $b = 2.5$ m**

$$\therefore \quad W = \frac{b + b_1}{2} H\rho = \frac{2.5 + 1}{2} \times 5 \times 20 = 175 \text{ kN/m}$$

Factor of safety against sliding is given by

$$F = \frac{\mu\, W}{P_a} = \frac{175 \tan \delta'}{75} = \frac{175}{75} \tan 30° = 1.35$$

This is less than the recommended value of 1.5. Hence increase the base width b to 3 m. Then F comes out to be 1.54.

18.5. MASONRY DAMS

Dams are structures constructed to retain water to the required height. In other words, dams are just like retaining walls where the material retained is water. However, since the height of water retained (or stored) is very much higher, dams are much more massive than the usual retaining walls. A *gravity dam* is the one which resists the external forces (major force being the water pressure) by way of its own weight. Major forces acting on a gravity dam are : *(i)* hydrostatic water pressure *(ii)* uplift pressure exerted by water seeping below the base, and *(iii)* self weight of the dam. If the dam is located in earthquake prone zone, earth quake forces should also be considered. Other forces of minor importance are wave pressure, ice pressure and silt pressure which will not be considered here.

Mode of failure: Criteria of safe design

Thus, a gravity dam is subjected to a number of vertical forces (*i.e.*, self weight and uplift pressure) and a number of horizontal forces (*i.e.*, hydrostatic water pressure and earth quake force etc.). These forces can be resolved into a net vertical force $\Sigma\, W$ and a net horizontal force $\Sigma\, H$. The resultant of these two will obviously strike the base at some eccentricity e, causing unequal normal pressure distribution below the base. In addition to this, the horizontal forces will try to overturn the dam, and this tendency is resisted by the restoring moments provided by the weight of the dam. Similarly, the net horizontal force ($\Sigma\, H$) will try to slide the dam along the foundation. Hence the modes of failure of a gravity dam are :

(i) Overturning *(ii)* Sliding

(iii) Compression or crushing *(iv)* Tension

1. Overturning **: The overturning of the dam section takes place when the resultant force at any section cuts the base of the dam downstream of the toe. On the other hand, if the resultant cuts the base within the body of the dam, there will be no overturning.**

For stability requirements, the dam must be safe against overturning. The factor of safety against overturning is defined as the ratio of righting moment to the overturning moments:

$$\text{F.S.} = \frac{\Sigma \text{ Righting moments}}{\Sigma \text{ Overturning moments}} = \frac{\Sigma M_R}{\Sigma M_O} \quad \text{...(18.34)}$$

Since the magnitude and direction of each and every force is known, their moments about the toe can be easily found and F.S. can be computed. The factor of safety against overturning should not be less than 1.5.

2. Sliding **: A dam will fail in sliding at its base, or at any other level, if the horizontal forces causing sliding are more than the resistance available to it at that level. The resistance against sliding may be due to friction alone, or due to friction and shear strength of the joint. Shear strength develops at the base if benched foundations are provided and at other joints if the joints are carefully laid so that a good bond develops. Shear strength also comes into play because of the interlocking of stones in masonry dams.**

If the shear strength is not taken into account, the factor of safety is known as *factor of safety against sliding*. The factor of safety against sliding is defined as the ratio of *actual coefficient of static friction* (μ) on the horizontal joint to the *sliding friction*. The *sliding factor* (S.F.) is the minimum coefficient of friction required to prevent sliding. If ΣH is the net horizontal force causing the sliding and $\Sigma (V - U)$ is the *net vertical force*, the sliding factor (tan θ) is given by :

$$\text{S.F.} = \tan\theta = \frac{\Sigma H}{\Sigma (V - U)} \quad \text{...(18.35)}$$

and the factor of safety against sliding is

$$\text{F.S.} = \frac{\mu}{\tan\theta} = \frac{\mu\, \Sigma (V - U)}{\Sigma H} \quad \text{...(18.36)}$$

The coefficient of friction μ varies from 0.65 to 0.75. The factor of safety against sliding should be greater than 1.

It is considered that a low gravity dam should be safe against sliding, considering friction alone. However, in large dams, shear strength of the joint should also be considered for an economical design. The factor of safety in that case is commonly known as the *shear friction factor* (S.F.F.) and is defined by the equation :

$$\text{S.F.F.} = \frac{\mu\, \Sigma (V - U) + bq}{\Sigma H} \quad \text{...(18.37)}$$

where q = shear strength of the joint (usually 1.4 N/mm²)

b = width of the joint of section

3. Compression or crushing **: In order to calculate the normal stress distribution at the base, or at any section, let ΣH be the total horizontal force and ΣW be the total vertical force ($= \Sigma (V - U)$) and let R be the resultant force, cutting the base at an eccentricity e from the centre of base width b (Fig. 18.18)**

Thus, direct stress $= \dfrac{\Sigma W}{b \times 1}$

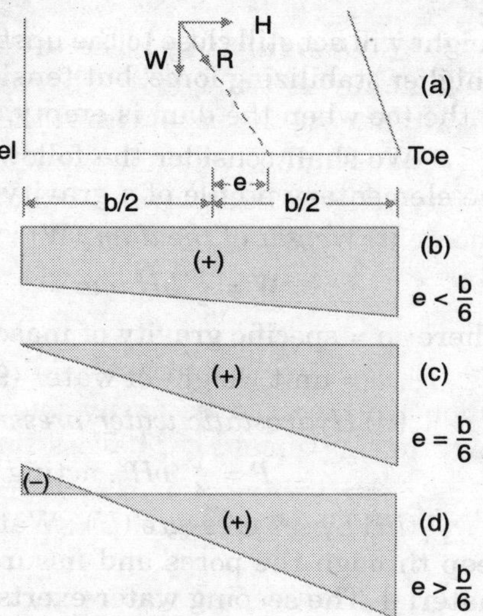

Fig. 18.18. Normal Stress Distribution at the Base

Bending stress

$$= \pm \frac{M}{Z} = \pm \frac{\Sigma W.e}{\frac{1}{6}b^2} = \pm \frac{6\,\Sigma W.e}{b^2}$$

Hence the *total normal stress* p_n is given by

$$p_n = \frac{\Sigma W}{b}\left(1 \pm \frac{6e}{b}\right) \quad \text{...(18.38)}$$

Thus, at toe, $$p_n = \frac{\Sigma W}{b}\left(1 + \frac{6e}{b}\right) \quad \text{...[18.38(a)]}$$

and, at heel, $$p_n = \frac{\Sigma W}{b}\left(1 - \frac{6e}{b}\right) \quad \text{...[18.38(b)]}$$

Fig. 18.18 (*b*) shows the normal stress distribution in general case when the pressures at both toe and heel are compressive. Evidently the maximum compressive stress occurs at the toe and for safety this should not be greater than the allowable compressive stress σ_c for the foundation material. Hence from *strength* point of view.

$$\frac{\Sigma W}{b}\left(1 + \frac{6e}{b}\right) \le \sigma_c \quad \text{...(18.39)}$$

When the eccentricity *e* is equal to *b*/6, we get

$$(p_n)_{toe} = \frac{\Sigma W}{b}\left(1 + \frac{6}{b} \times \frac{b}{6}\right) = \frac{2\,\Sigma W}{b} \quad \text{...(18.40)}$$

The corresponding stress at the heel in that circumstance will evidently be zero. Fig. 18.18 (*c*) shows the pressure distribution for this case.

4. Tension : The normal stress at the heel is given by

$$(p_n)_{heel} = \frac{\Sigma W}{b}\left(1 - \frac{6e}{b}\right) \quad \text{...[Eq. 18.38 (b)]}$$

It is evident that if $e > b/6$, the normal stress at the heel will be –ve or tensile. No tension should be permitted at any point of the dam under any circumstance. For no tension to develop, the eccentricity should be less than *b*/6. In other words, the resultant should always lie within the *middle third*.

Elementary profile of a gravity dam : Triangular profile

In the absence of any force other than the forces due to water, an elementary profile will be triangular in section, having zero width at the water level, where water pressure is zero, and a maximum base width *b*, where the maximum water pressure acts. Thus, the section of the elementary profile is of the same shape as the hydrostatic pressure distribution diagram. For the reservoir empty condition, a right angled triangular profile as shown in Fig. 18.19 will provide the maximum possible stabilizing force against overturning, without causing tension in the base. This is so because the weight of the dam acts at a distance of *b*/3 from the upstream face and is close to it. If any triangular profile, other than the right angled one, is provided, its weight will act still close to the upstream face to provide a higher stabilizing force, but tension will be developed at the toe when the dam is empty.

We shall consider the following forces acting on the elementary profile of a gravity dam.

(*i*) *Weight of the dam* (*W*) :

$$W = \frac{1}{2}\, bH\, \rho\, w$$

where ρ = specific gravity of masonry and

w = unit weight of water (9.81 kN/m³)

(*ii*) *Hydrostatic water pressures* (*P*)

$$P = \frac{1}{2}\, wH^2, \text{ acting } \frac{1}{3}\, H \text{ above the base.}$$

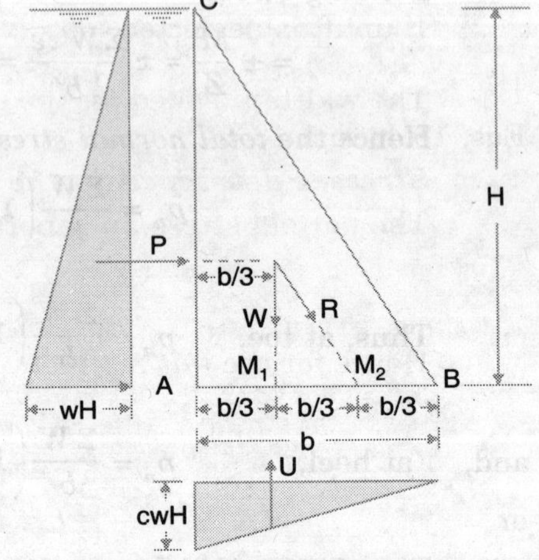

Fig. 18.19. Elementary Profile

(*iii*) *Uplift pressure* (*U*) : Water has a tendency to seep through the pores and fissures of the foundation material. The seeping water exerts pressure, known as uplift pressure, which must be accounted for in stability calculations. A portion of the weight of the dam will be supported on the upward pressure of water; hence net foundation reaction due to vertical force will reduce. The uplift pressure distribution, in absence of tail water, is shown in Fig. 18.19.

$$\therefore \qquad U = \frac{1}{2} \times cwb\text{H} \qquad \text{...(18.41)}$$

where c = uplift pressure intensity coefficient.

Base width of the elementary profile : The base width of the elementary profile is found under two criteria : (*i*) stress criterion and (*ii*) stability criterion.

(i) Stress criterion **:** When reservoir is empty the resultant should act at the inner third point (M_1) so that no tension is developed. For the reservoir full condition for no tension to develop, the resultant R must pass through the outer third point (M_2) shown in Fig. 18.19.

Taking the moment of all forces about M_2 and equating it to zero (since the moment of R about M_2 is zero), we get

$$\frac{1}{2}\, wH^2 \cdot \frac{H}{3} + \frac{1}{2}\, c\, w\, b\, H \,.\, \frac{b}{3} - \frac{1}{2}\, bH\, \rho\, w. \,\frac{b}{3} = 0$$

Multiplying all the terms by $6/wH$, we get :

$$H^2 + cb^2 - b^2 \rho = 0 \quad \text{or} \quad b^2\,(\rho - c) = H^2$$

Hence
$$b = \frac{H}{\sqrt{\rho - c}} \qquad \text{...(18.42)}$$

If uplift is not considered, $c = 0$

$$\therefore \qquad b = \frac{H}{\sqrt{\rho}} \qquad \text{...[18.42 (a)]}$$

(ii) Stability or sliding criterion **:** For no sliding to occur, horizontal force causing sliding should be balanced by the frictional forces opposing the same.

Hence
$$P = \mu\,(W - U)$$

$$\therefore \qquad \frac{1}{2}\, wH^2 = \mu \left(\frac{1}{2}\, bH\, \rho\, w - \frac{1}{2}\, cb\, wH\right)$$

From which
$$b = \frac{H}{\mu\,(\rho - c)} \qquad \text{...(18.43)}$$

If uplift is neglected, $b = \dfrac{H}{\mu\rho}$ ...[18.43 (a)]

The width provided for the elementary profile should be greater of the widths given by Eqs. 18.42 and 18.43.

Stresses developed in the elementary profile **:**

The normal stress, in general, is given by

$$p_n = \frac{\Sigma W}{b}\left(1 \pm \frac{6e}{b}\right), \quad \text{where } \Sigma W = (W - U) \text{ in this case and } e = b/6$$

Hence for the *full reservoir case*, we have

$$(p_n)_{toe} = \frac{W-U}{b}(1+1) = \frac{2(W-U)}{b}$$

or
$$(p_n)_{toe} = \frac{2}{b}\left[\tfrac{1}{2}bH\,\rho\,w - \tfrac{1}{2}cbwH\right] = wH\,(\rho - c) \qquad ...(19.44)$$

Also,
$$(p_n)_{heel} = \frac{(W-U)}{b}(1-1) = 0$$

Reservoir empty condition **:** When the reservoir is empty, the only force acting on the elementary profile will be its weight, acting through the first third point M_1. Hence, the maximum compressive stress at the heel will be $= \dfrac{W}{b}(1+1) = \dfrac{2W}{b}$, and the corresponding normal stress at the toe will be $\dfrac{W}{b}$ (1 – 1) = zero.

Rectangular profile

Consider a rectangular dam of width b and height H.

$$W = bH\,\rho\,w\ ; \quad P = \frac{wH^2}{2}\ ; \quad U = \tfrac{1}{2}\,cwbH.$$

Let the resultant pass through a point M, distant $\bar{x}$ from the toe B. Taking moments about the toe, we have.

$$\Sigma M = W\,.\,\frac{b}{2} - \frac{PH}{3} - U\,.\,\frac{2b}{3}$$

Also
$$\Sigma W = W - U$$

$$\therefore \quad \bar{x} = \frac{\Sigma M}{\Sigma W} = \frac{W\dfrac{b}{2} - \dfrac{PH}{3} - \dfrac{U\,.\,2b}{3}}{W - U}$$

Fig. 18.20. Rectangular Profile

Substituting the values of W, P and U, we get,

$$\bar{x} = \frac{bH\rho w\,.\,\dfrac{b}{2} - \dfrac{wH^3}{6} - \dfrac{2}{3}b\left(\tfrac{1}{2}\,cwbH\right)}{bH\rho w - \tfrac{1}{2}\,cwbH}$$

or
$$\bar{x} = \frac{\dfrac{b}{2}\left(bH\rho w - \tfrac{1}{2}\,cwbH\right) - \dfrac{wH^3}{6} - \left(\tfrac{1}{2}\,cwbH\right)\dfrac{b}{6}}{bH\rho w - \tfrac{1}{2}\,cwbH} = \frac{b}{2} - \frac{\dfrac{wH^3}{6} + \dfrac{1}{2}cwbH\,.\,\dfrac{b}{6}}{bH\rho w - \tfrac{1}{2}\,cwbH}$$

Hence $$e = \frac{b}{2} - \bar{x} = \frac{\frac{wH^3}{6} + \frac{1}{2} cwbH \cdot \frac{b}{6}}{bH\rho w - \frac{1}{2} cwbH} = \frac{\frac{H^2}{6} + \frac{cb^2}{12}}{b\rho - \frac{1}{2} cb}$$

For no tension to develop

$$e = \frac{b}{6} = \frac{\frac{H^2}{6} + \frac{cb^2}{12}}{b\,\rho - \frac{1}{2} cb}$$

or $$\frac{b^2}{6}\left(\rho - \frac{c}{2}\right) = \frac{H^2}{6} + \frac{cb^2}{12} \quad \text{or} \quad b^2 (\rho - c) = H^2$$

From which $$b = \frac{H}{\sqrt{\rho - c}} \qquad \text{...(18.45)}$$

If $c = 0$ (no uplift), $$b = \frac{H}{\sqrt{\rho}} \qquad \text{...[18.45 (a)]}$$

Also, the limiting height is given by, $H_{\lim} = b\sqrt{\rho - c}$ ...[18.45 (*b*)]

From stability considerations, providing a F.S. of 1.5 against sliding, we have

$$\mu (W - U) = 1.5\, P$$

or $$\mu (b\, H\rho w - \tfrac{1}{2} cwbH) = \frac{3}{2} \times \frac{wH^2}{2} \quad \text{or} \quad \mu b\left(\rho - \frac{c}{2}\right) = \frac{3}{4} H$$

From which $$b = \frac{3}{4} \frac{H}{\mu\left(\rho - \frac{c}{2}\right)} \qquad \text{...(18.46)}$$

If $c = 0$ (*i.e.*, no uplift) $$b = \frac{3}{4} \frac{H}{\mu\,\rho} \qquad \text{...[18.46 (a)]}$$

Also, limiting height is, $$H_{\lim} = \frac{4}{3}\, b\, \mu \left(\rho - \frac{c}{2}\right) \qquad \text{...(18.47)}$$

Neglecting uplift, $$H_{\lim} = \frac{4}{3}\, b \mu\, \rho \qquad \text{...[18.47 (a)]}$$

From *crushing point of view,* when $e = b/6$, we get

$$p_n = \frac{2(W - U)}{b} = \frac{2}{b}\left(bH\rho w - \tfrac{1}{2} cwbH\right)$$

$$\therefore \quad \sigma_c = p_n = 2\, wH\left(\rho - \tfrac{1}{2} c\right)$$

$$\therefore \quad H_{\lim} = \frac{\sigma_c}{2\, w(\rho - \frac{1}{2} c)} \qquad \text{...(18.48)}$$

If uplift is neglected, $$H_{\lim} = \frac{\sigma_c}{2\, w\, \rho} \qquad \text{...[18.48 (a)]}$$

Trapezoidal profile : Consider a trapezoidal profile with top width a and bottom width b, with upstream face vertical.

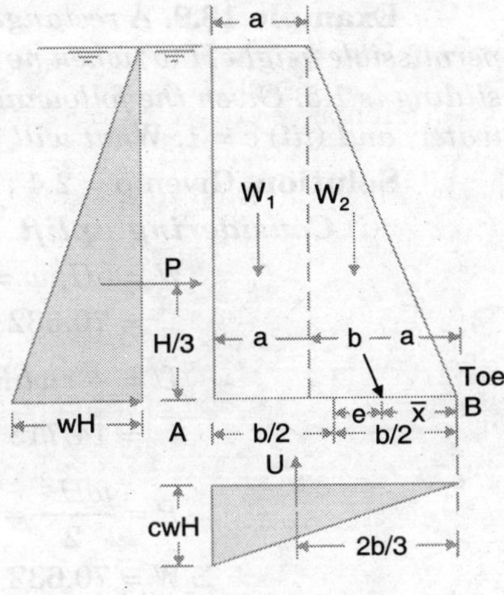

Fig. 18.21

W_1 = weight of rectangular portion = $aH\rho w$

W_2 = weight of triangular portion = $\frac{1}{2}$ $bH.\ \rho\ w$

$$P = \text{water pressure} = \frac{wH^2}{2}$$

$$C = \text{uplift pressure} = \tfrac{1}{2}\ cwbH$$

Let the resultant fall at $\bar{x}$ from the toe.

If ΣM is the sum of moments of all the forces about the toe, we have

$$\Sigma M = \bar{x}\ .\ \Sigma W \quad \text{or} \quad \bar{x} = \frac{\Sigma M}{\Sigma W}$$

$$\therefore \qquad \bar{x} = \frac{aH\rho w\left(b - \frac{a}{2}\right) + \frac{1}{2}\, bH\rho w \times \frac{2}{3}\,(b-a) - \frac{wH^3}{6} - \frac{1}{2}\, cwbH \times \frac{2}{3}\, b}{aH\rho w + \frac{1}{2}\, bH\rho w - \frac{1}{2}\, cwbH}$$

or
$$\bar{x} = \frac{a\rho\left(b - \frac{a}{2}\right) + \frac{1}{3}\, b\,\rho\,(b-a) - \frac{H^2}{6} - \frac{1}{3}\, cb^2}{a\,\rho + \frac{1}{2}\,\rho\, b - \frac{1}{2}\, cb} \qquad \text{...(18.49)}$$

$$\text{Eccentricity} \quad e = \frac{b}{2} - \bar{x} \quad \text{and} \quad p_n = \frac{\Sigma W}{b}\left(1 \pm \frac{6e}{b}\right).$$

Practical profile of a gravity dam

The elementary profile (*i.e.*, triangular profile) of a gravity dam is only a *theoretical profile.* However, such a profile is not possible in practice because of provision of (*i*) roadway at top, (*ii*) additional loads due to road way and (*iii*) free board. Free board is the margin provided between the top of the dam and the high flood level (H.F.L.) in the reservoir to prevent the splashing of the waves over the non-overflow section. Due to these provisions, the resultant forces of the weight of the dam and the water pressure falls outside the middle third of the base of the dam when the reservoir is full. To eliminate tension, some masonry is to be provided to the upstream side. Fig. 18.22 shows the dimensions of the practical profile of a gravity dam. The free board usually is kept equal to $1.5h_w$, where h_w is the height of waves. The top width equal to about 14% of h has been found to be economical.

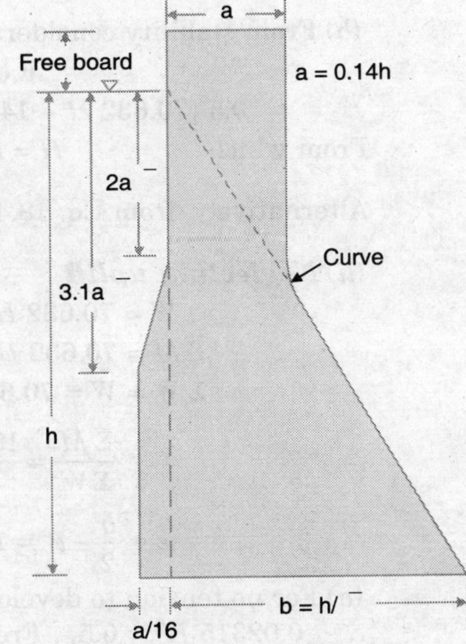

Fig. 18.22. Practical Profile of a Low Gravity Dam

Example 18.9. *A rectangular masonry dam is 3 m at the base. Compute the maximum permissible height H (a) when no tension is permissible, and (b) when the factor of safety against sliding is 1.5. Given the following : (i) μ = 0.5, (ii) density of masonry = 2.4 times the density of water, and (iii) c = 1. What will be the corresponding values of H if uplift is neglected ?*

Solution. Given $\rho = 2.4$; $w = 9.81$ kN/m^3

(i) ***Considering uplift***

$$W = bH\rho w = 3 \times 2.4 \times 9.81\, H$$

$$= 70.632\, H \text{ kN}$$

$$U = \tfrac{1}{2}\, cwbH = \tfrac{1}{2} \times 1 \times 9.81 \times 3\, H$$

$$= 14.715\, H \text{ kN}$$

$$P = \frac{wH^2}{2} = \frac{9.81\, H^2}{2} = 4.905\, H^2 \text{ kN}$$

$$\Sigma W = 70.632\, H - 14.715\, H = 55.917\, H$$

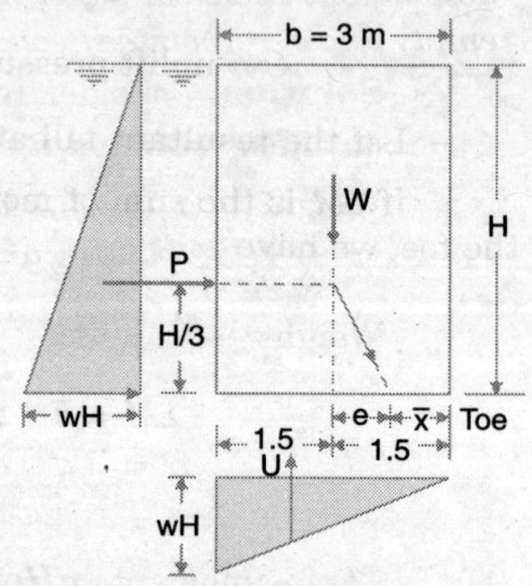

Fig. 18.23

Taking moments about the toe,

$$\Sigma M = (70.632\, H \times 1.5) - (14.715\, H \times 2) - 4.905\, H^3/3$$

$$= 76.518\, H - 1.635\, H^3$$

$$\therefore \quad \bar{x} = \frac{\Sigma M}{\Sigma W} = \frac{76.518\, H - 1.635\, H^3}{55.917\, H} = 1.368 - 0.02924\, H^2$$

(a) For no tension to develop, $e = b/6 = 3/6 = 0.5$ m

$$\therefore \quad e = 1.5 - \bar{x} = 0.132 + 0.02924\, H^2 = 0.5$$

From which , $H_{\text{lim}} \approx 3.55$ m

Alternatively, from Eq. 18.45 (b), $H_{\text{lim}} = b\sqrt{\rho - c} = 3\sqrt{2.4 - 1} = \mathbf{3.55\ m}$

(b) From stability consideration, providing a F.S. of 1.5 against sliding, we have

$$\mu\,(W - U) = 1.5\, P$$

$$\therefore \quad 0.5\,(70.632\, H - 14.715\, H) = 1.5 \times 4.905\, H^2 \quad \text{or} \quad 27.9585 = 7.3575\, H$$

From which $\quad H = H_{\text{lim}} = \mathbf{3.8\ m}$

Alternatively, from Eq. 18.47, $H_{\text{lim}} = \dfrac{4}{3}\, b\,\mu\left(\rho - \dfrac{c}{2}\right) = \dfrac{4}{3} \times 3 \times 0.5\,(2.4 - 0.5) = 3.8$ m.

(ii) ***Neglecting uplift***

$W = 70.632\, H$ and $P = 4.905\, H^2$, as before

$$\Sigma M = 70.632\, H \times 1.5 - 4.905\, H^3/3 = 105.948\, H - 1.635\, H^3.$$

$$\Sigma W = W = 70.632\, H$$

$$\therefore \quad \bar{x} = \frac{\Sigma M}{\Sigma W} = \frac{105.948\, H - 1.635\, H^3}{70.632\, H} = 1.5 - 0.02315\, H^3$$

$$\therefore \quad e = \frac{b}{2} - \bar{x} = 1.5 - (1.5 - 0.02315\, H^2) = 0.02315\, H^2$$

(a) For no tension to develop, $e = b/6 = 3/6 = 0.5$ m

$\therefore \quad 0.02315\, H^2 = 0.5$, From which $H = H_{\text{lim}} \approx \mathbf{4.65\ m}$

(*b*) From stability consideration, providing a F.S. of 1.5 against sliding,

$$\mu\, W = 1.5\, P$$

or $\quad 0.5 \times 70.632\, H = 1.5 \times 4.905\, H^2,\quad$ From which H = **4.8 m.**

Example 18.10. *A masonry dam 6 m high is 1.5 m wide at top and 4.5 m wide at the bottom, with vertical water face. Determine the normal stresses at the toe and heel for reservoir empty and reservoir full conditions. Take ρ = 2.4 and c = 1.*

Solution. Consider 1 m length of the dam.

(*a*) ***Reservoir empty :***

Weight $\quad W_1 = 1.5 \times 6 \times 1\,(2.4 \times 9.81) = 211.9$ kN

Weight $\quad W_2 = \frac{1}{2} \times 3 \times 6\,(2.4 \times 9.81) = 211.9$ kN

$$\Sigma W = W_1 + W_2 = 211.9 + 211.9 = 423.8 \text{ kN}$$

Taking moments about the toe (*B*) we have

$$\bar{x} = \frac{\Sigma M}{\Sigma W} = \frac{211.9 \times 3.75 + 211.9 \times 2}{423.8} = 2.875 \text{ m}$$

$$\therefore \quad e = \frac{b}{2} - \bar{x} = 2.25 - 2.875 = -0.625 \text{ m}$$

(*i.e.*, resultant falls to the left of middle point.)

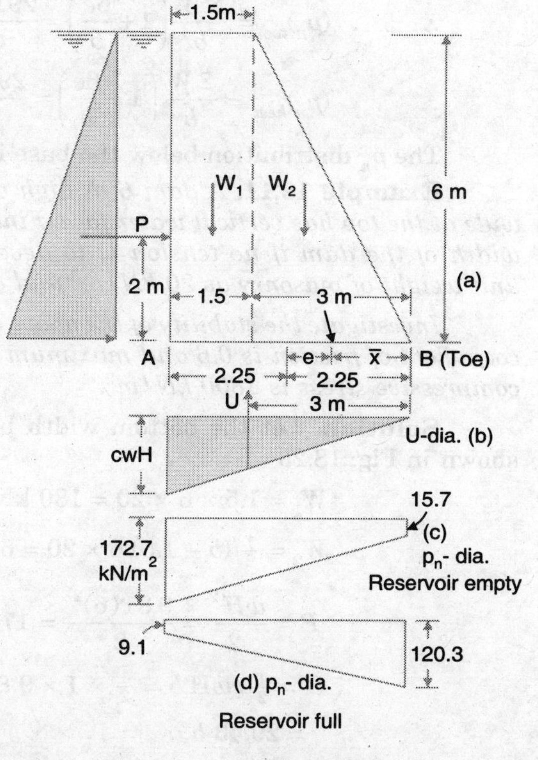

Fig. 18.24

$$\therefore \quad (p_n)_{toe} = \frac{\Sigma W}{b}\left(1 - \frac{6e}{b}\right) = \frac{423.8}{4.5}\left(1 - \frac{6 \times 0.625}{4.5}\right) = 15.7 \text{ kN/m}^2 \text{ (comp.)}$$

$$\therefore \quad (p_n)_{heel} = \frac{\Sigma W}{b}\left(1 + \frac{6e}{b}\right) = \frac{423.8}{4.5}\left(1 + \frac{6 \times 0.625}{4.5}\right) = 172.7 \text{ kN/m}^2 \text{ (comp.)}$$

The p_n distribution below the base is shown in Fig. 18.24 (*c*).

(*b*) ***Reservoir full :*** $\quad W_1 = 211.9$ kN and $W_2 = 211.9$ kN, as before.

$$P = \frac{wH^2}{2} = \frac{9.81\,(6)^2}{2} = 176.58 \text{ kN}$$

$$U = \tfrac{1}{2}\, cw\, H.b = \tfrac{1}{2} \times 1 \times 9.81 \times 6 \times 4.5 = 132.435 \text{ kN.}$$

$$\Sigma W = 211.9 + 211.9 - 132.435 = 291.365 \text{ kN}$$

Taking moments about the toe, we have

$$\bar{x} = \frac{\Sigma M}{\Sigma W} = \frac{211.9 \times 3.75 + 211.9 \times 2 - 132.453 \times 3 - 176.58 \times 2}{291.365} = 1.606 \text{ m}$$

$$\therefore \quad e = \frac{b}{2} - \bar{x} = 2.25 - 1.606 = 0.644 \text{ m}$$

$$\therefore \quad (p_n)_{toe} = \frac{\Sigma W}{b}\left(1 + \frac{6e}{b}\right) = \frac{291.365}{4.5}\left(1 + \frac{6 \times 0.644}{4.5}\right) = 120.3 \text{ kN/m}^2 \text{ (comp.)}$$

$$\therefore \quad (p_n)_{heel} = \frac{\Sigma W}{b}\left(1 - \frac{6e}{b}\right) = \frac{291.365}{4.5}\left(1 - \frac{6 \times 0.644}{4.5}\right) = 9.1 \text{ kN/m}^2 \text{ (comp.)}$$

The p_n distribution below the base is shown in Fig. 18.24 (*d*).

Example 18.11. *A dam 6 m high and 1.5 m wide at the top has vertical water face. Find the base width of the dam if no tension is to develop. Take unit weight of masonry as 20 kN/m³ and c = 1.*

Investigate the stability of the above dam if the coefficient of friction is 0.6 and maximum allowable compressive stress is 1800 kN/m².

Solution. Let the bottom width be *b* m, as shown in Fig. 18.25

$$W_1 = 1.5 \times 6 \times 20 = 180 \text{ kN}$$

$$W_2 = \tfrac{1}{2}\,(b - 1.5)\, 6 \times 20 = 60\,(b - 1.5)$$

$$P = \frac{wH^2}{2} = \frac{9.81\,(6)^2}{2} = 176.58 \text{ kN}$$

$$U = \tfrac{1}{2}\, cw\text{H}.b = \tfrac{1}{2} \times 1 \times 9.81 \times 6\, b$$

$$= 29.43\, b.$$

$$\Sigma W = 180 + 60\,(b - 1.5) - 29.43\, b$$

$$= 90 + 30.57\, b$$

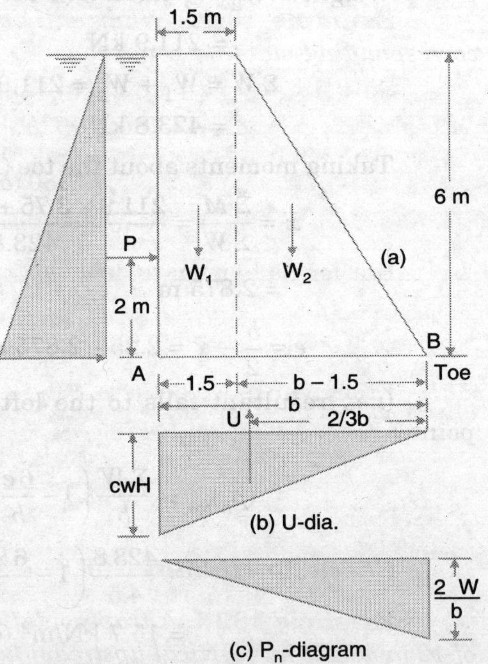

Fig. 18.25

Taking moments about the toe, we get

$$\bar{x} = \frac{\Sigma M}{\Sigma W} = \frac{180\,(b - 1.5 + 0.75) + 60(b - 1.5)\,\frac{2}{3}\,(b - 1.5) - 176.58 \times 2 - 29.43.\, b\, \frac{2}{3}\, b}{90 + 30.57\, b}$$

For no tension to develop, *e* should be equal to *b*/6

$$\therefore \quad \frac{b}{2} - \bar{x} = \frac{b}{6} \quad \text{or} \quad \bar{x} = \frac{b}{3}$$

Hence,
$$\frac{b}{3} = \frac{180\,(b - 0.75) + 40\,(b - 1.5)^2 - 353.16 - 19.62\, b^2}{90 + 30.57\, b}$$

From which, we get $b^2 - 2.944\, b - 39.074 = 0$

Hence $\qquad b \approx \mathbf{4.95\ m}$

Check for crushing

$$\Sigma W = 90 + 30.57\,b = 90 + 30.57 \times 4.95 = 241.322 \text{ kN}$$

$$(p_n)_{max.} = \frac{2\,\Sigma W}{b} = \frac{2(241.322)}{4.95} = 97.5 \text{ kN/m}^2 < 1800. \text{ OK.}$$

Check for sliding :

$$\text{F.S.} = \frac{\mu\,\Sigma W}{b} = \frac{0.6 \times 241.322}{176.58} = 0.82$$

Hence the dam is *unsafe* against sliding.

If a minimum factor of safety of 1 is required under the conditions of uplift, we have

$$\mu\,\Sigma W = P$$

$\therefore$ $0.6\,(90 + 30.57\,b) = 176.58$, From which $b = \mathbf{6.683\ m.}$

Example 18.12. *What will the base width if uplift is neglected ? Investigate the corresponding stability conditions.*

Solution. $W_1 = 180$ kN ; $W_2 = 60\,(b - 1.5)$; $P = 176.58$ kN

$$\Sigma W = 180 + 60\,(b - 1.5) = 90 + 60\,b$$

$$\bar{x} = \frac{\Sigma M}{\Sigma W} = \frac{180\,(b - 1.5 + 0.75) + 60\,(b - 1.5)\dfrac{2}{3}(b - 1.5) - 176.58 \times 2}{90 + 60\,b}$$

But for no tension to develop, $\bar{x} = b/3$

$$\frac{b}{3} = \frac{180\,(b - 0.75) + 40\,(b - 1.5)^2 - 353.16}{90 + 60\,b}$$

From which, we get $b^2 + 1.5\,b - 19.908 = 0$. Hence $b = \mathbf{3.774\ m.}$

Check for crushing : $\Sigma W = 90 + 60\,b = 90 + 60 \times 3.774 = 316.44$ kN

$$(p_n)_{max.} = \frac{2\,\Sigma W}{b} = \frac{2 \times 316.44}{3.774} = 167.7 \text{ kN/m}^2 < 1800. \text{ Hence safe.}$$

Check for sliding $$\text{F.S.} = \frac{\mu\,\Sigma W}{P} = \frac{0.6 \times 316.44}{176.58} = 1.076.$$

Example 18.13. *A trapezoidal dam section of 12 m height has vertical upstream face, with 3 m top width and 9 m bottom width. In order to increase the storage, it is proposed to raise the water height by erecting steel shuttering at the top of the dam, in line with the vertical face of the dam. Determine the maximum permissible height of shuttering so that no tension is developed at the upstream end of the base, under the reservoir full condition. Neglect foundation seepage. Take unit weight of masonry as 21 kN/m³.*

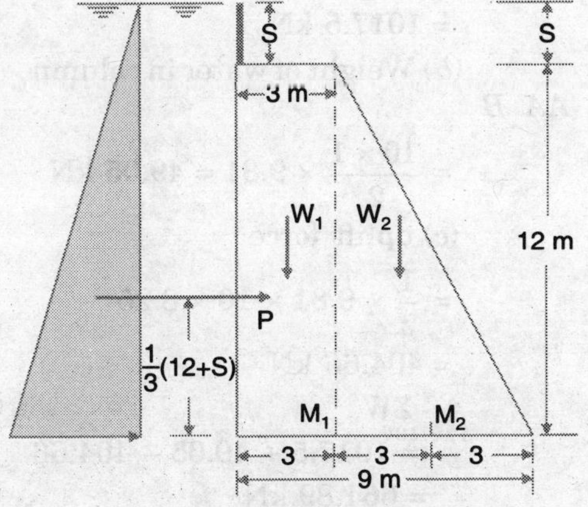

Fig. 18.26

Solution. (Fig. 18.26)

Let the height of shutters be S. In order that no tension is developed, the resultant must pass through the outer third point M_2, distant 3 m from the toe. Hence ΣM at point M_2 must be zero.

Now $W_1 = 3 \times 12 \times 21 = 756$ kN acting at 4.5 m from M_2

$W_2 = \frac{1}{2} \times 6 \times 12 \times 21 = 756$ kN acting at 1 m from M_2.

$$P = \frac{wH^2}{2} = \frac{9.81(12+S)^2}{2} = 4.905\,(12+S)^2 \text{ kN}$$

acting at $\frac{1}{3}(12 + S)$ m above the base.

Taking moments about M_2

$$\Sigma M = 756 \times 4.5 + 756 \times 1 - 4.905\,(12+S)^2 \times \tfrac{1}{3}\,(12+S) = 0$$

or $(12 + S)^3 = 2543$ or $(12 + S) = 2543^{\frac{1}{3}} = 13.65$

$\therefore$ $S = \mathbf{1.65\ m}$

Hence height of shuttering should not exceed 1.65 m.

Example 18.14. *A masonry dam 10 m high is trapezoidal in section with a top width of 1 m and bottom width of 8.25 m. The face exposed to water has a batter of 1:10. Test the stability of the dam. Find out the normal stresses at the toe and heel of the dam. Assume unit weight of masonry as 22 kN/m³ and permissible shear stress of joint = 1200 kN/m². Take c = 1 and μ = 0.75.*

Solution. Consider 1 m length of the dam. Assuming no free board, the dam is tested by considering only three forces, (*i*) Weight of dam (*ii*) Hydrostatic water pressure and (*iii*) uplift pressure

(*i*) ***Vertical forces :***

(*a*) Self weight of dam

$$= \frac{8.25 + 1}{2} \times 10 \times 1 \times 22$$

$= 1017.5$ kN

(*b*) Weight of water in column *AA′ B*

$$= \frac{10 \times 1}{2} \times 9.81 = 49.05 \text{ kN}$$

(*c*) uplift force

$$= \frac{1}{2} \times 9.81 \times 10 \times 8.25$$

$= 404.66$ kN

$\therefore$ ΣW

$= 1017.5 + 49.05 - 404.66$

$= 661.89$ kN

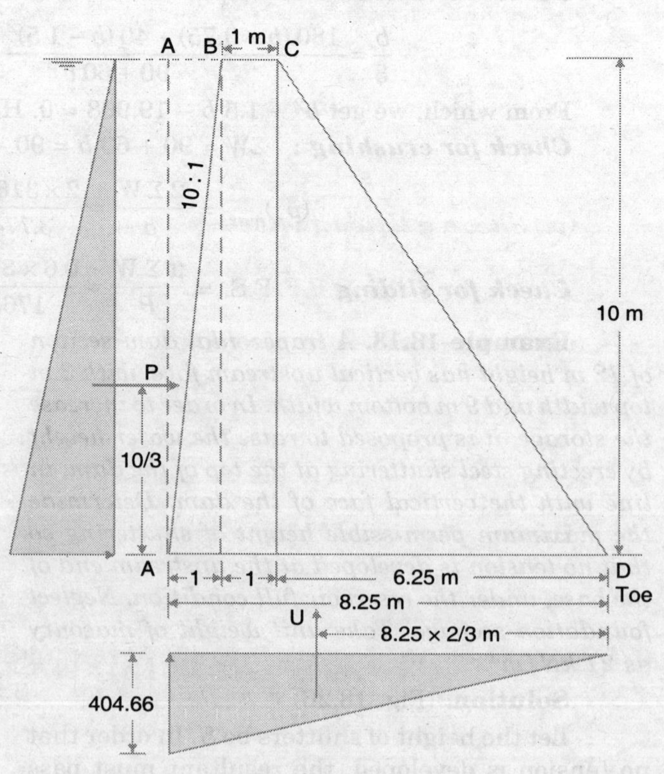

Fig. 18.27

(ii) ***Horizontal force***

$$P = \frac{wH^2}{2} = \frac{9.81\,(10)^2}{2} = 490.5 \text{ kN.}$$

(iii) ***Moment of forces about toe (D)***

(a) Due to self weight of the dam :

$$\text{Moment} = \left\{\left(\tfrac{1}{2} \times 1 \times 10 \times 22\right)\left(7.25 + \tfrac{1}{3}\right)\right\} + \{(1 \times 10 \times 22)\,(6.25 + 0.5)\}$$

$$+ \left\{\left(\tfrac{1}{2} \times 6.25 \times 10 \times 22\right)\left(\tfrac{2}{3} \times 6.25\right)\right\} = 5183.7 \text{ kN-m}$$

(b) Due to column of water $AA'B$:

$$\text{Moment} = \left(\tfrac{1}{2} \times 10 \times 1 \times 9.81\right)\left(8.25 - \tfrac{1}{3}\right) = 388.3 \text{ kN-m}$$

(c) Due to uplift pressure :

$$\text{Moment} = 404.66 \times \frac{2}{3} \times 8.25 = 2225.6 \text{ kN-m}$$

(d) Due to horizontal water pressure :

$$\text{Moment} = 490.5 \times \frac{10}{3} = 1635 \text{ kN-m}$$

$$\therefore \quad \Sigma M = 5183.7 + 388.3 - 2225.6 - 1635 = 1711.4 \text{ kN}$$

(iv) ***F.S. against overturning***

$$\text{F.S.} = \frac{\Sigma M_R}{\Sigma M_O} = \frac{5183.7 + 388.3}{2225.6 + 1635} = 1.44$$

(v) ***F.S. against sliding***

$$\text{F.S.} = \frac{\mu\, \Sigma W}{\Sigma H} = \frac{0.75 \times 661.89}{490.5} = 1.012 > 1$$

(vi) ***Shear friction factor***

$$\text{S.F.F.} = \frac{\mu\, \Sigma W + bq}{\Sigma H} = \frac{0.75 \times 661.89 + 8.25 \times 1200}{490.5} = 21.2$$

(vii) ***Normal stress at the base***

$$\bar{x} = \frac{\Sigma M}{\Sigma W} = \frac{1711.4}{661.89} = 2.586 \text{ m}$$

$$\therefore \quad e = \frac{b}{2} - \bar{x} = \frac{8.25}{2} - 2.586 = 1.539 \text{ m}; \quad \frac{b}{6} = \frac{8.26}{6} = 1.375$$

Hence $e > b/6$ which is not desirable.

$$\therefore \quad (p_n)_{toe} = \frac{\Sigma W}{b}\left(1 + \frac{6e}{b}\right) = \frac{661.89}{8.25}\left(1 + \frac{6 \times 1.539}{8.25}\right) = 170 \text{ kN/m}^2$$

$$(p_n)_{heel} = \frac{\Sigma W}{b}\left(1 - \frac{6e}{b}\right) = \frac{661.89}{8.25}\left(1 - \frac{6 \times 1.539}{8.25}\right) = -\,9.6 \text{ kN/m}^2$$

Hence tension is developed at the heel.

Example 18.15. *Design the practical profile of a gravity dam of stone masonry, given the following data :*

R.L. of base of dam	*= 1450 m*
R.L. of H.F.L.	*= 1480.5 m*
Specific gravity of masonry	*= 2.4*
Height of waves	*= 1 m.*

Solution. Height of waves, $h_w = 1$ m. $\therefore$ F.B. $= 1.5\, h_w = 1.5$ m

$\therefore$ R.L. to top of dam = 1480.5 + 1.5 = 1482 m

Hence height of dam = 1482 – 1450 = 32 m

The practical profile of the dam can be designed with respect to the recommendations marked in Fig. 18.22.

Depth of water, $h = 1480.5 - 1450 = 30.5$ m

Top width a can be determined on the following criteria :

(*i*) a = 14% of H, for economy

$= 0.14 \times 32 \approx 4.5$ m

(*ii*) a = width of roadway, if any.

However, keep $a = 4.5$ m

Base width of elementary profile

$$= \frac{h}{\sqrt{\rho}} = \frac{30.5}{\sqrt{2.4}} = 19.7 \text{ m}$$

$$\text{Upstream offset} = \frac{a}{16} = \frac{4.5}{16} \approx 0.3 \text{ m}$$

$\therefore$ Total base width $= 19.7 + 0.3$

$= 20$ m

Distance upto which the u/s slope is vertical, from the u/s water level

$$= 2a\sqrt{\rho} = 2 \times 4.5\sqrt{2.4} = 14 \text{ m}$$

Distance upto which the u/s slope is inclined, from the u/s water level

$$= 3.1\, a\sqrt{\rho} = 3.1 \times 4.5\sqrt{2.4} = 21.6 \text{ m}$$

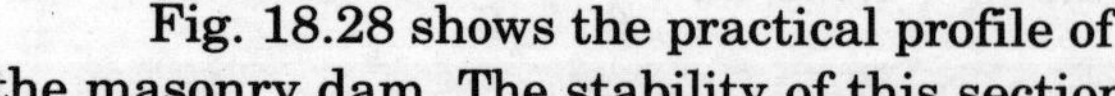

Fig. 18.28 shows the practical profile of the masonry dam. The stability of this section can now be tested as usual.

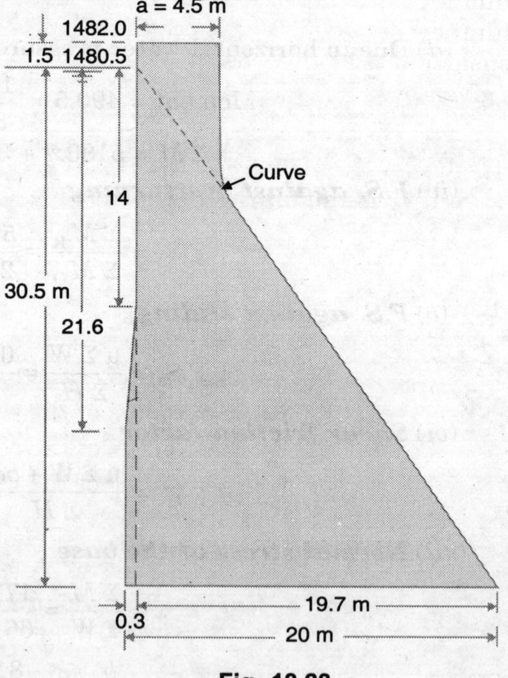

Fig. 18.28

18.6. MASONRY CHIMNEYS

Chimneys are used in almost all industries for the escape of flue gases to such a height that diffusion of gases takes place so that they do not contaminate the surroundings. The height of chimney is fixed on the basis of required draft. However, masonry chimneys are restricted to small heights only, since the section of the chimney becomes very heavy for large height, requiring heavy foundations. Masonry chimneys are generally built in bricks, with either the common bricks or radial bricks. Masonry chimneys are constructed in various shapes, such as circular, square, hexagonal or octagonal; circular shape is preferred over others. Chimney are lined from inside with fire bricks, atleast upto 20 to 25% of its height from the bottom.

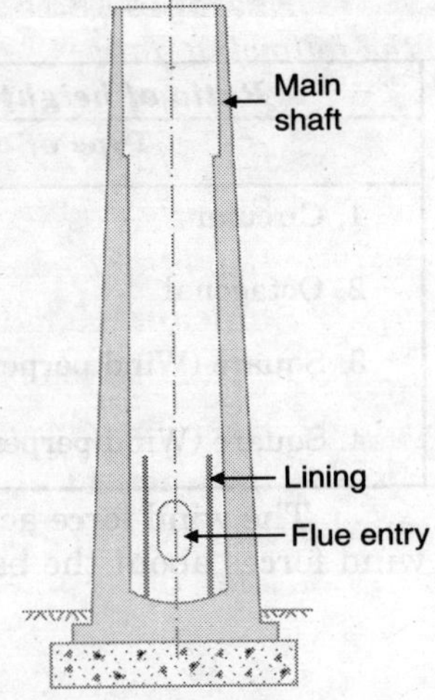

Fig. 18.29. Masonry Chimney

Forces acting on a chimney

Various forces that act on a chimney are :

1. Self weight of masonry.
2. Weight of brick lining.
3. Wind pressure.
4. Seismic force.

1. Self weight of chimney : As shown in Fig. 18.29, chimneys are tapered from top to bottom. The position of the chimney between two successive sections of height h may be considered to be the frustum of a hollow cone, the volume of which is given by,

$$V = \frac{h}{3}\left(A_1 + A_2 + \sqrt{A_1 A_2}\right) \quad \text{...(18.50)}$$

If γ_m is the unit weight of the masonry, the self weight of the chimney is given by

$$W_s = \gamma_m \frac{h}{3}\left(A_1 + A_2 + \sqrt{A_1 A_2}\right) \quad \text{...[18.50 (a)]}$$

where A_1 = area of cross-section at the bottom section = $\pi\,(R_1^{\,2} - r_1^{\,2})$

A_2 = area of cross-section at the top section = $\pi(R_2^{\,2} - r_2^{\,2})$

R_1 = outer radius of the cone at the bottom section

r_1 = inner radius of the cone at the bottom section

R_2 = outer radius of the cone at the top section

r_2 = inner radius of the cone at the top section

The stress f_s due to self weight is, $f_s = \dfrac{W_s}{A} = \dfrac{W_s}{\pi(R_1^{\,2} - r_1^{\,2})}$ ...(18.51)

2. Weight of lining : The weight of lining (W_l) is given by

$$W_l = \gamma_l \,.\, \pi\, d_l\, t_l\, h_l \quad \text{...(18.52)}$$

where γ_l = unit weight of lining material

d_l = mean diameter of lining

t_l = thickness of lining

h_l = height of lining

It is to be noted that the lining is supported directly at the base.

3. Wind force : Wind force on the chimneys can be determined on the basis of recommendations contained in IS 875-1987. Table 18.1 gives the values of shape factors as per IS 875. The wind force on chimney = shape factor × design wind pressure × projected area of the chimney.

TABLE 18.1. SHAPE FACTORS (IS : 875)

Ratio of height to base width	***0 to 4***	***4 to 8***	***8 and over***
Type of chimney		***Shape factor***	
1. Circular	0.7	0.7	0.7
2. Octagonal	0.8	0.9	1.0
3. Square (Wind perpendicular to diagonal)	0.8	0.9	1.0
4. Square (Wind perpendicular to face)	1.0	1.15	1.3

The wind force acting on the chimney has an overturning effect. If M_w is the moment of wind forces about the base, the compressive and tensile stresses at the base are given by

$$f_w = \pm \frac{M_w}{I} R_1 = + \frac{M_w}{Z} \qquad \text{...(18.53)}$$

where R_1 = the outer radius at the base section.

I = moment of inertia of the section at base = $\frac{\pi}{4}(R_1^4 - r_1^4)$

Hence the total stress due to self weight and wind is given by

$$f = f_s \pm f_w = \frac{W_s}{A} \pm \frac{M_w}{I} R_1 \qquad \text{...(18.54)}$$

The value of *maximum stress* (using + sign) should not exceed the allowable compressive stress for masonry. Also, the minimum stress (using – sign) should be compressive in nature, *i.e.,* no tension should develop at the base.

Example 18.16. *A circular chimney of brick masonry has uniform section throughout its height, having an inner diameter of 2 m and outer diameter of 3 m. Find the maximum permissible height of the chimney if no tension is to develop and if the allowable compressive stress in masonry is 1.5 N/mm². Take a uniform wind pressure intensity of 1.5 kN/m². Take unit weight of brick masonry as 22 kN/m³.*

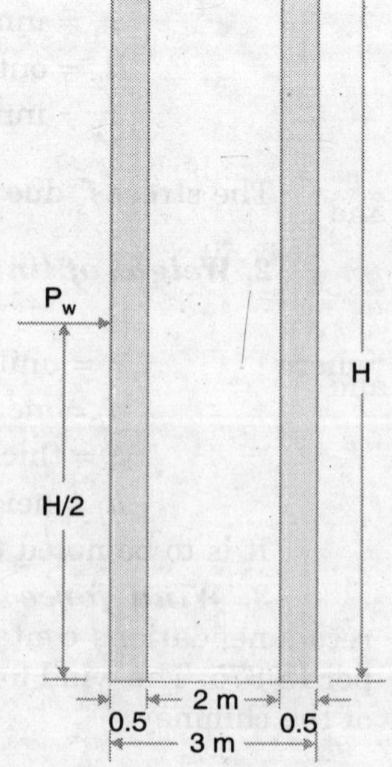

Fig. 18.30

Solution. Let H be the height of chimney

$$A = \frac{\pi}{4}(3^2 - 2^2) = \frac{5}{4}\pi \text{ m}^2$$

$\therefore$ Self weight of chimney,

$$W_s = \frac{5}{4}\pi H \times 22 = 27.5\,\pi H \text{ kN}$$

P_w = shape factor × wind pressure × projected area

$= 0.7 \times 1.5 \times 3H \times 3.15\, H$ kN

acting at $H/2$ from the base

$$M_w = 3.15\, H (H/2) = 1.575\, H^2 \text{ kN-m}$$

$$I = \frac{\pi}{64}(3^4 - 2^4) = \frac{65}{64}\pi \text{ m}^4$$

$$y_{max} = 1.5 \text{ m}$$

$$\therefore \quad f_{max.} = \frac{W_s}{A} + \frac{M_w}{I} y_{max.} = \frac{27.5 \pi H}{\frac{5}{4}\pi} + \frac{1.575 H^2}{\frac{65}{64}\pi} \times 1.5$$

$$= 22 H + 2.326 \frac{H^2}{\pi}$$

$$f_{min.} = \frac{W_s}{A} - \frac{M_w}{I} y_{max} = 22 H - 2.326 \frac{H^2}{\pi}$$

For no tension to develop, $f_{min} = 0 = 22 H - 2.326 \dfrac{H^2}{\pi}$

From which **H = 29.71 m.**

$$\therefore \quad f_{max} = 22 \times 29.71 + \frac{2.326 (29.71)^2}{\pi} = 653.71 + 653.71$$

$$= 1307.4 \text{ kN/m}^2 = 1.3074 \text{ N/mm}^2 < 1.5. \text{ Hence safe.}$$

Thus, maximum permissible height of chimney = 29.71 m.

Example 18.17. *Fig. 18.31 shows a hollow square chimney having uniform inside dimension of 1.6 m × 1.6 m and varying out side dimension at different heights. Check the stability of the chimney by taking uniform wind pressure intensity of 1.5 kN/m² throughout the height. Take unit weight of masonry as 22 kN/m³ and permissible compressive stress in brick masonry as 1.8 N/mm².*

Solution. Let us test the stability of the chimney under two conditions:

(*i*) wind blowing at right angles to the face,

and (*ii*) wind blowing at right angles to the diagonal.

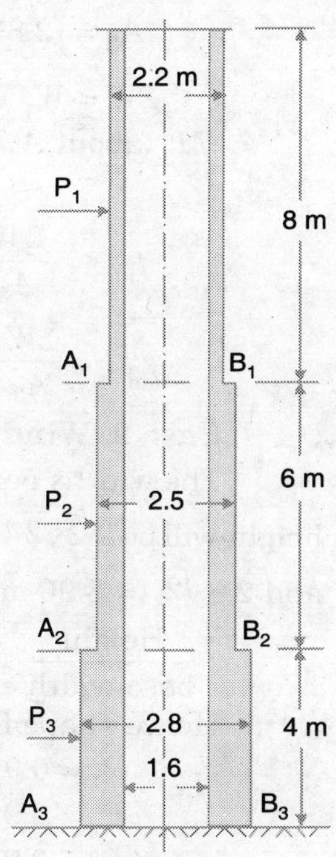

Fig. 18.31

Case 1. Wind blowing at right angles to the face.

$$\frac{\text{Height}}{\text{base}} \text{ ratio} = \frac{18}{2.8} = 6.43$$

∴ Shape factor = 1.15 (Table 18.1)

Wind forces P_1 on top 8 m height :

$$P_1 = 1.15 \times 1.5 \times 2.2 \times 8 = 30.36 \text{ kN}$$

Similarly, $P_2 = 1.15 \times 1.5 \times 2.5 \times 6 = 25.875$ kN

and $P_3 = 1.15 \times 1.5 \times 2.8 \times 4 = 19.32$ kN

The self weight of the three portions are :

$$W_1 = [(2.2)^2 - (1.6)^2] \times 8 \times 22 = 401.28 \text{ kN}$$

$$W_2 = [(2.5)^2 - (1.6)^2] \times 6 \times 22 = 487.08 \text{ kN}$$

and $W_3 = [(2.8)^2 - (1.6)^2] \times 4 \times 22 = 464.64$ kN

Check at section A_1B_1

$$A_1 = (2.2)^2 - (1.6)^2 = 2.28 \text{ m}^2$$

$$I_1 = \frac{1}{12} [(2.2)^4 - (1.6)^4] = 1.405 \text{ m}^4$$

$$y_{1, max.} = \frac{2.2}{2} = 1.1 \text{ m}$$

$$M_{w1} \text{ (about } A_1B_1) = P_1 \times \frac{8}{2} = 30.36 \times 4 = 121.44 \text{ kN-m}$$

$$f_{min.} = \frac{W_1}{A_1} - \frac{M_{w1}}{I_1} y_{1, max} = \frac{401.28}{2.28} - \frac{121.44}{1.405} \times 1.1$$

$= 176 - 95 = 81 \text{ kN/m}^2$ (safe)

$$f_{max} = \frac{W_1}{A_1} + \frac{M_{w1}}{I_1} y_{1,max} = 176 + 95 = 271 \text{ kN/m}^2$$

$= 0.271 \text{ N/mm}^2 < 1.8$ (Safe)

Fig. 18.32

Check at section A_2 B_2

$$A_2 = (2.5)^2 - (1.6)^2 = 3.69 \text{ m}^2$$

$$I_2 = \frac{1}{12}[(2.5)^4 - (1.6)^4] = 2.709 \text{ m}^4$$

$y_{2,\,max} = 1.25$ m

$\Sigma W = W_1 + W_2 = 401.28 + 487.08 = 888.36$ kN

$$M_{w2} \text{ (about } A_2B_2) = P_1\,(4 + 6) + P_2 \times \frac{6}{2} = 30.36 \times 10 + 25.875 \times 3 = 381.225 \text{ kN}$$

$$\therefore \quad f_{\min} = \frac{\Sigma W}{A_2} - \frac{M_{w2}}{I_2} y_{2,\,\max} = \frac{888.36}{3.69} - \frac{381.225}{2.709} \times 1.25 = 240.7 - 175.9 = 64.8 \text{ kN/m}^2$$

$$f_{\max} = \frac{\Sigma W}{A_2} + \frac{M_{w2}}{I_2} \cdot y_{2,\,\max} = 240.7 + 175.9 = 416.6 \text{ kN/m}^2 = 0.4166 \text{ N/mm}^2 < 1.8 \text{ (safe)}$$

Check at section A_3B_3

$$A_3 = (2.8)^2 - (1.6)^2 = 5.28 \text{ m}^2 \; ; I_3 = \frac{1}{12}[(2.8)^4 - (1.6)^4] = 4.576 \text{ m}^4 \; ; y_{3,\,\max} = 1.4 \text{ m.}$$

$\Sigma W = W_1 + W_2 + W_3 = 401.28 + 487.08 + 464.64 = 1353$ kN

M_3 (about A_3B_3) $= P_1 \times 14 + P_2 \times 7 + P_3 \times 2$

$= 30.36 \times 14 + 25.875 \times 7 + 19.32 \times 2 = 644.805$ kN-m

$$f_{\min} = \frac{\Sigma W}{A_3} - \frac{M_{w3}}{I_3} \cdot y_{3,\,\max} = \frac{1353}{5.28} - \frac{644.805}{4.576} \times 1.4 = 256.3 - 197.3 = 59 \text{ kN/m}^2$$

$$f_{\max} = \frac{\Sigma W}{A_3} + \frac{M_{w3}}{I_3} \cdot y_{3,\,\max} = 256.3 + 197.3 = 453.6 \text{ kN/m}^2 = 0.4536 \text{ N/mm}^2 < 1.8 \text{ (safe)}$$

Case 2. Wind blowing at right angles to the diagonal

The widths perpendicular to the directions of wind (*i.e.*, length of diagonals), at various height will be $2.2\sqrt{2}$ (= 3.111 m), $2.5\sqrt{2}$ (= 3.536 m) and $2.8\sqrt{2}$ (= 3.96 m) for the three heights.

$$\frac{\text{Height}}{\text{base width}} \text{ ratio} = \frac{18}{3.96} = 4.55.$$

Hence shape factor = 0.9

$\therefore \quad P_1 = 0.9 \times 1.5 \times 3.111 \times 8 = 33.6$ kN

$P_2 = 0.9 \times 1.5 \times 3.536 \times 6 = 28.64$ kN

$P_3 = 0.9 \times 1.5 \times 3.96 \times 4 = 21.384$ kN

Check at section A_1B_1

$A_1 = 2.28 \text{ m}^2$ (as before);

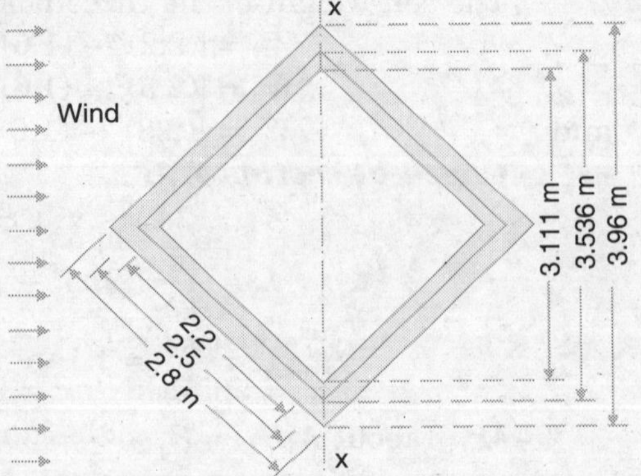

Fig. 18.33

$$I_1 = \frac{2}{12}\left[3.111\left(\frac{3.111}{2}\right)^3 - 2.263\left(\frac{2.263}{2}\right)^3\right] = 1.405 \text{ m}^4$$

$$y_{1,\,max} = \frac{3.111}{2} = 1.556 \text{ m}$$

$$M_{w1} = P_1 \times 4 = 33.6 \times 4 = 134.4 \text{ kN-m}$$

$$\therefore \quad f_{min} = \frac{W_1}{A_1} - \frac{M_{w1}}{I_1} y_{1,\,max} = \frac{401.28}{2.28} - \frac{134.4}{1.405} \times 1.556 = 176 - 148.8 = 27.2 \text{ kN/m}^2$$

$$f_{max} = \frac{W_1}{A_1} + \frac{M_{w1}}{I_1} . y_{1,\,max} = 176 + 148.8 = 324.8 \text{ kN/m}^2$$

$$= 0.325 \text{ N/mm}^2 < 1.8 \text{ safe.}$$

Check at section A_2B_2 **:** $A_2 = 3.69 \text{ m}^2$

$$I_2 = \frac{2}{12}\left[3.536\left(\frac{3.536}{2}\right)^3 - 2.263\left(\frac{2.263}{2}\right)^3\right] = 2.710 \text{ m}^4$$

$$y_{2,max} = \frac{3.536}{2} = 1.768 \text{ m}$$

$$M_{w2} = P_1 \times 10 + P_2 \times 3 = 33.6 \times 10 + 28.64 \times 3 = 421.9 \text{ kN-m}$$

$$\therefore \quad f_{min} = \frac{\Sigma W}{A_2} - \frac{M_{w2}}{I_2} . y_{2\,max} = \frac{888.36}{3.69} - \frac{421.9}{2.71} \times 1.768$$

$$= 240.7 - 275.3 = -34.6 \text{ kN/m}^2. \text{ (unsafe)}$$

$$f_{max.} = \frac{\Sigma W}{A_2} + \frac{M_{w2}}{I_2} . y_{2,\,max} = 240.7 + 275.3 = 516 \text{ kN/m}^2 = 0.516 \text{ N/mm}^2 < 1.8$$

Check at section A_3B_3 **:** $A_3 = 5.28 \text{ m}^2$

$$I_3 = \frac{2}{12}\left[3.96\left(\frac{3.96}{2}\right)^3 - 2.263\left(\frac{2.263}{2}\right)^3\right] = 4.576 \text{ m}^4$$

$$y_{3\,max} = \frac{3.96}{2} = 1.98 \text{ m}$$

$$M_{w3} = P_1 \times 14 + P_2 \times 7 + P_3 \times 2 = 33.6 \times 14 + 28.64 \times 7 + 21.384 \times 2$$

$$= 713.648 \text{ kN-m}$$

$$f_{min} = \frac{\Sigma W}{A_3} - \frac{M_{w3}}{I_3} . y_{3,\,max} = \frac{1353}{5.28} - \frac{713.648}{4.576} \times 1.98$$

$$= 256.3 - 308.8 = -52.5 \text{ kN/m}^2 \text{ (unsafe)}$$

$$f_{max.} = \frac{\Sigma W}{A_3} + \frac{M_{w3}}{I_3} \, y_{3,\,max.} = 256.3 + 308.8$$

$$= 565.1 \text{ kN/m}^2 = 0.565 \text{ N/mm}^2 < 1.8.$$

Example 18.18. *A free standing brick shaft hexagonal in cross-section is 50 m high and is subjected to a wind loading of 1.5 kN/m² on the total exposed area. But for the direction of wind as shown in Fig. 18.34 (a), the loading is to be modified by a coefficient 0.75 to allow for the shape. The distance between opposite external faces is 4 m at the base and 2.5 m at the top. The brick work is 1 m thick at the base and 0.25 m thick at the top and that the total weight of the shaft is 5000 kN. Determine the distribution of stress across the base.*

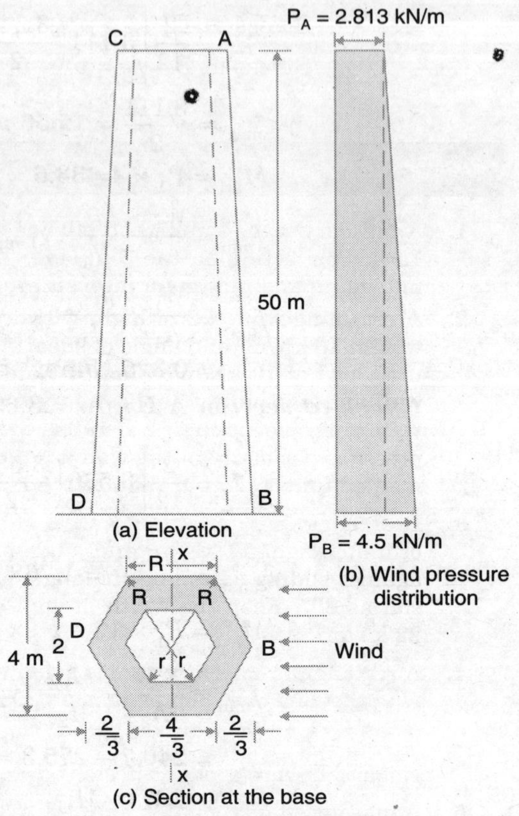

Fig 18.34

Solution. Wind pressure at base, per m height :

$$P_B = 0.75 \times 1.5 \times 4 \times 1 = 4.5 \text{ kN/m}$$

Wind pressure at top, per m height

$$P_A = 0.75 \times 1.5 \times 2.5 \times 1 = 2.813 \text{ kN/m}$$

The wind pressure distribution is shown in Fig. 18.34(*b*).

$$M_w \text{ about the base} = \left(2.813 \times 50 \times \frac{50}{2}\right) + \frac{1}{2}$$

$$(4.5 - 2.813) \times 50 \times \frac{50}{3} \approx 4219 \text{ kN-m}$$

For the base section [Fig. 18.34 (*c*)]

$$R = \frac{4}{\sqrt{3}} \; ; r = \frac{2}{\sqrt{3}}$$

$$I_{xx} \text{ of base} = 2\left[\left(\frac{1}{36} \times 4 \times \frac{8}{3\sqrt{3}}\right) + \left(\frac{1}{2} \times 4 \times \frac{2}{\sqrt{3}}\right) \times \left(\frac{4}{2\sqrt{3}} + \frac{2}{3\sqrt{3}}\right)^2 + \left(\frac{1}{3} \times 4 \times \frac{8}{3\sqrt{3}}\right)\right]$$

$$- 2\left[\left(\frac{1}{36} \times 2 \frac{1}{3\sqrt{3}}\right) + \left(\frac{1}{2} \times 2 \times \frac{1}{\sqrt{3}}\right) \times \left(\frac{1}{\sqrt{3}} + \frac{1}{3\sqrt{3}}\right)^2 + \left(\frac{1}{3} \times 2 \times \frac{1}{3\sqrt{3}}\right)\right]$$

$$= 2\left\{\frac{8}{27\sqrt{3}} + \frac{256}{27\sqrt{3}} + \frac{32}{9\sqrt{3}} - \frac{1}{54\sqrt{3}} - \frac{16}{27\sqrt{3}} - \frac{2}{9\sqrt{3}}\right\}$$

$$= \frac{2 \times 675}{54\sqrt{3}} \text{ m}^4 = \frac{25}{\sqrt{3}} \text{ m}^4.$$

$$\therefore \quad Z_{xx} \text{ of base} = \frac{25}{\sqrt{3}} \div \frac{4}{\sqrt{3}} = 6.25 \text{ m}^3$$

$$\therefore \quad p_b = \pm \frac{M_w}{Z_{xx}} = \pm \frac{4219}{6.25} \approx \pm 675 \text{ kN/m}^2$$

$$\text{Area of base,} \quad A = 6\left\{\frac{1}{2}\left(\frac{4}{\sqrt{3}} + \frac{2}{\sqrt{3}}\right) \times 1\right\} = \frac{6 \times 6}{2\sqrt{3}} = 6\sqrt{3} \text{ m}^2$$

$$\therefore \quad p_0 = \frac{W_s}{A} = \frac{5000}{6\sqrt{3}} \approx 481 \text{ kN/m}^2$$

$\therefore$ Stress at D, $f_D = p_0 + p_b = 481 + 675 =$ **1156 kN/m²** (comp.)

Stress at B, $f_B = p_0 - p_b = 481 - 675 = -$ **194 kN/m²** (*i.e.*, tension).

PROBLEMS

1. A masonry wall, 4 m high, 5 m wide and 1 m thick is subjected to a horizontal wind pressure of 1.5 kN/m² acting on the 5 m side. Taking the unit weight of masonry as 21 kN/m³, find the maximum and minimum intensities of stresses induced at the base of the wall.

2. A retaining wall, 4.2 m high, with a smooth vertical back, retains a dry sandy backfill of unit weight 18 kN/m³ and angle of shearing resistance of 30°. The backfill carries a uniformly distributed load of 10 kN/m². Find by Rankine's theory the total active pressure per metre length of wall and its point of application above the base. If the water table rises behind the back of the wall to an elevation of 2.1 m below the top of the wall, what is the change in the total active pressure per metre of the wall ? Assume no change in the angle of shearing resistance and submerged unit weight of sand as 9.5 kN/m³.

3. Design a gravity retaining wall, 6 m high, with vertical back to retain a dry cohesionless backfill of unit weight 17 kN/m³ and angle of shearing resistance of 24°. Find also the factor of safety against sliding assuming the angle of friction between the base of the wall and the foundation soil as 30°. Keep the width of wall at the top as 1.2 m. Take the unit weight of masonry as 22 kN/m³.

4. A rectangular masonry dam is 2.4 m wide at the base. Compute the maximum permissible height H of the dam (*a*) when no tension is permissible and (*b*) when the factor of safety against sliding is 1.5. Take $\mu = 0.5$ and unit weight of masonry equal to $2\frac{1}{2}$ times the unit weight of water. Neglect under-seepage effects.

5. A masonry dam 8 m high and 2 m wide at the top has vertical water face. Find the base width of the dam if no tension is to develop. Take unit weight of masonry as 22 kN/m³. Neglect uplift pressure. Also, investigate the stability of the dam section, taking $\mu = 0.7$ and maximum allowable compressive stress as 180 kN/m².

6. Solve problem 5, considering the uplift pressure. Take $c = 1$.

7. A hollow square chimney has inside dimensions of 1.5 m × 1.5 m at the top and 2 m × 2 m size at the bottom. It has uniform thickness of 0.5 m throughout its height of 16 m. Check the stability of the chimney as its base, for a uniform wind pressure intensity of 1.5 kN/m². Take unit weight of brick masonry as 20 kN/m³ and permissible compressive stress as 2 N/mm².

8. A masonry chimney is built so that the outside of a cross-section is a regular hexagon and the inside a circle. Show that if no tensile stress is to occur the resultant thrust at any horizontal cross-section due to dead weight and the wind must lie inside a hexagon inscribed in a circle of radius equal to

$$\frac{\left(\frac{5}{8}R^4 - \frac{\pi}{2\sqrt{3}}r^4\right)}{R\left(\frac{3\sqrt{3}}{2}R^2 - \pi r^2\right)}$$

as shown in Fig. 18.35.

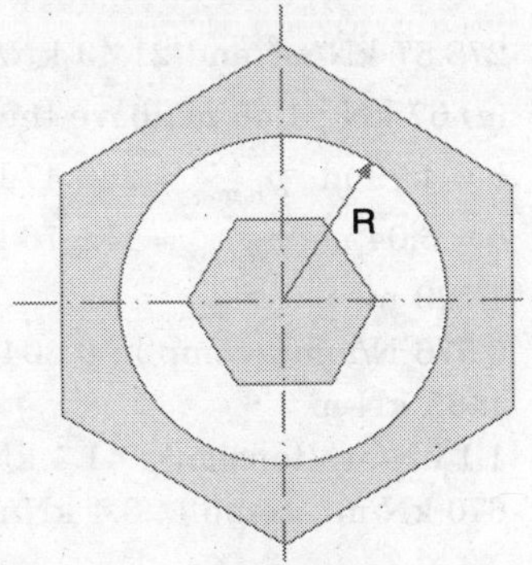

Fig. 18.35

9. Show that in order to prevent tensile stress being set up in masonry chimney the line of resultant thrust must no deviate from the centre line by more than $\dfrac{r_0^{\,2} + r_i^{\,2}}{4r_0}$, where r_0 and r_i are the outer and inner radii.

10. A square chimney 25 m high, having an opening of 1 m by 1 m is subjected to a horizontal wind pressure of 1.5 kN/m^2. Find the necessary thickness of brickwork at base if the density of masonry is 21 kN/m^3 and the maximum permissible stress on brick masonry is limited to 0.8 N/mm^2.

11. Find the maximum and minimum stress intensities at the base of a uniform circular chimney, 4 m external diameter and 2 m internal diameter 20 m high, subjected to horizontal wind pressure of 1.5 kN/m^2 of projected area. The density of masonry may be taken as 22 kN/m^3.

12. Calculate the bending moment at the base of chimney shaft 30 m high tapering from 3 m external diameter at base to 1.5 m external diameter at top. The horizontal wind pressure acting on the flat surface is 2.5 kN/m^2 and the coefficient of wind resistance is 0.66.

13. A 1 metre high brick chimney with three flues as shown in Fig. 18.36 is subjected to a wind pressure of 1.5 kN/m^2 acting normally to the longer side. The weight of brickwork is 20 kN/m^3. Calculate the maximum stresses on the windward and leeward sides. Would you consider the chimney safe ?

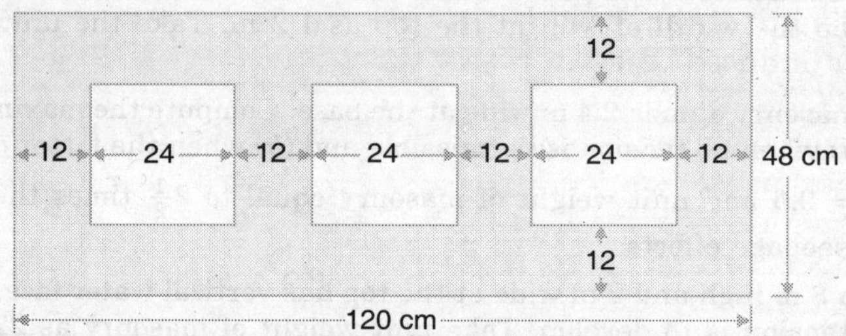

Fig. 18.36

14. A masonry dam trapezoidal in section with one face vertical is 2.5 m high, 5 m wide at top and 15 m wide at bottom. It retains water towards vertical face for the total depth of 24 m. If density of masonry is 20 kN/m^3 and that of water 10 kN/m^3, calculate the maximum and minimum stress intensities induced at the base.

ANSWERS

1. 278.57 kN/m^2 and 21.43 kN/m^2

2. (*a*) 67 kN ; 1.55 m above the base (*b*) Increase in pressure 15.72 kN

5. b = 4.79 m ; $p_{n.\max}$ = 249.47 kN/m^2 ; F.S. sliding = 1.33

6. b = 6.04 m ; $p_{n.\max}$ = 155.76 kN/m^2 ; F.S. against sliding = 1.05

10. 1.129 m.

11. 0.576 N/mm^2 (comp.) ; 0.304 N/mm^2 (comp.).

12. 1485 kN-m.

13. 1.1 kN/m^2 (tension) ; 41.1 kN/m^2 (comp.). Not safe.

14. 670 kN/m^2 (comp.) ; 3.4 kN/m^2 (tensile).

19

Columns

19.1. INTRODUCTION

If a short length of bar or block is subjected to a compressive force P, uniform compressive stress $p = P/A$ is induced. Such a compression member fails by *crushing* on increasing the value of force P. The behaviour of such a member has been discussed in chapter 1. Even if such a short compression member is subjected to eccentric load (or to a direct compressive force and a moment simultaneously), the failure is primarily by crushing. However, there remain the cases in which compression members of relatively longer length fail by *buckling* or by *lateral bending* even under a central load. Such members are commonly known as *columns*. While a *short compression member* undergoes negligible lateral deflection, *column is a compression member that is so slender compared to its length that under gradually increasing loads, it fails by buckling at loads considerably less than those required to cause failure by crushing.* Though there is no clear cut line of demarcation between a short compression member and a column, *a compression member is generally considered to be a column when its unsupported length is more than 10 times its least lateral dimension.*

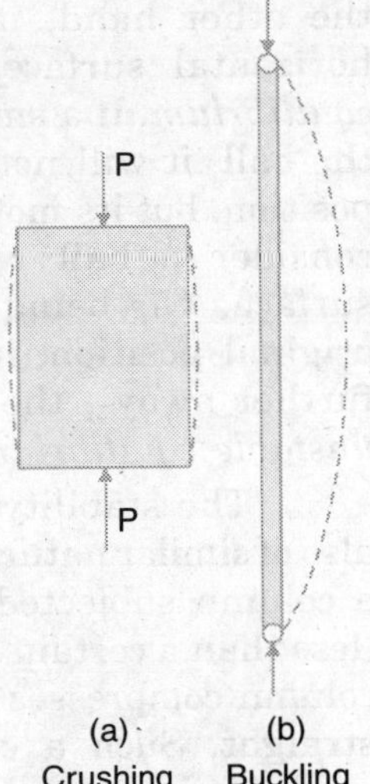

Fig. 19.1

Thus, a *compression member* is divided into three categories (*i*) *short compression members*, which fail primarily by crushing without buckling, (*ii*) *long columns* which fail by buckling or excessive lateral bending, and (*iii*) *intermediate columns* which fail by a combination of crushing and buckling. A compression member is classified by its length and least lateral dimension. Most of the practical cases of compression members fall under the second category. The *critical load*, which the member can carry before failure, depends upon (*i*) member dimensions (area, shape and length) and (*ii*) end conditions. For columns, the problem is one of determining the *critical load* which causes *elastic instability*.

Any structural member loaded axially in compression is called a *column* in the general sense of the word. It is variously termed as stanchion, post, strut, boom etc., depending upon its use as a particular member in a structure. *Columns* are ordinarily used in buildings, as vertical members, to carry loads of beams, slabs etc. *Stanchions* are steel columns made of rolled steel sections, commonly used in buildings. The term *post* is loosely used for a column ; in a truss bridge girder, end compression members are called *end posts*. The term *strut* is commonly used for compression member in a roof truss; it may either be in vertical position or in inclined position. The principal compression member in a crane is called a *boom*.

19.2. MODES OF FAILURE OF A COLUMN : ELASTIC INSTABILITY

A column subjected to axial load, may fail under the following modes :

(*i*) crushing

(*ii*) buckling

and (*iii*) mixed mode of buckling and crushing.

If a short length of a bar or column is subjected to a compressive force P, a uniform compressive stress p is induced, p being equal to P/A. As the value of P is increased, p will increase, and ultimately, failure will occur due to *crushing* or *yielding* of the material. In the case of *long columns* subjected to axial loading, *buckling* occurs due to *elastic instability*. Every such column has a *critical* load (P_{cr}) which causes elastic instability, due to which column suddenly fails due to excessive bending stresses on the section. Columns of intermediate length fail in the mixed mode of buckling and crushing. Most of the practical columns fail due to this mixed mode of failure.

Critical Equilibrium

Equilibrium of an elastic body may be (*i*) stable, (*ii*) neutral and (*iii*) unstable. Consider a ball resting on a concave surface [Fig. 19.2 (*a*)]. If the ball is given a small displacement from its position, it will return to its initial position. The ball is, therefore in *stable equilibrium*. On the other hand, if the ball is resting on a horizontal surface, it will be in a *neutral equilibrium*. If a small displacement is given to the ball, it will not return to its initial position, but its motion will stop. Lastly, consider a ball resting on a convex surface. On being displaced from its original position, it continues to move further away ; the ball is therefore in *unstable equilibrium*.

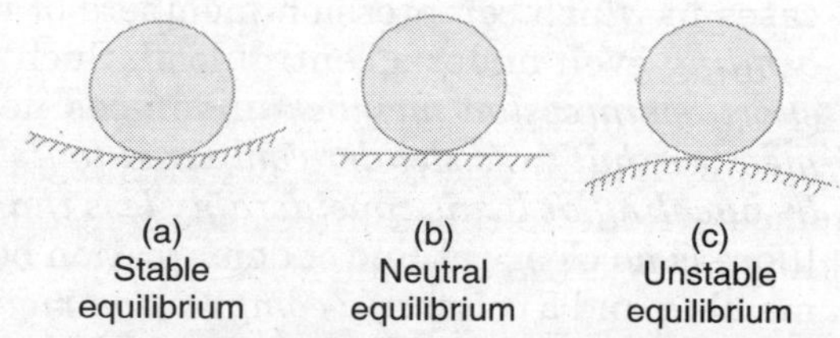

Fig. 19.2. Equilibrium Conditions

The stability of a long column is also of similar nature. Fig. 19.3 (*ai*) shows a column subjected to an axial force P (less than a certain critical load P_{cr}) ; the column compresses a little and remains straight. Such a column will be in a condition of *stable equilibrium*. Even if it is deflected slightly by some lateral load δQ and then the lateral load is removed,

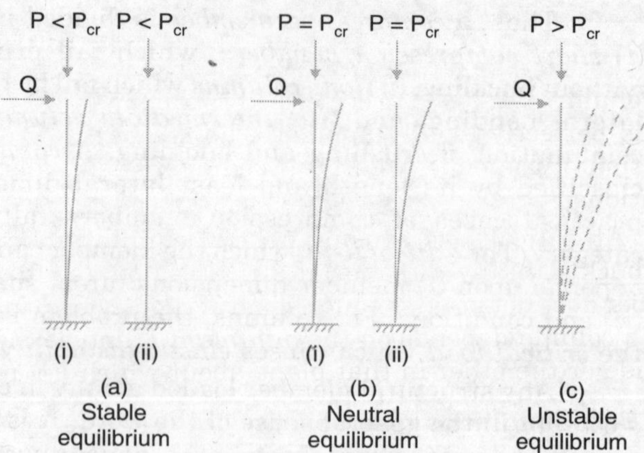

Fig. 19.3. Equilibrium Conditions of a Column

the column will become straight again and will assume the original configuration of equilibrium (Fig. 19.3 *aii*). The same cycle may be repeated with a force P increasing in magnitude until a critical load P_{cr} is reached at which the column remains in a slightly deformed position after the application and removal of lateral force δQ. This instantaneous condition is the state of *neutral equilibrium* [Fig. 19.3 (*b*)]. Finally if force P exceeds the critical one, the straight configuration of equilibrium of the column will become unstable as shown in Fig. 19.3 (*c*). At a load greater than the critical, a curvilinear configuration is stable and it can be shown by exact differential equations of elastic curve that a mere increase of 1.5% in P_{cr} causes a maximum sideways deflection equal to 22% of the length of the column, provided the material remains elastic. Such large deflections cannot be tolerated in engineering practice, since large stresses are induced in the column by the resulting eccentricity of the applied force and the column collapses. Therefore, *the determination of critical force P_{cr} is of utmost importance, since it represents the ultimate capacity of an ideal structure*. Investigations show that *instability* has been the cause of many accidents and structural damages.

Buckling of columns : Buckling stresses

The curved configuration of a column, under an axial load is called *buckled shape*. Due to this, the distribution of stress over the section will not be uniform, and the resulting eccentricity, however slight, will cause bending moment. This bending moment produces bending stresses which are referred as *buckling stresses* simply to prevent confusion with bending stresses produced by the eccentricity of applied loads. *Thus the buckling stresses and bending stresses are the same except that the former, by definition result from axial loads while the later from the applied eccentric load.* However, it should be clearly noted that buckling stresses are in addition to the direct compressive stress due to applied load.

In general, the buckling tendency of a column varies with the ratio of the length to least lateral dimension. The ratio is known as *slenderness ratio*. For tall slender columns, this ratio is large and if failure occurs, it will be entirely due to buckling (mode 2). When this ratio is very small, failure occurs due principally to *yielding* or crushing (mode 1). Between these extremes are the so called intermediate columns where failure will be due to combination of buckling and crushing (mode 3).

19.3. EULER'S THEORY : LONG COLUMNS

Let us take a column which is very long in proportion to its cross-sectional dimensions. The column is assumed to be perfectly straight and homogeneous in quality, and the compressive load is perfectly axially applied. The strength to resist buckling is greatly affected by the condition of the ends, whether fixed or free. We shall, therefore, take different cases of end conditions.

At critical load causing neutral equilibrium, stated in § 19.2, a compression member may buckle in any direction if its I is same about both the axes. Generally, a compression member does not possess equal flexural rigidity in all directions. The *significant* flexural rigidity EI of a column depends on the *minimum* I, and at the critical load a column buckles either to one side or the other in that plane about which I is minimum.

Ideal End Conditions of a Column

A column may have two types of end conditions (*i*) Position restraint, and (*ii*) Direction restraint.

An end has a *position restraint* when it is not free to change its position (or is not free to translate), but is free to rotate. The *hinged end* of a column is a typical example of position restraint. In the *direction restraint*, the end of the column is not free to change its direction. It may be free to change its position, or it may not be free to change its position. However, an end of a column may have both the position restraint as well as direction restraint.

Depending upon various *combinations of restraints*, there may be the following four cases of ends conditions (Fig. 19.4) :

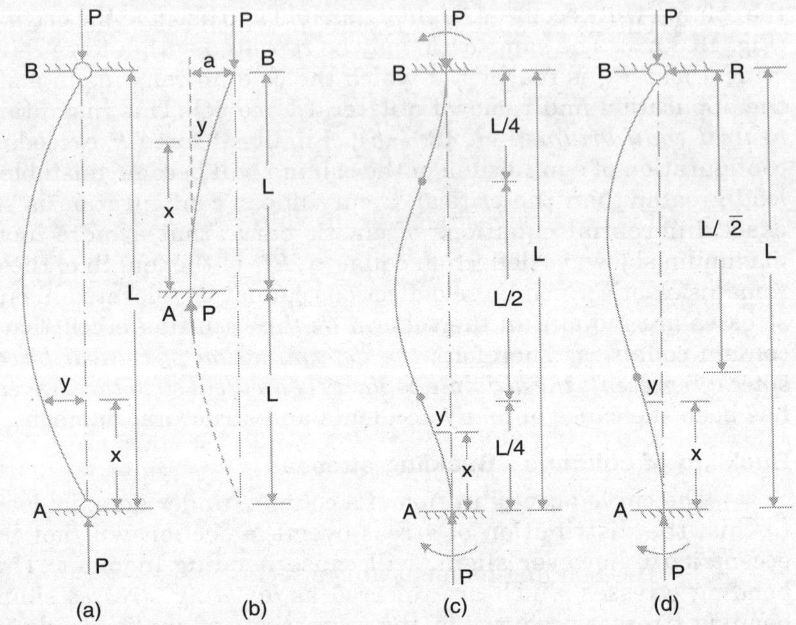

Fig. 19.4. Ideal End Conditions

***Case 1* :** Both ends of column hinged [Fig. 19.4 (*a*)] : *standard case.*

***Case 2* :** One end of column fixed and the other end free [Fig. 19.4 (*b*)].

***Case 3* :** Both ends of column fixed [Fig. 19.4 (*c*)].

***Case 4* :** One end of column fixed and the other end hinged [Fig. 19.4 (*d*)].

Case (*a*) : Both ends hinged

Consider a column with its ends free to rotate around frictionless pins. The buckled shape shown in Fig. 19.4 (*a*) is possible only at a critical or Euler load, as prior to this load, the column remains straight. The smallest force at which a buckled shape is possible is called the *critical force*.

Consider a section at a distance x from A and let y be its deflection from the centre line. We have, therefore,

$$EI\,\frac{d^2y}{dx^2} = -M_x = -Py.$$

Re-arranging the above equation, we get

$$\frac{d^2y}{dx^2} + \frac{P}{EI}\,y = 0.$$

This is a standard differential equation, the solution of which is

$$y = c_1 \cos x\sqrt{\frac{P}{EI}} + c_2 \sin x\sqrt{\frac{P}{EI}}$$

where c_1 and c_2 are constants of integration and can be found by applying end conditions.

At A, $x = 0$ and $y = 0$ $\quad\therefore\quad c_1 = 0$

At B, $x = L$ and $y = 0$ $\quad\therefore\quad 0 = c_2 \sin L\sqrt{\dfrac{P}{EI}}$

This is possible if c_2 is zero, in which case, the column has not bent at all, or if

$$\sin L\sqrt{\frac{P}{EI}} = 0.$$

$$\therefore \qquad L\sqrt{\frac{P}{EI}} = 0, \pi, 2\pi \ldots\ldots$$

Taking the least significant value (*i.e.*, π), we get

$$P = \frac{\pi^2 EI}{L^2}.$$

We shall call this *critical load* as *Euler's load* P_E.

Hence
$$P_E = \frac{\pi^2 EI}{L^2} \qquad \ldots(19.1)$$

Case (*b*) : One end fixed, other end free

Let end A be the direction fixed as well as position fixed and the end B be free. Let a be the deflection of the free end [Fig. 19.4 (*b*)].

At any distance x from A, let y be the deflection.

The bending moment at the section

$$= -P(a - y).$$

Hence
$$EI\frac{d^2y}{dx^2} = -M = P(a - y)$$

or
$$\frac{d^2y}{dx^2} + \frac{P}{EI}y = \frac{Pa}{EI}$$

The solution of the above differential equation is

$$y = c_1 \cos x\sqrt{\frac{P}{EI}} + c_2 \sin x\sqrt{\frac{P}{EI}} + a$$

At A, $x = 0$ and $y = 0$ $\therefore$ $c_1 = -a$

Also at A, $x = 0$ and $\frac{dy}{dx} = 0$

$$\frac{dy}{dx} = -c_1\sqrt{\frac{P}{EI}} \sin x\sqrt{\frac{P}{EI}} + c_2\sqrt{\frac{P}{EI}} \cos x\sqrt{\frac{P}{EI}}$$

$$\therefore \qquad \frac{dy}{dx} = 0 = c_2\sqrt{\frac{P}{EI}}$$

Hence $c_2 = 0$, since P is not zero.

Substituting the value of c_1 and c_2, we get

$$y = -a\cos x\sqrt{\frac{P}{EI}} + a = a\left(1 - \cos x\sqrt{\frac{P}{EI}}\right)$$

Again, at B, $x = L$ and $y = a$

$$a = a\left(1 - \cos L\sqrt{\frac{P}{EI}}\right) \quad \text{or} \quad \cos L\sqrt{\frac{P}{EI}} = 0$$

Hence $L\sqrt{\frac{P}{EI}} = \frac{\pi}{2}, \frac{3\pi}{2}, \frac{5\pi}{2}.$

Taking the least significant value, we get

$$L\sqrt{\frac{P}{EI}} = \frac{\pi}{2} \quad \text{or} \quad P = P_E = \frac{\pi^2 EI}{4L^2} \qquad ...(19.2)$$

Case (*c*) : Both ends fixed

Let both the ends A and B be direction fixed as well as position fixed. The deformed shape of the column has been shown in Fig. 19.4 (*c*). The column is convex for a length of $L/4$ from either ends and concave for the middle half. At any distance x from A, we have

$$EI\frac{d^2y}{dx^2} = M - Py$$

where M is fixed moment at the ends, as in the case of a fixed beam.

Re-arranging the above equation, we get

$$\frac{d^2y}{dx^2} + \frac{P}{EI}y = \frac{M}{EI}$$

The solution of the above differential equation is

$$y = c_1 \cos x\sqrt{\frac{P}{EI}} + c_2 \sin x\sqrt{\frac{P}{EI}} + \frac{M}{P} \qquad ...(1)$$

Differentiating it, we get

$$\frac{dy}{dx} = -c_1\sqrt{\frac{P}{EI}}\sin x\sqrt{\frac{P}{EI}} + c_2\sqrt{\frac{P}{EI}}\cos x\sqrt{\frac{P}{EI}} \qquad ...(2)$$

At A, $x = 0$, $\quad y = 0 \quad \therefore \quad c_1 = -\frac{M}{P}$

At A, $x = 0$, $\quad \frac{dy}{dx} = 0 \quad \therefore \quad c_2 = 0.$

Substituting the values of $\frac{M}{P}$ and c_2 in (1), we get

$$y = c_1 \cos x\sqrt{\frac{P}{EI}} - c_1 = c_1\left(\cos x\sqrt{\frac{P}{EI}} - 1\right)$$

At B, $x = L$ and $\quad y = 0 = c_1\left(\cos L\sqrt{\frac{P}{EI}} - 1\right)$

Either $c_1 = 0$ in which case the column has not bent at all or

$$\cos L\sqrt{\frac{P}{EI}} - 1 = 0 \quad \text{or} \quad \cos L\sqrt{\frac{P}{EI}} = 1$$

$$\therefore \quad L\sqrt{\frac{P}{EI}} = 0, 2\pi, 4\pi \qquad ...(3)$$

Also, at B, $x = L$ and $\frac{dy}{dx} = 0$. Hence from (2)

$$\therefore \quad 0 = -c_1\sqrt{\frac{P}{EI}}\sin L\sqrt{\frac{P}{EI}} \quad \text{or} \quad \sin L\sqrt{\frac{P}{EI}} = 0$$

$$\therefore \quad L\sqrt{\frac{P}{EI}} = 0, \pi, 2\pi, 3\pi \qquad ...(4)$$

The minimum significant value of $L\sqrt{\frac{P}{EI}}$ consistent both with (3) and (4) is 2π.

$$\therefore \quad L\sqrt{\frac{P}{EI}} = 2\pi \quad \text{or} \quad P = P_E = \frac{4\pi^2 EI}{L^2} \qquad ...(19.3)$$

Case (*d*) : One end fixed, the other hinged

Fig. 19.4 (*d*) represents the deflected shape of a column which is position and direction fixed at *A* and is hinged at *B*. Since end *B* is direction fixed, a bending moment *M* is induced at *A* and this involves the presence of a force *R* at right angles to *AB* to keep equilibrium. Hence, we have

$$EI\frac{d^2y}{dx^2} = -Py + R(L-x)$$

or
$$\frac{d^2y}{dx^2} + \frac{P}{EI}y = \frac{R}{EI}(L-x)$$

The solution of the above differential equation is

$$y = c_1 \cos x\sqrt{\frac{P}{EI}} + c_2 \sin x\sqrt{\frac{P}{EI}} + \frac{R}{P}(L-x) \qquad ...(1)$$

$$\therefore \quad \frac{dy}{dx} = -c_1\sqrt{\frac{P}{EI}}\sin x\sqrt{\frac{P}{EI}} + c_2\sqrt{\frac{P}{EI}}\cos x\sqrt{\frac{P}{EI}} - \frac{R}{P} \qquad ...(2)$$

At A, $x = 0$ and $y = 0$ $\therefore$ $c_1 = \frac{RL}{P}$

At A, $x = 0$, $\frac{dy}{dx} = 0 = c_2\sqrt{\frac{P}{EI}} - \frac{R}{P}$ or $c_2 = \frac{R}{P}\sqrt{\frac{EI}{P}}$.

Substituting the value of c_1 and c_2 in (1), we get

$$y = -\frac{RL}{P}\cos x\sqrt{\frac{P}{EI}} + \frac{R}{P}\sqrt{\frac{EI}{P}}\sin x\sqrt{\frac{P}{EI}} + \frac{R}{P}(L-x)$$

Now at B, $x = L$ and $y = 0$

$$\therefore \quad 0 = -\frac{RL}{P}\cos L\sqrt{\frac{P}{EI}} + \frac{R}{P}\sqrt{\frac{EI}{P}}\sin L\sqrt{\frac{P}{EI}}$$

or
$$\tan L\sqrt{\frac{P}{EI}} = L\sqrt{\frac{P}{EI}}$$

Thus the value of $L\sqrt{\frac{P}{EI}}$ has to be such that its tangent is equal to the angle itself. The smallest root of the above equation is therefore,

$$L\sqrt{\frac{P}{EI}} = 4.49 \text{ radians}$$

or
$$\frac{L^2P}{EI} = 20 = 2\pi^2 \text{ (approx.)}$$

or
$$P = P_E = \frac{2\pi^2 EI}{L^2} \qquad ...(19.4)$$

19.4. THE EQUIVALENT LENGTH

In all the four cases, we have deduced the Euler's crippling load. In general, Euler's crippling load can be expressed in the form $P_E = \frac{\pi^2 EI}{CL^2}$ in which C is a constant which takes into account the method of fixing the ends (*i.e.*, whether the ends are fixed or hinged or free).

For a column hinged at both the ends, represented in Fig. 19.4 (*a*), $C = 1$, and hence it will be called the ***standard case***. For case (*b*), $C = 4$; for case (*c*), $C = \frac{1}{4}$ and for case (*d*), $C = \frac{1}{2}$. In the place of using the actual length L of the column, we can use that length of the column which is *equivalent* to that of a pin jointed column. Thus

$$P_E = \frac{\pi^2 EI}{L_E{}^2} = \frac{\pi^2 EI}{l^2} \qquad \text{...(19.5)}$$

The above expression is applicable for all end conditions. L_E (or l) in the above formula is equal to $L\sqrt{C}$ and is known as *equivalent length* or *effective length*. The equivalent length of different columns is given below :

End conditions	*Both ends hinged*	*One end fixed other free*	*Both ends fixed*	*One end fixed other hinged*
Equivalent length L_E (or l) $= L\sqrt{C}$	L	$2L$	$\frac{L}{2}$	$\frac{L}{\sqrt{2}}$

The concept of *equivalent length* is useful in practical applications, since actual compression members are seldom truely hinged or completely fixed. In many cases, the equivalent length or effective length is estimated or is based on results of tests. For commonly used flat ended columns, the point of inflection at the critical load occurs somewhere between the two limiting cases represented in Fig. 19.4 (*a*) and Fig. 19.4 (*c*). In building and bridge practice, the equivalent length is taken equal to $\frac{3}{4}$ L for columns whose ends are riveted to other members, and equal to $\frac{7}{8}$ L for pinned ends, as the pins are known to develop some moment of resistance by friction.

It is to be noted that no factor of safety is included in the Euler's equations.

19.5. LIMITATIONS OF EULER'S FORMULA

The Euler's theory is based on the following assumptions :

(*i*) The axis of the strut is perfectly straight when unloaded.

(*ii*) The line of thrust coincides exactly with the unstrained axis of the strut.

(*iii*) The flexural rigidity EI is uniform.

(*iv*) The material is isotropic.

(*v*) The buckling value of $P = P_E$ is assumed to obtain for all degrees of flexure.

Usually, the first two assumptions are not fully realised in practice. The column may have initial curvature, distortion or crookedness. The theory, therefore, refers to an *ideal strut* and not to a real one.

It is interesting to note that no *strength property* of the material appears in the Euler's formulae, yet they determine the carrying capacity of a column. The only material property involved is the elastic modulus *E*, which physically represents the stiffness characteristic of the material. *The effect of the imperfection is to convert the strut problem from a problem of stability to a problem of stresses, of which Euler's theory takes no account.*

The critical stress p_E, which is defined as an average stress over the cross-section is given by

$$p_E = \frac{P_E}{A} = \frac{\pi^2 EI}{AL^2} = \frac{\pi^2 EAr^2}{AL^2} = \frac{\pi^2 Er^2}{L^2} \frac{\pi^2 E}{(L/r)^2} \quad \ldots[19.5\ (a)]$$

for the standard case, where *r* is the *radius of gyration*.

In the above expression, *L/r* is known as the *slenderness ratio.* The graphical interpretations of equation 19.5 is shown by the curve *ACB* in Fig. 19.5, where the critical column stress is plotted versus the slenderness ratio for mild steel ($E = 2 \times 10^5$ N/mm^2). It will be appreciated that the curve is entirely defined by the magnitude of *E* and is independent of its ultimate strength. A long column has greater value of *L/r* and loses its strength at very small compressive stress. This condition cannot be improved by taking a steel of higher strength, since the modulus of steel does not vary much with alloy and heat treatment, and remains practically constant.

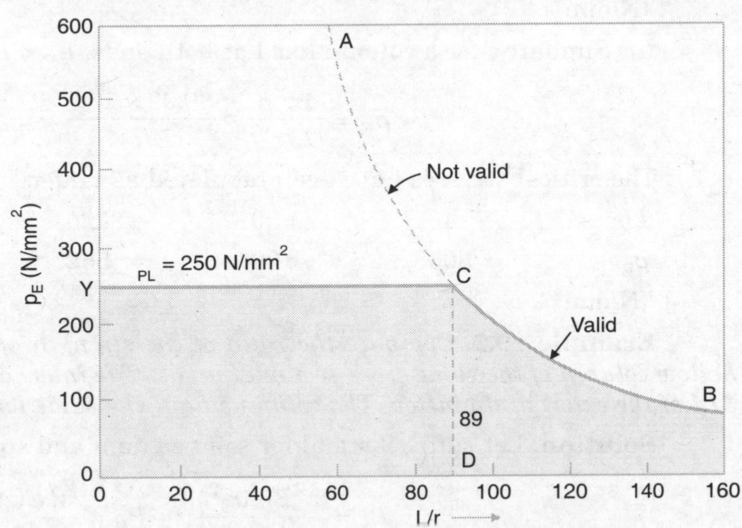

Fig. 19.5. Validity Limits of Euler's Formula

In Fig. 19.5, let *OY* represent the proportional limit (σ_{PL}) of the material. It is clear that the Euler formula cannot possibly apply if *L/r* is less than *OD* since for stress greater than *OY* the material becomes plastic and no longer follows Hooke's Law. Taking proportional limit (σ_{PL}) equal to 250 N/mm^2 for mild steel, we have $\frac{L}{r} = OD = \sqrt{\frac{\pi^2 \times 2 \times 10^5}{250}} = 89.$

Hence the Euler's formula is applicable only if L/r is greater than 89 for mild steel column hinged at both the ends. Hence the critical stress is given by solid line *YCB* of Fig. 19.5. The dotted portion of the curve in not valid.

Example 19.1. *Using Euler's formula, calculate the critical stresses for a series of struts having slenderness ratio of 40, 80, 120, 160 and 200 under the following conditions : (i) Both ends hinged, (ii) Both ends fixed. Take* $E = 2.05 \times 10^5\ N/mm^2$.

Solution. (*i*) $p_E = \dfrac{\pi^2 E \,.\, r^2}{L_E{}^2}$, where L_E = equivalent length.

For both ends hinged, $p_E = \dfrac{\pi E r^2}{L^2} = \dfrac{\pi^2 E}{(L/r)^2} = \dfrac{20.23 \times 10^5}{(L/r)^2}$

Substituting the value of L/r in the above formula, the critical stress can be calculated and tabulated as under :

$L/r \rightarrow$	40	80	120	160	200
$p_E \rightarrow$ (N/mm^2)	1264	316	140	79	50.6

(*ii*) Similarly, for a column fixed at both ends, $L_E = L/2$.

$$\therefore \quad p_E = \frac{4\pi^2 E}{(L/r)^2} = \frac{80.93 \times 10^5}{(L/r)^2}$$

The critical stresses have been tabulated as under :

$L/r \rightarrow$	40	80	120	160	200
$p_E \rightarrow$ (N/mm^2).	5058	1264	562	316	202.3

Example 19.2. *Compare the ratio of the strength of a solid steel column to that of a hollow column of the same cross-sectional areas. The internal diameter of the hollow column is 3/4 of the external diameter. The columns have the same length and are pinned at the ends.*

Solution. Let suffix *S* stand for solid column and suffix *H* stand for hollow column.

$$P_S = \frac{\pi^2 E(A r_S{}^2)}{L^2} \;;\quad P_H = \frac{\pi^2 E(A r_H{}^2)}{L^2} \quad \text{Hence} \quad \frac{P_H}{P_S} = \left(\frac{r_H}{r_S}\right)^2$$

Now, for solid section, $r_S{}^2 = \dfrac{\pi}{64} d^4 \times \dfrac{4}{\pi d^2} = \dfrac{d^2}{16}$

For hollow section, $r_H{}^2 = \dfrac{\dfrac{\pi}{64}(D^4 - D'^4)}{\dfrac{\pi}{4}(D^2 - D'^2)} = \left(\dfrac{D^2 + D'^2}{16}\right)$ ...(1)

where D and D' and external and internal diameters respectively.

$$\therefore \quad \frac{P_H}{P_S} = \frac{r_H^{\ 2}}{r_S^{\ 2}} = \frac{(D^2 + D'^2)}{d^2} = \frac{D^2 + \frac{9}{16}D^2}{d^2} = \frac{25}{16}\cdot\frac{D^2}{d^2} \qquad \ldots(2)$$

Since cross-sectional area of the both are equal, we have

$$A = \frac{\pi}{4}d^2 = \frac{\pi}{4}\left\{D^2 - \left(\frac{3}{4}D\right)^2\right\}$$

or

$$d^2 = \frac{7}{16}D^2 \quad \text{or} \quad \frac{D^2}{d^2} = \frac{16}{7}$$

$$\therefore \quad \frac{P_H}{P_S} = \frac{25}{16}\cdot\frac{D^2}{d^2} = \frac{25}{16}\times\frac{16}{7} = \mathbf{\frac{25}{7}}.$$

Example 19.3. *A straight bar of alloy 900 mm long and 12.7 mm × 4.8 mm in section is mounted in strut testing machine and loaded axially till it buckles. Assuming the Euler formula for pinned ends to apply, estimate the maximum central deflection before the material attains its yield point at 300 N/mm². Take E = 7.7 × 10⁴ N/mm².*

Solution. $$I_{min.} = \frac{1}{12}(12.7)(4.8)^3 = 117 \text{ mm}^4$$

$$P_E = \frac{\pi^2 EI}{L^2} = \frac{9.86\times7.7\times10^4}{900\times900}\times 117 = 109.7 \text{ N}$$

$M_{max.}$ at half length = $P.\,\delta$, where δ is the maximum central deflection.

$$\therefore \quad P.\,\delta = M = f_b\,.\,Z = f_b\left(\frac{117}{2.4}\right) = 48.75\,f_b$$

$$\therefore \quad f_b = \frac{P.\,\delta}{48.75} = \frac{109.7\times\delta}{48.75} = 2.25\,\delta \qquad \ldots(1)$$

Also, $$f_0 = \frac{P}{A} = \frac{109.7}{12.7\times4.8} = 1.80 \text{ N/mm}^2 \qquad \ldots(2)$$

But $$f_b + f_0 = 300$$

$$\therefore \quad 2.25\,\delta + 1.80 = 300$$

$$\therefore \quad \delta = \frac{300 - 1.8}{2.25} = \mathbf{132.5 \text{ mm.}}$$

Example 19.4. *A straight cylindrical bar is 15 mm diameter and 1.2 m long. It is freely supported at its two ends in a horizontal position and loaded at the centre with a concentrated load of 90 N. The central deflection is found to be 5 mm.*

If placed vertical and loaded along its axis, what load would cause it to buckle ? What is the ratio of the maximum stress in the two cases ?

Solution. $$\delta = 5 = \frac{WL^3}{48EI} = \frac{90\times1200}{48}\,\frac{L^2}{EI}$$

(where L, E and I are in mm units).

$$\therefore \quad \frac{EI}{L^2} = \frac{90 \times 1200}{48 \times 5} = 450 \qquad \ldots(1)$$

Now $\quad P_E = \dfrac{\pi^2 EI}{L^2} = \pi^2 \times 450 = 4441 \text{ N}$

As a beam, $\quad M_{max} = \dfrac{WL}{4} = \dfrac{90 \times 1200}{4} = 27000 \text{ N-mm}$

$$\therefore \quad f_b = \frac{M}{Z} = \frac{27000}{\frac{\pi}{32}(15)^3} = 81.5 \text{ N/mm}^2$$

As a strut, $\quad p_E = \dfrac{P_E}{A} = \dfrac{4441}{\frac{\pi}{4}(15)^2} = 25.13 \text{ N/mm}^2$

$$\therefore \quad \frac{f_b}{p_E} = \frac{81.5}{25.13} = \mathbf{3.24.}$$

19.6. INTERMEDIATE COLUMNS: EMPIRICAL FORMULAE

We have seen that Euler's formula is valid only for long columns, *i.e.,* for column having (L/r) ratio greater than a certain value for a particular material. Also, short columns fail in direct compression only. Euler's formula, though valid for (L/r) higher than 89 (for Mild Steel having $\sigma_{PL} = 250$ N/mm^2), and $L/r = 99$ (for mild steel having $\sigma_{PL} = 200$ N/mm^2) does not take into account the direct compressive stress, and hence gives correct results only for very long columns. For columns intermediate between very long and short ones, Euler's formula, though applicable, does not give correct results. For such columns, empirical formulae have been proposed by various workers. However, none of these have been accepted universally for intermediate columns, partly because of their departure from the stress strain relationship when the stresses exceed the proportional limit, and partly because of their intermediate mixture of direct and flexural stresse. We shall consider here the following empirical formulae,

1. Rankine's formula
2. Gorden's formula
3. Johnson's formulae.

19.7. RANKINE'S FORMULA

Rankine proposed an empirical formula for columns which cover all cases ranging from very short to very long struts. He proposed the relation

$$\frac{1}{P} = \frac{1}{P_C} + \frac{1}{P_E}$$

where $P_C = fc.A$ = ultimate load for a struct, and

$$P_E = \frac{\pi^2 EI}{L^2} = \text{Eulerian crippling load for the standard case.}$$

In the above relation, $1/P_C$ is constant for a material. For short columns, P_E is very large and hence $1/P_E$ is small in comparison to $1/P_C$, thus making the crippling load P approximately equal to P_C. For long columns, P_E is extremely small and hence $1/P_E$ is large as compared to $1/P_C$, thus making the crippling load P approximately equal to P_E. Thus the value of P obtained from the above relation covers all cases ranging from short to long columns or struts.

The above relation can be re-arranged as

$$P = \frac{P_C P_E}{P_E + P_C} = \frac{P_C}{1 + \dfrac{P_C}{P_E}} = \frac{f_C A}{1 + \dfrac{f_C A L^2}{\pi^2 EI}}$$

or

$$P = \frac{f_C . A}{1 + \left(\dfrac{f_C}{\pi^2 E}\right)\left(\dfrac{L}{r}\right)^2} = \frac{f_C . A}{1 + a\left(\dfrac{L}{r}\right)^2} \qquad \text{...(19.6)}$$

In the above equation, f_C is the crushing stress for the material and a is the Rankine's constant for the material. Though the formula is empirical, the value of the constant a is not calculated from the values of f_C, π^2 and E, but is determined experimentally.

The following table gives the values of f_C and a for the materials most commonly used for columns and struts.

S.No.	*Material*	f_C		*a for ends hinged*
		kg/cm²	*N/mm²*	
1.	Wrought Iron	2550	255	$\frac{1}{9000}$
2.	Cast iron	5670	567	$\frac{1}{1600}$
3.	Mild Steel	3300	330	$\frac{1}{7500}$
4.	Strong Timber	500	50	$\frac{1}{750}$

Equation 19.6 is the Rankine's formula for the standard case of two hinged columns. It is sometimes known as Rankine-Gordon formula. For columns with other end conditions, the value of the constant will be changed accordingly. However, since a is a constant for a particular material used as a hinged column, it is not convenient to remember the values of a for other end conditions. It is, therefore, better to modify the Rankine's formula and remember it in the form :

$$P = \frac{f_C A}{1 + a\left(\dfrac{L_E}{r}\right)^2} \qquad \text{...(19.7)}$$

in which a is the Rankine's constant for a particular material and for the standard case of column and L_E is the equivalent length of the column. Since equivalent length is generally known, it can be substituted in the above formula to get the Rankine's formula for a particular case of end conditions. Equations 19.6 and 19.7 do not include the factor of safety.

Another formula, similar to Eq. 19.7 was proposed by *Ritter.* His expression for an *average axial stress* is

$$\frac{P}{A} = \frac{f}{1 + a\left(\frac{L}{r}\right)^2} \quad \text{...[19.7 (a)]}$$

in which f is allowable stress and $a = \dfrac{f_C}{\pi^2 E}$. The only difference between Eqs. 19.7 and 19.8 is that the later includes the factor of safety.

Example 19.5. *Determine the section of a cast iron hollow cylindrical column 5 metre long with ends firmly built-in if it carries an axial load of 300 kN. The ratio of internal to external diameter is 3/4. Use factor of safety of 8. Take $f_C = 567$ N/mm² and Rankine's constant $a = 1/1600$.*

Solution. Let D = external diameter and d = internal diameter $= \frac{3}{4} D$

$$A = \frac{\pi}{4}(D^2 - d^2) = 0.343\, D^2 \; ; \quad r^2 = \frac{D^2 + d^2}{16} = \frac{D^2 + \frac{9}{16}D^2}{16} = \frac{25}{256} D^2$$

Since the factor of safety is 8, the crippling load = 300 × 8 = 2400 kN.

The modified Rankine's formula is

$$P = \frac{f_C A}{1 + a\left(\frac{L_E}{r}\right)^2}, \text{ where } L_E = \frac{L}{2}. \text{ For a column fixed at both ends.}$$

Hence $$2400 \times 1000 = \frac{567 \times 0.343 D^2}{1 + \frac{1}{1600}\left(\frac{5 \times 1000}{2}\right)^2 \left(\frac{256}{25 D^2}\right)}$$

or $$D^4 - 12340\, D^2 - 4.936 \times 10^8 = 0, \text{ on simplification.}$$

Hence $D^2 = 29286$ or D = **171.1 mm** and $d = 171.1 \times 0.75$ = **128.3 mm.**

Example 19.6. *Compare the crippling loads given by Euler's and Rankine's formula for a tubular steel strut 2.3 m long having outer and inner diameters 38 mm and 33 mm respectively, loaded through pin joints at each end. Take the yield stress as 335 N/mm², the Rankine's constant = 1/7500 and E = 0.205 × 10⁶ N/mm². For what length of strut of this cross-section does the Euler formula cease to apply ?*

Solution. $$I = \frac{\pi}{64}(38^4 - 33^4) = 4.51 \times 10^4 \text{ mm}^4 \; ; \; A = \frac{\pi}{4}(38^2 - 33^2) = 275 \text{ mm}^2$$

$$r = \sqrt{\frac{4.51 \times 10^4}{275}} = 12.8 \text{ mm} \; ; \; L = 2.3 \text{ m} = 2300 \text{ mm}$$

$$P_E = \frac{\pi^2 EI}{L^2} = \frac{\pi^2 \times 0.205 \times 10^6 \times 4.5 \times 10^4}{2300 \times 2300} = 17250 \text{ N} = \mathbf{17.25 \text{ kN.}}$$

$$P_R = \frac{f_C . A}{1 + a\left(\frac{L}{r}\right)^2} = \frac{335 \times 275}{1 + \frac{1}{7500}\left(\frac{2300}{12.8}\right)^2} = \frac{335 \times 275}{1 + 4.305} = 17370 \text{ N} = 17.37 \text{ kN}$$

$\therefore$ $$\frac{P_E}{P_R} = \frac{17.25}{17.37} = \mathbf{0.99} \text{ ; Also, } \frac{P_E}{A} = 335 \text{ N/mm}^2 = \frac{\pi^2 E r^2}{L^2}$$

Validity limit of $\frac{L}{r} = \sqrt{\frac{\pi^2 E}{335}} = \pi\sqrt{\frac{0.205 \times 10^6}{335}} = \mathbf{77.7.}$

Example 19.7. *A hollow cast iron column with fixed ends supports an axial load of 1000 kN. If the column is 5 m long and has an external diameter of 250 mm, find the thickness of metal required. Use Rankine's formula, taking a constant 1/6400 and assume a working stress of 80 N/mm^2.*

Solution. The Rankine's constant a for cast iron has empirical value of 1/1600. As the given value is 1/4 of this, it includes the effect of end conditions. Let D and d be external and internal diameters respectively.

$\therefore$ $$P = \frac{f_C . A}{1 + \frac{1}{6400}\left(\frac{L}{r}\right)^2}, \text{ where } L = \text{overall length.}$$

Now $$A = \frac{\pi}{4}(D^2 - d^2) \text{ and } r^2 = \frac{D^2 + d^2}{16}$$

$\therefore$ $$P = 1000000 = \frac{80 \times \frac{\pi}{4}(D^2 - d^2)}{1 + \frac{1}{6400}\frac{(500)^2 \times 16}{D^2 + d^2}} = \frac{62.83\,(D^2 - d^2)}{1 + \frac{625}{D^2 + d^2}}$$

or $$1000000 + \frac{625 \times 1000000}{D^2 + d^2} = 62.83\,(D^2 - d^2)$$

or $$10^6 (D^2 + d^2) + 625 \times 10^6 = 62.83\,(D^4 - d^4)$$

or $$625 \times 10^8 + 10^6 d^2 + 625 \times 10^6 = 2454 \times 10^8 - 62.83\, d^4$$

$\therefore$ $$d^4 + 15916\, d^2 - 29.01 \times 10^8 = 0$$

From which $$d^2 = 46488 \quad \text{and} \quad d = 215.6 \text{ mm}$$

$\therefore$ $$\text{Metal thickness} = \frac{250 - 215.6}{2} = \mathbf{17.19 \text{ mm.}}$$

19.8. GORDON'S FORMULA

Gordon proposed the following formula :

$$P = \frac{f_A}{1 + a' \dfrac{L^2}{b^2}} \qquad \text{...(19.8)}$$

in which a' = Gordon's constant $= \dfrac{ab^2}{r^2}$

b = Overall lesser dimension ; a = Rankine's constant.

Gordon's constant a' is not a mathematical constant like Rankine's but its value depends upon the properties of the material and the shape of the section.

19.9. JOHNSON'S FORMULA

Johnson proposed the following two formulae :

$$P = A\left(f - n\frac{L}{r}\right) \qquad \text{...(19.9)}$$

$$P = A\left(f - n'\frac{L^2}{r^2}\right) \qquad \text{...[19.9(a)]}$$

Equation 19.9 is a *straight line formula,* as a straight line will be obtained if P/A is plotted against L/r. The constant n depends on the material and upon the end conditions of the column, f is the allowable stress of the material and therefore, P is the safe load for the column.

For mild steel, f = 1125 kg/cm^2 and n = 5.625 for both ends hinged and n' = 4.219 for both ends fixed. Thus, Johnson's straight line formula becomes

$$P = A\left(1125 - 5.625\frac{L}{r}\right) \text{ kg for both ends hinged} \qquad \text{...[19.9 (b)]}$$

and

$$P = A\left(1125 - 4.219\frac{L}{r}\right) \text{ for both ends fixed} \qquad \text{...(19.10)}$$

Equation 19.9(a) is a *parabolic formula*, as a parabola will be obtained if P/A is plotted against L/r. The constant n' depends on the material and upon the end conditions of the column.

19.10. DESIGN OF COLUMNS

The procedure for designing the column is that of a trial and error, except for the circular section in which L/r is directly known in terms of diameter. However, for other sections, a suitable value of L/r is assumed to start with. Using any suitable formula, the value of P/A is calculated. Since P is known and P/A is also known, the area of a section A is known. Now, from the standard structural steel tables, a suitable section is chosen having the area of section A calculated above and having L/r *nearer* to the assumed value. The calculations are then repeated with the new value of L/r and a new area A is found.

19.11. LONG COLUMNS UNDER ECCENTRIC LOADING : SECANT FORMULA

(*a*) **Empirical Formula:** In chapter 18, we have seen that if a short column is subjected to an eccentric compressive load *P*, the maximum compressive stress is given by

$$f = f_0 + f_b = \frac{P}{A} + \frac{Pe}{Z} = \frac{P}{A} + \frac{Pe.y_c}{Ar^2}$$

where y_c = distance of the extreme compressive fibre from N.A.

Hence $$f = \frac{P}{A}\left(1 + \frac{ey_c}{r^2}\right)$$

If f = permissible stress for the material, the corresponding load is given by

$$P = \frac{f\,A}{\left(1 + \frac{ey_c}{r^2}\right)} \qquad \text{...(1)}$$

in which $\left(1 + \frac{ey_c}{r^2}\right)$ is a *reduction factor* which takes into account the *eccentricity* of loading.

For a long column loaded with axial loading, the safe load, by Rankine's formula, is given by

$$P = \frac{f.A}{1 + a\left(\frac{L}{r}\right)^2} \qquad \text{...(2)}$$

in which $\left\{1 + a\left(\frac{L}{r}\right)^2\right\}$ is a *reduction factor* which takes into account the *buckling* of the column.

Hence, for a long column loaded with eccentric loading, the modified Rankine's formula can be empirically stated as

$$P = \frac{f.A}{\left(1 + e\frac{y_c}{r^2}\right)\left(1 + a\,\frac{L^2}{r^2}\right)} \qquad \text{...(19.11)}$$

The above formula is for the standard case. For other cases, the constant a can be changed accordingly.

(*b*) **The Secant Formula:** We shall now investigate the problem from the first principles for the *standard case.*

At any section distant *x* from *A*, *y* is the deflection with reference to the line of action of *P* which is acting at an eccentricity *e* at the ends. The B.M. is + *Py*.

Hence $$EI\,\frac{d^2y}{dx^2} = -M = -Py$$

or $$\frac{d^2y}{dx^2} + \frac{P}{EI}\,y = 0$$

The solution of the above differential equation is

$$y = C_1 \sin x \sqrt{\frac{P}{EI}} + C_2 \cos x \sqrt{\frac{P}{EI}} \quad \text{...(1)}$$

At A, $x = 0$ and $y = e$

Hence, $C_2 = e$

Also, $\frac{dy}{dx} = C_1 \sqrt{\frac{P}{EI}} \cos x \sqrt{\frac{P}{EI}} - C_2 \sqrt{\frac{P}{EI}} \sin x \sqrt{\frac{P}{EI}}$

At the mid-height of the column,

$$\frac{dy}{dx} = 0 \quad \text{and} \quad x = L/2$$

$$0 = C_1 \sqrt{\frac{P}{EI}} \cos \frac{L}{2} \sqrt{\frac{P}{EI}} - e \sqrt{\frac{P}{EI}} \sin \frac{L}{2} \sqrt{\frac{P}{EI}}$$

or

$$C_1 = e \frac{\sin \frac{L}{2} \sqrt{\frac{P}{EI}}}{\cos \frac{L}{2} \sqrt{\frac{P}{EI}}}$$

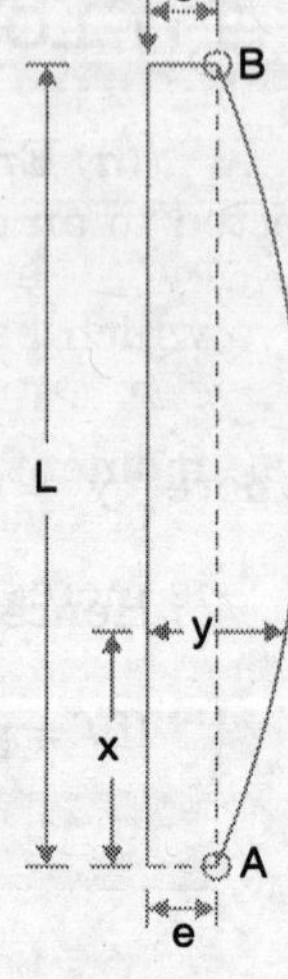

Fig. 19.6

Substituting the values of C_1 and C_2 in (1), we get

$$y = e \left[\frac{\sin \frac{L}{2} \sqrt{\frac{P}{EI}}}{\cos \frac{L}{2} \sqrt{\frac{P}{EI}}} \sin x \sqrt{\frac{P}{EI}} + \cos x \sqrt{\frac{P}{EI}} \right] \quad \text{...(2)}$$

At $x = L/2$, $y = y_{max}$.

$$\therefore \quad y_{max} = e \left[\frac{\sin^2 \frac{L}{2} \sqrt{\frac{P}{EI}}}{\cos \frac{L}{2} \sqrt{\frac{P}{EI}}} + \cos \frac{L}{2} \sqrt{\frac{P}{EI}} \right] = e \sec \frac{L}{2} \sqrt{\frac{P}{EI}} \quad \text{...(3)}$$

The largest bending moment M_{max} occurs at $x = L/2$, where y is maximum.

$$\therefore \quad M_{max} = Py_{max} = Pe \sec \frac{L}{2} \sqrt{\frac{P}{EI}}$$

The maximum compressive stress occurring in the column may be computed by usual formula, as

$$f = \frac{P}{A} + \frac{M}{Z}$$

or

$$f = \frac{P}{A} + \frac{Pe \sec . \frac{L}{2} \sqrt{\frac{P}{EI}}}{Z} = \frac{P}{A} \left(1 + \frac{ey_c}{r^2} \sec \frac{L}{2} \sqrt{\frac{P}{EI}} \right) \quad \text{...(19.12)}$$

This equation, is known as the *secant formula* for columns, since it contains a secant term.

In order to cover the cases for all end conditions, the secant formula can be written as

$$f = \frac{P}{A} + \frac{P.e \sec \dfrac{L_E}{2}\sqrt{\dfrac{P}{EI}}}{Z} = \frac{P}{A}\left[1 + \frac{ey_c}{r^2} \sec \frac{L_E}{2}\sqrt{\frac{P}{EI}}\right] \qquad \text{...(19.13)}$$

In equation 19.12, r is obtained from the value of I associated with the axis around which bending occurs.

It is interesting to note that in a short column with an eccentric load P, the bending moment at *all sections* is $P.e$ while in the case of *long column*, the maximum bending moment is increased by a factor $\sec \frac{L_E}{2}\sqrt{\frac{P}{EI}}$.

19.12. PERRY'S FORMULA FOR LONG COLUMN WITH ECCENTRIC LOAD

It will be noted that equation 19.12 is quite convenient to calculate the maximum value of stress, if P and e are given. However, if it is required to calculate the load P for a given stress and limit of eccentricity, Eq. 19.12 is not convenient, since secant function contains P itself. It is necessary, therefore, to put the above equation in a more workable form :

$$\sec \frac{L}{2}\sqrt{\frac{P}{EI}} = \sec \sqrt{\frac{PL^2}{4EI}} = \sec \frac{\pi}{2}\sqrt{\frac{P}{P_E}}$$

where P_E = Eulerian load = $\frac{\pi^2 EI}{L^2}$.

Prof. Perry found that $\sec \frac{\pi}{2}\sqrt{\frac{P}{P_E}} \simeq \frac{1.2\, P_E}{P_E - P}$

$$\therefore \qquad \sec \frac{L}{2}\sqrt{\frac{P}{EI}} \simeq \frac{1.2\, P_E}{P_E - P}$$

Substituting in equation 19.12, we get

$$f = \frac{P}{A}\left(1 + \frac{ey_c}{r^2} \cdot \frac{1.2\, P_E}{P_E - P}\right)$$

Writing $f_0 = \frac{P}{A}$ and $p_E = \frac{P_E}{A}$, we get

$$f = f_0\left(1 + \frac{ey_c}{r^2} \cdot \frac{1.2\, p_E}{p_E - f_0}\right)$$

$$\therefore \qquad \left(\frac{f}{f_0} - 1\right)\left(1 - \frac{f_0}{p_E}\right) = \frac{1.2 e y_c}{r^2} \qquad \text{...(19.14)}$$

Equation 19.14 is known as Prof. Perry's approximate formula from which f_0 and hence $P(= f_0 A)$ can be easily calculated.

Example 19.8. *A mild steel column is of hollow circular section with 120 mm as external diameter and 90 mm as internal diameter. The column is 3 m long, hinged at both the ends and has to carry a load of 80 kN at an eccentricity of 20 mm from the geometrical axis. Calculate the maximum and minimum intensities of stresses. Also, calculate the maximum permissible eccentricity so that no tension is induced anywhere in the section. Take E = 2.05 × 10⁵ N/mm².*

Solution. $A = \frac{\pi}{4}(120^2 - 90^2) = 4950 \text{ mm}^2$; $I = \frac{\pi}{64}(120^4 - 90^4) = 6.96 \times 10^6 \text{ mm}^4$

$$r^2 = \frac{I}{A} = \frac{6.96 \times 10^6}{4950} = 1406 \text{ mm}^2 \; ; \; Z = \frac{6.96 \times 10^6}{60} = 11.6 \times 10^4 \text{ mm}^3$$

$$M_{max} = P.\, e \sec \frac{L}{2}\sqrt{\frac{P}{EI}} = 80000 \times 20 \sec \frac{3000}{2}\sqrt{\frac{80000}{2.05 \times 10^5 \times 6.96 \times 10^6}}$$

$$= 1600000 \sec 0.355 = 1600000 \times 1.0666 = 1.706 \times 10^6 \text{ N-mm}$$

$$f_b = \frac{M}{Z} = \frac{1.706 \times 10^6}{11.6 \times 10^4} = 14.7 \text{ N/mm}^2 \text{ ; Also, } f_0 = \frac{P}{A} = \frac{80000}{4950} = 16.2 \text{ N/mm}^2$$

$\therefore$ $f_{max} = f_0 + f_b = 16.2 + 14.7 =$ **30.9 N/mm²** (comp.)

$f_{min} = f_0 - f_b = 16.2 - 14.7 =$ **1.5 N/mm²** (comp.)

For no tension $f_b = f_0$

$\therefore$ $14.7 \times \frac{e}{20} = 16.2$ or $e = \frac{16.2 \times 20}{14.7} =$ **22 mm.**

Example 19.9. *A mild steel column is of hollow circular section with 120 mm external diameter and 90 mm internal diameter. The column is 3 m long and is hinged at both the ends. Calculate the maximum permissible load with an eccentricity of 20 mm if the maximum compressive stress is limited to 80 N/mm². Take E = 2.05 × 10⁵ N/mm².*

Solution. As in Example 19.8,

$A = 4950 \text{ mm}^2$; $I = 6.96 \times 10^6 \text{ mm}^4$; $r^2 = 1406 \text{ mm}^2$; $Z = 11.6 \times 10^4 \text{ mm}^3$

$$P_E = \frac{\pi^2 EI}{L^2} = \frac{\pi^2 \times 2.05 \times 10^5 \times 6.96 \times 10^6}{3000 \times 3000} = 15.65 \times 10^5 \text{ N}$$

$$p_E = \frac{P_E}{A} = \frac{15.65 \times 10^5}{4950} = 316 \text{ N/mm}^2$$

$f = 80 \text{ N/mm}^2$ (given)

Substituting in Perry's formula (Eq. 19.14), we get

$$\left(\frac{80}{f_0} - 1\right)\left(1 - \frac{f_0}{316}\right) = \frac{1.2 \times 20 \times 60}{1406} = 1.024$$

or $(80 - f_0)(316 - f_0) = 1.024 \times 316 f_0$

or $f_0^2 - 719.6 f_0 + 25280 = 0$, From which $f_0 = 37 \text{ N/mm}^2$

$\therefore$ Safe load $P = f_0 A = 37 \times 4950 = 183330 \text{ N} =$ **183.83 kN.**

19.13. INDIAN STANDARD FORMULA: DESIGN FORMULA

(*a*) **IS Code formula as per IS 800–1962 :**

In Eq. 19.13 putting $f = f_{max}$, we have

$$f_{max} = \frac{P}{A}\left(1 + \frac{ey_c}{r^2} \sec \frac{L}{2r}\sqrt{\frac{P}{EA}}\right)$$

or

$$\frac{P}{A} = f_c = \frac{f_{max}}{1 + \frac{ey_c}{r^2} \sec \frac{L}{2r}\sqrt{\frac{f_c}{E}}}$$

Taking m = factor of safety, we get $f_c = m\sigma_{ac}$, where σ_{ac} = allowable stress in axial compression.

Taking $f_{max} = f_y$ = guaranteed minimum yield stress for the material, we get

$$\sigma_{ac} = \frac{f_y / m}{1 + \frac{ey_c}{r^2} \sec \frac{L}{2r} \sqrt{\frac{m\sigma_{ac}}{E}}}$$

Indian Standard Code of Practice (IS : 800–1962) recommended the value of factor $\frac{ey_c}{r^2}$ as 0.20. Hence for L/r between 0 to 150, the secant formula takes the following form :

$$\sigma_{ac} = \frac{f_y / m}{1 + 0.2 \sec\left(\frac{L}{r} \sqrt{\frac{m\sigma_{ac}}{4E}}\right)} \quad ...(19.15)$$

In the above Indian Standard formula, the factor of safety m can be taken as 1.68, and E may be taken equal to 2.047×10^5 N/mm^2 for steel conforming to IS : 226–1962.

For columns having $L/r \geq 160$, the allowable stress can be found from the following expression :

$$\sigma_{ac}' = \sigma_{ac}\left(1.2 - \frac{L}{800\, r}\right) \quad ...(19.16)$$

where σ_{ac}' = allowable stress in axial compression for $l/r \geq 160$.

(*b*) **IS Code formula as per IS 800–1984 : Merchant-Rankine formula**

The above formula (Eq. 19.15), suggested in earlier versions of the IS Code, was dropped in its 1984 version.

Indian Standard IS : 800–1984 adopted Merchant-Rankine formula for determining permissible stress :

$$\sigma_{ac} = \frac{f_{cc}.\, f_y}{m\left[(f_{cc})^n + (f_y)^n\right]^{1/n}} \quad ...(19.17)$$

where σ_{ac} = permissible stress in axial compression (N/mm^2)

f_{cc} = elastic critical stress in compression = $\frac{\pi^2 E}{\lambda^2}$

E = modulus of elasticity of steel = 2×10^5 N/mm^2

f_y = yield stress in steel (N/mm^2)

$\lambda = L/r$ = Slenderness ratio

m = factor of safety = 10/6

n = a factor, assumed as 1.4.

Adopting m = 10/6, we get

$$\sigma_{ac} = 0.6\, \frac{f_{cc} \cdot f_y}{\left[(f_{cc})^n + (f_y)^n\right]^{1/n}} \quad ...[19.17(a)]$$

According to IS : 800–1984, the direct stress in compression on the gross cross-sectional area of axially loaded compression member shall not exceed 0.6 f_y, nor the permissible value σ_{ac} calculated using the above formula.

Values of σ_{ac} for some of the Indian Standard structural steels are given in Table 19.1.

TABLE 19.1. PERMISSIBLE STRESS σ_{ac} (N/mm² or MPa) IN AXIAL COMPRESSION FOR STEELS WITH VARIOUS YIELD STRESS

$f_y \rightarrow$ $\lambda \downarrow$	220	230	240	250	260	280	300	320	340	360	380	400	420	450	480	510	540
10	132	138	144	150	156	168	180	192	204	215	227	239	251	269	287	305	323
20	131	137	142	148	154	166	177	189	201	212	224	235	246	263	280	297	314
30	128	134	140	145	151	162	172	183	194	204	215	225	236	251	266	280	295
40	124	129	134	139	145	154	164	174	183	192	201	210	218	231	243	255	267
50	118	123	127	132	136	145	153	161	168	176	183	190	197	207	216	225	233
60	111	115	118	122	126	133	139	146	152	158	163	168	173	180	187	193	199
70	102	106	109	112	115	120	125	130	135	139	142	147	150	155	160	164	168
80	93	96	98	101	103	107	111	115	118	121	124	127	129	133	136	139	141
90	85	87	88	90	92	95	98	101	103	105	108	109	111	114	116	118	119
100	76	78	79	80	82	84	86	88	90	92	93	94	96	97	99	100	101
110	68	69	71	72	73	74	76	77	79	80	81	82	83	84	85	86	87
120	61	62	63	64	64	66	67	67	69	70	71	71	72	73	73	74	75
130	55	55	56	57	57	58	59	60	61	61	62	62	63	63	64	64	65
140	49	50	50	51	51	52	53	53	54	54	54	55	55	56	56	56	57
150	44	45	45	45	46	46	47	47	48	48	48	49	49	49	49	50	50
160	40	40	41	41	41	42	42	42	43	43	43	43	43	44	44	44	44
170	36	36	37	37	37	37	38	38	38	38	39	39	39	39	39	39	39
180	33	33	33	33	33	34	34	34	34	35	35	35	35	35	35	35	35
190	30	30	30	30	30	30	31	31	31	31	31	31	32	32	32	32	32
200	27	27	28	28	28	28	28	28	28	28	28	28	28	28	28	28	28
210	25	25	25	25	25	25	26	26	26	26	26	26	26	26	26	26	26
220	23	23	23	23	23	23	23	24	24	24	24	24	24	24	24	24	24
230	21	21	21	21	21	21	21	21	22	22	22	22	22	22	22	22	22
240	20	20	20	20	20	20	20	20	20	20	20	20	20	20	20	20	20
250	18	18	18	18	18	18	18	18	18	19	19	19	19	19	19	19	19

19.14. PRACTICAL END CONDITIONS AND EFFECTIVE LENGTH FACTORS

The ideal end conditions discussed in the previous articles are only of theoretical importance, since in actual practice, none of these can be achieved. However, the concept of *equivalent length* or *effective length* developed earlier is useful. Table 19.2 gives Indian Standard recommendations (IS : 800–1984) for finding effective length in the given plane. The values of these tables may also be adopted where columns form part of framed structure. Refer Fig. 19.7 along with Table 19.2.

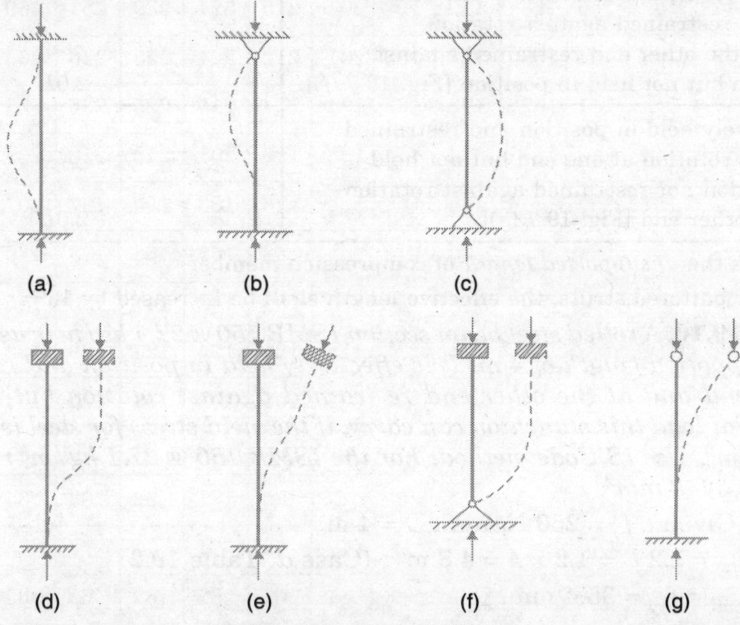

Fig. 19.7. End conditions

TABLE 19.2. EFFECTIVE LENGTHS OF COMPRESSION MEMBERS (IS : 800–1984)

Degree of end restraint of compression member	*Recommended value of effective length*
(*a*) Effectively held in position and restrained against rotation at both ends [Fig. 19.7 (*a*)].	$0.65L$
(*b*) Effectively held in position at both ends and restrained against rotation at one end [Fig. 19.7 (*b*)].	$0.80L$
(*c*) Effectively held in position at both ends, but not restrained against rotation [Fig. 19.7 (*c*)].	$1.00L$
(*d*) Effectively held in position and restrained against rotation at one end, and at the other end restrained against rotation but not held in position [Fig. 19.7 (*d*)].	$1.20L$

Degree of end restraint of compression member	*Recommended value of effective length*
(*e*) Effectively held in position and restrained against rotation at one end and at the other end partially restrained against rotation but not held in position [Fig. 19.7 (*e*)].	1.5L
(*f*) Effectively held in position at one end but not restrained against rotation, and at the other end restrained against rotation but not held in position [Fig. 19.7 (*f*)].	2.0L
(*g*) Effectively held in position and restrained against rotation at one end but not held in position nor restrained against rotation at the other end [Fig. 19.7 (*g*)].	2.0L

Note 1. L is the *unsupported length* of compression member.

Note 2. For battered struts, the effective length shall be increased by 10%.

Example 19.10. *A rolled steel beam section ISMB 150 @ 27.1 kg / m is used as a stanchion and has an unsupported length of 4 m. It is effectively held in position and restrained against rotation at one end and at the other end restrained against rotation but held in position. Determine the axial load this stanchion can carry, if the yield stress for steel is 250 N / mm² and E = 2 × 10⁵ N / mm². Use IS Code method. For the ISMB 150 @ 27.1 kg / m,* r_{xx} *= 65 mm,* r_{yy} *= 35.4 mm and a = 3448 mm².*

Solution. Given : $f_y = 250$ N/mm², $L = 4$ m.

$\therefore$ $L_E = 1.2\,L = 1.2 \times 4 = 4.8$ m (Case *d*, Table 19.2)

$r_{min} = r_{yy} = 35.4$ mm.

$$\lambda_{max} = \frac{L_E}{r_{min}} = \frac{4.8 \times 1000}{35.4} = 135.6\ ;\ f_{cc} = \frac{\pi^2 E}{\lambda^2} = \frac{\pi^2 (2.0 \times 10^5)}{(135.6)^2} = 107.35 \text{ N/mm}^2$$

$$\sigma_{ac} = 0.6\,\frac{f_{cc}.\ f_y}{[(f_{cc})^n + (f_y)^n]^{1/n}} = \frac{0.6 \times 107.35 \times 250}{[(107.35)^{1.4} + (250)^{1.4}]^{1/1.4}} = 53.22 \text{ N/mm}^2.$$

Alternatively from Table 19.1, for $f_y = 250$ and $\lambda = 135.6$, we have

$$\sigma_{ac} = 57 - (57 - 51) \times \frac{5.6}{10} = 53.64 \text{ N/mm}^2$$

$\therefore$ $P = a.\ \sigma_{ac} = 3448 \times 53.22 = 183503$ N = **183.503 kN.**

19.15. USE OF STRUCTURAL STEEL SECTIONS AS STRUT

Upto this stage, we have considered only those sections which are symmetrical. Usually, structural steel sections such as T-section, channel section and angle sections are used as struts, specially in framed structures. In these sections, both X-axis and Y-axis may not be symmetrical about the section. Three cases may arise : (*i*) both X-axis and Y-axis are symmetrical, such as in I-section [Fig. 19.8 (*a*)] (*ii*) only one of the axes (either Y-axis or X-axis) is the axis of symmetry, such as in T-section and channel section, [Fig. 19.8 (*b, c*)], and (*iii*) none of the axes are symmetrical, such as in an angle section [Fig. 19.8 (*d*)].

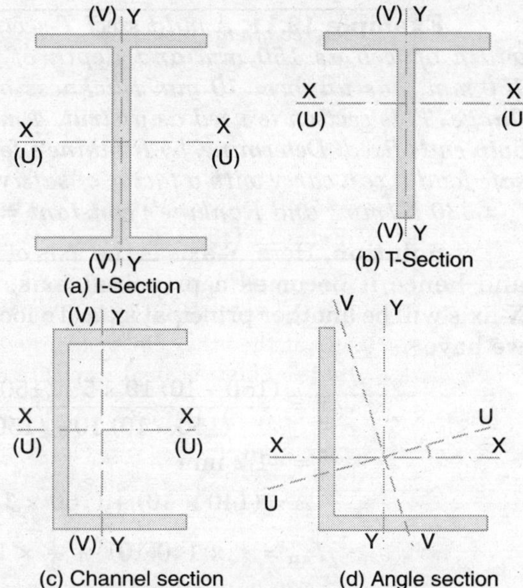

Fig. 19.8. Structural steel sections

In the case of I-section, both X-axis as well as Y-axis are the axes of symmetry. Hence the *principal axes* (U-axis and V-axis) will coincide with X-axis and Y-axis respectively. *Principal axes* of a section are defined as a pair of rectangular axes through the centre of gravity of a plane area such that the *product of inertia* is zero. If U-U and V-V are the principle axes, the product of inertia $\Sigma u.v.\delta a = 0$, where δa is an elementary area with co-ordinates u and v referred to the principal axes. If a plane area has an axis of symmetry, it is obviously a principal axis since the axis of symmetry has to satisfy the condition $\Sigma u.v.\delta a = 0$ about it. In that case, the other axis, perpendicular to the previous one, is also a principal axis. In I-section, both X-axis and Y-axis are symmetrical. Hence U-axis coincides with X-axis, while V-axis coincides with Y-axis. The minimum moment of inertia (I_{XX} or I_{YY}) is to be used in various column formulae.

In T-section, Y-axis is the axis of symmetry, and hence it is the principal axis (V-axis). The other principal axis (U-axis) will coincide with X-axis. Similarly, in channel section has X-axis is axis of symmetry. Hence U-axis will coincide with X-axis, while V-axis will coincide with Y-axis. Minimum of I_{XX} or I_{YY} is to be used.

In the case of an angle section, however, there is no axis of symmetry. Hence the principal axes U-axis and V-axis will be inclined to X-axis and Y-axis, as shown in Fig. 19.8 (*d*). For such a section, we have

$$I_{XX} = \Sigma y^2\, \delta a\ ;\ I_{YY} = \Sigma x^2.\delta a \text{ and } I_{XY} = \Sigma x.y.\delta a.$$

The moment of inertia I_{UU} and I_{VV} about the principal axes are given by the following equations :

$$I_{UU} = \frac{I_{XX} + I_{YY}}{2} + \sqrt{\left(\frac{I_{XX} - I_{YY}}{2}\right)^2 + I_{XY}^2} \qquad \ldots[19.18\ (a)]$$

$$I_{VV} = \frac{I_{XX} + I_{YY}}{2} - \sqrt{\left(\frac{I_{XY} - I_{YY}}{2}\right)^2 + I_{XY}^2} \qquad \ldots[19.18\ (b)]$$

Thus knowing I_{XX}, I_{YY} and I_{XY}, the principal moments of inertia I_{UU} and I_{VV} can be calculated from the above analytical expressions. *It should be noted that the moments of inertia of a section about its principal axes have maximum and minimum values respectively.* The minimum value should be used in the column formulae.

Example 19.11. *A mild steel T-section having width of web as 150 mm and depth of section as 150 mm, has uniform 10 mm thickness of web and flange. This section is used as a strut, 4 m long with both ends fixed. Determine, by Rankine's formula, the safe load it can carry with a factor of safety of 3. Take f_c = 330 N/mm² and Rankine's constant =1/7500.*

Fig. 19.9

Solution. Here, Y-axis is the axis of symmetry, and hence it becomes a principal axis. Therefore X-axis will be another principal axis. To locate X-axis, we have

$$\bar{x} = \frac{(150-10)\,10 \times 5 + (150 \times 10 \times 75)}{(150-10)\,10 + (150 \times 10)}$$

$$= 41.2 \text{ mm}$$

$$A = (140 \times 10) + (150 \times 10) = 2900 \text{ mm}^2$$

$$I_{AB} = \tfrac{1}{3} \times 140(10)^3 + \tfrac{1}{3} \times 10 \times (150)^3 \simeq 11.297 \times 10^6$$

$$\therefore \quad I_{XX} = I_{AB} - A\bar{x}^2 = 11.297 \times 10^6 - 2900(41.2)^2 = 6.374 \times 10^6 \text{ mm}^4$$

Also, $$I_{YY} = \frac{1}{12} \times 10(150)^3 + \frac{1}{12} \times 140(10)^3 = 2.824 \times 10^6 \text{ mm}^4$$

$$\therefore \quad I_{min} = I_{YY} = 2.824 \times 10^6 \text{ mm}^4.$$

Now $L = 4$ m.

Since the column is fixed at the ends, $L_E = \dfrac{L}{2} = \dfrac{4}{2} = 2 \text{ m} = 2000 \text{ mm}$

$$r = \sqrt{\frac{I}{A}} = \sqrt{\frac{2.824 \times 10^6}{2900}} = 31.2 \text{ mm}$$

$$P = \frac{f_c \cdot A}{1 + a\left(\dfrac{L_E}{r}\right)^2} = \frac{330 \times 2900}{1 + \dfrac{1}{7500}\left(\dfrac{2000}{31.2}\right)^2} = 618260 \text{ N} = 618.26 \text{ kN}$$

$$\therefore \quad P_{safe} = \frac{P}{\text{F.S.}} = \frac{618.26}{3} \simeq \mathbf{206\ kN.}$$

Example 19.12. *A steel column made of a 4 m long channel section, 300 mm × 100 mm, is fixed at both the ends. The thickness of flange is 11.6 mm while the thickness of web is 6.8 mm. Using Rankines formula, calculate the load it can carry with a factor of safety of 3. Take f_c = 330 N/mm² and Rankines constant = 1/7500.*

Solution. Since X-X axis is symmetrical, it becomes one of the principal axes. Hence Y-Y axis is other principal axis. Thus, I_{XX} and I_{YY} will be the principal moments of intertia. To locate the position of Y-Y axis, we have

$$\bar{y} = \frac{(2 \times 11.6 \times 100 \times 50) + (300 - 2 \times 11.6) \times 6.8 \times 3.4}{(2 \times 11.6 \times 100) + (300 - 2 \times 11.6) \times 6.8} = 29.13$$

$$A = (2 \times 11.6 \times 100) + (300 - 2 \times 11.6)\ 6.8$$

$$= 4202 \text{ mm}^2$$

$$I_{AB} = \tfrac{1}{3} \times 11.6\,(100)^3 \times 2 + \tfrac{1}{3}(300 - 2 \times 11.6)\,(6.8)^3$$

$$= 7762345 \text{ mm}^4$$

$$\therefore \quad I_{YY} = 7762345 - 4202(29.13)^2 = 4.197 \times 10^6 \text{ mm}^4$$

$$I_{XX} = \frac{1}{12}\,[BD^3 - bd^3]$$

$$= \frac{1}{12}\,[100(300)^3 - (100 - 6.8)(300 - 2 \times 11.6)^3]$$

$$= 60.285 \times 10^6 \text{ mm}^4$$

$$\therefore \quad I_{min} = I_{YY} = 4.197 \times 10^6 \text{ mm}^4$$

Now $L = 4$ m

Since the column is fixed at the ends,

$$L_E = \frac{L}{2} = \frac{4}{2} = 2 \text{ m} = 2000 \text{ mm.}$$

$$r = \sqrt{\frac{I}{A}} = \sqrt{\frac{4.197 \times 10^6}{4202}} = 31.6 \text{ mm}$$

$$P = \frac{f_c \cdot A}{1 + a\left(\dfrac{L_E}{r}\right)^2} = \frac{330 \times 4202}{1 + \dfrac{1}{7500}\left(\dfrac{2000}{31.6}\right)^2} = 903890 \text{ N}$$

$$\therefore \quad P_{safe} = \frac{P}{3} = \frac{903890}{3} = 301297 \simeq \mathbf{301.3\ kN.}$$

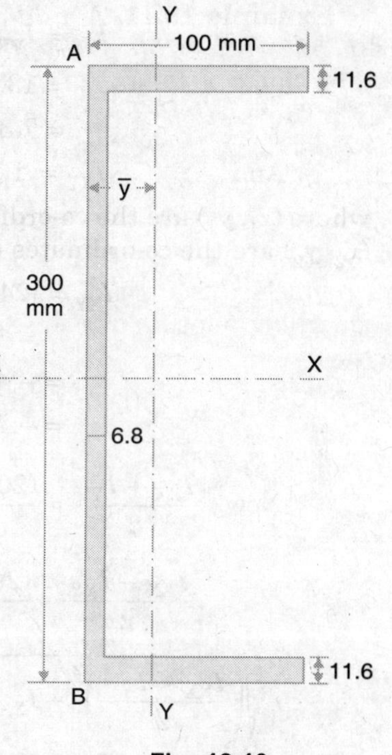

Fig. 19.10

Example 19.13. *An unequal angle, 60 mm × 40 mm × 6 mm is used as a strut for a length of 3 m. The strut may be considered as hinged at top and fixed at bottom. Using Euler's formula, calculate the safe load the column can carry, at a factor of safety of 2.0. Take E = 2 × 10⁵ N/mm².*

Solution. Both X-X axis as well as *Y-Y* axis are not symmetrical about the section. Hence these are not the principal axes. The moment of inertia I_{UU} and I_{VV} will have to be calculated :

$$A = A_1 + A_2 = (40 \times 6) + (54 \times 6) = 240 + 325 = 564 \text{ mm}^2$$

$$\bar{x} = \frac{(40 \times 6 \times 3) + (54 \times 6 \times 33)}{564} = 20.2 \text{ mm}$$

$$\bar{y} = \frac{(40 \times 6 \times 20) + (54 \times 6 \times 3)}{564} = 10.2 \text{ mm}$$

$$I_{PQ} = \left(\tfrac{1}{3} \times 6 \times 60^3\right) + \left(\tfrac{1}{3} \times 34 \times 6^3\right) = 43.44 \times 10^4 \text{ mm}^4$$

$$I_{XX} = I_{PQ} - A\bar{x}^2 = 43.44 \times 10^4 - 564(20.2)^2 = 20.34 \times 10^4 \text{ mm}^4$$

$$I_{PR} = \left(\tfrac{1}{3} \times 54 \times 6^3\right) + \left(\tfrac{1}{3} \times 6 \times 40^3\right) = 13.19 \times 10^4 \text{ mm}^4$$

$$I_{YY} = I_{PR} - A\bar{y}^2$$
$$= 13.19 \times 10^4 - 564(10.2)^2$$
$$= 7.33 \times 10^4 \text{ mm}^4$$

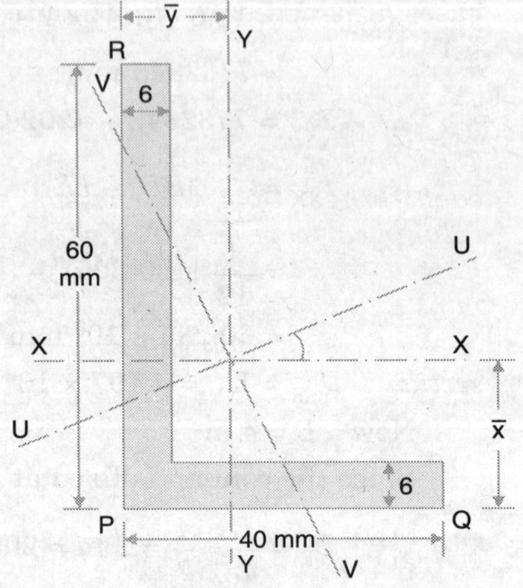

Fig. 19.11

And $I_{XY} = A_1.x_1.y_1 + A_2.x_2.y_2$

where (x_1, y_1) are the co-ordinates of C.G. of area A_1 and (x_2, y_2) are the co-ordinates of C.G. of area A_2.

$$\therefore \quad I_{XY} = \{240(20 - 10.2)(3 - 20.2)\} + \{324(33 - 20.2)(3 - 10.2)\}$$
$$= -4.05 \times 10^4 - 2.99 \times 10^4$$
$$= -7.04 \times 10^4 \text{ mm}^4$$

Now $$\frac{I_{XX} + I_{YY}}{2} = \frac{(20.34 + 7.33)10^4}{2} = 13.84 \times 10^4$$

$$\frac{I_{XX} - I_{YY}}{2} = \frac{(20.34 - 7.33)10^4}{2} = 6.5 \times 10^4$$

$$\sqrt{\left(\frac{I_{XX} - I_{YY}}{2}\right)^2 + I_{XX}^2} = \sqrt{(6.5 \times 10^4)^2 + (-7.04 \times 10^4)^2} = 9.58 \times 10^4$$

$$\therefore \quad I_{UU} = 13.84 \times 10^4 + 9.58 \times 10^4 = 23.42 \times 10^4 \text{ mm}^4 ;$$
$$I_{VV} = 13.84 \times 10^4 - 9.58 \times 10^4 = 4.26 \times 10^4 \text{ mm}^4.$$

Check : $I_{XY} + I_{YY} = I_{UU} + I_{VV}$

$$(20.34 \times 10^4 + 7.33 \times 10^4) = (23.42 \times 10^4 + 4.26 \times 10^4)$$

or $$27.67 \times 10^4 \simeq 27.68 \times 10^4$$

Hence $I_{min.} = I_{VV} = 4.26 \times 10^4 \text{ mm}^4$.

Now, from Euler's formula for column, fixed at base and hinged at top,

$$P_E = \frac{2\pi^2 EI}{L^2} = \frac{2\pi^2 \times 2 \times 10^5 \times 4.26 \times 10^4}{(3000)^2} = 18686 \text{ N}$$

$$\therefore \quad P_{safe} = \frac{18686}{2} = \mathbf{9686 \text{ N} = 9.686 \text{ kN}}$$

Note. It is essential to find I_{UU} and I_{VV} even if the angle has *equal legs,* since I_{VV} will be *minimum.*

19.16. BUILT-UP COLUMNS

A built-up column consists of either a rolled steel section with steel plates attached to its flanges, or of a combination of two or more rolled steel sections with or without steel plates. In most of the cases, these built-up sections are symmetrical about both the axes, and hence the principal axes coincide with X and Y-axes. In some cases, the built-up section has only one axis as the axis of symmetry ; in such cases also, the principal axes coincide with X and Y-axes. In either case, one has to find I_{xx} as well as I_{yy}, and choose I_{min} out of the two values. However, there may exist cases of built-up sections, which have no axis of symmetry. In such cases, it is necessary to compute I_{UU} and I_{VV} to find the minimum value of I, as explained in § 19.15.

Example 19.14. *Fig. 19.12 shows a built-up column section composed of ISHB 250 @ 54.7 kg/m with 400 × 16 mm plate attached to each flange. Compute the safe load which the column can carry, if it is 4 m long, with both ends hinged. Use IS Code formula, taking f_y = 250 N/mm² and E = 2 × 10⁵ N/mm².*

Given the following properties for the ISHB section :

$$I_{xx} = 7983.9 \times 10^4 \text{ mm}^4$$
$$I_{yy} = 2011.7 \times 10^4 \text{ mm}^4$$
$$a = 6971 \text{ mm}^2$$

Width of flange = 250 mm ;

Depth = 250 mm.

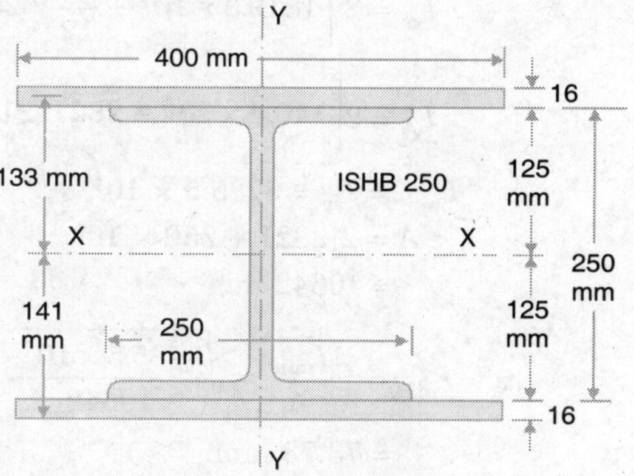

Fig. 19.12

Solution. $I_{xx} = 7983.9 \times 10^4 + 2\left[\frac{1}{12} \times 400(16)^3 + 400 \times 16(125+8)^2\right]$

$$= 30653 \times 10^4 \text{ mm}^4$$

$$I_{yy} = 2011.7 \times 10^4 + 2\left[\frac{1}{12} \times 16(400)^3\right] = 19078 \times 10^4 \text{ mm}^4$$

$$\therefore \quad I_{min} = I_{yy} = 19078 \times 10^4 \text{ mm}^4.$$

$$A = 6971 + 2 \times 400 \times 16 = 19771 \text{ mm}^2$$

$$\therefore \quad r_{min} = \sqrt{\frac{I_{min}}{A}} = \sqrt{\frac{19078 \times 10^4}{19771}} = 98.23 \text{ mm}$$

$$\therefore \quad L_E = L = 4 \text{ m} = 4000 \text{ mm}.$$

$$\therefore \quad \lambda = \frac{L_E}{r} = \frac{4000}{98.23} = 40.72$$

$$f_{cc} = \frac{\pi^2 E}{\lambda^2} = \frac{\pi^2 \times 2.0 \times 10^5}{(40.72)^2} = 1190.46 \text{ N/mm}^2$$

$$\sigma_{ac} = 0.6 \frac{f_{cc} \cdot f_y}{[(f_{cc})^n + (f_y)^n]^{1/n}} = \frac{0.6 \times 1190.46 \times 250}{[(1190.46)^{1.4} + (250)^{1.4}]^{1/1.4}} \simeq 139 \text{ N/mm}^2$$

$$\therefore \quad P = A.\ \sigma_{ac} = 1971 \times 139 = 273969 \text{ N} \simeq \mathbf{273.97\ kN}$$

Example 19.15. *A column section is made up of two channels ISMC 200 @ 22.1 kg/m, placed back to back, at a distance of 100 mm, with two flange plates of size 250 mm × 10 mm, arranged as shown in Fig. 19.13. The column is 5 m long, and has its both ends hinged. Determine the safe load which the column can take, using IS Code formula. Take f_y = 250 N/mm² and E = 2 × 10⁵ N/mm². Each channel has the following properties :*

$$I_{xx} = 1819.3 \times 10^4 \text{ mm}^2 ;\ I_{yy} = 140.4 \times 10^4 \text{ mm}^2 ;$$
$$a = 2821 \text{ mm}^2 ;\ h = 200 \text{ mm} ;\ b = 75 \text{ mm}.$$

Distance of centroid from web = c_{yy} = 21.7 mm.

Solution.

$$I_{xx} = 2\left[1819.3 \times 10^4 + \frac{1}{12} \times 250(10)^3 + 250 \times 10(105)^2\right] = 9155.3 \times 10^4 \text{ mm}^4$$

$$I_{yy} = 2\left[140.4 \times 10^4 + 2821(21.7 + 50)^2 + \frac{1}{12} \times 10 \times 250^3\right] = 5785.5 \times 10^4 \text{ mm}^4$$

$\therefore \quad I_{min} = I_{yy} = 5785.5 \times 10^4$

$A = 2[2821 + 250 \times 10]$

$= 10642 \text{ mm}^2$

$$\therefore \quad r_{min} = \sqrt{\frac{I_{min}}{A}} = \sqrt{\frac{5785.5 \times 10^4}{10642}}$$

$= 73.73$ mm.

$L_E = L = 5 \text{ m} = 5000$

$$\therefore \quad \lambda = \frac{L_E}{r} = \frac{5000}{73.73} = 67.81$$

$$\sigma_{ac} = 0.6 \frac{f_{cc} \cdot f_y}{[(f_{cc})^n + (f_y)^n]^{1/n}}$$

where
$$f_{cc} = \frac{\pi^2 E}{\lambda^2} = \frac{\pi^2 \times 2.1 \times 10^5}{(67.81)^2}$$

$= 450.71 \text{ N/mm}^2$

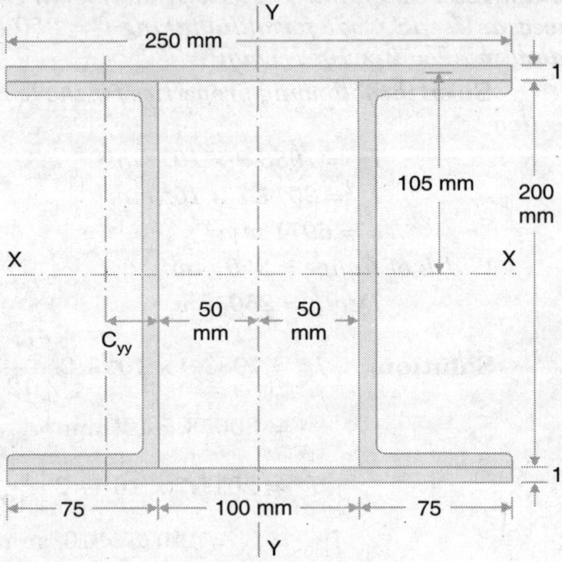

Fig. 19.13

$$\therefore \quad \sigma_{ac} = \frac{0.6 \times 450.71 \times 250}{[(450.71)^{1.4} + (250)^{1.4}]^{1/1.4}} = 115.71 \text{ N/mm}^2$$

$\therefore \quad P = A.\sigma_{ac} = 10642 \times 115.71 = 1231373 \text{ N} =$ **1231.37 kN.**

Example 19.16. *Given the built-up column section of Example 19.15 (Fig. 19.13). Determine the maximum eccentricity for a load of 600 kN from the Y-Y axis if the maximum permissible stress is 100 N/mm². Take f_y = 250 N/mm².*

Solution. Length of column = 5 m. $\therefore$ L_E = 5 m.

From Example 19.15, we have $I_{yy} = 5785.5 \times 10^4 \text{ mm}^4$; $A = 10642 \text{ mm}^2$

$$\therefore \quad Z_{yy} = \frac{I_{yy}}{B/2} = \frac{5785.5 \times 10^4}{125} = 462.84 \times 10^3 \text{ mm}^3$$

$$\frac{L_E}{2}\sqrt{\frac{P}{EI}} = \frac{5000}{2}\sqrt{\frac{600 \times 1000}{2 \times 10^5 \times 5785.5 \times 10^4}} = 0.5693 \text{ radians}$$

Now from Eq. 19.13 (secant formula for eccentric load),

$$f = \frac{P}{A} + \frac{P.e.\sec \frac{L_E}{2}\sqrt{\frac{P}{EI}}}{Z}$$

$$\therefore \quad 100 = \frac{600 \times 1000}{10642} + \frac{600 \times 1000\, e \sec 0.5693}{462.84 \times 10^3}$$

or $100 = 56.38 + 1.539e$

From which $e = \mathbf{28.34\ mm.}$

19.17. ADDITIONAL EXAMPLES USEFUL FOR COMPETITIVE EXAMINATIONS

Example 19.17. *A solid member has the cross-section as shown in Fig. 19.14. It is to be used as a column having 3.5 m effective length. Determine the value of d if the maximum allowable load is to be 400 kN. Take $f_y = 250\ N/mm^2$ and $E = 2 \times 10^5\ N/mm^2$.*

Solution.

$$A = 2 \times \tfrac{1}{2}(2d)\left(\frac{d}{2}\right) = d^2$$

$$I_{XX} = 2 \times \frac{1}{12}(2d)\left(\frac{d}{2}\right)^3 = \frac{d^4}{24}$$

$$I_{YY} = 2 \times \frac{1}{12}(d)(d)^3 = \frac{d^4}{6}$$

$$\therefore \quad I_{min} = I_{XX} = \frac{d^4}{24}.$$

Fig. 19.14

$$r_{min} = \sqrt{\frac{I_{min}}{A}} = \sqrt{\frac{d^4/24}{d^2}} = \frac{d}{\sqrt{24}}\ ;\ \lambda_{max} = \frac{L}{r_{min}} = \frac{3500}{d/\sqrt{24}} = \frac{17146.4}{d}$$

$$f_{cc} = \frac{\pi^2 E}{\lambda^2} = \frac{\pi^2 \times 2.0 \times 10^5}{(17146.4/d)^2} = 6.714 \times 10^{-3}\, d^2$$

Now $\quad P = 400 \times 10^3 = \sigma_{ac} \times A\ ;\ \therefore\ \sigma_{ac} = \dfrac{P}{A} = \dfrac{400 \times 10^3}{d^2}$

But $\quad \sigma_{ac} = \dfrac{0.6\, f_{cc} \cdot f_y}{\left[(f_{cc})^n + (f_y)^n\right]^{1/n}}$ or $\sigma_{ac}{}^n\,[(f_{cc})^n + (f_y)^n] = (0.6 \cdot f_y \cdot f_{cc})^n$

or $\quad \left(\dfrac{400 \times 10^3}{d^2}\right)^{1.4} [(6.714 \times 10^{-3}\, d^2)^{1.4} + (250)^{1.4}] = [0.6 \times 250 \times 6.714 \times 10^{-3}\, d^2]^{1.4}$

or $\quad 9.0736 \times 10^{-4}\, d^{2.8} + 2275.7 = 1.45 \times 10^{-8}\, d^{5.6}$

or $\quad d^{5.6} - 62569.4\, d^{2.8} - 1.5693 \times 10^{11} = 0.$

or $\quad x^2 - 62569.4\, x - 1.5693 \times 10^{11} = 0$ (where $x = d^{2.8}$) From which $x = 428658.$

$\therefore \quad d = x^{1/2.8} = (428658)^{1/2.8} = 102.7$ mm

Keep $\quad \boldsymbol{d} = \mathbf{103\ mm.}$

Check $\quad r_{min} = \dfrac{103}{\sqrt{24}} = 21.02\ ;\ \lambda = \dfrac{17146.4}{103} = 166.5$

$\therefore \quad \sigma_{ac} = 41 - (41 - 37)\dfrac{6.5}{10} = 38.4\ N/mm^2\ ;\ A = (103)^2$

$\therefore \quad P = (103)^2 \times 38.4 \times 10^{-3} = 407.4$ kN. Hence O.K.

Example 19.18. *A cast iron hollow column of 200 mm external diameter and 160 mm internal diameter is 4 m long. It is fixed at its both ends and subjected to an eccentric load of 150 kN. Determine the maximum eccentricity, in order that there is no tension any where is the section. Take E = 0.94 × 10^5 N/mm^2.*

Solution. $A = \frac{\pi}{4}(200^2 - 160^2) = 11309.7 \text{ mm}^2$

$$I = \frac{\pi}{64}(200^4 - 160^4) = 4637 \times 10^4 \text{ mm}^4 \; ; \; Z = \frac{4637 \times 10^4}{100} = 463700 \text{ mm}^3$$

For column with fixed ends, $L_E = \frac{L}{2} = \frac{4}{2} = 2$ m

$$\frac{L_E}{2}\sqrt{\frac{P}{EI}} = \frac{2000}{2}\sqrt{\frac{150 \times 10^3}{0.94 \times 10^5 \times 4637 \times 10^4}} = 0.1856 \text{ radian} = 10.629°$$

Total stress at any location $= p_0 \pm p_b$.

In order that there is no tension anywhere in the section, the direct stress (p_0) should be equal to the bending stress (p_b)

$$\therefore \quad \frac{P}{A} = \frac{P \cdot e \sec \frac{L_E}{2}\sqrt{\frac{P}{EI}}}{Z}.$$

or
$$\frac{150 \times 10^3}{11309.7} = \frac{150 \times 10^3 \; e \sec 10.629°}{463700} = \frac{150 \times 10^3 \times 1.01746\, e}{463700}$$

$$\therefore \quad e = \frac{463700}{11309.7 \times 1.01746} = \mathbf{40.3 \text{ mm}}.$$

Example 19.19. *A straight cylindrical bar is 20 mm diameter and 2 m long. It is freely supported at its two ends in a horizontal position and loaded at the centre with a concentrated load of 120 N. The central deflection is found to be 4 mm.*

If placed vertical and loaded along its axis, what load would cause it to buckle ? What is the ratio of the maximum stress in the two cases ?

Solution.

Now
$$y_c = 4 = \frac{W L^3}{48 EI} = \frac{120\,(2000)}{48} \times \frac{L^2}{E I}$$

(where *L*, *E* and *I* are in mm units)

$$\therefore \quad \frac{EI}{L^2} = \frac{120\,(2000)}{48 \times 4} = 1250 \qquad \ldots(1)$$

As a strut:
$$P_E = \frac{\pi^2 . E . I}{L^2} = \pi^2 \times 1250 = \mathbf{12337 \text{ N}} \qquad \ldots(2)$$

$$\therefore \quad f_E = \frac{P_E}{A} = \frac{12337}{\frac{\pi}{4}(20)^2} = 39.27 \text{ N/mm}^2$$

As a beam,
$$M_{max} = \frac{W L}{4} = \frac{120\,(2000)}{4} = 60000 \text{ N-mm}$$

$$\therefore \quad f_b = \frac{M}{Z} = \frac{60000}{\frac{\pi}{32}(20)^3} = 76.39 \text{ N/mm}^2$$

$$\therefore \quad \frac{f_b}{f_E} = \frac{76.39}{39.27} = \mathbf{1.945.}$$

Example 19.20. *Select the lightest wide flange joist that can be used as a column 5 m long to support an axial load of 500 kN, with a factor of safety of 3. Assume both ends hinged. Take $\sigma_{PL} = 200\ N/mm^2$ and $E = 2 \times 10^5\ N/mm^2$.*

Solution. For steel with proportional limit $\sigma_{PL} = 200$ N/mm², the limiting value of L/r is given by:

$$p_E = \sigma_{PL} = \frac{\pi^2 E I}{A L^2} = \frac{\pi^2 E r^2}{L^2} = \frac{\pi^2 E}{(L/r)^2}$$

$$\therefore \quad \left(\frac{L}{r}\right)_{lim.} = \sqrt{\frac{\pi^2 E}{\sigma_{PL}}} = \sqrt{\frac{\pi^2 \times 2 \times 10^5}{200}} = 99.35$$

Hence if $L/r \geq 99.35$, Euler's formula applies. *If $L/r < 99.35$, the limiting stress is the proportional limit.*

The specified working load (= 500 kN), when multiplied by the F.S. (= 3), gives the Euler's load. Hence $P_E = 500 \times 3 = 1500$ kN.

$$\therefore \quad I_{min} = \frac{P_E L^2}{E \pi^2} = \frac{(1500 \times 10^3)(5000)^2}{2 \times 10^5 (\pi^2)} = 1899.8 \times 10^4 \text{ mm}^4$$

Also, for slenderness ratio $L/r \geq 99.35$, we have

$$r_{min} \leq \frac{L}{99.35} = \frac{5000}{99.35} = 50.33 \text{ mm.}$$

This criteria requires that the section must have a least $I \geq 1899.8 \times 10^4$ mm⁴ and a least $r \leq 50.33$ mm. This is satisfied by choosing a section ISWB 500 having $I_{min} = I_{yy} = 2987.8 \times 10^4$ mm⁴ and $r_{min} = 49.6$ mm.

However, if the selection were based on the proportional limit, the section must have a minimum area $A_{min} = \dfrac{P_E}{\sigma_{PL}} = \dfrac{1500 \times 10^3}{200} = 7500$ mm² and r_{min} greater 50.33 mm (so that L/r is less than 99.35). These conditions are satisfied by a section ISWB 550 having $A = 14334$ mm² and $r_{min} = 51.1$ mm.

The lightest section is therefore ISWB 500.

Example 19.21. *A column section, shown in Fig. 19.15, consists of three plates, each of thickness t welded together. It carries an axial load of 1000 kN over an effective length of 3.6 m. Taking $f_y = 250\ N/mm^2$ and $E = 2 \times 10^5\ N/mm^2$, determine the thickness t.*

Solution. $$I_{XX} = \frac{1}{12} t (20\,t)^3 + 2\left[\frac{1}{12} \times (10\,t)\,t^3 + 10\,t^2 \left(10\,t + \frac{t}{2}\right)^2\right] = 2273.33\,t^4$$

$$I_{YY} = 2 \times \frac{1}{12} \times t\,(10\,t)^3 + \frac{1}{12}(20\,t)\,t^3 = 168.33\,t^4$$

$$A = 2\,[t\,(10\,t)] + 20\,t \times t = 40\,t^2$$

$$r_{min} = \sqrt{\frac{I_{min}}{A}} = \sqrt{\frac{168.33\,t^4}{40\,t^2}} = 2.0514\,t$$

$$\lambda = \frac{l}{r_{min}} = \frac{3600}{2.0514\,t} = \frac{1754.9}{t}$$

$$\sigma_{ac} = \frac{P}{A} = \frac{1000 \times 10^3}{40\,t^2} = \frac{25000}{t^2}$$

$$f_{cc} = \frac{\pi^2 E}{\lambda^2} = \frac{\pi^2 \times 2 \times 10^5 \times t^2}{(1754.9)^2} = 0.64095\,t^2$$

Now $\sigma_{ac} = \dfrac{0.6\,f_{cc}\,.\,f_y}{[f_{cc}^n + f_y^n]^{1/n}}$ where $n = 1.4$

$$\therefore \quad \frac{25000}{t^2} = \frac{0.6\,(0.64095 t^2)\,250}{[(0.64096\,t^2)^{1.4} + (250)^{1.4}]^{1/1.4}}$$

or $$\frac{260.03}{t^4} = \frac{1}{[(0.5365\,t^{2.8} + 2275.7]^{\frac{1}{1.4}}}$$

or $$2404.54\,[0.5365\,t^{2.8} + 2275.7] = t^{5.6}$$

or $$t^{5.6} - 1290\,t^{2.8} - 5.47 \times 10^6 = 0$$

or $$x^2 - 1290\,x - 5.47 \times 10^6 = 0 \text{ (where } x = t^{2.8})$$

From which $x = 3071.5 = t^{2.8}$

$$\therefore \quad t = (3071.5)^{\frac{1}{2.8}} = (3071.5)^{0.3571} \approx 17.6 \text{ mm}$$

Keep $t = 18$ mm

Check $\therefore$ $r_{min} = 2.0514\,t = 36.925$; $\lambda = \dfrac{3600}{36.925} = 97.50$

$$\therefore \quad \sigma_{ac} = 90 - (90 - 80)\,\frac{7.50}{10} = 82.5 \text{ N/mm}^2$$

$$A = 40\,t^2 = 40(18)^2 = 12960 \text{ mm}^2$$

$\therefore$ $P = A\,.\,\sigma_{ac} = 12960 \times 82.5 \times 10^{-3} = 1069.2$ kN. Hence O.K.

Fig. 19.15

Example 19.22. *A uniform bar of cross-sectional area A and flexural stiffness EI is heated so that its temperature varies linearly from t/2 at one end to t at the other end. One end is pin jointed to a rigid foundation. The other end is pin jointed so that it can slide in the direction of the length of the bar, the thermal expansion of which is resisted by a compression spring of stiffness k. If there is no load in the spring when t = 0, obtain an expression for the stress in the bar when it is heated and show that it buckles in flexure when*

$$t = \frac{4\,\pi^2\,I}{3\,\alpha\,L^2\,A}\left(1 + \frac{E\,A}{k\,L}\right)$$

where α = *coefficient of linear thermal expansion.*

Solution. Average temperature along the bar $\dfrac{1}{2}\left(\dfrac{1}{2}t + t\right) = \dfrac{3}{4}\,t$

$\therefore$ Thermal expansion of bar $= \Delta_t = \dfrac{3}{4}\alpha L t$ ...(1)

Let P = force exerted by the spring on the bar.

Then, compression produced in the bar $= \dfrac{PL}{AE}$ ...(2)

Also, compression of the spring $= \dfrac{P}{k}$ ...(3)

Now, net expansion of the bar = compression of spring

$$\therefore \quad \frac{3}{4}\alpha L t - \frac{PL}{AE} = \frac{P}{k}$$

From which
$$P = \frac{\frac{3}{4}\alpha L t}{\frac{L}{AE} + \frac{1}{k}} \quad ...(a)$$

and stress in the bar
$$= p = \frac{P}{A} = \frac{\frac{3}{4}\alpha L t}{\frac{L}{E} + \frac{A}{k}} \quad \textbf{(Ans.)} \quad ...(b)$$

Now, the bar will buckle when buckling load $P = \dfrac{\pi^2 EI}{L^2}$ ...(c)

Equating (*a*) and (*c*), we get
$$\frac{\pi^2 EI}{L^2} = \frac{\frac{3}{4}L\alpha t}{\frac{L}{AE} + \frac{1}{k}}$$

Rearranging, we get
$$t = \frac{4\pi^2 I}{3\alpha L^2 A}\left(1 + \frac{AE}{kL}\right).$$

Example 19.23. ***Partial fixing of ends:*** *A strut of length 2a has each end fixed in an elastic material which exerts a restraint μ per unit angular displacement. Prove that the critical load P is given by the equation.*

$$\mu\, n \tan na + P = 0 \text{ where } n^2 = \frac{P}{EI}$$

Such a strut, 2.5 m in length, has a theoretical critical load of 15 kN on the assumption of pinned ends. Determine the percentage increase in the critical load if the constraint offered at the ends is 170 N-m per degree of rotation. (U.L.)

Solution. Let M = restraining moment at each end

Then, as in the case of strut with fixed ends [Fig. 19.4(*c*)],

$$EI\frac{d^2y}{dx^2} = -P.y + M$$

or
$$\frac{d^2y}{dx^2} + \frac{P}{EI}y = \frac{M}{EI} \quad ...(1)$$

Putting $\dfrac{P}{E\,I} = n^2$, we get

$$\frac{d^2y}{dx^2} + n^2 y = \frac{M}{E\,I} \qquad \text{...[1 (a)]}$$

The general solution of the above equation is

$$y = A \sin n\,x + B \cos n\,x + \frac{M}{P} \qquad \text{...(2)}$$

when $x = 0,\ y = 0$. Hence $B = -\dfrac{M}{P}$ ...(a)

Also, from (2) $\dfrac{dy}{dx} = A\,n \cos n\,x - B\,n \sin n\,x$ ...(3)

At $x = 0$, $\dfrac{dy}{dx} = \left(\dfrac{dy}{dx}\right)_0$

$\therefore$ $\left(\dfrac{dy}{dx}\right)_0 = A\,n$

$\therefore$ $M = \mu \left(\dfrac{dy}{dx}\right)_0 = \mu\,A\,n$ or $A = \dfrac{M}{\mu\,N}$ ...(b)

Substituting the values of A and B, we get from (2)

$$y = \frac{M}{\mu\,N} \sin n\,x + \frac{M}{P}(1 - \cos n\,x) \qquad \text{...(4)}$$

At the centre, $x = \dfrac{L}{2} = a$ and $\dfrac{dy}{dx} = 0$

Hence from (3) $0 = \dfrac{M}{\mu} \cos n\,a + \dfrac{M\,n}{P} \sin n\,a$

or $\mu\,n \tan n\,a + P = 0$ (Proved) ...(c)

For pinned ends $\dfrac{\pi^2\,E\,I}{(2a)^2} = 15000$

From which $EI = \dfrac{15000\,(2500)^2}{\pi^2} = 9499 \times 10^6$ N-mm^2

$$n = \sqrt{\frac{P}{E\,I}} = \frac{\sqrt{P}}{\sqrt{9499 \times 10^6}} = \frac{\sqrt{P}}{97462}$$

and $n\,a = \dfrac{\sqrt{P}}{97462}\left(\dfrac{2500}{2}\right) = 0.0128\,\sqrt{P}$

Also, $\mu = 170$ N-m/degree of rotation $\dfrac{170 \times 10^3}{\pi/180}$ N-mm/rad $= 9740 \times 10^3$ N-mm/rad

Hence equation (c) can be written as

$$\tan(0.0128\sqrt{P}) = -\frac{P}{9740\times10^3\,\dfrac{\sqrt{P}}{97462}} = -0.01\sqrt{P}$$

The least solution of the above equation is for $\sqrt{P} = 166$. Hence $P = (166)^2 =$ **27500 N**

$\therefore$ % increase over the value for pinned ends $= \dfrac{27500-15000}{1500} =$ **83.3**

Example 19.24. ***Partial Restraint against lateral deflection :*** *A strut of length L is encastered at its lower ends ; its upper end is elastically supported against lateral deflection so that the resisting force is k times the end deflection. Show that the crippling load P is given by*

$$\frac{\tan\alpha L}{\alpha L} = 1-\frac{P}{kL},\ \text{where } \alpha^2 = \frac{P}{EI}.$$

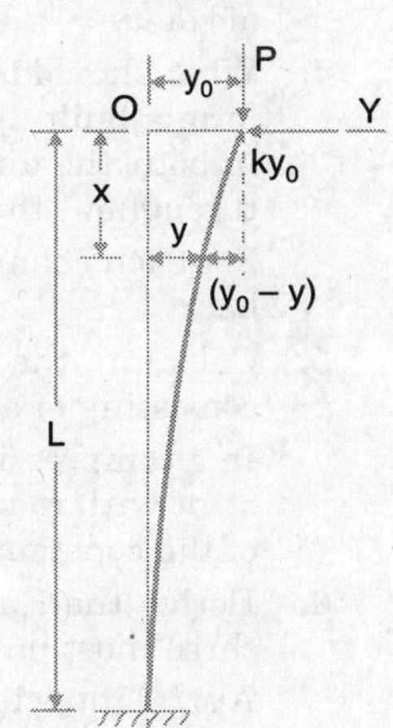

Fig. 19.16

Solution. Let us choose the origin at O, and the axes as shown in Fig. 19.16

Hence $\qquad EI\dfrac{d^2y}{dx^2} = P(y_0 - y) - k\,y_0\,x$

Taking $\dfrac{P}{EI} = \alpha^2$, we get

$$\frac{d^2y}{dx^2} + \alpha^2 y = \alpha^2 y_0 - ky_0\frac{x}{EI} \qquad ...(1)$$

The solution of the above equation is

$$y = A\sin\alpha x + B\cos\alpha x + y_0 - \frac{ky_0x}{P} \qquad ...(2)$$

At $\qquad x = 0,\ y = y_0 \quad$ Hence $B = 0$

Also, at $x = L,\ y = 0 \quad \therefore \quad A\sin\alpha L + y_0 - \dfrac{ky_0L}{P} = 0 \qquad ...(a)$

Also, at $x = L,\ \dfrac{dy}{dx} = 0 \quad \therefore \quad A\alpha\cos\alpha L - \dfrac{ky_0}{P} = 0 \qquad ...(b)$

From (b), we get $\qquad A = \dfrac{ky_0}{P\alpha\cos\alpha L}$

Substituting in (a), we get

$$\frac{ky_0\tan\alpha L}{P\alpha} + y_0 - \frac{ky_0L}{p} = 0$$

Multiplying by $\dfrac{P}{ky_0L}$ and rearranging, we get

$$\frac{\tan\alpha L}{\alpha L} = 1 - \frac{P}{kL}.$$

PROBLEMS

1. Calculate the length of a strut for 15 kN collapse load. The strut is made up of 2 mm thick mild steel pipe having its external diameter 4 cm and one end of the strut is fixed while the other is free. $E = 2.05 \times 10^5$ N/mm^2.
2. A R.S.J. 300 mm × 140 mm × 44.2 kg of 5 m length is used as stanchion with one end fixed and the other free. Find the safe axial load for the stanchion using Rankine's constant and a factor of safety of 3. $I_{xx} = 8603.6$ cm^4 ; $I_{yy} = 453.9$ cm^4 ; Area of section = 56.26 cm^2.
3. Find the safe axial load on a steel stanchion of 8 metre height having one end fixed and the other hinged which is built-up of two 250 mm × 100 mm standard channels placed 10 m apart back to back with two 300 mm × 10 mm plates riveted to each flange. For individual channel section, $I_{xx} = 3687.9$ cm^4 ; $I_{yy} = 298.4$ cm^4 ; $A = 35.65$ cm^2 ; distance of *C.G.* from base = 2.70 cm. Adopt factor of safety = 4. Use Rankine's formula.
4. What should be the distance between the back faces of two channels 250 mm × 100 mm placed to form a built-up column in order that under axial loads, equal resistance to buckling is provided in both the directions ? On both sides, the flanges are connected by a plate 360 mm × 10 mm throughout the length of the column.

   For each channel, $A = 35.65$ cm^2 ; $k_{xx} = 10.17$ cm; $k_{yy} = 2.89$ cm ; Point of C.G. = 2.70 cm from its back.
5. Calculate the crippling stresses, using the Euler formula, for a pin-ended strut 180 cm long consisting of a tube of 6.35 cm outside diameter and wall thickness 0.32 cm.

   In a compression test a short length of this tube failed at a load of 310 kN and when tested as a strut with rounded ends, 180 cm long, it failed at a load of 155 kN. Find from this data the value of the constant a in the Rankine formula. Take $E = 2.07 \times 10^5$ N/mm^2.
6. Derive the Euler formula for strut hinged at both ends. State clearly what you regard as the three most important assumptions made in deriving this formula.

   A straight cylindrical bar is 1.5 cm in diameter and 1.25 m long. It is freely supported at its two ends in a horizontal position and loaded at the centre with a concentrated load of 100 N. The central deflection is found to be 5 mm.

   If placed vertical and loaded along its axis, what load would cause it to buckle ? What is the ratio of the maximum stresses in the two cases ?

ANSWERS

1. 142 cm. **2.** 35.4 kN. **3.** 995 kN.

4. 19.6 cm. **5.** 288 N/mm^2 ; $\dfrac{1}{7100}$ **6.** 5120 N ; 3.25.

SECTION 4 SHAFTS, SPRINGS AND CYLINDERS

CHAPTERS

20

Torsion of Shafts

20.1. INTRODUCTION

In chapter 9, we have seen that in a beam, the bending moment is resisted by longitudinal tensile and compressive stresses which vary in intensity at different points in the depth of the beam. Thus, if a moment is applied in a vertical plane containing the longitudinal axis of the beam, it will set bending stresses in the beam. If on the contrary, the moment is applied in a vertical plane perpendicular to the longitudinal axis of the beam or a shaft, it will be subjected to a torque causing twist or torsion in the member. Members subjected simultaneously to torque and bending have also been treated in this chapter. The most common examples of torsion are line shafts transmitting torque or power as shown in Fig. 20.1.

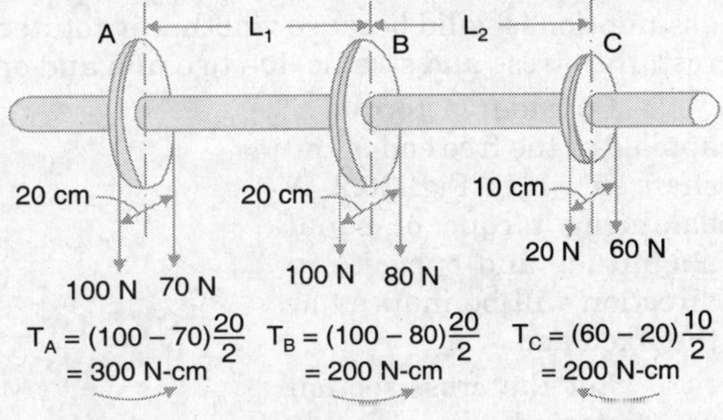

Fig. 20.1

To find internal torque in statically determinate members, only one equation of statics $\Sigma M_z = 0$ is required, where Z-axis is directed along the member. As in the case of determination of B.M., pass a plane at the desired section perpendicular to the longitudinal axis of the member and remove everything to either side of the cut. The internal or resisting torque necessary to maintain equilibrium of the isolated part is determined. Thus, the internal or resisting torque between A and B is 300 N-cm and between B and C is 500 N-cm.

The mathematical treatment of stress and deformations induced by torsion is too complicated, except in the cases of solid and hollow circular shafts. We shall, therefore, confine ourselves to these simple cases only.

Assumptions

We shall make the following assumptions in developing the equations for stresses and deformations due to torsion :

1. Plane normal sections of shaft remain plane after twisting *i.e.*, no warping or distortion of parallel planes normal to the axis of the member takes place.
2. Torsion is uniform along the rod *i.e.*, all normal cross-sections which are at the same axial distance suffer equal relative rotation.
3. Radii remain straight after torsion.
4. Stress is proportional to strain, *i.e.*, all the stresses are within the elastic limit.
5. The material is homogeneous and isotropic.

The stress set up at any point in a cross-section is one of pure shear. There is, thus, a state of simple shear; the two planes across which the stress is wholly tangential are (1) the plane containing the point and perpendicular to the axis, and (2) the plane through that point and the axis.

20.2. RELATION BETWEEN TWISTING MOMENT, TWIST AND SHEAR STRESS

For the purpose of developing the expressions for the torsional stress and strain, we shall assume that one end of the shaft is fixed and a moment is applied at the other end, the plane of application of moment being perpendicular to the longitudinal axis of the beam. This assumption is valid because whether it rotates at uniform speed to transmit the power or is at rest, the stress and strains due to equal and opposite couples at its ends will remain the same.

Consider a torque T applied at the free end of the shaft shown in Fig. 20.2. A balancing torque of equal magnitude and opposite in direction will be induced at the fixed end.

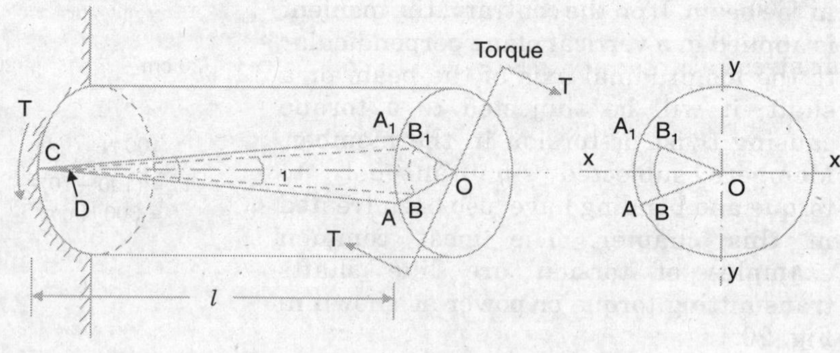

Fig. 20.2

Let the cross-section be distorted through an angle θ, shifting the radial line OA to OA_1. On the surface of the shaft, any line such as CA will be distorted to CA_1, through an angle ϕ.

Let f_s be the intensity of shear stress at the surface of the shaft. We have, therefore,

$$\phi = \frac{f_s}{N} \qquad \ldots(1)$$

Also

$$\phi = \frac{AA_1}{CA} = \frac{OA.\theta}{CA} = \frac{R\theta}{l} \qquad \ldots(2)$$

Hence $$\phi = \frac{f_s}{N} = \frac{R\theta}{l} \qquad ...(3)$$

or $$\frac{f_s}{R} = \frac{N\theta}{l} \qquad ...(4)$$

Similarly, a layer such as DB distant r from the axis of the shaft, will be distorted to a new position DB_1 through an angle ϕ_1 and, we have

$$\frac{q}{N} = \phi_1 = \frac{BB_1}{DB} = \frac{r\theta}{l}$$

or $$\frac{q}{r} = \frac{N\theta}{l} \qquad ...(5)$$

Hence, we have,

$$\frac{f_s}{R} = \frac{q}{r} = \frac{N\theta}{l}. \qquad ...(20.1)$$

Thus it can be stated that the *intensity of shear stress at any point in the cross-section of a shaft subjected to pure torsion is proportional to its distance from the centre.* This shear intensity is zero at the axis and increases linearly to maximum of f_s at the surface, as shown in Fig. 20.3. The longitudinal axis of the shaft is the neutral axis.

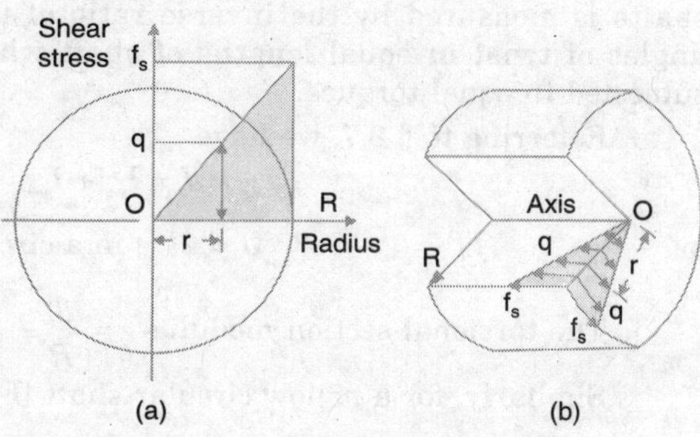

Fig. 20.3

Resisting Torque

From conditions of equilibrium, the external torque T must be balanced by resisting torque, *i.e.*, by moments of the tangential shearing stresses acting on any transverse section.

Consider an elementary concentric ring of radius r and thickness δr. Let the intensity of shear stress at this radius be q. We have $q = \frac{f_s}{R} r$ (from equation 20.1).

The resistance couple set up by this ring will be given by :

$$\delta T_r = q.\delta a.r = \frac{f_s}{R} r.\delta a.r$$

where δa – area of elementary ring = $2\pi r.\delta r$.

Hence the resistance set up by the whole shift is

$$T_r = \sum_O^R \delta T_r = \int_O^R \frac{f_s}{R} \delta a.r^2 = \frac{f_s}{R}\int_O^R \delta a.r^2$$

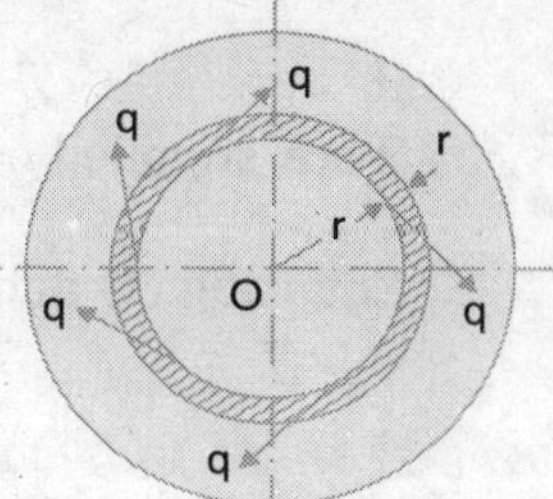

Fig. 20.4

Now $\int_O^R \delta a.r^2$ represents the *polar moment of inertia* J about the longitudinal axis of the shaft.

Hence, $$T_r = \frac{f_s}{R}.J \text{ or } \frac{T_r}{J} = \frac{f_s}{R} \qquad ...(20.2)$$

Connecting equations 20.1 and 20.2, we get

$$\frac{q}{r} = \frac{f_s}{R} = \frac{T_r}{J} = \frac{N\theta}{l} \qquad ...(20.3)$$

Equation 20.3 can be compared with equation 9.6 of bending. The expression q/r corresponds to f/y ; T_r/J corresponds to M/I and $N\theta/l$ corresponds to E/R. The expression J/R corresponds to $I/(d/2)$ and is called the *torsional section modulus*. The expression NJ corresponds to expression EI and will be called *torsional rigidity* of shaft. The relative stiffness of two shafts is measured by the inverse ratio of the angles of twist in equal lengths of shaft when subjected to equal torques.

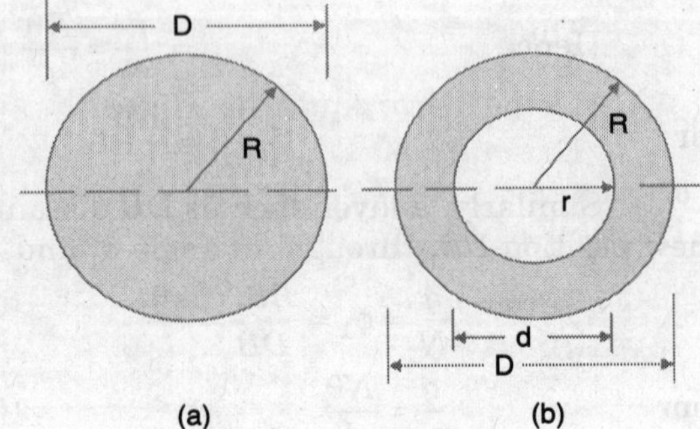

Fig. 20.5

Referring to § 8.7, we have

$$I_{zz} = J = I_{xx} + I_{yy}$$

or $$J = 2I_{xx}, \text{ in a circular section} = 2 \times \frac{\pi}{64} D^4 = \frac{\pi}{32} D^4$$

The torsional section modulus $= \dfrac{J}{R} = \dfrac{\pi}{16} D^3$

Similarly, for a hollow circular shaft [Fig. 20.5(*b*)]

$$J = \frac{\pi}{32} (D^4 - d^4) = \frac{\pi}{2} (R^4 - r^4).$$

20.3. DESIGN OF SHAFTS

In SI system, power (P) is measured in watts (W).

$$1 \ W = 1 \text{ Joule/sec} = 1 \text{ N-m/sec.}$$

Let the power be P kW.

Work done in transmitting P kilo watts $= 1 \times P \times 10^3$ N-m/sec.

$= P \times 60 \times 10^6$ N-mm/min. $= 60P$ kN-m/min. ...(1)

If T = torque in kN-m,

Work supplied = (average couple) × (angle turned through one minute (= $T.2\pi n$) ...(2)

Equating (1) and (2), we get

$$60P = 2\pi nT \quad \text{or} \quad P = \frac{2\pi nT}{60} \text{ kW} \qquad \text{...[20.4 (a)]}$$

and $$T = \frac{60P}{2\pi n} \text{ kN-m} \qquad \text{...[20.4 (b)]}$$

[It should be noted that 1 H.P. (metric) = 0.75 kW.]

The resistance torque (T_r) of the shaft must at least be equal to T above, so that maximum stress f_s is not exceeded.

Thus, $$T = T_r = f_s \cdot \frac{J}{R} \quad \text{or} \quad \frac{J}{R} = \frac{T}{f_s} \qquad \text{...(20.5)}$$

In the above equation, T is known from equation 20.4 (*b*) and f_s is the allowable stress. Hence the section modulus J/R can be determined. Knowing J/R, the diameter of the shaft can be calculated.

Thus, $$T = f_s \cdot \frac{J}{R} = f_s \frac{\pi}{16} d^3 \qquad \text{...(20.6)}$$

In the above equation, T must be substituted in N-mm units if f_s is in N/mm^2 units.

Example 20.1. *Calculate the maximum intensity of shear stress induced and the angle of twist in degrees for a length of 10 metres for a solid shaft of 100 mm diameter transmitting 112.5 kW at 150 R.P.M. Take N = 8.2 × 10^4 N/mm^2 for the material of the shaft.*

Solution. $$T = \frac{60P}{2\pi n} = \frac{60 \times 112.5}{2\pi \times 150} = 7.162 \text{ kN-m} = 7.162 \times 10^6 \text{ N-mm}$$

or $$f_s = \frac{16T}{\pi d^3} = \frac{16 \times 7.162 \times 10^6}{\pi\,(100)^3} = \mathbf{36.5\ N/mm^2}$$

Also, $$\theta = \frac{f_s}{R} \cdot \frac{l}{N} = \frac{36.5}{50} \times \frac{10 \times 1000}{8.2 \times 10^4} = \mathbf{0.089\ radian.}$$

Example 20.2. *A solid circular shaft of steel is 50 mm in diameter. Find the power transmitted at 120 R.P.M. if the permissible shear stress is 62.5 N/mm^2.*

Solution. $$T_r = f_s \frac{\pi}{16} d^3 = 62.5 \frac{\pi}{16} (50)^3 = 1.534 \times 10^6 \text{ N-mm} = 1.534 \text{ kN-m}$$

Power $$P = \frac{2\pi n T}{60} \text{ kW (where } T \text{ is in kN-m.)} = \frac{2\pi \times 120 \times 1.534}{60} = \mathbf{19.28\ kW.}$$

Example 20.3. *A solid shaft has to transmit 90 kW power at 160 R.P.M. Find the suitable diameter for the shaft if the maximum torque transmitted in each revolution exceeds the mean by 25%. Take allowable shear stress as 70 N/mm^2.*

Solution. The average torque

$$T_{av} = \frac{60P}{2\pi n} \text{ kN-m} = \frac{60 \times 90}{2\pi \times 160} = 5.37 \text{ kN-m}$$

$$T_{max} = 1.25\, T_{av} = 1.25 \times 5.37 = 6.71 \text{ kN-m} = 6.71 \times 10^6 \text{ N-mm}$$

$$d^3 = \frac{16T}{f_s \cdot \pi} = \frac{16 \times 6.71 \times 10^6}{70 \times \pi} = 488513 \text{ mm}^3$$

$$\therefore \quad d = 78.76 \text{ mm} \triangleq \mathbf{80\ mm.}$$

Example 20.4. *450 kW power has to be transmitted at 100 R.P.M. Find (a) the necessary diameter of a solid circular shaft, (b) the necessary diameter of hollow circular section, the inside diameter being 3/4 of the external diameter. Allowable shear stress = 75 N/mm^2 and the density of material = 77 kN/m^3.*

Solution. $$T = \frac{60P}{2\pi n} = \frac{60 \times 450}{2\pi \times 100} = 42.97 \text{ kN-m} = 42.97 \times 10^6 \text{ N-mm}$$

For a Solid Section $$d^3 = \frac{16T}{f_s \cdot \pi} = \frac{16 \times 42.97 \times 10^6}{75 \times \pi} \quad \text{or} \quad d \triangleq \mathbf{143\ mm}$$

For a hollow Circular Section, $d = \frac{3}{4} D$

$$\therefore \quad T = f_s \frac{\pi}{16}\left(\frac{D^4 - d^4}{D}\right) = \frac{f_s}{16} \pi D^3 \left(1 + \frac{9}{16}\right)\left(1 - \frac{9}{16}\right) = \frac{f_s\,\pi}{(16)^3} D^3 \times 175$$

$$\therefore \quad D^3 = \frac{(16)^3}{175 f_s\,\pi} T = \frac{(16)^3 \times 42.97 \times 10^6}{175 \times 75 \times \pi} = 4.269 \times 10^6$$

$\therefore$ D = **162 mm** and $d = \dfrac{3}{4} \times 162 =$ **121.5 mm.**

Weight of solid shaft per mm length is

$$W_1 = \frac{\pi}{4}(143)^2 \times 1\,\rho = \frac{\pi}{4}\rho\,(20449)$$

where ρ = unit weight of material per mm^3.

Weight of hollow shaft per mm length is

$$W_2 = \frac{\pi}{4}\left[(162)^2 - (121.5)^2\right] \times 1 \times \rho = \frac{\pi}{4}\rho\,(11482)$$

$\therefore$ % saving $= \dfrac{20449 - 11482}{20449} \times 100 =$ **43.9%**.

Example 20.5. *Select a suitable diameter of solid shaft of circular section to transmit 112.5 kW of power at 200 R.P.M. if the allowable shear stress is 75 N/mm³ and the allowable twist is 1° in a length of 3 metres. Take N = 0.082 × 10⁶ N/mm².*

Solution. $T = \dfrac{60P}{2\pi n} = \dfrac{60 \times 112.5}{2 \times \pi \times 200} = 5.372$ kN-m $= 5.372 \times 10^6$ N-mm

For Strength: $d^3 = \dfrac{16T}{f_s\,.\,\pi} = \dfrac{16 \times 5.732 \times 10^6}{75 \times \pi} = 0.3648 \times 10^6$

$\therefore$ d = 71.4 mm

For Stiffness: $\theta = \dfrac{Tl}{NJ} = \dfrac{32Tl}{N\pi d^4}$

$\therefore$ $d^4 = \dfrac{32Tl}{\pi N\theta}$ where $\theta = 1° = \dfrac{\pi}{180}$ radians

$\therefore$ $d^4 = \dfrac{32 \times 5.372 \times 10^6 \times 3000}{\pi \times 0.082 \times 10^6 \times 1} \times \dfrac{180}{\pi} = 11470 \times 10^4$

$\therefore$ d = 103.5 mm $\simeq$ **105 mm** (say).

This is greater than the diameter found from strength criteria. Thus a diameter of 105 mm will be suitable.

Example 20.6. *A hollow steel shaft 4 m long is to transmit 150 kW power at 150 R.P.M. The total angle of twist in this length is not to exceed 2.5° and the allowable shear stress is 60/mm². Determine the inside and outside diameters, if N = 0.082 × 10⁶ N/mm².*

Solution.

For stiffness: $\theta = \dfrac{Tl}{NJ} = \dfrac{Tl \times 32}{N\pi\,(D^4 - d^4)}$ or $D^4 - d^4 = \dfrac{Tl \times 32}{N\pi\,\theta}$ (where $\theta = \dfrac{2.5\pi}{180}$ radians) ...(1)

For strength: $T = f_s \dfrac{J}{R} = f_s \dfrac{\pi}{16}\left(\dfrac{D^4 - d^4}{D}\right)$ or $(D^4 - d^4) = \dfrac{16DT}{f_s\pi}$ ...(2)

Equating (1) and (2), we get $\dfrac{32Tl}{N\pi\theta} = \dfrac{16DT}{f_s\pi}$

or $$D = \frac{2lf_s}{N\theta} = \frac{2 \times 4000 \times 60}{0.082 \times 10^6 \times (2.5\pi/180)} = \mathbf{134\ mm}$$

Now, from (2), $D^4 - d^4 = \dfrac{16DT}{f_s \pi}$ where $T = \dfrac{60P}{2\pi n}$ kN-m $= \dfrac{60P}{2\pi n} \times 10^6$ N-mm

$$\therefore \quad D^4 - d^4 = \frac{16D}{f_s \pi} \times \frac{60P}{2\pi n} \times 10^6$$

or
$$(134)^4 - d^4 = \frac{16 \times 134}{60\pi} \times \frac{60 \times 150}{2\pi \times 150} \times 10^6 = 108.6 \times 10^6$$

$$\therefore \quad d = \mathbf{120.9\ mm.}$$

Example 20.7. *A solid shaft of aluminium of length 1.5 m and of 60 mm diameter is to be replaced by a tubular steel shaft of the same length and the same outside diameter, such that each of two shafts have the same angle of twist per unit torsional moment over the total length. Determine the inner diameter of the tubular steel shaft, if the modulus of rigidity of steel is three times that of aluminium.*

Solution. Given $L_s = L_A = 1.5$ m ; $D = 60$ mm ; $N_s = 3\ N_A$; $\theta_S = \theta_A$; $T_S = T_A$

Let d = inner diameter of steel shaft.

Polar moment of inertia of solid aluminium shaft $= J_A = \dfrac{\pi}{32} D^4 = \dfrac{\pi}{32} (64)^4$

Polar moment of inertia of hollow steel shaft $= J_S = \dfrac{\pi}{32} (D^4 - d^4) = \dfrac{\pi}{32} [(60)^4 - d^4]$

We have, $\dfrac{TL}{\theta} = NJ$

Since T, θ and L are the same for both the shafts, we have

$$N_A J_A = N_S J_S$$

or
$$N_A \times \frac{\pi}{32} (60)^4 = N_S \frac{\pi}{32} [(60)^4 - d^4]$$

or
$$(60)^4 = \frac{N_S}{N_A} [(60)^4 - d^4] = 3\ [(60)^4 - d^4]$$

or $\dfrac{12960000}{3} = 12960000 - d^4$ From which $d = \mathbf{54.22\ mm.}$

Example 20.8. *A solid shaft of 250 mm diameter has the same cross-sectional area as the hollow shaft of the same material with inside diameter of 200 mm. (a) Find the ratio of power transmitted by the two shafts for the same angular velocity, and (b) compare the angles of twist in equal lengths of these shafts, when stressed to the same intensity.*

Solution. Let the outside diameter of hollow shaft $= D$ mm

Area of cross-section of solid shaft $= A_S = \dfrac{\pi}{4} (250)^2$

Area of cross-section of hollow shaft $= A_H = \dfrac{\pi}{4} [D^2 - (200)^2]$

Since $A_S = A_H$, we have $\dfrac{\pi}{4} [D^2 - (200)^2] = \dfrac{\pi}{4} (250)^2$.

From which, we get $D = [(250)^2 + (200)^2]^{\frac{1}{2}} = 320.16$ mm

(a) Ratio of power transmitted

We know that $P = \dfrac{2\pi nT}{60}$

Shafts having same angular velocity will have same r.p.m. Hence ratio of H.P. transmitted will be equal to the ratio of torques transmitted. Hence if f_{sa} = allowable shear stress for the material, we have

$$\frac{P_S}{P_H} = \frac{T_S}{T_H} = \frac{\frac{\pi}{16} f_{sa} D'^3_S}{\frac{\pi}{16} f_{sa} \left[\frac{D^4 - d^4}{D}\right]} = \frac{(250)^3}{\frac{(320.16)^4 - (200)^4}{320.16}} = 0.56165$$

or $$\frac{P_H}{P_S} = \mathbf{1.78.}$$

(b) Angles of twist

Let θ_S = angles of twist for a solid shaft and θ_H be the angle of twist for the hollow shaft, for the same length.

In general, we have $\theta = \dfrac{f_S L}{R N}$, where R = outside radius

Now, for both shafts, f_S, N and L are the same

$$\therefore \quad \frac{\theta_S}{\theta_H} = \frac{R_H}{R_S} = \frac{D_H}{D_S} = \frac{320.16}{250} = \mathbf{1.2806.}$$

20.4. TORSION OF SHAFTS NOT CIRCULAR IN SECTION

The ordinary theory of torsion developed so far is true only for circular sections. For other shapes, the sections originally plane become warped on twisting. For non-circular sections, an exact study of stresses and strains produced by torsion is rather complicated and is beyond the scope of the present volume. An approximate analysis for square shafts is given in example 20.9.

Example 20.9. *A square shaft is required for transmitting power to a 200 kN overhead travelling crane. The load is lifted at a rate of 10 metres per minute. Taking the efficiency of crane gearing as 50%, calculate the necessary size of the shaft to run at 150 R.P.M. The twist must not exceed 1° in a length of 30 times the side of the square. Take N = 0.82 × 10⁵ N/mm².*

Solution. Let the side of the square shaft = S

Work done per minute to lift weight = 200 × 10 kN-m = 2000 × 10^6 N-mm.

Crane efficiency = 50%

$$\therefore \quad \text{Work to be supplied per minute} = \frac{2000 \times 10^6}{0.5} = 4 \times 10^9 \text{ N-mm}$$

$$\text{Torque, } T = \frac{\text{work supplied}}{2\pi n} = \frac{4 \times 10^9}{2\pi \times 150} = 4.24 \times 10^6 \text{ N-mm}$$

For a square shaft, we have

$$\theta = \frac{410Tl}{NS^4}, \text{ where } \theta = \text{angle of twist in degrees; } S = \text{side of square.}$$

$$\therefore \quad S^4 = \frac{410Tl}{N\theta} = \frac{410 \times 4.24 \times 10^6}{0.82 \times 10^5 \times 1} (30S) \quad \text{from which} \quad \mathbf{S = 86\ mm.}$$

20.5. KEYS AND COUPLINGS

Keys. A shaft rotating at a certain speed transmits power through pulley which is keyed to it. Thus, a key is also subjected to twisting moment and shearing stress is developed in it.

Let q_k = shear stress in the key.
r = distance of the key from the centre of the shaft.
l = length of the key;
b = width of the key.

Fig. 20.6

The moment of resistance of the key = $q_k\, r.l.b.$
This is equal to the torque transmitted.

Hence we have, $\quad T = q_k\, r.l.b \left(\text{where } T = \frac{\pi}{16} d^3 . f_s\right).$ ...[20.6(a)]

Coupling. Two sections or lengths of a shaft are connected through pulley which is keyed to it. Thus a coupling has to transmit the torque from one section to the other thereby subjected to shear stress.

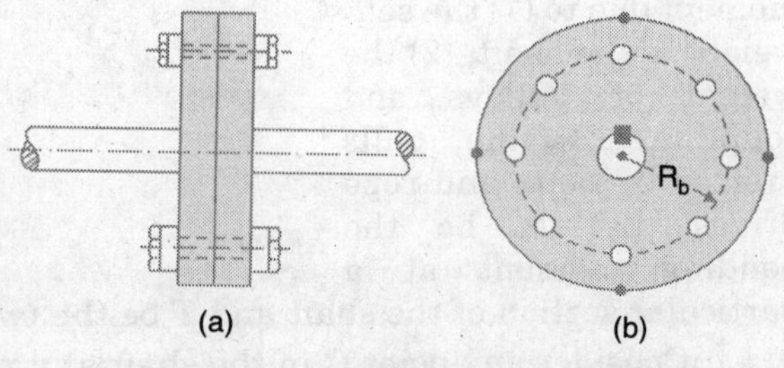

Fig. 20.7

Let n = number of bolts in the coupling
R_b = radius of the circle on which the bolts are arranged
d_b = diameter of each bolt
q_b = shear stress in each bolt.

The total moment transmitted by n bolts = $n\left(q_b \frac{\pi}{4} d_b{}^2\right) R_b$

This must be equal to the external torque.

$$\therefore \quad T = q_b\, n\, R_b \frac{\pi}{4} d_b{}^2, \left(\text{where } T = \frac{\pi}{16} d^3 . f_s\right). \quad ...(20.7)$$

Example 20.10. *A shaft of 100 mm diameter transmits 112.5 kW power at 180 R.P.M. A flanged coupling is keyed to the shaft by means of a key 100 mm long and 30 mm wide. The coupling has 6 bolts of 20 mm diameter symmetrically arranged along a bolt circle of 300 mm diameter. Calculate the shearing stresses in the shaft, the key and the bolts of the coupling.*

Solution. $\quad T = \frac{60P}{2\pi n} = \frac{60 \times 112.5}{2\pi \times 180} = 5.97 \text{ kN-m} = 5.97 \times 10^6 \text{ N-mm}.$

Now $f_s \dfrac{\pi}{16} d^3 = T = 5.97 \times 10^6$ or $f_s = \dfrac{5.97 \times 10^6 \times 16}{\pi (100)^3}$ = **30.4 N/mm²**

Again, for stress q_k in key, we have

$$q_k (100 \times 30) \times \frac{100}{2} = T = 5.97 \times 10^6 \quad \text{or} \quad q_k = \frac{5.97 \times 10^6}{15 \times 10^4} = \mathbf{39.8\ N/mm^2}$$

For the stress q_b in bolts, we have

$$q_b \times 6 \times \frac{\pi}{4} (20)^2 \times 150 = T = 5.97 \times 10^6$$

Hence, $q_b = \dfrac{5.97 \times 10^6}{6\pi \times 100 \times 150}$ = **21.1 N/mm².**

20.6. COMBINED BENDING AND TORSION : PRINCIPAL STRESSES

We have studied the effect of pure twisting moment on the shaft. A shaft transmitting power is also subjected to bending moment due to (1) the self-weight of the shaft, (2) the weight of pulleys and coupling, (3) the pulls exerted by belts and rope drives. Let M be the bending moment at a particular section of the shaft and T be the twisting moment.

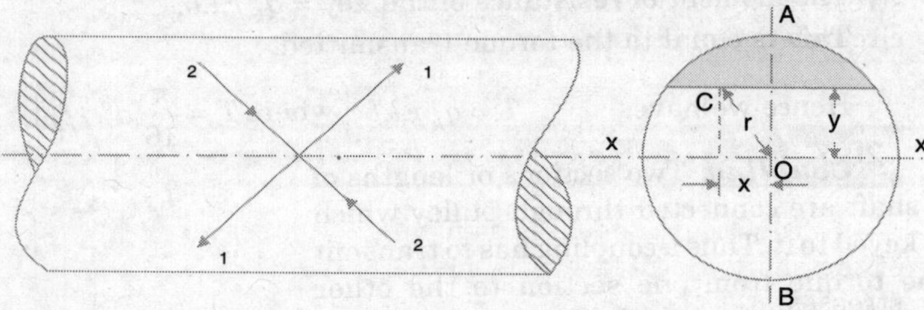

Fig. 20.8

Consider any point C in the shaft at a radial distance r from the centre and a distance y from the neutral axis. The shear stress due to torsion is given by

$$q = \frac{T}{J} r \qquad \ldots(1)$$

The bending stress at this point is given by

$$f = \frac{M}{I} y \qquad \ldots(2)$$

The shear stress resulting from bending forces, at the plane is given by

$$q' = \frac{F}{I.z} (A\bar{y}) \qquad \ldots(3)$$

Thus, at any point in the shaft, the component stresses are (1) the shear stress due to torsion, (2) the bending stress (tensile or compressive) and (3) the shear stress due to bending forces. In shafts which are not very long, the principal stresses will be maximum at the surface point (such as A and B). Thus, at A or B the stresses will be :

(1) The shear stress f_s due to torsion.

(2) The bending stress f_b due to bending.

The shear stress due to bending will be zero at point *A* or *B*.

Thus we have $$f_s = \frac{T}{J} R = \frac{16T}{\pi d^3} \quad \text{...(4)}$$

and $$f_b = \frac{M}{I} R = \frac{32M}{\pi d^3} \quad \text{...(5)}$$

And the principal stresses are

$$\sigma = \frac{f_b}{2} \pm \sqrt{\frac{f_b^2}{4} + f_s^2} = \frac{16}{\pi d^3} \{M \pm \sqrt{M^2 + T^2}\} \quad \text{...(20.8)}$$

The position of principal planes are given by

$$\tan 2\theta_p = \frac{2f_s}{f_b} = \frac{T}{M} \quad \text{...(20.9)}$$

It is to be noted that the principal stresses σ_1 and σ_2 will be of opposite sign.

In the case of a very short shaft, the effect of bending stress may be very small in comparison to the shear caused by bending forces. In such cases, the maximum principal stress may occur within the shaft. Usually, however, the maximum principal stress is at the circumference, as given by equation 20.8.

20.7. HYPOTHESIS OF ELASTIC FAILURE

To resist the bending moment and the twisting moment, the selection of suitable diameter of the shaft can be done on the basis of different hypothesis of elastic failure for compound stresses.

For the common two dimensional case, the hypothesis are as follows :

(1) **Principal Stress Hypothesis**

Let f = stress at elastic failure in simple direct stress. We have, thus

$$f_1 = \sigma_1 = \frac{16}{\pi d^3} \{M + \sqrt{M^2 + T^2}\} = \frac{32}{\pi d^3} M_e \quad \text{...(20.10)}$$

where $M_e = \frac{1}{2} \{M + \sqrt{M^2 + T^2}\}$ and is known as *equivalent bending moment*.

(2) **Maximum Principal Strain Hypothesis**

$$f_1 = \sigma_1 - \frac{1}{m}\sigma_2 = \frac{1}{2} f_b \left(1 - \frac{1}{m}\right) + \left(1 + \frac{1}{m}\right)\sqrt{\frac{f_b^2}{4} + f_s^2}$$

or $$f = \frac{16}{\pi d^3}\left[M\left(1 - \frac{1}{m}\right) + \left(1 + \frac{1}{m}\right)\sqrt{M^2 + T^2}\right] \quad \text{...(20.11)}$$

(3) **Maximum Shear Stress Hypothesis**

$$q_{max} = \frac{f}{2} = \frac{\sigma_1 - \sigma_2}{2} = \sqrt{\frac{f_b^2}{4} + f_s^2} = \frac{16}{\pi d^3}\sqrt{M^2 + T^2} = \frac{16}{\pi d^3} T_e \quad \text{...(20.12)}$$

In which $\sqrt{M^2 + T^2} = T_e$ = *equivalent twisting moment*, which alone would produce the same maximum shear stress.

(4) **Maximum Elastic Strain Energy Hypothesis**

$$f^2 = \sigma_1^2 + \sigma_2^2 - \frac{2\sigma_1\sigma_2}{m}$$

or $$f^2 = f_b^2 + 2f_s^2\left(1+\frac{1}{m}\right)$$

or $$f^2 = \left(\frac{32M}{\pi d^3}\right)^2 + 2\left(\frac{16T}{\pi d^3}\right)^2\left(1+\frac{1}{m}\right)$$

or $$f = \frac{16}{\pi d^3}\sqrt{4M^2 + \frac{2(m+1)}{m}T^2} \quad \text{...(20.13)}$$

(5) **Maximum Elastic Shear Strain Energy Hypothesis**

$$f^2 = \sigma_1^2 + \sigma_2^2 - \sigma_1\sigma_2 = f_b^2 + 3f_s^2$$

or $$f^2 = \left(\frac{32M}{\pi d^3}\right)^2 + 3\left(\frac{16T}{\pi d^3}\right)^2$$

$\therefore$ $$f = \frac{16}{\pi d^3}\sqrt{4M^2 + 3T^2} \quad \text{...(20.14)}$$

Example 20.11. *A shaft of 100 mm diameter is subjected to a bending moment of 10 kN-m and a twisting moment of 15 kN-m at a certain cross-section. Calculate the maximum direct stress induced in the section and specify the position of the plane on which it acts. Taking 1/m = 0.3, find the stress, which acting alone will produce the same maximum (i) strain, (ii) strain energy.*

Solution. $T = 15$ kN-m $= 15 \times 10^6$ N-mm

Shear stress $$f_s = \frac{16T}{\pi d^3} = \frac{16 \times 15 \times 10^6}{\pi (100)^3} = 76.4 \text{ N/mm}^2$$

Bending stress $$f_b = \frac{32M}{\pi d^3} = \frac{32 \times 10 \times 10^6}{\pi (10)^3} = 101.9 \text{ N/mm}^2.$$

$$\tan 2\theta_p = \frac{2f_s}{f_b} = \frac{T}{M} = \frac{15}{10} = 1.5; \quad \text{Hence } 2\theta_p = 56° \, 18'$$

$\theta_{p1} = 28° \, 9'$ or $61° \, 51'$ with the axis of the shaft

$\theta_{p2} = 118° \, 9'$ or $28° \, 9'$ with the axis of the shaft.

$$\sigma_1 = \frac{f_b}{2} + \sqrt{\frac{f_b^2}{4} + f_s^2} = \frac{101.9}{2} + \sqrt{\left(\frac{101.9}{2}\right)^2 + (76.4)^2} = \mathbf{142.5 \text{ N/mm}^2}.$$

$$\sigma_2 = \frac{f_b}{2} - \sqrt{\frac{f_b^2}{4} + f_s^2} = \frac{101.9}{2} - \sqrt{\left(\frac{101.9}{2}\right)^2 + (76.4)^2} = \mathbf{-40.6 \text{ N/mm}^2}.$$

(*i*) *For maximum strain,* we have

$$f = E.e = \sigma_1 - \frac{1}{m}\sigma_2 = 142.5 + (40.6 \times 0.3) = \mathbf{154.7 \text{ N/mm}^2}.$$

(*ii*) *For maximum strain energy,* we have

$$f^2 = \sigma_1^2 + \sigma_2^2 - \frac{2\sigma_1\sigma_2}{m} = (142.5)^2 + (-40.6)^2 + (2 \times 142.5 \times 40.6 \times 0.3)$$

$$= 2.54 \times 10^4$$

$\therefore$ $$f = \mathbf{159.4 \text{ N/mm}^2}.$$

Example 20.12. *A circular shaft is supported in bearings 3 metres apart and transmits 75 kW power at 150 R.P.M. At 1 metre from one-bearing, it carries a pulley exerting a transverse load of 30 kN on the shaft. Determine the suitable diameter for the shaft if (i) the maximum direct stress is not to exceed 100 N/mm², (ii) the maximum intensity of shear stress is not to exceed 50 N/mm², (iii) the stress which acting alone would produce the same maximum strain, is not to exceed 100 N/mm², (iv) the direct stress which acting alone would produce the same maximum strain energy is not to exceed 100 N/mm². Take 1/m = 0.25.*

Solution. $$T = \frac{60P}{2\pi n} = \frac{60 \times 75}{2\pi \times 150} = 4.77 \text{ kN-m} = 4.77 \times 10^6 \text{ N-mm}$$

$$M = \frac{Wab}{L} = \frac{30 \times 1 \times 2}{3} = 20 \text{ kN-m} = 20 \times 10^6 \text{ N-mm}$$

(i) Principal stress criterion

$$f = \frac{16}{\pi d^3} \times \left[M + \sqrt{M^2 + T^2} \right]$$

or $$d^3 = \frac{16}{\pi \times 100} \left[20 + \sqrt{(20)^2 + (4.77)^2} \right] \times 10^6 = 2.066 \times 10^6$$

∴ d = **127.3 mm.**

(ii) Maximum shear stress criterion

$$q_{max} = \frac{16}{\pi d^3} \sqrt{M^2 + T^2}$$

or $$d^3 = \frac{16}{\pi \times 50} \left[\sqrt{(20)^2 + (4.77)^2} \right] \times 10^6 = 2.094 \times 10^6$$

∴ d = **128 mm.**

(iii) Maximum strain criterion

$$f = \frac{16}{\pi d^3} \left[\left(1 - \frac{1}{m}\right) M + \left(1 + \frac{1}{m}\right) \sqrt{M^2 + T^2} \right]$$

∴ $$d^3 = \frac{16}{\pi \times 100} \left[(1 - 0.25)20 + (1 + 0.25) \sqrt{(2)^2 + (4.77)^2} \right] \times 10^6 = 2.075 \times 10^6$$

∴ d = **127.5 mm.**

(iv) Maximum strain energy criterion

$$f = \frac{16}{\pi d^3} \sqrt{4M^2 + \frac{2(m+1)}{m} T^2}$$

or $$d^3 = \frac{16 \times 10^6}{\pi \times 100} \sqrt{4(20)^2 + 2\,(1 + 0.25)\,(4.77)^2} = 2.073 \times 10^6$$

∴ d = **127.5 mm**

Hence adopt the diameter of shaft as **128 mm.**

20.8. EFFECT OF END THRUST

When a shaft is subjected to an end thrust in addition to the bending moment and torsion, constant compressive stress will be induced in the shaft. The total direct stress at any point in the shaft will be the resultant of direct stress due to bending and the direct stress due to the thrust. Thus, for point A of Fig. 20.6, the total direct compressive stress will be given by

$$f_c = \frac{M}{I} R + \frac{P}{\pi R^2} = f_b + f_0 \quad \ldots(20.15)$$

where P = end thrust and f_0 = direct stress due to thrust.

This value of f_c must be used in place of f_b in the equation 20.8 to calculate the compressive principal stress.

Similarly, the total direct tensile stress will be given by

$$f_t = f_b - f_0 \quad \ldots(20.16)$$

If f_t is positive, it must be used in place of f_b in equation 20.8 to calculate the tensile principal stress.

Example 20.13. *A propeller shaft of 200 mm external diameter and 100 mm internal diameter has to transmit 1350 kW power at 80 R.P.M. In addition, it is subjected to a B.M. of 10 kN-m and an end thrust of 100 kN. Find (i) the principal stresses and their planes, (ii) the maximum shear stress and its plane, (iii) the stress which acting alone would produce the same maximum strain. Take 1/m = 0.3.*

Solution. Area of cross-section $= \frac{\pi}{4}(200^2 - 100^2) = 2.356 \times 10^4 \text{ mm}^2$

Section modulus of shaft $= \frac{\pi}{32}\left(\frac{D^4 - d^4}{D}\right) = \frac{\pi}{32} \frac{(200)^4 - (100)^4}{200}$

$= 73.63 \times 10^4 \text{ mm}^3$

Torsional modulus of section $= \frac{\pi}{16} \frac{D^4 - d^4}{D} = 147.26 \times 10^4 \text{ mm}^3$

Now $f_0 = \frac{P}{A} = \frac{100 \times 1000}{2.356 \times 10^4} = 4.24 \text{ N/mm}^2$; $f_b = \frac{M}{Z} = \frac{10 \times 10^6}{73.63 \times 10^4} = 13.58 \text{ N/mm}^2$

$\therefore$ Max. comp. direct stress $= f_c = 4.24 + 13.58 = 17.82 \text{ N/mm}^2$

Torque $T = \frac{60P}{2\pi n} = \frac{60 \times 1350}{2\pi \times 80} = 161.1 \text{ kN-m} = 161.1 \times 10^6 \text{ N-mm.}$

$\therefore$ Max. shear stress, $f_s = \frac{161.1 \times 10^6}{147.26 \times 20^4} = 109.4 \text{ N/mm}^2$

(*i*) Principal stresses are :

$$\sigma_1 = \frac{17.82}{2} + \sqrt{\left(\frac{17.82}{2}\right)^2 + (109.4)^2} = \mathbf{118.7 \text{ N/mm}^2} \text{ (comp.)}$$

$$\sigma_2 = \frac{17.82}{2} - \sqrt{\left(\frac{17.82}{2}\right)^2 + (109.4)^2} = -\mathbf{100.9 \text{ N/mm}^2} \text{ (}i.e.\text{, tensile)}$$

(*ii*) Max. shear intensity $= \sqrt{\left(\frac{17.82}{2}\right)^2 + (109.4)^2} = \mathbf{109.8 \text{ N/mm}^2}$

(*iii*) Max. stress to give the maximum strain is

$$f = \sigma_1 - \frac{1}{m}\sigma_2 = 118.7 + 0.3 \times 100.9 \simeq \mathbf{149 \text{ N/mm}^2}.$$

20.9. TORSIONAL STRAIN ENERGY

We know that the elastic energy in a material subjected to uniform shear stress is equal to $q^2/2N$ per unit volume. In order to calculate the torsional strain energy for a solid circular shaft, consider an elementary ring for a radius r and thickness δr. The torsional strain energy of this ring for a length l is given by,

$$\delta U = \frac{q^2}{2N}(2\pi r \delta r)l \qquad \text{...(1)}$$

But $\dfrac{q}{r} = \dfrac{f_s}{R}$ or $q = \dfrac{f_s}{R} r$

$$\therefore \quad \delta U = \frac{f_s^{\,2} r^2}{2R^2 N}(2\pi r l dr) = \frac{f_s^{\,2} r^3}{2R^2 N} 2\pi l dr$$

Hence the total torsional strain energy stored in the shaft is

$$U = \frac{f_s^{\,2} 2\pi l}{2R^2 N}\int_0^R r^3 dr = \frac{f_s^{\,2}\pi l}{R^2 N}\cdot\frac{R^4}{4} = \frac{f_s^{\,2}}{4N}(\pi R^2 l)$$

$$= \frac{f_s^{\,2}}{4N} \times \text{volume of the shaft} \qquad \text{...(20.17)}$$

Similarly, for a hollow shaft of internal diameter d and external diameter D, the total torsional strain energy will be

$$U = \frac{f_s^{\,2}\pi l}{R^2 N}\int_{d/2}^{D/2} r^3 dr = \frac{f_s^{\,2}\pi l (D^4 - d^4)}{64 N R^2}$$

$$= \frac{f_s^{\,2}}{4N}\left(\frac{D^2 + d^2}{D^2}\right) \times \text{volume of the shaft} \qquad \text{...(20.18)}$$

For a very thin tube, D approaches d and hence

$$U \simeq \frac{f_s^{\,2}}{2N} \times \text{volume of the shaft} \qquad \text{...(20.19)}$$

If f_s in equations 20.17, 20.18 and 20.19 represent the intensity of shear stress at the elastic limit, the expressions for U may be called the *proof torsional resilience.*

Example 20.14. *A hollow shaft, subjected to pure torque, attains a maximum shear stress of f_s. Given that the strain energy per unit volume is $f_s^{\,2}/3N$, calculate the ratio of shaft diameters. Determine the actual diameters of such a shaft to transmit 5 MW at 120 r.p.m. when energy stored is 25000 Nm/m³ and N = 80000 N/mm².*

Solution. From Eq. 20.18, We have

$$\frac{U}{\text{Volume}} = \frac{f_s^{\,2}}{4N}\left(\frac{D^2 + d^2}{D^2}\right) \text{ for a hollow shaft}$$

But $$\frac{U}{\text{Volume}} = \frac{f_s^{\,2}}{3N} \text{ (Given)}$$

$$\therefore \quad \frac{f_s^{\,2}}{4N}\left(\frac{D^2 + d^2}{D^2}\right) = \frac{f_s^{\,2}}{3N}$$

or $$\frac{D^2 + d^2}{D^2} = \frac{4}{3}, \text{ from which } \frac{d}{D} = \sqrt{\frac{1}{3}} \quad \text{or} \quad \frac{D}{d} = \sqrt{3} = 1.732 \qquad ...(1)$$

If $$\frac{f_s^2}{3N} = 25000 \text{ N-m/m}^3 = 25000 \times 10^{-6} \text{ N-mm/mm}^3 \text{ (Given)}$$

$$f_s = (25000 \times 10^{-6} \times 3 \times 80000)^{1/2} = 77.46 \text{ N/mm}^2$$

Also $$T = \frac{60P}{2\pi n} = \frac{60 \times 5 \times 10^3}{2\pi\,(120)} = 397.89 \text{ kN-m} = 397.89 \times 10^6 \text{ N-mm}$$

Now $$f_s = T \cdot \frac{R}{J} = T \frac{D/2}{\frac{\pi}{32}(D^4 - d^4)} = \frac{16DT}{\pi(D^4 - d^4)} \qquad ...(2)$$

Substituting the values of f_s and T and noting that $\frac{d}{D} = \left(\frac{1}{3}\right)^{\frac{1}{2}}$, we get

$$77.46 = \frac{16\,D\,(397.89 \times 10^6)}{\pi D^4\left(1 - \frac{1}{9}\right)} = \frac{2280 \times 10^6}{D^3}$$

From which $D^3 = 29.43 \times 10^6$ or $D \simeq$ **309 mm** and $d = 309/\sqrt{3} \simeq$ **178 mm.**

20.10. SHAFTS IN SERIES AND PARALLEL

When two dissimilar shafts are connected together to form one shaft, the shaft is known as a *composite shaft*. If the driving torque is applied at one end, and the resisting torque at the other end, the shafts are said to be connected in *series* [Fig. 20.9(*a*)]. In such a case both the shafts carry the same torque T, and the total angle of twist at the fixed or resisting end is the sum of the separate angles of twist of the two shafts :

$$\theta = \theta_1 + \theta_2 = \frac{Tl_1}{N_1 J_1} + \frac{Tl_2}{N_2 J_2} \qquad ...(20.20)$$

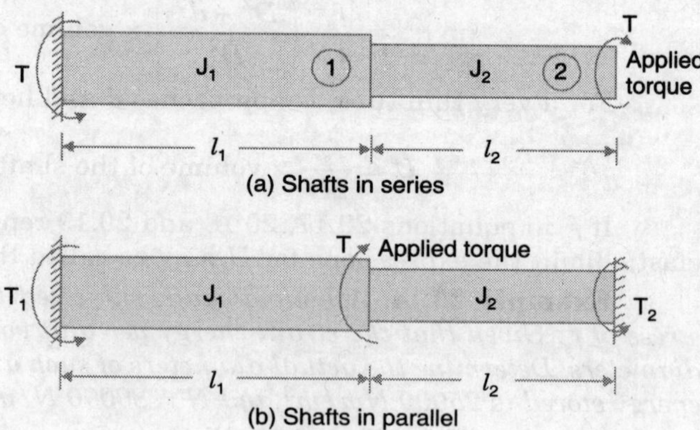

(a) Shafts in series

(b) Shafts in parallel

Fig. 20.9

If $N_1 = N_2 = N$ (*i.e.*, both shafts of the same material)

$$\theta = \frac{T}{N}\left(\frac{l_1}{J_1} + \frac{l_2}{J_2}\right) \qquad ...(20.21)$$

If the torque T is applied at the junction of the two shafts and the resisting torques T_1 and T_2 at their remote ends, the shafts are said to be connected in *parallel*. In such a case, the angle of twist is the same for both the shafts.

$$\theta_1 = \theta_2$$

$$\therefore \quad \frac{T_1 l_1}{N_1 J_1} = \frac{T_2 l_2}{N_2 J_2} \quad \text{or} \quad \frac{T_1}{T_2} = \frac{l_2}{l_1} \cdot \frac{N_1}{N_2} \cdot \frac{J_1}{J_2} \quad \text{...(20.22)}$$

If both the shafts are of the same material,

$$\frac{T_1}{T_2} = \frac{J_1}{J_2} \cdot \frac{l_2}{l_1} \quad \text{...(20.23)}$$

Example 20.15. *A 3 cm diameter circular steel shaft is provided with enlarged portions A and B as shown in Fig. 20.10. On to the enlarged portion, a steel tube of 1.5 mm thickness is shrunk. While the shrinking process is going on, the 3 cm shaft is held twisted by a couple of magnitude of 300 N-m. When the tube is firmly set on the shaft, this twisting couple is removed. Calculate what twisting couple is left on the shaft, the shaft and the tube being made of the same material.*

Solution. For the shaft, $J_1 = \frac{\pi}{32}(3)^4 = 7.95 \text{ cm}^4$;

For the tube, $J_2 = \frac{\pi}{32}\{(6.3)^4 - (6)^4\} = 27.4 \text{ cm}^4$

Initially, Let θ_1 = initial angle of twist in the shaft.

$$\therefore \quad \frac{N\theta_1}{l} = \frac{T_1}{J_1} = \frac{300 \times 100}{7.95} = 3780 \quad \text{or} \quad \theta_1 = \frac{3780 l}{N} \quad \text{...(1)}$$

Finally, Let $\theta_1{}'$ = Residual angle of twist in the shaft

$\theta_2{}'$ = Residual angle of twist in the tube

T = Residual twisting couple on the shaft

Fig. 20.10

$$T = \frac{N\theta_1{}' J_1}{l} = \frac{N\theta_2{}' J_2}{l}$$

$$\text{or} \quad \theta_1{}' = \theta_2{}' \frac{J_2}{J_1} = \theta_2{}' \frac{27.4}{7.95} = 3.45\,\theta_2{}' \quad \text{...(2)}$$

But from compatibility of twist, we have

$$\theta_1{}' + \theta_2{}' = \theta_1 \quad \text{...(3)}$$

$$\therefore \quad 3.45\theta_2{}' + \theta_2{}' = \theta_1 = \frac{3780 l}{N} \quad \text{or} \quad 4.45\theta_2{}' = \frac{3780 l}{N}$$

$$\therefore \quad 4.45 \frac{Tl}{NJ_2} = \frac{3780 l}{N}, \qquad \left(\text{since } \theta_2{}' = \frac{Tl}{NJ_2}\right)$$

$$\therefore \quad T = \frac{3780 J_2}{4.45} = \frac{3780 \times 27.4}{4.45} = 23200 \text{ N-cm} = \mathbf{232 \text{ N-m.}}$$

Example 20.16. *A compound shaft, shown in Fig. 20.11 consists of a 1 metre long aluminium bar securely joined to a 1 m length of brass bar, the diameter of each being 50 mm. Calculate the maximum torque that can be applied at the lower end if the allowable angle of twist is 1° and the allowable shear stress in aluminium is 75 N/mm² and that in brass is 50 N/mm². Take N for brass = 0.34 × 10⁵ N/mm² and that for aluminium as 0.27 × 10⁵ N/mm².*

Solution. Both the shafts are connected in series.

We will find the torque on the basis of the following three conditions. The minimum of the three will be the allowable torque.

(1) The torque T_1 to set up the working shear stress in brass.

(2) The torque T_2 to set up the working shear stress in aluminium.

(3) The torque T_3 to create 1° twist to entire shaft.

1 m Brass

1 m Aluminium

50 mm

Fig. 20.11

$$(1)\ T_1 = \frac{f_s J}{R} = \frac{50 \times \frac{\pi}{32}(50)^4}{25} = 1.228 \times 10^6 \text{ N-mm}$$

$$(2)\ T_2 = \frac{f_s J}{R} = \frac{75 \times \pi(50)^4}{32 \times 25} = 1.842 \times 10^6 \text{ N-mm}$$

$$(3)\ \theta = \frac{T_3 l_1}{N_1 J} + \frac{T_3 l_2}{N_2 J} = \frac{T_3 l}{J}\left(\frac{1}{N_1} + \frac{1}{N_2}\right)$$

$$= \frac{T_3 l}{J}\left(\frac{N_2 + N_1}{N_1 N_2}\right) \quad \text{where} \quad l_1 = l_2 = l = 1 \text{ m} = 1000 \text{ mm}$$

$$\therefore \quad T_3 = \frac{J\theta N_1 N_2}{(N_1 + N_2)l} = \frac{\pi \times (50)^4 \times 0.27 \times 0.34 \times 10^5 \times 10^5}{32(0.27 + 0.34)10^5 \times 1000} \times \frac{1 \times \pi}{180}$$

$= 1.612 \times 10^5$ N-mm = **0.1612 kN-m.**

T_3 is the minimum of the above three torques. Maximum allowable torque

= **1.612 × 10⁵ N-mm = 161.2 N-m.**

Example 20.17. *A steel shaft of 3 cm diameter and 1 metre long is rigidly fixed at the ends. A twisting moment of 600 N-m is applied at a distance of 250 mm from one end. Calculate (a) the fixing couples at the ends, (b) the maximum shear stress, (c) the angle of twist of the section where the twisting moment has been applied. Take N = 0.82 × 10⁵ N/mm².*

Solution. Let the reactive torques at the ends be T_1 and T_2 respectively.

For equilibrium, we have $T_1 + T_2 = T = 600000$ N-mm ...(1)

The other equation can be obtained from the consideration of the torsional deflection (*i.e.*, θ) of the length l_1 and l_2 at their junction.

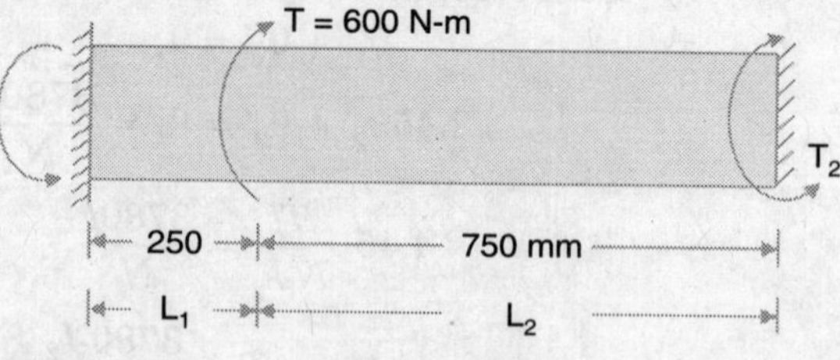

Fig. 20.12

Thus, we have, the compatibility equation,

$$\theta = \frac{T_1 l_1}{NJ} = \frac{T_2 l_2}{NJ} \quad \text{or} \quad T_1 l_1 = T_2 l_2$$

$$\therefore \quad T_1 = T_2 \frac{l_2}{l_1} = T_2 \frac{750}{250} = 3T_2 \qquad \ldots(2)$$

Substituting this in (1), we get $3T_2 + T_2 = 600000$

or $T_2 = 150000$ N-mm = **150 N-m** and $T_1 = 450000$ N-mm = **450 N-m**

The maximum shear stress will occur where T is maximum. Hence for the segment l_1, we have

$$f_s = \frac{16T_1}{\pi d^3} = \frac{16 \times 450000}{\pi (30)^3} = \mathbf{84.88\ N/mm^2}$$

Angle of twist $\quad \theta = \frac{f_s l_1}{RN} = \frac{84.88 \times 250}{15 \times 0.82 \times 10^5} = \mathbf{0.01725\ radians.}$

20.11. SHAFTS OF VARYING DIAMETER

Fig. 20.13 shows a shaft AB of length L, having diameter d_1 (or radius r_1) at one end A and diameter d_2 (or radius r_2) at the other end B, and is subjected to *uniform torque T*. Consider a strip of thickness dx at a distance x from the end A. The radius r_x at this section is given by

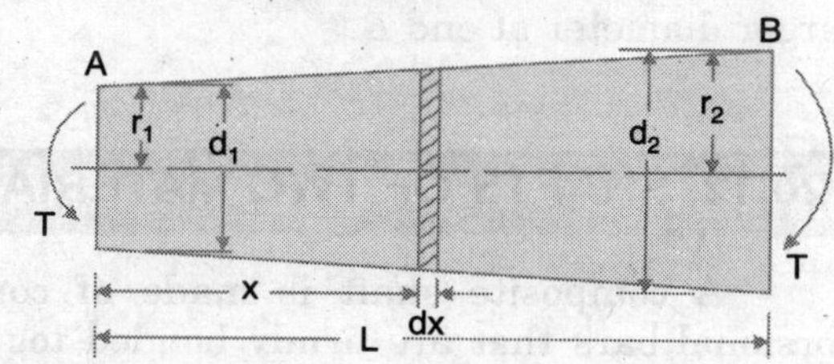

Fig. 20.13

$$r_x = r_1 + \frac{(r_2 - r_1)\,x}{L}$$

The polar moment of inertia of the elemental section is given by

$$J = \frac{\pi}{2} r_x^{\,4} = \frac{\pi}{2}\left[r_1 + \left(\frac{r_2 - r_1}{L}\right)x\right]^4$$

If $\delta\theta$ is the angle through which the shaded element of length δx is twisted, then

$$\delta\theta = \frac{TL}{NJ} = \frac{T\,\delta x}{N\dfrac{\pi}{2}\left[r_1 + \left(\dfrac{r_2 - r_1}{L}\right)x\right]^4}$$

Hence the total angle of twist for the whole shaft is

$$\theta = \sum_0^L \delta\theta = \frac{2T}{N\pi}\int_0^L \frac{dx}{\left[r_1 + (r_2 - r_1)\dfrac{x}{L}\right]^4} = \frac{2T}{N\pi}\left(-\frac{1}{3}\right)\left(\frac{L}{r_2 - r_1}\right)\left[\frac{1}{\left\{r_1 + \left(\dfrac{r_2 - r_1}{L}\right)x\right\}^3}\right]_0^L$$

or
$$\theta = \frac{2}{3}\frac{TL}{N\pi (r_2 - r_1)}\left[\frac{1}{r_1^{\,3}} - \frac{1}{r_2^{\,3}}\right] \qquad \text{...[20.24 (a)]}$$

or
$$\theta = \frac{32\,TL}{3N\pi (d_2 - d_1)}\left[\frac{1}{d_1^{\,3}} - \frac{1}{d_2^{\,3}}\right] \qquad \text{...[20.24 (b)]}$$

Another convenient form of expression for θ is as under

$$\theta = \frac{TL}{NJ_1}\left(\frac{\beta^2 + \beta + 1}{3\beta^3}\right) \qquad \text{...[20.24 (c)]}$$

where $\quad J_1 = \frac{\pi d_1^{\,4}}{32} \quad$ and $\quad \beta = \frac{d_2}{d_1}$

If $\beta = 1$ (*i.e.,* $d_2 = d_1 = d$) the above expression reduces to

$$\theta = \frac{TL}{NJ} = \frac{TL}{N\left(\frac{\pi d^4}{32}\right)} = \frac{32TL}{N \pi d^4} \qquad \text{...[20.24 (d)]}$$

which is the same expression as obtained for a shaft of uniform section.

If $\beta = 2$ (*i.e.,* $d_2 = 2d_1$), we get

$$\theta = \frac{7}{24}\frac{TL}{NJ_1} = \frac{7}{24}\frac{TL}{N\left(\frac{\pi d_1^4}{32}\right)} = \frac{28}{3}\frac{TL}{N\pi d_1^4} \qquad \text{...[20.24 (e)]}$$

This is much less than for the case of $\beta = 1$ because of increased stiffness due to the larger diameter at end B.

20.12. SHAFTS OF TWO MATERIALS: COMPOSITE SHAFT

A composite shaft is made of concentric circular torsional bars that are firmly bonded together to act as a single member. Fig. 20.14 shows such a composite shaft where a hollow tube and a core are securely bonded to act as a solid shaft. Such a shaft is statically indeterminate, since both the hollow tube and the solid core have different properties.

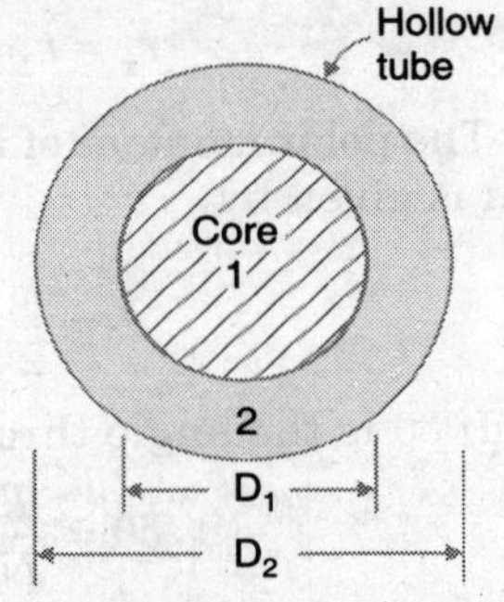

Fig. 20.14. Composite Shaft

Let us have suffix 1 for core and 2 for the hollow tube. Let the composite shaft be acted upon by a torque T. This is evidently resisted by the torque T_1 and T_2 developed in the core and tube respectively.

Hence from the condition of statical equilibrium,

$$T = T_1 + T_2 \qquad \text{...(1)}$$

Since both the parts are rigidly bonded together, they must rotate by the same amounts, *i.e.,* the angle of twist must be the same. Hence from the compatibility of rotations, we have

$$\theta = \frac{T_1 L}{N_1 J_1} = \frac{T_2 L}{N_2 J_2} \quad \text{or} \quad \frac{T_1}{T_2} = \frac{N_1}{N_2}\cdot\frac{J_1}{J_2} \qquad \text{...(2)}$$

From (1) and (2), we get

$$T_1 = T\left(\frac{N_1 J_1}{N_1 J_1 + N_2 J_2}\right) \quad \text{and} \quad T_2 = T\left(\frac{N_2 J_2}{N_1 J_1 + N_2 J_2}\right) \qquad \text{...(20.25)}$$

Hence $$\theta = \frac{TL}{N_1 J_1 + N_2 J_2} \qquad \text{...(20.26)}$$

The maximum shear stresses f_{s1} and f_{s2} in the core and tube are given by

$$f_{s1} = \frac{T_1 (D_1/2)}{J_1} \quad \text{and} \quad f_{s2} = \frac{T_2 (D_2/2)}{J_2} \qquad \text{...(20.27)}$$

Hence $$\frac{f_{s1}}{f_{s2}} = \frac{T_1 D_1}{T_2 D_2}\cdot\frac{J_2}{J_1} = \frac{N_1}{N_2}\cdot\frac{D_1}{D_2} \qquad \text{...(20.28)}$$

Example 20.18. *A solid circular shaft has a radius of 100 mm at one end and 115 mm at the other end, the length of the shaft being 2 m. Calculate the percentage error committed if θ is calculated on the basis of a mean radius of 120 mm.*

Solution. Given:

$r_1 = 100$ mm ($d_1 = 200$ mm) and $r_2 = 115$ mm ($d_2 = 230$ mm)

$L = 2$ m $= 2000$ mm

From Eq. 20.24 (*b*),

$$\theta = \frac{32\,TL}{3N\pi\,(d_2 - d_1)}\left[\frac{1}{d_1^3} - \frac{1}{d_2^3}\right] = \frac{32}{3}\cdot\frac{T}{N}\cdot\frac{2000}{\pi\,(230-200)}\left[\frac{1}{(200)^3} - \frac{1}{(230)^3}\right]$$

$$= 9.69 \times 10^{-6}\frac{T}{N} \text{ radian}$$

Alternatively, from Eq. 20.24 (*c*), $\theta = \dfrac{TL}{NJ_1}\left(\dfrac{\beta^2 + \beta + 1}{3\beta^3}\right)$

Here $\beta = \dfrac{d_2}{d_1} = \dfrac{230}{200} = 1.15$; $J_1 = \dfrac{\pi d_1^4}{32} = \dfrac{\pi}{32}\,(200)^4 = 157.08 \times 10^6$ mm⁴

$$\therefore \quad \theta = \frac{T}{N} \times \frac{2000}{157.08 \times 10^6}\left[\frac{(1.15)^2 + 1.15 + 1}{3\,(1.15)^3}\right] = 9.69 \times 10^{-6}\,\frac{T}{N} \text{ radian}$$

For a solid shaft average diameter $d = \dfrac{1}{2}\,(d_1 + d_2) = \dfrac{1}{2}\,(200 + 230) = 215$ mm

$$\theta = \frac{32\,TL}{N\pi d^4} = \frac{32T}{N} \times \frac{2000}{\pi\,(215)^4} = 9.534 \times 10^{-6}\,\frac{T}{N} \text{ radian}$$

$$\therefore \quad \text{Percentage error} = \frac{9.69 - 9.534}{9.69} \times 100 = \mathbf{1.61\%}.$$

Example 20.19. *A gun-metal sleeve is fixed securely to a steel shaft and the compound shaft is subjected to a torque. If the torque on the sleeve is twice that on the shaft, find the ratio of external diameter of sleeve to diameter of the shaft.*

If the limits of shearing stress in the gun-metal and steel are 45 and 80 N/mm² respectively, find the torque that may be transmitted by the compound shaft when the steel shaft diameter is 50 mm. Take $N_{steel} = 2.5\,N_{gm}$. (U.L.)

Solution. Let us use suffix *s* for steel shaft and suffix *g* for the gun metal sleeve. Also, let *d* be the external diameter of the steel shaft and *D* be the external diameter of the gun-metal sleeve. The internal diameter of the gun-metal sleeve is evidently equal to *d*.

Now $\dfrac{\theta}{L} = \dfrac{T}{JN}$, in general

Since both the steel shaft and gun-metal sleeve are securely fixed, θ/*L* is the same for both.

$$\therefore \quad \frac{T_s}{J_s N_s} = \frac{T_g}{J_g N_g} \quad \text{or} \quad \frac{J_g}{J_s} = \frac{T_g}{T_s}\cdot\frac{N_s}{N_g} = 2 \times 2.5 = 5.$$

$$\text{or} \quad \frac{\frac{\pi}{32}\,(D^4 - d^4)}{\frac{\pi}{32}\,d^4} = 5 \quad \text{or} \quad D^4 - d^4 = 5\,d^4$$

Giving $D/d = (6)^{\frac{1}{4}} = 1.5651$

Again $\dfrac{f_{sg}}{f_{ss}} = \dfrac{(TR/J)_g}{(TR/J)_s} = \dfrac{(TD/2J)_g}{(TD/2J)_s} = \left(\dfrac{T_g}{T_s}\right)\left(\dfrac{D}{d}\right)\left(\dfrac{J_s}{J_g}\right)$

or $\dfrac{f_{sg}}{f_{ss}} = \left(\dfrac{T_g}{T_s}\right)\left(\dfrac{D}{d}\right)\left(\dfrac{d^4}{D^4 - d^4}\right) = (2)\,(1.5651)\left(\dfrac{1}{5}\right) = 0.62604$

or $\dfrac{f_{ss}}{f_{sg}} \simeq 1.6$

If the maximum stress in gun-metal = 45, corresponding maximum steel in steel will be = 1.6 × 45 = 72 N/mm², which is less than permissible value of 80 N/mm².

Now torque carried by steel shaft $= \dfrac{J}{R} f_{ss} = \dfrac{\pi}{16} (50)^3 \times 72 \times 10^{-3} = 1767$ **N-m.**

Hence total torque of composite shaft = (2 + 1) × 1767 = 5301 N-m.

20.13. ADDITIONAL EXAMPLES USEFUL FOR COMPETITIVE EXAMINATIONS

Example 20.20. *Derive an expression for the allowable twisting moment for a thin walled tube. Also, derive an approximate expression for the strength weight ratio for such a tube. Take f_s as the allowable shear stress. It is assumed that the tube does not buckle.*

Solution. Thickness of the wall of the tube $= t = \dfrac{D-d}{2}$; outer radius of tube = R

where D = outer diameter = $2R$ and d = inner diameter = $2(R - t)$

$$\therefore \quad J = \frac{\pi}{32}(D^4 - d^4) = \frac{\pi}{32}[(2R)^4 - 2^4 (R-t)^4] = \frac{\pi}{2}[R^4 - (R-t)^4]$$

$$= \frac{\pi}{2}[4R^3 t - 6R^2t^2 + 6Rt^3 - t^4] = \frac{\pi}{2} R^4 \left[4\frac{t}{R} - 6\left(\frac{t}{R}\right)^2 + 6\left(\frac{t}{R}\right)^3 - \left(\frac{t}{R}\right)^4\right]$$

For a thin tube, t/R is quite small and hence higher power of t/R can be neglected.

$$\therefore \quad J \simeq 2\pi R^3 t \qquad \text{...(20.29)}$$

$$\text{Again,} \quad T = \frac{J}{R} \cdot f_s = 2\pi R^2 t f_s \qquad \text{...(20.30)}$$

$$\text{Also,} \quad A = \pi[R^2 - (R-t)^2] = \pi[2Rt - t^2] = \pi R^2 \left[\frac{2t}{R} - \left(\frac{t}{R}\right)^2\right]$$

Neglecting higher powers of t/R, we get $A \simeq 2\pi Rt$

Let γ be the unit weight of the material.

∴ **Weight of tube of length** $L = W = AL\gamma = 2\pi Rt L\gamma$

$$\therefore \quad \text{Strength-weight ratio} = \frac{T}{W} = \frac{2\pi R^2 t f_s}{2\pi Rt L\gamma} = \frac{\mathbf{Rf_s}}{\mathbf{L\gamma}} \qquad \text{...(20.31)}$$

which is the required expression.

Example 20.21. *A shaft tapers uniformly from a radius (r + a) at one end to (r – a) at the other. If it is under the action of an axial torque T and a = 0.1 r, find the percentage error in the angle of twist for a given length when calculated on the assumption of a constant radius r.* (U.L.)

Solution. Let L = length of the shaft

Change in radius $= (r + a) - (r - a) = 2a$, in a length L.

Consider a section at a distance x from the small end.

Radius at this section is, $r_x = r - a + \dfrac{2ax}{L}$

Angle of twist of a length δx is given by

$$\delta\theta = \frac{T \cdot \delta x}{NJ} \quad \left(\text{where } J = \frac{\pi}{32} d^4 = \frac{\pi}{32}(2r)^4 = \frac{\pi}{2} r^4, \text{ in general}\right)$$

$$\therefore \quad \delta\theta = \frac{T \times 2\,\delta x}{N\pi\,(r - a + 2ax/L)^4}$$

$$\therefore \quad \text{Total angle,} \quad \theta = \frac{2T}{N\pi}\int_0^L \frac{dx}{(r - a + 2ax/L)^4} = \frac{2T}{3N\pi}\frac{L}{2a}\left[\frac{1}{(r-a)^3} - \frac{1}{(r+a)^3}\right]$$

Substituting $a = 0.1\,r = r/10$, we get

$$\theta = \frac{10TL}{3N\pi\, r^4}\left[\frac{1000}{729} - \frac{1000}{1331}\right] = 2.065\frac{TL}{N\pi\, r^4} \quad \ldots(1)$$

For a shaft of uniform radius r,

$$\theta = \frac{TL}{NJ} = \frac{TL}{N\left(\dfrac{\pi}{2} r^4\right)} = \frac{2TL}{N\pi\, r^4} \quad \ldots(2)$$

$$\therefore \quad \text{Percentage error} = \frac{2.065 - 2}{2.065} \times 100 = \mathbf{3.15\%.}$$

Example 20.22. *A composite shaft consists of a steel rod of 80 mm diameter surrounded by a closely fitting tube of brass. Find the outside diameter of the brass tube when a torque of 2000 N-m is applied on the composite shaft and shared equally by the two materials. Take N for steel as 0.84 × 10⁵ N/mm² and N for brass as 0.42 × 10⁵ N/mm². Also, determine the common angles of twist in a length of 5 metres.*

Solution. Given : Outer diameter of steel rod = D_s = 80 mm

Let us use suffix s for steel rod and suffix b for brass tube

D_b = outer diameter of brass tube (to be determined)

d_b = inner diameter of brass tube = D_s = 80 mm

Torque T = 2000 N-m; $T_s = T_b$ = 2000/2 = 1000 N-m

(a) Outside diameter of brass tube

In general, $\dfrac{T}{J} = \dfrac{N\theta}{L}$ or $\theta = \dfrac{TL}{NJ}$

$$\therefore \quad \frac{T_b L_b}{N_b J_b} = \frac{T_s L_s}{N_s J_s} \quad (\text{But } T_b = T_s \text{ and } L_b = L_s)$$

$\therefore$ $$N_bJ_b = N_sJ_s \quad \text{or} \quad J_b = \frac{N_s}{N_b} \cdot J_s$$

or $$\frac{\pi}{32}(D_b^4 - 80^4) = \frac{0.84 \times 10^5}{0.42 \times 10^5} \times \frac{\pi}{32}(80)^4$$

or $$D_b^4 - 40.96 \times 10^6 = 2 \times 40.96 \times 10^6$$

or $$D_b^4 = 3 \times 40.96 \times 10^6, \text{ from which } D_b = \mathbf{105.3\ mm}$$

(b) *Common angle of twist*

$$\theta = \theta_s = \frac{T_sL_s}{N_sJ_s} = \frac{(1000 \times 10^3)(5000)}{0.84 \times 10^5 \times \frac{\pi}{32}(80)^4} = \mathbf{0.0148\ radian}$$

Check : $$\theta_b = \frac{T_bL_b}{N_bJ_b} = \frac{(1000 \times 10^3)(5000)}{0.42 \times 10^5 \times \frac{\pi}{32}\left[(105.3)^4 - (80)^4\right]} = \mathbf{0.0148\ radian}$$

Hence $\theta = \theta_b = \theta_s = \mathbf{0.0148\ radian = 0.847°}$.

Example 20.23. *A shaft ACB of 1 m length and 100 mm external diameter is bored, for a part of its length AC to 50 mm diameter and for the remaining length CB to 75 mm diameter bore. If the shear stress is not to exceed 76 N/mm², find the maximum power the shaft can transmit at a speed of 240 r.p.m.*

If the angle of twist in the length of 50 mm diameter bore is equal to that in the 75 mm diameter bore, find the length of the shaft that has been bored to 50 mm and 75 mm diameters.

Solution. Let us use suffix 1 for length *AC* and 2 for *CB*.

Fig. 20.15

(a) *Maximum power*

$$T = \frac{\pi}{16} f_s \left(\frac{D^4 - d^4}{D}\right),$$

Since *D* is the same for both the portions, safe torque will be governed by the greater value of *d*.

$\therefore$ $$T_{safe} = T_2 = \frac{\pi}{16} \times 76\left[\frac{100^4 - 75^4}{100}\right] = 10.2 \times 10^6 \text{ N-mm} = 10.2 \text{ kN-m}$$

Now $$P_{safe} = \frac{2\pi n T_{safe}}{60} = \frac{2\pi(240)(10.2)}{60} = \mathbf{256.4\ kW}$$

(b) *Individual lengths*

Since θ is the same for both the portions, and since *T* and *N* are also the same, we have

$$\frac{L_1}{J_1} = \frac{L_2}{J_2} \quad \therefore \quad \frac{L_1}{L_2} = \frac{J_1}{J_2} = \frac{\frac{\pi}{32}\left[100^4 - 50^4\right]}{\frac{\pi}{32}\left[100^4 - 75^4\right]} = 1.371$$

or $$L_1 = 1.371 L_2$$

Also, $$L_1 + L_2 = L = 1000$$

$\therefore$ $$1.371 L_2 + L_2 = 1000, \text{ from which } L_2 = \mathbf{421.8\ mm}$$

$\therefore$ $$L_1 = 1.371 \times 421.8 = \mathbf{578.2\ mm.}$$

Example 20.24. *A steel shaft ABCD having a total length of 3 m consists of the following different sections : AB is hollow having outside and inside diameters of 100 mm and 60 mm respectively, and BC and CD are solid, BC having a diameter of 100 mm and CD a diameter of 85 mm. If the angle of twist is the same for each section, determine the length of each section. Find the value of the applied torque and the total angle of twist, if maximum shear stress in the hollow portion is 50 N/mm² and N = 0.82 × 10⁵ N/mm².*

Solution. Given $L = 3$ m

(a) Length of individual portions

Total length $L = L_{ab} + L_{bc} + L_{cd} = 3000$ mm ...(1)

Since the angle of twist for each portion is the same, we have

$$\frac{TL_{ab}}{NJ_{ab}} = \frac{TL_{bc}}{NJ_{bc}} = \frac{TL_{cd}}{NJ_{cd}} \quad \text{or} \quad \frac{L_{ab}}{J_{ab}} = \frac{L_{bc}}{J_{bc}} = \frac{L_{cd}}{J_{cd}}$$

or

$$\frac{L_{ab}}{\frac{\pi}{32}(100^4 - 60^4)} = \frac{L_{bc}}{\frac{\pi}{32}(100)^4} = \frac{L_{cd}}{\frac{\pi}{32}(85)^4}$$

or

$$\frac{L_{ab}}{87040000} = \frac{L_{bc}}{100000000} = \frac{L_{cd}}{52200625}$$

$\therefore$
$$L_{bc} = \frac{100000000}{87040000} L_{ab} = 1.1489\, L_{ab} \quad ...(2)$$

and
$$L_{cd} = \frac{52200625}{87040000} L_{ab} = 0.5997\, L_{ab} \quad ...(3)$$

Substituting the values of L_{bc} and L_{cd} in (1), we get

$L_{ab} + 1.1489\, L_{ab} + 0.5997\, L_{ab} = 3000$ From which L_{ab} = **1091.5 mm**

Hence $L_{bc} = 1.1489 \times 1091.5$ = **1254 mm** and $L_{cd} = 0.5997 \times 1091.5$ = **654.5 mm.**

(b) Applied torque on the shaft

$$T_{ab} = \frac{\pi}{16}\frac{(100^4 - 60^4)}{100} \times 50 = 8.545 \times 10^6 \text{ N-mm} = 8.545 \text{ kN-m}$$

$$T_{bc} = \frac{\pi}{16}(100)^3 \times 50 = 9.817 \times 10^6 \text{ N-mm} = 9.817 \text{ kN-m}$$

$$T_{cd} = \frac{\pi}{16}(85)^3 \times 50 = 6.029 \times 10^6 \text{ N-mm} = 6.029 \text{ kN-m}$$

Hence applied torque will be the minimum of the three, *i.e.*, **6.029 kN-m.**

(c) Total angle of twist

$$\theta_{cd} = \frac{TL_{cd}}{J_{cd} \cdot N} = \frac{6.029 \times 10^6\ (654.4)}{\frac{\pi}{32}(85)^4 \times 0.82 \times 10^5} = 9.39 \times 10^{-3} \text{ radian}$$

$\therefore$ Total $\theta = \theta_{ab} + \theta_{bc} + \theta_{cd} = 3\,\theta_{cd} = 3 \times 9.39 \times 10^{-3}$ = **0.0282 radian.**

Example 20.25. *A shaft ABC [Fig. 20.16 (a)] of 60 mm diameter is driven at A by a motor that transmits 100 kW to the shaft at 20 Hz. The gears at B and C remove 60 kW and 40 kW respectively. Compute the maximum shear stress f_s in the shaft and the angle of twist (θ) between the ends A and C. Take N = 0.8 × 10⁵ N/mm².*

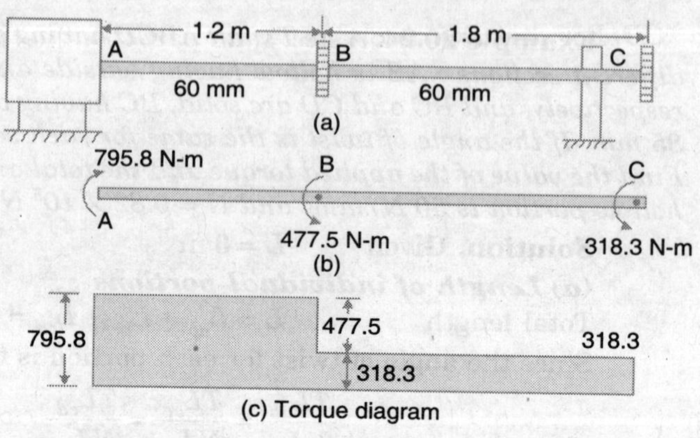

(c) Torque diagram

Fig. 20.16

Solution. Given $f = 20$ Hz

Now 1 Hz = 1 r.p.s. = 60 r.p.m.

$\therefore \quad n = 20 \times 60 = 1200$ r.p.m.

$$T = \frac{60\,P}{2\,\pi\,n} \text{ kN-m}$$

$$= \frac{60 \times 100}{2\,\pi\,(1200)}$$

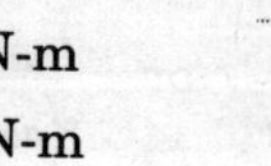

= 0.7958 kN-m = 795.8 N-m

$\therefore \quad T_B = 795.8 \times 0.6 = 477.5$ N-m

and $\quad T_C = 795.8 \times 0.4 = 318.3$ N-m

(a) Shear stress

$$\therefore \quad f_{sab} = \frac{16\,T_{ab}}{\pi\,d^3} = \frac{16\,(795.8 \times 10^3)}{\pi\,(60)^3} = 18.76 \text{ N/mm}^2$$

For *CB*,
$$f_{scb} = \frac{16\,T_{cb}}{\pi\,d^3} = \frac{16 \times 318.3 \times 10^3}{\pi\,(60)^3} = 7.51 \text{ N/mm}^2$$

$\therefore \quad f_{s,\,\max}$ = **18.76 N/mm²**, which occurs in part *AB*.

(b) Angles of twist

or
$$\theta_{ab} = \frac{T_{ab}\,L_{ab}}{NJ} = \frac{(795.8 \times 10)^3\,(1200)}{(0.8 \times 10^5)\,\dfrac{\pi}{32}\,(60)^4} = 0.00938 \text{ radian}$$

and
$$\theta_{bc} = \frac{T_{bc}\,L_{bc}}{NJ} = \frac{(318.3 \times 10^3)\,(1800)}{(0.8 \times 10^5)\,\dfrac{\pi}{32}\,(60)^4} = 0.00563 \text{ radian}$$

$\therefore$ Total angle of twist $\theta = \theta_{ab} + \theta_{bc} = 0.00938 + 0.00563 =$ **0.01501 radian = 0.86°**.

Example 20.26. *A solid circular bar with fixed ends is acted upon by two oppositely directed torques T as shown in Fig. 20.17(a). Find (a) the reactive torques at ends A and D, (b) the angle of twist at B, and (c) the angle of twist at the mid section M of the bar.*

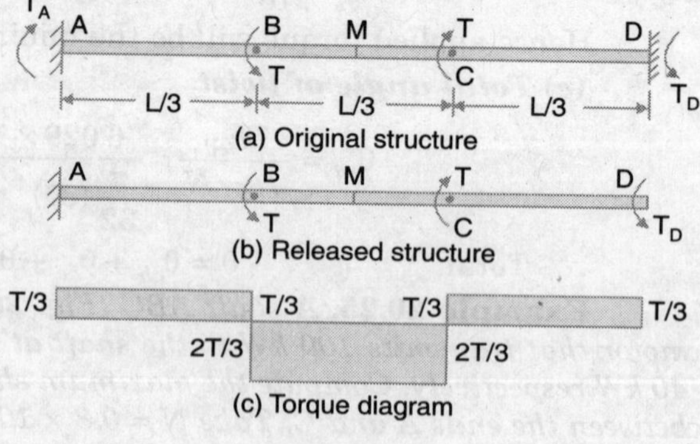

(c) Torque diagram

Fig. 20.17

Solution. *(a) Reactive torques at ends A and D.* Let T_A and T_D be the reactive torques, acting in the directions as marked in Fig. 20.17(*a*). From statics we have

$$T_A - T + T - T_D = 0$$

or
$$T_A = T_D \qquad \ldots(1)$$

In order to obtain another equation between T_A and T_D, select T_D as the redundant torque. Fig. 20.17(*b*) shows the released structure. The three torques *T* (at *B*), *T* (at *C*) and T_D (at *D*) act on this released structure. They produce an angle of twist θ_D at end *D*, which is equal to the algebraic sum of the angles of twist of the three parts *AB*, *BC* and

CD. For part DC, torque is T_D; for part CB, the torque is $(T_D - T)$, while for part A, the torque is equal to $T_D - T + T = T_D$.

$$\therefore \quad \theta_D = \theta_{AB} + \theta_{BC} + \theta_{CD} = \frac{T_{AB} \, . \, L_{AB}}{NJ} + \frac{T_{BC} \, . \, L_{BC}}{NJ} + \frac{T_{CD} \, L_{CD}}{NJ}$$

Since the angle of rotation at end D, in the original bar is equal to zero, we have $\theta_D = 0$.

$$\therefore \quad T_D\left(\frac{L}{3}\right) + (T_D - T)\left(\frac{L}{3}\right) + T_D\left(\frac{L}{3}\right) = 0 \quad \text{or} \quad T_D + (T_D - T) + T_D = 0$$

or $\quad 3 \times T_D = T$ from which $T_D = \dfrac{T}{3}$ **(Ans)**

Also, $\quad T_A = T_D = \dfrac{T}{3}$ **(Ans)**

Fig. 20.17 (c) shows the torque diagram.

(b) Angle of twist at B

$$\theta_B = \frac{T_{AB} \, . \, L_{AB}}{NJ} = \frac{\left(\frac{T}{3}\right)\left(\frac{L}{3}\right)}{NJ} = \frac{TL}{9\,NJ} \quad \textbf{(Ans)}$$

(c) Angle of twist at mid-point M

$$\theta_M = \frac{T_{AB} \, . \, L_{AB}}{NJ} + \frac{T_{BM} \, . \, L_{BM}}{NJ} = \frac{\left(\frac{T}{3}\right)\left(\frac{L}{3}\right)}{NJ} + \frac{\left(\frac{-2T}{3}\right)\left(\frac{L}{6}\right)}{NJ} = \frac{TL}{9\,NJ} - \frac{TL}{9\,NJ} = 0.$$

Example 20.27. *A solid circular shaft AB of total L is fixed against rotation at both ends, as shown in Fig. 20.18. The shaft has diameters d_a and d_b in parts AC and CB respectively. A torque T acts at section C. What should be the length a and b for the most economical design of the shaft ?*

Solution. Let T_A and T_B be two reactive torques at A and B, in the directions marked. Choosing T_B as the redundant structure, the released structure is shown in Fig. 20.18 (b).

From statics, $\quad T_A + T_B - T = 0 \quad$ or $\quad T_A + T_B = T \qquad$ ...(1)

For the most economical design, T_A must be equal to T_B. Hence from (1) $T_B = \dfrac{T}{2}$.

Portion CB is subjected to a torque T_B $(= T/2)$ while portion CA is subjected to a torque $T_B - T = \dfrac{T}{2} - T = -\dfrac{T}{2}$.

(b) Released structure

Fig. 20.18

$$\therefore \quad \theta_B = 0 = \frac{T_{AC} \, . \, L_{AC}}{N\,J_{AC}} + \frac{T_{CB} \, . \, L_{CB}}{N\,J_{CB}}$$

$$\therefore \quad \frac{(T_B - T) \, . \, a}{N \, \frac{\pi}{32} \, (d_a)^4} + \frac{T_B \, (L - a)}{N \, \frac{\pi}{32} \, (d_b)^4} = 0$$

or $$\frac{\left(-\frac{T}{2}\right) a}{(d_a)^4} + \frac{(T/2)\,(L - a)}{(d_b)^4} = 0$$

or $\quad a\,(d_b)^4 - L\,(d_a)^4 + a\,(d_a)^4 = 0 \quad$ or $\quad a\,(d_b{}^4 + d_a{}^4) = L(d_a{}^4)$

From which $\quad a = \dfrac{d_a{}^4 \, L}{d_a{}^4 + d_b{}^4}$ **(Ans).**

PROBLEMS

1. A solid circular shaft of steel is 6 cm in diameter and transmits 60 kW power at 140 R.P.M. Taking $N = 0.82 \times 10^5$ N/mm^2, calculate the maximum intensity of stress and the angle of twist in degrees for a length of 6 metres.

2. Find the power transmitted by a shaft of 10 cm diameter at 200 R.P.M. if the permissible shear stress is 65 N/mm^2.

3. A solid steel shaft has to transmit 75 kW power at 200 R.P.M. Taking the allowable shear stress as 70 N/mm^2, find a suitable diameter for the shaft if the maximum torque transmitted in each revolution exceeds the mean by 30%.

4. In the previous problem, find the suitable diameter of a hollow shaft the inside diameter being 0.7 of the outside diameter and hence calculate the percentage saving in weight of the shaft.

5. A shaft has to transmit 375 kW power at 80 R.P.M. The shaft must not be stressed beyond 75 N/mm^2 and must not twist more than 1° in a length of 4 m. Select a suitable diameter. Take $N = 0.82 \times 10^5$ N/mm^2.

6. A steel shaft of 5 cm diameter and 3.0 metre long is rigidly fixed at the ends. A twisting moment of 2 kN-m is applied to it at a distance of 1 m from one end. Calculate (*a*) the fixing couples at the ends, (*b*) the maximum shear stress, (*c*) the angle of twist of the section, where the twisting moment has been applied. Take $N = 0.82 \times 10^5$ N/mm^2.

7. Calculate the angle of twist for a shaft having diameter of 6 cm at one end and 7 cm at the other end in a length of 2 m. Also, find the % error committed in calculating θ, if it is calculated on the basis of an average diameter of 6.5 cm.

8. A shaft of 8 cm diameter transmits 150 kW power at 180 r.p.m. A flanged coupling is keyed to the shaft by means of a key 10 cm long and 3 cm wide. The coupling has 6 bolts of 2 cm diameter symmetrically arranged along a bolt circle of 20 cm diameter. Calculate the shear stress in the shaft, the key and the bolts of the coupling.

9. A shaft of 10 cm diameter is subjected to a bending moment of 8 kN-m and a twisting moment of 12 kN-m at a certain cross-section. Calculate the maximum direct stress induced in the section and specify the position of the plane on which it acts. Taking $1/m = 0.3$, find the stress which acting alone, will produce the same maximum (*i*) strain, (*ii*) strain energy.

10. A circular shaft supported in bearings 4 metre apart transmits 75 kW power at 120 r.p.m. A pulley provided at 1.5 metre from one bearing exerts a transverse load of 40 kN on the shaft. Determine a suitable diameter of the shaft if (*i*) the maximum direct stress is not to exceed 90 N/mm^2, (*ii*) the maximum intensity of shear stress is not to exceed 45 N/mm^2, (*iii*) the stress which, acting alone, would produce the same maximum strain is not to exceed 90 N/mm^2, (*iv*) the direct stress which acting alone would produce the same maximum strain energy, is not to exceed 90 N/mm^2. Take $1/m = 0.25$.

11. A propeller shaft 30 cm external diameter and 15 cm internal diameter transmit 1800 kW power at 100 r.p.m. There is at the same time a bending moment of 12 kN-m and an end thrust of 300 kN. Find (*i*) the principal stresses and their planes, (*ii*) the maximum shear stress, (*iii*) the stress which acting alone will produce the same maximum strain. Take $1/m = 0.3$.

12. A compound shaft 1.5 m long, fixed at one end is subjected to a torque of 15 kN-m at the extreme end and of 20 kN-m at the junction point as shown in Fig. 20.19. The shaft has a diameter of 12 cm for a length of 75 cm from the fixed end and a diameter of 9 cm for the remaining length. Determine (*a*) the maximum shearing stress in each of the portion of the shaft, (*b*) the angles of twist at the junction of the two sections and at the free end. N for steel = 0.82×10^5 N/mm^2.

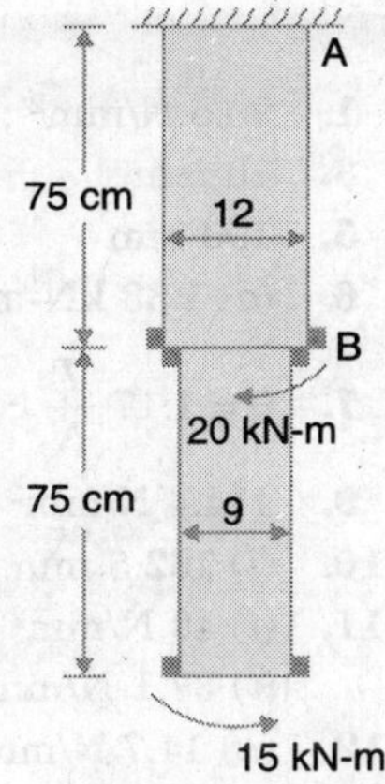

Fig. 20.19

13. A compound shaft is formed by surrounding a 5 cm diameter solid brass shaft by a steel tube having wall thickness of 6 mm. The metals are securely attached at their juncture. Determine the increase in the torque carrying capacity of the composite shaft over that of the brass shaft alone. For brass $N = 0.34 \times 10^5$ N/mm² ; N for steel = 0.82×10^5 N/mm². The allowable shear stress for brass = 60 N/mm² and for steel = 80 N/mm².

14. A horizontal circular shaft of diameter d and diametral moment of inertia I is subjected to bending moment $M \cos \theta$ in a vertical plane and to an axial twisting moment $M \sin \theta$. Show that the principal stresses at the ends of a vertical diameter are $\frac{1}{2} Mk (\cos \theta \pm 1)$, where $k = d/2I$. If the strain energy is the criterion of failure, show that

$$S = \frac{S_0\sqrt{2}}{\sqrt{\cos^2 \theta(1-\sigma)+(1+\sigma)}}, \text{ where } S = \text{max, shearing stress}$$

S_0 = max. shearing stress in the special case when $\theta = 0$ and σ = Poisson's ratio. *(U.L.)*

15. A solid shaft of diameter d is subjected to an axial thrust and an axial torque T. Show that the principal stresses at a point on surface of the shaft is given by $\frac{2P}{\pi d^2}\left(1 \pm \sqrt{1+\frac{64T^2}{P^2d^2}}\right)$.

16. Two circular shafts of the same material and length, one hollow and the other solid, are to have the same torsional strength. Show that if the hollow shaft has an external diameter equal to twice the internal diameter, the torsional stiffness of the hollow shaft will be about 20% greater than that of the solid shaft. By how much per cent will the weight of hollow shaft be less than that of the solid shaft ?

17. The centre line of a thin walled tube of circular section form three sides of a rectangle in which $AB = \frac{1}{2} BC = CD = a$. The tube is fixed in a horizontal plane, its ends A and D being encastered into a vertical wall, and it supports a weight W at the middle of BC. Assuming that the modulus of rigidity for the material of the tube is 2/5 of its Young's Modulus, show that the torque excerted on portions AB and DC is $\frac{Wa}{9}$ and that the deflection of the weight is $\frac{5Wa^3}{18\,EI}$, where I is the moment of inertia of the cross-section of the tube about its diameter.

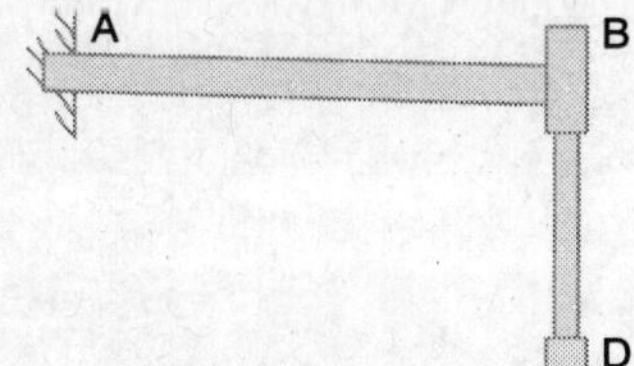

Fig. 20.20

18. A horizontal rod AB, of circular cross-section, is clamped rigidly at A and has a bracket BD fixed to its free end as shown in Fig. 20.20. The length $AB = 2BD$. If a weight is hung from D prove that the ratio of angle of bend to the angle of twist at the end B is equal to $\frac{1}{\sigma+1}$. Assume that the rod is not over-strained.

19. A solid cylindrical gun-metal shaft is reinforced by a cylindrical steel casing. The diameter of the gun-metal core is d_1 and the external diameter of the tight-fitting steel casing is d_2. Prove that the maximum shearing stresses in gun-metal and steel are respectively

$$\frac{16N_1d_1T}{\pi[N_2(d_2^4-d_1^4)+N_1d_1^4]} \text{ and } \frac{16N_2d_2T}{\pi[N_2(d_2^4-d_1^4)+N_1d_1^4]},$$ where N_1 and N_2 are the Moduli of rigidity for the gun-metal and steel respectively and T is axial torque applied to the composite shaft.

ANSWERS

1. 96.5 N/mm^2 ; 13.48° **2.** 267 kW

3. 70 mm **4.** D = 76.4 mm ; d = 53.4 mm ; 38%

5. 190 mm

6. (*a*) 1.33 kN-m ; 0.667 kN-m ; (*b*) 54.5 N/mm ; (*c*) 0.0265 radian

7. $\theta = 1.17 \dfrac{T}{N}$; 2.39% **8.** 79 N/mm^2 ; 66.3 N/mm^2 ; 42.3 N/mm^2

9. 114.4 N/mm^2 to 28° 9′ or 61° 51′ with the axis (*i*) 124.2 N/mm^2 ; 128.2 N/mm^2

10. (*i*) 162.5 mm, (*ii*) 163 mm, (*iii*) 162.1 mm, (*iv*) 163 mm

11. (*i*) 40 N/mm^2 (comp.) at 40° 41′ and 34.2 N/mm^2 (tensile at 130° 41′
(*ii*) 37.1 N/mm^2 (*iii*) 29.7 N/mm^2

12. (*a*) 14.7 N/mm^2 ; 105 N/mm^2 **13.** 90.6%

16. 21.7%.

21

Springs

21.1. INTRODUCTION

Spring may be regarded as a device for storing up energy in the form of resilience. They are used either as a storage of energy or else are used to absorb excess energy, depending upon the function of a particular spring under consideration. A spring in the clock-work, for example, is used to store work in the form of strain energy which is regained when the spring takes its original shape. A carriage spring, on the other hand, is used to absorb shocks, etc. and the effect of the blow on the carriage is greatly reduced. The best form of spring is that which absorbs greatest amount of energy for a given stress.

Strictly speaking, all forms of elastic bodies such as struts, ties, beams, etc. are springs as they have resilience, but they are too stiff to be used. We may divide springs chiefly into two categories :

(*i*) Torsion spring (*ii*) Bending spring.

A *torsion spring* is the one which is subjected to a twisting moment and the resilience is mainly due to torsion. Closely-coiled helical springs subjected to axial pull fall under this category. Such a spring is made of rod or wire in the form of a helix described on a right circular cylinder.

A *bending spring* is the one which is subjected to bending only and the resilience is mainly due to bending. The leaf or plate springs fall under this category.

An open coiled helical spring falls under both the categories.

Stiffness. The stiffness of a spring is defined as load required to produce unit deflection.

21.2. CLOSE-COILED HELICAL SPRING : AXIAL PULL

Close coiled helical spring is a *torsion spring*. It is assumed that this type of helical spring is so closely coiled that each turn is practically a plane at right angle to the axis of the helix and the stresses upon the material are almost of pure torsion. The bending couple is negligible in comparison with the torsional couple and the length (l) of the spring may be taken equal to $2\pi Rn$, where n = Number of turns and R is the mean radius of the coil.

We shall first consider the effect of axial pull on such a spring. Consider a spring having R as mean radius of coil and d as the diameter of the wire, as shown in Fig. 21.1. The spring will be equivalent to a shaft of diameter d and of length l such that $l = 2\pi nR$, where n is the total number of coils. The equivalent shaft has been shown in Fig. 21.2.

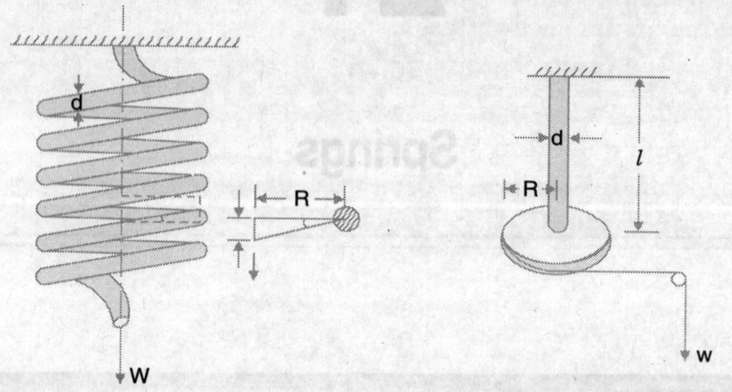

Fig. 21.1 **Fig. 21.2**

From Eq. 20.2, we have

$$\frac{T}{J} = \frac{N\theta}{l}, \text{ where } \theta = \text{angle of twist at the end}$$

or $$\theta = \frac{Tl}{NJ} = \frac{WRl}{N\frac{\pi}{32}d^4} = \frac{32WRl}{\pi d^4 N} \text{ radians} \qquad \text{...[21.1 (a)]}$$

or $$\theta = \frac{32WR(2\pi nR)}{N\pi d^4} = \frac{64WR^2n}{Nd^4} \text{ radians} \qquad \text{...[21.1 (b)]}$$

Thus, the free end will twist through an angle θ given by Eq. 21.1. Consquently, the free end will have an axial movement $R\theta$. If δ is the axial movement or deflection of the free end, we have

$$\delta = R\theta = \frac{32WR^2l}{\pi Nd^4} \text{ or } = \frac{64WR^3n}{Nd^4} \qquad \text{...(21.2)}$$

The deflection δ can also be obtained by equating the torsional strain energy to the work done by axial force and deflection, *viz.*,

$$\frac{1}{2}W\delta = \frac{1}{2}T\theta = \frac{1}{2}(WR)\frac{32WRl}{N\pi d^4}$$

or $$\delta = \frac{32WR^2l}{N\pi d^4} = \frac{64WR^3n}{Nd^4} \qquad \text{...[21.2(a)]}$$

The stiffness s of the spring will be given by

$$s = \frac{W}{\delta} = \frac{Nd^4}{64nR^3} \qquad \text{...(21.3)}$$

Any section distant R from the axis is subjected to (*a*) torsion $T = W \cdot R$ and (*b*) shear force W due to axial load. Hence shear stress is induced at the section, due to both T and W. Maximum intensity of shear stress due to T, at any section is given by

$$f_{s1} = \frac{T}{J} R = \frac{16T}{\pi d^3} = \frac{16WR}{\pi d^3} \quad ...[21.4\ (a)]$$

Also, shear stress due to W is given by

$$f_{s2} = \frac{4W}{\pi d^2} \quad ...[21.4\ (b)]$$

Hence total maximum shear stress is

$$(f_s)_{max} = \frac{16WR}{\pi d^3} + \frac{4W}{\pi d^2} = \frac{16WR}{\pi d^3}\left(1 + \frac{d}{4R}\right) \quad ...(21.4)$$

Springs of Square Section Wire :

If the spring is made of square section wire ($b \times b$), we have

$$\theta = \frac{410Tl}{Nb^4} \text{ degrees} = \frac{7.11Tl}{Nb^4} \text{ radians} = \frac{7.11\,WRl}{Nb^4} \text{ radians} \quad ...[21.5\ (a)]$$

$$\delta = R\theta = \frac{7.11WR^2 l}{Nb^4}, \text{ or } = \frac{44.7WR^3 n}{Nb^4} \quad ...[21.5\ (b)]$$

Stiffness, $$s = \frac{W}{\delta} = \frac{Nb^4}{44.7nR^3} \quad ...[21.5\ (c)]$$

The maximum deflection will be there when the spring 'shuts', *i.e.*, when the coils are touching each other. Hence $\delta = (L - nd)$, where L = axial length of the spring of circular section wire.

Example 21.1. ***A closely coiled helical spring made out of a 10 mm diameter steel rod has 10 complete coils, each of mean diameter of 80 mm. Calculate stress induced in the section of the rod, the deflection under the pull and the amount of energy stored in the spring during the extension, if it is subjected to an axial pull of 200 N. Take N (or G) = 0.84×10^5 N/mm^2.***

Solution. $T = WR = 200 \times 40 = 8000$ **N-mm**

But $$T = f_s \frac{\pi}{16} d^3 \text{ or } f_{s1} = \frac{16T}{\pi d^3} = \frac{16 \times 8000}{\pi (10)^3} = 40.74 \text{ N/mm}^2$$

Also, $$f_{s2} = \frac{4W}{\pi d^2} = \frac{4 \times 200}{\pi (10)^2} = 2.55 \text{ N/mm}^2$$

$\therefore$ $(f_s)_{max} = 40.74 + 2.55 = \mathbf{43.29\ N/mm^2}$

Deflection, $$\delta = R \,.\, \theta = \frac{64WR^3 n}{Nd^4} = \frac{64 \times 200 \times (40)^3 \times 10}{0.84 \times 10^5 (10)^4} = 9.75 \text{ mm.}$$

$\therefore$ **Energy stored** $= \frac{1}{2} W\delta = \frac{1}{2} \times 200 \times 9.75 = 975$ **mm-N = 0.975 m-N = 0.975 J.**

Example 21.2. ***A wagon weighing 25 kN is moving at 3 kilometres per hour. How many springs each of 24 coils will be required in a bufferstop to absorb the energy of motion during a compression of 200 mm ? The mean diameter of coils is 240 mm and the diameter of steel rod comprising the coil is 20 mm. Take N = 0.9×10^5 N/mm^2.***

Solution. $$v = 3 \text{ km/hour} = \frac{3 \times 1000 \times 100}{60 \times 60} = 83.4 \text{ cm/sec.}$$

$\therefore$ K.E. absorbed by buffer springs $= \dfrac{Wv^2}{2g} = \dfrac{25 \times 1000\,(83.4)^2}{2 \times 981} = 88.63 \times 10^3$ cm-N

$= 88.63 \times 10^4$ mm-N.

If W is the axial load, for each spring for a compression of δ, we have

$$\delta = \frac{64WR^3n}{Nd^4} \quad \text{or} \quad W = \frac{Nd^4\delta}{64R^3n} = \frac{0.9 \times 10^5 (20)^4 (200)}{64\,(120)^3 \times 24} = 1085 \text{ N}$$

Energy stored by one spring $= \frac{1}{2} W\delta = \frac{1}{2} \times 1085 \times 200 = 10.85 \times 10^4$ mm-N

$\therefore$ No. of springs required $= \dfrac{88.63 \times 10^4}{10.85 \times 10^4} = 8.17 \simeq 9$ (say).

Example 21.3. *A safety valve is 80 mm dia. to blow off at pressure of 1 N/mm² gauge. It is held by a close coiled compression spring of circular steel bar. The mean diameter is 160 mm and the initial compression is 20 mm. Find the diameter of steel bar and the number of convolutions necessary if the shear stress allowed is 75 N/mm². Take N = 0.86 × 10⁵ N/mm².*

Solution. W = total force = area of valve × pressure $= \dfrac{\pi}{4}(80)^2 \times 1 = 5026.5$ N

Now
$$f_s = 75 = \frac{16WR}{\pi d^3}\left(1 + \frac{d}{4R}\right) = \frac{16 \times 5026.5 \times 80}{\pi d^3}\left(1 + \frac{d}{4R}\right)$$

or
$$\frac{d^3}{\left(1 + \dfrac{d}{4R}\right)} = 27306$$

The above equation can be solved by trial. For this, neglecting $d/4R$ in the first trial, we get $d \simeq 30$ mm. This gives $d/4R = 30/4 \times 80 \simeq 0.1$.

$$\therefore \quad \frac{d^3}{1 + 0.1} = 27306 \text{ which gives } d = \mathbf{31\ mm.}$$

Let the number of convolutions be n.

$$\delta = \frac{64WR^3n}{Nd^4} \quad \text{or} \quad n = \frac{\delta\, Nd^4}{64WR^3} = \frac{20 \times 0.86 \times 10^5 (31)^4}{64 \times 5026.5 (80)^3} = 9.6 \simeq \mathbf{10.}$$

Example 21.4. *A weight of 200 N is dropped on to a helical compression spring made of 15 mm steel wire colsely coilded to a mean diameter of 150 mm with 24 coils. If the instantaneous compression is 100 mm, calculate the height of drop. Take N = 0.90 × 10⁵ N/mm².*

Solution. Let h = height of drop, in mm

W = equivalent gradually applied load in N to produce the same compression.

$$\delta = \frac{64WR^3n}{Nd^4} \quad \text{or} \quad W = \frac{Nd^4\delta}{64R^3n} = \frac{0.90 \times 10^5 (15)^4 (100)}{64\,(75)^3 (24)} = 707 \text{ N.}$$

Equating the energy supplied by the impact load to the energy stored, we have

$$P(h + \delta) = \frac{1}{2} W\delta$$

or $\quad 200(h + 100) = \frac{1}{2} \times 707 \times 100 = 35350$, from which $h =$ **76.8 mm.**

Example 21.5. *A closely coiled spring, made out of 10 mm square rod has 12 coils of 120 mm mean diameter. Find the maximum permissible axial load for this spring if the maximum shear stress is limited to 260 N/mm². Also, calculate the deflection under the load and the stiffness of the spring. Take $N = 0.84 \times 10^5$ N/mm².*

Solution. For a square section, we have

$$T = 0.208\, b^3 \,.\, f_s = 0.208\,(10)^3\,(260) = 54080 \text{ N-mm}$$

But $\quad T = WR \quad$ or $\quad W = \frac{T}{R} = \frac{54080}{60} = \mathbf{901.3\ N}$

$$\theta = \frac{7.11 \times 2\pi\, WR^2 n}{Nb^4} = \frac{7.11 \times 2 \times \pi \times 901.3\,(60)^2 \times 12}{0.84 \times 10^5\,(10)^4} = 2.07 \text{ radians}$$

$\therefore \quad \delta = R\theta = 60 \times 2.07 = \mathbf{124.2\ mm}$

Stiffness $\quad s = \frac{W}{\delta} = \frac{901.3}{124.2} = \mathbf{7.26\ N/mm.}$

Example 21.6. *Compare the resilience of a closely coiled helical spring of square section wire with that of a circular section if the volume of both the springs is the same. Prove that for equal weight a closely coiled circular section helical spring of given diameter is 1.36 times as strong and will absorb 1.62 times as much energy as square section.*

Solution. Let b = length of each side of the square ; d = diameter of circular section.

Work absorbed by the square spring

$$= \frac{1}{2}\, T \,.\, \theta = \frac{1}{2}\,(0.208\, f_s b^3)\, \frac{7.11\, Tl}{Nb^4}$$

$$= \frac{0.208\, f_s b^3}{2} \times \frac{7.11\, l}{Nb^4}\,(0.208\, f_s b^3) = \frac{0.154 f_s^{\,2} b^2 l}{N} = \frac{0.154 f_s^{\,2}}{N}\, V$$

Hence resilience $= \frac{0.154 f_s^{\,2}}{N}$ ...(1)

Similarly, for a solid circular section,

$$\text{Work absorbed} = \frac{1}{2}\, T \,.\, \theta = \frac{1}{2}\, T .\, \frac{Tl}{NJ} = \frac{1}{2}\left(f_s\, \frac{\pi d^3}{16}\right)^2 \frac{l}{N \frac{\pi}{32} d^4}$$

$$= \frac{f_s^{\,2}}{N} \times \frac{\pi d^2 l}{16} = \frac{1}{4}\, \frac{f_s^{\,2}}{N} \times \text{volume}$$

$\therefore \quad$ Resilience $= 0.25\, \frac{f_s^{\,2}}{N}$ ...(2)

$$\therefore \quad \frac{\text{Resilience of circular section}}{\text{Resilience of square section}} = \frac{0.25}{0.154} = \mathbf{1.62.}$$

That is, for equal weight, the circular section will absorb 1.62 times as much energy as of square section.

Again, for circular section, $T_1 = f_s\, \frac{\pi d^3}{16} = \frac{f_s d}{4} \times$ area

(Neglecting stress due to direct force)

For square section, $T_2 = 0.208\, f_s b^3 = 0.208\, f_s b \times \text{area}$

$$\therefore \quad \frac{T_1}{T_2} = \frac{d}{4 \times 0.208b} \qquad \text{...(3)}$$

But for equal weight, $\frac{\pi}{4}\, d^2 l = b^2 l$ or $d = b\sqrt{\frac{\pi}{4}} = 1.128b$

Hence $\frac{T_1}{T_2} = \frac{1.128\, b}{4 \times 0.208\, b} = \mathbf{1.36.}$

Thus, for equal weight, the circular section is 1.36 times as strong as a square section.

21.3. CLOSE COILED HELICAL SPRING : AXIAL COUPLE

If a close coiled helical spring fixed at one end is subjected at the other end to a twisting couple M about the central axis of the spring, tending to unwind or wind up the spring, the whole spring will be subjected to a bending moment equal to the torque M applied.

Assuming, as an approximation, that the coils having considerable initial curvature behave like a beam of initial zero curvature, we have

$$\textit{Bending moment } M = EI \times (\textit{change of curvature}) = EI\left(\frac{1}{R'} - \frac{1}{R}\right) \qquad \text{...(1)}$$

where R' is the new radius of curvature and R is the original radius of curvature, R' being less than R. That is, the *sense* of the B.M. M applied in the above equation is such that it increases the curvature. Let n be the original number of coils and n' be the resulting number of coils $(n' > n)$ after the application of M. Hence, we have

$$l = 2\pi R n = 2\pi R' n'$$

where $\frac{1}{R} = \frac{2\pi n}{l}$ and $\frac{1}{R'} = \frac{2\pi n'}{l}$

Hence from (1), $$M = EI\left\{\frac{2\pi n'}{l} - \frac{2\pi n}{l}\right\} = \frac{EI.2\pi}{l}\,(n' - n) \qquad \text{...(21.6)}$$

The total twist ϕ of the free end in radians is

$$\phi = 2\pi\,(n' - n) = \frac{Ml}{EI} \qquad \text{...(21.7)}$$

The above result could also be obtained from consideration of strain energy. Thus,

$$\tfrac{1}{2} M\phi = \frac{1}{2EI}\int_0^l M^2 dx = \frac{1}{2}\frac{M^2 l}{EI}$$

Hence $\phi = \frac{Ml}{EI}$, which is the same as Eq. 21.7.

The change in curvature or angle of bend per unit length is

$$\frac{d\phi}{dl} = \frac{1}{R'} - \frac{1}{R} = \frac{M}{EI} \qquad \text{...(21.8)}$$

For a solid circular section, we get from Eq. 21.7,

$$\phi = \frac{64Ml}{\pi E d^4} \text{ or } \frac{128MRn}{Ed^4} \text{ radians} \qquad \text{...[21.9 (a)]}$$

For a square section of side b, we get from Eq. 21.7,

$$\phi = \frac{12Ml}{Eb^4} = \frac{24\pi MRn}{Eb^4} \text{ radians} \qquad \text{...[21.9 (b)]}$$

For solid circular section, the maximum bending stress f_b is given by

$$f_b = \frac{M}{I}\,y = \frac{M}{Z} = \frac{32M}{\pi d^3} \qquad \text{...(21.10)}$$

The strain energy stored in the spring = $\frac{1}{2}M\phi$.

Also, in terms of allowable stress f, we have

$$U = \frac{f^2}{8E} \times \text{volume of spring, for circular section, and}$$

$$U = \frac{f^2}{8E} \times \text{volume of spring, for rectangular section.}$$

$$\text{Also, stiffness for torsion} = \frac{M}{\phi} = \frac{EI}{2\pi nR} \text{ (N/m per radian)} \qquad \text{...(21.11)}$$

Example 21.7. *A closely coiled helical spring is made of 10 mm diameter steel rod, the coils having 10 complete turns and a mean diameter of 80 mm. Calculate the increase in number of turns and the bending stress induced in the section, if it is subjected to an axial thrust of 10 N-m. Take E = 2.1 × 10⁵ N/mm².*

Also, calculate the torsional stiffness of the spring.

Solution. $I = \frac{\pi}{64}d^4 = \frac{\pi}{64}(10)^4 = 490.9 \text{ mm}^4 \; ; \; Z = \frac{I}{d/2} = \frac{490.9}{5} = 98.2 \text{ mm}^3$

$$M = 10 \text{ N-m} = 10000 \text{ N-mm} \; ; \; f_b = \frac{M}{Z} = \frac{10 \times 1000}{98.2} = \mathbf{101.9 \text{ N/mm}^2}$$

$$\phi = \frac{Ml}{EI} = \frac{M(2\pi nR)}{EI} = \frac{10000 \times 2\pi \times 10 \times 40}{2.1 \times 10^5 \times 490.9} = 0.2438 \text{ radians}$$

But $\phi = 2\pi(n' - n)$ or $n' - n = \frac{\phi}{2\pi} = \frac{0.2438}{2\pi} = 0.0388$

or $n' = 0.0388 + 10 = 10.0388$

Torsional stiffness of the spring

$$= \frac{M}{\phi} = \frac{EI}{2\pi nR} = \frac{2.1 \times 10^5 \times 490.9}{2\pi \times 10 \times 40} = 41 \times 10^3 \text{ N-mm/radian}$$

$$= \mathbf{41 \text{ N-m/radian.}}$$

21.4. OPEN COILED HELICAL SPRING : AXIAL FORCE

In the case of open coiled helical spring, the coils are not close together and the angle α cannot be treated as small. The bending couple, therefore, cannot be considered 'negligible' in comparison with the torsional couple. Thus, an open coiled helical spring falls under both the categories of springs *i.e.,* the torsion spring and the bending spring.

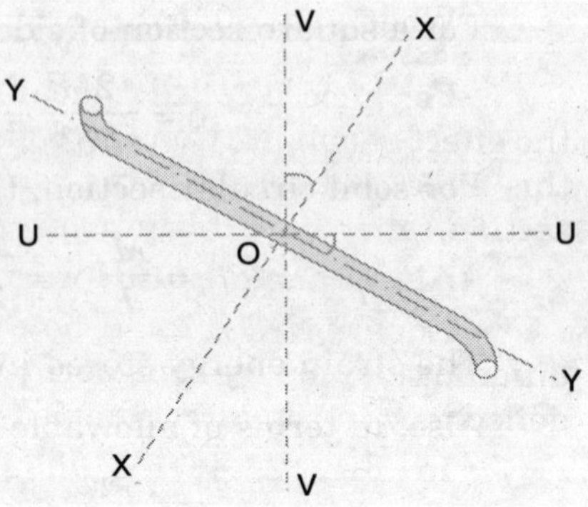

Fig. 21.3

In Fig. 21.3, let the centre line of spring at any point be inclined at an angle α to horizontal.

Let R = mean radius of each coil ;
n = total number of turns ;
l = length of spring ;
d = diameter of wire.

We have thus $l = 2\pi Rn \sec \alpha$ ...(21.12)

The axial force W exerts a moment WR about axis-UU. This moment can be resolved in two components (*a*) a moment T in the plane XX, causing a twisting action, and (*b*) a moment M normal to the plane XX, causing a bending action.

Thus, we have : $T = WR \cos \alpha$ and $M = WR \sin \alpha$ ...(21.13)

The angle of torsion θ, in the plane XX will be $\theta = \dfrac{Tl}{NJ}$

Work done against torsional stress $= \dfrac{1}{2} T\theta = \dfrac{T^2 l}{2NJ} = \dfrac{W^2R^2 l \cos^2 \alpha}{2NJ}$ ...(1)

Work done against bending stress $= \dfrac{M^2 l}{2EI} = \dfrac{W^2R^2 l \sin^2 \alpha}{2EI}$ ...(2)

$\therefore$ Total work done = (1) + (2) $= \dfrac{W^2R^2 l \cos^2 \alpha}{2NJ} + \dfrac{W^2R^2 l \sin^2 \alpha}{2EI}$ ...(3)

If δ is the axial extension, we have, work done $= \frac{1}{2} W\delta$ ...(4)

Equating (3) and (4),

$$\frac{1}{2} W\delta = \frac{W^2R^2 l \cos^2 \alpha}{2NJ} + \frac{W^2R^2 l \sin^2 \alpha}{2EI} = \frac{W^2R^2 l}{2}\left(\frac{\cos^2 \alpha}{NJ} + \frac{\sin^2 \alpha}{EI}\right)$$

$$\therefore \quad \delta = WR^2 l \left(\frac{\cos^2 \alpha}{NJ} + \frac{\sin^2 \alpha}{EI}\right) = W R^2.2\pi nR \sec \alpha \left(\frac{\cos^2 \alpha}{NJ} + \frac{\sin^2 \alpha}{EI}\right) \quad \text{...(21.14)}$$

or
$$\delta = 2\pi n W R^3 \sec \alpha \left(\frac{\cos^2 \alpha}{NJ} + \frac{\sin^2 \alpha}{EI}\right) \quad \text{...(5)}$$

But $J = 2I = \dfrac{\pi}{32} d^4$ (for circular section).

Hence
$$\delta = \frac{64WR^3 n \sec \alpha}{d^4}\left(\frac{\cos^2 \alpha}{N} + \frac{2 \sin^2 \alpha}{E}\right) \quad \text{...(6)}$$

Taking $\dfrac{E}{N} = 2.5$, we have

$$\delta = \frac{64WR^3 n \sec \alpha}{d^4 E} (2.5 \cos^2 \alpha + 2 \sin^2 \alpha) = \frac{64WR^3 n \sec \alpha}{d^4 N} (\cos^2 \alpha + 0.8 \sin^2 \alpha)$$

or
$$\delta = \frac{64WR^3 n \sec \alpha}{d^4 N} (1 - 0.2 \sin^2 \alpha) \quad \text{...(21.15)}$$

When $\alpha = 0$, $\delta = \dfrac{64WR^3n}{Nd^4}$, which is the same as equation 21.2.

For a close coiled helical spring of circular section, it can be seen from Eq. 21.15, that the effect of obliquity on the deflection for all values of α below 10° is about 1% and for $\alpha = 45°$, this effect is about 10% as compared to the case of a close coiled helical spring of circular section, where $\alpha = 0$. However, for other sections, the effect is greater.

Alternative Solution

Equation 21.14 can be obtained from a different manner by considering the *component* deflections in the directions *XX* and *YY*.

In Fig. 21.4 (*a*), *ab* represents the movement due to *twisting* per unit length of the coil, in the plane *XX*. Similarly in Fig. 21.4 (*b*), *a′b′* represents the movement due to *bending* per unit length of coil, in the plane *YY*.

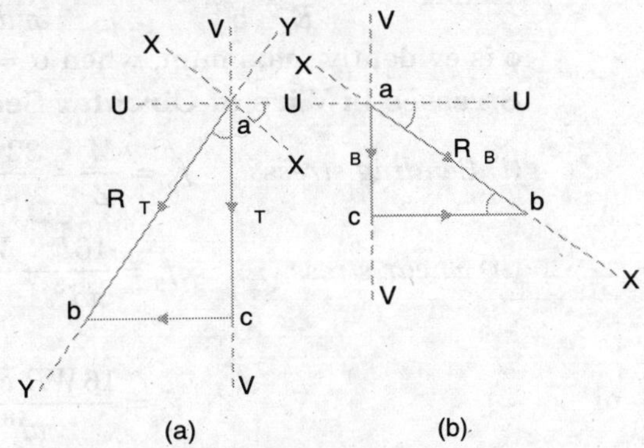

Fig. 21.4

If θ_T = angle of torsion, we have, $ab = R\theta_T$

The movement of *ab* is equivalent to a movement *ac* vertically downwards and a movement *cb* horizontally.

Thus, $ac = \delta_T = ab \cos\alpha = R\,\theta_T \cos\alpha$

and $b = ab \sin\alpha = R\theta_T \sin\alpha$ ...(7)

The movement *cb* will have winding action tending to *increase* the number of turns. Similarly, the bending deflection per unit length $= a'b' = R\theta_B$, where θ_B = angular change in the centre line of the spring due to bending.

The movement *a′b′* is equivalent to the movement *a′c′* vertically downwards and a movement *c′b′* horizontally.

Thus, $a'c' = \delta_B = a'b' \sin\alpha = R\theta_B \sin\alpha$ and $c'b' = a'b' \cos\alpha = R\theta_B \cos\alpha$ ...(8)

The movement *c′b′* will have unwinding action tending to *decrease* the number of turns.

But $\theta_T = \dfrac{T}{NJ}$; $\theta_B = \dfrac{M}{EI}$; $T = WR\cos\alpha$ and $M = WR\sin\alpha$.

Total vertical deflection per unit length $= \delta_T + \delta_B$

$$= R\theta_T \cos\alpha + R\theta_B \sin\alpha = \frac{RT\cos\alpha}{NJ} + \frac{RM\sin\alpha}{EI}$$

$$= \frac{WR^2\cos^2\alpha}{NJ} + \frac{WR^2\sin^2\alpha}{EI} = WR^2\left(\frac{\cos^2\alpha}{NJ} + \frac{\sin^2\alpha}{EI}\right)$$

Deflection for the length $l = WR^2 l\left(\dfrac{\cos^2\alpha}{NJ} + \dfrac{\sin^2\alpha}{EI}\right)$.

The above expression is the same as Eq. 21.14 obtained from the consideration of strain energy.

Again, angular winding up moment per unit length $= \dfrac{cb - c'b'}{R} = \theta_T \sin\alpha - \theta_B \cos\alpha$.

Total winding up in moment for length $l = \phi$

$$= (\theta_T \sin\alpha - \theta_B \cos\alpha)\, l$$

$$= WR\left(\frac{\sin\alpha\cos\alpha}{NJ} - \frac{\sin\alpha\cos\alpha}{EI}\right)l \quad \text{...(9)}$$

or $$\phi = \frac{32WRl\sin\alpha\cos\alpha}{\pi d^4 N}\left(1 - \frac{2N}{E}\right), \text{ for a solid circular section.}$$

Taking $$\frac{N}{E} = \frac{2}{5},\ \phi = \frac{32WRl\sin\alpha\cos\alpha}{5\pi d^4 N} \quad \text{...(21.16)}$$

ϕ is evidently maximum, when $\alpha = 45°$.

Stresses in Wire of Circular Section :

(i) Bending stress, $$f_b = \frac{M}{Z} = \frac{32WR\sin\alpha}{\pi d^3}.$$

(ii) Shear stress, $$f_s = \frac{16T}{\pi d^3} + \frac{W\sin\alpha}{\frac{\pi}{4}d^2} = \frac{16WR\cos\alpha}{\pi d^3} + \frac{4W\sin\alpha}{\pi d^2}$$

or $$f_s = \frac{16WR\cos\alpha}{\pi d^3}\left(1 + \frac{d\tan\alpha}{4R}\right).$$

In the above, the term $\frac{d\tan\alpha}{4R}$ is usually very small, its value ranging between 0.01 to 0.02. Hence this factor may be neglected.

In that case, $$f_s \simeq \frac{16WR\cos\alpha}{\pi d^3}.$$

(iii) Principal stresses σ_1 and σ_2 :

$$\sigma = \frac{f_b}{2} \pm \sqrt{\frac{f_b^2}{2} + f_s^2} = \frac{16WR}{\pi d^3}\left[\sin\alpha \pm \sqrt{\sin^2\alpha + \cos^2\alpha}\right]$$

$$= \frac{16WR}{\pi d^3}(\sin\alpha \pm 1) \quad \left(\text{Taking } f_s \simeq \frac{16WR\cos\alpha}{\pi d^3}\right) \quad \text{...(21.17)}$$

(iv) Maximum shear stress f_s' is given by

$$f_s' = \frac{\sigma_1 - \sigma_2}{2} = \frac{16WR}{\pi d^3} \quad \text{...(21.18)}$$

21.5. OPEN COILED HELICAL SPRING : AXIAL TORQUE

Let the spring be subjected to an axial torque M. M will be reckoned positive if it tends to increase ϕ (or to increase the curvature of the coils).

The resolved part of M about $XX = M\cos\alpha$, causing bending.

The resolved part of M about $YY = M\sin\alpha$, causing torsion.

From the equation of strain energy, we have

$$\frac{1}{2}M\phi = \frac{1}{2}\frac{M^2 l\cos^2\alpha}{EI} + \frac{1}{2}\frac{M^2 l\sin^2\alpha}{NJ}$$

or $$\phi = Ml\left(\frac{\cos^2\alpha}{EI} + \frac{\sin^2\alpha}{NJ}\right) = 2\pi\, RnM \,.\, \sec\alpha\left(\frac{\cos^2\alpha}{EI} + \frac{\sin^2\alpha}{NJ}\right) \qquad ...(21.19)$$

or $$\phi = \frac{64\,MR}{d^4}\, n\sec\alpha\left(\frac{2\cos^2\alpha}{E} + \frac{\sin^2\alpha}{N}\right)$$

When $\alpha = 0$, $\phi = \dfrac{128MRn}{Ed^4}$ radians which is the same as equation 21.9.

The axial extension caused by the couple M may be found by resolving the rotations as before. For the circular section, we get

$$\delta = MRl\sin\alpha\cos\alpha\left(\frac{1}{NJ} - \frac{1}{EI}\right) = \frac{32\,MRl}{\pi d^4}\sin\alpha\cos\alpha\left(\frac{1}{N} - \frac{2}{E}\right)$$

or $$\delta = \frac{64\,MR^2}{d^4}\, n\sin\alpha\left(\frac{1}{N} - \frac{2}{E}\right) \qquad ...(21.20)$$

Example 21.8. *An open coiled helical spring is made out of 15 mm diameter steel rod, the coils having 12 complete turns and a mean diameter of 100 mm, the angle of helix being 12°. Calculate the deflection under an axial load of 300 N and maximum intensities of direct and shear stresses induced in the section of the wire.*

If the axial load of 300 N is replaced by an axial torque of 8 N-m, calculate the angle of rotation about the axis of the coil and the axial deflection. Take $N = 0.84 \times 10^5$ N/mm² and $E = 2.1 \times 10^5$ N/mm².

Solution. (*a*) $T = WR\cos\alpha = 300 \times 50\cos 12° = 14672$ N-mm

$M = WR\sin\alpha = 300 \times 50\sin 12° = 3119$ N-mm

Deflection, $$\delta = \frac{64W\,R^3\, n\sec\alpha}{d^4}\left[\frac{\cos^2\alpha}{N} + \frac{2\sin^2\alpha}{E}\right]$$

$$= \frac{64 \times 300\,(50)^3 \times 12\sec 12°}{(15)^4} \times \left[\frac{\cos^2 12°}{0.84\times 10^5} + \frac{2\sin^2 12°}{2.1\times 10^5}\right] = 6.86 \text{ mm}$$

$$f_b = \frac{M}{Z} = \frac{3119}{\frac{\pi}{32}(15)^3} = \mathbf{9.41\ N/mm^2}$$

$$f_s = \frac{14672}{\frac{\pi}{16}(15)^3} = 22.1 \text{ N/mm}^2 \left(\text{Neglecting the term } \frac{d\tan\alpha}{4R}\right)$$

However, $$\frac{d\tan\alpha}{4R} = \frac{15\tan 12°}{4\times 50} = 0.0159$$

Actual $$f_s = 22.1\,(1 + 0.0159) = 22.5 \text{ N/mm}^2$$

Thus the effect of the term $\dfrac{d\tan\alpha}{4R}$ is 1.6% only, and this term therefore can safely be neglected.

$$\sigma = \frac{f_b}{2} \pm \sqrt{\frac{f_b{}^2}{4} + f_s{}^2} = \frac{9.41}{2} \pm \sqrt{\left(\frac{9.41}{2}\right)^2 + (22.1)^2} = 4.705 \pm 22.595$$

$\therefore$ σ_1 = **27.3 N/mm²** and σ_2 = **– 17.9 N/mm²**

Maximum intensity of shear stress $= f_s' = \sqrt{\frac{f_b{}^2}{4} + f_s{}^2} \simeq$ **22.6 N/mm².**

(*b*) If the axial load is replaced by an axial torque *M*, the angle of rotation is given by

$$\phi = \frac{64MR\,n \sec\alpha}{d^4}\left[\frac{2\cos^2\alpha}{E} + \frac{\sin^2\alpha}{N}\right]$$

$$= \frac{64 \times 8000 \times 50 \times 12 \sec 12°}{(15)^4}\left[\frac{2\cos^2 12°}{2.1 \times 10^5} + \frac{\sin^2 12°}{0.84 \times 10^5}\right] = 0.0597 \text{ radians} = 3° \ 24'$$

The deflection δ is given by

$$\delta = \frac{64MR^2 n \sin\alpha}{d^4}\left[\frac{1}{N} - \frac{2}{E}\right]$$

$$= \frac{64 \times 8000\,(50)^2 \times 12 \sin 12°}{(15)^4}\left[\frac{1}{0.84 \times 10^5} - \frac{2}{2.1 \times 10^5}\right] = \mathbf{15\ mm.}$$

21.6. CARRIAGE SPRINGS : LEAF SPRINGS (SEMI-ELLIPTIC SPRINGS)

A carriage spring or leaf spring (or *semi-elliptic spring*) consists of number of parallel strips of metal of same width, placed one above the other as shown in Fig. 21.5. The plates are bent to the same radius so that they contact only at their edges. When the load *W* is applied at the centre, the change of curvature of each plate is uniform and the same for all the plates, and the contact will continue to be at the ends only. Therefore, each plate may be regarded as supported at its point of contact with the one below it, the load transmitted at its overhanging end being *W*/2. The overhang $a = l/2n$, where n = number of plates. Suppose that all the plates except the one at top are cut longitudinally through the centre and placed as shown in Fig. 21.5 (*b*), they would make up the diamond shape '*equivalent plate*'. The width of the equivalent plate at the centre = $B = nb$, and the thickness will be t, where b is the width of each plate.

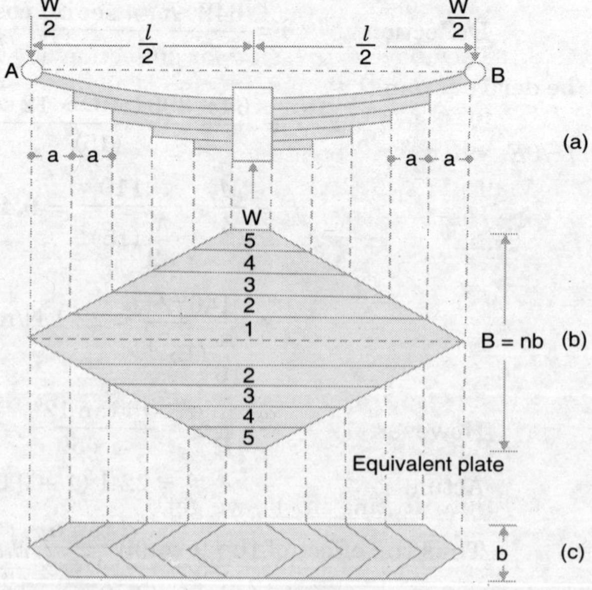

Fig. 21.5

In order to draw the B.M. diagram and determine the shape of the plates, consider a plate *AB* of uniform thickness *t*, loaded as shown in Fig. 21.6. The maximum bending moment will be *Wa*/2, and B.M. at any distance *x* from *A*, and between *A* and *C*, will be *Wx*/2. We have $\frac{1}{R} = \frac{M}{EI}$. Hence, in order that the radius of each plate after bending may remain constant, we must have *M*/*I* as constant.

The bending moment is constant from *C* to *D*, but is not constant from *A* to *C* and from *D* to *B*. Thus, in order to make *M*/*I* constant, *I* must vary from *A* to *C* and *D* to *B*. This can be done by making the ends triangular in plan, the thickness being constant.

Then, at any distance *x* from *A*, $I_x = \frac{b_x t^3}{12}$ and $M_x = \frac{Wx}{2}$

$$\therefore \quad \frac{M_x}{I_x} = \frac{W.x}{2} \times \frac{12}{b_x t^3} = \frac{6W}{t^3}\frac{x}{b_x} = \frac{6W}{t^3} \times \frac{a}{b} = \frac{6aW}{bt^3} = \text{constant.}$$

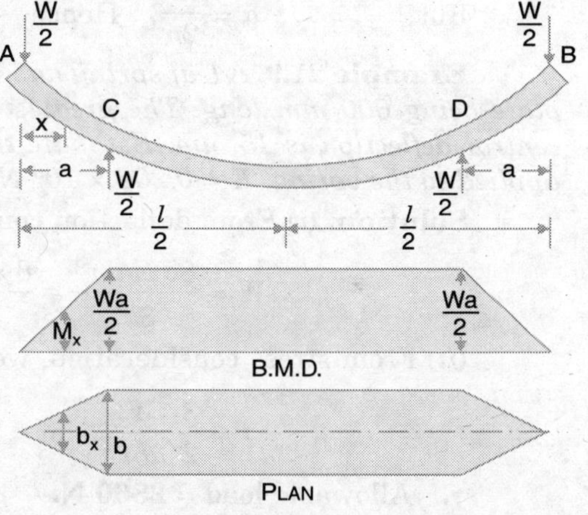

Fig. 21.6

The upward vertical deflection for the 'equivalent plate', which bends to a circular arc

$$= \delta = \frac{Ml^2}{8EI}, \text{ where } I = \frac{nbt^3}{12} \text{ (at the centre).}$$

$$\therefore \quad \delta = \frac{Ml^2}{8EI} = \frac{12Ml^2}{8Enbt^3} = \frac{3Ml^2}{2Enbt^3}$$

For a centrally loaded beam, $M = \frac{Wl}{4}$

$$\therefore \quad \delta = \frac{3Wl^3}{8Enbt^3} \quad \text{...(21.21)}$$

The above expression for deflection can also be derived from the consideration of strain energy.

We know that Resilience due to bending = $f^2/6E$, where f = bending stress.

Total work done is stressing = $\frac{f^2}{6E}$ × volume of the equivalent plate

$$= \frac{f^2}{6E}\left(lt\frac{nb}{2}\right)$$

$$\therefore \quad \frac{1}{2}W\delta = \frac{f^2 lntb}{12E} \quad \text{...(1)}$$

But $$f = \frac{M}{I} \times \frac{t}{2} = \frac{Wl}{(4nbt^3/12)} \cdot \frac{t}{2} = \frac{3}{2}\frac{Wl}{nbt^2} \quad \text{...(21.22)}$$

Substituting in (1), we get

$$\frac{1}{2}W\delta = \frac{9W^2l^2}{4n^2b^2t^4} \times \frac{lntb}{12E} = \frac{3W^2l^3}{16nbt^3E}$$

Hence $\delta = \frac{3Wl^3}{8Enbt^3}$, as before.

If δ_0 is the deflection corresponding to a load W_0 to make the plate flat, we have

$$\delta_0 = \frac{3W_0 l^3}{8Enbt^3} \quad \text{or} \quad W_0 = \frac{8Enbt^3\,\delta_0}{3l^3} \qquad ...(21.23)$$

In the above equation, W_0 is known as "*proof load*" or *test load.*

At the proof load, the maximum bending stress f induced in the plate will be given by

$$f = \frac{3W_0 l}{2nbt^2} = \frac{3l}{2nbt^2} \times \frac{8Enbt^3\,b\delta_0}{3l^3} \quad \text{or} \quad f = \frac{4Et\delta_0}{l^2} \qquad ...(21.24)$$

For a given material, f and E are prescribed. Therefore, equation 21.24 fixes proper relationship between the thickness and initial radius.

The deflection of the ends of the spring relative to the centre

$$= \delta' = \delta_0 - \delta = \frac{Ml^2}{8EI} = \frac{1}{8}\cdot\frac{Wa}{2}\frac{l^2}{E}\cdot\frac{12}{bt^3} = \frac{3}{4}\frac{Wal^2}{Ebt^3}$$

But $\quad a = \frac{l}{2n}$. Hence $\delta' = \delta_0 - \delta = \frac{3Wl^3}{8nEbt^3}$ ...(21.25)

Example 21.9. *A leaf spring has 12 plates each 50 mm wide and 5 mm thick, the longest plate being 600 mm long. The greatest bending stress is not to exceed 180 N/mm² and the central deflection is 15 mm. Estimate the magnitude of the greatest central load that can be applied to the spring. E = 0.206 × 10⁶ N/mm².*

Solution. (*i*) From deflection consideration, we have (Eq. 21.21)

$$\therefore \quad W = \frac{8\delta\,Enbt^3}{3l^3} = \frac{8 \times 15 \times 0.206 \times 10^6 \times 12 \times 50(5)^3}{3(600)^3} = 2860 \text{ N.}$$

(*ii*) From stress consideration, we have

$$f = \frac{3}{2}\frac{Wl}{nbt^2} \quad \text{or} \quad W = \frac{2fnbt^2}{3l} = \frac{2 \times 180 \times 12 \times 50 \times (5)^2}{3 \times 600} = 3000 \text{ N}$$

$\therefore$ Allowable load = **2860 N.**

Example 21.10. *A laminated steel spring, simply supported at the ends and centrally loaded with a span of 75 cm is required to carry a proof load as 7.5 kN, and the central deflection is not to exceed 50 mm. The bending stress must not be greater than 400 N/mm². Plates are available is multiple of 1 mm for thickness and in multiples of 3 mm for width.*

Determine suitable values for thickness, width and number of plates and the radius to which the plates should be formed.

Assume the width to be twelve times the thickness. Take E = 2 × 10⁵ N/mm².

Solution. From Eq. 21.21, we have

$$\delta = \frac{3Wl^3}{8Enbt^3} \quad \text{or} \quad nbt^3 = \frac{3Wl^3}{8E\delta} \qquad ...(1)$$

Also $\quad f = \frac{3}{2}\frac{Wl}{nbt^2} \quad \therefore \quad \text{or} \quad nbt^2 = \frac{3Wl}{2f}$ ...(2)

Dividing (1) by (2), we have

$$t = \frac{3Wl^3}{8E\delta}\,\frac{2f}{3Wl} = \frac{l^2 f}{4E\delta} = \frac{750 \times 750 \times 400}{4 \times 2 \times 10^5 \times 50} = 5.63 \text{ mm}$$

To the nearest 1 mm, $t = 6$ mm. Hence, $b = 12t = 12 \times 6 = 72$ mm

Also, from (2), $n = \dfrac{3Wl}{2fbt^2} = \dfrac{3 \times 7500 \times 750}{2 \times 400 \times 72\,(6)^2} = 8.14$

$\therefore$ Number of plates = 9 (say)

Hence from (2), the modified f is given by

$$f = \frac{3}{2}\,\frac{Wl}{nbt^2} = \frac{3 \times 7500 \times 750}{2 \times 9 \times 72\,(6)^2} = 361.7 \text{ N/mm}^2$$

$$\therefore \qquad R = \frac{E}{f} \,.\, y = \frac{2 \times 10^5}{361.7} \times 3 \simeq \mathbf{1660\ mm.}$$

21.7. QUARTER ELLIPTIC SPRINGS

Fig. 21.7 shows a quarter elliptic spring of unloaded length L, consisting of n plates of width b and thickness t.

Let W = load at the end of the spring tending to straighten it.

δ = corresponding deflection at the free end.

Now $\dfrac{1}{R} = \dfrac{M}{EI}$

But M = B.M. at the fixed end = WL and I = moment of inertia at fixed end

$$= n \times \frac{bt^3}{12}.$$

$$\therefore \quad \frac{1}{R} = \frac{12WL}{nbt^3E} \qquad \text{...(1)}$$

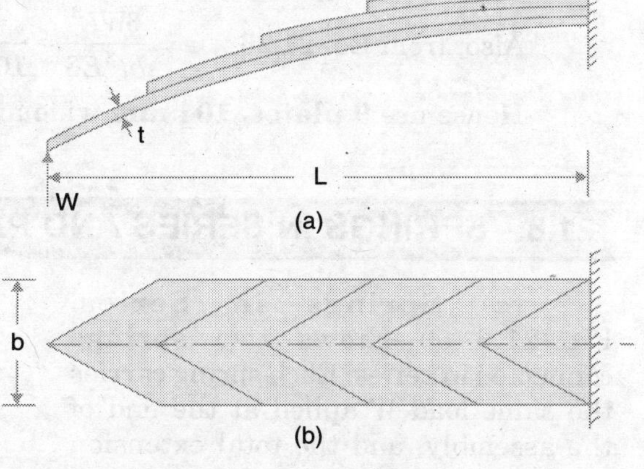

Fig. 21.7

Again, from the properties of a circle, $(2R - \delta)\,\delta = L.L.$

Neglecting the product of small quantities, $2R\delta \simeq L^2$

$$\text{Hence} \qquad \delta = \frac{L^2}{2R} = \frac{6WL^3}{nbt^3E} \qquad \text{...(21.26)}$$

$$\text{Maximum bending stress } f = \frac{M}{I}\,y = \frac{WL\,.\,\frac{1}{2}t}{\frac{1}{12}nbt^3} = \frac{6WL}{nbt^2} \qquad \text{...(21.27)}$$

If W_0 is the proof load and δ_0 is the free height of the spring, we have from Eq. 21.26,

$$W_0 = \frac{nbt^3E\delta_0}{6L^3}. \qquad \text{...(21.28)}$$

Example 21.11. *A 'quarter-elliptic' laminated spring is to be 80 cm long and is to be made of steel plates which are available in multiples of 4 mm in width and 1 mm in thickness. The spring is to carry a load of 10 kN and the bending stress must not exceed 320 N/mm². The maximum deflection must not be greater than 80 mm. Find suitable values of the number and size of plates required, taking the width as eight times the thickness. Take E = 2 × 10⁵ N/mm².*

Solution. From Eqs. 21.27 and 21.26, we have

$$f = \frac{6WL}{nbt^2} \quad \text{and} \quad \delta = \frac{6WL^3}{nbt^3 E}$$

$$\frac{f}{\delta} = \frac{Et}{L^2} \quad \text{or} \quad t = \frac{f}{\delta} \cdot \frac{L^2}{E} = \frac{320}{80} \times \frac{(800)^2}{2 \times 10^5} = 12.8 \text{ mm}$$

Since thickness available is in multiple of 1 mm, adopt t = **13 mm**

∴ Width of plate, $b = 8t = 13 \times 8 =$ **104 mm**. This is also in multiple of 4 mm.

Again, from Eq. 21.27, $n = \dfrac{6WL}{fbt^2} = \dfrac{6 \times 10 \times 1000\,(800)}{320\,(104)\,(13)^2} = 8.53$

Also, from Eq. 21.26, $n = \dfrac{6WL^3}{bt^3 E\delta} = \dfrac{6 \times 10000\,(800)^3}{104\,(13)^3\,(2 \times 10^5)\,(80)} = 8.4.$

Hence use **9 plates, 104 mm** wide and **13 mm** thick.

21.8. SPRINGS IN SERIES AND PARALLEL

(*a*) **Springs in Series.** Fig. 21.8 (*a*) shows two springs connected in series. Each spring carries the same load *W* aplied at the end of the assembly, and the total extension is the sum of extensions of the springs. Let s_1 and s_2 be the stiffness of the component springs and s be the stiffness of the composite spring.

Now, Total extension = sum of extensions of two springs

$$\therefore \quad \delta = \delta_1 + \delta_2$$

$$\therefore \quad \frac{W}{s} = \frac{W}{s_1} + \frac{W}{s_2}$$

$$\text{or} \quad \frac{1}{s} = \frac{1}{s_1} + \frac{1}{s_2} \qquad \text{...(21.29)}$$

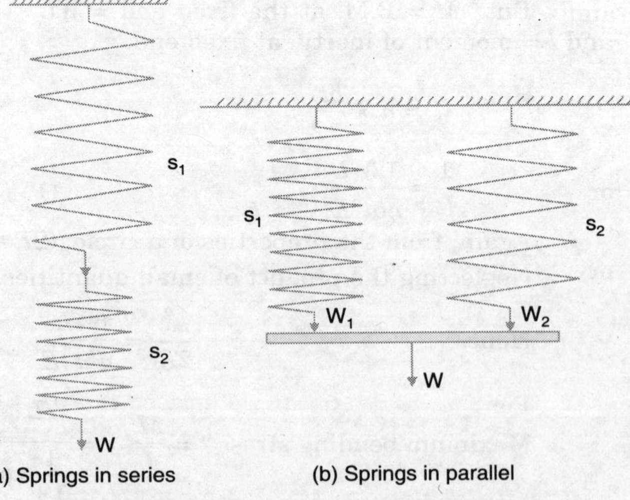

Fig. 21.8

(*b*) **Springs in Parallel.** Fig. 21.8 (*b*) shows two springs connected in parallel. The extension of both the springs is equal, while the sum of the load carried by each spring is equal to the external load.

$$\therefore \quad W = W_1 + W_2 \quad \text{or} \quad \delta.s = \delta.s_1 + \delta.s_2 \quad \therefore \quad s = s_1 + s_2 \qquad \text{...(21.30)}$$

where s is the stiffness of the composite spring.

Example 21.12. *A composite spring has two close-coiled helical steel springs connected in series ; each spring has 12 coils at a mean diameter of 20 mm. Find the diameter of the wire in one of the springs if the diameter in the other spring is 2.5 mm and the stiffness of the composite spring is 1 N/mm. Estimate the greatest load that can be carried by the composite spring and the corresponding extension for a maximum shearing stress of 180 N/mm². N = 0.82 × 10⁵ N/mm².*

Solution. (Fig. 21.9)

In general, $s = \frac{W}{\delta} = \frac{Nd^4}{8D^3n}$ or $\frac{1}{s} = \frac{8D^3n}{Nd^4}$

Let s = stiffness of composite spring = 1 N/mm

s_1 = stiffness of spring 1 ; s_2 = stiffness of spring 2

δ_1 = extension of spring 1 ; δ_2 = extension of spring 2

δ = total extension of the composite spring = $\delta_1 + \delta_2$.

Fig. 21.9

Now $\frac{1}{s} = \frac{\delta}{W}$; but $s = 1$ and $\delta = \delta_1 + \delta_2$

$$\therefore \quad \frac{1}{1} = \frac{\delta_1 + \delta_2}{W} = \frac{\delta_1}{W} + \frac{\delta_2}{W} = \frac{1}{s_1} + \frac{1}{s_2} \quad \text{...(1)}$$

Also,
$$\frac{1}{s_1} = \frac{8D^3n}{Nd_1^{\,4}} = \frac{8 \times 8000 \times 12}{N(2.5)^4} = \frac{1.97 \times 10^4}{N}$$

and
$$\frac{1}{s_2} = \frac{8D^3n}{Nd_2^{\,4}} = \frac{8 \times 8000 \times 12}{Nd_2^{\,4}} = \frac{7.68 \times 10^5}{Nd_2^{\,4}}$$

Hence from (1),
$$\frac{1.97 \times 10^4}{N} + \frac{7.68 \times 10^5}{Nd_2^{\,4}} = 1$$

or
$$\frac{7.68 \times 10^5}{d_2^{\,4}} = N - 1.97 \times 10^4 = (0.82 - 0.197) \times 10^5 = 0.623 \times 10^5$$

or
$$d_2^{\,4} = \frac{7.68}{0.623} = 12.327 \quad \text{or} \quad d_2 = 1.874 \text{ mm}$$

Now
$$f_s = \frac{16WR}{\pi d^3}\left(1 + \frac{d}{4R}\right) \quad \text{...(Eq. 21.4)}$$

For f_s to be maximum, d is to be minimum.

Hence $d_{min} = d_2 = 1.874$ mm

$$\therefore \quad 180 = \frac{16 \times W \times 10}{\pi (1.874)^3}\left(1 + \frac{1.874}{4 \times 10}\right) = 8.10\, W$$

$\therefore$ **W = 22.22 N**

$\therefore$ Total extension $\delta = s.W = 1 \times 22.22 =$ **22.22 mm.**

Example 21.13. *Fig. 21.10 shows a compound spring consisting of two co-axial close coiled helical springs. The details are as follows :*

	Mean dia. D (mm)	*Dia. of wire d (mm)*	*No. of turns n*	*Free length l (mm)*
Outer spring	60	6	10	120
Inner spring	40	5	8	100

If the spring is subjected to an axial load W = 600 N, determine for each spring (a) the change in length, (b) the amount of load carried, and (c) the maximum shearing stress. Take N = 0.18 × 10⁵ N/mm².

W

Fig. 21.10

Solution. Refer to the outer and inner springs as 1 and 2 respectively and use corresponding suffixes.

Stiffness $$s_1 = \frac{Nd_1^4}{64n_1R_1^3} = \frac{0.8 \times 10^5 (6)^4}{64 \times 10\,(30)^3} = 6 \text{ N/mm}$$

Stiffness $$s_2 = \frac{Nd_2^4}{64n_2R_2^3} = \frac{0.8 \times 10^5 (5)^4}{64 \times 8\,(20)^3} = 12.20 \text{ N/mm.}$$

Spring 1 is longer than spring 2 by 20 mm.

Hence to compress spring 1 by 20 mm, load taken by spring 1 is = 6 × 20 = 120 N.

The remaining load of (600 – 120) = 480 N will compress both the springs. Let the common compression be equal to y.

$$\therefore \quad \frac{480}{y} = s_1 + s_2 = 6 + 12.2 = 18.2 \quad \text{or} \quad y = \frac{480}{18.2} = \mathbf{26.37 \text{ mm}}$$

Hence total $\delta_1 = 20 + y = 20 + 26.37 =$ **46.37 mm** and $\delta_2 = y =$ **26.37 mm**

Also, $W_1 = s_1 \,.\, \delta_1 = 6 \times 46.37 =$ **278.2 N** ; $W_2 = s_2 \,.\, \delta_2 = 12.2 \times 26.37 =$ **321.8 N**

(**Check :** $W_1 + W_2 = 600$ N)

$$f_{s1} = \frac{16W_1R_1}{\pi d_1^3}\left(1 + \frac{d_1}{4R_1}\right) = \frac{16 \times 278.2 \times 30}{\pi (6)^3}\left(1 + \frac{6}{4 \times 30}\right) = \mathbf{206.6 \text{ N/mm}^2}$$

$$f_{s2} = \frac{16W_2R_2}{\pi d_2^3}\left(1 + \frac{d_2}{4R_2}\right) = \frac{16 \times 321.8 \times 20}{\pi (5)^3}\left(1 + \frac{5}{4 \times 20}\right) = \mathbf{278.6 \text{ N/mm}^2.}$$

21.9. FLAT SPIRAL SPRINGS

Flat spiral springs are commonly used in clock work mechanisms. It consists of a uniform thin strip of width b and thickness t, wound into a spiral in one plane, and *pinned* at its outer end O, as shown in Fig. 21.11. A spindle is attached to the centre of the spiral which is wound up by applying a torque (T) to the spindle. When the spring is wound up under the action of torque T, reactions R_x and R_y are induced at the pinned end O, in the horizontal and vertical directions, as marked in Fig. 21.11.

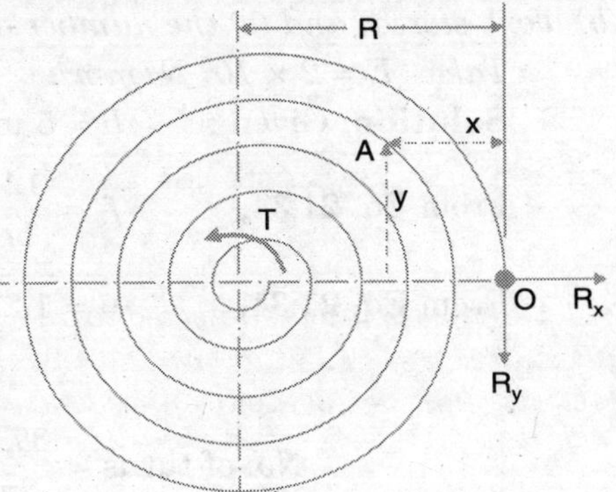

Fig. 21.11. Flat spiral spring

Now consider an element A of length ds of the spring having co-ordinates x and y, with respect to point O. The bending moment at the point is given by

$$M = R_y \,.\, x - R_x \,.\, y \qquad ...(1)$$

Also, by taking moments about the axis of the spindle, we have

$$T = R_y \,.\, R \quad \text{or} \quad R_y = T/R. \qquad ...(2)$$

where R = maximum radius of the spring

$$\therefore \quad \text{Strain energy of the element, } dU = \frac{M^2 \,.\, ds}{2EI} \qquad ...(3)$$

$$\therefore \quad \text{Total strain energy } U = \int \frac{(R_y x - R_x y)^2}{2EI}\, ds = \int \left[\frac{T}{R}x - R_x \,.\, y\right]^2 \frac{ds}{2EI} \qquad ...(4)$$

Since O is a fixed point, $\dfrac{\partial U}{\partial R_x} = 0$. Hence we get from (4)

$$R_x = \frac{\left(\dfrac{T}{R}\right) x\, y\, ds}{y^2 ds} = 0, \text{ since } y^2\, ds \text{ is zero by symmetry.}$$

Hence from (4),

$$U = \int \frac{T^2}{R^2} x^2 \frac{ds}{2EI} \qquad ...(5)$$

$$\therefore \qquad \theta = \frac{\partial U}{\partial T} = \frac{\left(\dfrac{2T}{R^2}\right) \int x^2 ds}{2EI} \qquad ...(6)$$

Treating the spiral as a uniform disc, $\int x^2\, ds \triangleq \left(\dfrac{R^2}{4} + R^2\right) L \triangleq 1.25\, R^2 L$, where L is the total length of the spiral.

$$\therefore \qquad \theta = \frac{\left(\dfrac{2T}{R^2}\right)(1.25\, R^2 L)}{2EI} = 1.25\, \frac{TL}{EI} \qquad ...(21.31)$$

Again, strain energy $\quad U = \dfrac{1}{2}\, T\, \theta = \dfrac{1.25\, T^2 L}{2EI} \qquad ...(7) \qquad ...(21.32)$

M_{max} (at the left hand end) $= R_y \,.\, 2R = \left(\dfrac{T}{R}\right) 2R = 2T. \qquad ...(21.33)$

Maximum stress $\quad f_s = \dfrac{M_{max}}{Z} = \dfrac{2T}{Z} = \dfrac{2T}{\left(\dfrac{bt^2}{6}\right)} = \dfrac{12T}{bt^2} \qquad ...(21.34)$

where $\quad b$ = width of the strip and t = thickness of the strip of the spring.

Example 21.14. *A flat spiral spring is 5 mm wide, 0.2 mm thick and 2 m long. If the maximum stress induced at the point of maximum B.M. is 600 N/mm², determine (a) torque (b) work stored, and (c) the number of turns to wind up the spring.*

Take $E = 2 \times 10^5$ *N/mm²*.

Solution. Given : $b = 5$ mm ; $t = 0.2$ mm ; $L = 2000$ mm ; $f_s = 600$ N/mm²

From Eq. 21.34, $f_s = \frac{12T}{bt^2}$ or $T = \frac{bt^2}{12} \cdot f_s = \frac{5(0.2)^2}{12} \times 600 = \mathbf{10\ N\text{-}mm}$

From Eq. 21.31, $\theta = 1.25 \frac{TL}{EI} = \frac{1.25 \times 10\,(2000)}{2 \times 10^5 \left(\frac{1}{12} \times 5 \times 0.2^3\right)} = 37.5$ radians

$\therefore$ No. of turns $= \frac{37.5}{2\pi} = \mathbf{5.97}$ turns of the spindle

Work stored in the spring $= \frac{1}{2} T\theta = \frac{1}{2} \times 10 \times 37.5 = \mathbf{187.5\ N\text{-}mm.}$

21.10. ADDITIONAL EXAMPLES USEFUL FOR COMPETITIVE EXAMINATIONS

Example 21.15. *A close-coiled helical spring is to have a stiffness of 900 N/m in compression with a maximum load of 45 N and a maximum shearing stress of 120 N/mm². The 'solid' length of the spring (i.e., coils touching) is 45 mm. Find the wire diameter, mean coil radius, and number of coils. Take N = 40000 N/mm².*

Solution. Given : $S = 900$ N/m $= 0.9$ N/mm ; $W = 45$ N ; $f_s = 120$ N/mm²;

$$S = \frac{Nd^4}{64R^3n} \quad \text{or} \quad d^4 = \frac{64\,R^3n}{N} S = \frac{64\,R^3n}{40000} \times 0.9 = 1.44 \times 10^{-3} R^3n \qquad \text{...(1)}$$

$$f_s = \frac{16\,WR}{\pi d^3} \quad \text{or} \quad R = \frac{\pi d^3 f_s}{16\,W} = \frac{\pi\,(120)\,d^3}{16 \times 45} = 0.5236\, d^3 \qquad \text{...(2)}$$

Also $n = \frac{\text{solid length}}{d} = \frac{45}{d}$ ...(3)

Substituting the value of R and n in (1), we get

$$d^4 = 1.44 \times 10^{-3}\,(0.5236\,d^3)^3 \left(\frac{45}{d}\right) = 0.0093\,d^8$$

$\therefore$ $d^4 = \frac{1}{0.0093} = 107.5$. From which $d = \mathbf{3.22\ mm}$

Hence from (2), $R = 0.5236\,(3.22)^3 = \mathbf{17.48\ mm}$

and, from (3), $n = 45/3.22 = 13.975 \simeq \mathbf{14.}$

Example 21.16. *The mean diameter of a close-coiled helical spring is 10 times the diameter of the wire and when 3000 N-m of work is done upon the spring, a compression of 300 mm is produced. Given that the maximum torsional shear stress under these conditions is not to exceed 500 N/mm² and N = 0.84 × 10⁵ N/mm², determine (a) the diameter of wire (b) the mean diameter of coils (c) number of free coils.*

Solution. Given $D = 10\,d$; $U = 3000$ N-m $= 3 \times 10^6$ N-mm ;

$\delta = 300$ mm ; $f_s = 500$ N/mm²

$$U = \frac{1}{2} W \cdot \delta \quad \text{or} \quad W = \frac{2U}{\delta} = \frac{2 \times 3 \times 10^6}{300} = 20 \times 10^3 \text{ N}$$

Now
$$f_s = \frac{16WR}{\pi d^3} = \frac{8WD}{\pi d^3} = \frac{8W(10\,d)}{\pi d^3} = \frac{80W}{\pi d^2}$$

$$\therefore \quad d^2 = \frac{80W}{\pi f_s} = \frac{80 \times 20 \times 10^3}{\pi\,(500)} = 1018.6 \text{ From which } \quad d = \mathbf{31.92\ mm}$$

Mean diameter of coils, $D = 10\,d =$ **319.2 mm** ; $R = 159.6$ mm

$$n = \frac{Nd^4\delta}{64\,WR^3} = \frac{0.84 \times 10^5\,(31.92)^4\,(300)}{64 \times 20 \times 10^3\,(159.6)^3} = \mathbf{5.03.}$$

Example 21.17. *In a compound helical spring, the inner spring is arranged within and concentric with the outer one, but is 9 mm shorter. The outer spring has ten coils of mean diameter 24 mm, and the wire diameter is 3 mm. Find the stiffness of the inner spring if an axial load of 150 N causes the outer one to compress 18 mm.*

If the radial clearance between the springs is 1.5 mm, find the wire diameter of the inner spring when it has eight coils. N = 77000 N/mm². (U.L.)

Solution. Since the inner spring is shorter than the outer spring by 9 mm, the outer spring compresses first, before both the springs start taking the load.

∴ Load taken by the outer spring to compress 18 mm is given by

$$W_1 = \frac{N\,d_1^4\delta_1}{64\,R_1^3 n_1} = \frac{77000\,(3)^4\,(18)}{64\,(12)^3(10)} = 101.5 \text{ N}$$

∴ Load taken by the inner spring $\quad W_2 = 150 - 101.5 = 48.5$ N

Compression of inner spring $\quad \delta_2 = 18 - 9 = 9$ mm

∴ Stiffness of inner spring $\quad S_2 = \dfrac{W_2}{\delta_2} = \dfrac{48.5}{9} =$ **5.39 N/mm**

Mean diameter of outer spring, $\quad D_1 = 24$ mm

∴ Mean diameter of inner spring, $D_2 = 24 - 3 - 2 \times 1.5 - d_2 = 18 - d_2$

Also, number of coils for inner spring, $n_2 = 8$

For inner spring,
$$S_2 = \frac{N\,d_2^4}{64\,n_2 R_2^3} = \frac{N\,d_2^4}{8\,n_2 D_2^3} = \frac{N\,d_2^4}{8n_2\,(18 - d_2)^3}$$

or
$$\frac{d_2^4}{(18 - d_2)^3} = \frac{8n_2 S_2}{N} = \frac{8 \times 8\,(5.39)}{77000} = \frac{1}{223.2}$$

The above equation is solved by trial and error. In the first trial, neglecting d_2 in comparison to 18, we get

$$d_2^4 = \frac{(18)^3}{223.3}, \text{ from which, we get } d_2 = 2.26 \text{ mm}$$

Second trial :
$$\frac{d_2^4}{(18 - 2.26)^3} = \frac{1}{223.2}, \text{ from which } d_2 = 2.04 \text{ mm}$$

Final trial :
$$\frac{d_2^4}{(18 - 2.04)^3} = \frac{1}{223.2}, \text{ from which } d_2 = \mathbf{2.06\ mm.}$$

Example 21.18. *A close-coiled helical spring, whose free length when not compressed is 150 mm, is required to absorb strain energy equal to 50 N-m when fully compressed with the coils in contact. The maximum torsional shear stress is limited to 140 N/mm². Assuming a mean coil diameter of 100 mm, find the diameter of steel wire required and, the number of coils. Take N = 0.8 × 10⁵ N/mm².*

Solution. Given : Uncompressed length = 150 mm ; solid length = nd

$U = 50$ N-m $= 50 \times 10^3$ N-mm ; $f_s = 140$ N/mm² ; $D = 100$ mm ; $R = 50$ mm

δ = uncompressed length – solid length = 150 – nd

Now $$U = \frac{f_s^2}{4N} \times V = \frac{f_s^2}{4N} \times \frac{\pi}{4} d^2 \, (2\pi Rn)$$

$$\therefore \quad 50 \times 10^3 = \frac{(140)^2}{4 \times 0.8 \times 10^5} \times \frac{\pi}{4} d^2 \, (2\pi \times 50)n$$

From which, we get $nd^2 = 3308.4$ ...(1)

Also, $$f_s = \frac{16WR}{\pi d^3} \quad \text{or} \quad \frac{W}{d^3} = \frac{f_s \pi}{16R} = \frac{140\pi}{16 \times 50} = 0.5498 \qquad \text{...(2)}$$

Again $$\delta = \frac{64\, WR^3 n}{Nd^4}$$

or $$(150 - nd) = \frac{64\, WR^3 n}{Nd}\left(\frac{W}{d^3}\right) = \frac{64\,(50)^3\, n}{0.8 \times 10^5 d} \times 0.5498$$

Substituting the value of $nd = \dfrac{3308.4}{d}$ and $n = \dfrac{3308.4}{d^2}$, we get

$$\left(150 - \frac{3308.4}{d}\right) = \frac{54.98}{d}\left(\frac{3308.4}{d^2}\right)$$

or $$150\, d^3 - 3308.4 d^2 - 181896 = 0 \quad \text{or} \quad d^3 - 22.06\, d^2 - 1212.6 = 0$$

Solving this equation by trial and error, we get d = **24.1 mm**

Hence from (1), $n = 3308.4/(24.1)^2 \approx$ **5.7.**

Example 21.19. *A stiff bar of negligible weight transmits a load W to a combination of 3 springs, as shown in Fig. 21.12. The three springs are made of the same material and out of rods of equal diameters. They are of the same length before loading. The number of coils in the three springs are 10, 12 and 15 respectively while the mean radii of the coils are in the proportion 1 : 1.2 : 1.4 respectively. Find the distance x such that the stiff bar remains horizontal after applying the load.*

Solution. Since the bar is rigid, and remains horizontal after the load P is applied, the deflection of each spring will be the same. Hence $\delta_1 = \delta_2 = \delta_3 = \delta$ (say)

Also, the springs are made of the same material and out of the rods of equal diameter, we have $N_1 = N_2 = N_3 = N$ and $d_1 = d_2 = d_3 = d$

Also : Given $n_1 = 10$; $n_2 = 12$ and $n_3 = 15$ and $R_1 : R_2 : R_3 = 1 : 1.2 : 1.4$. Hence $R_2 = 1.2\, R_1$ and $R_3 = 1.4\, R_1$.

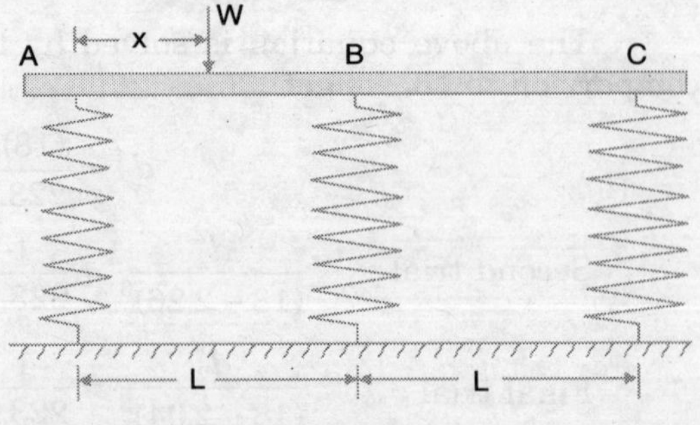

Fig. 21.12

Now $$W_1 = \frac{Nd^4\delta}{64\,R_1^3 n_1} = \frac{Nd^4\delta}{64\,R_1^3(10)} = \frac{Nd^4\delta}{640R_1^3} \quad ...(1)$$

$$W_2 = \frac{Nd^4\delta}{64\,R_2^3 n_2} = \frac{Nd^4\delta}{64\,(1.2\,R_1)^3(12)} = \frac{Nd^4\delta}{1327.1R_1^3} \quad ...(2)$$

and $$W_3 = \frac{Nd^4\delta}{64\,R_3^3 n_3} = \frac{Nd^4\delta}{64\,(1.4\,R_1)^3\,15} = \frac{Nd^4\delta}{2634.2R_1^3} \quad ...(3)$$

From (1) and (2), we get $W_2 = \frac{640}{1327.1}\,W_1 = 0.4823\,W_1$

From (1) and (3), we get $W_3 = \frac{640}{2634.2}\,W_1 = 0.2430\,W_1$

Now, taking moments about the line of action of W_1, we get

$W_2 \,.\, L + W_3\,(2L) = W \,.\, x$ or $0.4823\,W_1 L + 0.2430\,W_1\,(2L) = Wx$

$$\therefore \quad x = \frac{(0.4823\,W_1 + 0.4860\,W_1)L}{W} \quad ...(a)$$

But $$W = W_1 + W_2 + W_3 = W_1 + 0.4823\,W_1 + 0.2430\,W_1 = 1.7253\,W_1 \quad ...(b)$$

Hence from (*a*), we get $x = \frac{0.4823\,W_1 + 0.4860\,W_1}{1.7253\,W_1}\,L =$ **0.5612 *L*.**

Example 21.20. *A close-coiled helical spring of circular section extends 1 cm when subjected to an axial load W, and there is an angular rotation of 1 radian when a torque T is independently applied about the axis. If D is the mean coil diameter, show that*

$$\frac{T}{W} \approx \frac{D^2\,(1+\mu)}{4} \text{ where } \mu \text{ is the Poisson's ratio.}$$

Determine Poisson's ratio if D = 3 cm, a load of 55 N extends the spring 5.45 cm, and a torque of 300 N-mm produces an angular rotation of 60°.

Solution. We know that $$W = \frac{Nd^4\delta}{8D^3 n} \quad ...(1)$$

Also, we have $$\varphi = \frac{TL}{EI} = \frac{T\,(\pi\,Dn)}{E\,\frac{\pi}{64}\,d^4} = \frac{64\,TDn}{Ed^4} \text{ or } T = \frac{Ed^4\varphi}{64Dn} \quad ...(2)$$

Here, $\delta = \varphi = 1$. Hence $$\frac{T}{W} = \left(\frac{Ed^4}{64Dn}\right)\left(\frac{8D^3 n}{Nd^4}\right) = \frac{E}{N}\cdot\frac{D^2}{8}$$

But $\frac{E}{N} = 2\,(1+\mu)$

$$\therefore \quad \frac{T}{W} = \frac{D^2\,(1+\mu)}{4} \text{ (Proved)} \quad ...(3)$$

For the given data : $D = 3$ cm ; $W = 55$ N/ 5.45 = 10 092 N/cm

$$T = \frac{300}{600 \times \frac{\pi}{180}} = \frac{300 \times 3}{\pi} \text{ N-mm/radian} = \frac{90}{\pi} \text{ N-cm/radian}$$

Hence from (3), we get $(1 + \mu) = \frac{4T}{WD^2} = \frac{4 \times 90/\pi}{10.092\,(3)^2} = 1.262$

$\therefore \qquad \mu = \mathbf{0.262}$

Example 21.21. *An open coiled helical spring and a close-coiled helical spring are similar in every respect, except that the elongation of the open coiled spring is 2.5% greater than that of the close-coiled spring, under the same load. Determine the angle of inclination of the coils of the open coil springs. Take E / N = 2.5.*

Solution. Using suffix 1 for open coiled spring and 2 for close-coiled spring,

We have $\qquad \delta_1 = 1.025\, \delta_2$

$$\therefore \quad \frac{64WR^3.n}{d^4 \cos\alpha}\left[\frac{\cos^2\alpha}{N} + \frac{2\sin^2\alpha}{E}\right] = 1.025 \times \frac{64WR^3n}{Nd^4}$$

or $$\frac{1}{N\cos\alpha}\left[\cos^2\alpha + \frac{2N}{E}\sin^2\alpha\right] = \frac{1.025}{N}$$

or $$\cos^2\alpha + \frac{2}{2.5}(1 - \cos^2\alpha) = 1.025 \cos\alpha \quad \text{or} \quad \cos^2\alpha + 0.8\,(1 - \cos^2\alpha) = 1.025\cos\alpha$$

or $$0.2\cos^2\alpha - 1.025\cos\alpha + 0.8 = 0 \quad \text{or} \quad \cos^2\alpha - 5.125\cos\alpha + 4 = 0$$

$$\therefore \quad \cos\alpha = \frac{5.125 \pm \sqrt{(5.125)^2 - 4 \times 4 \times 1}}{2 \times 1} = 4.102 \text{ and } 0.9604$$

$\therefore \qquad \cos\alpha = 0.9604$ From which $\alpha = \mathbf{16.16°}$.

Example 21.22. *An open coiled helical spring of a given length of wire, 5 mm is diameter is coiled to a mean diameter of 40 mm, and is loaded with an axial load. If the spring is assumed to be a closely coiled, it would cause an error of 2.5% in axial deflection. Determine the pitch of the open-coiled spring. Take μ = 0.3.*

Solution. Given : $\quad d = 5$ mm ; $D = 40$ mm ; $R = 20$ mm ; $\mu = 0.3$

Now $$E = 2N(1 + \mu) \quad \text{or} \quad \frac{2N}{E} = \frac{1}{1+\mu} = \frac{1}{1+0.3} = 0.7692 \qquad \text{...(1)}$$

Let us use suffix 1 for open coiled spring and 2 for closed coiled spring.

We have $L_1 = 2\pi Rn \sec\alpha$ and $L_2 = 2\pi Rn$. Hence $L_1 > L_2$ and therefore $\delta_1 > \delta_2$

$$\therefore \quad \frac{\delta_1 - \delta_2}{\delta_1} = 0.025 \quad \text{or} \quad 1 - \frac{\delta_2}{\delta_1} = 0.025 \quad \text{or} \quad \delta_2 = 0.975\,\delta_1 \qquad \text{...(2)}$$

$$\therefore \quad \frac{64WR^3n}{Nd^4} = 0.975 \times \frac{64WR^3n}{d^4\cos\alpha}\left[\frac{\cos^2\alpha}{N} + \frac{2\sin^2\alpha}{E}\right]$$

or $$1 = \frac{0.975}{\cos\alpha}\left[\cos^2\alpha + \frac{2N}{E}\sin^2\alpha\right] = \frac{0.9575}{\cos\alpha}\left[\cos^2\alpha + 0.7692\,(1 - \cos^2\alpha)\right]$$

$$\therefore \quad 1.0256\cos\alpha = \cos^2\alpha + 0.7692 - 0.7692\cos^2\alpha$$

or $\qquad \cos^2 \alpha - 4.4437 \cos \alpha + 3.3328 = 0$

$\therefore \qquad \cos \alpha = \dfrac{4.4437 \pm \sqrt{(4.4437)^2 - 4 \times 3.3328}}{2} = 3.4883 \quad \text{or} \quad 0.9554$

$\therefore \qquad \cos \alpha = 0.9554 \quad \text{From which } \alpha = 17.17°$

Now, if p is the pitch, we have $\dfrac{p}{\pi D} = \tan \alpha$

$\therefore \qquad p = \pi D \tan \alpha = \pi (40) \tan 17.17° =$ **38.84 mm.**

Example 21.23. *An open coiled helical spring made of 4 mm dia. steel wire, has 8 coils of 40 mm diameter and 30 mm pitch. The spring is subjected to a winding torque about the axis of the spring, resulting in an increase in the number of coils by 0.8, the length remaining the same (i.e., 240 mm). Determine (a) the torque required, and (b) minimum elastic limit strength of steel to permit this amount of winding. Take $E = 2 \times 10^5$ N/mm² and $N = 0.83 \times 10^5$ N/mm².*

Solution. Given : $d = 4$ mm ; $n = 8$; $D = 40$ mm ; $R = 20$ mm ; $p = 30$ mm ; $n' - n = 0.8$

Vertical length of spring $= p \,.\, n = 30 \times 8 = 240$ mm.

The spring is subjected to a winding torque T. Due to this torque, the vertical length of the spring is bound to increase, *i.e.*, δ has to be a finite positive value. However, since the length does not increase (*i.e.*, since $\delta = 0$), a restraining force W exists there, due to which the resulting deflection is zero. In other words, the spring is subjected to (*a*) a winding torque T and (*b*) a restraining force W of opposite nature. Under the combined action of T and W, we have

$$\delta = \frac{64\, T R^2 n \sin \alpha}{d^4}\left[\frac{1}{N} - \frac{2}{E}\right] - \frac{64\, W R^3 n \sec \alpha}{d^4}\left[\frac{\cos^2 \alpha}{N} + \frac{2 \sin^2 \alpha}{E}\right] \qquad \text{...(1)}$$

and
$$\varphi = \frac{64\, T R n \sec \alpha}{d^4}\left[\frac{2\cos^2 \alpha}{E} + \frac{\sin^2 \alpha}{N}\right] - \frac{64\, W R^2 n \sin \alpha}{d^4}\left[\frac{1}{N} - \frac{2}{E}\right] \qquad \text{...(2)}$$

Now $\tan \alpha = \dfrac{p}{\pi D} = \dfrac{30}{\pi (D)}$, from which $\alpha = 13.427°$

$\therefore \quad \sin \alpha = 0.2322; \cos \alpha = 0.9727; \sec \alpha = 1.0281 \quad \sin^2 \alpha = 0.0539; \cos^2 \alpha = 0.9461$

Substituting the values in (1) and putting $\delta = 0$, we get

$$\frac{64\, T (20)^2 8 (0.2322)}{(4)^4}\left[\frac{1}{0.8 \times 10^5} - \frac{2}{2 \times 10^5}\right] - \frac{64\, W (20)^3 8 (1.0281)}{(4)^4}\left[\frac{0.9461}{0.83 \times 10^5} + \frac{2 \times 0.0539}{2 \times 10^5}\right] = 0$$

or
$$4.644 \times 10^{-4}\, T - 0.1964\, W = 0$$

From which $\quad W = 2.3648 \times 10^{-3}\, T \qquad$ ...(3)

Now angle of rotation $\varphi = 0.8$ coils $= 0.8\,(2\pi)$ radians $= 5.0265$ radians

Substituting $\varphi = 5.0265$ radians and $W = 2.3648 \times 10^{-3}\, T$ in Eq. (2), we get

$$5.0265 = \frac{64\, T (20)\, 8 (1.0281)}{(4)^4}\left[\frac{2 \times 0.9461}{2 \times 10^5} + \frac{0.0539}{0.83 \times 10^5}\right] - \frac{64 (2.3648 \times 10^{-3} T)(20)^2 8 (0.2322)}{(4)^4}\left[\frac{1}{0.83 \times 10^5} - \frac{2}{2 \times 10^5}\right]$$

or
$$5.0265 = 4.1578 \times 10^{-4}\, T - 8.9974 \times 10^{-7}\, T$$

From which, we get $T = \textbf{12116 N-mm} = \textbf{12.116 N-m}$

Now $$\frac{M}{I} = \frac{f_b}{y} = \frac{f_b}{d/2} \quad \text{or} \quad f_b = \frac{M}{I}\frac{d}{2} = \frac{T}{\frac{\pi}{64}d^4} \cdot \frac{d}{2} = \frac{32T}{\pi d^3}$$

(since the spring is subjected to a winding torque T only)

$$\therefore \qquad f_b = \frac{32 \times 12116}{\pi\,(4)^3} = \textbf{1928.2 N/mm}^2$$

PROBLEMS

1. A closely coiled spring is made out of 15 mm steel rod the coils having 12 complete turns and a mean diameter of 120 mm. It carries an axial pull of 400 N. Calculate the shear stress induced in the section of the rod. If $N = 0.85 \times 10^5$ N/mm^2, calculate the deflection under the pull and the amount of energy stored in the spring during the extension.
2. A truck weighing 30 kN and moving at 5 kilometre per hour has to be brought to rest by a buffer. Find how many springs, each of 18 coils, will be required to store the energy of motion during a compression of 20 cm. The spring is made out of 25 mm diameter steel rod coiled to a mean diameter of 24 cm. $N = 0.84 \times 10^5$ N/mm^2.
3. Find the maximum permissible axial load for a closely coiled spring made out of 10 mm square rod with 16 coils of 12 cm mean diameter. If the maximum shear stress is limited to 300 N/mm^2, calculate also the deflection under the load if $N = 0.84 \times 10^5$ N/mm^2.
4. A weight of 500 N is dropped on to a helical compression spring made of 20 mm wire closely coiled to a mean diameter of 12 cm with 20 coils. If the instantaneous compression is 60 mm, calculate the height of drop. $N = 0{\cdot}84 \times 10^5$ N/mm^2.
5. A weight of 300 N is dropped through a height of 50 cm to a closely coiled compression spring which instantaneously compresses to 10 cm under the impact. If the diameter of steel rod is 20 mm and the mean diameter of the coil is 12 cm, find the maximum instantaneous stress produced by the impact and the number of coils in the spring. Take $N = 0.84 \times 10^5$ N/mm^2.
6. A closely coiled helical spring is made out of 15 mm diameter steel rod, the coils having 12 complete turns and of mean diameter of 9 cm, Calculate the increase in the number of turns and the bending stress induced in the section, if it is subjected to an axial twist of 8 N-m. $E = 2.1 \times 10^5$ N/mm^2.
7. An open coil helical spring is made out of 10 mm diameter steel rod, the coils having 10 complete turns, and a mean diameter 80 mm, the angle of helix being 15°. Calculate the deflection under an axial load of 250 N and the maximum intensities of direct and shear stresses induced in the section of the wire. If the axial load of 250 N is replaced by an axial torque of 6 N-m, calculate the angle of rotation about axis of the coil and axial deflection. $N = 0.85 \times 10^5$ N/mm^2 and $E = 2.1 \times 10^5$ N/mm^2.
8. Deduce an expression for the extension of an open coiled helical spring carrying an axial load W. Take α as the inclination of the coils, d the diameter of the wire and R the mean radius of the coils.

   Find by what percentage the axial extension is under-estimated if the inclination of the coils is neglected for a spring in which $\alpha = 25°$, $E = 2.11 \times 10^5$ N/mm^2 ; $N = 0.808 \times 10^5$ N/mm^2.
9. An open coiled helical spring of round wire of diameter d has a mean coil's diameter D with n complete turns, the coil makes an angle α with plane perpendicular to the axis of the spring.

   If the modulus of elasticity is E and of rigidity N, derive an expression for the extension due to an axial load W in terms of the symbols given.

From this expression, derive the value of the extension of a close coiled spring.

If the close coiled spring formula is used in finding the extension of an open coiled spring, determine the maximum value of α if the error in the value of extension is not to exceed 1%. Assume $E = 2.5$ N.

10. Assuming without proof the formulae for the strength and stiffness of a round bar in torsion, derive an expression for the extension of a close coiled helical spring in terms of W, the mean coil diameter D, and the diameter of the wire section d, the number of active coils n and the modulus of rigidity N.

11. A carriage spring of length L having n plates of width b and thickness t, carries a central load W. Obtain expressions for (*i*) the maximum bending stress f in terms of W, L, n, b and t, (*ii*) the strain energy in terms of f, E and volume of spring.

12. A composite spring has two close coiled helical steel springs in series ; each spring has a mean coil diameter of 8 times the diameter of its wire. One spring has 20 coils and wire diameter of 2.5 mm, Find the diameter of the wire in the other spring if it has 15 coils and the stiffness of the composite spring is 2 N/mm.

Find the greatest axial load that can be applied to the spring and the corresponding extension for a maximum shearing stress of 300 N/mm². $N = 0.82 \times 10^5$ N/mm².

ANSWERS

1. $f_s = 36.2$ N/mm² ; $\delta = 15.6$ mm; $U = 3.12$ N-m **2.** 9 Nos.

3. $W = 1040$ N; $\delta = 191.4$ mm **4.** 115 mm

5. $f_s = 149$ N/mm² ; $n = 27$ **6.** $n' - n = 0.00834$; $f = 24.2$ N/mm²

7. $\delta = 12.45$ mm; $\sigma_1 = 64$ N/mm² ; $\sigma_2 = -37.6$ N/mm² ; $f_s = 50.8$ N/mm²; $\phi' = 8.8°$; $\Delta = 0.355$ mm

8. 5.43% **9.** $\alpha = 9° 54'$

12. $d_2 = 1.875$ mm; W = 48.8 N $\delta = 24.4$ mm.

22

Thin Cylindrical and Spherical Shells

22.1. INTRODUCTION

Shells whose walls are subjected to pressure from fluids, dry materials or gases are in frequent use in engineering. Steam boilers, reservoirs, tanks, working chambers of engines, etc., are the common examples of shells. If the thickness of the wall of the shell is less than 1/10 to 1/15 of the diameter of the shell, it is treated as a *thin shell* and the stress developed in their walls under *internal* pressure can easily be found. In such cases, it can be assumed with sufficient accuracy that the normal stresses (tensile or compressive) are uniformly distributed through the thickness of the wall. However, if the walls are thick, the stresses are no longer uniformly distributed and the problem becomes complex. In this chapter, we shall develop the equations for the stresses in the walls of a *thin shell*. The theory of *thick shell* has been discussed in Chapter 23.

22.2. THIN CYLINDRICAL SHELL : INTERNAL PRESSURE

Consider a thin seamless cylindrical shell of internal diameter d, wall thickness t and length l, subjected to an internal intensity of pressure p. Due to this, the following stresses are induced in the cylinder :

(*i*) Circumferential tensile stress (or hoop stress) f

(*ii*) Longitudinal (or axial) tensile stress f_0

(*iii*) Radial compressive stress which varies from a value p at the inner surface to the atmospheric pressure at the outer surface. It is required to find the intensity of circumferential stress (or hoop stress) f and intensity of longitudinal stress f_0.

Circumferential Stress *f*

Fig. 22.1 (*a*) shows the cross-section of cylindrical shell. Consider two elementary strips of the shell, each at an angle θ on either side of the vertical diameter and subtending an angle $\delta\theta$ at the centre. The normal force on each elementary ring of length l is

$$\delta P_n = p.r\delta\theta.l$$

The resultant of the two elementary forces δP_n will be

$$\delta P = 2\delta P_n.\cos\theta = 2pr.\delta\theta.l\cos\theta$$

in the vertical direction.

The total effect of the radial pressure on the upper or lower portion of the shell is known as the bursting force and is given by

$$P = \Sigma\delta P = \int_0^{\pi/2} 2prl\cos\theta\, d\theta$$

$$P = 2prl = pd.l$$

$$= p \times \text{projected area} \quad ...(22.1)$$

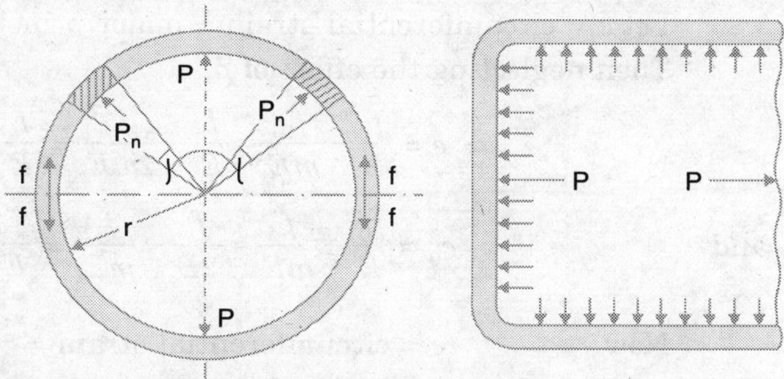

Fig. 22.1

If f = intensity of circumferential stress. We have, for equilibrium

$$f(2tl) = p.d.l \quad \text{or} \quad f = \frac{pd}{2.t} \quad ...(22.2)$$

However, if η is the efficiency of the longitudinal riveted joint,

$$f = \frac{p.d}{2t\eta} \quad ...(22.3)$$

Longitudinal Stress f_0

Fig. 22.1 (*b*) shows the longitudinal section of the cylindrical shell. The bursting force P due to the effect of the pressure on the end is given by

$$P = p\frac{\pi}{4}d^2$$

If f_0 = intensity of longitudinal stress, we have, for equilibrium

$$f_0(\pi d.t) = p.\frac{\pi}{4}d^2 \quad \text{or} \quad f_0 = \frac{p.d}{4t} \quad ...(22.4)$$

Thus, the intensity of longitudinal stress is one-half of the intensity of circumferential or hoop stress.

If η is the efficiency of circumferential riveted joint,

$$f_0 = \frac{pd}{4t\eta}. \quad ...(22.5)$$

22.3. CHANGE IN VOLUME OF CYLINDRICAL SHELL

We have seen that at any point in the metal of the cylinder, the circumferential stress f across the longitudinal section is $pd/2t$ and the longitudinal stress across the circumferential section is $pd/4t$. Since both these stresses are *wholly normal* and since the internal pressure p is also normal, we have at any point, three principal stresses : f, f_0 and p. However, p is generally very small in comparison to f and f_0 and is, therefore, neglected in calculating the strains.

The maximum value of the shear stress will evidently be

$$= \frac{f-(-p)}{2} = \frac{f+p}{2} \quad \text{or} \quad = \frac{f}{2}, \text{ if } p \text{ is neglected.}$$

Let e = circumferential strain = major principal strain and e_0 = longitudinal strain. Then neglecting the effect of p,

$$e = \frac{f}{E} - \frac{f_0}{mE} = \frac{f}{E} - \frac{f}{2mE} = \frac{f}{E}\left(1 - \frac{1}{2m}\right) = \frac{pd}{2tE}\left(1 - \frac{1}{2m}\right) \quad ...(22.6)$$

and

$$e_0 = \frac{f_0}{E} - \frac{f}{mE} = \frac{f}{2E} - \frac{f}{mE} = \frac{f}{E}\left(\frac{1}{2} - \frac{1}{m}\right) = \frac{pd}{2tE}\left(\frac{1}{2} - \frac{1}{m}\right) \quad ...(22.7)$$

Now

$$e = \text{circumferential strain} = \frac{\text{change in circumference}}{\text{original circumference}} = \frac{\pi\delta d}{\pi d} = \frac{\delta d}{d}$$

Hence $\delta d = ed$; Similarly, $\delta l = e_0 l$ ...(22.8)

Thus, from Eq. 22.8, the change in the dimensions of the cylinder can be found out.

For the volumetric strain, $V = \frac{\pi}{4} d^2 l$

$$\delta V = \frac{\pi}{4} d^2 \delta l + 2 \cdot \frac{\pi}{4} ld \cdot \delta d \; ; \quad \therefore \quad \frac{\delta V}{V} = e_v = \frac{\delta l}{l} + 2\frac{\delta d}{d}$$

or

$$e_v = e_0 + 2e$$

Substituting the value of e_0 and $2e$, we get

$$e_0 = \frac{pd}{2tE}\left(\frac{1}{2} - \frac{1}{m}\right) + \frac{pd}{tE}\left(1 - \frac{1}{2m}\right) = \frac{pd}{2tE}\left(\frac{5}{2} - \frac{2}{m}\right)$$

However, if p is also considered, we have

$$e = \frac{pd}{2tE}\left(1 - \frac{1}{2m}\right) + \frac{p}{mE}, \; p \text{ being compressive}$$

and

$$e_0 = \frac{pd}{2tE}\left(\frac{1}{2} - \frac{1}{m}\right) + \frac{p}{mE}$$

$$\therefore \quad e_v = e_0 + 2e = \frac{pd}{2tE}\left(\frac{5}{2} - \frac{2}{m}\right) + \frac{3p}{mE} \quad ...(22.9)$$

Thus, if the radial stress is considered, the volumetric strain is increased by $3p/mE$.

22.4. DESIGN OF THIN CYLINDRICAL SHELL

If f_a is the allowable tensile stress, the major principal stress f should be less than f_a.

Thus $\frac{pd}{2t} \le f_a$.

If it is required to design a cylinder of given diameter and internal pressure, we have

$$t \ge \frac{pd}{2f_a} \quad ...(22.10)$$

Example 22.1. *A seamless pipe of 1 metre diameter is to carry fluid under a pressure of 2 N/mm². Calculate the necessary thickness of metal, if maximum stress is not to exceed 80 N/mm².*

Solution. $t \geq \frac{pd}{2f_a} = \frac{2 \times 1000}{2 \times 80} = \mathbf{12.5\ mm.}$

Example 22.2. *A cylindrical shell is 3 m long, 1 metre internal diameter and 15 mm metal thickness. Calculate the maximum intensity of shear stress induced and also the changes in the dimensions of the shell if it is subjected to an internal pressure of 1.5 N/mm². Take E = 0.204 × 10⁶ N/mm² and 1/m = 0.3.*

Solution. Circumferential stress $f = \frac{pd}{2t} = \frac{1.5 \times 1000}{2 \times 15} = 50\ \text{N/mm}^2$ (tensile)

Longitudinal stress $f_0 = \frac{1}{2} f = 25\ \text{N/mm}^2$ (tensile).

Maximum shear stress at any point in the thickness of the metal

$$= \frac{f+p}{2} = \frac{50+1.5}{2} = 25.75\ \text{N/mm}^2$$

or

$$= \frac{f}{2} = \frac{50}{2} = 25\ \text{N/mm}^2, \text{ if } p \text{ is neglected.}$$

Again, circumferential strain $e = \frac{f}{E} - \frac{f_0}{mE} = \frac{1}{E}(50 - 25 \times 0.3) = \frac{42.5}{E}$ (increase)

$$\therefore \quad \delta d = e.d = \frac{42.5}{0.204 \times 10^6} \times 1000 = \mathbf{0.208\ mm} \text{ (increase)}$$

Longitudinal strain $e_0 = \frac{f_0}{E} - \frac{f}{mE} = \frac{1}{E}(25 - 0.3 \times 50) = \frac{10}{E}$ (increase)

$$\therefore \quad \delta l = e_0.l = \frac{10}{0.204 \times 10^6} \times 3000 = \mathbf{0.147\ mm} \text{ (increase)}$$

Now $\frac{\delta V}{V} = e_0 + 2e = \frac{10}{E} + \frac{2 \times 42.5}{E} = \frac{95}{E}$

$$\therefore \quad \delta V = \frac{95}{0.204 \times 10^6}\left[\frac{\pi}{4}(1000)^2 \times 3000\right] = \mathbf{1097 \times 10^3\ mm^3} \text{ (increase).}$$

Example 22.3. *A boiler of 2 m diameter is made of 20 mm thick mild steel plates. Taking the efficiency of the longitudinal riveted joint as 75%, calculate the permissible steam pressure in the boiler if the maximum tensile stress in the plate section through the rivets is not to exceed 100 N/mm². Calculate the circumferential stress in the solid plate-section at this pressure and also longitudinal stress in the plate-section through the rivets of a circumferential joint if the efficiency is 65%.*

Solution. $f = \frac{pd}{2t\eta}$ or $p = \frac{2ft\eta}{d} = \frac{2 \times 100 \times 20 \times 0.75}{200} = 1.5\ \text{N/mm}^2$

At this internal pressure, the circumferential stress across the solid plate section will be

$$f = \frac{pd}{2t} = \frac{1.5 \times 2000}{2 \times 20} = \mathbf{75\ N/mm^2}$$

Similarly, the longitudinal stress across a riveted circumferential will be

$$f_0 = \frac{pd}{4t\eta} = \frac{1.5 \times 2000}{4 \times 20 \times 0.65} = \mathbf{57.6\ N/mm^2.}$$

Example 22.4. *A copper tube of 50 mm internal diameter, 1 metre long and 1.25 mm thick has closed ends and is filled with water under pressure. Neglecting any distortion of the end plates, determine the alteration of pressure when an additional volume of 3 cubic centimetres of water is pumped into the tube.*

Modulus of elasticity for copper $= 1.03 \times 10^5$ *N/mm*^{2}.

Poissons ratio $= 0.3$

Bulk Modulus for water $= 2.1 \times 10^3$ *N/mm*^{2}.

Solution. Internal volume of cylinder $= V = \frac{\pi}{4}(50)^2 \times 1000 = 19.63 \times 10^5$ mm^3

Let p = alteration in pressure.

For the tube :

Hoop stress $f = \frac{pd}{2t} = \frac{50p}{2 \times 1.25} = 20\,p$ (tensile)

Longitudinal stress $= f_0 = \frac{pd}{4t} = 10\,p$ (tensile)

Volumetric strain $e_v = e_0 + 2e = \frac{1}{E}(10p - 0.3 \times 20p) + \frac{2}{E}(20p - 0.3 \times 10p) = \frac{38p}{E}$

Change in the volume of the tube $= \frac{38p}{E} \times V$ (increase) ...(1)

For water :

Volumetric strain $e_v' = \frac{p}{K}$; Change in volume of water $= \frac{p}{K} \times V$

The total space created in the tube = Volumetric change of the tube + Volumetric change of water

Hence $3 \times 1000\ (\text{mm}^3) = \frac{38p}{E} \times V + \frac{p}{K} \times V$

or $$3000 = pV\left(\frac{38}{E} + \frac{1}{K}\right) = 19.63 \times 10^5 p\left[\frac{38}{1.03 \times 10^5} + \frac{1}{2.1 \times 10^3}\right]$$

or $$3000 = 1659\,p \quad \text{or} \quad p = \frac{3000}{1659} = \mathbf{1.81\ N/mm^2}.$$

Example 22.5. *A thin cylinder of 100 mm internal diameter and wall thickness 2 mm has its ends closed by rigid plates and is then filled with water. When an external pull of 20 kN is applied to the ends, the water pressure, read by the gauge, is observed to fall by 0.075 N/mm*^{2}*. Neglecting any end effects due to plates, determine the value of Poisson's ratio for the metal. Take E for the metal* $= 2.1 \times 10^5$ *N/mm*^{2} *and bulk modulus of water* $= 2.17 \times 10^3$ *N/mm*^{2}.

Solution. Let μ = Poisson's ratio.

For the tube : $e_v = e_0 + 2e$

where e_0 and e are the corresponding changes in strain due to the axial load.

Change in longitudinal stress due to axial load

$$f_0 = \frac{20000 - \frac{\pi}{4}(100)^2 \times 0.075}{\pi d t} = \frac{20000 - 589}{\pi \times 100 \times 2} = 30.9\ \text{N/mm}^2\ \text{(tensile)}$$

Change in hoop stress due to fall in internal pressure

$$= f = \frac{pd}{2t} = \frac{0.075 \times 100}{2 \times 2} = 1.88 \text{ N/mm}^2$$

The above change is compressive, because p decreases.

$$\therefore \quad e = \frac{1}{E}(1.88 + 30.9\,\mu)\,(\text{decrease})\,;\; e_0 = \frac{1}{E}(30.9 + 1.88\,\mu) \qquad (\text{increase})$$

Now $\quad e_v = e_0 + 2e = \frac{1}{E}[(30.9 + 1.88\,\mu) - 2(1.88 + 30.9\,\mu)] = \frac{1}{E}[27.14 - 59.92\,\mu] \quad \ldots(1)$

For water :

$$e_v = \frac{p}{K} = \frac{0.075}{2.17 \times 10^3} \qquad \ldots(2)$$

Since no water escapes, we have

Volumetric strain of water = Volumetric strain of tube

$$\therefore \quad \frac{1}{2.1 \times 10^5}(27.14 - 59.92\mu) = \frac{0.075}{2.17 \times 10^3} \quad \text{From which } \mu = \mathbf{0.332.}$$

Example 22.6. *A steel tube having outside and inside diameters of 40 mm and 35 mm is firmly plugged at both the ends, leaving an internal length of 200 mm between the flat ends of the plugs. The plugs are designed so that water can be admitted to the inner space and also so that an axial pull can be applied to the tube. If the tube is subjected to an axial pull of 20 kN and in addition is filled with water at a gauge pressure of 2 N/mm², find the volume of water which will escape from the tube if the axial load is removed and the inner space opened to the atmosphere. Bulk modulus of water = 2.1 × 10³ N/mm² ; E for steel = 2.1 × 10⁵ N/mm². Take 1/m = 0.286.*

Solution. $\quad V = \text{internal volume} = \frac{\pi}{4}(35)^2 \times 200 = 19.24 \times 10^4 \text{ mm}^3$

Let $\quad P$ = initial pull = 20 kN = 20000 N ; p = initial water pressure = 2 N/mm²

Initially, the longitudinal stress $= f_0 = \frac{1}{A}\left(\frac{\pi}{4}d^2 p + P\right)$

$$= \frac{\frac{\pi}{4}(35)^2 \times 2 + 20000}{\frac{\pi}{4}(40^2 - 35^2)} = \frac{1924 + 20000}{294.4} = 74.5 \text{ N/mm}^2$$

and $\quad$ Hoop stress $= f = \frac{pd}{2t} = \frac{2 \times 35}{2(20 - 17.5)} = 14 \text{ N/mm}^2$ (tensile)

Transverse compressive stress = water pressure = 2 N/mm²

Longitudinal strain, $\quad e_0 = \frac{1}{E}(74.5 - 0.286 \times 14 + 2 \times 0.286) = \frac{71.07}{E}$

Hoop strain, $\quad e = \frac{1}{E}[14 - 0.286 \times 74.5 + 0.286 \times 2] = -\frac{6.74}{E}$

$\therefore$ Volumetric stain, $e_v = e_0 + 2e = \frac{71.07}{E} - \frac{2 \times 6.74}{E} = \frac{57.59}{E}$ (increase)

δV = increase is volume of tube caused by water pressure and external force on tube

$$= \frac{57.59}{2.1 \times 10^5} \times 19.24 \times 10^4 = 52.8 \text{ mm}^3$$

$\delta V'$ = compression of water under compressive stress p

$$= \frac{pV}{K} = \frac{2 \times 19.24 \times 10^4}{2.1 \times 10^3} = 183.2 \text{ mm}^3$$

$\therefore$ Volume of water escaping when both external load and internal pressure is removed

$$= \delta V + \delta V' = 52.8 + 183.2 = \mathbf{236\ mm^3}.$$

22.5. THIN SPHERICAL SHELL : INTERNAL PRESSURE

Let us now take a thin spherical shell of diameter d and thickness t, subjected to an internal pressure p. Consider the equilibrium of a hemisphere and its contents as shown in Fig. 22.2. The fluid pressure causes an upward force P given by

$$P = p \times \text{projected area} = p\left(\frac{\pi}{4} d^2\right)$$

If f = hoop or tangential stress of resistance, we have for equilibrium

$$f \pi d t = P = p\left(\frac{\pi}{4} d^2\right) \quad \text{or} \quad f = \frac{pd}{4t} \qquad \text{...(22.11)}$$

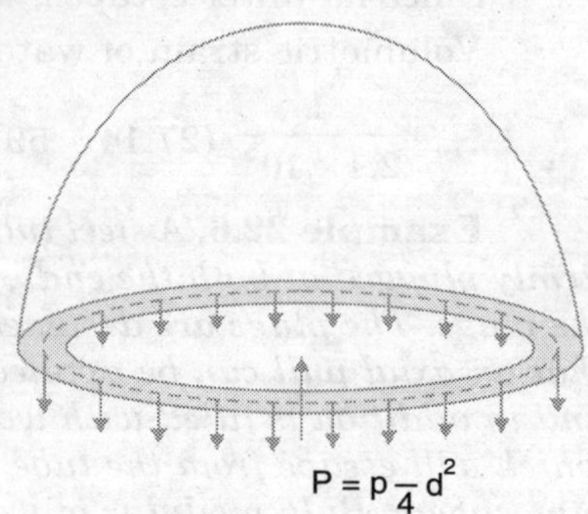

Fig. 22.2

However, if η = efficiency of joint, $f = \dfrac{pd}{4\eta t}$.

The stress f is the same in every direction tangential to the spherical shell. Thus,

$$f_0 = f = \frac{pd}{4t}.$$

At any point in the material, the three principal stresses are :

(1) $f = \dfrac{pd}{4t}$ (tensile) ; (2) $f_0 = f = \dfrac{pd}{4t}$ (tensile) and (3) p (compressive).

The third principal stress p is generally negligible. The shear stress is

$$= \frac{1}{2}\left(\frac{pd}{4t} + p\right) \text{ or} \simeq \frac{pd}{8t}, \text{ if } p \text{ is neglected.}$$

The hoop strain, $$e = \left(\frac{f}{E} - \frac{f}{mE}\right), \text{ if } p \text{ is neglected} = \frac{f}{E}\left(1 - \frac{1}{m}\right) = \frac{pd}{4tE}\left(1 - \frac{1}{m}\right) \qquad \text{...(22.12)}$$

$$\delta d = ed \;;\; V = \frac{\pi}{6} d^3 \text{ and } \delta V = 3.\frac{\pi}{6} d^2.\delta d$$

$$\therefore \quad \frac{\delta V}{V} = 3\frac{\delta d}{d} = 3e = \frac{3pd}{4tE}\left(1 - \frac{1}{m}\right) \qquad \text{...(22.13)}$$

However, if radial stress is considered,

$$e = \frac{pd}{4tE}\left(1-\frac{1}{m}\right)+\frac{p}{mE} \quad \text{and} \quad \frac{\delta V}{V} = e_v = \frac{3pd}{4tE}\left(1-\frac{1}{m}\right)+\frac{3p}{mE}.$$

Example 22.7. *A seamless spherical shell of 1 m internal diameter and 5 mm thick is filled with a fluid under pressure until its volume increases by 200 cubic centimetres. Calculate the pressure exerted by the fluid on the shell. Take E = 2.05 × 10⁵ N/mm² and 1/m = 0.3 for the material.*

Solution. Volume of the shell $= V = \frac{\pi}{6}\, d^3 = \frac{\pi}{6}\,(1000)^3 = 0.524 \times 10^9 \text{ mm}^3$

$$\frac{\delta V}{V} = \frac{200(10^3)}{0.524 \times 10^9} = 3.82 \times 10^{-4}$$

But $$\frac{\delta V}{V} = 3\,e \quad \text{or} \quad e = \frac{3.82 \times 10^{-4}}{3} = 1.273 \times 10^{-4}$$

Hence $$e = \frac{f}{E}\left(1-\frac{1}{m}\right) = 1.273 \times 10^{-4}$$

or $$f = \frac{1.273 \times 10^{-4}}{1-0.3} \times 2.05 \times 10^5 = 37.3 \text{ N/mm}^2$$

Now $$p = \frac{4tf}{d} = \frac{4 \times 5 \times 37.3}{1000} = \mathbf{0.75\ N/mm^2}.$$

Example 22.8. *Compare (a) the maximum tensile stresses and (b) the proportional increases in volume of a thin cylinder and a thin spherical shell having the same internal pressure and the* $\frac{\textit{diameter}}{\textit{thickness}}$ *ratio. Take 1/m = 0.3.*

Solution. Let suffix 1 be for cylinder and suffix 2 be for sphere.

$$f_1 = \frac{pd}{2t};\ f_2 = \frac{pd}{4t}; \quad \text{Hence } \frac{f_1}{f_2} = 2$$

$$e_{v1} = \frac{pd}{2tE}\left(\frac{5}{2}-\frac{2}{m}\right) = \frac{pd}{2tE}(2.5-0.6) = \frac{1.9pd}{2tE}$$

$$e_{v2} = \frac{3pd}{4tE}\left(1-\frac{1}{m}\right) = \frac{3pd}{4tE}(1-0.3) = \frac{2.1pd}{4tE}$$

$$\therefore \quad \frac{e_{v1}}{e_{v2}} = \frac{1.9pd}{2tE} \times \frac{4tE}{2.1pd} = \frac{7.6}{4.2} = \mathbf{1.81.}$$

22.6. THIN CYLINDER OR TUBE REINFORCED BY EXTERNAL WINDINGS

We have seen in the design of thin shells that the thickness of the shell depends on the magnitude of internal pressure. For high internal pressure, such as that in a gun tube, the increase in thickness of the tube will be uneconomical. There is, therefore, necessity of reinforcing the cylinder or tube by either providing a wire winding or providing a compound tube. The theory of compound tube has been discussed separately in § 23.5. Wire or strip of rectangular section is wound under tension, the tension in the wire being theoretically adjusted so as to finally give, under the internal pressure in the cylinder, a constant limiting stress or strain through the winding. The tube will thus be initially in a state of compression.

When the internal pressure is applied, the bursting force is jointly resisted by the shell and wire section. This gives one relation between the stresses set up in the shell and wire section. The other relation is obtained from the compatibility equation that the strain at the common surface of shell and wire is the same. The resultant stress in the shell or wire will then be equal to the stresses due to winding plus tresses due to the internal pressure.

Example 22.9. *A cast iron pipe of 200 mm internal diameter and 13 mm metal thickness is closely wound with a layer of 5 mm diameter steel wire under a tensile stress of 40 N/mm². Calculate the stresses set up in pipe and the steel wire, if water under a pressure of 3N/mm² is admitted into the pipe.*

For C.I., $E = 1.01 \times 10^5$ N/mm² and $1/m = 0.3$; For steel, $E = 2.04 \times 10^5$ N/mm².

Solution. ***Before water is admitted***

Tensile stress in wire = 40 N/mm²

In 1 mm length of pipe, there are $\left(\frac{1}{5} \times 2\right) = 0.4$ sections of wire, and total area of cross-section of wire per 1 mm length $= 0.4 \times \frac{\pi}{4} (5)^2 = 7.85$ mm².

Total compressive force on the pipe = 7.85 × 40 = 314.2 N/mm

Intensity of compressive stress on pipe $= \frac{314.2}{2 \times 13 \times 1} = 12.1$ N/mm².

After water is admitted

Let the resisting stress in the pipe be f_P and that in the wire be f_W.

Bursting force, per mm length of pipe $= pdl = 3 \times 200 \times 1 = 600$ N

Resisting force by the pipe per mm length $= f_P \times 1 \times 2t = f_P \times 1 \times 2 \times 13 = 26 f_P$

Resisting force of wire per mm length $= f_W \times 7.85$

Since total bursting force = total resisting force, we have

$$600 = 26 f_P + 7.85\, f_W \quad \ldots(1)$$

Also, from compatibility equation, $e_P = e_W$

But
$$e_P = \frac{f_P}{E_P} - \frac{1}{m}\frac{pd}{4t}\cdot\frac{1}{E_P} = \frac{1}{E_P}\left\{f_P - \frac{0.3 \times 3 \times 200}{4 \times 13}\right\} = \frac{f_P - 3.46}{E_P}$$

and
$$e_W = f_W / E_W$$

Hence
$$\frac{f_W}{E_W} = \frac{f_P - 3.46}{E_P}$$

or
$$f_W = \frac{2.04}{1.01}[f_P - 3.46] = 2.02\,(f_P - 3.46) \quad \ldots(2)$$

Substituting the value of f_W in (1), we get

$600 = 26 f_P + 15.86\, f_P - 54.86$ From which $f_P = 15.64$ N/mm² (tensile)

and $f_W = 2.02\,(15.64 - 3.46) = 24.6$ N/mm² (tensile)

Hence final stress in pipe = 15.64 – 12.1 = **3.54 N/mm²** (tensile) and final stress in wire = 24.6 + 40 = **64.6 N/mm²** (tensile).

Example 22.10. *A copper tube of outside diameter 38 mm and inside diameter 35.5 mm, is closely wound with steel wire 0.75 mm diameter. Estimate the tension at which the wire must have been wound if an internal gauge pressure of 20 kg / cm² produce a tensile circumferential stress of 70 kg / cm² in the copper tube. E for steel = 1.6 E for copper.*

Solution. Suffixes T and W refer to tube and wire respectively. Since $1/m$ is not given, it is assumed that the longitudinal expansion of the tube is constrained.

Before water is admitted :

Let the tensile stress in wire be f kg/cm².

In 1 cm length of the pipe, there are $\left(\frac{1}{0.075} \times 2\right)$ sections of wire, and the total area of

cross-section of wire per 1 cm length of wire $= \frac{2}{0.075} \times \frac{\pi}{4} (0.075)^2 = 0.118$ cm².

Total compressive force on the pipe $= 0.118\, f$ kg/cm²

Initial compressive stress on pipe $= \frac{0.118\, f}{(3.8 - 3.55)} = \frac{0.118\, f}{0.25} = 0.472\, f$ kg/cm² (comp.)

After water is admitted :

Let the resisting stress in the pipe be f_P and that in the wire be f_W.

Bursting force, per cm length of pipe $= pdl = 20 \times 3.55 \times 1 = 71$ kg-wt

Resisting force, per cm length of pipe $= f_T \times l \times 2t = f_T \times 1 \times (3.8 - 3.55) = 0.25\, f_T$

Resisting force per cm length of wire $= f_W\, (0.118)$

Hence, we have

$$71 = 0.25\, f_T + 0.118\, f_W \qquad \text{...(1)}$$

Also, from compatibility equation, $e_T = e_W$

Now, $e_T = f_T/E_T$ and $e_W = f_W/E_W$

The longitudinal stress has not been taken into account, since $1/m$ has not been given.

Hence,

$$\frac{f_T}{E_T} = \frac{f_W}{E_W} \quad \text{or} \quad f_W = f_T \frac{E_W}{E_T} = 1.6\, f_T \qquad \text{...(2)}$$

Substituting the value in (1), we get

$$f_T = \frac{71}{0.25 + (0.118 \times 1.6)} = \frac{71}{0.25 + 0.189} = 161.5 \text{ kg/cm}^2$$

and

$$f_W = 1.6 \times 161.5 = 259 \text{ kg/cm}^2.$$

Hence, the final circumferential stress in tube = $161.6 - 0.472\, f$.

But this has been given equal to 70 kg/cm²

$\therefore$ $161.5 - 0.472\, f = 70$ from which $f =$ **193.5 kg/cm²**

Hence the wire must be wound at a tensile stress of 193.5 kg/cm².

22.7. CYLINDRICAL SHELL WITH HEMISPHERICAL ENDS

Many times, hemispherical ends are used in cylindrical shells, to minimise the bending stresses in the cylinder when the ends are flat. Let t_1 be the thickness of the cylindrical shell and t_2 be the thickness of the hemispherical ends. *The thicknesses t_1 and t_2 are proportioned so that the radial expansion is the same in both cylinder and the hemisphere.* By doing so, the bending stresses are eliminated at the junction of the two parts. Let p be the internal pressure.

Hoop stress in cylindrical portion

$$f = \frac{pd}{2t_1}$$

Longitudinal stress in cylindrical portion, $f_0 = \dfrac{pd}{4t_1}$

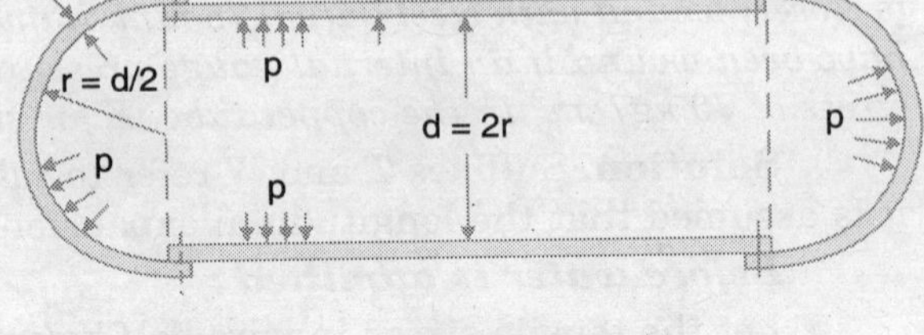

Fig. 22.3

∴ Hoop strain,

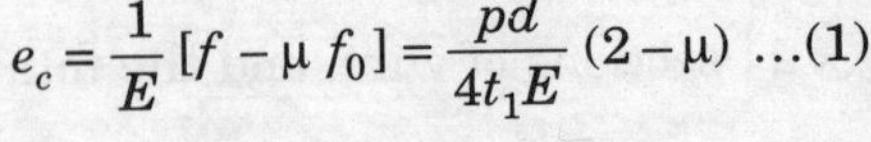

$$e_c = \frac{1}{E}[f - \mu f_0] = \frac{pd}{4t_1 E}(2 - \mu) \quad ...(1)$$

Hoop stress in spherical ends, $f = \dfrac{pd}{4t_2}$

∴ Hoop strain $\qquad e_s = \dfrac{1}{E}(f - \mu f) = \dfrac{pd}{4t_2 E}(1 - \mu) \qquad ...(2)$

For no distortion of the junction under pressure, $e_c = e_s$

$$\therefore \quad \frac{2-\mu}{t_1} = \frac{1-\mu}{t_2} \quad \text{or} \quad \frac{t_1}{t_2} = \frac{2-\mu}{1-\mu} \qquad ...(22.14)$$

For most metals, μ is approximately = 0.3.

$$\therefore \quad \frac{t_1}{t_2} = \frac{2-0.3}{1-0.3} = \frac{1.7}{0.7} = \mathbf{\frac{17}{7}}. \qquad ...[22.14\ (a)]$$

22.8. ADDITIONAL EXAMPLES USEFUL FOR COMPETITIVE EXAMINATIONS

Example 22.11. *A non-ferrous metal tube having a bore of 32 mm and a wall thickness of 1.6 mm has plugged ends. The effective length of the tube between the plugs is 500 mm and an internal fluid pressure 2 N/mm² is applied. An axial pull of 2000 N is also applied externally to the plugs. Determine when the forces are acting (a) the change in the internal diameter of the tube (b) change in the length. Take E = 104500 N/mm² and Poisson's ratio = 0.35.*

Solution. Given: $\quad d = 32$ mm; $t = 1.6$ mm; $L = 500$ mm; $p = 2$ N/mm² ; $P = 2000$ N

Method 1 : Method of super-position : *Both the effects are evaluated separately.*

(a) Strains due to internal pressure

$$f = \frac{pd}{2t} = \frac{2 \times 32}{2 \times 1.6} = 20 \text{ N/mm}^2 \text{ (tensile)} ; f_0 = \frac{pd}{4t} = \frac{f}{2} = 10 \text{ N/mm}^2 \text{ (tensile)}$$

$$e = \frac{1}{E}\left(f - \frac{f_0}{m}\right) = \frac{1}{104500}(20 - 10 \times 0.35) = \mathbf{1.579 \times 10^{-4}} \text{ (tensile)}$$

$$e_0 = \frac{1}{E}\left(f_0 - \frac{f}{m}\right) = \frac{1}{104500}(10 - 20 \times 0.35) = \mathbf{0.287 \times 10^{-4}} \text{ (tensile)}$$

(b) Strains due to external load

$$f_0' = \frac{P}{A} = \frac{2000}{\pi d t} = \frac{2000}{\pi(32)(1.6)} = 12.43 \text{ N/mm}^2$$

$$\therefore \quad e_0' = \frac{f_0'}{E} = \frac{12.43}{104500} = 1.189 \times 10^{-4} \text{ (tensile)}$$

$$e' = -\frac{f_0'}{m\,E} = -0.35 \times \frac{12.43}{104500} = -0.416 \times 10^{-4} \text{ (i.e., compressive)}$$

(c) *By super-position* $e_f = e + e' = (1.579 - 0.416)\,10^{-4} = 1.163 \times 10^{-4}$ (tensile)

$$\therefore \quad \delta d = e_f \,.\, d = 1.163 \times 10^{-4} \times 32 = \mathbf{0.00372\ mm} \text{ (increase)}$$

$$e_{of} = e_0 + e_0' = (0.287 + 1.189)\,10^{-4} = 1.476 \times 10^{-4} \text{ (tensile)}$$

$$\therefore \quad \delta L = e_{of} \,.\, L = 1.476 \times 10^{-4} \times 500 = \mathbf{0.0738\ mm} \text{ (increase)}$$

Method 2 : Direct Method

Since the hoop stress is not affected by the presence of axial load, $f = \dfrac{pd}{2t} = \dfrac{2 \times 32}{2 \times 1.6}$ $= 20$ N/mm^2 (tension). The axial stress is affected by the presence of axial load, and is not one half of the hoop stress. Consider the equilibrium of one of the plugs.

$$f_0(\pi\, d\, t) = p\frac{\pi}{4}d^2 + 2000 \text{ ; or } f_0\,(\pi \times 32 \times 1.6) = 2 \times \frac{\pi}{4}\,(32)^2 + 2000$$

From which $\quad f_0 = 22.43$ N/mm^2 (tension)

$$\therefore \quad e = \frac{1}{E}\left(f - \frac{f_0}{m}\right) = \frac{1}{104500}\,(20 - 0.35 \times 22.43) = 1.163 \times 10^{-4}$$

and $\quad \delta d = e\,.\,d = 1.163 \times 10^{-4} \times 32 = \mathbf{0.00372\ mm}$ (increase)

Also, $\quad e_0 = \dfrac{1}{E}\left(f_0 - \dfrac{f}{m}\right) = \dfrac{1}{104500}\,(22.43 - 0.35 \times 20) = 1.476 \times 10^{-4}$

and $\quad \delta L = e_0\,.\,L = 1.476 \times 10^{-4} \times 500 = \mathbf{0.0738\ mm}$ (increase).

Example 22.12. *A steel tube 1.27 mm thick and 100 mm internal diameter is plugged at each end to form a closed cylinder with internal length of 250 mm. The tube is completely filled with oil and subjected to a compressive load of 45 kN. Find (a) pressure in N/mm² produced in the oil, (b) the resulting circumferential stress in N/mm² in the tube wall. Take E = 2 × 10⁵ N/mm² and 1/m = 0.28 for steel and K = 2667 N/mm² for oil.*

Solution. Given : $\quad d = 100$ mm ; $t = 1.27$ mm; $L = 250$ mm; $P = 45$ kN.

Let $\quad p$ = internal pressure, in N/mm^2

f = circumferential stress, in N/mm^2

f_0 = longitudinal stress, in N/mm^2 (compressive)

Consider the equilibrium of the plugs

$$f_0(\pi\, d\, t) + p\frac{\pi}{4}d^2 = P = 45000$$

or $\quad f_0\,(\pi \times 100 \times 1.27) + p\dfrac{\pi}{4}\,(100)^2 = 45000$

or $\quad f_0 = 112.787 - 19.685\,p$ (compression) $\quad$ ...(1)

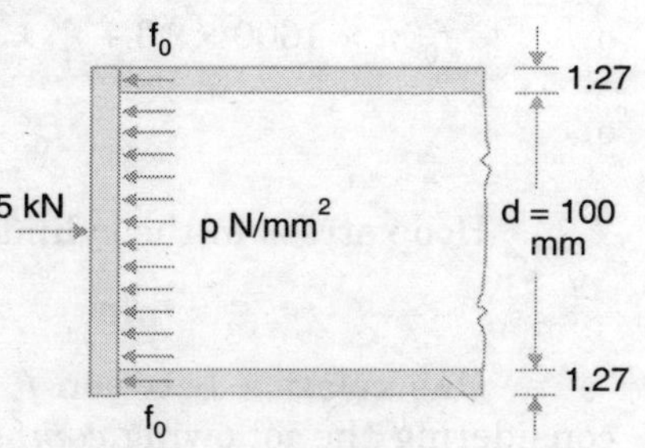

Fig. 22.4

Also, $\quad f = \dfrac{pd}{2t} = p\left(\dfrac{100}{2 \times 1.27}\right)$

$= 39.37\,p$ (tension) $\quad$ ...(2)

Now $\frac{\delta V}{V} = 2e + e_0$

The volumentric strain is negative, *i.e.*, compressive.

$$\therefore \quad -\frac{p}{K} = \frac{2}{E}\left[f - \frac{f_0}{m}\right] + \frac{1}{m}\left(f_0 - \frac{f}{m}\right) \quad \text{or} \quad -\frac{pE}{K} = 2\left(f - \frac{f_0}{m}\right) + \left(f_0 - \frac{f}{m}\right)$$

or $$-p \times \frac{2 \times 10^5}{2667} = 2(f - 0.28 f_0) + (f_0 - 0.28 f) = 1.72 f + 0.44 f_0 \quad ...(3)$$

Substituting the values of f and f_0 from (2) and (1), we get

$$-74.99\,p = 1.72\,(39.37\,p) - 0.44\,(112.787 - 19.685\,p) = 76.378\,p - 49.626$$

From which p = **0.328 N/mm²** ; $\therefore$ $f = 39.37\,p$ = **12.91 N/mm²** (tension)

Example 22.13. *A cylindrical tank is 1.6 m diameter, 2 m long and 10 mm thick. Its ends are flat and are joined by nine tie bars, each 30 mm diameter equally spaced. If the tie bars are initially stressed to 40 N/mm² and the tank is filled with water, find (a) the increase in capacity when the pressure is raised to 1.5 N/mm², and (b) the final stress in the tie bars. Take E = 2 × 10⁵ N/mm² and μ = 0.29.*

Solution. Given : d = 1.6 m; L = 2 m; t = 10 mm ; p = 1.5 N/mm²; φ = 30 mm

Initial Condition: The tie bars are under tensile stress (f_t) of 40 N/mm². Let f_0 be the longitudinal *compressive* stress in the cylinder walls (Fig. 22.5 *a*).

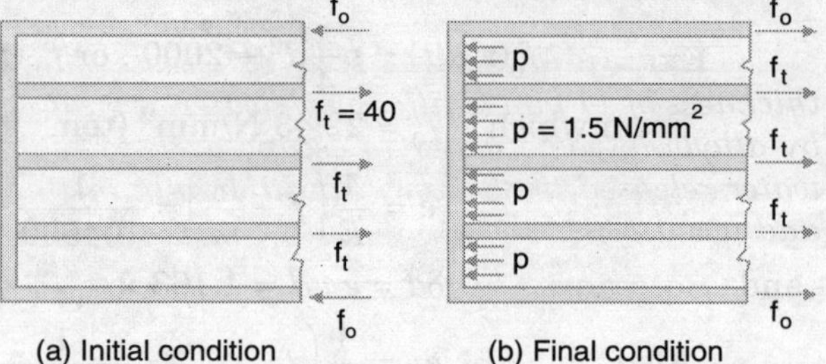

Fig. 22.5

From equilibrium conditions,

$$f_0\,(\pi\,d\,t) = f_t\,n\frac{\pi}{4}\varphi^2$$

or $$f_0\,(\pi \times 1600 \times 10) = 40 \times 9 \times \frac{\pi}{4}\,(30)^2 \quad \text{From which} \quad f_0 = 5.063 \text{ N/mm}^2$$

There is no hoop stress initially.

Final condition (Fig. 22.5 *b*)

Let f_t' be the final tensile stress in tie bars and f_0' be the final *tensile* longitudinal stress in the cylinder. From equilibrium conditions

$$f_0'\,\pi\,d\,t + f_t'\,n\,\frac{\pi}{4}\varphi^2 = p\left(\frac{\pi}{4}d^2 - n\,\frac{\pi}{4}\varphi^2\right)$$

or $$f_0'\,\pi \times 1600 \times 10 + f_t' \times 9 \times \frac{\pi}{4}\,(30)^2 = 1.5\left[\frac{\pi}{4}(1600)^2 - 9 \times \frac{\pi}{4}(30)^2\right]$$

or $$f_0' + 0.1266\,f_t' = 59.81 \quad ...(a)$$

Hoop stress in the cylinder $= \frac{pd}{2t} = \frac{1.5 \times 1600}{2 \times 10} = 120$ N/mm²

(This hoop stress is not affected by the tie bars)

One relation between f_t' and f_0' is given by Eq. (*a*). Another relation is obtained by considering the following *compatibility equation*:

The increase in Longitudinal strain for the bars = Increase in Longitudinal strain for Cylinder

or $$e_t = e_{oc}$$

where e_t = increase in longitudinal strain of tie bars = $\frac{f_t' - 40}{E}$

e_{oc} = increase in longitudinal strain of cylinder = final strain – initial strain

$$= \frac{1}{E}(f_0' - 0.29 \times 120) - \frac{1}{E}(-5.063)$$

$$\therefore \quad \frac{f_t' - 40}{E} = \frac{1}{E}(f_0' - 34.8 + 5.063) \quad \text{or} \quad f_t' = f_0' + 10.263 \qquad ...(b)$$

From Eqs (*a*) and (*b*), we get f_0' = 51.94 N/mm² and f_t' = **62.20 N/mm²**

Increase in capacity = (2 × increase in hoop strain + increase in longitudinal strain) × volume

$$= (2\,e + e_0)\,V = \frac{1}{2 \times 10^5}\,[2\,(120 - 0.29 \times 51.94 - 0.29 \times 5.063) + 62.20 - 40]$$

$$\times \frac{\pi}{4}(1600)^2 \times 2000 = \mathbf{4.61 \times 10^6\ mm^3.}$$

Example 22.14. *A spherical vessel having an inside diameter of 760 mm and a wall thickness of 11 mm is filled with water at gauge pressure of 5.6 N/mm². The pressure is lowered by allowing some water to escape and, to reduce the pressure to atmospheric, the volume of water released is 885 cm³. Find the bulk modulus of water. Take E for steel = 2 × 10⁵ N/mm² and μ = 0.286.*

Solution. Given : d = 760 mm; t = 11 mm; p = 5.6 N/mm²

$$V = \frac{4}{3}\pi r^3 = \frac{4}{3}\pi\left(\frac{760}{2}\right)^3 = 229.85 \times 10^6\ \text{mm}^3$$

Contraction of vessel

$$= \delta V_1 = \frac{3\,p\,d}{4\,t\,E}\left(1 - \frac{1}{m}\right)V = \frac{3 \times 5.6 \times 760}{4 \times 11 \times 2 \times 10^5}(1 - 0.286) \times 229.85 \times 10^6 = 238110\ \text{mm}^3$$

Expansion of water $= \delta V_2 = \frac{pV}{K} = \frac{5.6 \times 229.85 \times 10^6}{K} = \frac{1287.2 \times 10^6}{K}$

∴ Total $\delta V = 885 \times 10^3 = \delta V_1 + \delta V_2 = 238110 + \frac{1287.2 \times 10^6}{K}$

From which **K = 1990 N/mm².**

PROBLEMS

1. A spherical shell of 1 m diameter is subjected to an internal pressure of 1.2 N/mm². Taking the maximum allowable stress as 100 N/mm², find the necessary thickness of plate. Take the joint efficiency at 75%.
2. Derive an expression for the proportional increase in the capacity of a thin cylindrical shell subjected to an internal fluid pressure.

   Calculate the increase in the volume enclosed by a boiler shell, 3 m long and 1 m in diameter, when it is subjected to an internal pressure of 1.6 N/mm². The wall thickness is such that the maximum tensile stress in the shell is 20 N/mm² under this pressure. Take E = 2.05 × 10⁵ N/mm² and $1/m$ = 0.28.

3. **Derive a formula for the proportional increase of capacity of a thin spherical shell due to an internal pressure. Calculate the increase in volume of spherical shell, 1 m diameter and 12 mm thick, when it is subjected to an internal pressure of 1.6 N/mm². Take $E = 2.05 \times 10^5$ N/mm² and $1/m = 0.28$.**

4. **Show that when a thin walled spherical vessel of dimeter d and thickness t is subjected to an internal pressure p, the increase in volume is equal to $\dfrac{\pi p d^4}{8tE}\left(1-\dfrac{1}{m}\right)$.**

5. **A copper tube of inside diameter 5 cm and outside diameter 5.3 cm is closely wound with steel wire of diameter 0.7 mm. Find the tension at which the wire must be wound on the tube if an internal pressure of 1.3 N/mm² is required before the copper is subjected to tensile stress, the tube being free to expand or contract axially.**

For steel, $E = 2.04 \times 10^5$ N/mm² ; For copper, $E = 1.225 \times 10^5$ N/mm² ; and $1/m = 0.33$

ANSWERS

1. 0.4 cm. **2.** 446 cm³

3. 183 cm³. **4.** 408 kg/cm² ; 40.8 N/mm².

23

Thick Cylindrical and Spherical Shells

23.1. LAME'S THEORY

In the analysis of thin cylinders we had assumed that the circumferential stress f is constant throughout the thickness of the shell and that the radial stress p is negligible in comparison to f and f_0. In thick cylinder, however the circumferential stress no longer remains constant but varies along the thickness and the radial pressure p is no longer negligible since a thick cylinder is required to have heavy internal pressure. The problem of thick cylinder is somewhat of complex nature and was solved by Lame on the following assumptions :

(1) The material is homogeneous and isotropic.

(2) Plane sections of the cylinder perpendicular to the longitudinal axis remain plane under the pressure. *This assumption is nearly true at a considerable distance from the ends.* As a consequence of this assumption the longitudinal strain is the same at all points in the cylinder wall, *i.e.,* it is independent of the radius.

Hence, to satisfy the requirements of uniform longitudinal strain, we have

$$e_0 = \frac{1}{E}\left(f_0 - \frac{f_x}{m} + \frac{p_x}{m}\right) = \text{constant}$$

or

$$f_x - p_x = \text{constant} = 2A \text{ (say)} \qquad \ldots(23.1)$$

In the above expression, f_x is the circumferential stress (assumed tensile) and p_x is the radial pressure at any radius x (see Fig. 23.1).

Equation 23.1 is an important assumption on which thick cylinder theory is based.

Fig. 23.1 (a) shows a thick cylinder subjected to internal and external radial pressures. Consider an angular ring of the cylinder, of internal radius x and thickness δx. Let the internal radial pressure on this ring be p_x and external pressure $(p_x + \delta p_x)$. On any small element of this ring, f_x is circumferential stress. The condition for equilibrium of one-half of this ring are similar to those in the case of thin cylinder. Thus, we have

The bursting force = $(p_x 2xl) - (p_x + \delta p_x)\, 2(x + \delta x)l = 2l[-\,p_x\delta x - x\delta p_x - \delta x.\delta p_x]$

$= -\,2l(p_x\delta x + x\delta p_x)$, neglecting the products of small quantities.

The resisting force = $2f_x l\delta x$

For equilibrium, we have

$2lf_x\delta x = -\,2l(p_x\delta x + x\delta p_x)$

or $$f_x = -\,p_x - x\,\frac{\delta p_x}{\delta x}$$

In the limit when thickness of the element is reduced infinitely.

$$f_x + p_x + x\,\frac{dp_x}{dx} = 0 \quad ...(i)$$

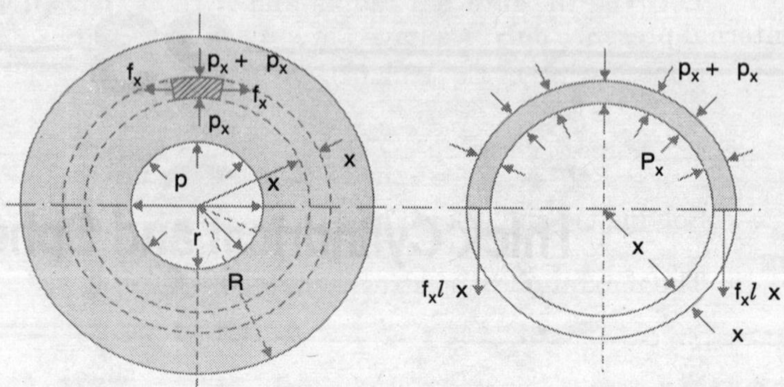

Fig. 23.1

Another relation is obtained from the assumption that the longitudinal strain is independent of x. Thus from equation 23.1, we have

$$f_x - p_x = 2A \qquad ...(ii)$$

Substituting the value of f_x in (i),

$$(p_x + 2A) + p_x + x\frac{dp_x}{dx} = 0 \quad \text{or} \quad \frac{dp_x}{dx} = -\frac{2(p_x + A)}{x}$$

or $$\frac{dp_x}{(p_x + A)} = -\,\frac{2dx}{x}$$

Integrating, $\log_e (p_x + A) = -\log_e x^2 + \log_e B$, where $\log_e B$ is a constant of integration.

$$\therefore \quad \log_e (p_x + A) = \log_e \frac{B}{x^2} \quad \text{or} \quad p_x = \frac{B}{x^2} - A$$

Hence, from (ii), $$f_x = \frac{B}{x^2} - A + 2A = \frac{B}{x^2} + A$$

Thus, we have, Lame's equations :

$$p_x = \frac{B}{x^2} - A \quad ...[23.2\ (a)] \quad \text{and} \quad f_x = \frac{B}{x^2} + A \quad ...[23.2\ (b)]$$

The constants A and B can be evaluated from the known internal and external radial pressure and radius. *It should be remembered that in the above equation, p_x is compressive and f_x is tensile.*

Longitudinal Stress (f_0) : The longitudinal stress f_0 can be found in the same manner, as found for the thin cylinder.

Thus, bursting force = $p_r\, \pi\, r^2$; Resisting force = $f_0\, \pi(R^2 - r^2)$

Equating the two, we get $f_0\, \pi\, (R^2 - r^2) = p_r\, \pi\, r^2$

From which $$f_0 = p_r\,\frac{r^2}{R^2 - r^2} \qquad ...[23.2\ (c)]$$

23.2. SPECIAL CASES

(1) INTERNAL PRESSURE = p_r ; EXTERNAL PRESSURE ZERO

Let r be the internal radius and R the external radius. If a cylinder is subjected to internal pressure only, we have : at $x = r, p = p_r$, and at $x = R, p = 0$.

Substituting in Eq. 23.2 (a), we get

$$p_r = \frac{B}{r^2} - A \quad \text{...(1)} \quad \text{and} \quad p_R = \frac{B}{R^2} - A = 0 \quad \text{...(2)}$$

Solving (1) and (2), we have $A = p_r \dfrac{r^2}{R^2 - r^2}$ and $B = p_r \dfrac{r^2 R^2}{R^2 - r^2}$

The maximum hoop stress will occur at $x = r$.

$$\therefore \quad f_r = p_r \frac{r^2 R^2}{(R^2 - r^2) r^2} + p_r \frac{r^2}{R^2 - r^2} = p_r \left(\frac{R^2 + r^2}{R^2 - r^2} \right) \quad \text{...(23.3)}$$

(2) EXTERNAL PRESSURE = p_R AND INTERNAL PRESSURE ZERO

At $x = R, p = p_R$ and at $x = r, p = p_r = 0$

Substituting in Eq. 23.2 (a), we have

$$p_R = \frac{B}{R^2} - A \text{ and } p_r = 0 = \frac{B}{r^2} - A \quad \text{...(1)}$$

Solving, we get $\quad B = -p_R \dfrac{R^2 r^2}{R^2 - r^2}$ and $A = -p_R \dfrac{R^2}{R^2 - r^2} \quad$ ...(2)

Hence the maximum hoop stress at radius r is

$$f_r = -p_R \left(\frac{2R^2}{R^2 - r^2} \right) \quad \text{...(23.4)}$$

The negative sign indicates that the hoop stress will be *compressive*.

(3) EXTERNAL PRESSURE = p_R AND INTERNAL PRESSURE = p_r

At $x = R, p = p_R$ and at $x = r, p = p_r$

Substituting in Eq. 23.2 (a), we have

$$p_R = \frac{B}{R^2} - A \quad \text{and} \quad p_r = \frac{B}{r^2} - A$$

Solving, we get $B = (p_r - p_R) \dfrac{r^2 R^2}{R^2 - r^2}$ and $A = \dfrac{p_r r^2 - p_R \,.\, R^2}{R^2 - r^2}$

The maximum hoop stress occurs at radius r and is given by

$$f_r = \frac{(p_r - p_R) R^2}{R^2 - r^2} + \frac{p_r r^2 - p_R R^2}{R^2 - r^2} = \frac{p_r R^2 - p_R R^2 + p_r r^2 - p_R R^2}{R^2 - r^2} = \frac{p_r (R^2 + r^2) - 2 p_R R^2}{R^2 - r^2}$$

...(23.5)

(4) SOLID CIRCULAR SHAFT HAVING EXTERNAL RADIAL PRESSURE

Let a solid rod or shaft of radius R be subjected to an external pressure p as shown in Fig. 23.2 (a).

$$p_x = \frac{B}{x^2} - A \quad \text{and} \quad f_x = \frac{B}{x^2} + A.$$

At $x = 0$, p_x and f_x become infinite, which suggest that $B = 0$. Hence the Lame equations reduce to

$$p_x = -A = p.$$

Thus, the intensity of radial pressure is constant everywhere and is equal to external pressure.

Again $\quad f_x = 0 + A = -p_x = -p$

The minus sign indicates that the pressure is compressive.

Thus the intensity of hoop pressure is constant everywhere and is compressive, its value being equal to the external pressure.

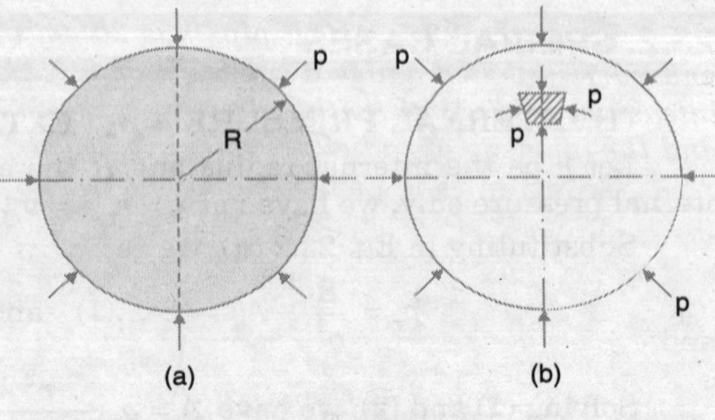

Fig. 23.2

The direction of stresses at any point in the material have been shown in Fig. 23.2 (*b*).

23.3. GRAPHICAL METHOD : LAME'S LINES

We have seen above that if the conditions of pressure etc., are known constants A and B can be evaluated algebraically and f_x and p_x found. The procedure is sometimes laborious. A more convenient way of finding the expressions for f_x and p_x is from the construction of Lame line (Fig. 23.3).

On any axis of reference, mark the points at distances $1/R^2$ and $1/r^2$ on either side of the point O. Taking the first case in which the cylinder is subjected to internal pressure only, we have $p = p_r$ at $x = r$. Hence, at the point G distant $1/r^2$ from O, erect a perpendicular GH of value p_r above the line (plotting compressive stress above the line and tensile stress below the line). Again at the point F distant $1/R^2$ from O, erect a perpendicular of zero value (since $p_R = 0$). Join H and F and prolong the line to the other side of O. Thus, the slope of the Lame line HF is known. At the point G' distant $1/r^2$, erect perpendicular $G'H'$ to cut the Lame line in H'. The ordinate $G'H'$ is the hoop stress f_r (tensile) at the radius r and is to be found out.

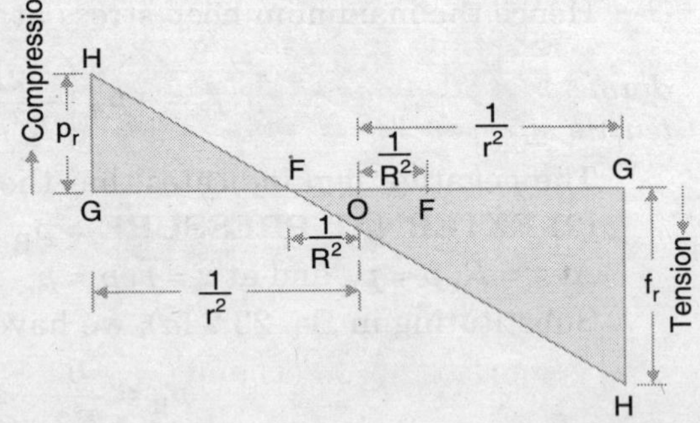

Fig. 23.3

By inspection of similar triangles, we have

$$\frac{HG}{GF} = \frac{H'G'}{G'F} \quad \text{or} \quad \frac{p_r}{\frac{1}{r^2} - \frac{1}{R^2}} = \frac{f_r}{\frac{1}{r^2} + \frac{1}{R^2}}$$

From which, $$f_r = p_r \frac{\frac{1}{R^2} + \frac{1}{r^2}}{\frac{1}{r^2} - \frac{1}{R^2}} = p_r \frac{R^2 + r^2}{R^2 - r^2},$$ which is same as equation 23.3.

The Lame line can similarly be constructed for other cases also.

Example 23.1. *A C.I. pipe has 20 cm internal diameter and 50 mm metal thickness, and carries water under a pressure of 5 N/mm². Calculate the maximum and minimum intensities of circumferential stress and sketch distribution of circumferential stress intensity and the intensity of radial pressure across the section.*

Solution. The Lame's equations are

$$p_x = \frac{B}{x^2} - A \qquad \text{...(1)}$$

and

$$f_x = \frac{B}{x^2} + A \qquad \text{...(2)}$$

At $x = 100$, $p_x = 5$ N/mm² ;

and at $x = 150$, $p_x = 0$.

Substituting in (1), we get

$$5 = \frac{B}{(100)^2} - A \quad \text{and} \quad 0 = \frac{B}{(150)^2} - A$$

From which $B = 90000$ and $A = 4$

Substituting in (1) and (2), we get

$$p_x = \frac{90000}{x^2} - 4 \quad \text{and} \quad f_x = \frac{90000}{x^2} + 4$$

At $x = 100$ mm, $p = \frac{90000}{10000} - 4 = 5$ N/mm² and $f = \frac{90000}{10000} + 4 = 13$ N/mm²

At $x = 150$ mm, $p = 0$ and $f = 8$ N/mm²

The variations of p and f are parabolic and have been shown in Fig. 23.4.

Fig. 23.4

Example 23.2. *Calculate the thickness of metal necessary for a cylindrical shell of internal diameter of 80 mm to withstand an internal pressure of 25 N/mm², if the maximum permissible tensile stress is 125 N/mm².*

Solution. The Lame's equations are

$$p_x = \frac{B}{x^2} - A \quad \text{...(1)} \quad \text{and} \quad f_x = \frac{B}{x^2} + A \quad \text{...(2)}$$

At $x = 40$ mm, $p_x = 25$ N/mm² and $f_{max} = 125$ N/mm²

Substituting in (1) and (2), we get

$$25 = \frac{B}{1600} - A \quad \text{and} \quad 125 = \frac{B}{1600} + A$$

Solving these, we get $B = (125 + 25) \times \frac{1600}{2} = 120000$ and $A = 50$

$$\therefore \quad p_x = \frac{120000}{x^2} - 50$$

At the outer side, $x = R$ and $p = 0$ Hence $0 = \frac{120000}{R^2} - 50$

or

$$R^2 = \frac{120000}{50} = 2400 \quad \text{or} \quad R \simeq 49 \text{ mm}$$

$\therefore$ Thickness of metal = 49 – 40 = **9 mm.**

Example 23.3. *Find the ratio of thickness to internal diameter of a tube subjected to internal pressure when the pressure is 5/8 of the value of the maximum permissible circumferential stress.*

Find the increase in internal diameter of such a tube of 100 mm internal diameter when the internal pressure is 90 N/mm². Take E = 2.05 × 10⁵ N/mm² and 1/m = 0.29. Neglect the longitudinal strain.

Solution. The Lame line for this case has been shown in Fig. 23.3 from which, we get

$$f_r = p_r \frac{R^2 + r^2}{R^2 - r^2}. \quad \text{But} \quad \frac{p_r}{f_r} = \frac{5}{8} \quad \text{(given)}$$

$$\therefore \quad \frac{R^2 + r^2}{R^2 - r^2} = \frac{f_r}{p_r} = \frac{8}{5} \quad \text{or} \quad 5R^2 + 5r^2 = 8R^2 - 8r^2$$

or $$3R^2 = 13r^2 \quad \text{or} \quad \frac{R}{r} = \sqrt{\frac{13}{3}} = 2.08$$

or $$\frac{r+t}{r} = \frac{\frac{d}{2} + t}{d/2} = \frac{d + 2t}{d} = 2.08 \quad \text{or} \quad 2.08d = d + 2t$$

From which $$\frac{t}{d} = \mathbf{0.54.}$$

Now, when $p_r = 90$, $f_r = \frac{8}{5} \times 90 = 144$ N/mm² (tensile)

Ignoring the longitudinal stress, we have

$$e_r = \frac{\delta d}{d} = \frac{1}{E}\left(f_r + \frac{p_r}{m}\right) = \frac{144 + 0.29 \times 90}{2.05 \times 10^5} = \frac{170.1}{2.05 \times 10^5}$$

$$\therefore \quad \delta d = \frac{170.1 \times 100}{2.05 \times 10^5} = \mathbf{0.083\ mm.}$$

23.4. DESIGN OF THICK CYLINDRICAL SHELL

We have seen that the maximum values of f and p occur at the inner radius r. Thus at any point at radius r, the three principal stresses are :

(*i*) The radial stress p_r (compressive)

(*ii*) Circumferential stress f_r (tensile)

$$f_r = p_r \frac{R^2 + r^2}{R^2 - r^2} = p_r \frac{K^2 + 1}{K^2 - 1}, \text{ where } K = \frac{R}{r}.$$

(*iii*) The longitudinal stress, $f_0 = \dfrac{p_r r^2}{R^2 - r^2} = \dfrac{p_r}{K^2 - 1}$ (tensile).

Out of three principal stresses, the circumferential stress is the largest. For a given internal pressure, the safe ratio of thickness of wall to bore of the tube can be found on the basis of any of the five theories of elastic failure.

(*i*) **Maximum Principal Stress Criteria**

Let f be the simple stress for elastic failure. Then,

$$p_r\left(\frac{K^2+1}{K^2-1}\right) \le f \qquad \text{...(23.6)}$$

(*ii*) **Maximum Principal Strain Criteria**

Maximum strain, $e_r = \dfrac{f_r}{E} + \dfrac{p_r}{mE} - \dfrac{f_0}{mE}$

$$\therefore \qquad f = e_r \,.\, E \ge f_r + \frac{p_r}{m} - \frac{f_0}{m}$$

or

$$f \ge p_r\left[\frac{K^2+1}{K^2-1} + \frac{1}{m} - \frac{1}{m}\cdot\frac{1}{K^2-1}\right] \ge p_r\left[\frac{K^2+1}{K^2-1} + \frac{1}{m}\cdot\frac{K^2-2}{K^2-1}\right] \qquad \text{...(23.7)}$$

However, if the shell does not carry longitudinal direct stress, we have

$$f = e_r \,.\, E \ge p_r\left(\frac{K^2+1}{K^2-1} + \frac{1}{m}\right) \qquad \text{...(23.8)}$$

(*iii*) **Maximum Shear Stress Criteria**

The maximum shear intensity $= \dfrac{f_r + p_r}{2} = \dfrac{p_r}{2}\left[\dfrac{K^2+1}{K^2-1} + 1\right] = \dfrac{p_r}{2}\,\dfrac{2K^2}{K^2-1}$

Hence

$$\frac{f}{2} \ge \frac{p_r}{2}\,\frac{2K^2}{K^2-1} \quad \text{or} \quad f \ge p_r\,\frac{2K^2}{K^2-1} \qquad \text{...(23.9)}$$

(*iv*) **Maximum Strain Energy.** We know that the strain energy U per unit volume under the three principal stresses are given by

$$U = \frac{1}{2E}\left\{\sigma_1^{\,2} + \sigma_2^{\,2} + \sigma_3^{\,2} - \frac{2}{m}(\sigma_1\sigma_2 + \sigma_2\sigma_3 + \sigma_3\sigma_1)\right\}$$

Substituting $\sigma_1 = f_r = p_r\,\dfrac{K^2+1}{K^2-1}$, $\sigma_2 = p_r$ and $\sigma_3 = f_0 = p_r\,\dfrac{1}{K^2-1}$, we get

$$U = \frac{p_r^{\,2}}{2E}\,\frac{\left\{2\left(1+\frac{1}{m}\right)K^4 + 3\left(1-\frac{2}{m}\right)\right\}}{(K^2-1)^2}$$

The energy for a single tensile stress, $U = \dfrac{f^2}{2E}$.

Hence we get the criteria

$$f \ge p_r\,\frac{\sqrt{2\left(1+\frac{1}{m}\right)K^4 + 3\left(1-\frac{1}{m}\right)}}{(K^2-1)} \qquad \text{...(23.10)}$$

Similarly, it can be shown that if f_0 is neglected, we get

$$f \ge p_r\,\frac{\sqrt{2\left(1+\frac{1}{m}\right)K^4 + 2\left(1-\frac{1}{m}\right)}}{(K^2-1)} \qquad \text{...(23.11)}$$

(*v*) **Maximum Shear Strain Energy.** Substituting the corresponding values σ_1, σ_2 and σ_3 in equation for U, it can be shown that the shear strain energy U per unit volume is

$\frac{p_r^2}{2E} \cdot \frac{3K^4}{(K^2 - 1)}$.

Hence $$f \geq p_r \frac{\sqrt{3K^4}}{(K^2 - 1)} \qquad ...(23.12)$$

However, if f_0 is neglected,

$$f \geq p_r \frac{\sqrt{3K^4 + 1}}{K^2 - 1}. \qquad ...(23.13)$$

Example 23.4. *A thick cylindrical shell is of 200 mm internal diameter and has to withstand an internal pressure of 25 N/mm². Calculate the thickness of metal necessary for the cylinder on the basis of* (i) *the maximum principal stress,* (ii) *the maximum principal strain,* (iii) *the maximum shear, and* (iv) *the maximum strain energy. Neglect the longitudinal direct stress. The permissible tensile stress of the shell is 125 N/mm² and 1/m = 0.3.*

Solution.

(i) Maximum principal stress hypothesis :

From Eq. 23.6

$$p_r \frac{K^2 + 1}{K^2 - 1} \leq f \quad \text{or} \quad \frac{K^2 + 1}{K^2 - 1} \leq \frac{125}{25} \leq 5$$

From which, $K^2 \geq 1.5$ or $K = 1.225$

$\therefore$ $R = K.r = 1.225 \times 100 = 122.5$ mm

$\therefore$ Metal thickness $= R - r = 122.5 - 100 =$ **22.5 mm.**

(ii) Maximum principal strain hypothesis :

From Eq. 23.8,

$$p\left(\frac{K^2 + 1}{K^2 - 1} + \frac{1}{m}\right) \leq f$$

or $$\frac{K^2 + 1}{K^2 - 1} + 0.3 \leq \frac{125}{25} \leq 5 \quad \text{or} \quad \frac{K^2 + 1}{K^2 - 1} = 4.7$$

From which $K^2 = 1.54$ or $K = 1.24$

$\therefore$ $R = K \,.\, r = 1.24 \times 100 = 124$ mm

Metal thickness $= 124 - 100 =$ **24 mm.**

(iii) Maximum shear stress hypothesis :

From Eq. 23.9,

$$\frac{p_r K^2}{K^2 - 1} \leq \frac{f}{2} \quad \text{or} \quad \frac{K^2}{K^2 - 1} \leq \frac{125}{25 \times 2} = 2.5$$

From which $K^2 = 1.67$ or $K = 1.29$

$\therefore$ $R = K.r = 1.29 \times 100 = 129$ mm

$\therefore$ Metal thickness $= 129 - 100 =$ **29 mm.**

(iv) Maximum strain energy hypothesis :

From Eq. 23.11,

$$f^2 \geq \frac{2p_r^2\left[K^4\left(1+\frac{1}{m}\right)+\left(1-\frac{1}{m}\right)\right]}{(K^2-1)^2}$$

or $(125)^2 (K^2 - 1)^2 = 2(25)^2 (1.3K^4 + 0.7)$ or $12.50 (K^4 + 1 - 2K^2) = 1.3K^4 + 0.7$

or $11.20K^4 = 25K^2 - 11.80 = 0$ or $K^4 - 2.23K^2 + 1.055 = 0$

or $(K^2 - 1.11)^2 = 0.175$ From which $K^2 = 1.528$ or $K = 1.235$

$\therefore$ $R = K \, . \, r = 1.235 \times 100 = 123.5$ mm

$\therefore$ Metal thickness = 123.5 – 100 = **23.5 mm.**

23.5. COMPOUND CYLINDERS

In the case of thin cylinders, we have seen that if it is wound with wire under tension, the whole shell will be in a state of initial compression and when the fluid is admitted, the resultant hoop stress will be less. The same can be done in the case of thick cylinders subjected to great internal pressure, so that the maximum hoop stress may be reduced. Another method, which is commonly used for the design of gun tube is to shrink one tube on to the other. By keeping proper initial difference in the common radii, the required shrinkage pressure at the junction can be developed. The initial shrinkage pressure p' will set initial compressive hoop stresses in inner tube and initial tensile hoop stresses in the outer tube. When the fluid is admitted, the internal pressure will be jointly resisted by the internal and external tubes. Due to fluid pressure, tensile hoop stresses will be set in the inner and outer tubes. The final hoop stresses in the tubes will be the algebraic sum of initial stresses and stresses due to fluid pressure. Thus, the final hoop stress will be more or less uniform. Fig. 23.5 shows the distribution of stresses in a compound cylinder.

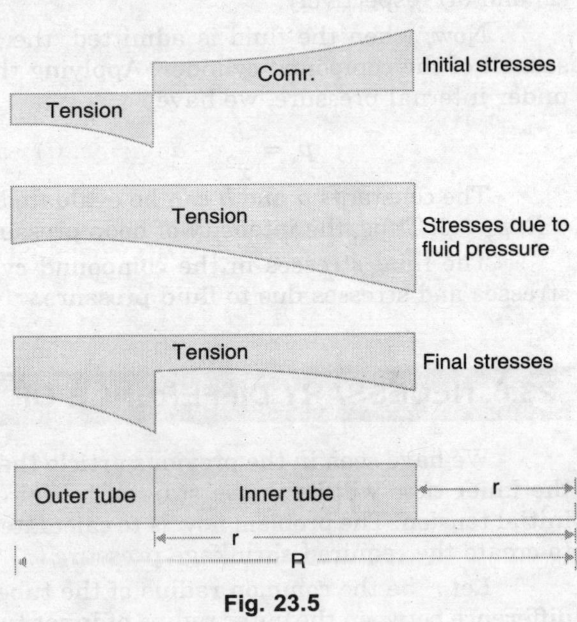

Fig. 23.5

Let r' be the radius at the common surface and p' the shrinkage pressure. The Lame's equations for the inner tube are

$$p_x = \frac{B}{x^2} - A \quad \ldots(1) \quad \text{and} \quad f_x = \frac{B}{x^2} + A \quad \ldots(2)$$

and for the outer tube, they are

$$p_x = \frac{B'}{x^2} - A' \quad ...(3) \quad \text{and} \quad f_x = \frac{B'}{x^2} + A' \quad ...(4)$$

In the above four equations, the four unknowns, A, B, A' and B' can be evaluated by applying the initial conditions, which are as follows :

For inner tube : Inner radius r and outer radius r' :

At $x = r, p = 0$ and at $x = r', p = p'$

$$\text{Hence} \quad \frac{B}{r^2} - A = 0 \quad ...(i) \quad \text{and} \quad \frac{B}{r'^2} - A = p' \quad ...(ii)$$

For outer tube : Inner radius r' and outer radius R :

At $x = R, p = 0$ and at $x = r', p = p'$

$$\text{Hence} \quad \frac{B'}{R^2} - A' = 0 \quad ...(iii) \quad \text{and} \quad \frac{B'}{r'^2} - A' = p' \quad ...(iv)$$

From Eqs. (*i*) to (*iv*), the four constants A, B, A' and B' can be known. Knowing the four constants, the initial hoop stresses in the inner and outer tubes can be found by using equation (2) and (4) respectively.

Now, when the fluid is admitted, the intensity of pressure is resisted jointly by the section of the compound cylinder. Applying the Lame's equations for the compound cylinder under internal pressure, we have

$$p_x = \frac{b}{x^2} - a \quad ...(i) \quad \text{and} \quad f_x = \frac{b}{x^2} + a \quad ...(ii)$$

The constants a and b can be evaluated by the conditions that at $x = r$, $p_x = p_r$ and at $x = R$, $p_x = 0$. Thus, the intensity of hoop pressure due to the fluid pressure can be calculated.

The final stresses in the compound cylinder will be the algebraic sum of the initial stresses and stresses due to fluid pressure.

23.6. NECESSARY DIFFERENCE OF RADII FOR SHRINKAGE

We have seen in the previous article that if p' is the shrinkage pressure at the junction, the inner tube will be in the state of initial compression and the outer tube in the state of initial tension. The problem now is to calculate the necessary difference of radii at the junction to create the required shrinkage pressure.

Let r' be the common radius of the tubes at junction after shrinking on. Let δr_1 be the difference between the outer radius of inner tube and r', and δr_2 be the difference between the inner radius of the outer tube and r'. If $\delta r'$ is the difference in the radii before shrinking on, we have

$$\delta r' = \delta r_1 + \delta r_2 \quad ...(1)$$

For the inner tube, the circumferential strains at the common radius r' is given by

$$\frac{\delta r_1}{r'} = \frac{1}{E}\left\{\left(\frac{B}{r'^2} + A\right) + \frac{p'}{m}\right\} \text{ compressive} \quad ...(2)$$

Similarly, for the outer tube ; the circumferential strain at the common radius r' is given by

$$\frac{\delta r_2}{r'} = \frac{1}{E}\left\{\left(\frac{B'}{r'^2} + A'\right) + \frac{p'}{m}\right\} \text{ (tensile)} \quad ...(3)$$

From (2) and (3), we have

$$\frac{\delta r_1 + \delta r_2}{r'} = \frac{\delta r'}{r'} = \frac{1}{E}\left\{\left(\frac{B'}{r'^2} + A'\right) - \left(\frac{B}{r'^2} + A\right)\right\} \qquad \ldots(23.14)$$

Thus, the original difference of radii at the junction, before shrinking on divided by r′ is equal to the algebraic difference of the hoop stresses for the two tubes at the junction divided by E.

Example 23.5. *A compound cylinder, formed by shrinking one tube on to another, is subjected to an internal pressure of 60 N/mm². Before the fluid is admitted, the internal and external diameter of the compound cylinder are 120 mm and 200 mm and the diameter at the junction is 160 mm. If after shrinking on, the radial pressure at the common surface is 8 N/mm², calculate the final stresses set up by the section.*

Solution. Given : $R = 100$ mm ; $r = 60$ mm ; $r' = 80$ mm

$p_r = 60$ N/mm² ; $p' = 8$ N/mm² ; $p_R = 0$

Let the Lame's equations for the inner tube be

$$p_x = \frac{B}{x^2} - A \quad \ldots(1) \; ; \quad f_x = \frac{B}{x^2} + A \qquad \ldots(2)$$

and for the outer tube be

$$p_x = \frac{B'}{x^2} - A' \quad \ldots(3) \quad \text{and} \quad f_x = \frac{B'}{x^2} + A' \qquad \ldots(4)$$

(a) Before the fluid is admitted : Inner tube

At $x = 60$ mm, $p = 0 \quad \therefore \quad \frac{B}{3600} - A = 0$

At $x = 80$ mm, $p = p' = 8$ N/mm² $\quad \therefore \quad \frac{B}{6400} - A = 8$

From which $B = -65829$ and $A = -18.29$

Hence the circumferential stress at any point in the inner tube will be given by

$$f_x = -\frac{65829}{x^2} - 18.29 \qquad \ldots(a)$$

The minus sign indicates that the stress will be wholly compressive.

At $x = 60$ mm, $f_{60} = -\frac{65829}{3600} - 18.29 = -18.29 - 18.29 = 36.58$ N/mm² (comp.)

At $x = 80$ mm, $f_{80} = \frac{65829}{6400} - 18.29 = -10.29 - 18.29 = 28.58$ N/mm² (comp.)

(b) Before the fluid is admitted : Outer tube

At $x = 100$ mm, $p_x = 0 \quad \therefore \quad \frac{B'}{10000} - A' = 0$

At $x = 80$ mm, $p_x = p' = 8 \quad \therefore \quad \frac{B'}{6400} - A' = 8$

From which $B' = 142222$ and $A' = 14.22$

Hence, the circumferential stress at any point in the outer tube will be given by :

$$f_x = \frac{142222}{x^2} + 14.22 \qquad \ldots(b)$$

At $x = 80$ mm, $f_{80} = \dfrac{142222}{6400} + 14.22 = 36.44$ N/mm² (tensile)

At $x = 100$ mm, $f_{100} = \dfrac{142222}{10000} + 14.22 = 28.44$ N/mm² (tensile).

(c) ***After the fluid is admitted : Compound tube***

Let the Lame's equation be $p_x = \dfrac{b}{x^2} - a$

At $x = 60$ mm, $p_x = 60$, $\therefore$ $60 = \dfrac{b}{3600} - a$

At $x = 100$ mm, $p_x = 0$, $\therefore$ $0 = \dfrac{b}{10000} - a$

Solving which, $b = 337500$ and $a = 33.75$

Hence, the hoop stress at any point in the compound tube is given by

$$f_x = \frac{337500}{x^2} + 33.75 \qquad \text{...(c)}$$

At $x = 60$ mm, $f_{60} = \dfrac{337500}{3600} + 33.75 = 127.5$ N/mm² (tensile)

At $x = 80$ mm, $f_{80} = \dfrac{337500}{6400} + 33.75 = 86.48$ N/mm² (tensile)

At $x = 100$ mm, $f_{100} = \dfrac{337500}{10000} + 33.75 = 67.5$ N/mm² (tensile)

The *final* hoop stresses at different points are tabulated below :

(+ for tension ; – for compression)

Hoop stress *(N/mm²)*	*Inner tube*		*Outer tube*	
	x = 60 mm	*x = 80 mm*	*x = 80 mm*	*x = 100 mm*
(*i*) Initially	– 36.58	– 28.58	+ 36.44	+ 28.44
(*ii*) Due to fluid pressure	+ 127.5	+ 86.48	+ 86.48	+ 67.5
Final	+ 90.92	+ 57.9	+ 122.92	+ 95.94

Example 23.6. *In example 23.5, calculate the necessary difference in the diameter of the two tubes at the common surface before shrinking on, so as to produce a radial pressure of 8 N/mm². Also calculate the minimum temperature to which the outer tube should be heated before it can be slipped on. Take $E = 2.05 \times 10^5$ N/mm² ; $\alpha = 6.2 \times 10^{-6}$ per °F and $1/m = 0.3$.*

Solution. We have, from equation 23.14,

$$\frac{\delta r'}{r'} = \frac{1}{E}\left\{\left(\frac{B'}{r'^2} + A'\right) - \left(\frac{B}{r'^2} + A\right)\right\}$$

$$= \frac{1}{E}\{36.44 - (-28.58)\} = \frac{65.02}{2.06 \times 10^5} = 3.15 \times 10^{-4}$$

$\therefore$ $\delta r' = 3.15 \times 10^{-4} \times 80 = 0.0252$ mm or $\delta d' =$ **0.0504 mm**

Now, $\pi d' \alpha \,.\, t = \pi \,.\, \delta \boldsymbol{d}'$, where $d' = 2 \times 80 = 160$ mm

$$\therefore \qquad t = \frac{\delta d'}{\alpha . d'} = \frac{0.0504}{160 \times 6.2 \times 10^{-6}} = \mathbf{50.8° \ F.}$$

Example 23.7. *A solid bar of uniform circular section is subjected to uniform radial pressure. Show that the stress at any point in a plane section parallel to the axis of the bar is compressive equal in magnitude to the radial stress.*

A steel bar of 50 mm diameter is pressed into a steel sleeve so that when assembled, the radial pressure is 14 N/mm² and the circumferential stress at the inside of the sleeve is 21 N/mm². Determine the increase of radial pressure when the bar is subjected to an axial compressive load of 70 kN. Poisson's Ratio = 0.304.

Solution. Let p be the uniform radial pressure on the solid bar.

The Lame's equations are

$$p = \frac{B}{x^2} - A \text{ (comp.) and } f = \frac{B}{x^2} + A \text{ (tensile)}$$

At $x = 0$, p and f are each infinite unless $B = 0$. Hence $B = 0$

$$\therefore \qquad p = -A \quad \text{or} \quad A = -p \quad \text{and} \quad f = +A = -p.$$

The minus sign shows that f is compressive. This holds for any point in the plane section at distance x from the centre. Hence the stress at any point is equal in the magnitude to the radial stress.

Let suffix 1 stand for sleeve and 2 for the bar.

For the sleeve

Initially, f = 21 N/mm² (tension) and p = 14 N/mm² (comp.)

R is unknown and end load = 0

Finally, when P = 70 kN (comp.)

$$\therefore \qquad f_0 = \frac{70000}{\frac{\pi}{4}(50)^2} = 35.67 \text{ N/mm}^2$$

Let $\quad \Delta f$ = change in hoop stress, when P is applied

$\quad \Delta p$ = change in radial stress, when P is applied.

Change in longitudinal strain of the rod $= \Delta e_0 = \dfrac{35.67}{E}$ (comp.)

Change in circumferential strain of the sleeve = Δe

$$\Delta e = \left(\frac{\Delta f}{E} + \frac{1}{mE} \Delta p\right) = \frac{1}{E}\left[\Delta f + \frac{\Delta p}{m}\right] \text{ expansion} \qquad \ldots(1)$$

For the rod

Change in circumferential strain for the rod $= \Delta e' = \dfrac{\Delta e_0}{m} + \dfrac{\Delta p}{mE} - \dfrac{\Delta p}{E}$

$$= \frac{1}{m} \cdot \frac{35.67}{E} + \frac{1}{m} \cdot \frac{\Delta p}{E} - \frac{\Delta p}{E} = \frac{1}{E}\left[\frac{35.67}{m} + \frac{\Delta p}{m} - \Delta p\right] \qquad \ldots(2)$$

Since the sleeve and shaft are in contact at common radius,

$$\Delta e = \Delta e'$$

$$\therefore \qquad \Delta f + \Delta p \, \frac{1}{m} = \frac{35.67}{m} + \frac{\Delta p}{m} - \Delta p \quad \text{or} \quad \Delta p = \frac{35.67}{m} - \Delta f \qquad \ldots(i)$$

This is one relation between Δp and Δf. To get another relation between Δp and Δf, construct the Lame's line for initial and final conditions, as shown in Fig. 23.6.

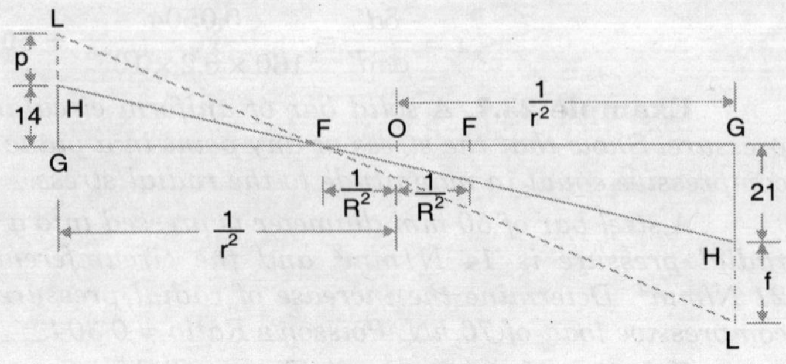

Fig. 23.6

Let r be the internal radius of sleeve (= external radius of shaft) and R be the external radius of sleeve. Mark point F and F' and G, G' as usual. At $x = r$, mark $p = GH = 14$ and $f = G'H' = 21$. Join H and H'. The solid line represents the initial conditions and the dotted line shows the final condition, in which Δp and Δf are the corresponding increase. By inspection from Fig. 23.6, we have

$$\frac{\Delta f}{\Delta p} = \frac{f}{p} = \frac{21}{14} = 1.5. \quad \text{Hence} \quad \Delta f = 1.5\, \Delta p \qquad \ldots(ii)$$

Substituting in (*i*), we get $\Delta p = \dfrac{35.67}{m} - 1.5\, \Delta p$

From which $\Delta p = \dfrac{35.67}{2.5} \times 0.304 = \mathbf{4.34\ N/mm^2}$.

Example 23.8. *Two thick steel cylinders A and B, closed at the ends, have the same conditions, the outer diameter of each being 1.6 times the inner diameter. The cylinder A is subject to internal fluid pressure only and B to external fluid pressure only. Find the ratio of the pressures on these cylinders :*

(*i*) *When the greatest circumferential stress has the same numerical value for each cylinder.*

(*ii*) *When the greatest circumferential strain has the same numerical value for each cylinder. Poisson's ratio = 0.304.*

Solution. Let p_a and p_b be the pressure in cylinders A and B respectively.

Condition (i) :

Due to internal pressure in A, the maximum hoop stress f_a will be at the inner radius r and will be tensile. Due to the external pressure in B, the maximum hoop stress f_b will also be at the inner radius r and will be compressive.

According to the given condition, f_a (tension) = f_b (compression)

For cylinder A : $f_a = p_a \dfrac{R^2 + r^2}{R^2 - r^2}$; For cylinder B : $f_b = p_b \dfrac{2R^2}{R^2 - r^2}$

But $f_a = f_b$

$$\therefore \quad p_a \left(\frac{R^2 + r^2}{R^2 - r^2}\right) = p_b \frac{2R^2}{R^2 - r^2}$$

or
$$\frac{p_a}{p_b} = \frac{2R^2}{R^2 + r^2} = \frac{2(1.6r)^2}{(1.6r)^2 + r^2} = \frac{2 \times 2.56}{2.56 + 1} = \mathbf{1.439}.$$

Condition (ii) :

According to the given condition, maximum hoop strain e_a for A = maximum hoop strain e_b for B. Neglecting the longitudinal strain e_a in the cylinder A,

$$e_a = \frac{f_a}{E} + \frac{1}{m}\frac{p_a}{E} \quad \text{...(expansion)}$$

For the cylinder B, maximum hoop strain, e_b occurs at the inner radius where p_r is zero. Hence,

$$e_b = \frac{f_b}{E} \text{ (compressive) ;} \quad \text{But } e_a = e_b$$

$$\therefore \quad f_a + \frac{p_a}{m} = f_b \quad \text{...(1)}$$

But for A, $$f_a = p_a \frac{R^2 + r^2}{R^2 - r^2} = p_a \frac{2.56r^2 + r^2}{2.56r^2 - r^2} = \frac{3.56}{1.56} p_a = 2.28\, p_a \quad \text{...(2)}$$

and for B, $$f_b = p_b \frac{2R^2}{R^2 - r^2} = p_b \frac{2 \times 2.56r^2}{2.56r^2 - r^2} = \frac{5.12}{1.56} p_b = 3.28\, p_b \quad \text{...(3)}$$

Substituting the values of f_a and f_b in (1), we get

$$2.28p_a + 0.304p_a = 3.28\, p_b \quad \text{or} \quad \frac{p_a}{p_b} = \frac{3.28}{2.584} = \mathbf{1.27}$$

However, if longitudinal strain e_0 for A is taken into account

$$e_0 = \frac{1}{E}\frac{\pi r^2 p_a}{\pi(R^2 - r^2)} = \frac{r^2 p_a}{E(2.56r^2 - r^2)} = \frac{p_a}{1.56E}$$

$$\therefore \quad e_a = \frac{1}{E}\left(f_a + \frac{p_a}{m} - \frac{1}{m}\frac{p_a}{1.56}\right) = \frac{1}{E}\left(f_a + \frac{0.358\, p_a}{m}\right)$$

and $$e_b = \frac{f_b}{E} \text{ as before.}$$

$$\therefore \quad f_a + \frac{0.358}{m} p_a = f_b$$

But $f_a = 2.28p_a$ and $f_b = 3.28p_b$ as before.

Hence $2.28p_a + 0.358 \times 0.304p_a = 3.28p_b$. or $2.389p_a = 3.28p_b$

$$\therefore \quad \frac{p_a}{p_b} = \frac{3.28}{2.389} = \mathbf{1.373.}$$

Example 23.9. *The external diameter of steel collar is 200 mm and the internal diameter decreases by 0.125 mm when shrunk on to a solid steel shaft of 125 mm diameter. Find the reduction in diameter of the shaft, the radial pressure between the collar and the shaft and hoop stress at the inner surface of the tube. Poisson's ratio = 0.304 and E = 2.05 × 10^5 N/mm^2.*

Solution. Let suffix 1 stand for the shaft and 2 for collar.

Due to the radial pressure at the junction, the shaft will have compressive stress throughout.

Let p = radial pressure at the junction ; f = hoop pressure at the junction

$$f = p\left(\frac{R^2 + r^2}{R^2 - r^2}\right) = p\,\frac{40000 + 15625}{40000 - 15625} = 2.282\, p$$

Now $$e_2 = \frac{1}{E}\left(f + \frac{p}{m}\right) \quad \text{But } e_2 = \frac{0.125}{125} = 1 \times 10^{-3} \text{ and } f = 2.282\, p$$

Hence $1 \times 10^{-3} = \dfrac{1}{E}\left(2.282p + \dfrac{p}{m}\right)$ or $p(2.282 + 0.304) = 2.05 \times 10^5 \times 10^{-3}$

From which $p =$ **79.3 N/mm² (comp.)**

$\therefore$ $f = 2.282\,p = 2.282 \times 79.3 = 180.9$ N/mm² (tension)

Again $e_1 = \dfrac{p}{E}\left(1 - \dfrac{1}{m}\right) = \dfrac{79.3}{2.05 \times 10^5}(1 - 0.304) = 2.692 \times 10^{-4}$

$\delta d_1 = d_1 . e_1 = 125 \times 2.692 \times 10^{-4} =$ **0.0336 mm.**

Example 23.10. *A compound cylinder is to be made by shrinking outer tube of 300 mm external diameter on to an inner tube of 150 mm internal diameter. Determine the common diameter at the junction if the greatest circumferential stress in the inner tube is to be two-thirds of the greatest circumferential stress in the outer tube.*

Solution. Let suffix 1 be for the outer tube and 2 for the inner tube.

Let R be the external radius of outer tube, r the inner radius of inner tube and r' the common radius. Let p' be the radial pressure at the junction.

The maximum circumferential stress f_2 for the inner tube will occur at the inner radius r. The maximum circumferential stress f_1 for the outer tube will occur at the radius r'.

For the outer tube, $f_1 = p'\dfrac{R^2 + r'^2}{R^2 - r'^2} = p'\dfrac{K+1}{K-1}$ where $K = \dfrac{R^2}{r'^2} = \dfrac{(150)^2}{r'^2} = \dfrac{22500}{r'^2}$

For the inner tube,

$$f_2 = p'\frac{2r'^2}{r'^2 - r^2} = p'\frac{2}{1 - \dfrac{(75)^2}{r'^2}} = p'\frac{8}{4 - \dfrac{(150)^2}{r'^2}} = p'\frac{8}{4 - K}$$

According to the given condition,

$$f_2 = \frac{2}{3}f_1 \quad \text{or} \quad p'\frac{8}{4-K} = \frac{2}{3}p'\left(\frac{K+1}{K-1}\right)$$

or $12(K - 1) = (K + 1)(4 - K)$ or $K^2 + 9K - 16 = 0$

from which $K = 1.52 = \dfrac{22500}{r'^2}$

$\therefore$ $r'^2 = \dfrac{22500}{1.52}$ or $r' = 121.7$ mm

Common diameter at junction = **243.4 mm.**

Example 23.11. *Find the ratio of thickness to internal diameter for a tube subjected to internal pressure when the ratio of the internal pressure to the greatest circumferential stress is 0.5. Find the alteration of thickness of metal in such a tube, 250 mm internal diameter, when the internal pressure is 75 N/mm². Take 1/m = 0.304 and E = 2.05 × 10⁵ N/mm².*

Solution. Let R = external radius, r = internal radius

$\therefore$ t = thickness of metal = $R - r$

Let $r = nt$

If p is the internal pressure and f is the greatest circumferential stress, we have

$$\frac{f}{p} = \frac{R^2 + r^2}{R^2 - r^2} = \frac{(r+t)^2 + r^2}{(R+r)(R-r)} = \frac{2r^2 + t^2 + 2rt}{(R+r)t} = \frac{2r^2 + 2rt + t^2}{(2r+t)t}$$

$$= \frac{2n^2t^2 + 2nt^2 + t^2}{(2nt + t)t} = \frac{2n^2 + 2n + 1}{2n + 1}$$

But $\dfrac{p}{f} = \dfrac{1}{2}$

$\therefore \quad \dfrac{2n + 1}{2n^2 + 2n + 1} = \dfrac{1}{2}$ or $2n^2 + 2n + 1 = 4n + 2$

or $2n^2 - 2n - 1 = 0$, from which $n = 1.366$

$\therefore \quad \dfrac{r}{t} = n = 1.366$

$\therefore \quad \dfrac{t}{r} = \dfrac{1}{1.366}$; or $\dfrac{t}{d} = \dfrac{1}{2.732} = \mathbf{0.366.}$

Change in thickness excluding longitudinal stress :

$$e_r = \text{strain at the inner radius} = \frac{1}{E}\left(f_r + \frac{p_r}{m}\right) = \frac{1}{E}\left(2p_r + \frac{p_r}{m}\right), \text{ since } f_r = 2p_r$$

$$= \frac{p_r}{E}\left(2 + \frac{1}{m}\right)$$

$$\therefore \quad \delta r = re_r = \frac{2.304r}{E}\,p_r \qquad \text{...(1)}$$

$$e_R = \text{strain in outer radius} = \frac{f_R}{E} \quad \text{...(2)} \quad \text{since } p_R = 0$$

Now, Lame's equations are $f = \dfrac{B}{x^2} + A$ and $p = \dfrac{B}{x^2} - A$

Applying these at the inner radius,

$f_r - p_r = 2A$ or $2p_r - p_r = 2A$

$\therefore \quad 2A = p_r$

Applying Lame's equations at the outer radius,

$f_R - p_R = 2A$ or $f_R - 0 = p_r$

$\therefore \quad f_R = p_r$

Substituting in (2), we get $\quad e_R = \dfrac{f_R}{E} = \dfrac{p_r}{E}$ ...(2)

$\therefore \quad R.e_R = \dfrac{p_r}{E}.R$

Change in thickness $= \Delta t = re_r - Re_R = \dfrac{2.304r}{E}\,p_r - R\dfrac{p_r}{E} = \dfrac{p_r}{E}\,(2.304r - R)$

Now when $d = 200$ mm, $t = 0.366 \times 200 = 73.2$ mm

$\therefore \quad D = 200 + (73.2 \times 2)$ or $R = 100 + 73.2 = 173.2$ mm

$\therefore \quad \Delta t = \dfrac{75}{2.05 \times 10^5}\,(2.304 \times 100 - 173.2) = 0.0209$ mm.

Change in thickness due to longitudinal stress :

$$f_0 = \frac{\pi r^2 p_r}{\pi(R^2 - r^2)} = \frac{75 \times 10000}{30000 - 10000} = 37.5 \text{ N/mm}$$

$\Delta t'$ = change in the thickness due to $f_0 = \dfrac{f_0}{mE} t = \dfrac{37.5 \times 0.304}{2.05 \times 10^5} \times 73.2$

$= 0.0047$ mm

Hence total change in thickness

$= \Delta t + \Delta t' = 0.0209 + 0.0047 =$ **0.0256 mm.**

23.7. THICK SPHERICAL SHELLS

Fig. 23.7 shows a thick spherical shell of inner radius r and outer radius R, subjected to internal pressure p.

Let p_x be the radial compressive stress at any radius x.

f_x = circumferential tensile stress, which is equal in all directions perpendicular to the radius.

The equation between the principal stresses, p_x and f_x may be formed by considering the forces on an elementary spherical shell of radius x and thickness δx and equating the bursting force and resisting force on the elementary spherical shell.

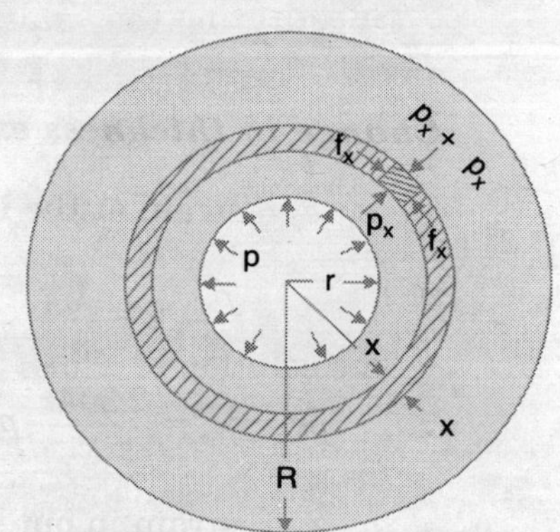

Fig. 23.7

Thus $\quad \pi x^2 p_x - \pi(x + \delta x)^2 (p_x + \delta p_x) = 2\pi x \delta x f_x$

Simplifying the above, and neglecting small quantities of second and higher order, we get

$$2f_x = -2p_x - x\frac{dp_x}{dx} \quad \text{or} \quad f_x = -p_x - \frac{x}{2}\frac{dp_x}{dx} \qquad \ldots[1\,(a)]$$

Differentiating,
$$\frac{df_x}{dx} = -\frac{dp_x}{dx} - \frac{1}{2}\left\{x\frac{d^2p_x}{dx^2} + \frac{dp_x}{dx}\right\} \qquad \ldots[1\,(b)]$$

The three principal stresses at any point in the elementary spherical shell are :

(*i*) The radial pressure p_x (compressive)

(*ii*) The hoop stress f_x (tensile)

(*iii*) The hoop stress $f_y = f_x$ (tensile), acting on a plane perpendicular to that of (*ii*).

The radial strain at any point is

$$e_x = \frac{p_x}{E} + \frac{2f_x}{mE} \text{ (compressive)} = -\frac{1}{E}\left(p_x + \frac{2f_x}{m}\right) \text{ (tensile)} \qquad \ldots(2)$$

The circumferential strain at any point is

$$e_y = \frac{f_x}{E} + \frac{p_x}{mE} - \frac{f_x}{mE} = \frac{1}{E}\left\{\frac{m-1}{m}f_x + \frac{p_x}{m}\right\} \text{ (tensile)} \qquad \ldots(3)$$

Let the radius x increase to $(x + u)$ owing to purely radial displacements of points.

The radial strain
$$e_x = \frac{\delta(x+u) - \delta x}{\delta x} = \frac{\delta u}{\delta x} \qquad \ldots(4)$$

Circumferential strain
$$e_y = \frac{(x+u)\delta\theta - x\delta\theta}{x\delta\theta} = \frac{u}{x} \qquad \ldots(5)$$

From (4) and (5), $e_x = \frac{\delta u}{\delta x} = \frac{\delta}{\delta x}(xe_y) = e_y + x\frac{\delta e_y}{\delta x}$

Substituting the value of e_x and e_y from (2) and (3), we get

$$-\frac{1}{E}\left(p_x + \frac{2f_x}{m}\right) = \frac{1}{E}\left\{\frac{m-1}{m}f_x + \frac{p_x}{m}\right\} + \frac{x}{E}\left\{\frac{m-1}{m}\frac{\delta f_x}{\delta x} + \frac{1}{m}\frac{\delta p_x}{\delta x}\right\}$$

Simplying and re-arranging, we get

$$(m+1)(p_x + f_x) + (m-1)\,x\frac{df_x}{dx} + x\frac{dp_x}{dx} = 0$$

Substituting the values of f_x and $\frac{df_x}{dx}$ from (1), we get

$$(m+1)\left(p_x - p_x - \frac{x}{2}\frac{dp_x}{dx}\right) + (m-1)\,x\left[-\frac{dp_x}{dx} - \frac{1}{2}\left\{x\frac{d^2p_x}{dx^2} + \frac{dp_x}{dx}\right\}\right] + x\frac{dp_x}{dx} = 0$$

Simplifying, we get $\frac{d^2p_x}{dx^2} + \frac{4}{x}\frac{dp_x}{dx} = 0$

The solution of which $\left(\text{as an equation in } \frac{dp_x}{dx}\right)$ is

$$x^4\frac{dp_x}{dx} = \text{constant} = -6B \text{ (say)} \quad \text{or} \quad \frac{dp_x}{dx} = -\frac{6B}{x^4} \qquad ...(6)$$

which on integration gives

$$p_x = \frac{2B}{x^3} + A, \text{ where } A \text{ is constant.} \qquad ...(23.15)$$

Again, from [1 (*a*)], $f_x = -p - \frac{x}{2}\frac{dp_x}{dx} = -\frac{2B}{x^3} - A - \frac{x}{2}\left(-\frac{6B}{x^4}\right) = -\frac{2B}{x^3} - A + \frac{3B}{x^3}$

or

$$f_x = \frac{B}{x^3} - A \qquad ...(23.16)$$

Hence p_x and f_x are given by Eqs. 23.15 and 23.16 respectivily.

Applying the given conditions,

At $x = r$, $p_x = p$; At $x = R$, $p_x = 0$

$\therefore$ $p = \frac{2B}{r^3} + A$ and $0 = \frac{2B}{R^3} + A$

From which $B = p\frac{R^3r^3}{2(R^3 - r^3)}$ and $A = -\frac{pr^3}{R^3 - r^3}$.

Example 23.12. *A spherical shell of 80 mm internal diameter has to withstand an internal pressure of 30 N/mm². Calculate the thickness of the shell if the maximum permissible tensile stress is 80 N/mm².*

Solution. $p_x = \frac{2B}{x^3} + A$ and $f_x = \frac{B}{x^3} - A$

At $x = 40$ mm, $p = 30$ N/mm^2 ; $\therefore$ $\frac{2B}{64000} + A = 30$ ...(1)

At $x = 40$ mm, $f = 80$ N/mm^2 $\therefore$ $\frac{B}{64000} - A = 80$ ...(2)

From (1) and (2), we get $B = (80 + 30)\dfrac{64000}{3} = 2.347 \times 10^6$

and $$A = \frac{B}{64000} - 80 = \frac{2.347 \times 10^6}{64000} - 80 = 36.67 - 80 = -43.33$$

Hence $$p_x = \frac{4.694 \times 10^6}{x^3} - 43.33$$

At $x = R$, $$p_x = 0 = \frac{4.694 \times 10^6}{R^3} - 43.33 \quad \text{or} \quad R^3 = \frac{4.694 \times 10^6}{43.33} = 108308$$

$\therefore$ $R = 47.7$ mm

$\therefore$ Thickness of shell $= 47.7 - 40 =$ **7.7 mm.**

23.8. ADDITIONAL EXAMPLES USEFUL FOR COMPETITIVE EXAMINATIONS

Example 23.13. *Determine the maximum allowable internal pressure to which a cylinder with an internal diameter of 200 mm and external diameter of 300 mm may be subjected if the cylinder is to have a factor of safety of 3. Use maximum shear stress theory of failure. Yield point stress determined by actual test for the material is 300 N/mm². What will be the internal pressure if the maximum principal stress theory of failure is used ?*

Solution. Given $r = 100$ mm ; $R = 150$ mm ; F.S. = 3 ; $f_y = 300$ N/mm²

$$f_{s,\,max}\ \text{(allowable)} = \frac{f_y}{2 \times \text{F.S.}} = \frac{300}{2 \times 3} = 50 \text{ N/mm}^2$$

Lame's equations are : $p_x = \dfrac{B}{x^2} - A$ and $f_x = \dfrac{B}{x^2} + A$

At $x = r = 100$ mm, $$p_x = p = \frac{B}{(100)^2} - A \quad \text{...(i)}$$

At $x = R = 150$ mm, $$p_x = 0 = \frac{B}{150^2} - A \quad \text{...(ii)}$$

From (*ii*), $A = \dfrac{B}{22500}$. Substituting in (*i*), we get $B = 18000\,p$. Hence $A = \dfrac{18000\,P}{22500} = 0.8\,p$

Hence $$f_x = \frac{18000\,p}{x^2} + 0.8\,p \quad \text{...(iii)}$$

At $x = r$, $$f_x = f_{x,\,max} = f_r = \frac{18000\,p}{(100)^2} + 0.8\,p = 2.6\,p \text{ (tensile)}$$

Also $$p_r = p \text{ (compressive)} \quad \therefore \quad f_{s,\,max} = \frac{2.6\,p - (-p)}{2} = 1.8\,p \quad \text{...(iv)}$$

Hence we have $1.8\,p = 50$; from which $p = 50/1.8 =$ **27.78 N/mm²**

If maximum principal stress theory is used, $\dfrac{f_y}{\text{F.S.}} = f_r$

or $300/3 = 2.6\,p$, from which $p =$ **38.46 N/mm²**

Comparing the results obtained by the two theories of failure, we observe that the safe pressure is **27.78 N/mm²**, obtained by the shear stress theory of failure.

Example 23.14. *The maximum stress permitted in a thick cylinder, radii 100 mm and 150 mm is 25 N/mm². The external pressure is 7.5 N/mm². What internal pressure can be applied ? Plot curves showing the variation of hoop and radial stresses through the material.*

Solution. Given $R = 150$ mm ; $r = 100$ mm ; $f_{max} = 25$ N/mm² ; $p_R = 7.5$ N/mm²

Lame's equations are $p_x = \dfrac{B}{x^2} - A$ and $f_x = \dfrac{B}{x^2} + A$

At $x = R = 150$ mm, $p_x = p_R = 7.5 = \dfrac{B}{(150)^2} - A$ ...(1)

Also, at $x = r = 100$ mm, $f_r = f_{max} = 25 = \dfrac{B}{(100)^2} + A$ ...(2)

From (1), $A = \dfrac{B}{22500} - 7.5$; Hence from (2), $25 = \dfrac{B}{10000} + \dfrac{B}{22500} - 7.5$

From which $B = 225000$. Hence $A = \dfrac{225000}{22500} - 7.5 = 2.5$

$$\therefore \quad f_x - p_x = \left(\frac{B}{x^2} + A\right) - \left(\frac{B}{x^2} - A\right) = 2A = 5$$

Thus, there is a constant difference of 5 N/mm² between the hoop and radial stresses.

At $r = 100$ mm

$$p_{100} = \frac{225000}{(100)^2} - 2.5 = 20 \text{ N/mm}^2$$

At $x = 125$ mm

$$p_{125} = \frac{225000}{(125)^2} - 2.5 = 11.9 \text{ N/mm}^2$$

$$f_{125} = \frac{225000}{(125)^2} + 2.5 = 16.9 \text{ N/mm}^2$$

At $x = 150$ mm, $p_R = 7.5$

$$f_{150} = \frac{225000}{(150)^2} + 2.5 = 12.5 \text{ N/mm}^2$$

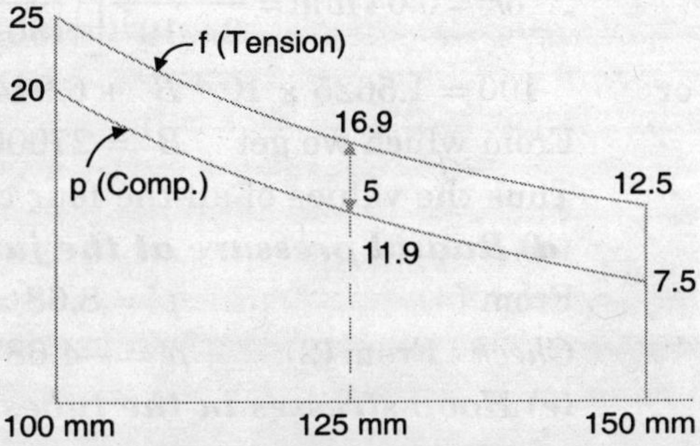

25
f (Tension)
20
16.9
p (Comp.)
5
12.5
11.9
7.5
100 mm
125 mm
150 mm

Fig. 23.8

Fig. 23.8 shows the graphical representation of stress variation.

Example 23.15. *A steel tube 240 mm external diameter is to be shrunk on another steel tube of 80 mm internal diameter. After shrinking, the diameter at the junction is 100 mm. Before shrinking on, the difference of diameter at the junction was 0.08 mm.*

Calculate the radial pressure at the junction and hoop stress developed in the two tubes after shrinking on. Take E as 200 kN/mm².

Solution. Given $R = 120$ mm; $r = 40$ mm; $r' = 800$ mm; $\delta r' = 0.04$ mm

(a) For the inner tube, before the fluid is admitted :

$$p_x = \frac{B}{x^2} - A$$

At $x = r = 40$ mm, $p_x = 0 = \dfrac{B}{(40)^2} - A$, From which $A = \dfrac{B}{1600}$ ...(1)

At $x = r' = 80$ mm, $p_x = p' = \dfrac{B}{(80)^2} - A = \dfrac{B}{6400} - \dfrac{B}{1600} = -\dfrac{3}{6400} B = -4.6875 \times 10^{-4} B$ ...(2)

(b) For the outer tube, before the fluid is admitted : $p_x = \dfrac{B'}{x^2} - A'$

At $x = R = 120$ mm, $\quad p_x = 0 = \dfrac{B'}{(120)^2} - A'$, from which $A' = \dfrac{B'}{14400}$ ...(3)

At $x = r' = 80$ mm, $\quad p_x = p' = \dfrac{B'}{(80)^2} - A' = \dfrac{B'}{6400} - \dfrac{B'}{14400} = 8.68 \times 10^{-5} B'$ ...(4)

Equating (2) and (4), we get $-4.6875 \times 10^{-4} B = 8.68 \times 10^{-5} B'$

From which $\quad B = -0.1852 B'$ ...(*a*)

Also, from (1), $\quad A = \dfrac{B}{1600} = -1.157 \times 10^{-4} B'$ ...(*b*)

and from (3) $\quad A' = \dfrac{B'}{14400} = 6.944 \times 10^{-5} B'$ ...(*c*)

Thus the values of the three constants *A*, *B* and *A′* are known in terms of *B′*.

(c) Difference in radii :

$$\frac{\delta r'}{r'} = \frac{1}{E}\left[\left(\frac{B'}{r'^2} + A'\right) - \left(\frac{B}{r'^2} + A\right)\right]$$

$$\therefore\ \delta r' = 0.04 \text{ mm} = \frac{80}{2 \times 10^5}\left[\left\{\frac{B'}{(80)^2} + 6.944 \times 10^{-5} B'\right\} - \left\{\frac{-0.1852 B'}{(80)^2} - 1.157 \times 10^{-4} B'\right\}\right]$$

or $\quad 100 = 1.5625 \times 10^{-4} B' + 6.944 \times 10^{-5} B' + 2.894 \times 10^{-5} B' + 1.157 \times 10^{-4} B'$

From which we get $\quad B' = 270000$. Hence $A' = 18.75$; $A = -31.24$ and $B = -50000$

Thus the values of all the four constants are know.

(d) Radial pressure at the junction :

From (4), $\quad p' = 8.68 \times 10^{-5} B' = 8.68 \times 10^{-5} \times 270000 =$ **23.44 N/mm²** (comp.)

Check : From (2), $\quad p' = -4.6875 \times 10^{-4} B = -4.6875 \times 10^{-4} \times (-50000) = 23.44$ N/mm²

(e) Hoop stresses in the tubes :

For the inner tube, $\quad f_x = \dfrac{B}{x^2} + A$

At $x = r = 40$ mm; $\quad f_r = \dfrac{-50000}{(40)^2} - 31.24 = -62.49$ N/mm² = **62.49 N/mm²** (comp.)

At $x = r' = 80$ mm $\quad f_r' = \dfrac{50000}{(80)^2} - 31.24 = 7.81 - 31.24 = -39.05$ N/mm²

$= $ **39.05 N/mm²** (comp.)

For the outer tube, $\quad f_x = \dfrac{B'}{x^2} + A'$

At $x = r' = 80$ mm $\quad f_r' = \dfrac{270000}{(80)^2} + 18.75 = 42.19 + 18.75 =$ **60.94 N/mm²** (tensile)

At $x = R = 120$ mm $\quad f_R = \dfrac{270000}{(120)^2} + 18.75 =$ **37.5 N/mm²** (tensile).

PROBLEMS

1. A steel sleeve is pressed on to a solid steel shaft of 50 mm diameter. The radial pressure between the shaft and the sleeve is 175 N/mm^2 and the hoop stress at the inner surface of the sleeve is 421 N/mm^2. If an axial compressive load of 54.5 kN is now applied to the shaft, determine the alteration of radial pressure. Take $1/m$ = 0.304.
2. A steel shaft 38 mm diameter is to be encased in a bronze sleeve 57 mm outside diameter which is to be forced into position and, before forcing on, the inside diameter of the sleeve is smaller than the diameter of the shaft, the difference in these diameters being 0.051 mm.

   Find, due to forcing on,

   (*a*) the radial pressure between the shaft and the sleeve,

   (*b*) the maximum hoop stress in the sleeve,

   (*c*) the change in outside diameter of the sleeve,

   For steel, $E = 2.08 \times 10^5$ N/mm^2 ; $1/m$ = 0.29.

   For bronze, $E = 1.25 \times 10^5$ N/mm^2 ; $1/m$ = 0.34.
3. State the formula used in thick cylinder calculations giving the relationship between (*a*) the hoop stress and the radius, (*b*) the radial stress and the radius, and from these obtain the relationship, in terms of inside and outside diameter between the maximum hoop stress and internal pressure in the case of a cylinder subjected to internal pressure only.

   If a cylinder of internal diameter d, wall thickness t and subjected to internal pressure only, is assumed to be a thin shell, what is the greatest value of the ratio t/d if the error in the estimated maximum hoop stress is not to exceed 5%.
4. A cast iron cylinder has outside and inside diameter of 203 mm and 127 mm. If the ultimate tensile strength of the cast iron is 157.5 N/mm^2, find according to each of the following theories of failure the internal pressure which would cause rupture :

   (*a*) Maximum principal stress theory,

   (*b*) Maximum strain theory,

   (*c*) Maximum strain energy theory.

   Assume no longitudinal stress in the cylinder. Which of the results obtained do you consider should be applied to this case ? Take $1/m$ = 0.25.
5. A steel cylinder of outside diameter of 305 mm and inside diameter 254 mm is shrunk on to one having diameter 254 mm and 203 mm, the interference fit being such that under internal pressure p, the inner tensile stress in both cylinders = 86 N/mm^2.

   Find the initial difference in nominal 254 mm diameter and the value of p if $E = 2.11 \times 10^5$ N/mm^2.

ANSWERS

1. 2.39 N/mm^2.
2. (*a*) 49.45 N/mm^2 ; (*b*) 128.5 N/mm^2 ; (*c*) 0.0364 cm.
3. $f = p\left(\dfrac{D^2 + d^2}{D^2 - d^2}\right)$; $\dfrac{t}{d} = 0.05025$.
4. (*a*) 6.9 N/mm^2 ; (*b*) 6.23 N/mm^2 ; (*c*) 5.81 N/mm^2. The lowest value given by (*c*) should be applied.
5. 2.68×10^{-2} mm ; p = 37.75 N/mm^2.

   Find, due to this alone,

   For bronze, $E = 1.28 \times 10^5$ N/mm², $1/m = 0.34$

1. [illegible] N/mm²

2. (a) [illegible] N/mm² (b) [illegible] N/mm² (c) 0.0048 mm

[illegible]

6. [illegible] mm; [illegible] N/mm²

SECTION 5 MISCELLANEOUS TOPICS

CHAPTERS

24

Stresses in Statically Determinate Frames

24.1. GENERAL

A framework is defined as an assemblage of bars which is able to resist geometrical distortion under any system of applied loads. Such framework can either be statically determinate or statically indeterminate. A *statically determinate* frame is one in which the axial forces in the members can be obtained by the application of the equations of static equilibrium. For a plane frame, there are three such equations (*i.e.*, $\Sigma V = 0$, $\Sigma H = 0$ and $\Sigma M = 0$), and for a space frame there are six such equations. A *statically indeterminate frame* is one in which the axial forces in the members cannot be obtained by the application of the equations of static equilibrium alone, and the deformation of the structure is to be taken into account for the analysis.

Perfect Frame

Frames may be classified under three heads.

(*i*) Perfect frame (*ii*) Deficient frame, and (*iii*) Redundant frame.

A *perfect frame* is the one which is composed of a number of members just sufficient to keep the frame in equilibrium. A simplest type of a perfect frame is a triangular frame shown in Fig. 24.1 (*a i*) consisting of three members, so that when the frame is loaded at its joints (or nodes), the shape of the frame is not distorted. Fig. 24.1 (*a ii*) and (*a iii*) show other types of perfect frames, obtained by adding two additional members meeting at each new joint, thus resulting in additional triangular shapes.

Thus, we observe that starting with a simple triangular frame consisting of three members and three joints, every additional joint would require two additional members. Hence the relation between number of the members and number of joints in a perfect frame can be expressed by the relation.

$$n = 2j - 3 \qquad \text{...(24.1)}$$

where n = number of members and j is the number of joints.

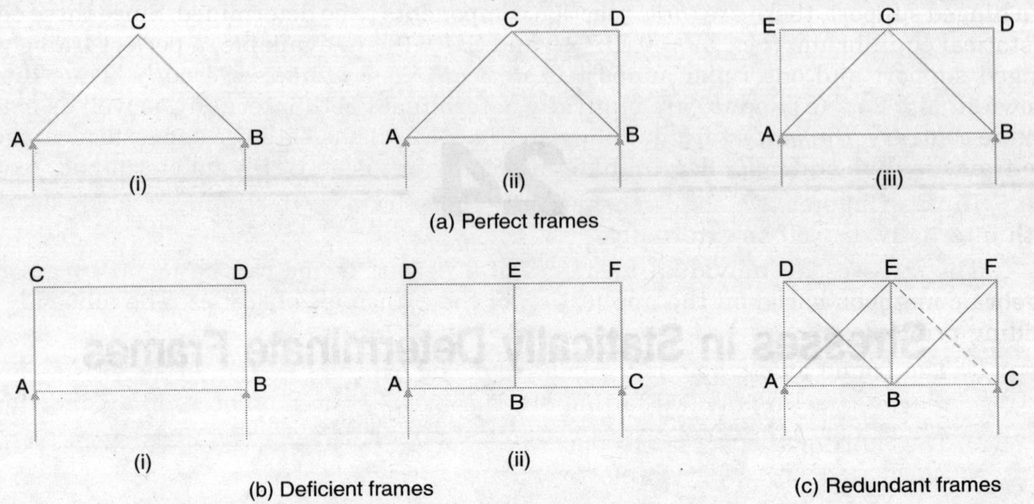

Fig. 24.1

If the number of members in a frame is *less* than the one given by Eq. 24.1, the frame is known as a *deficient frame* [Fig. 24.1 (*b*)]. Such frames are *unstable*. On the contrary, if the number of members in a frame is *more* than the one given by Eq. 24.1, it is known as *redundant frame* [Fig. 24.1 (*c*)]. Such frames are a *statically indeterminate*.

Types of supports in a perfect frame

A perfect frame may have two types of supports:

(*i*) Free support or roller support, and (*ii*) hinged support

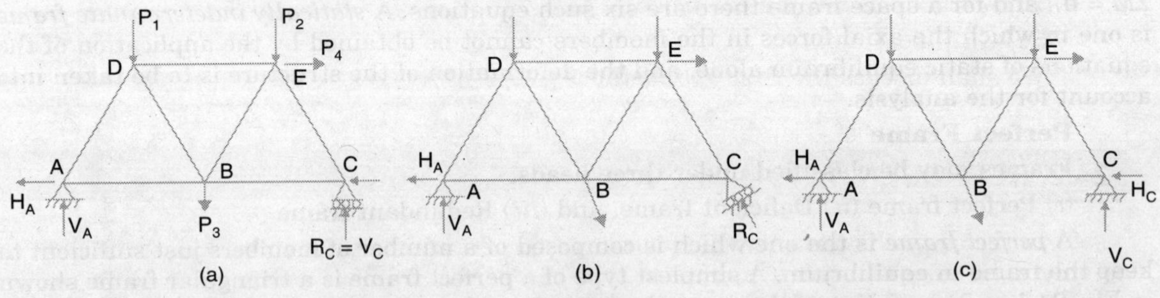

Fig. 24.2

A *roller support* (or simple support) is the one along which the *translation* is possible along its base. The reaction at a roller support is always perpendicular to the roller base along which movement takes place. Thus, in Figs. 24.2 (*a*) and (*b*), the support at C is a roller support, where the reaction R_C is perpendicular to the roller base.

A *hinged support* is the one in which only rotation is possible and there is no translation. At such a support, the direction of reaction depends upon the load system of the truss. Such a reaction can always be resolved in horizontal and vertical directions. The support at A in Figs. 24.2 (*a*) and (*b*) is a hinged support in which the reaction (R_A) is in inclined direction.

Thus, we find that a roller (or simple) support has only one *reaction component*, while a hinged support has two reaction components (*i.e.*, H and V), for general system of loading. Thus, in a perfect frame having one support as roller (or simple) support and the other support as a hinged support, there are *three* reaction components to be determined. Since three equations of statical equilibrium (*i.e.*, $\Sigma V = 0$ $\Sigma H = 0$ and $\Sigma M = 0$) are available, a perfect frame with one hinged support and one roller support is *statically determinate externally*. Thus the frames shown in Fig. 24.2 (*a*) and (*b*) are statically determinate both internally as well as externally. On the contrary, the perfect frame shown in Fig. 24.2 (*c*) has both the supports as hinged ones; the frame is thus statically determinate internally but is statically indeterminate externally.

In this Chapter, we shall consider only those frames which are statically determinate both internally as well as externally.

The stresses in individual members of a perfect frame can be found by graphical or algebraic methods based on the application of the principles of statics. The following are the guiding principles :

(1) The frame as a whole is a rigid body. The external system of forces, *i.e.,* loads and reactions acting upon it form by themselves a system of forces in equilibrium.

(2) The pulls or thrusts of the several members meeting in any joint form a system of concurrent forces in equilibrium.

(3) Any portion of structure may be taken as a rigid body held in equlibrium by the external forces acting upon it together with forces exerted upon it, through members, by remainder of the structure.

(4) The framework can resist geometrical distortion.

Methods of Solving Statically Determinate Frames

The main methods used for the analysis of statically determinate frames are :

(*i*) Methods of joints

(*ii*) Methods of sections

(*iii*) Methods of tension-coefficients

(*iv*) Graphical method.

We shall discuss the first two methods in this Chapter.

24.2. REPRESENTATION OF AXIAL FORCES IN MEMBERS

In the case of frames, external forces/loads are applied at the joints only. This gives rise to only *axial forces* in the members of the frame. These axial forces, also commonly called *stresses*, may be either tensile or compressive in nature. Fig. 24.3 (*a i*) shows a member *AB* of a frame, subjected to an axial compressive force *P* at each end *A* and *B*. The *internal reactive compressive forces*, also commonly called *stress*, will act in opposite directions and will be represented by a set of arrows pointing outwards at *A* and *B*, as shown in Fig. 24.3 (*a ii*). Similarly, Fig. 24.3 (*b i*) shows a members *AB* subjected to axial tensile force *P* at ends/joints *A* and *B*. The *internal reactive tensile forces* will act in apposite directions and will be represented by a set of arrows pointing *inwards* at *A* and *B*, as shown in Fig. 24.3 (*b ii*). *Thus a compressive force/stress*

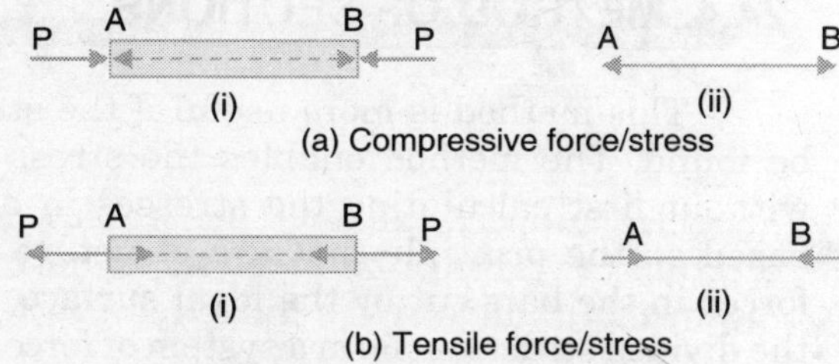

Fig. 24.3. Representation of Forces/stresses in members of frame

is represented by a set of arrows pointing outwards at the ends and a tensile force is represented by a set of arros pointing inwards from the ends.

24.3. METHOD OF JOINTS

When two or more number of members are meeting at one joint and if there are only two unknown forces, the two unknown forces can be determined by considering the equilibrium of the components, in two perpendicular directions, of the forces acting on the joint. Since the joint is in equilibrium, the sum of the components in each direction must be zero. This yields two equations from which two unknown forces can be obtained.

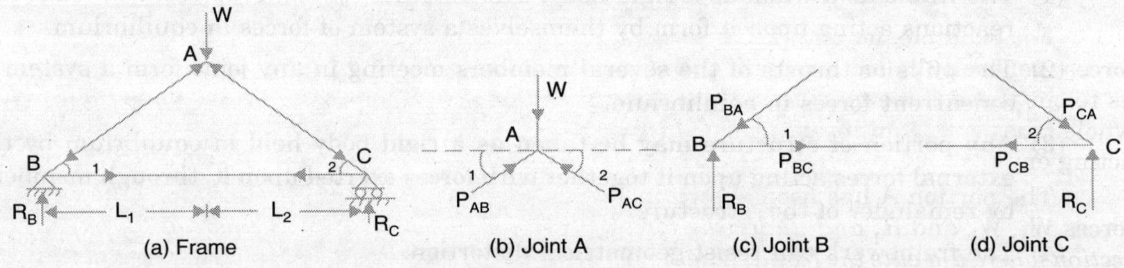

Fig. 24.4

Let us take simple example of a triangular frame shown in Fig. 24.4 (*a*). An external load *W* is acting at joint *A*. The forces in the members *AB* and *AC* can be found by considering the equilibrium of the joint *A* [Fig. 24.4 (*b*)]. Reaction $R_B = WL_2/L$ (↑) and $R_C = WL_1/L$ (↑)

Resolving horizontally all the forces at joint *A*,

$$P_{AB} \cos \theta_1 = P_{AC} \cos \theta_2 \quad \text{...(1)}$$

Resolving vertically, we get

$$P_{AB} \sin \theta_1 + P_{AC} \sin \theta_2 = W \quad \text{...(2)}$$

where P_{AB} and P_{AC} are the forces or stresses in *AB* and *AC* respectively.

Solving equations (1) and (2), P_{AB} and P_{AC} can be easily calculated.

The force or stress in *AB* can also be calculated by considering the equilibrium of joint *B*, where $P_{BA} \sin \theta_1 = R_B$ (compressive). Similarly, P_{CA} can also found considering the equilibrium of joint *C*. Thus $P_{CA} \sin \theta_2 = R_C$ (compressive).

24.4. METHOD OF SECTIONS

This method is more useful if the stresses in a limited number of members only are to be found. The method enables the stress in a member of a simple frame to be calculated without first calculating the stresses in a great number of other members. The method is based on the principle that if a structure is divided by an ideal surface into two parts, the forces in the bars cut by the ideal surface, together with the external forces on either part of the divided structure, form a system of forces in equilibrium. The equations of static equilibrium can be applied to *any one* portion of the frame.

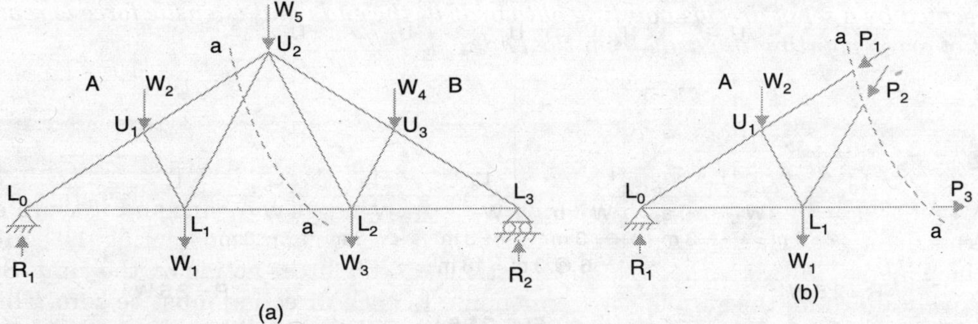

Fig. 24.5

For example, let us take the case of a frame shown in Fig. 24.5 (*a*). In order to find the forces (generally called stresses) in the members U_1U_2, L_1U_2 and L_1 L_2, pass a section *a-a* so as to cut these three members, separating the whole truss into two parts *A* and *B*. Since the whole truss is in equilibrium, each portion *A* or *B* is in equilibrium under the system of forces acting on it.

The portion *A* has been shown in Fig. 24.5 (*b*). It is in equilibrium under the external forces W_1, W_2 and R_1 *and under the forces* P_1, P_2 *and* P_3 *exerted by the portion B on A at the section where a-a cuts the frame.* These six forces keep the portion *A* of the truss in equilibrium. The equations of static equilibrium for the portion *A* are :

(*i*) The algebraic sum of the resolved parts of these above forces in any direction must be zero.

(*ii*) The algebraic sum of the moment of the forces about any point must be zero.

Thus, to find the force P_1 in member U_1U_2, take the moment of the forces about the point L_1. The other two unknown forces P_2 and P_3 pass through this point. Hence the unknown force P_1 can be easily found from moment equation. To find the force P_2, take the moment of all the force about point L_0 through which P_1 and P_3 pass. Similarly, to find the force P_3, take the moment of all the forces about the point U_2, where P_1 and P_2 meet. Thus, the forces P_1, P_2 and P_3, which are either pulls or pushes exerted by the portion *B* (and hence pull or pushes in the members U_1U_2, L_1U_2 and L_1L_2), can be found.

The method is effective only if it cuts three members. To evaluate the pull or push in any one of these members, the moments are taken about the point of intersection of other two forces.

The method of joints and method of sections will now be applied to some of common frames.

24.5. PRATT TRUSS WITH PARALLEL CHORDS

Let us consider a Pratt truss with parallel chords, loaded as shown in Fig. 24.6. We shall apply both the methods to find the forces in all the members of the truss. The reaction *R* at each support is equal to 2.5 *W*.

To find the forces in L_0L_1 and L_0U_1, consider joint L_0. We have thus,

$$P_{L_0U_1} \sin \theta = R = 2.5\ W$$

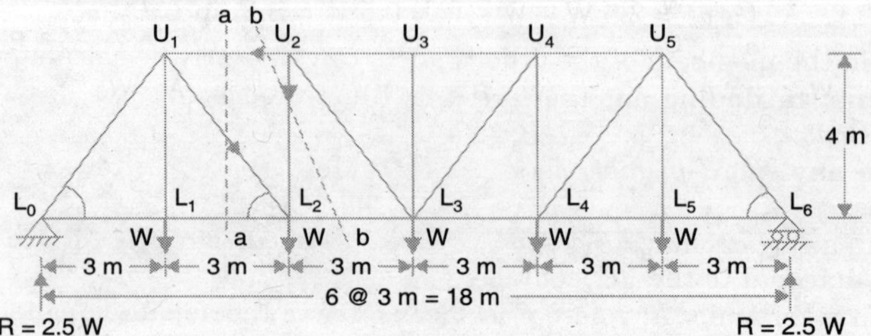

Fig. 24.6

$$\therefore \qquad P_{L_0U_1} = \frac{2.5\,W}{\sin\theta} \text{ where } P_{L_0U_1} \text{ is the force in member } L_0U_1$$

$$\sin\theta = \frac{4}{\sqrt{3^2 + 4^2}} = \frac{4}{5} = 0.8\ ;\ \cos\theta = \frac{3}{5} = 0.6$$

$$P_{L_0U_1} = \frac{2.5\,W}{0.8} = 3.125\,W \text{ (compressive)}$$

Also, $\quad P_{L_0L_1} = P_{L_0U_1} \cos\theta = 3.125\,W \times 0.6 = 1.875\,W$ (tension)

Considering joint L_1, $P_{L_1U_1} = W$ (tension)

To find the forces in the members of panel L_1L_2, pass a section *a-a*, to cut the member U_1U_2, U_1L_2 and L_1L_2.

Consider the equilibrium of the portion to the left of the section *a.a*. Thus, for $P_{U_1U_2}$, take moments about L_2. The moments of the external forces = $(2.5 \times 6 - 1 \times 3)\,W = 12\,W$ in the clockwise direction.

The direction of $P_{U_1U_2}$ must be such as to provide anticlock-wise moment for equilibrium of portion under consideration.

Thus $\quad P_{U_1U_2}\,(4) = 12\,W \quad$ or $\quad P_{U_1U_2} = \dfrac{12\,W}{4} = 3\,W.$

Since $P_{U_1U_2}$ pushes the portion of the truss to the left of section *a-a*, $P_{U_1U_2}$ is compressive.

Similarly, for $P_{L_1L_2}$ take moments about U_1. The moment of the external forces = $2.5 \times 3W = 7.5\,W$, in the clockwise direction.

The direction of $P_{L_1L_2}$ must be such so as produce anticlock-wise moment, for equilibrium of the portion under consideration.

Thus, $\quad P_{L_1L_2}\,(4) = 7.5\,W \quad$ or $\quad P_{L_1L_2} = \dfrac{7.5\,W}{4} = 1.875\,W$ (pull or tension)

To find $P_{U_1L_2}$, resolve the forces vertically.

Thus $\quad P_{U_1L_2} \sin\theta = R - W = 2.5 - W = 1.5\,W = F \quad$ where $\quad F = $ S.F. in panel L_1L_2.

Hence $\quad P_{U_1L_2} = F \operatorname{cosec}\theta = \dfrac{1.5\,W}{0.8} = 1.875\,W.$

Since $P_{U_1L_2}$ *pulls* the left portion of the truss, $P_{U_1L_2}$ is *tensile*.

Similarly, to find the forces in the members of the panel L_2L_3 pass a section *b-b* and consider the equilibrium of the portion to the left of *b-b*. $P_{U_2L_2}$ can be found by resolving the forces vertically.

Thus $\qquad P_{U_2L_2} = 2.5\ W - W - W = 0.5\ W = \text{S.F. in } L_2L_3.$

The forces in the members of the other panels can similarly be calculated.

Conclusions. In finding out the forces as illustrated above, we draw the following important *conclusions*:

(1) Force in any *chord member* is given by B.M. at the opposite joint divided by the height of the truss.

(2) Forces in any *diagonal* = F cosec θ, where F is shear force in the panel and θ is the inclination of the diagonal to the horizontal.

(3) Stress in a vertical = F, where F is shear force in the adjoining panel.

(4) The *direction* of the force in any cut member under consideration will be such as to either balance the moments of all the external forces about one point or to balance the resolved parts of all the external forces in one particular (say vertical) direction. If the *left* portion of the truss is under consideration for equilibrium, the *direction* of the force in any cut section will be marked by an arrow to the *right* of the cutting section. For example, the arrow in U_1U_2 to the right section *a-a* represents a compressive force exerted by the right portion of the truss on to the left portion and hence member U_1U_2 will be in compression. Similarly, the arrow in U_1L_2 to the right of section represents a tensile force exerted by the right portion of the truss on the left and hence member U_1L_2 will be in tension.

The stresses in all members of the truss have been tabulated in Table 24.1. The direction of the forces have been marked in Fig. 24.7 with proper arrows.

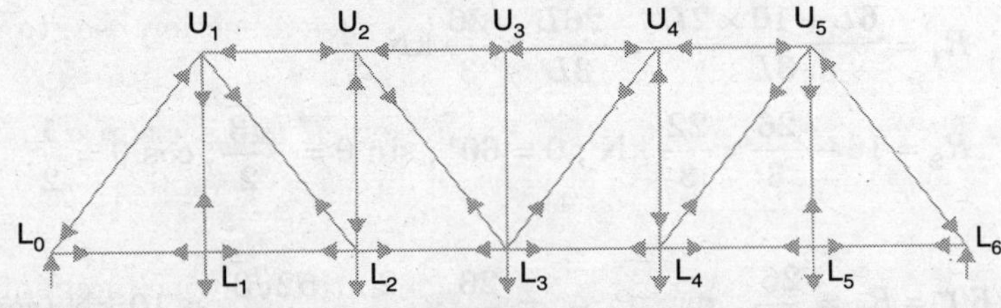

Fig. 24.7

TABLE 24.1. PRATT TRUSS WITH PARALLEL CHORDS

+ For compression — For tension

Member	*Force/Stress*	*Member*	*Force/Stress*
L_0L_1	– 1.875 W	U_5L_6	+ 3.125 W
L_1L_2	– 1.875 W	U_1L_1	– W
L_2L_3	– 3 W	U_1L_2	– 1.875 W
L_3L_4	– 3 W	U_2L_2	+ 0.5 W
L_4L_5	– 1.875 W	U_2L_3	– 0.625 W
L_5L_6	– 1.875 W	U_3L_3	0
L_0U_1	+ 3.125 W	U_4L_3	– 0.625 W
U_1U_2	+ 3 W	U_4L_4	+ 0.5 W
U_2U_3	+ 3.375 W	U_5L_4	– 1.875 W
U_3U_4	+ 3.375 W	U_5L_5	– W
U_4U_5	+ 3 W		

24.6. WARREN GIRDER : LOADS AT LOWER JOINTS

Let us now take a Warren girder of equilateral triangles, having 3 bays of length L each, loaded at lower joints as shown in Fig. 24.8.

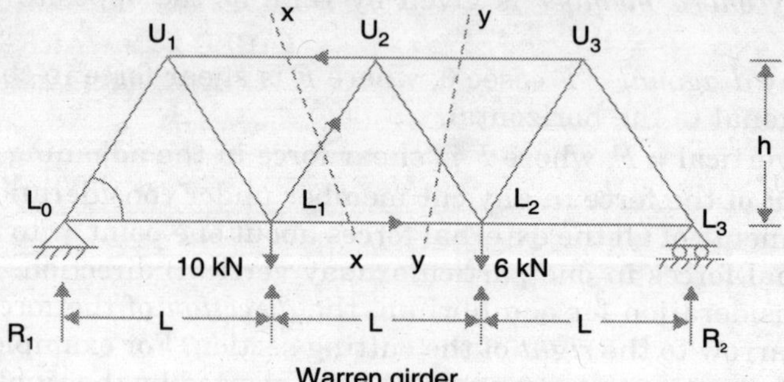

Fig. 24.8

Height of girder $= h = L\dfrac{\sqrt{3}}{2} = 0.866L$

To find R_1, take moment about L_3

$$\therefore \quad R_1 = \frac{6L + 10 \times 2L}{3L} = \frac{26L}{3L} = \frac{26}{3} \text{ kN}$$

$$R_2 = 16 - \frac{26}{3} = \frac{22}{3} \text{ kN} \; ; \; \theta = 60^\circ \; ; \; \sin\theta = \frac{\sqrt{3}}{2} ; \; \cos\theta = \frac{1}{2}.$$

***Joint L_0* :**

$$P_{L_0U_1} \sin 60^\circ = R_1 = \frac{26}{3} \quad \text{or} \quad P_{L_0U_1} = \frac{26}{3} \times \frac{2}{\sqrt{3}} = \frac{52\sqrt{3}}{9} \simeq 10 \text{ kN (compression)}$$

$$\therefore \quad P_{L_0L_1} = P_{L_0U_1} \cos 60^\circ = \frac{10}{2} = 5 \text{ kN (tension)}$$

***Joint U_1* :** Since L_0U_1 and U_1L_1 have same inclination with vertical,

$$P_{U_1L_1} = -P_{U_1L_0} \quad \text{or} \quad P_{U_1L_1} = 10 \text{ kN (tension).}$$

To find forces in all the members of panel L_1L_2, pass a section x-x, and consider the equilibrium of the left portion of the truss.

$$P_{U_1U_2} = \frac{M_{L1}}{h} = \frac{R_1 L}{L\sqrt{3}/2} = \frac{26}{3} L \times \frac{2}{\sqrt{3}L} = \frac{52\sqrt{3}}{9} \simeq 10 \text{ kN (comp.)}$$

$$P_{L_1L_2} = \frac{M_{U2}}{h} = \left(\frac{26}{3} \times \frac{3}{2} L - \frac{10L}{2}\right) \frac{2}{\sqrt{3}L} = (13 - 5)\frac{2}{\sqrt{3}} = 9.23 \text{ kN (tension)}$$

$$P_{L_1U_2} = \frac{F_{L_1L_2}}{\sin 60^\circ} = \frac{R_1 - 10}{\sin 60^\circ}\left(\frac{26}{3} - 10\right)\frac{2}{\sqrt{3}} = -1.54 = 1.54 \text{ kN (tension)}$$

Joint U_2 : Since L_1U_2 and L_2U_2 have the same inclination,

$$P_{U_2L_2} = -P_{U_2L_1} = 1.54 \text{ kN (comp.)}$$

Similarly, the forces in other members can be found out. The forces in all members have been tabulated in Table 24.2 and the directions of the forces in all the members have been marked in Fig. 24.9.

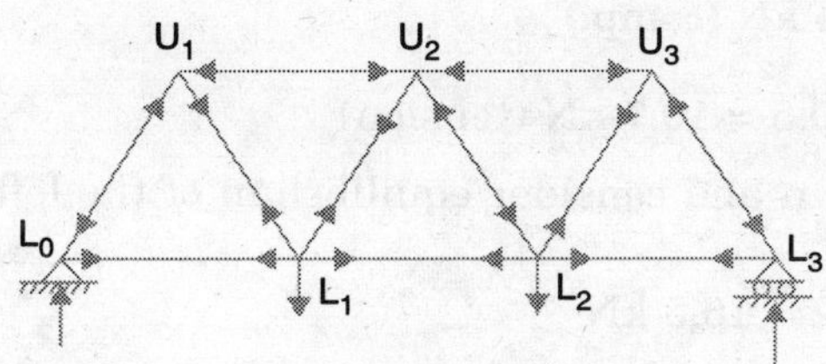

Fig. 24.9

TABLE 24.2 WARREN GIRDER : LOADS AT LOWER CHORD

+ Compression − Tension

Member	*Stress (kN)*	*Member*	*Stress (kN)*
L_0L_1	− 5	L_0U_1	+ 10
L_1L_2	− 9.23	U_3L_3	+ 8.45
L_2L_3	− 4.23	U_1L_1	− 10
U_1U_2	+ 10	U_2L_1	− 1.54
U_2U_3	+ 8.45	U_2L_2	+ 1.54
		U_3L_2	− 8.45

24.7. WARREN GIRDER : LOADS AT UPPER AND LOWER JOINTS

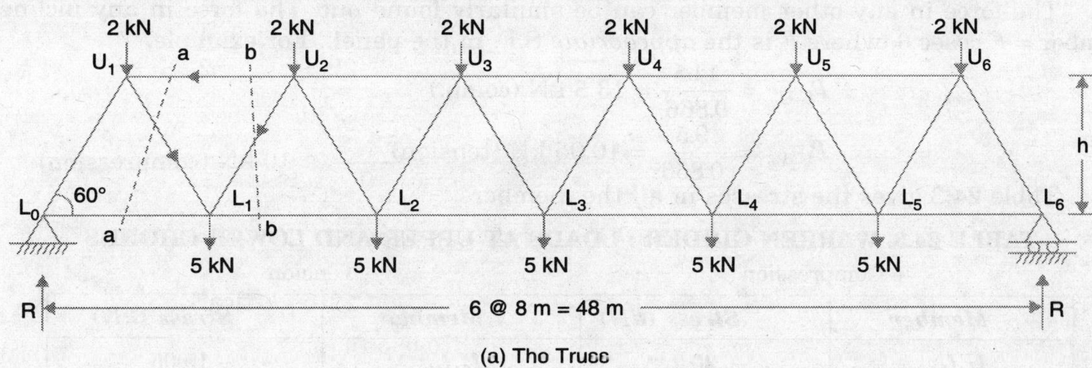

(a) Tho Trucc

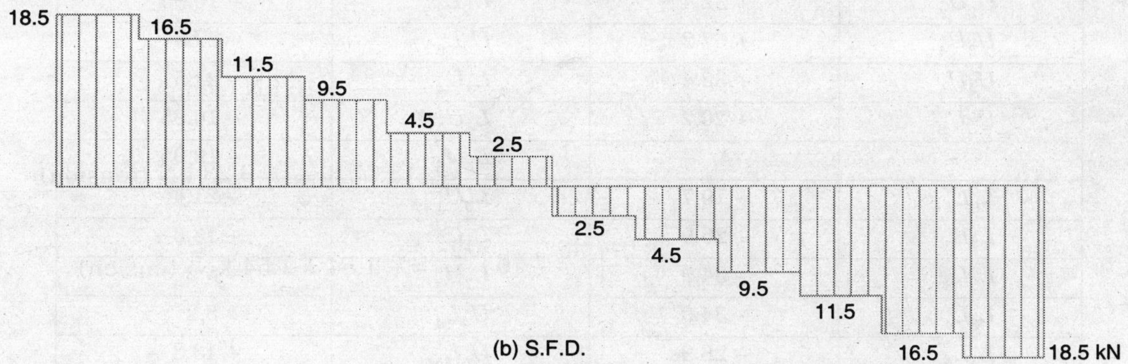

(b) S.F.D.

Fig. 24.10

Fig. 24.10 shows a Warren girder loaded at both upper and lower joints. There are 6 panels, each 8 m long.

Height h of the girder $= \dfrac{8 \times \sqrt{3}}{2} = 4\sqrt{3}$ m $= 6.93$ m

Reactions $R = 2 \times 3 + 5 \times \dfrac{5}{2} = 18.5$ kN

$$\theta = 60^\circ : \sin\theta = \frac{\sqrt{3}}{2} = 0.866 \; ; \cos\theta = \frac{1}{2} = 0.5.$$

***Joint L_0* :** $P_{L_0U_1} = \dfrac{18.5}{\sin 60^\circ} = \dfrac{18.5}{0.866} = 21.4$ kN (comp.)

$P_{L_0L_1} = P_{L_0U_1} \cos 60^\circ = 21.4 \times 0.5 = 10.7$ kN (tension)

***Panel U_1U_2* :** Pass sections *a-a* and *b-b* as shown and consider equilibrium of the left portion every time.

$$P_{U_1L_1} \sin 60^\circ = \text{S.F.} = R - 2 = 18.5 - 2 = 16.5 \text{ kN}$$

$$\therefore \quad P_{U_1L_1} = \frac{16.5}{0.866} = 19.05 \text{ kN (tension)}$$

The only difference between this case and that of § 24.5 is that here there are two values of S.F., one for the left hand half of the panel and other for the right hand half the panel. The S.F. diagram has also been plotted in Fig. 24.10 (*b*).

$$P_{U_1U_2} = \frac{M_{L1}}{h} = \frac{(18.5 \times 8) - (2 \times 4)}{6.93} = \frac{140}{6.93} = 20.2 \text{ kN (comp.)}$$

The force in any other member can be similarly found out. The force in any inclined member = F cosec θ, where F is the *appropriate* S.F. in the panel. For example,

$$P_{L_1U_2} = \frac{11.5}{0.866} = 13.3 \text{ kN (comp.)}$$

$$P_{U_2L_2} = \frac{9.5}{0.866} = 10.95 \text{ kN (tension)}$$

Table 24.3 gives the stresses in all the members.

TABLE 24.3. WARREN GIRDER : LOADS AT UPPER AND LOWER CHORDS

+ Compression – Tension

Member	***Stress (kN)***	***Member***	***Stress (kN)***
U_1U_2	+ 20.2	U_1L_1	– 19.05
U_2U_3	+ 32.4	U_2L_2	– 10.95
U_3U_4	+ 67.2	U_3L_3	– 2.85
U_4U_5	+ 32.4	L_3U_4	– 2.85
U_5U_6	+ 20.2	L_4U_5	– 10.95
		L_5U_6	– 19.05
L_0L_1	– 10.7	L_0U_1	+ 21.4
L_1L_2	– 26.9	L_1U_2	+ 13.3
L_2L_3	– 34.9	L_2U_3	+ 5.2
L_3L_4	– 34.9	U_4L_4	+ 5.2
L_4L_5	– 26.9	U_5L_5	+ 13.3
L_5L_6	– 10.7	U_6L_6	+ 21.4

24.8. PRATT TRUSS WITH INCLINED CHORDS

Fig. 24.11 shows a Pratt truss with inclined chords, loaded at lower points.

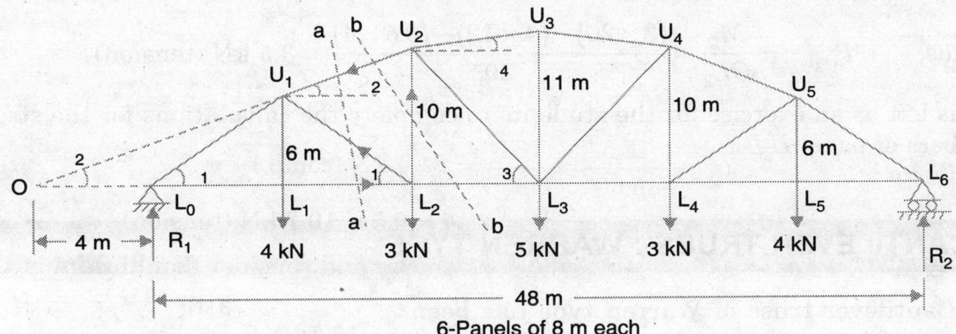

Fig. 24.11

Reactions $R_1 = R_2 = 9.5$ kN ;

$$\sin\theta_1 = \frac{6}{\sqrt{36+64}} = \frac{6}{10} = 0.6 \text{ ; } \cos\theta_1 = \frac{8}{10} = 0.8 \text{ ;}$$

$$\tan\theta_2 = \frac{10-6}{8} = 0.5 \text{ ; } \theta_2 = 26^\circ\ 32 \text{ ; } \sin\theta_2 = 0.446 \text{ ; } \cos\theta_2 = 0.893.$$

$$\sin\theta_3 = \frac{10}{\sqrt{100+64}} = \frac{10}{12.8} = 0.781 \text{ ; } \cos\theta_3 = \frac{8}{12.8} = 0.625$$

$$\sin\theta_4 = \frac{11-10}{\sqrt{(11-10)^2+64}} = \frac{1}{\sqrt{65}} = \frac{1}{8.06} = 0.124 \text{ ; } \cos\theta_4 = \frac{8}{8.06} = 0.992.$$

Panel L_0L_1

$$P_{L_0U_1}\sin\theta_1 = 9.5. \text{ Hence } P_{L_0U_1} = \frac{9.5}{0.6} = 15.85 \text{ kN (comp.)}$$

$$P_{L_0L_1} = P_{L_0U_1}\cos\theta_1 = 15.85 \times 0.8 = 12.65 \text{ kN (tension)}$$

$$P_{U_1L_1} = 4 \text{ kN (tension)}$$

Panel L_1L_2 **:** Pass a section *a-a* as shown,

$$P_{L_1L_2} = \frac{M_{U1}}{6} = \frac{9.5\times 8}{6} = \frac{76}{6} = 12.67 \text{ kN (tension)}$$

Produce U_2U_1 to meet the bottom chord line at O as shown.

$$OL_2 = \frac{U_2L_2}{\tan\theta_2} = \frac{10}{0.5} = 20 \text{ m. Hence } OL_0 = 20 - 2\times 8 = 4 \text{ m.}$$

To find $P_{U_1L_2}$ take moments at O.

$$\text{Thus,}\quad P_{U_1L_2} = \frac{M_0}{OL_2\sin\theta_1} = \frac{(4\times 12)-(9.5\times 4)}{20\times 0.6} = \frac{48-38}{12} = 0.834 \text{ kN (comp).}$$

Also, $$P_{U_1U_2} = \frac{M_{L_2}}{OL_2 \sin \theta_2} = \frac{(9.5 \times 16) - (4 \times 8)}{20 \times 0.446} = 13.45 \text{ kN (compression)}$$

To find $P_{U_2L_2}$ pass section *b-b* as shown and take moments about *O*.

Thus, $$P_{U_2L_2} = \frac{M_0}{OL_2} = \frac{(3 \times 20) + (4 \times 12) - (9.5 \times 4)}{20} = 3.5 \text{ kN (tension)}$$

It is left as an exercise for the students to complete the calculations for the stresses in the members of panel L_2L_3.

24.9. CANTILEVER TRUSS : WARREN TYPE

A Cantilever truss of Warren type has been shown in Fig. 24.12.

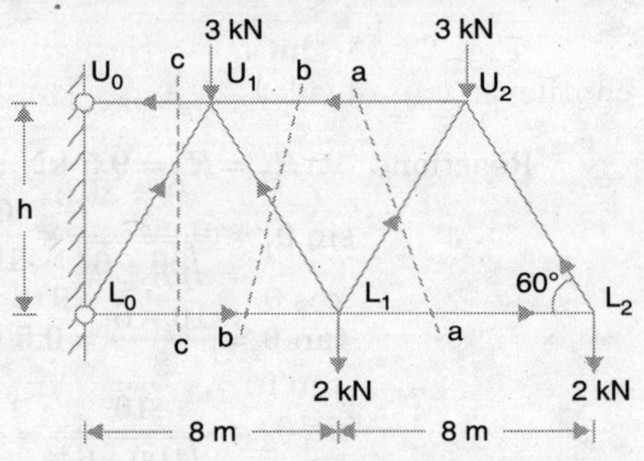

Fig. 24.12

$$\sin \theta = \sin 60° = \frac{\sqrt{3}}{2} = 0.866 \;;$$

$\cos 60° = 0.5$; Height $h = \sqrt{64 - 16} = 6.93$ m.

***Panel L_1L_2* :**

$$P_{L_2U_2} = \frac{2}{\sin 60°} = \frac{2}{0.866} = 2.31 \text{ kN (tension)}$$

$$P_{L_2L_1} = P_{L_2U_2} \cos 60° = 2.31 \times 0.5 = 1.155 \text{ (comp.)}$$

Pass a section *a-a* and consider the equilibrium of right hand portion.

$$P_{L_1U_2} = \frac{F_{L_1L_2}}{\sin 60°} = \frac{3+2}{0.866} = 5.77 \text{ kN (compression).}$$

***Panel U_1U_2* :**

$$P_{U_1U_2} = \frac{M_{L_1}}{6.93} = \frac{(2 \times 8) + (3 \times 4)}{6.93} = 4.04 \text{ kN (tension)}$$

Pass a section *b-b* and consider the equilibrium of right hand portion.

$$P_{U_1L_1} = \frac{F_{L_0L_1}}{\sin 60°} = \frac{2+2+3}{0.866} = 8.08 \text{ kN (tension)}$$

***Panel L_0L_1* :**

$$P_{L_0L_1} = \frac{M_{U_1}}{6.93} = \frac{(3 \times 8) + (2 \times 12) + (2 \times 4)}{6.93} = 8.08 \text{ kN (compression).}$$

Pass a section *c-c* and consider the equilibrium of the right hand portion.

$$P_{L_0U_1} = \frac{\text{S.F.}}{\sin 60°} = \frac{3+3+2+2}{0.866} = 11.54 \text{ kN (compression)}$$

$$P_{U_0U_1} = \frac{M_{L_0}}{6.93} = \frac{(2 \times 16) + (2 \times 8) + (3 \times 12) + (4 \times 3)}{6.93} = 13.85 \text{ kN (tension).}$$

The stresses have been tabulated in Table 24.4.

TABLE 24.4. WARREN TYPE CANTILEVER TRUSS

+ Compression — Tension

Member	*Stress (kN)*	*Member*	*Stress (kN)*
U_0U_1	– 13.85	L_0L_1	+ 8.08
U_1U_2	– 4.04	L_1L_2	+ 1.155
U_1L_1	– 8.08	L_1U_2	+ 5.77
U_2L_2	– 2.31	L_0U_1	+ 11.54

24.10. CANTILEVER TRUSS : TRIANGULAR TYPE

Fig. 24.13 shows a triangular type cantilever truss loaded on upper joints.

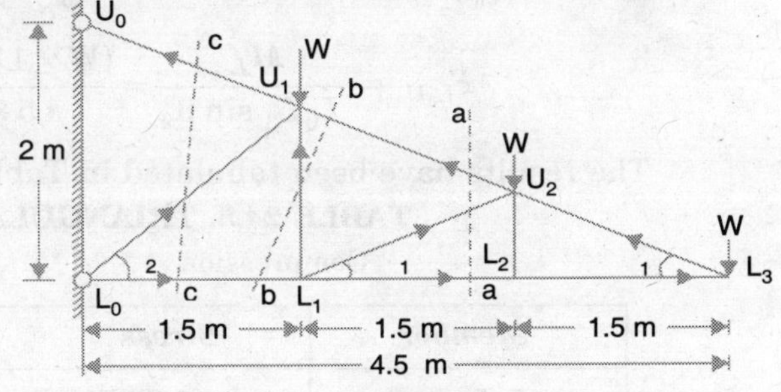

Fig. 24.13

$$\sin \theta_1 = \frac{2}{\sqrt{4 + 20.3}} = \frac{2}{4.9} = 0.407\ ;$$

$$\cos \theta_1 = \frac{4.5}{4.9} = 0.918.$$

$$\tan \theta_1 = \frac{2}{4.5} = 0.444$$

$$U_2L_2 = \frac{2}{3}\ \text{m}\ ;\ U_1L_1 = \frac{4}{3}\ \text{m}.$$

$$\tan \theta_2 = \frac{U_1L_1}{L_0L_1} = \frac{1.333}{1.5} = 0.89 \quad \therefore \quad \theta_2 = 41^\circ\ 40'$$

$$\sin \theta_2 = 0.665.$$

Panel L_2L_3 :

$$P_{L_3U_2} = \frac{W}{\sin \theta_1} = 2.45\ W \text{ (tension)}$$

$$P_{L_3L_2} = P_{L_3U_2} \cos \theta_1 = 2.45\ W \times 0.918 = 2.25\ W \text{ (comp.)}$$

Since there is no vertical load at L_2, $P_{L_2U_2} = 0$.

Panel L_1L_2 :

Pass a section *a-a* and consider the equilibrium of the right hand portion.

$$P_{U_1U_2} = \frac{M_{L_1}}{L_1L_3 \sin \theta_1} = \frac{(W \times 3) + (W \times 1.5)}{3 \times 0.407} = 3.68\ W \text{ (tension)}.$$

$$P_{L_1L_2} = \frac{M_{U_2}}{U_2L_2} = \frac{1.5\ W \times 3}{2} = 2.25\ W \text{ (comp.)}$$

To find $P_{L_1U_2}$ take moments about L_3.

$$P_{L_1U_2} = \frac{M_{L_3}}{L_1L_3 \sin \theta_1} = \frac{W \times 1.5}{3 \times 0.407} = 1.225\ W \text{ (comp.)}$$

Panel L_0L_1 :

To find $P_{U_1L_1}$ pass a section *b-b* and take moments about L_3.

$$P_{U_1L_1} = \frac{M_{L_3}}{L_1L_3} = \frac{W \times 1.5}{3} = 0.5\ W \text{ (tension)}$$

Pass a section *c-c* and consider the equilibrium of right hand portion.

$$P_{L_0L_1} = \frac{M_{U_1}}{U_1L_1} = \frac{(W \times 3) + (W \times 1.5)}{4/3} = 3.37\ W \text{ (comp.)}$$

$$P_{U_0U_1} = \frac{M_{L_0}}{L_0L_3 \sin \theta_1} = \frac{(W \times 4.5) + (W \times 3) + (W \times 1.5)}{4.4 \times 0.407} = 4.91\ W \text{ (tension)}$$

To find $P_{L_0U_1}$ take moment about L_3.

$$P_{L_0U_1} = \frac{M_{L_3}}{L_0L_3 \sin \theta_2} = \frac{(W \times 1.5) + W \times 3}{4.5 \times 0.665} = 1.5\ W \text{ (comp.)}$$

The results have been tabulated in Table 24.5.

TABLE 24.5. TRIANGULAR CANTILEVER TRUSS

+ Compression − Tension

Member	***Stress***	***Member***	***Stress***
L_0L_1	+ 3.37 *W*	U_1L_1	− 0.5 *W*
L_1L_2	+ 2.25 *W*	U_2L_2	0
L_2L_3	+ 2.25 *W*	U_1L_0	+1.5 *W*
U_0U_1	− 4.91 *W*	U_2L_1	+ 1.225 *W*
U_1U_2	− 3.68 *W*		
U_2L_3	− 2.45 *W*		

24.11. K-TRUSS

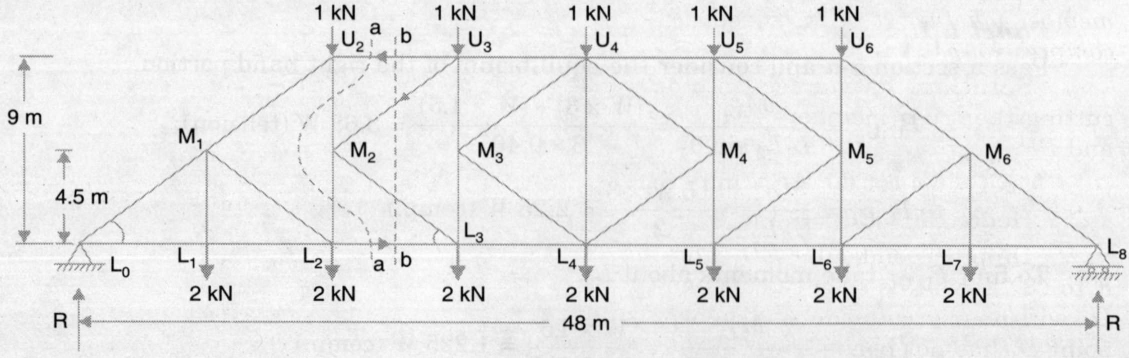

Fig. 24.14

Let us now analyse a K-truss as shown in Fig. 24.14.

We shall investigate the stresses in the members of panel L_2L_3 only.

Pass sections a-a and b-b as shown and consider the equilibrium of left portion every time.

$$\sin\theta = \frac{4.5}{\sqrt{(4.5)^2 + 6^2}} = 0.6 \text{ ; Reaction } R = \frac{19}{2} = 9.5 \text{ kN.}$$

Section a-a :

$$P_{U_2U_3} = \frac{M_{L_2}}{9} = \frac{(9.5 \times 12) - (2 \times 6)}{9} = \frac{102}{9} = 11.35 \text{ (comp.)}$$

Similarly, $P_{L_2L_3} = \dfrac{M_{U_2}}{9} = \dfrac{102}{9} = 11.35$ kN (tension).

Section b-b :

Consider the equilibrium of joint M_2. Since inclination of M_2U_3 and M_2L_3 is the same, we get $P_{M_2U_3} = P_{M_2L_3}$.

Again considering the equilibrium of the portion to left of the section b-b, we have

$P_{M_2U_3} \sin\theta + P_{M_2L_3} \sin\theta$ = shear force in panel L_2L_3.

$$\therefore \quad 2P_{M_2U_3} \sin\theta = 2P_{M_2L_3} \sin\theta = 11.35 - 2 - 1 - 2 = 6.35.$$

$$\therefore \quad P_{M_2U_3} = P_{M_2L_3} = \frac{6.35}{2\sin\theta} = \frac{6.35}{2 \times 0.6}$$

$$\therefore \quad P_{M_2U_3} = 5.29 \text{ kN (comp.) ; } P_{M_2L_3} = 5.29 \text{ kN (tension).}$$

It is left as an exercise for the students to calculate forces in other members.

24.12. ADDITIONAL EXAMPLES USEFUL FOR COMPETITIVE EXAMINATIONS

Example 24.1. *The roof truss shown in Fig. 24.15 and loaded as indicated, is supported on a roller bearing at B and pin jointed at A. Find the forces in DC, EC and EF by method of sections, stating for each member whether it is in tension or compression.*

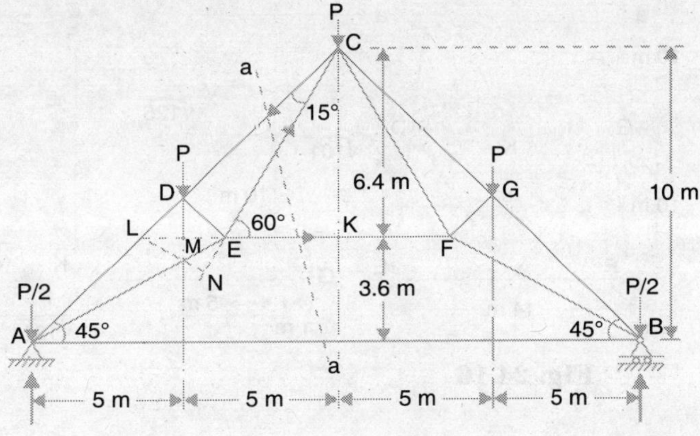

Fig. 24.15

Solution. Pass a section a-a cutting the truss members DC, EC and EF.

$EK = 6.4 \cot 60° = 3.7$ m ;

Hence $EF = EC = 7.4$ m.

Drop perpendicular ED_1 from E to CD. It will be seen that D and D_1 will nearly coincide and hence point D_1 has not been shown in the diagram.

$\therefore \quad ED_1 \simeq ED = EC \sin 15° = 7.40 \times 0.259 = 1.915$ m.

Pass a section *a-a*

For finding P_{CD}, take moments about *E*. Consider the equilibrium of the portion to the left of section *a-a*.

$$\therefore \quad P_{CD} = \frac{M_E}{ED} = \frac{1.5P(10-3.70) - P(5-3.70)}{1.915} = 4.265P \text{ (compression).}$$

For finding P_{EF} take moments about *C*.

$$\therefore \quad P_{EF} = \frac{M_C}{6.4} = \frac{(1.5P \times 10) - (P \times 5)}{6.4} = 1.562P \text{ (tension).}$$

For finding P_{CE}, produce *CD* and *FE* to meet in *L*. From *L* drop perpendicular *LN* on *CE* produced.

$$LE = DE\sqrt{2} = 1.951\sqrt{2} = 2.71 \text{ m}$$

$$LN = LE \sin 60° = 2.71 \sin 60° = 2.35 \text{ m}$$

$$LM = LK - MK = 6.4 - 5 = 1.4 \text{ m.}$$

Taking moments about *L*, we get

$$P_{CE} = \frac{M_L}{LN} = \frac{1.5P(5-1.4) + P \times 1.4}{2.35} = 2.902P \text{ (tension).}$$

Example 24.2. *Fig. 24.16 shows a symmetrical structure carrying vertical loads of 12 kN at A and B and a horizontal load of 6 kN at B. Determine the forces in the four members HG, HK, CK and CD.*

Solution.

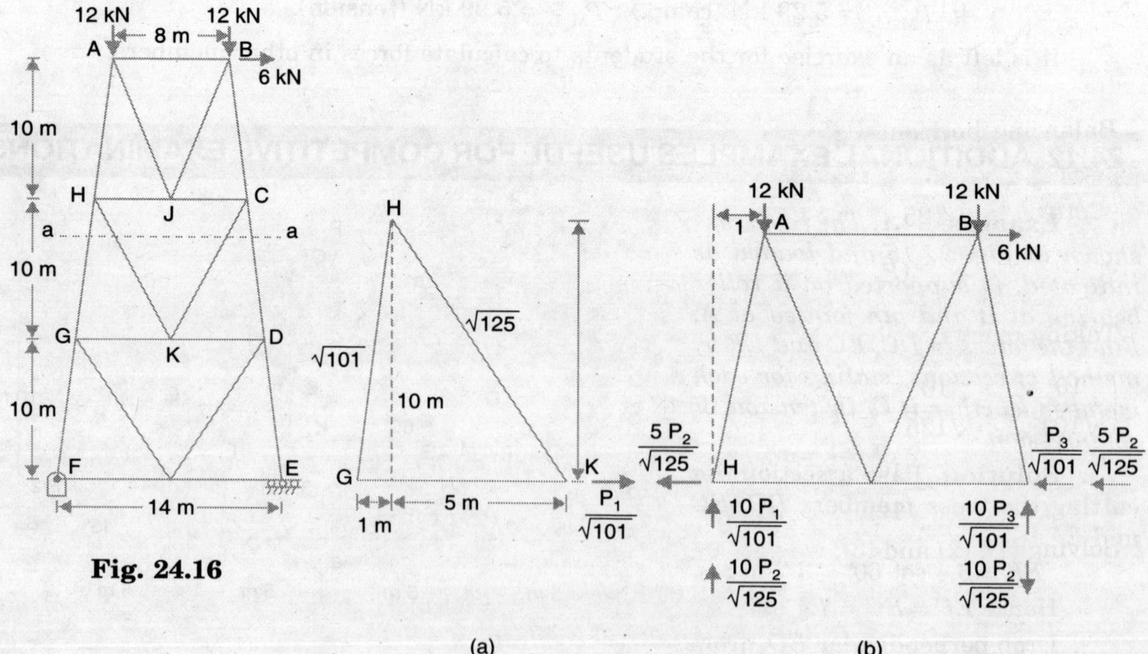

Fig. 24.16

Fig. 24.17

Let P_1 = force in *GH* (assumed compressive)
P_2 = force in *HK* (assumed compressive)
P_3 = force in *DC* (assumed compressive)

At the joint *K*, the vertical components of the forces in *HK* and *CK* must balance and hence the forces in these members are equal. Thus, force in *CK* = P_2 (assumed tensile).

The horizontal and vertical components of P_1, P_2 and P_3 are as follows [see Fig. 24.17 (*a*)]

$$P_{1H} = \frac{P_1 \times 1}{\sqrt{101}} \rightarrow ;\ P_{1V} = \frac{10}{\sqrt{101}} P_1 \uparrow \text{ (For member } GH)$$

$$P_{2H} = \frac{5}{\sqrt{125}} P_2 \leftarrow ;\ P_{2V} = \frac{10}{\sqrt{125}} P_2 \uparrow \text{ (For member } HK)$$

$$P_{2H} = \frac{5P_2}{\sqrt{125}} \leftarrow ;\ P_{2V} = \frac{10P_2}{\sqrt{125}} \downarrow \text{ (For member } CK)$$

$$P_{3H} = \frac{P_3}{\sqrt{101}} \leftarrow ;\ H_{2V} = \frac{10}{\sqrt{101}} P_3 \uparrow \text{ (For member } DC)$$

Pass a section *a-a*, as shown in Fig. 24.16, and consider the equilibrium of the upper portion. Fig. 24.17 (*b*) shows the upper portion replacing P_1, P_2 and P_3 by their components.

Balancing vertically

$$\frac{10}{\sqrt{101}} P_1 + \frac{10}{\sqrt{101}} P_3 + \frac{10P_2}{\sqrt{125}} - \frac{10P_2}{\sqrt{125}} = 12 + 12$$

$$\therefore \quad P_1 + P_3 = \frac{\sqrt{101}}{101} \times 24 \qquad \text{...(1)}$$

Balancing horizontally,

$$\frac{1}{\sqrt{101}} P_3 + \frac{5}{\sqrt{125}} P_2 + \frac{5}{\sqrt{125}} P_2 - \frac{1}{\sqrt{101}} \text{P}_1 = 6$$

$$(P_3 - P_1) \frac{1}{\sqrt{101}} + \frac{10}{\sqrt{125}} P_2 = 6 \qquad \text{...(2)}$$

Taking moments about H

$$\left(\frac{10}{\sqrt{101}} P_3 - \frac{10}{\sqrt{125}} P_2 \right) 10 = (12 \times 1) + (12 + 9) + (6 \times 10)$$

$$\frac{10}{\sqrt{125}} P_2 = \frac{10}{\sqrt{101}} P_3 - 18 \qquad \text{...(3)}$$

Solving (1), (2) and (3), we get

$P_1 = P_{GH}$ = 2.01 kN (comp.)
$P_2 = P_{KH}$ = 4.472 kN (comp.)
$P_2 = P_{KC}$ = 4.472 kN (tension.)
$P_3 = P_{DC}$ = 22.11 kN (comp.)

Example 24.3. *Determine the forces in the members of the frame shown in Fig. 24.18.*

Solution.

(a) Geometry and reactions

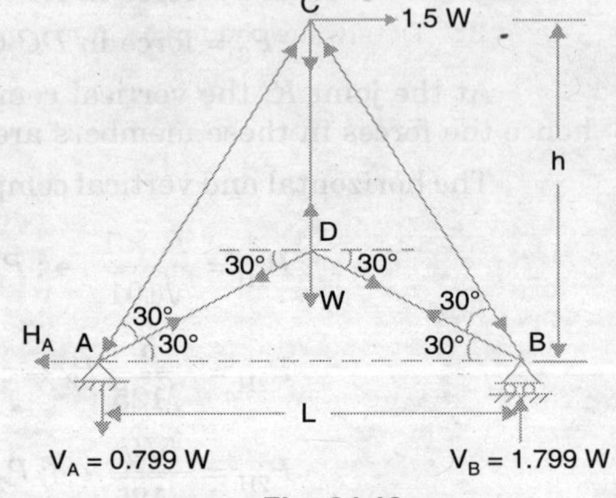

Fig. 24.18

$$h = \frac{L}{2} \tan 60^\circ = 0.866\, L$$

$$V_A = \frac{1}{L}\left(W \times \frac{L}{2} - 1.5\, W \times 0.866 L\right)$$

$$= -\,0.799\, W \text{ (i.e., 0.799 W } \downarrow)$$

$$V_B = \frac{1}{L}\left(W \times \frac{L}{2} + 1.5\, W \times 0.866 L\right)$$

$$= 1.799\, W\, (\uparrow)$$

Since $\Sigma H = 0$, we get $H_A = 1.5\, W\, (\leftarrow)$.

(b) Joint B: By inspection, P_{BC} will be compressive and P_{BD} will be tensile.

Hence $P_{BC} \cos 60^\circ = P_{BD} \cos 30^\circ$, from which

$$P_{BD} = P_{BC} \cdot \frac{\cos 60^\circ}{\cos 30^\circ} = 0.5774\, P_{BC} \qquad \ldots(i)$$

Also, $P_{BD} \sin 30^\circ + V_B = P_{BC} \sin 60^\circ$

or $(0.5774\, P_{BC}) \times 0.5 + 1.799\, W = 0.866\, P_{BC}$

From which $P_{BC} = 3.116\, W$ (comp.)

Hence $P_{BD} = 0.5774 \times 3.116\, W = 1.799\, W$ (tension)

(c) Joint A: By inspection P_{AC} will be compressive while P_{AD} will be tensile.

Resolving vertically, $P_{AC} \sin 60^\circ + V_A = P_{AD} \sin 30^\circ$,

$$\therefore \quad P_{AD} = \frac{1}{\sin 30^\circ}\,[P_{AC} \sin 60^\circ + V_A]$$

$$= \frac{1}{0.5}\,[0.866\, P_{AC} + 0.799\, W] = 1.732\, P_{AC} + 1.598\, W$$

Resolving horizontally,

$$P_{AC} \cos 60^\circ + H_A = P_{AD} \cos 30^\circ$$

or $0.5\, P_{AC} + 1.5\, W = 0.866\,(1.732\, P_{AC} + 1.598\, W)$

From which, $P_{AC} = 0.116\, W$ (comp.)

and $P_{AD} = 1.799\, W$ (tension)

(d) Joint D: By inspection, P_{DC} will tensile.

Resolving vertically, $P_{DC} = P_{DA} \sin 30 + P_{DB} \sin 30^\circ + W$

or $P_{DC} = 2 \times 1.799\, W \times 0.5 + W = 2.799\, W$ (tension)

PROBLEMS

1. Determine the forces in the members of the truss loaded as shown in Fig. 24.19.

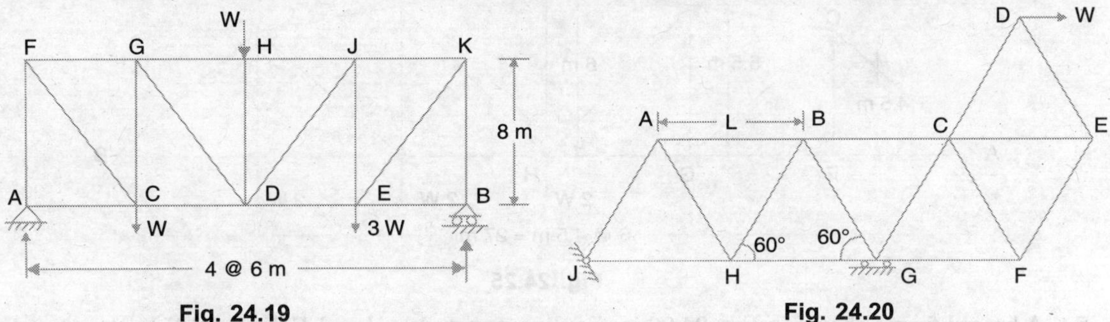

Fig. 24.19

Fig. 24.20

2. A pin jointed structure shown in Fig. 24.20 is pinned to an abutment at J and rests on rollers at G. A load W is applied horizontally at D. Determine the force in all the members.
3. A crane structure is shown in Fig. 24.21. The length of the member AD is 4 m ; all other members are 2 m long. Determine the forces in all the members due to a vertical load of 10 kN at the point A.

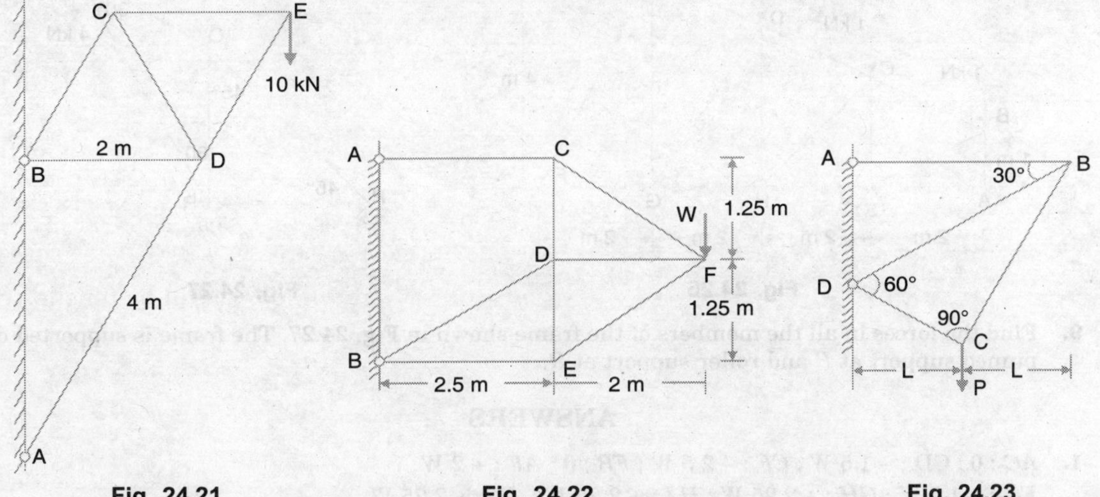

Fig. 24.21

Fig. 24.22

Fig. 24.23

4. Determine the forces in members of the frame loaded as shown in Fig. 24.22.
5. Determine the forces in the members of the frame loaded as shown in Fig. 24.23.
6. A uniformly distributed load of 3 kN per metre run is carried at the panel points of the lower beam of the girder shown in Fig. 24.24. Find the forces in members 1, 2, 3 and 4.

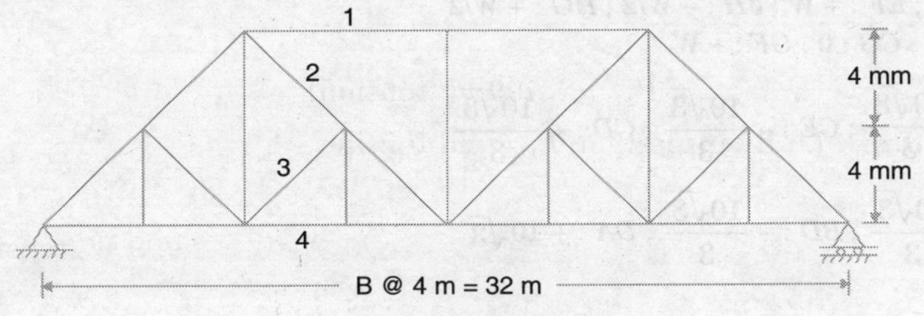

Fig. 24.24

7. Find by calculation, forces in the members *CD*, *DG* and *GH* of the frame loaded as shown in Fig. 24.25.

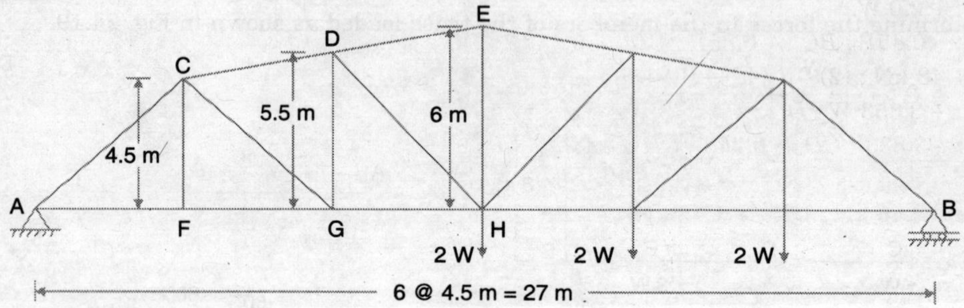

Fig. 24.25

8. A braced frame shown in Fig. 24.26 is simply supported at *J* and *F*. Four 1 kN loads act at *B*, *C*, *D* and *E*. Determine the forces in the members *CJ*, *CD*, *CH* and *JH*. (*U.L.*)

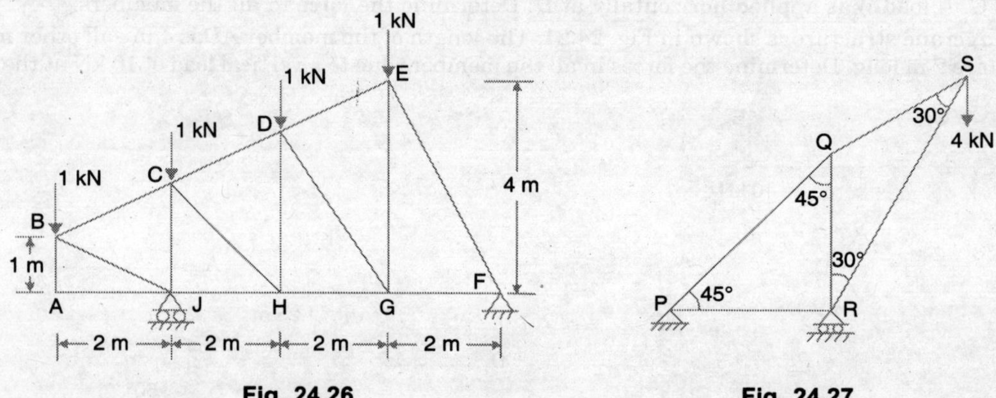

Fig. 24.26 **Fig. 24.27**

9. Find the forces in all the members of the frame shown in Fig. 24.27. The frame is supported on a pinned support at *P* and roller support at *R*.

ANSWERS

1. *AC* : 0 ; CD : – 1.5 W ; *CF* : – 2.5 W ; *FB* : 0 ; *AF* : + 2 W
FG : + 1.5 W ; *GH* : + 2.25 W ; *HJ* : + 2.25 W ; *JK* : + 2.25 W
KB : + 3 W ; *FC* : – 2.5 W ; *GC* : + W ; *GD* : – 1.25 W
HD : + W ; *JD* : 0 ; *JE* : 0 ; *KE* : – 3.75 W.

2. *AB* : – W ; *BC* : – 2W ; *CE* : – W ; *CD* : – W
AJ : – W ; *AH* : + W ; *BH* : – W ; *BG* : + W
CF : – W ; *EF* : + W ; *JH* : – W/2 ; *HG* : + W/2
DE : + W ; *CG* : 0 ; *GF* : + W.

3. $DE : +\dfrac{20\sqrt{3}}{3}$; $CE : -\dfrac{10\sqrt{3}}{3}$; $CD : +\dfrac{10\sqrt{3}}{3}$

$CB : -\dfrac{10\sqrt{3}}{3}$; $BD : -\dfrac{10\sqrt{3}}{3}$; $DA : +10\sqrt{3}$.

4. AC : – 1.8 W ; CE : – 2.12 W ; FE : – 0.236 W : BE : – 0.2 W
CD : + 1.125 W ; DE : + 0.127 W ; DB : + 2.23 W
DF : + 2.0 W.
5. AB : – 0.85P ; BC : – 0.85P ; CD : – 0.517P ; DB : + 1.5P.
6. (1) + 48 kN ; (2) – 17 kN ; (3) + 8.5 kN ; (4) – 42 kN.
7. CD : + 3.353 W ; DG : + 1.273 W ; GH : – 3.273 W.
8. CJ : + 2.833 ; CD, + 0.248 ; CH : – 1.725 ; JH : + 1
9. PQ : – 4.88 kN ; QS : – 4 kN ; SR : + 6.9 kN ;
RP : + 3.46 kN ; QR : + 1.48 kN.

25

Method of Tension Coefficients

25.1. INTRODUCTION

In Chapter 24, we have studied two methods of analysis of plane frames : (*i*) method of joints, and (*ii*) method of sections. We can also find out the forces in the members of plane frame, by the *graphical method*. We now introduce a more general method, called the '*method of tension coefficients*' which is equally applicable to both plane frames as well as space frames. The method of *tension coefficients*, first introduced by Prof. R.V. Southwell, is in effect a neat and systematic presentation of the 'method of joints'. The method is particularly useful to space frames in which other methods prove to be cumbersome and tedious. In this chapter, we shall introduce the method, and illustrate its application for plane frames. Its application for the *space frames* has been illustrated in Chapter 26.

25.2. TENSION COEFFICIENT

The *tension coefficient* for a member of a frame is defined as the *pull* or *tension* in that member divided by its length. Thus,

$$t = \frac{T}{L} \qquad \text{...(25.1)}$$

where t is the tension coefficient for the member, T is the pull in the member and L is its length.

Consider a member AB of a pin-jointed perfect frame in equilibrium, under a given system of external forces (or reactions) acting at the joints. Let T_{AB} be the resulting pull in the member.

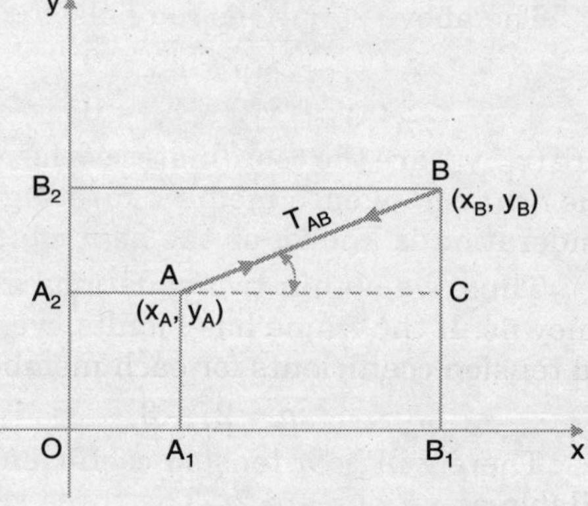

Fig. 25.1

Let (x_A, y_A) be the coordinates of A and (x_B, y_B) be the coordinates of joints B, referred to suitable axes of reference.

Component of pull T_{AB} in x-direction is

$$T_{AB} \cdot \cos\theta = T_{AB} \cdot \frac{AC}{AB} = T_{AB} \frac{(x_B - x_A)}{L_{AB}} = t_{AB}(x_B - x_A) \quad ...(1)$$

where t_{AB} = tension coefficient for $AB = \dfrac{T_{AB}}{L_{AB}}$

Similarly, component of T_{AB} in y-direction is

$$T_{AB} \cdot \sin\theta = T_{AB} \cdot \frac{CB}{AB} = T_{AB} \frac{(y_B - y_A)}{L_{AB}} = t_{AB}(y_B - y_A) \quad ...(2)$$

Thus, we observe that the component of forces in the members along the x and y directions can very easily be expressed in terms of *tension coefficients* and the coordinates of the ends of the members with reference to the chosen reference axes.

The length of the member AB is obviously given by

$$L_{AB} = \sqrt{(x_B - x_A)^2 + (y_B - y_A)^2} \quad ...(25.2)$$

A member which is in *compression* will have a *negative tension coefficient.*

25.3. ANALYSIS OF PLANE FRAMES

Let us now apply the method of tension coefficients for the analysis of plane frame. Let us consider a joint A, where a number of members AB, AC, AD,...... etc. are meeting. Let P_A be the external force acting at the joint A, and let X_A and Y_A be the components of this force P_A, in x and y directions. Since the joint is in equilibrium under the external force P_A and the *pulls* in the members meeting at joint A, the algebraic sum of *resolved* parts of these forces in any direction must be equal to zero.

Resolving the forces in x and y directions, and equating the algebraic sum of these forces in each direction to zero, we get

$$t_{AB}(x_B - x_A) + t_{AC}(x_C - x_A) + t_{AD}(x_D - x_A) ...... + X_A = 0 \quad ...[25.3\ (a)]$$

and

$$t_{AB}(y_B - y_A) + t_{AC}(y_C - y_A) + t_{AD}(y_D - y_A) ...... + Y_A = 0 \quad ...[25.3\ (b)]$$

The above equations may also be written in *compact* form as under :

$$\Sigma\, t\,(x_F - x_N) + X_A = 0 \quad ...[25.4\ (a)]$$

and

$$\Sigma\, t\,(y_F - y_N) + Y_A = 0 \quad ...[25.4\ (b)]$$

where (x_F, y_F) are the coordinates of the *far end* of each member, and (x_N, y_N) are the coordinates of the *near end* of each member meeting at the joint. The end of the member at the joint under consideration is known as the *near end.*

Thus, we obtain two equations at each joint, in which *tension coefficients* are the only unknowns. If the frame has j joints, we will have $2j$ such equations, the solution of which will yield tension coefficients for each member. If there are n members in a perfect frame, we have

$$n = 2j - 3,$$

There will be n tension coefficients, for which we require only $(2j - 3)$ equations, while available equations are $2j$. The three surplus equations will be useful either for determining the external reactions acting on the frame or for applying the check.

In order to apply the method, it is assumed that all the members are in tension. In the final solution, a member in compression will then automatically get *negative tension coefficient*. However, in setting the equations, utmost care must be taken in assigning correct sign to relevant terms. The term will be taken as positive or negative according as they tend to move the joint in the positive or negative directions of the axes of reference. The origin should be so selected that the coordinates of various joints can be written easily.

The method will now be illustrated with the help of few examples.

Example 25.1. *A plane frame consists of two members AB and CB, hinged at A and C to the wall, as shown in Fig. 25.2. Determine the forces in the two members due a vertical force P applied at joint P.*

Solution. Let us take the origin at joint C, and Cx and Cy be the axes of reference. The coordinates of the three joints are

$$C(0, 0)\ ;\ B(2, 0)\ ;\ A(0, 1.5).$$

There are only two members and therefore, there will be only two tension coefficients. Let us therefore take joint B and set two equations at that joint, assuming that every member is in state of tension, exerting a pull on the joint, though in the present case member BA will be in tension while member BC will be in compression. The tension coefficient for BC will automatically workout to be negative.

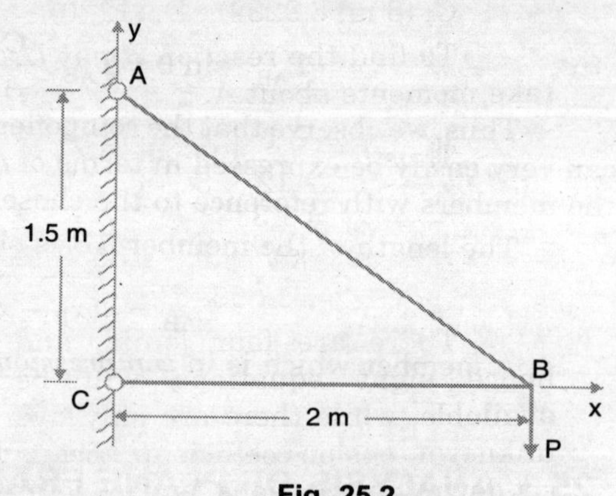

Fig. 25.2

Length $L_{BA} = \sqrt{(0-2)^2 + (1.5-0)^2} = 2.5$ m ; $L_{BC} = 2$ m (given).

At the joint B, we have the following two equations in x and y directions.

$$t_{BA}\,(x_A - x_B) + t_{BC}\,(x_C - x_B) + 0 = 0$$

and
$$t_{BA}\,(y_A - y_B) + t_{BC}\,(y_C - y_B) - P = 0$$

(Negative sign has been placed before P since force P acts in the negative y-direction)

Substituting the values, we get

$$t_{BA}\,(0-2) + t_{BC}\,(0-2) = 0 \quad \text{or} \quad t_{BA} + t_{BC} = 0 \qquad ...(1)$$

and
$$t_{BA}\,(1.5-0) + t_{BC}\,(0-0) - P = 0 \quad \text{or} \quad 1.5t_{BA} = P \qquad ...(2)$$

Solving (1) and (2), we get

$$t_{BA} = \frac{P}{1.5} \quad \text{and} \quad t_{BC} = -\,t_{BA} = -\frac{P}{1.5}.$$

Minus sign suggests that member BC will be in compression.

∴ Force in member $BA = T_{BA} = t_{BA} \,.\, L_{BA}$

$$= \frac{P}{1.5} \times 2.5 = \mathbf{1.6667}\boldsymbol{P} \text{ (tension).}$$

and Force in member $BC = T_{BC} = t_{BC} \times L_{BC} = -\frac{P}{1.5} \times 2$

$$= -\,1.3333P = \mathbf{1.3333\ P} \text{ (comp.).}$$

Example 25.2. *A truss, shown in Fig. 25.3 is loaded with two point loads of 2P and P at joints B and C. Determine the forces in all the members.*

Solution. Given : $L_{AB} = 8$ m : $L_{BC} = 4$ m ; $L_{CD} = 8$ m and $L_{AD} = 12$ m.

Let us keep the origin at *A*, with positive reference directions as shown in Fig. 25.3.

Now $AE = AB \cos 60° = 8 \cos 60° = 4 \text{ m} = FD$

$BE = AB \sin 60° = 8 \sin 60° = 6.9282 \text{ m}$; $ED = 12 - 4 = 8$ m.

$\therefore$ $BD = (BE^2 + ED^2)^{1/2} = \sqrt{(6.9282)^2 + (8)^2} = 10.583$ m.

The coordinates of various points are as under :

A, (0, 0) ; *B* (4, 6.9282) ;
C, (8 m, 6.9282) ; *D*, (12, 0).

To find the reaction R_D at *D*, take moments about *A*.

Thus $R_D = \frac{1}{12}[2P \times 4 + P \times 8]$
$= 1.3333\ P$;

$\therefore$ $R_A = 2P + P - 1.3333\ P$
$= 1.6667\ P.$

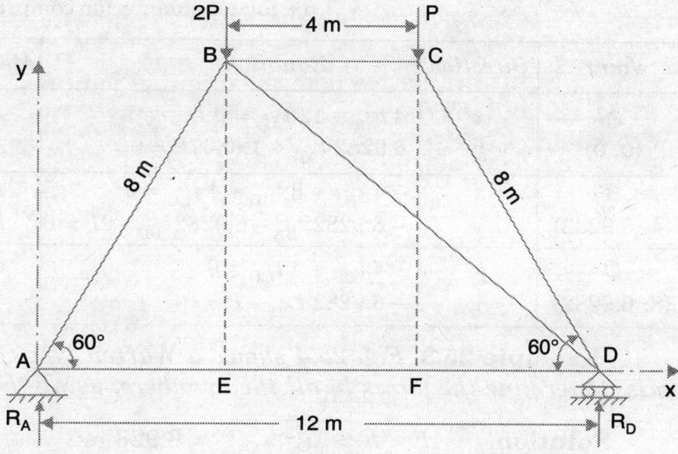

Fig. 25.3

There are four joints, and hence eight equations will be available, while there are only five unknown tension coefficients. Let us set down the equations for each joint assuming that every member is in state of tension, exerting a pull at the joint, though actually all members except *AD* are in compression.

Joint A : $t_{AB}(x_B - x_A) + t_{AD}(x_D - x_A) = 0$ and $t_{AB}(y_B - y_A) + t_{AD}(y_D - y_A) + R_A = 0$

Substituting the values, we get

$4\ t_{AB} + 12\ t_{AD} = 0$ ...(1) and $6.9282\ t_{AB} + 1.6667P = 0$ ...(2)

Solving, we get $t_{AB} = -0.2406P$ and $t_{AD} = +0.0802P$.

$\therefore$ $T_{AB} = t_{AB} \cdot L_{AB} = -0.2406 \times 8P = -1.9246P$ (*i.e.*, compression)

$T_{AD} = t_{AD} \cdot L_{AD} = +0.0802P \times 12 = +0.9624P$ (tension)

Joint B : $t_{BA}(x_A - x_B) + t_{BD}(x_D - x_B) + t_{BC}(x_C - x_B) = 0$

and $t_{BA}(y_A - y_B) + t_{BD}(y_D - y_B) + t_{BC}(y_C - y_B) - 2P = 0$

Substituting the values, we get

$-4\ t_{BA} + 8\ t_{BD} + 4\ t_{BC} = 0$ ...(3) and $-6.9282\ t_{BA} - 6.9282\ t_{BD} - 2P = 0$ ...(4)

From (4): $t_{BA} + t_{BD} = -0.2887P$; But $t_{BA} = t_{AB} = -0.2406P$.

$\therefore$ $t_{BD} = -0.2887P + 0.2406\ P = -0.0481P$.

Hence from (3), $-t_{BA} + 2t_{BD} + t_{BC} = 0$ or $0.2406P - 2 \times 0.0481P + t_{BC} = 0$.

From which $t_{BC} = -0.1444P$.

Hence $T_{BC} = -0.1444P \times 4 = -0.5776P$ (*i.e.*, compression)

and $T_{BD} = -0.0481P \times 10.583 = -0.509P$ (*i.e.*, compression).

Joint C : $t_{CB}(x_B - x_C) + t_{CD}(x_D - x_C) = 0$ and $t_{CB}(y_B - y_C) + t_{CD}(y_D - y_C) - P = 0$.

Substituting the values, we get

$-4\ t_{CB} + 4\ t_{CD} = 0$ ...(5)

and $-6.9282\ t_{CD} - P = 0$ ...(6)

Solving, we get $t_{CD} = -0.1444P$.

and $t_{CB} = t_{CD} = 0.1444P$ (which checks the earlier result).

$\therefore$ $T_{CD} = -0.1444P \times 8 = -1.1552P$ (*i.e.*, compression)

Thus the values of forces in all the members have been obtained. Additional two equations available at joint D can be used to check the values to T_{DB} and T_{DA} obtained earlier.

Sometimes, it is preferable to arrange all the computations in a tabular form shown below.

(+ for tension ; – for compression)

Joint	*Direction*	*Equation*	*Member*	*t*	*L(m)*	*T*
A	*x*	$4\,t_{AB} + 12\,t_{AD} = 0$	*AB*	$-0.2406P$	8	$-1.9246P$
(0, 0)	*y*	$6.9282\,t_{AB} + 1.6667P = 0$	*AD*	$+0.0802P$	12	$+0.9624P$
B	*x*	$-4\,t_{BA} + 8\,t_{BD} + 4\,t_{BC} = 0$	*BC*	$-0.1444P$	4	$-0.5776P$
(4, 6.9282)	*y*	$-6.9282\,t_{BA} - 6.9282\,t_{BD} - 2P = 0$	*BD*	$-0.0481P$	10.583	$-0.509P$
C	*x*	$-4\,t_{CB} + 4\,t_{CD} = 0$	*CD*	$-0.1444P$	8	$-0.1552P$
(8, 6.9282)	*y*	$-6.9282\,t_{CD} - P = 0$				

Example 25.3. *Fig. 25.4 shows a Warren type cantilever truss alongwith the imposed loads. Determine the forces in all the members, using the method of tension coefficients.*

Solution. $AF = h = \sqrt{8^2 - 4^2} = 6.928$ m.

Let us take origin at F, and select x-axis and y-axis along FD and FA respectively. The coordinates of various points are as under :

Joint	*x*	*y*
F	0	0
A	0	6.928
B	4	6.928
C	12	6.928
D	16	0
E	8	0

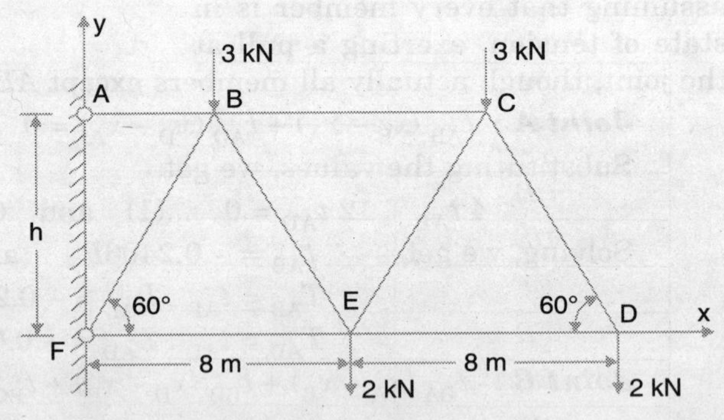

Fig. 25.4

Joint D : The two equations are

$$t_{DE}\,(x_E - x_D) + t_{DC}\,(x_C - x_D) + X_D = 0 \qquad ...(i)$$

$$t_{DE}\,(y_E - y_D) + t_{DC}\,(y_C - y_D) + Y_D = 0 \qquad ...(ii)$$

Substituting the values, we obtain

$$-8\,t_{DE} - 4\,t_{DC} = 0 \quad ...(1) \quad \text{and} \quad 6.928\,t_{DC} - 2 = 0 \qquad ...(2)$$

From (2), $t_{DC} = +0.289$; ...(*a*) and from (1), $t_{DE} = -\dfrac{1}{2}\,t_{DC} = -0.144$ ...(*b*)

Joint C : The two equation are

$$t_{CD}\,(x_D - x_C) + t_{CE}\,(x_E - x_C) + t_{CB}\,(x_B - x_C) + X_C = 0 \qquad ...(iii)$$

$$t_{CD}\,(y_D - y_C) + t_{CE}\,(y_E - y_C) + t_{CB}\,(y_B - y_C) + Y_C = 0 \qquad ...(iv)$$

Substituting the values, we get

$$4\,t_{CD} - 4\,t_{CE} - 8\,t_{CB} = 0 \quad ...(3) \quad \text{and} \quad -6.928\,t_{CD} - 6.928\,t_{CE} - 3 = 0 \qquad ...(4)$$

From (4), $t_{CE} = -t_{CD} - 0.433 = -0.289 - 0.433 = -0.722$ ...(*c*)

From (3), $t_{CB} = 0.5\,t_{CD} - 0.5\,t_{CE} = (0.5 \times 0.289) + (0.5 \times 0.722) = +0.5055$ ...(*d*)

Joint E : The two equations are

$$t_{EC}(x_C - x_E) + t_{ED}(x_D - x_E) + t_{EB}(x_B - x_E) + t_{EF}(x_F - x_E) + X_E = 0 \qquad ...(v)$$

$$t_{EC}(y_C - y_E) + t_{ED}(y_D - y_E) + t_{EB}(y_B - y_E) + t_{EF}(y_F - y_E) + Y_E = 0 \qquad ...(vi)$$

Substituting the values, we get

$$4\,t_{EC} + 8\,t_{ED} - 4\,t_{EB} - 8\,t_{EF} = 0 \quad ...(5) \quad \text{and} \quad 6.928\,t_{EC} + 6.928\,t_{EB} - 2 = 0 \qquad ...(6)$$

From (6), $t_{EB} = 0.2887 - t_{EC} = 0.2887 + 0.722 = +1.011$ ...(*e*)

From (5), $t_{EF} = 0.5\,t_{EC} + t_{ED} - 0.5\,t_{EB} = (-0.5 \times 0.722) + (-0.144) - (0.5 \times 1.011)$
$= -1.0105.$

Joint B : The two equations are

$$t_{BC}(x_C - x_B) + t_{BE}(x_E - x_B) + t_{BF}(x_F - x_B) + t_{BA}(x_A - x_B) + X_B = 0$$

$$t_{BC}(y_C - y_B) + t_{BE}(y_E - y_B) + t_{BF}(y_F - y_B) + t_{BA}(y_A - y_B) + Y_B = 0$$

Substituting the values, we get

$$8\,t_{BC} + 4t_{BE} - 4\,t_{BF} - 4t_{BA} = 0 \quad ...(7) \quad \text{and} \quad -6.928t_{BE} - 6.928t_{BF} - 3 = 0 \qquad ...(8)$$

From (8), $t_{BF} = -t_{BE} - 0.433 = -1.011 - 0.433 \simeq -1.444$ ...(*f*)

From (7), $t_{BA} = 2\,t_{BC} + t_{BE} - t_{BF} = (2 \times 0.5055) + 1.011 + 1.444 = +3.466.$

The values of tension coefficients and forces in various members are tabulated below.

(+ for tension ; – for compression)

Member	*Length (m)*	*t*	*T (kN)*
AB	4	+ 3.466	+ 13.864
BC	8	+ 0.5055	+ 4.044
CD	8	+ 0.289	+ 2.312
DE	8	– 0.144	–1.152
EF	8	– 1.0105	– 8.084
FB	8	– 1.444	– 11.552
EC	8	– 0.722	– 5.776
EB	8	+ 1.011	+ 8.088

PROBLEMS

1. Using the method tension coefficients, find forces in the members of the triangular type cantilever truss shown in Fig. 24.13.

2. Using the method of tension coefficients, determine the forces in the members of the crane structure shown in Fig. 24.21.

3. Using the method of tension coefficients, determine the forces in the members of the frame shown in Fig. 24.22.

4. Using the method of tension coefficients, determine the forces in the members of the frame shown in Fig. 24.23.

5. Find the forces in all the member of the frame shown in Fig. 24.27. The frame is supported on a pinned support at *P* and roller support at *R*.

26

Space Frames

26.1. INTRODUCTION

A *space frame* or *space truss* is a three dimensional assemblage of line members, each member being joined at its ends to the foundation or to other members by frictionless ball-and-socket joints. The simplest space frame consists of six members joined to form a tetrahedron [Fig. 26.1 (*a*)]. By beginning with six members forming a tetrahedron, a *stable space frame* can be constructed by successive addition of three new members and a joint. One such frame is shown in [Fig. 26.1 (*b*)].

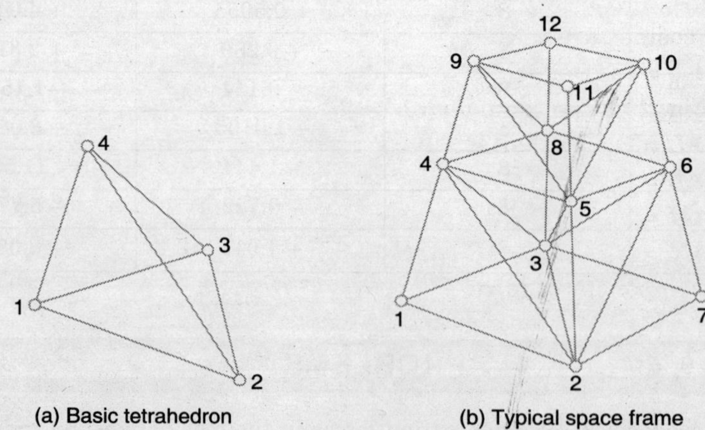

(a) Basic tetrahedron

(b) Typical space frame

Fig. 26.1. Simple space frames.

In order to form a stable (or rigid) space frame, a sufficient number of members have to be arranged in a suitable manner explained above, starting with a basic tetrahedron. The original tetrahedron consists of *six members* and *four joints*. Since for each additional joint there are three additional members, the relationship between the number of members (n) and number of joints (j) is given by

$$n - 6 = 3\,(j - 4) \quad \text{or} \quad n = 3j - 6 \qquad \text{...(26.1)}$$

The above equation gives the *minimum* number of members required to build a *stable* truss or frame. If a truss or frame has less than this number of members, it cannot be stable. If it has more members, then the truss or frame will be termed *internally statically indeterminate*.

26.2. METHOD OF TENSION COEFFICIENTS APPLIED TO SPACE FRAMES

The method of tension coefficients, explained in the previous chapter, can be extended to space frames by writing the equation of equilibrium of forces at each joint, in each direction. However, in space frame, there are three axes of reference ; x-axis, y-axis and z-axis, the first two axes being considered in a horizontal plane and z-axis in a vertical plane. Thus, there will be *three equations* in place of two equations used for plane frames.

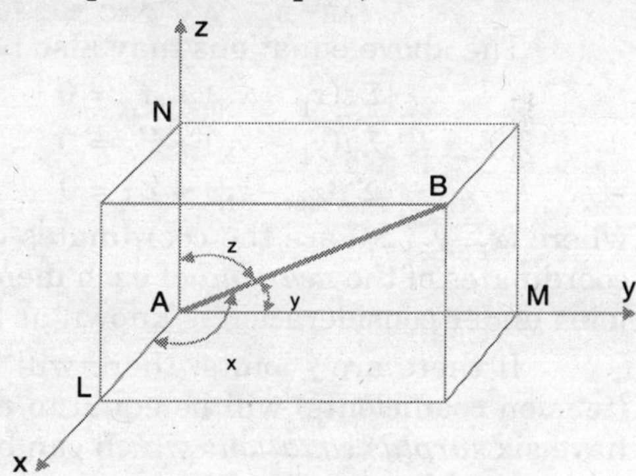

Fig. 26.2

Let us consider a member AB of a space frame, carrying a tensile force T_{AB}. Let (x_A, y_A, z_A) be coordinates of point A, and (x_B, y_B, z_B) be the coordinates of point B, referred to the three axes x, y, z of reference. Let the line AB be inclined at angles θ_x, θ_y, θ_z respectively to the three reference axes. Evidently.

$$L_{AB} = [(x_B - x_A)^2 + (y_B - y_A)^2 + (z_B - z_A)^2]^{1/2} \qquad \text{...(26.2)}$$

The direction cosines are :

$$l_{AB} = \cos\theta_x \; ; m_{AB} = \cos\theta_y \text{ and } n_{AB} = \cos\theta_z \qquad \text{...(26.3)}$$

As shown in Fig. 26.2 the components of AB along x, y and z axes are :

$$AL = (x_B - x_A) \; ; AM = (y_B - y_A) \; ; AN = (z_B - z_A)$$

Hence,

$$AL = (x_B - x_A) = AB\cos\theta_x = l_{AB} \,.\, L_{AB}$$

$$AM = (y_B - y_A) = AB\cos\theta_y = m_{AB} \,.\, L_{AB}$$

$$AN = (z_B - z_A) = AB\cos\theta_z = n_{AB} \,.\, L_{AB}$$

The resolved component of force T_{AB} along x-direction is

$$T_{AB}\cos\theta_x = T_{AB}\frac{(x_B - x_A)}{L_{AB}} = t_{AB}(x_B - x_A) \qquad \text{...}(a)$$

where t_{AB} = tension coefficient for AB.

Similarly, the resolved component of force T_{AB} along y and z directions are

$$T_{AB}\cos\theta_y = T_{AB}\frac{(y_B - y_A)}{L_{AB}} = t_{AB}(y_B - y_A) \qquad \text{...}(b)$$

$$T_{AB}\cos\theta_z = T_{AB}\frac{(z_B - z_A)}{L_{AB}} = t_{AB}(z_B - z_A) \qquad \text{...}(c)$$

Let P_A be the external force at joint A, and let X_A, Y_A and Z_A be the resolved components of this force in x, y and z directions respectively. Let AB, AC, AD, ... be the members meeting at the joint A. Since the joint is in equilibrium under the external force P_A and the pulls in the members meeting at the joint A, the algebraic sum of *resolved* parts of these forces in any direction must be equal to zero.

Hence resolving the forces in x, y and z directions and equating the algebraic sum of these forces in each direction to zero, we get the following three equations for joint A :

$$\left.\begin{aligned} t_{AB}(x_B - x_A) + t_{AC}(x_C - x_A) + t_{AD}(x_D - x_A) \text{ ......} X_A &= 0 \\ t_{AB}(y_B - y_A) + t_{AC}(y_C - y_A) + t_{AD}(y_D - y_A) \text{ ......} Y_A &= 0 \\ t_{AB}(z_B - z_A) + t_{AC}(z_C - z_A) + t_{AD}(z_D - z_A) \text{ ......} Z_A &= 0 \end{aligned}\right\} \quad \text{...(26.4)}$$

The above equations may also be written in compact form as under :

$$\left.\begin{aligned} \Sigma\, t(x_F - x_N) + X_A &= 0 \\ \Sigma\, t(y_F - y_N) + Y_A &= 0 \\ \Sigma\, t(z_F - z_N) + Z_A &= 0 \end{aligned}\right\} \quad \text{...(26.5)}$$

where (x_F, y_F, z_F) are the coordinates of the *far end* of each member and (x_N, y_N, z_N) are the coordinates of the *near end* of each member meeting at the joint. The end of the member at the joint under consideration is known as the *near end*.

If there are j joints, there will be $3j$ such equations, while the number of unknowns (tension coefficients) will be equal to number of members n, where $n = 3j - 6$. Thus, we will have six *surplus equations* which can be utilised either for determining the external reactions or for verifying the results.

The method will now be illustrated with the help of few examples.

26.3. ILLUSTRATIVE EXAMPLES

Example 26.1. *A pair of shear leg has length of each leg as 5 m, and the distance between their feet is 4 m. The line joining the feet of the legs is 7 m from the foot of the guy rope. If the length of the guy rope is 10 m, find the thrust in each leg and the pull in the guy rope when a load of 100 kN is suspended from the head.*

Solution. Let the origin pass through the foot of the guy rope, and let the y-axis pass through the mid-point C of the line AB joining the feet of the two legs, as shown in Fig. 26.3 along with the directions of x and z axes. H is the head of the shear leg. Let the guy rope make an angle θ with the y-axis.

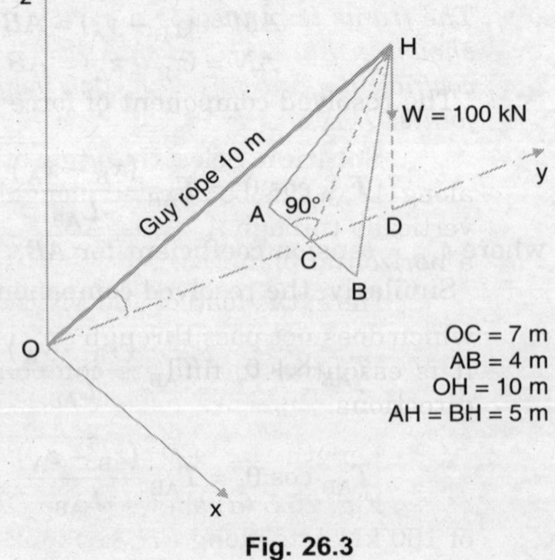

Fig. 26.3

Now $CH = \sqrt{AH^2 - AC^2} = \sqrt{5^2 - 2^2}$

$= \sqrt{21} = 4.5826$ m

From $\Delta\, OCH$, $\cos\theta = \dfrac{OH^2 + OC^2 - CH^2}{2\, OC.OH}$

$$= \frac{10^2 + 7^2 - 21}{2 \times 7 \times 10} = 0.9143$$

$\theta = 23.9°$; $\sin\theta = 0.4051$

From H, drop a perpendicular HD on y-axis.

$\therefore \quad OD = OH \cos\theta = 10 \times 0.9143 = 9.143$ m ;

$HD = OH \sin\theta = 10 \times 0.4051 = 4.051$ m.

$\therefore \quad CD = OD - OC = 9.143 - 7 = 2.143$ m.

Hence the co-ordinates of various points are as under :

Point	*Co-ordinates*		
	x	*y*	*z*
O	0	0	0
H	0	+ 9.143	+ 4.051
A	– 2	+ 7	0
B	+ 2	+ 7	0

At the head H, the three equations are as under :

$$t_{HA}(x_A - x_H) + t_{HB}(x_B - x_H) + t_{HO}(x_O - x_H) = 0 \quad ...(i)$$

$$t_{HA}(y_A - y_H) + t_{HB}(y_B - y_H) + t_{HO}(y_O - y_H) = 0 \quad ...(ii)$$

$$t_{HA}(z_A - z_H) + t_{HB}(z_B - z_H) + t_{HO}(z_O - z_H) + Z_H = 0 \quad ...(iii)$$

Substituting the values, we get

$$-2\, t_{HA} + 2\, t_{HB} = 0 \quad ...(1)$$

$$-2.143\, t_{HA} - 2.143\, t_{HB} - 9.143\, t_{HO} = 0 \quad ...(2)$$

and $$-4.051\,(t_{HA} + t_{HB} + t_{HO}) - 100 = 0 \quad ...(3)$$

From (1), $t_{HA} = t_{HB}$ (as expected)

From (2), $-(2.143 + 2.143)\, t_{HA} = 9.143\, t_{HO}$

From which $t_{HA} = -2.1332\, t_{HO}$

From (3), $-4.051\,(-2.1332 - 2.1332 + 1)\, t_{HO} = 100$

From which $t_{HO} = 7.5573$

$\therefore \quad t_{HA} = t_{HB} = -2.1332 \times 7.5573 = -16.1212.$

Hence *pull* in guy rope $= t_{HO} \times L_{HO} = 7.5573 \times 10 \simeq$ **75.75 kN**.

Force in each shear leg $= -16.1212 \times 5 \simeq -80.61$ kN.

$\therefore$ *Thrust* in each leg = **80.61 kN**.

Example 26.2. *A space frame consists of six members : AF, BE, BF, FE, EC and FD. The frame is pinned to a vertical wall at ABCD in such a way that ABCD form a square as shown in Fig. 26.4. Also, ABEF is a rectangle in a horizontal plane. Using method of tension coefficients, find forces in each member due to a load of 100 kN applied at E acting towards the joints D.*

Solution. Select the origin at A. Let x-axis be directed along AF, y-axis be directed along AB and z-axis be diected vertically through A. Thus, $ABEF$ is in the x-y plane which a horizontal plane.

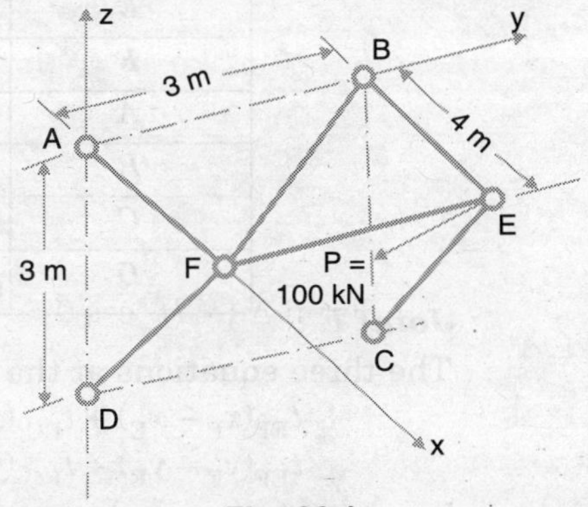

Fig. 26.4

Here, the load of 100 kN is in an inclined direction which does not pass through any of the three axes. Hence it is essential to find its components along x, y and z directions.

Length $\quad FD = \sqrt{3^2 + 4^2} = 5$ m.

Fig. 26.5 (a) shows plane $DFEC$, in which the load of 100 kN acts along ED. Evidently, angle $\theta = \tan^{-1}(3/5)$, from which $\theta = 30.964°$, $\sin\theta = 0.5145$ and $\cos\theta = 0.8575$.

∴ Resolved component of the force along EF (*i.e.*, along y-direction)

$$= 100 \sin\theta = 100 \times 0.5145 = 51.45 \text{ kN}$$

Resolved component of the force along EC

$$= 100 \cos\theta = 100 \times 0.8575 = 85.75 \text{ kN}.$$

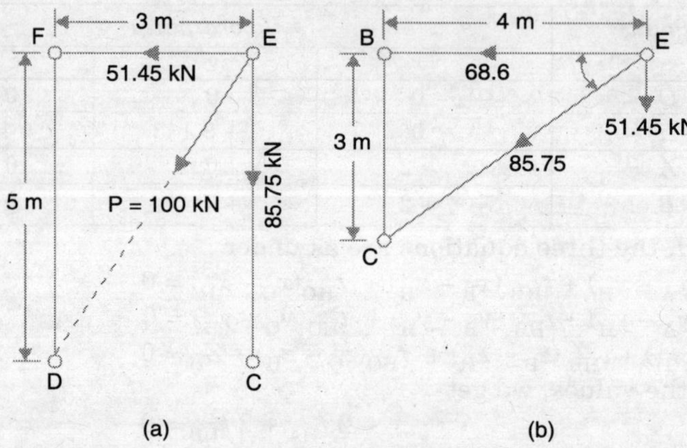

Fig. 26.5

Fig. 26.5 (*b*) shows plane BEC, in which BC is vertical, BE is horizontal, and force in EC is 85.75 kN.

$$\cos\alpha = \frac{BE}{EC} = \frac{4}{5}\ ;\ \sin\alpha = \frac{BC}{EC} = \frac{3}{5}.$$

∴ Resolved component of force in EC, along EB (*i.e.*, along x-direction).

$$= 85.75 \times \frac{4}{5} = 68.6 \text{ kN}$$

Resolved component of force in EC, along vertical direction

$$= 85.75 \times \frac{3}{5} = 51.45 \text{ kN}$$

Thus, resolved components of force P along the three directions are

$$P_x = -68.6 \text{ kN}\ ;\ P_y = -51.45 \text{ kN}\ ;\ P_Z = -51.45 \text{ kN}\ ;$$

The coordinates of various points are as under :

Point	x	y	z
E	4	3	0
F	4	0	0
A	0	0	0
B	0	3	0
C	0	3	– 3
D	0	0	– 3

Joint E :

The three equations at the joint E are :

$$t_{EF}(x_F - x_E) + t_{EC}(x_C - x_E) + t_{EB}(x_B - x_E) + P_x = 0 \quad ...(i)$$

$$t_{EF}(y_F - y_E) + t_{EC}(y_C - y_E) + t_{EB}(y_B - y_E) + P_y = 0 \quad ...(ii)$$

$$t_{EF}(z_F - z_E) + t_{EC}(z_C - z_E) + t_{EB}(z_B - z_E) + P_z = 0 \quad ...(iii)$$

Substituting the values, we get

$$-4\,t_{EC} - 4\,t_{EB} - 68.6 = 0 \qquad ...(1)$$

$$-3\,t_{EF} - 51.45 = 0 \qquad ...(2)$$

$$-3\,t_{EC} - 51.45 = 0 \qquad ...(3)$$

From (2), $t_{EF} = -17.15$

From (3), $t_{EC} = -17.15$

From (1), $t_{EB} = 0$

$\therefore$ $T_{EF} = -17.15 \times 3 = \mathbf{-51.45\ kN}$

$T_{EC} = -17.15 \times 5 = \mathbf{-85.75\ kN}$

$T_{EB} = 0$ m.

Joint F **:** The three equations are

$$t_{FA}(x_A - x_F) + t_{FB}(x_B - x_F) + t_{FD}(x_D - x_F) + t_{FE}(x_E - x_F) = 0$$

$$t_{FA}(y_A - y_F) + t_{FB}(y_B - y_F) + t_{FD}(y_D - y_F) + t_{FE}(y_E - y_F) = 0$$

$$t_{FA}(z_A - z_F) + t_{FB}(z_B - z_F) + t_{FD}(z_D - z_F) + t_{FE}(z_E - z_F) = 0$$

Substituting the values, we get

$$-4\,t_{FA} - 4\,t_{FB} - 4\,t_{FD} = 0 \qquad ...(4)$$

$$3\,t_{FB} + 3\,t_{FE} = 0 \qquad ...(5)$$

$$-3\,t_{FD} = 0 \qquad ...(6)$$

From (6), $t_{FD} = 0$

From (5), $t_{FB} = -t_{FE} = -t_{EF} = +17.15$

From (4), $-4\,t_{FA} - 4 \times 17.15 = 0$ or $t_{FA} = -17.15$

Hence $T_{FD} = 0$; $T_{FB} = +17.15 \times 5 = \mathbf{+85.75\ kN}$

$T_{FA} = -17.15 \times 4 = -\mathbf{68.60\ kN}$.

Example 26.3. *The feet of a tripod resting on a smooth ground are tied by horizontal bars forming a triangle BCD, as shown in Fig. 26.6 (a), where E is the mid-point of CD and F is the mid-point of BE. The apex A [Fig. 26.6 (b)] of the tripod is 3 m vertical above point F. Determine the forces in all the members due to a load of 100 kN suspended from apex A.*

Solution.

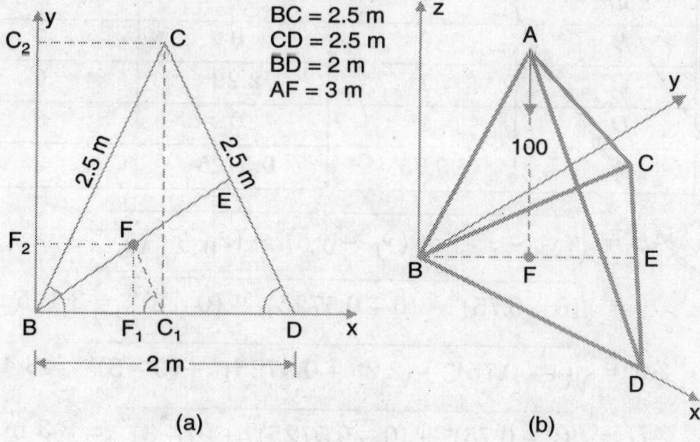

Fig. 26.6

Fig. 26.6 (a) shows the plan of the base triangle BCD in the horizontal plane (x-y plane). Let B be the origin, and let x-axis be directed along BD ; y-axis will be perpendicular to x-axis, while z-axis will be directed vertically upwards.

Drop CC_1 perpendicular to BD. Obviously, $BC_1 = C_1D = 1$ m.

Also, $$CC_1 = \sqrt{(2.5)^2 - (1)^2} = 2.29 \text{ m.}$$

$$\cos \alpha = \frac{1}{2.5} = 0.4 \; ; \; \sin \alpha = \frac{2.29}{2.5} = 0.916.$$

Since F is the mid-point of BE and C_1 is the mid-point of BD, FC_1 will be parallel to ED and will be equal to half of ED.

$$\therefore \quad FC_1 = \frac{1}{2} ED = \frac{1}{4} CD = \frac{1}{4} \times 2.5 = 0.625 \text{ m}$$

From F, drop perpendiculars FF_1 and FF_2 on x and y-axes.

From ΔFF_1C_1, $$FF_1 = FC_1 \sin \alpha = 0.625 \times 0.916 = 0.5725 \text{ m}$$
$$F_1C_1 = FC_1 \cos \alpha = 0.625 \times 0.4 = 0.25 \text{ m}$$
$$\therefore \quad BF_1 = 1 - 0.25 = 0.75.$$

Since point A is vertically above F, the x and y coordinates of point A will be the same as that of point F.

Alternatively, the coordinates of point F may be found as under :

(i) For E : $$x_E = \frac{x_C + x_D}{2} = \frac{1+2}{2} = 1.5 \; ; \; y_E = \frac{y_C + y_D}{2} = \frac{2.29 + 0}{2} = 1.145$$
$$z_E = \frac{z_C + z_D}{2} = \frac{0+0}{2} = 0$$

(ii) For F : $$x_F = \frac{x_B + x_E}{2} = \frac{0 + 1.5}{2} = 0.75 \; ; \; y_F = \frac{y_B + y_E}{2} = \frac{0 + 1.145}{2} = 0.5725$$
$$z_F = \frac{z_B + z_E}{2} = \frac{0+0}{2} = 0.$$

Hence the co-ordinates of various points are as under :

Point	*x*	*y*	*z*
B	0	0	0
C	1	2.29	0
D	2	0	0
A	0.75	0.5725	3

Length $$AB = \sqrt{(x_B - x_A)^2 + (y_B - y_A)^2 + (z_B - z_A)^2}$$
$$= \sqrt{(0 - 0.75)^2 + (0 - 0.5725)^2 + (0 - 3)^2} = 3.145 \text{ m}$$

Length $$AC = \sqrt{(1 - 0.75)^2 + (2.29 - 0.5725)^2 + (0 - 3)^2} = 3.466 \text{ m}$$

Length $$AD = \sqrt{(2 - 0.75)^2 + (0 - 0.5725)^2 + (0 - 3)^2} = 3.3 \text{ m.}$$

Joint A : The three equations are

$$t_{AB}(x_B - x_A) + t_{AC}(x_C - x_A) + t_{AD}(x_D - x_A) + X_A = 0 \quad \text{...(i)}$$

$$t_{AB}(y_B - y_A) + t_{AC}(y_C - y_A) + t_{AD}(y_D - y_A) + Y_A = 0 \quad ...(ii)$$

$$t_{AB}(z_B - z_A) + t_{AC}(z_C - z_A) + t_{AD}(z_D - z_A) + Z_A = 0 \quad ...(iii)$$

Substituting the values, we get

$$-0.75\, t_{AB} + 0.25\, t_{AC} + 1.25\, t_{AD} = 0 \quad ...(1)$$

$$-0.5725\, t_{AB} + 1.7175\, t_{AC} - 0.5725\, t_{AD} = 0 \quad ...(2)$$

$$-3\,(t_{AB} + t_{AC} + t_{AD}) - 100 = 0 \quad ...(3)$$

Solving (1), (2) and (3), we get

$$t_{AB} = -16.667\ ;\ t_{AC} = -8.333,\ t_{AD} = -8.333.$$

Joint B **: Because of *smooth* ground, the reactions at *B*, *C* and *D* will be wholly *vertical*. The three equations are**

$$t_{BA}(x_A - x_B) + t_{BC}(x_C - x_B) + t_{BD}(x_D - x_B) + X_B = 0 \quad ...(iv)$$

$$t_{BA}(y_A - y_B) + t_{BC}(y_C - y_B) + t_{BD}(y_D - y_B) + Y_B = 0 \quad ...(v)$$

$$t_{BA}(z_A - z_B) + t_{BC}(z_C - z_B) + t_{BD}(z_D - z_B) + Z_B = 0 \quad ...(vi)$$

Substituting the values, we get

$$0.75\, t_{BA} + 1\, t_{BC} + 2\, t_{BD} = 0 \quad ...(4)$$

$$0.5725\, t_{BA} + 2.29\, t_{BC} = 0 \quad ...(5)$$

$$3\, t_{BA} + R_B = 0 \quad ...(6)$$

From (6), $R_B = -3\, t_{BA} = 3 \times 16.667 = \mathbf{50\ kN}$

From (5), $t_{BC} = -0.25\, t_{BA} = \mathbf{+\,4.167\ kN}$

From (4), $t_{BD} = \frac{1}{2}[0.75 \times 16.667 - 4.167] = \mathbf{4.167\ kN}$

Joint D **:**

$$t_{DA}(x_A - x_D) + t_{DB}(x_B - x_D) + t_{DC}(x_C - x_D) + X_D = 0 \quad ...(vii)$$

$$t_{DA}(y_A - y_D) + t_{DB}(y_B - y_D) + t_{DC}(y_C - y_D) + Y_D = 0 \quad ...(viii)$$

$$t_{DA}(z_A - z_D) + t_{DB}(z_B - z_D) + t_{DC}(z_C - z_D) + Z_D = 0 \quad ...(ix)$$

Substituting the values

$$1.25\, t_{DA} + 2\, t_{DB} + t_{DC} = 0 \quad ...(7)$$

$$0.5725\, t_{DA} + 2.29\, t_{DC} = 0 \quad ...(8)$$

$$3\, t_{DA} + R_D = 0 \quad ...(9)$$

From (9), $R_D = -3\, t_{DA} = 3 \times 8.333 = \mathbf{25\ kN}$

From (8), $t_{DC} = -\frac{0.5725}{2.29}\, t_{DA} = +\,2.083$ kN

From (7), $t_{DB} = \frac{1}{2}(t_{DC} + 1.25\, t_{DA}) = -\frac{1}{2}(2.083 - 1.25 \times 8.333) = 4.167$ (check).

Hence the values of tension coefficients and forces in all the six members are as under.

Member	*t*	*L*	*T(kN)*
AB	– 16.667	3.145	– 52.42
AC	– 8.333	3.466	– 28.88
AD	– 8.333	3.20	– 27.50
BC	+ 4.167	2.5	+ 10.42
CD	+ 2.083	2.5	+ 5.21
DB	+ 4.167	2	+ 8.33

Example 26.4. *A space frame shown in Fig. 26.7 is supported at A, B, C and D in a horizontal plane, through ball joints. The member EF is horizontal, and is at a height of 3 m above the base. The loads at the joints E and F, shown in the figure act in a horizontal plane. Find the forces in all the members of the frame.*

Solution. Let the origin be at B, with x and y axes along BC and BA respectively, and let z-axis be directed vertically. The coordinates of various points are as under :

Point	*x*	*y*	*z*
B	0	0	0
C	7	0	0
D	7	6	0
A	0	6	0
E	2	3	3
F	5	3	3

Fig. 26.7

$L_{AE} = \sqrt{(2-0)^2 + (3-6)^2 + (3-0)^2}$

$= 4.6904 \text{ m} = L_{BE} = L_{CF} = L_{DF}$

$L_{BF} = \sqrt{(5-0)^2 + (3-0)^2 + (3-0)^2} = 6.5574 \text{ m}$

$L_{EF} = 3$ m (given).

Joint E **:** The three equations at joint E as follows :

$$t_{EA}(x_A - x_E) + t_{EB}(x_B - x_E) + t_{EF}(x_F - x_E) + X_E = 0 \quad ...(i)$$
$$t_{EA}(y_A - y_E) + t_{EB}(y_B - y_E) + t_{EF}(y_F - y_E) + Y_E = 0 \quad ...(ii)$$
$$t_{EA}(z_A - z_E) + t_{EB}(z_B - z_E) + t_{EF}(z_F - z_E) + Z_E = 0 \quad ...(iii)$$

Substituting the values, we obtain

$$-2\, t_{EA} - 2\, t_{EB} + 3\, t_{EF} + 5 = 0 \quad ...(1)$$
$$3\, t_{EA} - 3\, t_{EB} + 10 = 0 \quad ...(2)$$
$$-3\, t_{EA} - 3\, t_{EB} = 0 \quad ...(3)$$

From (3), $\quad t_{EA} = -\, t_{EB}$

From (2), $\quad t_{EA} + t_{EA} = -\dfrac{10}{3} \quad$ or $\quad t_{EA} = -\dfrac{5}{3} = -\,1.667 \quad ...(a)$

$\therefore \quad t_{EB} = +\dfrac{5}{3} = +\,1.667 \quad ...(b)$

From (1), $\left(2 \times \dfrac{5}{3}\right) - \left(2 \times \dfrac{5}{3}\right) + 3t_{EF} + 5 = 0$ From which $t_{EF} = -\dfrac{5}{3} = -\,1.667 \quad ...(c)$

Joint F **:** The three equation are

$$t_{FD}(x_D - x_F) + t_{FC}(x_C - x_F) + t_{FB}(x_B - x_F) + t_{FE}(x_E - x_F) + X_F = 0 \quad ...(iv)$$
$$t_{FD}(y_D - y_F) + t_{FC}(y_C - y_F) + t_{FB}(y_B - y_F) + t_{FE}(y_E - y_F) + Y_F = 0 \quad ...(v)$$
$$t_{FD}(z_D - z_F) + t_{FC}(z_C - z_F) + t_{FB}(z_B - z_F) + t_{FE}(z_E - z_F) + Z_F = 0 \quad ...(vi)$$

Substituting the values, we get

$$2\,t_{FD} + 2\,t_{FC} - 5\,t_{FB} - 3\,t_{FE} = 0 \quad ...(4)$$

$$3\,t_{FD} - 3\,t_{FC} - 3\,t_{FB} + 15 = 0 \quad ...(5)$$

$$-\,3\,t_{FD} - 3\,t_{FC} - 3\,t_{FB} = 0 \quad ...(6)$$

From (4), $t_{ED} + t_{FC} - 2.5\,t_{FB} = -\,2.5$ ...(*d*)

From (5), $t_{FD} - t_{FC} - t_{FB} = -\,5$ ...(*e*)

From (6), $t_{FD} + t_{FC} + t_{FB} = 0$ ...(*f*)

From (*e*) and (*f*), $t_{FD} = -\,2.5$; **From (*d*) and (*e*),** $t_{FB} = +\,0.7143$

From (*f*), $t_{FC} = +\,1.7857$.

Hence the tension coefficients and forces in the various members will be as shown in the table below.

Member	*Length (m)*	*t*	*T (kN)*
EA	4.6904	– 1.6667	– 7.817
EB	4.6904	+ 1.6667	+ 7.817
EF	3.0	– 1.6667	+ 5.0
FC	4.6904	+ 1.7857	+ 8.376
FD	4.6904	– 2.50	– 11.726
FB	6.5574	+ 0.7143	+ 4.684

PROBLEMS

1. Find forces in all members of the space frame shown in Fig. 26.8. Take *AB* = 4 m, *AD* = 5 m and *AF* = 6 m.

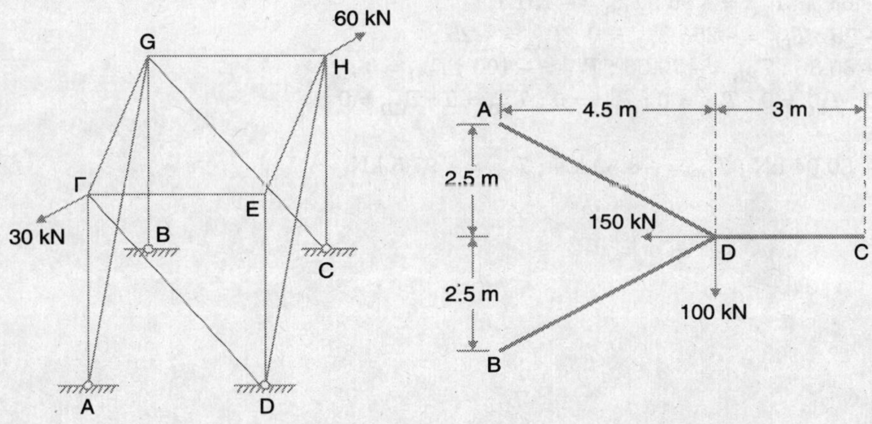

Fig. 26.8 **Fig. 26.9**

2. Fig. 26.9 shows the plan of a tripod, the feet *A*, *B* and *C* being in the same horizontal plane and the apex *D* being 3.75 m above the plane. Horizontal loads of 100 kN and 150 kN applied at *D* in the directions shown. Find the forces in the members assuming that all joint are pin-joints.

3. A frame padestal shown in Fig. 26.10 is simply supported at *B* and has two reaction supports at *A* and *C*. Determine the reaction and forces in all the members, due to horizontal force 100 kN acting at *F*. Take *AB* = *BC* = 4 m.

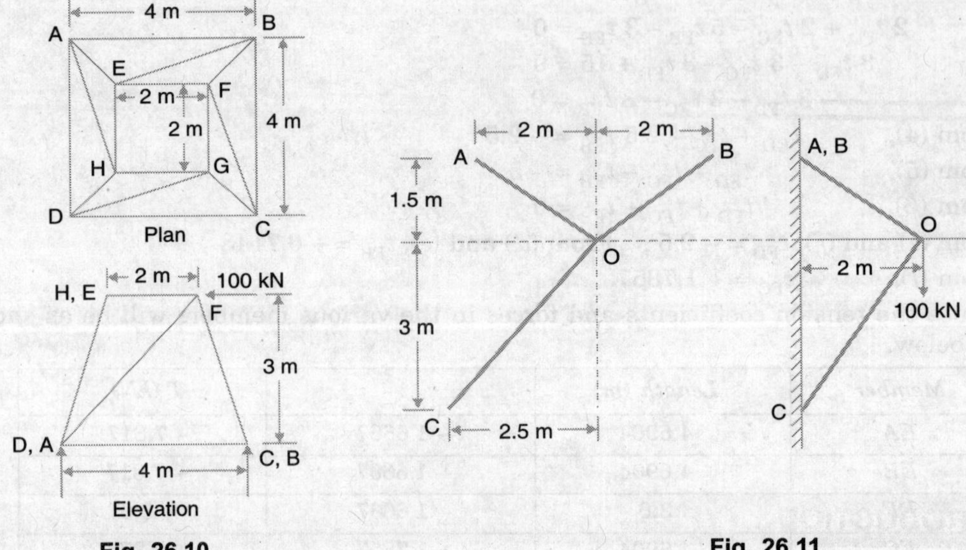

Fig. 26.10 **Fig. 26.11**

4. **Fig. 26.11 shows two views of a tripod bracket. All connections are pinned. Find the forces in magnitude and nature in the three members due to a vertical load of 100 kN acting at *O*.**

(*Based on U.L.*)

ANSWERS

1. $T_{EA} = 0$; $T_{FD} = +37.5$; $T_{FE} = 0$; $T_{FG} = +18.74$; $T_{GA} = -30$
$T_{GB} = +28.11$; $T_{GC} = -60$; $T_{GH} = +46.85$; $T_{HE} = 0$
$T_{HD} = +40$; $T_{HC} = -37.48$; $T_{ED} = 0$.

2. $T_{DA} = +63.7$; $T_{DC} = +96$; $T_{DB} = -191.1$.

3. $T_{AB} = -75$; $T_{BC} = -25$; $T_{CD} = 0$; $T_{DA} = +25$;
$T_{EA} = +35.36$; $T_{EB} = -79.06$; $T_{EF} = -100$; $T_{EH} = 0$;
$T_{EB} = 0$; $T_{FC} = 0$; $T_{FG} = 0$; $T_{GC} = 0$; $T_{GD} = 0$; $T_{HD} = 0$;
$T_{HA} = 0$.

4. $T_{OA} = +80.04$ kN ; $T_{OB} = -8.89$ kN ; $T_{OC} = -97.5$ kN.

27

Riveted Joints

27.1. INTRODUCTION : TYPES OF RIVETED CONNECTIONS

Riveting is a method of joining together structural members by inserting ductile metal pins, called rivets, into holes of members to be connected and forming a head at the end of each rivet to prevent the members from coming apart. It is apparent that the strength of the original members is reduced by providing a riveted joint. A riveted joint may be classified according to (*a*) arrangement of rivets and plates, (*b*) mode of load transmission, and (*c*) nature and location of load with respect to the rivet group.

(*a*) Classification According to Arrangements of Rivets and Plates

The common types of riveted joint for joining plates, etc., are shown in Figs. 27.1 to 27.4.

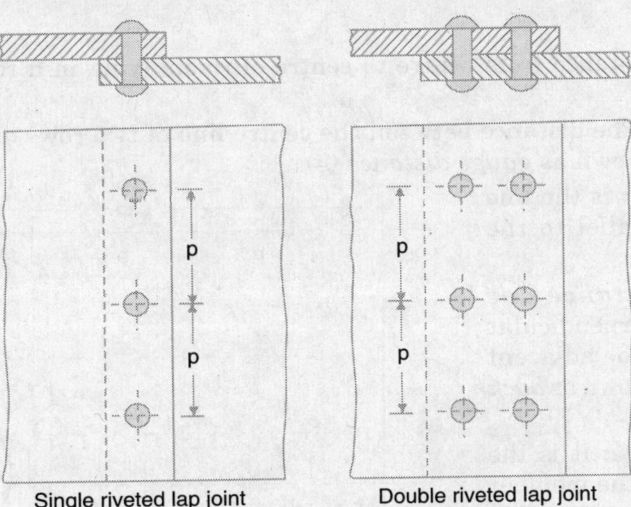

Single riveted lap joint

Fig. 27.1

Double riveted lap joint

Fig. 27.2

Fig. 27.1 shows a *single riveted lap joint* in which the plates overlap. Fig. 27.2 shows a *double riveted lap joint*. The lap joint has the disadvantage that the line of pull is such as to cause bending stresses tending to distort the joint.

Fig. 27.3 and 27.4 show *single riveted butt joint* and *double riveted butt joint* respectively, in which the edges of the plates come flush, and cover plates are used to join them. In the *single cover butt joint*, there are bending stresses developed, tending to distort the joint. This possibility is completely eliminated by using a *double cover butt joint*.

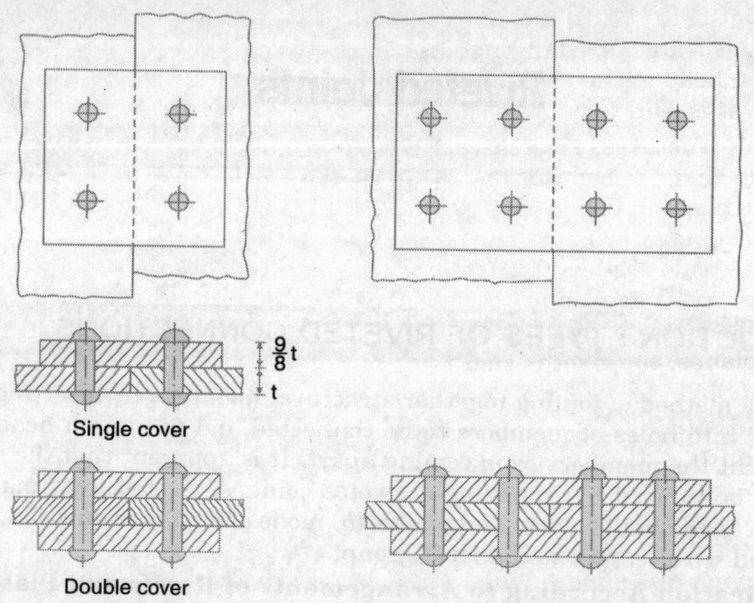

Fig. 27.3. Single Riveted Butt Joint **Fig. 27.4.** Double Riveted Butt Joint

Definitions

1. *Pitch*: The distance from centre to centre of rivets lying in a row is termed as the *pitch* (p).

2. *Back pitch*: The distance between the centre line of two rows of rivets is called the *back pitch*. It is also known as *gauge distance* (g).

3. *Gauge line*: It is the line of rivets which is parallel to the direction of stress.

4. *Gauge distance or 'gauge' (g)*: It is the perpendicular distance between two adjacent gauge lines. It is also known as *back pitch*.

5. *Edge distance*: It is the distance of the edge of the member or cover plates from the centre line of extreme rivet hole.

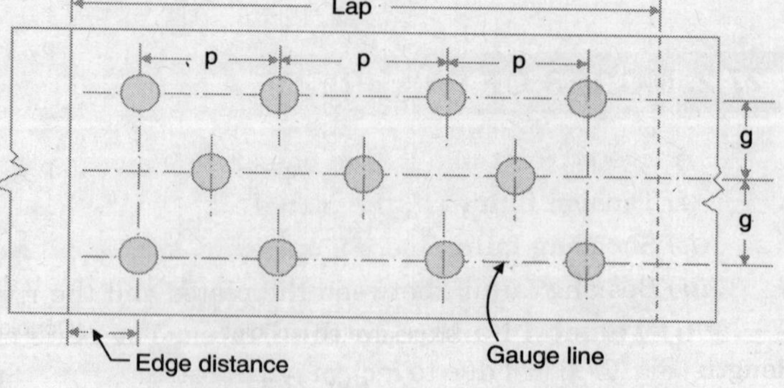

Fig. 27.5. Definition sketch

6. *Lap*: It is the distance normal to the joint between edges of the overlapping plates in a lap joint or between the joint and the end of cover plates in a butt joint.

(*b*) Classification According to the Mode of Load Transmission. Riveted connections can also be classified according to the *mode* of the load transmission by the rivets. If the load is transmitted through bearing between the plate and the shank of the rivet producing shear in the rivet (Fig. 27.6), the rivet is said to be in *shear*. If the load is transmitted through bearing between the plate and the head of the rivet, producing tension in the rivet, the rivet is said to be in *tension* (Fig. 27.7).

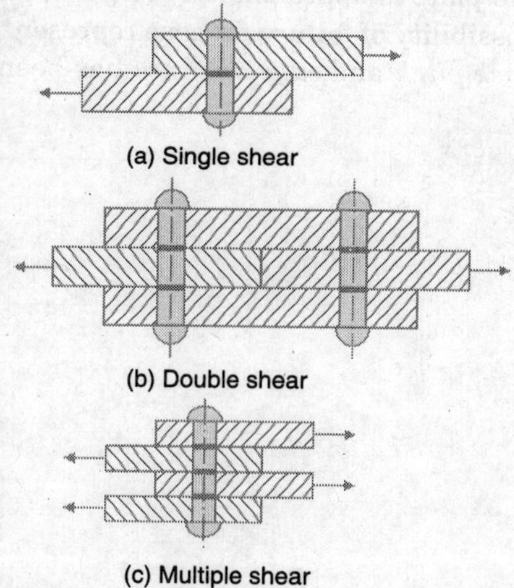

Fig. 27.6

Mostly in structures, riveted connections are to have the rivets in shear. When the load is transmitted by shear in only one plane in the rivet, the rivet is said to be in *single shear* [Fig. 27.6 (*a*)]. When the loading on the rivet is such as to have the load transmitted by shear in two planes, the rivet is said to be in *double shear* [Fig. 27.6 (*b*)]. When the load is transmitted by shear in more than two planes, the rivet is said to be in *multiple shear* [Fig. 27.6 (*c*)].

(*c*) Classification According to the Nature and Location of Load. A riveted connection can also be classified according to the nature and location of load with respect to the rivet group. When the load passes through the centroid of rivet cross-sectional areas, the connection is said to be carrying *direct load*. When the load does not pass through the centroid of rivet group, it is said to be an *eccentric load connection*. When the load transmitted consists of a torque or moment, it is said to be a *moment connection*. Sometimes, in any ordinary beam-to-column connection, there is usually a shear force at the end of the beam and there also exists a certain amount of end moment between the beam and the column. Such a connection is called *shear moment connection*.

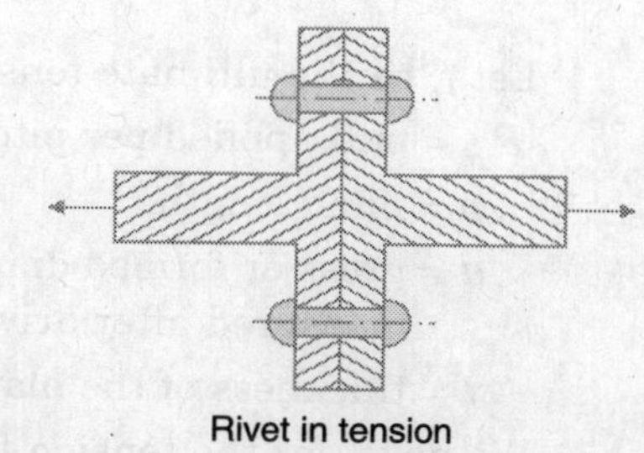
Rivet in tension

Fig. 27.7

27.2. MODES OF FAILURE OF A RIVETED JOINT

A riveted joint may fail in one of the following ways :

(*i*) Tension failure in the plates.

(*ii*) Shearing failure across one or more planes and shearing failure of the rivet.

(*iii*) Bearing failure between the plates and the rivets.

(*i*) Tension Failure in the Plates. The plate can fail by tearing off across the pitch length [Fig. 27.8 (*a*)] due to lack of tensile strength on a section along a row of rivets. The plate may also tear along a diagonal [Fig. 27.8 (*b*)]. However, this type of failure is unlikely to happen if the back pitch is at least 13 times the rivet diameter. Lastly, a plate may tear between the rivet hole and the edge of the plate [Fig. 27.8 (*c*) and (*d*)]. These types of the failure are unlikely to occur if the distance between the centre of the rivet hole and the edge of the plate is approximately twice the diameter of the rivet. However, we shall investigate the possibility of failure for case represented in [Fig. 27.8 (*a*)] only, as it is assumed that sufficient *back pitch* and *edge distance* has been provided.

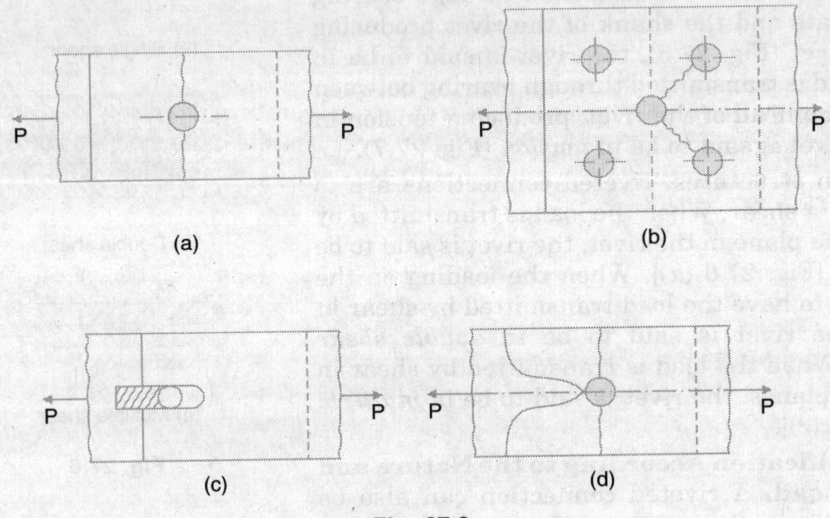

Fig. 27.8

Let f_t be the ultimate tensile strength for the plate material.

P_{ut} = pull applied per pitch length for tension failure.

p = pitch length.

d = gross or formed diameter of rivet. (It is the diameter of the rivet on the hole, measured after driving, and hence is equal to the diameter of the hole itself)

t = thickness of the plate.

We have, for the tension failure of the plate.

$$P_{ut} = f_t \times \text{(resisting section)} = f_t\,(p - d)\,t \qquad ...(27.1)$$

***(ii)* Shearing Failure of the Rivet.** In a riveted joint, the rivet may itself fail in shear. The tendency is to cut through the rivet across the section lying in the plane between the plates it connects [Fig. 27.9 (*a*) and (*b*)].

(a) Single shear (b) Double shear

Fig. 27.9

In analysing this possible manner of failure, one must always note whether a rivet acts in single shear or double shear. In the later case, two cross-sectional areas of the same rivet resist the applied force. The shearing stress is assumed to be uniformly distributed over the cross-section of a rivet.

Let P_{us} = pull required per pitch length for shear failure.

f_s = ultimate shear strength of rivet material.

Resisting area of rivet section = $\frac{\pi}{4}d^2$, in *single shear* and $2\frac{\pi}{4}d^2$ in *double shear*. Hence, we have

$$\left.\begin{aligned} P_{us} &= \frac{\pi}{4}d^2 f_s \text{ for single shear} \\ \text{and} \quad P_{us} &= \frac{\pi}{2}d^2 f_s \text{ for double shear} \end{aligned}\right\} \qquad ...(27.2)$$

(*iii*) Bearing Failure of Plate or Rivet. A riveted joint may fail if a rivet crushes the material of the plate against which it bears or if the rivet itself is deformed by plate acting on it [Fig. 27.10 (*a*)]. In calculating the resistance to bearing, it is assumed that the bearing stress is uniform [Fig. 27.10 (*b*)].

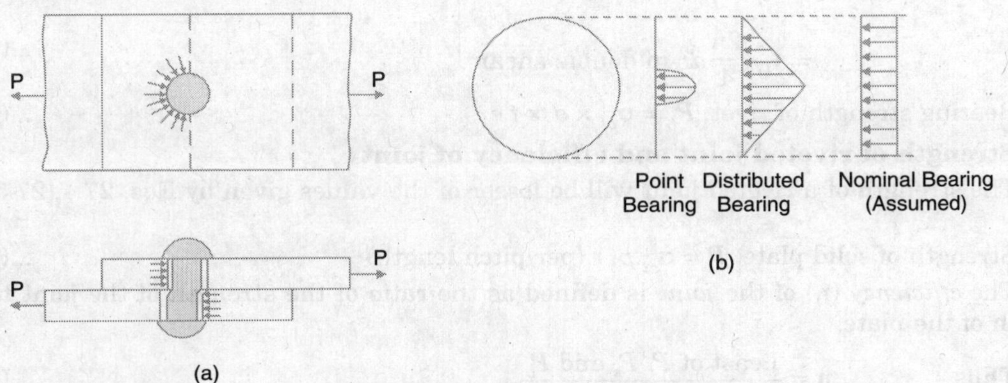

Fig. 27.10

Let P_{ub} = pull per pitch length for bearing failure.

f_b = ultimate crushing strength of rivet material.

$\therefore \quad P_{ub}$ = intensity of radial pressure × projected area = $f_b\ d.t$ ...(27.3)

Eqs. 27.1, 27.2 and 27.3 are applicable for single riveted joint. If however, there are *n rows* of rivets per pitch length, we have :

$$P_{ut} = (p - d)\ t.\ f_t \qquad ...[27.1(a)]$$

$$P_{us} = n\frac{\pi}{4}d^2 f_s \text{ for single shear} \qquad ...[27.2(a)]$$

or

$$= 2n\ \frac{\pi}{4}d^2 f_s \text{ for double shear}$$

and

$$P_{ub} = n\ .\ d\ t\ f_b \qquad ...[27.3(a)]$$

27.3. STRENGTH OF RIVETED JOINT

In the previous section, we have computed the values of pull required for the failure of a riveted joint, using ultimate strength values for the material. However, same expressions can be used to compute the *strength* of a riveted joint, by using the *permissible values of the stresses* in the materials, as shown below.

1. Strength of plate

Let σ_{at} = permissible stress in plate in axial tension.

Strength of plate, P_t, per pitch length is given by

$$P_t = \sigma_{at}\,(p - d)t \qquad ...(27.4)$$

2. Strength of rivet or rivet value

The strength of rivet, commonly known as *rivet value* (*R*) is the smaller of (*i*) shearing strength of rivet, and (*ii*) bearing strength of rivet.

Let τ_{vf} = permissible shearing stress in rivet

σ_{pf} = allowable bearing stress in rivet.

∴ Shearing strength of rivet,

$$\left.\begin{aligned} P_s &= \tau_{vf}\,.\,\frac{\pi}{4}d^2 \text{ in single shear} \\ \text{or} \quad &= \tau_{vf}\,\frac{2\pi}{4}d^2 \text{ in double shear} \end{aligned}\right\} \qquad ...(27.5)$$

Bearing strength of rivet, $P_b = \sigma_{pf} \times d \times t$ ...(27.6)

Strength of riveted joint and efficiency of joint

The strength of a riveted joint will be lesser of the values given by Eqs. 27.4, 27.5 and 27.6.

Strength of solid plate, $P = \sigma_{at}.p.\,t$ (per pitch length) ...(27.7)

The *efficiency* (η) of the *joint* is defined as the ratio of the strength of the joint to the strength of the plate.

$$\text{Thus,} \qquad \eta = \frac{\text{Least of } P_t\ P_s \text{ and } P_b}{P} \qquad ...(27.8)$$

It should be noted that Eqs. 27.4, 27.5 and 27.6 give the strength of a single riveted joint. However, if there are *n* rows of rivets per pitch length, we have

$$P_t = \sigma_{at}\,(p - d)t \qquad ...[27.4\ (a)]$$

$$P_s = n.\,\tau_{vf}\,.\,\frac{\pi}{4}\,d^2 \text{ (in single shear)} \qquad ...[27.5\ (a)]$$

$$\text{or} \qquad = n\,\tau_{vf}\,\frac{2\pi}{4}\,d^2 \text{ (in double shear)}$$

$$\text{and} \qquad P_b = n\,\sigma_{pf}\,.\,d.t \qquad ...[27.6\ (a)]$$

27.4. DESIGN OF RIVETED JOINT : AXIAL LOAD

When the line of action of the load coincides with the centre of gravity of rivet areas, the design of connection is based on the *nominal stress*, which must not exceed the allowable stress. The following assumptions, which are nearly correct for load approaching the ultimate, are made in computing the nominal stresses.

Assumption :

1. Initial tensile stress in the rivet is neglected.
2. The frictional resistance of slip between the plates in neglected.
3. The plates are rigid.

4. The rivet fills the hole completely.
5. Deformation of the plates under the load is neglected.
6. Shearing deformation of the rivets is assumed proportional to the shearing stress.
7. Shearing stress in the rivet is assumed to be uniformly distributed over the rivet cross-section.
8. Unit shearing stress in all the rivets of a joint is uniform.
9. Tensile stress concentration due to rivet holes in the plates are neglected.
10. Bearing stress between rivets and plates is assumed to be uniformly distributed over the nominal contact surface between the rivets and plates.
11. Bending of rivets is neglected.

1. Pitch of Rivets. The pitch of rivets in a riveted joint should be such that under a pull P_t, at the permissible stresses in the rivet for shearing and bearing are not exceeded.

From Eq. 27.4, $P_t = \sigma_{at} (p - d)t$

Strength of rivet = Rivet value $R = P_s$ or P_b whichever is less.

$$\therefore \quad P_t \le R \quad \text{or} \quad \sigma_{at} (p - d)\, t \le R \qquad \text{...(27.7)}$$

From Eq. 27.7, *pitch* (p) can be determined. If Eq. 27.7 is fulfilled, the strength of joint will be P_t and will be lesser than (or at the most equal to) P_s or P_b. The efficiency of the joint designed in this manner will be given by

$$\eta = \frac{P_t}{P} = \frac{(p-d)t\,\sigma_{at}}{p.t\,\sigma_{at}} = \frac{p-d}{p} \qquad \text{...(27.8)}$$

IS 800-1984 lays down the following specifications for the pitch of rivets :

(a) Minimum pitch. The distance between centres of rivets should be not less than 2.5 times the nominal diameter of rivet.

(b) Maximum Pitch

(*i*) The distance between centres of any two adjacent rivets (including tacking rivets) shall not exceed 32 t or 300 mm, whichever is less, where t is the thickness of the thinner outside plate.

(*ii*) The distance between centres of two adjacent rivets, in a line lying in the direction of stress, shall not exceed 16 t or 200 mm, whichever is less in tension members, and 12 t or 200 mm, whichever is less in compression members. In the case of compression members in which forces are transferred through butting faces, this distance shall not exceed 4.5 times the diameter of the rivets for a distance from the abutting faces equal to 1.5 times the width of the member.

(*iii*) The distance between centres of any two consecutive rivets in a line adjacent and parallel to an edge of an outside plate shall not exceed (100 mm + 4 t) or 200 mm, whichever is less in compression or tension members.

(*iv*) When rivets are staggered at equal intervals and the gauge does not exceed 75 mm, the distances specified in (*ii*) and (*iii*) between centres of rivets, may be increased by 50 percent.

2. Edge distance

(*a*) The minimum distance from the centre of any hole to the edge of a plate shall not be less than that given in Table 27.1.

(*b*) Where two or more parts are connected together, a line of rivets shall be provided at a distance of not more than 37 mm + 4 t from the nearest edge, where t is the thickness in mm of the thinner outside plate. In the case of work not exposed to weather, this may be increased to 12 t.

TABLE 27.1. EDGE DISTANCE OF HOLES

Nominal dia. of rivet (mm)	***Dia. of hole (mm)***	***Distance to sheared or hand flame cut hole (mm)***	***Distance to rolled, machine flame cut, sawn or planed hole (mm)***
(*1*)	(*2*)	(*3*)	(*4*)
12 or below	13.5 or below	19	17
14	15.5	25	22
16	17.5	29	25
18	19.5	32	29
20	21.5	32	29
22	23.5	38	32
24	25.5	44	38
27	29.0	51	44
30	32.0	57	51
33	35.0	57	51

3. Diameter or rivet. *Nominal diameter* of rivet is the diameter of unheated rivet; it is the *stated diameter* of the rivet, available in the market. The diameter of rivet for a given plate thickness is generally chosen from the *Unwin's formula* :

$$d = 6.04\sqrt{t} \qquad \text{...(27.9)}$$

where t = thickness of plate in mm, and d = dia. of rivet in mm.

The diameter of the rivet so found is rounded off to the available size of rivets which are manufactured in nominal diameters of 12, 14, 16, 18, 20, 22, 24, 27, 30, 33, 36, 39, 42 and 48 mm. Commonly used sizes for structural steel works are 16, 18, 20 and 22 mm.

For computation of strength etc., the *formed diameter* (equal of the diameter of hole) is used.

4. Tacking Rivets. Tacking rivets are generally used when a structural member consists of two angles (or such other sections), so that the two act as one unit. These rivets are also known as *stitch rivets*. As per IS : 800–1984, when the maximum distance between centres of two adjacent rivets exceeds the specified maximum pitch, taking rivets not subjected to calculated stress shall be used.

The taking rivets shall have a pitch in line not exceeding 32 times the thickness of outside plate or 300 mm, whichever is less. Where the plates are exposed to the weather, the pitch in the line shall not exceed 16 times the thickness of outside plate or 200 mm, whichever is less. In both cases, the lines of rivets shall not be apart at a distance greater than these pitches. All these requirements shall apply to compression members generally, subject to the stipulation in the Code affecting the design the construction of compression members.

In tension member composed of two flats, angles, channels or tees in contact back-to-back or separated back-to-back by a distance not exceeding the aggregate thickness of the connected parts, with solid distance pieces where the parts are separated, shall be provided at pitch in line out exceeding 1000 mm.

For compression members, the tacking rivets shall be as a pitch in line not exceeding 600 mm.

5. Rivets through packing. Number of rivets carrying calculated shear through a packing shall be increased above the number required by normal calculation by 2.5 percent for each 2.0 mm thickness of packing except that, for packing having a thickness of 6 mm or less, no increase need be made. For double shear connections packed on both sides, the number of additional rivets required shall be determined from the thickness of the thicker packing. The additional rivets should preferably be placed in an extension of the packing.

27.5. WORKING STRESSES IN RIVETS

1. The working stresses (or maximum permissible stresses) in mild steel shop rivets as per IS : 800–1984 are given in Table 27.2.

TABLE 27.2. MAXIMUM PERMISSIBLE STRESSES IN RIVETS

Type of Rivet	*Axial Tension, σ_{tf} N/mm² (MPa)*	*Shear, τ_{vf} N/mm² (MPa)*	*Bearing, σ_{pf} N/mm² (MPa)*
1. Power driven rivets	100	100	300
2. Hand driven rivets	80	80	250

2. The permissible stress in a high tensile steel rivet shall be those given in Table 27.2 *multiplied* by the ratio of tensile strength of the rivet material to the tensile strength as specified in IS : 1149-1973.

Note. For field rivets the permissible stresses shall be reduced by 10 percent.

3. The calculated bearing stress of a rivet on the parts connected by it shall not exceed : (*a*) the value of f_y for hand driven rivets, and (*b*) the value $1.2\,f_y$ for power driven rivets, where f_y is the yield stress of the connected parts.

Where the end distance of a rivet (that is, the edge distance in the direction in which it bears) is less than a limit of twice the effective diameter of the rivet, the permissible bearing stress of that rivet on the connected parts shall be reduced in the ratio of the actual and the distance to that limit.

4. **Combined shear and tension.** Rivets and bolts subjected to both shear and axial tension shall be so proportioned that the calculated shear and axial tension do not exceed the allowable stresses τ_{vf} and σ_{tf} and the expression $\left(\frac{\tau_{vf}, cal.}{\tau_{vf}} + \frac{\sigma_{tf}, cal.}{\sigma_{tf}}\right)$ does not exceed 1.4.

Example 27.1. *Determine the rivet value of 20 mm diameter rivets connecting 12 mm thick plates, if it is in (a) single shear, (b) double shear. The permissible stresses for rivet in shear and bearing are 80 N/mm² and 250 N/mm² and for plate in bearing is 250 N/mm².*

Solution. Gross diameter of rivet, $d = 20 + 1.5 = 21.5$ mm.

Gross area of rivet = $\frac{\pi}{4} d^2 = \frac{\pi}{4}$ (21.5)^2 = 363.05 mm^2.

Bearing strength of rivet,

$$P_b = \sigma_{pf} \times d \times t = 250 \times 21.5 \times 12 = 64500 \text{ N} \quad ...(i)$$

Strength of rivet in single shear,

$$P_{s1} = \tau_{vf} \cdot \frac{\pi}{4} d^2 = 80 \times 363.05 = 29044 \text{ N} \quad ...(ii)$$

Strength of rivet in double shear,

$$P_{s2} = \tau_{vf} \cdot 2 \times \frac{\pi}{4} d^2 = 2 \times 29044 = 58088 \text{ N} \quad ...(iii)$$

Hence, Rivet value in single shear = smaller of P_b and P_{s1} = **29044 N.**

Rivet value in double shear = smaller of P_b and P_{s2} = **58088 N.**

Example 27.2. *A double riveted double cover butt joint in plates 16 mm thick is made with 20 mm rivets at 80 mm pitch. Calcutate the pull per pitch length at which the joint will fail and also its efficiency. Take f_t = 480 N/mm^2, f_b = 760 N/mm^2 and f_s = 380 N/mm^2.*

Solution. Gross diameter or formed diameter of rivet = 20 + 1.5 = 21.5 mm.

For the tension failure in the plates,

$$P_{ut} = f_t (p - d)\, t = 480\,(80 - 21.5)\, 16 \times 10^{-3} = 449.28 \text{ kN.}$$

Since there are two rivets in one pitch length, and each rivet is in double shear, we have

$$P_{us} = 2\left(2 \times \frac{\pi}{4} d^2\right) f_s = \pi\,(21.5)^2 \times 380 \times 10^{-3} \simeq 551.84 \text{ kN.}$$

Also $\quad P_{ub} = 2 \cdot d \cdot t \cdot f_b = 2 \times 21.5 \times 16 \times 760 \times 10^{-3} = 522.88$ kN.

The joint will thus fail at a pull of 449.28 kN, the plates giving way be tearing off.

∴ Strength of the joint = 449.28 kN.

Strength of plate = $P_u = p \cdot d \cdot f_t = 80 \times 16 \times 480 \times 10^{-3} = 614.4$ kN.

$$\therefore \quad \eta = \frac{\text{Strength of joint}}{\text{Strength of solid plate}} = \frac{449.28}{614.4} \times 100 = \mathbf{73.12\%}.$$

Example 27.3. *A single riveted lap joint is used to connect 12 mm thick plates, by providing 20 mm diameter rivets at 50 mm pitch. Determine the strength of the joint and joint efficiency. Take working stress in shear in rivets = 80 N/mm^2, working stress in bearing in rivets = 250 N/mm^2 and working stress in axial tension in plates = 0.6 f_y, where f_y = 260 N/mm^2.*

Solution. Formed diameter of rivets = 20 + 1.5 = 21.5 mm

$$\sigma_{at} = 0.6\, f_y = 0.6 \times 260 = 156 \text{ N/mm}^2$$

Strength of plate in tension, per pitch length,

$$P_t = \sigma_{at}\,(p - d)\, t = 156\,(50 - 21.5) \times 12 = 53352 \text{ N} \quad ...(1)$$

Strength of rivet in single shear

$$P_t = \tau_{vf} \times \frac{\pi}{4} d^2 = 80 \times \frac{\pi}{4}\ (21.5)^2 = 29044 \text{ N} \quad ...(2)$$

Strength rivet in bearing

$$P_b = \sigma_{pf} \times d \times t = 250 \times 21.5 \times 12 = 64500 \text{ N} \quad ...(3)$$

∴ Strength of joint = minimum of the above three values = **29044 N.**

Strength of solid plate,

$$P = \sigma_{at}\, pt = 156 \times 50 \times 12 = 93600 \text{ N} \quad ...(4)$$

$$\therefore \quad \text{Joint efficiency} = \frac{\text{Least of } P_t, P_s \text{ and } P_b}{P} = \frac{29044}{93600} \times 100 = \mathbf{31.03\%}.$$

Example 27.4. *Determine the load which can be transmitted per pitch length of a double cover but joint connected by 24 mm diameter shop rivets at 100 mm pitch. The thickness of main plates and cover plates are 16 mm and 12 mm respectively. Take allowable tensile strength of plates equal to 150 N/mm², allowable shear stress in rivets equal to 100 N/mm² and allowable stress in bearing for rivets equal to 300 N/mm². Also, determine the efficiency of the joint.*

Solution. Given: $\sigma_{pf} = 300$ N/mm², $\tau_{vf} = 100$ N/mm² and $\sigma_{at} = 150$ N/mm².

Gross diameter of rivets = 24 + 1.5 = 25.5 mm.

(*i*) *Strength of rivet in double shear*

$$P_s = \tau_{vf} \,.\, 2\,\frac{\pi}{4} d^2 = 100 \times 2 \times \frac{\pi}{4} (25.5)^2 = 102141 \text{ N} \quad ...(i)$$

(*ii*) *Strength of rivet in bearing.* As the total thickness of cover plates is more than the thickness of the main plate, the strength of the rivet will be found for bearing on main plates.

$$P_b = \sigma_{pf} \times d \times t = 300 \times 25.5 \times 16 = 122400 \text{ N} \quad ...(ii)$$

(*iii*) *Strength of plate, per pitch length*

$$P_t = \sigma_{at}\, (p - d) = 150\,(100 - 25.5)\, 16 = 178800 \text{ N} \quad ...(iii)$$

∴ Strength of joint per pitch length = minimum of (*i*), (*ii*) and (*iii*) = 102141 N.

∴ Load which can be transmitted = **102141 N.**

Strength of solid plate $= \sigma_{at} \,.\, pt = 150 \times 100 \times 16 = 240000$ N

$$\therefore \quad \eta \text{ of joint} = \frac{102141}{240000} \times 100 = \mathbf{42.56\%}.$$

Example 27.5. *A double cover butt joint is used to connect plates of 12 mm thick. Design the riveted joint and determine its efficiency. Use power driven rivets and take permissible stresses as per IS : 800–1984. Take permissible axial tension in plate = 0.6 f_y, where f_y = 260 N/mm².*

Solution. The diameter of the rivet is found on the basis of Unwin's formula :

$$d = 6.04\sqrt{t} = 6.04\sqrt{12} = 20.9 \text{ mm}.$$

Since Unwin's formula gives slightly higher values, adopt nominal diameter of rivet = 20 mm. Gross diameter of rivet = 20 + 1.5 = 21.5 mm.

The permissible stresses for shop rivets, as per IS : 800–1984 are as follows :

$$\tau_{vf} = 100 \text{ N/mm}^2 \,;\, \sigma_{pf} = 300 \text{ N/mm}^2 \,;\, \sigma_{at} = 0.6\, f_{fy} = 0.6 \times 260 = 156 \text{ N/mm}^2.$$

Let p be the pitch of the rivets.

Strength of rivet in double shear

$$P_s = \tau_{vf} \,.\, 2\,\frac{\pi}{4} d^2 = 100 \times 2 \times \frac{\pi}{4} (21.5)^2 = 72610 \text{ N}.$$

Strength of rivet in bearing on main plate

$$P_b = \sigma_{pf} \,.\, d \,.\, t \,. = 300 \times 21.5 \times 12 = 77400 \text{ N}.$$

∴ *Rivet value* = 72610 N ...(*i*)

Strength of plate per pitch length = $\sigma_{at}(p - d)\,t$

$= 156\,(p - 21.5)12 = 1872\,(p - 21.5)$ N ...(*ii*)

Equating this to the rivet value, we get

$1872(p - 21.5) = 72610$, from which $p = 60.3$ mm. Adopt $p = 60$ mm.

Adopt thickness of each cover plate = $\frac{5}{8}$ × thickness of main plate = $\frac{5}{8} \times 12 = 7.5$ mm.

Keep thickness = 8 mm

Efficiency of joint = $\frac{p-d}{p} \times 100 = \frac{60-21.5}{60} \times 100 =$ **64.2%.**

27.6. RIVETED JOINTS IN STRUCTURAL STEEL WORK

In the case of a roof truss or bridge truss, the load (P) to be carried by the member is always known. The number of the rivets (n) required to connect the member to the other member is given by

$$n = \frac{P}{\text{strength of one rivet}} = \frac{P}{R} \qquad \text{...(27.10)}$$

where P is the pull or push carried by the member.

In the case of a tension member carrying pull, the arrangement of the rivets found above on each side of the joint is of utmost importance, since it will directly determine the width of the flat. Rivets in a joint may be arranged in two forms :

(*a*) Chain riveting (Fig. 27.11)

(*b*) Diamond riveting (Fig. 27.12)

Figs. 27.11 and 27.12 shows to alternative arrangements for a case when $n = 6$.

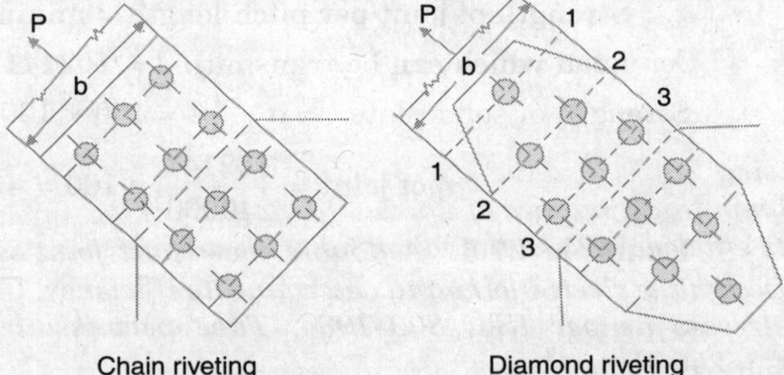

Fig. 27.11 Fig. 27.12

In chain riveting shown in Fig. 27.11 for a case when $n = 6$, the flat is weakened by three rivets holes. The width b of the flat in this case must, therefore, be such that.

$$P = \sigma_{at}(b - 3d)\,t \quad \text{or} \quad b = \frac{P}{t\,\sigma_{at}} + 3d \qquad \text{...(27.11)}$$

In the case of diamond riveting shown in Fig. 27.12, the flat is weakened at the section 1-1 by one rivet hole only and the width of the flat will be given by

$$P = \sigma_{at}(b - d)\,t \quad \text{or} \quad b = \left(\frac{P}{t\,\sigma_{at}} + d\right) \qquad \text{...(27.12)}$$

which is less by $2d$ than the previous case. The saving in the width by this amount is of utmost importance in the case of long bridge diagonals. The diamond riveting is, therefore, used in preference to the chain riveting. It can be shown that the joint gets successively stronger at section 2-2, 3-3, etc. Thus, the strength of joint at section 2-2 is

$$P_2 = \sigma_{at}(b - 2d)\, t + \text{(strength of one rivet in front).}$$

Similarly, the strength of the joint at section 3-3 is

$$P_3 = \sigma_{at}(b - 3d)\, t + \text{(strength of three rivets in front)}$$

Out of P, P_1, P_2 and P_3, force P is usually the least. Therefore, the efficiency of joint is given by

$$\eta = \frac{(b - d)\, t\, \sigma_{at}}{b\, t\, \sigma_{at}} = \frac{b - d}{b} \qquad \text{...(27.13)}$$

27.7. RIVETED JOINTS IN CYLINDRICAL AND SPHERICAL SHELLS

In the case of *cylindrical shells,* hoop stress is $\frac{pD}{2t}$, where p is the internal pressure, D is the diameter of the shell and t is the thickness of the plate. If the shell is riveted longitudinally, the efficiency of the joint should also be taken into account while finding the thickness. Similarly, in the case of spherical shells, hoop stress is $\frac{pD}{4t}$. Hence,

$$\text{For cylindrical shells,} \quad t = \frac{pD}{2\sigma_{at} \cdot \eta} \qquad \text{...(27.14)}$$

$$\text{For spherical shells,} \quad t = \frac{pD}{4\sigma_{at} \cdot \eta} \qquad \text{...(27.15)}$$

Example 27.6. *Fig. 27.13 shows the joint of a boiler shell made of 12 mm thick plates using 20 mm diameter rivets at a pitch of 20 mm. The two cover plates of the butt joint are 8 mm thick, but are of unequal length. Determine the strength of the joint per pitch length and its efficiency. Take the following values of permissible stresses :*

(i) σ_{at} *= 150 N/mm², (ii)* σ_{pf} *= 250 N/mm² and (iii)* τ_{vf} *= 80 N/mm².*

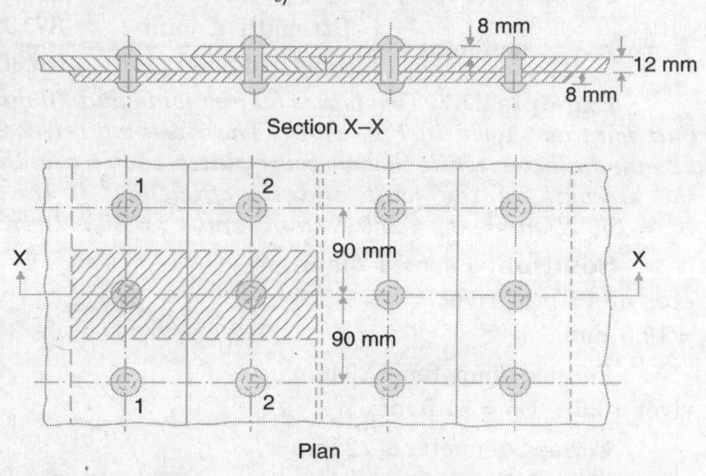

Fig. 27.13

Solution. Here we observe that rivets in row 1 are in single shear, and they bear against 8 mm plate. However, rivets in row 2 are in double shear, and they bear against 12 mm main plate because the total thickness of the two cover plates is more than that of the main plate.

Formed diameter of rivet = 20 + 1.5 = 21.5 mm.

(a) Rivet value in section 1-1

Strength of rivet in single shear

$$= \tau_{vf} \cdot \frac{\pi}{4} d^2 = 80 \times \frac{\pi}{4} (21.5)^2$$

$$= 29044 \text{ N.}$$

Strength of rivet in bearing on 8 mm plate = $\sigma_{pf} \times d \times t = 250 \times 21.5 \times 8 = 43000$ N.

$\therefore$ Rivet value in section 1-1 = $R_1 = 29044$ N.

(b) Rivet value is section 2-2

Strength of rivet in double shear = $\tau_{vf} \times 2\frac{\pi}{4}d^2 = 2 \times 29044 = 58088$ N.

Strength of rivet in bearing on 12 mm plate = $\sigma_{pf} \times d \times t = 250 \times 21.5 \times 12 = 64500$.

$\therefore$ Rivet value in section 2-2 = $R_2 = 58088$ N.

(c) Strength of joint

Consider one pitch length of the joint, as shown hatched in Fig. 27.13. We will consider three possible chances of failure : (*i*) failure of rivets at section 1-1 and 2-2, (*ii*) tearing of main plate at section 1-1 and (*iii*) tearing of cover plates at section 2-2.

(*i*) *Strength of joint on the basis of failure of rivets* :

Strength at section 1-1 and 2-2 = $R_1 + R_2 = 29044 + 58088 = 87132$ N ...(*i*)

(*ii*) *Strength of plates at section 1-1*

At section 1-1, main plate is more strong than the cover plate. Hence strength of the joint on the basis of strength of plate

$$= \sigma_{at}(p - d)\,t = 150(90 - 21.5) \times 12 = 123300 \text{ N} \quad ...(ii)$$

(*iii*) *Strength of cover plates at section 2-2*

At section 2-2, the total thickness of two cover plates is more than the main plate. Hence strength of the joint on the basis of strength of cover plates

$$= \sigma_{at}(p - d)\,t = 150(90 - 21.5)2 \times 8 = 164400 \text{ N} \quad ...(iii)$$

(*iv*) *Strength of main plate at section 2-2*

Strength of joint at 2-2 = strength of main plate at 2-2 + strength of rivets at 1-1

$$= \sigma_{at}(p - d)\,t + R_1 = 150(90 - 21.5)12 + 29044 = 152344 \text{ N} \quad ...(iv)$$

$\therefore$ Strength of joint per pitch length = least of (*i*), (*ii*), (*iii*) and (*iv*) = 87132 N.

Strength of solid plate = $\sigma_{at} \,.\, p \,.\, t = 150 \times 90 \times 12 = 162000$ N.

$$\therefore \quad \eta \text{ of joint} = \frac{\text{Strength of joint}}{\text{Strength of solid plate}} = \frac{87132}{162000} \times 100 = \mathbf{53.79\%}.$$

Example 27.7. *Two plates 80 mm wide and 10 mm thick are joined with a triple riveted butt joint as shown in Fig. 27.14. The rivets at 1-1, 2-2 and 3-3 are of 18 mm, 20 mm and 22 mm diameter, while the two cover plates, each 6 mm thick, are of unequal lengths. Determine the strength of the joint and its efficiency, if the allowable values of stresses are :* $\sigma_{at} = 150\ N/mm^2$, $\sigma_{pf} = 250\ N/mm^2$ *and* $\tau_{vf} = 80\ N/mm^2$.

Solution. Formed diameter of 18 mm rivet = 18 + 1.5 = 19.5 mm

Formed diameter of 20 mm rivet = 20 + 1.5 = 21.5 mm

Formed diameter of 22 mm rivet = 22 + 1.5 = 23.5 mm

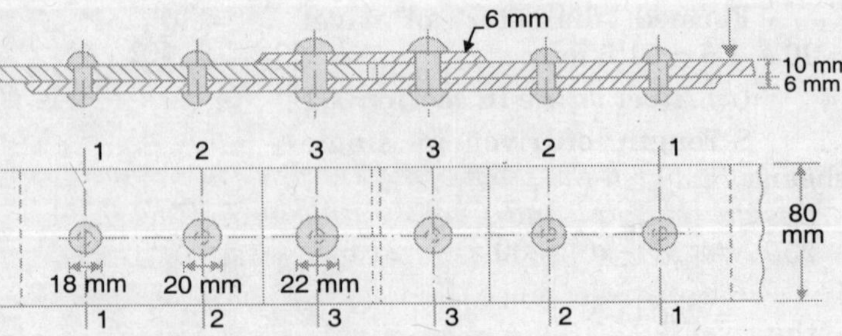

Fig. 27.14

1. *Rivet value of 18 mm rivet*. The rivet is in single shear.

Strength of rivet in single shear $= 80 \times \frac{\pi}{4}(19.5)^2 = 23892$ N.

Strength of rivet in bearing on 6 mm plate $= 250 \times 19.5 \times 6 = 29250$ N.

$\therefore$ Rivet value of 18 mm rivet $= R_1 = 23892$ N.

2. *Rivet value of 20 mm rivet.* The rivet is in single shear.

Strength of rivet in single shear $= 80 \times \frac{\pi}{4}(21.5)^2 = 29044$ N.

Strength of rivet in bearing on 6 mm plate $= 250 \times 21.5 \times 6 = 32250$ N.

$\therefore$ Rivet value of 20 mm rivet $= R_2 = 29044$ N.

3. *Rivet value of 22 mm rivet.* The rivet is in double shear. Also, since the total thickness of cover plates is more than the main plate, the bearing strength of the rivet will be on the basis of 10 mm plate.

$\therefore$ Strength of rivet in double shear $= 80 \times 2 \times \frac{\pi}{4}(23.5)^2 \simeq 69398$ N.

Strength of rivet in bearing on 10 mm plate $= 250 \times 23.5 \times 10 = 58750$ N.

$\therefore$ Rivet value of 22 mm rivet $= R_3 = 58750$ N.

(*a*) Strength of joint on the basis of rivet values

$$= R_1 + R_2 + R_3 = 23892 + 29044 + 58750 = 111686 \text{ N} \quad ...(i)$$

(*b*) Strength of joint on the basis of failure of main plate at section 1-1

$$= \sigma_{at}(p - d)t = 150(80 - 19.5) \times 10 = 90750 \text{ N} \quad ...(ii)$$

(*c*) Strength of joint on the basis of failure of main plate at section 2-2

= tearing of main plate at section 2-2 + rivet value of rivets at 1-1

$$= \sigma_{at}(p - d)t + R_1 = 150(80 - 21.5)10 + 23892 = 111642 \text{ N} \quad ...(iii)$$

(*d*) Strength of joint on the basis of failure of main plate at section 3-3

= tearing of main plate at section 3-3 + Rivet value of rivets at 2-2 + rivet value of rivets at 1-1

$$= \sigma_{at}(p - d)t + R_2 + R_1.$$
$$= 150(80 - 23.5)10 + 29044 + 23892 = 137989 \text{ N} \quad ...(iv)$$

(*e*) Strength of joint on the basis of failure of cover plates at 3-3

$$= \sigma_{at}(p - d)t = 150(80 - 23.5)(6 + 6) = 101700 \text{ N} \quad ...(v)$$

$\therefore$ *Strength of joint* = Least of (*i*), (*ii*), (*iii*), (*iv*) and (*v*)

= 90750 N.

Strength of solid plate $= \sigma_{at} \cdot p \cdot t = 150 \times 80 \times 10 = 120000$ N.

$\therefore$ η of joint $= \frac{90750}{120000} \times 100 =$ **75.6%.**

Example 27.8. *Two plates 12 mm and 10 mm thick are joined by a triple riveted lap joint, in which the pitch of the central row of rivets is half the pitch of rivets in the outer rows. Design the joint and find its efficiency. Take σ_{at} = 150 N/mm², τ_{vf} = 80 N/mm² and σ_{pf} = 250 N/mm².*

Solution. Since the Unwin's formula always gives slightly higher value of diameter of the rivet, we will use smaller of the thickness of the two plates.

$$\therefore \quad d = 6.04\sqrt{t} = 6.04\sqrt{10} = 19.1 \text{ mm}$$

Use 20 mm rivets. Gross diameter of rivets = 20 + 1.5 = 21.5 mm.

(*i*) **Rivet value**

Strength of rivets in single shear

$$= \tau_{vf} \cdot \frac{\pi}{4} d^2 = 80 \times \frac{\pi}{4} (21.5)^2$$
$$= 29044 \text{ N}$$

Strength of rivets in bearing on 10 mm plate

$$= \sigma_{pf} \cdot d \cdot t = 250 \times 21.5 \times 10$$
$$= 53750 \text{ N}$$

$\therefore$ Rivet value $R = 29044$ N.

(*ii*) Strength of thinner plate per pitch length, along section 1-1

$$\sigma_{at}(p - d)\, t = 150\,(p - 21.5)10$$
$$= 1500\,p - 32250.$$

(*iii*) Strength of thinner plate per pitch length, along section 2-2

$$= \sigma_{at}(p - 2d)\, t + R$$
$$= 150\,(p - 2 \times 21.5)\, 10 + 29044$$
$$= 1500\,p - 35456.$$

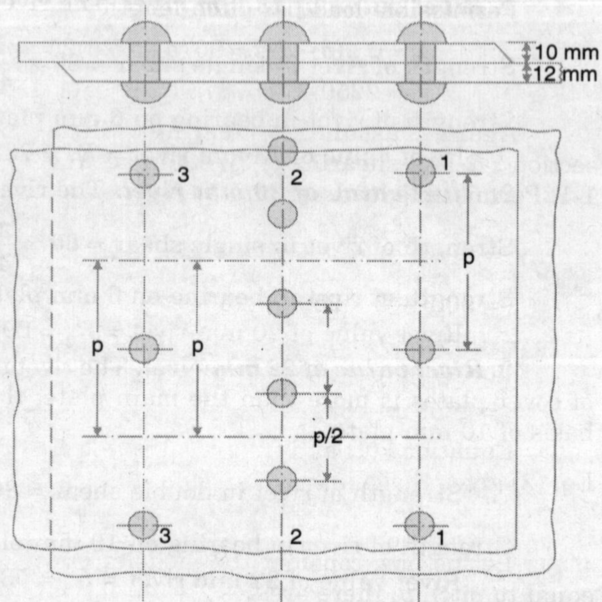

Fig. 27.15

Hence section 2-2 is weaker, along which the strength of plate is (1500 p – 35456). For maximum joint efficiency, the strength of plate per pitch length should be equal to strength of rivets per pitch length.

$$\therefore \quad 1500\,p - 35456 = 4R = 4 \times 29044$$

From which, p = 101 mm.

Minimum permissible pitch = 2.5 × 21.5 = 53.8 mm.

Maximum permissible pitch = 32 t = 32 × 10 = 320 mm.

Keep pitch equal to 54 mm for the inner row and 2 × 54 = 108 mm for the outer rows.

$\therefore$ Strength of joint = 1500p – 35456 = 1500 × 108 – 35456 = 126544 N or equal to 4R (= 4 × 29044 = 116176 N) whichever is less.

$\therefore$ Strength of joint = 116176 N

Strength of solid plate = $\sigma_{at}\, pt$ = 150 × 108 × 10 = 162000 N

$$\therefore \quad \text{Efficiency of joint} = \frac{116176}{162000} \times 100 = \mathbf{71.7\%}.$$

Example 27.9. *Two plates 25 mm thick are connected by a treble riveted butt joint with two cover straps. The pitch of rivets in the outer row is twice the pitch of those in other rows and the diameter of rivet is 24 mm. Taking the resistance of rivets in double shear equal to 1.75 times their resistance in single shear, determine p for equal tearing and shearing resistance. Also determine the efficiency of the joint.*

Assume $\sigma_{at} = 90$ *N/mm*^{2} *and* $\tau_{vf} = 60$ *N/mm*^{2}. ***(U.P.S.C. 1967)***

Solution. Formed diameter of rivet = d = 24 + 1.5 = 25.5 mm.

Consider strip width of joint equal to pitch p.

Let us assume tearing along section 1-1

Permissible load per pitch length

$$= \sigma_{at}(p-d)\,t = 90\,(p-25.5)25$$

$$= 2250\,p - 57375 \text{ N} \quad ...(1)$$

Again, assuming tearing along section 2-2 and shearing of rivets in section 1-1, Permissible load per pitch length

$$= \sigma_{at}\,(p - 2d)\,t + \tau_{vf} \times 1.75 \times \frac{\pi}{4} d^2$$

$$= 90(p - 2 \times 25.5) \times 25$$

$$+ 60 \times 1.75 \times \frac{\pi}{4}(25.5)^2$$

$$= 2250\,p - 61126 \quad ...(2)$$

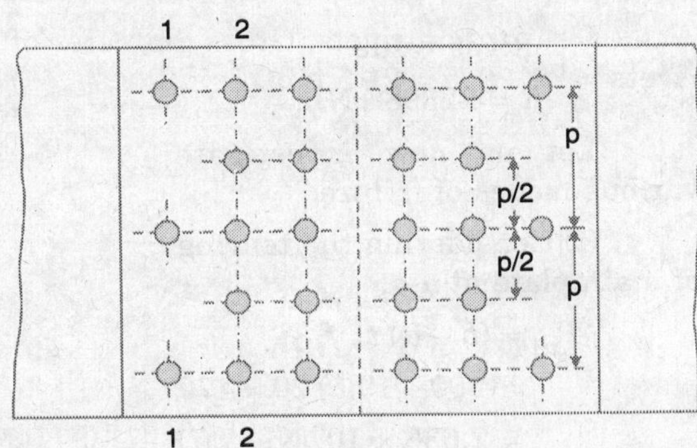

Fig. 27.16

Equations (1) and (2) gives two modes of failures due to tearing of plates, out of which Eq. (2) gives the smaller load. Hence permissible load, per pitch length, in tearing

$$= 2250\,p - 61126 \quad ...(a)$$

Let us now consider the mode of failure due to shearing of rivets. In the strip of joint equal to pitch p, there are 2 + 2 + 1 = 5 rivets. Each rivet is in double shear.

Hence permissible load, per pitch length,

$$= 5 \times \tau_{vf} \times 1.75 \times \frac{\pi}{4} d^2 = 5 \times 60 \times 1.75 \times \frac{\pi}{4}\,(25.5)^2$$

$$= 268120 \text{ N} \quad ...(b)$$

Equating (a) and (b), we get

$$2250\,p - 61126 = 268120$$

From which, p = **146.3 mm.**

Permissible load which can be carried by solid plate

$$= \sigma_{at}\,.\,p\,.\,t = 90 \times 146.3 \times 25 = 329175 \text{ N}$$

$$\therefore \quad \eta \text{ of the joint} = \frac{268120}{329175} \times 100 = \mathbf{81.5\%}.$$

Example 27.10. *A butt joint is used to connect two 20 mm thick plates as shown in Fig. 27.17. The diameter of rivets is 18 mm. Determine the efficiency of the joint. Take $f_t = 470\ N/mm^2$, $f_s = 375\ N/mm^2$ and $f_b = 750\ N/mm^2$.*

Solution. Formed diameter of the rivets = 18 + 1.5 = 19.5 mm.

Each rivet is in double shear.

Strength of each rivet in double shear

$$= P_{us} = 2 \times \frac{\pi}{4} d^2 \times f_s = 2 \times \frac{\pi}{4}\,(19.5)^2 \times 375 = 223986 \text{ N}$$

Strength of each rivet in bearing

$$= P_{ub} = f_b \times d \times t = 750 \times 19.5 \times 20 = 292500 \text{ N}$$

∴ Rivet value

$R = 223986$ N.

Let us now investigate various modes of failure.

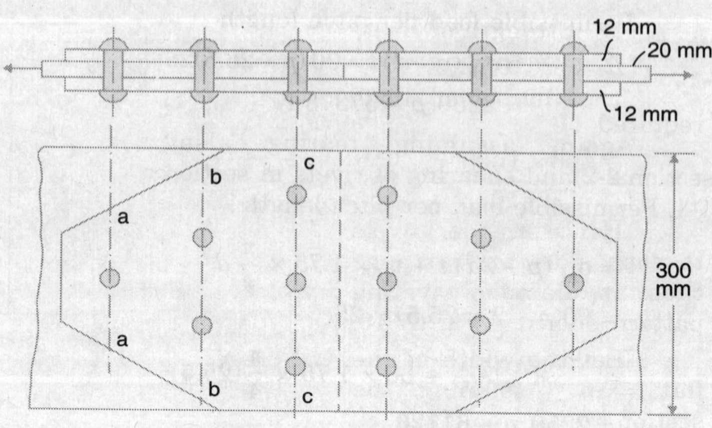

Fig. 27.17

(*i*) Considering the tearing of main plate at *a-a*,

$$P_{u1} = (b - d)\,t\,.\,f_t$$
$$= (300 - 19.5)\,20 \times 470$$
$$= 2.636 \times 10^6 \text{ N} \quad ...(i)$$

(*ii*) Considering tearing of main plate at *b-b* and shearing of rivet at *a-a*,

$$P_{u2} = (b - 2d)\,t\,.\,f_t + P_{us}$$
$$= (300 - 2 \times 19.5)\,20 \times 470 + 223986 = 2.677 \times 10^6 \text{ N} \quad ...(ii)$$

(*iii*) Considering tearing of main plate at *c-c* and shearing of rivets at *a* and *b*,

$$P_{u3} = (b - 3d)\,t\,f_t + 3\,P_{us}$$
$$= (300 - 3 \times 19.5)\,20 \times 470 + 3 \times 223986 = 2.27 \times 10^6 \text{ N} \quad ...(iii)$$

(*iv*) Considering failure by shearing of all the rivets,

$$P_{u4} = 6\,P_{us} = 6 \times 223986 = 1.344 \times 10^6 \text{ N} \quad ...(iv)$$

(*v*) Considering failure of bearing surfaces,

$$P_{u5} = 6\,P_{ub} = 6 \times 292500 = 1.755 \times 10^6 \text{ N} \quad ...(v)$$

(*vi*) Considering tearing of cover plates at *c-c*,

$$P_{u6} = (b - 3d)\,2t\,.\,f_t$$
$$= (300 - 3 \times 19.5)\,2 \times 12 \times 470 = 2.724 \times 10^6 \text{ N} \quad ...(vi)$$

The most likely mode of failure is (*iv*) in which the joint fails by shearing of the rivets, the strength of joint being = 1.344×10^6 N.

Strength of solid plate = $b\,.\,t.\,f_t = 300 \times 20 \times 470 = 2.82 \times 10^6$ N.

$$\therefore \quad \eta \text{ of joint} = \frac{1.344 \times 10^6}{2.82 \times 10^6} \times 100 = \mathbf{47.7\%}.$$

Example 27.11. *The diagonal of a bridge truss is made of 16 mm thick flat and has to transmit a pull of 600 kN. The diagonal is to be connected to 16 mm thick gusset plate by a double cover butt joint with 20 mm rivets. Calculate the number of rivets and width of flat required. Take permissible stresses as follows :* $\sigma_{at} = 150$ *N/mm*^{2} ; $\tau_{vf} = 100$ *N/mm*^{2} *and* $\sigma_{pf} = 300$ *N/mm*^{2}*. Sketch the joint and calculate the efficiency of the joint. Also, determine (i) the actual stresses induced in the flat and the rivers ; and (ii) thickness of cover plates.*

Solution. Formed diameter of rivet = 20 + 1.5 = 21.5 mm.

The rivets are in double shear. Strength of rivet in double shear is

$$P_s = 2 \times 100 \times \frac{\pi}{4}\,(21.5)^2 = 72610 \text{ N}.$$

Strength or rivet in bearing against main plate,

$$P_b = 21.5 \times 16 \times 300 = 103200 \text{ N}.$$

∴ Strength of rivet,

$R = 72610$ N.

∴ Number of rivets required

$$= \frac{600 \times 1000}{72610} = 8.26.$$

Hence provide 9 rivets to each side, and arrange them in diamond riveting pattern, shown in Fig. 27.18.

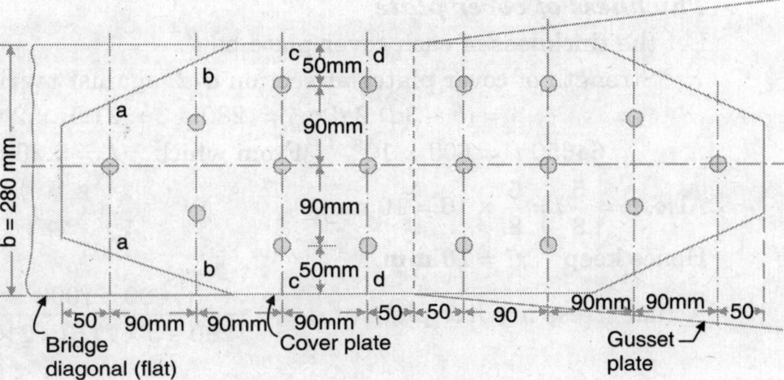

Fig. 27.18

Let the width of the flat = b. Assuming the section to be weakened by one rivet hole only,

$$P_t = (b - d)\, t\, \sigma_{at} = (b - 21.5)\, 16 \times 150$$

Equating this to external load, we get

$$(b - 21.5)\, 16 \times 150 = 600 \times 10^3$$

From which, $b = 271.5$ mm. Keep $b = 280$ mm

$$\therefore \qquad \eta = \frac{b-d}{b} = \frac{280 - 21.5}{280} \times 100 = \mathbf{92.3\%}.$$

Provide pitch and back pitch = 90 mm. Provide edge distance = 50 mm.

Actual Stresses in Rivets

$$f_s = \frac{600 \times 1000}{2 \times \frac{\pi}{4}(21.5)^2 \times 9} = 91.81 \text{ N/mm}^2$$

$$f_b = \frac{600 \times 1000}{21.5 \times 16 \times 9} = 193.8 \text{ N/mm}^2.$$

Actual Stresses in Flat

At section *a-a*, $P = f_1(b - d)\, t$

$$\therefore \qquad f_1 = \frac{600000}{(280 - 21.5)\, 16} = 145 \text{ N/mm}^2$$

At section *b-b*, $P = f_2(b - 2d)\, t + \dfrac{P}{n}$

$$\therefore \qquad 600000 = f_2(280 - 2 \times 21.5) \times 16 + \frac{600000}{9} \text{ From which, } f_2 = 140.6 \text{ N/mm}^2.$$

At section *c-c*, $P = f_3(b - 3d)\, t + \dfrac{3P}{n}$

$$\therefore \qquad 600000 = f_3(280 - 3 \times 21.5)\, 16 + \frac{3 \times 600000}{9} \quad \text{or} \quad f_3 = 116 \text{ N/mm}^2$$

At section *d-d* $P = f_4(b - 3d)\, t + \dfrac{6P}{n}$

$$\therefore \qquad 600000 = f_4(280 - 3 \times 21.5)\, 16 + \frac{6 \times 600000}{9} \quad \text{or} \quad f_4 = 58 \text{ N/mm}^2.$$

Thickness of cover plate

Let the thickness of each cover plate be t'.

$\therefore$ Strength of cover plates at section d-d, against tearing

$$= (b - 3d)\, 2\, t'\, \sigma_{at} = (280 - 3 \times 21.5) \times 2\, t' \times 150 = 64650\, t'.$$

$\therefore$ $64650\, t' = 600 \times 10^3$ From which $t' = 9.28$ mm ;

Also, $t' = \dfrac{5}{8} t = \dfrac{5}{8} \times 16 = 10$ mm.

Hence keep t' = **10 mm.**

Actual stress in cover plate at d-d $= \dfrac{600 \times 1000}{(280 - 3 \times 21.5) \times 2 \times 10} = \mathbf{139.2\ N/mm^2}$.

27.8. ECCENTRIC LOAD CONNECTION AND MOMENT CONNECTION

Fig. 27.19 shows an eccentric load connection, the load P acting at a distance e from the centre of gravity of rivet shear area. Introducing equal and opposite forces P at the C.G., the external loading is equivalent to

(*i*) A direct load P acting at the C.G. of the group.

(*ii*) A couple $M = P.e$.

Let f_1 = intensity of shear stress in each rivet, due to direct load

Then $f_1 = \dfrac{P}{\Sigma A}$ where ΣA is the sum of the rivet areas.

If A is the area of cross-section of any particular rivet, the force in the rivet, due to direct load is given by

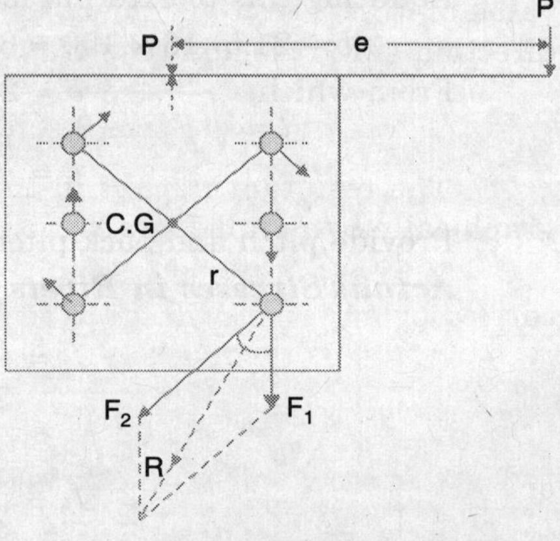

Fig. 27.19

$$F_1 = f_1 A = \frac{PA}{\Sigma A} \qquad \text{...[27.16 (a)]}$$

If all the rivets are of equal area A each, the equation 27.16 (a) reduces to

$$F_1 = \frac{PA}{nA} = \frac{P}{n} \qquad \text{...[27.16 (b)]}$$

Again, let f_2 = intensity of shear stress in each rivet due to the couple M. In order to find the stress f_2 due to couple, it is assumed that (*i*) the rivets are perfectly elastic and (*ii*) the unit stress on each rivet is proportional to its radius vector and acts in the direction perpendicular to its radius vector.

Thus, $f_2 \propto r$, where r = distance of any rivet from C.G.

or $$f_2 = kr \qquad \text{...(1)}$$

If F_2 is the force in the rivet, we have

$$F_2 = Af_2 = Akr \qquad \text{...(2)}$$

The moment of resistance of the rivet is = $F_2 r = Akr^2$

Total moment of resistance of the rivet group

$= \Sigma Akr^2 = k\Sigma Ar^2$

This must be equal to the couple $P.e$.

$$\therefore \quad k\Sigma Ar^2 = P.e \quad \text{or} \quad k = \frac{P.e}{\Sigma Ar^2} \qquad \text{...(3)}$$

Substituting the value of k in (1), we get

$$f_2 = \frac{P.e.r}{\Sigma Ar^2} \quad \text{and} \quad F_2 = \frac{P.eA.r}{\Sigma Ar^2} = \frac{M.A.r}{\Sigma A.r^2} \qquad \text{...[27.17 (a)]}$$

If all the rivets are of equal area

$$F_2 = \frac{M.r}{\Sigma r^2} \qquad \text{...[27.17 (b)]}$$

In order to find the resultant force R on a rivet, it must be noted that F_1 acts in the direction of P while P_2 acts in a direction perpendicular to the radius vector. The maximum value of R in a rivet group will be for the rivet in which F_1 and the resolved part of F_2 in the direction of F_1, are additive. If F_2 makes angle θ with the direction of F_1, we have

$$R = \sqrt{(F_2 \cos\theta + F_1)^2 + (F_2 \sin\theta)^2} = \sqrt{F_1^2 + F_2^2 + 2F_1F_2 \cos\theta} \qquad \text{...(27.18)}$$

The resultant force R in the heavily loaded rivet or any other rivet can also be found graphically by parallelogram of forces.

If in the place of eccentric load, a pure moment M acts on the rivet group, the load carried by any rivet will be given by

$$R = \frac{M.r.A}{\Sigma Ar^2}, \text{ in general} \qquad \text{...[Eq. 27.17(a)]}$$

or $$R = \frac{M.r}{\Sigma r^2} \text{ for the particular case when all the rivets have the same area.}$$

...[Eq. 27.17(b)]

The direction of R in this case will be perpendicular to the radius vector.

Example 27.12. *A load of 100 kN is carried by a bracket riveted to the flange plate of a stanchion as shown in Fig. 27.20. Each rivet is of 24 mm diameter. Calculate the maximum shear intensity induced in any rivet.*

Solution. Formed diameter of each rivet

$= 24 + 1.5 = 25.5$ mm

$M = P.e = 100000 \times 200 = 2 \times 10^7$ N-mm

Since area of section of all the rivets is the same

$$F_1 = \frac{P}{n} = \frac{100000}{8} = 12500 \text{ N}$$

$\Sigma r^2 = 4(40^2 + 120^2) + 4(40^2 + 40^2)$

$= 64000 + 12800$

$= 76800 \text{ mm}^2$

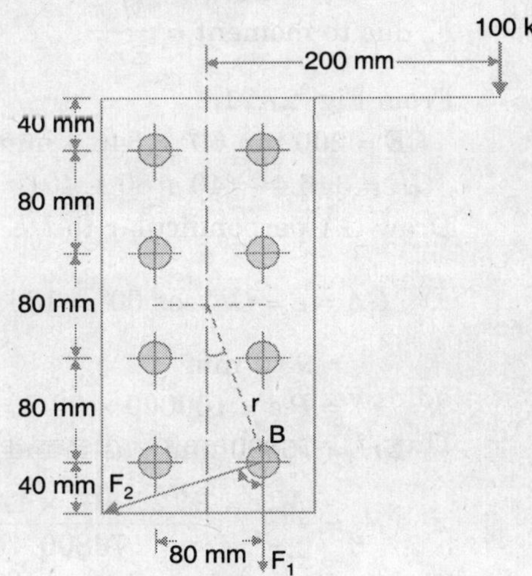

Fig. 27.20

The rivet B will be the heavily loaded rivet for which

$$r = \sqrt{(40)^2 + (120)^2} = 126.5 \text{ mm}$$

$$\therefore \quad F_2 = \frac{M.r}{\Sigma r^2} = \frac{2 \times 10^7 \times 126.5}{76800} = 32940 \text{ N}$$

$$\sin \theta = \frac{40}{126.5} \; ; \; \cos \theta = \frac{120}{126.5}$$

$$\therefore \quad F_2 \sin \theta = 32940 \times \frac{40}{126.5} = 10416 \text{ N} \; ; \; F_2 \cos \theta = 32940 \times \frac{120}{126.5} = 31248 \text{ N}$$

$$\therefore \quad R = \sqrt{(F_1 + F_2 \sin \theta)^2 + (F_2 \cos \theta)^2} = \sqrt{(12500 + 10416)^2 + (31248)^2}$$

$$= 38750 \text{ N.}$$

If the rivet is in single shear

$$f_s = \frac{38750}{\frac{\pi}{4}(25.5)^2} = \mathbf{75.9 \ N/mm^2}.$$

Example 27.13. *Solve example 27.12 if the load of 100 kN acts at an inclination of 60° to the horizontal.*

Solution. Apply equal and opposite force at C.G., in the direction of the given load of 100 kN. The given system of loading is thus equivalent to

(*i*) A direct load P acting through the C.G. of the rivet group and in the direction of the applied load, and

(*ii*) a moment $M = P.e$

$$F_1 \text{ due to direct load} = \frac{P}{n} = \frac{100000}{8} = 12500$$

in the direction of P.

$$F_2 \text{ due to moment} = \frac{M.r}{\Sigma r^2}.$$

From Fig. 27.21,

$CE = 200 \tan 60° = 346.4$ mm ;

$GE = 346.4 - (40 + 80 + 40) = 186.4$ mm.

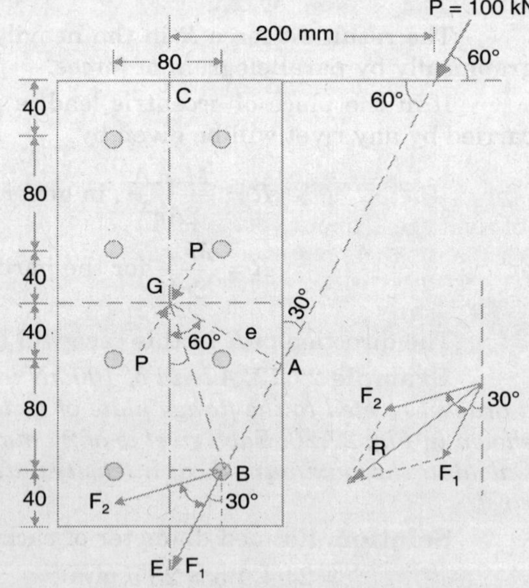

Fig. 27.21

Draw GA perpendicular to DE

$$\therefore \quad GA = e = GE \cos 60° = 186.4 \times \frac{1}{2}$$

$$= 93.2 \text{ mm}$$

$$\therefore \quad M = P.e = 100000 \times 93.2 = 93.2 \times 10^5 \text{ N/mm}$$

$\Sigma r^2 = 76800$ mm^2, as found is example 27.12 and $r = 126.5$ mm.

$$\therefore \quad F_2 = \frac{M.r}{\Sigma r^2} = \frac{93.2 \times 10^5 \times 126.5}{76800} = 15350 \text{ N.}$$

The maximum F_2 will occur in the rivet marked B. The resultant load $R = 26000$ N (found graphically). Alternatively, $\alpha = \sin^{-1}(40/126.5) = 18.43°$

$\therefore$ Inclination of F_2 with vertical = 90° – α = 90° – 18.43° = 71.57°.

Inclination of F_1 with vertical = 90° – 60° = 30°

$\therefore$ Angle θ between F_1 and F_2 = 71.57° – 30° = 41.57°.

$$\therefore \qquad R = \sqrt{F_1^2 + F_2^2 + 2F_1F_2 \cos\theta}$$

$$\sqrt{(12500)^2 + (15350)^2 + 2(12500 \times 15350)\cos 41.57^\circ} = 26057 \text{ N.}$$

$$\text{Maximum shear intensity} = \frac{R}{\text{area of rivet}} = \frac{26057}{\frac{\pi}{4}(25.5)^2} = \mathbf{51.02\ N/mm^2}.$$

Example 27.14. *Calculate the shearing stress in the rivets B and C for the connection shown in Fig. 27.22. Rivets A and B have 14 mm diameter and C has a diameter of 22 mm.*

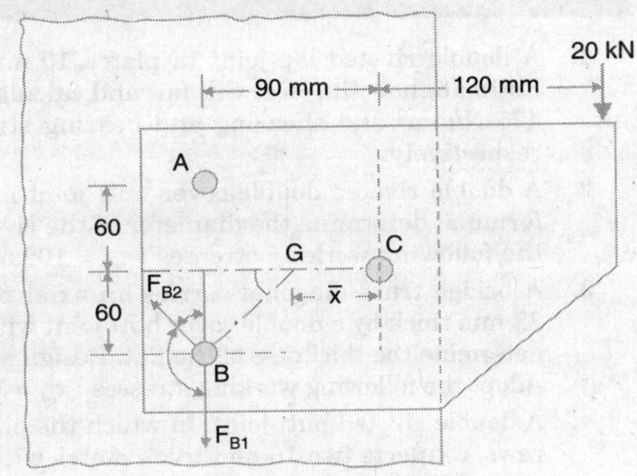

Fig. 27.22

Solution. Formed dia. of rivets *A* and *B* = 15.5 mm

Area of section of rivets *A* and *B*

$$= \frac{\pi}{4}(15.5)^2 = 188.7 \text{ mm}^2$$

Formed dia. of rivet *C*

$$= 22 + 1.5 = 23.5 \text{ mm}$$

Area of section of rivet *C*

$$= \frac{\pi}{4}(23.5)^2 = 433.7 \text{ mm}^2$$

To find the position of C.G., take moments of rivet areas about vertical line passing through *C*.

$$\text{Thus} \qquad \bar{x} = \frac{2 \times 188.7 \times 90}{(2 \times 188.7) + 433.7} = 41.9 \text{ mm}$$

$$\Sigma A = (2 \times 188.7) + 433.7 = 811.1 \text{ mm}^2$$

$$M = P.e. = 20000(120 + 41.9) = 3238000 \text{ N-mm.}$$

In the rivet *B*,

$$F_{B1} = \frac{P.A}{\Sigma A} = \frac{20000 \times 188.7}{811.1} = 4653 \text{ N}$$

Similarly, in rivet *C*

$$F_{C1} = \frac{P.A}{\Sigma A} = \frac{20000 \times 433.7}{811.1} = 10690 \text{ N}$$

$$\Sigma Ar^2 = 2 \times 188.7(48.1^2 + 60^2) + 433.7(41.9)^2 = 2993200$$

$$F_{B2} = \frac{M.r.A}{\Sigma Ar^2} = \frac{3238000 \times 188.7\sqrt{(48.1)^2 + 60^2}}{2993200} = 15700 \text{ N}$$

$$\text{and} \qquad F_{C2} = \frac{M.r.A}{\Sigma A.r^2} = \frac{3238000 \times 41.9 \times 433.7}{2993200} = 19660 \text{ N}$$

$$R_C = 10690 + 19660 = 30350 \text{ N}$$

$$\alpha = \tan^{-1}\frac{60}{90 - 41.9} = 51.28^\circ\ ;\ \theta = 180^\circ - 51.28^\circ = 128.72^\circ.$$

$$\therefore \quad R_B = \sqrt{F_{B1}^2 + F_{B2}^2 + 2F_{B1}.F_{B2}\cos\theta}$$

$$= \sqrt{(4653)^2 + (15700)^2 + 2 \times 4653 \times 15700 \cos 128.72^\circ} = 13295 \text{ N.}$$

$$\therefore \quad f_B = \frac{R_B}{188.7} = \frac{13295}{188.7} = \mathbf{70.45\ N/mm^2.}$$

$$f_C = \frac{R_C}{433.7} = \frac{30350}{433.7} = \mathbf{69.98\ N/mm^2.}$$

PROBLEMS

1. A double riveted lap joint in plates 10 mm thick is made with 16 mm rivets at 60 mm pitch. Estimate how the joint will fail and calculate its efficiency if the tearing strength of the plates is 475 N/mm² and shearing and bearing strength of the rivets are 380 N/mm² and 750 N/mm² respectively.

2. A double riveted double cover butt joint is used to connect plates 12 mm thick. Using Unwin's formula, determine the diameter of the rivet, revet value, pitch and efficiency of the joint. Adopt the following working stresses : τ_{vf} = 102.5 N/mm² ; σ_{pf} = 236 N/mm² and σ_{at} = 150 N/mm².

3. A bridge truss diagonal carries an axial pull of 500 kN. It is to be connected to a gusset-plate 22 mm thick by a double cover butt joint with 22 mm rivets. If the width of the flat tie bar is 25 cm, determine the thickness of the flat. Design an economical joint. Determine the efficiency of the joint. Adopt the following working stresses : τ_{vf} = 102.5 N/mm² ; σ_{pf} = 236 N/mm² and σ_{at} = 150 N/mm².

4. A double riveted butt joint, in which the pitch of rivet in the outer rows is twice that in the inner rows, connects two 16 mm thick plates with two cover plates each 12 mm thick. The diameter of rivets is 22 mm. Determine the pitches of rivets in the two rows if the working stresses are not to exceed the following limits.

Tensile stress in plates : 100 N/mm².

Shear stress in rivets : 75 N/mm².

Bearing stress in rivets and plates : 150 N/mm².

Make a fully dimensioned sketch of the joint.

5. Explain the basis of calculations usually made for determining the distribution of load amongst the individual rivets of a group. A tie bar is attached to a gusset plate by four rivets arranged at the corners as shown in Fig. 27.23 the pitch of the rivet being *p*. The pull is applied symmetrically. If the rivet *D* is now removed, the pull being maintained at its former value, calculate by what percentage the load on each of the rivets is increased ?

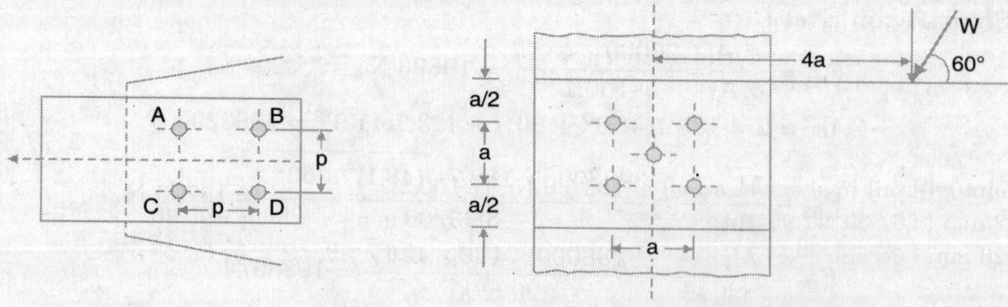

Fig. 27.23 **Fig. 27.24**

6. State the assumptions usually made in estimating the load carried by each rivet in a riveted joint which is subjected to both shear and bending in the plane of the joint.

   In the arrangement shown in Fig. 27.24 the five rivets are symmetrically arranged at the corners and the centre of the square of side *a*. Find in terms of *W*, the local transmitted by the most heavily loaded rivet.

7. A riveted steel bracket connection has 22 mm diameter rivets 12 in number arranged as shown in Fig. 27.25. Determine the load *P* so that allowable stress in the extremely loaded rivet is just reached. Take safe permissible stress in bearing in rivet = 236 N/mm^2. Safe permissible stress in shearing in rivet = 102.5 N/mm^2.

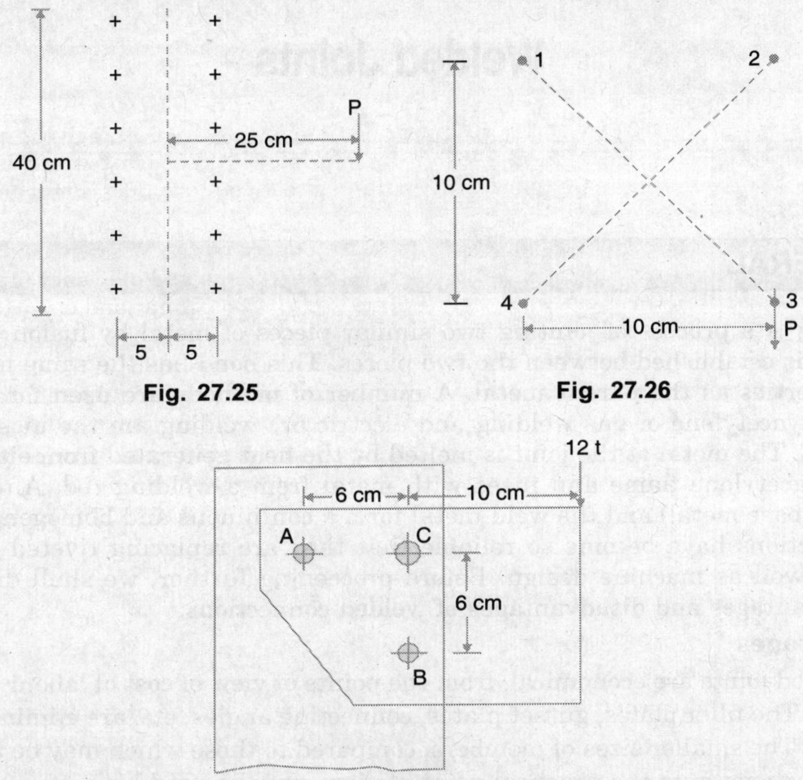

Fig. 27.25

Fig. 27.26

Fig. 27.27

8. If the worst rivet in the system shown in Fig. 27.26 may be stressed to 94.5 N/mm^2, calculate the safe value for the eccentric load *P*. The rivets are 27 mm in diameter and in single shear.

9. Calculate the maximum load carried by any rivet shown in Fig. 27.27. Rivets *A* and *B* are of 2 cm^2 cross-sectional area and rivet *C* of 4 cm^2 area.

ANSWERS

1. Joint will fail in shear at a pull of 182700 N ; η = 59.2%
2. 22 mm ; 66,600 N ; 95 mm ; 75.2%
3. t = 16 mm ; n = 6 ; η = 90.8%
4. 128 mm : 64 mm
5. A : 17.6% ; B : 21.2% ; C : 67.5%
6. 1.24 W
7. 186 kN
8. 158.1 kN
9. 15650 kg is C.

28

Welded Joints

28.1. GENERAL

Welding is a process of jointing two similar pieces of metal by fusion or pressure. A metallic bond is established between the two pieces. This bond has the same mechanical and physical properties as the parent metal. A number of methods are used for the process of fusion. The oxyacetylene or gas welding and electric arc welding are the most important of these methods. The metal at the joint is melted by the heat generated from either an electric arc or an oxyacetylene flame and fuses with metal from a welding rod. After cooling, the parent metal (base metal) and the weld metal form a continuous and homogeneous joint. The welded connections have become so reliable that they are replacing riveted joints, both in structural as well as machine design. Before proceeding further, we shall discuss here, in short, the advantages and disadvantages of welded connections.

Advantages

1. Welded joints are economical, from the points of view of cost of labour and materials, both. The filler plates, gusset plates, connecting angles, etc. are eliminated in welded joint. The smaller sizes of members, compared to those which may be used in riveted connections from the practical point of view, may be used here.
2. The efficiency of the welded joint is 100% as compared to an efficiency of 75 to 100% in case of riveted joints.
3. The fabrication of a complicated structure is easier by connection as in case of a circular steel pipe. The alterations or additions in existing structure are facilitated by it.
4. The welding provides very rigid joints. This is in keeping with the modern trend of providing rigid frames.
5. The noise associated with the riveting work is a source of a great nuisance. This is avoided in welding operation.
6. When riveting is done in populated localities, safety precautions to protect the public from the flying rivets has to be taken. No such precautions are necessary in case of welding operations.

7. The welded structures look more pleasing in comparison to the riveted ones.

Disadvantages

Not withstanding the advantages narrated above, the welded connections have a number of disadvantages in comparison to the riveted connections. The same have been narrated below :

1. No provision for expansion and contraction is kept in welded connection and, therefore there is possibility of cracks developing in such structures.
2. Due to uneven heating and cooling of the members during welding the members may distort resulting in additional stresses.
3. The inspection of welding work is more difficult and costlier than the riveting work. The welding work requires a skilled person, while, a semi-skilled person can do the riveting work.
4. On account of extreme heat, fatique may take place.

28.2. TYPES OF WELDS

In structural practice, the following types of welds are used :

1. Butt weld or groove weld.
2. Fillet weld.
3. Plug or slot weld.

1. Butt Weld or Groove Weld. When the joining members are to be jointed in such a way that they form a T or butt against each other, butt weld is used.

The butt weld is usually made convex on either sides. This extra area is called *reinforcement*. The reinforcement varies from 1 to 3 mm.

The common types of butt welds have been shown in Fig. 28.1. The square butt joints shown in Fig. 28.1 (*a*) and (*b*) are used for thickness less than 8 mm. The effective thickness of the weld, called *throat thickness*, is less than the thickness T of the plates jointed. It is taken as $\frac{3}{4}$ T. In single V-butt joint [Fig. 28.1 (*c*)], the throat thickness is taken at $\frac{3}{4}$ T. The butt welds of Fig. 28.1 (*d*) and (*e*) are fully effective. In Fig. 28.1 (*e*), a backing strip has been used. As a rule, in single-V, single-U, single-J butt welds, with backing strips, and in which the welding has

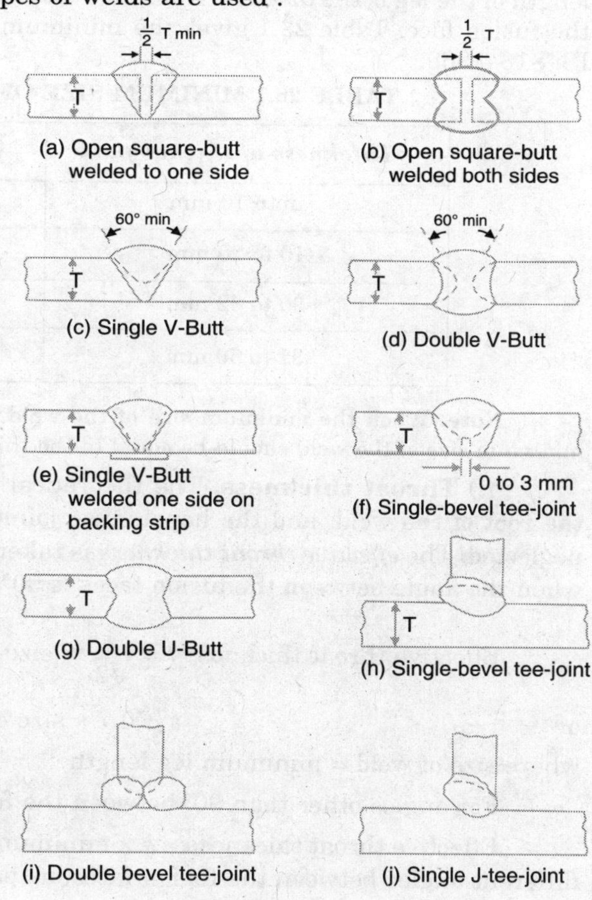

Fig. 28.1. Butt welds

been done from one side, full penetration is not possible, and the effective throat thickness is taken equal to $\frac{3}{4}$ T.

In all the above mentioned welds with backing strips and in case of double-V, double-U and double-J butt welds, full penetration is possible and the effective thickness of the throat is taken equal to the thickness of the plates joined. Whenever two plates of different thickness are joined, the thickness of the thinner plate must be taken into account.

2. Fillet Welds. When the lapped plates are to be jointed, fillet welds are used. These are generally of right angled triangle shape. The outer surface is generally made *convex*, as shown in Fig. 28.2. A filled weld is specified by the following.

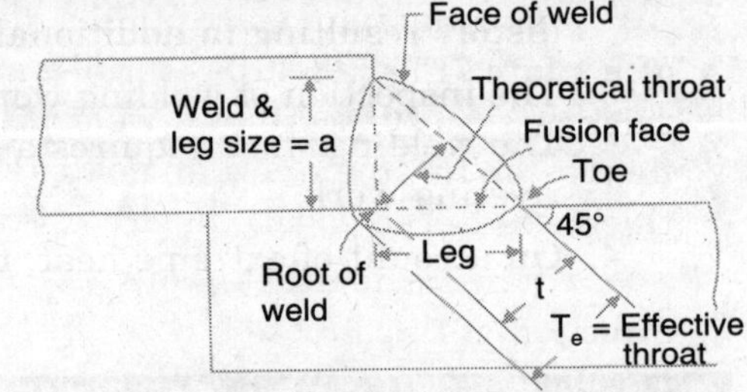

Fig. 28.2. Fillet weld terminology

(*i*) Size of weld.

(*ii*) Throat thickness.

(*iii*) Length of weld.

(*i*) **Size of weld :** The sides containing the right angle of the fillet are called *legs*. The *size* of the weld is specified by *minimum leg length*. The length of the leg is the distance from the root of the weld to the toe of the weld, measured along the fusion face. Table 28.1 gives the minimum size of single run fillet weld, as specified by IS: 816–1969.

TABLE 28.1 MINIMUM SIZE OF SINGLE RUN FILLET WELD

Thickness of thicker part	*Min. size*
upto 10 mm	3 mm
10 to 20 mm	5 mm
20 to 32 mm	6 mm
32 to 50 mm	8 mm (first run) ; 10 mm (min.)

Note. When the minimum size of the weld is greater than the thickness of thinner part, the minimum size of the weld should be equal to the thickness of thinner part.

(*ii*) **Throat thickness:** The theoretical throat is the perpendicular distance between the root of the weld, and the hopotenuse joining the two ends of the legs. Reinforcement is neglected. The *effective throat thickness* is taken equal to the theoretical throat thickness, and when the angle between the fusion faces is 90° (as is generally the case), we have :

$$\text{Effective throat thickness } t = \frac{1}{\sqrt{2}} \times \text{size of weld}$$

or

$$t \simeq 0.7 \times \text{size of weld.} \qquad \text{...(28.1)}$$

where size of weld = minimum leg length

For angles other than 90° between the fusion faces,

Effective throat thickness = k × minimum leg length. Table 28.2 gives the values of k for different angles between the fusion faces, as per IS : 816–1969.

TABLE 28.2 VALUES OF *k*

Angle	*60° to 90°*	*91° to 100°*	*101° to 106°*	*107° to 113°*	*114° to 120°*
k	0.7	0.65	0.60	0.55	0.50

It may be noted that a fillet weld is not used for jointing parts if the angle between the fusion faces is less than 60° or greater than 120°.

The maximum size of fillet weld at the square edge of a plate [Fig. 28.3(*a*)] is 1.5 mm less than the plate thickness and in case of a weld at the rounded edges of flanges or the toe of an angles is kept three fourths the thickness of the edge [Fig. 28.3(*b*)].

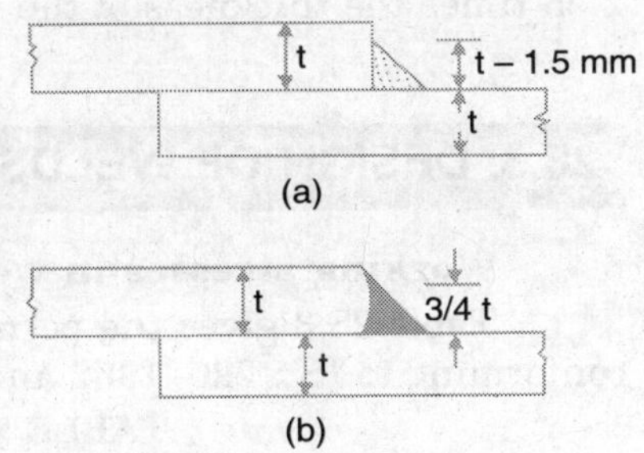

Fig. 28.3. Maximum size of weld

When the fillet weld is placed parallel to the direction of the forces on both the sides of the member, it is called *side fillet weld*. When the weld is placed at the end of the member, such that it is perpendicular to the direction of the force, it is called *end fillet weld*. If the axis of the weld is inclined to the direction of force, it is known as *diagonal fillet weld*.

(*iii*) **Effective length of weld:** The effective length of the weld is taken as overall length minus twice the weld size. The effective length should not be less than four times the size of the weld, otherwise the weld size must be taken as one fourth of its effective length. If only the side welds are used, the length of each side fillet weld must not be less than the perpendicular distance between the two. When the ends are returned, as shown in Fig. 28.4 (*b*), the ends should be carried continuous for a distance not less than twice the size of the weld, specially when the joint is subjected to tensile force. *Only the effective length is shown on the drawing and the additional length* (*i.e.*, 2 × *weld size*) *is provided by the welder.*

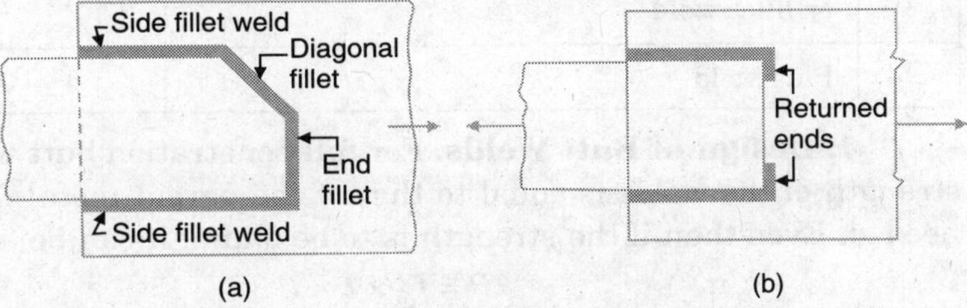

Fig. 28.4

3. Plug or Slot Weld. Whenever sufficient space is not available for providing the necessary length of the fillet weld, the plug or slot weld is provided to strengthen the connection. The slot or plug weld is also used for equalising the stress in plates and to prevent buckling in case of wide plates.

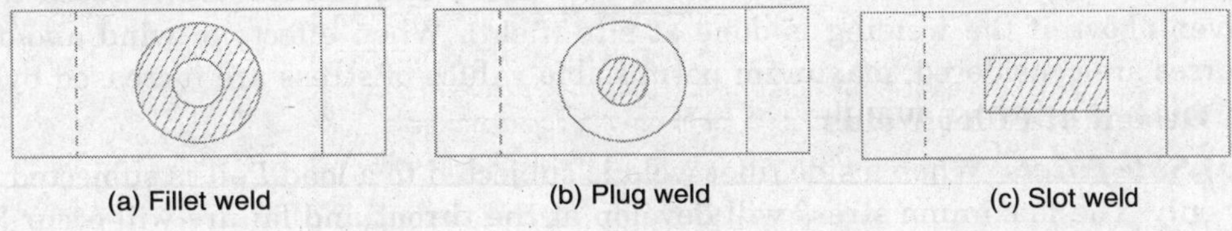

Fig. 28.5

A circular or a slotted hole is made into one of the members to be jointed. The weld metal is then filled in the hole. Otherwise a fillet weld is provided along the edge of the hole [Fig. 28.5 (*a*)].

The minimum diameter and the width of the hole should not be less than the thickness of the part containing the hole plus 8 mm, the maximum diameter and width being limited to 2.25 times the thickness of the plate punched.

28.3. DESIGN OF WELDS

Working stresses in welds

Table 28.3 gives the permissible stresses in welds as per IS : 816–1969, for mild steel conforming to IS : 226–1962 and electrodes conforming to IS : 814–1974.

TABLE 28.3 WORKING STRESSES IN WELDS

S.No.	*Kind of stress*	*Max. permissible values (converted in SI units)*	
1.	Tension or compression stress through throat of butt weld	$0.6\ f_y$	150 N/mm^2
2.	Bending stress in tension or compression	$0.66\ f_y$	165 N/mm^2
3.	Shear on section through throat of butt or fillet weld	$0.44\ f_y$	110 N/mm^2
4.	Plug weld	$0.44\ f_y$	100 N/mm^2

1. Design of Butt Welds. For full penetration butt welds [Fig. 28.1 (*c*), (*d*), (*e*), (*g*)] the strength of the welds is equal to that of the parent metal. In such cases, no calculations are needed. Even then if the strength is to be found, it can be found by the following formula :

$$P = f_w \times t \times L \qquad \text{...(28.2)}$$

where P = strength of weld.

t = effective thickness, including reinforcement.

L = length of the weld.

f_w = working stress in tension or compression.

For full penetration welds, the effective thickness is equal to the thickness of the thinner plate. If the full penetration weld has not been used, the effective thickness must be determined. It must be noted here that a butt weld is generally subjected to either tensile or compressive stress.

Maximum permissible value of stresses for shear and tension are reduced to 80% of those given above if the welding is done at site (field). When effects of wind and/or earthquake forces are considered, maximum permissible values of stress are increased by 20%.

2. Design of Fillet Welds

(a) Side fillets. When a side fillet weld is subjected to a load P, it is subjected to shear stresses only. The maximum stress will develop at the throat and failure will occur by shear along the throat. If we assume a uniform distribution of shear stress,

$$P = \text{Stress} \times \text{Area}$$

$$\text{Area} = \text{Throat thickness} \times \text{Length} = 0.707 \times h \times L.$$

$$\therefore \quad P = 0.707 \times h \times L \times \tau_{vp} \quad \text{...(28.3)}$$

where τ_{vp} = permissible maximum shear stress in weld.

(b) End fillets. If a load P is applied to an end filled weld as shown in Fig. 28.6 it will be subjected of different types of stresses on different planes. The plane DB will be subjected to tensile stress and the plane AC to shearing stress. In both the cases the stress will be numerically equal to P/hL. The section AD is subjected to tensile and shear stresses both, each equal to

$$\frac{P/\sqrt{2}}{Lh/\sqrt{2}} = \frac{P}{Lh}.$$

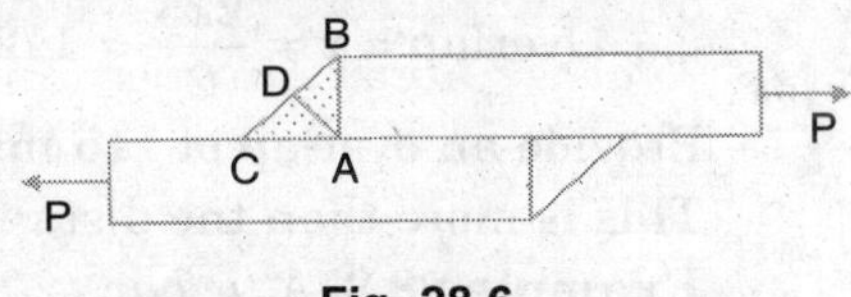

Fig. 28.6

It can be proved that the maximum tension on this section is $\frac{1.06P}{Lh}$ and the maximum shear stress is $\frac{1.12\,P}{Lh}$. As the material is weaker in shear, we shall consider the shear stress only. It may be noted that the shear stress developed in this case is less than that developed in case of sides fillets. So for convenience in design, no distinction is made between the side fillets and end fillets.

3. Plug or Slot Welds. The strength of a plug or slot weld is equal to the cross-sectional area of the plug or slot, multiplied by the allowable stress.

If fillet weld is provided in the plug or slot, its effective length is taken equal to the average length of the throat of the fillet. Generally, it is taken equal to the length of a line running parallel to the vertical lag of the weld at one fourth the leg dimension h from it. For a hole of diameter d the lenght of the fillet is $\pi\left(d - \frac{h}{2}\right)$. The maximum length is limited to 10 times the thickness of the material.

Example 28.1. *Find the safe load that can be transmitted by the fillet weld shown in Fig. 28.7. The size of the weld is 8 mm, take the safe stress in the weld as 110 N/mm².*

Solution. Effective length of weld,

$$L = 80 + 60 + 80 = 220 \text{ mm}$$

Size of weld, $\quad h = 8$ mm

Throat thickness, $\quad t = 0.7\,h = 0.7 \times 8 = 5.6$ mm

$$\therefore \quad \text{Safe load } P = t \,.\, L \,.\, \tau_{vp} = 5.6 \times 220 \times 110$$

$$= 135520 \text{ N} = \mathbf{135.32 \text{ kN}}.$$

80 mm
60
P
P

Fig. 28.7

Example 28.2. *A 100 mm × 10 mm plate is to be welded to another plate 150 mm × 10 mm by the fillet welding on three sides as shown in Fig. 28.8. The size of the weld is 6 mm. Find out the necessary overlap of the plate if the smaller plate is to develop full strength. The allowable stress in plates 150 N/mm² and allowable stress in weld as 100 N/mm².*

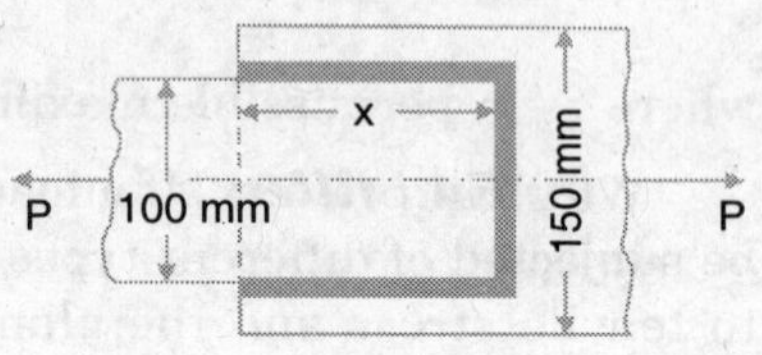

Fig. 28.8

Solution. The total load taken by the smaller plate

$= 100 \times 10 \times 150 = 15000$ N = 150 kN.

Throat thickness $= 0.7\ h = 0.7 \times 6 = 4.2$ mm

Allowable load per linear mm $= 4.2 \times 110 = 462$ N

$\therefore$ Total length of weld required $= \dfrac{150000}{462} = 325$ mm

The length of end fillet = 100 mm

$\therefore$ Length to be provided in side fillet

$= 325 - 100 = 225$ mm

$\therefore$ Overlap $= x = \dfrac{225}{2} = 112.5$ m

Provide an overlap of 115 mm.

This is more than the distance of 100 mm between two side fillets.

Example 28.3. *A tie bar 120 mm × 10 mm is to be connected to the other of size 120 mm × 14 mm. If the tie bars are to be loaded by 160 kN, find out the size of the end fillets such that the stresses in both the end fillets are same.*

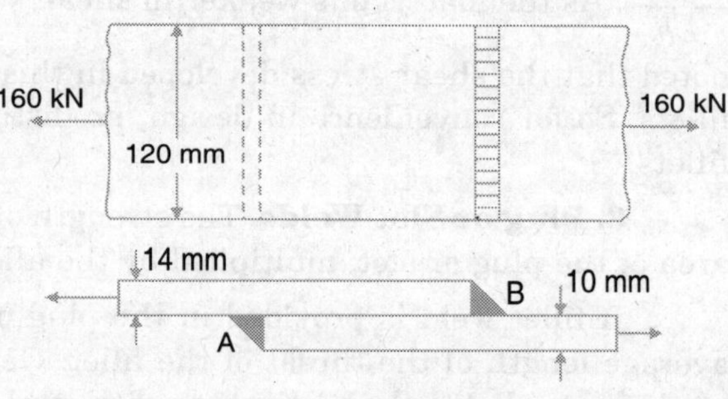

Fig. 28.9

Solution. The portion of plates between the welds stretch by the same amount. Therefore, the strain and hence the stress in both the plates are same. The force carried by each plate will be proportional to its thickness. Thus if the 10 mm thick plate carries P_1 force, the 14 mm thick plate will carry $1.4P_1$. Therefore, to keep the stresses same in both the end welds, we must keep the size of the welds in proportion to the thickness of the respective plates.

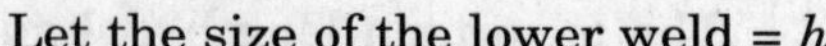

Let the size of the lower weld = h

$\therefore$ Size of the upper weld = $1.4\ h$

The length of the welds in each case = 120 mm

Strength of lower weld $= 0.7 \times h \times 120 \times 110 = 9240\ h$ N

Strength of upper weld $= 0.7 \times 1.4\ h \times 120 \times 110 = 12936$ N

Total load to be carried by the tie bar = 160000 N

$\therefore$ $9240\ h + 12936\ h = 160000$, from which $h = 7.22$ mm.

We shall keep $h = 7.5$ mm. Therefore, the size of the lower weld = 7.5 mm and the size of the upper weld $= 1.4\ h = 1.4 \times 7.5 = 10.5$ mm.

The maximum allowable size of lower weld $= t - 1.5 = 10 - 1.5 = 8.5 < 7.5$ mm.

Similarly, the size of upper weld is lesser than the maximum allowable.

Example 28.4. *A tension member consisting of two channel sections 200 mm × 75 mm @ 22.1 kg/m back to back is to be connected to gusset plate. Design the welded joint for the condition that the section is loaded to its full strength. Take A = 2821 sq. mm, thickness of the flange = 11.4 mm and the thickness of the web = 6.1 mm, permissible stresses in the weld equal to 110 N/mm² and permissible stress in the section in axial tension = 150 N/mm².*

Solution. In case of rolled section, the size of the weld is taken of three fourth of the thickness.

∴ Maximum size of the weld

$$= \frac{3}{4} \times 6.1 = 4.6 \text{ mm.}$$

We shall provide 4 mm weld.

Strength of weld per lineal cm

$$= 0.7 \times 4 \times 110 = 308 \text{ N}$$

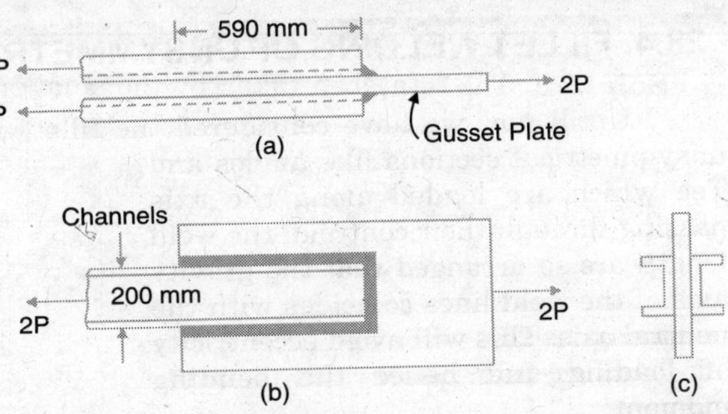

Fig. 28.10

The load to be carried by each channel

$$= 2821 \times 150 = 423150 \text{ N}$$

Total length of weld required for one channel

$$= \frac{423150}{308} \simeq 1380 \text{ mm}$$

Length of the end weld = 200 cm

Therefore, length of side welds = 1380 – 200 = 1180 mm

$$\therefore \quad \text{Overlap} = \frac{1180}{2} = 590 \text{ mm.}$$

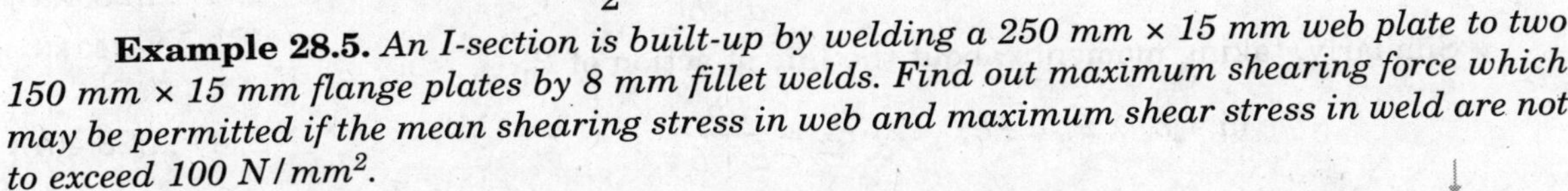

Example 28.5. *An I-section is built-up by welding a 250 mm × 15 mm web plate to two 150 mm × 15 mm flange plates by 8 mm fillet welds. Find out maximum shearing force which may be permitted if the mean shearing stress in web and maximum shear stress in weld are not to exceed 100 N/mm².*

Solution. The moment of inertia of the section about xx is,

$$I_{xx} = \frac{1}{12}\left[150 \times 280^3 - 135 \times 250^3\right] = 98.62 \times 10^6 \text{ mm}^4$$

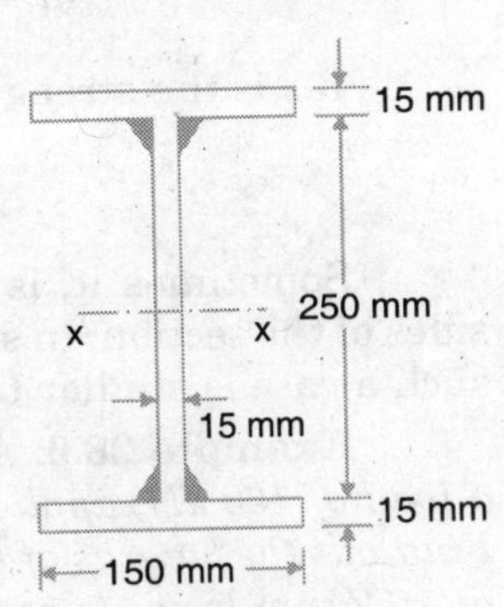

Fig. 28.11

Shear stress q at the section passing through welds is given by

$$q = \frac{F}{I_{xx}\, b}\,(A\bar{y})$$

where b = Effective thickness = 2 × throat thickness

$$= 2 \times 8 \times 0.7 = 11.2 \text{ mm.}$$

$A\bar{y}$ = Moment of the flange area about xx

$$= 150 \times 15(125 + 7.5) = 298125 \text{ mm}^3$$

q = Allowable shear stress = 100 N/mm²

$$\therefore \quad 100 = \frac{F}{98.62 \times 10^6 \times 11.2}[298125]$$

from which F = 370497 N

The maximum shear force limited on the web

$$= 250 \times 15 \times 100 = 375000 \text{ N}$$

∴ The max. allowable shearing force = 370497 N ≃ **37.05 kN.**

28.4. FILLET WELDING OF UNSYMMETRICAL SECTIONS : AXIALLY LOADED

Uptill now we have considered the fillet welding of symmetrical sections. In case of unsymmetrical sections like angles and Tee which are loaded along the axis passing through their centroid, the weld length are so arranged that the gravity axis of the weld lines coincides with the neutral axis. This will avoid eccentricity of loading and hence the bending moment.

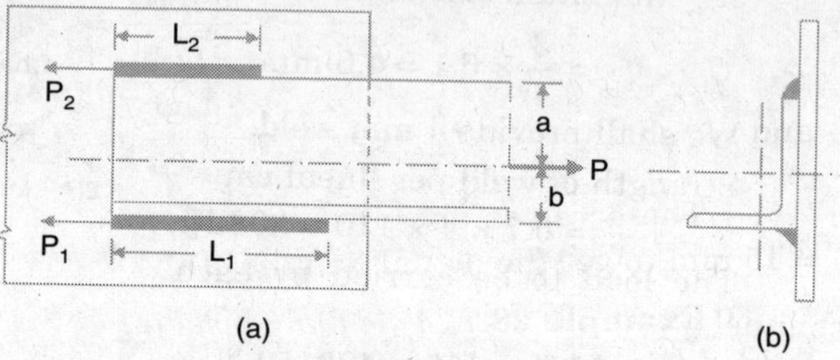

Fig. 28.12

Let us consider an angle section subjected to load P, welded to a gusset plate as shown in Fig. 28.12.

Let L_1 and L_2 be the required lengths of welds on the two faces, the P_1 and P_2 be the resisting forces exerted by the respective welds. These are assumed to act along the edges of the angle.

Taking moments about the line of action of P_2 we obtain,

$$(a + b) \times P_1 = Pa, \quad \therefore \quad P_1 = \frac{Pa}{a+b} \qquad ...(1) ...[28.4(a)]$$

Similarly, taking moments about the line of action of P_1,

$$(a + b) \times P_2 = Pb, \quad \therefore \quad P_2 = \frac{Pb}{a+b} \qquad ...(1) ...[28.4(b)]$$

If s is the strength of the weld per unit length of the weld,

$$L_1 = \frac{P_1}{s} = \frac{Pa}{s(a+b)} \quad \text{and} \quad L_2 = \frac{Pb}{s(a+b)} \qquad ...(28.5)$$

Sometimes it, is not possible to accommodate the required length of the weld on the sides of the section. In such cases, the end fillets are also provided. The procedure for analysing such a case is similar to the one described above.

Example 28.6. *An equal angle 65 mm × 65 mm @ 9.4 kg/m of thickness 10 mm carries a load of 160 kN, applied along its centroidal axis. The angle is to be welded to a gusset plate. Find out the lengths of the side fillet welds required at the heel and the toe of angle. Its C.G. is at 19.7 mm from its heel. Take permissible stress in weld equal to 110 N/mm².*

Solution. Taking moments about the line of action of P_2

$$65P_1 = 160 \times 1000 \times 45.3$$

$$\therefore \quad P_1 = 111508 \text{ N}$$

$$\therefore \quad P_2 = P - P_1$$

$$= 160000 - 111508$$

$$= 48492 \text{ N.}$$

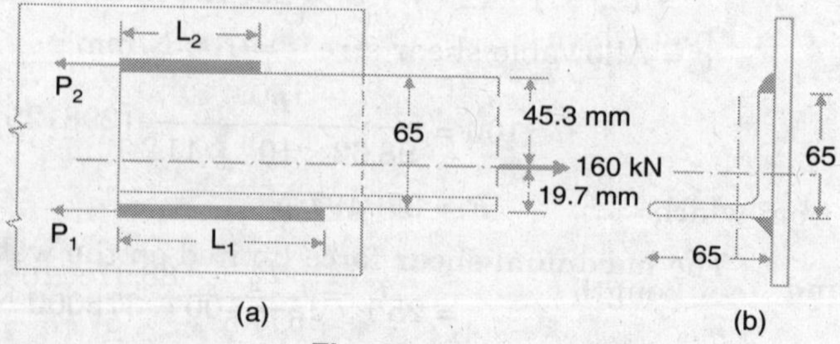

Fig. 28.13

The maximum size of the weld for a rounded edge at the toe of the angle is three-fourths of the thickness,

i.e., $\frac{3}{4} \times 10 = 7.5$ mm.

Strength of weld per lineal mm

$$= 0.7 \times 7.5 \times 110 = 577.5 \text{ N}$$

$$\therefore \quad L_1 = \frac{P_1}{577.5} = \frac{111508}{577.5} = 193 \text{ mm}$$

and

$$L_2 = \frac{P_2}{577.5} = \frac{48492}{577.5} = 84 \text{ mm.}$$

These values are effective and must be increased by twice the weld size, *i.e.*, 2 × 7.5 = 15 mm to get the actual lengths of the welds.

Example 28.7. *A tie bar consisting of a single angle 60 mm × 60 mm × 10 mm is to be welded to a gusset plate. The tie bar carries a load of 150 kN along its centroidal axis. Design the joint if both the side fillets and end fillets are to be provided. The centroidal axis lies at 18.5 mm from the heel of the angle. Take permissible stress in the weld equal to 110 N/mm².*

Solution. The maximum size of the fillet weld at the end, along the square edge of the angle will be 1.5 mm less than the thickness of the angle. Therefore, the maximum size of the end fillet

$$= 10 - 1.5 = 8.5 \text{ mm.}$$

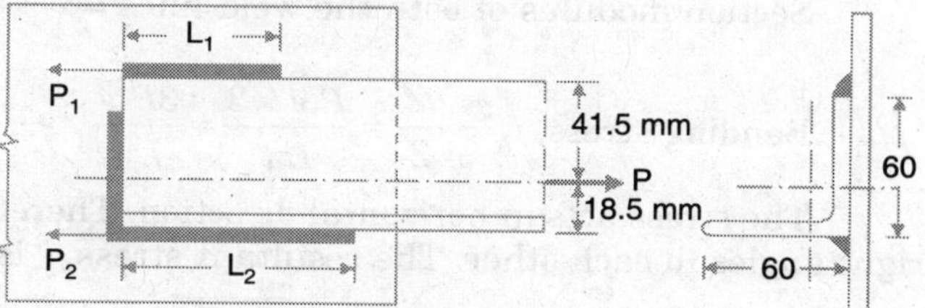

Fig. 28.14

The maximum size of the side fillets, along the rounded edge

$$= \frac{3}{4} \times 10 = 7.5 \text{ mm}$$

We shall provide 7.5 mm weld throughout.

Strength of the weld per cm length = 0.7 × 7.5 × 110 = 577.5 N

The end weld will be placed symmetrical about the line of action of the load in order to avoid eccentricity. The maximum length of the end weld is, therefore, equal to 2 × 18.5 = 37 mm.

The strength of the end weld = 577.5 × 37 = 21368 N = 21.368 kN.

Taking moments about the line of action of force P_1, we get

$$60\, P_2 = 150 \times 41.5 - 21.368 \times 41.5$$

From which

$$P_2 = 88.97 \text{ kN}$$

$$\therefore \quad P_1 = 150 - 88.97 - 21.368 = 39.662 \text{ kN}$$

$$\therefore \text{ Length} \quad L_1 = \frac{P_1}{577.5} = \frac{39.662 \times 1000}{677.5} = 69 \text{ mm}$$

and

$$\text{Length} \quad L_2 = \frac{P_2}{577.5} = \frac{88.97 \times 1000}{577.5} = 154 \text{ mm.}$$

Twice the size of the weld, *i.e.*, 2 × 7.5 = 15 mm must be added to above lengths to get the actual lengths of the side fillets.

28.5. WELDED JOINT SUBJECTED TO BENDING MOMENT

If the load P acting on a welded joint does not pass through the centroid of weld lines, it will subject the welded joint to bending moments in addition to the direct load. Consider I-beam welded to the flange of a column and subjected to a load P at eccentricity e, shown in Fig. 28.15.

Let L be the length of the weld and t be the throat thickness. The joint is subjected to a load P and a moment $P \times e$.

Fig. 28.15

The area resisting direct stress $= 2 \times L \times t$.

$$\therefore \quad \text{Direct throat stress } f_d = \frac{P}{2Lt} \qquad \text{...(28.6)}$$

It will act in the direction of the load, *i.e.*, vertically in this case.

$$\text{Section modulus of both the weld lines} = \frac{2 \times t \times L^2}{6} = \frac{L^2 t}{3}$$

$$\text{Bending stress, } f_b = \frac{M}{Z} = \frac{P.e \times 3}{L^2 t} = \frac{3P.e}{L^2 t} \qquad \text{...(28.7)}$$

The stress acts in horizontal direction. Therefore, the direct and bending stresses act at right angles to each other. The resultant stress f_r is given by

$$f_r = \sqrt{f_d^2 + f_b^2}\,. \qquad \text{...(28.8)}$$

For preliminary design, f_b can be assumed to be equal to f_r or τ_{vp}.

$$\text{In that case } f_b = \tau_{vp} = \frac{3\,P.e}{t.L^2}\,.$$

$$\text{From which} \qquad L = \sqrt{\frac{3\,P.e}{t.\tau_{vp}}} \qquad \text{...(28.9)}$$

Example 28.8. *A bracket consisting of an I-section is connected to the flange of a vertical column by means of two side fillets 250 mm deep and 8 mm thick, as shown in Fig. 28.15. The bracket carries a load of 160 kN at an eccentricity of 60 mm. Calculate the throat stress in the weld.*

Solution. Throat thickness = 0.7 × 8 = 5.6 mm

$$\text{Direct stress} \qquad f_d = \frac{160000}{2 \times 250 \times 5.6} = 57.14 \text{ N/mm}^2$$

Bending moment $= 160000 \times 60 = 96 \times 10^5$ N-mm

Z_{xx} of the weld lines $= 2 \times \frac{1}{6} \times 5.6\,(250)^2 = 116667 \text{ mm}^2$.

$$\text{Bending stress} \qquad f_b = \frac{M}{Z_{xx}} = \frac{96 \times 10^5}{116667} = 82.29 \text{ N/mm}^3$$

$$\text{Resultant stress} \qquad f = \sqrt{f_d^2 + f_b^2} = \sqrt{(57.14)^2 + (82.29)^2} = \mathbf{100.2\ N/mm^2}.$$

Example 28.9. *A bracket carrying a load of 120 kN is connected to a column by means of two horizontal fillet welds, each 150 mm long and 10 mm thick. The load acts at 80 mm from the face of the column. Find the throat stress.*

Solution. Fig. 28.16 shows the arrangement of the bracket.

Throat thickness = 0.7 × 10 = 7 mm

Direct force = 120 kN

$$\text{Direct stress } f_d = \frac{120 \times 1000}{2 \times 150 \times 7}$$

$= 57.14$ N/mm², acting in vertical direction.

Fig. 28.16

Moment $= 120000 \times 80 = 96 \times 10^5$ N-mm

The forces due to bending in the two welds will form a resisting couple

$= 150 \times 7 \times f_b \times 120 = 1.26 \times 10^5 f_b$

∴ $1.26 \times 10^5 f_b = 96 \times 10^5$

From which $f_b = 76.19$ N/mm², acting in vertical direction.

∴ Resultant stress $= \sqrt{(76.19)^2 + (57.14)^2} =$ **95.2 N/mm².**

28.6. WELDED JOINT SUBJECTED TO TORSION

If a bracket carrying an eccentrically applied load P is connected to the column as shown in Fig. 28.17 the welds will be subjected to torsion. The difference between this sort of bracket connection and that shown in Fig. 28.15 must be noted. In this case the moment is acting in a plane containing the welds, while in previous case, the moment was acting in a plane perpendicular to the welds. The method of analysis for welded joint subjected to torsion is similar to that for riveted joints.

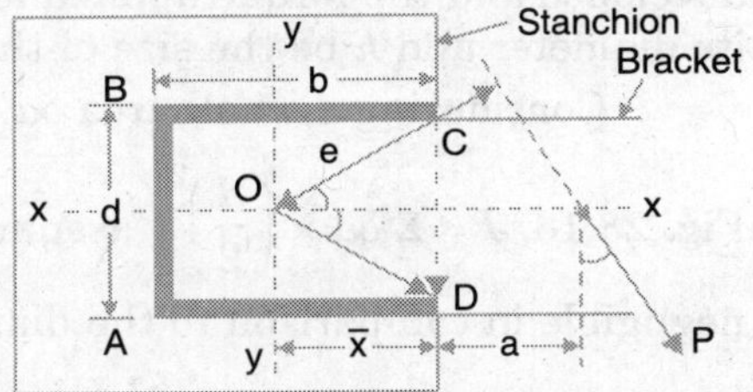

Fig. 28.17

Let the load be transmitted to the stanchion by means of welds of throat thickness t on top, bottom and one side of the bracket.

Total throat area $= t\,(d + 2b)$

$$\text{Direct throat stress} \quad f_d = \frac{P}{t(d + 2b)}.$$

The direct stress will act in the direction of load P.

To find out the torsional stress in the welds, the position of centroid of the welds must be determined. Let $\bar{x}$ be the distance of centroid from the face of the stanchion CD.

$$\bar{x} = \frac{2b \times t \times \dfrac{b}{2} + d \times t \times b}{2b \times t + d \times t} = \frac{b^2 + bd}{2b + d} \quad \text{...(28.10)}$$

$$\therefore \quad \text{Eccentricity } e = (a + \bar{x}) \cos \theta \quad \text{...(28.11)}$$

$$\text{Twisting moment} \quad T = P \times e \quad \text{...(28.12)}$$

It will be assumed that the stresses developed due to torsion at any point of the weld will be proportional to its distance from the centroid and that it will act at right angles to the radius vector.

If T is the twisting moment and f_t is the stress in weld caused due to twisting.

$$f_t = \frac{T \cdot r}{J} \qquad \text{...(28.13)}$$

where J is the polar moment of inertia of the weld sections about the centroid O.

$$J = I_{zz} = I_{xx} = I_{yy}$$

Here
$$I_{xx} = 2 \times bt \times \frac{d^2}{4} + \frac{td^3}{12}$$

and
$$I_{yy} = \frac{2tb^3}{12} + 2bt\left(\bar{x} - \frac{b}{2}\right)^2 + td\,(b - \bar{x})^2$$

The maximum stress will occur at C and D. The direct stress will act in the direction of the load and the stress due to torsion will act at right angles to the direction of the radius vector. The resultant of the two will give the total stress at the point.

Example 28.10. *A circular shaft of diameter 120 mm is welded to a rigid plate by a fillet weld of size 6 mm. If a torque of 8 kN-m applied to the shaft, find the maximum stress in the weld.*

Solution. Given: T = 8 kN-m = 8×10^6 N-mm

We shall first derive relation between the stresses developed and the torque applied for such a shaft. Let d be its diameter and h be the size of the weld.

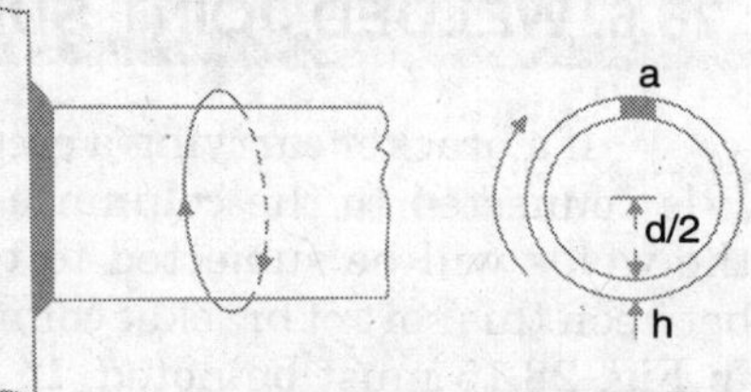

Fig. 28.18

Considering a small area δa of the weld as shown in Fig. 28.18, $J = \Sigma\, \delta a \times \left(\frac{d}{2}\right)^2$ (assuming the size of the weld negligible in comparison to the diameter of the shaft).

$$\therefore \quad J = \left(\frac{d^2}{4}\right) \Sigma\, \delta a = \frac{d^2}{4}(\pi\, dt) = \frac{\pi d^3}{4} t$$

where t = throat thickness = $0.7h$ = 0.7×6 = 4.2 mm

$$\Sigma\, \delta a = \pi d \times \left(\frac{h}{\sqrt{2}}\right)$$

$$\therefore \quad J = \frac{d^2}{4} \times \frac{\pi \cdot dh}{\sqrt{2}} = \frac{\pi d^3 h}{4\sqrt{2}}$$

Torsional shear stress

$$f_t = \frac{T \cdot r}{J} = T \cdot \frac{d}{2} \times \frac{4}{\pi d^3 t} = \frac{2T}{\pi d^2 t} \qquad \text{...(28.14)}$$

Substituting the values, we get

$$f_t = \frac{2 \times 8 \times 10^6}{\pi (120)^2 \times 4.2} = \mathbf{84.2\ N/mm^2.}$$

Example 28.11. *A bracket is subjected to a load of 100 kN as shown in Fig. 28.19. A bracket is welded to a stanchion by means of three lines of weld on three sides as indicated in the figure. Find out size of the welds so that the load is carried safely.*

Solution. Let f be the throat thickness of the weld.

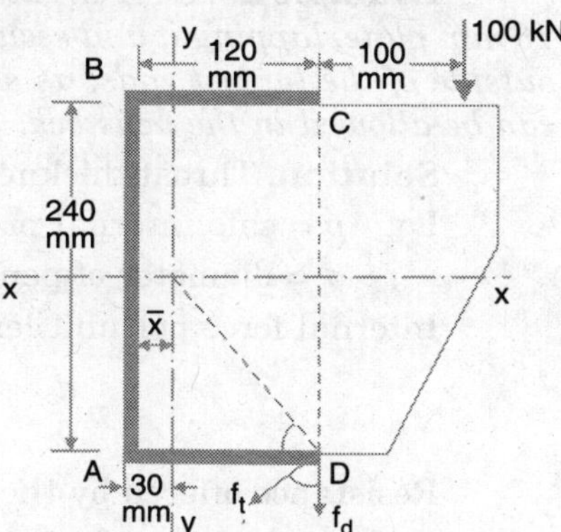

Fig. 28.19

If $\bar{x}$ is the distance of centroid of weld are from AB,

$$\bar{x} \simeq \frac{2 \times 120\, t \times 60}{2 \times 120\, t + 240\, t} = 30 \text{ mm}$$

$\therefore$ Eccentricity of the load

$= 100 + 120 - 30 = 190$ mm

$$I_{xx} = \frac{t\,(240)^3}{12} + 2 \times 120\, t\,(120)^2$$

$= 460.8 \times 10^4\, t \text{ mm}^4.$

$$I_{yy} = 240 \times t(30)^2 + 2 \times \frac{1}{12}\, t(120)^3$$

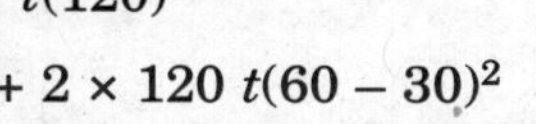

$$+ 2 \times 120\, t(60 - 30)^2$$

$= 72 \times 10^4\, t \text{ mm}^4$

$\therefore$ $J = I_{zz} = I_{xx} + I_{yy} = 460.8 \times 10^4\, t + 72 \times 10^4\, t = 532.8 \times 10^4\, t \text{ mm}^4.$

Area $= 2 \times 120 \times t + 240\, t = 480\, t \text{ mm}^2$

The maximum stress due to torsion will occur either at C or D.

Length of radius vector for C or D $= \sqrt{(120)^2 + (90)^2} = 150$ mm.

Maximum stress due to torsion

$$f_t = \frac{T.r}{J} = \frac{100000 \times 190 \times 150}{532.8 \times 10^4\, t} = \frac{534.9}{t} \text{ N/mm}^2.$$

$$\text{Direct stress, } f_d = \frac{P}{A} = \frac{10000}{480\, t} = \frac{208.3}{t} \text{ N/mm}^2$$

The angle between the stresses is θ, as shown in Fig. 28.19.

$$\cos\theta = \frac{90}{150} = 0.6$$

$\therefore$ Resultant stress

$$f_r = \sqrt{f_t^2 + f_d^2 + 2 f_t \cdot f_d \cos\theta}$$

$$= \sqrt{\left(\frac{534.9}{t}\right)^2 + \left(\frac{208.3}{t}\right)^2 + 2\left(\frac{534.9}{t}\right)\left(\frac{208.3}{t}\right) \times 0.6} = \frac{680.6}{t} \text{ N/mm}^2$$

Allowable shear stress = 110 N/mm²

$$\therefore \quad 110 = \frac{680.6}{t}, \text{ from which } t = 6.19 \text{ mm}$$

$$\text{Size of weld} = \frac{6.19}{0.7} = 8.84 \text{ mm}$$

Hence provide 9 mm size weld.

28.7 ADDITIONAL EXAMPLES USEFUL FOR COMPETITIVE EXAMINATIONS

Example 28.12. *A circular penstock of mild steel, of 1.6 m diameter, is fabricated from 16 mm plate, lapping it and securing it by fillet welds of 12 mm size, provided on the inside and outside of the lapped ends, as shown in Fig. 28.20. Determine the safe internal pressure that can be allowed in the penstock. Take safe stress in weld equal to 100 N/mm².*

Solution. Throat thickness $t = 0.7h = 0.7 \times 12 = 8.4$ mm.

Let p = safe internal pressure (N/mm²)

d = diameter of penstock (mm) = 1600 mm (given)

Internal force per unit length, causing bursting of pipe

$$= p\frac{d}{2} \quad \text{...(1)}$$

Resistance offered by the weld, per unit length

$$= 2\,t\,.\tau_{vp} \quad \text{...(2)}$$

Equating the two, we get

$$p\frac{d}{2} = 2\,t\,.\,\tau_{vp}.$$

$$\therefore \quad p = \frac{4\,t\,\tau_{vp}}{d} \quad \text{...(28.15)}$$

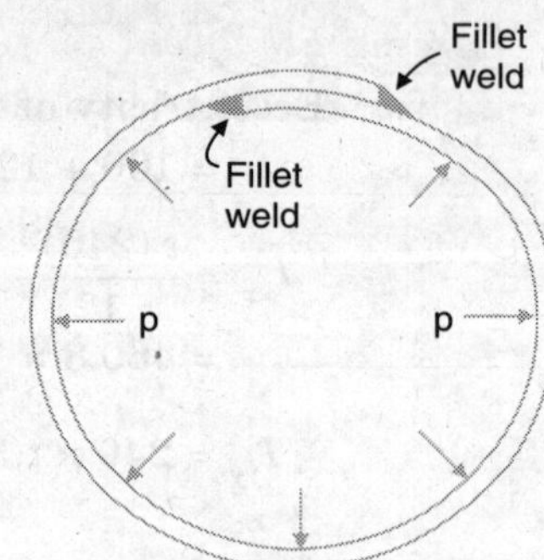

Fig. 28.20

Substituting the given values, we get

$$p = \frac{4 \times 8.4 \times 110}{1600} = \mathbf{2.31\ N/mm^2}.$$

Example 28.13. *Two plates 160 mm × 10 mm and 160 mm × 12 mm are to be butt welded. Calculate the strength of the welded joint in tension, if (a) a single-V butt weld, (b) a double-V butt welded is used to connect them. Take permissible stress in weld as 150 N/mm².*

(a) Single-V butt joint

Effective throat thickness

$$= \frac{5}{8} \times 10 = 6.25 \text{ mm.}$$

Hence, $\quad P = l\,.\,t.\,\tau_{vp} = 160 \times 6.25 \times 150 = 150000 \text{ N} = \mathbf{150\ kN}.$

(b) Double-V butt joint

In the case of double-V butt joint, complete penetration of weld takes place. Hence effective throat thickness = thickness of thinner plate jointed = 10 mm.

$$\therefore \quad P = l\,t\,\tau_{vp}$$

$$= 160 \times 10 \times 150$$

$$= 240000 \text{ N} = \mathbf{240\ kN}.$$

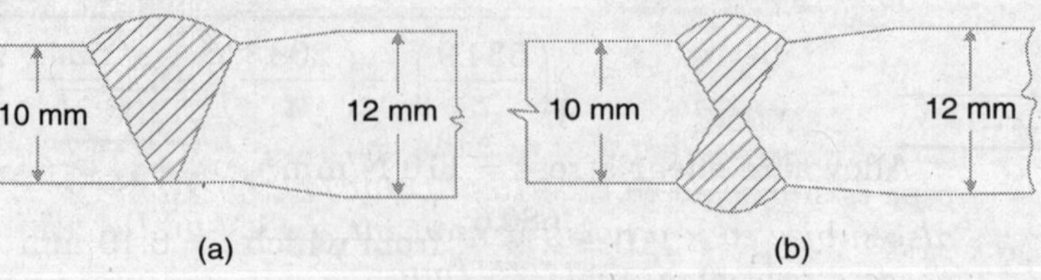

Fig. 28.21

The joints are shown in Fig. 28.21.

Example. 28.14. *A plate bracket, carrying a load of 100 kN at an eccentricity of 120 mm, is connected to the face of a steel stanchion by fillet welds on both the sides of the plate, as shown in Fig. 28.22. Determine*

(a) the size of the fillet weld.

(b) If 8 mm fillet weld is used, determine the depth of the bracket.

(c) If 8 mm fillet weld is used with a bracket of 250 mm depth, calculate the resulting stress in the weld.

Solution. (*a*) Here, d = 300 mm. Let t be the throat thickness.

Vertical shear stress :

$$f_a = \frac{P}{2dt} = \frac{100 \times 1000}{2 \times 300\, t} = \frac{166.67}{t} \text{ N/mm}^2 \quad ...(1)$$

Horizontal shear stress :

$$f_b = \frac{3\,P.e}{td^2} = \frac{3 \times 100 \times 1000 \times 120}{t\,(300)^2} = \frac{400}{t} \text{ N/mm}^2 \quad ...(2)$$

$$\therefore \quad f_r = \sqrt{f_a^2 + f_b^2} = \sqrt{\left(\frac{166.67}{t}\right)^2 + \left(\frac{400}{t}\right)^2} = \frac{433.3}{t} \text{ N/mm}^2.$$

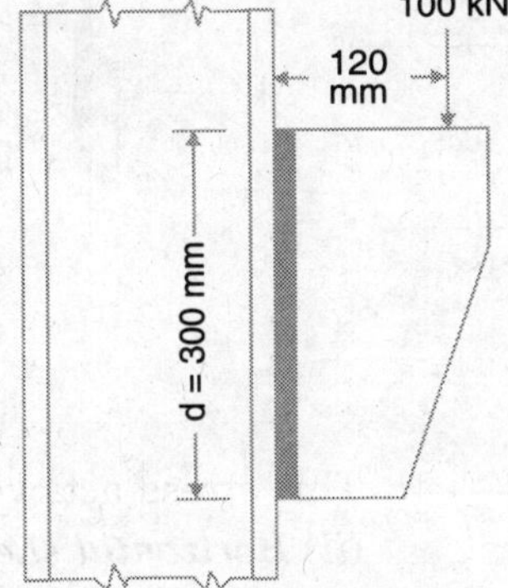

Fig. 28.22

Equating this to permissible shear stress τ_{vp} = 110 N/mm²,

we get $$t = \frac{433.3}{110} = 3.94 \text{ mm}$$

$\therefore$ Size of weld $= \dfrac{t}{0.7} = \dfrac{3.97}{0.7}$ = 5.6 mm. Hence provide 6 mm size weld.

(*b*) If size of weld is 8 mm, we have $t = 0.7 \times 8 = 5.6$ mm

Hence from Eq. 28.9

$$d = \sqrt{\frac{3\,P.e}{t\,\tau_{vp}}} = \sqrt{\frac{3 \times 100 \times 1000 \times 120}{5.6 \times 110}} \approx \mathbf{242\ mm}$$

(*c*) Given: size of weld = 8 mm and d = 250 mm.

$$\therefore \quad t = 0.7 \times 8 = 5.6 \text{ mm}$$

$$f_a = \frac{P}{2dt} = \frac{100 \times 1000}{2 \times 250 \times 5.6} = 35.71 \text{ N/mm}^2$$

$$f_b = \frac{3\,P.e}{td^2} = \frac{3 \times 100 \times 1000 \times 120}{5.6\,(250)^2} = 102.85 \text{ N/mm}^2$$

$$\therefore \quad f_r = \sqrt{f_a^2 + f_b^2} = \sqrt{(35.71)^2 + (102.85)^2} = \mathbf{108.9\ N/mm^2} < 110 \text{ N/mm}^2.$$

Example 28.15. *A circular steel pipe 120 mm external diameter and 100 mm internal diameter is welded to a rectangular plate 12 mm thick by fillet weld around the perimeter. The pipe is subjected to a vertical point load of 12 kN acting at 500 mm from the welded end. It is also subjected to a twisting moment of 4 kN-m. Determine the size of the weld, taking permissible shear stress equal to 110 N/mm².*

Solution. The weld section is subjected to *three* stresses.

(*i*) Vertical shearing stress f_a due to vertical force P = 12 kN

(*ii*) Horizontal shearing stress f_b due to moment

$M = P\,.\,e = 12 \times 10^3 \times 500 = 6 \times 10^6$ N-mm.

(*iii*) Torsional shearing stress f_t due to torsional moment $T = 4$ kN-m $= 4 \times 10^6$ N-mm.

Let t be the throat thickness of the fillet weld.

(*i*) *Vertical shear stress f_a due to* S.F. *P*

$$f_a = \frac{P}{L.t} = \frac{P}{\pi.d.t} = \frac{12000}{\pi \times 120 \times t} = \frac{31.8}{t} \text{ N/mm}^2$$

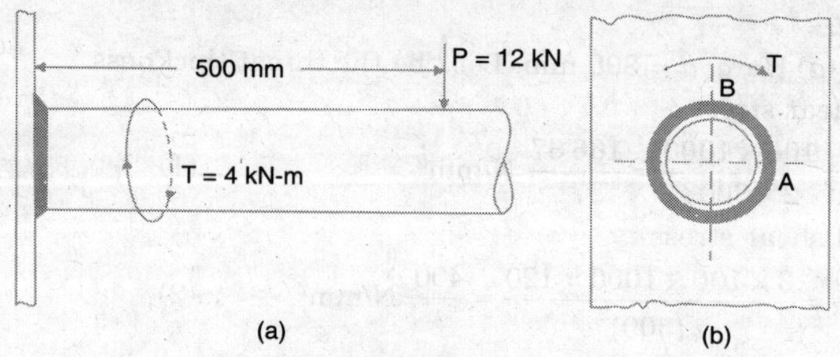

Fig. 28.23

This stress acts vertically downwards, at all sections of the weld.

(*ii*) *Horizontal shear stress f_b due to $M = P \,.\, e = 6 \times 10^6$* N-mm

$$f_b = \frac{M}{I}.y$$

where $\quad I \approx \dfrac{\pi d^3}{8} t = \dfrac{\pi(120)^3}{8} t = 678584\,t$ and $y = \dfrac{d}{2} = 60$ mm.

$$\therefore \quad f_b = \frac{6 \times 10^6}{678584t} \times 60 = \frac{530.5}{t} \text{ N/mm}^2$$

This stress acts perpendicular to the plane of the weld.

(*iii*) *Torsional shear stress f_t due to $T = 4$* kN-m.

$$f_t = \frac{T}{J}.r$$

where $\quad J \approx \dfrac{\pi d^3}{4} t = \dfrac{\pi(120)^3}{4} t = 1357168\,t$ and $r = \dfrac{d}{2} = 60$ mm

$$\therefore \quad f_t = \frac{4 \times 10^6}{1357168t} \times 60 = \frac{176.8}{t} \text{ N/mm}^2.$$

This stress acts in the plane of weld, along tangent (*i.e.*, perpendicular to the radius vector).

Let us now consider the combined effect of the three.

At point A: f_b is zero while f_a and f_t are additive since both act vertically downwards. Hence the resultant stress is

$$f_{rA} = f_a + f_t = \frac{31.8}{t} + \frac{176.8}{t} = \frac{208.6}{t} \text{ N/mm}^2 \qquad \text{...}(i)$$

At point B : f_a and f_t act at right angles to each other, in the same plane (*i.e.*, the plane of the weld) while f_b attains its maximum value and acts perpendicular to the plane of the weld.

Hence the resultant stress is

$$f_{rB} = \sqrt{f_a^2 + f_t^2 + f_b^2} = \sqrt{\left(\frac{31.8}{t}\right)^2 + \left(\frac{176.8}{t}\right)^2 + \left(\frac{530.5}{t}\right)^2} = \frac{560.1}{t} \text{ N/mm}^2 \qquad \text{...(ii)}$$

Thus maximum stress is induced at point B, which is the critical point. The stress at this point should not exceed permissible stress τ_{vp} = 110 N/mm².

$$\therefore \qquad \frac{560.1}{t} \le 110, \quad \text{from which} \quad t \ge \frac{560.1}{110} = 5.09$$

$$\text{Hence size of weld} \qquad = \frac{t}{0.7} = \frac{5.09}{0.7} = 7.27 \text{ mm}$$

As per IS 816–1969, for the thickness = 12 mm of the thicker part, minimum size of weld is 5 mm. Hence provide **7.5 mm** size fillet weld.

PROBLEMS

1. A 150 mm × 15 plate is welded to other plate by two side welds 12 cm each and end fillet of 100 mm length. Find the safe axial load to which this joint may be subjected if the size of the weld is 7 mm.

2. A 100 mm × 10 plate is welded to other by means of two end fillets and two side fillets of 8 mm as shown in Fig. 28.24. If the plate is loaded to its full strength, find out the required overlap length.

Fig. 28.24

3. An equal angle 75 mm × 75 mm @ 11.0 kg/m is subjected to a load for 180 kN, whose line of action passes through the centroid of the section, which is 22.2 mm from the heel. This angle is to be welded to a gusset plate. If the size of the weld is to be 8 mm, find the length of the size fillet welds.

4. An I-section is made up of a 200 mm × 10 mm thick web plate welded to two flange plates 120 mm × 10 mm thick by means of fillet welds to size 6 mm. Calculate the maximum shear force which this section can resist.

5. Fig. 28.25 shows a 10 mm angle bracket 100 mm wide welded to the flange of a steel stanchion. It carries a vertical load of 240 kN. The connection consists of continuous 10 mm weld extending along the top and both sides and returned at the bottom of the bracket. Treating the 240 kN load as a vertical shear load (*i.e.*, neglecting bending moment), calculate the depth of bracket, taking 110 N/mm² as the working stress in the transverse weld and 79 N/mm² in the longitudinal weld.

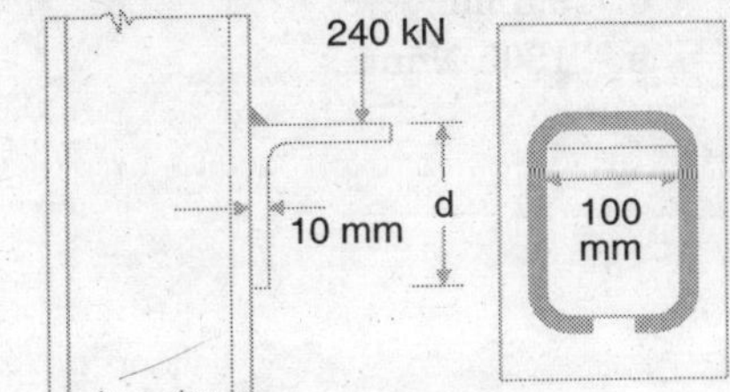

Fig. 28.25

6. An I-section bracket carrying 120 kN load, is connected to by a column as shown in Fig. 28.26 mean of two side fillet welds 200 mm deep. The load is eccentric by 70 mm. Calculate the size of the fillet weld.

7. A bracket consisting of a Tee-section 150 mm × 150 mm and 10 mm is connected to a column as shown in Fig. 28.26. The bracket carries 150 kN load at 80 mm eccentricity. If the size of the weld is 6 mm, find out the maximum throat stress.

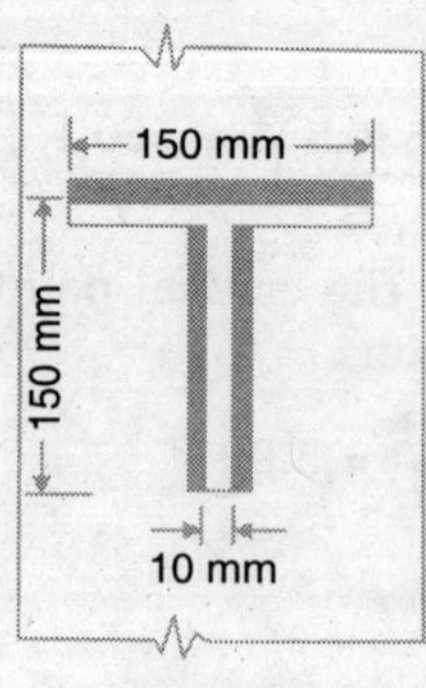

Fig. 28.26

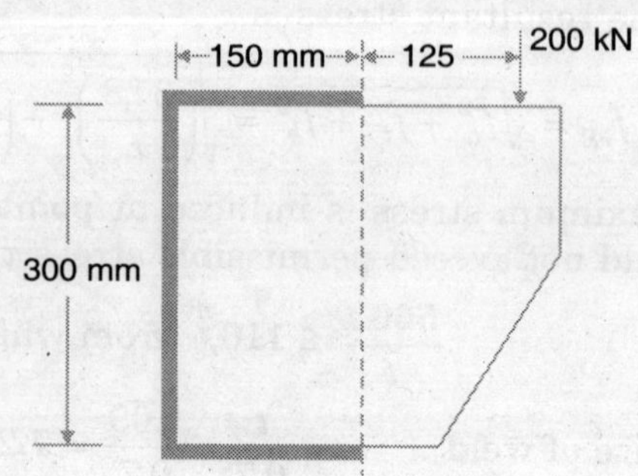

Fig. 28.27

8. The bracket shown in Fig. 28.27 is welded to a stanchion by side fillet welds on three sides indicated by heavy lines. Calculate the maximum forces per inch of weld metal when the bracket carries the load of 200 kN acting as shown.

9. A bracket is welded to a stanchion by fillet welds, having a throat thickness of 9 mm and a load of 180 kN is applied in the plane of the bracket as shown in Fig. 28.28. The weld extends round three sides and has the given dimensions. Determine the maximum stress on the throat of the weld.

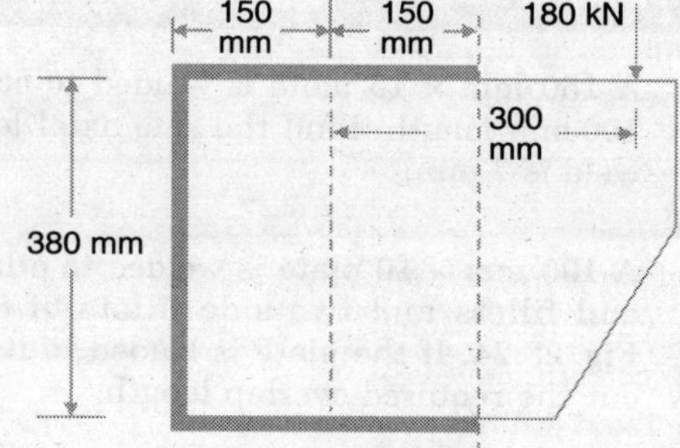

Fig. 28.28

ANSWERS

1. 172.5 kN
2. 72.5 mm
3. 218 mm and 91.5 mm
5. 160 mm
6. 9.5 mm
8. 1100 N/mm^2
9. 1800 N/mm^2.

Index

(The numbers refer to pages)

S

T

U

V

W

Y